Elementary Algebra

Graphs & Authentic Applications

Third Edition

Jay Lehmann

College of San Mateo

Pearson

330 Hudson Street, NY NY 10013

Director, Courseware Portfolio Management: Michael Hirsch
Courseware Portfolio Manager: Rachel Ross
Courseware Portfolio Management Assistant: Shannon Bushee
Content Producer: Tamela Ambush
Managing Producer: Karen Wernholm
Producer: Shana Siegmund
Manager, Courseware QA: Mary Durnwald
Associate Content Producer, TestGen: Sneh Singh
Manager, Content Development: Robert Carroll
Product Marketing Manager: Alicia Frankel
Marketing Assistant: Brooke Imbornone
Field Marketing Managers: Jennifer Crum and Lauren Schur
Senior Author Support/Technology Specialist: Joe Vetere
Manager, Rights and Permissions: Gina Cheselka
Manufacturing Buyer: Carol Melville, LSC Communications
Text Design and Composition: SPi Global
Cover Design: Jenny Willingham
Cover Image: Paseven/Shutterstock

Library of Congress Cataloging-in-Publication Data

Names: Lehmann, Jay, author.
Title: Elementary algebra : graphs & authentic applications / Jay Lehmann,
    College of San Mateo.
Description: Third edition. | Boston : Pearson Education, Inc., 2018. |
    Includes index.
Identifiers: LCCN 2017052377 | ISBN 9780134756998
Subjects: LCSH: Algebra—Textbooks.
Classification: LCC QA152.3 .L45 2018 | DDC 512.9—dc23
LC record available at https://lccn.loc.gov/2017052377

1 17

ISBN-10:   0-13-475699-1
ISBN-13: 978-0-13-475699-8

Contents

Total Profit for Top-100-
Grossing Concert Tours (p. 37)

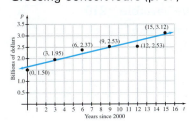

1

Presidential Election Voter
Turnout (p. 89)

Year	Percent of Eligible Voters Who Voted
1980	59.2
1984	59.9
1988	57.4
1992	61.9
1996	54.2
2000	54.7
2004	63.8
2008	63.6
2012	57.5
2016	60.0

2

Percentages of Adult Internet
Users Who Use Social
Networking Sites (p. 170)

Year	Percent
2010	60
2011	65
2012	67
2013	73
2014	74
2015	76

3

Revenues from Social
Network Gaming and
Mobile Games (pp. 241–242)

Year	Revenue (billions of dollars)
2009	5.4
2010	7.0
2011	7.5
2012	6.7
2013	9.0
2014	9.9
2015	11.2

4 SIMPLIFYING EXPRESSIONS AND SOLVING EQUATIONS   183

Total College Student
Discretionary Spending
(pp. 266–267)

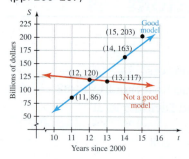

5 LINEAR EQUATIONS IN TWO VARIABLES AND LINEAR INEQUALITIES IN ONE VARIABLE   249

Numbers of Part-Time and
Full-Time Instructors (p. 307)

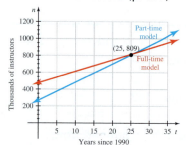

6 SYSTEMS OF LINEAR EQUATIONS AND SYSTEMS OF LINEAR INEQUALITIES   303

Contents v

Percentages of 10th Graders Who Have Used Marijuana in the Past 30 Days (p. 377)

Year	Percent
2008	13.8
2009	15.9
2010	16.7
2011	17.6
2012	17.0
2013	18.0
2014	16.6
2015	14.8

7 POLYNOMIALS AND PROPERTIES OF EXPONENTS   365

Worldwide iPhone Sales (p. 474)

Year	Sales (millions)
2007	1.4
2009	20.7
2011	72.3
2013	150.3
2015	231.2

8 FACTORING POLYNOMIALS AND SOLVING POLYNOMIAL EQUATIONS   434

Percentages of Lawyers Who Are Women (pp. 521–522)

Year	Percent
1970	5
1980	14
1990	24
2000	30
2010	33
2017	35

9 SOLVING QUADRATIC EQUATIONS   483

Numbers of Households
Burglarized (p. 541)

Year	Number of Households Burglarized (millions)	Total Number of Households (millions)
2011	1.63	118.7
2012	1.56	121.1
2013	1.43	122.5
2014	1.27	123.2
2015	1.13	124.6

Numbers of 3D Movie Screens
(p. 623)

Year	Number of Screens (thousands)
2010	14.7
2011	25.6
2012	33.1
2013	36.8
2014	38.4

Tracing a Curve (p. A-2)

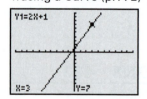

Preface

"The question of common sense is always, 'What is it good for?'—a question which would abolish the rose and be answered triumphantly by the cabbage."

—James Russell Lowell

These words seem to suggest that poet and editor James Russell Lowell (1819–1891) took Elementary Algebra. How many times have your students asked, "What is it good for?" After years of responding "You'll find out in the next course," I began an ongoing quest to develop a more satisfying and substantial response to my students' query.

This ongoing quest has led me to author three algebra texts and, most recently, a new stat prep text, *A Pathway to Introductory Statistics*. I have a passion for using authentic data, centered around a curve-fitting approach to help students learn in context.

Curve-Fitting Approach   Although there are many ways to center an Elementary Algebra course around authentic applications, I chose a curve-fitting approach for several reasons. A curve-fitting approach

- allows great flexibility in choosing interesting, authentic, current situations to model.
- emphasizes key concepts and skills in a natural, substantial way.
- deepens students' understanding of equations in two variables because it requires students to describe these equations graphically, numerically, symbolically, and verbally.
- unifies the many diverse topics of a typical Elementary Algebra course.

There is yet one more reason I chose a curve-fitting approach. An Elementary Algebra course is meant to prepare some students for a Calculus STEM track and others for Statistics, Liberal Arts Math, and so on (non-STEM tracks). This is a great challenge because Calculus, Statistics, and so on are vastly different courses not only in content but also in the type of problem solving they require. Teaching algebra with curve fitting empowers instructors to prepare students for all types of content and problem solving.

To fit a curve to data, students learn the following four-step modeling process:

1. Examine the data set to determine which type of model, if any, to use.
2. Find an equation of the model.
3. Verify that the model fits the data.
4. Use the model to make estimates and predictions.

This four-step process weaves together topics that are crucial to the course. Students must notice numerical patterns from data displayed in tables, recognize graphical patterns in scatterplots, find equations of models, graph models, and solve equations.

Not only does curve fitting foster cohesiveness within chapters, but it also creates a parallel theme for each set of chapters that introduces and discusses a new type of model. This structure enhances students' abilities to observe similarities and differences among fundamental models such as linear models, quadratic models, rational models, and radical models.

Curve fitting serves as a portal for students to see the usefulness of mathematics so they become fully engaged in the class. Once involved, students are more receptive to all aspects of the course.

NEW TO THE THIRD EDITION

Students will benefit from the following changes to the third edition of *Elementary Algebra: Graphs and Authentic Applications*:

- In previous editions, all authentic data sets in the print text were represented by similar, yet generic (inauthentic), data sets in MyLab Math to provide algorithmically-generated similar exercises for students completing homework in MyLab Math. However, in the new edition, where possible, MyLab Math exercises maintain the authenticity of the data. This has been accomplished by sampling from a large data set to generate six authentic data sets that inherit the same trend.

- *MyLab Math Exercises*: The number of skill, modeling, and conceptual exercises in MyLab Math has been increased to fully capture the spirit of the print textbook. In fact, for the first time ever, Related Review exercises (described later in the preface) will be assignable in MyLab Math.

- *Large Data Sets*: Many students who use this textbook will not perform regression analysis in their careers, but some *will* work with large data sets. Such work will also help prepare students to take Statistics. With this in mind, new exercises that involve large data sets have been sprinkled throughout the textbook. They directly follow the heading "Large Data Sets." The data sets consist of as many as thousands of rows and tens of columns of data.

- DATA *Downloadable Data Sets*: To support the appropriate use of technology when completing exercises and labs, data sets that consist of 16 or more data values can now be downloaded as Excel files at MyLab Math and at the Pearson Downloadable Student Resources for Math and Statistics website: http://www.pearsonhighered.com/mathstatsresources. These data sets in MyLab Math can also be opened in StatCrunch. Exercises that involve such data sets are flagged in the print textbook by the icon DATA.

- *Augmented Data Sets*: To make the data sets as current and relevant as possible, 162 data sets in examples and exercises have been augmented to include values for recent years.

- *New Data Sets*: 212 data sets in examples and exercises have been replaced with more compelling and contemporary topics such as immigration, national health care, and trust in the mass media.

- *Climate Change Labs*: All eight Climate Change labs have been updated to address the latest data and political events concerning this incredibly important global issue.

- *Graphing Calculator Instructions*: Appendix A, which consists of TI-83/TI-84 graphing calculator instructions, was available only online in the previous edition. To make the appendix more accessible to students, it is now included in the textbook.

- *StatCrunch Instructions*: Some departments that require StatCrunch for their Statistics courses introduce StatCrunch in their Elementary Algebra courses. To support such departments, Appendix B, which contains StatCrunch instructions, has been added to the textbook.

- *Section Opener Explorations*: Explorations that can be used at the start of a section have been moved from the preceding section to the current section. The new placement will visually remind instructors to assign such explorations and make it easier for students to access them.

- *Statistics Terminology*: To better support students who will take Statistics, the terminology has been improved: The words *scattergram*, *independent variable*, and *dependent variable* have been replaced with *scatterplot*, *explanatory variable*, and *response variable*.

- *Graphing Linear Equations and Linear Models*: The technique of graphing equations of the forms $x = a$ and $y = b$ has been moved from Section 3.2 to Section 3.1. This way, all equations of the forms $y = mx + b$ and $x = a$ are now contained within one section (3.1), and Section 3.2 is now devoted to unit analysis and graphing linear models.

- *Color*: More color has been used to enhance connections between equations, graphs, tables, and coordinates of ordered pairs.

CONTINUED FROM THE SECOND EDITION

Unique Organization   Many college students who take Elementary Algebra had significant difficulties with the equivalent courses in high school. These students face a greater challenge in the college courses because they must complete the course in one semester, rather than two. Instead of presenting the material in the "same old way," this textbook provides a unique organization that will better aid students in succeeding. Three key aspects of the organization—providing the "big picture," early graphing, and spiraling of concepts—are described in the following paragraphs.

Providing the Big Picture   The text uses modeling to provide the "big picture" before going into details. For example, Chapter 1 gives an overview of linear modeling, which is the main theme of Chapters 1–6, and Section 7.2 provides an overview of quadratic modeling, which is a major focus of Chapters 7–9. Using modeling to provide the big picture not only is good pedagogy, but also sets the tone that this course will be different, interesting, alive, and relevant, inviting students' creativity into the classroom.

Early Graphing   In Chapter 3, students learn to graph linear equations only in the forms $y = mx + b$ and $x = a$. This way, they can focus on the fundamental concepts of slope and y-intercept. As many professors have reported, students do exceedingly well in Chapter 3. This early-graphing organization postpones simplifying expressions and solving equations, buying students a bit more time to find their "sea legs" before moving on to the more challenging symbolic manipulation work in Chapter 4. By the time that students reach Section 5.1, they are ready to graph equations that are not in slope-intercept form, but can be put into it.

The early-graphing approach also enables students to solve equations graphically as well as symbolically. Most of Chapters 4–11 include exercises that reinforce the connection between graphing and solving equations or systems of equations.

Spiraling of Concepts   If a concept is never revisited, students may not retain it. Fortunately, curve fitting naturally revisits concepts as students' tool bag of models grows. In each modeling section, exercises require students to compare the implications of using the various types of equations to model authentic situations. In addition, students' retention of key concepts can be enhanced in Chapters 5–11 by completing two special types of exercises, **Related Review** and **Expressions, Equations, and Graphs**. These types of exercises are described in greater detail later in the preface.

Modeling Exercises   To give this third edition a current and lively feel, the vast majority of the hundreds of modeling exercises in the text have been updated or replaced. Most of the application exercises contain tables of data, but some describe data in paragraph form to give students practice in picking out relevant information and defining variables. Both types of applications are excellent preparation for subsequent courses (especially Statistics).

Group Explorations   All sections of this text contain one or two explorations that support student investigation of a concept. Instructors can use explorations as collaborative activities during class time or as part of homework assignments. The "Section Opener" explorations are meant to have students discover the section's concepts at the start of class. The other explorations are designed to have students apply concepts they have learned in the section in new ways. Both types of explorations can empower students to become active explorers of mathematics and open the door to the wonder and beauty of the subject.

Taking It to the Lab Sections   Laboratory assignments have been included at the end of most chapters to deepen students' understanding of concepts and the scientific method. These labs reinforce the idea that mathematics is useful. They are also an excellent avenue for more in-depth writing assignments.

Some of the labs are about climate change and have been written at a higher reading level than the rest of the text in order to give students a sense of what it is like to perform research. Students will find that by carefully reading (and possibly rereading) the background information, they can comprehend the information and apply concepts they have learned in the course to make estimates and predictions about this compelling, current, and authentic situation.

Balanced Extensive Homework Sections   Most exercise sets contain a large number of modeling, skill, and conceptual exercises to allow professors maximum flexibility in setting assignments.

Related Review   These exercises (in every section of Chapters 5–11) relate current concepts to previously learned concepts. Such exercises assist students in seeing the "big picture" of the course. This exercise type is now also assignable in MyLab Math.

Expressions, Equations, and Graphs   These exercises (in every section of Chapters 5–11) help students gain a solid understanding of these three core concepts, including how to distinguish among them.

Technology   The text assumes students have access to technology such as the TI-83 or TI-84 graphing calculator, Excel, or StatCrunch. Technology of this sort allows students to construct scatterplots and check the fit of a model quickly and accurately. It also empowers students to verify their results from Homework exercises and efficiently explore mathematical concepts in the Group Explorations.

The text supports instructors in holding students accountable for all aspects of the course without the aid of technology, including finding equations of linear models. (Linear regression equations are included in the Answers section because it can be difficult or impossible to anticipate which points a student will choose in trying to find a reasonable equation.)

Appendix A: Using a TI-83 or TI-84 Graphing Calculator   Appendix A contains step-by-step instructions for using the TI-83 and TI-84 graphing calculators. A subset of this appendix can serve as a tutorial early in the course. In addition, when the text requires a new calculator skill, students are referred to the appropriate section in Appendix A.

Appendix B: Using StatCrunch   Appendix B contains step-by-step instructions for using StatCrunch. The appendix describes how to enter data and construct scatterplots.

Exposition   If students can't make sense of the prose, it doesn't matter how precise it is. One of my top goals is to write descriptions that are straightforward, accessible, clear, and rigorous.

Tips for Success   Many sections close with tips that are intended to help students succeed in the course. A complete listing of these tips is included in the Index.

Warnings   These are discussions (flagged by the margin entry "WARNING") that address students' common misunderstandings about key concepts and help students avoid such misunderstandings.

Chapter Opener   Each chapter begins with a description of an authentic situation that can be modeled by the concepts discussed in the chapter.

GETTING IN TOUCH

I would love to hear from you and would greatly appreciate receiving your comments regarding this text. If you have any questions, please ask them, and I will respond.

Thank you for your interest in preserving the rose.

Jay Lehmann
MathNerdJay@aol.com

Resources for Success
Get the Most Out of MyLab Math

for *Elementary Algebra,* Third Edition, by Jay Lehmann

When it comes to developmental math, one size does not fit all. Jay Lehmann's *Elementary Algebra* offers market-leading content written by an author-educator, tightly integrated with the #1 choice in digital learning—MyLab Math. MyLab Math courses can be tailored to the needs of instructors and students, while weaving the author's voice and unique approach into all elements of the course. Learning mathematical concepts through authentic data comes through from the text to the MyLab course seamlessly.

Take advantage of the following resources to get the most out of your MyLab Math course.

Conceptual Understanding and Motivation

New! Large Data Sets in exercises and explorations get students accustomed to working with as many as thousands of rows of data. Data sets that involve 16 or more values are available for download to support the appropriate use of technology. Noted with a **DATA** icon, these exercises are ideal for using technology, like StatCrunch or Excel, to analyze the data and synthesize concepts. In today's age of "big data," it's important for students to see how technology can efficiently and accurately help when working with large data sets.

Large Data Sets

47. **DATA** Access the data about airborne times and distances of Delta Airlines flights, which are available at MyLab Math and at the Pearson Downloadable Student Resources for Math & Stats website. Let T be the airborne time (in minutes) and D be the distance (in miles) for a flight.
   a. Construct a scatterplot of the data.
   b. Give a possible reason why the scatterplot consists of vertically aligned clumps of data points.
   c. On the basis of just the scatterplot, guess whether Delta offers more *routes* that are less than 1000 miles or greater than 1000 miles. On the basis of just the scatterplot, why is it not possible to be sure?

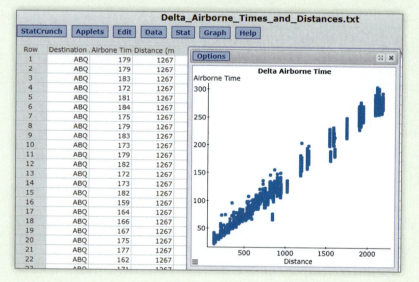

New! StatCrunch is a web-based statistical software available from within the MyLab Math course that students can use to easily analyze data sets from exercises and the text. Through StatCrunch users can access tens of thousands of shared data sets, create and conduct online surveys, perform complex analyses using the powerful statistical software, and generate compelling reports.

pearson.com/mylab/math

New! Select exercises with authentic data have been carefully revised to retain authentic data values, even when regenerating algorithmically. Oftentimes students sacrifice working with real-world data when they regenerate exercises with new values in MyLab Math. In this revision, the author has taken special care to ensure that many exercises' algorithmic versions of the question still ask the student to work with actual data pulled from real-world situations.

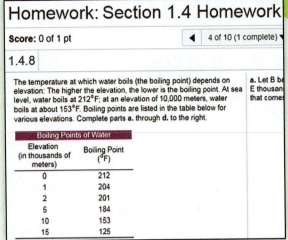

Updated! The video program provides students with extra help for each objective of the textbook. The videos highlight key examples, and a modern interface allows easy navigation. Videos have been updated to reflect all changes in the current edition.

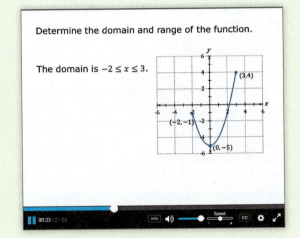

Personalized Learning and Preparedness

New! Skill Builder exercises offer just-in-time additional adaptive practice. The adaptive engine tracks student performance and delivers questions to each individual that adapt to his or her level of understanding. This new feature allows instructors to assign fewer questions for homework, allowing students to complete as many or as few questions needed.

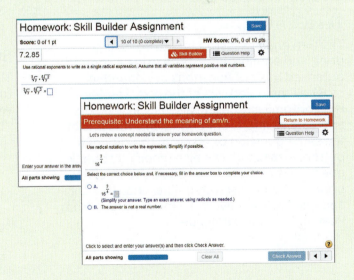

Enhanced Sample Assignments make course set-up easier by giving instructors a starting point for each section and chapter. Homework assignments have been carefully curated for this specific text and include a thoughtful mix of question types. Find these sample assignments in the Assignment Manager, under Copy and Assign Sample Assignments.

Resources for Success

Instructor Resources

The following instructor resources are available to download from the Instructor Resource Center at www.pearson.com, or in your MyLab Math course.

Instructor's Resource Manual

This manual, written by the author, contains suggestions for pacing the course and creating homework assignments. It discusses how to incorporate technology and how to structure project assignments. The manual also contains section-by-section suggestions for presenting lectures and for undertaking the explorations in the text.

Power Points

These fully editable lecture slides include definitions, key concepts, and examples for use in a lecture setting and are available for each section of the text.

Instructor's Solutions Manual

This manual includes complete solutions to the even-numbered exercises in the text.

TestGen

TestGen enables instructors to build, edit, print, and administer tests by using a computerized bank of questions developed to cover all the objectives of the text. TestGen is algorithmically based, allowing instructors to create multiple, but equivalent, versions of the same question or test with the click of a button. Instructors can also modify test-bank questions or add new questions. Tests can be printed or administered online. The software and test bank are available for download from Pearson's online catalogue.

Student Resources

New! Concepts and Explorations Notebook: Working with Authentic Data

This new compelling resource for students correlates to the text and provides students with opportunities to dig into data and solve problems using pencil and paper. The workbook includes:

- Explorations that offer collaborative activities to support discovery of key concepts.
- Modeling exercises with authentic data that give students more practice on this multifaceted concept, that can be sometimes hard to fully accomplish through MyLab Math.
- Projects that can be either open-ended or more guided, and ask students to dig deeper into a data set and think critically.
- Graphing exercises that ask students to practice graphing on their own, beyond what they do in MyLab Math.
- Mini-Essay questions that prompt students to think conceptually, also beyond what they do in MyLab Math!

Student's Solutions Manual

This manual contains the complete solutions to the odd-numbered exercises in the Homework sections of the text.

pearson.com/mylab/math

To the Student

You are about to embark on an exciting journey. In this course, you will learn not only more about algebra but also how to apply algebra to describe and make predictions about authentic situations. "Authentic situations" might make you think twice, but this just means situations that are *really* happening in the world. This text contains data that describe hundreds of these situations. Most of the data have been collected from recent publications, so the information is current and of interest to the general public. There is data about profit from concerts, success in school, climate change, sports, and so on. I hope it interests you too.

Working with authentic data will make mathematics more meaningful. While working with data about authentic situations, you will learn mathematical concepts more easily because they will be connected to familiar contexts. And you will see that almost any situation can be viewed mathematically. That vision will help you understand the situation and make estimates and/or predictions.

Many of the problems you will explore in this course involve data collected in a scientific experiment, survey, or census. The practical way to deal with such data sets is to use technology. So, a graphing calculator or computer system is probably required.

Analyzing authentic situations is a lifelong skill. We are living in the "age of data." In addition to working with data sets in this text, your instructor may assign some of the labs. Here you will collect data through experiment or research. This will give you a more complete picture of how you can use the approaches presented in this text in everyday life, and likely in your lifelong careers. Being able to work with and understand data can lead to higher-paying jobs and success.

Hands-on explorations are rewarding and fun. This text contains explorations with step-by-step instructions that will lead you to *discover* concepts, rather than hear or read about them. Because discovering a concept is exciting, it is more likely to leave a lasting impression on you. Also, as you progress through the explorations, your ability to make intuitive leaps will improve, as will your confidence in doing mathematics. Over the years, students have remarked to me time and time again that they never dreamed that learning math could be so much fun.

Jay has a wide variety of interests. He is pictured here playing with his rock band, The Procrastinistas. (Photo courtesy of Rick Gilbert)

This text contains special features to help you succeed. Many sections contain a Tips for Success feature. These tips are meant to inspire you to try new strategies to help you succeed in this course and future courses. If you browse through all the tips early in the course, you can take advantage of as many of them as you wish. Then, as you progress through the text, you'll be reminded of your favorite strategies. A complete listing of Tips for Success is included in the Subject Index.

Other special features that can support you include Warnings, which can help you avoid common misunderstandings; Key Points summaries, which can help you review and retain concepts and skills addressed in the chapter you have just read; Related Review exercises, which can help you understand current concepts in the context of previously learned concepts; and Expressions, Equations, and Graphs exercises, which can help you understand and distinguish among these three core concepts.

Feel free to contact me. It is my pleasure to read and respond to e-mails from students who are using my text. If you have any questions or comments about the text, feel free to contact me.

Jay Lehmann
MathNerdJay@aol.com

Acknowledgments

Writing a modeling text is an endurance run I couldn't have completed without the dedicated assistance of many people. First, I'm greatly indebted to Keri, my wife, who yet again served as an irreplaceable sounding board for the countless decisions that went into creating this book.

I acknowledge several people at Pearson Education. I'm very grateful to Editor-in-Chief Michael Hirsch, who has shared in my vision for this text and made significant investments to make that vision happen. The book has been greatly enhanced through the support of Senior Acquisitions Editor Rachel Ross, who made a multitude of contributions, including assembling an incredible team to develop and produce this text. The team includes Content Producer Tamela Ambush, who handled countless tasks to support me in preparing the manuscript for production, leading to a significantly better book.

Heartfelt thanks goes to Project Managers Thomas Russell, and Erin Hernandez who both orchestrated the many aspects of production.

And thanks to Rick Gilbert for the awesome photograph in the "To the Student" section of the author performing with his band, The Procrastinistas, at the Hotel Utah in San Francisco.

I thank these reviewers, whose thoughtful, detailed comments helped me sculpt this text into its current form:

Scott Adamson, *Chandler-Gilbert Community College*
Thomas Adamson, *Phoenix College*
Ken Anderson, *Chemeketa Community College*
Gwen Autin, *Southeastern Louisiana University*
Mona Baarson, *Jackson Community College*
Sam Bazzi, *Henry Ford Community College*
Joel Berman, *Valencia Community College—East*
Nancy Brien, *Middle Tennessee State University*
Ronnie Brown, *University of Baltimore*
Barbara Burke, *Hawaii Pacific University*
Laurie Burton, *Western Oregon University*
Paula Castagna, *Fresno City College*
James Cohen, *Los Medanos College*
Jeff Cohen, *El Camino College*
Joseph DeGuzman, *Norco College*
Cynthia Ellis, *Purdue University, Fort Wayne*
Junko Forbes, *El Camino College*
William P. Fox, *Francis Marion University*
Cathy Gardner, *Grand Valley State University*
James Gray, *Tacoma Community College*
Kathryn M. Gundersen, *Three Rivers Community College*
Miriam Harris-Botzum, *Lehigh Carbon Community College*
Stephanie Haynes, *Davis & Elkins College*
Rick Hough, *Skyline College*
Tracey Hoy, *College of Lake County*
Denise Hum, *Cañada College*
Evan Innerst, *Cañada College*
Judy Kasabian, *El Camino College*
Peter Kay, *Western Illinois University*
Charles Klein, *De Anza College*
Julianne M. Labbiento, *Lehigh Carbon Community College*
Jason Malozzi, *Lehigh Carbon Community College*
Debra Martin, *Purdue University, Fort Wayne*
Diane Mathios, *De Anza College*
Jim Matovina, *Community College of Southern Nevada*

Jane E. Mays, *Grand Valley State University*
Scott McDaniel, *Middle Tennessee State University*
Tim Merzenich, *Chemeketa Community College*
R. Messer, *American River College*
Jason L. Miner, *Santa Barbara City College*
Nolan Mitchell, *Chemeketa Community College*
Camille Moreno, *Cosumnes River College*
Lisa Mosser, *Jackson Community College*
Ellen Musen, *Brookdale Community College*
Charlie Naffziger, *Central Oregon Community College*
Chris Nord, *Chemeketa Community College*
Donna Marie Norman, *Jefferson Community College*
Denise Nunley, *Glendale Community College*
Karen D. Pain, *Palm Beach State College*
Ernest Palmer, *Grand Valley State University*
Ellen Rebold, *Brookdale Community College*
Jody Rooney, *Jackson Community College*
James Ryan, *State Center Community College District, Clovis*
Barbara Savage, *Roxbury Community College*
Ned Schillow, *Lehigh Carbon Community College*
Ingrid Scott, *Montgomery College*
Kathy Self, *Georgia Military College—Milledgeville*
David Shellabarger, *Lane Community College*
Laura Smallwood, *Chandler-Gilbert Community College*
John Szeto, *Southeastern Louisiana University*
Janet Teeguarden, *Ivy Tech Community College*
Lorna TenEyck, *Chemeketa Community College*
Cindy Vanderlaan, *Purdue University, Fort Wayne*
Lenove Vest, *Lower Columbia College*
Ollie Vignes, *Southeastern Louisiana University*
Linda Wagner, *Purdue University, Fort Wayne*
Karen Wiechelman, *University of Louisiana at Lafayette*
Robin Williams, *Palomar College*
Lisa Winch, *Kalamazoo Valley Community College*
Cathy Zucco-Teveloff, *Rider University*

Index of Applications

Introduction to Modeling

1

Think about the last concert you attended. What was the ticket price? Was it worth it? The total profit for the top-100-grossing concert tours has increased greatly (see Table 1). In Example 2 of Section 1.4, we will predict when the total profit will be $3.6 billion.

In this course, we will discuss how to describe the relationship between two quantities that occur in an authentic situation. For example, we will describe the percentage of Americans who use wearable devices such as fitness trackers and smartwatches for various age groups. In Chapters 1–6, we will focus on how to use (straight) lines to describe authentic situations. In Chapters 7–11, we will discuss other types of *curves* that can be used to describe authentic situations.

Table 1 Total Profit for Top-100-Grossing Concert Tours

Year	Total Profit (billion dollars)
2000	1.50
2003	1.95
2006	2.37
2009	2.53
2012	2.53
2015	3.12

Source: *Pollstar*

1.1 Variables and Constants

Objectives

» Describe the meaning of *variable* and *constant*.

» Describe the meaning of *counting numbers, integers, rational numbers, irrational numbers, real numbers, positive numbers,* and *negative numbers*.

» Use a number line to describe numbers.

» Describe a number as a decimal number.

» Graph data.

» Find the average (or mean) of a group of numbers.

» Describe a concept or procedure.

In this section, we will work with *variables* and *constants*, two extremely important building blocks of algebra. We will also discuss various types of numbers and how to describe numbers visually.

Variables

In arithmetic, we work with numbers. In algebra, we work with *variables* as well as numbers.

Definition Variable

A **variable** is a symbol that represents a quantity that can vary.

For example, we can define h to be the height (in feet) of a specific child. Height is a quantity that varies: As time passes, the child's height will increase. So, h is a variable. When we say $h = 4$, we mean the child's height is 4 feet.

We will discuss other roles of a variable in Sections 2.1 and 4.3.

"The Ramsey account? Should be done any minute..."

Example 1   Using a Variable to Represent a Quantity

1. Let s be a car's speed (in miles per hour). What is the meaning of $s = 60$?
2. Let n be the number of employees (in millions) who work from home at least half the time. For the year 2015, $n = 3.7$ (Source: *Global Workplace Analytics*). What does that mean in this situation?
3. Let t be the number of years since 2015. What is the meaning of $t = 4$?

Solution

1. The speed of the car is 60 miles per hour.
2. In 2015, 3.7 million people worked from home at least half the time.
3. $2015 + 4 = 2019$; so, $t = 4$ represents the year 2019.

There are many benefits to using variables. For example, in Problem 2 of Example 1, we found that the simple equation "$n = 3.7$" means the same thing as the wordy sentence "3.7 million people worked from home at least half the time." Variables can help us describe some situations with a small amount of writing.

In Problem 3 of Example 1, we described the year 2019 by using $t = 4$. So, our definition of t allows us to use smaller numbers to describe various years—an approach that will be helpful throughout the course.

We will see other benefits of variables as we proceed through the course.

Example 2   Using a Variable to Represent a Quantity

Choose a symbol to represent the given quantity. Explain why the symbol is a variable. Give two numbers that the variable can represent and two numbers that it cannot represent.

1. The weight (in pounds) of a baby at birth
2. The number of people who live in a two-bedroom house

Solution

1. Let w be the weight (in pounds) of a baby at birth. The weight of a baby at birth can vary, so w is a variable. For example, w can represent the numbers 6 and 8 because babies can weigh 6 or 8 pounds at birth. The variable w does not represent 0 or 300 because babies cannot weigh 0 or 300 pounds at birth!
2. Let n be the number of people who live in a two-bedroom house. The number of people who live in a two-bedroom house can vary, so n is a variable. For example, n can represent the numbers 2 and 3 because 2 or 3 people can live in a two-bedroom house. The variable n cannot represent the numbers 5000 or $\frac{1}{2}$ because 5000 people cannot live in a two-bedroom house and half of a person doesn't make sense.

In Problem 1 of Example 2, we stated that the units of w are pounds. Without stating the units of w, "$w = 10$" could mean the baby's weight was 10 ounces, 10 pounds, or 10 tons! In defining a variable, it is important to describe the variable's units.

Constants

A variable is a symbol that represents a quantity that can vary. When we use a symbol to represent a quantity that does *not* vary, we call that symbol a *constant*. So, 2, 0, 4.8, and π are constants. The constant π is approximately equal to 3.14.

Definition   Constant

A **constant** is a symbol that represents a specific number (a quantity that does *not* vary).

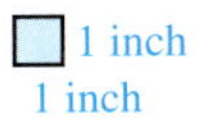

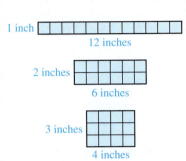

Figure 1 One square inch

In Example 3, we will compare the meanings of a variable and a constant while we consider the widths, lengths, and areas of some rectangles. The **area** (in square inches) of a flat surface is the number of square inches that it takes to cover the surface (see Fig. 1). The area of a rectangle is equal to the rectangle's length times its width.

Example 3   Comparing Constants and Variables

A rectangle has an area of 12 square inches. Let W be the width (in inches), L be the length (in inches), and A be the area (in square inches).

1. Sketch three possible rectangles of area 12 square inches.
2. Which of the symbols W, L, and A are variables? Explain.
3. Which of the symbols W, L, and A are constants? Explain.

Solution

1. We sketch three rectangles for which the width times the length is equal to 12 square inches (see Fig. 2).
2. The symbols W and L are variables because they represent quantities that vary.
3. The symbol A is a constant because in this problem the area does not vary—the area is always 12 square inches.

Figure 2 Three possible rectangles of area 12 square inches

Counting Numbers

When we describe people, it often helps to describe them in terms of certain categories, such as gender, ethnicity, and employment. In mathematics, it helps to describe numbers in terms of categories, too. We begin by describing the *counting numbers*, which are the numbers 1, 2, 3, 4, 5, and so on.

> **Definition**   Counting numbers (natural numbers)
>
> The **counting numbers**, or **natural numbers**, are the numbers
> $$1, 2, 3, 4, 5, \ldots$$

The three dots mean that the pattern of the numbers shown continues without ending. In this case, the pattern continues with 6, 7, 8, and so on. When a list of numbers goes on forever, we say that there are an *infinite* number of numbers.

Integers

Next, we describe the *integers*, which include the counting numbers and other numbers.

> **Definition**   Integers
>
> The **integers** are the numbers
> $$\ldots, -3, -2, -1, 0, 1, 2, 3, \ldots$$

The three dots on both sides mean that the pattern of the numbers shown continues without ending in both directions. In this case, the pattern continues with $-4, -5, -6$, and so on, and with 4, 5, 6, and so on.

If you write a check for more money than is in your checking account, you will have a negative balance. The balance -60 dollars is an integer.

The **positive integers** are the numbers $1, 2, 3, \ldots$. The **negative integers** are the numbers $-1, -2, -3, \ldots$. The integer 0 is neither positive nor negative. So, the integers consist of the counting numbers (which are positive integers), the negative integers, and 0.

The Number Line

We can visualize numbers on a *number line* (see Fig. 3).

Each point (location) on the number line represents a number. The numbers increase from left to right. We refer to the distance between two consecutive integers on the number line as 1 *unit* (see Fig. 3).

Figure 3 The number line

Example 4   Graphing Integers on a Number Line

Draw dots on a number line to represent the integers between -2 and 3, inclusive.

Solution

The integers between -2 and 3, inclusive, are $-2, -1, 0, 1, 2,$ and 3. "Inclusive" means to include the first and last numbers, which in this case are -2 and 3. We sketch a number line and draw dots at the appropriate locations for the numbers $-2, -1, 0, 1, 2,$ and 3 (see Fig. 4).

Figure 4 Graphing the numbers $-2, -1,$ 0, 1, 2, and 3

When we draw dots on a number line, we say that we are "plotting points" or "graphing numbers."

In Example 4, we worked with the integers between -2 and 3, inclusive: $-2, -1, 0, 1, 2,$ and 3. Here are the integers between -2 and 3: $-1, 0, 1,$ and 2. We did not include -2 or 3 because the word "inclusive" was not used. When working with such problems, it is important to check whether the word "inclusive" is used.

WARNING

Rational Numbers

For a fraction $\dfrac{n}{d}$, we call n the **numerator** and d the **denominator**. The dash between the numerator and the denominator is the **fraction bar**:

$$\text{Numerator} \longrightarrow \frac{n}{d} \longleftarrow \text{Fraction bar}$$
$$\text{Denominator} \longrightarrow$$

A fraction can be used to describe a part of a whole. For example, consider the meaning of $\dfrac{5}{8}$ of a pizza. If we divide the pizza into 8 slices of equal area, 5 of the slices make up $\dfrac{5}{8}$ of the pizza (see Fig. 5).

The number $\dfrac{5}{8}$ is called a *rational number*.

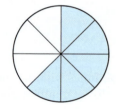

Figure 5 $\dfrac{5}{8}$ of a pizza

Definition   Rational numbers

The **rational numbers** are the numbers that can be written in the form $\dfrac{n}{d}$, where n and d are integers and d is nonzero.

We specify that d is nonzero because, as we shall see later, division by zero does not make sense.

Here are some examples of rational numbers:

$$\frac{3}{7} \qquad \frac{-2}{5} \qquad 4 = \frac{4}{1}$$

Rational numbers include all the integers because any integer n can be written as $\dfrac{n}{1}$.

Irrational Numbers

There are numbers represented on the number line that are *not* rational. These numbers are called **irrational numbers**. An irrational number *cannot* be written in the form $\dfrac{n}{d}$, where n and d are integers and d is nonzero. The number $\sqrt{2}$ is the number greater than zero that we multiply by itself to get 2. The number $\sqrt{2}$ is an irrational number. Here are some more examples of irrational numbers:

$$\pi \qquad \sqrt{3} \qquad \sqrt{5}$$

We know that $\sqrt{9} = 3 = \dfrac{3}{1}$ because $3 \times 3 = 9$. So, $\sqrt{9}$ is rational (not irrational).

Decimals

The list price of an Xbox One console is \$299.99, which is a decimal number.

Any rational number or irrational number can be written as a decimal number.

A rational number can be written as a decimal number that either terminates or repeats:

$$\dfrac{3}{4} = \underbrace{0.75}_{\text{terminates}} \qquad \dfrac{3}{11} = \underbrace{0.27272727\ldots}_{\text{repeats}}$$

We can use an overbar to write the repeating decimal $0.272727\ldots = 0.\overline{27}$.

An irrational number can be written as a decimal number that neither terminates nor repeats. It is impossible to write all the digits of an irrational number, but we can approximate the number by rounding. For example, earlier we *approximated* π by rounding to the second decimal place: $\pi \approx 3.14$.

Real Numbers

Recall that each point on the number line represents a number. We call all the numbers represented by all the points on the number line the *real numbers*.

Definition   Real numbers

The **real numbers** are all the numbers represented on the number line.

The real numbers are made up of the rational numbers and the irrational numbers. Here are some real numbers:

$$-1.8 \qquad -1 \qquad -\dfrac{7}{10} \qquad 0 \qquad 0.4 \qquad \dfrac{6}{5} \qquad \pi$$

We graph these real numbers in Fig. 6.

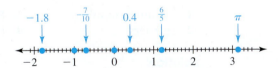

Figure 6 Graphing the real numbers $-1.8,\ -1,\ -\dfrac{7}{10},\ 0,\ 0.4,\ \dfrac{6}{5},$ and π

We use an arrow in labeling each point that does not fall on a labeled tick mark.

Example 5   Graphing Real Numbers on a Number Line

Graph the number on a number line.

1. $-\dfrac{7}{4}$ **2.** 2.3

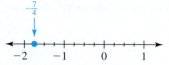

Figure 7 Graphing the number $-\frac{7}{4}$

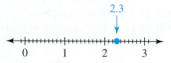

Figure 8 Graphing the number 2.3

Solution

1. We draw a number line so that the distance between tick marks is $\frac{1}{4}$ unit (see Fig. 7). To graph $-\frac{7}{4}$, we draw a dot at the seventh tick mark to the left of 0.

2. We draw a number line so that the distance between tick marks is $0.1 = \frac{1}{10}$ unit (see Fig. 8). To graph 2.3, we draw a dot at the third tick mark to the right of 2.

Figure 9 illustrates how the various types of numbers we have discussed so far are related. In particular, it shows that every counting number is an integer, every integer is a rational number, and every rational number is a real number. It also shows that irrational numbers are the real numbers that are not rational.

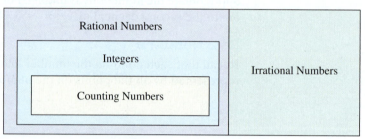

All numbers shown here are real numbers.

Figure 9 The real numbers

Example 6   Identifying Types of Numbers

Consider the following numbers:

$$-8 \qquad -2.56 \qquad 0 \qquad \frac{3}{5} \qquad 2 \qquad \sqrt{13} \qquad 98$$

Which of these numbers are the given type of number?

1. Counting numbers   3. Rational numbers   5. Real numbers
2. Integers   4. Irrational numbers

Solution

1. The counting numbers are 2 and 98.
2. The integers are $-8, 0, 2,$ and 98.
3. The rational numbers are the numbers $-8, -2.56, 0, \frac{3}{5}, 2,$ and 98.
4. The irrational number is $\sqrt{13}$.
5. The real numbers are all seven of the numbers.

Graphing Data

Data are values of quantities that describe authentic situations. For example, the following heights of six people, all in inches, are data: 64, 71, 75, 68, 71, and 69. We often can get a better sense of data by graphing than by just looking at the data values.

Example 7   Graphing Data

Over the course of a semester, a student took five quizzes. Here are the points he earned on the quizzes, in chronological order: 0, 4, 7, 9, 10. Let q be the number of points earned by the student on a quiz.

1. Graph the student's scores on a number line.
2. Did the quiz scores increase, decrease, stay approximately constant, or none of these?
3. Did the *increases* between the successive quiz scores increase, decrease, stay approximately constant, or none of these?

Solution

1. We sketch a number line and write "q" to the right of the number line and the units "Points" underneath the number line (see Fig. 10). Then we graph the numbers 0, 4, 7, 9, and 10.

Figure 10 Graphing the quiz scores

2. From the opening paragraph, we know that the quiz scores increased. (From the graph alone, we cannot tell that the quiz scores increased because the order in which the quizzes were taken is not indicated.)
3. As we look from left to right at the points plotted on the graph, we see that the distance between adjacent points decreases. This means that the increases between successive quiz scores decreased. That is, the jump from 0 to 4 is greater than the jump from 4 to 7, and so on.

Average

It is often helpful to use a single number to represent a group of numbers. One such number is the *average* (or *mean*).

> **Definition   Average, mean**
>
> To find the **average** (or **mean**) of a group of numbers, we divide the sum of the numbers by the number of numbers in the group.

For example, to find the average of the quiz scores included in Example 7, we first add the scores: $0 + 4 + 7 + 9 + 10 = 30$. We then divide the total, 30, by the number of quiz scores, 5:

$$30 \div 5 = 6 \text{ points}$$

So, the average quiz score is 6 points.

In general, the average of a group of numbers estimates the center of the numbers graphed on a number line. Figure 11 illustrates this concept for the quiz data.

Figure 11 The average quiz score, 6 points, estimates the center of the graphed quiz data

WARNING   Always include the units of an average. For example, we say that the average quiz score is *6 points*, not 6.

Averages can be especially helpful in comparing two data sets. For instance, a student whose average quiz score is 9 points would have performed much better, in general, than a student whose average quiz score is 3 points.

Example 8   Graphing Data and the Mean

The numbers (in thousands) of charter schools in the United States for the years 2009, 2010, 2011, 2012, 2013, and 2014 are 4.7, 5.0, 5.3, 5.7, 6.1, and 6.5, respectively (Source: *National Center for Education Statistics*). Let n be the number (in thousands) of charter schools in a given year.

1. Graph the data. Find the average of the data values, and indicate it on the graph.
2. Did the number of charter schools increase, decrease, stay approximately constant, or none of these from 2009 to 2014, inclusive? Explain.
3. Did the *increases* in the number of charter schools increase, decrease, stay approximately constant, or none of these from 2009 to 2014, inclusive? Explain.

Solution

1. We sketch a number line and write "n" to the right of the number line and the units "Thousands of charter schools" underneath the number line (see Fig. 12). Because the data values are between 4.7 and 6.5, inclusive, we write the numbers 4.5, 5.0, 5.5, 6.0, and 6.5, equally spaced on the number line. Then we graph the numbers 4.7, 5.0, 5.3, 5.7, 6.1, and 6.5.

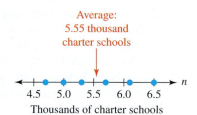

Figure 12 Graphing the data

To find the average, we divide the sum of the data values, 4.7 + 5.0 + 5.3 + 5.7 + 6.1 + 6.5 = 33.3, by 6: 33.3 ÷ 6 = 5.55 thousand charter schools. We indicate the average, 5.55 thousand charter schools, in Fig. 12.

2. From the opening paragraph, we know that the number of charter schools is increasing. (From the graph alone, we cannot tell that the number of charter schools is increasing because the years are not indicated.)
3. By reading the opening paragraph of this example and viewing the almost equal spacing between dots on the graph, we see that the increases in the number of charter schools were approximately constant.

In Fig. 12, we wrote the numbers 4.5, 5.0, 5.5, 6.0, and 6.5 on the number line. **When we write numbers on a number line, they should increase by a fixed amount and be equally spaced.**

There are two limitations with graphing the data in Example 8 on a *single* number line. One limitation is that the years are not indicated. The other limitation is that it is not clear what to do when the number of charter schools for two of the years is the same. We will address both limitations in Section 1.2.

Positive and Negative Numbers

The **negative numbers** are the real numbers less than 0, and the **positive numbers** are the real numbers greater than 0 (see Fig. 13).

Figure 13 The location of the negative numbers and the positive numbers on the number line

Some examples of negative numbers are -13, -5.2, $-\dfrac{3}{4}$, and $-\sqrt{2}$. Some examples of positive numbers are 13, 5.2, $\dfrac{3}{4}$, and π. As we discussed earlier, the number 0 is neither positive nor negative.

We say that the *sign* of a negative number is negative and that the *sign* of a positive number is positive. To include zero, we define the **nonnegative numbers** as the positive numbers together with 0. Likewise, we define the **nonpositive numbers** as the negative numbers together with 0.

Example 9   Graphing a Negative Quantity

A person bounces several checks and, as a result, is charged service fees. If b is the balance (in dollars) of the checking account, what value of b means the person owes $50? Graph the number on a number line.

Solution

Because the person *owes* money, the value of b is negative: $b = -50$. We graph -50 on a number line in Fig. 14.

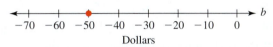

Dollars

Figure 14 Graphing the number $b = -50$

Describing a Concept or Procedure

In some homework exercises, you will be asked to describe, in general, a concept or procedure.

Guidelines on Writing a Good Response

- Create an example that illustrates the concept or outlines the procedure. Looking at examples or exercises may jump-start you into creating your own example.
- Using complete sentences and correct terminology, describe the key ideas or steps of your example. You can review the text for ideas, but write your description in your own words.
- Describe also the concept or the procedure in general without referring to your example. It may help to reflect on several examples and what they all have in common.
- In some cases, it will be helpful to point out the similarities and the differences between the concept or the procedure and other concepts or procedures.
- Describe the benefits of knowing the concept or the procedure.
- If you have described the steps in a procedure, explain why it is permissible to follow these steps.
- Clarify any common misunderstandings about the concept, or discuss how to avoid making common mistakes when following the procedure.

Example 10   Responding to a General Question about a Concept

Describe the meaning of "variable."

Solution

Let t be the number of hours that a person works in a department store. The symbol t is an example of a variable because the value of t can vary. In general, a variable is a symbol that stands for an amount that can vary. A symbol that stands for an amount that does *not* vary is called a constant.

There are many benefits to using variables. We can use a variable to concisely describe a quantity; using the earlier definition of t, we see that the equation $t = 8$ means a person works in a department store for 8 hours. By using a variable, we can also use smaller numbers to describe various years.

In defining a variable, it is important to describe its units.

▶

◈ Group Exploration

Reasonable values of a variable

1. Let u be the number of units (credits or hours) a student is currently taking at your college. You may replace the word "units" with "credits" or "hours" if appropriate.
   a. Which of the following values of u are reasonable in this situation? Explain.

      i. $u = 15$ iv. $u = 15.5$
      ii. $u = -5$ v. $u = 15.1$
      iii. $u = 200$ vi. $u = 0$

   b. Describe all the real numbers that are reasonable values of u. Use a number line, a list of numbers, words, or some other way to describe these numbers.

2. A few months ago, a person bought a Porsche 911 Carrera Turbo for $160,700. It has a 17.7-gallon fuel tank.

Let g be the amount of gasoline (in gallons) that is in the tank.
   a. Which of the following values of g are reasonable in this situation? Explain.

      i. $g = 7$ iv. $g = 17.7$
      ii. $g = 19$ v. $g = 0$
      iii. $g = -4$ vi. $g = 10.392$

   b. Describe all the real numbers that are reasonable values of g.

3. The legal capacity of a club is 180 people. Let n be the number of people who are at the club. You may assume that the number of people in the club never exceeds the legal limit. Describe all the reasonable values of n.

Tips for Success   Take Notes

It is always a good idea to take notes during classroom activities. Not only will you have something to refer to later when doing the homework, but also you will have something to help you prepare for tests. In addition, taking notes makes you become even more involved with the material, which means you will probably increase both your understanding and your retention of it.

◈ Homework 1.1

For extra help ▶ **MyLab Math**   ▦ Watch the videos in MyLab Math

Respond to the questions in Exercises 1–12 by using complete sentences.

1. Let n be the number (in thousands) of fans who attend a Kendrick Lamar concert. What does $n = 25$ mean in this situation?

2. Let n be the number of home runs hit by Chris Davis in a season. For 2015, the value of n is 47 (Source: *Major League Baseball*). What does $n = 47$ mean in this situation?

3. Let n be the number (in millions) of Americans who have a smartphone. For 2016, the value of n is 199 (Source: *comScore*). What does $n = 199$ mean in this situation?

4. Let p be the percentage of children ages 5–18 who participate in organized physical activity. The value of p is about 60 for 2017 (Source: *Statistical Brain Research Institute*). What does $p = 60$ mean in this situation?

5. Let s be the annual iPad® sales (in millions). The value of s is 54.9 for 2015 (Source: *Apple*). What does $s = 54.9$ mean in this situation?

6. Let p be the percentage of American workers who are in a union. For 2015, the value of p is 11.1 (Source: *Bureau of Labor Statistics*). What does $p = 11.1$ mean in this situation?

7. Let p be BP's annual profit (in billions of dollars). For 2015, the value of p is -6.5 (Source: *Reuters*). What does $p = -6.5$ mean in this situation?

8. Let T be the temperature (in degrees Fahrenheit). What does $T = -10$ mean in this situation?

9. Let t be the number of years since 2010. What does $t = 9$ mean in this situation?

10. Let t be the number of years since 2005. What does $t = 13$ mean in this situation?

11. Let t be the number of years since 2005. What does $t = -3$ mean in this situation?

12. Let t be the number of years since 2010. What does $t = -2$ mean in this situation?

For Exercises 13–20, choose a variable name for the given quantity. Give two numbers that the variable can represent and two numbers that it cannot represent.

13. The height (in inches) of a person

14. The amount of time (in hours) that a student prepares for an exam

15. The price (in dollars) of a video game

16. The number of students enrolled in an algebra class

17. The total time (in hours) a person works in a week

18. The temperature (in degrees Fahrenheit) in an oven

19. The annual salary (in thousands of dollars) of a person

20. The value (in thousands of dollars) of a new home

21. A rectangle has an area of 24 square inches. Let W be the width (in inches), L be the length (in inches), and A be the area (in square inches).
    a. Sketch three possible rectangles of area 24 square inches.
    b. Which of the symbols W, L, and A are variables? Explain.
    c. Which of the symbols W, L, and A are constants? Explain.

22. A rectangle has an area of 36 square feet. Let W be the width (in feet), L be the length (in feet), and A be the area (in square feet).
    a. Sketch three possible rectangles of area 36 square feet.
    b. Which of the symbols W, L, and A are variables? Explain.
    c. Which of the symbols W, L, and A are constants? Explain.

23. The *perimeter* of a rectangle is the sum of the lengths of all four sides. A rectangle has a perimeter of 20 feet. Let W be the width, L be the length, and P be the perimeter, all with units in feet.
    a. Sketch three possible rectangles of perimeter 20 feet.
    b. Which of the symbols W, L, and P are variables? Explain.
    c. Which of the symbols W, L, and P are constants? Explain.

24. The *perimeter* of a rectangle is the sum of the lengths of all four sides. A rectangle has a perimeter of 16 inches. Let W be the width, L be the length, and P be the perimeter, all with units in inches.
    a. Sketch three possible rectangles of perimeter 16 inches.
    b. Which of the symbols W, L, and P are variables? Explain.
    c. Which of the symbols W, L, and P are constants? Explain.

25. The length of a rectangle is 3 inches more than the width. Let W be the width (in inches), L be the length (in inches), and A be the area (in square inches).
    a. Sketch three possible rectangles of length 3 inches more than the width.
    b. Which of the symbols W, L, and A are variables? Explain.
    c. Which of the symbols W, L, and A are constants? Explain.

26. The length of a rectangle is twice the width. Let W be the width, L be the length, and P be the perimeter, all with units in inches. [**Hint:** *Twice* means to multiply by 2.]
    a. Sketch three possible rectangles in which the length is twice the width.
    b. Which of the symbols $W, L,$ and P are variables? Explain.
    c. Which of the symbols W, L, and P are constants? Explain.

27. The width of a rectangle is 2 yards. Let W be the width, L be the length, and P be the perimeter, all with units in yards.
    a. Sketch three possible rectangles of width 2 yards.
    b. Which of the symbols W, L, and P are variables? Explain.
    c. Which of the symbols W, L, and P are constants? Explain.

28. The width of a rectangle is 5 centimeters. Let W be the width (in centimeters), L be the length (in centimeters), and A be the area (in square centimeters).
    a. Sketch three possible rectangles of width 5 centimeters.
    b. Which of the symbols W, L, and A are variables? Explain.
    c. Which of the symbols W, L, and A are constants? Explain.

Graph all the given numbers on one number line.

29. $5, -2, 0, -3, 4, -1$

30. $-4, 1, -6, 2, 7, -3$

31. $-\dfrac{2}{3}, -1, \dfrac{7}{3}, 1, -\dfrac{5}{3}, 2$

32. $\dfrac{1}{4}, 0, -2, -\dfrac{5}{4}, \dfrac{9}{4}, 1$

33. $2.8, 3.6, 0.4, 2.5, 1.1, 1.8$

34. $3, 1.5, 2.3, 0.9, 2.7, 3.4$

35. $-2, 3.1, 1.2, -1.8, 0.5, 1$

36. $1, 0.2, -2.4, -0.7, 1.9, -1$

Graph the numbers on a number line.

37. Counting numbers between 3 and 8

38. Counting numbers between 1 and 5

39. Integers between -2 and 2, inclusive

40. Integers between -4 and 4, inclusive

41. Integers between -1 and 4, inclusive

42. Integers between -6 and 3, inclusive

43. Negative integers between -4 and 4

44. Positive integers between -4 and 4

For Exercises 45–50, consider the following numbers:

$$-9.7 \qquad -4 \qquad 0 \qquad \frac{3}{5} \qquad \sqrt{7} \qquad 3 \qquad \pi \qquad 356$$

Which of these numbers are the given type of number?

45. Counting numbers

46. Integers

47. Negative integers

48. Rational numbers

49. Irrational numbers

50. Real numbers

Give three examples of the following types of numbers.

51. Negative integers

52. Positive integers

53. Negative integers less than -7

54. Negative integers greater than -5

55. Integers that are not counting numbers

56. Rational numbers that are not integers

57. Rational numbers between 1 and 2

58. Irrational numbers between 1 and 10

59. Real numbers between -3 and -2

60. Real numbers that are not rational numbers

Use points on a number line to describe the given values of a variable. Find the average of the values, and indicate it on the number line.

61. A student goes to a college for six semesters. Here are the numbers of units (credits or hours) taken per semester: 10, 12, 6, 9, 15, 14. Let u be the number of units taken in one semester.

62. During the summer, a student visits a music website five times. Here are the numbers of songs downloaded: 2, 0, 1, 5, 4. Let n be the number of songs downloaded in one visit to the website.

63. The percentages of disposable personal annual income spent on food for various years are 16%, 14%, 13%, 11%, and 10%. Let p be the percentage of disposable personal annual income spent on food.

64. The percentages of airline flights that are on time for various years are 79%, 82%, 75%, 77%, and 76%. Let p be the percentage of flights in a year that are on time.

For Exercises 65–68, use points on a number line to describe the given values of a variable.

65. The average annual lost time (in hours) due to traffic congestion on highways for various years is 22, 30, 24, 27, and 32. Let L be the average annual lost time (in hours) due to traffic congestion.

66. The U.S. annual per-person consumption of sports drinks (in gallons) for various years is 1.9, 2.5, 2.1, 2.3, and 2.2. Let c be the per-person consumption (in gallons per year) of sports drinks in a year.

67. The low temperatures (in degrees Fahrenheit) for three days in December in Chicago are 5°F above zero, 4°F below zero, and 6°F below zero. Let F be the low temperature (in degrees Fahrenheit) for one day.

68. Here are a company's annual profits and losses for various years: loss of $5 million, profit of $3 million, and loss of $8 million. Let p be the company's annual profit (in millions of dollars).

69. The revenue (in billions of dollars) of Apple in the years 2011, 2012, 2013, and 2014 is 109, 156, 171, and 183, respectively (Source: *Apple*). Let r be the annual revenue (in billions of dollars) of Apple.
   a. Use points on a number line to describe the given values of r. Find the average of the values, and indicate it on the number line.
   b. Did the annual revenue increase, decrease, stay approximately constant, or none of these from 2011 to 2014, inclusive?
   c. Did the *increases* in the annual revenue increase, decrease, stay approximately constant, or none of these from 2011 to 2014, inclusive? Explain.

70. Sales (in millions) of cars in the years 2011, 2012, 2013, and 2014 are 6.1, 7.2, 7.6, and 7.7, respectively (Source: *WardsAuto*). Let s be the annual car sales (in millons).
   a. Use points on a number line to describe the given values of s. Find the average of the values, and indicate it on the number line.
   b. Did annual car sales increase, decrease, stay approximately constant, or none of these from 2011 to 2014, inclusive?
   c. Did the *increases* in annual car sales increase, decrease, stay approximately constant, or none of these from 2011 to 2014, inclusive? Explain.

71. The number (in thousands) of craft beer breweries (brewpubs, microbreweries, and regional craft breweries) in the years 2009, 2010, 2011, 2012, 2013, and 2014 is 1.6, 1.7, 2.0, 2.4, 2.9, and 3.7, respectively (Source: *Brewers Association*). Let n be the number (in thousands) of craft beer breweries.
   a. Use points on a number line to describe the given values of n. Find the average of the values, and indicate it on the number line. Round to the second decimal place.
   b. Did the number of craft beer breweries increase, decrease, stay approximately constant, or none of these from 2009 to 2014, inclusive?
   c. Did the *increases* in the number of craft beer breweries increase, decrease, stay approximately constant, or none of these from 2009 to 2014, inclusive? Explain.

72. The number of cities worldwide where Uber operates in the years 2010, 2011, 2012, 2013, and 2014 is 1, 7, 19, 67, and 276, respectively (Source: *Uber*). Let n be the number of cities where Uber operates.
   a. Use points on a number line to describe the given values of n. Find the average of the values, and indicate it on the number line.
   b. Did the number of cities where Uber operates increase, decrease, stay approximately constant, or none of these from 2010 to 2014, inclusive?
   c. Did the *increases* in the number of cities where Uber operates increase, decrease, stay approximately constant, or none of these from 2010 to 2014, inclusive? Explain.

Concepts

73. Let T be the temperature in degrees Fahrenheit.
   a. What value of T represents the temperature that is 5°F below zero?
   b. A student says that T represents only positive numbers and zero because there is no negative sign. Is the student correct? Explain.

74. A student says the integers between 2 and 5 are the numbers 2, 3, 4, and 5. Is the student correct? Explain.

75. a. Find the average of each pair of numbers. Then plot the two given numbers and their average on a number line.
  i. 7, 9   **ii.** 1, 5   **iii.** 2, 8
  b. What patterns do you notice in your work in part (a)?
  c. How many real numbers are there between 0 and 1? Explain. [**Hint:** Use the concept of average to help you show that you can keep finding more numbers.]

76. We can describe how far apart two numbers are on the number line. For example, the numbers 3 and 7 are 4 units apart. How far apart are two consecutive integers on the number line? How far apart are two consecutive even integers? How far apart are two consecutive odd integers?

77. How is a variable different from a constant?

78. The average of a student's four test scores is 90 points. If the student takes another test and the average of all five tests is 90 points, what is the student's score on the fifth test? Explain.

79. List the various types of numbers discussed in this section and describe the meanings of each type. (See page 9 for guidelines on writing a good response.)

80. Describe how to graph a negative quantity. (See page 9 for guidelines on writing a good response.)

1.2 Scatterplots

Objectives

» Describe the meaning of *ordered pair, coordinate,* and *coordinate system.*

» Construct scatterplots.

» Describe the meaning of *explanatory variable* and *response variable.*

» Read bar graphs.

» Plot points on a coordinate system.

In Section 1.1, we used a single number line to describe values of a quantity. In this section, we will use a pair of number lines to describe two quantities that are related. For instance, in Example 3, we will compare the average price of a Super Bowl ticket with the Super Bowl number.

The Coordinate System

A number line is convenient for displaying values of a variable. However, there are limitations to using a *single* number line. For example, suppose a student earns the following points (listed in chronological order) from taking five quizzes: 7, 6, 9, 8, 9. We let q stand for the number of points earned by the student on a quiz. We use a number line to graph the scores in Fig. 15.

Figure 15 Graphing the scores

There are two limitations with using one number line to graph these scores. First, the graph does not show that there were two scores of 9 points each. Second, the graph does not show which was the first score, the second score, and so on.

There is a way to address both limitations. To begin, we let n be the quiz number. For the first quiz, $n = 1$. For the second quiz, $n = 2$, and so on. Next, we organize the values of the variables n and q (the quiz score) in Table 2.

The "1" and "7" in the first row of Table 2 indicate that when $n = 1$, $q = 7$. This means the student's score on the first quiz was 7 points. If we agree to write the quiz number first and the quiz score second, we can use the ordered pair $(1, 7)$ to mean that when $n = 1$, $q = 7$. An **ordered pair** is a pair of numbers (written in parentheses and separated by a comma) for which the order of the numbers is meaningful. We call each of the numbers in an ordered pair a **coordinate**. For $(1, 7)$ in this situation, we call 1 the *n-coordinate* and 7 the *q-coordinate*.

The ordered pair $(2, 6)$ indicates that when $n = 2$, $q = 6$. This means the student's score on the second quiz was 6 points, which agrees with the second row of Table 2.

We call pairs of numbers such as $(3, 9)$ ordered pairs because the order in which the numbers appear matters: The ordered pair $(3, 9)$ means the student's score on the third quiz was 9 points, whereas the ordered pair $(9, 3)$ means the student's score on the ninth quiz was 3 points.

Table 2 Values of n and q

n	q
1	7
2	6
3	9
4	8
5	9

We graph the ordered pairs by using *two* number lines, which are called **axes** (singular: **axis**). To start, we draw a horizontal number line called the *n*-axis and a vertical number line called the *q*-axis (see Fig. 16). We refer to such a pair of axes as a **coordinate system**. The **origin** is the intersection point of the axes. The axes divide the coordinate system into four regions called **quadrants**, which we call Quadrants I, II, III, and IV. The quadrants do not include the axes.

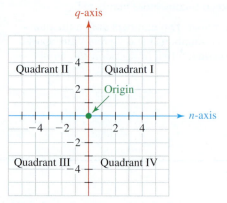

Figure 16 Coordinate system

Constructing Scatterplots

Next, we plot the ordered pair $(3, 9)$ shown in the third row of Table 2. To do so, we start at the origin, look 3 units to the right and 9 units up, and then draw a dot (see Fig. 17). In Fig. 18, we plot all the ordered pairs listed in Table 2.

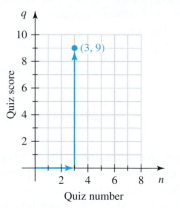

Figure 17 Plot (3, 9)

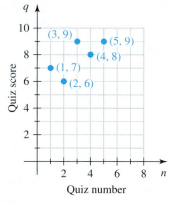

Figure 18 Plot the ordered pairs from Table 2

Note that we have addressed both of the limitations of using just one number line. The coordinate system in Fig. 18 shows that there are two scores of 9, and it also shows which score is from the first quiz, the second quiz, and so on.

As we look at the plotted points in Fig. 18 from left to right, the points, in general, go upward. This means the quiz scores, in general, are increasing.

A graph of plotted ordered pairs, such as the graph in Fig. 18, is called a **scatterplot**.

Explanatory and Response Variables

The quiz number affects the student's score. For example, if a certain quiz is more difficult than the others, the student may earn a lower score on that quiz than on the other quizzes. Because *n* affects (explains) *q*, we call *n* the *explanatory variable*. We call *q* the *response variable* because *q* is affected by (responds to) *n*.

Definition Explanatory and response variables

Assume that an authentic situation can be described by using the variables t and p, and assume that t affects (explains) p. Then

- We call t the **explanatory variable** (or **independent variable**).
- We call p the **response variable** (or **dependent variable**).

Example 1 Identifying Explanatory and Response Variables

For each situation, identify the explanatory variable and the response variable.

1. Let L be the loudness (in decibels) of the sound produced by a foghorn that is d miles away from you.
2. A car is traveling at speed s (in miles per hour) on a dry asphalt road, and the brakes are suddenly applied. Let d be the stopping distance (in feet).

Solution

1. The farther you are from the foghorn, the softer the sound you will hear. So, the distance d affects (explains) the loudness L. Thus, d is the explanatory variable and L is the response variable. (Notice that loudness does not affect the distance.)
2. The greater the traveling speed, the greater the stopping distance will be. So, the traveling speed s affects (explains) the stopping distance d. Thus, s is the explanatory variable and d is the response variable. (Notice that the stopping distance does not affect the traveling speed.)

For an ordered pair (a, b), we write the value of the explanatory variable in the first (left) position and the value of the response variable in the second (right) position. Note that we did just that for the quiz score application. That is, we listed values of the explanatory variable n in the first position and values of the response variable q in the second position. For example, the ordered pair $(4, 8)$ means that when $n = 4$, $q = 8$. In other words, the student's score on the fourth quiz is 8 points.

Example 2 Determining the Meaning of an Ordered Pair

1. Let n be the total number (in thousands) of Subway® restaurants at t years since 2010. What does the ordered pair $(5, 43.9)$ mean in this situation (Source: *Subway*)?
2. Let p be a runner's pulse rate (in beats per minute) when his speed is s miles per hour. What does the ordered pair $(10, 160)$ mean in this situation?

Solution

1. The year affects (explains) the total number of restaurants. So t is the explanatory variable and n is the response variable. The ordered pair $(5, 43.9)$ means that $t = 5$ and $n = 43.9$. There were 43.9 thousand restaurants in $2010 + 5 = 2015$.
2. The runner's speed affects (explains) his pulse rate. So s is the explanatory variable and p is the response variable. The ordered pair $(10, 160)$ means that $s = 10$ and $p = 160$. When the runner's speed is 10 miles per hour, his pulse rate is 160 beats per minute.

Table 3 Values of n and q

n	q
1	7
2	6
3	9
4	8
5	9

For tables of ordered pairs, we list the values of the explanatory variable in the first (left) column and the values of the response variable in the second (right) column. For example, in Table 3 the values of n are in the first column and the values of q are in the second column.

For coordinate systems, we describe the values of the explanatory variable with the horizontal axis and the values of the response variable with the vertical axis. For example, in Fig. 18, the horizontal axis is the n-axis and the vertical axis is the q-axis.

Columns of Tables and Axes of Coordinate Systems

Assume that an authentic situation can be described by using two variables. Then

- For tables, the values of the explanatory variable are listed in the first column and the values of the response variable are listed in the second column (see Table 4).
- For coordinate systems, the values of the explanatory variable are described by the horizontal axis and the values of the response variable are described by the vertical axis (see Fig. 19).

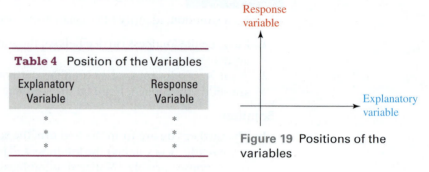

Table 4 Position of the Variables

Explanatory Variable	Response Variable
*	*
*	*
*	*

Figure 19 Positions of the variables

More about Scatterplots

Table 5 Average Ticket Prices for the Super Bowl

Super Bowl Number	Average Ticket Price (dollars)
1 (I)	12
5 (V)	15
10 (X)	20
15 (XV)	40
20 (XX)	75
25 (XXV)	150
30 (XXX)	250
35 (XXXV)	325
40 (XL)	550
45 (XLV)	875
50 (L)	1750

Source: *NFL.com*

Example 3   Constructing a Scatterplot

The average ticket prices for the Super Bowl are shown in Table 5 for various years. Let p be the average ticket price (in dollars) and n be the Super Bowl number.

1. Draw a scatterplot of the data.
2. For the Super Bowls described in Table 5, which Super Bowl had the highest average ticket price? What was that price?
3. Describe any patterns you see in the prices.

Solution

1. A scatterplot of the data is shown in Fig. 20. It makes sense to think of n as the explanatory variable because the Super Bowl number affects (explains) the average ticket price (and not the other way around). So, we let the horizontal axis be the

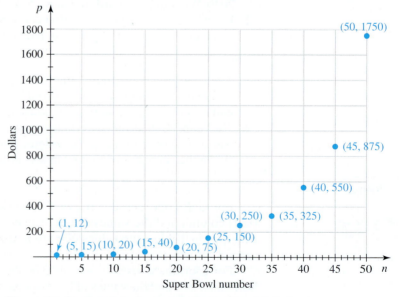

Figure 20 Scatterplot of average Super Bowl ticket prices

n-axis. Note that we write the variable names "*n*" and "*p*" and the units "Super Bowl number" and "Dollars" on the appropriate axes.

Recall from Section 1.1 that when we write numbers on an axis, they should increase by a fixed amount and be equally spaced. Because the Super Bowl numbers are between 1 and 50, inclusive, we write the numbers $5, 10, 15, \ldots, 50$ equally spaced on the *n*-axis. Because the prices shown in Table 5 are between \$12 and \$1750, inclusive, we write the numbers $200, 400, 600, \ldots, 1800$ on the *p*-axis.

The ordered pair $(1, 12)$ indicates that the average ticket price was \$12 for Super Bowl I.

2. From Table 5 and the scatterplot in Fig. 20, we see that the highest average ticket price was \$1750 in Super Bowl L.

3. From Table 5 and the scatterplot in Fig. 20, we see that average ticket prices have increased. We also see that from Super Bowl I to Super Bowl XXX, inclusive, the *increases* in average ticket prices have increased. That is, as we look from left to right at the first seven points plotted on the coordinate system, we see that the vertical distance between adjacent points increases.

▶

In Example 3, constructing a scatterplot helped us notice some patterns. Such observations will help us make predictions about authentic situations later in this course.

Example 4   Constructing a Scatterplot with Age Groups

The percentages of adults who own a home are listed in Table 6 for various age groups.

Table 6 Percentages of Adults Who Own a Home

Age Group (years)	Age Used to Represent Age Group (years)	Percent
20–24	22.0	19
25–29	27.0	38
30–34	32.0	46
35–44	39.5	51
45–54	49.5	54
55–64	59.5	57
65–74	69.5	60

Source: *American Community Survey*

Let *p* be the percentage of adults who own a home when they are at age *a* years.

1. Construct a scatterplot of the data.
2. Describe any patterns in the percentages of adults who own a home.

Solution

1. A look at the first row of Table 6 suggests that we use $a = 22.0$ to represent the age group from 20 years to 24 years. The age 22.0 years is the average of the ages 20 years and 24 years. (Try it.) Likewise, we will use 27.0 to represent the age group from 25 years to 29 years and so on.

A scatterplot of the data is shown in Fig. 21. It makes sense to think of the variable *a* as the explanatory variable because age affects (explains) the percentage of adults who own a home (and not the other way around). So, we let the horizontal axis be the *a*-axis. Note that we write the variable names "*a*" and "*p*" and the units "Age in years" and "Percent" on the appropriate axes.

The ordered pair $(22.0, 19)$ indicates that, for the age group from 20 years to 24 years, 19% of adults own a home.

Because the ages used to represent age groups are between 22.0 years and 69.5 years, inclusive, we write the numbers $10, 20, 30, \ldots, 80$ equally spaced on the a-axis. Because the percents shown in Table 6 are between 19% and 60%, inclusive, we write the numbers 20, 40, and 60 equally spaced on the p-axis.

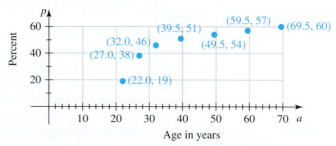

Figure 21 Scatterplot of homeowner data

2. From Table 6 and the scatterplot in Fig. 21, we see that the percentage of adults who own a home increases as their age increases. We also see that until age 49.5 years, the increase in the percentage decreases: from one row of Table 6 to the next, the percentage increases, but by less than it did in previous rows. In fact, the increase in the percentage from age 32 years to 39.5 years is less than the increase in the percentage from age 27 years to 32 years, even though the age span 32–39.5 years is longer than the age span 27–32 years. Likewise, the increase in the percentage from age 39.5 years to 49.5 years is less than the increase in the percentage from age 32 years to 39.5 years, even though the age span 39.5–49.5 years is longer than the age span 32–39.5 years.

In Example 5, we will define a variable to represent time.

Example 5   Defining a Variable for Time

Let t be the number of years since 1990. Find the values of t that represent the years 1990, 1996, 2005, 2010, and 2017.

Solution
We can represent 1990 by $t = 0$ because 1990 is 0 years after 1990. We can represent 1996 by $t = 6$ because 1996 is 6 years after 1990. We list the value of t for each of the years 1990, 1996, 2005, 2010, and 2017 in Table 7.

Table 7 Values of t

Year	Years since 1990 t
1990	0
1996	6
2005	15
2010	20
2017	27

because

$1990 - 1990 = 0$
$1996 - 1990 = 6$
$2005 - 1990 = 15$
$2010 - 1990 = 20$
$2017 - 1990 = 27$

The values of t in Table 7 are much smaller numbers than the years they represent. When working with authentic situations, we will often perform calculations that involve years. Using definitions similar to the one in Example 5 will enable us to perform those calculations with smaller numbers. It is also easier to label the axes of a coordinate system with smaller numbers.

Table 8 Life Expectancies

Year of Birth	Life Expectancy (years)
1980	73.7
1985	74.7
1990	75.4
1995	75.8
2000	76.9
2005	77.4
2010	78.3
2015	79.3

Source: *U.S. Census Bureau*

Table 9 Values of t and L

t (years since 1980)	L (years of life)
0	73.7
5	74.7
10	75.4
15	75.8
20	76.9
25	77.4
30	78.3
35	79.3

Example 6	Constructing a Scatterplot with Zigzag Lines on an Axis

A *life expectancy* is a prediction of how long a person will live. Table 8 shows life expectancies at birth for Americans in various years. Let L be the life expectancy at birth (in years) for an American born t years after 1980.

1. Construct a scatterplot of the data.
2. Describe any patterns in life expectancies of Americans.

Solution

1. First, we list the values of t and L in Table 9. For example, $t = 0$ represents 1980 because 1980 is 0 years after 1980. Also, $t = 5$ represents 1985 because 1985 is 5 years after 1980.

   A scatterplot of the data is shown in Fig. 22. Because L is the response variable, we let the vertical axis be the L-axis. We use zigzag lines on the L-axis to indicate that the part of the axis between 0 and about 71 is not displayed. This is done so we can show a clear view of the data points without having to make the coordinate system exceedingly tall.

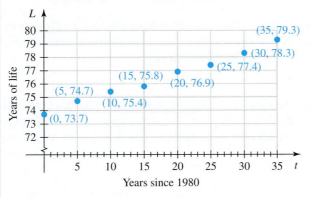

Figure 22 Life expectancy scatterplot

2. From Table 8 and Fig. 22, we see that the life expectancy of Americans has been increasing fairly steadily since 1980.

Using zigzag lines on the L-axis in Fig. 22 helped us have a clear view of the data points. To see how poor a view we have without the zigzag lines, see Fig. 23.

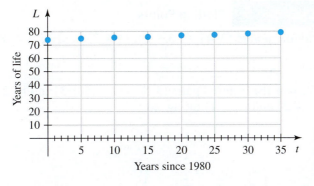

Figure 23 Poor view of life expectancy scatterplot (Fig. 22), without zigzag lines

Without the zigzag lines, it is very difficult to plot the data points at the correct height. Also, in Section 1.3, we will begin to estimate coordinates of points by reading graphs, and it would be very difficult to do so accurately by using the graph in Fig. 23. In fact, it is hard to tell that the heights of the points increase from left to right.

Bar Graphs

A **bar graph** is a diagram with two axes that we can use to compare measurements of two or more items (see Fig. 24). Along one axis we list the items, and along the other axis we mark tick marks, write numbers, and write the units of the measurements. We use a bar to indicate the measurement of each item.

Example 7   Reading a Bar Graph

The U.S. revenues (in millions of dollars) of some Broadway-based movie musicals (excluding remakes and sequels) are illustrated in the bar graph in Fig. 24 (Source: *Box Office Mojo*).

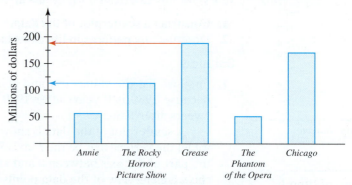

Figure 24 Bar graph of lifetime U.S. revenues of some Broadway-based movie musicals

1. Estimate the U.S. revenue of *The Rocky Horror Picture Show*.
2. Which Broadway-based movie musical listed in the bar graph has had the largest U.S. revenue? What is that revenue?

Solution

1. We look at the top of the bar for *The Rocky Horror Picture Show* and then look to the left at the vertical axis. It appears the U.S. revenue is about $113 million.
2. We identify the tallest bar, which is the one for *Grease*. We look at the top of the bar and then look to the left at the vertical axis. It appears the U.S. revenue is about $189 million.

Plotting Points on a Coordinate System

When we plot points that are not being used to describe authentic situations, we call the horizontal axis the *x-axis* and the vertical axis the *y-axis*. Then x is the explanatory variable and y is the response variable. The ordered pair $(6, 3)$ means $x = 6$ and $y = 3$. So, the x-coordinate is 6 and the y-coordinate is 3.

Example 8   Plotting Points

Plot the points $(3, 4)$, $(-5, -3)$, $(-4, 2)$, and $(5, -4)$ on a coordinate system.

Solution

We plot the ordered pairs $(3, 4)$ and $(-5, -3)$ in Fig. 25, and we plot the ordered pairs $(-4, 2)$ and $(5, -4)$ in Fig. 26.

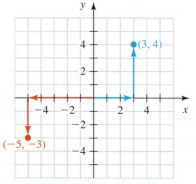

Figure 25 Plotting the ordered pairs $(3, 4)$ and $(-5, -3)$

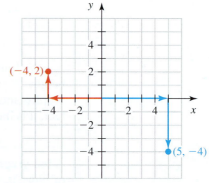

Figure 26 Plotting the ordered pairs $(-4, 2)$ and $(5, -4)$

◆ Group Exploration

Analyzing points below, above, and on a line containing points with equal coordinates

The scatterplot in Fig. 27 compares the scores of Test 2 and Test 3 for 35 calculus students taught by the author. Each dot represents a pair of scores for exactly one student.

Test 2 Scores versus Test 3 Scores

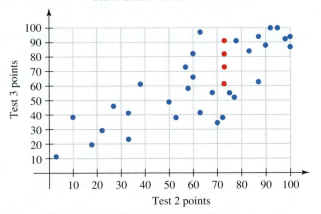

Figure 27 Scores of Test 2 and Test 3

1. Do the heights of points in the scatterplot generally increase or decrease from left to right? What does that mean in this situation?

2. On the scatterplot, imagine a line that goes through the points $(20,20)$, $(40,40)$, $(60,60)$, $(80,80)$, and $(100,100)$.

Or use a strand of thread, a stretched rubber band, or the edge of a clear ruler to form the line. Which data points lie on the line? What do they mean in this situation?

3. Give the coordinates of three points that lie above the line. For your first point, determine whether the corresponding student scored higher on Test 2 or Test 3. Is the same true for your other two points? the other points above the line?

4. Give the coordinates of three points that lie below the line. For your first point, determine whether the corresponding student scored higher on Test 2 or Test 3. Is the same true for your other two points? the other points below the line?

5. Two students have equal total scores for Test 2. The data point for one student is above the line, and the data point for the other student is below the line. Which student shows more promise? Explain.

6. A student who scored 73 points on Test 2 was absent for Test 3. Predict what the student's Test 3 score would have been by computing the average Test 3 score for the students (represented by the red dots) who scored 73 points on Test 2.

Tips for Success   Study Time

For each hour of class time, study for at least two hours outside class. If your math background is weak, you may need to spend more time studying.

One way to study is to do what you are doing now: Read the text. Class time is a great opportunity to be introduced to new concepts and to see how they fit together with previously learned ones. However, there is usually not enough time to address details as well as a textbook can. In this way, a textbook can serve as a supplement to what you learn in class.

◆ Homework 1.2

For extra help ▶ **MyLab Math**   Watch the videos in MyLab Math

For Exercises 1–16, plot the given points in a coordinate system.

1. $(5, 1)$ **2.** $(2, 3)$ **3.** $(4, -2)$

4. $(3, -4)$ **5.** $(-5, 4)$ **6.** $(-1, 3)$

7. $(-3, -6)$ **8.** $(-5, -2)$ **9.** $(0, 2)$

10. $(0, -4)$ **11.** $(-3, 0)$ **12.** $(1, 0)$

13. $(2.5, -4.5)$ **14.** $(-3.5, 1.5)$

15. $(-1.3, -3.9)$ **16.** $(-2.4, -4.1)$

17. What is the x-coordinate of the ordered pair $(2, -4)$?

18. What is the y-coordinate of the ordered pair $(2, -4)$?

For Exercises 19–28, identify the explanatory variable and the response variable.

19. Let n be the number of hours a student studies for a quiz, and let s be the student's score (in points) on the quiz.

20. Let t be the number of years a person has worked for a company, and let s be the person's salary (in dollars).

21. Let h be the height (in inches) of a girl, and let a be the age (in years) of the girl.

22. Let p be the percentage of colleges that would accept a student whose grade point average (GPA) is g points.

23. Let T be the tuition (in dollars) for enrolling in c credits (units or hours) of classes.

24. Let p be the percentage of men at age a years who have gray hair.

25. Let A be the floor area (in square feet) of a classroom, and let n be the number of students who can comfortably fit into the classroom.

26. A person cooks a potato in an oven for an hour and then removes the potato and allows it to cool. Let t be the number of minutes since the potato was removed from the oven, and let F be the temperature (in degrees Fahrenheit) of the potato.

27. Let t be the number of seconds after a baseball is hit upward, and let h be the baseball's height (in feet).

28. Let p be the percentage of people at age a years who own a computer.

For Exercises 29–36, describe what the given ordered pair represents.

29. Let n be the average number of magazine subscriptions sold per week by a telemarketer who works t hours per week. What does the ordered pair $(32, 43)$ mean in this situation?

30. Let c be the total cost (in dollars) of buying n pens. What does the ordered pair $(5, 10)$ mean in this situation?

31. Let p be the percentage of U.S. consumers who binge watch an average of n episodes of television at a time in 2016. What does the ordered pair $(5, 70)$ mean in this situation (Source: *Deloitte*)?

32. Let p be the percentage of Internet users who use n social networking sites. What does the ordered pair $(4, 5)$ mean in this situation (Source: *Pew Research Center*)?

33. Let b be the annual defense spending (in billions of dollars) at t years since 2000. What does the ordered pair $(15, 598.5)$ mean in this situation (Source: *International Institute for Strategic Studies*)?

34. Let n be the number (in millions) of ads Google blocked because they were viewed as rude, dishonest, or dangerous in the year that is t years since 2010. What does the ordered pair $(5, 780)$ mean in this situation (Source: *Fortune*)?

35. Let p be the percentage of Americans who feel that a college education is very important at t years since 2015. What does the ordered pair $(-2, 70)$ mean in this situation (Source: *Gallup*)?

36. Let p be the percentage of Americans who are satisfied with the size and influence of major corporations at t years since 2015. What does the ordered pair $(-1, 35)$ mean in this situation (Source: *Gallup*)?

37. Construct a scatterplot of the ordered pairs listed in Table 10.

Table 10 Some Ordered Pairs

x	y
2	5
7	9
11	10
14	9
16	5

38. Construct a scatterplot of the ordered pairs listed in Table 11.

Table 11 Some Ordered Pairs

x	y
−5	2
−4	3
−1	5
4	9
11	15

39. Find the coordinates of points A, B, C, D, E, and F shown in Fig. 28.

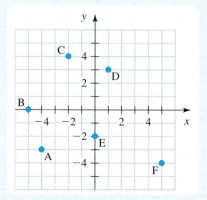

Figure 28 Exercise 39

40. Find the coordinates of points A, B, C, D, E, and F shown in Fig. 29.

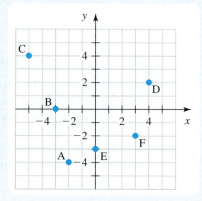

Figure 29 Exercise 40

41. The number of pages in each of the books in the Harry Potter series is listed in Table 12.

Table 12 Numbers of Pages in the Books of the Harry Potter Series

Book Number	Number of Pages
1	309
2	341
3	435
4	734
5	870
6	652
7	784

Let p be the number of pages and b be the corresponding book number.

a. Construct a scatterplot of the data.
b. Which book has the greatest number of pages?
c. From which book to the next did the number of pages increase the most? Explain how you can tell this by inspecting your scatterplot.

42. The average life spans of various denominations of bills are shown in Table 13. For example, the average life span of a $1 bill is 5.8 years before it is taken out of circulation due to wear and tear.

Table 13 Average Life Spans of Denominations of Bills

Value of Bill (dollars)	Average Life Span (years)
1	5.8
5	5.5
10	4.5
20	7.9
50	8.5
100	15.0

Source: *Federal Reserve System*

Let L be the average life span (in years) of a bill that is worth d dollars.
a. Construct a scatterplot of the data.
b. Explain why it makes sense that the average life span of a $100 bill is greater than the average life span of a $10 bill.
c. Each year, many more $10 bills are printed than $100 bills. Give at least two reasons why this makes sense.

43. Tiger Woods's golf tournament earnings and number of wins are shown in Table 14 for various years.

Table 14 Tiger Woods's Tournament Earnings and Numbers of Wins

Year	Tournament Earnings (millions of dollars)	Wins
2006	9.9	8
2007	10.9	7
2008	5.8	4
2009	10.5	6
2010	1.3	0
2011	0.7	0
2012	6.1	3
2013	8.6	5
2014	0.1	0
2015	0.4	0

Source: *PGA Tour*

Let E be Tiger Woods's tournament earnings (in millions of dollars) in the year that is t years since 2000. For example, $t = 6$ represents 2006, and $t = 7$ represents 2007.
a. Construct a scatterplot of the data.
b. For which of the years shown in Table 14 were Woods's tournament earnings the least? What were those earnings?
c. For which of the years shown in Table 14 were Woods's tournament earnings the greatest? What were those earnings?
d. For the years shown in Table 14, did Woods have his greatest tournament earnings in the same year as his greatest number of wins? Give two reasons why this could be possible.

44. The average price per barrel of crude oil and the total fuel cost for the global airline industry are shown in Table 15 for various years.

Table 15 Average Price per Barrel of Crude Oil and Total Fuel Cost for the Global Airline Industry

Year	Average Price per Barrel of Crude Oil (dollars)	Total Airline Fuel Cost (billions of dollars)
2011	111	191
2012	112	228
2013	109	230
2014	100	226
2015	54	181

Source: *IATA*

Let C be the total airline fuel cost (in billions of dollars) in the year that is t years since 2010. For example, $t = 1$ represents 2011, and $t = 2$ represents 2012.
a. Construct a scatterplot of the data.
b. For which of the years shown in Table 15 was the total airline fuel cost the least? What was that cost?
c. For which of the years shown in Table 15 was the total airline fuel cost the greatest? What was that cost?
d. For the years shown in Table 15, was the total fuel cost the greatest in the same year that the average price per barrel was the greatest? Explain why this is possible.

45. The numbers of firearm discoveries at U.S. airports are shown in Table 16 for various years.

Table 16 Numbers of Firearm Discoveries at U.S. Airports

Year	Number of Firearm Discoveries
2010	1123
2011	1320
2012	1556
2013	1813
2014	2212
2015	2653

Source: *Transportation Security Administration*

Let n be the number of firearm discoveries at U.S. airports in the year that is t years since 2010.
a. Construct a scatterplot of the data.
b. Did the number of firearm discoveries increase, decrease, stay approximately constant, or none of these?
c. Did the *increases* in the number of firearm discoveries increase, decrease, stay approximately constant, or none of these? Explain.
d. Give at least two possible reasons why the number of firearm discoveries has increased.

46. The average U.S. hourly pay is shown in Table 17 for various years.

Table 17 Average U.S. Hourly Pay

Year	Average U.S. Hourly Pay (dollars)
1995	11.67
2000	14.03
2005	16.15
2010	19.06
2015	21.05

Source: *St. Louis Federal Reserve Bank*

Let p be the average U.S. hourly pay (in dollars) at t years since 1990.

a. Construct a scatterplot of the data.

b. Did the average U.S. hourly pay increase, decrease, stay approximately constant, or none of these? Explain.

c. Did the five-year *increase* in the average U.S. hourly pay increase, decrease, stay approximately constant, or none of these? Explain.

47. The number of automobile accidents per 1000 licensed drivers per year are shown in Table 18 for various age groups.

Table 18 Automobile Accidents

Age Group (years)	Age Used to Represent Age Group (years)	Accident Rate (number of accidents per 1000 licensed drivers per year)
16–20	18.0	111.4
21–24	22.5	83.7
25–34	29.5	62.4
35–44	39.5	50.3
45–54	49.5	43.2
55–64	59.5	35.8
65–74	69.5	26.9
over 74	80.0	24.9

Source: *National Highway Traffic Safety Administration*

Let r be the automobile accident rate (number of accidents per 1000 licensed drivers per year) for licensed drivers at age a years.

a. Construct a scatterplot of the data.

b. Which age group shown in Table 18 has the lowest accident rate?

c. Which age group shown in Table 18 has the highest accident rate?

d. As the age of drivers increases, does the accident rate increase, decrease, approximately stay the same, or none of these?

e. Many states put limits on teenage driving. For example, some states do not allow 16-year-old drivers to drive at night. Some states require parental supervision at all times. Why do you think these regulations were adopted?

48. The percentages of Americans of various age groups who are ordering more takeout food than they did two years ago are shown in Table 19.

Table 19 Percentages of Americans Who Are Ordering More Takeout Food Than They Did Two Years Ago

Age Group (years)	Age Used to Represent Age Group (years)	Percent
18–24	21.0	34
25–34	29.5	31
35–44	39.5	27
45–54	49.5	17
55–64	59.5	15
over 64	70.0	7

Source: *National Restaurant Association Survey*

Let p be the percentage of Americans at age a years who are ordering more takeout food than they did two years ago.

a. Construct a scatterplot of the data.

b. Which of the points in your scatterplot is highest? What does that mean in this situation?

c. Which of the points in your scatterplot is lowest? What does that mean in this situation?

d. Do the heights of the points in your scatterplot increase or decrease from left to right? What does that mean in this situation?

49. Several inventions are listed in Table 20, along with the years they were invented and how long it took for one-quarter of the U.S. population to use them ("mass use").

Table 20 Number of Years until Inventions Reached Mass Use

Invention	Year Invented	Years until Mass Use
Electricity	1873	46
Telephone	1876	35
Gasoline-Powered Automobile	1886	55
Radio	1897	31
Television	1923	29
Microwave Oven	1953	36
VCR	1965	13
Personal Computer	1975	16
Mobile Phone	1985	11
CD Player	1985	8
World Wide Web	1991	7
DVD Player	1997	5

Source: *Newsweek*

Let M be the number of years elapsed until an invention reached mass use if it was invented at t years since 1870.

a. Construct a scatterplot of the data.

b. Compare the time it took to reach mass use for recent inventions versus earlier inventions. In your opinion, why did this happen?

c. Does the datum for the microwave oven fit the pattern you described in part (b)? Explain.

d. For a while after the microwave oven was invented, many people feared it would cause radiation poisoning, blindness, or impotence. Discuss the impact of these fears in terms of your response to part (c).

e. Explain why the datum for the gasoline-powered automobile does not fit the pattern you described in part (b). Why do you think this happened?

50. The percentages of adults of various age groups who approve of single men raising children on their own are shown in Table 21.

Table 21 Percentages of Adults Who Approve of Single Men Raising Children on Their Own

Age Group (years)	Age Used to Represent Age Group (years)	Percent
18–34	26.0	81
35–44	39.5	73
45–54	49.5	73
55–64	59.5	66
over 64	70.0	47

Source: *Taylor Nelson Sofres*

Let p be the percentage of adults at age a years who approve of single men raising children on their own.

a. Construct a scatterplot of the data.

b. Which age group shown in Table 21 has the most faith in single men raising children on their own?

c. Which age group shown in Table 21 has the least faith in single men raising children on their own?

d. A student says that the data show that young adults who approve of single men raising children on their own change their minds about this issue as they grow older. What would you tell the student?

51. The average starting salaries for employees with a bachelor's degree are illustrated in the bar graph in Fig. 30 for various fields of study (Source: *National Association of Colleges and Employers*).

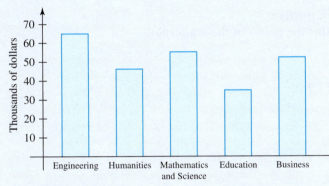

Figure 30 Exercise 51

a. For which field shown is the average starting salary the highest? What is that salary?

b. For which field shown is the average starting salary the lowest? What is that salary?

c. Estimate the average starting salary for employees with a mathematics or science degree.

52. In baseball, a grand slam is a home run with the bases loaded. The top six numbers of career grand slams for major league baseball players are illustrated in Fig. 31 (Source: *MLB.com*).

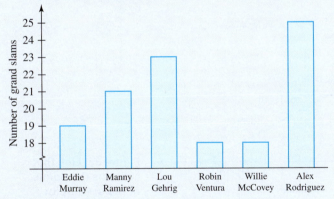

Figure 31 Exercise 52

a. Estimate Robin Ventura's number of career grand slams.

b. Which player holds the record for the greatest number of career grand slams? What is that number of grand slams?

c. Which player hit exactly 21 grand slams?

Concepts

53. List five ordered pairs whose *x*-coordinate is 3. Then construct a scatterplot of the ordered pairs. What do you notice about the arrangement of the points in your scatterplot? Explain why this makes sense.

54. List five ordered pairs whose *y*-coordinate is 2. Then construct a scatterplot of the ordered pairs. What do you notice about the arrangement of the points in your scatterplot? Explain why this makes sense.

55. The points where the sides of a triangle, rectangle, or any other polygon meet are called *vertices*. A square has vertices at $(2, 1)$ and $(2, 5)$. How many possible positions are there for the other two vertices? Find the coordinates of the vertices for two of these possible positions.

56. The points where the sides of a triangle, rectangle, or any other polygon meet are called *vertices*. A rectangle has vertices at $(2, 4)$ and $(7, 4)$. How many possible positions are there for the other two vertices? Find the coordinates of the vertices for four of these possible positions.

57. Describe the signs of the *x*-coordinate and the *y*-coordinate for a point that lies in the given quadrant.
   a. Quadrant I
   b. Quadrant II
   c. Quadrant III
   d. Quadrant IV

58. Describe all points on a coordinate system whose *x*-coordinates are 0.

59. Describe all points on a coordinate system whose *y*-coordinates are 0.

60. Compare a number line with a coordinate system. When is it useful to describe data by a number line? When is it useful to describe data by a coordinate system?

61. Compare the meaning of *explanatory variable* with the meaning of *response variable*.

1.3 Exact Linear Relationships

In this section, we will use a scatterplot to help us sketch a line that describes an authentic situation, such as the descent of a hot-air balloon. We will then use the line to make estimates and predictions about the situation.

Linear Models

Example 1   Using a Line to Describe an Authentic Situation

A person lowers her hot-air balloon by gradually releasing air from the balloon. Let a be the balloon's altitude (in feet) above the ground after she has released air from the balloon for t minutes. Values of t and a are listed in Table 22.

1. Construct a scatterplot of the data.
2. Draw the line that contains the points of the scatterplot.

Solution

1. We draw a scatterplot in Fig. 32.

Table 22 Altitudes of a Hot-Air Balloon

Time (minutes) t	Altitude (feet) a
0	2200
2	1800
4	1400
6	1000
8	600

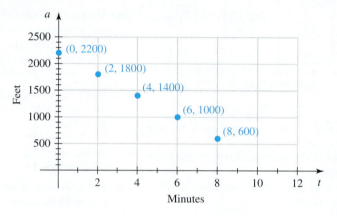

Figure 32 Scatterplot of balloon data

2. In mathematics, a "line" means a *straight* line. In Fig. 33, we draw the line that contains the data points shown in Fig. 32.

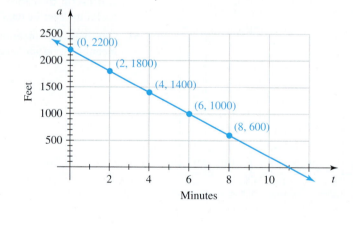

Figure 33 Line that contains the points of the scatterplot

The scatterplot in Fig. 32 accurately describes the altitude of the balloon at 0, 2, 4, 6, and 8 minutes. But it does not describe the altitude at other times.

If we imagine the line in Fig. 33 to be made up of points, then we can use the line to describe the altitudes at 0, 2, 4, 6, and 8 minutes accurately. We can also use the line to estimate the altitude for other times between 0 and 8 minutes. These results will be good estimates, provided that the altitude of the balloon declined steadily.

In addition, we can use the line to predict altitudes for times a little after 8 minutes. However, these predictions will be accurate only if the altitude of the balloon continued to decline steadily.

If the altitudes of the balloon over a span of time are described accurately by a line, we say that time and altitude (and the variables t and a) are *linearly related* for that span of time.

> **Definition   Linearly related**
>
> If two quantities of an authentic situation are described accurately by a line, then the quantities (and the variables representing those quantities) are **linearly related**.

The process of choosing a line to represent the relationship between balloon altitudes and time is an example of modeling.

> **Definition   Model**
>
> A **model** is a mathematical description of an authentic situation. We say the description *models* the situation.

We call the line in Fig. 33 a *linear model*. In Chapters 7–11, we will discuss other types of models. The term "model" is being used in much the same way as it is used in "airplane model." Just as an airplane designer can use the behavior of an airplane model in a wind tunnel to predict the behavior of an actual airplane, a linear model can be used to predict what might happen in a situation in which two variables are linearly related.

> **Definition   Linear model**
>
> A **linear model** is a nonvertical line that describes the relationship between two quantities in an authentic situation.

Using a Linear Model to Make Estimates and Predictions

Example 2   Making Estimates and Predictions

1. Use the linear model shown in Fig. 33 to estimate the balloon's altitude when air has been released for 5 minutes.
2. Use the linear model to predict when the balloon's altitude is 400 feet.

Solution

1. To estimate the altitude of the balloon when air has been released for 5 minutes, we locate the point on the linear model where the t-coordinate is 5 (see Fig. 34). The a-coordinate of that point is 1200. So, *according to the model*, the altitude is 1200 feet when air has been released for 5 minutes.

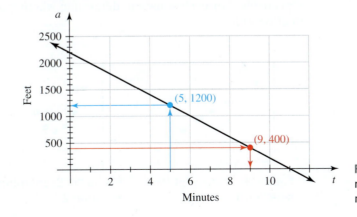

Figure 34 Using the linear model to make an estimate and a prediction

Table 23 Altitudes of
a Balloon

Time (minutes) t	Altitude (feet) a
0	2200
2	1800
4	1400
6	1000
8	600

We verify our work by checking that our result is consistent with the values in Table 23. Because the altitude is 1400 feet after 4 minutes and 1000 feet after 6 minutes, it makes sense that the altitude would be between 1000 and 1400 feet when air has been released for 5 minutes, which checks with our result of 1200 feet.

2. To find when the balloon's altitude is 400 feet, we locate the point on the linear model where the a-coordinate is 400 (see Fig. 34). The t-coordinate of that point is 9. So, *according to the linear model*, the altitude is 400 feet when air has been released for 9 minutes.

   Again, we verify our work by checking that our result is consistent with the values in Table 23. Because the altitude is 600 feet when air has been released for 8 minutes, it follows that air would have been released for more than 8 minutes for the altitude to be less than 600 feet, which checks with our result of 9 minutes.

Input and Output

In Example 2, we found that when the value of the explanatory variable t is 5, the corresponding value of the response variable a is 1200. We say the *input* 5 leads to the *output* 1200. The blue arrows in Fig. 34 show the action of the input $t = 5$ leading to the output $a = 1200$.

> **Definition**   Input, output
>
> An **input** is a permitted value of the *explanatory* variable that leads to at least one **output**, which is a permitted value of the *response* variable.

For a value to be permitted, it must make physical sense and be defined. For instance, in Example 2, the value -50 is not a permitted value of the variable a because it does not make sense for the balloon's altitude to be -50 feet. Later in the course we will discuss values that are not permitted for mathematical reasons.

Sometimes we will go "backward," from an output back to an input. For instance, in Example 2, we found that the output $a = 400$ originates from the input $t = 9$. The red arrows in Fig. 34 show the action of going backward from the output $a = 400$ to the input $t = 9$.

When to Use a Line to Model Data

Next, we will discuss how to determine whether an authentic situation can be described well by a linear model.

| **Example 3** | Deciding Whether to Use a Line to Model Data |

Consider the scatterplots of data for situations 1, 2, and 3 shown in Figs. 35, 36, and 37, respectively. For each situation, determine whether a linear model would describe the situation well.

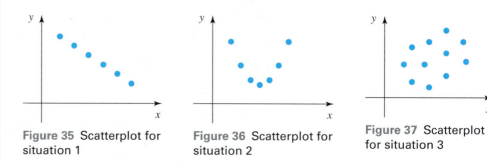

Figure 35 Scatterplot for situation 1

Figure 36 Scatterplot for situation 2

Figure 37 Scatterplot for situation 3

Solution

It appears that the data points for situation 1 lie on a line; a linear model would describe situation 1 well. The data points for situation 2 do not lie close to one line; a linear model would not describe situation 2. (In Chapters 7–9, we will discuss a type of nonlinear model that would describe situation 2 well.) The data points for situation 3 do not lie near a line; a linear model would not describe this situation.

▶

We construct a scatterplot of data to determine whether the data points lie on a line. If the points lie on a line, then we draw the line and use it to make estimates and predictions.

Intercepts of a Line

Consider the line sketched in Fig. 38. The line intersects the x-axis at the point $(6, 0)$. The point $(6, 0)$ is called the *x-intercept*. Also, the line intersects the y-axis at the point $(0, 3)$. The point $(0, 3)$ is called the *y-intercept*.

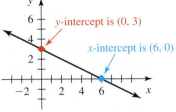

Figure 38 Intercepts of a line

Definition   Intercepts of a line

An **intercept** of a line is any point where the line and an axis (or axes) of a coordinate system intersect. There are two types of intercepts of a line sketched on a coordinate system with an x-axis and a y-axis:

- An **x-intercept** of a line is a point where the line and the x-axis intersect (see Fig. 39). The y-coordinate of an x-intercept is 0.

- A **y-intercept** of a line is a point where the line and the y-axis intersect (see Fig. 39). The x-coordinate of a y-intercept is 0.

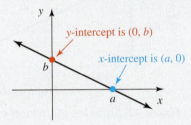

Figure 39 Intercepts of a line

Example 4   Finding Intercepts and Coordinates

Refer to Fig. 40 for the following problems.

1. Find the x-intercept of the line.
2. Find the y-intercept of the line.
3. Find y when $x = -6$.
4. Find x when $y = -3$.

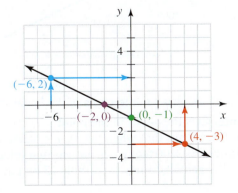

Figure 40 Problems 1–4 of Example 4

Solution

1. The line and the *x*-axis intersect at $(-2, 0)$. So, the *x*-intercept is $(-2, 0)$.
2. The line and the *y*-axis intersect at $(0, -1)$. So, the *y*-intercept is $(0, -1)$.
3. The blue arrows in Fig. 40 show that the input $x = -6$ leads to the output $y = 2$. So, $y = 2$ when $x = -6$.
4. The red arrows in Fig. 40 show that the output $y = -3$ originates from the input $x = 4$. So, $x = 4$ when $y = -3$.

Intercepts of a Linear Model

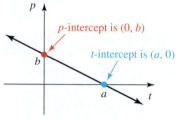

Figure 41 Intercepts of a linear model

Suppose that a linear model describes the relationship between two variables *t* and *p*, where *t* affects (explains) *p*. Then the *t-intercept* is a point where the line and the *t*-axis intersect, and the *p-intercept* is a point where the line and the *p*-axis intersect (see Fig. 41).

Example 5   Finding Intercepts of a Linear Model

1. The linear model from Examples 1 and 2 is shown in Fig. 42. Find the *t*-intercept of the model. What does it mean in this situation?
2. Find the *a*-intercept of the model. What does it mean in this situation?

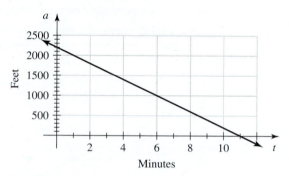

Figure 42 Altitudes of balloon model

Solution

1. The line intersects the *t*-axis at the point $(11, 0)$. See Fig. 43. So, the *t*-intercept is $(11, 0)$. This means that when $t = 11$, $a = 0$. The model predicts that the balloon reached the ground 11 minutes after air began to be released from the balloon.

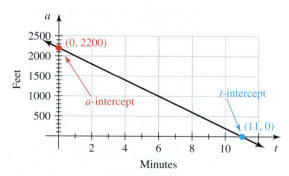

Figure 43 Intercepts of the linear model

2. The line intersects the *a*-axis at the point $(0, 2200)$. See Fig. 43. So, the *a*-intercept is $(0, 2200)$. This means that when $t = 0$, $a = 2200$. The model estimates that the balloon's altitude was 2200 feet when air was first released from the balloon. In fact, this estimate is the actual altitude (see Table 23 on page 28).

Table 24 Profits from Selling CDs

Sales (number of CDs) n	Profit (dollars) P
50	−1150
100	−800
150	−450
300	600
350	950

Example 6   Using a Linear Model to Make Estimates

An underground rock band manufactures 500 CDs of its original music and tries to sell the CDs at the band's concerts. Let P be the profit (in dollars) from selling a total of n CDs. Some values of n and P are listed in Table 24.

1. Construct a scatterplot of the data. Then draw a reasonable model.
2. Estimate the band's profit from selling all 500 CDs.
3. If the band loses $975, estimate how many CDs will have been sold.
4. Estimate the P-intercept of the model. What does it mean in this situation?
5. Estimate the n-intercept of the model. What does it mean in this situation?

Solution

1. The scatterplot is shown in Fig. 44 (the black points). Because the points lie on a line, we use the line as a model.
2. The blue arrows show that the input $n = 500$ leads to the output $P = 2000$. So, if the band sells 500 CDs, the profit will be $2000.
3. A negative value of P represents a loss. The red arrows show that the output $P = -975$ originates from the input $n = 75$. So, if the band sells 75 CDs, it will lose $975.

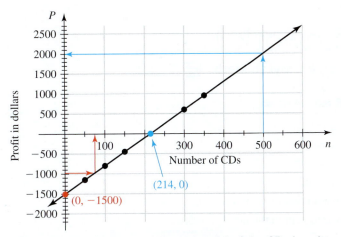

Figure 44 Scatterplot and linear model of the CD situation

4. The linear model and the P-axis intersect at the point $(0, -1500)$. This means that the band will lose $1500 if no CDs are sold.
5. The linear model and the n-axis intersect at about the point $(214, 0)$. This means that the band will have a profit of about 0 dollars (and hence break even) if 214 CDs are sold.

It is impossible to do a perfect job of sketching models and estimating coordinates of points. Your results for homework exercises will likely be different from the answers provided near the end of this textbook. However, if you do a careful job, your results should be close to those in the book.

◆ Group Exploration

Section Opener: Linear modeling

An airplane is beginning to descend. Let a be the altitude (in thousands of feet) of the airplane at t minutes since it began its descent. Some pairs of values of t and a are shown in Table 25.

1. Construct a scatterplot of the data.
2. Draw a line that contains the points in your scatterplot. We call the line a *linear model*.

Table 25 Altitudes of an Airplane

Time (minutes) t	Altitude (thousands of feet) a
0	24
5	20
10	16
15	12
20	8

3. Use your line to estimate the altitude of the airplane 8 minutes after it began its descent. [**Hint:** On the line, locate the point whose t-coordinate is 8.]

4. Use your line to estimate when the airplane reached an altitude of 10 thousand feet.

5. What was the altitude of the airplane when it began its descent?

6. Use your line to estimate when the airplane reached the ground.

Tips for Success   Get in Touch with Classmates

It is wise to exchange phone numbers and e-mail addresses with some classmates. If you have to miss class, then you have someone to contact to find out what you missed and what homework was assigned.

Homework 1.3

For extra help ▶ MyLab Math   Watch the videos in MyLab Math

For Exercises 1–6, refer to Fig. 45.

1. Find y when $x = -2$.
2. Find y when $x = 4$.
3. Find x when $y = -2$.
4. Find x when $y = 4$.
5. What is the x-intercept of the line?
6. What is the y-intercept of the line?

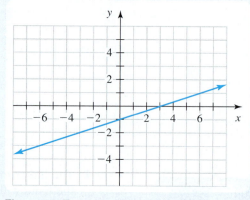

Figure 46 Exercises 7–12

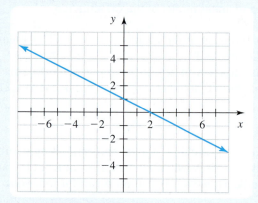

Figure 45 Exercises 1–6

For Exercises 7–12, refer to Fig. 46.

7. Find y when $x = -3$.
8. Find y when $x = 6$.
9. Find x when $y = -3$.
10. Find x when $y = 0$.
11. What is the y-intercept of the line?
12. What is the x-intercept of the line?

13. Some ordered pairs are listed in Table 26.
    a. Construct a scatterplot of the points shown in Table 26.
    b. Draw a line that contains the points in your scatterplot.
    c. Find y when $x = 3$.
    d. Find x when $y = 6$.
    e. What is the y-intercept of your line?
    f. What is the x-intercept of your line?

Table 26 Some Ordered Pairs

x	y
2	20
4	16
6	12
8	8
10	4

14. Some ordered pairs are listed in Table 27.
    a. Construct a scatterplot of the points shown in Table 27.
    b. Draw a line that contains the points in your scatterplot.

c. Find y when $x = 4$.

d. Find x when $y = 17$.

e. What is the y-intercept of your line?

f. What is the x-intercept of your line?

Table 27 Some Ordered Pairs

x	y
−6	4
−2	8
2	12
6	16
10	20

15. Water is steadily pumped out of a flooded basement. Let v be the volume of water (in thousands of gallons) that remains in the basement t hours after water began to be pumped. A linear model is shown in Fig. 47.

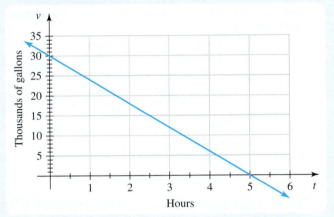

Figure 47 Exercise 15

a. How much water is in the basement after 2 hours of pumping?

b. After how many hours of pumping will 5 thousand gallons remain in the basement?

c. How much water was in the basement before any water was pumped out?

d. After how many hours of pumping will all the water be pumped out of the basement?

16. Let B be the balance (in dollars) of a student's checking account at t months since the student opened the account. A linear model is shown in Fig. 48.

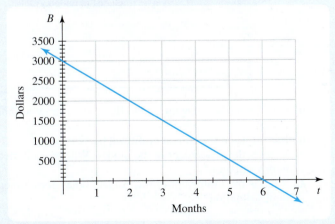

Figure 48 Exercise 16

a. What was the balance 3 months after the student opened the account?

b. When was the balance $500?

c. What is the B-intercept of the model? What does it mean in this situation?

d. What is the t-intercept of the model? What does it mean in this situation?

17. A scatterplot for a situation is graphed in Fig. 49. Is there a line that is a reasonable model of this situation? If yes, sketch the line. If no, explain why not.

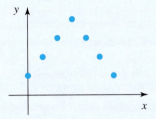

Figure 49 Exercise 17

18. A scatterplot for a situation is graphed in Fig. 50. Is there a line that is a reasonable model of this situation? If yes, sketch the line. If no, explain why not.

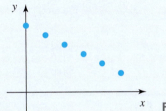

Figure 50 Exercise 18

19. Some ordered pairs are listed in Table 28.

a. Construct a scatterplot of the data shown in Table 28.

b. Is there a linear relationship between x and y? Explain.

Table 28 Some Ordered Pairs

x	y
0	10
5	5
8	2
10	1
12	2
13	5
14	10

20. Some ordered pairs are listed in Table 29.

a. Construct a scatterplot of the data shown in Table 29.

b. Is there a linear relationship between x and y? Explain.

Table 29 Some Ordered Pairs

x	y
2	1
3	6
4	9
5	10
6	9
7	6
8	1

21. Let d be the distance traveled (in miles) after a student has driven for t hours (not counting pit stops). Some pairs of values of t and d are shown in Table 30.

Table 30 Times and Distances for a Car

t (hours)	d (miles)
0	0
1	60
2	120
3	180
4	240

   a. Construct a scatterplot of the data. Then draw a linear model.
   b. Estimate how far the student has traveled in 2.5 hours.
   c. Estimate how long it took the student to travel 210 miles.

22. A student works part time at the college bookstore. Let p be the student's pay (in dollars) for working t hours. Some pairs of values of t and p are shown in Table 31.

Table 31 Pay for Working t Hours

t (hours)	p (dollars)
0	0
5	60
10	120
15	180
20	240

   a. Construct a scatterplot of the data. Then draw a linear model.
   b. Estimate the student's pay for working 7 hours.
   c. Estimate the number of hours the student must work to earn $156.

23. Let E be a college's enrollment (in thousands of students) at t years since the college began. Some pairs of values of t and E are shown in Table 32.

Table 32 Ages and Enrollments of a College

t (years)	E (thousands of students)
0	5
1	7
2	9
3	11
4	13

   a. Construct a scatterplot of the data. Then draw a linear model.
   b. Predict the enrollment when it has been 6 years since the college opened.
   c. Predict when the enrollment will reach 19 thousand students.

24. Let s be a person's salary (in thousands of dollars) after he has worked t years at a company. Some pairs of values of t and s are shown in Table 33.

Table 33 Years Worked and Salary

t (years)	s (thousands of dollars)
0	20
2	24
4	28
6	32
8	36

   a. Construct a scatterplot of the data. Then draw a linear model.
   b. Estimate the person's salary after he has worked 5 years at the company.
   c. Estimate when the person's salary was $34 thousand.
   d. What is the s-intercept of the model? What does it mean in this situation?

25. Let v be the value (in dollars) of a company's stock at t years since 2010. Some pairs of values of t and v are shown in Table 34.

Table 34 Values of a Stock

t (years)	v (dollars)
1	28
2	24
4	16
6	8
7	4

   a. Construct a scatterplot of the data. Then draw a linear model.
   b. Estimate when the value of the stock was $12.
   c. What is the t-intercept of the model? What does it mean in this situation?
   d. What is the v-intercept of the model? What does it mean in this situation?

26. Let p be the annual profit (in millions of dollars) of a company at t years since 2010. Some pairs of values of t and p are shown in Table 35.

Table 35 Annual Profits of a Company

t (years)	p (millions of dollars)
1	20
2	18
3	16
5	12
6	10

   a. Construct a scatterplot of the data. Then draw a linear model.
   b. Predict when the annual profit will be $2 million.
   c. What is the p-intercept of the model? What does it mean in this situation?
   d. What is the t-intercept of the model? What does it mean in this situation?

27. Let g be the number of gallons of gasoline that remain in a car's gasoline tank after the car has been driven d miles since the tank was filled. Some pairs of values of d and g are shown in Table 36.

Table 36 Miles Traveled and Gallons of Gasoline

d (miles)	g (gallons)
40	11
80	9
120	7
160	5
200	3
240	1

   a. Construct a scatterplot of the data. Then draw a linear model.
   b. Estimate how much gasoline is in the tank after the driver has gone 140 miles since last filling up.
   c. Estimate the number of miles driven since the tank was last filled if 2 gallons of gasoline remain in the tank.
   d. Find the d-intercept of the model. What does it mean in this situation?
   e. Find the g-intercept of the model. What does it mean in this situation?

28. Let v be the value (in thousands of dollars) of a car when it is t years old. Some pairs of values of t and v are listed in Table 37.

Table 37 Ages and Values of a Car

t (years)	v (thousands of dollars)
1	18
3	14
5	10
7	6
9	2

   a. Construct a scatterplot of the data. Then draw a linear model.
   b. Estimate the age of the car when it is worth $4 thousand.
   c. Estimate the value of the car when it is 6 years old.
   d. What is the v-intercept of the model? What does it mean in this situation?
   e. What is the t-intercept of the model? What does it mean in this situation?

29. Let r be the annual revenue (in millions of dollars) of a company at t years since 2010. Some pairs of values of t and r are shown in Table 38.

Table 38 Annual Revenues of a Company

t (years)	r (millions of dollars)
0	8
1	11
3	17
5	23
6	26

   a. Construct a scatterplot of the data. Then draw a linear model.
   b. Predict the revenue in 2020.
   c. Estimate when the annual revenue was $14 million.
   d. What is the r-intercept of the model? What does it mean in this situation?

30. Let v be the value (in dollars) of a company's stock at t years since 2010. Some pairs of values of t and v are shown in Table 39.

Table 39 Values of a Stock

t (years)	v (dollars)
0	15
2	19
4	23
5	25
6	27

   a. Construct a scatterplot of the data. Then draw a linear model.
   b. Estimate the value of the stock in 2013.
   c. Predict when the value of the stock will be $35.
   d. What is the v-intercept of the model? What does it mean in this situation?

31. Let a be the altitude (in thousands of feet) of an airplane at t minutes since the airplane began its descent. Some pairs of values of t and a are shown in Table 40.

Table 40 Altitudes of an Airplane

t (minutes)	a (thousands of feet)
0	36
5	30
10	24
15	18
20	12

   a. Construct a scatterplot of the data. Then draw a linear model.
   b. Use your model to estimate the airplane's altitude 12 minutes after it began its descent.
   c. Use your model to estimate when the airplane will reach the ground.
   d. Assume that your line does a good job of modeling the airplane's descent up until the last 2 thousand feet, at which point the airplane then descends at a slower rate than before. Is the estimate you made in part (c) an underestimate or an overestimate? Explain.

32. Let a be the altitude (in feet) of a hot-air balloon after the air in the balloon is released for t minutes. Some pairs of values of t and a are shown in Table 41.

Table 41 Altitudes of a Hot-Air Balloon

t (minutes)	a (feet)
0	1800
1	1600
3	1200
4	1000
6	600

a. Construct a scatterplot of the data. Then draw a linear model.
b. Estimate the balloon's altitude after air has been released for 5 minutes.
c. Estimate when the balloon will reach the ground.
d. Assume that your line does a good job of modeling the balloon's descent up until the last 400 feet, at which point the balloon then descends at a faster rate than before. Is your estimate in part (c) an underestimate or an overestimate? Explain.

Concepts

33. Let n be the number (in millions) of U.S. passports issued in the year that is t years since 2005. Some pairs of values of t and n are shown in Table 42.

Table 42 Numbers of U.S. Passports Issued

t (years)	n (millions)
2	18.4
4	14.2
6	12.6
8	13.5
10	15.6
11	18.7

Source: *U.S. Department of State*

a. Construct a scatterplot of the data.
b. Are the variables t and n linearly related? Explain.

34. Let p be the percentage of flights that are delayed at t years since 2000. Some pairs of values of t and p are shown in Table 43.

Table 43 Percentages of Flights That Are Delayed

t (years)	p (percent)
7	24
8	25
9	20
10	19
11	20
12	15
13	17
14	25
15	20
16	16

Source: *Bureau of Transportation Statistics*

a. Construct a scatterplot of the data.
b. Are the variables t and p linearly related? Explain.

35. A student says the y-intercept of the ordered pair $(2, 5)$ is 5. Is the student correct? Explain.

36. A student says the x-intercept of the ordered pair $(-3, 4)$ is -3. Is the student correct? Explain.

37. A student says the x-intercept of a line is $(0, 2)$. Is the student correct? Explain.

38. A student says the y-intercept of a line is $(5, 0)$. Is the student correct? Explain.

39. A student says the x-intercept of a line is 5. Is the student correct? Explain.

40. Are there any lines for which the x-intercept is the same point as the y-intercept? If yes, sketch such a line, and what is that point? If no, explain why not.

41. Sketch three distinct lines that all have the same x-intercept.

42. Sketch three distinct lines that all have the same y-intercept.

43. a. Sketch a nonvertical line in a coordinate system. Find any outputs for the given input. State how many outputs there are for that single input.
    i. the input 2
    ii. the input 4
    iii. the input -3
   b. For your line, a single input leads to how many outputs? Explain.
   c. For *any* nonvertical line, a single input leads to how many outputs? Explain.

44. Explain why the x-coordinate of a y-intercept of a line is 0.

45. Give an example of a linear model other than the ones in the textbook.

46. In your own words, describe the meaning of *linear model*. (See page 9 for guidelines on writing a good response.)

47. Describe how to find a linear model of a situation and how to use the model to make estimates and predictions. (See page 9 for guidelines on writing a good response.)

1.4 Approximate Linear Relationships

Objectives

» Describe the meaning of *approximately linearly related*.

» Use a linear model to make estimates and predictions.

» Find errors in estimations.

» Identify when *model break-down* occurs.

In this section, we will use a line to model a situation in which data points lie close to the line, but not necessarily on the line.

Modeling When Variables Are Approximately Linearly Related

Example 1 Using a Line to Model Data

The total profit for the top-100-grossing concert tours are shown in Table 44 for various years. Let p be the total profit (in billions of dollars) at t years since 2000. Construct a scatterplot of the data, and draw a line that comes close to the points of the scatterplot.

Solution

First, we list values of t and p in Table 45. For example, $t = 0$ represents 2000 because 2000 is 0 years after 2000; and $t = 3$ represents 2003 because 2003 is 3 years after 2000.

Next, we construct a scatterplot in Fig. 51. It makes sense to think of t as the explanatory variable, so we let the horizontal axis be the t-axis. Because p is the response variable, the vertical axis is the p-axis.

Table 44 Total Profit for Top-100-Grossing Concert Tours

Year	Total Profit (billions of dollars)
2000	1.50
2003	1.95
2006	2.37
2009	2.53
2012	2.53
2015	3.12

Source: *Pollstar*

Table 45 Using Values of t to Stand for the Years

Number of Years since 2000 t	Total Profit (billions of dollars) p
0	1.50
3	1.95
6	2.37
9	2.53
12	2.53
15	3.12

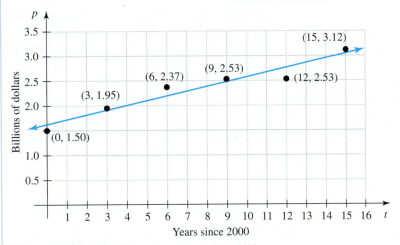

Figure 51 Total-profit scatterplot and model

Then we sketch a line that comes close to points of the scatterplot (see Fig. 51). The line needn't contain any of the points, but it should come close to all of them. Figure 52 shows that many such lines are possible.

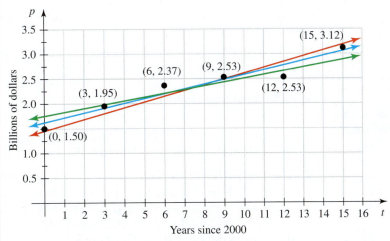

Figure 52 A few of the many reasonable linear models

In Fig. 51, we sketched a line that describes the total-profit situation. However, this description is not exact. For example, the line does not describe exactly what happened in the years 2000, 2003, 2006, 2009, 2012, and 2015 because the line does not contain any of the data points. However, the line does come pretty close to these data points, so it suggests pretty good approximations for those years.

If the points in a scatterplot of data lie close to (or on) a line, we say the relevant variables are **approximately linearly related**. For the concert-tour situation, the variables t and p are approximately linearly related.

Using a Linear Model to Make Estimates and Predictions

Because all the total-profit data points lie close to our linear model in Fig. 51, it seems reasonable that data points for the years between 2000 and 2015 that are not shown in Table 44 might also lie close to the line. Similarly, it is reasonable that data points for at least a few years before 2000 and for at least a few years after 2015 might also lie near the line.

| Example 2 | Making Estimates and a Prediction |

1. Use the linear model shown in Fig. 51 to estimate the total profit in 2006.
2. Use the linear model to estimate the total profit in 2012.
3. Use the model to predict in which year the total profit will be \$3.6 billion.

Solution

1. The year 2006 corresponds to $t = 6$ because $2006 - 2000 = 6$. The blue arrows in Fig. 53 show that the input $t = 6$ leads to the approximate output $p = 2.2$. So, according to the model, the total profit in 2006 was approximately \$2.2 billion.

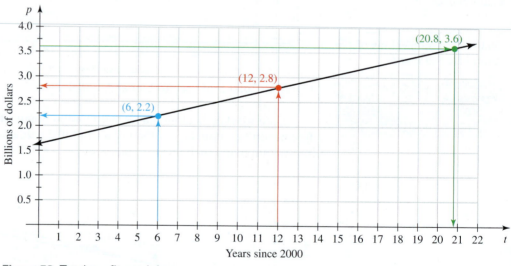

Figure 53 Total-profit model

Table 46 Total Profit

Year	Total Profit (billions of dollars)
2000	1.50
2003	1.95
2006	2.37
2009	2.53
2012	2.53
2015	3.12

The actual total profit in 2006 was \$2.37 billion (see Table 46). We will compute the error in the estimate in Example 3.

2. The year 2012 corresponds to $t = 12$ because $2012 - 2000 = 12$. The red arrows in Fig. 53 show that the input $t = 12$ leads to the approximate output $p = 2.8$. So according to the model, the approximate total profit in 2012 was \$2.8 billion.

The actual total profit in 2012 was \$2.53 billion (see Table 46). We will compute the error in the estimate in Example 3.

3. The green arrows in Fig. 53 show that the output $p = 3.6$ originates from the approximate input $t = 20.8 \approx 21$. So, according to the linear model, the total profit in about $2000 + 21 = 2021$ will be $3.6 billion.

Because the total profit in 2015 was $3.12 billion and total profit has been increasing, it follows that the model would predict that the total profit sometime *after* 2015 would be $3.6 billion, which checks with our result of 2021.

We construct a scatterplot of data to determine whether the relevant variables are approximately linearly related. If so, we draw a line that comes close to the data points and use the line to make estimates and predictions.

WARNING   It is a common error to try to find a line that contains the greatest number of points. However, our goal is to find a line that comes close to *all* the data points. For example, even though model 1 in Fig. 54 does not contain any of the data points shown, it fits the complete group of data points much better than does model 2, which contains three data points.

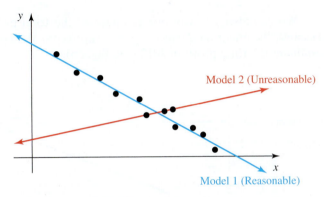

Figure 54 Comparing the fit of two models

Errors in Estimations

WARNING   It is also a common error to confuse the meaning of data points and points that lie on a linear model. Data points are accurate descriptions of an authentic situation. Points on a model may or may not be accurate descriptions.

For example, the data in Table 46 are accurate values for the total profits for the given years. For the linear model in Fig. 51 (page 37), some points on the line describe the situation well, but some points on the line do not. The advantage of using the linear model is that we can estimate the total profit for years other than those in Table 46.

The **error** in an estimate is the amount by which the estimate differs from the actual value. For an overestimate, the error is positive. For an underestimate, the error is negative. If the estimate is equal to the actual value, then the error is 0.

Example 3   Calculating Errors

1. In Example 2, we estimated that the total profit was $2.2 billion in 2006. The actual total profit was $2.37 billion (see Table 46). Calculate the error in the estimate.
2. In Example 2, we estimated that the total profit was $2.8 billion in 2012. The actual total profit was $2.53 billion (see Table 46). Calculate the error in the estimate.

Solution

1. Because $2.37 - 2.2 = 0.17$ and we underestimated the actual total profit, the error is -0.17 billion dollars.
2. Because $2.8 - 2.53 = 0.27$ and we overestimated the actual total profit, the error is 0.27 billion dollars.

By viewing the total-profit scatterplot and model in the same coordinate system, we can see why our estimate of the total profit in 2006 is an underestimate. Because the linear model at $(6, 2.2)$ is *below* the data point $(6, 2.37)$, the model *underestimates* the total profit in 2006 (see Fig. 55).

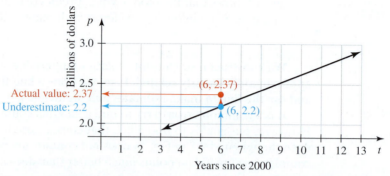

Figure 55 Comparing the data point and the model for 2006

We can also see why our estimate of the total profit in 2012 is an overestimate. Because the linear model at $(12, 2.8)$ is *above* the data point $(12, 2.53)$, the model *overestimates* the total profit in 2012 (see Fig. 56).

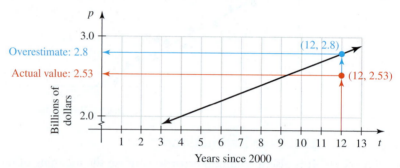

Figure 56 Comparing the data point and the model for 2012

Underestimates and Overestimates

Suppose that an explanatory variable t and a response variable p are approximately linearly related. For a given data point (a, b),

- If a linear model is below the data point, the model underestimates the value of p when $t = a$.
- If a linear model is above the data point, the model overestimates the value of p when $t = a$.

Model Breakdown

If a model gives an estimate that is not reasonably close to the actual value, we say that *model breakdown* has occurred. We will see another type of model breakdown in Example 4.

Example 4   Intercepts of a Linear Model; Model Breakdown

The percentages of drivers who were stopped by police are shown in Table 47 for various ages.

Table 47 Percentage of Drivers Stopped by Police

Age Group (years)	Age Used to Represent Age Group (years)	Percent
16–19	17.5	11.0
20–29	24.5	13.0
30–39	34.5	9.8
40–49	44.5	8.4
50–59	54.5	7.0
over 59	65.0	4.1

Source: *Bureau of Justice Statistics*

1. Let p be the percentage of drivers at age a years who were stopped by police. Find a linear model that describes the relationship between a and p.
2. Estimate at what age 12% of drivers were stopped by police.
3. Find the p-intercept of the model. What does the p-intercept mean in this situation?
4. Find the a-intercept of the linear model. What does the a-intercept mean in this situation?
5. Use the model to estimate the percentage of 110-year-old drivers who were stopped by police.

Solution

1. We construct a scatterplot (see Fig. 57). It appears a and p are approximately linearly related, so we sketch a line that comes close to the data points.

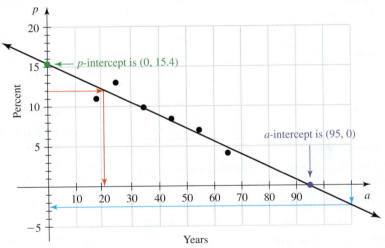

Figure 57 Driver scatterplot and model

2. The red arrows in Fig. 57 show that the output $p = 12$ originates from the approximate input $a = 21$. According to the model, 12% of 21-year-old drivers were stopped by police.
3. The approximate p-intercept is $(0, 15.4)$, or $p = 15.4$, when $a = 0$. According to the model, 15.4% of 0-year-old drivers were stopped by police. This estimate doesn't make sense.
4. The approximate a-intercept is $(95, 0)$, or $p = 0$, when $a = 95$. According to the model, not one 95-year-old driver was stopped by police. However, one would expect that at least one 95-year-old driver was stopped by police.
5. The blue arrows in Fig. 57 show that the input $a = 110$ leads to the approximate output $p = -2.5$. So, according to the model, -2.5% of 110-year-old drivers were stopped by police. This estimate doesn't make sense.

In Problem 5 in Example 4, we found that when $a = 110, p = -2.5$. The model estimates -2.5% of 110-year-old drivers were stopped by police, which does not make sense. So, $a = 110$ is *not* an input and $p = -2.5$ is *not* an output. This is an example of model breakdown.

> **Definition**   Model breakdown
>
> When a model yields a prediction that does not make sense or an estimate that is not a good approximation, we say that **model breakdown** has occurred.

If you are using a model to make an estimate or a prediction and model breakdown occurs, you should describe the result, state that model breakdown occurred, and explain why you know it has occurred.

In Section 1.3, we discussed why your estimates and predictions in the homework exercises will likely be different from the answers given near the end of this textbook. Here we note another reason this will likely happen: If two variables are approximately linearly related, there will be many reasonable linear models to choose from. However, if you do a careful job, your results should be close to those in the textbook.

 # Group Exploration

Section Opener: Approximately linearly related variables

The percentages of CEOs in Fortune 500 companies who are women are shown in Table 48 for various years.

Table 48 Percentages of CEOs in Fortune 500 Companies Who Are Women

Year	Percent
2003	1.4
2006	2.0
2009	3.0
2012	3.6
2015	4.0

Source: *Catalyst*

Let p be the percentage of CEOs who are women at t years since 2000.

1. Construct a scatterplot of the data.
2. Draw a (straight) line that comes close to all the data points in your scatterplot.
3. Use your line to estimate the percentage of CEOs in Fortune 500 companies who were women in 2013. On the basis of that percentage, how many female CEOs were there? There are 500 companies in the Fortune 500. Assume there is one CEO in each of these companies.
4. Use your line to estimate when 2.5% of CEOs in Fortune 500 companies were women.
5. What is the p-intercept of the linear model? What does it mean in this situation?
6. In 2014, 26 of the 500 Fortune companies had female CEOs. Did the percentage of CEOs who were women increase each year, according to the model? Is this actually true?

Tips for Success   Practice Exams

When studying for an exam (or quiz), try creating your own exam to take for practice. To create your exam, select several homework exercises from each section on which you will be tested. Choose a variety of exercises that address concepts your instructor has emphasized. Your test should include many exercises that are moderately difficult and some that are challenging. Completing such a practice test will help you reflect on important concepts and pin down what types of problems you need to study more.

It is a good idea to work on the practice exam for a predetermined period. Doing so will help you get used to a timed exam, build your confidence, and lower your anxiety about the real exam.

If you are studying with another student, you can each create a test and then take each other's test. Or you can create a test together and each take it separately.

Homework 1.4

For extra help ▶ **MyLab Math**   📺 Watch the videos in MyLab Math

1. Some ordered pairs are listed in Table 49.

Table 49 Some Ordered Pairs

x	y
1	13
3	9
5	8
7	4
9	2

a. Construct a scatterplot of the points shown in Table 49.
b. Are the variables x and y linearly related, approximately linearly related, or neither?
c. Draw a line that comes close to the points in your scatterplot.
d. Which point on your line has x-coordinate 8?
e. Which point on your line has y-coordinate 6?
f. What is the y-intercept of your line?
g. What is the x-intercept of your line?

2. Some ordered pairs are listed in Table 50.

Table 50 Some Ordered Pairs

x	y
−8	−5
−5	−2
−2	0
0	5
3	7

a. Construct a scatterplot of the points shown in Table 50.
b. Are the variables x and y linearly related, approximately linearly related, or neither?
c. Draw a line that comes close to the points in your scatterplot.
d. Which point on your line has x-coordinate −1?
e. Which point on your line has y-coordinate −3?
f. What is the y-intercept of your line?
g. What is the x-intercept of your line?

3. The scatterplot in Fig. 58 compares the prices of hot dogs and soft drinks at all the Major League Baseball (MLB) stadiums (Source: *Team Marketing Report*).

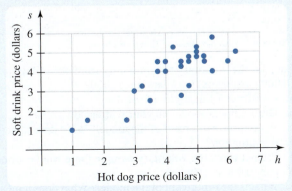

Figure 58 Exercise 3

Let h be the price of a hot dog and s be the price of a soft drink, both in dollars.
a. Are the variables h and s linearly related, approximately linearly related, or neither?
b. Trace the scatterplot in Fig. 58 carefully onto a piece of paper. Then, on the same scatterplot, draw a linear model.
c. Estimate the price of a soft drink at a stadium that charges $4 for a hot dog.
d. Estimate the price of a hot dog at a stadium that charges $5 for a soft drink.
e. Does the line go up or down from left to right? Explain why that makes sense in this situation.

4. The scatterplot in Fig. 59 compares carbohydrates and calories for 29 pizzas made by six of the leading pizza companies (Sources: *Domino's, Little Caesar's, Papa John's, Pizza Hut, DiGiorno Frozen, Kashi Frozen*).

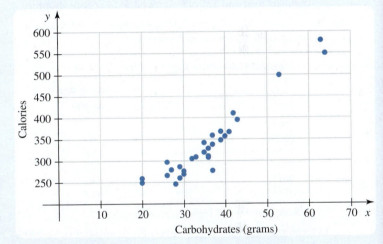

Figure 59 Exercise 4

Let x be the amount of carbohydrates (in grams) and y be the number of calories.
a. Are the variables x and y linearly related, approximately linearly related, or neither?
b. Trace the scatterplot in Fig. 59 carefully onto a piece of paper. Then, on the same scatterplot, draw a linear model.
c. Estimate the number of calories for a pizza with 30 carbohydrates.
d. Estimate the amount of carbohydrates for a pizza with 450 calories.
e. Does the line go up or down from left to right? Explain why that makes sense in this situation.

5. The amounts consumers plan to spend on back-to-school shopping are shown in Table 51 for various years.

Table 51 Amounts Consumers Plan to Spend on Back-to-School Shopping

Year	Amount (billions of dollars)
2007	18.4
2009	17.4
2011	22.8
2013	26.7
2015	24.9
2016	27.3

Source: *Euromonitor*

Let a be the amount (in billions of dollars) consumers plan to spend on back-to-school shopping at t years since 2005. For example, $t = 2$ represents 2007, and $t = 4$ represents 2009.
a. Construct a scatterplot of the data.
b. Draw a linear model on your scatterplot.
c. Use your model to predict the amount consumers planned to spend on back-to-school shopping in 2014.
d. Use your model to estimate in which year consumers planned to spend $21 billion on back-to-school shopping.

6. The average ages of light vehicles in the United States are shown in Table 52 for various years.

Table 52 Average Ages of Light Vehicles

Year	Average Age (years)
2000	8.9
2003	9.7
2006	9.9
2009	10.3
2012	11.2
2015	11.4

Source: *Bureau of Transportation Statistics*

Let a be the average age (in years) of light vehicles since 2000. For example, $t = 0$ represents 2000, and $t = 3$ represents 2003.
a. Construct a scatterplot of the data.
b. Draw a linear model on your scatterplot.
c. Use your model to estimate the average age of light vehicles in 2010.
d. Use your model to estimate when the average age of light vehicles was 9.3 years.

7. The numbers of animal and plant species in the world that are listed as threatened (critically endangered, endangered, or vulnerable) are shown in Table 53 for various years.

Table 53 Numbers of Threatened Species

Year	Number of Species Listed (thousands)
2000	11.0
2003	12.3
2006	16.1
2009	17.3
2012	20.2
2015	23.3
2016	23.9

Source: *IUCN*

Let n be the number (in thousands) of species that are listed as threatened at t years since 2000.
a. Construct a scatterplot of the data.
b. Are the variables t and n linearly related, approximately linearly related, or neither? Explain.
c. Draw a linear model on your scatterplot.
d. Estimate when 22 thousand species were listed.
e. Predict the number of species that will be listed in 2021.

8. The percentages of Americans who say there should be a ban on the possession of handguns are shown in Table 54 for various years.

Table 54 Percentages of Americans Who Say There Should Be a Ban on the Possession of Handguns

Year	Percent
1992	43
1995	38
2000	36
2005	35
2010	28
2015	27

Source: *Gallup*

Let p be the percentage of Americans who say there should be a ban on the possession of handguns at t years since 1990.
a. Construct a scatterplot of the data.
b. Are the variables linearly related, approximately linearly related, or neither? Explain.
c. Draw a linear model on your scatterplot.
d. Use your model to estimate when 31% of Americans said there should be a ban on the possession of handguns.
e. Use your model to predict the percentage of Americans in 2021 who will say there should be a ban on the possession of handguns.

9. If there are too many ticketed passengers for a flight, a person can volunteer to be "bumped" onto another flight. The voluntary bumping rates for large U.S. airlines (number of bumps per 10,000 passengers, January through September) are shown in Table 55 for various years.

Table 55 Voluntary Bumping Rates of U.S. Airlines

Year	Voluntary Bumping Rate (number of bumps per 10,000 passengers)
2003	15
2006	11
2009	12
2012	9
2015	8

Source: *U.S. Department of Transportation*

Let r be the voluntary bumping rate (number of bumps per 10,000 passengers) at t years since 2000.
a. Construct a scatterplot of the data.
b. Draw a linear model on your scatterplot.
c. What is the r-intercept of the model? What does it mean in this situation?
d. Predict when the voluntary bumping rate will be 4 bumps per 10,000 passengers.
e. What is the t-intercept of the model? What does it mean in this situation?

10. The death rate from heart disease in the United States has decreased greatly since 1960 (see Table 56).

Table 56 Death Rates Due to Heart Disease

Year	Death Rate (number of deaths per 100,000 people)
1960	559
1970	492
1980	409
1990	317
2000	253
2010	177
2015	169

Source: *U.S. Center for Health Statistics*

Let r be the death rate (number of deaths per 100,000 people) from heart disease for the year that is t years since 1960.
a. Construct a scatterplot of the data.
b. Draw a linear model on your scatterplot.
c. What is the t-intercept of the model? What does it mean in this situation?
d. Predict the death rate from heart disease in 2020. The U.S. population will be about 335 million in that year. Predict the number of Americans who will die of heart disease in 2020.

11. The percentages of Americans who went to the movies at least once in the past year are shown in Table 57 for various age groups.

Table 57 Percentages of Americans Who Go to the Movies

Age Group (years)	Age Used to Represent Age Group (years)	Percent
18–24	21.0	76
25–34	29.5	69
35–44	39.5	68
45–54	49.5	60
55–64	59.5	51
65–74	69.5	44
over 74	80.0	31

Source: *U.S. National Endowment for the Arts*

Let p be the percentage of Americans at age a years who go to movies.
a. Construct a scatterplot of the data.
b. Draw a linear model on your scatterplot.
c. Estimate what percentage of Americans at age 19 go to the movies.
d. At what age do half of Americans go to the movies?
e. What is the a-intercept of the model? What does it mean in this situation?

12. The percentages of elementary school students who participated in an after-school arts activity at least once a month are shown in Table 58 for various household incomes.

Table 58 Percentages of Elementary School Students Who Participate in an After-School Arts Activity

Income Group (thousands of dollars)	Income Used to Represent Income Group (thousands of dollars)	Percent
under 15	7.5	5.7
15–30	22.5	9.3
30–50	40.0	13.6
50–75	62.5	20.3
over 75	100.0	29.8

Source: *Department of Education*

Let p be the percentage of elementary school students from households with income d (in thousands of dollars) who participate in an after-school arts activity at least once a month.
a. Construct a scatterplot of the data.
b. Draw a linear model on your scatterplot.
c. Estimate the percentage of elementary school students from households with income $50 thousand who participate in an after-school arts activity.
d. Estimate the income of households from which 25% of elementary school students participate in an after-school arts activity.
e. What is the p-intercept of the model? What does it mean in this situation?

13. The average salaries of public school teachers are shown in Table 59 for various years.

Table 59 Average Salaries of Public School Teachers

Year	Average Salary (in thousands of dollars)
1990	31
1995	37
2000	42
2005	48
2010	55
2014	57

Source: *National Center for Education Statistics*

Let s be the average salary (in thousands of dollars) at t years since 1990.
a. Construct a scatterplot of the data.
b. Draw a linear model on your scatterplot.
c. What is the s-intercept of the model? What does it mean in this situation?
d. Predict the average salary in 2022.
e. Predict when the average salary will reach $65 thousand.

14. The numbers of police fatalities are shown in Table 60 for various years.

Table 60 Numbers of Police Fatalities

Year	Number of Fatalities
1994	179
1997	173
2000	162
2003	150
2006	156
2009	125
2012	131
2015	123

Source: *National Law Enforcement Officers Memorial*

Let n be the number of police fatalities in the year that is t years since 1990.
  a. Construct a scatterplot of the data.
  b. Draw a linear model on your scatterplot.
  c. What is the n-intercept? What does it mean in this situation?
  d. Use your model to predict the number of police fatalities in 2020.
  e. Use your model to predict in which year there will be 100 police fatalities.

15. The loudness of sound can be measured by using a *decibel scale.* Some examples of sounds at various sound levels are listed in Table 61.

Table 61 Examples of Sound Levels

Sound Level (decibels)	Example
0	Faintest sound heard by humans
27	Bedroom at night
50	Dishwasher next room
71	Vacuum cleaner at 10 ft
82	Garbage disposal at 3 ft
98	Inside subway train
111	Rock band

Source: Caltrans Noise Manual, *M. M. Hatano*

The sound level of music from a Pioneer MT-2000® stereo-CD-receiver system is controlled by the system's volume number. The sound levels of music for various volume numbers are shown in Table 62.

Table 62 Sound Levels of Music Played by a Stereo

Volume Number	Sound Level (decibels)
6	60
8	66
10	69
12	74
14	78
16	82
18	86
20	90

Source: *J. Lehmann*

Let S be the sound level (in decibels) for a volume number n.
  a. Construct a scatterplot of the data in Table 62.
  b. Draw a linear model on your scatterplot.
  c. Use your model to estimate the sound level when the volume number is 19.
  d. Use your model to estimate for what volume number the sound level is comparable to that of a vacuum cleaner at 10 ft (see Table 61).
  e. What is the S-intercept? What does it mean in this situation?

16. DATA. Five human subjects were injected with various concentrations of LSD, and with each concentration they solved as many simple arithmetic problems that they could in three minutes. Their work was recorded as the percentage of the number of problems they could do before being injected with LSD. The LSD concentrations and the average test percentages are shown in Table 63.

Table 63 LSD Concentrations and Average Test Percentages

LSD Concentration (nanograms per milliliter of plasma)	Average Test Percentage
1.17	78.93
2.97	58.20
3.26	67.47
4.69	37.47
5.83	45.65
6.00	32.92
6.41	29.97

Source: Correlation of Performance Test Scores with "Tissue Concentration" of Lysergic Acid Diethylamide in Human Subjects. *Wagner et al.*

Let p be the average test percentage when the LSD concentration was c nanograms per milliliter of plasma.
  a. Construct a scatterplot of the data.
  b. Draw a linear model on your scatterplot.
  c. Use your model to estimate what the average test percentage would be for an LSD concentration of 4 nanograms per milliliter of plasma.
  d. What is the c-intercept? What does it mean in this situation?
  e. What is the p-intercept? What does it mean in this situation? Has model breakdown occurred? Explain.

17. The numbers of ride-related injuries at fixed-site amusement parks are shown in Table 64 for various years.

Table 64 Numbers of Ride-Related Injuries at Fixed-Site Amusement Parks

Years	Number of Injuries (thousands)
2003	2.0
2004	1.6
2005	1.8
2006	1.8
2007	1.7
2008	1.5
2009	1.2
2010	1.3
2011	1.2
2012	1.4
2013	1.4
2014	1.2

Source: *National Safety Council*

Let n be the number (in thousands) of ride-related injuries at fixed-site amusement parks in the year that is t years since 2000. A scatterplot of the data and a linear model are sketched in Fig. 60.

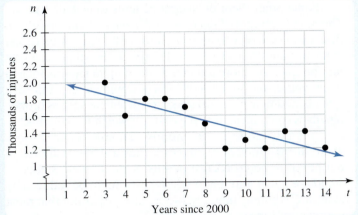

Figure 60 Exercises 17 and 18

a. Use the linear model to estimate the number of ride-related injuries in 2004.
b. What was the actual number of ride-related injuries in 2004?
c. Is the result you found in part (a) an underestimate or an overestimate? Explain how you can tell this from the graph of the scatterplot and the sketch of the model. Calculate the error in the estimate.

18. Refer to Exercise 17, including Table 64 and Fig. 60.
   a. Use the linear model to estimate the number of ride-related injuries in 2013.
   b. What was the actual number of ride-related injuries in 2013?
   c. Is the result you found in part (a) an underestimate or an overestimate? Explain how you can tell this from the graph of the scatterplot and the sketch of the model. Calculate the error in the estimate.

Large Data Sets

19. **DATA** Access the data about cruise ships, which are available at MyLab Math and at the Pearson Downloadable Student Resources for Math & Stats website. Let S be the crew size (in hundreds) for a cruise ship with length L hundred feet.
   a. Construct a scatterplot of the data. Print your scatterplot.
   b. Draw a linear model on your scatterplot.
   c. Find the L-intercept. What does it mean in this situation?
   d. Estimate the length of a cruise ship with 6 hundred crew members.
   e. Estimate the crew size for a 1000-foot-long cruise ship. [**Hint:** The units of L are *hundreds* of feet.]
   f. Estimate the crew size for two 500-foot-long cruise ships. Is your result less than, equal to, or greater than the result you found in part (e)?
   g. A student says that two 500-foot-long ships should require the same number of crew members as a 1000-foot-long ship because $500 + 500 = 1000$. What would you tell the student?

20. **DATA** Access the data about gas mileages of cars, which are available at MyLab Math and at the Pearson Downloadable Student Resources for Math & Stats website. Let H be the highway gas mileage and C be the city gas mileage, both in miles per gallon.
   a. Construct a scatterplot of the data. Let H be the explanatory variable and C be the response variable. Print your scatterplot.
   b. For cars with highway gas mileages between 12 and 36 miles per gallon, draw a linear model on your scatterplot.
   c. Estimate the city gas mileage for a car with highway gas mileage 30 miles per gallon.
   d. Estimate the highway gas mileage for a car with city gas mileage 14 miles per gallon.
   e. Find the H-intercept. What does it mean in this situation?
   f. Some of the data points in your scatterplot represent gas mileages of more than one car. On the basis of this fact, explain why you might not have drawn the best linear model.
   g. Now consider *all* the data points in your scatterplot. For data points that are far from your model, are they mostly above the model or below the model? What type of car do most of these data points represent?

Concepts

21. Suppose that an explanatory variable t and a response variable p are approximately linearly related. If a data point (c, d) is below a linear model, does the model underestimate or overestimate the value of p when $t = c$? Explain.

22. Suppose that an explanatory variable t and a response variable p are approximately linearly related. If a linear model is below a data point (c, d), does the model underestimate or overestimate the value of p when $t = c$? Explain.

23. When modeling a situation in which the variables are approximately linearly related, different students may all do good work yet not get the same results. Draw a scatterplot and at least two reasonable linear models to show how this is possible.

24. Which is more desirable, finding a linear model that contains several, but not all, data points or finding a linear model that does not contain any data points but comes close to all data points? Include in your discussion some sketches of scatterplots and linear models.

25. Compare the meaning of *linearly related* with that of *approximately linearly related*.

26. A student comes up with a shortcut for modeling a situation. Instead of plotting all the given data points, the student plots only two of the data points and draws a line that contains the two chosen points. Give an example to illustrate what can go wrong with this shortcut.

27. A person collects data by doing research. If the data points lie exactly on a line, will all points on the line describe the situation exactly? Explain.

28. Let t be the number of years since 2015. A student says $t = -2$ is an example of model breakdown because time can't be negative. Is the student correct? Explain.

Taking It to the Lab

Climate Change Lab

Many scientists are greatly concerned that the average temperature of the surface of Earth has increased since 1900 (see Table 65).

DATA. Table 65 Average Surface Temperatures of Earth

Year	Average Temperature (degrees Fahrenheit)	Year	Average Temperature (degrees Fahrenheit)
1900	57.1	1960	57.2
1905	56.8	1965	57.0
1910	56.6	1970	57.3
1915	57.0	1975	57.1
1920	56.9	1980	57.6
1925	56.9	1985	57.3
1930	57.1	1990	57.8
1935	57.0	1995	57.9
1940	57.3	2000	57.8
1945	57.3	2005	58.3
1950	56.9	2010	58.3
1955	57.0	2015	59.0

Source: *NASA–GISS*

Although it may not seem that the average temperatures shown in Table 65 have increased much, many scientists believe an increase as small as 3.6°F could be a dangerous climate change.[*] Global warming would cause the extinction of plants and animals, lead to severe water shortages, create more extreme weather events, increase the number of heat-related illnesses and deaths, and melt glaciers, which would raise ocean levels and, thus, submerge coastlands.

Despite these alarming predictions, not all experts are concerned. Robert Mendelsohn, an environmental economist at Yale University, argues that "global warming will increase agricultural production in the northern half of the United States" and that "the southern half will be able to maintain its current level of production." Even from a global perspective, he believes, the benefits of global warming will offset the damages.[†]

To test such theories, Peter S. Curtis, an ecologist at Ohio State University, and colleagues ran experiments that showed that increased carbon dioxide levels do in fact increase plant growth. However, the nutritional value of the produce was lower—so much lower that the increase in growth did not make up for the decrease in nutrition.[‡]

Some other theories suggest benefits to global warming. The scientific report *Impacts of a Warming Arctic* points out that it will be easier to extract oil from the Arctic due to less extensive and thinner sea ice. However, the study also says that oil spills are more difficult to clean up in icy seas than in open waters and that many species would suffer from such spills. In addition, potential structural problems such as broken pipelines could mean that the costs outweigh the benefits.[§]

A study done by economist Thomas Gale Moore at Stanford University suggests that a 4.5°F increase in temperature could reduce deaths in the United States by 40,000 per year and that medical costs might be reduced by at least $20 billion annually. Moore also points out that most people prefer warmer climates.[¶]

Most scientists do not share Moore's perspective. On a global scale, they believe warmer climates could increase the spread of diseases such as malaria and, thus, increase the global death rate and medical costs.

Although there may be some relatively small and short-lived benefits to global warming, the vast majority of scientists agree that global warming is already taking its toll and that further warming would bring catastrophic results.

Glaciologists report that, over the past century, glaciers around the globe have been melting.[‖] Biologists note that many species throughout the world have changed their habitats in search of cooler climates.[**] One species, the golden toad, was not able to migrate and as a result has become extinct due to heat stress.[††]

Looking to the future, an international study, the most comprehensive analysis of its kind, predicts that 15% to 37% of all species of plants and animals—well over a million species—will become extinct by 2050.[***] Klaus Toepfer, head of the United Nations Environment Programme (UNEP), said, "If one million species become extinct . . . it is not just the plant and animal kingdoms and the beauty of the planet that will suffer. Billions of people, especially in the developing world, will suffer too as they rely on nature for such essential goods and services as food, shelter and medicines."[†††]

[*]"Meeting the Climate Challenge: Recommendations of the International Climate Change Taskforce," The Institute for Public Policy Research/The Center for American Progress/The Australian Institute, January 2005.

[†]From *The Impact of Climate Change on the United States Economy*, R. Mendelsohn and J. Neumann (eds.), Cambridge University Press.

[‡]L. M. Jablonski, X. Wang, and P. S. Curtis, "Plant Reproduction under Elevated CO_2 Conditions," *New Phytologist* (156):9–26, 2002.

[§]Report of the Arctic Climate Impact Assessment, Cambridge University Press, 2004.

[¶]"In sickness or in health: The Kyoto Protocol versus global warming," Hoover Institution, Stanford University, August 2000.

[‖]National Snow and Ice Data Center, 2003.

[**]T. L. Root et al., "Fingerprints of global warming on wild animals and plants," January 2, 2003, *Nature*, 421:57–60.

[††]J. A. Pounds et al., "Biological response to climate change on a tropical mountain," 1999, *Nature (London)*, 398(6728):611–615.

[***]C. D. Thomas et al., "Extinction risk from climate change," January 8, 2004, *Nature*, 427:145–148.

[†††]Reported by UNEP (United Nations Environment Programme), January 8, 2004.

Analyzing the Situation

1. Discuss those theories that describe the benefits of global warming and whether they are likely correct.

2. What are some possible costs of global warming? In your opinion, do the possible costs outweigh the possible benefits? Explain.

3. Let A be the average surface temperature of Earth (in degrees Fahrenheit) at t years since 1900. *Carefully* construct a scatterplot of the data in Table 65.

4. On the basis of your scatterplot, in what year is it first clear that global warming is occurring? Explain. Also, explain why it makes sense that the first World Climate Conference convened in 1979.

5. Are the variables t and A approximately linearly related for the years 1900–2015? Are the variables approximately linearly related for the years 1965–2010? Explain.

6. Draw a line that comes close to the points in your scatterplot for the years 1965–2010.

7. Use your linear model to estimate the average global temperature in 2015. Is your result an underestimate or an overestimate? Describe what *might* be starting to happen.

Volume Lab

In this lab, you will explore the relationship between the volume of some water in a cylinder and the height of the water. Check with your instructor whether you should collect your own data or use the data listed in Table 66.

Table 66 Heights of Water in a Cylinder with Radius 4.45 Centimeters

Height (centimeters)	Volume (ounces)
0	0
0.9	2
1.9	4
2.9	6
3.8	8
4.8	10
5.7	12

Source: *J. Lehmann*

Materials

You will need the following items:

- A "perfect" cylinder (the diameter of the top should equal the diameter of the base) that can hold at least 8 ounces of water
- At least 8 ounces of water
- A $\frac{1}{4}$-cup measuring cup
- A ruler

Recording of Data

Pour $\frac{1}{4}$ cup (2 ounces) of water into the cylinder, and measure the height of the water, using units of centimeters. Then continue adding $\frac{1}{4}$ cup of water and measuring the height after you have added each $\frac{1}{4}$ cup until there is at least 8 ounces of water in the cylinder. Also, measure the height of the cylinder in units of centimeters.

Analyzing the Data

1. Display your data in a table similar to Table 66. If you are using the data in Table 66, the height of the cylinder is 12 centimeters.

2. Let V be the volume of water (in ounces) in the cylinder when the height is h centimeters. Assume that h is the explanatory variable. Construct a scatterplot of the data.

3. Draw a linear model on your scatterplot.

4. What is the V-intercept of your model? What does it mean in this situation?

5. Use the model to estimate the volume of water when the height of the water is 3 centimeters.

6. Use the model to estimate the height of 7 ounces of water in the cylinder.

7. What is the height of the cylinder? Use this height and the model to estimate the maximum amount of water that the cylinder can hold.

8. Indicate on your graph of the model where model breakdown occurs. Also, describe in words when model breakdown occurs.

Linear Graphing Lab: Topic of Your Choice

Your objective in this lab is to use a linear model to describe some authentic situation. Choose a situation that has not been discussed in this text. Your first task will be to find some data. Almanacs, newspapers, magazines, and scientific journals are good resources. You may want to try searching on the Internet. Or you can conduct an experiment. Choose something that interests you!

Analyzing the Situation

1. What two quantities did you explore? Define variables for the quantities. Include units in your definitions.

2. Which variable is the explanatory variable? Which variable is the response variable? Explain.

3. Describe how you found your data. If you conducted an experiment, provide a careful description with specific details of how you ran your experiment. If you didn't conduct an experiment, state the source of your data.

4. Include a table of your data.

5. Construct a scatterplot of your data. (If your data are not approximately linear, find some data that are.)

6. Draw a linear model on your scatterplot.

7. Choose a value for your explanatory variable. On the basis of your chosen value for the explanatory variable, use your model to find a value for your response variable. Describe what your result means in the situation you are modeling.

8. Choose a value for your response variable. On the basis of your chosen value for the response variable, use your model to find a value for your explanatory variable. Describe what your result means in the situation you are modeling.

9. Comment on your lab experience.
   a. For example, you might address whether this lab was enjoyable, insightful, and so on.
   b. Were you surprised by any of your findings? If so, which ones?
   c. How would you improve your process for this lab if you were to do it again?
   d. How would you improve your process if you had more time and money?

 # Chapter Summary

Key Points of Chapter 1

Section 1.1   Variables and Constants

Variable	A **variable** is a symbol that represents a quantity that can vary.
Constant	A **constant** is a symbol that represents a specific number (a quantity that does *not* vary).
Counting numbers or natural numbers	The **counting numbers**, or **natural numbers**, are the numbers $1, 2, 3, 4, 5, \ldots$.
Integers	The **integers** are the numbers $\ldots, -3, -2, -1, 0, 1, 2, 3, \ldots$.
Rational numbers	The **rational numbers** are the numbers that can be written in the form $\frac{n}{d}$, where n and d are integers and d is nonzero.
Real numbers	The **real numbers** are all the numbers represented on the number line.
Irrational numbers	The **irrational numbers** are the real numbers that are not rational.
Average or mean	To find the **average** (or **mean**) of a group of numbers, we divide the sum of the numbers by the number of numbers in the group.
Negative numbers and positive numbers	The **negative numbers** are the real numbers less than 0, and the **positive numbers** are the real numbers greater than 0.

Section 1.2   Scatterplots

Scatterplot	A **scatterplot** is a graph of plotted ordered pairs.
Identifying explanatory and response variables	Assume that an authentic situation can be described by using the variables t and p, and assume that t affects (explains) p. Then • We call t the **explanatory variable** (or **independent variable**). • We call p the **response variable** (or **dependent variable**). • For an **ordered pair** (a, b), we write the value of the explanatory variable in the first (left) position and the value of the response variable in the second (right) position.
Columns of tables and axes of coordinate systems	Assume that an authentic situation can be described by using two variables. Then • For tables, the values of the explanatory variable are listed in the first column and the values of the response variable are listed in the second column. • For coordinate systems, the values of the explanatory variable are described by the horizontal axis and the values of the response variable are described by the vertical axis.

Section 1.3   Exact Linear Relationships

Linearly related	If two quantities of an authentic situation are described accurately by a line, then the quantities (and the variables representing those quantities) are **linearly related**.
Model	A **model** is a mathematical description of an authentic situation.

Section 1.3 Exact Linear Relationships (*Continued*)

Linear model	A **linear model** is a nonvertical line that describes the relationship between two quantities in an authentic situation.
Input and output	An **input** is a permitted value of the *explanatory* variable that leads to at least one **output**, which is a permitted value of the *response* variable.
Determine whether data points lie on a line	We construct a scatterplot of data to determine whether the data points lie on a line. If the points lie on a line, then we draw the line and use it to make estimates and predictions.
Intercepts of a line	An **intercept** of a line is any point where the line and an axis (or axes) of a coordinate system intersect. There are two types of intercepts of a line sketched on a coordinate system with an *x*-axis and a *y*-axis: • An ***x*-intercept** of a line is a point where the line and the *x*-axis intersect. The *y*-coordinate of an *x*-intercept is 0. • A ***y*-intercept** of a line is a point where the line and the *y*-axis intersect. The *x*-coordinate of a *y*-intercept is 0.

Section 1.4 Approximate Linear Relationships

Approximately linearly related	If the points in a scatterplot of data lie close to (or on) a line, we say that the relevant variables are **approximately linearly related**.
Determine whether variables are approximately linearly related	We construct a scatterplot of data to determine whether the relevant variables are approximately linearly related. If so, we draw a line that comes close to the data points and use the line to make estimates and predictions.
Underestimates and overestimates	Suppose that an explanatory variable t and a response variable p are approximately linearly related. For a given data point (a, b), • If a linear model is below the data point, the model underestimates the value of p when $t = a$. • If a linear model is above the data point, the model overestimates the value of p when $t = a$.
Model breakdown	When a model yields a prediction that does not make sense or an estimate that is not a good approximation, we say that **model breakdown** has occurred.

Chapter 1 Review Exercises

1. Let B be the total box office gross (in billions of dollars) from U.S. and Canada movie theaters. For 2015, the value of B is 11.13 (Source: *Box Office Mojo*). What does that mean in this situation?

2. Let t be the number of years since 1995. What does $t = 26$ represent?

3. Choose a variable name for the percentage of students who are full-time students at a college. Give two numbers that the variable can represent and two numbers that it cannot represent.

4. A rectangle has a perimeter of 40 inches. Let W be the width, L be the length, and P be the perimeter, all with units in inches.
   a. Sketch three possible rectangles with a perimeter of 40 inches.
   b. Which of the symbols W, L, and P are variables? Explain.
   c. Which of the symbols W, L, and P are constants? Explain.

5. Graph the numbers -2, $-\frac{3}{2}$, 0, 1, $\frac{5}{2}$, and 3 on a number line.

6. Graph the negative integers between -5 and 5 on a number line.

7. Here are a company's profits and losses for various years: profit of $2 million, loss of $4 million, loss of $1 million, profit of $3 million. Let p be the profit (in millions of dollars). Use points on a number line to describe the profits and losses of the company.

8. Plot the points $(2, 4)$, $(-3, -1)$, $(5, -2)$, and $(-4, 5)$ in a coordinate system.

9. What is the y-coordinate of the ordered pair $(3, -6)$?

10. What is the x-coordinate of the ordered pair $(-4, -7)$?

For Exercises 11 and 12, identify the explanatory variable and the response variable.

11. Let p be the percentage of Americans at age a years who own a home.

12. Let a be the average salary (in dollars) for a person with t years of education.

13. Let n be the total number of U.S. billionaires at t years since 2010. What does the ordered pair $(6, 540)$ mean in this situation (Source: *Forbes*)?

14. Let r be the annual revenue (in billions of dollars) from ADHD drugs at t years since 2000. What does the ordered pair $(14, 11)$ mean in this situation (Source: *IBISWorld*)?

15. Construct a scatterplot of the ordered pairs listed in Table 67.

Table 67 Some Ordered Pairs

x	y
−5	−2
−3	4
0	1
2	3
4	−5

16. The average gas mileages of cars are shown in Table 68 for various years.

Table 68 Average Gas Mileages of Cars

Year	Average Gas Mileage (miles per gallon)
1980	15
1990	19
2000	20
2010	22
2014	21

Source: *Federal Highway Administration*

Let g be the average gas mileage (in miles per gallon) of cars at t years since 1980.
a. Construct a scatterplot of the data.
b. For which of the years shown in Table 68 was the average gas mileage of cars the highest?
c. For which of the years shown in Table 68 was the average gas mileage of cars the lowest?

17. The countries with the top six percentages of electricity generated by nuclear power are shown in the bar graph in Fig. 61 (Source: *International Atomic Energy Agency*).

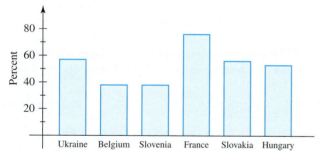

Figure 61 Exercise 17

a. Which country generates the largest percentage of its electricity by nuclear power? Estimate that percentage.
b. Of the countries included in Fig. 61, which two countries generate the smallest percentage of their electricity by nuclear power? Estimate that percentage.
c. Estimate the percentage of Ukraine's electricity that is generated by nuclear power.

For Exercises 18–23, refer to Fig. 62.

18. Find y when $x = -2$.
19. Find y when $x = 6$.
20. Find x when $y = -4$.
21. Find x when $y = 1$.

22. What is the y-intercept of the line?
23. What is the x-intercept of the line?

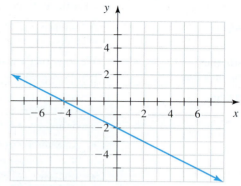

Figure 62 Exercises 18–23

24. Some ordered pairs are listed in Table 69.
a. Construct a scatterplot of the points shown in Table 69.
b. Draw a line that contains the points in your scatterplot.
c. Find y when $x = 11$.
d. Find x when $y = 5$.
e. What is the x-intercept of your line?
f. What is the y-intercept of your line?

Table 69 Some Ordered Pairs

x	y
1	11
2	10
5	7
8	4
9	3

25. Some ordered pairs are listed in Table 70.

Table 70 Some Ordered Pairs

x	y
1	10
2	8
3	6
4	4
5	2
6	0

a. Construct a scatterplot of the points shown in Table 70.
b. Are the variables x and y linearly related, approximately linearly related, or neither?

26. Let p be the annual profit (in millions of dollars) of a company at t years since 2010. Some pairs of values of t and p are shown in Table 71.

Table 71 Profits of a Company

t (years)	p (millions of dollars)
1	20
3	16
4	14
6	10
7	8

a. Construct a scatterplot of the data. Then draw a linear model.
b. Predict the profit in 2020.
c. Estimate when the profit was $18 million.
d. What is the *p*-intercept of the model? What does it mean in this situation?
e. What is the *t*-intercept of the model? What does it mean in this situation?

27. What is the *y*-coordinate of an *x*-intercept of a line?
28. Some ordered pairs are listed in Table 72.

Table 72 Some Ordered Pairs

x	y
1	23
3	17
6	10
8	9
9	4

a. Construct a scatterplot of the points shown in Table 72.
b. Are the variables *x* and *y* linearly related, approximately linearly related, or neither?
c. Draw a linear model on your scatterplot.
d. Which point on your line has *x*-coordinate 5?
e. Which point on your line has *y*-coordinate 20?
f. What is the *y*-intercept of your line?
g. What is the *x*-intercept of your line?

29. The percentages of American adults who are obese are shown in Table 73 for various years.

Table 73 Percentages of American Adults Who Are Obese

Year	Percent
2004	32.2
2006	34.3
2008	33.7
2010	35.7
2012	34.9
2014	37.7

Source: *National Center for Health Statistics*

Let *p* be the percentage of American adults who are obese at *t* years since 2000.
a. Construct a scatterplot of the data.
b. Draw a linear model on your scatterplot.

c. Use your model to predict the percentage of American adults who will be obese in 2021.
d. Use your model to predict when 41% of Americans will be obese.
e. The data in Table 73 were obtained by in-house clinical measurements. In a different study, Gallup used self-reported heights and weights to estimate the percentages of Americans who are obese. Why does it make sense that Gallop's estimate of 27.7% for 2014 is less than the result of 37.7% shown in Table 73?

30. Willie Mays, with all-around talent, was one of the greatest baseball players of all time. Mays's statistics on stolen bases from 1956 to 1963 are shown in Table 74.

Table 74 Willie Mays: Numbers of Stolen Bases

Year	Number of Stolen Bases
1956	40
1957	38
1958	31
1959	27
1960	25
1961	18
1962	18
1963	8

Source: *The Sports Encyclopedia: Baseball 2004, D. S. Neft et al., 2004, St. Martin's Press, NY.*

Let *n* be Willie Mays's number of stolen bases in the year that is *t* years since 1955.
a. Construct a scatterplot of the data.
b. Draw a linear model on your scatterplot.
c. Find the *n*-intercept of the model. What does it mean in this situation?
d. Find the *t*-intercept of the model. What does it mean in this situation?
e. In 1955, Mays stole 24 bases. Does your linear model underestimate or overestimate his number of stolen bases in that year? Has model breakdown occurred? Explain.
f. In 1971, Mays stole 23 bases. Does your linear model underestimate or overestimate his number of stolen bases in that year? Has model breakdown occurred? Explain.

Chapter 1 Test

1. A rectangle has an area of 36 square feet. Let *W* be the width (in feet), *L* be the length (in feet), and *A* be the area (in square feet).
   a. Sketch three possible rectangles of area 36 square feet.
   b. Which of the symbols *W*, *L*, and *A* are variables? Explain.
   c. Which of the symbols *W*, *L*, and *A* are constants? Explain.

2. Graph the integers between −4 and 2, inclusive, on a number line.

3. The low temperatures (in degrees Fahrenheit) for four days in January in Indianapolis, Indiana, are 5°F below zero,

7°F above zero, 2°F above zero, and 3°F below zero. Let *F* be the low temperature (in degrees Fahrenheit) for any one day. Use points on a number line to describe the given values of *F*.

4. The number of electric cars (in thousands) in use in the United States for various years is 4.5, 5.2, 7.0, 8.7, and 10.4. Let *n* be the number of electric cars (in thousands) in use. Use points on a number line to describe the given values of *n*. Find the average of the values and indicate it on the number line.

5. Let *c* be the total cost (in dollars) of *n* tickets to a hip-hop concert. What is the response variable? Explain.

6. Let s be the salary (in millions of dollars) of Joe Mauer of the Minnesota Twins in the year that is t years since 2010. What does the ordered pair $(6, 23)$ mean in this situation (Source: *Spotrac*)?

7. The percentages of Americans of various age groups who are without health insurance are shown in Table 75.

Table 75 Percentages of Americans Who Are Uninsured

Age Group (years)	Age Used to Represent Age Group (years)	Percent
0–17	8.5	5.5
18–25	21.5	14.8
26–34	30.0	18.5
35–64	49.5	10.7
over 64	70.0	1.6

Source: *Gallup*

Let p be the percentage of Americans who are without health insurance at age a years.
a. Construct a scatterplot of the data.
b. Which point in your scatterplot is highest? What does that mean in this situation?
c. Which point in your scatterplot is lowest? What does that mean in this situation?

For Exercises 8–11, refer to Fig. 63.

8. Find y when $x = -4$.

9. Find x when $y = 1$.

10. What is the y-intercept of the line?

11. What is the x-intercept of the line?

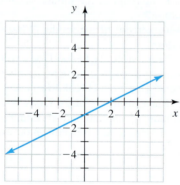

Figure 63 Exercises 8–11

12. Let s be a person's salary (in thousands of dollars) after she has worked t years at a company. Some pairs of values of t and s are shown in Table 76.

Table 76 Years Worked and Salary

t (years)	s (thousands of dollars)
0	21
2	25
3	27
5	31
6	33

a. Construct a scatterplot of the data. Then draw a linear model.
b. Estimate the person's salary after she has worked 4 years at the company.
c. Estimate when the person's salary will be $35 thousand.
d. What is the s-intercept of the model? What does it mean in this situation?

13. Describe, in your own words, the meaning of *linear model*.

14. *Space debris* is the collection of old satellites, spent rocket stages, mission-related trash, and the breakdown of these objects, all orbiting Earth. The debris can do serious damage to spacecraft. A 3-millimeter-long object would collide with energy similar to being hit by a bullet, and a 10-centimeter-long object would be equivalent to a large bomb (Source: *Aerospace*). The numbers of space debris are shown in Table 77 for various years.

Table 77 Numbers of Space Debris

Year	Number of Space Debris (thousands)
1960	0.02
1970	2.6
1980	5.5
1990	7.5
2000	9.8
2011	16.0

Source: *NASA Orbital Debris Program Office*

Let n be the number of space debris (in thousands) at t years since 1960.
a. Construct a scatterplot of the data.
b. Draw a linear model on your scatterplot.
c. Use your model to estimate when there were 9 thousand space debris.
d. Use your model to estimate the number of space debris in 2010.
e. In addition to the typical debris released between 2000 and 2010, two unusual events occurred: the Chinese military exploded a dead satellite in 2007 and a Russian spacecraft collided with a satellite in 2009. Estimate the total number of debris created by the two events. The actual number of debris in 2010 was 16.0 thousand.

15. Describe the meaning of *model breakdown*. Give an example of it. If we use a model to make an estimate and the result is not equal to the actual value, has model breakdown necessarily occurred? Explain.

Operations and Expressions

The number of AIDS deaths decreased from 17,679 deaths in 2003 to 14,561 deaths in 2007, increased to 17,774 deaths in 2009, and decreased to 12,963 deaths in 2013 (see Table 1). Why do you think the decrease, the increase, and then the decrease occurred? In Example 9 in Section 2.4, we will calculate how the number of deaths has changed in various years.

In this chapter, we will describe authentic quantities by using *expressions*. We will discuss how to perform operations and in which order we should perform them, which will help us use expressions to find values of authentic quantities. We will also discuss how to use subtraction and division to compare quantities pertaining to an authentic situation, such as the average movie ticket prices for various years.

Table 1 Numbers of AIDS Deaths in the United States

Year	Number of AIDS Deaths
2003	17,679
2004	17,154
2005	16,823
2006	15,564
2007	14,561
2008	16,084
2009	17,774
2010	14,399
2011	14,122
2012	13,984
2013	12,963

Source: *Centers for Disease Control and Prevention*

2.1 Expressions

Objectives

» Describe the meaning of *expression* and of *evaluate an expression.*

» Use expressions to describe authentic quantities.

» Evaluate expressions.

» Translate English phrases into and from mathematical expressions.

» Describe two roles of a variable.

» Evaluate expressions with more than one variable.

In this section, we will work with expressions—a very important concept in algebra.

Expressions

Addition, subtraction, multiplication, and division are examples of *operations*. In arithmetic, you performed operations with numbers. Because variables represent numbers, we can perform operations with variables, too.

Example 1   Using Operations with Variables and Numbers

Each employee at a small company receives a $500 bonus at the end of the year. For each employee's annual salary shown, find the employee's annual salary plus bonus.

1. $28,000   **2.** $32,000   **3.** *s* dollars

Solution

1. The employee's annual salary plus bonus is $28{,}000 + 500 = 28{,}500$ dollars.
2. The employee's annual salary plus bonus is $32{,}000 + 500 = 32{,}500$ dollars.
3. In Problems 1 and 2, we added the annual salary and \$500, the bonus, to find the results. So, the employee's annual salary plus bonus (in dollars) is $s + 500$.

▶

In Example 1, we took s to be an employee's annual salary and $s + 500$ to be the employee's annual salary plus bonus. We call s and $s + 500$ *expressions*.

Definition   Expression

An **expression** is a constant, a variable, or a combination of constants, variables, operation symbols, and grouping symbols, such as parentheses.

Here are some more examples of expressions:

$$t + 6 \qquad \pi \qquad L + W - 9 \qquad y \qquad 4 \qquad 5 \div (x + 2)$$

In Example 1, we used a variable to represent a quantity from an authentic situation. Sometimes we use variables to represent numbers in a math problem that is not being used to describe an authentic situation. In this case, we often use x for the variable. For example, we could let x represent a number. In this case, x could be *any* number.

To avoid confusing the multiplication symbol $\times$ and the variable name x, we use $\cdot$ or no operation symbol to indicate multiplication. For example, each of the following expressions describes multiplying 2 by 3:

$$2 \cdot 3 \qquad 2(3) \qquad (2)3 \qquad (2)(3)$$

And each of the following expressions describes multiplying 2 by k:

$$2 \cdot k \qquad 2k \qquad 2(k) \qquad (2)k \qquad (2)(k)$$

Using Expressions to Describe Authentic Quantities

We can use expressions to describe authentic quantities. In Example 2, we will find such an expression by noticing a pattern as we calculate values of a quantity.

Example 2   Describing a Quantity

Hot dogs sell for \$2 apiece at the Atlanta Falcons' Mercedes-Benz Stadium. Find the total cost of buying the given number of hot dogs.

1. 3 hot dogs   2. 5 hot dogs   3. 8 hot dogs   4. n hot dogs

Solution

1. Three hot dogs cost $3(2) = 6$ dollars.
2. Five hot dogs cost $5(2) = 10$ dollars.
3. Eight hot dogs cost $8(2) = 16$ dollars.
4. In Problems 1–3, we found the total cost by multiplying the number of hot dogs by 2, the cost (in dollars) per hot dog. So, if there are n hot dogs, the total cost (in dollars) is $n(2)$.

▶

In each of Examples 3 and 4, we will use a table to help us find an expression that describes an authentic quantity.

Table 2 Original and New Test Scores

Original Score (points)	New Score (points)
60	60 + 5
70	70 + 5
80	80 + 5
90	90 + 5
s	s + 5

Example 3   Using a Table to Find an Expression

An instructor adds 5 points to each student's test score. Find the new scores if the original scores are 60 points, 70 points, 80 points, and 90 points. Show the arithmetic to help you see the pattern. Organize the calculations in a table, and include an expression that stands for the new score if the original score is s points.

Solution

First, we construct Table 2. From the last row of the table, we see that the expression $s + 5$ represents the new score (in points) for a test with an original score of s points.

Table 3 Driving Times and Distances

Driving Time (hours)	Distance (miles)
1	$75 \cdot 1$
2	$75 \cdot 2$
3	$75 \cdot 3$
4	$75 \cdot 4$
t	$75 \cdot t$

Example 4   Using a Table to Find an Expression

A person drives at a constant speed of 75 miles per hour. Find the distance (in miles) traveled in 1, 2, 3, and 4 hours of driving at that speed. Show the arithmetic to help you see the pattern. Organize the calculations in a table, and include an expression that stands for the distance (in miles) traveled in t hours.

Solution

First, we construct Table 3. From the last row of the table, we see that the expression $75t$ represents the distance traveled (in miles) in t hours.

Evaluating Expressions

In Example 4, we used $75t$ to describe the distance traveled (in miles) in t hours. This means if the driving time is 5 hours, the distance traveled is $75(5) = 375$ miles. To find the distance, we substituted 5 for t. We say we have *evaluated* the expression $75t$ for $t = 5$.

Definition   Evaluate an expression

We **evaluate an expression** by substituting a number for each variable in the expression and then calculating the result. If a variable appears more than once in the expression, the same number is substituted for that variable each time.

When we evaluate an expression, it is good practice to use parentheses each time a number is substituted for a variable. For example, here we evaluate $5x$ for $x = 3$:

$$5(3) = 15$$

This strategy will be especially helpful when we evaluate an expression for a negative number, which we will begin to do in Section 2.3.

Example 5   Evaluating Expressions

1. In Example 1, we used s to represent an employee's annual salary (in dollars) and $s + 500$ to represent the employee's annual salary plus bonus (in dollars). Evaluate $s + 500$ for $s = 40{,}000$, and describe the meaning of the result.
2. In Example 2, we used n to represent the number of hot dogs bought and $n(2)$ to represent the total cost (in dollars) of n hot dogs. Evaluate $n(2)$ for $n = 4$, and describe the meaning of the result.

Solution

1. We substitute $40{,}000$ for s in $s + 500$:

$$(40{,}000) + 500 = 40{,}500$$

So, the annual salary plus bonus is \$40,500.

2. We substitute 4 for n in $n(2)$:

$$(4)(2) = 8$$

So, the total cost of 4 hot dogs is \$8.

Translating English Phrases into and from Expressions

In order to use mathematics to find results for authentic situations, we must translate from English into mathematics, or vice versa. To do this, the following definitions are helpful:

Definition   Product, factor, and quotient

Let a and b be numbers. Then

- The **product** of a and b is ab. We call a and b **factors** of ab.
- The **quotient** of a and b is $a \div b$, where b is not zero.

For example, because $6 \cdot 3 = 18$, the number 18 is the product of 6 and 3 and the numbers 6 and 3 are factors of 18. The quotient of 6 and 3 is $6 \div 3 = 2$.

Here are some examples of English phrases or sentences and mathematical expressions that have the same meaning:

Operation	English Phrase or Sentence	Mathematical Expression
Addition	A number plus 3	$x + 3$
	The sum of a number and 3	$x + 3$
	The total of a number and 3	$x + 3$
	Add a number and 3.	$x + 3$
	3 more than a number	$x + 3$
	A number increased by 3	$x + 3$
Subtraction	A number minus 3	$x - 3$
	The difference of a number and 3	$x - 3$
	Subtract 3 from a number.	$x - 3$
	3 less than a number	$x - 3$
	A number decreased by 3	$x - 3$
Multiplication	Multiply 3 by a number.	$3x$
	3 times a number	$3x$
	The product of 3 and a number	$3x$
	Twice a number	$2x$
	One-third of a number	$\dfrac{1}{3}x$
Division	Divide a number by 3.	$x \div 3$
	The quotient of a number and 3	$x \div 3$
	The ratio of a number to 3	$x \div 3$

WARNING   To subtract 2 from 5, we write $5 - 2$, not $2 - 5$. For example, suppose you have \$5 and you take \$2 from the \$5. Then you have $5 - 2 = 3$ dollars left. So, subtracting 2 from 5 is $5 - 2$.

> ### Example 6   Translating from English into Mathematics
>
> Let x be a number.
>
> 1. Translate the English phrase "The product of 2 and the number" into an expression.
> 2. Evaluate the result you found in Problem 1 for $x = 3$.
> 3. Evaluate the result you found in Problem 1 for $x = 7$.
>
> #### Solution
>
> 1. The expression is $2x$.   2. $2(3) = 6$ 3. $2(7) = 14$

> ### Example 7   Translating from English into Mathematics
>
> Let x be a number. Translate the English phrase or sentence into an expression. Then evaluate the expression for $x = 6$.
>
> 1. The quotient of the number and 3   2. Subtract the number from 8.
>
> #### Solution
>
> 1. The expression is $x \div 3$. Next, we evaluate $x \div 3$ for $x = 6$:
> $$(6) \div 3 = 2$$
> 2. The expression is $8 - x$. Next, we evaluate $8 - x$ for $x = 6$:
> $$8 - (6) = 2$$

> ### Example 8   Translating from Mathematics into English
>
> Let x be a number. Translate the expression into an English phrase.
>
> 1. $6 - x$ 2. $8x$
>
> #### Solution
>
> 1. The difference of 6 and the number
> 2. The product of 8 and the number

Roles of a Variable

In Chapter 1, we used a variable to represent a quantity that can vary. In Example 6, we used a variable for another reason: In the expression $2x$, the variable x is used as a *placeholder* for a number to be substituted for x. First, we substituted 3 for x. Then we substituted 7 for x.

> **Roles of a Variable**
>
> Here are two roles of a variable:*
>
> 1. A variable can represent a quantity that can vary.
> 2. In an expression, a variable is a placeholder for a number.

Sometimes a variable serves both roles. For example, consider the variable s in Problem 1 in Example 5. Recall that s represents an employee's annual salary, which can vary over time or from employee to employee. The variable s also serves as a placeholder for a number in the expression $s + 500$, which describes the employee's annual salary plus bonus (in dollars).

---

*We will discuss one more role of a variable in Section 4.3.

Figure 1 The length L and width W of a rectangle

Expressions with More Than One Variable

An expression may contain more than one variable. For example, let W be the width (in feet) and let L be the length (in feet) of a rectangle (see Fig. 1).

Recall that the area of a rectangle is equal to the length times the width of the rectangle, so the area (in square feet) is equal to the expression LW. We can evaluate the expression LW for $L = 4$ and $W = 3$:

$$(4)(3) = 12$$

So, a 3-foot by 4-foot rectangle has an area of 12 square feet.

Note the power of algebra in that the expression LW *concisely* tells us how to find the area of *any* rectangle, no matter what its dimensions are.

Example 9   Evaluating an Expression in Two Variables

If it takes a student T minutes to complete a test that has n questions, then $T \div n$ is the average time (in minutes) taken to respond to one question. Evaluate $T \div n$ for $T = 48$ and $n = 16$. What does the result mean in this situation?

Solution

We substitute 48 for T and 16 for n in the expression $T \div n$ and then calculate the result:

$$48 \div 16 = 3$$

If it takes a student 48 minutes to respond to 16 questions, the average response time is 3 minutes per question.

▶

Example 10   Translating from English into Mathematics

Write the phrase as a mathematical expression, and then evaluate the result for $x = 8$ and $y = 4$.

1. The sum of x and y **2.** The quotient of x and y

Solution

1. The expression is $x + y$. Next, we evaluate $x + y$ for $x = 8$ and $y = 4$:

$$(8) + (4) = 12$$

2. The expression is $x \div y$. Next, we evaluate $x \div y$ for $x = 8$ and $y = 4$:

$$(8) \div (4) = 2$$

▶

◆ Group Exploration ─────────────────────────

Section Opener: Expressions

1. An instructor adds 3 bonus points to each student's test score. For the given original test score, find the final test score.
  a. 74 points
  b. 88 points
  c. r points [**Hint:** Think about how you got your results for parts (a) and (b).]

2. A student gets paid $8 per hour for working at a music store. For the given number of hours worked, find the total amount of money earned (in dollars).
  a. 5 hours   **b.** 9 hours   **c.** t hours

3. In a lottery, some people won $200, which they will share equally. For the given number of people, find each person's share (in dollars).
  a. 4 people   **b.** 10 people   **c.** n people

Group Exploration

Expressions used to describe a quantity

Consider the expression $x + 2$. Suppose that a child has grown 2 inches within the last year. We could define x to be the child's height (in inches) last year, and then $x + 2$ would be the child's current height (in inches).

Describe a situation in which x represents a meaningful quantity and the expression given describes another meaningful quantity.

1. $x + 3$

2. $x - 4$

3. $3x$

4. $x \div 2$

For each of the four expressions, evaluate it for a reasonable value of x and describe the meaning of the result.

Tips for Success   Make Good Use of This Text

You can get more out of this course by making good use of the text. Before class, consider previewing the material for 10 minutes. You can do this by reading the objectives and the boxed statements. Even if what you read doesn't make much sense to you, previewing will flag key concepts that you can focus on during class time.

After class, read the relevant section(s). When looking at each example, figure out how it goes from one step to the next.

Then begin working on the homework assignment. If you have difficulty with an exercise, locate a similar example to help guide you. You may need to seek outside help for more challenging exercises. If you needed to look at examples or get outside help for a large number of exercises, then it is important that you keep doing additional exercises until you are self-sufficient. After all, unless your instructor allows open-book tests or collaborative tests, you won't be able to read the text or seek help from others during an exam.

Homework 2.1

For extra help ▶   **MyLab Math**   Watch the videos in MyLab Math

For Exercises 1–12, evaluate the expression for $x = 6$.

1. $x + 2$
2. $5 + x$
3. $9 - x$
4. $x - 4$
5. $7x$
6. $x(9)$
7. $x \div 3$
8. $30 \div x$
9. $x + x$
10. $x - x$
11. $x \cdot x$
12. $x \div x$

13. If a person buys n albums by Isaiah Rashad on iTunes, the total cost is $10.99n$ dollars. Evaluate $10.99n$ for $n = 4$. What does your result mean in this situation?

14. If a student earns a total of T points on five tests, then $T \div 5$ is the student's average test score (in points). If a student earns a total of 440 points on five tests, what is the student's average test score?

15. For the period 2008–2014, if U is the average daily price (in dollars) of hotels in urban centers, then $U - 65.25$ is approximately the average daily price (in dollars) of hotels at airports (Source: *STR*). The average daily price of hotels in urban centers was \$167.94 in 2014. Estimate the average daily price of hotels at airports in 2014.

16. For the period 2010–2016, if F is the number (in thousands) of Ford employees, then $F + 17.3$ is approximately the number (in thousands) of General Motors employees (Sources: *Ford;*

General Motors). There were about 60.9 thousand Ford employees in 2016. Estimate the number of General Motors employees in 2016.

17. Each student at a community college pays a student services fee of \$20.

a. Complete Table 4 to help find an expression that describes the total cost (in dollars) of tuition plus the services fee if a student pays t dollars for tuition. Show the arithmetic to help you see a pattern.

Table 4 Tuition and Total Cost

Tuition (dollars)	Total Cost (dollars)
400	
401	
402	
403	
t	

b. Evaluate the expression you found in part (a) for $t = 417$. What does your result mean in this situation?

18. A person is driving 5 miles per hour over the speed limit.
   a. Complete Table 5 to help find an expression that describes the driving speed (in miles per hour) if the speed limit is s miles per hour. Show the arithmetic to help you see a pattern.

Table 5 Speed Limit and Driving Speed

Speed Limit (miles per hour)	Driving Speed (miles per hour)
35	
40	
45	
50	
s	

   b. Evaluate the expression you found in part (a) for $s = 65$. What does your result mean in this situation?

19. On December 30, 2016, the closing value of each share of Twitter stock was $16.30 (Source: *Google Finance*).
   a. Complete Table 6 to help find an expression that describes the total value (in dollars) of n shares of the stock. Show the arithmetic to help you see a pattern.

Table 6 Number of Shares and Total Value

Number of Shares	Total Value (dollars)
1	
2	
3	
4	
n	

   b. Evaluate the expression you found in part (a) for $n = 7$. What does your result mean in this situation?

20. A pair of Nike men's Elite Basketball Crew socks sells for $4.05.
   a. Complete Table 7 to help find an expression that describes the total cost (in dollars) of n pairs of socks. Show the arithmetic to help you see a pattern.

Table 7 Number of Pairs and Total Cost

Number of Pairs	Total Cost (dollars)
1	
2	
3	
4	
n	

   b. Evaluate the expression you found in part (a) for $n = 9$. What does your result mean in this situation?

21. The tuition rate at Community College of Philadelphia for a Philadelphia resident is $153 per credit hour (unit or hour).
   a. Complete Table 8 to help find an expression that describes the total cost (in dollars) of enrolling in n credit hours of classes. Show the arithmetic to help you see a pattern.
   b. Evaluate the expression you found in part (a) for $n = 15$. What does your result mean in this situation?

Table 8 Number of Credit Hours and Total Cost

Number of Credit Hours	Total Cost (dollars)
1	
2	
3	
4	
n	

22. The length of a rectangular garden is 20 feet.
   a. Complete Table 9 to help find an expression that describes the area (in square feet) of the rectangle if the width is w feet. Show the arithmetic to help you see a pattern.

Table 9 Width and Area

Width (feet)	Area (square feet)
1	
2	
3	
4	
w	

   b. Evaluate the expression you found in part (a) for $w = 10$. What does your result mean in this situation?

Let x be a number. Translate the English phrase or sentence into a mathematical expression. Then evaluate the expression for x = 8.

23. The number plus 4

24. 8 minus the number

25. The quotient of the number and 2

26. Add 6 and the number.

27. Subtract 5 from the number.

28. 15 more than the number

29. The product of 7 and the number

30. The difference of the number and 7

31. 16 divided by the number

32. Multiply the number by 5.

Let x be a number. Translate the expression into an English phrase.

33. $x \div 2$ **34.** $6 \div x$ **35.** $7 - x$

36. $x - 2$ **37.** $x + 5$ **38.** $4 + x$

39. $9x$ **40.** $x(5)$ **41.** $x - 7$

42. $x + 3$ **43.** $x(2)$ **44.** $x \div 5$

Evaluate the expression for x = 6 and y = 3.

45. $x + y$ **46.** $y + x$ **47.** $x - y$

48. xy **49.** yx **50.** $x \div y$

For Exercises 51–54, translate the phrase into a mathematical expression. Then evaluate the expression for x = 9 and y = 3.

51. The product of x and y **52.** The sum of x and y

53. The difference of x and y **54.** The quotient of x and y

55. If a car travels at a constant speed of r miles per hour for t hours, it will travel rt miles. Evaluate rt for $r = 62$ and $t = 3$. What does your result mean in this situation?

56. Let b be the balance (in dollars) of a checking account. If a check is written for d dollars, then the new balance (in dollars) is $b - d$. Evaluate $b - d$ for $b = 3758$ and $d = 994$. What does your result mean in this situation?

57. If a car can travel m miles on g gallons of gasoline, then the car's gas mileage is $m \div g$ miles per gallon. Evaluate $m \div g$ for $m = 240$ and $g = 12$. What does your result mean in this situation?

58. If T is the total cost (in dollars) for n students to go on a ski trip, then $T \div n$ is the cost (in dollars) per student. Evaluate $T \div n$ for $T = 9000$ and $n = 20$. What does your result mean in this situation?

59. Let C be the total cost (in dollars) of manufacturing some computers and R be the total revenue (in dollars) from selling the computers. Then $R - C$ is the total profit (in dollars). If the total cost of manufacturing some computers is \$315,000 and the total revenue from selling the computers is \$485,000, what is the total profit?

60. For the period 1992–2009, if E is the average verbal SAT score (in points) for a certain year, then the average math SAT score (in points) for that year is approximately $E + t$, where t is the number of years since 1992. The average verbal SAT score was 501 points in 2009. Estimate the average math SAT score in 2009.

Concepts

61. A person gets paid $5t$ dollars for t hours of work.
  a. Evaluate $5t$ for $t = 1$, $t = 2$, $t = 3$, and $t = 4$. Describe the meaning of these results.
  b. Refer to the results you found in part (a) to determine how much the person gets paid per hour. Explain.
  c. Compare the result you found in part (b) with the expression $5t$. What do you notice?

62. The total price of n loaves of bread is $3n$ dollars.
  a. Evaluate $3n$ for $n = 1$, $n = 2$, $n = 3$, and $n = 4$. Describe the meaning of these results.
  b. Refer to the results you found in part (a) to determine the cost per loaf of bread. Explain.
  c. Compare the result you found in part (b) with the expression $3n$. What do you notice?

63. A person drives $50t$ miles in t hours.
  a. Evaluate $50t$ for $t = 1$, $t = 2$, $t = 3$, and $t = 4$. Describe the meaning of your results.
  b. Refer to the results you found in part (a) to determine at what speed the person is traveling. Explain.
  c. Compare the result you found in part (b) with the expression $50t$. What do you notice?

64. An elevator rises $2t$ yards in t seconds.
  a. Evaluate $2t$ for $t = 1$, $t = 2$, $t = 3$, and $t = 4$. Describe the meaning of your results.
  b. Refer to the results you found in part (a) to determine at what speed the elevator is rising. Explain.
  c. Compare the result you found in part (b) with the expression $2t$. What do you notice?

65. Compare the meaning of *variable* with the meaning of *expression*. (See page 9 for guidelines on writing a good response.)

66. Give an example of an expression containing a variable, and then evaluate it three times to get three different results.

67. Give an example of a variable that is used to represent a quantity that varies.

68. Give an example of a variable that is used as a placeholder for a number in an expression.

69. Describe an authentic situation for the expression $8x$. Include a definition for the variable x in your description.

70. Describe an authentic situation for the expression $200 \div x$. Include a definition for the variable x in your description.

2.2 Operations with Fractions

Objectives

» Describe the meaning of a fraction.

» Explain why division by zero is undefined.

» Describe the rules for $a \cdot 1$, $\frac{a}{1}$, and $\frac{a}{a}$.

» Perform operations with fractions.

» Find the prime factorization of a number.

» Simplify fractions.

In this section, we will perform operations with fractions, which are used in numerous fields, including music, social science, business, engineering, and medicine.

Meaning of a Fraction

A fraction can be used to describe a part of a whole. For example, consider the meaning of $\frac{3}{4}$ of a pizza. If we divide the pizza into 4 slices of equal area, 3 of the slices make up $\frac{3}{4}$ of the pizza (see Fig. 2).

Even though the orange region in Fig. 3 is 1 of 3 parts, it is *not* $\frac{1}{3}$ of the pizza because the 3 parts do not have equal area. The orange region *is* equal to $\frac{1}{2}$ of the pizza because it is 1 of 2 parts of equal area that make up the pizza (see Fig. 4).

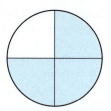

Figure 2 $\frac{3}{4}$ of a pizza

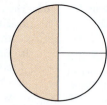

Figure 3 The 3 parts do not have equal area

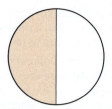

Figure 4 The 2 parts have equal area, so the orange part is $\frac{1}{2}$ of the pizza

The fraction $\frac{a}{b}$ means $a \div b$. For example, $\frac{8}{4} = 8 \div 4 = 2$. So 8 quarters of pizza make 2 pizzas with 4 slices each (see Fig. 5).

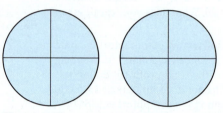

Figure 5 The 8 quarters of pizza make 2 pizzas

Division by Zero

We can think of division in terms of repeated subtraction. For example, $17 \div 5$ is equal to 3 with a remainder of 2 (try it). This means if we subtract 5 from 17 three times, the result is 2 (the remainder):

$$17 - 5 = 12, \qquad 12 - 5 = 7, \qquad 7 - 5 = 2$$

Note that the remainder, 2, is less than the divisor, 5.

As a matter of fact, the remainder must always be less than the divisor. This rule will help us see that division by 0 is undefined. For example, consider $8 \div 0$. No matter how many times we subtract 0 from 8, the result is always 8:

$$8 - 0 = 8, \qquad 8 - 0 = 8, \qquad 8 - 0 = 8, \qquad \text{and so on}$$

If $8 \div 0$ is defined, the remainder would have to be the repeated result 8. Because the remainder must be less than the divisor, it is implied that 8 is less than 0, which is false. So, $8 \div 0$ is undefined. In fact, any number divided by 0 is undefined.

Division by Zero

The fraction $\frac{a}{b}$ is undefined if $b = 0$. Division by 0 is undefined.

For example, $\frac{6}{0}$ is undefined. If you use a calculator to divide by 0, the screen will likely display "Error," "ERR:," "E," or "ERR: Divide by 0" to indicate that division by 0 is undefined.

WARNING   However, the fraction $\frac{0}{6}$ *is* defined. In fact, $\frac{0}{6} = 0$. For example, if a person eats zero sixths of a pizza, this means that the person didn't eat any pizza.

Rules for $a \cdot 1$, $\frac{a}{1}$, and $\frac{a}{a}$

The products $4 \cdot 1 = 4$, $5 \cdot 1 = 5$, and $8 \cdot 1 = 8$ suggest the following property:

Multiplying a Number by 1

$$a \cdot 1 = a$$

In words: A number multiplied by 1 is that same number.

When we write statements such as $a \cdot 1 = a$, we mean if we evaluate $a \cdot 1$ and a for *any* value of a in both expressions, the results will be equal. We say that the expressions $a \cdot 1$ and a are **equivalent expressions**.

The quotients $\frac{4}{1} = 4 \div 1 = 4, \frac{5}{1} = 5 \div 1 = 5,$ and $\frac{8}{1} = 8 \div 1 = 8$ suggest the following property:

Dividing a Number by 1

$$\frac{a}{1} = a$$

In words: A number divided by 1 is that same number.

Finally, the quotients $\frac{4}{4} = 4 \div 4 = 1, \frac{5}{5} = 5 \div 5 = 1,$ and $\frac{8}{8} = 8 \div 8 = 1$ suggest the following property:

Dividing a Nonzero Number by Itself

If a is nonzero, then

$$\frac{a}{a} = 1$$

In words: A nonzero number divided by itself is 1.

The properties $a \cdot 1 = a$, $\frac{a}{1} = a$, and $\frac{a}{a} = 1$ (where a is nonzero) will help us when we work with fractions.

Multiplication of Fractions

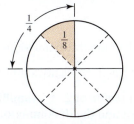

Figure 6 $\frac{1}{2}$ of $\frac{1}{4}$ of a pizza is $\frac{1}{8}$ of a pizza

Figure 6 illustrates that $\frac{1}{2}$ of $\frac{1}{4}$ of a pizza is $\frac{1}{8}$ of a pizza. We can calculate this result by finding the product $\frac{1}{2} \cdot \frac{1}{4}$:

$$\frac{1}{2} \cdot \frac{1}{4} = \frac{1 \cdot 1}{2 \cdot 4} = \frac{1}{8}$$

Multiplying Fractions

If b and d are nonzero, then

$$\frac{a}{b} \cdot \frac{c}{d} = \frac{ac}{bd}$$

In words: To multiply two fractions, write the numerators as a product and write the denominators as a product.

Example 1   Finding the Product of Two Fractions

Find the product $\frac{2}{5} \cdot \frac{3}{7}$.

Solution

$$\frac{2}{5} \cdot \frac{3}{7} = \frac{2 \cdot 3}{5 \cdot 7}$$ *Write numerators and denominators as products:* $\frac{a}{b} \cdot \frac{c}{d} = \frac{ac}{bd}$

$$= \frac{6}{35}$$ *Find products.*

Prime Factorization

When we work with fractions, it can sometimes help to work with prime numbers.

> **Definition   Prime number**
>
> A **prime number**, or **prime**, is any counting number larger than 1 whose only positive factors are itself and 1.

Here are the first ten primes:

$$2, 3, 5, 7, 11, 13, 17, 19, 23, 29$$

Sometimes when we work with fractions, it is helpful to write a number as a product of primes. We call this product the **prime factorization** of the number.

Example 2   Writing a Number as a Product of Primes

Write 54 as a product of primes.

Solution

$$54 = 6 \cdot 9 \qquad \text{\textit{Write 54 as a product of two numbers.}}$$
$$= 2 \cdot 3 \cdot 3 \cdot 3 \qquad \text{\textit{Find prime factorizations of 6 and 9.}}$$

The prime factorization of 54 is $2 \cdot 3 \cdot 3 \cdot 3$.

Simplifying Fractions

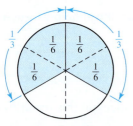

Figure 7 $\dfrac{4}{6} = \dfrac{2}{3}$

Figure 7 illustrates that $\dfrac{4}{6} = \dfrac{2}{3}$. We say $\dfrac{2}{3}$ is *simplified* because the numerator and denominator do not have positive factors other than 1 in common. The fraction $\dfrac{4}{6}$ is not simplified because the numerator and the denominator have a common factor of 2. To **simplify** a fraction, we write it as an equal fraction in which the numerator and the denominator do not have any common positive factors other than 1.

Example 3   Simplifying Fractions

Simplify.

1. $\dfrac{4}{6}$ **2.** $\dfrac{30}{42}$

Solution

1. We begin to simplify $\dfrac{4}{6}$ by finding the prime factorizations of the numerator 4 and the denominator 6:

$$\frac{4}{6} = \frac{2 \cdot 2}{2 \cdot 3} = \frac{2}{2} \cdot \frac{2}{3} = 1 \cdot \frac{2}{3} = \frac{2}{3}$$

Our result matches with what we found in Fig. 7. By performing long division or using a calculator, we can check that both fractions are equal to the repeating decimal $0.\overline{6}$.

2. We begin to simplify $\dfrac{30}{42}$ by finding the prime factorizations of the numerator, 30, and the denominator, 42:

$$\frac{30}{42} = \frac{2 \cdot 3 \cdot 5}{2 \cdot 3 \cdot 7} = \frac{2 \cdot 3}{2 \cdot 3} \cdot \frac{5}{7} = 1 \cdot \frac{5}{7} = \frac{5}{7}$$

Simplifying fractions can make it easier to work out certain problems. Also, if two fractions are simplified, it is easy to tell whether they are equal. **If the result of an exercise is a fraction, simplify it.**

Simplifying a Fraction

To simplify a fraction,

1. Find the prime factorizations of the numerator and denominator.

2. Find an equal fraction in which the numerator and the denominator do not have common positive factors other than 1 by using the property

$$\frac{ab}{ac} = \frac{a}{a} \cdot \frac{b}{c} = 1 \cdot \frac{b}{c} = \frac{b}{c}$$

where a and c are nonzero.

In Example 4, we will multiply two fractions and simplify the result.

Example 4   Finding the Product of Two Fractions

Find the product $\frac{8}{9} \cdot \frac{15}{4}$.

Solution

$$\frac{8}{9} \cdot \frac{15}{4} = \frac{8 \cdot 15}{9 \cdot 4}$$ *Write numerators and denominators as products:* $\frac{a}{b} \cdot \frac{c}{d} = \frac{ac}{bd}$

$$= \frac{2 \cdot 2 \cdot 2 \cdot 3 \cdot 5}{3 \cdot 3 \cdot 2 \cdot 2}$$ *Find prime factorizations.*

$$= \frac{2 \cdot 5}{3}$$ *Simplify:* $\frac{2 \cdot 2 \cdot 3}{2 \cdot 2 \cdot 3} = 1$

$$= \frac{10}{3}$$ *Multiply.*

Division of Fractions

The **reciprocal** of $\frac{a}{b}$ is $\frac{b}{a}$. For example, the reciprocal of $\frac{3}{8}$ is $\frac{8}{3}$. We will need to find the reciprocal of a fraction when we divide two fractions.

Dividing Fractions

If $b, c,$ and d are nonzero, then

$$\frac{a}{b} \div \frac{c}{d} = \frac{a}{b} \cdot \frac{d}{c}$$

In words: To divide by a fraction, multiply by its reciprocal.

Example 5   Finding the Quotient of Two Fractions

Find the quotient $\frac{3}{4} \div \frac{1}{8}$.

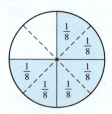

Figure 8 $\frac{3}{4}$ of a pizza divided into slices of size $\frac{1}{8}$ of the pizza gives 6 slices of pizza

Press $\boxed{(}$ $\boxed{3}$ $\boxed{÷}$ $\boxed{4}$ $\boxed{)}$ $\boxed{÷}$ $\boxed{(}$ $\boxed{1}$ $\boxed{÷}$ $\boxed{8}$ $\boxed{)}$
$\boxed{\text{ENTER}}$.

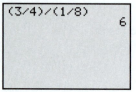

Figure 9 Verify the work

Solution

$$\frac{3}{4} \div \frac{1}{8} = \frac{3}{4} \cdot \frac{8}{1} \qquad \textit{Multiply by reciprocal of } \frac{1}{8}, \textit{ which is } \frac{8}{1}; \frac{a}{b} \div \frac{c}{d} = \frac{a}{b} \cdot \frac{d}{c}$$

$$= \frac{3 \cdot 8}{4 \cdot 1} \qquad \textit{Write numerators and denominators as products.}$$

$$= \frac{3 \cdot 2 \cdot 2 \cdot 2}{2 \cdot 2 \cdot 1} \qquad \textit{Find prime factorizations.}$$

$$= \frac{3 \cdot 2}{1} \qquad \textit{Simplify: } \frac{2 \cdot 2}{2 \cdot 2} = 1$$

$$= 6 \qquad \frac{a}{1} = a$$

Our result makes sense because $\frac{3}{4}$ of a pizza divided into slices, each of size $\frac{1}{8}$ of the pizza, gives 6 slices (see Fig. 8).

We use a graphing calculator to check our work in Example 5 (see Fig. 9).

When you use a calculator to check work with fractions, it is good practice to enclose each fraction in parentheses. You will see the importance of using parentheses when we discuss the order of operations in Section 2.6.

To find the reciprocal of 6, we use the fact $6 = \frac{6}{1}$. So, the reciprocal of 6 is $\frac{1}{6}$.

Example 6   Evaluating an Expression

Evaluate $\frac{a}{b} \div c$ for $a = 21, b = 2,$ and $c = 3$.

Solution

We substitute $a = 21, b = 2,$ and $c = 3$ into the expression $\frac{a}{b} \div c$:

$$\frac{(21)}{(2)} \div (3) = \frac{21}{2} \div \frac{3}{1} \qquad \textit{Write 3 as a fraction: } 3 = \frac{3}{1}$$

$$= \frac{21}{2} \cdot \frac{1}{3} \qquad \textit{Multiply by reciprocal of } \frac{3}{1}, \textit{ which is } \frac{1}{3}; \frac{a}{b} \div \frac{c}{d} = \frac{a}{b} \cdot \frac{d}{c}$$

$$= \frac{21 \cdot 1}{2 \cdot 3} \qquad \textit{Write numerators and denominators as products.}$$

$$= \frac{3 \cdot 7 \cdot 1}{2 \cdot 3} \qquad \textit{Find prime factorizations.}$$

$$= \frac{7}{2} \qquad \textit{Simplify: } \frac{3}{3} = 1$$

In Example 6, the result is $\frac{7}{2}$, which is an improper fraction (that is, the numerator is larger than the denominator). For *nonmodeling* exercises, if a fractional result is in improper form, we will leave it in that form. For *modeling* exercises, if a result is in improper form, we will write it as a mixed number. For example, we say that a car trip takes $3\frac{1}{2}$ hours rather than $\frac{7}{2}$ hours.

Addition of Fractions

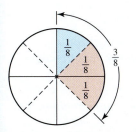

Figure 10 $\frac{1}{8}$ pizza plus $\frac{2}{8}$ pizza is $\frac{3}{8}$ pizza

Figure 10 illustrates that $\frac{1}{8}$ of a pizza plus $\frac{2}{8}$ of a pizza is equal to $\frac{3}{8}$ of a pizza. This illustration suggests that, to find the sum $\frac{1}{8} + \frac{2}{8}$, we add the numerators 1 and 2 and write the result, 3, over the common denominator, 8:

$$\frac{1}{8} + \frac{2}{8} = \frac{1+2}{8}$$
$$= \frac{3}{8}$$

Adding Fractions with the Same Denominator

If b is nonzero, then

$$\frac{a}{b} + \frac{c}{b} = \frac{a+c}{b}$$

In words: To add two fractions with the same denominator, add the numerators and write the result above the common denominator.

Example 7   Adding Fractions with the Same Denominator

Find the sum $\frac{4}{15} + \frac{6}{15}$.

Solution

$$\frac{4}{15} + \frac{6}{15} = \frac{4+6}{15} \qquad \textit{Write numerators as a sum and keep common denominator: } \frac{a}{b} + \frac{c}{b} = \frac{a+c}{b}$$
$$= \frac{10}{15} \qquad \textit{Find sum.}$$
$$= \frac{2}{3} \qquad \textit{Simplify.}$$

Least Common Denominators

To find the sum $\frac{1}{4} + \frac{5}{6}$, in which the denominators of the fractions are different, we find an equal sum of fractions in which the denominators are equal. First, we list the multiples of 4 and the multiples of 6:

Multiples of 4:   4, 8, 12, 16, 20, 24, 28, 32, 36, . . .
Multiples of 6:   6, 12, 18, 24, 30, 36, 42, 48, 54, . . .

Common multiples of 4 and 6 are

$$12, 24, 36, \ldots$$

Note that 12 is the least (lowest) number in the list. We call it the least common multiple of 4 and 6. The **least common multiple (LCM)** of a group of numbers is the smallest number that is a multiple of *all* the numbers in the group.

To find the sum $\frac{1}{4} + \frac{5}{6}$, we use the fact $\frac{a}{a} = 1$, where a is nonzero, to write an equal sum of fractions in which each denominator is equal to the LCM, 12:

$$\frac{1}{4} + \frac{5}{6} = \frac{1}{4} \cdot 1 + \frac{5}{6} \cdot 1 \qquad a = a \cdot 1$$

$$= \frac{1}{4} \cdot \frac{3}{3} + \frac{5}{6} \cdot \frac{2}{2} \qquad 1 = \frac{a}{a}$$

$$= \frac{3}{12} + \frac{10}{12} \qquad \text{Multiply numerators and multiply denominators:}$$
$$\frac{a}{b} \cdot \frac{c}{d} = \frac{ac}{bd}$$

$$= \frac{13}{12} \qquad \text{Add numerators and keep common denominator:}$$
$$\frac{a}{b} + \frac{c}{b} = \frac{a + c}{b}$$

We also call 12 the least common denominator of $\frac{1}{4}$ and $\frac{5}{6}$. The **least common denominator** (**LCD**) of a group of fractions is the LCM of the denominators of all the fractions.

Example 8   Adding Fractions with Different Denominators

Find the sum $\frac{5}{8} + \frac{5}{6}$.

Solution

We list multiples of 8 and multiples of 6:

Multiples of 8:   8, 16, 24, 32, 40, 48, . . .
Multiples of 6:   6, 12, 18, 24, 30, 36, . . .

The LCD is 24. We write an equal sum of fractions in which each denominator is 24:

$$\frac{5}{8} + \frac{5}{6} = \frac{5}{8} \cdot \frac{3}{3} + \frac{5}{6} \cdot \frac{4}{4} \qquad \textit{LCD is 24.}$$

$$= \frac{15}{24} + \frac{20}{24} \qquad \text{Multiply numerators and multiply denominators:}$$
$$\frac{a}{b} \cdot \frac{c}{d} = \frac{ac}{bd}$$

$$= \frac{35}{24} \qquad \text{Add numerators and keep common denominator:}$$
$$\frac{a}{b} + \frac{c}{b} = \frac{a + c}{b}$$

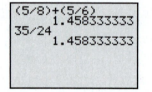

Figure 11 Verify the work

▶   We use a graphing calculator to verify the work (see Fig. 11).

Subtraction of Fractions

The rule for subtracting two fractions with the same denominator is similar to the rule for adding such fractions, except we subtract the numerators.

Subtracting Fractions with the Same Denominator

If b is nonzero, then

$$\frac{a}{b} - \frac{c}{b} = \frac{a - c}{b}$$

In words: To subtract two fractions with the same denominator, subtract the numerators and write the result above the common denominator.

Example 9   Subtracting Fractions with the Same Denominator

Find the difference $\frac{5}{8} - \frac{3}{8}$.

Solution

$$\frac{5}{8} - \frac{3}{8} = \frac{5-3}{8}$$ *Write numerators as a difference and keep common denominator:* $\frac{a}{b} - \frac{c}{b} = \frac{a-c}{b}$

$$= \frac{2}{8}$$ *Find difference.*

$$= \frac{1}{4}$$ *Simplify.*

▶

Subtracting fractions with different denominators is similar to adding them. The first step is to rewrite each fraction so that each denominator is the LCD.

Example 10   Subtracting Fractions with Different Denominators

Find the difference $\frac{8}{9} - \frac{3}{5}$.

Solution

We list the multiples of 9 and the multiples of 5:

Multiples of 9:   9, 18, 27, 36, 45, 54, 63, 72, 81, . . .
Multiples of 5:   5, 10, 15, 20, 25, 30, 35, 40, 45, . . .

The LCD is 45. We now rewrite each fraction with the denominator 45:

$$\frac{8}{9} - \frac{3}{5} = \frac{8}{9} \cdot \frac{5}{5} - \frac{3}{5} \cdot \frac{9}{9}$$ *LCD is 45.*

$$= \frac{40}{45} - \frac{27}{45}$$ *Multiply numerators and multiply denominators:* $\frac{a}{b} \cdot \frac{c}{d} = \frac{ac}{bd}$

$$= \frac{13}{45}$$ *Subtract numerators and keep common denominator:* $\frac{a}{b} - \frac{c}{b} = \frac{a-c}{b}$

▶

Adding (or Subtracting) Fractions with Different Denominators

To add (or subtract) two fractions with different denominators, use the fact that $\frac{a}{a} = 1$, where a is nonzero, to write an equal sum (or difference) of fractions for which each denominator is the LCD.

◈ Group Exploration

Illustrations of simplifying fractions and operations with fractions

Draw a picture of a pizza to show that the true statement makes sense. [**Hint:** See Figs. 6, 7, 8, and 10.]

1. $\frac{6}{8} = \frac{3}{4}$

2. $\frac{5}{8} + \frac{2}{8} = \frac{7}{8}$

3. $\frac{5}{6} - \frac{4}{6} = \frac{1}{6}$

4. $\frac{1}{2} \cdot \frac{1}{3} = \frac{1}{6}$

5. $\frac{2}{3} \div \frac{1}{6} = 4$

Tips for Success   Review Your Notes as Soon as Possible

How often do you get confused by class notes you wrote earlier the same day, even though the class activities made sense to you? If this happens a lot, review your notes as soon after class as possible. Even reviewing your notes for just a few minutes between classes will help. This will increase your likelihood of remembering what you learned in class and will give you the opportunity to add new comments to your notes while the class experience is still fresh in your mind.

Homework 2.2

For extra help ▶   **MyLab Math**   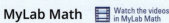 Watch the videos in MyLab Math

1. What is the denominator of $\frac{3}{7}$?

2. What is the numerator of $\frac{2}{5}$?

Write the number as a product of primes.

3. 20   **4.** 18   **5.** 36   **6.** 24

7. 45   **8.** 27   **9.** 78   **10.** 105

Simplify. Then use a calculator to check your work.

11. $\frac{6}{8}$ **12.** $\frac{10}{14}$ **13.** $\frac{3}{12}$ **14.** $\frac{7}{28}$

15. $\frac{18}{30}$ **16.** $\frac{27}{54}$ **17.** $\frac{20}{50}$ **18.** $\frac{49}{63}$

19. $\frac{5}{25}$ **20.** $\frac{9}{81}$ **21.** $\frac{20}{24}$ **22.** $\frac{15}{18}$

Perform the indicated operation. Then use a calculator to check your work.

23. $\frac{1}{3} \cdot \frac{2}{5}$ **24.** $\frac{6}{7} \cdot \frac{4}{9}$ **25.** $\frac{4}{5} \cdot \frac{3}{8}$ **26.** $\frac{2}{3} \cdot \frac{5}{6}$

27. $\frac{5}{21} \cdot 7$ **28.** $\frac{5}{12} \cdot 2$ **29.** $\frac{5}{8} \div \frac{3}{4}$ **30.** $\frac{7}{12} \div \frac{2}{3}$

31. $\frac{8}{9} \div \frac{4}{3}$ **32.** $\frac{4}{7} \div \frac{8}{3}$ **33.** $\frac{2}{3} \div 5$ **34.** $\frac{4}{9} \div 2$

35. $\frac{2}{7} + \frac{3}{7}$ **36.** $\frac{5}{9} + \frac{2}{9}$ **37.** $\frac{5}{8} + \frac{1}{8}$ **38.** $\frac{2}{15} + \frac{8}{15}$

39. $\frac{4}{5} - \frac{3}{5}$ **40.** $\frac{5}{7} - \frac{2}{7}$ **41.** $\frac{11}{12} - \frac{7}{12}$ **42.** $\frac{13}{18} - \frac{9}{18}$

43. $\frac{1}{4} + \frac{1}{2}$ **44.** $\frac{1}{3} + \frac{5}{9}$ **45.** $\frac{5}{6} + \frac{3}{4}$ **46.** $\frac{3}{8} + \frac{1}{6}$

47. $4 + \frac{2}{3}$ **48.** $2 + \frac{3}{7}$ **49.** $\frac{7}{9} - \frac{2}{3}$ **50.** $\frac{3}{4} - \frac{1}{2}$

51. $\frac{5}{9} - \frac{2}{7}$ **52.** $\frac{5}{6} - \frac{4}{7}$ **53.** $3 - \frac{4}{5}$ **54.** $1 - \frac{9}{7}$

Perform the indicated operation. If the fraction is undefined, say so. Then use a calculator to check your work.

55. $\frac{3172}{3172}$ **56.** $\frac{62}{62}$ **57.** $\frac{599}{1}$ **58.** $\frac{215}{1}$

59. $\frac{842}{0}$ **60.** $\frac{713}{0}$ **61.** $\frac{0}{621}$ **62.** $\frac{0}{798}$

63. $\frac{824}{631} \cdot \frac{631}{824}$ **64.** $\frac{173}{190} \cdot \frac{190}{173}$

65. $\frac{544}{293} - \frac{544}{293}$ **66.** $\frac{345}{917} - \frac{345}{917}$

Evaluate the given expression for w = 4, x = 3, y = 5, and z = 12.

67. $\frac{w}{z}$ **68.** $\frac{z}{x}$ **69.** $\frac{x}{w} \div \frac{y}{z}$

70. $\frac{y}{z} \cdot \frac{w}{x}$ **71.** $\frac{x}{w} - \frac{y}{z}$ **72.** $\frac{y}{x} + \frac{y}{z}$

Use a calculator to compute. Round the result to two decimal places.

73. $\frac{19}{97} \cdot \frac{65}{74}$ **74.** $\frac{67}{71} \cdot \frac{381}{399}$ **75.** $\frac{684}{795} \div \frac{24}{37}$

76. $\frac{149}{215} \div \frac{31}{52}$ **77.** $\frac{89}{102} - \frac{59}{133}$ **78.** $\frac{614}{701} + \frac{391}{400}$

For Exercises 79 and 80, draw a picture of a pizza to show that the true statement makes sense.

79. $\frac{2}{8} = \frac{1}{4}$ **80.** $\frac{1}{4} + \frac{2}{4} = \frac{3}{4}$

81. A rectangular plot of land has a length of $\frac{2}{5}$ mile and a width of $\frac{1}{4}$ mile. What is the area of this plot?

82. A rectangular picture has a width of $\frac{2}{3}$ foot and a length of $\frac{3}{4}$ foot. What is the perimeter of this picture?

83. For an elementary algebra course, total course points are calculated by adding points earned on homework assignments, quizzes, tests, and the final exam. If the total of scores on tests is worth $\frac{1}{2}$ of the course points and the final exam score is worth $\frac{1}{4}$ of the course points, what fraction of the course points comes from homework assignments and quizzes?

84. A family spends $\frac{1}{3}$ of its income for the mortgage and $\frac{1}{6}$ of its income for food. What fraction of its income remains?

For Exercises 85 and 86, let x be a number. Translate the expression into an English phrase.

85. $\dfrac{x}{3}$

86. $\dfrac{5}{x}$

87. Some friends pay a total of $19 for a pizza. Each of the n friends pays an equal share of the cost. Complete Table 10 to help find an expression that describes the cost (in dollars) per person. Show the arithmetic to help you see a pattern.

Table 10 Cost per Person for the Pizza

Number of People	Cost per Person (dollars)
2	
3	
4	
5	
n	

88. A tutor charges $45 for a tutoring session that lasts for t hours. Complete Table 11 to help find an expression that describes the cost (in dollars) per hour. Show the arithmetic to help you see a pattern.

Table 11 Cost per Hour for the Session

Total Time (hours)	Cost per Hour (dollars per hour)
2	
3	
4	
5	
t	

Concepts

89. a. Perform the indicated operation.

   i. $\dfrac{5}{6} \cdot \dfrac{2}{3}$ **ii.** $\dfrac{5}{6} \div \dfrac{2}{3}$

   iii. $\dfrac{5}{6} + \dfrac{2}{3}$ **iv.** $\dfrac{5}{6} - \dfrac{2}{3}$

  b. Compare the methods you used to perform the operations in part (a). Describe how the methods are similar and how they are different.

90. a. Find each product.

   i. $\dfrac{2}{3} \cdot \dfrac{3}{2}$ **ii.** $\dfrac{4}{7} \cdot \dfrac{7}{4}$ **iii.** $\dfrac{1}{6} \cdot \dfrac{6}{1}$

  b. On the basis of the results you found in part (a), use words to describe a property of a fraction and its reciprocal. Then describe the property in terms of variables.

91. A student tries to find the product $\dfrac{1}{2} \cdot \dfrac{1}{3}$:

$$\frac{1}{2} \cdot \frac{1}{3} = \left(\frac{1}{2} \cdot \frac{3}{3}\right) \cdot \left(\frac{1}{3} \cdot \frac{2}{2}\right)$$

$$= \frac{3}{6} \cdot \frac{2}{6}$$

$$= \frac{6}{36}$$

$$= \frac{1}{6}$$

What would you tell the student?

92. A student tries to find the sum $\dfrac{2}{3} + \dfrac{5}{6}$:

$$\frac{2}{3} + \frac{5}{6} = \frac{2+5}{3+6} = \frac{7}{9}$$

Describe any errors. Then find the sum correctly.

93. A student tries to find the product $2 \cdot \dfrac{3}{5}$:

$$2 \cdot \frac{3}{5} = \frac{6}{10} = \frac{3}{5}$$

Describe any errors. Then find the product correctly.

94. A student tries to find the product $3 \cdot \dfrac{7}{2}$:

$$3 \cdot \frac{7}{2} = \frac{7}{3 \cdot 2} = \frac{7}{6}$$

Describe any errors. Then find the product correctly.

95. Greenskeepers have been mowing golf putting surfaces progressively lower over the past half century (see Table 12).

Table 12 Grass Heights on Golf Putting Surfaces

Decade	Year Used to Represent Decade	Grass Height (inches)
1950s	1955	$\dfrac{1}{4}$
1960s	1965	$\dfrac{7}{32}$
1970s	1975	$\dfrac{3}{16}$
1980s	1985	$\dfrac{5}{32}$
1990s	1995	$\dfrac{1}{8}$
2000s	2005	$\dfrac{1}{10}$

Source: *Golf Course Superintendents Association of America*

Let h be the grass height (in inches) at t years since 1950.

  a. Construct a scatterplot of the data.

  b. Draw a linear model on your scatterplot.

  c. Predict the grass height in 2025.

  d. Predict when the grass height will be $\dfrac{1}{16}$ inch.

  e. Find the t-intercept. What does it mean in this situation?

96. a. Use a number line to find the distance between 0 and each number.

   i. 4   **ii.** -3 **iii.** -6

  b. *Without* using a number line, describe in general how you can find the distance between 0 and a number on a number line.

97. Explore how to add two negative numbers:

  a. Use a calculator to find each sum of two negative numbers.

   i. $-1 + (-5)$ **ii.** $-6 + (-2)$ **iii.** $-3 + (-4)$

  b. What pattern do you notice? If you do not see a pattern, continue finding sums of any two negative numbers until you do.

  c. Without using a calculator, find the sum $-4 + (-5)$. Then use a calculator to check your work.

  d. State a general rule for how to add two negative numbers.

98. Explore how to add two numbers with different signs:
  a. First, consider the case in which the positive number is farther from 0 on the number line than the negative number is. Use a calculator to find the following sums:
    i. $5 + (-2)$ **ii.** $7 + (-1)$ **iii.** $8 + (-3)$
  b. What pattern do you notice in your work from part (a)? If you do not see a pattern, continue finding similar sums until you do.
  c. Now consider the case in which the positive number is closer to 0 on the number line than the negative number is. Use a calculator to find the following sums:
    i. $2 + (-5)$ **ii.** $1 + (-7)$ **iii.** $3 + (-8)$
  d. What pattern do you notice in your work from part (c)?

  e. Now consider the case in which the two numbers are the same distance from 0 on the number line, but on opposite sides of 0. Use a calculator to find the following sums:
    i. $4 + (-4)$ **ii.** $7 + (-7)$ **iii.** $9 + (-9)$
  f. What pattern do you notice in your work from part (e)?
  g. Without using a calculator, find each sum. Then use a calculator to check your work.
    i. $6 + (-4)$ **ii.** $3 + (-7)$ **iii.** $6 + (-6)$
  h. State a general rule for how to add two numbers with different signs.

99. Why is division by 0 undefined?

100. Explain why we do not add the denominators of two fractions when we add the fractions.

2.3 Absolute Value and Adding Real Numbers

Objectives

» Find the opposite of a number.

» Find the opposite of the opposite of a number.

» Find the absolute value of a number.

» Add real numbers by thinking in terms of money, the number line, and absolute value.

» Add real numbers pertaining to authentic situations.

» Find an expression to model an authentic quantity.

In this section, our main objective is to add real numbers.

The Opposite of a Number

Note that in Fig. 12 both the numbers -3 and 3 are 3 units from 0 on the number line, but they are on opposite sides of 0. We say that -3 is the *opposite* of 3, that 3 is the *opposite* of -3, and that -3 and 3 are *opposites*.

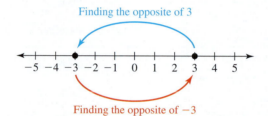

Figure 12 Finding the opposite of 3 and the opposite of -3

Two numbers are called **opposites** of each other if they are the same distance from 0 on the number line, but are on different sides of 0. We find the opposite of a number by writing a negative sign in front of the number. For example, the opposite of 3 is -3 (see Fig. 12).

The Opposite of the Opposite of a Number

Now consider this true statement:

The opposite of -3 is 3 (see Fig. 12).

In symbols, we write

The opposite of -3 is equal to   3.

$$-\quad(-3)\quad=\quad3$$

Here are some more examples of finding the opposite of a negative number:

$$-(-2) = 2$$
$$-(-7) = 7$$

We use a graphing calculator to find $-(-7)$. See Fig. 13. We use the button $\boxed{-}$ for subtraction and the button $\boxed{(-)}$ for negative numbers and for taking opposites.

Press $\boxed{(-)}\boxed{(}\boxed{(-)}\boxed{7}\boxed{)}\boxed{\text{ENTER}}$.

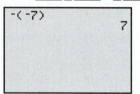

Figure 13 Calculating $-(-7)$

WARNING   If we use the subtraction button $\boxed{-}$ to try to find $-(-7)$, a T1-83 or T1-84 calculator will pull up the previous answer, and after we press ENTER, it will display an error message.

We can view $-(-7)$ as finding the opposite of -7 or as finding the opposite of the opposite of 7.

Finding the Opposite of the Opposite of a Number

$$-(-a) = a$$

In words: The opposite of the opposite of a number is equal to that same number.

We use parentheses to separate two opposite symbols or an operation symbol and an opposite symbol.

Example 1   Finding Opposites

Find the opposite.

1. $-(-5)$ **2.** $-(-(-8))$

Solution

1. $-(-5) = 5$ $-(-a) = a$
2. $-(-(-8)) = -(8)$ $-(-a) = a$
                 $= -8$ *Write without parentheses.*

Absolute Value

The *absolute value* of a number a, written $|a|$, is the distance the number a is from 0 on the number line.

Definition   Absolute value

The **absolute value** of a number is the distance the number is from 0 on the number line.

So $|-3| = 3$ because -3 is a distance of 3 units from 0, and $|3| = 3$ because 3 is a distance of 3 units from 0 (see Fig. 14).

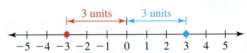

Figure 14 Both -3 and 3 are a distance of 3 units from 0

Press [2nd] [0] [ENTER] [(−)] [3] [)] [ENTER].

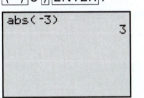

Figure 15 Calculating $|-3|$

On a graphing calculator, "abs" stands for absolute value. We find $|-3|$ in Fig. 15.

Example 2   Finding Absolute Values of Numbers

Calculate.

1. $|2|$ **2.** $|-2|$
3. $-|2|$ **4.** $-|-2|$

Solution

1. $|2| = 2$ because 2 is a distance of 2 units from 0 (see Fig. 16).

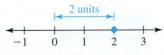

Figure 16 $|2| = 2$

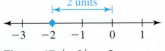

Figure 17 $|-2| = 2$

2. $|-2| = 2$ because -2 is a distance of 2 units from 0 (see Fig. 17).

3. $-|2| = -(2)$ $|2| = 2$
       $= -2$ *Write without parentheses.*

4. $-|-2| = -(2)$ $|-2| = 2$
       $= -2$ *Write without parentheses.*

We use a graphing calculator to check that $-|-2| = -2$ (see Fig. 18).

Press $\boxed{(-)}$ $\boxed{\text{2nd}}$ $\boxed{0}$ $\boxed{\text{ENTER}}$
$\boxed{(-)}$ $\boxed{2}$ $\boxed{)}$ $\boxed{\text{ENTER}}$.

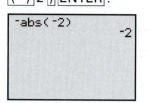

Figure 18 Check that
$-|-2| = -2$

Addition of Two Numbers with the Same Sign

Thinking about credit card balances or the number line can help us see how to add numbers with the same sign.

Example 3   Finding the Sum of Two Numbers with the Same Sign

1. A person has a credit card balance of $0. If she uses her credit card to make two purchases, one for $2 and one for $5, what is the new balance?
2. Write a sum that is related to the computation in Problem 1.
3. Use a number line to illustrate the sum found in Problem 2.

Solution

1. By making purchases for $2 and $5, the person now owes $7. So, the new balance is -7 dollars.
2. Here is the sum:

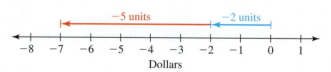

3. Using the number line, imagine moving 2 units to the left of 0 and then 5 more units to the left. Figure 19 illustrates that $-2 + (-5) = -7$.

Figure 19 Illustration of $-2 + (-5) = -7$

In Example 3, we found that $-2 + (-5) = -7$. To get this result, we added the debts of 2 and 5 to get a total debt of 7. Note that 2 and 5 are the absolute values of -2 and -5. Note also that the result, -7, of the original sum has the same sign as both -2 and -5. These observations suggest the following procedure:

Adding Two Numbers with the Same Sign

To add two numbers with the same sign,

1. Add the absolute values of the numbers.
2. The sum of the original numbers has the same sign as the sign of the original numbers.

Example 4   Finding the Sum of Two Numbers with the Same Sign

Find the sum.

1. $-3 + (-6)$

2. $-\dfrac{1}{5} + \left(-\dfrac{3}{5}\right)$

Solution

1. First, we add the absolute values of the numbers -3 and -6: $3 + 6 = 9$. Because both -3 and -6 are negative, their sum is negative. So, $-3 + (-6) = -9$.

2. By adding the absolute values of the fractions, we have $\frac{1}{5} + \frac{3}{5} = \frac{4}{5}$. Because both original fractions are negative, their sum is negative. So,

$$-\frac{1}{5} + \left(-\frac{3}{5}\right) = -\frac{4}{5}$$

Press $\boxed{(-)}\,3\,\boxed{+}\,\boxed{(}\,\boxed{(-)}$
$6\,\boxed{)}\,\boxed{\text{ENTER}}$.

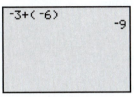

Figure 20 Calculating $-3 + (-6)$

After completing an exercise in this section's homework by hand, you can use a graphing calculator to check your work. For example, we check our work for $-3 + (-6)$, Problem 1 in Example 4 (see Fig. 20).

Addition of Two Numbers with Different Signs

Thinking about the number line or exchanges of money can also help us see how to add numbers with different signs.

| Example 5 | Finding the Sum of Two Numbers with Different Signs |

1. A brother owes his sister \$5. If he then pays her back \$2, how much does he still owe her?
2. Write a sum that is related to your work in Problem 1.
3. Use a number line to illustrate the sum you found in Problem 2.

Solution

1. By owing his sister \$5 and paying her back \$2, the brother now owes his sister \$3.
2. Here's the sum:

$$\underbrace{-5}_{\text{Owe \$5}} \; + \; \underbrace{2}_{\text{Pay back \$2}} \; = \; \underbrace{-3}_{\text{Now owe \$3}}$$

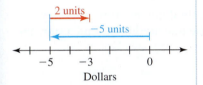

Figure 21 Illustration of $-5 + 2 = -3$

3. Using the number line, imagine moving 5 units to the left of 0 and then 2 units to the right of -5. Figure 21 illustrates that $-5 + 2 = -3$.

In Problem 2 in Example 5, we found that $-5 + 2 = -3$. We can get this result by first finding the difference of 5 and 2:

$$5 - 2 = 3$$

We can think of this operation as lowering a debt of \$5 by \$2 to get a debt of \$3, so the result is -3. Note that the result, -3, has the same sign as -5, which has a larger absolute value than 2. These observations suggest the following procedure:

Adding Two Numbers with Different Signs

To add two numbers with different signs,

1. Find the absolute values of the numbers. Then subtract the smaller absolute value from the larger absolute value.

2. The sum of the original numbers has the same sign as the original number with the larger absolute value.

Example 6   Finding the Sum of Two Numbers with Different Signs

Find the sum.

1. $-4 + 7$

2. $3 + (-9)$

3. $-\dfrac{5}{6} + \dfrac{2}{3}$

Solution

1. First, we find that $7 - 4 = 3$. Because 7 has a larger absolute value than -4, and because 7 is positive, the sum is positive: $-4 + 7 = 3$.
2. First, we find that $9 - 3 = 6$. Because -9 has a larger absolute value than 3, and because -9 is negative, the sum is negative: $3 + (-9) = -6$.
3. First, we write the fractions so that the denominators are the same.

$$-\frac{5}{6} + \frac{2}{3} = -\frac{5}{6} + \frac{2}{3} \cdot \frac{2}{2} \qquad \textit{LCD is 6.}$$
$$= -\frac{5}{6} + \frac{4}{6} \qquad \textit{Multiply numerators and multiply denominators: } \frac{a}{b} \cdot \frac{c}{d} = \frac{ac}{bd}$$

Next, we subtract the smaller absolute value from the larger absolute value:

$$\frac{5}{6} - \frac{4}{6} = \frac{1}{6}$$

Because $-\dfrac{5}{6}$ has a larger absolute value than the fraction $\dfrac{4}{6}$, and because $-\dfrac{5}{6}$ is negative, the sum is negative:

$$-\frac{5}{6} + \frac{4}{6} = -\frac{1}{6}$$

▶

We have discussed three ways to add real numbers: thinking in terms of money, the number line, and absolute value. **It is good practice to use one method to find a sum and then use another method (or a calculator) as a check.**

Example 7   Translating from English into Mathematics

Let x be a number. Translate the phrase "the sum of -4 and the number" into a mathematical expression. Then evaluate the expression for $x = -2$.

Solution

The expression is $-4 + x$. We substitute -2 for x in the expression $-4 + x$ and then find the sum:

$$-4 + (-2) = -6$$

▶

Adding Real Numbers in Authentic Situations

Knowing how to add real numbers is a useful skill when you work with quantities that can be negative, such as balances of checking accounts and temperature readings.

Example 8   Applications of Adding Real Numbers

1. A person bounces several checks and is charged service fees such that the balance of the checking account is -90.75 dollars. If the person then deposits 300 dollars, what is the balance?
2. Three hours ago, the temperature was $-11°F$. If the temperature has increased by $5°F$ in the last 3 hours, what is the current temperature?

Solution

1. The balance is $-90.75 + 300$ dollars. To find this sum, we first find the difference $300 - 90.75 = 209.25$. Because 300 has a larger absolute value than -90.75 and because 300 is positive, the sum is positive: $-90.75 + 300 = 209.25$. So, the balance is $209.25.

2. The temperature is $-11 + 5$ degrees Fahrenheit. To find this sum, we first find the difference $11 - 5 = 6$. Because -11 has a larger absolute value than 5 and because -11 is negative, the sum is negative: $-11 + 5 = -6$. So, the current temperature is $-6°F$.

Modeling with Expressions

In Example 9, we will use an expression to describe an authentic quantity.

Example 9   Finding and Evaluating an Expression

A person fills up his 15-gallon gasoline tank. Let c be the amount of gasoline (in gallons) consumed after the tank has been filled up.

1. Use a table to help find an expression that describes the amount of gasoline (in gallons) that remains in the tank.
2. Evaluate the expression found in Problem 1 for $c = 7$. What does the result mean in this situation?

Table 13 Gasoline Remaining

Gasoline Consumed (gallons)	Gasoline Remaining (gallons)
1	$15 + (-1)$
2	$15 + (-2)$
3	$15 + (-3)$
4	$15 + (-4)$
c	$15 + (-c)$

Solution

1. First, we construct Table 13. We show the arithmetic to help us see the pattern. From the last row of the table, we see that the expression $15 + (-c)$ represents the amount of gasoline (in gallons) that remains.
2. We evaluate $15 + (-c)$ for $c = 7$:

$$15 + (-(7)) = 15 + (-7) = 8$$

If 7 gallons of gasoline are consumed, 8 gallons of gasoline will remain.

◈ Group Exploration

Adding a number and its opposite

1. Evaluate $a + (-a)$ for the given values of a.

   a. $a = 2$ **b.** $a = 3$ **c.** $a = 5$

2. Evaluate $a + (-a)$ for the given values of a.
   [**Hint:** Use $-(-a) = a$.]

   a. $a = -2$ **b.** $a = -3$ **c.** $a = -5$

3. What do the results you found in Problems 1 and 2 suggest about $a + (-a)$?

Tips for Success   Affirmations

Do you have difficulty with math? If so, do you ever tell yourself (or others) that you are not good at it? This is called *negative self-talk*. The more you say this, the more likely your subconscious will believe it—and you *will* do poorly in math.

You can counteract years of negative self-talk by telling yourself with conviction that you are good at math.

It might seem strange to state that something is true that hasn't happened yet, but it works! Such statements are called *affirmations*.

There are four guiding principles for getting the most out of saying affirmations:

1. Say affirmations that imply the desired event is currently happening. For example, say

"I am good at algebra," not "I will be good at algebra."

2. Say that your desired result is continuing to improve. For example, say

"I am good at algebra, and I continue to improve at it."

3. Say affirmations in the positive rather than in the negative. For example, say

"I attend each class," not "I don't cut classes."

4. Say affirmations with conviction.

If you would like to learn more about affirmations, the book *Creative Visualization* (Bantam Books, 1985), by Shakti Gawain, is an excellent resource.

Homework 2.3

For extra help ▶   **MyLab Math**   Watch the videos in MyLab Math

Compute by hand. Then use a calculator to check your work.

1. $-(-4)$ 2. $-(-9)$ 3. $-(-(-7))$
4. $-(-(-2))$ 5. $|3|$ 6. $|6|$
7. $|-8|$ 8. $|-1|$ 9. $-|4|$
10. $-|5|$ 11. $-|-7|$ 12. $-|-9|$

Find the sum by hand. Then use a calculator to check your work.

13. $2 + (-7)$ 14. $5 + (-3)$ 15. $-1 + (-4)$
16. $-3 + (-2)$ 17. $7 + (-5)$ 18. $6 + (-9)$
19. $-8 + 5$ 20. $-3 + 4$ 21. $-7 + (-3)$
22. $-9 + (-5)$ 23. $4 + (-7)$ 24. $8 + (-2)$
25. $1 + (-1)$ 26. $8 + (-8)$ 27. $-4 + 4$
28. $-7 + 7$ 29. $12 + (-25)$ 30. $17 + (-14)$
31. $-39 + 17$ 32. $-89 + 57$
33. $-246 + (-899)$ 34. $-347 + (-594)$
35. $25,371 + (-25,371)$ 36. $127,512 + (-127,512)$
37. $-4.1 + (-2.6)$ 38. $-3.7 + (-9.9)$
39. $-5 + 0.2$ 40. $-0.3 + 7$
41. $2.6 + (-99.9)$ 42. $37.05 + (-19.26)$
43. $\dfrac{5}{7} + \left(-\dfrac{3}{7}\right)$ 44. $\dfrac{2}{5} + \left(-\dfrac{1}{5}\right)$
45. $-\dfrac{5}{8} + \dfrac{3}{8}$ 46. $-\dfrac{5}{6} + \dfrac{1}{6}$
47. $-\dfrac{1}{4} + \left(-\dfrac{1}{2}\right)$ 48. $-\dfrac{2}{3} + \left(-\dfrac{5}{6}\right)$
49. $\dfrac{5}{6} + \left(-\dfrac{1}{4}\right)$ 50. $\dfrac{2}{3} + \left(-\dfrac{3}{4}\right)$

Use a calculator to find the sum. Round the result to two decimal places.

51. $-325.89 + 6547.29$ 52. $-7498.34 + 6435.28$

53. $-17,835.69 + (-79,735.45)$
54. $-38,487.26 + (-83,205.87)$
55. $-\dfrac{34}{983} + \left(-\dfrac{19}{251}\right)$ 56. $-\dfrac{937}{642} + \left(-\dfrac{825}{983}\right)$

Evaluate the expression for $a = -4$, $b = 3$, and $c = -2$.

57. $a + b$ 58. $b + a$ 59. $a + c$ 60. $b + c$

For Exercises 61–64, let x be a number. Translate the English phrase into a mathematical expression. Then evaluate the expression for $x = -6$.

61. 2 more than the number
62. The number increased by 3
63. The sum of -5 and the number
64. The number plus -8
65. A person bounces several checks and is charged service fees such that the balance of the checking account is -75 dollars. If the person then deposits $250, what is the balance?
66. A person bounces several checks and is charged service fees such that the balance of the checking account is -112.50 dollars. If the person then deposits $170, what is the balance?
67. A check register is shown in Table 14. Find the final balance of the checking account.

Table 14 Check Register

Check No.	Date	Description of Transaction	Payment	Deposit	Balance
					−89.00
	7/18	Transfer		300.00	
3021	7/22	State Farm	91.22		
3022	7/22	MCI	44.26		
	7/31	Paycheck		870.00	

68. A check register is shown in Table 15. Find the final balance of the checking account.

Table 15 Check Register

Check No.	Date	Description of Transaction	Payment	Deposit	Balance
					−135.00
	2/31	Paycheck		549.00	
253	3/2	FedEx Office	10.74		
	3/3	ATM	21.50		
254	3/7	Barnes & Noble	17.19		

69. A person has a credit card balance of −5471 dollars. If she sends a check to the credit card company for $2600, what is the new balance?

70. A person has a credit card balance of −2739 dollars. If he sends a check to the credit card company for $530, what is the new balance?

71. A student has a credit card balance of −3496 dollars. If he sends a check to the credit card company for $2500 and then uses his credit card to purchase a bicycle for $613 and a helmet for $24, what is the new balance?

72. A student has a credit card balance of −873 dollars. If she sends a check to the credit card company for $500 and then uses the card to buy a tennis racquet for $249 and a tennis outfit for $87, what is the new balance?

73. Three hours ago, it was −5°F. If the temperature has increased by 9°F in the last 3 hours, what is the current temperature?

74. Four hours ago, it was −12°F. If the temperature has increased by 8°F in the last 4 hours, what is the current temperature?

75. A person just lost 20 pounds on a diet.
  a. Complete Table 16 to help find an expression that describes the person's current weight (in pounds) if the person's weight before the diet was B pounds. Show the arithmetic to help you see a pattern.

Table 16 Weights before and after the Diet

Weight before Diet (pounds)	Weight after Diet (pounds)
160	
165	
170	
175	
B	

  b. Evaluate the expression you found in part (a) for $B = 169$. What does your result mean in this situation?

76. An electronics store is offering a weekend sale of $35 off the retail price of any of its LCD televisions.
  a. Complete Table 17 to help find an expression that describes the sale price (in dollars) if the retail price is r dollars. Show the arithmetic to help you see a pattern.

Table 17 Retail and Sale Prices

Retail Price (dollars)	Sale Price (dollars)
350	
400	
450	
500	
r	

  b. Evaluate the expression you found in part (a) for $r = 270$. What does your result mean in this situation?

77. The balance in a person's checking account is −80 dollars.
  a. Complete Table 18 to help find an expression that describes the new balance (in dollars) if the person deposits d dollars. Show the arithmetic to help you see a pattern.

Table 18 Deposits and New Balances

Deposit (dollars)	New Balance (dollars)
50	
100	
150	
200	
d	

  b. Evaluate the expression you found in part (a) for $d = 125$. What does your result mean in this situation?

78. One hour ago, the temperature was −2°F.
  a. Complete Table 19 to help find an expression that describes the current temperature (in degrees Fahrenheit) if the temperature decreased by x degrees Fahrenheit in the past hour. Show the arithmetic to help you see a pattern.

Table 19 Decreases in Temperature and Current Temperatures

Decrease in Temperature (degrees Fahrenheit)	Current Temperature (degrees Fahrenheit)
1	
2	
3	
4	
x	

  b. Evaluate the expression you found in part (a) for $x = 7$. What does your result mean in this situation?

Concepts

79. If a is negative and b is negative, what can you say about the sign of $a + b$? Use a number line to show this property.

80. If a is positive, b is negative, and b is larger in absolute value than a, what can you say about the sign of $a + b$? Use a number line to show this property.

81. If $a + b = 0$, what can you say about a and b?

82. If $a + b$ is positive, what can you say about a and b?

83. **a.** Evaluate $-a$ for $a = -3$.
   b. Evaluate $-a$ for $a = -4$.
   c. Evaluate $-a$ for $a = -6$.
   d. A student says that $-a$ represents only negative numbers because $-a$ has a negative sign. Is the student correct? Explain.

84. **a.** Evaluate $a + b$ for $a = -2$ and $b = 5$.
   b. Evaluate $b + a$ for $a = -2$ and $b = 5$.
   c. Compare the results you found in parts (a) and (b).
   d. Evaluate $a + b$ and $b + a$ for $a = -4$ and $b = -9$, and then compare the results.
   e. Evaluate $a + b$ and $b + a$ for values of your choosing for a and b, and then compare the results.
   f. Is the statement $a + b = b + a$ true for all numbers a and b? Explain.

2.4 Change in a Quantity and Subtracting Real Numbers

Objectives

» Find the change in a quantity.

» Subtract real numbers.

» Find a change in elevation.

» Determine the sign of the change for an increasing or decreasing quantity.

In this section, we will discuss how to use subtraction to compute how much a quantity has changed. For example, we can compute the increase in a population of wolves or the decrease in the percentage of eligible voters who voted in an election.

Change in a Quantity

If the value of a stock increases from $5 to $8, we say the value *changed* by $3. Finding the change in a quantity is a very important concept in mathematics and has many applications. A company is extremely focused on the change in its profits. During an operation, a surgeon keeps a close eye on the change in a patient's blood pressure. You probably care deeply about a change in your GPA.

> **Example 1**   Finding the Change in a Quantity
>
> **1.** If a student's CD collection increases from 2 CDs to 7 CDs, find the change in the number of CDs.
>
> **2.** Write a difference that is related to the computation in Problem 1.

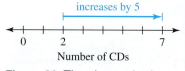

Figure 22 The change in the number of CDs is 5 CDs

Solution

1. If the number of CDs increases from 2 CDs to 7 CDs, then the change in the number of CDs is 5 CDs (see Fig. 22).

2. Here is the difference:

Change in the number of CDs		Ending number of CDs		Beginning number of CDs
5	=	7	−	2

In Example 1, we found the change in the number of CDs by finding the difference of the ending number of CDs and the beginning number of CDs.

> **Change in a Quantity**
>
> The change in a quantity is the ending amount minus the beginning amount:
>
> Change in the quantity = Ending amount − Beginning amount

In Example 2, we will find the changes in a quantity from one year to the next.

Table 20 Average Movie Ticket Prices

Years	Average Price (dollars)
2010	7.89
2011	7.93
2012	7.96
2013	8.13
2014	8.17
2015	8.43

Source: *National Association of Theater Owners*

Example 2   Finding Changes in a Quantity

Average movie ticket prices are shown in Table 20 for various years.

1. Find the change in the average movie ticket price from 2010 to 2011.
2. For the period 2010–2015, find each of the changes in the average movie ticket price from one year to the next.
3. From which year to the next did the average price increase the most?

Solution

1. We find the difference of the average price in 2011 (ending) and the average price in 2010 (beginning):

Ending average price (in dollars)   Beginning average price (in dollars)   Change in average price (in dollars)

$$7.93 \quad - \quad 7.89 \quad = \quad 0.04$$

So, the average ticket price from 2010 to 2011 changed by $0.04.

2. The changes in the average ticket price from one year to the next are listed in Table 21. The changes were found by computing the differences similar to the one found in Problem 1.

Table 21 Changes in Average Movie Ticket Prices from Year to Year

Years	Average Price (dollars)	Change in Average Price (dollars)
2010	7.89	—
2011	7.93	$7.93 - 7.89 = 0.04$
2012	7.96	$7.96 - 7.93 = 0.03$
2013	8.13	$8.13 - 7.96 = 0.17$
2014	8.17	$8.17 - 8.13 = 0.04$
2015	8.43	$8.43 - 8.17 = 0.26$

Source: *National Association of Theater Owners*

3. The average ticket price changed by $0.26 from 2014 to 2015, the greatest increase from any year to the next.

Subtraction of Real Numbers

Exploring the change in a quantity can help us see how to subtract real numbers.

Example 3   Finding the Difference of Two Real Numbers

1. A college's enrollment decreases from 7 thousand students to 2 thousand students. What is the change in the enrollment?
2. Write a difference that is related to the computation in Problem 1.

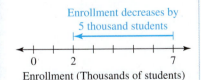

Enrollment decreases by 5 thousand students

Enrollment (Thousands of students)

Figure 23 Enrollment decreases from 7 thousand students to 2 thousand students

Solution

1. Because the enrollment has decreased from 7 thousand students to 2 thousand students, the change is −5 thousand students. The change is negative because the enrollment is decreasing (see Fig. 23).
2. The change in the enrollment is the difference of the ending enrollment and the beginning enrollment:

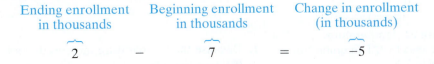

Ending enrollment in thousands   Beginning enrollment in thousands   Change in enrollment (in thousands)

$$2 \quad - \quad 7 \quad = \quad -5$$

In Example 3, we found that

$$2 - 7 = -5$$

Note that $2 + (-7)$ gives the same result:

$$2 + (-7) = -5$$

This means

$$2 - 7 = 2 + (-7)$$

which suggests that subtracting a number is the same as adding the opposite of that number:

Subtract 7.   Add the opposite of 7.
$$2 - 7 = 2 + (-7)$$

Subtracting a Real Number

$$a - b = a + (-b)$$

In words: To subtract a number, add its opposite.

To subtract real numbers, we first write the difference as a related sum and then find the sum.

Example 4   Finding Differences of Real Numbers

Find the difference.

1. $4 - 6$

2. $\dfrac{2}{9} - \dfrac{5}{9}$

Solution

1.   Subtract 6.   Add the opposite of 6.
$$4 - 6 \quad = \quad 4 + (-6) = -2 \quad a - b = a + (-b)$$

2. $\dfrac{2}{9} - \dfrac{5}{9} = \dfrac{2}{9} + \left(-\dfrac{5}{9}\right)$ *Add the opposite of* $\dfrac{5}{9}$*:* $a - b = a + (-b)$

$$= -\dfrac{3}{9}$$

$$= -\dfrac{1}{3} \qquad \textit{Simplify.}$$

Considering the change in a quantity can also help us see how to subtract a negative number.

Example 5   Subtracting a Negative Number

1. The temperature increases from $-2°F$ to $7°F$. Find the change in temperature.
2. Write a difference that is related to the work in Problem 1.
3. Find the difference obtained in Problem 2 by using the rule $a - b = a + (-b)$.

Solution

1. Because the temperature increased from $-2°F$ to $7°F$, the change in temperature is $9°F$ (see Fig. 24).

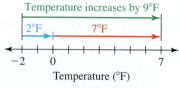

Figure 24 Temperature increases by 9°F in going from −2°F to 7°F

2. The change in temperature is the ending temperature minus the beginning temperature:

Ending temperature (°F)		Beginning temperature (°F)		Change in temperature (°F)
7	−	(−2)	=	9

3.

$$\overbrace{7 - (-2)}^{\text{Subtract } -2.} = \overbrace{7 + 2}^{\substack{\text{Add the opposite} \\ \text{of } -2. \text{ (So, add 2.)}}} \qquad a - b = a + (-b)$$
$$= 9 \qquad\qquad\qquad \textit{Add.}$$

Note that to find the difference $7 - (-2)$, we add 2 and 7. This makes sense because, in going from −2°F to 0°F, the temperature increases by 2°F, and in continuing from 0°F to 7°F, the temperature increases by another 7°F (see Fig. 24).

We use a calculator to check our work in Example 5 (see Fig. 25). Recall from Section 2.3 that we use the button $\boxed{-}$ for subtraction and the button $\boxed{(-)}$ for negative numbers and for taking opposites.

It is good practice to do homework exercises first by hand and then by using a calculator to check your hand results.

Press 7 $\boxed{-}$ $\boxed{(}$ $\boxed{(-)}$ 2 $\boxed{)}$ $\boxed{\text{ENTER}}$.

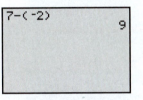

```
7-(-2)
              9
```

Figure 25 Calculating $7 - (-2)$

Example 6   Subtracting a Negative Number

Find the difference.

1. $4 - (-6)$ **2.** $-9 - (-3)$

Solution

1.

$$\overbrace{4 - (-6)}^{\text{Subtract } -6.} = \overbrace{4 + 6}^{\substack{\text{Add the opposite} \\ \text{of } -6. \text{ (So, add 6.)}}} \qquad a - b = a + (-b)$$
$$= 10 \qquad\qquad\qquad \textit{Add.}$$

2.

$$\overbrace{-9 - (-3)}^{\text{Subtract } -3.} = \overbrace{-9 + 3}^{\substack{\text{Add the opposite} \\ \text{of } -3. \text{ (So, add 3.)}}} \qquad a - b = a + (-b)$$
$$= -6 \qquad\qquad\qquad \textit{Add.}$$

Example 7   Translating from English into Mathematics

Translate the phrase "the difference of a and b" into a mathematical expression. Then evaluate the expression for $a = 3$ and $b = -7$.

Solution

The expression is $a - b$. We substitute 3 for a and −7 for b in the expression $a - b$ and then find the difference:

$$(3) - (-7) = 3 + 7 \qquad a - b = a + (-b)$$
$$= 10 \qquad \textit{Add.}$$

Figure 26 Elevations of 200 feet and −200 feet

Change in Elevation

In Example 8, you will work with *elevation*. An object that has a *positive* elevation of 200 feet is 200 feet *above* sea level (see Fig. 26). An object that has a *negative* elevation of −200 feet is 200 feet *below* sea level (see Fig. 26).

Example 8   Finding a Change in Elevation

The Golden Gate Bridge has two towers that support the two main cables of the bridge (see Fig. 27). The top of each tower is at an elevation of 746 feet, and the foot of each tower is at an elevation of −136 feet (136 feet below sea level). Find the height of each tower.

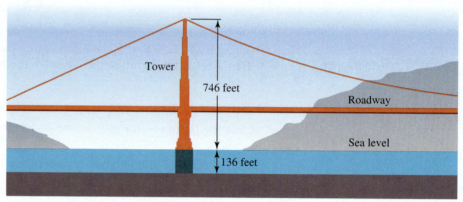

Figure 27 Golden Gate Bridge

Solution

We can find the height of each tower by computing the change in elevation from the bottom of each tower to the top:

$$
\underbrace{746}_{\substack{\text{Top elevation} \\ \text{(in feet)}}} - \underbrace{(-136)}_{\substack{\text{Bottom elevation} \\ \text{(in feet)}}} = 746 + 136 \quad a - b = a + (-b)
$$

$$
= 882 \quad \text{Add.}
$$

So, the height of each tower is 882 feet.

Changes of Increasing and Decreasing Quantities

An increasing quantity has a positive change. For instance, in Example 5, the temperature *increased* from −2°F to 7°F and the change in temperature was *positive* (9°F).

A decreasing quantity has a negative change. For instance, in Example 3, the college's enrollment *decreased* from 7 thousand students to 2 thousand students and the change in enrollment was *negative* (−5 thousand students).

Changes of Increasing and Decreasing Quantities

- An increasing quantity has a positive change.
- A decreasing quantity has a negative change.

In Example 9, we will consider the meaning of a quantity with a positive or negative change.

Table 22 Numbers of AIDS Deaths in the United States

Year	Number of AIDS Deaths
2003	17,679
2004	17,154
2005	16,823
2006	15,564
2007	14,561
2008	16,084
2009	17,774
2010	14,399
2011	14,122
2012	13,984
2013	12,963

Source: *U.S. Centers for Disease Control and Prevention*

Example 9   Finding Changes in Quantities

The numbers of AIDS deaths in the United States are shown in Table 22 for various years.

1. Find the change in the number of AIDS deaths from 2003 to 2004. What does your result mean in terms of the number of AIDS deaths?
2. Find the change in the number of AIDS deaths from 2008 to 2009. What does your result mean in terms of the number of AIDS deaths?
3. Find the change in the number of AIDS deaths from one year to the next, beginning in 2003.
4. From what year(s) to the next did the number of AIDS deaths increase the most? What is the change in the number of deaths?
5. From what year(s) to the next did the number of AIDS deaths decrease the most? What is the change in the number of deaths?

Solution

1. The change is $17{,}154 - 17{,}679 = -525$ deaths. So, the number of AIDS deaths from 2003 to 2004 decreased by 525 deaths.
2. The change is $17{,}774 - 16{,}084 = 1690$ deaths. So, the number of AIDS deaths from 2008 to 2009 increased by 1690 deaths.
3. The changes in number of AIDS deaths from one year to the next are listed in Table 23. The changes were found by computing differences similar to those found in Problems 1 and 2.

Table 23 Changes in AIDS Deaths from Year to Year

Year	Number of AIDS Deaths	Change in the Number of AIDS Deaths from Previous Year to Current Year
2003	17,679	—
2004	17,154	$17{,}154 - 17{,}679 = -525$
2005	16,823	$16{,}823 - 17{,}154 = -331$
2006	15,564	$15{,}564 - 16{,}823 = -1259$
2007	14,561	$14{,}561 - 15{,}564 = -1003$
2008	16,084	$16{,}084 - 14{,}561 = 1523$
2009	17,774	$17{,}774 - 16{,}084 = 1690$
2010	14,399	$14{,}399 - 17{,}774 = -3375$
2011	14,122	$14{,}122 - 14{,}399 = -277$
2012	13,984	$13{,}984 - 14{,}122 = -138$
2013	12,963	$12{,}963 - 13{,}984 = -1021$

4. The number of AIDS deaths increased the most from 2008 to 2009. The change in the number of deaths is 1690 deaths.
5. The number of AIDS deaths decreased the most from 2009 to 2010. The change in the number of deaths is -3375 deaths.

In Example 10, we will find an expression that describes the change in a quantity.

Example 10   Finding an Expression Describing Change

In 2017, a financial planner had 132 clients. Let n be the number of clients in 2018.

1. Use a table to help find an expression that describes the change in the number of clients from 2017 to 2018.
2. Evaluate the expression found in Problem 1 for $n = 115$. What does the result mean in this situation?

Table 24 Changes in Number of Clients

Number of Clients in 2018	Change in the Number of Clients from 2017 to 2018
133	$133 - 132$
134	$134 - 132$
135	$135 - 132$
136	$136 - 132$
n	$n - 132$

Solution

1. First, we construct Table 24. We show the arithmetic to help us see the pattern. From the last row of the table, we see that the expression $n - 132$ represents the change in the number of clients.
2. We evaluate $n - 132$ for $n = 115$:

$$(115) - 132 = 115 + (-132) = -17$$

So, there was a change of -17 clients from 2017 to 2018. In other words, the financial planner's client base decreased by 17 clients.

▶

 Group Exploration

Section Opener: Subtracting numbers

1. Find the difference $6 - 4$ and the sum $6 + (-4)$, and compare your results.
2. Find the difference $7 - 3$ and the sum $7 + (-3)$, and compare your results.
3. Find the difference $9 - 2$ and the sum $9 + (-2)$, and compare your results.

4. In Problems 1–3, for each difference, there is a related sum that gives the same result. Write $a - b$ as a sum.
5. In Problem 4, you wrote $a - b$ as a sum. Use this method to find the given difference.
   a. $8 - 3$ **b.** $2 - 5$ **c.** $-4 - 3$

Tips for Success   Math Journal

Do you tend to make the same mistakes repeatedly throughout a math course? If so, it might help to keep a journal in which you list errors you have made on assignments, quizzes, and tests. For each error you list, include the correct solution, as well as a description of the concept needed to solve the problem correctly. You can review this journal from time to time to help you avoid making these errors.

 Homework 2.4

For extra help ▶   **MyLab Math**   Watch the videos in MyLab Math

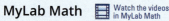

Find the difference by hand. Then use a calculator to check your work.

1. $6 - 8$
2. $3 - 7$
3. $-1 - 5$
4. $-3 - 9$
5. $2 - (-7)$
6. $5 - (-1)$
7. $-3 - (-2)$
8. $-7 - (-3)$
9. $4 - 7$
10. $-4 - 7$
11. $4 - (-7)$
12. $-4 - (-7)$
13. $-3 - 3$
14. $-7 - 7$
15. $-54 - 25$
16. $-100 - 257$
17. $381 - (-39)$
18. $-1939 - (-352)$
19. $2.5 - 7.9$
20. $5.8 - 3.7$
21. $-6.5 - 4.8$
22. $-1.7 - 7.4$
23. $3.8 - (-1.9)$
24. $3.1 - (-3.1)$
25. $13.6 - (-2.38)$
26. $-159.24 - (-7.8)$

27. $-\dfrac{1}{3} - \dfrac{2}{3}$
28. $-\dfrac{1}{5} - \dfrac{4}{5}$
29. $-\dfrac{1}{8} - \left(-\dfrac{5}{8}\right)$
30. $-\dfrac{4}{9} - \left(-\dfrac{7}{9}\right)$
31. $\dfrac{1}{2} - \left(-\dfrac{1}{4}\right)$
32. $\dfrac{5}{12} - \left(-\dfrac{1}{6}\right)$
33. $-\dfrac{1}{6} - \dfrac{3}{8}$
34. $-\dfrac{2}{3} - \dfrac{2}{5}$

Perform the indicated operation by hand. Then use a calculator to check your work.

35. $-5 + 7$
36. $-3 + 9$
37. $-6 - (-4)$
38. $-4 - (-3)$
39. $\dfrac{3}{8} - \dfrac{5}{8}$
40. $-\dfrac{5}{6} + \dfrac{1}{6}$
41. $-4.9 - (-2.2)$
42. $-6.4 + 3.5$
43. $-2 + (-5)$
44. $-5 + (-8)$
45. $10 - 12$
46. $5 - 9$

For Exercises 47–52, use a calculator to perform the indicated operation. Round the result to two decimal places.

47. $-234.913 - 2893.26$

48. $-6178.39 - 52.387$

49. $29{,}643.52 - (-83{,}284.39)$

50. $83{,}451.6 - (-408.549)$

51. $-\dfrac{17}{89} - \dfrac{51}{67}$

52. $-\dfrac{49}{56} - \dfrac{85}{97}$

53. Three hours ago, the temperature was $7°F$. If the temperature has decreased by $19°F$ in the last 3 hours, what is the current temperature?

54. Four hours ago, the temperature was $-12°F$. If the temperature has increased by $18°F$ in the last 4 hours, what is the current temperature?

55. Three hours ago, the temperature was $-4°F$. Now the temperature is $7°F$. What is the change in temperature for the past 3 hours?

56. Four hours ago, the temperature was $-2°F$. Now the temperature is $-13°F$. What is the change in temperature for the past 4 hours?

57. Two hours ago, the temperature was $8°F$. The temperature is now $-4°F$.
   a. What is the change in temperature for the past 2 hours?
   b. Estimate the change in temperature for the past hour.
   c. Explain why your estimate in part (b) may not be the actual change in temperature for the past hour.

58. Three hours ago, the temperature was $-6°F$. The temperature is now $9°F$.
   a. What is the change in temperature for the past 3 hours?
   b. Estimate the change in temperature for the past hour.
   c. Explain why your estimate in part (b) may not be the actual change in temperature for the past hour.

59. The lowest elevation in the United States is at Death Valley, California (-282 feet), and the highest elevation is at the top of Denali (formerly Mount McKinley), Alaska (20,320 feet). Find the change in elevation from Death Valley to the top of Denali.

60. The lowest elevation on (dry) land in the world is at the edge of the Dead Sea, along the Israel–Jordan border (-1312 feet), and the highest elevation is at the top of Mount Everest, along the Nepal–Tibet border (29,035 feet). Find the change in the elevation from the Dead Sea to Mount Everest.

61. DATA The U.S. presidential election in 2000 was the closest presidential race in the electoral vote since 1876. Yet, only a little over half of eligible voters chose to cast a vote (see Table 25).

Table 25 Presidential Election Voter Turnout

Year	Percent of Eligible Voters Who Voted
1980	59.2
1984	59.9
1988	57.4
1992	61.9
1996	54.2
2000	54.7
2004	63.8
2008	63.6
2012	57.5
2016	60.0

Source: *U.S. Census Bureau, Current Population Study*

a. For the years listed in Table 25, find the changes in percent turnout from one presidential election to the next.
b. What was the greatest increase in percent turnout?
c. In 1993, in an attempt to increase the number of eligible voters, a "motor voter" law was passed that made voter registration a part of the process of applying for a driver's license. As a result, about 11 million new voters were registered. Compare the change in percent turnout between 1992 and 1996 with other changes you found in part (a). On the basis of the information in Table 25 alone, we cannot know for sure, but does it seem that many of these 11 million people voted? Explain.

62. In the 1930s, the gray wolf was hunted to near extinction across the western United States. In 1995, 14 wolves were reintroduced to Yellowstone National Park. In the following year, 17 more wolves were released into the park. The wolf population in Yellowstone National Park is shown in Table 26 for various years.

Table 26 Wolf Population

Year	Population
2008	124
2009	96
2010	97
2011	98
2012	83
2013	95
2014	104

Source: *National Park Service, Yellowstone National Park*

a. For the years listed in Table 26, find the changes in population from each year to the next.
b. From what year(s) to the next did the population increase the most? What is the change in population?
c. From what year(s) to the next did the population decrease the most? What is the change in population?
d. From 2013 to 2014, the change in the population was 9 wolves. Does that mean that there were 9 births? Explain.

63. The changes in Toyota Prius® car sales (in thousands of cars) from one year to the next are shown in Table 27.

Table 27 Changes in Toyota Prius Sales

Years	Changes in Sales (thousands of cars)
2008–2009	-19
2009–2010	1
2010–2011	-4
2011–2012	87
2012–2013	-2
2013–2014	-28
2014–2015	-14

Source: *Toyota*

a. If 159 thousand cars were sold in 2008, what were the sales in 2015?
b. During which period(s) were sales increasing?
c. During which period(s) were sales decreasing?

64. The changes in the number of Patriot Groups (conspiracy-minded antigovernment groups) from one year to the next are shown in Table 28.

Table 28 Changes in the Number of Patriot Groups

Year	Change in the Number of Patriot Groups
2009–2010	312
2010–2011	450
2011–2012	86
2012–2013	−264
2013–2014	−222
2014–2015	124

Source: *Southern Poverty Law Center*

a. If there were 512 Patriot Groups in 2009, how many Patriot Groups were there in 2015?

b. During which period(s) were the number of Patriot Groups increasing?

c. During which period(s) were the number of Patriot Groups decreasing?

65. A student scored 87 points on the first exam of the semester.

a. Complete Table 29 to help find an expression that describes the change in score (in points) from the first exam to the second exam if the student scored p points on the second exam. Show the arithmetic to help you see a pattern.

Table 29 Scores on the Second Exam and Changes in Scores

Score on the Second Exam (points)	Change in Score (points)
80	
85	
90	
95	
p	

b. Evaluate the expression you found in part (a) for $p = 81$. What does your result mean in this situation?

66. A year ago, the value of a stock was $35.

a. Complete Table 30 to help find an expression that describes the change in the stock's value (in dollars) if the current value is x dollars. Show the arithmetic to help you see a pattern.

Table 30 Current Values and Changes in Values

Current Value (dollars)	Change in Value (dollars)
30	
35	
40	
45	
x	

b. Evaluate the expression you found in part (a) for $x = 44$. What does your result mean in this situation?

67. Last year the enrollment at a college was 24,500 students.

a. Complete Table 31 to help find an expression that describes the current enrollment if the *change* in enrollment in the past year is c students. Show the arithmetic to help you see a pattern.

Table 31 Changes in Enrollments and Current Enrollments

Change in Enrollment	Current Enrollment
100	
200	
300	
400	
c	

b. Evaluate the expression you found in part (a) for $c = -700$. What does your result mean in this situation?

68. Last year there were 820 deer in a state park.

a. Complete Table 32 to help find an expression that describes the current deer population if the *change* in population in the past year is c deer. Show the arithmetic to help you see a pattern.

Table 32 Changes in Population and Current Populations

Change in Population	Current Population
10	
20	
30	
40	
c	

b. Evaluate the expression you found in part (a) for $c = -25$. What does your result mean in this situation?

Evaluate the expression for $a = -5$, $b = 2$, and $c = -7$.

69. $a + b$ **70.** $a + c$ **71.** $a - b$

72. $c - a$ **73.** $b - c$ **74.** $b - a$

Let x be a number. Translate the English phrase or sentence into a mathematical expression. Then evaluate the expression for $x = -5$.

75. −3 minus the number

76. The number decreased by 4

77. 8 less than the number

78. Subtract 5 from the number.

79. Subtract −2 from the number.

80. The difference of the number and −6

Concepts

81. A student tries to find the difference $7 - (-5)$:

$$7 - (-5) = 7 - 5 = 2$$

Describe any errors. Then find the difference correctly.

82. A student tries to find the difference $2 - 6$:

$$2 - 6 = 6 - 2 = 4$$

Describe any errors. Then find the difference correctly.

83. A quantity increases from amount a to amount b.
   a. Find the change in the quantity.
      i. $a = 3, b = 5$
      ii. $a = 1, b = 9$
      iii. $a = 2, b = 7$
   b. By referring to your work in part (a), explain why it makes sense that if a quantity increased, then the change will be positive.

84. A quantity decreases from amount a to amount b.
   a. Find the change in the quantity.
      i. $a = 8, b = 2$
      ii. $a = 9, b = 3$
      iii. $a = 5, b = 1$
   b. By referring to your work in part (a), explain why it makes sense that if a quantity decreased, then the change will be negative.

85. a. Use a calculator to find each product of two numbers with different signs:
      i. $-2(5)$
      ii. $-4(6)$
      iii. $-7(9)$
   b. What pattern do you notice? If you do not see a pattern, continue finding products of two numbers with different signs until you do.
   c. Without using a calculator, find the product $-3(7)$. Then use a calculator to check your work.
   d. State a general rule for how to multiply two numbers with different signs.

86. a. Use a calculator to find each product of two negative numbers.
      i. $-2(-5)$
      ii. $-4(-6)$
      iii. $-7(-9)$
   b. What pattern do you notice? If you do not see a pattern, continue finding products of two negative numbers until you do.
   c. Without using a calculator, find the product $-3(-7)$. Then use a calculator to check your work.
   d. State a general rule for how to multiply two negative numbers.

87. a. Evaluate $a - b$ for $a = 8$ and $b = 5$.
   b. Evaluate $b - a$ for $a = 8$ and $b = 5$.
   c. Compare the results you found in parts (a) and (b).
   d. Evaluate $a - b$ and $b - a$ for $a = -2$ and $b = 4$, and compare your results.
   e. Evaluate $a - b$ and $b - a$ for values of your choosing for a and b, where a does not equal b. Compare your results.
   f. From your work on parts (a) through (e), what connection do you notice between $a - b$ and $b - a$, where a does not equal b?

88. If the temperature increases from $3°F$ to $5°F$, we find the change in temperature (in degrees Fahrenheit) by performing a subtraction:
$$5 - 3 = 2$$
If the temperature increases from $-3°F$ to $5°F$, we find the change in temperature (in degrees Fahrenheit) by eventually performing an addition:
$$5 - (-3) = 5 + 3 = 8$$
Explain why it makes sense that in the first situation we subtract and in the second we eventually add. [**Hint:** How far is -3 from 0 on the number line? How far is 5 from 0 on the number line?]

89. If x is positive and y is negative, find the sign of $x - y$, if possible. If it is impossible to find the sign, explain why.

90. If x and y are both negative, find the sign of $x - y$, if possible. If it is impossible to find the sign, explain why.

2.5 Ratios, Percents, and Multiplying and Dividing Real Numbers

Objectives

» Find the ratio of two quantities.

» Describe the meaning of *percent*.

» Convert percentages to and from decimal numbers.

» Find the percentage of a quantity.

» Multiply and divide real numbers.

» Determine which fractions with negative signs are equal to each other.

In this section, we will use ratios and percents to describe various quantities. Ratios and percents are important tools used in many fields, including business, political science, journalism, and psychology. We will also discuss how to multiply and divide real numbers.

The Ratio of Two Quantities

Recall from Section 2.1 that the ratio of a to b is the quotient $a \div b$. Usually, we write the ratio of a to b as the fraction $\dfrac{a}{b}$ or as $a : b$.

We can use a ratio to compare two quantities. For example, if a person has 6 cats and 2 dogs, then the ratio of cats to dogs is

$$\frac{6 \text{ cats}}{2 \text{ dogs}} = \frac{3 \text{ cats}}{1 \text{ dog}}$$

We say there are "3 cats to 1 dog." This means there are 3 cats per dog. Or we can say there are 3 times as many cats as dogs.

The ratio of 3 cats to 1 dog is an example of a unit ratio. A **unit ratio** is a ratio written as $\dfrac{a}{b}$ with $b = 1$ or as $a : b$ with $b = 1$.

Example 1   Finding a Unit Ratio

In the academic year 2016–2017, the average annual charge for tuition and fees was $9650 at public four-year colleges and $3520 at public two-year colleges (Source: *College Board*). Find the unit ratio of the average annual charge at public four-year colleges to the average annual charge at public two-year colleges. What does the result mean?

Solution

We divide the average annual charge at public four-year colleges by the average annual charge at public two-year colleges:

$$\text{public four-year colleges} \longrightarrow \frac{\$9650}{\$3520} \approx \frac{2.74}{1} \longleftarrow \text{public two-year colleges}$$

So, the average annual charge for tuition and fees at public four-year colleges is about 2.74 times the average annual charge for tuition and fees at public two-year colleges.

Example 2   Comparing Ratios

The **median** of a group of numbers is the number in the middle (or the average of the two numbers in the middle) when the numbers are listed in order from smallest to largest. The median sales prices of existing homes and the median household incomes in 2014 are shown in Table 33 for four regions of the United States.

Table 33 Median Sales Prices of Existing Homes and Median Household Incomes

Region	Median Sales Price of Existing Homes (dollars)	Median Household Income (dollars)
Northeast	402,800	59,210
Midwest	269,700	54,267
South	257,700	49,655
West	333,900	57,688

Sources: *U.S. Census; National Association of Realtors®*

1. For the Northeast, find the unit ratio of the median sales price of existing homes to the median household income. What does the result mean?
2. For each of the four regions, find the unit ratio of the median sales price of existing homes to the median household income. Taking into account the median household income of each region, list the regions in order of affordability of existing homes, from greatest to least.

Solution

1. We divide the median sales price of existing homes in the Northeast by the median household income in the Northeast:

$$\text{Median sales price of existing homes} \longrightarrow \frac{\$402,800}{\$59,210} \approx \frac{6.80}{1} \longleftarrow \text{Median household income}$$

So, the median sales price of an existing home in the Northeast is about 6.80 times the median household income in that region.
2. We find the unit ratios for each region by dividing the region's median sales price of existing homes by the region's median household income (see Table 34).

Table 34 Unit Ratios of Median Sales Prices of Existing Homes to Median Household Incomes

Region	Median Sales Price of Existing Homes (dollars)	Median Household Income (dollars)	Unit Ratio of Median Sales Price of Existing Homes to Median Household Income
Northeast	402,800	59,210	$\frac{402{,}800}{59{,}210} \approx 6.80$
Midwest	269,700	54,267	$\frac{269{,}700}{54{,}267} \approx 4.97$
South	257,700	49,655	$\frac{257{,}700}{49{,}655} \approx 5.19$
West	333,900	57,688	$\frac{333{,}900}{57{,}688} \approx 5.79$

The lower the unit ratio, the more affordable the existing homes are in the region. So, the regions, in order of affordability of existing homes, from greatest to least, are Midwest, South, West, Northeast.

Meaning of Percent

Suppose there are 53 women in a class of 100 students. Then the ratio of the number of women to the total number of students is $\frac{53}{100}$. We say that 53% of the students are women.

Definition Percent

Percent means "for each hundred": $a\% = \dfrac{a}{100}$

For example, 37% means 37 for each 100 (the ratio $\frac{37}{100}$, or the unit ratio $\frac{0.37}{1}$). In Fig. 28, the area of the shaded region is 37% of the area of the large square because 37 of 100 parts of equal area are shaded.

Converting Percentages to and from Decimal Numbers

Because 37% is the ratio $\frac{37}{100}$, 37% is 37 hundredths, or 0.37:

$$37\% = \frac{37}{100} = 0.\underbrace{\overbrace{3}^{\substack{\text{tenths} \\ \text{place}}} \; \overbrace{7}^{\substack{\text{hundredths} \\ \text{place}}}}_{\text{37 hundredths}}$$

So, to write 37% as a decimal number, first we remove the percent symbol. Then we divide 37 by 100, which is equivalent to moving the decimal point two places to the left:

$$37\% = 37.0\% = 0.37$$

two places to the left

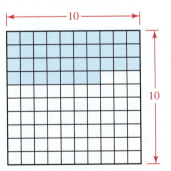

Figure 28 The area of the shaded region is 37% of the area of the large square

To write 0.37 as a percentage, first we multiply 0.37 by 100, which is equivalent to moving the decimal point two places to the right. Then we insert a percent symbol:

$$0.37 = 37.0\% = 37\%$$

two places to the right

Converting Percentages to and from Decimal Numbers

- To write a percentage as a decimal number, remove the percent symbol and divide the number by 100 (move the decimal point two places to the left).
- To write a decimal number as a percentage, multiply the number by 100 (move the decimal point two places to the right) and insert a percent symbol.

Example 3   Converting Percentages and Decimal Numbers

Write each percentage as a decimal number, and write each decimal number as a percentage.

1. 86%   **2.** 5%   **3.** 0.125

Solution

1. To write 86% as a decimal number, we remove the percent symbol and move the decimal point two places to the left:

$$86\% = 86.0\% = 0.86$$

two places to the left

2. To write 5% as a decimal number, we remove the percent symbol and move the decimal point two places to the left, using 0 in the tenths place as a placeholder:

$$5\% = 5.0\% = 0.05$$

two places to the left

3. To write 0.125 as a percentage, we move the decimal point two places to the right and insert a percent symbol:

$$0.125 = 12.5\%$$

two places to the right

▶

WARNING   From Problem 2 in Example 3, we see 5% is *not* equal to 0.5. Rather, 5% is equal to 0.05. Remember to move the decimal point *two* places to the left, using 0 in the tenths place as a placeholder.

Percentage of a Quantity

How do we find the percentage of a quantity? For example, consider 75% of 4. That is the same as $\frac{75}{100}$ of 4. To find a fraction of a number, we *multiply* the fraction by that number:

$$\frac{75}{100} \text{ of } 4 = \frac{75}{100} \cdot 4 = \frac{3}{4} \cdot \frac{4}{1} = \frac{3}{1} = 3$$

So, using decimal notation, we find 75% of 4 by *multiplying* 0.75 by 4:

$$75\% \text{ of } 4 = 0.75(4) = 3$$

To see whether our result makes sense, we first form a large square made up of 4 medium-size squares of equal area (see Fig. 29). To find 75% of the 4 squares, we divide the large square into 100 small squares of equal area and shade 75 of them. The shaded region contains 3 of the 4 medium-size squares, which checks with our earlier computations.

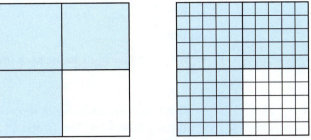

Figure 29 75% of 4 medium-size squares is made up of 75 small squares (in blue), or 3 medium-size squares (in blue)

Finding the Percentage of a Quantity

To find the percentage of a quantity, multiply the decimal form of the percentage and the quantity.

Example 4   Finding the Percentage of a Quantity

1. A person buys a Fender American Standard Jazz Bass for $1500 at Guitar Center in Indianapolis, Indiana, which has a sales tax of 7%. How much money is the sales tax?
2. At the University of Oregon, 75.8% of 24,935 applicants were accepted in the fall semester of 2016 (Source: *University of Oregon*). How many applicants were accepted?

Solution

1. $0.07(1500) = 105$; so, the sales tax is $105.
2. $0.758(24,935) = 18,900.73 \approx 18,901$; so, 18,901 applicants were accepted.

▶

By definition, 100% means 100 for each 100. In other words, 100% of a quantity is *all* the quantity. For example, 100% of 21 guitars is 21 guitars.

One Hundred Percent of a Quantity

One hundred percent of a quantity is *all* the quantity.

We will continue to work with ratios and percents as we discuss how to multiply and divide real numbers.

Multiplication of Two Numbers with Different Signs

We can think of multiplication as repeated addition. For example, 3(5) is equal to the sum of three 5s:

$$3(5) = 5 + 5 + 5 = 15$$

Also, 3(5) is equal to the sum of five 3s:

$$3(5) = 3 + 3 + 3 + 3 + 3 = 15$$

We can use the idea of repeated addition to help us find the product of two numbers with different signs.

Example 5   Finding the Product of Two Numbers with Different Signs

Find the product.

1. $4(-2)$ **2.** $(-6)(3)$

Solution

1. We write $4(-2)$ as the sum of four -2s:

$$4(-2) = (-2) + (-2) + (-2) + (-2) = -8$$

This result makes sense in terms of money. If you borrow $2 from a friend four times, you will owe the friend $8.

2. We write $(-6)(3)$ as the sum of three -6s:

$$(-6)(3) = (-6) + (-6) + (-6) = -18$$

In Example 5, we found the products of two numbers with different signs. Note that both results were negative:

$$\overbrace{4(-2)}^{\text{Different signs}} = \overbrace{-8}^{\text{Negative}}$$

$$\overbrace{(-6)(3)}^{\text{Different signs}} = \overbrace{-18}^{\text{Negative}}$$

Multiplying Two Numbers with Different Signs

The product of two numbers that have different signs is negative.

Example 6   Finding the Product of Two Numbers with Different Signs

Find the product.

1. $7(-4)$ **2.** $(-0.2)(0.3)$

Solution

1. Because the signs of 7 and -4 are different, their product is negative: $7(-4) = -28$.

2. Because the signs of -0.2 and 0.3 are different, their product is negative: $(-0.2)(0.3) = -0.06$.

Multiplication of Two Numbers with the Same Sign

We have discussed how to multiply numbers with different signs. What if the signs are the same? To begin this investigation, consider the following pattern:

$$3(-5) = -15$$
$$2(-5) = -10$$
This factor decreases by 1. $\qquad$ $1(-5) = -5$ $\qquad$ The product increases by 5.
$$0(-5) = 0$$

It turns out that this pattern continues. So, we have

This factor decreases by 1. ───┤ $$\begin{aligned} -1(-5) &= 5 \\ -2(-5) &= 10 \\ -3(-5) &= 15 \end{aligned}$$ ├── The product increases by 5.

Note that for each of the last three computations, the product of the two negative numbers is positive. This is, in fact, always true. Here we find another product of two negative numbers:

$$\overbrace{(-7)(-9)}^{\text{Same signs}} = \overbrace{63}^{\text{Positive}}$$

Multiplying Two Numbers with the Same Sign

The product of two numbers that have the same sign is positive.

Example 7   Finding the Product of Two Numbers with the Same Sign

Find the product.

1. $-5(-6)$

2. $\left(-\dfrac{3}{2}\right)\left(-\dfrac{5}{7}\right)$

Solution

1. Because -5 and -6 have the same sign, their product is positive: $-5(-6) = 30$.

2. Because $-\dfrac{3}{2}$ and $-\dfrac{5}{7}$ have the same sign, their product is positive:

$$\left(-\frac{3}{2}\right)\left(-\frac{5}{7}\right) = \frac{15}{14}$$

In Fig. 30, we show a multiplication table for some specific numbers. In Fig. 31, we summarize the multiplication sign rules for all nonzero real numbers.

·	4	−4
2	8	−8
−2	−8	8

·	+	−
+	+	−
−	−	+

Figure 30 Multiplication table for 2, −2, 4, and −4

Figure 31 Multiplication table for all nonzero real numbers

Example 8   Multiplying Real Numbers in an Authentic Situation

A person's credit card balance is -2340 dollars. If the person pays off 30% of the balance, what is the new balance?

Solution

If the person pays off 30% of the balance, then $100\% - 30\% = 70\%$ of the balance remains. We find 70% of -2340:

$$0.70(-2340) = -1638$$

The new balance is -1638 dollars.

Division of Real Numbers

We can get an idea of how to divide real numbers by writing multiplications as related divisions. For example, consider this statement:

$$2 \cdot 3 = 6 \text{ implies that } 6 \div 3 = 2.$$

We now write a similar statement for $(-2)(-3)$:

$$(-2)(-3) = 6 \text{ implies that } \overbrace{6 \div (-3)}^{\text{Different signs}} = \overbrace{-2}^{\text{Negative}}.$$

This statement suggests that the quotient of two numbers with different signs is negative. Now consider the following statement:

$$(2)(-3) = -6 \text{ implies that } \overbrace{-6 \div (-3)}^{\text{Same signs}} = \overbrace{2}^{\text{Positive}}.$$

This statement suggests that the quotient of two numbers with the same sign is positive. Both statements suggest that the sign rules for dividing real numbers are similar to those for multiplying real numbers.

> **Multiplying or Dividing Real Numbers**
>
> The product or quotient of two numbers that have different signs is negative. The product or quotient of two numbers that have the same sign is positive.

Example 9   Finding Quotients of Real Numbers

Find the quotient.

1. $-10 \div 2$

2. $-\dfrac{1}{6} \div \left(-\dfrac{3}{5}\right)$

Solution

1. Because -10 and 2 have different signs, the quotient is negative: $-10 \div 2 = -5$. This makes sense in terms of money. If we divide a debt of \$10 by 2, the result is a debt of \$5.

2. The quotient of two negative numbers is positive. To find the result, we divide the absolute value of the fractions:

$$\frac{1}{6} \div \frac{3}{5} = \frac{1}{6} \cdot \frac{5}{3} \quad \textit{Multiply by reciprocal of } \frac{3}{5}, \textit{ which is } \frac{5}{3}; \frac{a}{b} \div \frac{c}{d} = \frac{a}{b} \cdot \frac{d}{c}$$

$$= \frac{5}{18} \quad \textit{Multiply numerators and multiply denominators.}$$

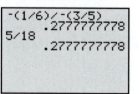

Press $\boxed{(-)}\boxed{(}\boxed{(}\,1\,\boxed{\div}\,6\,\boxed{)}\boxed{)}\,\boxed{\div}\,\boxed{(-)}$
$\boxed{(}\,3\,\boxed{\div}\,5\,\boxed{)}\,\boxed{\text{ENTER}}$.

```
-(1/6)/-(3/5)
           .2777777778
5/18
           .2777777778
```

Figure 32 Verify the work

We use a graphing calculator to check our work for Problem 2 in Example 9 (see Fig. 32).

Example 10   Application of a Ratio of Two Real Numbers

A person has credit card balances of -3950 dollars on a Visa® account and -1225 dollars on a MasterCard® account.

1. Find the unit ratio of the Visa balance to the MasterCard balance.
2. If the person wishes to pay off both accounts gradually in the same amount of time, describe how the result in Problem 1 can help guide the person in making his next payment.

Solution

1. We divide the Visa balance by the MasterCard balance:

$$\frac{-3950}{-1225} \approx \frac{3.22}{1}$$

So, the Visa balance is about 3.22 times the MasterCard balance.

2. For each \$1 the person pays to his MasterCard account, he should pay about \$3.22 to his Visa account. (The ratio will need to be recalculated each month to take into account recent purchases, cash advances, and so on, as well as possible differences in interest rates on the cards.)

Equal Fractions with Negative Signs

Notice that $\dfrac{-6}{2} = -3$, $\dfrac{6}{-2} = -3$, and $-\dfrac{6}{2} = -3$. So we can write

$$\frac{-6}{2} = \frac{6}{-2} = -\frac{6}{2}$$

The positions of the negative signs in the three equal expressions suggest the following property.

Equal Fractions with Negative Signs

If $b \neq 0$, then

$$\frac{-a}{b} = \frac{a}{-b} = -\frac{a}{b}$$

If the result of a computation is a negative fraction, we write the result in the form $-\dfrac{a}{b}$ rather than $\dfrac{-a}{b}$ or $\dfrac{a}{-b}$.

Example 11   Adding Fractions with Negative Signs

Find the sum $\dfrac{3}{-5} + \dfrac{1}{5}$.

Solution

$$\frac{3}{-5} + \frac{1}{5} = \frac{-3}{5} + \frac{1}{5} \qquad \frac{a}{-b} = \frac{-a}{b}$$

$$= \frac{-3 + 1}{5} \qquad \text{\textit{Write numerators as a sum and keep common}}$$
$$\text{\textit{denominator: }} \frac{a}{b} + \frac{c}{b} = \frac{a + c}{b}.$$

$$= \frac{-2}{5} \qquad \text{\textit{Find sum.}}$$

$$= -\frac{2}{5} \qquad \frac{-a}{b} = -\frac{a}{b}$$

Group Exploration

Section Opener: Finding the product of a positive number and a negative number

We can think of the multiplication of two counting numbers as a repeated addition. For example, we can think of 4(3) as adding four 3s:

$$4(3) = 3 + 3 + 3 + 3 = 12$$

We can use this idea to help us find the product of a positive number and a negative number.

1. Write each of the products that follow as a repeated sum. Then find the sum.

   a. $3(-2)$ b. $5(-4)$ c. $7(-1)$

2. Are the results you found in Problem 1 positive or negative? What can you say about the product of a positive number and a negative number? If you are not sure, try some more multiplications.

3. Explain why the observation you made in Problem 2 makes sense. [**Hint:** What can you say about a negative number plus a negative number?]

Tips for Success   Use Your Instructor's Office Hours

Helping students during office hours is part of an instructor's job. Keep in mind that your instructor wants you to succeed and hopes you take advantage of all opportunities to learn.

It is a good idea to come prepared to office visits. For example, if you are having trouble with a concept, attempt some related exercises and bring your work so your instructor can see where you are having difficulty. If you miss a class, it is helpful to read the material, borrow class notes, and try to complete assigned exercises before visiting your instructor, so you get the most out of the visit.

 Homework 2.5

For extra help ▶   **MyLab Math**   Watch the videos in MyLab Math

For Exercises 1–10, write the percentage as a decimal number or write the decimal number as a percentage, as appropriate.

1. 63%   2. 91%   3. 9%   4. 4%   5. 0.08

6. 0.01   7. 7.3%   8. 3.8%   9. 0.052   10. 0.089

11. Find 35% of 7000 cars.

12. Find 67% of 4500 computers.

13. A person buys a refrigerator for $229.99 (not including sales tax) at Cummins Appliance in Baltimore, Maryland, which has a sales tax of 6%. How much money is the sales tax?

14. A person buys groceries for $125.35 (not including grocery tax) in Harmons Grocery in Salt Lake City, Utah, which has a grocery tax of 3%. How much money is the grocery tax?

15. At the University of Arkansas, 52% of 22,159 undergraduates were female in spring semester, 2017 (Source: *University of Arkansas*). How many female undergraduates were there?

16. At the University of Iowa, 26% of 23,357 undergraduates lived on campus in spring semester, 2017 (Source: *University of Iowa*). How many undergraduates lived on campus?

Perform the indicated operation by hand. Then use a calculator to check your work.

17. $-2(6)$ 18. $-5(4)$ 19. $-3(-6)$

20. $-8(-9)$ 21. $1(-1)$ 22. $5(-2)$

23. $-40 \div 5$ 24. $-63 \div 7$ 25. $25 \div (-5)$

26. $24 \div (-3)$ 27. $-56 \div (-7)$ 28. $-1 \div (-1)$

29. $-15(-37)$ 30. $-124(-29)$ 31. $936 \div (-24)$

32. $1008 \div (-21)$ 33. $-0.2(-0.4)$ 34. $-0.3(-0.3)$

35. $2.5(-0.39)$ 36. $3.7(-5.24)$ 37. $-0.06 \div 0.2$

38. $-0.12 \div 0.3$ 39. $\dfrac{36}{-4}$ 40. $\dfrac{-9}{3}$

41. $\dfrac{-32}{-8}$ 42. $\dfrac{-72}{-8}$ 43. $\dfrac{1}{2}\left(-\dfrac{1}{5}\right)$

44. $\dfrac{1}{3}\left(-\dfrac{7}{5}\right)$ 45. $\left(-\dfrac{4}{9}\right)\left(-\dfrac{3}{20}\right)$ 46. $\left(-\dfrac{7}{25}\right)\left(-\dfrac{5}{21}\right)$

47. $-\dfrac{3}{4} \div \dfrac{7}{6}$ 48. $-\dfrac{5}{7} \div \dfrac{15}{8}$

49. $-\dfrac{24}{35} \div \left(-\dfrac{16}{25}\right)$ 50. $-\dfrac{3}{8} \div \left(-\dfrac{9}{20}\right)$

Perform the indicated operation by hand. Then use a calculator to check your work.

51. $6 + (-9)$ 52. $-9 + (-4)$ 53. $-39 \div (-3)$

54. $-49 \div 7$ 55. $4 - (-2)$ 56. $-2 - 7$

57. $10(-10)$ **58.** $-5(-9)$ **59.** $-\dfrac{3}{4}+\dfrac{1}{2}$

60. $-\dfrac{8}{3}+\left(-\dfrac{5}{9}\right)$ **61.** $\left(-\dfrac{10}{7}\right)\left(-\dfrac{14}{15}\right)$ **62.** $\dfrac{9}{2}\left(-\dfrac{4}{21}\right)$

63. $\dfrac{3}{4}-\dfrac{5}{3}$ **64.** $-\dfrac{3}{8}-\left(-\dfrac{1}{10}\right)$

65. $-\dfrac{3}{8}\div\dfrac{5}{6}$ **66.** $-\dfrac{22}{9}\div\left(-\dfrac{33}{18}\right)$

Simplify.

67. $\dfrac{-16}{20}$ **68.** $\dfrac{-15}{35}$

69. $\dfrac{-18}{-24}$ **70.** $\dfrac{-35}{-21}$

Perform the indicated operation by hand. Then use a calculator to check your work.

71. $\dfrac{3}{-4}+\dfrac{1}{4}$ **72.** $\dfrac{5}{-6}+\dfrac{1}{6}$ **73.** $\dfrac{4}{7}-\left(\dfrac{3}{-7}\right)$

74. $\dfrac{2}{3}-\left(\dfrac{1}{-3}\right)$ **75.** $\dfrac{5}{6}+\dfrac{7}{-8}$ **76.** $\dfrac{1}{4}+\dfrac{5}{-6}$

Use a calculator to perform the indicated operation. Round the result to two decimal places.

77. $-26.87(-381.572)$ **78.** $-489.2(-8.39)$

79. $222.045\div(-32.76)$ **80.** $64.958\div(-3.716)$

81. $-\dfrac{11}{18}\left(-\dfrac{15}{19}\right)$ **82.** $-\dfrac{169}{175}\left(-\dfrac{64}{71}\right)$

83. $-\dfrac{59}{13}\div\dfrac{27}{48}$ **84.** $-\dfrac{75}{22}\div\dfrac{13}{48}$

Evaluate the expression for $a=-6$, $b=4$, and $c=-8$.

85. ab **86.** ac **87.** $\dfrac{a}{b}$ **88.** $\dfrac{b}{a}$

89. $-ac$ **90.** $-bc$ **91.** $-\dfrac{b}{c}$ **92.** $-\dfrac{a}{c}$

Let w be a number. Translate the English phrase into a mathematical expression. Then evaluate the expression for $w=-8$.

93. The quotient of the number and 2

94. The number divided by 4

95. The product of the number and -5

96. -2 times the number

For Exercises 97 and 98, write the ratio as a fraction.

97. The ratio of 6 to 8

98. The ratio of 9 to 15

99. The 1776-foot-tall One World Trade Center at the site where New York's World Trade Center once stood is the tallest building in the United States. The John Hancock Tower in Boston is 790 feet tall. Find the unit ratio of the height of One World Trade Center to the height of the John Hancock Tower. What does your result mean in this situation?

100. The number of nuclear bomb tests since 1945 in the United States is 1032 tests and in China is 45 tests (Source: *SIPRI*). Find the unit ratio of the number of nuclear bomb tests since 1945 in the United States to that in China. What does your result mean in this situation?

101. There were 540 U.S. billionaires in 2016 and 412 U.S. billionaires in 2011 (Source: *Forbes*). Find the unit ratio of the number of U.S. billionaires in 2016 to the number of U.S. billionaires in 2011. What does your result mean in this situation?

102. In December 2016, the average number of viewers was 4.79 million viewers per day for the TV show *Today* and 3.77 million viewers for the competing show *CBS This Morning* (Source: *Nielsen*). Find the unit ratio of the average number of viewers per day of *Today* to the average number of viewers per day of *CBS This Morning*. What does your result mean in this situation?

103. A recipe for roasted red-pepper pasta calls for 4 red bell peppers and 5 black olives. Calculate the given unit ratio. What does your result mean in this situation?
  a. The unit ratio of the number of red bell peppers to the number of black olives
  b. The unit ratio of the number of black olives to the number of red bell peppers

104. A recipe for beef stroganoff calls for 2 cups of sliced mushrooms and 4 cups of cooked noodles. Calculate the given unit ratio. What does your result mean in this situation?
  a. The unit ratio of the number of cups of sliced mushrooms to the number of cups of cooked noodles
  b. The unit ratio of the number of cups of cooked noodles to the number of cups of sliced mushrooms

105. The *full-time equivalent enrollment* (FTE enrollment) at a college is the number of full-time students it would take for their total credits (units or hours) to equal the total credits in which both part-time and full-time students combined are enrolled in one semester. The number of *full-time equivalent (FTE) faculty* is the number of full-time faculty it would take to teach all the courses that are taught by both part-time and full-time faculty combined. The FTE enrollments and the numbers of FTE faculty are shown in Table 35 for various colleges.

Table 35 FTE Enrollments and Numbers of FTE Faculty

College	FTE Enrollment	Number of FTE Faculty
Butler University	4834.0	425.0
St. Olaf College	3019.0	253.0
Stonehill College	2382.6	196.3
University of Massachusetts Amherst	24,936.7	1379.0
Texas A&M University	48,619.0	2114.0

Sources: *Butler University, St. Olaf College, Stonehill College, University of Massachusetts, Texas A&M University*

  a. Find the unit ratio of FTE enrollment at Texas A&M University to the FTE enrollment at St. Olaf College. What does your result mean in this situation?
  b. Find the unit ratio of the number of FTE faculty at the University of Massachusetts Amherst to the number of FTE faculty at Butler University. What does your result mean in this situation?
  c. Find the unit ratio of FTE enrollment to the number of FTE faculty at each of the colleges listed in Table 35.

d. Which college listed in Table 35 has the largest ratio of FTE enrollment to the number of FTE faculty? Which has the smallest?

e. A person believes that the ratio of FTE enrollment to the number of FTE faculty is lower at Stonehill College than at St. Olaf College because Stonehill College has the lower FTE enrollment. Is that person correct? Explain.

106. The 2016 populations and land areas are shown in Table 36 for various states.

Table 36 Populations and Land Areas

State	Population	Land Area (square miles)
Alaska	741,894	571,951
California	39,250,017	155,959
Michigan	9,928,300	56,804
New Jersey	8,944,469	7417
New York	19,745,289	47,214

Sources: *U.S. Census Bureau; Infoplease*

a. Find the unit ratio of New York's population to New Jersey's population. What does your result mean in this situation?

b. Find the unit ratio of Alaska's land area to California's land area. What does your result mean in this situation?

c. The unit ratio of population to land area is called the *population density*. Find the population density of each state listed in Table 36.

d. Which state listed in Table 36 has the greatest population density? Which has the least?

e. A person believes that Michigan has a greater population density than New Jersey because Michigan's population is more than New Jersey's population. Is that person correct? Explain.

107. A person has credit card balances of −4360 dollars on a Discover® account and −1825 dollars on a MasterCard® account.

a. Find the unit ratio of the Discover balance to the MasterCard balance.

b. If the person wishes to pay off both accounts gradually in the same amount of time, describe how the result you found in part (a) can help guide the person in making her next payment.

108. A person has credit card balances of −6810 dollars on a Visa® account and −2950 dollars on a Sears® account.

a. Find the unit ratio of the Visa balance to the Sears balance.

b. If the person wishes to pay off both accounts gradually in the same amount of time, describe how the result you found in part (a) can help guide the person in making his next payment.

109. A person's credit card balance is −3720 dollars. If the person pays off 15% of the balance, what is the new balance?

110. A person's credit card balance is −1590 dollars. If the person pays off 35% of the balance, what is the new balance?

111. A student has zero balance on a credit card. The student uses the credit card to buy 12.3 gallons of gasoline at a cost of $2.40 per gallon. What is the new balance?

112. A person has zero balance on a credit card. The person uses the credit card to buy three lamps at a cost of $89.50 per lamp. What is the new balance?

Concepts

113. **a.** Find the sum $-2 + (-4)$.
b. Find the product $-2(-4)$.
c. Consider the following statements:
   - Two negative numbers make a positive.
   - A negative number times a negative number is equal to a positive number.

   Which statement is clearer? Explain.
d. Compare the sign rule for adding two negative numbers with the sign rule for multiplying two negative numbers.

114. **a.** Find the sum $-2 + 3$.
b. Find the product $-2(3)$.
c. Consider the following statements:
   - A negative number times a positive number is equal to a negative number.
   - A negative and a positive make a negative.

   Which statement is clearer? Explain.
d. Compare the sign rule for adding numbers with different signs with the sign rule for multiplying numbers with different signs.

115. Which of the following fractions are equal? (There may be more than one pair of answers.)

$$\frac{a}{b} \quad \frac{-a}{b} \quad \frac{a}{-b} \quad -\frac{a}{b} \quad \frac{-a}{-b} \quad -\frac{-a}{-b}$$

116. **a.** Is $\dfrac{12}{-4}$ positive or negative? Explain.
b. If a is positive and b is negative, is $\dfrac{a}{b}$ positive or negative? Explain.
c. A student says that $\dfrac{a}{b}$ is positive because it has no negative signs. Is the student correct? Explain.

117. Discuss in terms of repeated addition why it makes sense that $3(-6)$ is negative.

118. Discuss in terms of repeated addition why it makes sense that $(-4)(5)$ is negative.

119. If ab is negative, what can you say about a or b?

120. If ab is positive, what can you say about a or b?

121. If $ab = 0$, what can you say about a or b?

122. If $\dfrac{a}{b}$ is negative, what can you say about a or b?

123. **a.** Perform the indicated operations in $(8 \div 2) \cdot 4$ by first doing the division and then doing the multiplication.
b. Perform the indicated operations in $8 \div (2 \cdot 4)$ by first doing the multiplication and then doing the division.
c. Compare the results you found in parts (a) and (b). Does it matter in which order we multiply and divide? Explain.

124. **a.** Find each product.
   i. $(-1)(-1)$
   ii. $(-1)(-1)(-1)$
   iii. $(-1)(-1)(-1)(-1)$
   iv. $(-1)(-1)(-1)(-1)(-1)$
b. Describe any patterns that you notice in the results you found in part (a). If you don't see a pattern, find some more products of −1s until you do.
c. Find the product:

$$\underbrace{(-1)(-1)(-1) \cdots (-1)}_{524 \text{ factors of } -1}$$

d. Find the product:

$$\underbrace{(-1)(-1)(-1)\cdots(-1)}_{\text{847 factors of }-1}$$

125. Some students believe $2x$ is greater than x for *all* real numbers.
  a. Find three values of x where $2x$ is greater than x.
  b. Find three values of x where $2x$ is less than x.
  c. Find one value of x where $2x$ is equal to x.

126. Some students believe $\dfrac{2}{x}$ is less than 2 for *all* real numbers.
  a. Find three values of x where $\dfrac{2}{x}$ is less than 2.
  b. Find three values of x where $\dfrac{2}{x}$ is greater than 2.
  c. Find one value of x where $\dfrac{2}{x}$ is equal to 2.

2.6 Exponents and Order of Operations

Objectives

» Describe the meaning of *exponent*.

» Use the rules for order of operations to perform computations.

» Use the rules for order of operations to evaluate expressions.

» Use the rules for order of operations to make predictions.

In this section, we will discuss an operation called *exponentiation*. We will also discuss the order in which we should perform various operations.

Exponents

The notation x^2 stands for $x \cdot x$. So, $7^2 = 7 \cdot 7 = 49$. The notation x^3 stands for $x \cdot x \cdot x$. So, $2^3 = 2 \cdot 2 \cdot 2 = 8$.

Definition   Exponent

For any counting number n,

$$x^n = \underbrace{x \cdot x \cdot x \cdot \;\cdots\; \cdot x}_{n \text{ factors of } x}$$

We refer to x^n as the **power**, the **nth power of x**, or **x raised to the nth power**. We call x the **base** and n the **exponent**.

The expression 2^5 is a power. It is the 5th power of 2, or 2 raised to the 5th power. For 2^5, the base is 2 and the exponent is 5. Here, we label the base and the exponent of 2^5 and compute the power:

$$\overset{\text{Exponent}}{2^5} = \underbrace{2 \cdot 2 \cdot 2 \cdot 2 \cdot 2}_{\text{5 factors of 2}} = 32$$

Base

When we calculate a power, we say that we are performing **exponentiation**.

Notice that the notation x^1 stands for one factor of x, so $x^1 = x$.

Two powers of x have specific names. We refer to x^2 as the **square of x** or **x squared**. We refer to x^3 as the **cube of x** or **x cubed**.

For an expression of the form $-x^n$, we calculate x^n before taking the opposite. For example,

$$-3^4 = -\left(3^4\right) = -(3 \cdot 3 \cdot 3 \cdot 3) = -81$$

For -3^4, the base is 3. If we want the base to be -3, we enclose -3 in parentheses:

$$(-3)^4 = (-3)(-3)(-3)(-3) = 81$$

We use a graphing calculator to check both computations (see Fig. 33).

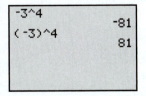

Figure 33 Compute -3^4 and $(-3)^4$

Example 1	Calculating Expressions That Have Exponents

Perform the exponentiation.

1. 5^4 **2.** -3^2 **3.** $(-3)^2$ **4.** $\left(\dfrac{4}{5}\right)^3$

Solution

1. $5^4 = 5 \cdot 5 \cdot 5 \cdot 5 = 625$ *The base is 5.*

2. $-3^2 = -(3 \cdot 3) = -9$ *The base is 3.*

3. $(-3)^2 = (-3)(-3) = 9$ *The base is -3.*

4. $\left(\dfrac{4}{5}\right)^3 = \dfrac{4}{5} \cdot \dfrac{4}{5} \cdot \dfrac{4}{5} = \dfrac{64}{125}$ *The base is $\dfrac{4}{5}$.*

Order of Operations

We can establish the order of operations by using *grouping symbols*, such as parentheses ( ), absolute-value symbols | |, and fraction bars. We do operations that lie within grouping symbols before we perform other operations.

But does it matter in which order we perform operations? From the following calculations, it is clear that it *does* matter:

$$(3 + 2) \cdot 4 = 5 \cdot 4 = 20 \quad \textit{First add; then multiply.}$$
$$3 + (2 \cdot 4) = 3 + 8 = 11 \quad \textit{First multiply; then add.}$$

With a fraction such as $\dfrac{7 + 3}{3 - 1}$, the following use of parentheses is assumed:

$$\frac{7 + 3}{3 - 1} = \frac{(7 + 3)}{(3 - 1)} = \frac{10}{2} = 5$$

So, we compute both the numerator and the denominator before we divide.

Example 2	Performing Operations

Perform the indicated operations.

1. $(7 + 2)(3 - 8)$ **2.** $\dfrac{11 - 3}{1 - 5}$

Solution

1. $(7 + 2)(3 - 8) = (9)(-5) = -45$

2. $\dfrac{11 - 3}{1 - 5} = \dfrac{8}{-4} = -2$

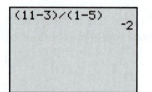

Figure 34 Verify the result

We use a graphing calculator to check our result (see Fig. 34).

For an expression such as $3 + 2 \cdot 4$, where grouping symbols do not specify the order of operations for all operations in the expression, there is an understood order of operations.

Order of Operations

We perform operations in the following order:

1. First, perform operations within parentheses or other grouping symbols, starting with the innermost group.

2. Then perform exponentiations.

3. Next, perform multiplications and divisions, going from left to right.

4. Last, perform additions and subtractions, going from left to right.

Example 3	Performing Operations

Perform the indicated operations.

1. $9 - 8 \div 4$ **2.** $10 \div 5 + (6 - 4) \cdot 5$ **3.** $2 - [7 + 4(3 - 5)]$

Solution

1. $9 - 8 \div 4 = 9 - 2$ *Divide before subtracting.*
$ = 7$ *Subtract.*

2. $10 \div 5 + (6 - 4) \cdot 5 = 10 \div 5 + 2 \cdot 5$ *Subtract within parentheses.*
$ = 2 + 2 \cdot 5$ *Divide, because the division is to the left of the multiplication.*
$ = 2 + 10$ *Multiply before adding.*
$ = 12$ *Add.*

We use a graphing calculator to verify our result (see Fig. 35).

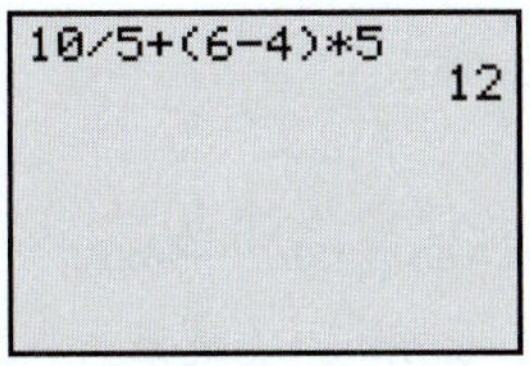

Figure 35 Verify the result

```
10/5+(6-4)*5
            12
```

3. $2 - [7 + 4(3 - 5)] = 2 - [7 + 4(-2)]$ *Subtract within innermost parentheses.*
$ = 2 - [7 + (-8)]$ *Multiply.*
$ = 2 - [-1]$ *Add.*
$ = 2 + 1$ *Simplify.*
$ = 3$ *Add.*

There is a connection between the order of operations and the strengths of the operations. We explore the strengths of exponentiation, multiplication, and addition by performing these operations on a pair of 10s:

Operation	Computation with 10s
Exponentiation	$10^{10} = 10,000,000,000$
Multiplication	$10 \cdot 10 = 100$
Addition	$10 + 10 = 20$

Exponentiation is much more powerful than multiplication, which in turn is more powerful than addition. Because dividing by a number is the same as multiplying by the reciprocal of the number, division is as powerful as multiplication. Because subtracting a number is the same as adding the opposite of the number, subtraction is as powerful as addition. Here is a summary of the strengths of the operations:

Operation	Strength of Operation
Exponentiation	Most Powerful
Multiplication and Division	Next Most Powerful
Addition and Subtraction	Weakest

Order of Operations and the Strengths of Operations

After we have performed operations in parentheses, the order of operations goes from the most powerful operation, exponentiation, to the next-most-powerful operations, multiplication and division, to the weakest operations, addition and subtraction.

Knowing the relationship between the order of operations and the strengths of the operations will likely help you remember the order of operations.

Example 4   Performing Operations

Perform the indicated operations.

1. $3 + 2^3 + 4(5)$

2. $7 + (2 - 6)^3 - 8 \div (-2)$

Solution

1. $3 + 2^3 + 4(5) = 3 + 8 + 4(5)$ *Perform exponentiation first:* $2^3 = 2 \cdot 2 \cdot 2 = 8$

$\qquad\qquad\qquad\quad = 3 + 8 + 20$ *Multiply before adding.*

$\qquad\qquad\qquad\quad = 31$ *Add.*

2.

$7 + (2 - 6)^3 - 8 \div (-2) = 7 + (-4)^3 - 8 \div (-2)$ *Work within parentheses first.*

$\qquad\qquad\qquad\qquad\quad = 7 + (-64) - 8 \div (-2)$ *Perform exponentiation:* $(-4)^3 = (-4) \cdot (-4) \cdot (-4) =$

$\qquad\qquad\qquad\qquad\quad = 7 + (-64) - (-4)$ *Divide.*

$\qquad\qquad\qquad\qquad\quad = 7 - 64 + 4$ *Simplify.*

$\qquad\qquad\qquad\qquad\quad = -57 + 4$ *Subtract.*

$\qquad\qquad\qquad\qquad\quad = -53$ *Add.*

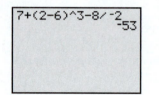

Figure 36 Verify the work

We use a graphing calculator to verify our work (see Fig. 36).

Expressions

Now that we know the order of operations, we can evaluate expressions that involve more than one operation.

Example 5   Evaluating an Expression

Evaluate $\dfrac{a - b}{c - d}$ for $a = 4, b = -2, c = -5,$ and $d = 4.$

Solution

We begin by substituting 4 for a, -2 for b, -5 for c, and 4 for d:

$$\frac{(4) - (-2)}{(-5) - (4)} = \frac{6}{-9}\qquad \textit{Subtract.}$$

$$= -\frac{6}{9}\qquad \frac{a}{-b} = -\frac{a}{b}$$

$$= -\frac{2}{3}\qquad \textit{Simplify.}$$

Example 6   Evaluating Expressions

1. Evaluate $4x^2$ for $x = -3$.
2. Evaluate $b^2 - 4ac$ for $a = 5, b = -3,$ and $c = -6$.

Solution

1. We begin by substituting -3 for x in the expression $4x^2$:

$$4(-3)^2 = 4(9)\qquad \textit{Perform exponentiation first.}$$

$$= 36\qquad \textit{Multiply.}$$

2. We begin by substituting 5 for a, -3 for b, and -6 for c in the expression $b^2 - 4ac$:

$$(-3)^2 - 4(5)(-6) = 9 - 4(5)(-6) \quad \textit{Perform exponentiation first.}$$
$$= 9 - (-120) \quad \textit{Multiply.}$$
$$= 9 + 120 \quad \textit{Simplify.}$$
$$= 129 \quad \textit{Add.}$$

▶

Example 7   Translating from English into Mathematics

Let x be a number. Translate the phrase "6 minus the product of -5 and the number" into a mathematical expression. Then evaluate the expression for $x = -3$.

Solution

First, we translate the given phrase into an expression:

$$\underbrace{\text{Six minus}}_{6\ -} \quad \underbrace{\begin{array}{c}\text{the product of } -5 \\ \text{and the number}\end{array}}_{(-5)x}$$

Then we substitute -3 for x in the expression $6 - (-5)x$ and perform the indicated operations:

$$6 - (-5)(-3) = 6 - 15 \quad \textit{Multiply before subtracting.}$$
$$= -9 \quad \textit{Subtract.}$$

▶

Making Predictions

In Example 8, we will find an expression, which we will use to make a prediction.

Example 8   Using a Table to Find an Expression

The number of nursing degrees awarded in 2014 was 206 thousand and has increased by about 11.2 thousand degrees per year since then (Source: *Integrated Postsecondary Education Data System*).

1. Use a table to help find an expression that stands for the number of nursing degrees (in thousands) awarded in the year that is t years since 2014.
2. Evaluate the expression you found in Problem 1 for $t = 7$. What does your result mean in this situation?

Solution

1. We construct Table 37. We show the arithmetic to help us see a pattern. From the last row of the table, we see that the number of nursing degrees awarded in the year that is t years since 2014 can be represented by the expression $11.2t + 206$.

Table 37 Number of Nursing Degrees

Years since 2014	Number of Nursing Degrees (thousands)
0	$11.2 \cdot 0 + 206$
1	$11.2 \cdot 1 + 206$
2	$11.2 \cdot 2 + 206$
3	$11.2 \cdot 3 + 206$
4	$11.2 \cdot 4 + 206$
t	$11.2 \cdot t + 206$

2. We substitute 7 for t in the expression $11.2t + 206$:

$$11.2(7) + 206 = 284.4$$

The number of nursing degrees that will be awarded in $2014 + 7 = 2021$ will be 284.4 thousand degrees, according to the model.

In Section 2.5, we found the percentage of a quantity. Now we will find the result of a percentage increase or percentage decrease in a quantity.

Example 9   Solving a Percentage Problem

Bicycle sales were 18.0 million bicycles in 2014, and they decreased by 3.3% in 2015 (Source: *National Bicycle Dealers Association*). What were bicycle sales in 2015?

Solution

The decrease in sales from 2014 to 2015 was $0.033(18.0)$ million bicycles. To find the sales (in millions of bicycles) in 2015, we subtract $0.033(18.0)$ from 18.0:

2014 sales (in millions)		Decrease in sales (in millions)		2015 sales (in millions)
18.0	−	0.033(18.0)	=	17.406

The sales in 2015 were about 17.4 million bicycles.

 Group Exploration

Section Opener: Order of operations

If an expression has more than one operation, we can use parentheses to indicate which operation to do first.

1. Perform the indicated operations in $(2 + 3) \cdot 4$ by first doing the addition and then doing the multiplication.

2. Perform the indicated operations in $2 + (3 \cdot 4)$ by first doing the multiplication and then doing the addition.

3. Compare the results you found in Problems 1 and 2. Does it matter in which order we add and multiply?

For each object, determine whether the area is $(2 + 3) \cdot 4$ or $2 + (3 \cdot 4)$. Explain.

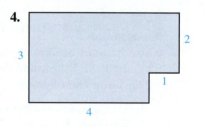

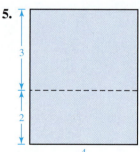

Tips for Success   Study in a Test Environment

Do you ever feel that you understand your homework assignments, yet you perform poorly on quizzes and tests? If so, you may not be studying enough to be ready to solve problems *in a test environment*. For example, although it is a good idea to refer to your lecture notes when you are stumped on a homework exercise, you must continue to solve similar exercises until you can solve them *without* referring to your lecture notes (unless your instructor uses open-notebook tests). The same idea applies to getting help from someone, referring to examples in the text, looking up answers in the back of the text, or any other form of support.

   Getting support to help you learn math is a great idea. Just make sure you spend the last part of your study time completing exercises without such support. One way to complete your study time would be to make up a practice quiz or test for you to do in a given amount of time.

Homework 2.6

For extra help ▶ **MyLab Math**   Watch the videos in MyLab Math

Perform the exponentiation by hand. Then use a calculator to check your work.

1. 4^3 2. 3^4 3. 2^5 4. 5^3
5. -8^2 6. -7^2 7. $(-8)^2$ 8. $(-7)^2$
9. $\left(\frac{6}{7}\right)^2$ 10. $\left(\frac{3}{5}\right)^3$

Perform the indicated operations by hand. Then use a calculator to check your work.

11. $3\cdot(5-1)$ 12. $8\cdot(2-6)$
13. $(2-5)(9-3)$ 14. $(2+8)(3-8)$
15. $4-(3-8)+1$ 16. $-6-(4-7)+5$
17. $\frac{1-10}{1+2}$ 18. $\frac{3-9}{2-4}$
19. $\frac{2-(-3)}{5-7}$ 20. $\frac{4-7}{-3-(-1)}$
21. $\frac{4-(-6)}{-7-8}$ 22. $\frac{1-9}{2-(-4)}$
23. $6+8\div2$ 24. $2-3\cdot5$
25. $-5-4\cdot3$ 26. $1+9\cdot(-4)$
27. $20\div(-2)\cdot5$ 28. $-16\div(-4)\cdot2$
29. $-9-4+3$ 30. $3-7+1$
31. $7-3(5+2)$ 32. $2-4(9-6)$
33. $3(5-1)-4(-2)$ 34. $2(2-5)+10\div5$
35. $15\div3-(2-7)(2)$ 36. $6(2+3)-5\cdot7$
37. $5-[4-7(5-9)]$ 38. $-3-[6+2(4-8)]$
39. $3[-2(1+6)-5]$ 40. $-2[4(3-8)+1]$
41. $\frac{7}{8}-\frac{3}{4}\cdot\frac{1}{2}$ 42. $\frac{5}{6}+\frac{2}{3}\div\frac{2}{5}$
43. $2+5^2$ 44. $8-3^2$
45. $-3(4)^2$ 46. $8(-2)^3$
47. $\frac{2^4}{4^2}$ 48. $\frac{5^2}{2^5}$
49. 4^3-3^4 50. 5^2+2^5
51. $45\div3^2$ 52. $-20\div2^2$
53. $(-1)^2-(-1)^3$ 54. $4^3-(-4)^3$
55. $-5(3)^2+4$ 56. $4(-2)^2-3$
57. $-4(-1)^2-2(-1)+5$ 58. $2(-4)^2+3(-4)-7$
59. $\frac{9-6^2}{12+3^2}$ 60. $\frac{10-(-2)^2}{3^3}$
61. $8-(9-5)^2-1$ 62. $4+(3-6)^2-2$
63. $8^2+2(4-8)^2\div(-2)$ 64. $(9-7)^2\cdot(-3)-2^4$

Use a calculator to perform the indicated operations. Round the result to two decimal places.

65. $13.28-35.2(17.9)+9.43\div2.75$
66. $8.53\div5.26+24.91-78.3(45.3)$

67. $5.82-3.16^3\div4.29$ 68. $1.98+8.22^5\cdot5.29$
69. $\frac{(25.36)(-3.42)-17.89}{33.26+45.32}$ 70. $\frac{(53.25)+99.83}{31.28-(6.31)(89.11)}$

Evaluate the expression for $a=-2$, $b=-4$, and $c=3$.

71. $a+bc$ 72. $ac-b$
73. $ac-b\div a$ 74. $c\div a+abc$
75. a^2-c^2 76. c^2-b^2
77. b^2-4ac 78. $-cb^2+a^2$
79. $\frac{-b-c^2}{2a}$ 80. $\frac{a^2-b}{c^2-b}$

Evaluate $\frac{a-b}{c-d}$ for the given values of a, b, c, and d.

81. $a=3, b=-10, c=-6, d=1$
82. $a=5, b=-1, c=-4, d=7$
83. $a=-3, b=7, c=1, d=-3$
84. $a=-2, b=6, c=5, d=-1$
85. $a=-8, b=-2, c=-15, d=-5$
86. $a=-3, b=-5, c=-8, d=-3$

Evaluate the expression for $x=-3$.

87. $-3x^2$ 88. $5x^2$
89. $-x^2+x$ 90. $-4x^2+4$
91. $2x^2-3x+5$ 92. $4x^2+x-2$

For Exercises 93–96, let x be a number. Translate the English phrase or sentence into a mathematical expression. Then evaluate the expression for $x=-4$.

93. 5 more than the product of −6 and the number
94. −3 minus the quotient of 8 and the number
95. Subtract 3 from the quotient of the number and −2.
96. The number plus the product of −5 and the number
97. Congressional pay in 2000 was $141.3 thousand, and it increased by approximately $3.6 thousand per year for the period 2000–2009 (Source: *Bureau of Labor Statistics*).
   a. Complete Table 38 to help find an expression that stands for the congressional pay (in thousands of dollars) at t years since 2000. Show the arithmetic to help you see a pattern.

Table 38 Congressional Pay

Years since 2000	Congressional Pay (thousands of dollars)
0	
1	
2	
3	
4	
t	

b. Evaluate the expression you found in part (a) for $t = 7$. What does your result mean in this situation?

c. Congress has had a pay freeze since 2009. If the congressional pay were increased by 1% in 2017, what would be the pay in that year?

98. The number of countries requiring picture warnings on cigarette packages was 70 countries in 2014 and has increased by approximately 8.8 countries per year since then (Source: *Canadian Cancer Society*).

a. Complete Table 39 to help find an expression that stands for the number of countries requiring picture warnings on cigarette packages at t years since 2014. Show the arithmetic to help you see a pattern.

Table 39 Numbers of Countries Requiring Picture Warnings on Cigarette Packages

Years since 2014	Number of Countries
0	
1	
2	
3	
4	
t	

b. Evaluate the expression you found in part (a) for $t = 8$. What does your result mean in this situation?

99. The population of Gary, Indiana, was about 152 thousand in 1980 and has decreased by about 2.1 thousand per year since then (Source: *U.S. Census Bureau*).

a. Complete Table 40 to help find an expression that stands for the population of Gary (in thousands) at t years since 1980. Show the arithmetic to help you see a pattern.

Table 40 Populations of Gary

Years since 1980	Population (thousands)
0	
1	
2	
3	
4	
t	

b. Evaluate the expression you found in part (a) for $t = 42$. What does your result mean in this situation?

100. The percentage of companies offering traditional benefit plans was 85% in 1990 and has decreased by approximately 1.1 percentage points per year since then (Source: *National Compensation Survey*).

a. Complete Table 41 to help find an expression that stands for the percentage of companies offering traditional benefit plans at t years since 1990. Show the arithmetic to help you see a pattern.

b. Evaluate the expression you found in part (a) for $t = 31$. What does your result mean in this situation?

Table 41 Percentages of Companies Offering Traditional Benefit Plans

Years since 1990	Percent
0	
1	
2	
3	
4	
t	

101. The number of terrorist attacks in the United States was 16 attacks in 2013, and it increased by 18.75% in 2014 (Source: *Global Terrorism Database*). How many terrorist attacks were there in 2014?

102. Digital ad revenue was $49.5 billion in 2014, and it increased by 20.4% in 2015 (Source: *IAB/PwC*). What was the digital ad revenue in 2015?

103. The number of Americans who were affected by plastic-bag bans was 18.3 million people in 2014, and it increased by 168% in 2015 (Source: *EPI*). How many Americans were affected by plastic-bag bans in 2015?

104. Nike's North American revenue (in billions of dollars) was $13.7 billion in 2015, and it increased by 8% in 2016 (Source: *Nike*). What was the revenue in 2016?

105. The number of subscribers of Weight Watchers was 3.7 million subscribers in the first quarter of 2014, and it decreased by 29.7% in the first quarter of 2015 (Source: *Weight Watchers*). What was the number of subscribers in the first quarter of 2015?

106. The number of Kmart stores was 2165 stores in 2000, and it decreased by 59.2% in 2016 (Source: *Kmart*). What was the number of stores in 2016?

107. The number of people who rode a bicycle at least six times per year was 43.1 million people in 2000, and it decreased by 17.4% in 2014 (Source: *National Bicycle Dealers Association*). What was the number of people who rode a bicycle at least six times per year in 2014?

108. The number of people participating in the Supplemental Nutrition Assistance Program (formerly known as food stamps) was 46.5 million people in 2015, and it decreased by 2.3% in 2016 (Source: *USDA*). How many people were in the program in 2016?

109. If a cube has sides of length s feet, then the volume of the cube is s^3 cubic feet. Find the volume of a cubic box with sides of length 2 feet.

110. If the radius of a sphere is r inches, then the volume of the sphere is $\frac{4}{3}\pi r^3$ cubic inches. Find the volume of a sphere with a radius of 3 inches.

Concepts

111. A student tries to perform the indicated operations in $2(3)^2 + 2(3) + 1$:

$$2(3)^2 + 2(3) + 1 = 6^2 + 2(3) + 1$$
$$= 36 + 2(3) + 1$$
$$= 36 + 6 + 1$$
$$= 43$$

Describe any errors. Then perform the operations correctly.

112. A student tries to evaluate $x^2 + 4x + 5$ for $x = -3$:

$$-3^2 + 4(-3) + 5 = -9 - 12 + 5$$
$$= -21 + 5$$
$$= -16$$

Describe any errors. Then evaluate the expression correctly.

113. A student thinks that $-4^2 = 16$ because a negative number times a negative number is equal to a positive number. Is the student correct? Explain.

114. A student tries to perform the indicated operations in $16 \div 2 \cdot 4$:

$$16 \div 2 \cdot 4 = 16 \div 8 = 2$$

Describe any errors. Then perform the operations correctly.

115. In Problem 2 of Example 2, we performed the indicated operations in $\dfrac{11 - 3}{1 - 5}$ by the following steps:

$$\frac{11 - 3}{1 - 5} = \frac{8}{-4} = -2$$

a. Perform the indicated operations in $(11 - 3) \div (1 - 5)$.
b. Perform the indicated operations in $11 - 3 \div 1 - 5$.
c. In using a calculator to simplify $\dfrac{11 - 3}{1 - 5}$, a student presses the following buttons:

$$11 \;\boxed{-}\; 3 \;\boxed{\div}\; 1 \;\boxed{-}\; 5$$

The result of this calculation is 3 rather than -2. Describe any errors. Then perform the operations correctly.

116. a. Perform the indicated operations in $(12 \div 3) \cdot 2$.
b. Perform the indicated operations in $12 \div (3 \cdot 2)$.
c. For the expression $12 \div 3 \cdot 2$, does it matter whether we multiply and divide from left to right or multiply and divide from right to left? Explain.
d. Perform the indicated operations in $12 \div 3 \cdot 2$.

117. a. Evaluate $(ab)c$ for $a = 2, b = 3$, and $c = 4$.
b. Evaluate $a(bc)$ for $a = 2, b = 3$, and $c = 4$.

c. Compare the results you found in parts (a) and (b).
d. Evaluate $(ab)c$ and $a(bc)$ for $a = 4, b = -2$, and $c = 5$, and then compare the results.
e. Evaluate $(ab)c$ and $a(bc)$ for a, b, and c values of your choosing, and then compare the results.
f. Do you think that the statement $(ab)c = a(bc)$ is true for all values of a, b, and c? Explain.
g. What does the statement $(ab)c = a(bc)$ imply about the order of operations?

118. a. Evaluate $(a + b) + c$ for $a = 2, b = 3$, and $c = 4$.
b. Evaluate $a + (b + c)$ for $a = 2, b = 3$, and $c = 4$.
c. Compare the results you found in parts (a) and (b).
d. Evaluate $(a + b) + c$ and $a + (b + c)$ for $a = 4, b = -2$, and $c = 5$, and then compare the results.
e. Evaluate $(a + b) + c$ and $a + (b + c)$ for a, b, and c values of your choosing, and then compare the results.
f. Do you think that the statement $(a + b) + c = a + (b + c)$ is true for all values of a, b, and c? Explain.
g. What does the statement $(a + b) + c = a + (b + c)$ imply about the order of operations?

119. a. Evaluate x^2 for x equal to $-2, -1, 0, 1$, and 2.
b. Which of the following describes the results you found in part (a)?

negative, nonpositive, positive, nonnegative

c. Describe x^2 for any real number x.

120. a. Compute without using a calculator.

   i. $(-1)^2$ **ii.** $(-1)^3$
   iii. $(-1)^4$ **iv.** $(-1)^5$
   v. $(-1)^{87}$ **vi.** $(-1)^{596}$

b. For what counting-number values of n is $(-1)^n$ equal to 1? Explain.
c. For what counting-number values of n is $(-1)^n$ equal to -1? Explain.

121. Some students believe x^2 is greater than x for *all* real numbers.
a. Find three values of x where x^2 is greater than x.
b. Find three values of x where x^2 is less than x.
c. Find two values of x where x^2 is equal to x.

122. Describe the order of operations in your own words.

Taking It to the Lab

Climate Change Lab (continued from Chapter 1)

Given the consensus among most scientists that global warming is occurring and that it could have catastrophic results, scientists have been searching for the cause of the warming (see Table 42).

Most scientists believe global warming is largely the result of carbon dioxide emissions from the burning of fossil fuels such as oil, coal, and natural gas. Carbon dioxide emissions in the United States and in the world have increased greatly since 1950 (see Table 43).

Not everyone agrees carbon dioxide emissions cause global warming, however. For example, Jonathan Adler, professor of environmental law at Case Western Reserve School of Law, argues that two-thirds of the temperature increase occurred in the first half of the 20th century, yet most of the carbon dioxide emissions occurred in the

DATA **Table 42** Average Surface Temperatures of Earth

Year	Average Temperature (degrees Fahrenheit)	Year	Average Temperature (degrees Fahrenheit)
1900	57.1	1960	57.2
1905	56.8	1965	57.0
1910	56.6	1970	57.3
1915	57.0	1975	57.1
1920	56.9	1980	57.6
1925	56.9	1985	57.3
1930	57.1	1990	57.8
1935	57.0	1995	57.9
1940	57.3	2000	57.8
1945	57.3	2005	58.3
1950	56.9	2010	58.3
1955	57.0	2015	59.0

Source: *NASA–GISS*

DATA **Table 43** Carbon Dioxide Emissions from Burning of Fossil Fuels

| Year | Carbon Dioxide Emissions (billions of metric tons) | |
	United States	World
1950	2.4	5.8
1955	2.7	7.2
1960	2.9	9.4
1965	3.5	11.2
1970	4.3	14.7
1975	4.4	16.5
1980	4.8	19.1
1985	4.6	19.4
1990	5.0	21.6
1995	5.3	22.2
2000	5.9	23.8
2005	6.0	28.4
2010	5.6	30.6
2014	5.4	35.7

Source: *U.S. Department of Energy*

second half of the century. He concludes that there is no link between carbon dioxide emissions and global warming. Adler believes global warming could be due to slight variations in the sun's output, combined with fluctuations in Earth's orbit.[*]

Offering another explanation for global warming, Enric Pallé and his colleagues at the Big Bear Solar Observatory and the California Institute of Technology conducted a 2004 study that suggests the global warming that took place from 1984 to 2000 could have been due to reduced cloud cover.[†] Other scientists have raised objections to the study's findings.

Although there have been many dissenters along the way, the vast majority of scientists now believe in the warming–emissions connection. Even in 1997 there was strong enough agreement to motivate the United Nations to negotiate a treaty called the Kyoto Protocol. The treaty's goal was to reduce annual greenhouse gas emissions to about 5% to 7% below 1990 levels by 2012. In November 2004, Russia cast the deciding vote to ratify the protocol, which took effect on February 16, 2005.

To create some flexibility in the treaty's requirements, a country that had exceeded its emissions limit could buy emissions credits from a country that was below its emissions limit. Also, a country could receive emissions credits by financing a project to help lower emissions in another country.

The treaty was legally binding for the 128 countries that ratified it. The United States declined to ratify the Kyoto Protocol, saying that reducing emissions to the point called for by the treaty would cripple the U.S. economy. If it had ratified the treaty, the United States would have had to reduce its 2012 greenhouse gas emissions to 7% below its emissions level in 1990.

Instead of ratifying the treaty, in February 2002, the Bush administration adopted a voluntary program, called the Global Climate Change Initiative (GCCI), that includes tax incentives to motivate companies to reduce their emissions. The Bush administration set a goal of lowering the carbon intensity in 2012 to 18% below its level in 2002. *Carbon intensity* is defined as the ratio of annual carbon dioxide emissions to annual economic output.

Critics of the GCCI plan point out that if the U.S. economy is improving, then it is possible for carbon intensity to decrease *even though annual carbon dioxide emissions continue to increase*. In fact, even though annual carbon dioxide emissions have increased during each of the past three decades, carbon intensity has declined by 18%, 23%, and 16% in those decades! However, the projected decrease for 2002–2012 was only 13%, so modest efforts would have had to be implemented to reach the goal of 18%.[‡]

Critics such as the Earth Policy Institute and the Pew Center on Global Climate Change also say the voluntary plan is unrealistic because businesses will make modest efforts to reduce emissions when the economy is doing well and will make little or no effort when the economy is doing poorly. These critics argue the United States would have been more likely to respond to the Kyoto Protocol because it was a legally binding, yet flexible, treaty.

The Obama administration spent $5.9 billion on GCCI from 2010 to 2016, helping developing countries do the following: adapt to climate change, develop clean energy infrastructures, and stop cutting down forests. The Congressional Research Service raised pros and cons of continuing such actions, including questioning whether such monies would

---

[*]"Global Warming—Hot Problem or Hot Air?" April 1998, *The Freeman*, 48(4), The Foundation for Economic Education, Inc.

[†]"Changes in Earth's Reflectance over the Past Two Decades," May 28, 2004, *Science* 304: 1299–1301.

[‡]"Early Release of the Annual Energy Outlook 2003" (November 2002), Energy Information Administration; and "Projected Greenhouse Gas Emissions," May 2002, *U.S. Climate Action Report 2002*, pp. 70–80, U.S. Department of State, Washington, DC.

be better spent in meeting domestic challenges, especially when the U.S. economy was struggling.[§]

All countries except the United States, Nicaragua, and Syria have either signed or ratified the Paris Climate Accord, which aims to limit the increase of global average temperature to well below 2°C above pre-industrial levels, to have global emissions peak as soon as possible, and to undertake rapid reductions thereafter.[#]

Although President Obama ratified the accord in 2016, President Trump withdrew from the accord in 2017, claiming that it would do little to reduce global warming, is unfair to the United States, and would hurt the U.S. economy.[##]

Analyzing the Situation

1. **a.** Find the change in the average global temperature from 1900 to 1950.
    b. Find the change in the average global temperature from 1950 to 2000.
    c. Recall that Adler believes there is no link between carbon dioxide emissions and global warming. What is his argument? Is his reasoning correct? Explain.

2. Let c be U.S. carbon dioxide emissions (in billions of metric tons) in the year that is t years since 1950. Construct a careful scatterplot of the data.

3. Draw a linear model on your scatterplot. Use your model to estimate U.S. carbon dioxide emissions in 2012.

4. Under the Kyoto Protocol, what would have been the largest quantity of U.S. carbon dioxide emissions allowed in 2012? Find the difference between this quantity and your estimate in Problem 3.

5. The numbers in the list 20, 22, 24, 26, 28 are increasing. Decide whether the ratios in the following list are increasing, decreasing, or neither:

$$\frac{20}{1}, \frac{22}{2}, \frac{24}{4}, \frac{26}{8}, \frac{28}{16}$$

6. Explain why your work in Problem 5 illustrates how it is possible for U.S. carbon intensity to decrease while U.S. annual carbon dioxide emissions increase. [**Hint:** Recall that carbon intensity is defined as the *ratio* of annual carbon dioxide emissions to annual economic output.]

7. Which seeks to lower U.S. carbon dioxide emissions more, the GCCI plan or the Paris Climate Accord? Explain. Taking into account the degree to which American companies would respond to either policy, do you think U.S. carbon dioxide emissions would be reduced more by the GCCI plan or by the Paris Climate Accord? Explain.

Stocks Lab

Imagine that you have $5000 that you plan to invest in five stocks for one week. In this lab, you will explore some possible outcomes of that investment.[**]

Collecting the Data

Use a newspaper or the Internet to select five stocks. For each stock, record the company name and the call letters of the stock. Here are some examples:

Twitter has the call letters TWTR.
Coca-Cola has the call letters KO.
EMC corporation has the call letters EMC.

Record the value of one share (the beginning share price) of each of the five stocks. Also, record how you distribute your $5000 investment. For example, you may invest all $5000 in one of the five stocks or $1000 in each of the five stocks, or you may opt for some other distribution of the money. The sum of your investments should equal or be close to $5000 by buying whole amounts of stock. Even if you do not invest money in some of the stocks, still record the information about all five stocks.

After one week, record the new value of one share (the ending share price) of each of the five stocks.

Analyzing the Data

1. Complete Table 44. The *profit* from a stock is the money you collect from the stock, minus the money you invested in the stock. What is the total profit from your $5000 investment?

2. For each of the five stocks, construct a bar graph displaying the beginning and ending values of one share. You can use the same axes for several bar graphs if the scaling is convenient.

Table 44 Five Stocks' Performances

Call Letters of Stock	Investment in Stock (dollars)	Beginning Share Price (dollars)	Number of Shares	Ending Share Price (dollars)	Money Collected from Stock (dollars)	Profit from Stock (dollars)

[§]Richard K. Lattanzio, "The Global Climate Change Initiative (GCCI): Budget Authority and Request. FY2010-FY2013," Congressional Research Service, *CRS Report for Congress*. March 15, 2012.

[#]Center for Climate and Energy Solutions, "Outcomes of the U.N. Climate Change Conference in Paris," December 2015.

[##] "Statement by President Trump on the Paris Climate Accord," www.whitehouse.gov, June 1, 2017.

[**]Lab suggested by Jim Ryan, State Center Community College District, Clovis Center, Clovis, CA.

3. Find the change in the share price of each of the five stocks. Which share price had the greatest change? The least? Explain how you can illustrate these changes on your bar graphs.

4. The *percent change* of the value of a stock can be found by dividing the change in value of the stock by the beginning value of the stock and then converting the decimal result into percent form (by multiplying by 100). For example, suppose that a stock's value increases from $7 to $9. The change in value of this stock is its ending value minus its beginning value: $9 - 7 = 2$ dollars.

Here we find the percent change in value:

$$\text{Percent change} = \frac{\text{Change in value}}{\text{Beginning value}} \cdot 100$$

$$= \frac{9 - 7}{7} \cdot 100 = \frac{2}{7} \cdot 100 \approx 28.57$$

So, the percent change is about 28.57%.

Now find the percent change in value for each of your five stocks. Which stock had the greatest percent change? The least? Explain how you can at least approximately compare the percent changes by viewing your bar graphs.

5. Among your five stocks, is there a pair of stocks for which one stock has the greater change but the other has the greater percent change? If yes, use these stocks to respond to the questions in parts (a)–(c). If no, then use the following values of fictional stocks A and B:

Stock A	Stock B
Increased from $4 to $5	Increased from $20 to $23

a. Find the profit earned from investing all the $5000 in the stock with the larger change in value.
b. Find the profit earned from investing all the $5000 in the stock with the larger percent change in value.
c. Which is the better measure of the growth of a stock, change in value or percent change in value? Explain.

6. a. Find the profit earned from each of the following scenarios:

    i. You invest the $5000 in the best-performing stock among the five stocks.
    ii. You invest the $5000 in the worst-performing stock among the five stocks.
    iii. You invest the $5000 by investing $1000 in each of the five stocks.

   b. Describe the benefits and drawbacks to investing your money in a number of stocks rather than in just one stock.

Chapter Summary

Key Points of Chapter 2

Section 2.1   Expressions

Expression	An **expression** is a constant, a variable, or a combination of constants, variables, operation symbols, and grouping symbols, such as parentheses.
Evaluate an expression	We **evaluate an expression** by substituting a number for each variable in the expression and then calculating the result. If a variable appears more than once in the expression, the same number is substituted for that variable each time.

Section 2.2   Operations with Fractions

Division by 0	The fraction $\frac{a}{b}$ is undefined if $b = 0$. Division by 0 is undefined.
Simplify a fraction	To simplify a fraction, 1. Find the prime factorizations of the numerator and denominator. 2. Find an equal fraction in which the numerator and the denominator do not have common positive factors other than 1 by using the property $$\frac{ab}{ac} = \frac{a}{a} \cdot \frac{b}{c} = 1 \cdot \frac{b}{c} = \frac{b}{c}$$ where a and c are nonzero.

Section 2.2   Operations with Fractions (*Continued*)

Simplify results	If the result of an exercise is a fraction, simplify it.
Multiplying fractions	$\dfrac{a}{b} \cdot \dfrac{c}{d} = \dfrac{ac}{bd}$, where b and d are nonzero.
Dividing fractions	$\dfrac{a}{b} \div \dfrac{c}{d} = \dfrac{a}{b} \cdot \dfrac{d}{c}$, where b, c, and d are nonzero.
Adding fractions	$\dfrac{a}{b} + \dfrac{c}{b} = \dfrac{a + c}{b}$, where b is nonzero.
Subtracting fractions	$\dfrac{a}{b} - \dfrac{c}{b} = \dfrac{a - c}{b}$, where b is nonzero.
How to add or subtract two fractions with different denominators	To add (or subtract) two fractions with different denominators, use the fact that $\dfrac{a}{a} = 1$, where a is nonzero, to write an equal sum (or difference) of fractions for which each denominator is the LCD.

Section 2.3   Absolute Value and Adding Real Numbers

The opposite of the opposite of a number	$-(-a) = a$
Absolute value	The **absolute value** of a number is the distance that the number is from 0 on the number line.
Adding two numbers with the same sign	To add two numbers with the same sign, **1.** Add the absolute values of the numbers. **2.** The sum of the original numbers has the same sign as the sign of the original numbers.
Adding two numbers with different signs	To add two numbers with different signs, **1.** Find the absolute values of the numbers. Then subtract the smaller absolute value from the larger absolute value. **2.** The sum of the original numbers has the same sign as the original number with the larger absolute value.

Section 2.4   Change in a Quantity and Subtracting Real Numbers

Change in a quantity	The change in a quantity is the ending amount minus the beginning amount: Change in the quantity = Ending amount − Beginning amount
Subtracting a number	To subtract a number, add its opposite: $a - b = a + (-b)$
Increasing quantity	An increasing quantity has a positive change.
Decreasing quantity	A decreasing quantity has a negative change.

Section 2.5   Ratios, Percents, and Multiplying and Dividing Real Numbers

Unit ratio	A **unit ratio** is a ratio written as $\dfrac{a}{b}$ with $b = 1$ or as $a : b$ with $b = 1$.
Percent	**Percent** means "for each hundred": $a\% = \dfrac{a}{100}$.
Writing a percentage as a decimal number	To write a percentage as a decimal number, remove the percent symbol and divide the number by 100 (move the decimal point two places to the left).
Writing a decimal number as a percentage	To write a decimal number as a percentage, multiply the number by 100 (move the decimal point two places to the right) and insert a percent symbol.
Finding the percentage of a quantity	To find the percentage of a quantity, multiply the decimal form of the percentage and the quantity.
One hundred percent	One hundred percent of a quantity is *all* the quantity.
The product or quotient of two numbers	The product or quotient of two numbers that have different signs is negative. The product or quotient of two numbers that have the same sign is positive.
Equal fractions with negative signs	If $b \neq 0$, then $\dfrac{-a}{b} = \dfrac{a}{-b} = -\dfrac{a}{b}$.

Section 2.6 Exponents and Order of Operations

Exponent	For any counting number n, $$x^n = \underbrace{x \cdot x \cdot x \cdot \cdots \cdot x}_{n \text{ factors of } x}$$ We refer to x^n as the **power**, the **nth power of x**, or **x raised to the nth power**. We call x the **base** and n the **exponent**.
Finding $-x^n$	For an expression of the form $-x^n$, we calculate x^n before taking the opposite.
Order of operations	We perform operations in the following order: 1. First, perform operations within parentheses or other grouping symbols, starting with the innermost group. 2. Then perform exponentiations. 3. Next, perform multiplications and divisions, going from left to right. 4. Last, perform additions and subtractions, going from left to right.
Order of operations and the strengths of operations	After we have performed operations in parentheses, the order of operations goes from the most powerful operation, exponentiation, to the next-most-powerful operations, multiplication and division, to the weakest operations, addition and subtraction.

Chapter 2 Review Exercises

Perform the indicated operations. Then use a calculator to check your work.

1. $8 + (-2)$

2. $(-5) + (-7)$

3. $6 - 9$

4. $8 - (-2)$

5. $8(-2)$

6. $8 \div (-2)$

7. $-24 \div (10 - 2)$

8. $(2 - 6)(5 - 8)$

9. $\dfrac{7 - 2}{2 - 7}$

10. $\dfrac{2 - 8}{3 - (-1)}$

11. $\dfrac{3 - 5(-6)}{-2 - 1}$

12. $3(-5) + 2$

13. $-4 + 2(-6)$

14. $2 - 12 \div 2$

15. $8 \div (-2) \cdot 5$

16. $4 - 6(7 - 2)$

17. $2(4 - 7) - (8 - 2)$

18. $-2(3 - 6) + 18 \div (-9)$

19. $-14 \div (-7) - 3(1 - 5)$

20. $-5 - [3 + 2(1 - 7)]$

21. $4.2 - (-6.7)$

22. $\dfrac{4}{9}\left(-\dfrac{3}{10}\right)$

23. $\left(-\dfrac{8}{15}\right) \div \left(-\dfrac{16}{25}\right)$

24. $\dfrac{5}{9} - \left(-\dfrac{2}{9}\right)$

25. $-\dfrac{5}{6} + \dfrac{7}{8}$

26. $\dfrac{-5}{2} - \dfrac{7}{-3}$

27. $(-9)^2$

28. -9^2

29. 2^4

30. $\left(\dfrac{3}{4}\right)^3$

31. $-6(3)^2$

32. $24 \div 2^3$

33. $(-2)^3 - 4(-2)$

34. $\dfrac{2^3}{3 + 3^2}$

35. $\dfrac{17 - (-3)^2}{5 - 4^2}$

36. $-3(2)^2 - 4(2) + 1$

37. $24 \div (3 - 5)^3$

38. $7^2 - 3(2 - 5)^2 \div (-3)$

Simplify.

39. $\dfrac{-18}{-24}$

40. $\dfrac{-28}{35}$

For Exercises 41 and 42, use a calculator to compute. Round the result to two decimal places.

41. $-5.7 + 2.3^4 \div (-9.4)$

42. $\dfrac{3.5(17.4) - 872.3}{54.2 \div 8.4 - 65.3}$

43. A rectangle has a width of $\dfrac{1}{4}$ yard and a length of $\dfrac{5}{6}$ yard. What is the perimeter of the rectangle?

44. A student has a credit card balance of -4789 dollars. If he sends a check to the credit card company for \$800 and then uses the card to buy a textbook for \$102.99 and a notebook for \$3.50, what is the new balance?

45. An airplane drops from 32,500 feet to 27,800 feet. Find the change in altitude.

46. Three hours ago, the temperature was 4°F. The temperature is now −8°F.
  a. What is the change in temperature for the past 3 hours?
  b. Estimate the change in temperature for the past hour.
  c. Explain why your estimate in part (b) may not be the actual change in temperature for the past hour.

47. Contributions from individuals and political action committees to major party nominees of presidential elections are shown in Table 45 for various years.

Table 45 Contributions from Individuals and Political Action Committees to Major Party Nominees of Presidential Elections

Year	Contribution (millions of dollars)	
	Democratic Nominee	Republican Nominee
2000	631	923
2004	824	981
2008	962	920
2012	1070	1022

Source: *Center for Responsive Politics*

a. Find the change in the annual contributions to the Democratic nominee from 2000 to 2004.

b. Find the change in the annual contributions to the Republican nominee from 2004 to 2008.

c. Over which 4-year period was there the greatest change in annual contributions to the Democratic nominee? What was that change?

d. Over which 4-year period was there the greatest change in annual contributions to the Republican nominee? What was that change?

48. Female teen cell-phone users sent or received an average of 79 texts per day. Male teen cell-phone users sent or received an average of 56 texts per day (Source: *Pew Research Center*). Find the unit ratio of the average daily number of texts sent or received by female teens to the average daily number of texts sent or received by male teens. What does your result mean?

For Exercises 49 and 50, write the percentage as a decimal number.

49. 75% **50.** 2.9%

51. In a survey of 1.2 million Americans, 15% of the people said Five Guys has the best hamburger (Source: *YouGov*). How many of those surveyed said Five Guys has the best hamburger?

52. In a survey of 68,300 people aged at least 12 years, 7.3% of the people had used marijuana in the past month (Source: *National Institute on Drug Abuse*). How many of those surveyed used marijuana in the past month?

53. A person's credit card balance is -5493 dollars. If the person pays off 20% of the balance, what is the new balance?

Evaluate the expression for $a = 2$, $b = -5$, $c = -4$, and $d = 10$.

54. $ac + c \div a$

55. $b^2 - 4ac$

56. $a(b - c)$

57. $\dfrac{-b - c^2}{2a}$

58. $2c^2 - 5c + 3$

59. $\dfrac{a - b}{c - d}$

For Exercises 60–63, let x be a number. Translate the English phrase into a mathematical expression. Then evaluate the expression for $x = -3$.

60. 5 more than the number

61. The number subtracted from -7

62. 2 minus the product of the number and 4

63. 1 plus the quotient of -24 and the number

64. If T is the total cost (in dollars) for a team to join a softball league and there are n players on the team, then $T \div n$ is the cost (in dollars) per player. Evaluate $T \div n$ for $T = 650$ and $n = 13$. What does your result mean in this situation?

65. A basement is flooded with 400 cubic feet of water. Each hour, 50 cubic feet of water is pumped out of the basement.

a. Complete Table 46 to help find an expression that stands for the volume (in cubic feet) of water in the basement after water has been pumped out for t hours. Show the arithmetic to help you see a pattern.

Table 46 Volumes of Water

Time (hours)	Volume of Water (cubic feet)
0	
1	
2	
3	
4	
t	

b. Evaluate the expression that you found in part (a) for $t = 7$. What does your result mean in this situation?

66. The number of Americans who flew to Europe in 2011 was 10.8 million, and it increased 10.2% by 2014 (Source: *U.S. Department of Commerce*). How many Americans flew to Europe in 2014?

Chapter 2 Test

For Exercises 1–14, perform the indicated operations by hand.

1. $-8 - 5$

2. $-7(-9)$

3. $-3 + 9 \div (-3)$

4. $(4 - 2)(3 - 7)$

5. $\dfrac{4 - 7}{-1 - 5}$

6. $5 - (2 - 10) \div (-4)$

7. $-20 \div 5 - (2 - 9)(-3)$

8. $0.4(-0.2)$

9. $-\dfrac{27}{10} \div \dfrac{18}{75}$

10. $-\dfrac{3}{10} + \dfrac{5}{8}$

11. 3^4

12. -4^2

13. $7 + 2^3 - 3^2$

14. $1 - (3 - 7)^2 + 10 \div (-5)$

15. Simplify $\dfrac{84}{-16}$.

16. Two hours ago, the temperature was 5°F. If the temperature has decreased by 9°F in the last 2 hours, what is the current temperature?

17. The chances of being audited by the Internal Revenue Service (IRS) are shown in Table 47 for various years.

Table 47 Tax Audit Rates

Year	Tax Audit Rate (number of audits per 1000 tax returns)
2003	6.5
2005	9.7
2007	10.3
2009	10.3
2011	11.1
2013	9.6
2015	8.4

Source: *Internal Revenue Service*

a. Find the change in the tax audit rate from 2003 to 2005.
b. Find the change in the tax audit rate from 2013 to 2015.
c. For which 2-year period was the change in the tax audit rate the most? What was that change?

18. The number of background checks conducted for gun purchases was 8.4 million checks in 2000 and 23 million checks in 2015 (Source: *FBI*). Find the unit ratio of the number of background checks in 2015 to the number of background checks in 2000. What does your result mean?

Evaluate the expression for a = −6, b = −2, c = 5, and d = −1.

19. $ac - \dfrac{a}{b}$

20. $\dfrac{a - b}{c - d}$

21. $a + b^3 + c^2$

22. $b^2 - 4ac$

For Exercises 23 and 24, let x be a number. Translate the English phrase into a mathematical expression. Then evaluate the expression for x = −5.

23. Twice the number minus the product of 3 and the number

24. 6 subtracted from the quotient of −10 and the number

25. U.S. Postal Service first-class mail volume was 63.8 billion pieces in 2014. The first-class mail volume has decreased by about 2.1 billion pieces per year since 2014 (Source: *U.S. Postal Service*).
a. Complete Table 48 to help find an expression that stands for the U.S. Postal Service first-class mail volume (in billions of pieces) in the year that is *t* years since 2014. Show the arithmetic to help you see a pattern.

Table 48 U.S. Postal Service First-Class Mail Volume

Years since 2014	First-Class Mail Volume (billions of pieces)
0	
1	
2	
3	
4	
t	

b. Evaluate the expression that you found in part (a) for *t* = 7. What does your result mean in this situation?

26. The number of bats broken in Major League Baseball games was 2400 bats in 2008, and it decreased by 63% in 2014 (Source: *Major League Baseball*). How many bats were broken in 2014?

Cumulative Review of Chapters 1 and 2

1. A rectangle has a perimeter of 36 inches. Let *W* be the width, *L* be the length, and *P* be the perimeter, all with units in inches.
a. Sketch three possible rectangles with a perimeter of 36 inches.
b. Which of the symbols *W*, *L*, and *P* are variables? Explain.
c. Which of the symbols *W*, *L*, and *P* are constants? Explain.

2. Graph the integers between −2 and 3, inclusive, on a number line.

3. Here are the changes (in dollars) in Microsoft's closing stock prices from one day to the next during the week of December 5, 2016: decrease of 0.67, increase of 1.82, decrease of 0.36, and increase of 0.96. Let *C* be the change of the stock's closing price (in dollars) from one day to the next. Use points on a number line to describe the changes in value of the stock.

4. What is the *x*-coordinate of the ordered pair $(-5, 3)$?

5. A person takes a bath. Let *V* be the volume (in gallons) of water in the bathtub at *t* minutes after the person pulls out the plug from the drain. Identify the explanatory variable and the response variable.

6. The revenues of ExxonMobile are shown in Table 49 for various years.

Table 49 Revenues of ExxonMobile

Years	Revenue (billions of dollars)
2009	302
2010	370
2011	467
2012	452
2013	421
2014	394
2015	259

Source: *ExxonMobile*

Let *r* be the annual revenue (in billions of dollars) at *t* years since 2005.
a. Construct a scatterplot of the data.
b. For which year shown in Table 49 was the revenue the most?
c. For which year shown in Table 49 was the revenue the least?
d. From which year to the next did the annual revenue decrease the most? What was the change in annual revenue?
e. From which year to the next did the annual revenue increase the most? What was the change in annual revenue?

For Exercises 7–10, refer to Fig. 37.

7. Find y when $x = -4$.   **8.** Find x when $y = 1$.

9. Find the y-intercept.   **10.** Find the x-intercept.

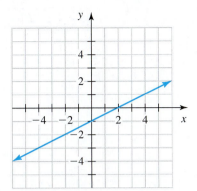

Figure 37 Exercises 7–10

11. A person is laid off from work. Let B be the balance (in thousands of dollars) in her checking account at t months since she was laid off. Some pairs of values of t and B are shown in Table 50.

Table 50 Balances in a Checking Account

t (months)	B (thousands of dollars)
2	16
3	14
5	10
6	8
8	4

a. Construct a scatterplot of the data. Then draw a linear model.

b. What was the balance 4 months after the person was laid off?

c. When was the balance $6 thousand?

d. What is the B-intercept? What does it mean in this situation?

e. What is the t-intercept? What does it mean in this situation?

12. The unemployment rates in January are shown in Table 51 for various years.

Table 51 Unemployment Rates

Year	Unemployment Rate (percent)
2010	9.8
2011	9.1
2012	8.3
2013	8.0
2014	6.6
2015	5.7
2016	4.9

Source: *U.S. Bureau of Labor Statistics*

Let r be the unemployment rate (in percent) at t years since 2010.

a. Construct a scatterplot of the data. Then draw a linear model.

b. Use the model to estimate the unemployment rate in 2013. Compute the error in your estimate.

c. Find the r-intercept of the model. What does it mean in this situation?

d. Find the t-intercept of the model. What does it mean in this situation?

For Exercises 13–18, perform the indicated operations by hand. Then use a calculator to check your work.

13. $\dfrac{3(-8) + 15}{2 - 7(2)}$ **14.** $-4(3) + 6 - 20 \div (-10)$

15. $\left(-\dfrac{14}{15}\right) \div \left(-\dfrac{35}{27}\right)$ **16.** $\dfrac{3}{8} - \dfrac{5}{6}$

17. $4 - (7 - 9)^4 + 20 \div (-4)$ **18.** $\dfrac{5 - 3^2}{4^2 + 2}$

19. In 2016, the low temperature in Chicago was 20°F on December 17 and −5°F on December 18. What is the change in low temperature?

20. A student has a credit card balance of −2692 dollars. If he sends a check to the credit card company for $850 and then uses the card to buy some gasoline for $23, what is the new balance?

21. Evaluate $\dfrac{a - b}{c - d}$ for $a = 1, b = -4, c = -3$, and $d = 7$.

22. Evaluate $b^2 - 4ac$ for $a = 2, b = -3$, and $c = -5$.

For Exercises 23 and 24, let x be a number. Translate the English phrase into a mathematical expression. Then evaluate the expression for x = −4.

23. The number minus the quotient of −12 and the number

24. 7 more than the product of −2 and the number

25. A stock was worth $42 last year. Let v be the value (in dollars) of the stock today. The expression

$$\frac{100(v - 42)}{42}$$

represents the percent growth of the investment. Evaluate the expression for $v = 45$. Round your result to the second decimal place. What does your result mean in this situation?

26. The enrollment at DeVry University was 67 thousand students in 2013 and has decreased by about 11.3 thousand students per year since then (Source: *DeVry Education Group*).

a. Complete Table 52 to help find an expression that stands for DeVry University's enrollment (in thousands of students) at t years since 2013. Show the arithmetic to help you see a pattern.

Table 52 DeVry University's Enrollment

Years since 2013	Enrollment (thousands of students)
0	
1	
2	
3	
4	
t	

b. Evaluate the expression that you found in part (a) for $t = 5$. What does your result mean in this situation?

3

Using Slope to Graph Linear Equations

Do you use social networking sites such as Facebook? The percentage of adult Internet users who use social networking sites increased to 76% in 2015 (see Table 1). In Exercise 27 of Homework 3.5, you will describe how quickly the percentage has increased over time.

In Chapter 1, we used a line to describe the relationship between two quantities that are linearly related. In this chapter, we will discuss how to describe such a relationship by a symbolic statement called an *equation*. We will discuss how to use an equation to sketch a line. We will also describe the steepness of a line and how that steepness is related to how quickly one quantity changes in relation to another, such as how quickly banks' average overdraft fee has increased over time.

Table 1 Percentages of Adult Internet Users Who Use Social Networking Sites

Year	Percent
2010	60
2011	65
2012	67
2013	73
2014	74
2015	76

Source: *Pew Research Center*

3.1 Graphing Equations of the Forms $y = mx + b$ and $x = a$

Objectives

» For an equation in two variables, describe the meaning of *solution, satisfy,* and *solution set.*
» Describe the meaning of the *graph* of an equation.
» Graph equations of the form $y = mx + b$.
» Describe the meaning of b in equations of the form $y = mx + b$.
» Graph equations of the form $y = b$ and $x = a$.
» Describe the Rule of Four for solutions of equations.
» Compare finding outputs using an equation with using a graph.

In this section, we will work with equations. An **equation** consists of an equality sign "=" with expressions on both sides. Here are some examples of equations:

$$y = 3x - 5, \qquad 2x - 4y = 8, \qquad x = 5$$

In Sections 1.3 and 1.4, we used lines to model data. It turns out we can describe any line by an equation.

Solutions, Satisfying Equations, and Solution Sets

Consider the equation $y = x + 4$. Let's find y when $x = 3$:

$$y = x + 4 \qquad \textit{Original equation}$$
$$y = 3 + 4 \qquad \textit{Substitute 3 for x.}$$
$$ = 7 \qquad \textit{Add.}$$

So, $y = 7$ when $x = 3$. Recall from Section 1.2 that the ordered-pair notation $(3, 7)$ is shorthand for saying that when $x = 3$, $y = 7$.

For the equation $y = x + 4$, we found that $y = 7$ when $x = 3$. This means the equation $y = x + 4$ becomes a true statement when we substitute 3 for x and 7 for y:

$$y = x + 4 \quad \textit{Original equation}$$
$$7 \overset{?}{=} 3 + 4 \quad \textit{Substitute 3 for x and 7 for y.}$$
$$7 \overset{?}{=} 7 \quad \textit{Add.}$$
$$\text{true}$$

We say that $(3, 7)$ is a *solution* of the equation $y = x + 4$ and that $(3, 7)$ *satisfies* the equation $y = x + 4$.

A *set* is a container. Much as an egg carton contains eggs, a *solution set* contains solutions.

> **Definition**   *Solution*, *satisfy*, and *solution set* of an equation in two variables
>
> An ordered pair (a, b) is a **solution** of an equation in terms of x and y if the equation becomes a true statement when a is substituted for x and b is substituted for y. We say (a, b) **satisfies** the equation. The **solution set** of an equation is the set of all solutions of the equation.

Example 1   Identifying Solutions of an Equation

1. Is $(2, 1)$ a solution of $y = 3x - 5$?
2. Is $(4, 9)$ a solution of $y = 3x - 5$?

Solution

1. We substitute 2 for x and 1 for y in the equation $y = 3x - 5$:

$$y = 3x - 5 \quad \textit{Original equation}$$
$$1 \overset{?}{=} 3(2) - 5 \quad \textit{Substitute 2 for x and 1 for y.}$$
$$1 \overset{?}{=} 6 - 5 \quad \textit{Multiply before subtracting.}$$
$$1 \overset{?}{=} 1 \quad \textit{Subtract.}$$
$$\text{true}$$

So, $(2, 1)$ is a solution of $y = 3x - 5$.

2. We substitute 4 for x and 9 for y in the equation $y = 3x - 5$:

$$y = 3x - 5 \quad \textit{Original equation}$$
$$9 \overset{?}{=} 3(4) - 5 \quad \textit{Substitute 4 for x and 9 for y.}$$
$$9 \overset{?}{=} 12 - 5 \quad \textit{Multiply before subtracting.}$$
$$9 \overset{?}{=} 7 \quad \textit{Subtract.}$$
$$\text{false}$$

So, $(4, 9)$ is *not* a solution of $y = 3x - 5$.

Definition of Graph

Next we will learn how to *graph* an equation. As a first step, we will plot some solutions of an equation in Example 2.

Example 2   Plotting Some Solutions of an Equation

Find five solutions of $y = 2x - 1$, and plot them in the same coordinate system.

Solution

To find solutions, we are free to choose *any* values we'd like to substitute for x, but it's a good idea to pick integers close to or equal to 0 so the solutions are easy to plot. For example, here we substitute 0, 1, and 2 for x:

$$y = 2(0) - 1 \qquad y = 2(1) - 1 \qquad y = 2(2) - 1$$
$$= 0 - 1 \qquad\qquad = 2 - 1 \qquad\qquad = 4 - 1$$
$$= -1 \qquad\qquad = 1 \qquad\qquad = 3$$

Solution: $(0, -1)$ Solution: $(1, 1)$ Solution: $(2, 3)$

Next we substitute -2 and -1 for x:

$$y = 2(-2) - 1 \qquad y = 2(-1) - 1$$
$$y = -4 - 1 \qquad\qquad y = -2 - 1$$
$$y = -5 \qquad\qquad\quad y = -3$$

Solution: $(-2, -5)$ Solution: $(-1, -3)$

We organize our findings in Table 2. In Fig. 1, we plot the five solutions.

Table 2 Solutions of $y = 2x - 1$

x	y
-2	$2(-2) - 1 = -5$
-1	$2(-1) - 1 = -3$
0	$2(0) - 1 = -1$
1	$2(1) - 1 = 1$
2	$2(2) - 1 = 3$

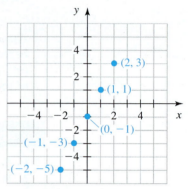

Figure 1 Five solutions of $y = 2x - 1$

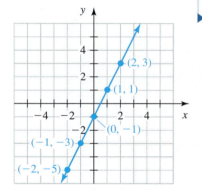

Figure 2 The line contains solutions of $y = 2x - 1$ found in Example 2

In Example 2, we plotted five solutions of $y = 2x - 1$. Note that a line contains these five points (see Fig. 2). It turns out *every* point on the line represents a solution of the equation $y = 2x - 1$. For example, in Fig. 3, we see that the point $(3, 5)$ lies on the line, and we can show that the ordered pair $(3, 5)$ does satisfy the equation $y = 2x - 1$:

$$y = 2x - 1 \qquad \textit{Original equation}$$
$$5 \stackrel{?}{=} 2(3) - 1 \qquad \textit{Substitute 3 for x and 5 for y.}$$
$$5 \stackrel{?}{=} 6 - 1 \qquad \textit{Multiply before subtracting.}$$
$$5 \stackrel{?}{=} 5 \qquad \textit{Add.}$$
$$\text{true}$$

It also turns out points that do not lie on the line represent ordered pairs that do *not* satisfy the equation. For example, by Fig. 3, we see that the point $(4, 2)$ does not lie on the line, and we can show that the ordered pair $(4, 2)$ does not satisfy the equation $y = 2x - 1$:

$$y = 2x - 1 \qquad \textit{Original equation}$$
$$2 \stackrel{?}{=} 2(4) - 1 \qquad \textit{Substitute 4 for x and 2 for y.}$$
$$2 \stackrel{?}{=} 8 - 1 \qquad \textit{Multiply before subtracting.}$$
$$2 \stackrel{?}{=} 7 \qquad \textit{Subtract.}$$
$$\text{false}$$

We call the line in Fig. 3 the *graph* of the equation $y = 2x - 1$.

Figure 3 $(3, 5)$ lies on the line, but $(4, 2)$ does not lie on the line

Definition   Graph

The **graph** of an equation in two variables is the set of points that correspond to all solutions of the equation.

The graph of an equation in two variables is a visual description of the solutions of the equation. Every point on the graph represents a solution of the equation. Every point *not* on the graph represents an ordered pair that is *not* a solution.

Graphs of Equations of the Form $y = mx + b$

Directly after Example 2, we found that the graph of the equation $y = 2x - 1$ is a line. The equation $y = 2x - 1$, or $y = 2x + (-1)$, is of the form $y = mx + b$, where $m = 2$ and $b = -1$. It turns out for any equation of the form $y = mx + b$, where m and b are constants, the graph is a line.

Graph of $y = mx + b$

The graph of an equation of the form $y = mx + b$, where m and b are constants, is a line.

Here are some equations whose graphs are lines:

$$y = 3x + 7, \qquad y = -4x - 5, \qquad y = -4x, \qquad y = x + 3, \qquad y = 2$$

The equation $y = -4x$ is of the form $y = mx + b$ because we can write it as $y = -4x + 0$. The equation $y = 2$ is of the form $y = mx + b$ because we can write it as $y = 0x + 2$.

Table 3 Solutions of $y = -2x + 3$

x	y
0	$-2(0) + 3 = 3$
1	$-2(1) + 3 = 1$
2	$-2(2) + 3 = -1$

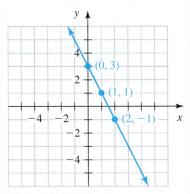

Figure 4 Graph of $y = -2x + 3$

Example 3   Graphing an Equation

Sketch the graph of $y = -2x + 3$. Also, find the y-intercept.

Solution

Because $y = -2x + 3$ is of the form $y = mx + b$, the graph is a line. Although we can sketch a line from as few as two points, we plot a third point as a check. If the third point is not in line with the other two, then we know that we have computed or plotted at least one of the solutions incorrectly.

To begin, we calculate three solutions of $y = -2x + 3$ in Table 3. We use 0, 1, and 2 as values of x because they correspond to points that are easy to plot. Then we plot the three points and sketch the line through them (see Fig. 4).

We use ZStandard followed by ZSquare to verify our graph (see Fig. 5). See Appendix A.3, A.4, and A.6 for graphing calculator instructions.

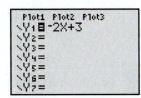 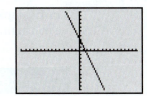

Figure 5 Graph of $y = -2x + 3$

From Table 3 and Fig. 4, we see that the y-intercept is $(0, 3)$.

Table 4 Solutions of $y = \dfrac{3}{2}x - 2$

x	y
0	$\dfrac{3}{2}(0) - 2 = -2$
2	$\dfrac{3}{2}(2) - 2 = 1$
4	$\dfrac{3}{2}(4) - 2 = 4$

Example 4   Graphing an Equation

Sketch the graph of $y = \dfrac{3}{2}x - 2$. Also, find the y-intercept.

Solution

In Table 4, we use 0 and multiples of 2 as values of x to avoid fractional values of y. Because

$$y = \frac{3}{2}x - 2$$

is of the form $y = mx + b$, the graph is a line. We plot the points that correspond to the solutions we found and sketch the line that contains them in Fig. 6.

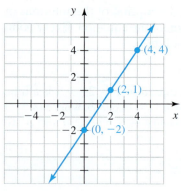

Figure 6 Graph of $y = \dfrac{3}{2}x - 2$

We use ZDecimal to verify our graph (see Fig. 7). See Appendix A.3, A.4, and A.6 for graphing calculator instructions.

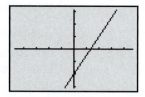

Figure 7 Graph of $y = \dfrac{3}{2}x - 2$

From Table 4 and Fig. 6, we see that the y-intercept is $(0, -2)$.

The Meaning of b for an Equation of the Form $y = mx + b$

Recall from Section 1.3 that the x-coordinate of a y-intercept is 0. For a line with an equation of the form $y = mx + b$, the y-intercept is $(0, b)$. For instance, in Example 3, we found that the line $y = -2x + 3$ has y-intercept $(0, 3)$. In Example 4, we found that the line $y = \dfrac{3}{2}x - 2$ has y-intercept $(0, -2)$.

Now consider any equation of the form $y = mx + b$. Substituting 0 for x gives

$$y = m(0) + b = 0 + b = b$$

which shows that the y-intercept is $(0, b)$.

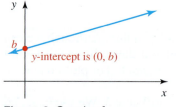

Figure 8 Graph of $y = mx + b$ has y-intercept $(0, b)$

---

y-Intercept of the Graph of $y = mx + b$
The graph of an equation of the form $y = mx + b$ has y-intercept $(0, b)$. See Fig. 8.

For the line $y = -5x + 9$, the y-intercept is $(0, 9)$, and for the line $y = 8x - 4$, the y-intercept is $(0, -4)$.

Graphing Equations of the Form $y = b$ and $x = a$

In Example 5, we will explore how to graph an equation of the form $y = b$, where b is a constant.

Table 5 Solutions of $y = 3$

x	y
-2	3
-1	3
0	3
1	3
2	3

Example 5   Graphing an Equation of the Form $y = b$

Sketch the graph of $y = 3$.

Solution

Note that y must be 3, but x can have any value. Some solutions of $y = 3$ are listed in Table 5. We plot the corresponding points and sketch the line through them (see Fig. 9). The graph of $y = 3$ is a horizontal line.

We can use a graphing calculator to verify our graph (see Fig. 10).

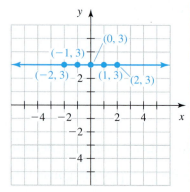

Figure 9 Graph of $y = 3$

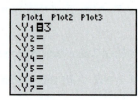
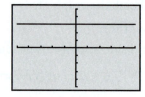

Figure 10 The graph of $y = 3$

In Example 5, we saw that the graph of the equation $y = 3$ is a horizontal line. Any equation that can be written in the form $y = b$, where b is a constant, has a horizontal line as its graph.

Table 6 Solutions of $x = 2$

x	y
2	-2
2	-1
2	0
2	1
2	2

Example 6   Graphing an Equation of the Form $x = a$

Sketch the graph of $x = 2$.

Solution

Note that x must be 2, but y can have any value. Some solutions of $x = 2$ are listed in Table 6. We plot the corresponding points and sketch the line through them (see Fig. 11). The graph of $x = 2$ is a vertical line.

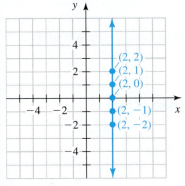

Figure 11 Graph of $x = 2$

In Example 6, we saw that the graph of the equation $x = 2$ is a vertical line. Any equation that can be written in the form $x = a$, where a is a constant, has a vertical line as its graph.

Equations for Horizontal and Vertical Lines

If a and b are constants, then

- The graph of $y = b$ is a horizontal line (see Fig. 12).
- The graph of $x = a$ is a vertical line (see Fig. 13).

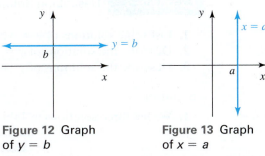

Figure 12 Graph of $y = b$

Figure 13 Graph of $x = a$

For example, the graphs of the equations $y = 5$ and $y = -2$ are horizontal lines. The graphs of the equations $x = 6$ and $x = -4$ are vertical lines.

In Example 6, we graphed the equation $x = 2$, where we assumed we were describing the relationship between the *two* variables, x and y (see Fig. 11). If there were just *one* variable, x, then we would graph the equation $x = 2$ using a number line to plot a single dot at 2 (see Fig. 14).

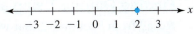

Figure 14 Graph of $x = 2$ when there is just one variable

When graphing equations throughout the rest of the course, if it is not clear whether there are one or two variables, you should assume there are two variables.

Equations Whose Graphs Are Lines

If an equation can be put into the form

$$y = mx + b \quad \text{or} \quad x = a$$

where m, a, and b are constants, then the graph of the equation is a line. We call such an equation a **linear equation in two variables**.

Any equation that can be put into the form $x = a$ has a vertical line as its graph. Any equation that can be put into the form $y = mx + b$ has a nonvertical line as its graph.

Rule of Four for Equations

We can describe the solutions of an equation in two variables in four ways. For instance, in Example 4, we described the solutions of the equation $y = \dfrac{3}{2}x - 2$ by using the equation and a graph (see Fig. 6 on page 124). We also described some of the solutions by using a table (see Table 4 on page 123). Finally, we can describe the solutions verbally: For each solution, the y-coordinate is three-halves of the x-coordinate minus 2.

Rule of Four for Solutions of an Equation

We can describe some of or all the solutions of an equation in two variables with:

1. an equation,   **2.** a table,

3. a graph, or   **4.** words.

These four ways to describe solutions are known as the **Rule of Four**.

Example 7   Describing Solutions by Using the Rule of Four

1. List some solutions of $y = 4x$ by using a table.
2. Describe the solutions of $y = 4x$ by using a graph.
3. Describe the solutions of $y = 4x$ by using words.

Solution

1. We list three solutions in Table 7.

Table 7 Solutions of $y = 4x$

x	y
-1	$4(-1) = -4$
0	$4(0) = 0$
1	$4(1) = 4$

2. We plot the solutions listed in Table 7 and sketch the line through them (see Fig. 15).
3. For each solution, the y-coordinate is four times the x-coordinate.

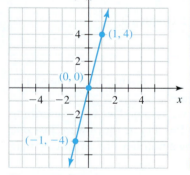

Figure 15 Graph of $y = 4x$

Finding Outputs Using an Equation versus Using a Graph

Recall from Section 1.2 that when we are not describing an authentic situation, we use x as the *explanatory variable* and y as the *response variable*. Recall from Section 1.3 that an *input* is a permitted value of the explanatory variable that leads to at least one *output*, which is a permitted value of the response variable.

In Chapter 1, we used a sketched line to see how an input leads to an output. It will often be easier and more efficient to use an *equation* of a line to perform such a task.

For example, in Fig. 2 on page 122, we graphed the equation $y = 2x - 1$. From the blue arrows in Fig. 16, we see that the input $x = 3$ leads to the output $y = 5$.

To use the equation, we substitute 3 for x in $y = 2x - 1$:

$$y = 2(3) - 1 = 5$$

Figure 16 The input $x = 3$ leads to the output $y = 5$

This work also shows that the input $x = 3$ leads to the output $y = 5$.

Group Exploration

Solutions of an equation

Consider the equation $y = x - 3$.

1. Use the coordinate system in Fig. 17 to sketch a graph of $y = x - 3$.

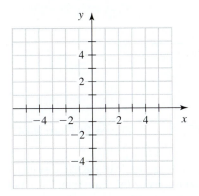

Figure 17 Problem 1

2. Pick three points that lie on the graph of $y = x - 3$. Do the coordinates of these points satisfy the equation $y = x - 3$?

3. Pick three points that do not lie on the graph of $y = x - 3$. Do the coordinates of these points satisfy the equation $y = x - 3$?

4. Which ordered pairs satisfy the equation $y = x - 3$? There are too many to list, but describe them in words. [**Hint:** You should say something about the points that do or do not lie on the line.]

5. The graph of an equation is sketched in Fig. 18. Which of the points A, B, C, D, E, and F represent ordered pairs that satisfy the equation?

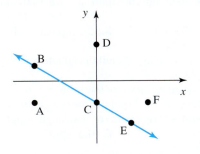

Figure 18 Problem 5

Tips for Success   Use the Graphing Calculator Appendix

Remember that Appendix A contains step-by-step graphing calculator instructions for many tools that are helpful for this course. Appendix A includes Appendix A.23, which describes how to respond to various error messages.

 # Homework 3.1

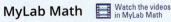

For extra help ▶ **MyLab Math**   🎬 Watch the videos in MyLab Math

Which of the given ordered pairs satisfy the given equation?

1. $y = 2x - 4$ $(-3, -10), (1, -3), (2, 0)$

2. $y = 4x - 12$ $(-2, -20), (1, -8), (3, 0)$

3. $y = -3x + 7$ $(-1, 4), (0, 7), (4, -5)$

4. $y = -5x + 8$ $(-2, 3), (0, 8), (3, -7)$

Find the y-intercept. Also, graph the equation by hand. Use a graphing calculator to verify your work.

5. $y = x + 2$
6. $y = x + 4$
7. $y = x - 4$
8. $y = x - 6$
9. $y = 2x$
10. $y = 5x$
11. $y = -3x$
12. $y = -4x$
13. $y = x$
14. $y = -x$
15. $y = \frac{1}{3}x$
16. $y = \frac{1}{2}x$

17. $y = -\frac{5}{3}x$
18. $y = -\frac{3}{2}x$
19. $y = 2x + 1$
20. $y = 3x + 2$
21. $y = 5x - 3$
22. $y = 4x - 1$
23. $y = -3x + 5$
24. $y = -2x + 4$
25. $y = -2x - 3$
26. $y = -4x - 2$
27. $y = \frac{1}{2}x - 3$
28. $y = \frac{1}{3}x - 2$
29. $y = -\frac{2}{3}x + 1$
30. $y = -\frac{4}{3}x + 5$

Graph the equation by hand.

31. $x = 3$
32. $x = 6$
33. $y = 1$
34. $y = 5$
35. $y = -2$
36. $y = -4$
37. $x = -1$
38. $x = -3$
39. $x = 0$
40. $y = 0$

Concepts

41. Recall that we can describe some of or all the solutions of an equation in two variables with an equation, a table, a graph, or words.
  a. Describe three solutions of $y = 2x - 3$ by using a table.
  b. Describe the solutions of $y = 2x - 3$ by using a graph.
  c. Describe the solutions of $y = 2x - 3$ by using words.

42. Recall that we can describe some of or all the solutions of an equation in two variables with an equation, a table, a graph, or words.
  a. Describe three solutions of $y = -4x + 5$ by using a table.
  b. Describe the solutions of $y = -4x + 5$ by using a graph.
  c. Describe the solutions of $y = -4x + 5$ by using words.

43. a. For the equation $y = 3x + 1$, find all outputs for the given input. State how many outputs there are for that single input.
  i. the input $x = 2$ **ii.** the input $x = 4$
  iii. the input $x = -2$
  b. For $y = 3x + 1$, how many outputs originate from any single input? Explain.
  c. Give an example of an equation of the form $y = mx + b$. Using your equation, find all outputs for the given input. State how many outputs there are for that single input.
  i. the input $x = 3$ **ii.** the input $x = 5$
  iii. the input $x = -3$
  d. For your equation, how many outputs originate from any single input? Explain.
  e. For *any* equation of the form $y = mx + b$, how many outputs originate from a single input? Explain.

44. a. Graph $y = 2x - 4$ by hand.
  b. For the equation $y = 2x - 4$, find all outputs for the given input. Explain by using arrows on your graph in part (a). State how many outputs there are for that single input.
  i. the input $x = 3$ **ii.** the input $x = 4$
  iii. the input $x = 5$
  c. For $y = 2x - 4$, how many outputs originate from any single input? Explain in terms of drawing arrows.
  d. Give an example of an equation of the form $y = mx + b$. Graph your equation by hand.
  e. Using your equation, find all outputs for the given input. Explain by using arrows on your graph in part (d). State how many outputs there are for that single input.
  i. the input $x = 1$ **ii.** the input $x = 3$
  iii. the input $x = -2$
  f. For your equation, how many outputs originate from any single input? Explain in terms of drawing arrows.
  g. For *any* equation of the form $y = mx + b$, how many outputs originate from a single input? Explain in terms of drawing arrows.

45. a. Graph the equation by hand. Find all x-intercepts and y-intercepts.
  i. $y = 3x$ **ii.** $y = -2x$ **iii.** $y = \dfrac{2}{5}x$
  b. What are the intercepts of the graph of an equation of the form $y = mx$, where $m \neq 0$?

46. Find the intersection point of the lines $y = 4x$ and $y = -5x$. Try to do this without graphing.

47. The graph of an equation is sketched in Fig. 19. Construct a table of ordered-pair solutions of this equation. Include at least five ordered pairs.

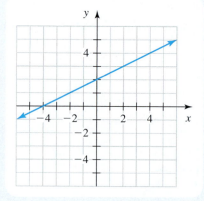

Figure 19 Exercise 47

48. The graph of an equation is sketched in Fig. 20. Construct a table of ordered-pair solutions of this equation. Your table should contain at least five ordered pairs.

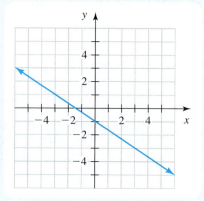

Figure 20 Exercise 48

For Exercises 49–56, refer to the graph sketched in Fig. 21.

49. Find y when $x = -4$.   **50.** Find y when $x = 0$.
51. Find y when $x = 2$.   **52.** Find y when $x = -2$.
53. Find x when $y = -1$.   **54.** Find x when $y = 0$.
55. Find x when $y = 2$.   **56.** Find x when $y = 3$.

Figure 21
Exercises 49–56

57. The graph of an equation is sketched in Fig. 22. Which of the points A, B, C, D, E, and F represent ordered pairs that satisfy the equation?

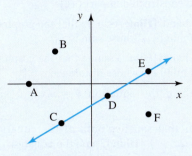

Figure 22 Exercise 57

58. The graphs of equations 1 and 2 are sketched in Fig. 23.

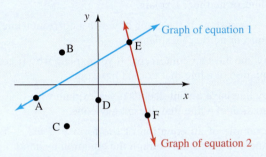

Figure 23 Exercise 58

For each part, decide which one or more of the points A, B, C, D, E, and F represent ordered pairs that
a. satisfy equation 1.　　**b.** satisfy equation 2.
c. satisfy both equations.　**d.** do not satisfy either equation.

59. Find a solution of $y = x + 2$ that lies in Quadrant II. How many solutions of this equation are in Quadrant II?

60. Find a solution of $y = x - 4$ that lies in Quadrant III. How many solutions of this equation are in Quadrant III?

61. Find an equation of a line that contains the points listed in Table 8. [**Hint:** For each point, what number can be added to the x-coordinate to get the y-coordinate?]

Table 8 Points on a Line
(Exercise 61)

x	y
0	3
1	4
2	5
3	6
4	7

62. Find an equation of a line that contains the points listed in Table 9. [**Hint:** For each point, what number can be subtracted from the x-coordinate to get the y-coordinate?]

Table 9 Points on a Line
(Exercise 62)

x	y
0	−1
1	0
2	1
3	2
4	3

63. Find an equation of a line that contains the points listed in Table 10.

Table 10 Points on a Line
(Exercise 63)

x	y
0	0
1	1
2	2
3	3
4	4

64. Find an equation of a line that contains the points listed in Table 11.

Table 11 Points on a Line
(Exercise 64)

x	y
0	0
1	−1
2	−2
3	−3
4	−4

65. Find an equation of the line sketched in Fig. 24.

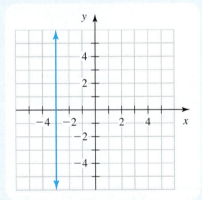

Figure 24 Exercise 65

66. Find an equation of the line sketched in Fig. 25.

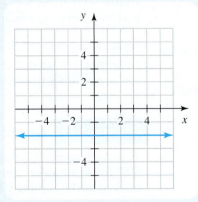

Figure 25 Exercise 66

67. The graph of an equation is sketched in Fig. 26.
a. Construct a table of ordered-pair solutions of this equation. Your table should contain at least five ordered pairs.
b. Find an equation of the line. [**Hint:** Recognize a pattern from the table you constructed in part (a).]

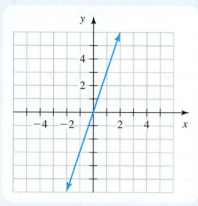

Figure 26 Exercise 67

68. The graph of an equation is sketched in Fig. 27.

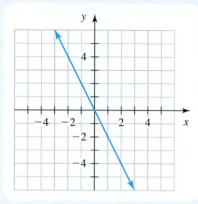

Figure 27 Exercise 68

a. Construct a table of ordered-pair solutions of this equation. Your table should contain at least five ordered pairs.
b. Find an equation of the line. [**Hint:** Recognize a pattern from the table you constructed in part (a).]

69. Graph $x + y = 5$ by hand. [**Hint:** Assume that the graph is a line. Think of pairs of numbers whose sum is 5.]

70. Graph $x - y = 5$ by hand. [**Hint:** Assume that the graph is a line. Think of pairs of numbers whose difference is 5.]

71. Why does the graph of the equation $y = mx + b$ have y-intercept $(0, b)$?

72. Find an ordered pair that satisfies both of the equations $y = x + 1$ and $y = -2x + 4$. [**Hint:** Graph the equations.]

73. Assume a is a constant. Is the graph of $x = a$ a horizontal line, a vertical line, or neither? Explain.

74. Assume b is a constant. Is the graph of $y = b$ a horizontal line, a vertical line, or neither? Explain.

75. Does every line have an x-intercept? If yes, explain why. If no, give an equation of a line that doesn't have one.

76. Does every line have a y-intercept? If yes, explain why. If no, give an equation of a line that doesn't have one.

77. Give an example of an equation of the form $y = mx + b$, where m and b are constants. Graph the equation by hand.

78. Give an example of an equation of the form $x = a$, where a is a constant. Graph the equation by hand.

79. Describe how to graph an equation of the form $y = mx + b$. (See page 9 for guidelines on writing a good response.)

80. In your own words, describe the Rule of Four for equations. (See page 9 for guidelines on writing a good response.)

3.2 Graphing Linear Models; Unit Analysis

Objectives

» Graph a linear model.

» Describe the Rule of Four for linear models.

» Perform a unit analysis of a linear model.

In this section, we will graph equations that describe authentic situations and use such graphs to make predictions. We will find the units of both sides of such equations.

Graphing a Linear Model

In Section 1.3, we defined a linear model as a nonvertical line that describes two quantities in an authentic situation. An equation of such a line is also called a *linear model*.

Example 1   Using a Linear Model to Make Predictions

A person earns a starting salary of $32 thousand at a company. Each year, he receives a $2 thousand raise. Let s be the person's salary (in thousands of dollars) after he has worked at the company for t years.

1. Use a table to help find an equation for t and s.
2. Substitute 8 for t in the equation. What does the result mean in this situation?
3. Graph the equation.

Table 12 Years at Company and Salaries

Years at Company t	Salary (thousands of dollars) s
0	$2 \cdot 0 + 32$
1	$2 \cdot 1 + 32$
2	$2 \cdot 2 + 32$
3	$2 \cdot 3 + 32$
4	$2 \cdot 4 + 32$
t	$2 \cdot t + 32$

Table 13 Years at Company and Salaries

t	s
0	$2(0) + 32 = 32$
1	$2(1) + 32 = 34$
2	$2(2) + 32 = 36$
3	$2(3) + 32 = 38$
4	$2(4) + 32 = 40$

4. What is the s-intercept? What does it mean in this situation?
5. When will the salary be $42 thousand?

Solution

1. We construct Table 12. From the last row of the table, we see that the salary s (in thousands of dollars) can be represented by $2t + 32$. So, $s = 2t + 32$.
2. We substitute 8 for t in the equation $s = 2t + 32$:

$$s = 2(8) + 32 = 48$$

So, the person's salary is $48 thousand after he has worked at the company for 8 years.

3. In Table 13, we substitute values for t in the equation $s = 2t + 32$ to find the corresponding values for s. Then, we plot the points and sketch a line that contains the points (see Fig. 28).
4. The model $s = 2t + 32$ is of the form $s = mt + b$, where $b = 32$. So, the s-intercept is $(0, 32)$. We can also find the s-intercept from Table 13 and Fig. 28. The s-intercept being $(0, 32)$ means that the starting salary is $32 thousand.
5. The red arrows in Fig. 29 show that the output $s = 42$ originates from the input $t = 5$. So, the person's salary will be $42 thousand after he has worked at the company for 5 years.

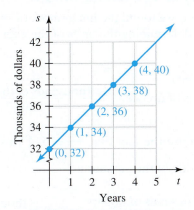

Figure 28 Salary model

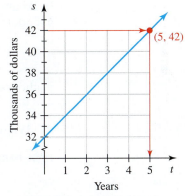

Figure 29 Using the salary model to make a prediction

We use a graphing calculator to verify our work (see Fig. 30). See Appendix A.4, A.5, and A.7 for graphing calculator instructions.

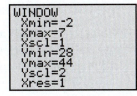
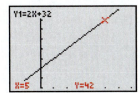

Figure 30 Verify the work

We can add two numbers in either order and get the same result. For example, $2 + 5 = 7$ and $5 + 2 = 7$. We can also multiply two numbers in either order and get the same result. For example, $3 \cdot 8 = 24$ and $8 \cdot 3 = 24$. Depending on how we had arranged the arithmetic in Table 12, we could have found any one of the following equations:

$$s = 2t + 32, \qquad s = 32 + 2t, \qquad s = t(2) + 32, \qquad s = 32 + t(2)$$

All these equations describe the same relationship between t and s and have the same graph.

Rule of Four for Linear Models

Just as we can use equations, tables, graphs, and words to describe solutions of equations in two variables, we can use these four ways to describe linear models. For instance, in Example 1, we first described a linear model by using words:

> A person earns a starting salary of $32 thousand at a company. Each year, he receives a $2 thousand raise.

We also described the model by using the equation $s = 2t + 32$, a table (see Table 13), and a graph (see Fig. 28).

Rule of Four for Linear Models

We can describe a linear model with

1. an equation,   **2.** a table,

3. a graph, or   **4.** words.

Unit Analysis of a Linear Model

In Example 1, we found the linear model $s = 2t + 32$. We can perform a *unit analysis* of a model by determining the units of the expressions on both sides of the equation:

$$\underset{\text{thousands of dollars}}{s} \quad = \quad \underset{\dfrac{\text{thousands of dollars}}{\text{year}}}{2} \quad \cdot \quad \underset{\text{years}}{t} \quad + \quad \underset{\text{thousands of dollars}}{32}$$

We can use the fact that $\dfrac{\text{years}}{\text{year}} = 1$ to simplify the units of the expression on the right-hand side of the equation:

$$\frac{\text{thousands of dollars}}{\text{year}} \cdot \text{years} + \frac{\text{thousands}}{\text{of dollars}} = \frac{\text{thousands}}{\text{of dollars}} + \frac{\text{thousands}}{\text{of dollars}}$$

So, the units of the expressions on both sides of the equation are thousands of dollars, which suggests that our equation is correct.

Definition   Unit analysis

We perform a **unit analysis** of a model's equation by determining the units of the expressions on both sides of the equation. The units of the expressions on both sides of the equation should be the same.

"I exercised, drank plenty of fluids, and avoided eating anything that even remotely had flavor."

"How'd you lose the weight?"

We can perform a unit analysis of a model's equation to help verify the equation.

Example 2   Using a Linear Model to Make a Prediction

Before starting a diet, a person weighs 160 pounds. While on the diet, she loses 3 pounds per month. Let w be the person's weight (in pounds) after she has been on the diet for t months.

1. Use a table to help find an equation for t and w.
2. Perform a unit analysis of the equation you found in Problem 1.
3. Predict the person's weight after she has been on the diet for 6 months.

Table 14 Numbers of Months and Weights

Time on Diet (months)	Weight (pounds)
t	w
0	$160 - 3 \cdot 0$
1	$160 - 3 \cdot 1$
2	$160 - 3 \cdot 2$
3	$160 - 3 \cdot 3$
4	$160 - 3 \cdot 4$
t	$160 - 3 \cdot t$

Solution

1. We construct Table 14. From the bottom row of the table, we see that we can model the situation by using the equation $w = 160 - 3t$.
2. Here is a unit analysis of the equation $w = 160 - 3t$:

$$\underbrace{w}_{\text{pounds}} = \underbrace{160}_{\text{pounds}} - \underbrace{3}_{\frac{\text{pounds}}{\text{month}}} \cdot \underbrace{t}_{\text{months}}$$

We can use the fact that $\dfrac{\text{months}}{\text{month}} = 1$ to simplify the units of the expression on the right-hand side of the equation:

$$\text{pounds} - \frac{\text{pounds}}{\text{month}} \cdot \text{months} = \text{pounds} - \text{pounds}$$

So, the units on both sides of the equation are pounds, which suggests that our equation is correct.

3. We substitute 6 for t in the equation $w = 160 - 3t$:

$$w = 160 - 3(6) = 142$$

The person will weigh 142 pounds after she has been on the diet for 6 months.

▶

 Group Exploration ───────────────────────────────

Section Opener: Finding an equation of a linear model

The value of a stock was \$5 in 2015. Each year the value increases by \$2. Let v be the value (in dollars) at t years since 2015.

1. Complete Table 15 to help find an equation for t and v. Show the arithmetic to help you see a pattern.

2. Graph the equation you found in Problem 1 by hand.

3. Use the equation you found in Problem 1 to predict the stock's value in 2021. Verify your result by using arrows on the graph you sketched in Problem 2.

4. Use the graph you sketched in Problem 2 to predict when the stock's value will be \$15. Explain by using arrows on your graph.

Table 15 Years and Values of a Stock

Years since 2015	Value (dollars)
t	v
0	
1	
2	
3	
4	
t	

Tips for Success   Study with a Classmate

It can be helpful to meet with a classmate and discuss what happened in class that day. Not only can you ask questions *of* each other, but you will learn just as much by explaining concepts *to* each other. Explaining a concept to someone else forces you to clarify your own understanding of the concept.

Homework 3.2

For extra help ▶   **MyLab Math**   📺 Watch the videos in MyLab Math

1. A person pays a $3 cover charge to hear the band Little Muddy. Let T be the total cost (in dollars) of the cover charge plus d dollars spent on drinks.

a. Complete Table 16 to help find an equation that describes the relationship between d and T. Show the arithmetic to help you see a pattern.

Table 16 Drink Cost and Total Cost

Drink Cost (dollars) d	Total Cost (dollars) T
2	
3	
4	
5	
d	

b. Perform a unit analysis of the equation you found in part (a).
c. Graph the equation by hand.
d. What is the T-intercept of the linear model? What does it mean in this situation?
e. If $10 is spent on drinks, find the total cost of the cover charge and drinks. Explain by using arrows on the graph you sketched in part (c).

2. A company offers a $2 mail-in rebate on a shaver. The *retail price* of a shaver is the price paid at the store (not including the $2 rebate). The *net price* is the price of the shaver, taking into account the money saved by the rebate. Let n be the net price (in dollars) of a shaver whose retail price is r dollars.

a. Complete Table 17 to help find an equation that describes the relationship between r and n. Show the arithmetic to help you see a pattern.

Table 17 Retail Price and Net Price

Retail Price (dollars) r	Net Price (dollars) n
4	
5	
6	
7	
r	

b. Perform a unit analysis of the equation you found in part (a).
c. Graph the equation by hand.
d. If the net price is $6, what is the retail price? Explain by using arrows on the graph you sketched in part (c).

3. Chemeketa Community College charged $94 per credit (unit or hour) for tuition in spring term 2017. Let T be the total cost (in dollars) of tuition for enrolling in c credits of classes.

a. Complete Table 18 to help find an equation that describes the relationship between c and T. Show the arithmetic to help you see a pattern.

Table 18 Numbers of Credits and Total Costs

Number of Credits c	Total Cost (dollars) T
3	
6	
9	
12	
c	

b. Perform a unit analysis of the equation you found in part (a).
c. Graph the equation by hand.
d. What is the total cost of tuition for 15 credits of classes? Explain by using arrows on the graph you sketched in part (c).

4. The average gas mileage of a Toyota Camry Hybrid XLE® is 40 miles per gallon. Let g be the number of gallons used to drive d miles.

a. Complete Table 19 to help find an equation that describes the relationship between d and g. Show the arithmetic to help you see a pattern.

Table 19 Distances and Gasoline Consumed

Distance (miles) d	Gasoline Consumed (gallons) g
40	
80	
120	
160	
d	

b. Perform a unit analysis of the equation you found in part (a).
c. Graph the equation by hand.
d. How many gallons are used to travel 100 miles? Explain by using arrows on the graph you sketched in part (c).

5. A person earns a starting salary of $24 thousand at a company. Each year, she receives a $3 thousand raise. Let s be the salary (in thousands of dollars) after she has worked at the company for t years.

a. Complete Table 20 to help find an equation for t and s. Show the arithmetic to help you see a pattern.
b. Perform a unit analysis of the equation you found in part (a).
c. Graph the equation by hand.
d. What is the s-intercept? What does it mean in this situation?
e. When will the person's salary be $42 thousand? Explain by using arrows on the graph you sketched in part (c).

Table 20 Years at Company and Salaries

Time at Company (years) t	Salary (thousands of dollars) s
0	
1	
2	
3	
4	
t	

6. The U.S. revenue from products that carry a gluten-free label was $23 billion in 2014 and has increased by about $3 billion per year since then (Source: *Nielsen*). Let r be the annual revenue (in billions of dollars) at t years since 2014.

    a. Complete Table 21 to help find an equation for t and r. Show the arithmetic to help you see a pattern.

Table 21 Years and Revenues from Products That Carry a Gluten-Free Label

Years since 2014 t	Revenue (billions of dollars) r
0	
1	
2	
3	
4	
t	

    b. Perform a unit analysis of the equation you found in part (a).
    c. Graph the equation by hand.
    d. Predict when the annual revenue will be $41 billion. Explain by using arrows on the graph you sketched in part (c).

7. Sales of Nintendo handheld games were 80 million in 2014 and have decreased by about 12 million per year since then (Source: *Nintendo*). Let s be the annual sales (in millions of games) at t years since 2014.

    a. Complete Table 22 to help find an equation for t and s. Show the arithmetic to help you see a pattern.

Table 22 Years and Nintendo Handheld Game Sales

Years since 2014 t	Sales (millions of games) s
0	
1	
2	
3	
4	
t	

    b. Perform a unit analysis of the equation you found in part (a).
    c. Graph the equation by hand.
    d. What is the s-intercept? What does it mean in this situation?

    e. Predict when annual sales will be 20 million games. Explain by using arrows on the graph you sketched in part (c).

8. Revenue from Ivory bar soap was $94 million in 2013 and has decreased by about $3 million per year since then (Source: *Euromonitor*). Let r be the annual revenue (in millions of dollars) at t years since 2013.

    a. Complete Table 23 to help find an equation for t and r. Show the arithmetic to help you see a pattern.

Table 23 Years and Revenues from Ivory Bar Soap

Years since 2013 t	Revenue (millions of dollars) r
0	
1	
2	
3	
4	
t	

    b. Perform a unit analysis of the equation you found in part (a).
    c. Graph the equation by hand.
    d. What is the r-intercept? What does it mean in this situation?
    e. Predict when the annual revenue will be $73 million. Explain by using arrows on the graph you sketched in part (c).

9. To make fudgelike brownies, a person bakes a brownie mix for 5 minutes less than the baking time suggested on the box. Let r be the suggested baking time (in minutes) and a be the actual baking time (in minutes).

    a. Find an equation that describes the relationship between r and a. Assume that a is the response variable. [**Hint:** If you have trouble finding the equation, construct a table of values for r and a.]
    b. Perform a unit analysis of the equation you found in part (a).
    c. Graph the equation by hand.
    d. If the actual baking time is 23 minutes, what is the baking time suggested on the box? Explain by using arrows on the graph you sketched in part (c).

10. A person pays $4 for parking at an arts-and-crafts fair. Let T be the total cost (in dollars) of parking plus v dollars spent on a vase.

    a. Find an equation that describes the relationship between v and T. [**Hint:** If you have trouble finding the equation, construct a table of values for v and T.]
    b. Perform a unit analysis of the equation you found in part (a).
    c. Graph the equation by hand.
    d. If the person spends $25 on the vase, find the total cost of parking and the vase. Explain by using arrows on the graph you sketched in part (c).

11. A person drives at a constant speed of 60 miles per hour. Let d be the distance (in miles) traveled after the person has driven for t hours.

    a. Find an equation that describes the relationship between t and d. [**Hint:** If you have trouble finding the equation, construct a table of values for t and d.]
    b. Perform a unit analysis of the equation you found in part (a).
    c. Graph the equation by hand.
    d. What is the d-intercept? What does it mean in this situation?

e. After how many hours will the person have traveled 150 miles? Explain by using arrows on the graph you sketched in part (c).

12. The San Francisco Bike Hut charges $8 per hour to rent a road bike. Let C be the total cost (in dollars) of renting a road bike for t hours.
  a. Find an equation that describes the relationship between t and C. [**Hint:** If you have trouble finding the equation, construct a table of values for t and C.]
  b. Perform a unit analysis of the equation you found in part (a).
  c. Graph the equation by hand.
  d. If a person paid $20 for renting a road bike, how long did the person ride the bike? Explain by using arrows on the graph you sketched in part (c).

13. For spring semester 2017, Cosumnes River College charged $46 per unit (credit or hour) for tuition. All students paid a $1 student representation fee each semester. Students who drove to school paid a $35 parking fee each semester. Let T be the total one-semester cost (in dollars) of tuition and fees for a student who drove to school and took u units of classes.
  a. Find an equation for u and T. [**Hint:** If you have trouble finding the equation, construct a table of values for u and T.]
  b. Perform a unit analysis of the equation you found in part (a).
  c. What is the total one-semester cost of tuition and fees for a student who drove to school and took 15 units of classes?

14. For spring semester 2017, undergraduates taking 15 hours (units or credits) of courses at Southeastern Louisiana University paid $2888.60 for tuition. Students could rent textbooks for $45 per course. Let T be the total one-semester cost (in dollars) of tuition and renting textbooks for an undergraduate who took n courses for a total of 15 hours.
  a. Find an equation for n and T. [**Hint:** If you have trouble finding the equation, construct a table of values for n and T.]
  b. Perform a unit analysis of the equation you found in part (a).
  c. What was the total one-semester cost of tuition and renting textbooks for an undergraduate who took five 3-hour courses?

15. The pressure in a bike tire is 62 pounds per square inch (psi). A person uses a bike pump to add air to the tire. The pressure increases by about 3 psi with each pump.
  a. Find an equation of a linear model to describe the situation. Explain what your variables represent.
  b. Perform a unit analysis of the equation you found in part (a).
  c. What is the pressure after 17 pumps?

16. Just before some bad publicity was released about a company, the company's stock was worth $95. Now that the bad publicity has been released, the value of the stock has been declining by $4 per week.
  a. Find an equation of a linear model to describe the situation. Explain what your variables represent.
  b. Perform a unit analysis of the equation you found in part (a).
  c. What is the value of the stock 14 weeks after the bad publicity was released?

17. A person fills up his car's 11-gallon gasoline tank and then drives at a constant 60 miles per hour. For each hour of driving, the car uses 2 gallons of gasoline. Let g be the amount of gasoline (in gallons) in the tank after the person has driven for t hours.
  a. Find an equation for t and g.
  b. Perform a unit analysis of the equation you found in part (a).
  c. Graph the equation by hand.
  d. For how long can the person drive before refueling if he wants to refuel when 1 gallon of gasoline is left in the tank? Explain by using arrows on the graph you sketched in part (c).
  e. Estimate the amount of gasoline in the tank after 8 hours of driving. Explain by using arrows on the graph you sketched in part (c). [**Hint:** Remember, if you think that model breakdown occurs, say so, say where, and explain why.]

18. When a small airplane begins to descend, its altitude is 4 thousand feet. The airplane descends $\frac{1}{2}$ thousand feet each minute. Let h be the altitude (in thousands of feet) of the airplane t minutes after it has begun to descend.
  a. Find an equation for t and h.
  b. Perform a unit analysis of the equation you found in part (a).
  c. Graph the equation by hand.
  d. When will the airplane reach an altitude of 1 thousand feet? Explain by using arrows on the graph you sketched in part (c).
  e. Predict the airplane's altitude 11 minutes after the airplane has begun its descent. Explain by using arrows on the graph you sketched in part (c). [**Hint:** Remember, if you think that model breakdown occurs, say so, say where, and explain why.]

19. *Fair-trade coffee* guarantees farmers a minimum fair price for their crops. Sales of fair-trade coffee were 167.1 million pounds in 2014 and have increased by about 13.6 million pounds per year since then (Source: *Fair Trade USA*). Let s be the annual sales (in millions of pounds) of fair-trade coffee at t years since 2014. Recall that we can describe a linear model using an equation, a table, a graph, or words.
  a. Use a table of values of t and s to describe the situation.
  b. Use an equation to describe the situation.
  c. Use a graph to describe the situation.

20. The number of albums sold was 257 million in 2014 and has decreased by about 25 million albums per year since then (Source: *Nielsen Music*). Let n be the number of albums (in millions) sold in the year that is t years since 2014. Recall that we can describe a linear model using an equation, a table, a graph, or words.
  a. Use a table of values of t and n to describe the situation.
  b. Use an equation to describe the situation.
  c. Use a graph to describe the situation.

21. A person has had a constant 45 novels since 2016. Let n be the number of novels that the person owns at t years since 2016. Find an equation that models the number of novels owned by the person for 2016 and thereafter. Also, graph the equation by hand.

22. A small company has had a constant 35 employees since 2015. Let n be the number of employees at t years since 2015. Find an equation that models the number of employees for 2015 and thereafter. Also, graph the equation by hand.

23. Let n be the number (in thousands) of independent CD and record stores at t years since 2010. A reasonable linear model is shown in Fig. 31 (Source: *National Arts Index*).

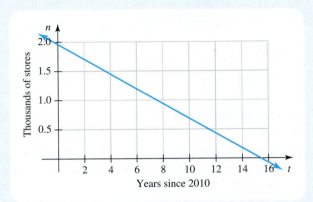

Figure 31 Exercise 23

a. Estimate when there were 1.5 thousand independent CD and record stores.
b. Predict the number of independent CD and record stores in 2022.
c. What is the n-intercept? What does it mean in this situation?
d. What is the t-intercept? What does it mean in this situation?

24. Let v be the total value (in billions of dollars) of unused gift cards at t years since 2010. A reasonable linear model is shown in Fig. 32 (Source: *CEB TowerGroup*).

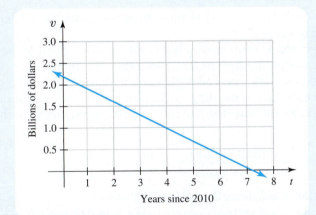

Figure 32 Exercise 24

a. Estimate in which year the total value of unused gift cards was $1 billion.
b. Estimate the total value of unused gift cards in 2016.
c. What is the t-intercept? What does it mean in this situation?
d. What is the v-intercept? What does it mean in this situation?

Concepts

25. The points $(1, 2)$ and $(5, 2)$ are plotted in Fig. 33.
a. In going from point $(1, 2)$ to point $(5, 2)$, find the change
   i. in the x-coordinate.   **ii.** in the y-coordinate.
b. In going from point $(5, 2)$ to point $(1, 2)$, find the change
   i. in the x-coordinate.   **ii.** in the y-coordinate.

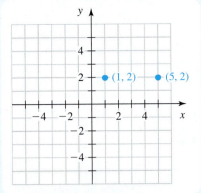

Figure 33 Exercise 25

26. The points $(3, -4)$ and $(3, 2)$ are plotted in Fig. 34.
a. In going from point $(3, -4)$ to point $(3, 2)$, find the change
   i. in the x-coordinate.   **ii.** in the y-coordinate.
b. In going from point $(3, 2)$ to point $(3, -4)$, find the change
   i. in the x-coordinate.   **ii.** in the y-coordinate.

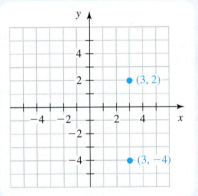

Figure 34 Exercise 26

27. The points $(-3, 1)$ and $(2, 4)$ are plotted in Fig. 35.
a. In going from point $(-3, 1)$ to point $(2, 4)$, find the change
   i. in the x-coordinate.   **ii.** in the y-coordinate.
b. In going from point $(2, 4)$ to point $(-3, 1)$, find the change
   i. in the x-coordinate.   **ii.** in the y-coordinate.

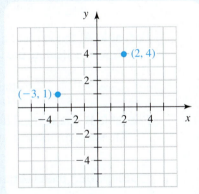

Figure 35 Exercise 27

28. The points $(1, 4)$ and $(5, -3)$ are plotted in Fig. 36.
a. In going from point $(1, 4)$ to point $(5, -3)$, find the change
   i. in the x-coordinate.   **ii.** in the y-coordinate.
b. In going from point $(5, -3)$ to point $(1, 4)$, find the change
   i. in the x-coordinate.   **ii.** in the y-coordinate.

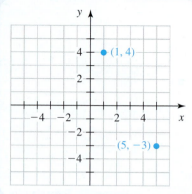

Figure 36 Exercise 28

29. a. Use a graphing calculator to graph the equations $y = x, y = 2x$, and $y = 3x$ in the same viewing window.
   b. List the lines $y = x, y = 2x$, and $y = 3x$ in order of steepness, from least to greatest.

c. What pattern does your list in part (b) suggest?
d. Give an equation of a line that is steeper than the lines $y = x, y = 2x$, and $y = 3x$.

30. a. Use a graphing calculator to graph the pair of equations in the same viewing window.
   i. $y = 2x + 1, y = 2x + 3$ **ii.** $y = 3x - 2, y = 3x - 5$
   iii. $y = -2x + 4, y = -2x + 3$
   b. Describe what pattern your work in part (a) suggests.
   c. Find an equation whose graph is parallel to the line $y = 4x - 5$.

31. In your own words, describe the Rule of Four for linear models.

32. Describe how to perform a unit analysis of a model's equation. Explain why a unit analysis can be used to help verify the equation.

3.3 Slope of a Line

Objectives

» Use a ratio to compare the steepness of two objects.

» Describe the meaning of, and how to calculate, the *slope* of a nonvertical line.

» Determine the sign of the slope of an increasing line and of a decreasing line.

» Explain why the slope of a horizontal line is zero and that the slope of a vertical line is undefined.

How do we measure the steepness of an object such as a ladder? In this section, we will discuss the *slope* of a line. This key concept has many applications in business, engineering, nursing, surveying, physics, social science, mathematics, and many other fields.

Comparing the Steepness of Two Objects

Consider the sketch of two cables (guy wires) running from the ground to a telephone pole in Fig. 37. Which cable is steeper?

Cable B is steeper than cable A, even though both cables reach a point at the same height on the building. To measure the steepness of each cable, we compare the *vertical* distance from the base of the pole to the top end of the cable with the *horizontal* distance from the bottom end of the cable to the building. In Section 2.5, we used a unit ratio to compare two quantities. Here, we calculate the unit ratio of vertical distance to horizontal distance for cable A:

$$\text{Cable A:} \quad \frac{\text{vertical distance}}{\text{horizontal distance}} = \frac{16 \text{ feet}}{8 \text{ feet}} = \frac{2}{1}$$

For cable A, the vertical distance is 2 times the horizontal distance.
   Now we calculate the unit ratio of vertical distance to horizontal distance for cable B:

$$\text{Cable B:} \quad \frac{\text{vertical distance}}{\text{horizontal distance}} = \frac{16 \text{ feet}}{4 \text{ feet}} = \frac{4}{1}$$

For cable B, the vertical distance is 4 times the horizontal distance.
   These calculations confirm cable B is steeper than cable A.

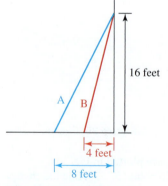

Figure 37 Cable A and cable B running from the ground to a telephone pole

Comparing the Steepness of Two Objects
To compare the steepness of two objects, compute the unit ratio $$\frac{\text{vertical distance}}{\text{horizontal distance}}$$ for each object. The object with the larger ratio is the steeper object.

Example 1   Comparing the Steepness of Two Objects

A portion of road A climbs steadily for 120 feet over a horizontal distance of 4250 feet. A portion of road B climbs steadily for 95 feet over a horizontal distance of 2875 feet. Which road is steeper? Explain.

Solution

Figure 38 shows sketches of the two roads, but the horizontal distances and vertical distances are not drawn to scale.

Figure 38 Roads A and B

Here, we calculate the unit ratio of the vertical distance to the horizontal distance for each road:

$$\text{Road A:} \quad \frac{\text{vertical distance}}{\text{horizontal distance}} = \frac{120 \text{ feet}}{4250 \text{ feet}} \approx \frac{0.028}{1}$$

$$\text{Road B:} \quad \frac{\text{vertical distance}}{\text{horizontal distance}} = \frac{95 \text{ feet}}{2875 \text{ feet}} \approx \frac{0.033}{1}$$

Road B is a little steeper than road A because road B's ratio of vertical distance to horizontal distance is greater than road A's.

The **grade** of a road is the ratio of the vertical distance to the horizontal distance, written as a percentage. Recall from Section 2.5 that to write a decimal number as a percentage, we move the decimal point two places to the right and insert the percent symbol. In Example 1, the grade of road A is about 2.8% and the grade of road B is about 3.3%.

Finding a Line's Slope

To measure the steepness (also called *slope*) of a nonvertical line, we will also use a ratio, but we will work with *changes* in a quantity, which can be negative, rather than with distances, which are always positive.

Consider the line that contains the points $(4, 2)$ and $(6, 5)$ sketched in Fig. 39. To go from point $(4, 2)$ to point $(6, 5)$, we look 2 units to the right and then look 3 units up. So, the horizontal change, called the *run*, is 2 and the vertical change, called the *rise*, is 3. The slope of the line is the ratio of the rise to the run:

$$\text{Slope} = \frac{\text{vertical change}}{\text{horizontal change}} = \frac{\text{rise}}{\text{run}} = \frac{3}{2}$$

In general, for any line, the **rise** is the vertical change and the **run** is the horizontal change in going from one point on the line to another point on the line. We use the letter m to represent the slope.

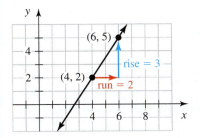

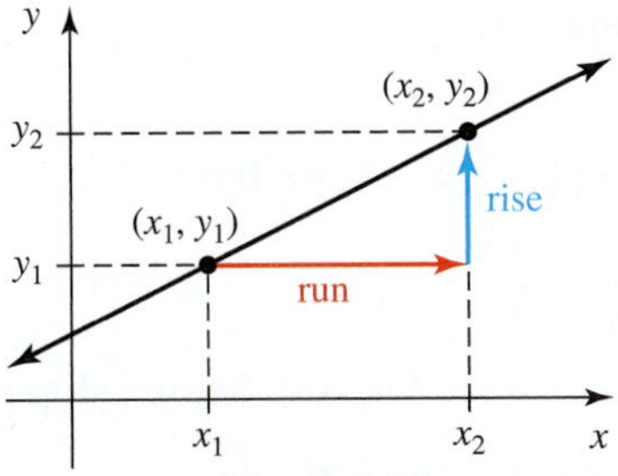

Figure 39 The slope is $\dfrac{3}{2}$

Figure 40 The rise and run between any two points

Definition   Slope of a nonvertical line

$$m = \text{slope} = \frac{\text{vertical change}}{\text{horizontal change}} = \frac{\text{rise}}{\text{run}}$$

In words: The **slope** of a nonvertical line is equal to the ratio of the rise to the run (in going from one point on the line to another point on the line). See Fig. 40.

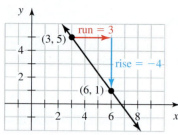

Figure 41 The slope is

$$\frac{-4}{3} = -\frac{4}{3}$$

Example 2   Finding the Slope of a Line

Find the slope of the line that contains the points $(3, 5)$ and $(6, 1)$.

Solution

We begin by plotting the points $(3, 5)$ and $(6, 1)$ and sketching the line that contains them (see Fig. 41). To go from point $(3, 5)$ to point $(6, 1)$, we must look 3 units to the right and then 4 units down. So, the run is 3 and the rise is -4. The slope of the line is the ratio of the rise to the run:

$$\text{Slope} = \frac{\text{vertical change}}{\text{horizontal change}} = \frac{\text{rise}}{\text{run}} = \frac{-4}{3} = -\frac{4}{3}$$

In Example 2, we found that, from point $(3, 5)$ to point $(6, 1)$, the run is 3. We can calculate this run by computing the change in the x-coordinates. Recall from Section 2.4 that we can find the change in a quantity by computing the ending amount minus the beginning amount:

$$\text{Change in } x\text{-coordinates} = \frac{x\text{-coordinate}}{\text{of ending point}} - \frac{x\text{-coordinate}}{\text{of beginning point}} = 6 - 3 = 3$$

In Example 2, we also found that, from point $(3, 5)$ to point $(6, 1)$, the rise is -4. We can calculate this rise by computing the change in the y-coordinates:

$$\text{Change in } y\text{-coordinates} = \frac{y\text{-coordinate}}{\text{of ending point}} - \frac{y\text{-coordinate}}{\text{of beginning point}} = 1 - 5 = -4$$

How do we calculate the slope of *any* nonvertical line if we are given two points on the line? First, we use the subscript 1 to identify x_1 and y_1 as the coordinates of the first point, (x_1, y_1). Likewise, we identify x_2 and y_2 as the coordinates of the second point, (x_2, y_2). When we look from point (x_1, y_1) to point (x_2, y_2), the run is the difference $x_2 - x_1$ and the rise is the difference $y_2 - y_1$ (see Fig. 42).

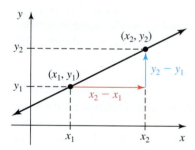

Figure 42 The slope of the line

is $m = \dfrac{y_2 - y_1}{x_2 - x_1}$

Calculating Slope

Let (x_1, y_1) and (x_2, y_2) be two distinct points of a nonvertical line. The slope of the line is

$$m = \frac{\text{vertical change}}{\text{horizontal change}} = \frac{y_2 - y_1}{x_2 - x_1}$$

(see Fig. 42).

A **formula** is an equation that contains two or more variables. We will refer to the equation $m = \dfrac{y_2 - y_1}{x_2 - x_1}$ as the **slope formula**.

Example 3   Finding the Slope of a Line

Find the slope of the line that contains the points $(2, 3)$ and $(8, 5)$.

Solution

Using the slope formula with $(x_1, y_1) = (2, 3)$ and $(x_2, y_2) = (8, 5)$, we have

$$m = \frac{y_2 - y_1}{x_2 - x_1} = \frac{5 - 3}{8 - 2} = \frac{2}{6} = \frac{1}{3}$$

By plotting points, we find that if the run is 6, then the rise is 2 (see Fig. 43). So, the slope is $m = \dfrac{\text{rise}}{\text{run}} = \dfrac{2}{6} = \dfrac{1}{3}$, which is the same as our result from using the formula.

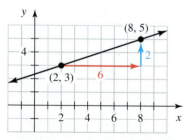

Figure 43 The slope is $\dfrac{2}{6} = \dfrac{1}{3}$

In Example 3, we calculated the slope of a line for $(x_1, y_1) = (2, 3)$ and $(x_2, y_2) = (8, 5)$. Here, we switch the roles of the two points to find the slope when $(x_1, y_1) = (8, 5)$ and $(x_2, y_2) = (2, 3)$:

$$m = \frac{y_2 - y_1}{x_2 - x_1} = \frac{3 - 5}{2 - 8} = \frac{-2}{-6} = \frac{1}{3}$$

The result is the same as that in Example 3. In general, when we use the slope formula with two points on a line, it doesn't matter which point we choose to be first, (x_1, y_1), and which point we choose to be second, (x_2, y_2).

WARNING   It is a common error to make incorrect substitutions into the slope formula. Carefully consider why the middle and right-hand formulas are incorrect:

Correct	**Incorrect**	**Incorrect**
$m = \dfrac{y_2 - y_1}{x_2 - x_1}$	$m = \dfrac{y_2 - y_1}{x_1 - x_2}$	$m = \dfrac{x_2 - x_1}{y_2 - y_1}$

Example 4   Finding the Slope of a Line

Find the slope of the line that contains the points $(3, 4)$ and $(7, 2)$.

Solution

Using the formula for slope with $(x_1, y_1) = (3, 4)$ and $(x_2, y_2) = (7, 2)$, we have

$$m = \frac{y_2 - y_1}{x_2 - x_1} = \frac{2 - 4}{7 - 3} = \frac{-2}{4} = -\frac{1}{2}$$

By plotting points, we find that when the run is 4, the rise is -2 (see Fig. 44). So, the slope is $\frac{-2}{4} = -\frac{1}{2}$, which is the same as our result from using the formula.

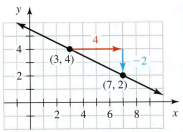

Figure 44 The slope is $\dfrac{-2}{4} = -\dfrac{1}{2}$

Slopes of Increasing and Decreasing Lines

Because the line in Fig. 45 goes upward from left to right, we say the line (and the graph) is **increasing**. A sign analysis of the rise and run in Fig. 45 shows that the slope of the line is positive.

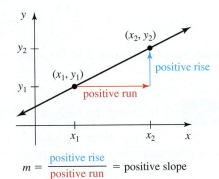

$$m = \frac{\text{positive rise}}{\text{positive run}} = \text{positive slope}$$

Figure 45 Increasing lines have positive slope

Because the line in Fig. 46 goes downward from left to right, we say the line (and the graph) is **decreasing**. A sign analysis of the rise and run in Fig. 46 shows that the slope of the line is negative.

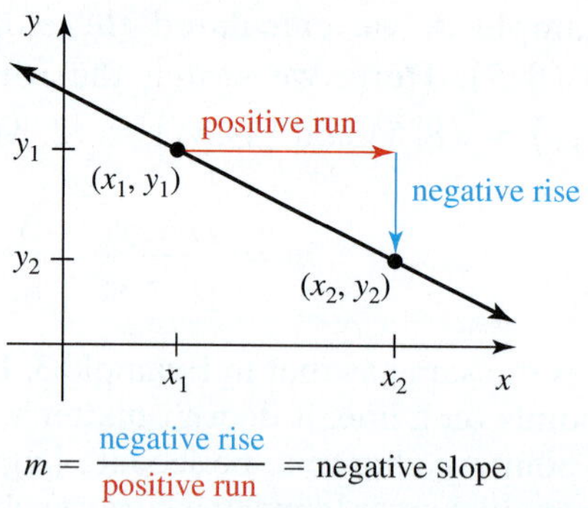

$$m = \frac{\text{negative rise}}{\text{positive run}} = \text{negative slope}$$

Figure 46 Decreasing lines have negative slope

Slopes of Increasing or Decreasing Lines

- An increasing line has positive slope (see Fig. 45).
- A decreasing line has negative slope (see Fig. 46).

When we compute the slope of two points that have negative coordinates, it can help to write first

$$\frac{(\ ) - (\ )}{(\ ) - (\ )}$$

and then insert the coordinates of the two points into the appropriate parentheses.

Example 5   Finding the Slope of a Line

Find the slope of the line that contains the points $(-5, 2)$ and $(3, -4)$.

Solution

$$m = \frac{(-4) - (2)}{(3) - (-5)} = \frac{-4 - 2}{3 + 5} = \frac{-6}{8} = -\frac{6}{8} = -\frac{3}{4}$$

Because the slope is negative, the line is decreasing.

▶

Example 6   Finding the Slope of a Line

Find the approximate slope of the line that contains the points $(-4.9, -3.5)$ and $(-2.3, 5.8)$. Round the result to the second decimal place.

Solution

$$m = \frac{(5.8) - (-3.5)}{(-2.3) - (-4.9)} = \frac{9.3}{2.6} \approx 3.58$$

So, the slope is approximately 3.58. Because the slope is positive, the line is increasing.

▶

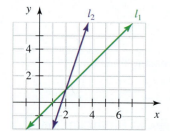

Figure 47 Find the slopes of the two lines

Example 7   Comparing the Slopes of Two Lines

Find the slopes of the two lines sketched in Fig. 47. Which line has the greater slope? Explain why this makes sense in terms of the steepness of a line.

Solution

In Fig. 48, we see that for line l_1, if the run is 1, the rise is 1. We calculate the slope of line l_1:

$$\text{Slope of line } l_1 = \frac{\text{rise}}{\text{run}} = \frac{1}{1} = 1$$

In Fig. 49, we see that for line l_2, if the run is 1, the rise is 3. We calculate the slope of line l_2:

$$\text{Slope of line } l_2 = \frac{\text{rise}}{\text{run}} = \frac{3}{1} = 3$$

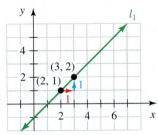

Figure 48 Line with lesser slope

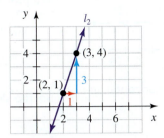

Figure 49 Line with greater slope

The slope of line l_2 is greater than the slope of line l_1, which is what we would expect because line l_2 is steeper than line l_1.

In Fig. 50, we show three decreasing lines and three increasing lines and their slopes. In Fig. 51, we show the same lines and the absolute values of their slopes.

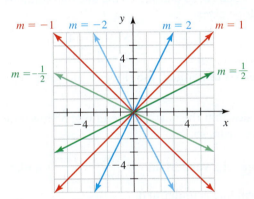

Figure 50 Slopes of some lines

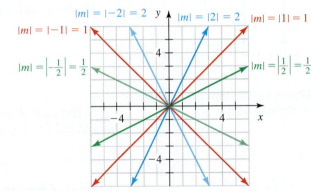

Figure 51 Absolute values of slopes of some lines

Of the six lines in Fig. 51, the steepest (blue) lines have the largest absolute value of slope (2). The next steepest (red) lines have the next largest absolute value of slope (1). The least steep (green) lines have the least absolute value of slope $\left(\dfrac{1}{2}\right)$.

> **Measuring the Steepness of a Line**
>
> The absolute value of the slope of a line measures the steepness of the line. The steeper the line, the larger the absolute value of its slope will be.

Slopes of Horizontal and Vertical Lines

So far, we have discussed slopes of lines that are increasing or decreasing. What about the slope of a horizontal line or the slope of a vertical line?

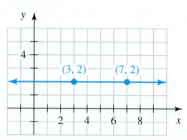

Figure 52 The horizontal line has zero slope

Example 8   Finding the Slope of a Horizontal Line

Find the slope of the line that contains the points $(3, 2)$ and $(7, 2)$.

Solution

We plot the points $(3, 2)$ and $(7, 2)$ and sketch the line that contains the points (see Fig. 52).

The formula for slope gives

$$m = \frac{2 - 2}{7 - 3} = \frac{0}{4} = 0$$

So, the slope of this horizontal line is zero.

In Example 8, we saw that the horizontal line in Fig. 52 has a slope equal to zero. *Any* horizontal line has no rise, so

$$\text{Slope of a horizontal line} = \frac{\text{rise}}{\text{run}} = \frac{0}{\text{run}} = 0$$

The slope of any horizontal line is zero.

Example 9   Finding the Slope of a Vertical Line

Find the slope of the line that contains the points $(3, 1)$ and $(3, 5)$.

Solution

We plot the points $(3, 1)$ and $(3, 5)$ and sketch the line that contains the points (see Fig. 53).

The formula for slope gives

$$m = \frac{5 - 1}{3 - 3} = \frac{4}{0}$$

Because division by zero is undefined, the slope of the vertical line is *undefined*.

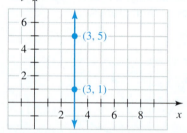

Figure 53 The vertical line has undefined slope

In Example 9, we saw that the vertical line in Fig. 53 has undefined slope. Note that *any* vertical line has zero run, so

$$\text{Slope of a vertical line} = \frac{\text{rise}}{\text{run}} = \left. \frac{\text{rise}}{0} \right\} \text{undefined}$$

The slope of any vertical line is undefined.

Slopes of Horizontal and Vertical Lines

- A horizontal line has a slope equal to zero (see Fig. 54).
- A vertical line has undefined slope (see Fig. 55).

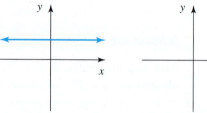

Figure 54 Horizontal lines have a slope equal to zero

Figure 55 Vertical lines have undefined slope

 **Group Exploration**

Section Opener: Graphical significance of *m* for $y = mx$

1. Use ZDecimal to graph these equations of the form $y = mx$ in order, and describe what you observe:

$$y = x, \quad y = 2x, \quad y = 3x, \quad y = 4x$$

2. Give an example of an equation of the form $y = mx$ whose graph is a line steeper than the lines you sketched in Problem 1.

3. Use ZDecimal to graph these equations in order, and describe what you observe:

$$y = -x, \quad y = -2x, \quad y = -3x, \quad y = -4x$$

4. Describe the graph of $y = mx$ in the following situations:
   a. *m* is a large positive number.
   b. *m* is a positive number near zero.
   c. *m* is a negative number near zero.
   d. *m* is less than -10.
   e. $m = 0$

5. Describe what you have learned in this exploration.

 Group Exploration

For a line, rise over run is constant

1. A line is sketched in Fig. 56. Plot the points $(-2, -5)$, $(1, 1)$, and $(3, 5)$. (Plotted correctly, these points will lie on the line.)

2. Using the points $(-2, -5)$ and $(1, 1)$, find the slope of the line.

3. Using the points $(1, 1)$ and $(3, 5)$, find the slope of the line.

4. Using the points $(-2, -5)$ and $(3, 5)$, find the slope of the line.

5. Using two other points of your choice, find the slope of the line.

6. What do you notice about the slopes you have calculated? Does it matter which two points on a line are used to find the slope of the line?

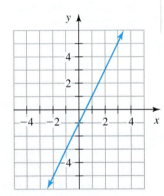

Figure 56 Problems 1–5

Tips for Success   Make Changes

If you have not had passing scores on tests and quizzes during the first part of this course, it is time to evaluate what the problem is, what changes you should make, and whether you can commit to making those changes.

Sometimes students must change how they study for a course. For example, Rosie did poorly on exams and quizzes for the first third of the course. It was not clear why she was not passing the course because she had good attendance, was actively involved in classroom work, and was doing the homework assignments. Suddenly, Rosie started getting As on every quiz and test. What had happened? Rosie said, "I figured out that, to do well in this course, it was not enough for me to do just the exercises that were assigned. So now I do a lot of extra exercises from each section."

Homework 3.3

For extra help ▶ **MyLab Math** Watch the videos in MyLab Math

1. A portion of road A climbs steadily for 210 feet over a horizontal distance of 3500 feet. A portion of road B climbs steadily for 275 feet over a horizontal distance of 5000 feet. Which road is steeper? Explain.

2. While taking off, airplane A climbs steadily for 4800 feet over a horizontal distance of 9000 feet. Airplane B climbs steadily for 5800 feet over a horizontal distance of 10,500 feet. Which plane is climbing at a greater incline? Explain.

3. Ski run A declines steadily for 80 yards over a horizontal distance of 400 yards. Ski run B declines steadily for 90 yards over a horizontal distance of 600 yards. Which ski run is steeper? Explain.

4. Ski run A declines steadily for 90 yards over a horizontal distance of 500 yards. Ski run B declines steadily for 130 yards over a horizontal distance of 650 yards. Which ski run is steeper? Explain.

Plot the two given points, and then sketch the line that contains the points. Find the run and rise in going from the first point listed to the second point listed. Find the slope of the line.

5. $(2, 3)$ and $(4, 6)$

6. $(1, 4)$ and $(6, 5)$

7. $(3, 6)$ and $(5, 2)$

8. $(2, 5)$ and $(6, 3)$

9. $(-4, 1)$ and $(2, 5)$

10. $(3, -4)$ and $(5, 2)$

11. $(-4, -2)$ and $(-2, -6)$

12. $(-3, -2)$ and $(-1, -4)$

Use the slope formula to find the slope of the line that passes through the two given points. State whether the line is increasing, decreasing, horizontal, or vertical.

13. $(1, 5)$ and $(3, 9)$

14. $(2, 3)$ and $(5, 12)$

15. $(3, 10)$ and $(5, 2)$

16. $(5, 8)$ and $(7, 2)$

17. $(2, 1)$ and $(8, 4)$

18. $(3, 2)$ and $(7, 4)$

19. $(2, 5)$ and $(8, 3)$

20. $(1, 7)$ and $(9, 5)$

21. $(-2, 4)$ and $(3, -1)$

22. $(-3, 4)$ and $(1, -2)$

23. $(5, -2)$ and $(9, -4)$

24. $(2, -3)$ and $(8, -6)$

25. $(-7, -1)$ and $(-2, 9)$

26. $(-6, -8)$ and $(-4, 2)$

27. $(-6, -9)$ and $(-2, -3)$

28. $(-5, -2)$ and $(-1, -3)$

29. $(6, -1)$ and $(-4, 7)$

30. $(4, -5)$ and $(-2, 10)$

31. $(-2, -11)$ and $(7, -5)$

32. $(-3, -1)$ and $(9, -3)$

33. $(0, 0)$ and $(4, -2)$

34. $(-6, -9)$ and $(0, 0)$

35. $(3, 5)$ and $(7, 5)$

36. $(-4, -6)$ and $(3, -6)$

37. $(-3, -1)$ and $(-3, -2)$

38. $(4, 2)$ and $(4, 7)$

For Exercises 39–46, find the approximate slope of the line that contains the two given points. Round your result to the second decimal place. State whether the line is increasing, decreasing, horizontal, or vertical.

39. $(-3.2, 5.1)$ and $(-2.8, 1.4)$

40. $(-1.9, 4.8)$ and $(-3.1, 5.5)$

41. $(4.9, -2.7)$ and $(6.3, -1.1)$

42. $(9.7, -6.8)$ and $(4.5, -2.7)$

43. $(-4.97, -3.25)$ and $(-9.64, -2.27)$

44. $(-3.22, -8.54)$ and $(-7.29, -6.13)$

45. $(-2.45, -6.71)$ and $(4.88, -1.53)$

46. $(-3.99, -2.49)$ and $(1.06, -3.76)$

47. Find the slope of the line sketched in Fig. 57.

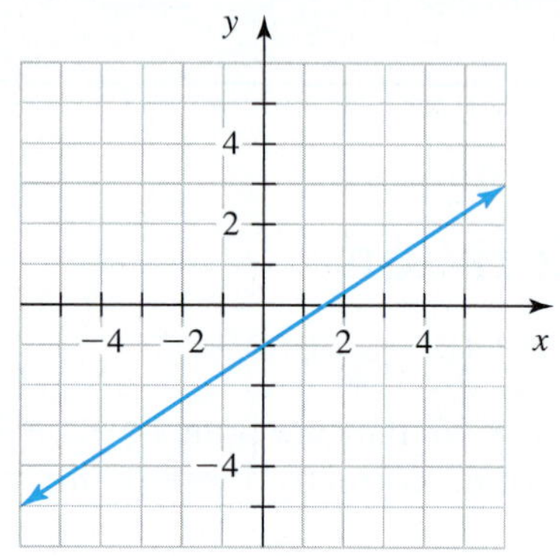

Figure 57 Exercise 47

48. Find the slope of the line sketched in Fig. 58.

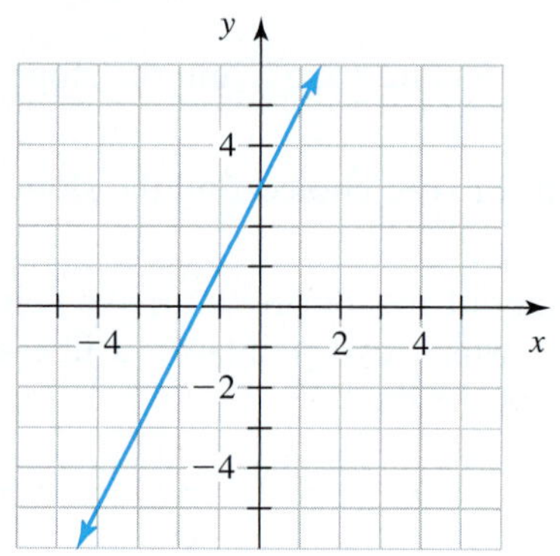

Figure 58 Exercise 48

49. Find the slope of the line sketched in Fig. 59.

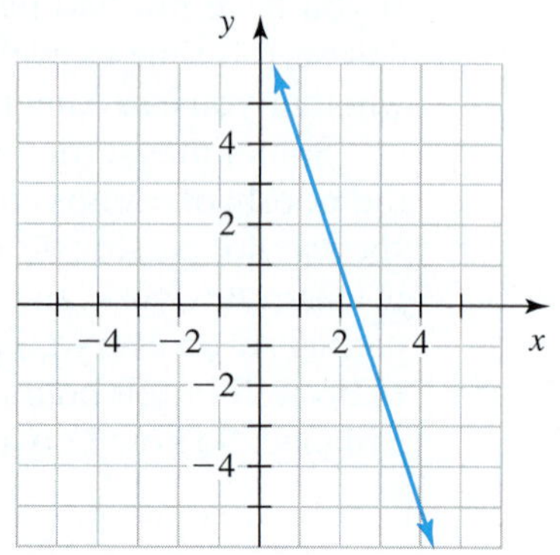

Figure 59 Exercise 49

50. Find the slope of the line sketched in Fig. 60.

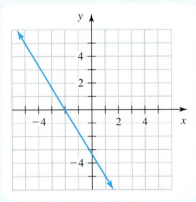

Figure 60 Exercise 50

Concepts

51. For each line sketched in Fig. 61, determine whether the line's slope is positive, negative, zero, or undefined.

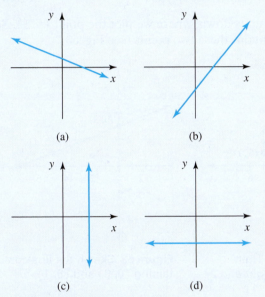

Figure 61 Exercise 51

52. Find the slope of the line with x-intercept $(-3, 0)$ and y-intercept $(0, 4)$.

Sketch a line that meets the given description. Find the slope of the line.

53. An increasing line that is nearly horizontal

54. A decreasing line that is nearly horizontal

55. A decreasing line that is nearly vertical

56. An increasing line that is nearly vertical

For Exercises 57–60, sketch a line that meets the given description.

57. The slope is a large positive number.

58. The slope is a positive number near zero.

59. The slope is a negative number near zero.

60. The slope is less than -5.

61. A student tries to find the slope of the line that contains the points $(1, 3)$ and $(4, 7)$:

$$\frac{4 - 1}{7 - 3} = \frac{3}{4}$$

Describe any errors. Then find the slope correctly.

62. A student tries to find the slope of the line that contains the points $(6, 4)$ and $(3, 9)$:

$$\frac{9 - 4}{6 - 3} = \frac{5}{3}$$

Describe any errors. Then find the slope correctly.

63. A student tries to find the slope of the line that contains the points $(-1, -5)$ and $(3, 8)$:

$$\frac{8 - 5}{3 - 1} = \frac{3}{2}$$

Describe any errors. Then find the slope correctly.

64. A student tries to find the slope of the line that contains the points $(2, 5)$ and $(6, 7)$:

$$\frac{7 - 5}{6 - 2} = \frac{2}{4}$$

Describe any errors. Then find the slope correctly.

65. Sketch a line with a slope of 2 and another line with a slope of 3. Does the steeper line have the greater slope?

66. Sketch a line with a slope of 1 and another line with a slope of $\frac{1}{2}$. Does the steeper line have the greater slope?

67. **a.** Sketch a line with a slope of -2 and another line with a slope of -3.
b. Does the steeper line have the greater slope?
c. Find the absolute value of the slope of each line. Does the steeper line have the greater absolute value of the slope?
d. Explain why the absolute value of the slope of a line is useful for comparing the steepness of lines.

68. **a.** Sketch a line with a slope of -2 and another line with a slope of 2.
b. Is the line with a slope of 2 steeper than the line with a slope of -2?
c. Find the absolute value of the slope of each line. Compare the results.
d. Explain why the absolute value of the slope of a line is useful for comparing the steepness of lines.

69. A line contains the points $(2, 1)$ and $(3, 4)$. Find three more points that lie on the line. [**Hint:** Find the slope of the line. Then use the slope and a point on the line to help you find other points on the line.]

70. A line contains the points $(1, 5)$ and $(4, 3)$. Find three more points that lie on the line. [**Hint:** See Exercise 69.]

71. Explain why the slope of a vertical line is undefined.

72. Explain why the slope of a horizontal line is 0.

73. Explain why the slope of an increasing line is positive.

74. Explain why the slope of a decreasing line is negative.

75. Both ladder A and ladder B are leaning against a building. Ladder A reaches a higher point on the building than ladder B does. Can we conclude ladder A is steeper than ladder B? Explain why this situation suggests we must take into account both rise and run when we measure the steepness of a line (or a ladder). Draw sketches of rises and runs of lines to illustrate your point.

76. Explain how to find the slope of a line.

3.4 Using Slope to Graph Linear Equations

Objectives

» Use the slope and the y-intercept of a line to sketch the line.

» Describe the meaning of m for an equation of the form $y = mx + b$.

» Graph an equation of the form $y = mx + b$ by using the line's slope and y-intercept.

» Find an equation of a line from its graph.

» Graph an equation of a linear model by using the model's slope and y-intercept.

» Describe the relationship between slopes of parallel lines.

» Describe the relationship between slopes of perpendicular lines.

In Sections 3.1 and 3.2, we graphed a linear equation in two variables by first finding solutions of the equation. In this section, we will discuss a more efficient way to graph such equations.

Using the Slope and the y-Intercept to Sketch a Line

We can use the slope and the y-intercept of a line to sketch the line.

Example 1   Sketching a Line

Sketch the line that has slope $m = -\dfrac{2}{5}$ and y-intercept $(0, 3)$.

Solution

Slope tells us how to go from one point on a line to another point on the line. So, we begin by plotting the y-intercept, $(0, 3)$. The slope is $-\dfrac{2}{5} = \dfrac{-2}{5} = \dfrac{\text{rise}}{\text{run}}$. From $(0, 3)$, we count 5 units to the right and 2 units down, where we plot the point $(5, 1)$. See Fig. 62. We then sketch the line that contains these two points (see Fig. 63).

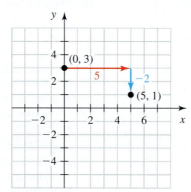

Figure 62 Plot $(0, 3)$. Then count 5 units to the right and 2 units down to plot $(5, 1)$

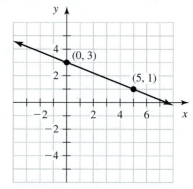

Figure 63 Sketch the line containing $(0, 3)$ and $(5, 1)$

Because the slope is negative, we can verify our work by checking that the line is decreasing.

The Meaning of m for $y = mx + b$

In Example 1, we saw that if we know the slope and the y-intercept of a line, we can sketch that line. Next, we will discuss how to determine the slope and y-intercept of a line from the line's equation.

Example 2   Finding the Slope of a Line

Find the slope of the line $y = 2x + 1$.

Solution

We list some solutions in Table 24 and sketch the graph of the equation in Fig. 64.
    If the run is 1, the rise is 2 (see Fig. 64). So, the slope is

$$m = \frac{\text{rise}}{\text{run}} = \frac{2}{1} = 2$$

Table 24 Solutions of $y = 2x + 1$

x	y
0	$2(0) + 1 = 1$
1	$2(1) + 1 = 3$
2	$2(2) + 1 = 5$

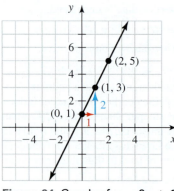

Figure 64 Graph of $y = 2x + 1$

In Example 2, we found that the line $y = 2x + 1$ has slope 2. Note that 2 is also the number multiplied by x in the equation $y = 2x + 1$. This observation suggests a general property about a linear equation of the form $y = mx + b$.

Finding the Slope and y-Intercept from a Linear Equation

For a linear equation of the form $y = mx + b$,

- the slope of the line is m and
- the y-intercept of the line is $(0, b)$.

We say that this equation is in **slope–intercept form**.

For example, the equation $y = -4x + 5$ is in slope–intercept form with $m = -4$ and $b = 5$. The graph of this equation is a line with slope -4 and y-intercept $(0, 5)$. The line $y = 8x - 2$ has slope 8 and y-intercept $(0, -2)$.

Graphing Equations of the Form $y = mx + b$

In Example 3, we will graph an equation in slope–intercept form.

Example 3   Graphing an Equation

Sketch the graph of $y = 3x - 4$.

Solution

Note that the y-intercept is $(0, -4)$ and the slope is $3 = \dfrac{3}{1} = \dfrac{\text{rise}}{\text{run}}$. To graph the line:

1. Plot the y-intercept $(0, -4)$.
2. From $(0, -4)$, count 1 unit to the right and 3 units up to plot a second point, which we see by inspection is $(1, -1)$. See Fig. 65.
3. Sketch the line that contains these two points (see Fig. 66).

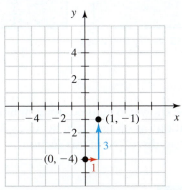

Figure 65 Plot $(0, -4)$. Then count 1 unit to the right and 3 units up to plot $(1, -1)$

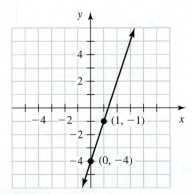

Figure 66 Sketch the line containing $(0, -4)$ and $(1, -1)$

Recall from Section 3.1 that every point on the graph of an equation represents a solution of the equation. We can verify our result by checking that both $(0, -4)$ and $(1, -1)$ are solutions of $y = 3x - 4$:

Check that $(0, -4)$ is a solution   **Check that $(1, -1)$ is a solution**

$$y = 3x - 4 \qquad\qquad\qquad y = 3x - 4$$
$$-4 \stackrel{?}{=} 3(0) - 4 \qquad\qquad -1 \stackrel{?}{=} 3(1) - 4$$
$$-4 \stackrel{?}{=} 0 - 4 \qquad\qquad\quad -1 \stackrel{?}{=} 3 - 4$$
$$-4 \stackrel{?}{=} -4 \qquad\qquad\qquad -1 \stackrel{?}{=} -1$$
$$\text{true} \qquad\qquad\qquad\qquad \text{true}$$

We check two ordered pairs (rather than just one) because two points determine a line.

Graphing an Equation in Slope–Intercept Form

To graph an equation of the form $y = mx + b$,

1. Plot the y-intercept $(0, b)$.

2. Use $m = \dfrac{\text{rise}}{\text{run}}$ to plot a second point. For example, if $m = \dfrac{2}{3}$, then count 3 units to the right (from the y-intercept) and 2 units up to plot another point.

3. Sketch the line that passes through the two plotted points.

Example 4   Graphing an Equation

Sketch the graph of $y = -\dfrac{3}{4}x + 4$.

Solution

The y-intercept is $(0, 4)$, and the slope is $-\dfrac{3}{4} = \dfrac{-3}{4} = \dfrac{\text{rise}}{\text{run}}$. To graph the line:

1. Plot the y-intercept $(0, 4)$.
2. From $(0, 4)$, count 4 units to the right and 3 units down to plot a second point, which we see by inspection is $(4, 1)$. See Fig. 67.
3. Sketch the line that contains these two points (see Fig. 68).

We use a graphing calculator to verify our work (see Fig. 69).

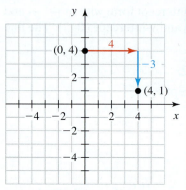

Figure 67 Plot $(0, 4)$. Then count 4 units to the right and 3 units down to plot $(4, 1)$

 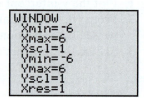 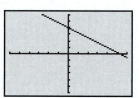

Figure 69 Verify the work

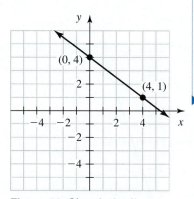

Figure 68 Sketch the line containing $(0, 4)$ and $(4, 1)$

We now know two methods for graphing a linear equation: We can first find solutions of the equation (as discussed in Sections 3.1 and 3.2), or we can first find the slope and y-intercept (as discussed in this section).

Example 5   Interpreting the Signs of m and b

Graph an equation of the form $y = mx + b$ where m is negative and b is positive.

Solution

We should sketch a decreasing line because the slope m is negative. We should draw a line whose y-intercept, $(0, b)$, is above the origin because b is positive. We sketch such a line in Fig. 70.

There is nothing special about the line in Fig. 70—any line that is decreasing and has its y-intercept above the origin will do.

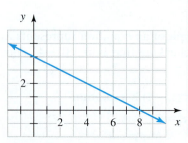

Figure 70 A decreasing line with y-intercept above the origin

Finding an Equation of a Line from a Graph

Given the slope and y-intercept of a nonvertical line, we can find an equation of the line.

Example 6   Finding an Equation of a Line from Its Slope and y-Intercept

Find an equation of the line that has slope $\frac{2}{3}$ and y-intercept $(0, -5)$.

Solution

To find an equation, we substitute $\frac{2}{3}$ for m and -5 for b in the equation $y = mx + b$:

$$y = \frac{2}{3}x + (-5)$$

$$y = \frac{2}{3}x - 5$$

Given an equation in slope–intercept form, $y = mx + b$, we can find the graph of the equation. We can also go backward: Given the graph, we can find an equation for it.

Example 7   Finding an Equation of a Line from Its Graph

Find an equation of the line sketched in Fig. 71.

Solution

From Fig. 72, we see that the y-intercept of the line is $(0, -1)$. We also see that if the run is 3, then the rise is -2. So, the slope is $\frac{-2}{3} = -\frac{2}{3}$. By substituting $-\frac{2}{3}$ for m and -1 for b in the equation $y = mx + b$, we have $y = -\frac{2}{3}x - 1$.

We can verify our equation by checking that both $(0, -1)$ and $(3, -3)$ satisfy the equation. Or we can use a graphing calculator to verify our work (see Fig. 73).

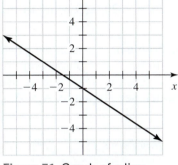

Figure 71 Graph of a line

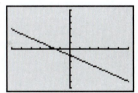

Figure 73 Verify the work

Figure 72 Finding the slope of the line

Finding an Equation of a Line from a Graph

To find an equation of a line from a graph,

1. Determine the slope m and the y-intercept $(0, b)$ from the graph.

2. Substitute your values for m and b into the equation $y = mx + b$.

Graphing an Equation of a Linear Model

We can graph an equation of a linear model by using the model's slope and y-intercept.

Table 25 Percentages of CEOs Who Are Women

Year	Percent
2006	2.0
2008	2.4
2010	3.0
2012	3.6
2015	4.0

Source: *Catalyst*

Example 8   Graphing a Model's Equation

The percentages of CEOs who are women are shown in Table 25 for various years. Let p be the percentage of CEOs who are women at t years since 2000. A reasonable model is

$$p = 0.23t + 0.6$$

1. Graph the model.
2. Predict when 5.2% of CEOs will be women.

Solution

1. The p-intercept is $(0, 0.6)$, and the slope is $0.23 = \dfrac{0.23}{1} = \dfrac{\text{rise}}{\text{run}}$. It will be easier to graph the model if we multiply the slope by $\dfrac{10}{10} = 1$ so the rise and run are larger:

$$0.23 = \frac{0.23}{1} \cdot \frac{10}{10} = \frac{2.3}{10} = \frac{\text{rise}}{\text{run}}$$

To graph the model, we first plot the p-intercept $(0, 0.6)$. From $(0, 0.6)$, we count 10 units to the right and 2.3 units up, where we plot the point $(10, 2.9)$. See Fig. 74. We then sketch the line that contains the two points.

Instead of using the slope to find the point $(10, 2.9)$, we could have substituted 10 for t in the equation $p = 0.23t + 0.6$ and solved for p:

$$p = 0.23(10) + 0.6 = 2.9$$

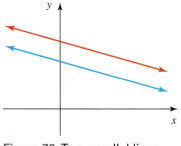

Figure 74 Female CEO model

So, the point $(10, 2.9)$ is a point on the linear model.

2. The red arrows in Fig. 74 show that the output $p = 5.2$ originates from the input $t = 20$. So, 5.2% of CEOs will be women in $2000 + 20 = 2020$, according to the model.

To use a graphing calculator to verify our work in Problems 1 and 2, we press $\boxed{\text{WINDOW}}$ and set Xmin to be -10 (for 1990), set Xmax to be 25 (for 2025), and use ZoomFit to set the values for Ymin and Ymax automatically (see Fig. 75). Then we use TRACE to check that $(20, 5.2)$ is a point on the linear model. See Appendix A.5, A.6, and A.7 for graphing calculator instructions.

 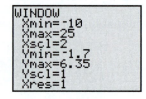 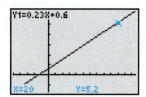

Figure 75 Verify the work

Slopes of Parallel Lines

Parallel lines are lines in the same plane that never intersect (see Fig. 76). How do the slopes of two parallel lines compare?

Figure 76 Two parallel lines

| Example 9 | Find the Slopes of Two Parallel Lines |

Find the slopes of the parallel lines l_1 and l_2 sketched in Fig. 77.

Solution

For both lines, if the run is 2, the rise is 1 (see Fig. 78). So, the slope of both parallel lines is

$$m = \frac{\text{rise}}{\text{run}} = \frac{1}{2}$$

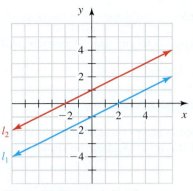

Figure 77 Two parallel lines

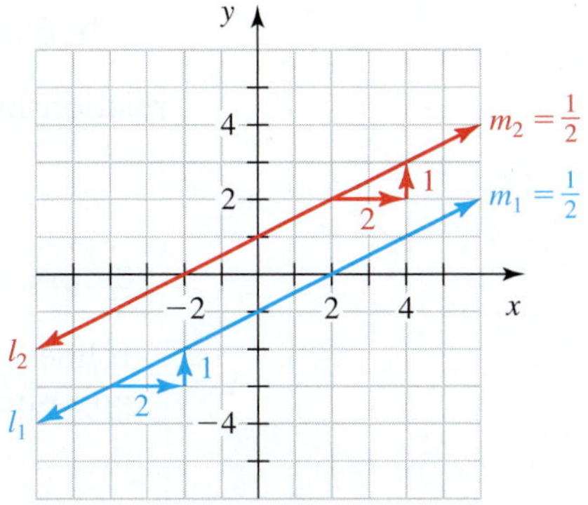

Figure 78 Calculate the slopes of the parallel lines

It makes sense that parallel nonvertical lines have equal slope because parallel lines have the same steepness.

Slopes of Parallel Lines

If lines l_1 and l_2 are parallel nonvertical lines, then the slopes of the lines are equal:

$$m_1 = m_2$$

Also, if two distinct lines have equal slope, then the lines are parallel.

Slopes of Perpendicular Lines

Two lines are called **perpendicular** if they intersect at a 90° angle (see Fig. 79). How do the slopes of two perpendicular lines compare?

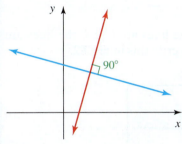

Figure 79 Two perpendicular lines

Example 10   Find the Slopes of Two Perpendicular Lines

Find the slopes of the perpendicular lines l_1 and l_2 sketched in Fig. 80.

Solution

From Fig. 81 we see that the slope of line l_1 is

$$m_1 = \frac{2}{3}$$

and the slope of line l_2 is

$$m_2 = \frac{-3}{2} = -\frac{3}{2}$$

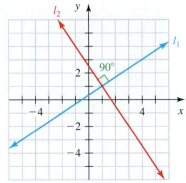

Figure 80 Two perpendicular lines

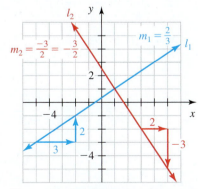

Figure 81 Calculate the slopes of the perpendicular lines

In Example 10, the slope $-\dfrac{3}{2}$ is the opposite of the reciprocal of the slope $\dfrac{2}{3}$. A similar relationship applies to any pair of perpendicular nonvertical lines.

Slopes of Perpendicular Lines

If lines l_1 and l_2 are perpendicular nonvertical lines, then the slope of one line is the opposite of the reciprocal of the slope of the other line:

$$m_2 = -\frac{1}{m_1}$$

Also, if the slope of one line is the opposite of the reciprocal of another line's slope, then the lines are perpendicular.

Example 11   Identifying Parallel Lines and Perpendicular Lines

Determine whether the pair of lines is parallel, perpendicular, or neither.

1. $y = 3x - 5$ and $y = 3x + 2$
2. $y = \dfrac{8}{5}x - 3$ and $y = -\dfrac{5}{8}x + 6$

Solution

1. The slope of each line is 3. Because the slopes of the lines are equal, the lines are parallel. We use ZStandard followed by ZSquare to verify this in Fig. 82.

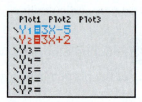 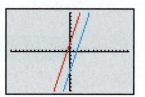

Figure 82 The lines appear to be parallel

2. The slopes of the lines are $\dfrac{8}{5}$ and $-\dfrac{5}{8}$. Because the slope $-\dfrac{5}{8}$ is the opposite of the reciprocal of the slope $\dfrac{8}{5}$, the lines are perpendicular. We use ZStandard followed by ZSquare to verify this in Fig. 83.

 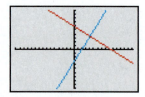

Figure 83 The lines appear to be perpendicular

Group Exploration

Section Opener: The meaning of *m* in the equation *y* = *mx* + *b*

1. **a.** Carefully sketch a graph of the line $y = 2x - 1$.

   b. Using the formula $m = \dfrac{\text{rise}}{\text{run}}$, find the slope of the line you sketched.

   c. What number is multiplied by x in the equation $y = 2x - 1$? How does it compare with the slope you found in part (b)?

2. **a.** Carefully sketch a graph of the line $y = -3x + 5$.

   b. Using the formula $m = \dfrac{\text{rise}}{\text{run}}$, find the slope of the line you sketched.

   c. What number is multiplied by x in the equation $y = -3x + 5$? How does it compare with the slope you found in part (b)?

3. Describe what you have learned in this exploration so far.

4. Without graphing, determine the slope of each line.

   a. $y = 4x - 7$ **b.** $y = -2x + 4$

   c. $y = \dfrac{2}{5}x - 3$ **d.** $y = x - 2$

   e. $y = 3$

 Group Exploration

Drawing lines with various slopes

1. On a graphing calculator, graph a group of lines (a *family of lines*) to make a starburst like the one in Fig. 84. List the equations of your lines.

2. On a graphing calculator, graph a family of lines to make a starburst like the one in Fig. 85. List the equations of your lines.

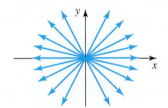

Figure 84 A starburst

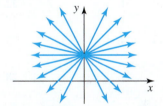

Figure 85 Another starburst

3. Summarize what you have learned about slope from this exploration, this section, and Section 3.3.

Tips for Success   Visualize

To help prepare themselves mentally and physically for competition, many exceptional athletes visualize themselves performing well at their event many times throughout their training period. For example, a runner training for the 100-meter dash might imagine getting set in the starting blocks, taking off right after the gun goes off, being in front of the other runners, and so on, right up until the moment of breaking the tape at the finish line.

In an experiment, three groups of basketball players were used to test the effectiveness of visualization. The first group warmed up by shooting baskets before a game. The second group visualized shooting baskets, but did not shoot any baskets during the pregame warm-up. The third group did not warm up or visualize before the game. The visualization group not only outperformed the group that did not warm up, but also did better than the group that warmed up by shooting baskets!

You can do visualizations, too. Visualize doing all the things you feel you need to do to succeed in the course. If you do this regularly, you will have better follow-through with what you intend to do. You will also feel more confident about succeeding.

Homework 3.4

For extra help ▶ **MyLab Math**   Watch the videos in MyLab Math

Sketch the line that has the given slope and contains the given point.

1. $m = \frac{2}{3}, (0, 1)$ **2.** $m = \frac{3}{5}, (0, 2)$

3. $m = -\frac{5}{2}, (0, 4)$ **4.** $m = -\frac{3}{4}, (0, -2)$

5. $m = -\frac{3}{2}, (0, 0)$ **6.** $m = -\frac{1}{2}, (0, 0)$

7. $m = 2, (0, 1)$ **8.** $m = 4, (0, -3)$

9. $m = -3, (0, -2)$ **10.** $m = -5, (0, 4)$

11. $m = -1, (0, 3)$ **12.** $m = 1, (0 - 2)$

13. $m = 0, (4, -5)$ **14.** $m = 0, (6, 3)$

15. m is undefined, $(2, -1)$ **16.** m is undefined, $(-1, -3)$

Determine the slope and the y-intercept. Then use the slope and the y-intercept to graph the equation by hand. Verify your work with a graphing calculator.

17. $y = \frac{2}{3}x - 1$ **18.** $y = \frac{1}{5}x + 2$ **19.** $y = -\frac{1}{3}x + 4$

20. $y = -\frac{3}{2}x - 1$ **21.** $y = \frac{4}{3}x + 2$ **22.** $y = \frac{5}{2}x - 3$

23. $y = -\frac{4}{5}x - 1$ **24.** $y = -\frac{1}{4}x + 2$ **25.** $y = \frac{1}{2}x$

26. $y = \frac{1}{4}x$ **27.** $y = -\frac{5}{3}x$ **28.** $y = -\frac{4}{5}x$

For Exercises 29–46, determine the slope and the y-intercept. Then use the slope and the y-intercept to graph the equation by hand. Verify your work by checking that two points on your line satisfy the given equation.

29. $y = 4x - 2$ **30.** $y = 2x - 4$ **31.** $y = -2x + 4$

32. $y = -3x + 5$ **33.** $y = -4x - 1$ **34.** $y = -2x - 2$

35. $y = x + 1$ **36.** $y = x - 4$ **37.** $y = -x + 3$

38. $y = -x + 2$ **39.** $y = -3x$ **40.** $y = 4x$

41. $y = x$ **42.** $y = -x$ **43.** $y = -3$

44. $y = -2$ **45.** $y = 0$ **46.** $y = 1$

47. The numbers of 7-Eleven stores worldwide are shown in Table 26 for various years.

Table 26 Numbers of 7-Eleven Stores Worldwide

Year	Number of Stores (in thousands)
2008	34
2010	38
2012	45
2014	52
2016	59

Source: *7-Eleven*

Let n be the number (in thousands) of 7-Eleven stores at t years since 2005. A reasonable model of this situation is

$$n = 3.2t + 23.2$$

a. Graph the model by hand.
b. Predict when there will be 75 thousand stores. Explain by using arrows on the graph you sketched in part (a).

48. The average monthly employee contribution required to cover a family in an employer-sponsored health plan has increased greatly since 2004 (see Table 27).

Table 27 Costs to Cover a Family in an Employer-Sponsored Health Plan

Year	Average Monthly Employee Contribution (dollars)
2004	222
2006	248
2008	280
2010	333
2012	360
2014	402
2015	413

Source: *Kaiser Family Foundation*

Let a be the average monthly employee contribution (in dollars) required to cover a family in an employer-sponsored health plan at t years since 2000. A reasonable model of the situation is

$$a = 18t + 144$$

a. Graph the model by hand.
b. Use your graph to predict when the average monthly employee contribution will be $500. Explain by using arrows on the graph you sketched in part (a).

49. The numbers of motor vehicle crash deaths per 100,000 people are shown in Table 28 for various years.

Table 28 Numbers of Motor Vehicle Crash Deaths per 100,000 People

Year	Deaths per 100,000 People
1975	20.6
1985	18.4
1995	15.9
2005	14.7
2014	10.2

Source: *Insurance Institute for Highway Safety*

Let d be the number of motor vehicle crash deaths per 100,000 people at t years since 1970. A reasonable model of the situation is

$$d = -0.25t + 22$$

a. Graph the model by hand.
b. Predict when there will be 9.5 motor vehicle crash deaths per 100,000 people. Explain by using arrows on the graph you sketched in part (a).

50. The number of injuries on farms to people under age 20 has dropped significantly since 2001 (see Table 29).

Table 29 Numbers of Injuries on Farms to People under Age 20

Year	Number of Injuries (in thousands)
2001	29
2004	28
2006	23
2009	16
2012	14
2014	12

Source: *National Institute for Occupational Safety and Health*

Let n be the number (in thousands) of injuries on farms to people under age 20 in the year that is t years since 2000. A reasonable model of this situation is

$$n = -1.5t + 31$$

a. Graph the model by hand.
b. Predict in which year 3 thousand people under age 20 will be injured on farms. Explain by using arrows on the graph you sketched in part (a).

51. Graphs of four linear equations are shown in Fig. 86. State whether m and b are positive, negative, zero, or undefined for the $y = mx + b$ form of each equation.

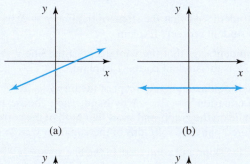

Figure 86 Exercise 51

52. Graphs of four linear equations are shown in Fig. 87. State the signs of the constants m and b for the $y = mx + b$ form of each equation.

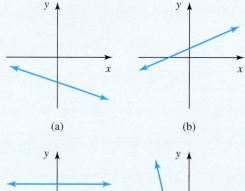

Figure 87 Exercise 52

Graph by hand an equation of the form $y = mx + b$ that meets the given criteria for m and b. Also, find an equation of each graph.

53. m is positive and b is positive
54. m is positive and b is negative
55. m is negative and b is negative
56. m is negative and $b = 0$
57. $m = 0$ and b is negative
58. $m = 0$ and $b = 0$

For Exercises 59–64, find an equation of a line that has the given slope and contains the given point.

59. $m = 3$, $(0, -4)$ 60. $m = -2$, $(0, 5)$

61. $m = -\dfrac{6}{5}$, $(0, 3)$ 62. $m = \dfrac{3}{4}$, $(0, -2)$

63. $m = -\dfrac{2}{7}$, $(0, 0)$ 64. $m = 0$, $(0, -1)$

65. Find an equation of the line sketched in Fig. 88.

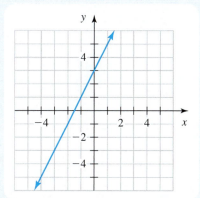

Figure 88 Exercise 65

66. Find an equation of the line sketched in Fig. 89.

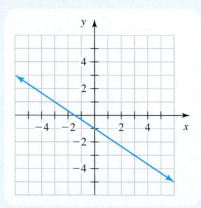

Figure 89 Exercise 66

For Exercises 67–78, determine whether the pair of lines is parallel, perpendicular, or neither. Explain.

67. $y = \dfrac{2}{5}x + 1$ and $y = -\dfrac{5}{2}x - 3$

68. $y = \dfrac{3}{7}x + 5$ and $y = -\dfrac{3}{7}x - 6$

69. $y = 3x - 1$ and $y = -3x + 2$

70. $y = 2x - 1$ and $y = 2x + 6$

71. $y = -4x + 2$ and $y = -4x + 3$

72. $y = -7x + 5$ and $y = \dfrac{1}{7}x - 8$

73. $y = \dfrac{2}{3}x - 1$ and $y = \dfrac{3}{2}x + 3$

74. $y = -\dfrac{4}{11}x + 2$ and $y = \dfrac{11}{4}x + 5$

75. $y = 2$ and $y = -4$ **76.** $x = -2$ and $y = 1$

77. $x = 0$ and $y = 0$ **78.** $x = -2$ and $x = 4$

79. Are the lines sketched in Fig. 90 parallel? Explain.

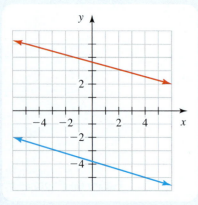

Figure 90 Exercise 79

80. Are the lines sketched in Fig. 91 perpendicular? Explain.

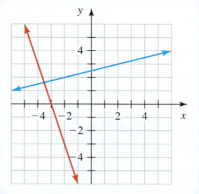

Figure 91 Exercise 80

For Exercises 81–86, refer to Fig. 92.

81. Find y when $x = -3$.   **82.** Find x when $y = -3$.

83. Find x when $y = 0$.   **84.** Find y when $x = 0$.

85. Find the slope of the line.   **86.** Find an equation of the line.

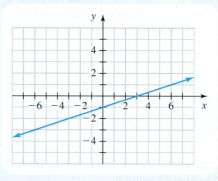

Figure 92
Exercises 81–86

Concepts

87. A student says that the slope of the line $y = 2x + 1$ is $2x$. Is the student correct? Explain.

88. A student says that the y-intercept of the line $y = 3x - 2$ is $(0, 2)$. Is the student correct? Explain.

89. Use the slope and y-intercept of the line $y = 2x - 1$ to graph the equation $y = 2x - 1$ by hand. Then choose two points that lie on the graph, and show that both of the corresponding ordered pairs are solutions of the equation $y = 2x - 1$.

90. Use the slope and y-intercept of the line $y = -2x + 3$ to graph the equation $y = -2x + 3$ by hand. Then choose two points that lie on the graph, and show that both of the corresponding ordered pairs are solutions of the equation $y = -2x + 3$.

91. Recall that we can describe some of or all the solutions of an equation in two variables with an equation, a table, a graph, or words.

   a. Describe the solutions of $y = \dfrac{1}{2}x + 2$ by using a graph.

   b. Describe three solutions of $y = \dfrac{1}{2}x + 2$ by using a table.

   c. Describe the solutions of $y = \dfrac{1}{2}x + 2$ by using words.

92. Recall that we can describe some of or all the solutions of an equation in two variables with an equation, a table, a graph, or words.

   a. Describe the solutions of $y = \frac{1}{3}x - 1$ by using a graph.

   b. Describe three solutions of $y = \frac{1}{3}x - 1$ by using a table.

   c. Describe the solutions of $y = \frac{1}{3}x - 1$ by using words.

93. A student tries to graph the equation $y = -3x + 1$ (see Fig. 93).

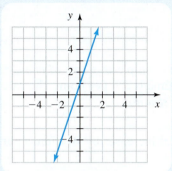

Figure 93 Exercise 93

Choose two points that lie on the line, and show that at least one of these points is not a solution of $y = -3x + 1$. Explain why your work shows that Fig. 93 is incorrect. Then sketch the correct graph.

94. A student tries to graph the equation $y = \frac{3}{2}x - 1$ (see Fig. 94).

Choose two points that lie on the line, and show that at least one of these points is not a solution of $y = \frac{3}{2}x - 1$. Explain why your work shows that Fig. 94 is incorrect. Then sketch the correct graph.

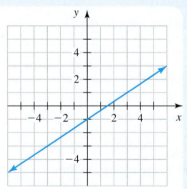

Figure 94 Exercise 94

95. A line passes through the point $(-1, 1)$ and has a slope of 2.
   a. Sketch the line.
   b. Find an equation of the line. [**Hint:** Refer to the line you sketched in part (a) to find b in $y = mx + b$.]

96. A line passes through the point $(1, 2)$ and has a slope of -3.
   a. Sketch the line.
   b. Find an equation of the line. [**Hint:** Refer to the line you sketched in part (a) to find b in $y = mx + b$.]

97. a. Find the slope of each line: $y = 3$, $y = -5$, and $y = 0$.
   b. Find the slope of the graph of any linear equation of the form $y = k$, where k is a constant.

98. a. Find the slope of each line: $x = 2$, $x = -4$, and $x = 0$.
   b. Find the slope of the graph of any equation of the form $x = k$, where k is a constant.

99. Graphs of the equations $y = mx + b$ and $y = kx + c$ (where m, b, k, and c are constants) are sketched in Fig. 95.

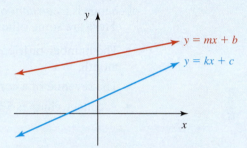

Figure 95
Exercise 99

   a. Which is greater, m or k? Explain.
   b. Which is greater, b or c? Explain.

100. Explain why distinct lines with equal slopes are parallel.

101. Create an equation of the form $y = mx + b$. Find two points on the graph of your equation by substituting two values for x. Use the two points to calculate the slope of the line and compare the result with your chosen value of m.

102. Describe how to use the method discussed in this section to graph an equation of the form $y = mx + b$.

3.5 Rate of Change

Objectives

» Calculate the rate of change of a quantity.

» Explain why slope is a rate of change.

» Use rate of change to help find an equation of a linear model.

» Describe the slope addition property.

In this section, we will describe how quickly a quantity changes in relation to another quantity. For example, we can describe how quickly the tuition of a college increases in relation to time or how quickly the altitude of an airplane declines in relation to time.

Calculating Rate of Change

Suppose the temperature increased *steadily* by 6°F in the past 3 hours. We can compute how much the temperature changed *per hour* by finding the unit ratio of the change in temperature (6°F) to the change in time (3 hours):

$$\frac{6°F}{3 \text{ hours}} = \frac{2°F}{1 \text{ hour}}$$

So, the temperature increased by 2°F per hour. This is an example of a *rate of change*. We say the rate of change of temperature with respect to time is 2°F per hour. The rate of change is a *constant* because the temperature increased *steadily*.

Here are some other examples of rates of change:

- The number of friends on a student's Facebook account increases by 10 friends per month.
- The revenue of a company decreases by $3 million per year.
- A college charges $70 per unit (hour or credit).

Example 1   Finding Rates of Change

1. An airplane's altitude increases steadily by 10,000 feet over a 5-minute period. Find the rate of change of altitude with respect to time.
2. The temperature increased steadily from 60°F at 6 A.M. to 72°F at 10 A.M. Find the rate of change of temperature with respect to time.

Solution

1. We find the unit ratio of the change in altitude (10,000 feet) to the change in time (5 minutes):

$$\frac{\text{change in altitude}}{\text{change in time}} = \frac{10,000 \text{ feet}}{5 \text{ minutes}} \qquad \textit{Find ratio.}$$

$$= \frac{2000 \text{ feet}}{1 \text{ minute}} \qquad \textit{Find unit ratio.}$$

So, the airplane climbed at a rate of 2000 feet per minute.

2. We find the unit ratio of the change in temperature to the change in time:

$$\frac{\text{change in temperature}}{\text{change in time}} = \frac{72°F - 60°F}{10:00 - 6:00} \qquad \textit{Change in a quantity is ending amount minus beginning amount.}$$

$$= \frac{12°F}{4 \text{ hours}} \qquad \textit{Subtract.}$$

$$= \frac{3°F}{1 \text{ hour}} \qquad \textit{Find unit ratio.}$$

So, the temperature increased 3°F per hour.

Formula for Rate of Change

Suppose that a quantity y changes steadily from y_1 to y_2 as a quantity x changes steadily from x_1 to x_2. Then the **rate of change** of y with respect to x is the ratio of the change in y to the change in x:

$$\frac{\text{change in } y}{\text{change in } x} = \frac{y_2 - y_1}{x_2 - x_1}$$

We often refer to rate of change *with respect to time* simply as "rate of change."

Example 2   Finding Rates of Change

1. Worldwide PC shipments declined approximately steadily from 365 million PCs in 2011 to 289 million PCs in 2015 (Source: *Gartner*). Find the approximate rate of change of worldwide PC shipments.
2. The total cost of 12 karate classes and an enrollment fee is $158. The total cost of 20 karate classes and the same enrollment fee is $230. The charge per class is the same, regardless of the number of classes for which you pay. Find the rate of change of the total cost with respect to the number of classes.

Solution

1. $\dfrac{\text{change in PC shipments}}{\text{change in time}}$

    $= \dfrac{289 \text{ million PCs} - 365 \text{ million PCs}}{\text{year } 2015 - \text{year } 2011}$ *Change in a quantity is ending amount minus beginning amount.*

    $= \dfrac{-76 \text{ million PCs}}{4 \text{ years}}$ *Subtract.*

    $= \dfrac{-19 \text{ million PCs}}{1 \text{ year}}$ *Find unit ratio.*

The rate of change of worldwide PC shipments was about −19 million PCs per year. Shipments had a yearly decline of about 19 million PCs. Our result is an approximation because worldwide PC shipments declined *approximately* steadily.

2. To be consistent in finding the signs of the changes, we assume that the number of classes increases from 12 to 20 and that the total cost increases from $158 to $230:

    $\dfrac{\text{change in total cost}}{\text{change in number of classes}} = \dfrac{230 \text{ dollars} - 158 \text{ dollars}}{20 \text{ classes} - 12 \text{ classes}}$ *Change in a quantity is ending amount minus beginning amount.*

    $= \dfrac{72 \text{ dollars}}{8 \text{ classes}}$ *Subtract.*

    $= \dfrac{9 \text{ dollars}}{1 \text{ class}}$ *Find unit ratio.*

The rate of change of the total cost with respect to the number of classes is $9 per class. So, the cost of each class is $9. Our result is exact because it is given that the charge per class is the same.

From our work in Example 2, we can see a connection between the sign of a rate of change and whether the response variable increases or decreases as the explanatory variable increases. In Problem 2, the rate of change was *positive* because the total cost *increases* (as the number of classes increases). In Problem 1, the approximate rate of change was *negative* because the number of PC shipments *decreased* (as time increased).

Signs of Rate of Change

Suppose that a quantity t affects (explains) a quantity p. Then

- If p increases steadily as t increases steadily, then the rate of change of p with respect to t is positive.
- If p decreases steadily as t increases steadily, then the rate of change of p with respect to t is negative.

Slope Is a Rate of Change

You may have noticed that the expression

$$\frac{y_2 - y_1}{x_2 - x_1}$$

we have been using to calculate rate of change is the same expression we use to calculate the slope of a line. In other words, the slope of a linear model is a rate of change. We will explore this important concept in Example 3.

Example 3   Comparing Slope with a Rate of Change

Suppose that a student travels at a constant rate on a road trip. Let d be the distance (in miles) that the student can drive in t hours. Some values of t and d are shown in Table 30.

1. Construct a scatterplot. Then draw a linear model.
2. Find the slope of the linear model.
3. Find the rate of change of the distance traveled in each given period. Compare each result with the slope of the linear model.
   a. From $t = 3$ to $t = 4$ b. From $t = 0$ to $t = 5$

Table 30 Times and Distances

Time (hours) t	Distance (miles) d
0	0
1	50
2	100
3	150
4	200
5	250

Solution

1. We draw a scatterplot and then draw a line that contains the data points (see Fig. 96).

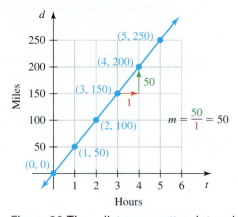

Figure 96 Time-distance scatterplot and model

2. Recall from Section 3.3 that the formula for slope is $m = \dfrac{y_2 - y_1}{x_2 - x_1}$. So, with the variables t and d, we have

$$\frac{d_2 - d_1}{t_2 - t_1}$$

We arbitrarily use the points $(3, 150)$ and $(4, 200)$ to calculate the slope of the linear model:

$$\frac{200 - 150}{4 - 3} = \frac{50}{1} = 50$$

So, the slope is 50. This checks with the calculation shown in Fig. 96.

3. a. First we calculate the rate of change of distance traveled from $t = 3$ to $t = 4$:

$$\frac{\text{change in distance}}{\text{change in time}} = \frac{200 \text{ miles} - 150 \text{ miles}}{4 \text{ hours} - 3 \text{ hours}} \qquad \textit{Change in a quantity is ending amount minus beginning amount.}$$

$$= \frac{50 \text{ miles}}{1 \text{ hour}} \qquad \textit{Subtract.}$$

$$= 50 \text{ miles per hour} \qquad \textit{Divide.}$$

The rate of change (50 miles per hour) is equal to the slope (50).

b. Now we calculate the rate of change of distance traveled from $t = 0$ to $t = 5$:

$$\frac{\text{change in distance}}{\text{change in time}} = \frac{250 \text{ miles} - 0 \text{ miles}}{5 \text{ hours} - 0 \text{ hours}} \qquad \textit{Change in a quantity is ending amount minus beginning amount.}$$

$$= \frac{250 \text{ miles}}{5 \text{ hours}} \qquad \textit{Subtract.}$$

$$= \frac{50 \text{ miles}}{1 \text{ hour}} \qquad \textit{Find unit ratio.}$$

$$= 50 \text{ miles per hour} \qquad \textit{Divide.}$$

The rate of change (50 miles per hour) is equal to the slope (50).

In Example 3, we found that the time t and the distance traveled d are linearly related. We also found that the slope of the linear model is equal to the rate of change of distance traveled with respect to time.

Slope Is a Rate of Change

If there is a linear relationship between the quantities t and p, and if t affects (explains) p, then the slope of the linear model is equal to the rate of change of p with respect to t.

In Problem 3 of Example 3, we calculated the same rate of change (50 miles per hour) for two different periods. In fact, the rate of change is 50 miles per hour for *any* period within the first 5 hours. This makes sense because the rate of change is equal to the slope of the line (50), which is a constant.

Constant Rate of Change

Suppose that a quantity t affects (explains) a quantity p. Then

* If there is a linear relationship between t and p, then the rate of change of p with respect to t is constant.
* If the rate of change of p with respect to t is constant, then there is a linear relationship between t and p.

Finding an Equation of a Linear Model

We can use what we have learned about rate of change to help us find an equation of a linear model.

Example 4   Finding a Model

In 2010, a college's enrollment was 20 thousand students. Each year, the enrollment increases by 2 thousand students. Let E be the enrollment (in thousands of students) at t years since 2010.

1. Is there a linear relationship between t and E? Explain.
2. Find the E-intercept of a linear model. What does it mean in this situation?
3. Find the slope of the linear model. What does it mean in this situation?
4. Find an equation of the linear model.

Solution

1. Because the rate of change of enrollment per year is a *constant* 2 thousand students per year, the variables t and E are linearly related.
2. We list some values of t and E in Table 31. We plot the corresponding points and sketch the line that contains the points in Fig. 97.

Table 31 College Enrollments

Years since 2000 t	Enrollment (thousands of students) E
0	20
1	22
2	24
3	26
4	28
5	30

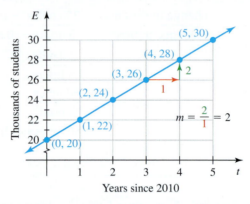

Figure 97 Enrollment scatterplot and model

From the table and the graph, we see that the E-intercept is $(0, 20)$. This means that the enrollment was 20 thousand students in the year 2010.

3. The rate of change of enrollment is 2 thousand students per year. So, the slope of the linear model is 2. This checks with the calculation shown in Fig. 97.
4. An equation of a line can be written in slope–intercept form, $y = mx + b$. Using t and E, we have $E = mt + b$. Because the slope is 2 and the E-intercept is $(0, 20)$, we have $E = 2t + 20$.

   To check our work with a graphing calculator, we begin by entering our model (see Fig. 98). Then we check that the entries in the graphing calculator table in Fig. 99 equal the entries in Table 31. We also check that the graph of our equation contains the points of the scatterplot of the data (see Fig. 100).

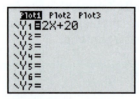

Figure 98 Enter the model

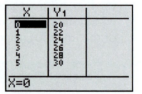

Figure 99 Use a table to verify the model

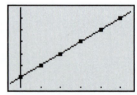

Figure 100 Use a graph to verify the model

For graphing calculator instructions on constructing tables, see Appendix A.13. For instructions on graphing an equation with a scatterplot, see Appendix A.8 and A.10.

In Example 4, we found the model $E = 2t + 20$. Here is the connection between parts of the equation and the situation:

$$E = \underbrace{2}_{\substack{\text{rate of} \\ \text{change of} \\ \text{enrollment}}} \cdot \; t \; + \underbrace{20}_{\substack{\text{enrollment at} \\ t = 0}}$$

Example 5   Finding a Model

Elevation affects (explains) the temperature at which water boils (the *boiling point*): the higher you are, the lower the boiling point is. At sea level (elevation 0), water boils at 212°F. The boiling point declines by 5.9°F for each thousand-meter increase in elevation (Source: *Thermodynamics, an Engineering Approach by Cengal & Boles*). Let B be the boiling point (in degrees Fahrenheit) at an elevation of E thousand meters.

1. Is there a linear relationship between E and B? If so, find the slope.
2. Find an equation of the model.
3. Perform a unit analysis of the equation.
4. Denali, the highest mountain in the United States, reaches 6.194 thousand meters. What is the boiling point of water at the peak?

Solution

1. Because the elevation affects (explains) the boiling point, we will consider the rate of change of the boiling point with respect to elevation. Because this rate of change is a *constant* −5.9°F per thousand meters, the variables E and B are linearly related and the slope is −5.9.
2. We want an equation of the form $B = mE + b$. Because $B = 212$ when $E = 0$, the B-intercept is $(0, 212)$. Recall that $m = -5.9$, so an equation is $B = -5.9E + 212$.
3. Here is a unit analysis of the equation $B = -5.9E + 212$:

$$\underbrace{B}_{\text{degrees Fahrenheit}} = \underbrace{-5.9}_{\frac{\text{degrees Fahrenheit}}{\text{thousand meters}}} \cdot \underbrace{E}_{\text{thousand meters}} + \underbrace{212}_{\text{degrees Fahrenheit}}$$

We can use the fact that $\dfrac{\text{thousand meters}}{\text{thousand meters}} = 1$ to simplify the units of the expression on the right-hand side of the equation:

$$\frac{\text{degrees Fahrenheit}}{\text{thousand meters}} \cdot \text{thousand meters} + \frac{\text{degrees}}{\text{Fahrenheit}} = \frac{\text{degrees}}{\text{Fahrenheit}} + \frac{\text{degrees}}{\text{Fahrenheit}}$$

So, the units of the expressions on both sides of the equation are degrees Fahrenheit, which suggests that our equation is correct.

4. To find the boiling point, we substitute the input 6.194 for E in the equation $B = -5.9E + 212$:

$$B = -5.9(6.194) + 212 \quad \textit{Substitute 6.194 for E.}$$
$$\approx 175.46 \qquad\qquad \textit{Perform indicated operations.}$$

So, the boiling point is about 175°F at the peak of Denali.

▶

Example 6   Analyzing a Model

Banks' average overdraft fees are shown in Table 32 for various years. Let F be the average overdraft fee (in dollars) at t years since 2000. A model of the situation is

$$F = 0.63t + 23.61$$

1. Use a graphing calculator to draw a scatterplot and the model in the same viewing window. Check whether the line comes close to the data points.
2. What is the slope of the model? What does it mean in this situation?
3. Find the rates of change in average overdraft fee from one year in Table 32 to the next one listed. Compare the rates of change with the result found in Problem 2.
4. What is the F-intercept? What does it mean in this situation?
5. Predict the average overdraft fee in 2021.

Table 32 Banks' Average Overdraft Fees

Year	Fee (dollars)
2000	23.74
2004	25.81
2008	28.95
2012	31.26
2015	33.07

Source: *Bankrate*

Solution

1. We draw the scatterplot and the model in the same viewing window (see Fig. 101). For graphing calculator instructions on drawing scatterplots and models, see Appendix A.8 and A.10.

 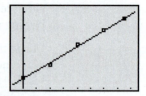

Figure 101 Fee scatterplot and model

The line comes close to the data points, so the model is a reasonable one.

2. The slope is 0.63 because $F = 0.63t + 23.61$ is of the form $y = mx + b$ and $m = 0.63$. According to the model, the average overdraft fee is increasing by 0.63 dollar per year.

3. The rates of change of average overdraft fee are shown in Table 33. All the rates of change are fairly close to 0.63 dollar per year.

Table 33 Rates of Change of Average Overdraft Fee

Year	Fee (dollars)	Rate of Change of Fee from Previous Year (dollars per year)
2000	23.74	—
2004	25.81	$(25.81 - 23.74) \div (2004 - 2000) \approx 0.52$
2008	28.95	$(28.95 - 25.81) \div (2008 - 2004) \approx 0.79$
2012	31.26	$(31.26 - 28.95) \div (2012 - 2008) \approx 0.58$
2015	33.07	$(33.07 - 31.26) \div (2015 - 2012) \approx 0.60$

4. The F-intercept is $(0, 23.61)$ because $F = 0.63t + 23.61$ is of the form $y = mx + b$ and $b = 23.61$. According to the model, the average overdraft fee was \$23.61 in 2000.

5. We substitute the input 21 for t in the equation $F = 0.63t + 23.61$:

$$F = 0.63(21) + 23.61 = 36.84$$

According to the model, the average overdraft fee will be about \$36.84 in 2021.

▶

WARNING   A common error in describing the meaning of the slope of a model is vagueness. For example, a description such as

> The slope means that it is increasing.

neither specifies the quantity that is increasing nor the rate of increase. The following statement includes the missing information:

> The slope of 0.63 means the average overdraft fee is increasing by 0.63 dollar per year.

Slope Addition Property

So far, we have discussed rate of change for linear models. Now we will explore rate of change for linear equations that are not used as models.

Table 34 Some Solutions of $y = 2x + 1$

x	y
0	1
1	3
2	5
3	7
4	9

Example 7   Interpreting Slope as a Rate of Change

What is the slope of the line $y = 2x + 1$? Interpret the slope as a rate of change.

Solution

The slope is 2. So, the rate of change of y with respect to x is 2. This means that if the value of x increases by 1, the value of y increases by 2 (see Table 34).

▶

Table 35 Some Solutions of $y = -3x + 8$

x	y
0	8
1	5
2	2
3	-1
4	-4

Example 8   Interpreting Slope as a Rate of Change

What is the slope of the line $y = -3x + 8$? Interpret the slope as a rate of change.

Solution

The slope is -3. So, the rate of change of y with respect to x is -3. This means that if the value of x increases by 1, the value of y decreases by 3 (see Table 35).

▶

Our observations made in Examples 7 and 8 suggest yet another way to think about slope: the **slope addition property**.

Slope Addition Property

For a linear equation of the form $y = mx + b$, if the value of x increases by 1, then the value of y changes by the slope m. In other words, if an input increases by 1, the output changes by the slope.

For the line $y = 4x - 9$, the slope is 4. If the value of x increases by 1, then the value of y changes by 4. For the line $y = -5x - 2$, the slope is -5. If the value of x increases by 1, then the value of y changes by -5.

Example 9   Identifying Possible Linear Equations

Some solutions of four equations are listed in Table 36. Which of the equations could be linear?

Table 36 Some Solutions of Four Equations

Equation 1		Equation 2		Equation 3		Equation 4	
x	y	x	y	x	y	x	y
1	23	4	12	0	3	50	8
2	20	5	17	1	6	51	8
3	17	6	22	2	12	52	8
4	14	7	27	3	24	53	8
5	11	8	32	4	48	54	8

Solution

1. Equation 1 could be linear: Each time the value of x increases by 1, the value of y changes by -3.
2. Equation 2 could be linear: Each time the value of x increases by 1, the value of y changes by 5.
3. Equation 3 is not linear: Each time the value of x increases by 1, the value of y does not change by the same value.
4. Equation 4 could be $y = 8$, which is a linear equation: Each time the value of x increases by 1, the value of y changes by 0.

▶

Table 37 Points on a Line

x	y
0	15
1	9
2	3
3	-3
4	-9

Example 10   Finding an Equation of a Line

A line contains the points listed in Table 37. Find an equation of the line.

Solution

For $y = mx + b$, the graph has y-intercept $(0, b)$. From Table 37, we see that the y-intercept is $(0, 15)$, so $b = 15$. As the value of x increases by 1, the value of y changes by -6. By the slope addition property, we know that $m = -6$. Therefore, an equation of the line is $y = -6x + 15$.

We use a graphing calculator table to verify our work in Fig. 102.

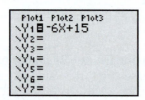 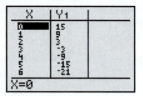

Figure 102 Verify the equation

Group Exploration

Averaging rates of change

Google's worldwide revenues are shown in Table 38 for various years.

1. Find the approximate rate of change of the annual revenue from 2011 to 2015. [**Hint:** Use the 2011 and 2015 dates in Table 38.]

2. Find the rate of change of annual revenue from each year to the next, beginning in 2011. [**Hint:** You should find four results.]

Table 38 Google's Worldwide Revenues

Years	Revenue (billions of dollars)
2011	38
2012	50
2013	56
2014	66
2015	75

Source: *Google*

3. Find the average of the four rates of change you found in Problem 2. [**Hint:** Divide the sum of the four rates of change by 4.]

4. Compare the results you found in Problems 1 and 3.

5. The following expression is an example of a *telescoping sum*:

$$(50 - 38) + (56 - 50) + (66 - 56) + (75 - 66)$$

Explain why the above sum is equal to $75 - 38$.

6. Use the result you found in Problem 5 to help explain why the following statement is true:

$$\frac{\frac{50-38}{2012-2011} + \frac{56-50}{2013-2012} + \frac{66-56}{2014-2013} + \frac{75-66}{2015-2014}}{4}$$

$$= \frac{75 - 38}{2015 - 2011}$$

Also, explain how this statement is related to your comparison in Problem 4.

7. Describe what you have learned from doing this exploration.

Tips for Success   Get the Most Out of Working Exercises

If you work an exercise by referring to a similar example in your notebook or in the text, it is a good idea to try the exercise again without referring to your source of help. If you need to refer to your source of help to solve the exercise a second time, consider trying the exercise a third time without help. When you finally complete the exercise without help, reflect on which concepts you used to work the exercise, where you had difficulty, and what the key idea was that opened the door of understanding for you.

A similar strategy can be used in getting help from a student, an instructor, or a tutor.

If this sounds like a lot of work, it is! But it is well worth it. Although it is important to complete each assignment, it is also important to learn as much as possible while progressing through it.

Homework 3.5

For extra help ▶ **MyLab Math**   ▦ Watch the videos in MyLab Math

1. The volume of water in a swimming pool increases steadily by 2400 gallons in an 8-hour period. Find the rate of change of the volume of water.

2. The height of a ti plant increases steadily by 48 inches in 12 years. Find the rate of change of the ti plant's height.

3. An airplane's altitude declines steadily by 24,750 feet over a 15-minute period. Find the rate of change of the airplane's altitude.

4. The temperature decreases steadily by 15°F over a 3-hour period. Find the rate of change of temperature.

For Exercises 5–10, round your result to the second decimal place as needed.

5. The number of shark attacks in the United States increased approximately steadily from 18 attacks in 1989 to 53 attacks in 2016 (Source: *University of Florida*). Find the approximate rate of change of shark attacks.

6. Target's revenue increased approximately steadily from $61.5 billion in 2007 to $73.8 billion in 2016 (Source: *Target*). Find the approximate rate of change of revenue.

7. The number of skier/snowboarder visits to U.S. slopes west of the Rockies has decreased approximately steadily from 12.2 million visits in 2011 to 7.0 million visits in 2015 (Source: *National Ski Areas Association*). Find the approximate rate of change of the number of visits.

8. The number of workers earning minimum wage decreased approximately steadily from 18 million workers in 2010 to 9 million workers in 2015 (Source: *U.S. Department of Labor*). Find the approximate rate of change of the number of workers earning minimum wage.

9. The percentage of Americans who said they'd like to lose weight decreased approximately steadily from 59% in 2011 to 51% to 2014 (Source: *Gallup*). Find the approximate rate of change of the percentage of Americans who said they'd like to lose weight.

10. The percentage of all e-mail that is spam decreased approximately steadily from 84.9% in 2010 to 54.1% in 2015 (Source: *Trustware*). Find the approximate rate of change of this percentage.

11. In-district students at Triton College pay $1017 for 9 credit hours (units) of classes and $1356 for 12 credit hours of classes (Source: *Triton College*). Find the approximate rate of change of the total cost of classes with respect to the number of credit hours of classes.

12. In San Francisco, the median price of a two-bedroom house is $1,635,000, and that of a four-bedroom house is $2,500,000 (Source: *Trulia*). Find the approximate rate of change of median price with respect to the number of bedrooms.

13. In order for a family living in New York to qualify for free health insurance with the program Child Health Plus, the family's annual income must be less than a maximum level. In 2016, the maximum level of a three-person family was $32,244 and the maximum level of a seven-person family was $58,764 (Source: *New York State Department of Health*). Find the approximate rate of change of maximum income level with respect to family size.

14. A person stacks some cups of uniform shape and size (one placed inside the next). The height of 3 stacked cups is 17.5 centimeters and of 5 stacked cups is 23.0 centimeters. Find the rate of change of the height of the stacked cups with respect to the number of cups.

15. A company's revenue in 2010 was $3 million. Each year its revenue increases by $4 million. Let r be the annual revenue (in millions of dollars) at t years since 2010.
    a. Is there a linear relationship between t and r? Explain. If the relationship is linear, find the slope and describe what it means in this situation.
    b. Recall that we can describe a linear model using an equation, a table, a graph, or words.
       i. Use an equation to describe the situation.
       ii. Use a table of values of t and r to describe the situation.
       iii. Use a graph to describe the situation.

16. As of February 1, a garage band knows 5 songs. Each week, the band members learn 2 more songs. Let n be the number of songs that the band knows at t weeks since February 1.
    a. Is there a linear relationship between t and n? Explain. If the relationship is linear, find the slope and describe what it means in this situation.
    b. Recall that we can describe a linear model using an equation, a table, a graph, or words.
       i. Use an equation to describe the situation.
       ii. Use a table of values of t and n to describe the situation.
       iii. Use a graph to describe the situation.

17. In 2014, the number of drive-in movie sites in the United States was 348. The number has been decreasing by about 12 sites per year since then (Source: *National Association of Theatre Owners*). Let n be the number of drive-in movie sites at t years since 2014.
    a. Is there an approximate linear relationship between t and n? Explain. If the relationship is approximately linear, find the slope and describe what it means in this situation.
    b. What is the n-intercept of the model? What does it mean in this situation?
    c. Find an equation of the model.
    d. Perform a unit analysis of the equation you found in part (c).
    e. Predict the number of sites in 2021.

18. Total spending for Father's Day gifts was $12.5 billion in 2014 and has increased by about $0.5 billion per year since then (Source: *National Retail Federation*). Let s be the annual total spending (in billions of dollars) for Father's Day gifts at t years since 2014.
    a. Is there an approximate linear relationship between t and s? Explain. If the relationship is approximately linear, find the slope and describe what it means in this situation.
    b. What is the s-intercept of the model? What does it mean in this situation?
    c. Find an equation of the model.
    d. Perform a unit analysis of the equation you found in part (c).
    e. Predict the total spending in 2020.

19. A student's savings account has a balance of $4700 on September 1. Each month, the balance declines by $650. Let B be the balance (in dollars) at t months since September 1.
   a. Find the slope of the linear model that describes this situation. What does it mean in this situation?
   b. What is the B-intercept of the model? What does it mean in this situation?
   c. Find an equation of the model.
   d. Perform a unit analysis of the equation you found in part (c).
   e. Find the balance on March 1 (6 months after September 1).

20. A person owns a propane-gas barbecue grill with a tank that holds 5 gallons of propane. The person always sets the temperature to 350°F, which uses 0.125 gallon of propane per hour. Let g be the number of gallons of propane that remain in the tank after t hours of cooking since the tank was last filled.
   a. Find the slope of the linear model that describes this situation. What does it mean in this situation?
   b. What is the g-intercept of the model? What does it mean in this situation?
   c. Find an equation of the model.
   d. The person fills the propane tank and then uses the grill for 3 hours. How much propane remains in the tank?
   e. Estimate the amount of propane that will remain in the tank after 50 hours of cooking since the tank was last filled.

21. For the spring semester 2017, part-time students at Centenary College paid $592 per credit (unit or hour) for tuition and paid a mandatory part-time student fee of $16 per semester.
   a. Find an equation of a linear model to describe the situation. Explain what your variables represent.
   b. Perform a unit analysis of the equation you found in part (a).
   c. What is the slope of your model? What does it mean in this situation?
   d. What was the total one-semester cost of tuition plus part-time student fee for 9 credits of classes?

22. For the spring semester 2017, California residents paid an enrollment fee of $46 per unit (credit or hour) at Santa Barbara City College. Students were also required to pay a $19 health fee each semester.
   a. Find an equation of a linear model to describe the situation. Explain what your variables represent.
   b. Perform a unit analysis of the equation you found in part (a).
   c. What is the slope of your model? What does it mean in this situation?
   d. What was the total one-semester cost of enrollment plus health fee for 15 units of classes?

23. A person drives her Toyota Prius® on a road trip. At the start of the trip, she fills up the 11.9-gallon tank with gasoline. During the trip, the car uses about 0.02 gallon of gasoline per mile. Let G be the number of gallons of gasoline remaining in the tank after driving d miles.
   a. What is the slope of the linear model that describes this situation? What does it mean in this situation?
   b. What is the G-intercept of the model? What does it mean in this situation?
   c. Find an equation of the model.
   d. Perform a unit analysis of the equation you found in part (c).
   e. If the person drives 525 miles before she refuels the car, how much gasoline is required to fill up the tank?

24. Although the United States and Great Britain use the Fahrenheit (°F) temperature scale, most countries use the Celsius (°C) scale. The temperature reading 0°C is equivalent to the Fahrenheit reading 32°F. An increase of 1°C is equivalent to an increase of 1.8°F. Let F be the Fahrenheit reading that is equivalent to a Celsius reading of C degrees. Assume that F is the response variable.
   a. Find the slope of a linear model that describes this situation. What does it mean in this situation?
   b. What is the F-intercept of the model? What does it mean in this situation?
   c. Find an equation of the model.
   d. Perform a unit analysis of the equation you found in part (c).
   e. If the temperature is 30°C, what is the Fahrenheit reading?

25. Let s be worldwide digital camera annual sales (in millions of cameras) at t years since 2010. A reasonable model is $s = -21.1t + 130.6$ (Source: *Camera and Imaging Products Association*).
   a. What is the slope? What does it mean in this situation?
   b. What is the s-intercept? What does it mean in this situation?
   c. Estimate the sales in 2016.

26. Let n be the number of volunteer firefighters who died on duty in the year that is t years since 2014. A reasonable model of the number of volunteer firefighters who have died on duty is $n = -1.97t + 40$ (Source: *National Fire Protection Agency*).
   a. What is the slope? What does it mean in this situation?
   b. What is the n-intercept? What does it mean in this situation?

27. The percentages of adult Internet users who use social networking sites are shown in Table 39 for various years.

Table 39 Percentages of Adult Internet Users Who Use Social Networking Sites

Year	Percent
2010	60
2011	65
2012	67
2013	73
2014	74
2015	76

Source: *Pew Research Center*

Let p be the percentage of adult Internet users who use social networking sites at t years since 2010. A model of the situation is $p = 3.23t + 61.10$.
   a. Use a graphing calculator to draw a scatterplot and the model in the same viewing window. Does the line come close to the data points?
   b. What is the slope? What does it mean in this situation?
   c. Find the rate of change of the percentage of adult Internet users who use social networking sites for each of the periods 2010–2012, 2012–2014, and 2014–2015. Compare the rate of the change of each of the three periods with the result you found in part (b).
   d. What is the p-intercept? What does it mean in this situation?
   e. Use the model to predict the percentage of adult Internet users who will use social networking sites in 2020.

28. The numbers of bird species involved in airline bird strikes are shown in Table 40 for various years.

Table 40 Numbers of Bird Species Involved in Airline Bird Strikes

Year	Number of Bird Species
2008	230
2009	259
2010	266
2011	282
2012	297
2013	314
2014	330

Source: *Federal Aviation Administration*

Let n be the number of bird species involved in airline bird strikes at t years since 2000. A model of the situation is $n = 15.75t + 109.32$.

a. Use a graphing calculator to draw a scatterplot and the model in the same viewing window. Does the line come close to the data points?

b. What is the slope? What does it mean in this situation?

c. What is the approximate rate of change of the number of bird species involved in airline bird strikes for each of the periods 2008–2010, 2010–2012, and 2012–2014. Compare the approximate rate of change of each of the three periods with the result you found in part (b).

d. Predict the number of bird species that will be involved in airline bird strikes in 2021.

e. Give two possible reasons why the entries in Table 40 for the number of bird species involved in airline strikes increase.

29. In an attempt to raise the low graduation rates of college football and basketball players, Division I institutions require that prospective athletes must meet new grade point average (GPA) standards (based on a maximum of 4.0) to play as freshmen, and enrolled athletes must meet these new standards to keep playing (see Table 41).

Table 41 Core GPAs Needed to Qualify to Play, for Given SAT Scores

SAT Score	Qualifying Core GPA
620	3.0
700	2.8
780	2.6
860	2.4
940	2.2
1020	2.0

Source: *NCAA*

Let G be the qualifying core GPA for an SAT score of s points. A model of the situation is $G = -0.0025s + 4.55$.

a. Use a graphing calculator to draw a scatterplot and the model in the same viewing window. Does the line come close to the data points?

b. If an athlete's SAT score is 400 points, the lowest possible score, estimate the student's qualifying core GPA. The actual qualifying core GPA is 3.55. Compute the error in the estimate.

c. What is the slope? What does it mean in this situation?

d. What is the G-intercept? What does it mean in this situation?

e. Does an athlete with a core GPA of 3.2 and an SAT score of 500 points meet the standards? Explain.

30. The average selling prices for a Nissan Altima® of various ages are shown in Table 42.

Table 42 Average Selling Prices of Nissan Altimas

Age (years)	Average Price (dollars)
1	17,963
2	15,954
3	14,423
4	12,216
5	10,509

Source: *AutoTrader®*

Let p be the average price (in dollars) of a Nissan Altima of age a years. A model of the situation is $p = -1864.6a + 19,806.8$.

a. Use a graphing calculator to draw a scatterplot and the model in the same viewing window. Does the line come close to the data points?

b. What is the slope? What does it mean in this situation?

c. Estimate the average price of an 8-year-old Nissan Altima.

31. A person is on a car trip. Let d be the distance (in miles) traveled in t hours of driving. A reasonable model is shown in Fig. 103.

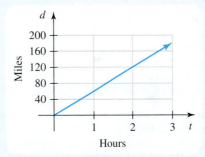

Figure 103 Exercise 31

a. Is the car traveling at a constant speed according to the model? Explain.

b. What is the speed of the car?

32. A person is on an airplane. Let d be the distance (in miles) traveled in t hours. A reasonable model is shown in Fig. 104.

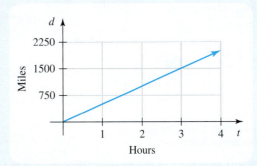

Figure 104 Exercise 32

a. Is the airplane traveling at a constant speed according to the model? Explain.

b. What is the speed of the airplane?

33. A person is on a car trip. Let V be the volume (in gallons) of gasoline that remains in the gasoline tank after t hours of driving. A reasonable model is shown in Fig. 105.

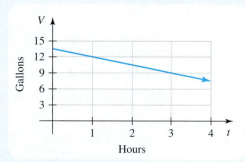

Figure 105 Exercise 33

a. Is the rate of change of gasoline remaining in the tank constant according to the model? Explain.

b. What is the rate of change of gasoline remaining in the tank?

34. Let F be the temperature (in degrees Fahrenheit) at t hours after noon. A reasonable model is shown in Fig. 106.

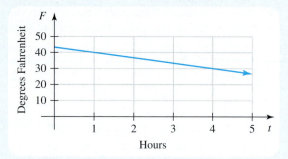

Figure 106 Exercise 34

a. Is the rate of change of temperature constant according to the model? Explain.

b. What is the rate of change of temperature?

35. The scatterplot and the model in Fig. 107 describe the relationship between 2013 U.S. revenues and 2013 worldwide revenues, both in millions of dollars, of the 15 worldwide best-selling video games.

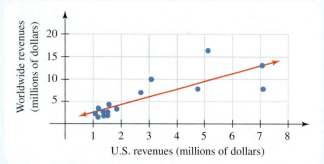

Figure 107 Exercise 35
(**Source:** *VGChartz*)

What is the slope of the model? What does it mean in this situation?

36. Many appliances such as televisions and electronic gadgets such as cell phones contain rare earth metals. The scatterplot and the model in Fig. 108 compare the annual revenue (in billions of dollars) of U.S. electronics and appliance stores with annual worldwide rare earth mining (in thousands of metric tons).

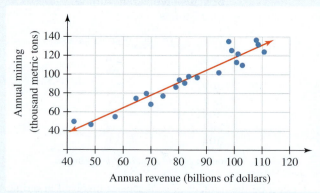

Figure 108 Exercise 36
(**Source:** *U.S Geological Survey*)

What is the slope of the model? What does it mean in this situation?

37. The scatterplot and the model in Fig. 109 compare the maximum weights (in pounds) with the average life expectancies (in years) for 15 dog breeds.

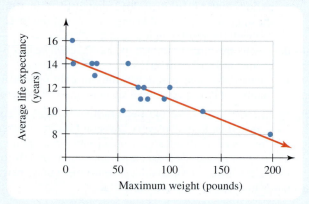

Figure 109 Exercise 37
(**Source:** *Find The Best*)

What is the slope of the model? What does it mean in this situation?

38. The scatterplot and the model in Fig. 110 compare the annual amounts (in millions of pounds) of herbicides used with the numbers (in millions) of bee colonies in the United States for years between 1939 and 2013, inclusive.

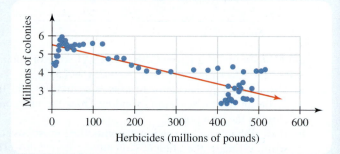

Figure 110 Exercise 38
(**Source:** *U.S. Department of Agriculture*)

What is the slope of the model? What does it mean in this situation?

39. A math tutor charges $30 per hour. Let c be the total charge (in dollars) for t hours of tutoring. Are t and c linearly related? If so, find the slope and describe what it means in this situation.

40. Regular gasoline sells for $2.75 per gallon at a certain Shell station in Charlestown, Massachusetts. Let T be the total charge (in dollars) for g gallons of gas. Are g and T linearly related? If so, find the slope and describe what it means in this situation.

41. Some solutions of four equations are listed in Table 43. Which of the equations could be linear?

Table 43 Solutions of Four Equations

Equation 1		Equation 2		Equation 3		Equation 4	
x	y	x	y	x	y	x	y
0	2	0	5	2	5	5	20
1	5	1	6	3	3	6	19
2	8	2	8	4	1	7	17
3	11	3	9	5	−1	8	14
4	14	4	11	6	−3	9	10

42. Some solutions of four equations are listed in Table 44. Which of the equations could be linear?

Table 44 Solutions of Four Equations

Equation 1		Equation 2		Equation 3		Equation 4	
x	y	x	y	x	y	x	y
0	2	0	35	4	−7	23	25
1	4	1	31	5	−3	24	23
2	8	2	27	6	1	25	20
3	16	3	23	7	5	26	18
4	32	4	19	8	9	27	15

43. Some solutions of four linear equations are listed in Table 45. Complete the table.

Table 45 Solutions of Four Linear Equations

Equation 1		Equation 2		Equation 3		Equation 4	
x	y	x	y	x	y	x	y
0	3	0	99	21	16	43	
1	8	1	92	22		44	
2		2		23	12	45	23
3		3		24		46	
4		4		25		47	29

44. Some solutions of four linear equations are listed in Table 46. Complete the table.

Table 46 Solutions of Four Linear Equations

Equation 1		Equation 2		Equation 3		Equation 4	
x	y	x	y	x	y	x	y
0	100	0	2	13	32	75	4
1	94	1	9	14		76	
2		2		15		77	
3		3		16	26	78	
4		4		17		79	16

45. Four sets of points are described in Table 47. For each set, find an equation of a line that contains the points.

Table 47 Solutions of Four Equations

Set 1		Set 2		Set 3		Set 4	
x	y	x	y	x	y	x	y
0	5	0	20	0	21	0	9
1	7	1	17	1	29	1	4
2	9	2	14	2	37	2	−1
3	11	3	11	3	45	3	−6
4	13	4	8	4	53	4	−11

46. Four sets of points are described in Table 48. For each set, find an equation of a line that contains the points.

Table 48 Solutions of Four Equations

Set 1		Set 2		Set 3		Set 4	
x	y	x	y	x	y	x	y
0	22	0	3	0	15	0	−12
1	19	1	7	1	6	1	−10
2	16	2	11	2	−3	2	−8
3	13	3	15	3	−12	3	−6
4	10	4	19	4	−21	4	−4

Large Data Sets

47. **DATA** Access the data about airborne times and distances of Delta Airlines flights, which are available at MyLab Math and at the Pearson Downloadable Student Resources for Math & Stats website. Let T be the airborne time (in minutes) and D be the distance (in miles) for a flight.
  a. Construct a scatterplot of the data.
  b. Give a possible reason why the scatterplot consists of vertically aligned clumps of data points.
  c. On the basis of just the scatterplot, guess whether Delta offers more *routes* that are less than 1000 miles or greater than 1000 miles. On the basis of just the scatterplot, why is it not possible to be sure?
  d. Print your scatterplot, and draw a linear model.
  e. Estimate the slope of the linear model. What does it mean in this situation?
  f. Use the reciprocal of the slope to help you estimate the speed of the airplanes in units of miles per *hour*.
  g. For the distance 849 miles, there is a clump of data points that lie below the other data points for that distance. Use the coordinates of the lowest data point in the clump to estimate the speed (in miles per hour) of the airplane represented by that data point. Explain why an error was likely made in recording the coordinates of the data point (as well as the rest of the data points in the clump).

48. **DATA** Access the data about heights and weights of the 2014 draft for NBA basketball teams, which are available at MyLab Math and at the Pearson Downloadable Student Resources for Math & Stats website. Let H be the height (in inches) and W be the weight (in pounds) of a player.
  a. Construct a scatterplot of the data.
  b. Print your scatterplot, and draw a linear model.
  c. Estimate the slope of the linear model. What does it mean in this situation?

d. Estimate the weight of an 80-inch-tall player.

e. Estimate the height of a 200-pound player.

f. At 75 inches in height, Marcus Smart weighed 227 pounds. At 82 inches in height, Cameron Bairstow weighed 252 pounds. On the basis of the linear model, which of the two players was heavier for his height? Explain.

Concepts

49. For the equation $y = 7x + 2$, describe the change in the value of y as the value of x is increased by 1.

50. For the equation $y = -9x - 5$, describe the change in the value of y as the value of x is increased by 1.

51. For the equation $y = -6x + 40$, use a table of solutions to show that if the value of x is increased by 1, then the value of y changes by the slope.

52. For the equation $y = 8x + 1$, use a table of solutions to show that if the value of x is increased by 1, then the value of y changes by the slope.

53. Recall that we can describe some of or all the solutions of an equation in two variables with an equation, a table, a graph, or words. Describe the slope of the line $y = 3x - 4$ in each of these four ways:

a. Explain how you can determine the slope of the line $y = 3x - 4$ from the equation.

b. Describe the slope of the line $y = 3x - 4$ in terms of "run" and "rise."

c. Describe the slope of the line $y = 3x - 4$ in terms of the slope addition property.

d. Describe the meaning of the slope of the line $y = 3x - 4$ in your own words.

54. Give an example of a linear model other than those in the textbook. What does the slope of the model mean in the situation?

55. For a linear equation of the form $y = mx + b$, by how much does the value of y change if the value of x increases by 2? Explain.

56. In your own words, describe the slope addition property.

57. Explain why slope is a rate of change.

Taking It to the Lab

Climate Change Lab (continued from Chapter 2)

Scientists estimate that the average global temperature has increased by about 1°F in the past century (see Table 49). Recall from the Climate Change Lab in Chapter 1 that many scientists believe an increase as small as 3.6°F could be dangerous. Because the planet's average temperature increased by about 1°F in the past century, it might seem it would take about another two-and-a-half centuries for Earth to reach a dangerous temperature level.

DATA ▼ Table 49 Average Surface Temperatures of Earth

Year	Average Temperature (degrees Fahrenheit)	Year	Average Temperature (degrees Fahrenheit)
1900	57.1	1960	57.2
1905	56.8	1965	57.0
1910	56.6	1970	57.3
1915	57.0	1975	57.1
1920	56.9	1980	57.6
1925	56.9	1985	57.3
1930	57.1	1990	57.8
1935	57.0	1995	57.9
1940	57.3	2000	57.8
1945	57.3	2005	58.3
1950	56.9	2010	58.3
1955	57.0	2015	59.0

Source: *NASA–GISS*

However, the Intergovernmental Panel on Climate Change (IPCC), composed of hundreds of scientists around the world, predicted that Earth's average temperature will rise 2.5°F to 10.4°F in the coming century.[*] Scientists from the United States and Europe have predicted that there is a 9-out-of-10 chance the global average temperature will increase 3°F to 9°F, with a range of 4°F to 7°F most likely.[†]

One result of the last century of global warming is that glaciers are receding and ocean levels are rising. For example, NASA scientist Bill Krabill estimates that Greenland's ice cap, the world's second largest, may be losing ice at a rate of 50 cubic kilometers per year.[‡] The ice cap contained 2.85 million cubic kilometers of ice in 2000. Climatologist Jonathan Gregory of the University of Reading, in the United Kingdom, believes that by 2050 the ice cap may start an irreversible runaway melting. A total meltdown could take 1000 years.[§]

Jonathan Overpeck, director of the Institute for the Study of Planet Earth at the University of Arizona, believes that Greenland's ice cap could melt completely in as little

---

[*]IPCC, 2001: Summary for Policymakers.

[†]T. M. L. Wigley and S. C. B. Raper, "Interpretation of high projections for global-mean warming," *Science* 293 (5529):451–454, July 20, 2001.

[‡]W. Krabill et al., "Greenland ice sheet: High-elevation balance and peripheral thinning," *Science* 289:428–430, 2000.

[§]J. M. Gregory et al., "Climatology: Threatened loss of the Greenland ice sheet," *Nature* (April 8, 2004):426–616.

as 150 years and that a partial melting of Greenland's ice cap by 2100 could raise ocean levels by more than 1 meter.[¶] Coastal engineers estimate that a 1-meter rise would translate into a loss of about 100 meters of land.[‖] In some areas of the world, land loss could be much greater.[**] For example, most of Florida is less than 1 meter above sea level.

Most scientists predict that global warming will also cause the extinction of plants and animals, bring severe water shortages, create extreme weather events, and increase the number of heat-related illnesses and deaths.

Analyzing the Situation

1. Let A be the average global temperature (in degrees Fahrenheit) at t years since 1900. Refer to the scatterplot that you constructed in Problem 3 of the Climate Change Lab of Chapter 1. (If you did not construct this scatterplot earlier, construct a *careful* one now, using the data listed in Table 49.) Does it appear that Earth's average temperature increased much from 1900 to 1965? Explain.

2. **a.** Estimate the average global temperature from 1900 to 1965. [**Hint:** Divide the sum of the temperatures by the number of temperature readings.]

   b. Estimate the average global temperature from 1990 to 2000. [**Hint:** Divide the sum of the temperatures by the number of temperature readings.]

   c. Use the results you found in parts (a) and (b) to estimate the change in Earth's average temperature in the past century. Compare your result with the scientists' estimate of 1°F.

3. In Problem 5 of the Climate Change Lab of Chapter 1, you showed that the variables t and A are approximately linearly related from 1965 to 2010. (If you didn't do this earlier, do so now.) Find the approximate rate of change of the average global temperature from 1965 to 2010.

4. Use the approximate rate of change found in Problem 3 to predict the change in average temperature for the coming century. Does your result fall within the 2.5–10.4°F IPCC range? Does it fall within either the 3–9°F or 4–7°F range predicted by the team of scientists from the United States and Europe? If yes, which one(s)?

5. In Problem 7 of the Climate Change Lab of Chapter 1, you showed that the linear temperature model for the years 1965–2010 underestimates the average global temperature in 2015. (If you did not do this earlier, do so now.) What *might* that mean about the results you found in Problem 4?

6. Explain why, by using terminology such as "9 out of 10" and "most likely," scientists have been able to predict narrower ranges of increase in global average temperatures in the coming century. Explain how this terminology, coupled with narrower ranges, will help policymakers better understand the risks involved in various courses of action or inaction.

7. Let I be the amount of ice (in cubic kilometers) in Greenland's ice cap at t years since 2000. Assuming that Greenland's ice cap continues to melt at the current rate, find an equation that models the situation.

8. **a.** Use your model to predict the amount of ice in Greenland's ice cap in 1000 years. If a total meltdown occurs in 1000 years, as predicted by Gregory, what must happen to the rate of melting?

   b. If a total meltdown occurs in 150 years, as predicted by Overpeck, what must happen to the rate of melting?

Workout Lab

In this lab, you will explore your walking or running speed. Check with your instructor about whether you should collect your own data or use the data listed in Table 50.

DATA. Table 50 Times for Walking 440-Yard Laps

Lap Number	Distance (yards)	Time (seconds)
0	0	0
1	440	217
2	880	436
3	1320	656
4	1760	878
5	2200	1095
6	2640	1308

Source: *J. Lehmann*

Materials

You will need the following items:

- A timing device
- A pencil or pen
- A small pad of paper

Preparation

Locate a running track on which you can walk or run. For most tracks, one lap is 440 yards; so, four laps are 1760 yards, or 1 mile. You may select some other type of route, provided that you know the distance of one lap and you can easily complete six laps. Or map out a route in your neighborhood and estimate the distance by measuring it with the odometer in a car.

Recording of Data

Start your timing device and begin walking or running. Complete six laps of your course. Each time you complete

---

[¶]American Geophysical Union meeting, October 2002.

[‖]National Oceanic and Atmospheric Administration.

[**]R. J. Nicholls and F. M. J. Hoozemans, "Vulnerability to sea-level rise with reference to the Mediterranean region," *Medcoast 95*, Vol. II, October 1995.

a lap, record the *total* elapsed time. It will be easier to have a friend record the times for you. You may go slowly or quickly, but try to move at a constant speed throughout this experiment.

Analyzing the Data

1. Describe your route and the distance of one lap.

2. Use a table to describe the six total elapsed times for the six laps.

3. Let d be the distance (in yards) after you have walked or run for t seconds. Throughout this lab, treat t as the explanatory variable and d as the response variable. Show the six pairs of values of t and d in a table.

4. Construct a scatterplot for the variables t and d.

5. For each of the six laps, calculate your approximate speed. Did you move at a steady rate, slow down, speed up, or engage in a combination of these?

6. Explain how your six approximate speeds are related to the position of the points in your scatterplot.

7. Find the average of the six approximate speeds you found in Problem 5. Compare this result with the approximate speed for the entire workout.

8. What is your approximate speed for the entire workout, in units of miles per hour?

9. Use the result you found in Problem 8 to predict how long it would take you to walk or run 2 miles.

10. Use the result you found in Problem 8 to predict how long it would take you to walk or run a marathon, which is 26.2 miles long. Has model breakdown occurred? Explain.

Balloon Lab

In this lab, you will explore how long it takes for air to be released from a balloon when it is inflated with various amounts of air. Check with your instructor about whether you should collect your own data or use the data listed in Table 51.

DATA Table 51 Average Release Times for a Balloon Inflated with a Single Breath Having a Volume of 20 Ounces

Number of Breaths	Volume (ounces)	Average Release Time (seconds)
0	0	0
5	100	1.9
10	200	3.5
15	300	6.1
20	400	6.0
25	500	7.7
30	600	10.6

Source: *J. Lehmann*

Materials

You will need one helper and the following items:

- A balloon
- A timing device
- A bucket, sink, or bathtub
- Water to fill the bucket, sink, or bathtub
- A transparent 1-cup or larger measuring cup (a larger cup is more convenient)

Preparation

Inflate and deflate the balloon fully several times to stretch it out. Each time you inflate the balloon, practice blowing into it with uniform-size breaths. There is no need to breathe deeply into the balloon. Medium-size breaths will work well for this lab.

Recording of Data

Perform the following tasks:

1. Count how many medium-size breaths it takes to fill the balloon.

2. For each trial of this experiment, you will time how long it takes for the balloon to release all the air inside it. Run three trials for each of six different volumes. Decide for which volumes you will have trials. For example, if it takes 30 breaths to fill the balloon, you could run three trials for each of the volumes of 5, 10, 15, 20, 25, and 30 breaths. To run a trial, first fill the balloon to the desired volume. Then begin timing as you release the balloon. Stop timing when the balloon is deflated completely. Record the time and the corresponding volume (in number of breaths).

3. To find the volume of a medium-size breath, fill a bucket, sink, or bathtub with water. Have one person fully submerge the measuring cup, so that there is no air in it, and then turn the cup upside down while it is still under water. The cup should not rest on the bottom of the container. Next, have a second person blow once into the balloon and then carefully release the air into the submerged cup. The air from the balloon will displace the water in the cup. The volume of air in the balloon will likely be more than what can fit in the cup, so the second person will have to stop releasing air from the balloon, and then the first person can empty out the air by resubmerging the cup. The second person should continue releasing air from the balloon into the cup until the balloon is empty. Depending on the size of the breath and the cup, the first person may have to resubmerge the cup several times. Then compute the volume of a medium-size breath by measuring the amount of air in the cup and taking into account how many times you filled the cup with air.

Analyzing the Data

1. For each of the volumes of 5, 10, 15, 20, 25, and 30 breaths, compute the average of the three release

times. Then record the numbers of breaths and the average release times in columns 1 and 3 of a table similar to Table 51. Use the volume of one breath to help you find the entries for the second column of the table.

2. Let T be the time (in seconds) it takes for the balloon to deflate completely when the initial volume is n ounces. Construct a scatterplot of the data.

3. Draw a linear model on your scatterplot.

4. What is the T-intercept of the linear model? What does it mean in this situation? If model breakdown occurs, draw a better model.

5. Find the slope of the linear model. What does it mean in this situation?

6. Find an equation of the model.

7. Use the model to estimate the release time for a volume other than any of the volumes you used for the trials.

8. As you inflated the balloon, did it take about the same amount of effort to breathe in each time, or did it get progressively easier or harder? Thinking about how that effort is related to release times, does it suggest that the data points should lie close to a line or to a curve that bends? Explain. Do the (actual) data points support your theory? Explain.

 # Chapter Summary

Key Points of Chapter 3

Section 3.1 Graphing Equations of the Forms $y = mx + b$ and $x = a$

Solution, satisfy, and solution set of an equation	An ordered pair (a, b) is a **solution** of an equation in terms of x and y if the equation becomes a true statement when a is substituted for x and b is substituted for y. We say (a, b) **satisfies** the equation. The **solution set** of an equation is the set of all solutions of the equation.
Graph	The **graph** of an equation in two variables is the set of points that correspond to all solutions of the equation.
Graph of $y = mx + b$	The graph of an equation of the form $y = mx + b$, where m and b are constants, is a line.
y-intercept of the graph of $y = mx + b$	The graph of an equation of the form $y = mx + b$ has y-intercept $(0, b)$.
Equations for horizontal and vertical lines	If a and b are constants, then • The graph of $y = b$ is a horizontal line. • The graph of $x = a$ is a vertical line.
Equations whose graphs are lines	If an equation can be put into the form $y = mx + b$ or $x = a$, where $m, a,$ and b are constants, then the graph of the equation is a line. We call such an equation a **linear equation in two variables**.
Rule of Four for solutions of an equation	We can describe some of or all the solutions of an equation in two variables with an equation, a table, a graph, or words. These four ways to describe solutions are known as the **Rule of Four**.

Section 3.2 Graphing Linear Models; Unit Analysis

Rule of Four for linear models	We can describe a linear model with an equation, a table, a graph, or words. These four ways to describe authentic situations are known as the **Rule of Four**.
Unit analysis	We perform a **unit analysis** of a model's equation by determining the units of the expressions on both sides of the equation. The units of the expressions on both sides of the equation should be the same.

Section 3.3 Slope of a Line

Comparing the steepness of two objects	To compare the steepness of two objects, compute the unit ratio $$\frac{\text{vertical distance}}{\text{horizontal distance}}$$ for each object. The object with the larger ratio is the steeper object.
Slope of a nonvertical line	Let (x_1, y_1) and (x_2, y_2) be two distinct points of a nonvertical line. Then the **slope** of the line is $$m = \frac{\text{vertical change}}{\text{horizontal change}} = \frac{\text{rise}}{\text{run}} = \frac{y_2 - y_1}{x_2 - x_1}$$
Slopes of increasing and decreasing lines	An increasing line has positive slope. A decreasing line has negative slope.
Measuring the steepness of a line	The absolute value of the slope of a line measures the steepness of the line. The steeper the line, the larger the absolute value of its slope will be.
Slopes of horizontal and vertical lines	A horizontal line has a slope equal to zero. A vertical line has undefined slope.

Section 3.4 Using Slope to Graph Linear Equations

Slope and y-intercept of a linear equation of the form $y = mx + b$; slope–intercept form	For a linear equation of the form $y = mx + b$, • the slope of the line is m and • the y-intercept of the line is $(0, b)$. We say that this equation is in **slope–intercept form**.
Using slope to graph an equation of the form $y = mx + b$	To graph an equation of the form $y = mx + b$, 1. Plot the y-intercept $(0, b)$. 2. Use $m = \dfrac{\text{rise}}{\text{run}}$ to plot a second point. 3. Sketch the line that passes through the two plotted points.
Finding an equation of a line from a graph	To find an equation of a line from a graph, 1. Determine the slope m and the y-intercept $(0, b)$ from the graph. 2. Substitute your values for m and b into the equation $y = mx + b$.
Slopes of parallel lines	If lines l_1 and l_2 are parallel nonvertical lines, then the slopes of the lines are equal: $$m_1 = m_2$$ Also, if two distinct lines have equal slope, then the lines are parallel.
Slopes of perpendicular lines	If lines l_1 and l_2 are perpendicular nonvertical lines, then the slope of one line is the opposite of the reciprocal of the slope of the other line: $$m_2 = -\frac{1}{m_1}$$ Also, if the slope of one line is the opposite of the reciprocal of another line's slope, then the lines are perpendicular.

Section 3.5 Rate of Change

Rate of change	Suppose that a quantity y changes steadily from y_1 to y_2 as a quantity x changes steadily from x_1 to x_2. Then the **rate of change** of y with respect to x is the ratio of the change in y to the change in x: $$\frac{\text{change in } y}{\text{change in } x} = \frac{y_2 - y_1}{x_2 - x_1}$$

Section 3.5 Rate of Change (*Continued*)

Signs of rate of change	Suppose that a quantity t affects (explains) a quantity p. Then • If p increases steadily as t increases steadily, then the rate of change of p with respect to t is positive. • If p decreases steadily as t increases steadily, then the rate of change of p with respect to t is negative.
Slope is a rate of change	If there is a linear relationship between the quantities t and p, and if t affects (explains) p, then the slope of the linear model is equal to the rate of change of p with respect to t.
Constant rate of change	Suppose that a quantity t affects (explains) a quantity p. Then • If there is a linear relationship between t and p, then the rate of change of p with respect to t is constant. • If the rate of change of p with respect to t is constant, then there is a linear relationship between t and p.
Slope addition property	For a linear equation of the form $y = mx + b$, if the value of x increases by 1, then the value of y changes by the slope m. In other words, if an input increases by 1, the output changes by the slope.

Chapter 3 Review Exercises

1. Which of the ordered pairs $(-3, 9)$, $(1, 2)$, and $(4, -5)$ satisfy the equation $y = -2x + 3$?

For Exercises 2–7, refer to the graph sketched in Fig. 111.

2. Find y when $x = 2$.

3. Find y when $x = -2$.

4. Find y when $x = 0$.

5. Find x when $y = -3$.

6. Find x when $y = -4$.

7. Find x when $y = 0$.

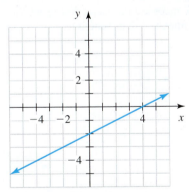

Figure 111 Exercises 2–7

8. While taking off, airplane A climbs steadily for 6500 feet over a horizontal distance of 12,700 feet. Airplane B climbs steadily for 7400 feet over a horizontal distance of 15,600 feet. Which plane is climbing at a greater incline? Explain.

Find the slope of the line passing through the two given points. State whether the line is increasing, decreasing, horizontal, or vertical.

9. $(-3, 1)$ and $(2, 11)$

10. $(-2, -4)$ and $(1, -7)$

11. $(4, -3)$ and $(8, -1)$

12. $(-6, 0)$ and $(0, -3)$

13. $(-5, 5)$ and $(2, -2)$

14. $(-10, -3)$ and $(-4, -5)$

15. $(-5, 2)$ and $(3, -7)$

16. $(-4, -1)$ and $(2, -5)$

17. $(-4, 7)$ and $(-4, -3)$

18. $(-5, 2)$ and $(-1, 2)$

For Exercises 19 and 20, find the approximate slope of the line that contains the two given points. Round your result to the second decimal place. State whether the line is increasing, decreasing, horizontal, or vertical.

19. $(5.4, 7.9)$ and $(8.3, -2.6)$

20. $(-8.74, -2.38)$ and $(-1.16, 4.77)$

21. Sketch a line whose slope is a negative number near zero.

Sketch the line that has the given slope and contains the given point.

22. $m = 3$, $(0, -4)$

23. $m = \dfrac{4}{3}$, $(0, 1)$

24. $m = 0$, $(2, -3)$

Determine the slope and the y-intercept. Use them to graph the equation by hand. Use a graphing calculator to verify your work.

25. $y = \dfrac{3}{4}x - 1$

26. $y = -\dfrac{1}{2}x + 3$

27. $y = -\dfrac{2}{5}x - 1$

28. $y = \dfrac{2}{3}x$

29. $y = -4x$

30. $y = 2x - 4$

31. $y = -3x + 1$

32. $y = x + 2$

33. $y = -5$

For Exercises 34 and 35, graph the equation by hand.

34. $x = -3$

35. $y = 2$

36. Recall that we can describe some of or all the solutions of an equation in two variables with an equation, a table, a graph, or words.

    a. Describe three solutions of $y = -2x + 1$ by using a table.

    b. Describe the solutions of $y = -2x + 1$ by using a graph.

    c. Describe the solutions of $y = -2x + 1$ by using words.

37. When a certain person loses his job, the balance in his checking account is $19 thousand. While the person is unemployed, the balance declines by $3 thousand per month. Let B be the balance (in thousands of dollars) in the checking account after t months of unemployment.
   a. Find an equation for t and B.
   b. Graph by hand the equation you found in part (a).
   c. What is the B-intercept? What does it mean in this situation?
   d. How long has the person been unemployed if the balance is $4 thousand? Explain by using arrows on the graph you sketched in part (b).

38. Find an equation of the line sketched in Fig. 112.

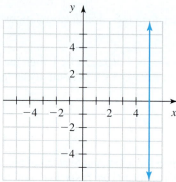

Figure 112 Exercise 38

39. Find an equation of the line sketched in Fig. 113.

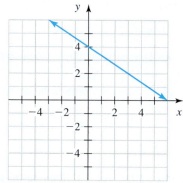

Figure 113 Exercise 39

For Exercises 40–43, determine whether the pair of lines is parallel, perpendicular, or neither. Explain.

40. $y = 3x - 2$ and $y = \frac{1}{3}x + 6$

41. $y = \frac{4}{7}x + 1$ and $y = -\frac{7}{4}x - 5$

42. $x = -2$ and $y = 5$

43. $x = -4$ and $x = 1$

44. The temperature declines steadily by 6°F over a 4-hour period. Find the rate of change of temperature.

45. The number of nuns in the United States decreased approximately steadily from 79.8 thousand in 2000 to 50.0 thousand in 2015 (Source: *The Official Catholic Directory*). Find the approximate rate of change of the number of nuns. Round your result to the second decimal place.

46. In Nashville, Tennessee, taxis charge a flat fee of $3 plus $2 per mile (Source: *Metropolitan Government of Nashville and*

Davidson County). Let c be the total charge (in dollars) for a trip of d miles.
   a. Find an equation for d and c.
   b. Perform a unit analysis of the equation you found in part (a).
   c. What is the cost of a 17-mile trip?

47. The IRS budget was $11.3 billion in 2014 and has decreased by about $0.3 billion per year since then (Source: *IRS*). Let A be the annual budget (in billions of dollars) at t years since 2014.
   a. Find the slope of the linear model that describes the situation. What does it mean in this situation?
   b. Find an equation of the model.
   c. Predict the budget in 2022.

48. The total amount Americans spent on Father's Day was $12.5 billion in 2014 and has increased by about $0.5 billion per year since then (Source: *National Retail Federation*). Let A be the total amount (in billions of dollars) Americans spent on Father's Day at t years since 2014.
   a. Find the slope of the linear model that describes the situation. What does it mean in this situation?
   b. What is the A-intercept of the model? What does it mean in this situation?
   c. Find an equation of the model.
   d. Predict the total amount Americans will spend on Father's Day in 2021.

49. The suggested retail price for a TI-84 Plus® graphing calculator is $119.99. Let C be the total cost (in dollars) of purchasing n of these calculators at the retail price. Are n and C linearly related? If so, what is the slope? What does the slope mean in this situation?

50. The percentages of American adults who are confident they will retire ahead of their schedule are described by the scatterplot and the model in Fig. 114 for various incomes (in thousands of dollars). Let p be the percentage of American adults with income I thousand dollars who are confident they will retire ahead of their schedule.

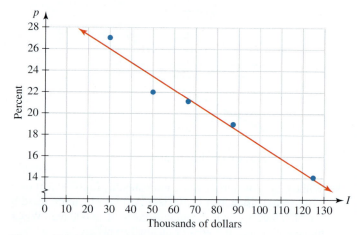

Figure 114 Exercise 50
(**Source:** *Country Financial*)

   a. Use the model to estimate the percentage of Americans with incomes of $60 thousand who are confident they will retire ahead of their schedule.
   b. Use the model to estimate at what income 16% of American adults are confident they will retire ahead of their schedule.
   c. What is the slope? What does it mean in this situation?

51. Some solutions of four equations are listed in Table 52. Which of the equations could be linear?

Table 52 Solutions of Four Equations

Equation 1		Equation 2		Equation 3		Equation 4	
x	y	x	y	x	y	x	y
0	17	0	2	3	4	1	−8
1	14	1	6	4	4	2	−3
2	11	2	9	5	4	3	2
3	8	3	11	6	4	4	7
4	5	4	12	7	4	5	12

52. Some solutions of four linear equations are listed in Table 53. Complete the table.

Table 53 Solutions of Four Linear Equations

Equation 1		Equation 2		Equation 3		Equation 4	
x	y	x	y	x	y	x	y
0	50	0	12	61	25	26	−4
1	41	1	16	62		27	
2		2		63		28	
3		3		64	19	29	
4		4		65		30	8

53. For the equation $y = -6x + 39$, describe the change in the value of y as the value of x is increased by 1.

Chapter 3 Test

For Exercises 1–4, refer to the graph sketched in Fig. 115.

1. Find y when $x = -3$.

2. Find x when $y = -1$.

3. Find the y-intercept.

4. Estimate the x-intercept.

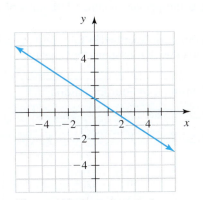

Figure 115 Exercises 1–4

5. Ski run A declines steadily for 115 yards over a horizontal distance of 580 yards. Ski run B declines steadily for 150 yards over a horizontal distance of 675 yards. Which ski run is steeper? Explain.

For Exercises 6–9, find the slope of the line passing through the two given points. State whether the line is increasing, decreasing, horizontal, or vertical.

6. $(3, -8)$ and $(5, -2)$

7. $(-4, -1)$ and $(2, -4)$

8. $(-5, 4)$ and $(1, 4)$

9. $(-2, -7)$ and $(-2, 3)$

10. Find the approximate slope of the line that contains the points $(-5.99, -3.27)$ and $(2.83, 8.12)$. Round your result to the second decimal place. State whether the line is increasing, decreasing, horizontal, or vertical.

11. Sketch the line that has a slope of $\frac{2}{5}$ and contains the point $(0, -3)$.

For Exercises 12–16, determine the slope and the y-intercept. Use them to graph the equation by hand.

12. $y = -\dfrac{3}{2}x + 2$

13. $y = \dfrac{5}{6}x$

14. $y = 3x - 4$

15. $y = 2$

16. $y = -2x + 3$

17. Find an equation of the line sketched in Fig. 116.

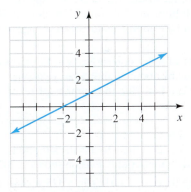

Figure 116 Exercise 17

18. A used car is worth $17 thousand. Each year, the car's value will decline by $2 thousand. Let v be the car's value (in thousands of dollars) at t years from now.

  a. Find an equation for t and v.

  b. Graph by hand the equation you found in part (a).

  c. What is the v-intercept? What does it mean in this situation?

  d. When will the value of the car be $5 thousand dollars? Explain by using arrows on the graph you sketched in part (b).

19. Graphs of four equations are shown in Fig. 117. State whether m and b are positive, negative, zero, or undefined for the $y = mx + b$ form of each equation.

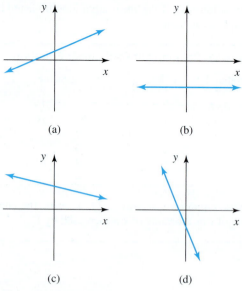

(a)   (b)

(c)   (d)

Figure 117 Exercise 19

For Exercises 20 and 21, determine whether the pair of lines is parallel, perpendicular, or neither. Explain.

20. $y = \dfrac{2}{5}x + 3$ and $y = \dfrac{5}{2}x - 7$

21. $y = -3x + 8$ and $y = -3x - 1$

22. The temperature decreases steadily by 12°F in a 5-hour period. Find the rate of change of temperature.

23. The U.S. revenue from cereal decreased approximately steadily from \$9.4 billion in 2012 to \$8.6 billion in 2016 (Source: *Nielsen*). Find the approximate rate of change of annual cereal revenue.

24. The number of television shows was 409 shows in 2015 and has increased by about 30 shows per year since then (Source: *FX Networks*). Let n be the number of television shows at t years since 2015.
  a. Find the slope of the model of the situation. What does it mean in this situation?
  b. Find an equation that models the situation.
  c. Predict the number of television shows in 2021.

25. The cooking times of a turkey in a 325°F oven are shown in Table 54 for various weights.

Table 54 Cooking Times of a Turkey in a 325°F Oven

Weight Group (pounds)	Weight Used to Represent Weight Group (pounds)	Cooking-Time Group (hours)	Time Used to Represent Cooking-Time Group (hours)
6–8	7	3.0–3.5	3.25
8–12	10	3.5–4.5	4
12–16	14	4.5–5.5	5
16–20	18	5.5–6.5	6
20–24	22	6.5–7.0	6.75

Source: *About, Inc.*

Let T be the cooking time (in hours) of a turkey that weighs w pounds. A model of the situation is $T = 0.24w + 1.64$.
  a. Use a graphing calculator to draw a scatterplot and the model in the same viewing window. Does the line come close to the data points?
  b. What is the slope? What does it mean in this situation?
  c. What is the T-intercept? What does it mean in this situation?
  d. Estimate the cooking time of a 19-pound turkey.

26. A person goes out for a run. Let d be the distance (in miles) traveled in t minutes of running. The graph in Fig. 118 describes the relationship between t and d.

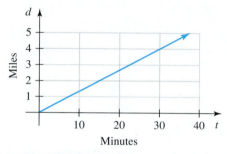

Figure 118 Exercise 26

  a. Is the person running at a constant speed according to the model? Explain.
  b. How fast is the person running? Use units of miles per hour.

27. Four sets of points are described in Table 55. For each set, find an equation of a line that contains the points.

Table 55 Solutions of Four Equations

Set 1		Set 2		Set 3		Set 4	
x	y	x	y	x	y	x	y
0	25	0	2	0	12	0	47
1	22	1	6	1	7	1	53
2	19	2	10	2	2	2	59
3	16	3	14	3	-3	3	65
4	13	4	18	4	-8	4	71

28. For the equation $y = 3x - 8$, describe the change in the value of y as the value of x is increased by 1.

Simplifying Expressions and Solving Equations

4

What was your average grade in high school? The percentage of freshmen at Boise State University whose GPA in high school was at least 3.5 increased from 37.8% in 2009 to 46.5% in 2015 (see Table 1). Do you think the increase is due to students learning more, teachers lowering their standards, or some other reason? In Exercise 51 of Homework 4.4, you will estimate when half of all Boise State University freshmen earned a GPA of at least 3.5 in high school.

In this chapter, we will discuss how to write expressions more simply. We will also discuss how to find solutions of equations in one variable. Once we have learned these skills, we will be able to make predictions efficiently about the explanatory variable (or the response variable) in authentic situations, such as the number of Starbucks stores in various years.

Table 1 Freshmen at Boise State University Whose GPA in High School Was at Least 3.5

Year	Percent
2009	37.8
2010	38.2
2011	39.9
2012	44.6
2013	43.8
2014	45.4
2015	46.5

Source: *Boise State University*

4.1 Simplifying Expressions

Objectives

» Describe the *commutative, associative,* and *distributive* laws.

» Compare the distributive and associative laws for multiplication.

» Describe the meaning of *equivalent expressions.*

» Simplify expressions.

» Subtract two expressions.

» Show that a statement is false.

In this section, we will write expressions more simply.

Commutative Laws

Consider the equations $2 + 6 = 6 + 2$ and $2 \cdot 6 = 6 \cdot 2$. These true statements suggest the *commutative laws.*

Commutative Laws for Addition and Multiplication

Commutative law for addition: $a + b = b + a$
Commutative law for multiplication: $ab = ba$

In words: We can add two numbers in either order and get the same result, and we can multiply two numbers in either order and get the same result.

Example 1	Using the Commutative Law for Addition

Use the commutative law for addition to write the expression in another form.

1. $3 + x$ **2.** $5 + 3w$

Solution

1. $3 + x = x + 3$ *Commutative law for addition: $a + b = b + a$*
2. $5 + 3w = 3w + 5$ *Commutative law for addition: $a + b = b + a$*

| Example 2 | Using the Commutative Law for Multiplication |

Use the commutative law for multiplication to write the expression in another form.

1. cd **2.** $-7x$ **3.** $k \cdot 8 + 2$

Solution

1. $cd = dc$ *Commutative law for multiplication: $ab = ba$*
2. $-7x = x(-7)$ *Commutative law for multiplication: $ab = ba$*
3. $k \cdot 8 + 2 = 8k + 2$ *Commutative law for multiplication: $ab = ba$*

▶

WARNING   There is no commutative law for subtraction. For example, $7 - 3 = 4$ but $3 - 7 = -4$. And there is no commutative law for division. For example, $3 \div 1 = 3$, but $1 \div 3 = \dfrac{1}{3}$.

A **term** is a constant, a variable, or a product of a constant and one or more variables raised to powers. Here are some terms:

$$4x \qquad -7 \qquad y \qquad -5xy^2 \qquad 97x^6y^3$$

The expression $7xy + 4x - 5y + 3$ has four terms: $7xy$, $4x$, $5y$, and 3. Note that, in the expression, the terms are separated by addition and subtraction symbols.

Variable terms are terms that contain variables. **Constant terms** are terms that do not contain variables. For example, $-5x$ is a variable term, and 8 is a constant term. We usually write a sum of both variable and constant terms with the variable terms to the left of the constant terms. So, for $5 - 8x$, we write

$$5 - 8x = 5 + (-8x) \qquad a - b = a + (-b)$$
$$= -8x + 5 \qquad \textit{Commutative law for addition: } a + b = b + a$$

Associative Laws

As we discussed in Section 2.6, we work from left to right when we perform additions and subtractions. However, if a sum is of the form $a + b + c$, we can get the same result by performing the addition on the left first or the one on the right first. For example,

$$(4 + 2) + 3 = 6 + 3 = 9$$
$$4 + (2 + 3) = 4 + 5 = 9$$

A similar law is true for an expression of the form abc:

$$(4 \cdot 2) \cdot 3 = 8 \cdot 3 = 24$$
$$4 \cdot (2 \cdot 3) = 4 \cdot 6 = 24$$

These examples suggest the *associative laws*.

Associative Laws for Addition and Multiplication

Associative law for addition: $a + (b + c) = (a + b) + c$
Associative law for multiplication: $a(bc) = (ab)c$

In words: For an expression $a + b + c$, we get the same result by performing the addition on the right first or the one on the left first. For an expression abc, we get the same result by performing the multiplication on the right first or the one on the left first.

It is important to know the difference between the associative laws and the commutative laws. The associative laws change the order of *operations*, whereas the commutative laws change the order of *terms* or even *expressions*.

For example, consider the expression $(3 + x) + 1$. By the associative law of addition, we can change the order of the additions:

$$(3 + x) + 1 = 3 + (x + 1)$$

By the commutative law of addition, we can change the order of the terms 3 and x:

$$(3 + x) + 1 = (x + 3) + 1$$

We can also use the commutative law to change the order of the expressions $x + 3$ and 1:

$$(x + 3) + 1 = 1 + (x + 3)$$

Example 3   Using the Associative Laws

Use an associative law to write the expression in another form.

1. $(x + 2) + y$ **2.** $w + (9 + k)$ **3.** $3(mp)$ **4.** $(wx)y$

Solution

1. $(x + 2) + y = x + (2 + y)$ *Associative law for addition:* $(a + b) + c = a + (b + c)$
2. $w + (9 + k) = (w + 9) + k$ *Associative law for addition:* $a + (b + c) = (a + b) + c$
3. $3(mp) = (3m)p$ *Associative law for multiplication:* $a(bc) = (ab)c$
4. $(wx)y = w(xy)$ *Associative law for multiplication:* $(ab)c = a(bc)$

▶

We can use a combination of the commutative law and the associative law, both for addition, to write the terms of $a + b + c$ in different orders:

$$a + c + b \qquad b + a + c \qquad b + c + a \qquad c + a + b \qquad c + b + a$$

We **rearrange the terms** of an expression by writing the terms in a different order.

Likewise, we **rearrange the factors** of an expression by writing the factors in a different order. Here we rearrange the factors of abc to get the following products:

$$acb \qquad bac \qquad bca \qquad cab \qquad cba$$

Example 4   Rearranging Terms

Rearrange the terms of $9 - 2x - 5 + 3$ so that the numbers can be added.

Solution

$$
\begin{aligned}
9 - 2x - 5 + 3 &= 9 + (-2x) + (-5) + 3 \quad && a - b = a + (-b) \\
&= -2x + 9 + (-5) + 3 \quad && \textit{Rearrange terms.} \\
&= -2x + 7 \quad && \textit{Add constant terms.}
\end{aligned}
$$

▶

It is helpful to practice problems such as the one in Example 4 until you can do each problem in one step.

Example 5   Using the Commutative and Associative Laws

Use the commutative and associative laws to remove the parentheses. Also, multiply and add numbers when possible.

1. $-3(8x)$ **2.** $(5 + 7p) + 8$

Solution

1. $\begin{aligned}[t] -3(8x) &= (-3 \cdot 8)x \quad && \textit{Associative law for multiplication: } a(bc) = (ab)c \\ &= -24x \quad && \textit{Multiply.} \end{aligned}$

2. $\begin{aligned}[t] (5 + 7p) + 8 &= (7p + 5) + 8 \quad && \textit{Commutative law for addition: } a + b = b + a \\ &= 7p + (5 + 8) \quad && \textit{Associative law for addition: } (a + b) + c = a + (b + c) \\ &= 7p + 13 \quad && \textit{Add.} \end{aligned}$

▶

Distributive Law

For the expression $2(3 + 4)$, we perform the addition before the multiplication. However, compare the result with that for computing $2 \cdot 3 + 2 \cdot 4$:

$$2(3 + 4) = 2(7) = 14$$
$$2 \cdot 3 + 2 \cdot 4 = 6 + 8 = 14$$

Because both results are equal to 14, we can write

$$2(3 + 4) = 2 \cdot 3 + 2 \cdot 4$$

The blue curves indicate what numbers we multiply by 2. We say that we *distribute* the 2 to both the 3 and the 4. This example suggests the **distributive law**.

Distributive Law

$$a(b + c) = ab + ac$$

In words: To find $a(b + c)$, distribute a to both b and c.

Example 6   Using the Distributive Law

Find the product.

1. $3(x + 5)$ **2.** $5(2x + 4y)$

Solution

1.
$$3(x + 5) = 3x + 3 \cdot 5 \quad \textit{Distributive law: } a(b + c) = ab + ac$$
$$= 3x + 15 \quad \textit{Multiply.}$$

2.
$$5(2x + 4y) = 5 \cdot 2x + 5 \cdot 4y \quad \textit{Distributive law: } a(b + c) = ab + ac$$
$$= 10x + 20y \quad \textit{Multiply.}$$

WARNING   In Problem 1 of Example 6, we found that $3(x + 5) = 3x + 15$. In applying the distributive law to an expression such as $3(x + 5)$, remember to distribute the 3 to *every* term in the parentheses. For example, the expression $3x + 5$ is an incorrect result.

Because subtracting a number is the same as adding the opposite of the number, it is also true that

$$a(b - c) = ab - ac$$

We can use the distributive law and the commutative law for multiplication to show that we can "distribute from the right" as well as from the left (see Exercises 115 and 116):

$$(a + b)c = ac + bc$$
$$(a - b)c = ac - bc$$

Example 7   Using the Distributive Law

Find the product.

1. $4(3x - 2y)$ **2.** $(x - 6)(3)$ **3.** $-2(6 + t)$ **4.** $-5(w - 3)$

Solution

1.
$$4(3x - 2y) = 4 \cdot 3x - 4 \cdot 2y \quad \textit{Distributive law: } a(b - c) = ab - ac$$
$$= 12x - 8y \quad \textit{Multiply.}$$

2. $(x - 6)(3) = x \cdot 3 - 6 \cdot 3$ *Distributive law: $(a - b)c = ac - bc$*
$= 3x - 18$ *Commutative law for multiplication: $ab = ba$; multiply.*

3. $-2(6 + t) = -2(6) + (-2t)$ *Distributive law: $a(b + c) = ab + ac$*
$= -12 + (-2t)$ *Multiply.*
$= -2t + (-12)$ *Commutative law for addition: $a + b = b + a$*
$= -2t - 12$ *$a + (-b) = a - b$*

4. $-5(w - 3) = -5w - (-5)(3)$ *Distributive law: $a(b - c) = ab - ac$*
$= -5w + 15$ *Multiply.*

It is extremely helpful to practice problems such as those in Example 7 until you can do them in one step.

We can also use the distributive law when there are more than two terms inside the parentheses.

Example 8 **Distributive Law**

Find the product $3(2t - 5w + 4)$.

Solution

$3(2t - 5w + 4) = 3 \cdot 2t - 3 \cdot 5w + 3 \cdot 4$ *Distributive law*
$= 6t - 15w + 12$ *Multiply.*

The products $-1 \cdot 2 = -2$, $-1 \cdot 5 = -5$, and $-1 \cdot 6 = -6$ suggest the following property:

Multiplying a Number by -1

$$-1a = -a$$

In words: -1 times a number is equal to the opposite of the number.

To remove parentheses in $-(x + 3)$, we use the fact that $-a = -1a$ to write

$-(x + 3) = -1(x + 3)$ *$-a = -1a$*
$= -1x + (-1)3$ *Distributive law: $a(b + c) = ab + ac$*
$= -x - 3$ *$-1a = -a$*

Example 9 **Removing Parentheses**

Remove parentheses in $-(x - 4y + 7)$.

Solution

$-(x - 4y + 7) = -1(x - 4y + 7)$ *$-a = -1a$*
$= -1x - (-1)4y + (-1)7$ *Distributive law*
$= -x + 4y - 7$ *$-1a = -a$*

Comparing the Distributive Law with the Associative Law for Multiplication

It is helpful to compare the distributive law with the associative law for multiplication to avoid confusing the two laws. For the expression $a(b + c)$, we distribute a to both of the terms b and c:

$$a(b + c) = ab + ac \quad \textit{Distributive law}$$

WARNING   For the expression $a(bc)$, we do *not* distribute the a to both of the factors b and c. Rather, we distribute a to just *one* of the factors:

$$a(bc) = (ab)c \quad \textit{Associative law for multiplication}$$

Equivalent Expressions

In Problem 1 of Example 6, we found that $3(x + 5) = 3x + 15$. In Example 10, we will explore the meaning of this statement.

Example 10   Evaluating Expressions

Evaluate both of the expressions $3(x + 5)$ and $3x + 15$ for the given values of x.

1. $x = 2$ **2.** $x = 4$
3. $x = 0, x = 1, x = 2, x = 3, x = 4, x = 5,$ and $x = 6$

Solution

1. First, we evaluate $3(x + 5)$ for $x = 2$:

$$3(2 + 5) = 3(7) = 21$$

Then we evaluate $3x + 15$ for $x = 2$:

$$3(2) + 15 = 6 + 15 = 21$$

Both results are equal to 21.

2. First, we evaluate $3(x + 5)$ for $x = 4$:

$$3(4 + 5) = 3(9) = 27$$

Then we evaluate $3x + 15$ for $x = 4$:

$$3(4) + 15 = 12 + 15 = 27$$

Both results are equal to 27.

3. We use a graphing calculator to construct a table for the equations $y = 3(x + 5)$ and $y = 3x + 15$ (see Fig. 1). See Appendix A.14 for graphing calculator instructions.

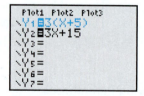 

Figure 1 Evaluate the expressions $3(x + 5)$ and $3x + 15$ for several values of x

For each value of x in the table, the values of y for the two equations are equal. So for each of these values of x, the two results of evaluating the two expressions are equal.

In Example 10, we found that for each value that we used to evaluate the expressions $3(x + 5)$ and $3x + 15$, the two results were equal.

Definition   Equivalent expressions

Two or more expressions are **equivalent expressions** if, when each variable is evaluated for *any* real number (for which all the expressions are defined), the expressions all give equal results.

Simplifying an Expression

We **simplify** an expression by using the laws of operations to remove any parentheses and to rearrange terms so that we can add any constant terms. We will discuss other ways to simplify expressions throughout this text. The result of simplifying an expression is a **simplified expression**, which is equivalent to the original expression.

Simplifying an expression often makes it easier to evaluate the expression and to graph an equation that contains the expression. We will see other benefits of simplifying expressions throughout the text.

Example 11   Simplifying Expressions

Simplify.

1. $4(2t - 3) - 7$ **2.** $5 - 6(x - 2)$ **3.** $1 - 2(5w + 3k - 6)$

Solution

1. $4(2t - 3) - 7 = 8t - 12 - 7$ Distributive law: $a(b - c) = ab - ac$
$\qquad\qquad\qquad = 8t - 19$ Subtract constant terms.

2. $5 - 6(x - 2) = 5 - 6x + 12$ Distributive law: $a(b - c) = ab - ac$
$\qquad\qquad\quad = -6x + 5 + 12$ Rearrange terms.
$\qquad\qquad\quad = -6x + 17$ Add constant terms.

We use a graphing calculator table to verify our work (see Fig. 2). For each value of x in the table, the two results of evaluating the two expressions are equal; this is strong evidence that the expressions are equivalent.

Figure 2 Verify the work

3. $1 - 2(5w + 3k - 6) = 1 - 10w - 6k + 12$ Distributive law
$\qquad\qquad\qquad\quad = -10w - 6k + 1 + 12$ Rearrange terms.
$\qquad\qquad\qquad\quad = -10w - 6k + 13$ Add.

After we simplify an expression, it is wise to use a graphing calculator table to verify that the result and the original expression are indeed equivalent.

Example 12   Translating from English into Mathematics

Let x be a number. Translate the phrase "1, plus 3 times the difference of twice the number and 4" into an expression. Then simplify the expression.

Solution

The expression is

$$\underbrace{1}_{\text{1, plus}} + \underbrace{3 \cdot}_{\text{3 times}} \underbrace{(2x - 4)}_{\substack{\text{the difference of twice} \\ \text{the number and 4}}}$$

twice the number

Next, we simplify the expression:

$$
\begin{aligned}
1 + 3 \cdot (2x - 4) &= 1 + 6x - 12 && \text{\textit{Distributive law: } } a(b - c) = ab - ac \\
&= 6x + 1 - 12 && \text{\textit{Rearrange terms.}} \\
&= 6x - 11 && \text{\textit{Subtract constant terms.}}
\end{aligned}
$$

Subtracting Two Expressions

Note that $a - 1b = a - b$. We can use this fact, written $a - b = a - 1b$, to help us subtract two expressions.

Example 13   Simplifying an Expression

Simplify $7 - (x + 9)$.

Solution

$$
\begin{aligned}
7 - (x + 9) &= 7 - 1(x + 9) && a - b = a - 1b \\
&= 7 - 1x + (-1)9 && \text{\textit{Distributive law: } } a(b + c) = ab + ac \\
&= 7 - x - 9 && -1a = -a \\
&= -x - 2 && \text{\textit{Rearrange terms; subtract constant terms.}}
\end{aligned}
$$

We use a graphing calculator table to verify our work (see Fig. 3).

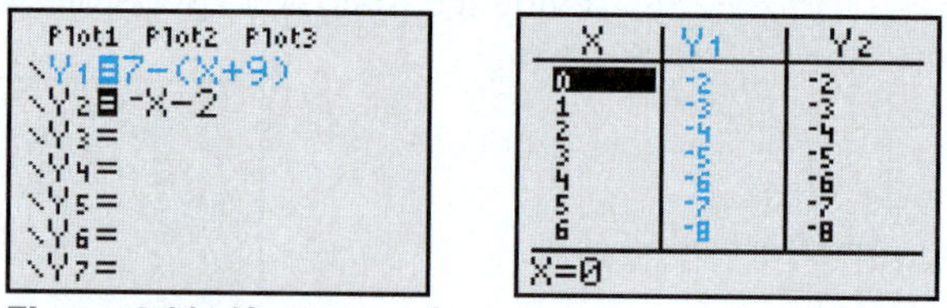

Figure 3 Verify the work

False Statements

We have discussed the commutative laws for addition and multiplication. Example 14 will show there are not similar laws for subtraction and division. **To show that an equation is false, we need find only *one* set of numbers that, when substituted for any variables, gives a result of the form $a = b$, where a and b are different real numbers.**

Example 14   Showing That a Statement Is False

Show that the statement is false.

1. $a - b = b - a$ **2.** $a \div b = b \div a$

Solution

1. We substitute the arbitrary number 5 for a and the arbitrary number 2 for b in the statement $a - b = b - a$:

$$a - b = b - a$$
$$5 - 2 \overset{?}{=} 2 - 5 \qquad \textit{Substitute 5 for a and 2 for b.}$$
$$3 \overset{?}{=} -3 \qquad \textit{Subtract.}$$
$$\text{false}$$

So, the statement $a - b = b - a$ is false.

2. We substitute the arbitrary number 6 for a and the arbitrary number 3 for b in the statement $a \div b = b \div a$:

$$a \div b = b \div a$$
$$6 \div 3 \overset{?}{=} 3 \div 6 \qquad \textit{Substitute 6 for a and 3 for b.}$$
$$2 \overset{?}{=} \frac{3}{6} \qquad \textit{Divide; } a \div b = \frac{a}{b}.$$
$$2 \overset{?}{=} \frac{1}{2} \qquad \textit{Simplify.}$$
$$\text{false}$$

So, the statement $a \div b = b \div a$ is false.

▶

In Example 14, we showed that there is no commutative law for either subtraction or division. In Exercises 111 and 112, you will show that there is no associative law for either subtraction or division as well.

 Group Exploration

Section Opener: Laws of operations

1. **a.** Evaluate $a(b + c)$ for $a = 2, b = 3$, and $c = 5$.
   b. Evaluate $ab + ac$ for $a = 2, b = 3$, and $c = 5$.
   c. Compare the results you found in parts (a) and (b).
   d. Evaluate both $a(b + c)$ and $ab + ac$ for $a = 4, b = 2$, and $c = 6$, and then compare the results.
   e. Evaluate both $a(b + c)$ and $ab + ac$ for values of your choosing for a, b, and c, and then compare the results.
   f. Make an educated guess as to whether the statement $a(b + c) = ab + ac$ is true for all numbers $a, b,$ and c.

2. Evaluate both $a(bc)$ and $(ab)(ac)$ for values of your choosing for $a, b,$ and c. Is the statement $a(bc) = (ab)(ac)$ true for all numbers $a, b,$ and c?

3. Evaluate both $a(bc)$ and $(ab)c$ for values of your choosing for $a, b,$ and c. Make an educated guess as to whether the statement $a(bc) = (ab)c$ is true for all numbers $a, b,$ and c.

4. Evaluate both $a + b$ and $b + a$ for values of your choosing for a and b. Make an educated guess as to whether the statement $a + b = b + a$ is true for all numbers a and b.

5. What are the main points of this exploration?

 Group Exploration

Equivalent expressions

1. Evaluate both $3(x + 2)$ and $x + 6$ for $x = 0$.

2. The results you found in Problem 1 should be equal. If not, check your work. If the results are equal, does that mean that $3(x + 2)$ and $x + 6$ are equivalent? Explain.

3. Evaluate both $3(x + 2)$ and $x + 6$ for $x = 5$. Compare your results, and explain what your comparison tells you.

4. Explain why it is good practice to evaluate expressions several times, each time for a different value of x, to be more confident that the expressions are equivalent.

5. Simplify $3(x + 2)$. Then evaluate both $3(x + 2)$ and your simplified result for several values of x to check that your work is correct.

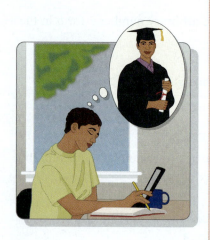

Tips for Success   Desire and Faith

To accomplish anything worthwhile, including succeeding in this course, requires substantial effort *and* faith that you will succeed. The following quote describes what it takes to achieve a goal:

> *"The secret of making something work in your lives is, first of all, the deep desire to make it work. Then the faith and belief that it can work. Then to hold that clear definite vision in your consciousness and see it working out step by step without one thought of doubt or disbelief."*

> —From *Footprints on the Path* by Eileen Caddy, © 1976, 1991.

Your deep desire to succeed in this course might be to earn a degree so you can earn more money. Or perhaps you will be motivated to learn algebra for the love of learning or to experience setting a goal and reaching it. Your faith and belief can come from knowing you, your instructor, and your college will do everything possible to ensure your success. To hold your vision of success "without one thought of doubt or disbelief" is a tall order, but the more you look for ways to succeed rather than feel discouraged, the better are your chances of success.

Homework 4.1

For extra help ▶   **MyLab Math**   Watch the videos in MyLab Math

Use the commutative law for addition to write the expression in another form.

1. $5 + x$ **2.** $x + 8$ **3.** $2p + 7$ **4.** $6 + 8w$

Use the commutative law for multiplication to write the expression in another form.

5. xy **6.** pw **7.** $-2x$

8. $-9x$ **9.** $15 + 4m$ **10.** $7n + 12$

Use an associative law to write the expression in another form.

11. $(x + 4) + y$ **12.** $(8 + x) + y$

13. $(4b)c$ **14.** $(7p)w$

15. $x + (y + 3)$ **16.** $9 + (k + d)$

17. $a(bc)$ **18.** $k(pw)$

Use the commutative and associative laws to simplify the expression. Use a graphing calculator table to verify your work.

19. $2(5x)$ **20.** $3(6x)$

21. $(p + 4) + 3$ **22.** $(k + 8) + 1$

23. $-4(-9x)$ **24.** $-5(-7x)$

25. $2 + (3b + 9)$ **26.** $1 + (8m + 4)$

27. $\frac{1}{2}(-8x)$ **28.** $\frac{2}{3}(-12x)$

29. $7\left(\frac{x}{4}\right)$ **30.** $6\left(\frac{x}{5}\right)$

Simplify.

31. $5 - 3x + 4 - 7$ **32.** $2 - 5x - 6 + 1$

33. $-4 + 2k + 8p - 3$ **34.** $5 - 3k + 2p - 9$

35. $x - \frac{9}{7} + y - \frac{5}{7}$ **36.** $x - \frac{5}{3} - y + \frac{2}{3}$

Use the distributive law to simplify the expression. Use a graphing calculator to verify your work when possible.

37. $3(x + 9)$ **38.** $6(x + 7)$

39. $(x - 5)2$ **40.** $(x - 8)4$

41. $-2(t + 5)$ **42.** $-4(w + 3)$

43. $-5(6 - 2x)$ **44.** $-8(5 - 3x)$

45. $(4x + 7)(-6)$ **46.** $(8x + 2)(-3)$

47. $2(3x - 5y)$ **48.** $4(2x - 8y)$

49. $-5(5x + 3y - 8)$ **50.** $-6(2x - 4y + 5)$

51. $-0.3(x + 0.2)$ **52.** $-0.5(x + 0.4)$

53. $\frac{1}{2}(4x + 8)$ **54.** $\frac{1}{3}(6x - 9)$

55. $-\frac{3}{5}(45 - 35x)$ **56.** $-\frac{5}{6}(30 + 12x)$

Simplify. Use a graphing calculator to verify your work when possible.

57. $3 + 2(x + 1)$ **58.** $1 + 5(x + 3)$

59. $4(a + 3) + 7$ **60.** $2(t + 4) + 1$

61. $-3(4x - 2) + 3$

62. $-5(2x - 3) + 7$

63. $4 - 3(3a - 5)$

64. $-2 - 6(2p - 3)$

65. $-3.7 + 4.2(2.5x - 8.3)$

66. $6.4 + 3.9(6.1x - 4.4)$

67. $-(t + 2)$

68. $-(a + 5)$

69. $-(8x - 9y)$

70. $-(4x - 3y)$

71. $-(5x + 8y - 1)$

72. $-(3x - 7y + 4)$

73. $-(3x + 2) + 5$

74. $-(4x + 1) + 3$

75. $8 - (x + 3)$

76. $4 - (x + 1)$

77. $-2 - (2t - 5)$

78. $-3 - (4a - 2)$

79. $7 - 2(4x - 7y + 5)$

80. $3 - 6(2x + 3y - 9)$

Simplify the expression. Then evaluate both the expression and your simplified result for x = 2 to check your work. Finally, evaluate both expressions for a value of x of your choosing (other than 2) to be even more confident that your work is correct.

81. $2(4x)$

82. $3(5x)$

83. $5(x - 7)$

84. $4(x - 3)$

85. $5 - 3(x + 4)$

86. $1 - 6(x + 3)$

For Exercises 87–94, let x be a number. Translate the English phrase into a mathematical expression. Then simplify the expression.

87. 3 times the sum of the number and 2

88. -5 times the difference of the number and 7

89. -4 times the difference of twice the number and 5

90. 2 times the sum of twice the number and 6

91. 5, minus 2 times the difference of the number and 4

92. 6, minus 3 times the sum of the number and 1

93. -2, plus 7 times the sum of twice the number and 1

94. 3, minus 2 times the difference of twice the number and 5

Concepts

95. a. Simplify $2(x - 3) + 4$.
b. Evaluate $2(x - 3) + 4$ for $x = 5$.
c. Evaluate the result you found in part (a) for $x = 5$.
d. Compare the results you found in parts (b) and (c). If the results are equal, what does this tell you? If the results are not equal, what does that tell you?

96. a. Simplify $7 - 4(x + 2)$.
b. Evaluate $7 - 4(x + 2)$ for $x = 3$.
c. Evaluate the result you found in part (a) for $x = 3$.
d. Compare the results you found in parts (b) and (c). If the results are equal, what does this tell you? If the results are not equal, what does that tell you?

97. A student tries to simplify $3(x + 4)$:
$$3(x + 4) = 3x + 4$$
Evaluate both $3(x + 4)$ and $3x + 4$ for $x = 2$, and compare your results. What does this comparison tell you?

98. A student tries to simplify $-2(x - 5)$:
$$-2(x - 5) = -2x - 10$$
Evaluate both $-2(x - 5)$ and $-2x - 10$ for $x = 3$ and compare your results. What does this comparison tell you?

99. It is a common error to write $a(bc) = (ab)(ac)$.
a. Evaluate both $a(bc)$ and $(ab)(ac)$ for $a = 2, b = 3$, and $c = 4$.
b. Explain why the results you found in part (a) show that the statement $a(bc) = (ab)(ac)$ is false.
c. Write a true statement that involves $a(bc)$.

100. It is a common error to write $a(b + c) = ab + c$.
a. Evaluate both $a(b + c)$ and $ab + c$ for $a = 2, b = 3$, and $c = 4$.
b. Explain why the results you found in part (a) show that the statement $a(b + c) = ab + c$ is false.
c. Write a true statement that involves $a(b + c)$.

For Exercises 101–106, simplify the right-hand side of the equation. Then graph the equation by hand. Finally, use graphing calculator graphs to verify your work.

101. $y = 2(x - 3)$

102. $y = -4(x + 2)$

103. $y = -3(x + 1)$

104. $y = -2(x - 2)$

105. $y = -2(3x)$

106. $y = 4(2x)$

107. A student works 7 hours on Monday, 3 hours on Tuesday, and 6 hours on Wednesday. The student earns $10 per hour. Show two ways to compute the student's total earnings. Use the distributive law to explain why both methods give the same result.

108. When we write $-(x + 5)$ as $-x - 5$, what *number* have we distributed? Show the missing steps.

109. Give three examples of expressions that are equivalent to the expression $2x - 6$.

110. Give an example of two expressions that are not equivalent, but that happen to give the same result when they are evaluated for $x = 0$.

111. Show that the statement $a - (b - c) = (a - b) - c$ is false by substituting 7 for a, 5 for b, and 1 for c. Explain why your work shows that the statement is false and that there is no associative law for subtraction.

112. Show that the statement $a \div (b \div c) = (a \div b) \div c$ is false by substituting 8 for a, 4 for b, and 2 for c. Explain why your work shows that the statement is false and why there is no associative law for division.

113. Explain what law was used in each step to rearrange terms of $a + b + c$ as follows:
$$\begin{aligned} a + b + c &= (a + b) + c \quad \text{\textit{Add from left to right.}} \\ &= (b + a) + c \\ &= b + (a + c) \\ &= b + (c + a) \\ &= (b + c) + a \\ &= b + c + a \quad \text{\textit{Add from left to right.}} \end{aligned}$$

114. Explain what law was used in each step to rearrange factors of abc as follows:
$$\begin{aligned} abc &= (ab)c \quad \text{\textit{Multiply from left to right.}} \\ &= a(bc) \\ &= a(cb) \end{aligned}$$

$$= (ac)b$$
$$= (ca)b$$
$$= cab \qquad \textit{Multiply from left to right.}$$

115. Explain what law was used in each step to show that $(a + b)c = ac + bc$:

$$(a + b)c = c(a + b)$$
$$= ca + cb$$
$$= ac + bc$$

116. Use the distributive law $a(b - c) = ab - ac$ and other laws to show that $(a - b)c = ac - bc$. [**Hint:** See Exercise 115.]

117. In Section 2.5, we learned that the product of two numbers with the same sign is positive. Explain what property or law was used in each step to show why $(-2)(-3) = 6$:

$$(-2)(-3) = (-1 \cdot 2)(-3)$$
$$= -1(2 \cdot (-3))$$
$$= -1(-6)$$
$$= -(-6)$$
$$= 6$$

118. Describe the meaning of *equivalent expressions*. (See page 9 for guidelines on writing a good response.)

119. Give an example of two equivalent expressions.

4.2 Simplifying More Expressions

Objectives

» Combine like terms.

» Simplify expressions.

» Locate errors in simplifying expressions by evaluating expressions.

In Section 4.1, we discussed several ways to simplify an expression. Now we will discuss yet another technique used to simplify an expression: *combining like terms*.

Combining Like Terms

The **coefficient** of a term is the constant factor of the term. For example, the coefficient of the term $-3x$ is -3. Because $x = 1 \cdot x$, the coefficient of the term x is 1. Because $-x = -1x$, the coefficient of the term $-x$ is -1. For the term 4, the coefficient is 4.

Like terms are either constant terms or variable terms that contain the same variable(s) raised to exactly the same power(s). For example, 5 and 9 are like terms; so are $3x$ and $8x$. Also, $2y^3$, $5y^3$, and $-6y^3$ are like terms.

If terms are not like terms, we say that they are **unlike terms**. For example, $3x$ and $8y$ are unlike terms because $3x$ contains an x but $8y$ does not and because $8y$ contains a y but $3x$ does not. The terms $4x^2$ and $7x^3$ are unlike terms because the exponents of x are different.

For a sum of like terms, such as $2x + 4x$, we can use the distributive law to write the sum as one term:

$$2x + 4x = (2 + 4)x \quad \textit{Distributive law: } ac + bc = (a + b)c$$
$$= 6x \qquad\quad \textit{Add.}$$

When we write a sum or difference of like terms as one term, we say we have **combined like terms**.

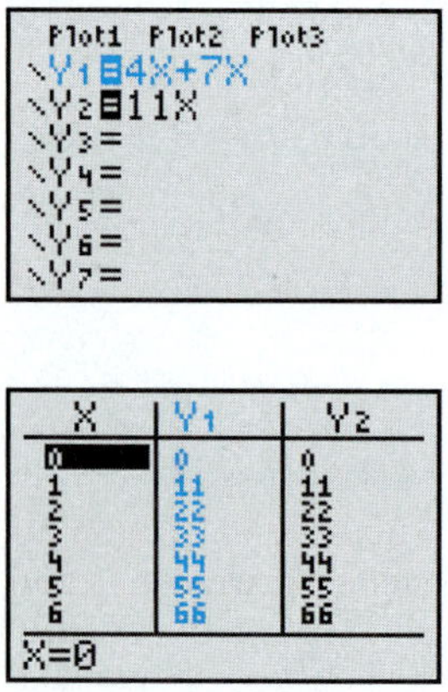

Figure 4 Verify the work

Example 1   Combining Two Like Terms

Combine like terms.

1. $4x + 7x$ **2.** $8x - 5x$ **3.** $6a + a$

Solution

1. $4x + 7x = (4 + 7)x \quad \textit{Distributive law: } ac + bc = (a + b)c$
           $= 11x \qquad\quad \textit{Add.}$

We use a graphing calculator table to verify our work (see Fig. 4).

2. $8x - 5x = (8 - 5)x \quad \textit{Distributive law: } ac - bc = (a - b)c$
           $= 3x \qquad\quad \textit{Subtract.}$

3. $6a + a = 6a + 1a \qquad a = 1a$
        $= (6 + 1)a \quad \textit{Distributive law: } ac + bc = (a + b)c$
        $= 7a \qquad\quad \textit{Add.}$

In Problem 1 of Example 1, we found that $4x + 7x = 11x$. We can find the sum $4x + 7x$ in one step by adding the coefficients of $4x$ and $7x$ (that is, 4 and 7) to get the coefficient of the result, 11:

$$4x + 7x = 11x$$

$$4 + 7 = 11$$

In Problem 2 of Example 1, we found that $8x - 5x = 3x$. Note that if we write $8x - 5x = 8x + (-5x)$, then we can add the coefficients of $8x$ and $-5x$ (that is, 8 and -5) to get the coefficient of the result, 3:

$$8x - 5x = 3x$$

$$8 + (-5) = 3$$

Combining Like Terms

To combine like terms, add the coefficients of the terms and keep the same variable factors.

Simplifying an Expression

We simplify an expression by removing parentheses and combining like terms.

Example 2   Simplifying Expressions

Simplify.

1. $2x + 5y - 6x + 2 + 3y + 7$ **2.** $-2(x + 5y) - 3(2x - 4y)$
3. $4(x - 2) - (x + 3)$

Solution

1. $2x + 5y - 6x + 2 + 3y + 7 = 2x - 6x + 5y + 3y + 2 + 7$ *Rearrange terms.*
$$= -4x + 8y + 9 \qquad\qquad\qquad \textit{Combine like terms.}$$

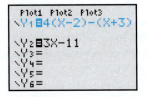

2. $-2(x + 5y) - 3(2x - 4y) = -2x - 10y - 6x + 12y$ *Distributive law*
$$= -2x - 6x - 10y + 12y \quad \textit{Rearrange terms.}$$
$$= -8x + 2y \qquad\qquad\qquad \textit{Combine like terms.}$$

3. We write $4(x - 2) - (x + 3)$ as $4(x - 2) - 1(x + 3)$ so that we can distribute the -1:

$$4(x - 2) - (x + 3) = 4(x - 2) - 1(x + 3) \quad a - b = a - 1b$$
$$= 4x - 8 - 1x - 3 \qquad \textit{Distributive law}$$
$$= 4x - 1x - 8 - 3 \qquad \textit{Rearrange terms.}$$
$$= 3x - 11 \qquad\qquad\quad \textit{Combine like terms.}$$

We use a graphing calculator table to verify the work (see Fig. 5).

Figure 5 Verify the work

Example 3   Translating from English into Mathematics

Let x be a number. Translate the phrase "3 times the number, minus 4, plus twice the number" into an expression. Then simplify the expression.

Solution

The expression is

3 times
the number,   minus 4,   plus   twice the
number

$$3x \qquad -4 \qquad + \qquad 2x$$

Next, we simplify the expression:

$$3x - 4 + 2x = 3x + 2x - 4 \quad \text{\textit{Rearrange terms.}}$$
$$= 5x - 4 \quad \text{\textit{Combine like terms.}}$$

▶

In Example 3, we translated an English phrase into a mathematical expression. In Example 4, we will translate a mathematical expression into an English phrase.

Example 4   Translating from Mathematics into English

Let x be a number. Translate the expression $x - 5 \cdot (x - 2)$ into an English phrase. Then simplify the expression.

Solution

Here is the translation:

$$x \; - \qquad\qquad 5 \cdot \qquad\qquad (x \; - \; 2)$$

the number,   5 times   the difference of
minus   the number and 2

One of many possible correct translations is "the number, minus 5 times the difference of the number and 2." Next, we simplify the expression:

$$x - 5 \cdot (x - 2) = x - 5x + 10 \quad \text{\textit{Distributive law}}$$
$$= -4x + 10 \quad \text{\textit{Combine like terms.}}$$

▶

Locate Errors in Simplifying Expressions

So far, we have used a graphing calculator table to verify our work in simplifying an expression. If we determine that we have made an error, we still have to find it to correct it. In Example 5, we will discuss how to pinpoint the step(s) in which any errors have been made.

Example 5   Locating an Error

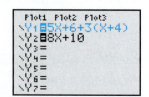

Consider the following incorrect work:

$$5x + 6 + 3(x + 4) = 5x + 6 + 3x + 4 \quad \text{\textit{Incorrect}}$$
$$= 5x + 3x + 6 + 4$$
$$= 8x + 10$$

1. Use a graphing calculator table to show that the work is incorrect.
2. Pinpoint where the error was made by evaluating the expression in each step for $x = 2$.

Solution

1. From Fig. 6, we can tell that the expressions $5x + 6 + 3(x + 4)$ and $8x + 10$ are *not* equivalent because the second and third columns of the table are different. So, an error has been made.
2. We evaluate each expression for $x = 2$:

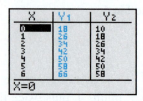

Figure 6 The work is incorrect

Expression	Evaluate the Expression for $x = 2$
$5x + 6 + 3(x + 4)$	$5(2) + 6 + 3(2 + 4) = 34$
$= 5x + 6 + 3x + 4$	$5(2) + 6 + 3(2) + 4 = 26$
$= 5x + 3x + 6 + 4$	$5(2) + 3(2) + 6 + 4 = 26$
$= 8x + 10$	$8(2) + 10 = 26$

Because the result of evaluating $5x + 6 + 3(x + 4)$ for $x = 2$ is different from the result of evaluating $5x + 6 + 3x + 4$ for $x = 2$, we conclude that an error was made in writing $5x + 6 + 3(x + 4) = 5x + 6 + 3x + 4$. (Note that the right-hand side of the equation *should* be $5x + 6 + 3x + 12$.)

Locating an Error in Simplifying an Expression

After simplifying an expression, use a graphing calculator table to check whether the result is equivalent to the original expression. If the expressions are not equivalent, an error was made. To pinpoint the error, evaluate all the expressions by using the same number(s) and find the pair of consecutive expressions for which the results are different.

 ## Group Exploration

Laws of operations

For the work shown, carefully describe any errors.

1. $3(x + 5) + 4x = 3x + 5 + 4x$
$= 3x + 4x + 5$
$= 7x + 5$

2. $2(xy) = (2x)(2y)$
$= 2(2)xy$
$= 4xy$

3. $5 + x - 3 = 5 + 3 - x$
$= 8 - x$
$= x - 8$

Tips for Success   **Read Your Response**

After writing a response to a question, read your response to make sure that it says what you intended it to say. Reading your response aloud, if you are in a place where you feel comfortable doing so, can help.

Homework 4.2

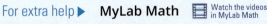

For extra help ▶ **MyLab Math**   Watch the videos in MyLab Math

Simplify. Use a graphing calculator table to verify your work when possible.

1. $2x + 5x$

2. $3x + 6x$

3. $9x - 4x$

4. $6x - 5x$

5. $-8w - 5w$

6. $-4p + 7p$

7. $-t + 5t$

8. $a + 3a$

9. $6.6x - 7.1x$

10. $-4.5x - 2.9x$

11. $\frac{2}{3}x + \frac{5}{3}x$

12. $\frac{9}{5}x - \frac{2}{5}x$

13. $2 + 4x - 5 - 7x$

14. $-8x - 1 + 3x - 4$

15. $-3p + 2 + p - 9$ **16.** $7 - w - 10 - 2w$

17. $3y + 5x - 2y - 2x + 1$ **18.** $5 - 3x + 7y + 6x$

19. $-4.6x + 3.9y + 2.1 - 5.3x - 2.8y$

20. $4.7 - 3.5y + 8.8x - 6.2y + 1.9x$

21. $-3(a - 5) + 2a$ **22.** $5(t - 8) - 6t$

23. $5.2(8.3x + 4.9) - 2.4$ **24.** $-3.8(2.7x - 5.5) - 8.4$

25. $4(3a - 2b) - 5a$ **26.** $-2(8m + 4n) - 3n$

27. $8 - 2(x + 3) + x$ **28.** $3 - 4(x + 1) - x$

29. $6x - (4x - 3y) - 5y$ **30.** $8x - (3x + 7y) - 2y$

31. $2t - 3(5t + 2) + 1$ **32.** $3a - 2(3a + 4) + 5$

33. $6 - 2(x + 3y) + 2y$ **34.** $4 - 5(x + 2y) + 7y$

35. $-3(x - 2) - 5(x + 4)$ **36.** $-2(x - 3) - 7(x + 2)$

37. $6(2x - 3y) - 4(9x + 5y)$

38. $2(7x - 5y) - 5(3x + y)$

39. $-(x - 1) - (1 - x)$

40. $-(6x - 7) - (7 - 6x)$

41. $2x - 5y - 3(2x - 4y + 7)$

42. $4x + 3y - 2(5x + 2y + 8)$

43. $5(2x - 4y) - (3x - 7y + 2)$

44. $3(4x - 3y) - (9x + 2y - 5)$

45. $\frac{2}{7}(a + 1) - \frac{4}{7}(a - 1)$ **46.** $\frac{3}{5}(t - 1) + \frac{1}{5}(t + 1)$

47. $5x - \frac{1}{2}(4x + 6)$ **48.** $7x - \frac{1}{3}(6x + 9)$

For Exercises 49–56, let x be a number. Translate the English phrase into a mathematical expression. Then simplify the expression.

49. The number plus the product of 5 and the number

50. The number minus the product of the number and 3

51. 4 times the difference of the number and 2

52. -6 times the sum of the number and 4

53. The number, plus 3 times the difference of the number and 7

54. The number, minus 5 times the sum of the number and 2

55. Twice the number, minus 4 times the sum of the number and 6

56. Twice the number, plus 9 times the difference of the number and 4

For Exercises 57–64, let x be a number. Translate the expression into an English phrase. Then simplify the expression.

57. $2x + 6x$ **58.** $3x - 8x$

59. $7(x - 5)$ **60.** $-2(x + 4)$

61. $x + 5(x + 1)$ **62.** $x - 8(x - 3)$

63. $2x - 3(x - 9)$ **64.** $2x + 6(x - 4)$

65. Find the sum of $3x - 7$ and $5x + 2$.

66. Find the sum of $6x - 3$ and $8x - 9$.

67. Find the difference of $4x + 8$ and $7x - 1$.

68. Find the difference of $5x - 9$ and $2x + 6$.

Simplify. Then evaluate both the expression and your result for x = 4 to check your work. Finally, evaluate both expressions for an x-value of your choosing (other than 4) to be even more confident that your work is correct.

69. $-2x + 5 - 3 + 7x$ **70.** $4x - 8 - 2 + x$

71. $4(x + 2) - (x - 3)$ **72.** $-(x - 7) + 4(x + 2)$

Concepts

73. a. Simplify $3(x + 4) + 5x$.
 b. Evaluate $3(x + 4) + 5x$ for $x = 2$.
 c. Evaluate the result you found in part (a) for $x = 2$.
 d. Compare the results you found in parts (b) and (c). If the results are equal, what does this tell you? If the results are not equal, what does that tell you?

74. a. Simplify $4(x - 2) - 3(x + 1)$.
 b. Evaluate $4(x - 2) - 3(x + 1)$ for $x = 3$.
 c. Evaluate the result you found in part (a) for $x = 3$.
 d. Compare the results you found in parts (b) and (c). If the results are equal, what does this tell you? If the results are not equal, what does that tell you?

75. Which of the following expressions are equivalent?

$$-2x - 3 \qquad -2(x - 3) \qquad 2(3 - x) \qquad -2x - 6$$
$$-3(x - 2) + x \qquad -2x + 6$$

76. Which of the following expressions are equivalent?

$$-(2x - 3) \qquad 3 - 2x \qquad -(3 - 2x) \qquad -2x + 3$$
$$-(3x - 2) + x + 1 \qquad -(x - 1) - (x - 2)$$

77. A student incorrectly simplifies $4(x - 3) + 5x - 1$:

Expression	Evaluate the Expression for $x = 2$
$4(x - 3) + 5x - 1$	
$= 4x - 3 + 5x - 1$	
$= 4x + 5x - 3 - 1$	
$= 9x - 4$	

Evaluate each of the four expressions for $x = 2$ to pinpoint where the error was made. Then simplify $4(x - 3) + 5x - 1$ correctly.

78. A student incorrectly simplifies $2(x + 1) + x - 4$:

Expression	Evaluate the Expression for $x = 3$
$2(x + 1) + x - 4$	
$= 2x + 2 + x - 4$	
$= 2x + x + 2 - 4$	
$= 2x - 2$	

Evaluate each of the four expressions for $x = 3$ to pinpoint where the error was made. Then simplify $2(x + 1) + x - 4$ correctly.

79. Give three examples of expressions equivalent to the expression $2(x - 5) + 3(x + 1)$.

80. Give three examples of expressions equivalent to x.

For Exercises 81–86, simplify the right-hand side of the equation. Then graph the equation by hand. Finally, use graphing calculator graphs to verify your work.

81. $y = 3x - 5x$

82. $y = x + 2x$

83. $y = 9x - 4 - 7x$

84. $y = 4x + 3 - 6x$

85. $y = 4(2x - 1) - 5x$

86. $y = -2(3x - 2) + 5x$

87. a. Pick a number. Next, subtract 3. Then double the result. Then subtract the original number. Finally, add 6. Compare your result with the original number.

b. Pick another number, and follow the instructions in part (a), including comparing your result with the original number.

c. Let x be a number, and follow the instructions in part (a). Your result should be an expression. Simplify the expression to explain the observations you made in parts (a) and (b).

88. Give an example of combining like terms. Then use the distributive law to show why we can do this.

89. Describe how to simplify an expression.

4.3 Solving Linear Equations in One Variable

Objectives

» Describe a *linear equation in one variable*.

» For an equation in one variable, describe *satisfy*, *solution*, *solution set*, and *solve*.

» Describe the three roles of variables.

» Describe *equivalent equations*.

» Apply the *addition property of equality*.

» Apply the *multiplication property of equality*.

» Solve a linear equation in one variable.

» Solve a percentage problem.

» Use a graph or a table to solve a linear equation in one variable.

Linear Equation in One Variable

In Sections 4.1 and 4.2, we simplified expressions. In this section and Section 4.4, we will use that skill to help us work with equations. So far, we have discussed how to graph linear equations in *two* variables, such as

$$y = x + 2 \qquad y = -5x \qquad y = 4x - 7 \qquad y = 2x - 6$$

In this section and Section 4.4, we will work with *linear equations in one variable*, such as

$$0 = x + 2 \qquad 9 = -5x \qquad 4x - 7 = 3 \qquad x + 1 = 2x - 6$$

We will see in this section and in Section 4.4 that we can put each of these four equations in the form $mx + b = 0$, where m and b are constants and $m \neq 0$.

> **Definition Linear equation in one variable**
>
> A **linear equation in one variable** is an equation that can be put into the form
>
> $$mx + b = 0$$
>
> where m and b are constants and $m \neq 0$.

Working with linear equations in one variable will enable us to make efficient estimates and predictions about authentic situations.

Meaning of Satisfy, Solution, Solution Set, and Solve

Consider the linear equation

$$x + 1 = 6$$

This equation becomes a false statement if we substitute 2 for x:

$$x + 1 = 6 \qquad \textit{Original equation}$$
$$(2) + 1 \overset{?}{=} 6 \qquad \textit{Substitute 2 for x.}$$
$$3 \overset{?}{=} 6 \qquad \textit{Add.}$$
$$\text{false}$$

However, the equation $x + 1 = 6$ becomes a true statement if we substitute 5 for x:

$$x + 1 = 6 \qquad \textit{Original equation}$$
$$(5) + 1 \overset{?}{=} 6 \qquad \textit{Substitute 5 for x.}$$
$$6 \overset{?}{=} 6 \qquad \textit{Add.}$$
$$\text{true}$$

We say 5 *satisfies* the equation $x + 1 = 6$ and 5 is a *solution* of the equation. In fact, 5 is the only solution of $x + 1 = 6$ because 5 is the only number that, when increased by 1, is equal to 6. We call the set containing only this number the *solution set* of the equation.

> **Definition**   *Solution, satisfy, solution set*, and *solve* for an equation in one variable
>
> A number is a **solution** of an equation in one variable if the equation becomes a true statement when the number is substituted for the variable. We say the number **satisfies** the equation. The set of all solutions of the equation is called the **solution set** of the equation. We **solve** the equation by finding its solution set.

Example 1   Identifying Solutions of an Equation

1. Is 3 a solution of the equation $5(x - 1) = 10 + 2x$?
2. Is 5 a solution of the equation $5(x - 1) = 10 + 2x$?

Solution

1. We begin by substituting 3 for x in $5(x - 1) = 10 + 2x$:

$$5(x - 1) = 10 + 2x \qquad \textit{Original equation}$$
$$5(3 - 1) \stackrel{?}{=} 10 + 2(3) \qquad \textit{Substitute 3 for x.}$$
$$5(2) \stackrel{?}{=} 10 + 6 \qquad \textit{Simplify.}$$
$$10 \stackrel{?}{=} 16 \qquad \textit{Simplify.}$$
$$\text{false}$$

So, 3 is not a solution of the equation $5(x - 1) = 10 + 2x$.

2. We begin by substituting 5 for x in $5(x - 1) = 10 + 2x$:

$$5(x - 1) = 10 + 2x \qquad \textit{Original equation}$$
$$5(5 - 1) \stackrel{?}{=} 10 + 2(5) \qquad \textit{Substitute 5 for x.}$$
$$5(4) \stackrel{?}{=} 10 + 10 \qquad \textit{Simplify.}$$
$$20 \stackrel{?}{=} 20 \qquad \textit{Simplify.}$$
$$\text{true}$$

So, 5 is a solution of the equation $5(x - 1) = 10 + 2x$.

Roles of a Variable

In Section 2.1, we discussed two roles of a variable. In one role, a variable represents a quantity that can vary. For example, if we let s represent the speed of an airplane, then the variable s can vary between 0 and 2200 miles per hour. In another role, a variable can serve as a placeholder in an expression. For example, the variable x is a placeholder for a number in the expression $2x + 5$.

A variable can be used in yet another way: In an equation, a variable is used to represent any number that is a solution of the equation. For instance, in Example 1, the variable x is used to represent any number that satisfies the equation $5(x - 1) = 10 + 2x$. In Problem 2 of Example 1, we found that 5 is such a number.

> **Roles of a Variable**
>
> Here are three roles of a variable:
>
> 1. A variable represents a quantity that can vary.
> 2. In an expression, a variable is a placeholder for a number.
> 3. In an equation, a variable represents any number that is a solution of the equation.

In this section we emphasize the third role of a variable.

Equivalent Equations

Consider the equation $x = 2$. We add 5 to both sides of the equation:

$$x = 2 \qquad \text{\textit{Original equation}}$$
$$x + 5 = 2 + 5 \qquad \text{\textit{Add 5 to both sides.}}$$
$$x + 5 = 7 \qquad \text{\textit{Add.}}$$

Note that 2 satisfies all three equations:

Equation	Does 2 satisfy the equation?	
$x = 2$	$(2) \overset{?}{=} 2$	true
$x + 5 = 2 + 5$	$(2) + 5 \overset{?}{=} 2 + 5$	true
$x + 5 = 7$	$(2) + 5 \overset{?}{=} 7$	true

In fact, 2 is the *only* number that satisfies any of these equations. So, the equations $x = 2$, $x + 5 = 2 + 5$, and $x + 5 = 7$ have the same solution set. We say that the three equations are *equivalent*.

> **Definition   Equivalent equations**
>
> **Equivalent equations** are equations that have the same solution set.

Addition Property of Equality

The fact that the equations $x = 2$ and $x + 5 = 2 + 5$ have the same solution set suggests that adding a number to both sides of an equation does not change an equation's solution set. This property is called the **addition property of equality**.

> **Addition Property of Equality**
>
> If A and B are expressions and c is a number, then the equations $A = B$ and $A + c = B + c$ are equivalent.

To solve an equation in one variable, x, we can sometimes use the addition property of equality to get x alone on one side of the equation. Then we can identify solutions of the equation. For example, for the equation $x = 3$, we can see the solution is 3.

Example 2   Solving an Equation by Adding a Number to Both Sides

Solve $x - 2 = 3$.

Solution

To get x alone on the left side, we undo the subtraction of 2 by adding 2 to *both* sides:

$$x - 2 = 3 \qquad \text{\textit{Original equation}}$$
$$x - 2 + 2 = 3 + 2 \qquad \text{\textit{Addition property of equality: Add 2 to both sides.}}$$
$$x + 0 = 5 \qquad \text{\textit{Simplify: } -a + a = 0}$$
$$x = 5 \qquad \text{\textit{Simplify: } a + 0 = a}$$

Next, we check that 5 satisfies the original equation, $x - 2 = 3$:

$$x - 2 = 3 \qquad \text{\textit{Original equation}}$$
$$(5) - 2 \overset{?}{=} 3 \qquad \text{\textit{Substitute 5 for x.}}$$
$$3 \overset{?}{=} 3 \qquad \text{\textit{Subtract.}}$$
$$\text{true}$$

So, the solution is 5.

After solving an equation, check that all your results satisfy the equation.

In Example 2, we worked with the equations $x - 2 = 3$, $x - 2 + 2 = 3 + 2$, $x + 0 = 5$, and $x = 5$. Although we could find the solution set with any of these equivalent equations, it is easiest to determine the solution set from the equation $x = 5$, which has the variable alone on the left side. **Our strategy in solving linear equations in one variable will be to use properties to get the variable alone on one side of the equation.**

Assume that A and B are expressions and that c is a number. Because subtracting a number is the same as adding the opposite of the number, the addition property of equality implies that the equations $A = B$ and $A - c = B - c$ are equivalent.

Example 3   Solving an Equation by Subtracting a Number from Both Sides

Solve $x + 4 = 6$.

Solution

To get x alone on the left side, we undo the addition of 4 by subtracting 4 from *both* sides:

$$x + 4 = 6 \qquad \textit{Original equation}$$
$$x + 4 - 4 = 6 - 4 \qquad \textit{Subtract 4 from both sides.}$$
$$x = 2 \qquad \textit{Simplify: } a + 0 = a$$

Next, we check that 2 satisfies the original equation, $x + 4 = 6$:

$$x + 4 = 6 \qquad \textit{Original equation}$$
$$(2) + 4 \stackrel{?}{=} 6 \qquad \textit{Substitute 2 for x.}$$
$$6 \stackrel{?}{=} 6 \qquad \textit{Add.}$$
$$\text{true}$$

The solution is 2.

▶

Multiplication Property of Equality

Consider the equation $x = 3$. We multiply both sides of $x = 3$ by 7:

$$x = 3 \qquad \textit{Original equation}$$
$$7 \cdot x = 7 \cdot 3 \qquad \textit{Multiply both sides by 7.}$$
$$7x = 21 \qquad \textit{Multiply.}$$

Note that 3 satisfies all three equations:

Equation	Does 3 satisfy the equation?	
$x = 3$	$(3) \stackrel{?}{=} 3$	*true*
$7 \cdot x = 7 \cdot 3$	$7 \cdot (3) \stackrel{?}{=} 7 \cdot 3$	*true*
$7x = 21$	$7(3) \stackrel{?}{=} 21$	*true*

In fact, 3 is the *only* number that satisfies any of these equations. So, the equations $x = 3$, $7 \cdot x = 7 \cdot 3$, and $7x = 21$ are equivalent. The fact that the equations $x = 3$ and $7 \cdot x = 7 \cdot 3$ are equivalent suggests the **multiplication property of equality**.

Multiplication Property of Equality

If A and B are expressions and c is a nonzero number, then the equations $A = B$ and $Ac = Bc$ are equivalent.*

---

*If $c = 0$, the equations may not be equivalent. For example, $x = 5$ and $x \cdot 0 = 5 \cdot 0$ (or $0 = 0$) are not equivalent. The only solution of $x = 5$ is 5, but the material in Section 4.4 will help us see that the solution set of $0 = 0$ is the set of all real numbers.

The multiplication property of equality is often helpful in solving equations that contain fractions. It is useful to know that a fraction times its reciprocal is equal to 1. For example,

$$\frac{2}{7} \cdot \frac{7}{2} = \frac{14}{14} = 1$$

Solving a Linear Equation in One Variable

In Example 4, we will use the multiplication property of equality to solve an equation.

| **Example 4** | Solving an Equation by Multiplying Both Sides by a Number |

Solve $\frac{4}{5}x = 8$.

Solution

To get x alone on the left side, we multiply *both* sides by the reciprocal of $\frac{4}{5}$, which is $\frac{5}{4}$:

$$\frac{4}{5}x = 8 \qquad \textit{Original equation}$$

$$\frac{5}{4} \cdot \frac{4}{5}x = \frac{5}{4} \cdot 8 \qquad \textit{Multiplication property of equality: Multiply both sides by } \frac{5}{4}.$$

$$1x = 10 \qquad \textit{Simplify.}$$

$$x = 10 \qquad \textit{1a = a}$$

The solution is 10. We use a graphing calculator table to check that, when 10 is substituted for x in the expression $\frac{4}{5}x$, the result is 8 (see Fig. 7). For graphing calculator instructions on using "Ask" in a table, see Appendix A.15.

  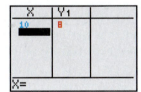

Figure 7 Verify the work

Assume that A and B are expressions and c is a nonzero number. Because dividing by a nonzero number is the same as multiplying by its reciprocal, the multiplication property of equality implies the equations $A = B$ and $\frac{A}{c} = \frac{B}{c}$ are equivalent.

| **Example 5** | Solving an Equation by Dividing Both Sides by a Number |

Solve $-12 = -3t$.

Solution

The variable term $-3t$ is on the right-hand side this time. To get t alone on this side, we undo the multiplication by -3 by dividing *both* sides by -3:

$$-12 = -3t \qquad \textit{Original equation}$$

$$\frac{-12}{-3} = \frac{-3t}{-3} \qquad \textit{Divide both sides by } -3.$$

$$4 = t \qquad \textit{Simplify.}$$

The solution is 4. We can check that 4 satisfies the original equation (try it).

The last step of Example 5 is $4 = t$. Note that the equation $t = 4$ is equivalent to $4 = t$ and that we can see from either equation that the solution is 4.

Example 6   Solving an Equation by Multiplying Both Sides by -1

Solve $-w = 5$.

Solution

Recall that $-w = -1w$. To get w alone on the left side, we undo the multiplication by -1 by dividing both sides by -1:

$$-w = 5 \qquad \textit{Original equation}$$
$$-1w = 5 \qquad \textit{$-w = -1w$}$$
$$\frac{-1w}{-1} = \frac{5}{-1} \qquad \textit{Divide both sides by -1.}$$
$$w = -5 \qquad \textit{Divide.}$$

Next, we check that -5 satisfies the original equation, $-w = 5$:

$$-w = 5 \qquad \textit{Original equation}$$
$$-(-5) \overset{?}{=} 5 \qquad \textit{Substitute -5 for w.}$$
$$5 \overset{?}{=} 5 \qquad \textit{$-(-a) = a$}$$
$$\text{true}$$

The solution is -5.

Example 7   Solving an Equation by Multiplying Both Sides by a Number

Solve $\dfrac{2x}{3} = \dfrac{5}{6}$.

Solution

Because $\dfrac{2}{3}x = \dfrac{2}{3} \cdot \dfrac{x}{1} = \dfrac{2x}{3}$, we first write $\dfrac{2x}{3}$ as $\dfrac{2}{3}x$. Then we multiply both sides by the reciprocal of $\dfrac{2}{3}$, which is $\dfrac{3}{2}$, to get x alone on the left side:

$$\frac{2x}{3} = \frac{5}{6} \qquad \textit{Original equation}$$
$$\frac{2}{3}x = \frac{5}{6} \qquad \textit{Write $\frac{2x}{3}$ as $\frac{2}{3}x$.}$$
$$\frac{3}{2} \cdot \frac{2}{3}x = \frac{3}{2} \cdot \frac{5}{6} \qquad \textit{Multiply both sides by $\frac{3}{2}$.}$$
$$1x = \frac{3 \cdot 5}{2 \cdot 2 \cdot 3} \qquad \textit{Simplify; $\frac{a}{b} \cdot \frac{c}{d} = \frac{ac}{bd}$}$$
$$x = \frac{5}{4} \qquad \textit{Simplify: $\frac{3}{3} = 1$}$$

The solution is $\dfrac{5}{4}$. We use a graphing calculator table to check that, when $\dfrac{5}{4} = 1.25$ is substituted for x in the expression $\dfrac{2x}{3}$, the result is $\dfrac{5}{6} \approx 0.83333$ (see Fig. 8).

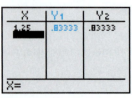

Figure 8 Verify the work

Solving Percentage Problems

Recall that to find the percentage of a quantity, we multiply the decimal form of the percentage and the quantity (Section 2.5).

| Example 8 | Solving a Percentage Problem |

A person deposits 28% of her weekly earnings into her savings account. If she deposits $448 into her savings account each week, how much is her weekly pay?

Solution

We let p be the weekly pay (in dollars). Because 28% of the weekly pay is equal to $448, we write

$$\underbrace{28\% \text{ of weekly pay}}_{0.28p} \quad \overset{\text{is equal to}}{=} \quad \overset{448}{448}$$

Now we solve the equation:

$$0.28p = 448$$
$$\frac{0.28p}{0.28} = \frac{448}{0.28} \qquad \textit{Divide both sides by 0.28.}$$
$$p = 1600 \qquad \textit{Simplify; divide.}$$

The weekly pay is $1600.

In most application problems in this text so far, variable names and their definitions have been provided. In Example 8, a key step was to create the variable name p and define it.

Using Graphing to Solve an Equation in One Variable

In Example 2, we showed that the solution of the equation $x - 2 = 3$ is 5. How can we use graphing to solve this equation? Here are three steps:

Step 1. We set y equal to the left side, $x - 2$, to form the equation $y = x - 2$, and we set y equal to the right side, 3, to form the equation $y = 3$. Then we graph the two equations $y = x - 2$ and $y = 3$ in the same coordinate system (see Fig. 9).

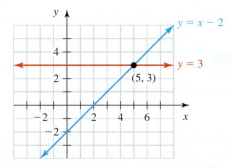

Figure 9 The intersection point is $(5, 3)$

Step 2. We find the intersection point of the two graphs, which is the point $(5, 3)$.

Step 3. The solution of the original equation $x - 2 = 3$ is the x-coordinate of the intersection point $(5, 3)$. So, the solution is 5.

In general, the solutions are all the x-coordinates of any intersection points.

Using Graphing to Solve an Equation in One Variable

To use graphing to solve an equation $A = B$ in one variable, x, where A and B are expressions,

1. Graph the equations $y = A$ and $y = B$ in the same coordinate system. (For example, if the original equation is $5x - 9 = 3x + 7$, then we would graph the equations $y = 5x - 9$ and $y = 3x + 7$.)

2. Find all intersection points.

3. The x-coordinates of those intersection points are the solutions of the equation $A = B$.

Example 9   Solving an Equation in One Variable by Graphing

The graphs of $y = \frac{3}{2}x + 1$, $y = 4$, and $y = -5$ are shown in Fig. 10. Use these graphs to solve the given equations.

1. $\frac{3}{2}x + 1 = 4$ **2.** $\frac{3}{2}x + 1 = -5$

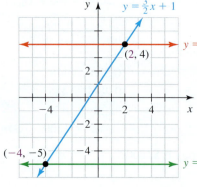

Figure 10 Solving $\frac{3}{2}x + 1 = 4$ and $\frac{3}{2}x + 1 = -5$

Solution

1. The graphs of $y = \frac{3}{2}x + 1$ and $y = 4$ intersect only at the point $(2, 4)$. The intersection point, $(2, 4)$, has x-coordinate 2. So, 2 is the solution of $\frac{3}{2}x + 1 = 4$. We can verify our work by checking that 2 satisfies the equation (try it).

2. The graphs of $y = \frac{3}{2}x + 1$ and $y = -5$ intersect only at the point $(-4, -5)$. The intersection point, $(-4, -5)$, has x-coordinate -4. So, -4 is the solution of $\frac{3}{2}x + 1 = -5$. We can verify our work by checking that -4 satisfies the equation (try it).

Example 10   Solving an Equation in One Variable by Graphing

Use "intersect" on a graphing calculator to solve the equation $-2x + 4 = x - 5$.

Solution

We use a graphing calculator to graph the equations $y = -2x + 4$ and $y = x - 5$ in the same coordinate system and then use "intersect" to find the intersection point, which turns out to be $(3, -2)$. See Fig. 11. See Appendix A.17 for graphing calculator instructions.

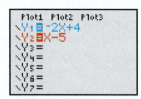 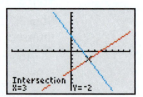

Figure 11 Using "intersect" to solve an equation in one variable

The intersection point, $(3, -2)$, has x-coordinate 3. So, the solution of the original equation is 3. We can verify our work by checking that 3 satisfies the equation $-2x + 4 = x - 5$ (try it).

Using a Table to Solve an Equation in One Variable

We can use a table of solutions of an equation in two variables to help us solve an equation in one variable.

Example 11   Solving an Equation in One Variable by Using a Table

Some solutions of $y = 3x - 5$ are shown in Table 2. Use the table to solve the equation $1 = 3x - 5$.

Table 2 Solutions of $y = 3x - 5$

x	y
−1	−8
0	−5
1	−2
2	1
3	4

Solution

If we substitute 1 for y in the equation $y = 3x - 5$, the result is the equation $1 = 3x - 5$, which is what we are trying to solve. From Table 2, we see that the output $y = 1$ origi-nates from the input $x = 2$. The ordered pair $(2, 1)$ satisfies the equation $y = 3x - 5$:

$$1 = 3(2) - 5$$

This means that 2 is a solution of the equation $1 = 3x - 5$.

▶

We have solved an equation by getting the variable alone on one side of the equation, by using graphing, and by using a table. All three methods give the same results.

Remember that **whether adding, subtracting, multiplying, or dividing, what is done to one side of the equation must also be done to the other side of the equation.**

Group Exploration

Locating an error in solving an equation

A student tries to solve the equation $x - 3 = 5$:

$$x - 3 = 5$$
$$x - 3 - 3 = 5 - 3$$
$$x - 0 = 2$$
$$x = 2$$

Substitute 2 for x in each of the four equations to determine which equations have 2 as a solution. Explain how your work shows that the student made an error and how your work helps you pinpoint the step in which the error was made.

Tips for Success   **Review Material**

At various times throughout this course, you can improve your understanding of alge-bra by reviewing material that you have learned so far. Your review should include solving problems, redoing explorations, and reexamining concepts and techniques from previous sections.

Homework 4.3

For extra help ▶   **MyLab Math**   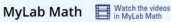 Watch the videos in MyLab Math

Determine whether 2 is a solution of the equation.

1. $3x + 1 = 7$

2. $-2x + 5 = 4$

3. $5(2x - 1) = 0$

4. $-3(x - 2) = 0$

5. $12 - x = 2(4x - 3)$

6. $5 - x = 3(2x - 1)$

For Exercises 7–50, solve. Verify that your result satisfies the equation.

7. $x - 3 = 2$

8. $x - 4 = 9$

9. $x + 6 = -8$

10. $x + 1 = -5$

11. $t - 9 = 15$

12. $a - 5 = 12$

13. $x + 11 = -17$

14. $x + 18 = -13$

15. $-5 = x - 2$

16. $-3 = x - 4$　　**17.** $x - 3 = 0$　　**18.** $x - 5 = 0$

19. $6r = 18$　　**20.** $4w = 24$　　**21.** $-3x = 12$

22. $-5x = 20$　　**23.** $15 = 3x$　　**24.** $24 = 2x$

25. $6x = 8$　　**26.** $4x = 6$　　**27.** $-10x = -12$

28. $-14x = -6$　　**29.** $-2x = 0$　　**30.** $-5x = 0$

31. $\dfrac{1}{3}t = 5$　　**32.** $\dfrac{1}{2}w = 4$　　**33.** $-\dfrac{2}{7}x = -3$

34. $-\dfrac{4}{9}x = -5$　　**35.** $-9 = \dfrac{3x}{4}$　　**36.** $-8 = \dfrac{2x}{5}$

37. $\dfrac{2}{5}p = -\dfrac{4}{3}$　　**38.** $\dfrac{4}{7}b = -\dfrac{5}{21}$　　**39.** $-\dfrac{3x}{8} = -\dfrac{9}{4}$

40. $-\dfrac{5x}{4} = -\dfrac{15}{8}$　　**41.** $-x = 3$　　**42.** $-x = 2$

43. $-\dfrac{1}{2} = -x$　　**44.** $-\dfrac{3}{4} = -x$　　**45.** $x + 4.3 = -6.8$

46. $x + 7.5 = -2.8$　　　**47.** $25.17 = x - 16.59$

48. $5.27 = x - 28.85$　　　**49.** $-3.7r = -8.51$

50. $-2.9w = 13.34$

51. In Chicago, a person must pay 9.5% of the price of a couch in sales tax. If the sales tax is $77.90, what is the price of the couch?

52. A person tips a waiter 18% of the bill. If the tip is $4.50, how much is the bill?

53. A student earns 85% of the points on a test. If the student earns 34 points, how many points is the test worth?

54. A book agent charges 15% to represent an author. If the agent earns $4.5 thousand from a book deal, how much does the author earn?

55. In 2014, there were 358 thousand students who earned bachelor's degrees in business, which was 19.2% of the students who earned bachelor's degrees in any field in that year (Source: *U.S. Center for Education Statistics*). How many students earned bachelor's degrees in 2014?

56. In 2015, the revenue from streaming music was $2.9 billion, which was 43.3% of the revenue from all digital music in that year (Source: *International Federation of the Phonographic Industry*). What was the revenue from all digital music in 2015?

57. During the period 1995–2015, the worldwide damage from geophysical events such as earthquakes was $763 billion, which was 29% of the total cost from worldwide natural disasters (Source: *United Nations Office for Disaster Risk Reduction*). What was the total cost from worldwide natural disasters during the period 1995–2015?

58. In 2015, there were 221 thousand new cases of lung and bronchus cancer, which was 13.3% of the new cases of all types of cancer in that year (Source: *American Cancer Society*). How many new cases of cancer were there in 2015?

For Exercises 59–62, use the graph of $y = -\dfrac{1}{2}x + 1$, shown in Fig. 12, to solve the given equation.

59. $-\dfrac{1}{2}x + 1 = 3$　　　　**60.** $-\dfrac{1}{2}x + 1 = 2$

61. $-\dfrac{1}{2}x + 1 = -1$　　　**62.** $-\dfrac{1}{2}x + 1 = -2$

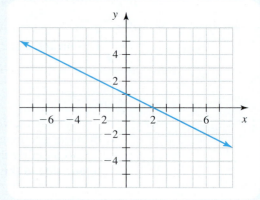

Figure 12 Exercises 59–62

Use "intersect" on a graphing calculator to solve the equation.

63. $x + 2 = 7$　　　　**64.** $x - 3 = 4$

65. $2x - 3 = 5$　　　**66.** $-3x + 5 = -7$

67. $-4(x - 1) = -8$　　**68.** $2(x + 5) = 6$

69. $\dfrac{2}{3}t - \dfrac{3}{2} = -\dfrac{7}{2}$　　**70.** $\dfrac{5}{2}w - \dfrac{5}{3} = \dfrac{10}{3}$

For Exercises 71–74, solve the given equation by referring to the solutions of $y = 5x - 3$ shown in Table 3.

71. $5x - 3 = 12$　　　**72.** $5x - 3 = 7$

73. $5x - 3 = -13$　　　**74.** $5x - 3 = -8$

Table 3 Exercises 71–74

x	y
-3	-18
-2	-13
-1	-8
0	-3
1	2
2	7
3	12

*Solve. [**Hint:** Combine like terms on the left side.]*

75. $2x + 5x = 14$　　　**76.** $3x + x = 20$

77. $4x - 5x = -2$　　　**78.** $2x - 8x = -4$

Concepts

79. A student tries to solve the equation $2x = 10$:

$$2x = 10$$
$$2x - 2 = 10 - 2$$
$$x = 8$$

Describe any errors. Then solve the equation correctly.

80. A student tries to solve the equation $x + 6 = 9$:

$$x + 6 = 9$$
$$x + 6 - 6 = 9$$
$$x + 0 = 9$$
$$x = 9$$

Describe any errors. Then solve the equation correctly.

81. A student solves the equation $4x = 12$:

$$4x = 12$$
$$\frac{4x}{4} = \frac{12}{4}$$
$$x = 3$$

Show that 3 satisfies each of the three equations.

82. A student solves the equation $x - 4 = 7$:

$$x - 4 = 7$$
$$x - 4 + 4 = 7 + 4$$
$$x + 0 = 11$$
$$x = 11$$

Show that 11 satisfies each of the four equations.

83. A student tries to solve the equation $x + 2 = 7$:

$$x + 2 = 7$$
$$x + 2 - 7 = 7 - 7$$
$$x - 5 = 0$$
$$x - 5 + 5 = 0 + 5$$
$$x = 5$$

Is the work correct? If yes, show a way to solve the equation in fewer steps. If no, describe the student's error(s) and then solve the equation correctly.

84. A student tries to solve the equation $\frac{3}{7}x = 2$:

$$\frac{3}{7}x = 2$$
$$7 \cdot \frac{3}{7}x = 7 \cdot 2$$
$$\frac{7}{1} \cdot \frac{3}{7}x = 14$$
$$3x = 14$$
$$\frac{3x}{3} = \frac{14}{3}$$
$$x = \frac{14}{3}$$

Is the work correct? If yes, show a way to solve the equation in fewer steps. If no, describe the student's error(s) and then solve the equation correctly.

85. Give an example of an equation for which it is helpful to add 7 to both sides. Then solve the equation.

86. Give an example of an equation for which it is helpful to subtract 7 from both sides. Then solve the equation.

87. Give an example of an equation for which it is helpful to multiply both sides by 7. Then solve the equation.

88. Give an example of an equation for which it is helpful to divide both sides by 7. Then solve the equation.

89. Are the equations $\frac{x}{3} = 2$ and $x - 1 = 4$ equivalent?

90. Are the equations $2x = 10$ and $x + 2 = 7$ equivalent?

91. Give three examples of equations that have 4 as their solution.

92. Give three examples of equations that have -2 as their solution.

93. Give an example of three equations that are equivalent. Explain.

94. Give an example of two equations that are not equivalent. Explain.

95. Are the equations $\frac{x - 5}{2} = \frac{3}{x + 1}$ and $\frac{x - 5}{2} + 6 = \frac{3}{x + 1} + 6$ equivalent? Explain.

96. Are the equations $x(x + 1) = 6$ and $2x(x + 1) = 12$ equivalent? Explain.

97. **a.** Solve $x + 2 = 7$.
   b. Solve $x + 5 = 9$.
   c. Solve $x + b = k$, where b and k are constants. [**Hint:** Refer to parts (a) and (b). The solution will be in terms of b and k.]

98. **a.** Solve $2x = 7$.
   b. Solve $5x = 9$.
   c. Solve $mx = p$, where m and p are constants and m is nonzero. [**Hint:** Refer to parts (a) and (b). The solution will be in terms of m and p.]

99. Describe in your own words what the addition property of equality means.

100. Describe in your own words what the multiplication property of equality means.

101. Explain why it is not necessary to state a subtraction property of equality: If A and B are expressions and c is a number, then the equations $A = B$ and $A - c = B - c$ are equivalent.

102. Explain why it is not necessary to state a division property of equality: If A and B are expressions and c is a nonzero number, then the equations $A = B$ and $\frac{A}{c} = \frac{B}{c}$ are equivalent.

4.4 Solving More Linear Equations in One Variable

Objectives

» Solve a linear equation in one variable.

» Make estimates and predictions by solving equations.

» Translate English sentences into and from mathematical equations.

» Use a graph or tables to solve a linear equation in one variable.

» Describe the meaning of *conditional equation*, *inconsistent equation*, and *identity* and how to solve these types of equations.

» Locate errors in solving linear equations.

In this section, we will solve more complicated linear equations in one variable than those in Section 4.3. We will solve such equations to help us make predictions about authentic situations.

Solving Linear Equations in One Variable

In Section 4.3, we used either the addition property of equality or the multiplication property of equality to solve a linear equation. In this section, we will use a combination of these properties to solve linear equations.

Example 1   Using the Addition and Multiplication Properties of Equality

Solve $2x - 3 = 5$.

Solution

We begin by adding 3 to both sides to get $2x$ alone on the left side:

$$
\begin{aligned}
2x - 3 &= 5 && \textit{Original equation} \\
2x - 3 + 3 &= 5 + 3 && \textit{Add 3 to both sides.} \\
2x &= 8 && \textit{Combine like terms.} \\
\frac{2x}{2} &= \frac{8}{2} && \textit{Divide both sides by 2.} \\
x &= 4 && \textit{Simplify.}
\end{aligned}
$$

Next, we check that 4 satisfies the equation $2x - 3 = 5$;

$$
\begin{aligned}
2x - 3 &= 5 && \textit{Original equation} \\
2(4) - 3 &\stackrel{?}{=} 5 && \textit{Substitute 4 for x.} \\
8 - 3 &\stackrel{?}{=} 5 && \textit{Multiply.} \\
5 &\stackrel{?}{=} 5 && \textit{Subtract.} \\
&\quad \text{true}
\end{aligned}
$$

So, the solution is 4.

▶

It is wise to check that a result (or the results) of solving an equation does indeed satisfy the equation.

In Example 1, we used the addition property of equality so that a variable term was alone on one side of the equation and a constant term was on the other side. Then we used the multiplication property of equality to get the variable alone on one side of the equation.

We will do the same thing in Example 2, but first we will simplify one side of the equation.

Example 2   Combining Like Terms to Help Solve an Equation

Solve $4x - 7x + 2 = 17$.

Solution

First, we combine like terms on the left side:

$$
\begin{aligned}
4x - 7x + 2 &= 17 && \textit{Original equation} \\
-3x + 2 &= 17 && \textit{Combine like terms.}
\end{aligned}
$$

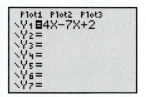

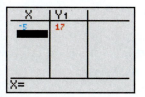

Figure 13 Verify the work

$$-3x + 2 - 2 = 17 - 2 \qquad \textit{Subtract 2 from both sides.}$$
$$-3x = 15 \qquad \textit{Combine like terms.}$$
$$\frac{-3x}{-3} = \frac{15}{-3} \qquad \textit{Divide both sides by } -3.$$
$$x = -5 \qquad \textit{Simplify.}$$

The solution is -5. We use a graphing calculator table to check that if -5 is substituted for x in the expression $4x - 7x + 2$, the result is 17 (see Fig. 13).

▶

In Example 3, we will solve an equation that contains both variable terms and constant terms on each side of the equation. To solve such an equation, we first use the addition property of equality to write the variable terms on one side of the equation and constant terms on the other side.

Example 3   Solving an Equation with Variable Terms on Both Sides

Solve $-x + 7 = 2x - 2$.

Solution

First, we subtract $2x$ from both sides to get all the variable terms on the left side:*

$$-x + 7 = 2x - 2 \qquad \textit{Original equation}$$
$$-x + 7 - 2x = 2x - 2 - 2x \qquad \textit{Subtract 2x from both sides.}$$
$$-3x + 7 = -2 \qquad \textit{Combine like terms.}$$
$$-3x + 7 - 7 = -2 - 7 \qquad \textit{Subtract 7 from both sides.}$$
$$-3x = -9 \qquad \textit{Combine like terms.}$$
$$\frac{-3x}{-3} = \frac{-9}{-3} \qquad \textit{Divide both sides by } -3.$$
$$x = 3 \qquad \textit{Simplify.}$$

The solution is 3. We use a graphing calculator table to check that if 3 is substituted for x in the expressions $-x + 7$ and $2x - 2$, the two results are equal (see Fig. 14).

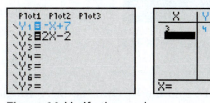

Figure 14 Verify the work

▶

When possible, we simplify each side of an equation before using properties such as the addition property of equality or the multiplication property of equality.

---

*The addition property of equality implies that we can add a number to both sides of an equation without changing the equation's solution set. We can also add (or subtract) a variable term of the form mx, where m is a constant, to both sides of an equation without changing the equation's solution set.

Example 4   Using the Distributive Law to Help Solve an Equation

Solve $-2(a - 3) + 4 = 4(a - 2)$.

Solution

First, we use the distributive law on each side:

$$
\begin{array}{ll}
-2(a - 3) + 4 = 4(a - 2) & \textit{Original equation} \\
-2a + 6 + 4 = 4a - 8 & \textit{Distributive law} \\
-2a + 10 = 4a - 8 & \textit{Combine like terms.} \\
-2a + 10 - 4a = 4a - 8 - 4a & \textit{Subtract 4a from both sides.} \\
-6a + 10 = -8 & \textit{Combine like terms.} \\
-6a + 10 - 10 = -8 - 10 & \textit{Subtract 10 from both sides.} \\
-6a = -18 & \textit{Combine like terms.} \\
\dfrac{-6a}{-6} = \dfrac{-18}{-6} & \textit{Divide both sides by } -6. \\
a = 3 & \textit{Simplify; divide.}
\end{array}
$$

Next, we check that 3 satisfies $-2(a - 3) + 4 = 4(a - 2)$:

$$
\begin{array}{ll}
-2(a - 3) + 4 = 4(a - 2) & \textit{Original equation} \\
-2(3 - 3) + 4 \overset{?}{=} 4(3 - 2) & \textit{Substitute 3 for a.} \\
-2(0) + 4 \overset{?}{=} 4(1) & \textit{Simplify.} \\
4 \overset{?}{=} 4 & \textit{Simplify.} \\
\text{true}
\end{array}
$$

So, 3 is the solution.

▶

A key step in solving an equation that contains fractions is to multiply both sides of the equation by the LCD so there are no fractions on either side of the equation. An equation that is "cleared of fractions" will be easier to solve than the original equation.

Example 5   Solving an Equation That Contains Fractions

Solve $\dfrac{2}{3}x + \dfrac{1}{6} = \dfrac{3}{4}$.

Solution

To find the LCD of the three fractions in the equation, we list the multiples of 3, the multiples of 6, and the multiples of 4:

$$
\begin{array}{ll}
\text{Multiples of 3:} & 3, 6, 9, 12, 15, 18, 21, \ldots \\
\text{Multiples of 6:} & 6, 12, 18, 24, 30, 36, 42, \ldots \\
\text{Multiples of 4:} & 4, 8, 12, 16, 20, 24, 28, \ldots
\end{array}
$$

The LCD is 12. Next, we multiply both sides of $\dfrac{2}{3}x + \dfrac{1}{6} = \dfrac{3}{4}$ by 12 to clear the equation of fractions:

$$
\dfrac{2}{3}x + \dfrac{1}{6} = \dfrac{3}{4} \qquad \textit{Original equation}
$$

$$
12 \cdot \left(\dfrac{2}{3}x + \dfrac{1}{6} \right) = 12 \cdot \dfrac{3}{4} \qquad \textit{Multiply both sides by the LCD, 12.}
$$

$$12 \cdot \frac{2}{3}x + 12 \cdot \frac{1}{6} = 12 \cdot \frac{3}{4} \qquad \textit{Distributive law}$$

$$8x + 2 = 9 \qquad \textit{Simplify.}$$

$$8x + 2 - 2 = 9 - 2 \qquad \textit{Subtract 2 from both sides.}$$

$$8x = 7 \qquad \textit{Combine like terms.}$$

$$\frac{8x}{8} = \frac{7}{8} \qquad \textit{Divide both sides by 8.}$$

$$x = \frac{7}{8} \qquad \textit{Simplify.}$$

The solution is $\frac{7}{8}$. We check our work by using ZDecimal to graph the equations $y = \frac{2}{3}x + \frac{1}{6}$ and $y = \frac{3}{4}$ in the same coordinate system and by using "intersect" to find the intersection point, $(0.875, 0.75)$. See Fig. 15. So, the solution of the original equation is $\frac{7}{8} = 0.875$, which checks.

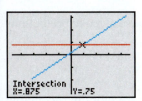

Figure 15 Verify the work

Example 6   Solving an Equation That Contains Fractions

Solve $\dfrac{3x - 1}{2} = \dfrac{4x + 2}{3}$.

Solution

We multiply both sides of $\dfrac{3x - 1}{2} = \dfrac{4x + 2}{3}$ by the LCD, 6, to clear the equation of fractions:

$$\frac{3x - 1}{2} = \frac{4x + 2}{3} \qquad \textit{Original equation}$$

$$6 \cdot \frac{3x - 1}{2} = 6 \cdot \frac{4x + 2}{3} \qquad \textit{Multiply both sides by the LCD, 6.}$$

$$3(3x - 1) = 2(4x + 2) \qquad \textit{Simplify.}$$

$$9x - 3 = 8x + 4 \qquad \textit{Distributive law}$$

$$9x - 3 - 8x = 8x + 4 - 8x \qquad \textit{Subtract 8x from both sides.}$$

$$x - 3 = 4 \qquad \textit{Combine like terms.}$$

$$x - 3 + 3 = 4 + 3 \qquad \textit{Add 3 to both sides.}$$

$$x = 7 \qquad \textit{Combine like terms.}$$

We can use "intersect" on a graphing calculator to check our work.

To solve some true-to-life problems, we will need to solve equations that contain decimal numbers. When solving such problems, we round results to two decimal places.

Example 7   Solving an Equation That Contains Decimal Numbers

Solve $2.71t = -3.4(5.9t - 4.8)$. Round any solutions to two decimal places.

Solution

$$2.71t = -3.4(5.9t - 4.8) \qquad \textit{Original equation}$$

$$2.71t = -20.06t + 16.32 \qquad \textit{Distributive law}$$

$$2.71t + 20.06t = -20.06t + 16.32 + 20.06t \quad \textcolor{blue}{\textit{Add 20.06t to both sides.}}$$

$$22.77t = 16.32 \qquad\qquad \textcolor{blue}{\textit{Combine like terms.}}$$

$$\frac{22.77t}{22.77} = \frac{16.32}{22.77} \qquad \textcolor{blue}{\textit{Divide both sides by 22.77.}}$$

$$t \approx 0.72 \qquad\qquad \textcolor{blue}{\textit{Round right-hand side to}}$$
$$\textcolor{blue}{\textit{two decimal places.}}$$

The solution is approximately 0.72. We can check that 0.72 approximately satisfies the original equation (try it).

Making Predictions by Solving Equations

Now that we know how to solve linear equations in one variable, we can make predictions about the explanatory variable of a linear model.

Example 8   Using a Model to Make Predictions

The percentage of Americans who believe gay/lesbian relations are morally acceptable was 58% in 2014 and has increased by about 1.5 percentage points per year since then (Source: *Gallup*). Let p be the percentage of Americans who believe gay/lesbian relations are morally acceptable at t years since 2014.

1. Find a model of the situation.
2. Predict what percentage of Americans will believe gay/lesbian relations are morally acceptable in 2021.
3. Predict when 66% of Americans will believe gay/lesbian relations are morally acceptable.

Solution

1. Because the percentage is increasing by about a *constant* 1.5 percentage points per year, the variables t and p are approximately linearly related. We want an equation of the form $p = mt + b$. Because the slope is 1.5 and the p-intercept is $(0, 58)$, a reasonable model is

$$p = 1.5t + 58$$

2. We substitute the input 7 for t in the equation $p = 1.5t + 58$ and solve for p:

$$p = 1.5(7) + 58 = 68.5$$

About 69% of Americans will believe gay/lesbian relations are morally acceptable in 2021, according to the model.

3. We substitute 66 for p in the equation $p = 1.5t + 58$ and solve for t:

$$66 = 1.5t + 58 \qquad\qquad \textcolor{blue}{\textit{Substitute 66 for p.}}$$
$$66 - 58 = 1.5t + 58 - 58 \qquad \textcolor{blue}{\textit{Subtract 58 from both sides.}}$$
$$8 = 1.5t \qquad\qquad \textcolor{blue}{\textit{Combine like terms.}}$$
$$\frac{8}{1.5} = \frac{1.5t}{1.5} \qquad\qquad \textcolor{blue}{\textit{Divide both sides by 1.5.}}$$
$$5.33 \approx t$$

In $2014 + 5.33 \approx 2019$, 66% of Americans will believe gay/lesbian relations are morally acceptable, according to the model. We verify our work in Problems 2 and 3 by using a graphing calculator table (see Fig. 16).

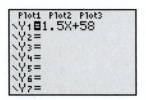

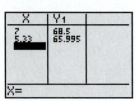

Figure 16 Verify the work

Example 9   Making a Prediction

The average weekly time Americans ages 18–24 years spend watching traditional television was 18.1 hours in 2015 and has decreased by about 2.1 hours per year since then (Source: *Nielsen*). Predict when the average weekly time will be 7.5 hours.

Solution

We let w be the average weekly time (in hours) Americans ages 18–24 years spend watching traditional television at t years since 2015. Because the average weekly viewing time decreased by about a *constant* 2.1 hours per year, the variables t and w are approximately linearly related. We want an equation of the form $w = mt + b$. Because the slope is -2.1 and the w-intercept is $(0, 18.1)$, a reasonable model is

$$w = -2.1t + 18.1$$

To find when the average weekly time will be 7.5 hours, we substitute 7.5 for w in the equation $w = -2.1t + 18.1$ and solve for t:

$$7.5 = -2.1t + 18.1 \qquad \textit{Substitute 7.5 for w.}$$
$$7.5 - 18.1 = -2.1t + 18.1 - 18.1 \qquad \textit{Subtract 18.1 from both sides.}$$
$$-10.6 = -2.1t \qquad \textit{Combine like terms.}$$
$$\frac{-10.6}{-2.1} = \frac{-2.1t}{-2.1} \qquad \textit{Divide both sides by -2.1.}$$
$$5.05 \approx t \qquad \textit{Simplify.}$$

In $2015 + 5.05 \approx 2020$, Americans ages 18–24 will spend 7.5 hours per week watching traditional television, on average, according to the model.

Example 10   Solving a Percentage Problem

In 2014, worldwide sales of digital cameras were 43.4 million cameras, down 31% from 2013 (Source: *Camera & Imaging Products Association*). What were the sales in 2013?

Solution

We let s be the worldwide sales (in millions of digital cameras) in 2013. Because the sales decreased by 31%, we write

Millions of cameras sold in 2013		Decrease in sales (in millions of cameras)		Millions of cameras sold in 2014
s	$-$	$0.31s$	$=$	43.4

Now we solve the equation:

$$s - 0.31s = 43.4$$
$$0.69s = 43.4 \qquad \textit{Combine like terms.}$$
$$\frac{0.69s}{0.69} = \frac{43.4}{0.69} \qquad \textit{Divide both sides by 0.69.}$$
$$s \approx 62.90$$

Worldwide sales in 2013 were about 62.90 million digital cameras.

Translating Sentences into and from Equations

Let x represent a number. Here are some sentences that have the same meaning as $x = 4$:

- The number is 4.
- The number is equal to 4.
- The number is the same as 4.

Example 11   Translating an English Sentence into an Equation

5, minus 2 times the sum of a number and 3 is 11. What is the number?

Solution

Let x be the number. Next, we translate the given information into an equation:

$$\underbrace{5\ -}_{\text{5, minus}}\ \ \underbrace{2\ \cdot}_{\text{2 times}}\ \ \underbrace{(x\ +\ 3)}_{\substack{\text{the sum of a}\\\text{number and 3}}}\ \ \underbrace{=\ 11}_{\text{is 11.}}$$

Then we solve the equation.

$$
\begin{aligned}
5 - 2(x + 3) &= 11 && \textit{Original equation}\\
5 - 2x - 6 &= 11 && \textit{Distributive law}\\
-2x - 1 &= 11 && \textit{Combine like terms.}\\
-2x - 1 + 1 &= 11 + 1 && \textit{Add 1 to both sides.}\\
-2x &= 12 && \textit{Combine like terms.}\\
\frac{-2x}{-2} &= \frac{12}{-2} && \textit{Divide both sides by } -2.\\
x &= -6 && \textit{Simplify.}
\end{aligned}
$$

So, the number is -6.

▶

In Example 11, we translated an English sentence into a mathematical equation, solved the equation, and then translated the result into an English sentence.

Example 12   Translating an Equation into an English Sentence

Let x be a number. Translate the equation $7x - 1 = 4x + 5$ into an English sentence.

Solution

$$\underbrace{7x}_{\substack{\text{7 times}\\\text{the number,}}}\qquad \underbrace{-\ 1}_{\text{minus 1,}}\qquad \underbrace{=}_{\substack{\text{is equal}\\\text{to}}}\qquad \underbrace{4x}_{\substack{\text{4 times}\\\text{the number,}}}\qquad \underbrace{+\ 5}_{\text{plus 5.}}$$

One of many correct translations is "7 times the number, minus 1 is equal to 4 times the number, plus 5." The solution of the equation is 2 (try it).

▶

Using Graphing to Solve an Equation in One Variable

In Section 4.3, we used graphing to solve some linear equations in one variable. We can use graphing to solve more complicated linear equations in one variable.

Example 13 Solving an Equation in One Variable by Graphing

The graphs of $y = -\frac{1}{2}x - 1$ and $y = -\frac{5}{4}x + 2$ are shown in Fig. 17. Use these graphs to solve the equation $-\frac{1}{2}x - 1 = -\frac{5}{4}x + 2$.

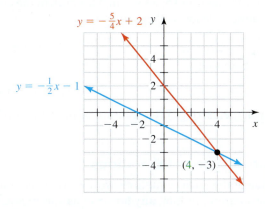

Figure 17 The intersection point is $(4, -3)$

Solution

The two lines intersect only at $(4, -3)$, whose x-coordinate is 4. So, 4 is the solution of the equation $-\frac{1}{2}x - 1 = -\frac{5}{4}x + 2$.

▶

Example 14 Solving an Equation in One Variable by Graphing

Use "intersect" on a graphing calculator to solve the equation $-\frac{3}{5}x + \frac{5}{2} = \frac{2}{3}x + \frac{16}{3}$, with the solution rounded to the second decimal place.

Solution

We use a graphing calculator to graph the equations $y = -\frac{3}{5}x + \frac{5}{2}$ and $y = \frac{2}{3}x + \frac{16}{3}$ on the same coordinate system and then use "intersect" to find the approximate intersection point, $(-2.24, 3.84)$. See Fig. 18. See Appendix A.17 for graphing calculator instructions.

The approximate intersection point $(-2.24, 3.84)$ has x-coordinate -2.24. So, the approximate solution of the original equation is -2.24.

▶

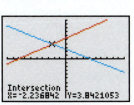

Figure 18 Using "intersect" to solve an equation in one variable

Using Tables to Solve an Equation in One Variable

Recall from Section 4.3 that we can also solve equations by using tables.

Example 15 Solving an Equation in One Variable by Using Tables

The solutions of $y = 2x - 5$ and $y = -4x + 13$ are shown in Tables 4 and 5, respectively. Use the tables to solve the equation $2x - 5 = -4x + 13$.

Table 4 Solutions of $y = 2x - 5$		Table 5 Solutions of $y = -4x + 13$	
x	y	x	y
0	−5	0	13
1	−3	1	9
2	−1	2	5
3	1	3	1
4	3	4	−3

Solution

From Tables 4 and 5, we see that for both of the equations $y = 2x - 5$ and $y = -4x + 13$, the input 3 leads to the output 1:

$$2(3) - 5 = 1 \quad \text{and} \quad -4(3) + 13 = 1$$

It follows that

$$2(3) - 5 = -4(3) + 13$$

which means that 3 is a solution of the equation $2x - 5 = -4x + 13$.

▶

Three Types of Equations

Each of the linear equations we solved in Section 4.3 and in this section so far has exactly one solution. In fact, **any linear equation in one variable has exactly one solution**. (You will show this in the second Exploration.)

A linear equation in one variable is an example of a conditional equation. A **conditional equation** is sometimes true and sometimes false, depending on which values are substituted for the variable(s). There are two other types of equations: inconsistent equations and identities.

If an equation does not have any solutions, we call the equation **inconsistent** and say that the solution set is the **empty set**. For example, the equation $x = x + 1$ is inconsistent because no number is 1 more than itself. Here we subtract x from both sides of the equation:

$$x = x + 1 \qquad \textit{Original equation}$$
$$x - x = x + 1 - x \quad \textit{Subtract x from both sides.}$$
$$0 = 1 \qquad \textit{Combine like terms.}$$
$$\text{false}$$

When we apply the usual steps to solve an inconsistent equation, the result is always a false statement.

An equation that is true for all permissible values of the variable(s) it contains is called an **identity**. For example, $x + 5 = x + 2 + 3$ is an identity because it is true for all real numbers. Here we add the numbers 2 and 3 and then subtract x from both sides:

$$x + 5 = x + 2 + 3 \quad \textit{Original equation}$$
$$x + 5 = x + 5 \qquad \textit{Add.}$$
$$x + 5 - x = x + 5 - x \quad \textit{Subtract x from both sides.}$$
$$5 = 5 \qquad \textit{Combine like terms.}$$
$$\text{true}$$

When we apply the usual steps to solve an identity, the result is always a true statement of the form $a = a$, where a is a constant.

Example 16   Solving Three Types of Equations

Solve the equation. State whether the equation is a conditional equation, an inconsistent equation, or an identity.

1. $2x + 1 = 9$ **2.** $3x + 2 = 3x + 7$ **3.** $2(x + 4) + 1 = 2x + 9$

Solution

1.
$$2x + 1 = 9 \qquad \textit{Original equation}$$
$$2x + 1 - 1 = 9 - 1 \qquad \textit{Subtract 1 from both sides.}$$
$$2x = 8 \qquad \textit{Combine like terms.}$$
$$\frac{2x}{2} = \frac{8}{2} \qquad \textit{Divide both sides by 2.}$$
$$x = 4 \qquad \textit{Simplify.}$$

The number 4 is the only solution, so the equation is conditional.

2.
$$3x + 2 = 3x + 7 \qquad \textit{Original equation}$$
$$3x + 2 - 3x = 3x + 7 - 3x \qquad \textit{Subtract 3x from both sides.}$$
$$2 = 7 \qquad \textit{Combine like terms.}$$
$$\text{false}$$

Because $2 = 7$ is a false statement, we conclude that the original equation is inconsistent and its solution set is the empty set.

3.
$$2(x + 4) + 1 = 2x + 9 \qquad \textit{Original equation}$$
$$2x + 8 + 1 = 2x + 9 \qquad \textit{Distributive law}$$
$$2x + 9 = 2x + 9 \qquad \textit{Combine like terms.}$$
$$2x + 9 - 2x = 2x + 9 - 2x \qquad \textit{Subtract 2x from both sides.}$$
$$9 = 9 \qquad \textit{Combine like terms.}$$
$$\text{true}$$

Because $9 = 9$ is a true statement (of the form $a = a$), we conclude that the original equation is an identity and its solution set is the set of all real numbers.

Locating Errors in Solving Linear Equations in One Variable

In Example 17, we will check whether a result satisfies equations. Doing this will help us locate an error.

Example 17 Locating an Error

Consider the following incorrect work in an attempt to solve the equation $2x + 1 = 5$:

$$2x + 1 = 5$$
$$2x + 1 - 1 = 5 - 1$$
$$2x = 4$$
$$\frac{2x}{2} = 4 \qquad \textit{Incorrect}$$
$$x = 4$$

1. Show that the work is incorrect by substituting the result 4 for x in the original equation.
2. Pinpoint the error by substituting 4 for x in all five equations.

Solution

1. We substitute 4 for x in the equation $2x + 1 = 5$:

$$2x + 1 = 5 \qquad \textit{Original equation}$$
$$2(4) + 1 \stackrel{?}{=} 5 \qquad \textit{Substitute 4 for x.}$$
$$9 \stackrel{?}{=} 5 \qquad \textit{Simplify.}$$
$$\text{false}$$

Because 4 does not satisfy the original equation $2x + 1 = 5$, we know that an error was made in solving that equation.

2. We substitute 4 for x in all five equations and note which pair of consecutive equations gives a false statement followed by a true statement.

Equation	Does 4 satisfy the equation?	
$2x + 1 = 5$	$2(4) + 1 \overset{?}{=} 5$	*false*
$2x + 1 - 1 = 5 - 1$	$2(4) + 1 - 1 \overset{?}{=} 5 - 1$	*false*
$2x = 4$	$2(4) \overset{?}{=} 4$	*false*
$\dfrac{2x}{2} = 4$	$\dfrac{2(4)}{2} \overset{?}{=} 4$	*true*
$x = 4$	$(4) \overset{?}{=} 4$	*true*

Because substituting 4 for x in $2x = 4$ gives a false statement, but substituting 4 for x in the next equation, $\dfrac{2x}{2} = 4$, gives a true statement, we know an error was made when only the left side of the equation $2x = 4$ was divided by 2:

$$2x = 4 \quad \textit{Third equation of the work}$$

$$\frac{2x}{2} = 4 \quad \textit{Incorrect}$$

Instead, *both* sides of the equation $2x = 4$ should be divided by 2:

$$2x = 4 \quad \textit{Third equation of the work}$$

$$\frac{2x}{2} = \frac{4}{2} \quad \textit{Divide both sides by 2.}$$

$$x = 2 \quad \textit{Simplify.}$$

So, the solution is 2, not 4. We can check that 2 satisfies the original equation $2x + 1 = 5$ (try it).

▶

Locating an Error in Solving a Linear Equation

If the result of an attempt to solve a linear equation in one variable does not satisfy the equation, an error was made. To pinpoint the error, substitute the result into all the equations in the work and find which pair of consecutive equations gives a false statement followed by a true statement.

◆ Group Exploration

Section Opener: Solving linear equations

1. Solve $x + 3 = 11$.

2. Solve $2x = 8$.

3. Solve $2x + 3 = 11$. [**Hint:** First, subtract 3 from both sides.]

4. Solve $6x - 4x + 3 = 11$. [**Hint:** First, combine like terms on the left side of the equation.]

5. Solve the equation $2x + 4 + 3x = 14$.

Group Exploration

Any linear equation in one variable has exactly one solution.

1. Solve $7x + 5 = 0$.

2. Solve $4x + 3 = 0$.

3. Solve $5x + 2 = 0$.

4. Solve $mx + b = 0$, where m and b are constants and $m \neq 0$. [**Hint:** Perform steps similar to those you followed for Problems 1–3.]

5. Describe a linear equation in one variable.

6. Does using the addition property of equality or the multiplication property of equality change an equation's solution set? Explain.

7. Explain why your responses to Problems 4–6 show that any linear equation in one variable has exactly one solution.

8. Use the result you found in Problem 4 to solve $8x + 3 = 0$ in one step.

Tips for Success   Choose a Good Time and Place to Study

To improve your effectiveness at studying, consider taking stock of when and where you are best able to study. Tracy, a student who lives in a sometimes distracting household, completes her assignments at the campus library just after she attends her classes. Gerome, a morning person, gets up early so that he can study before classes. Being consistent in the time and location for studying can help, too.

Homework 4.4

For extra help ▶   **MyLab Math**   ▦ Watch the videos in MyLab Math

Solve. Use a graphing calculator to verify your result.

1. $3x - 2 = 13$ 2. $5x - 1 = 9$ 3. $-4x + 6 = 26$

4. $-2x + 7 = 23$ 5. $-5 = 6x + 3$ 6. $-7 = 4x - 1$

7. $8 - x = -4$ 8. $2 - x = -9$ 9. $2x + 6 - 7x = -4$

10. $4x + 3 - 9x = -22$ 11. $5x + 4 = 3x + 16$

12. $7x + 5 = 4x + 17$ 13. $-3r - 1 = 2r + 24$

14. $-6w - 3 = 4w + 17$ 15. $9 - x - 5 = 2x - x$

16. $8 - 2x - 2 = 3x + x$ 17. $2(x + 3) = 5x - 3$

18. $-3(x - 4) = 2x + 2$

19. $1 - 3(5b - 2) = 4 - (7b + 3)$

20. $3 - 4(3p + 2) = 7 - (9p - 1)$

21. $4x = 3(2x - 1) + 5$

22. $2x = 5(2x + 9) + 3$

23. $3(4x - 5) - (2x + 3) = 2(x - 4)$

24. $-2(5x + 3) - (4x - 1) = 5(x + 2)$

25. $\dfrac{x}{2} - \dfrac{3}{4} = \dfrac{1}{2}$ 26. $\dfrac{x}{9} - \dfrac{1}{3} = \dfrac{2}{9}$

27. $\dfrac{5x}{6} + \dfrac{2}{3} = 2$ 28. $\dfrac{3x}{8} - \dfrac{1}{2} = 1$

29. $\dfrac{5}{6}k = \dfrac{3}{4}k + \dfrac{1}{2}$ 30. $\dfrac{3}{8}t = \dfrac{5}{6}t - \dfrac{1}{4}$

31. $\dfrac{7}{12}x - \dfrac{5}{3} = \dfrac{7}{4} + \dfrac{5}{6}x$ 32. $\dfrac{5}{4} + \dfrac{9}{2}x = \dfrac{3}{8}x - \dfrac{1}{4}$

33. $\dfrac{4}{3}x - 2 = 3x + \dfrac{5}{2}$ 34. $\dfrac{2}{5}x - 4 = 2x - \dfrac{3}{4}$

35. $\dfrac{3(x - 4)}{5} = -2x$ 36. $\dfrac{5(x - 2)}{3} = -4x$

37. $\dfrac{4x + 3}{5} = \dfrac{2x - 1}{3}$ 38. $\dfrac{3x + 2}{2} = \dfrac{6x - 3}{5}$

39. $\dfrac{4m - 5}{2} - \dfrac{3m + 1}{3} = \dfrac{5}{6}$

40. $\dfrac{2p + 4}{3} - \dfrac{5p - 7}{6} = \dfrac{11}{12}$

For Exercises 41–46, solve. Round the result to the second decimal place. Use a graphing calculator to verify your work.

41. $0.3x + 0.2 = 0.7$

42. $0.6x - 0.1 = 0.4$

43. $5.27x - 6.35 = 2.71x + 9.89$

44. $8.25x - 17.56 + 4.38x = 25.86$

45. $0.4x - 1.6(2.5 - x) = 3.1(x - 5.4) - 11.3$

46. $3.2x + 0.5(7.3 - x) = 4.7 - 6.4(x - 2.1)$

47. The average weight of a turkey was 30.4 pounds in 2014 and has increased by about 0.4 pound per year since then (Source: *U.S. Department of Agriculture*). Let *w* be the average weight (in pounds) of a turkey at *t* years since 2014.
  a. Find an equation of a model to describe the situation.
  b. Predict the average weight of a turkey in 2021.
  c. Predict when the average weight of a turkey will be 32.8 pounds.

48. The average salary for a hockey player in the National Hockey League (NHL) was $2.58 million in 2014 and has increased by about $0.23 million per year since then (Source: *National Hockey League Players Association*). Let *s* be the average salary (in millions of dollars) at *t* years since 2014.
  a. Find an equation of a linear model to describe the situation.
  b. Predict the average salary in 2021.
  c. The average salary of Major League Baseball players was $4.4 million in 2016. Predict when the average salary of NHL hockey players will reach that level.

49. Martha Stewart went to prison for selling nearly 4000 shares of ImClone Systems stock because she had inside information that the company was about to announce bad news that would result in the stock's value plummeting. Surprisingly, the value of a share of Martha Stewart Living Omnimedia stock increased by about $3.36 per month after she was sentenced. The value of a share of her stock was $8.16 just before she was sentenced (Source: *Bloomberg Financial Markets*). Let *v* be the value (in dollars) of a share of the stock at *t* months since Stewart was sentenced.
  a. Find an equation of a model to describe the situation.
  b. Estimate the value of a share of the stock 3 months after the sentencing, which is about when Stewart reported to prison.
  c. Estimate when the value of a share of the stock reached $28.32.

50. The number of fires was 1.30 million fires in 2014 and has decreased by about 0.04 million fires per year since then (Source: *National Fire Protection Association*). Let *n* be the number (in millions) of fires in the year that is *t* years since 2014.
  a. Find an equation of a model to describe the situation.
  b. Predict the number of fires in 2020.
  c. Predict in which year there will be 1 million fires.

51. The percentages of freshmen at Boise State University whose GPA in high school was at least 3.5 are shown in Table 6 for various years.

Table 6 Freshmen at Boise State University Whose GPA in High School Was at Least 3.5

Year	Percent
2009	37.8
2010	38.2
2011	39.9
2012	44.6
2013	43.8
2014	45.4
2015	46.5

Source: *Boise State University*

Let *p* be the percentage of freshmen at Boise State University at *t* years since 2000 who earned at least a 3.5 GPA in high school. A model of the situation is $p = 1.59t + 23.29$.

  a. Use a graphing calculator to draw a scatterplot and the model in the same viewing window. Does the line come close to the points?
  b. What is the slope? What does it mean in this situation?
  c. Predict the percentage of freshmen at Boise State University in 2020 who will have earned at least a 3.5 GPA in high school.
  d. Estimate when half the freshmen at Boise State University earned at least a 3.5 GPA in high school.
  e. Give at least two explanations of why the percentage of freshmen at Boise State University who have earned at least a 3.5 GPA is increasing.

52. The average selling prices of a home sold in San Bruno, California, are shown in Table 7 for various square footages.

Table 7 Average Selling Prices of Homes

Number of Square Feet	Number Used to Represent Square Feet	Average Selling Price (thousands of dollars)
500–1000	750	632
1001–1500	1250	733
1501–2000	1750	814
2001–2500	2250	894
2501–3000	2750	1025

Source: *Green Banker*

Let *p* be the average selling price (in thousands of dollars) of a home measuring *s* square feet. A model of the situation is $p = 0.19s + 488.15$.
  a. Use a graphing calculator to draw a scatterplot and the model in the same viewing window. Does the line come close to the data points?
  b. What is the slope? What does it mean in this situation?
  c. Estimate the square footage for which the average selling price is $950 thousand.

53. For a single passenger, Chicago taxis charge $3.25 plus $2.25 for each mile traveled (Source: *City of Chicago*).
   a. Find an equation of a linear model to describe the situation. Explain what your variables represent.
   b. If a person paid $32.50 for a cab fare in Chicago, how far was the ride?

54. Seattle taxis charge $2.60 plus $2.70 for each mile traveled (Source: *City of Seattle*).
   a. Find an equation of a linear model to describe the situation. Explain what your variables represent.
   b. If a person paid $45.80 for a cab fare in Seattle, how far was the ride?

55. In 2016, a one-year-old Ford Explorer was worth about $31,681, with a depreciation of about $3281 per year (Source: *AutoTrader.com*).
   a. Find an equation of a linear model to describe the situation. Explain what your variables represent.
   b. Estimate when the car will be worth $12,000.

56. In 2016, a one-year-old Nissan Pathfinder was worth about $23,929, with a depreciation of about $1672 per year (Source: *AutoTrader.com*).
   a. Find an equation of a linear model to describe the situation. Explain what your variables represent.
   b. Estimate when the car will be worth $4000.

57. A student has $62 set aside for an Xbox One S console, which sells for $299 (Source: *Microsoft*). If the student saves $20 each week, how long will it take until he can afford the console?

58. An employee's starting salary is $29,200. If the employee receives a raise of $1700 per year, after how many years will the employee's salary be $43,000?

59. In 2015, the percentage of workers who were in a union was 11.1% and has since decreased by about 0.13 percentage point per year (Source: *Bureau of Labor Statistics*). Predict when the percentage will be 10.3%.

60. Album sales in the United States were 241.4 million albums in 2015 and have decreased by about 25.6 million albums per year (Source: *Nielsen Music*). Predict when the annual sales will be 100 million albums.

61. In 2015, total U.S. VISA debit card purchases were $1.37 trillion, up 7.8% from 2014 (Source: *The Nilson Report*). What were the total U.S. VISA debit card purchases in 2014?

62. Spending on prescription drugs in the United States was $425 billion in 2015, up 12.2% from 2014 (Source: *IMS Health*). What was the spending in 2014?

63. In 2015, the average per-person annual consumption of cheese was 35.3 pounds, up 7% from 2010 (Source: *USDA*). What was the average per-person annual consumption of cheese in 2010?

64. In 2015, the number of trips taken by Americans on public transportation was 10.6 billion trips, up 12.8% from 2010 (Source: *American Public Transportation Association*). What was the number of trips taken by Americans in 2010?

65. Viacom's revenue was $13.3 billion in 2015, down 11.3% from 2014 (Source: *Viacom*). What was the revenue in 2014?

66. In 2015, Bank of America revenue was $82.5 billion, down 3.1% from 2014 (Source: *Bank of America*). What was the revenue in 2014?

67. In 2017, the average office space per worker was 151 square feet, down 32.9% from 2010 (Source: *CoreNet Global*). What was the average office space per worker in 2010?

68. The average selling price of Samsung's smartphones was $343 in 2015, down 14.3% from 2011 (Source: *IDC*). What was the average selling price in 2011?

For Exercises 69–76, find the number.

69. 3 more than the product of 5 and a number is 18.

70. 4 more than the product of 3 and a number is 11.

71. 3 times the difference of a number and 2 is −18.

72. 6 times the sum of a number and 2 is −12.

73. 3 subtracted from a number is equal to twice the number plus 1.

74. 4, plus 7 times a number is equal to 5 times the number, minus 3.

75. 1, minus 4 times the sum of a number and 5 is 9.

76. 3, plus 5 times the difference of a number and 2 is 18.

Let x be a number. Translate the given equation into an English sentence. Then solve the equation.

77. $2x - 3 = 7$

78. $3x + 8 = 2$

79. $6x - 3 = 8x - 4$

80. $2x + 9 = 7x + 12$

81. $2(x - 4) = 10$

82. $3(x + 6) = 12$

83. $4 - 7(x + 1) = 2$

84. $9 + 5(x - 3) = 4$

For Exercises 85–88, solve the given equation by referring to the solutions of $y = -3x + 7$ and $y = 5x + 15$ shown in Tables 8 and 9, respectively.

85. $-3x + 7 = 5x + 15$

86. $-3x + 7 = 4$

87. $5x + 15 = 5$

88. $5x + 15 = 25$

Table 8 Solutions of $y = -3x + 7$		Table 9 Solutions of $y = 5x + 15$	
x	y	x	y
−2	13	−2	5
−1	10	−1	10
0	7	0	15
1	4	1	20
2	1	2	25

Use "intersect" on a graphing calculator to solve the equation. Round the solution to the second decimal place.

89. $-4x + 8 = 2x - 9$

90. $-2x - 7 = x + 1$

91. $2.5x - 6.4 = -1.7x + 8.1$

92. $-1.5x - 9.3 = 3.1x + 2.1$

93. $\dfrac{23}{75}x - \dfrac{99}{38} = -\dfrac{52}{89}x - \dfrac{67}{9}$

94. $-\dfrac{54}{35}x + \dfrac{26}{21} = \dfrac{67}{95}x - \dfrac{72}{31}$

For Exercises 95–100, solve the given equation by referring to the graphs of $y = -\dfrac{3}{2}x + 2$ and $y = \dfrac{1}{2}x - 2$ shown in Fig. 19.

95. $-\dfrac{3}{2}x + 2 = \dfrac{1}{2}x - 2$

96. $-\dfrac{3}{2}x + 2 = -4$

97. $-\frac{3}{2}x + 2 = 5$

98. $\frac{1}{2}x - 2 = -3$

99. $\frac{1}{2}x - 2 = 0$

100. $-\frac{3}{2}x + 2 = 2$

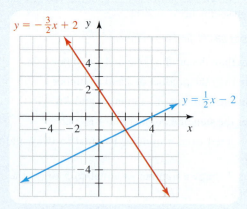

Figure 19 Exercises 95–100

For Exercises 101–106, solve the given equation by referring to the graphs of $y = -\frac{1}{3}x + \frac{5}{3}$ and $y = \frac{3}{2}x + \frac{7}{2}$ shown in Fig. 20.

101. $-\frac{1}{3}x + \frac{5}{3} = \frac{3}{2}x + \frac{7}{2}$

102. $-\frac{1}{3}x + \frac{5}{3} = 3$

103. $-\frac{1}{3}x + \frac{5}{3} = 1$

104. $\frac{3}{2}x + \frac{7}{2} = -4$

105. $\frac{3}{2}x + \frac{7}{2} = -1$

106. $-\frac{1}{3}x + \frac{5}{3} = 0$

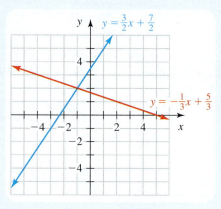

Figure 20 Exercises 101–106

Solve the equation. State whether the equation is a conditional equation, an inconsistent equation, or an identity.

107. $3x + 4x = 7x$

108. $9x - 3x = 6x$

109. $4x - 5 - 2x = 2x - 1$

110. $3x + 8x - 7 = 15x - 3 - 4x$

111. $3k + 10 - 5k = 4k - 2$

112. $5 - r + 8r = 3r + 1$

113. $2(x + 3) - 2 = 2x + 4$

114. $5(x - 2) + 1 = 5x - 9$

115. $5(2x - 3) - 4x = 3(2x - 1) + 6$

116. $4(3x + 2) - 6x = 2(3x - 4) + 1$

Large Data Sets

117. **DATA** Access the data about NFL football player ages and levels of experience, which are available at MyLab Math and at the Pearson Downloadable Student Resources for Math & Stats website. Let E be the experience (in years) of a player whose age is a years.
  a. Construct a scatterplot that describes the relationship between a and E. Print the scatterplot.
  b. Graph the equation $E = 0.90a - 19.94$ on the scatterplot. Does the line fit the data well?
  c. Estimate the level of experience of a 30-year-old player.
  d. Estimate the age of a player who has 11 years experience.
  e. Estimate the slope of the model. What does it mean in this situation?
  f. An athlete must be out of high school for at least 3 years to be eligible to play in the NFL. Identify any data points that must have at least one coordinate that is incorrect.

118. **DATA** Access the data about adults who exercise and adults who are obese, which are available at MyLab Math and at the Pearson Downloadable Student Resources for Math & Stats website. Let y be the percentage of adults who are obese and x be the percentage of adults who exercise for each of District of Columbia, Guam, Puerto Rico, Virgin Islands, and the 50 states, except Alaska.
  a. Construct a scatterplot that describes the relationship between x and y. Print the scatterplot.
  b. Graph the equation $y = -0.52x + 40.38$ on the scatterplot. Does the line fit the data well?
  c. Estimate the percentage of adults of a state who are obese if 30% of the adults exercise.
  d. Estimate the percentage of adults of a state who exercise if 30% of the adults are obese.
  e. Estimate the slope of the model. What does it mean in this situation?
  f. Find the y-intercept. What does it mean in this situation?

Concepts

119. A student incorrectly solves $3(x - 3) = 12$:

$$3(x - 3) = 12$$
$$3x - 9 = 12$$
$$3x - 9 - 9 = 12 - 9$$
$$3x = 3$$
$$\frac{3x}{3} = \frac{3}{3}$$
$$x = 1$$

Substitute 1 for x in each of the six equations, and explain how the results help you pinpoint the error. Describe the error. Then solve the equation correctly.

120. A student incorrectly solves $-2x + 1 = 3$:

$$-2x + 1 = 3$$
$$-2x + 1 - 1 = 3 - 1$$
$$-2x = 2$$
$$\frac{-2x}{-2} = 2$$
$$x = 2$$

Substitute 2 for x in each of the five equations, and explain how the results help you pinpoint the error. Describe the error. Then solve the equation correctly.

121. A student tries to solve $2(x - 5) = x - 3$:

$$2(x - 5) = x - 3$$
$$2x - 10 = x - 3$$
$$2x - 10 + 10 = x - 3 + 10$$
$$2x = x + 7$$
$$\frac{2x}{2} = \frac{x + 7}{2}$$
$$x = \frac{x + 7}{2}$$

Describe any errors. Then solve the equation correctly.

122. A student tries to solve the equation $\frac{1}{2}x + 3 = \frac{5}{2}$:

$$\frac{1}{2}x + 3 = \frac{5}{2}$$
$$2 \cdot \frac{1}{2}x + 3 = 2 \cdot \frac{5}{2}$$
$$x + 3 = 5$$
$$x + 3 - 3 = 5 - 3$$
$$x = 2$$

Describe any errors. Then solve the equation correctly.

123. Two students try to solve the equation $5x + 3 = 3x + 7$:

Student A

$$5x + 3 = 3x + 7$$
$$5x + 3 - 3x = 3x + 7 - 3x$$
$$2x + 3 = 7$$
$$2x + 3 - 3 = 7 - 3$$
$$2x = 4$$
$$\frac{2x}{2} = \frac{4}{2}$$
$$x = 2$$

Student B

$$5x + 3 = 3x + 7$$
$$5x + 3 - 3 = 3x + 7 - 3$$
$$5x = 3x + 4$$
$$5x - 3x = 3x + 4 - 3x$$
$$2x = 4$$
$$\frac{2x}{2} = \frac{4}{2}$$
$$x = 2$$

Compare the methods of solving the equation. Is one method better than the other? Explain.

124. Two students try to solve the equation $2x + 4 = 0$:

Student 1

$$2x + 4 = 0$$
$$2x + 4 + 6 = 0 + 6$$
$$2x + 10 = 6$$
$$2x + 10 - 10 = 6 - 10$$
$$2x = -4$$
$$2x \cdot 3 = -4 \cdot 3$$
$$6x = -12$$
$$\frac{6x}{6} = \frac{-12}{6}$$
$$x = -2$$

Student 2

$$2x + 4 = 0$$
$$2x + 4 - 4 = 0 - 4$$
$$2x = -4$$
$$\frac{2x}{2} = \frac{-4}{2}$$
$$x = -2$$

Did either student, both students, or neither student solve the equation correctly? Explain.

125. Solve $5(x - 2) = 2x - 1$. Show that your result satisfies each of the equations in your work.

126. Solve $3(x + 1) + 2 = x + 7$. Show that your result satisfies each of the equations in your work.

127. Consider the following equations:

$$x = 3$$
$$x + 2 = 3 + 2$$
$$x + 2 = 5$$
$$3(x + 2) = 3 \cdot 5$$
$$3(x + 2) = 15$$
$$3(x + 2) + 1 = 15 + 1$$
$$3(x + 2) + 1 = 16$$

What is the solution of the equation $3(x + 2) + 1 = 16$? What is the solution of the equation $3(x + 2) = 15$? Explain how you can respond to these questions without doing any work besides the work already included in this problem.

128. If an equation contains fractions, why is it helpful to multiply both sides of the equation by the LCD?

129. Consider the equation $7x + 4 = mx + b$. For what values of m and b is the following true?
   a. There is exactly one solution.
   b. The solution set is the set of all real numbers.
   c. The solution set is the empty set.

4.5 Comparing Expressions and Equations

Objectives

» Compare the meanings of *simplifying expressions* and *solving equations*.

» Translate English phrases and sentences into mathematical expressions and equations.

In this section, we compare the meanings of mathematical expressions and equations.

Comparing Expressions and Equations

First, we define a *linear expression in one variable*.

Definition   Linear expression in one variable

A **linear expression in one variable** is an expression that can be put into the form

$$mx + b$$

where m and b are constants and $m \neq 0$.

Next, we recall the definition of a linear equation from Section 4.3: A *linear equation in one variable* is an equation that can be put into the form $mx + b = 0$, where m and b are constants and $m \neq 0$.

For example, $4x + 7$ is a linear expression in one variable and $4x + 7 = 0$ is a linear equation in one variable. Also, the expression $2(x - 5) + 3x$ is linear because it can be put into the form $mx + b$, where $m \neq 0$. (Try it.) The equation $2(x - 5) + 3x = 8x - 4$ is linear because it can be put into the form $mx + b = 0$, where $m \neq 0$. (Try it.)

Example 1   Simplifying an Expression and Solving an Equation

1. Simplify $5(x + 2) - 2x$.
2. Solve $5(x + 2) - 2x = 2x + 14$.
3. Use a graphing calculator table to check the work in Problem 1.
4. Use a graphing calculator table to check the work in Problem 2.

Solution

1. $5(x + 2) - 2x = 5x + 10 - 2x$ *Distributive law*
   $$= 3x + 10 \qquad \text{Combine like terms.}$$
   The simplified expression is $3x + 10$.

2. $5(x + 2) - 2x = 2x + 14$ *Original equation*
   $5x + 10 - 2x = 2x + 14$ *Distributive law*
   $3x + 10 = 2x + 14$ *Combine like terms.*
   $3x + 10 - 2x = 2x + 14 - 2x$ *Subtract 2x from both sides.*
   $x + 10 = 14$ *Combine like terms.*
   $x + 10 - 10 = 14 - 10$ *Subtract 10 from both sides.*
   $x = 4$ *Combine like terms.*

   The solution is 4.

3. We use a graphing calculator to find a table for $y = 5(x + 2) - 2x$ and $y = 3x + 10$, using the original expression and the simplified result, respectively (see Fig. 21). For each value of x in the table, the values of y for both expressions are equal. So, we are confident that the two expressions are equivalent.

Figure 21 Verify the work

4. We use a graphing calculator table to check that if 4 is substituted for x in the expressions $5(x + 2) - 2x$ and $2x + 14$ (both sides of the original equation), the two results are equal (see Fig. 22).

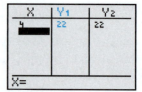

Figure 22 Verify the work

Note that if any other number besides 4 is substituted for x in the expressions $5(x + 2) - 2x$ and $2x + 14$, the two results will *not* be equal. That is because 4 is the *only* solution of the equation $5(x + 2) - 2x = 2x + 14$.

From our work in Example 1, we can make two important observations: First, the result of simplifying the *expression* $5(x + 2) - 2x$ is $3x + 10$, an *expression*; second, the result of solving the *equation* $5(x + 2) - 2x = 2x + 14$ is 4, a *number*.

Results of Simplifying an Expression and Solving an Equation

The result of simplifying an expression is an expression. The result of solving a linear equation in one variable is a number.

Example 2 Identifying an Error in Solving an Equation

A student tries to solve the equation $2(x - 3) = x$:

$$2(x - 3) = x$$
$$2x - 6 = x$$
$$2x - 6 + 6 = x + 6$$
$$2x = x + 6$$
$$x = \frac{x + 6}{2}$$

What would you tell the student?

Solution

The student found an equation that is equivalent to the original equation, but the student did not find the *solution* of the original equation. The solution of a linear equation is a number, not an expression that contains a variable. So, the solution cannot be $\frac{x + 6}{2}$.

Notice that x still appears on both sides of the equation. By getting x alone on one side of the equation, we can show that the solution is 6. (Try it.)

In Example 1, we simplified $5(x + 2) - 2x$ to $3x + 10$. So, if we use the same number to evaluate both expressions, the two results will be equal. This implies that *any* number satisfies the equation $5(x + 2) - 2x = 3x + 10$. For example, in Fig. 21 on page 226, we see that the numbers 0, 1, 2, 3, 4, 5, and 6 are solutions of this equation. Because any number satisfies the equation, the equation is an identity.

We also found that 4 is the solution of the linear equation $5(x + 2) - 2x = 2x + 14$. This means 4 is the *only* number that satisfies that equation. In general, a linear equation in one variable has exactly one solution.

Comparing Simplifying an Expression with Solving an Equation

- If an expression A in one variable is simplified to an expression B, then every real number (for which both expressions are defined) is a solution of the equation $A = B$. In other words, the equation $A = B$ is an identity.
- Exactly one number is a solution of a linear equation in one variable.

Example 3 Comparing Expressions and Equations

1. The simplified form of $4(2a - 5) - 2(3a + 4)$ is $2a - 28$. What does this mean?
2. The solution of $4(2a - 5) - 2(3a + 4) = 2$ is 15. What does this mean?

Solution

1. This means the linear expressions $4(2a - 5) - 2(3a + 4)$ and $2a - 28$ are equivalent. It also means every real number is a solution of the equation $4(2a - 5) - 2(3a + 4) = 2a - 28$. So, the equation is an identity.
2. This means 15 is the only number that satisfies the linear equation $4(2a - 5) - 2(3a + 4) = 2$. (Try it.)

Example 4   Comparing Expressions and Equations

1. Simplify $\frac{2}{5}x + \frac{1}{3}x$.

2. Solve $\frac{2}{5}x + \frac{1}{3}x = \frac{4}{5}$.

3. Compare the first steps of your work in Problems 1 and 2.

Solution

1. $\frac{2}{5}x + \frac{1}{3}x = \frac{3}{3} \cdot \frac{2}{5}x + \frac{5}{5} \cdot \frac{1}{3}x$ *The LCD of $\frac{2}{5}$ and $\frac{1}{3}$ is 15.*

$$= \frac{6}{15}x + \frac{5}{15}x$$ *Multiply numerators and multiply denominators:*
$$\frac{a}{b} \cdot \frac{c}{d} = \frac{ac}{bd}$$

$$= \frac{11}{15}x$$ *Combine like terms.*

2. $$\frac{2}{5}x + \frac{1}{3}x = \frac{4}{5}$$ *Original equation*

$$15 \cdot \left(\frac{2}{5}x + \frac{1}{3}x \right) = 15 \cdot \frac{4}{5}$$ *Multiply both sides by 15.*

$$15 \cdot \frac{2}{5}x + 15 \cdot \frac{1}{3}x = 15 \cdot \frac{4}{5}$$ *Distributive law*

$$6x + 5x = 12$$ *Simplify.*

$$11x = 12$$ *Combine like terms.*

$$\frac{11x}{11} = \frac{12}{11}$$ *Divide both sides by 11.*

$$x = \frac{12}{11}$$ *Simplify.*

3. To start simplifying the expression in Problem 1, we multiplied the term $\frac{2}{5}x$ by $\frac{3}{3} = 1$ and multiplied the term $\frac{1}{3}x$ by $\frac{5}{5} = 1$. To start solving the equation in Problem 2, we multiplied both sides of the equation by 15, which is *not* equal to 1.

Problem 3 in Example 4 suggests the following comparison of multiplying an expression by a number and multiplying both sides of an equation by a number:

Multiplying Expressions and Both Sides of Equations by Numbers

In simplifying an expression, the only number that we can multiply the expression or part of it by is 1. In solving an equation, we can multiply both sides of the equation by *any* number except 0.

By the property $a \cdot 1 = a$, we know multiplying an expression by 1 gives an equivalent expression. By the multiplication property of equality, we know multiplying both sides of an equation by *any* nonzero number gives an equivalent equation.

Translating from English into Mathematics

In Section 2.1, we discussed how to translate an English phrase or sentence into a mathematical expression, and in Section 4.4, we discussed how to translate an English sentence into an equation. We use an expression to represent a quantity, and we use an equation to state that one quantity is equal to another.

| Example 5 | Translating from English into Mathematics |

Let x be a number. Translate the following into an expression or an equation, as appropriate, and then simplify the expression or solve the equation:

1. 5, minus 3 times the sum of the number and 4
2. The product of 6 and the number is equal to the number subtracted from 2.

Solution

1. The phrase describes a single quantity. So, we translate the phrase into an expression:

$$
\underbrace{5\ -}_{\text{5, minus}}\quad \underbrace{3\ \cdot}_{\text{3 times}}\quad \underbrace{(x+4)}_{\substack{\text{the sum of the}\\\text{number and 4}}}
$$

Next, we simplify the expression:

$$
5 - 3\cdot(x+4) = 5 - 3x - 12 \qquad \textit{Distributive law}
$$
$$
= -3x - 7 \qquad \textit{Combine like terms.}
$$

2. The sentence states that one quantity is equal to another. So, we translate the sentence into an equation:

$$
\underbrace{6x}_{\substack{\text{The product of 6}\\\text{and the number}}}\quad \underbrace{=}_{\text{is equal to}}\quad \underbrace{2 - x}_{\substack{\text{the number subtracted}\\\text{from 2.}}}
$$

Next, we solve the equation:

$$
6x = 2 - x \qquad \textit{Original equation}
$$
$$
6x + x = 2 - x + x \qquad \textit{Add x to both sides.}
$$
$$
7x = 2 \qquad \textit{Combine like terms.}
$$
$$
\frac{7x}{7} = \frac{2}{7} \qquad \textit{Divide both sides by 7.}
$$
$$
x = \frac{2}{7} \qquad \textit{Simplify.}
$$

So, the solution is $\dfrac{2}{7}$.

Group Exploration

Section Opener: Comparing expressions and equations

1. Simplify the expression $2(x + 3)$.
2. Solve the equation $2(x + 3) = 0$.
3. Compare the results you found in Problems 1 and 2. Also, explain how to substitute values for x to check your work in the problems.
4. Simplify the expression $3x + 4x$.

5. Solve the equation $3x + 4x = 15 + 2x$.
6. Compare the results you found in Problems 4 and 5. Also, explain how to substitute values for x to check your work in the problems.
7. In general, compare the results of simplifying an expression with the results of solving an equation.

Group Exploration

Simplifying versus solving

1. Two students try to solve the equation $2 = \dfrac{x}{2} + \dfrac{x}{3}$:

Student A

$$2 = \frac{x}{2} + \frac{x}{3}$$

$$6 \cdot 2 = 6\left(\frac{x}{2} + \frac{x}{3}\right)$$

$$12 = 6 \cdot \frac{x}{2} + 6 \cdot \frac{x}{3}$$

$$12 = 3x + 2x$$

$$12 = 5x$$

$$\frac{12}{5} = \frac{5x}{5}$$

$$\frac{12}{5} = x$$

Student B

$$2 = \frac{x}{2} + \frac{x}{3}$$

$$= \frac{x}{2} + \frac{x}{3} - 2$$

$$= \frac{x}{2} \cdot \frac{3}{3} + \frac{x}{3} \cdot \frac{2}{2} - 2$$

$$= \frac{3x}{6} + \frac{2x}{6} - 2$$

$$= \frac{3x + 2x}{6} - 2$$

$$= \frac{5x}{6} - 2$$

Did either student, both students, or neither student solve the equation correctly? Explain.

2. Three students try to simplify the expression $\dfrac{x}{2} + \dfrac{x}{3}$:

Student C

$$\frac{x}{2} + \frac{x}{3} = 6\left(\frac{x}{2} + \frac{x}{3}\right)$$

$$= 6 \cdot \frac{x}{2} + 6 \cdot \frac{x}{3}$$

$$= 3x + 2x$$

$$= 5x$$

Student D

$$\frac{x}{2} + \frac{x}{3} = \frac{3}{3} \cdot \frac{x}{2} + \frac{2}{2} \cdot \frac{x}{3}$$

$$= \frac{3x}{6} + \frac{2x}{6}$$

$$= \frac{3x + 2x}{6}$$

$$= \frac{5x}{6}$$

Student E

$$\frac{x}{2} + \frac{x}{3} = 0$$

$$6\left(\frac{x}{2} + \frac{x}{3}\right) = 6 \cdot 0$$

$$6 \cdot \frac{x}{2} + 6 \cdot \frac{x}{3} = 0$$

$$3x + 2x = 0$$

$$5x = 0$$

$$\frac{5x}{5} = \frac{0}{5}$$

$$x = 0$$

Which students, if any, simplified the expression correctly? Explain.

Homework 4.5

Is the following a linear expression or a linear equation?

1. $3x + 7x = 8$

2. $2x + 4x - 7$

3. $3x + 7x$

4. $2x + 4x - 7 = 5$

5. $4 - 2(x - 9)$

6. $3(x - 1) = 7$

7. $4 - 2(x - 9) = 5$

8. $3(x - 1)$

Simplify the expression or solve the equation, as appropriate. Use a graphing calculator to verify your work.

9. $3x + 4x = 14$

10. $2x - 7x = 15$

11. $3x + 4x$

12. $2x - 7x$

13. $b - 5(b - 1)$

14. $-6k + 2(k + 4)$

15. $b - 5(b - 1) = 0$

16. $-6k + 2(k + 4) = 0$

17. $3(3x - 5) + 2(5x + 4) = 0$

18. $-4(2x + 1) - 2(3x - 6) = 0$

19. $3(3x - 5) + 2(5x + 4)$

20. $-4(2x + 1) - 2(3x - 6)$

21. $3(x - 2) - (7x + 2) = 4(3x + 1)$

22. $5(2x + 3) - (3x - 8) - 2(4x - 5)$

23. $3(x - 2) - (7x + 2) - 4(3x + 1)$

24. $5(2x + 3) - (3x - 8) = 2(4x - 5)$

25. $7.2p - 4.5 - 1.3p$

26. $8.3t + 9.2 - 7.7t$

27. $7.2p - 4.5 - 1.3p = 20.5 - 6.6p$

28. $8.3t + 9.2 - 7.7t = 2.9 - 1.2t$

29. $-3.5(x - 8) - 2.6(x - 2.8) = 13.93$

30. $-4.8(x + 3) + 6.5(x - 1.2)$

31. $-3.5(x - 8) - 2.6(x - 2.8)$

32. $-4.8(x + 3) + 6.5(x - 1.2) = -25.09$

33. $-\dfrac{6w}{8}$

34. $-\dfrac{15t}{10}$

35. $-\dfrac{6w}{8} = \dfrac{3}{2}$

36. $-\dfrac{15t}{10} = -\dfrac{5}{4}$

37. $\dfrac{5x}{6} + \dfrac{1}{2} - \dfrac{3x}{4}$

38. $\dfrac{2x}{3} - \dfrac{5x}{2} - 1 = 0$

39. $\dfrac{5x}{6} + \dfrac{1}{2} - \dfrac{3x}{4} = 0$

40. $\dfrac{2x}{3} - \dfrac{5x}{2} - 1$

41. $\dfrac{7}{2}x - \dfrac{5}{6} = \dfrac{1}{3} + \dfrac{3}{4}x$

42. $\dfrac{3}{5}x + 2x - \dfrac{5}{2} - \dfrac{7}{10}x$

43. $\dfrac{7}{2}x - \dfrac{5}{6} - \dfrac{1}{3} + \dfrac{3}{4}x$

44. $\dfrac{3}{5}x + 2x = \dfrac{5}{2} - \dfrac{7}{10}x$

For Exercises 45–56, let x be a number. Translate each of the following into an expression or an equation, as appropriate, and then simplify the expression or solve the equation.

45. The sum of 3 and twice the number is -10.

46. The difference of 3 times the number and 5 is -12.

47. 4, minus 6 times the difference of the number and 2

48. 6, plus 3 times the sum of the number and 8

49. The product of -9 and the number is equal to the difference of the number and 5.

50. The sum of the number and 4 is equal to the number subtracted from 1.

51. The quotient of the number and 2 is equal to 3 times the difference of the number and 5.

52. The quotient of the number and 3 is equal to 2 plus the quotient of the number and 6.

53. The number plus the product of the number and 6

54. The number minus the product of the number and -4

55. The number plus the quotient of the number and 2

56. 4 minus the quotient of the number and 5

Concepts

For Exercises 57 and 58, give an example of each. Then simplify or solve, as appropriate.

57. Linear expression

58. Linear equation in one variable

59. Two students try to solve the equation $\dfrac{3}{4}x - \dfrac{5}{6} = \dfrac{1}{3}$:

Student A

$$\frac{3}{4}x - \frac{5}{6} = \frac{1}{3}$$

$$\frac{3}{4}x = \frac{1}{3} + \frac{5}{6}$$

$$\frac{3}{3} \cdot \frac{3}{4}x = \frac{4}{4} \cdot \frac{1}{3} + \frac{2}{2} \cdot \frac{5}{6}$$

$$\frac{9}{12}x = \frac{4}{12} + \frac{10}{12}$$

$$\frac{9}{12}x = \frac{14}{12}$$

$$\frac{3}{4}x = \frac{7}{6}$$

$$\frac{4}{3} \cdot \frac{3}{4}x = \frac{4}{3} \cdot \frac{7}{6}$$

$$1x = \frac{2 \cdot 2 \cdot 7}{3 \cdot 2 \cdot 3}$$

$$x = \frac{14}{9}$$

Student B

$$\frac{3}{4}x - \frac{5}{6} = \frac{1}{3}$$

$$12\left(\frac{3}{4}x - \frac{5}{6}\right) = 12 \cdot \frac{1}{3}$$

$$12 \cdot \frac{3}{4}x - 12 \cdot \frac{5}{6} = 4$$

$$9x - 10 = 4$$

$$9x = 14$$

$$x = \frac{14}{9}$$

Compare the methods of solving the equation. Is one method better than the other? Explain.

60. A student believes that 5 is the solution (and the only solution) of the equation $-2(x + 4) = -2x - 8$ because 5 satisfies that equation:

$$-2(5 + 4) \stackrel{?}{=} -2(5) - 8$$
$$-18 \stackrel{?}{=} -18$$
$$\text{true}$$

Is the student correct? Explain.

61. A student believes that $x + 6$ is the solution of an equation. Is the student correct? Explain.

62. A student tries to simplify an expression. The student writes, "The solution of the expression is 5." Is the student correct? Explain.

63. A student tries to simplify the expression $\frac{1}{4}x + \frac{1}{3}x$:

$$\frac{1}{4}x + \frac{1}{3}x = 12\left(\frac{1}{4}x + \frac{1}{3}x\right)$$
$$= 12 \cdot \frac{1}{4}x + 12 \cdot \frac{1}{3}x$$
$$= 3x + 4x$$
$$= 7x$$

Describe any errors. Then simplify the expression correctly.

64. A student tries to simplify the expression $3x + 5x - 16$:

$$3x + 5x - 16 = 0$$
$$8x - 16 = 0$$
$$8x = 16$$
$$x = 2$$

Describe any errors. Then simplify the expression correctly.

65. The simplified form of $7 + 2(x + 3)$ is $2x + 13$. What does this mean?

66. The solution of $7 + 2(x + 3) = 19$ is 3. What does this mean?

67. A student tries to simplify $2(5 + x)$:

$$2(5 + x) = 10 + x$$

To check the work, the student evaluates $2(5 + x)$ and $10 + x$, using 0 for x:

Evaluate $2(5 + x)$ for $x = 0$: $2(5 + 0) = 10$

Evaluate $10 + x$ for $x = 0$: $10 + 0 = 10$

Because the results of evaluating the expressions are equal, the student decides that the work is correct. What would you tell the student about checking the work?

68. A student evaluates both of the expressions $3x + 1$ and $21 - 2x$, using 4 for x:

Evaluate $3x + 1$ for $x = 4$: $3(4) + 1 = 13$

Evaluate $21 - 2x$ for $x = 4$: $21 - 2(4) = 13$

Because the results of evaluating the expressions are equal, the student concludes that the expressions are equivalent. Is the student correct? Explain.

69. Give three examples of an equation whose solution is 5.

70. Give three examples of an expression equivalent to the expression 5.

71. When simplifying an expression, what is the only number we can multiply it by? Why?

72. When solving an equation, why can we multiply both sides by *any* nonzero number?

73. Explain the meaning of simplifying an expression. (See page 9 for guidelines on writing a good response.)

74. Explain the meaning of solving an equation. (See page 9 for guidelines on writing a good response.)

4.6 Formulas

Objectives

» Determine area and perimeter formulas of a rectangle and a total-value formula.

» Use formulas to solve various types of problems.

» Translate an English sentence into a formula.

» Solve a formula for a variable.

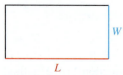

Figure 23 The length *L* and width *W* of a rectangle

Recall from Section 3.3 that a **formula** is an equation that contains two or more variables. We can use formulas to find quantities such as the area and perimeter of rectangular objects, the total value of a group of objects, and the average test score of a group of tests.

Area Formula of a Rectangle

Recall from Section 1.1 that the area of a rectangle is equal to the length times the width. We can write $A = LW$, where A is the area, L is the length, and W is the width (see Fig. 23). The equation $A = LW$ is a formula.

Example 1   Using the Area Formula of a Rectangle

An architect is designing a school building so that each classroom contains 35 student desks. Fire codes require that there be 18 square feet per student desk. If there is enough room for classrooms of length 28 feet, what is the smallest width permitted by the fire codes?

Solution

To find the area of the floor of one room, we multiply the area needed per student desk by the number of student desks: $18(35) = 630$ square feet. Next, we must find the width of a rectangle that has an area of 630 square feet and a length of 28 feet (see Fig. 24).

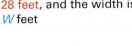

28 feet

Figure 24 The length is 28 feet, and the width is W feet

To do so, we substitute 630 for A and 28 for L in the formula $A = LW$, where the units of L and W are feet and the units of A are square feet. Then we solve for W:

$$A = LW \qquad \textit{Area formula of a rectangle}$$
$$630 = (28)W \qquad \textit{Substitute 630 for A and 28 for L.}$$
$$\frac{630}{28} = \frac{28W}{28} \qquad \textit{Divide both sides by 28.}$$
$$22.5 = W \qquad \textit{Simplify.}$$

The smallest width permitted is 22.5 feet.

To check, we find the product of the length and width: $28(22.5) = 630$, which is equal to the area (in square feet).

▶

In Example 1, we substituted values for A and L in the formula $A = LW$ and then solved for the variable W. In general, **to find a single value of a variable in a formula, we often substitute numbers for all the other variables and then solve for that one variable.**

Perimeter Formula of a Rectangle

The **perimeter** of a polygon, which is a geometrical object such as a triangle, rectangle, or trapezoid, is the total distance around the object. For example, consider a rectangle with length L and width W (see Fig. 25).

The perimeter P of a rectangle is equal to the sum of the lengths of the four sides:

$$P = L + W + L + W$$

Combining like terms on the right-hand side of the formula gives

$$P = 2L + 2W$$

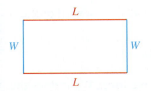

Figure 25 The length L and width W of a rectangle

Area and Perimeter of a Rectangle

For a rectangle with length L, width W, area A, and perimeter P,

- $A = LW$
- $P = 2L + 2W$

37.5 feet

Figure 26 The length is 37.5 feet, and the width is W feet

Example 2 Using the Perimeter Formula of a Rectangle

A landscaper has budgeted enough money for 100 feet of fencing to enclose a rectangular garden. If the length of the garden is to run the full length of the property, which is 37.5 feet, what will be the width of the garden (see Fig. 26)?

Solution

Because the 100-foot-long fencing is on the border of the garden, the perimeter of the garden is 100 feet. So, we substitute 100 for P and 37.5 for L in the formula $P = 2L + 2W$, where the units of P, L, and W are feet. Then we solve for W:

$$P = 2L + 2W \qquad \textit{Perimeter formula of a rectangle}$$
$$100 = 2(37.5) + 2W \qquad \textit{Substitute 100 for P and 37.5 for L.}$$
$$100 = 75 + 2W \qquad \textit{Simplify.}$$
$$100 - 75 = 75 + 2W - 75 \qquad \textit{Subtract 75 from both sides.}$$
$$25 = 2W \qquad \textit{Combine like terms.}$$
$$\frac{25}{2} = \frac{2W}{2} \qquad \textit{Divide both sides by 2.}$$
$$12.5 = W \qquad \textit{Write } \frac{25}{2} \textit{ as a decimal number.}$$

The width of the garden will be 12.5 feet.

To check, we find twice the length of 37.5 feet plus twice the width of 12.5 feet: $2(37.5) + 2(12.5) = 100$ feet, which is equal to the perimeter.

▶

When describing an authentic quantity, we use a decimal form rather than a fractional form. For instance, in Example 2 we say that the width is 12.5 feet rather than $\frac{25}{2}$ feet.

Total-Value Formula

Three quarters are worth $25 \cdot 3 = 75$ cents. Note that we found the total value of the quarters by multiplying the value of one quarter (25 cents) by the number of quarters (3).

Total-Value Formula

If n objects each have value v, then their total value T is given by

$$T = vn$$

In words: The total value is equal to the value of one object times the number of objects.

Example 3   Total-Value Formula

I'm beginning to understand why our tickets were half off.

On September 20, 2016, Chicago Cubs baseball tickets for upper-deck reserved outfield seating at Wrigley Field cost $11 each.

1. Find a formula of the total cost T (in dollars) of n of these tickets.
2. Perform a unit analysis of the formula found in Problem 1.
3. Substitute 187 for T in the formula from Problem 1, and solve for n. What does the result mean in this situation?

Solution

1. We substitute 11 for v in the formula $T = vn$:

$$T = 11n$$

2. Here is a unit analysis of the formula $T = 11n$:

$$\underbrace{T}_{\text{dollars}} = \underbrace{11}_{\dfrac{\text{dollars}}{\text{ticket}}} \quad \underbrace{n}_{\text{number of tickets}}$$

   The units of the expressions on both sides of the equation are dollars, which checks.

3. We substitute 187 for T in the formula $T = 11n$ and solve for n:

$$T = 11n \qquad \textit{Total-value formula}$$
$$187 = 11n \qquad \textit{Substitute 187 for T.}$$
$$17 = n \qquad \textit{Divide both sides by 11.}$$

The total cost of 17 upper-deck reserved outfield tickets at Wrigley Field is $187.

Translating from English into a Mathematics Formula

In Example 4, we will translate some information into a formula and then use the formula to find several quantities.

Example 4   Translating from English into Mathematics

A Celsius temperature reading is equal to $\frac{5}{9}$ times the difference of the Fahrenheit temperature reading and 32.

1. Write a formula of the Celsius temperature in terms of the Fahrenheit temperature.
2. Convert 50°F to the equivalent Celsius temperature.
3. Use a graphing calculator to convert 50°F, 59°F, 68°F, 77°F, and 86°F to the equivalent Celsius temperatures.

Solution

1. We let C be the Celsius reading and F be the equivalent Fahrenheit reading. Next, we translate the given information into a formula:

$$
\underbrace{\text{A Celsius reading}}_{C} \quad \underbrace{\text{is equal to}}_{=} \quad \underbrace{\dfrac{5}{9}}_{\frac{5}{9}} \quad \underbrace{\text{times}}_{\cdot} \quad \overbrace{\text{the difference of the Fahrenheit reading and 32.}}^{(F - 32)}
$$

So, the formula is $C = \dfrac{5}{9}(F - 32)$.

2. We substitute 50 for F in the formula $C = \dfrac{5}{9}(F - 32)$ and solve for C:

$$
C = \frac{5}{9}(50 - 32) = 10
$$

So, $10°$C is equivalent to $50°$F.

3. We use a graphing calculator table to substitute the inputs 50, 59, 68, 77, and 86 for F in the formula $C = \dfrac{5}{9}(F - 32)$. See Fig. 27. See Appendix A.13 for graphing calculator instructions.

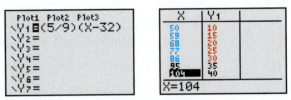

Figure 27 Finding Celsius temperatures

By viewing the second column of the graphing calculator table, we see that the values of C are 10, 15, 20, 25, and 30. So, $50°$F, $59°$F, $68°$F, $77°$F, and $86°$F are equivalent to $10°$C, $15°$C, $20°$C, $25°$C, and $30°$C, respectively.

Solving a Formula for a Variable

In each of Examples 1–3, we substituted values for all but one variable in a formula and then solved the equation for the remaining variable. Sometimes it will be better to reverse the two steps. That is, we can first solve a formula for a variable and then substitute values for the other variables.

Example 5 Solving a Formula for One of Its Variables

1. Solve the formula $A = LW$ for W.
2. Substitute 630 for A and 28 for L in the formula found in Problem 1.

Solution

1. We can solve the formula $A = LW$ for W in much the same way that we solved the equation $630 = (28)W$ in Example 1:

$630 = (28)W$	$A = LW$ *Area formula*
$\dfrac{630}{28} = \dfrac{28W}{28}$	$\dfrac{A}{L} = \dfrac{LW}{L}$ *Divide both sides by L.*
$22.5 = W$	$\dfrac{A}{L} = W$ *Simplify.*

The result is $\dfrac{A}{L} = W$. By switching the sides of the equation, we have $W = \dfrac{A}{L}$. We are done because W is alone on one side of the formula and does not appear on the other side.

2. We substitute 630 for A and 28 for L in the formula $W = \dfrac{A}{L}$:

$$W = \frac{630}{28} = 22.5$$

So, the width is 22.5 feet. This is the same result we found in Example 1.

▶

In Problem 1 of Example 5, the result of solving the formula $A = LW$ for W is the formula $W = \dfrac{A}{L}$. This means that the formulas give the same result when we substitute values for all but one variable and solve for the remaining variable. For example, we found the same value (22.5) for W in Example 1 and Problem 2 of Example 5. In general, **solving for a variable in a formula will not change the relationship between the variables in the formula.**

Example 6   Solving a Formula for One of Its Variables

1. Solve the formula $P = 2L + 2W$ for W.
2. A total of 100 feet of fencing will be used to enclose a rectangular garden. Use the formula $P = 2L + 2W$ or the formula found in Problem 1 to find the widths of the garden for the following lengths (in feet): 30, 32, 34, 36, 38, and 40.

Solution

1. We can solve the formula $P = 2L + 2W$ for W in much the same way that we solved the equation $100 = 75 + 2W$ in Example 2:

$100 = 75 + 2W$	$P = 2L + 2W$	*Perimeter formula*
$100 - 75 = 75 + 2W - 75$	$P - 2L = 2L + 2W - 2L$	*Subtract 2L from both sides.*
$25 = 2W$	$P - 2L = 2W$	*Combine like terms.*
$\dfrac{25}{2} = \dfrac{2W}{2}$	$\dfrac{P - 2L}{2} = \dfrac{2W}{2}$	*Divide both sides by 2.*
$\dfrac{25}{2} = W$	$\dfrac{P - 2L}{2} = W$	*Simplify.*

By switching the sides of the equation $\dfrac{P - 2L}{2} = W$, we have $W = \dfrac{P - 2L}{2}$.

2. To find the widths, we will use the formula $W = \dfrac{P - 2L}{2}$ because W is alone on one side of the equation. First, we substitute 100 for P:

$$W = \frac{100 - 2L}{2}$$

Then we use a graphing calculator table to substitute 30, 32, 34, 36, 38, and 40 for L in the formula $W = \dfrac{100 - 2L}{2}$ (see Fig. 28). By viewing the second column of the graphing calculator table, we see that the values of W are 20, 18, 16, 14, 12, and 10. So, for the lengths (in feet) 30, 32, 34, 36, 38, and 40, the widths (in feet) are 20, 18, 16, 14, 12, and 10, respectively.

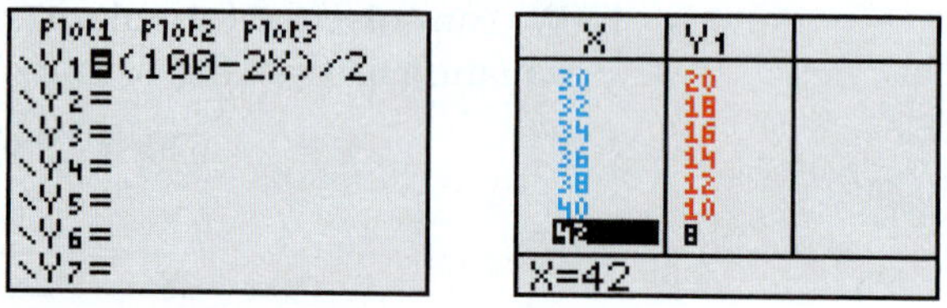

Figure 28 Find the six widths

In Example 2, we found the width of a rectangle by substituting 100 for P and 37.5 for L in the formula $P = 2L + 2W$ and then solving for W. This was efficient because we wanted to find only one width. But to find the widths corresponding to the six lengths in Example 6, substituting before solving would require solving *six* linear equations! Solving the formula $P = 2L + 2W$ for W and then constructing a table is much more efficient (see Example 6). In general, **to find several values of a variable in a formula, we usually solve the formula for that variable before we make any substitutions.**

The equation $2x - 5y = 10$ is a formula because it is an equation that contains two (or more) variables. In Example 7, we will solve this equation for y.

Example 7   Writing a Linear Equation in Slope–Intercept Form

Write the equation $2x - 5y = 10$ in slope–intercept form $(y = mx + b)$.

Solution

We solve $2x - 5y = 10$ for y:

$$2x - 5y = 10 \quad \textit{Original equation}$$
$$2x - 5y - 2x = 10 - 2x \quad \textit{Subtract 2x from both sides.}$$
$$-5y = -2x + 10 \quad \textit{Combine like terms; rearrange terms.}$$
$$\frac{-5y}{-5} = \frac{-2x}{-5} + \frac{10}{-5} \quad \textit{Divide both sides by -5.}$$
$$y = \frac{2}{5}x - 2 \quad \textit{Simplify.}$$

In Section 5.1, we will perform work similar to our work in Example 7 to help us graph linear equations.

Example 8   Solving a Formula for One of Its Variables

1. Solve the Fahrenheit–Celsius model $C = \frac{5}{9}(F - 32)$ for F.

2. Use a graphing calculator to convert $10°C, 15°C, 20°C, 25°C,$ and $30°C$ to the equivalent Fahrenheit temperatures.

Solution

1.
$$C = \frac{5}{9}(F - 32) \quad \textit{Celsius formula}$$
$$\frac{9}{5} \cdot C = \frac{9}{5} \cdot \frac{5}{9}(F - 32) \quad \textit{Multiply both sides by } \frac{9}{5}.$$
$$\frac{9}{5}C = F - 32 \quad \textit{Simplify.}$$
$$\frac{9}{5}C + 32 = F - 32 + 32 \quad \textit{Add 32 to both sides.}$$
$$\frac{9}{5}C + 32 = F \quad \textit{Combine like terms.}$$

The result is $F = \frac{9}{5}C + 32$.

2. We use a graphing calculator table to substitute $10, 15, 20, 25,$ and 30 for C in the formula $F = \frac{9}{5}C + 32$ (see Fig. 29). By viewing the second column of the graphing calculator table, we see that the values of F are $50, 59, 68, 77,$ and 86. So, $10°C, 15°C, 20°C, 25°C,$ and $30°C$ are equivalent to $50°F, 59°F, 68°F, 77°F,$ and $86°F$, respectively.

Figure 29 Finding Fahrenheit temperatures

Example 9 Solving a Linear Model for One of Its Variables

The average fee banks charged noncustomers to use their ATMs was $2.88 in 2015 and has increased by about $0.12 per year since then (Source: *Bankrate*®). Let F be the average ATM fee (in dollars) at t years since 2015.

1. Find an equation of a linear model to describe the situation.
2. Solve the equation found in Problem 1 for t.
3. Estimate when the average ATM fee was $3.
4. Use a graphing calculator table to predict in which years the average ATM fee will be $3.20, $3.40, $3.60, $3.80, and $4.00.

Solution

1. Because the average ATM fee increases by about $0.12 each year, we can model the situation by using a linear model with slope 0.12. Because the average ATM fee was $2.88 in 2015, the F-intercept is $(0, 2.88)$. So, a reasonable model is

$$F = 0.12t + 2.88$$

2.
$$F = 0.12t + 2.88 \qquad \text{\textit{Original equation}}$$
$$F - 2.88 = 0.12t + 2.88 - 2.88 \qquad \text{\textit{Subtract 2.88 from both sides.}}$$
$$F - 2.88 = 0.12t \qquad \text{\textit{Combine like terms.}}$$
$$\frac{F - 2.88}{0.12} = \frac{0.12t}{0.12} \qquad \text{\textit{Divide both sides by 0.12.}}$$
$$\frac{F - 2.88}{0.12} = t \qquad \text{\textit{Simplify.}}$$

The result is the formula $t = \dfrac{F - 2.88}{0.12}$.

3. To find the year, we use the formula $t = \dfrac{F - 2.88}{0.12}$ because t is alone on one side of the equation. So, we substitute 3 for F in $t = \dfrac{F - 2.88}{0.12}$:

$$t = \frac{3 - 2.88}{0.12} = 1$$

The average ATM fee was $3 in $2015 + 1 = 2016$, according to the model.

4. We use a graphing calculator table to substitute $3.20, 3.40, 3.60, 3.80$, and 4.00 for F in the equation $t = \dfrac{F - 2.88}{0.12}$ (see Fig. 30). By viewing the second column of the graphing calculator table, we see that the approximate values of t (rounded to the ones place) are $3, 4, 6, 8$, and 9. The model predicts that the average ATM fee will be $3.20, $3.40, $3.60, $3.80, and $4.00 in the years 2018, 2019, 2021, 2023, and 2024, respectively.

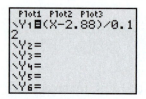

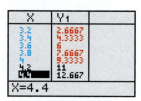

Figure 30 Finding the years

◆ **Group Exploration**

Section Opener: Solving a formula for one of its variables

1. **a.** Solve the equation $x + 2 = 6$.
   b. Solve the equation $x + 4 = 9$.
   c. Solve the equation $x + b = c$ for x. [**Hint:** Refer to your work in parts (a) and (b).]
2. **a.** Solve the equation $2x = 7$.
   b. Solve the equation $5x = 3$.
   c. Solve the equation $mx = b$ for x. [**Hint:** Refer to your work in parts (a) and (b).]
3. **a.** Solve the equation $3 + 5y = 7$.
   b. Solve the equation $2 + 7y = 8$.
   c. Solve the equation $ax + by = c$ for y.

Homework 4.6

 For extra help ▶   **MyLab Math**    Watch the videos in MyLab Math

Find a formula of the perimeter P of the polygon.

1. See Fig. 31.

2. See Fig. 32.

Figure 31 Exercise 1

Figure 32 Exercise 2

3. See Fig. 33.

4. See Fig. 34.

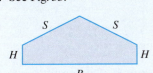

Figure 33 Exercise 3

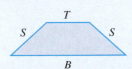

Figure 34 Exercise 4

5. See Fig. 35.

6. See Fig. 36.

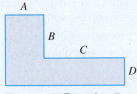

Figure 35 Exercise 5

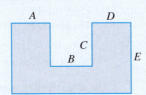

Figure 36 Exercise 6

For Exercises 7–16, substitute the given values for the variables and then solve the formula for the remaining variable. Round approximate results to the second decimal place.

7. $P = VI$; $P = 20$, $I = 4$ (power in an electrical circuit)

8. $PV = nRT$; $P = 0.97$, $V = 2.85$, $R = 0.082$, $T = 295$ (pressure, volume, number of moles, and temperature of a gas)

9. $A = \frac{1}{2}BH$; $A = 6$, $H = 3$ (area of a triangle)

10. $U = -\frac{GmM}{r}$; $U = -50$, $G = 9.8$, $M = 7$, $r = 2$ (gravitational potential energy)

11. $v = gt + v_0$; $v = 80$, $g = 32.2$, $v_0 = 20$ (speed of a projectile)

12. $A = P + Prt$; $A = 850$, $P = 500$, $r = 0.1$ (simple interest)

13. $S = 2WL + 2WH + 2LH$; $S = 52$, $W = 2$, $H = 4$ (surface area of a rectangular box)

14. $S = 2WL + 2WH + 2LH$; $S = 76$, $L = 2$, $H = 4$ (surface area of a rectangular box)

15. $A = \frac{a + b + c}{3}$; $A = 5$, $a = 2$, $c = 6$ (average length of a side of a triangle)

16. $\frac{x}{a} + \frac{y}{a} = 1$; $a = 2$, $y = 6$ (equation of a line)

17. A rectangular carpet has an area of 116 square feet and a width of 8 feet. Find the length of the carpet.

18. A rectangular floor has an area of 207 square feet and a length of 18 feet. Find the width of the floor.

19. A rectangle has a perimeter of 52 inches and a width of 10 inches. Find the length of the rectangle.

20. A rectangle has a perimeter of 88 inches and a length of 25 inches. Find the width of the rectangle.

21. A portable volleyball court includes a 177-foot cord to use as the rectangular boundary of the official-size court. The official width of a volleyball court is 29.5 feet. What is the official length of the court?

22. A photographer plans to use 55 inches of wood to make a rectangular frame for a photo. If the width of the frame is to be 12.5 inches, what will its length be?

23. A rectangle has a width of 3 inches and a length of x inches.
   a. Find a formula of the area A (in square inches) of the rectangle.
   b. Find a formula of the perimeter P (in inches) of the rectangle.
   c. Perform a unit analysis of the formula you found in part (b).

24. A rectangle has a width of x feet and a length of 7 feet.
   a. Find a formula of the area A (in square feet) of the rectangle.
   b. Find a formula of the perimeter P (in feet) of the rectangle.
   c. Perform a unit analysis of the formula you found in part (b).

25. A landscaper is going to dig a rectangular garden and enclose the garden with 40 feet of fencing. He can plant one flower per square foot of ground.
  a. Give three examples of rectangular gardens that could be enclosed with 40 feet of fencing. State the length and width of each garden.
  b. Find the area of each of the three gardens you described in part (a).
  c. Which of the gardens you described in part (a) would hold the greatest number of flowers? Explain.

26. A landscaper is going to dig a small rectangular garden that has an area of 36 square feet.
  a. Give three examples of rectangular gardens that have an area of 36 square feet. State the length and width of each garden.
  b. Find the perimeter of each of the three gardens you described in part (a).
  c. The landscaper plans to enclose the garden with fencing. Which of the gardens you described in part (a) would require the least amount of fencing? How much fencing is that?

27. a. What is the value (in cents) of 3 dimes?
  b. What is the value (in cents) of 4 dimes?
  c. Find a formula of the total value T (in cents) of d dimes.
  d. Perform a unit analysis of the formula you found in part (c).

28. a. What is the value (in dollars) of 3 nickels?
  b. What is the value (in dollars) of 4 nickels?
  c. Find a formula of the total value T (in dollars) of n nickels.
  d. Perform a unit analysis of the formula you found in part (c).

29. Find a formula of the total sales T (in dollars) of selling n guitars at $725 per guitar.

30. Find a formula of the total sales T (in dollars) of selling n gallons of gasoline at $3.95 per gallon.

31. Wye Oak played at the Republic in New Orleans, Louisiana, on September 21, 2016. Tickets purchased in advance sold for $15 each.
  a. Assuming all the tickets were purchased in advance, find a formula of the total sales T (in dollars) if x people bought tickets.
  b. Substitute 30,000 for T in the formula you found in part (a), and solve for x. What does your result mean in this situation?

32. T-Mobile charges $70 per month for unlimited texting, talking, and data.
  a. Find a formula of the total charge T (in dollars) for m months of service.
  b. Substitute $980 for T in the formula you found in part (a), and solve for m. What does your result mean in this situation?

33. Baseball tickets for field-level seating at Yankee Stadium in the Bronx ranged from $77 to $250 in 2016.
  a. Find a formula of the total cost C (in dollars) of k tickets that sold for $77 per ticket.
  b. Find a formula of the total cost E (in dollars) of n tickets that sold for $250 per ticket.

c. Find a formula of the total cost T (in dollars) of k tickets that sold for $77 per ticket and n tickets that sold for $250 per ticket.
  d. Use the formula you found in part (c) to find the value of n when $k = 65$ and $T = 27,505$. What does your result mean in this situation?

34. Football season tickets at University of Illinois, Urbana-Champaign ranged from $119 to $339 in 2017.
  a. Find a formula of the total cost C (in dollars) of k season tickets that sold for $119 per ticket.
  b. Find a formula of the total cost E (in dollars) of n season tickets that sold for $339 per ticket.
  c. Find a formula of the total cost T (in dollars) of k season tickets that sold for $119 per ticket and n season tickets that sold for $339 per ticket.
  d. Use the formula you found in part (c) to find the value of n when $k = 40$ and $T = \$26,795$. What does your result mean in this situation?

35. One cubic foot is shown in Fig. 37. The *volume*, in cubic feet, of an object is the number of cubic feet that it takes to fill that object. Let L be the length, W be the width, and H be the height, all in feet, of a rectangular box (see Fig. 38). The volume V (in cubic feet) of the rectangular box is equal to the length times the width times the height of the box.

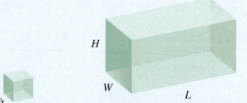

1 ft
1 ft   1 ft
Figure 37 One cubic foot

H
W
L
Figure 38 The length L, width W, and height H of a rectangular box

  a. Write a formula of the volume of the rectangular box.
  b. Find the height of the rectangular box if the volume is 48 cubic feet, the length is 3 feet, and the width is 2 feet.

36. Let B be the length of the base, T be the length of the top, and H be the height, all in inches, of a trapezoid (see Fig. 39).

The area A (in square inches) of the trapezoid is equal to $\frac{1}{2}$ of the height times the sum of the lengths of the base and top.

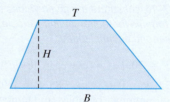

T
H
B
Figure 39 The base B, top T, and height H of a trapezoid

  a. Write a formula of the area of the trapezoid.
  b. Find the base length of the trapezoid if the area is 20 square inches, the height is 4 inches, and the top length is 3 inches.

37. The interest I (in dollars) earned from a simple-interest bank account is equal to the money P (in dollars) invested, times the decimal annual interest rate r, times the number of years t of the investment.
   a. Write a formula of I.
   b. Find the interest earned from investing $5000 in a 4% simple-interest account for 3 years. [**Hint:** If the interest rate is 4%, then the decimal interest rate is 0.04.]
   c. The balance B (in dollars) of a bank account is equal to the original money invested plus the interest. Find a formula of B in terms of P, r, and t. [**Hint:** Build on the formula you found in part (a).]
   d. Find the balance after 4 years in a 5% simple-interest account in which $2000 was originally invested.

38. The force F (in newtons) needed to lift an object is equal to the product of the object's mass m (in kilograms) and the constant a (in meters per second squared), which is the object's acceleration due to gravity.
   a. Write a formula of the force needed to lift an object.
   b. Describe the unit newton (N) in terms of kilograms (kg), meters (m), and seconds (s). [**Hint:** Perform a unit analysis of the right-hand side of the formula you found in part (a).]
   c. If a force of 25 newtons is required to lift an object and the constant a is 9.8 meters per second squared, find the mass of the object.

39. The average test score A (in points) of five test scores t_1, t_2, t_3, t_4, and t_5, all in points, can be computed by using the formula

$$A = \frac{t_1 + t_2 + t_3 + t_4 + t_5}{5}$$

If a student has test scores of 74, 81, 79, and 84, all in points, what score does she need on the fifth test so that her five-test average is 80 points?

40. The average test score A (in points) of four test scores t_1, t_2, t_3, and t_4, all in points, can be computed by using the formula

$$A = \frac{t_1 + t_2 + t_3 + t_4}{4}$$

If a student has test scores of 87, 92, and 86, all in points, what score does he need on the fourth test so that his four-test average is 90 points?

Solve the formula for the specified variable.

41. $A = LW$, for W

42. $P = VI$, for I

43. $PV = nRT$, for T

44. $PV = nRT$, for n

45. $U = -\dfrac{GmM}{r}$, for M

46. $U = -\dfrac{GmM}{r}$, for m

47. $A = \dfrac{1}{2}BH$, for B

48. $A = \dfrac{1}{2}BH$, for H

49. $v = gt + v_0$, for t

50. $y = mx + b$, for x

51. $A = P + Prt$, for r

52. $A = P + Prt$, for t

53. $A = \dfrac{a + b + c}{3}$, for b

54. $A = \dfrac{a + b + c}{3}$, for c

55. $y - k = m(x - h)$, for x

56. $y - k = m(x - h)$, for h

57. $\dfrac{x}{a} + \dfrac{y}{a} = 1$, for y

58. $\dfrac{x}{a} + \dfrac{y}{a} = 1$, for x

For Exercises 59–66, write the linear equation in slope–intercept form.

59. $3x + 4y = 16$

60. $2x + 3y = 24$

61. $2x + 4y - 8 = 0$

62. $7x + 3y - 21 = 0$

63. $5x - 2y = 6$

64. $4x - 5y = 30$

65. $-3x - 7y = 5$

66. $-5x - 9y = 2$

67. The numbers of worldwide smartphone subscriptions are shown in Table 10 for various years.

Table 10 Worldwide Smartphone Subscriptions

Year	Number of Subscriptions (billions)
2010	0.3
2011	0.8
2012	1.3
2013	1.7
2014	2.1
2015	2.6

Source: *Informa*

Let n be the number (in billions) of worldwide smartphone subscriptions at t years since 2010. A model of the situation is $n = 0.45t + 0.34$.
   a. Use a graphing calculator to draw a scatterplot and the model in the same viewing window. Does the line come close to the data points?
   b. Solve the equation $n = 0.45t + 0.34$ for t.
   c. Use the equation you found in part (b) to estimate when there were 3 billion worldwide smartphone subscriptions.
   d. Use a graphing calculator table to predict in which years the number of worldwide smartphone subscriptions will be 3.5, 4, 4.5, 5, and 5.5, all in billions.

68. Revenues in the United States from social network gaming and mobile games are shown in Table 11 for various years.

Table 11 Revenues from Social Network Gaming and Mobile Games

Year	Revenue (billions of dollars)
2009	5.4
2010	7.0
2011	7.5
2012	6.7
2013	9.0
2014	9.9
2015	11.2

Sources: *Entertainment Software Association*; *NPD Group*

Let r be the annual revenue (in billions of dollars) from social network gaming and mobile games at t years since 2000. A model of the situation is $r = 0.88t - 2.49$.
   a. Use a graphing calculator to draw a scatterplot and the model in the same viewing window. Does the line come close to the data points?
   b. Solve the equation $r = 0.88t - 2.49$ for t.
   c. Use the equation you found in part (b) to estimate when the revenue was $12 billion.

d. Use a graphing calculator table to predict in which years the revenue will be 13, 14, 15, 16, and 17, all in billions of dollars.

69. Sales of Kia® automobiles were 626 thousand automobiles in 2015 and have increased by about 44.6 thousand automobiles per year since then (Source: *Kia*). Let s be the annual sales (in thousands of automobiles) at t years since 2015.
  a. Find an equation of a linear model to describe the situation.
  b. Solve the equation you found in part (a) for t.
  c. Use the equation you found in part (b) to predict in which year Kia's automobile sales will be 700 thousand automobiles.
  d. Use a graphing calculator table to predict in which years Kia's automobile sales will be 750, 800, 850, 900, and 950, all in thousands of automobiles.

70. The percentage of adults who use social networking sites was 65% in 2015 and has increased by about 3.9 percentage points per year since then (Source: *Pew Research Center*). Let p be the percentage of adults who use social networking sites at t years since 2015.
  a. Find an equation of a model to describe the situation.
  b. Solve the equation you found in part (a) for t.
  c. Use the equation you found in part (b) to estimate when 70% of adults used social networking sites.
  d. Use a graphing calculator table to predict when the percentage of adults who use social networking sites will be 75%, 80%, 85%, 90%, and 95%.

71. The percentage of people who say they try to avoid snacking entirely was 39.1% in 2015 and has decreased by about 1.4 percentage points per year since then (Source: *The NPD Group*). Let p be the percentage of people who say they try to avoid snacking entirely at t years since 2015.
  a. Find an equation of a model to describe the situation.
  b. Use the equation you found in part (a) to estimate when 31% of people said they try to avoid snacking entirely.
  c. Solve the equation you found in part (a) for t.
  d. Use the equation you found in part (c) to estimate when 31% of people said they try to avoid snacking entirely.
  e. Compare the results you found in parts (b) and (d). Which equation was easier to use? Explain.
  f. Which equation is easier to use to predict the percentage of people in 2022 who will say they try to avoid snacking entirely? Explain. Then find that percentage.

72. Orange production in the United States was 148.6 million boxes in 2015 and has decreased by about 20.7 million boxes per year since then (Source: *USDA*). Let p be U.S. annual orange production (in millions of boxes) at t years since 2015.
  a. Find an equation of a model to describe the situation.
  b. Use the equation you found in part (a) to estimate when U.S. annual orange production was 100 million boxes.
  c. Solve the equation you found in part (a) for t.
  d. Use the equation you found in part (c) to estimate when U.S. annual orange production was 100 million boxes.
  e. Compare the results you found in parts (b) and (d). Which equation was easier to use? Explain.
  f. Which equation is easier to use to predict U.S. orange production in 2018? Explain. Then find that production.

73. A person travels d miles at a constant speed s (in miles per hour) for t hours.
  a. Complete Table 12 to help find a formula that describes the relationship among s, t, and d. Show the arithmetic to help you see a pattern.

Table 12 Speed, Time, and Distance

Speed (miles per hour) s	Time (hours) t	Distance (miles) d
50	4	
70	3	
65	2	
55	5	
s	t	

  b. Perform a unit analysis of the formula you found in part (a).
  c. Solve the formula you found in part (a) for t.
  d. Use the formula you found in part (c) to find the value of t when $d = 315$ and $s = 70$. What does your result mean in this situation?
  e. A student plans to drive from South Louisiana Community College in Lafayette, Louisiana, to Hot Springs, Arkansas. The speed limit is 75 mph in Louisiana and 70 mph in Arkansas. The trip involves 254 miles of travel in Louisiana followed by 152 miles of travel in Arkansas. If the student drives at the posted speed limits, estimate the driving time.

Concepts

74. a. Solve the equation $mx + b = 0$, where $m \neq 0$, for x.
  b. Explain why a linear equation in one variable has exactly one solution.

75. An object travels d miles at a constant speed s (in miles per hour) for t hours. A student believes that the formula of s is

$$s = dt \quad \textit{Incorrect}$$

Perform a unit analysis to show that the formula is incorrect.

76. A student believes that a formula of the total value T (in dollars) of n objects, each worth v dollars, is

$$n = Tv \quad \textit{Incorrect}$$

Perform a unit analysis to show that the formula is incorrect.

77. a. How does doubling the width and doubling the length of a rectangle affect the perimeter of the rectangle?
  b. How does doubling the width and doubling the length of a rectangle affect the area of the rectangle?

78. a. How does increasing the length of a rectangle by 1 inch affect the perimeter (in inches) of the rectangle?
  b. How does increasing the length of a rectangle by 1 inch affect the area (in square inches) of the rectangle?

79. For a rectangle of length L, width W, and perimeter P, explain why the perimeter is given by $P = 2L + 2W$.

80. If n objects each have value v, then their total value T is given by $T = vn$. Give an example of this.

81. In what cases would you first solve a formula for a variable and then make substitutions for any other variables, and in what cases would you first make substitutions for all but one variable and then solve for the remaining variable?

82. Give an example of an equation in one variable and an example of a formula. How are these equations different?

Chapter Summary

Key Points of Chapter 4

Section 4.1   Simplifying Expressions

Commutative law for addition	$a + b = b + a$
Commutative law for multiplication	$ab = ba$
Term	A **term** is a constant, a variable, or a product of a constant and one or more variables raised to powers.
Associative law for addition	$a + (b + c) = (a + b) + c$
Associative law for multiplication	$a(bc) = (ab)c$
Distributive law	$a(b + c) = ab + ac$
Multiplying a number by -1	$-1a = -a$
Equivalent expressions	Two or more expressions are **equivalent expressions** if, when each variable is evaluated for *any* real number (for which all the expressions are defined), the expressions all give equal results.
Showing that an equation is false	To show that an equation is false, we need find only *one* set of numbers that, when substituted for any variables, gives a result of the form $a = b$, where a and b are different real numbers.

Section 4.2   Simplifying More Expressions

Like terms	**Like terms** are either constant terms or variable terms that contain the same variable(s) raised to exactly the same power(s).
Combining like terms	To combine like terms, add the coefficients of the terms and keep the same variable factors.
Simplifying an expression	We simplify an expression by removing parentheses and combining like terms.
Locating an error in simplifying an expression	After simplifying an expression, use a graphing calculator table to check whether the result is equivalent to the original expression. If the expressions are not equivalent, an error was made. To pinpoint the error, evaluate all the expressions by using the same number(s) and find the pair of consecutive expressions for which the results are different.

Section 4.3   Solving Linear Equations in One Variable

Linear equation in one variable	A **linear equation in one variable** is an equation that can be put into the form $mx + b = 0$, where m and b are constants and $m \neq 0$.
Solution, satisfy, solution set, and solve for an equation in one variable	A number is a **solution** of an equation in one variable if the equation becomes a true statement when the number is substituted for the variable. We say that the number **satisfies** the equation. The set of all solutions of the equation is called the **solution set** of the equation. We **solve** the equation by finding its solution set.
Roles of a variable	Here are three roles of a variable: 1. A variable represents a quantity that can vary. 2. In an expression, a variable is a placeholder for a number. 3. In an equation, a variable represents any number that is a solution of the equation.
Equivalent equations	**Equivalent equations** are equations that have the same solution set.
Addition property of equality	If A and B are expressions and c is a number, then the equations $A = B$ and $A + c = B + c$ are equivalent.
Checking results after solving an equation	After solving an equation, check that all your results satisfy the equation.

Section 4.3   Solving Linear Equations in One Variable (*Continued*)

Multiplication property of equality	If A and B are expressions and c is a nonzero number, then the equations $A = B$ and $Ac = Bc$ are equivalent.
Using graphing to solve an equation in one variable	To use graphing to solve an equation $A = B$ in one variable, x, where A and B are expressions, **1.** Graph the equations $y = A$ and $y = B$ in the same coordinate system. **2.** Find all intersection points. **3.** The x-coordinates of those intersection points are the solutions of the equation $A = B$.

Section 4.4   Solving More Linear Equations in One Variable

Solving an equation that contains fractions	A key step in solving an equation that contains fractions is to multiply both sides of the equation by the LCD so that there are no fractions on either side of the equation.
Number of solutions	Any linear equation in one variable has exactly one solution.
Conditional equation, inconsistent equation, and identity	There are three types of equations: **1.** A **conditional equation** is sometimes true and sometimes false, depending on which values are substituted for the variable(s). **2.** If an equation does not have any solutions, we call the equation **inconsistent** and say that the solution is the **empty set**. When we apply the usual steps to solve an inconsistent equation, the result is always a false statement. **3.** An equation that is true for all permissible values of the variable(s) it contains is called an **identity**. When we apply the usual steps to solve an identity, the result is always a true statement of the form $a = a$, where a is a constant.
Locating an error in solving a linear equation	If the result of an attempt to solve a linear equation in one variable does not satisfy the equation, an error was made. To pinpoint the error, substitute the result into all the equations in the work and find which pair of consecutive equations gives a false statement followed by a true statement.

Section 4.5   Comparing Expressions and Equations

Linear expression in one variable	A **linear expression in one variable** is an expression that can be put into the form $mx + b$, where m and b are constants and $m \neq 0$.
Results of simplifying an expression and solving an equation	The result of simplifying an expression is an expression. The result of solving a linear equation in one variable is a number.
Connection between simplifying an expression and an identity	If an expression A in one variable is simplified to an expression B, then every real number (for which both expressions are defined) is a solution of the equation $A = B$. In other words, the equation $A = B$ is an identity.
Multiplying expressions and both sides of equations by numbers	In simplifying an expression, the only number that we can multiply the expression or part of it by is 1. In solving an equation, we can multiply both sides of the equation by *any* number except 0.

Section 4.6   Formulas

Formula	A **formula** is an equation that contains two or more variables.
Finding a single value of a variable in a formula	To find a single value of a variable in a formula, we often substitute numbers for all the other variables and then solve for that one variable.
Perimeter	The **perimeter** of a polygon, which is a geometrical object such as a triangle, rectangle, or trapezoid, is the total distance around the object.
Area and perimeter of a rectangle	For a rectangle with length L, width W, area A, and perimeter P, • $A = LW$ **Area formula** • $P = 2L + 2W$ **Perimeter formula**
Total-value formula	If n objects each have value v, then their total value T is given by $T = vn$.
Finding several values of a variable in a formula	To find several values of a variable in a formula, we usually solve the formula for that variable before making any substitutions.

Chapter 4 Review Exercises

1. Use the commutative law for addition to write the expression $5w + 9$ in another form.

2. Use the commutative law for multiplication to write the expression $8 + wp$ in another form.

Use an associative law to write the expression in another form.

3. $2 + (k + y)$ 4. $(bx)w$

Simplify. Use a graphing calculator table to verify your work when possible.

5. $-5(4x)$ 6. $-3(8x + 4)$

7. $\dfrac{4}{5}(15y - 35)$ 8. $-(3x - 6y - 8)$

9. $\dfrac{2}{9}x + \dfrac{5}{9}x$ 10. $5a + 2 - 13b - a + 4b$

11. $-5y - 3(4x + y) - 6x$ 12. $-2.6(3.1x + 4.5) - 8.5$

13. $-(2m - 4) - (3m + 8)$ 14. $4(3a - 7b) - 3(5a + 4b)$

For Exercises 15 and 16, let x be a number. Translate the English phrase into a mathematical expression. Then simplify the expression.

15. -4 times the difference of the number and 7

16. -7, plus 2 times the sum of the number and 8

17. Use values for $a, b,$ and c to show that $a(b + c) = ab + c$ is false. Then write a true statement that involves $a(b + c)$.

18. Give three examples of expressions that are equivalent to $3x - 9$.

19. Which of the following expressions are equivalent?
$$-5x - 20 \quad -5(x - 4) \quad 5(4 - x) \quad -5x - 4$$
$$-2(x - 10) - 3x \quad -5x + 20$$

For Exercises 20 and 21, graph the equation by hand.

20. $y = 2x + 3 - 4x$ 21. $y = -3(x - 2)$

22. Determine whether 3 is a solution of the linear equation $2 - 5x = -3(4x - 7)$.

For Exercises 23–33, solve. Use a graphing calculator to verify your result.

23. $a + 5 = 12$ 24. $-4x = 20$

25. $-p = -3$ 26. $-\dfrac{7}{3}a = 14$

27. $4.5x - 17.2 = -5.05$ 28. $5x - 9x + 3 = 17$

29. $8m - 3 - m = 2 - 4m$ 30. $8x = -7(2x - 3) + x$

31. $6(4x - 1) - 3(2x + 5) = 2(5x - 3)$

32. $\dfrac{w}{8} - \dfrac{3}{4} = \dfrac{5}{6}$ 33. $\dfrac{3p - 4}{2} = \dfrac{5p + 2}{4} + \dfrac{7}{6}$

34. A student tries to solve the equation $x - 5 = 2$:
$$x - 5 = 2$$
$$x - 5 + 5 = 2$$
$$x + 0 = 2$$
$$x = 2$$

Describe any errors. Then solve the equation correctly.

35. Give three examples of equations that have the solution -6.

36. Solve $-2.5(3.8x - 1.9) = 83.7$. Round the result to two decimal places.

37. The number of gun seizures by the New York Police Department was 1231 in 2015 and has decreased by about 74 gun seizures per year since then (Source: *NYPD*). Let n be the number of gun seizures in the year that is t years since 2015.
   a. Find an equation of a model.
   b. Predict in which year there will be 700 gun seizures.
   c. Predict the number of gun seizures in 2020.

For Exercises 38 and 39, find the number.

38. 4 times the difference of a number and 6 is 15.

39. 2, minus 3 times the sum of a number and 8 is 95.

40. Use "intersect" on a graphing calculator to solve the equation $\dfrac{1}{2}x + \dfrac{5}{3} = -\dfrac{2}{3}x - \dfrac{1}{4}$. Round the solution to the second decimal place.

For Exercises 41–44, solve the given equation by referring to the solutions of $y = -2x + 17$ and $y = 5x - 4$ shown in Tables 13 and 14.

41. $-2x + 17 = 5x - 4$ 42. $-2x + 17 = 15$

43. $5x - 4 = 6$ 44. $5x - 4 = -4$

Table 13 Solutions of $y = -2x + 17$	
x	y
0	17
1	15
2	13
3	11
4	9

Table 14 Solutions of $y = 5x - 4$	
x	y
0	-4
1	1
2	6
3	11
4	16

For Exercises 45–47, solve the equation. State whether the equation is a conditional equation, an inconsistent equation, or an identity.

45. $7x - 4 + 3x = 2 + 10x - 6$

46. $6(2x - 3) - (5x + 2) = -2(4x - 1)$

47. $2(x - 5) + 3 = 2x - 4$

48. A student incorrectly solves $4(x - 5) = 28$:
$$4(x - 5) = 28$$
$$4x - 20 = 28$$
$$4x - 20 - 20 = 28 - 20$$
$$4x = 8$$
$$\dfrac{4x}{4} = \dfrac{8}{4}$$
$$x = 2$$

Substitute 2 for x in each of the six equations, and explain how the results help you pinpoint the error. Describe the error. Then solve the equation correctly.

Is the following a linear expression or a linear equation?

49. $8 - 3(x + 5)$ 50. $8 - 3(x + 5) = 4x$

For Exercises 51–58, simplify the expression or solve the equation, as appropriate.

51. $6t - 8t$ 52. $0.1 + 0.5a - 0.3a = 0.7$

53. $6t - 8t = 10$

54. $0.1 + 0.5a - 0.3a$

55. $9(2p - 5) - 3(7p + 3)$

56. $\dfrac{5}{6}r - \dfrac{3}{4} = \dfrac{1}{6} + \dfrac{7}{2}r$

57. $9(2p - 5) - 3(7p + 3) = 0$

58. $\dfrac{5}{6}r - \dfrac{3}{4} - \dfrac{1}{6} + \dfrac{7}{2}r$

59. A student is trying to simplify an expression. The student writes, "The solution of the expression is 2." Is the student correct? Explain.

60. A student tries to simplify the expression $\dfrac{2}{3}x + \dfrac{7}{5}$:

$$\frac{2}{3}x + \frac{7}{5} = 15\left(\frac{2}{3}x + \frac{7}{5}\right)$$
$$= 15 \cdot \frac{2}{3}x + 15 \cdot \frac{7}{5}$$
$$= 10x + 21$$

Describe any errors. Then simplify the expression correctly.

For Exercises 61 and 62, let x be a number. Translate the English phrase or sentence into an expression or an equation. Then simplify the expression or solve the equation, as appropriate.

61. 4 times the difference of 6 and the number is 17.

62. The number minus the quotient of the number and 2

63. Find a formula of the perimeter P of the polygon shown in Fig. 40.

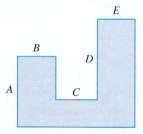

Figure 40 Exercise 63

64. a. Find a formula of the total cost T (in dollars) of n tickets that sell for $15 per ticket and w tickets that sell for $25 per ticket.

b. Use the formula you found in part (a) to find the value of w when $n = 370$ and $T = 11{,}050$. What does your result mean in this situation?

For Exercises 65–68, solve the formula for the specified variable.

65. $C = 2\pi r$, for r

66. $P = a + b + c$, for c

67. $3x - 6y = 18$, for y

68. $A = \dfrac{1}{2}H(B + T)$, for T

69. The number of Starbucks stores worldwide was 23.0 thousand in 2015 and has increased by about 1.5 thousand stores per year since then (Source: *Starbucks*). Let n be the number (in thousands) of Starbucks stores worldwide at t years since 2015.

   a. Find an equation of a model to describe the situation.
   b. Solve the equation you found in part (a) for t.
   c. Use the equation you found in part (a) to predict when there will be 32 thousand Starbucks stores.
   d. Use the equation you found in part (b) to predict when there will be 32 thousand Starbucks stores.
   e. Compare the results you found in parts (c) and (d). Which equation was easier to use? Explain.
   f. Which equation is easier to use to predict the number of Starbucks stores in 2022? Explain. Then find that number of stores.

70. The average salary of public elementary school teachers was $57,230 in 2015 and has increased by about $840 per year since then (Source: *National Education Association*). Predict when the average salary will be $63,000.

71. In 2014, the total attendance at Seaworld in Orlando, Florida, was 4.68 million, down 8% from 2013 (Source: *Seaworld*). What was the total attendance in 2013?

Chapter 4 Test

1. Use the commutative law for addition to write the expression $4 + 3p$ in another form.

2. Use an associative law to write the expression $3(xy)$ in another form.

For Exercises 3–6, simplify.

3. $-\dfrac{2}{3}(6x - 9)$

4. $9.36 - 2.4(1.7x + 3.5)$

5. $-5(2w - 7) - 3(4w - 6)$

6. $-(3a + 7b) - (8a - 4b + 2)$

7. A student incorrectly simplifies $3(x - 2) - (5x + 4)$:

Expression	Evaluate the Expression for $x = 3$
$3(x - 2) - (5x + 4)$	
$= 3x - 6 - 5x + 4$	
$= 3x - 5x - 6 + 4$	
$= -2x - 2$	

Evaluate each of the four expressions for $x = 3$ to pinpoint where the error was made. Then simplify $3(x - 2) - (5x + 4)$ correctly.

8. Graph $y = -2(x + 1)$ by hand.

For Exercises 9–14, solve.

9. $6x - 3 = 19$

10. $\dfrac{3}{5}x = 6$

11. $9a - 5 = 8a + 2$

12. $8 - 2(3t - 1) = 7t$

13. $3(2x - 5) - 2(7x + 9) = 49$

14. $\dfrac{7}{8}x + \dfrac{3}{10} = \dfrac{1}{4}x - \dfrac{1}{2}$

15. Solve $8.21x = 3.9(4.4x - 2.7)$. Round your result to two decimal places.

16. 4, minus 2 times the sum of a number and 7 is 54. Find the number.

For Exercises 17 and 18, simplify the expression or solve the equation, as appropriate.

17. $9(3x + 2) - (4x - 6)$ **18.** $9(3x + 2) - (4x - 6) = x$

19. A student believes that $x - 3$ is the solution of an equation. Is the student correct? Explain.

20. Give three examples of an expression equivalent to the expression 4.

For Exercises 21 and 22, let x be a number. Translate the following into an expression or an equation, as appropriate. Then simplify the expression or solve the equation.

21. 5 times the difference of the number and 2 is 29.

22. 2, plus 4 times the sum of 3 and the number

For Exercises 23–26, solve the given equation by referring to the graphs of $y = \frac{3}{2}x - 4$ and $y = \frac{1}{2}x - 2$ shown in Fig. 41.

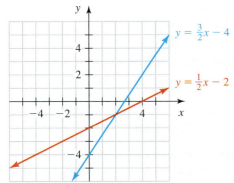

Figure 41 Exercises 23–26

23. $\frac{3}{2}x - 4 = \frac{1}{2}x - 2$ **24.** $\frac{3}{2}x - 4 = 2$

25. $\frac{1}{2}x - 2 = -3$ **26.** $\frac{1}{2}x - 2 = 0$

27. The number of U.S. patent applications was 630 thousand in 2015 and increases by about 22.8 thousand per year since then (Source: *U.S. Patent and Trademark Office*). Let n be the number of patent applications (in thousands) during the year that is t years since 2015.
   a. Find an equation of a model.
   b. Predict the number of patent applications in 2021.
   c. Predict when there will be 750 thousand patent applications.

28. A person has $8350 in a savings account. If the person withdraws $125 per month, after how many months will there be $6975 in the account?

29. In 2015, the number of complaints of online crime was 288 thousand complaints, up 7.1% from 2014 (Source: *Internet Crime Complaint Center*). What was the number of complaints in 2014?

30. A person plans to enclose a rectangular garden with fencing. If the width of the garden is 8 feet, and if 52 feet of fencing is required to enclose the garden, what is the length of the garden?

31. Solve the formula $A = \dfrac{a + b}{2}$ for the variable a.

Cumulative Review of Chapters 1–4

1. Choose a variable name for the number of pages in a book. Give two numbers that the variable can represent and two numbers that it cannot represent.

2. Graph the numbers -3, $-\frac{5}{2}$, 1, and $\frac{3}{2}$ on one number line.

3. Let n be the average number (in millions) of unique monthly visitors to the website YouTube in the year that is t years since 2010. What does the ordered pair $(5, 170.7)$ represent (Source: *YouTube*)?

For Exercises 4–7, perform the indicated operations by hand. Then use a calculator to check your work.

4. $4 + 3(-2)$

5. $-8 \div 4 - 2(7 - 10)$

6. $\dfrac{15}{8} \cdot \left(-\dfrac{4}{25}\right)$

7. $\left(-\dfrac{3}{10}\right) + \left(-\dfrac{7}{8}\right)$

8. Simplify $\dfrac{27}{-45}$.

9. A rectangle has a width of $\frac{3}{4}$ foot and a length of $\frac{5}{6}$ foot. What is the perimeter of the rectangle?

10. A student scores 92 points on one test and 85 points on the next test. What is the change in the scores?

11. Evaluate the expression $a(b - c)$ for $a = -3$, $b = 5$, and $c = -4$.

For Exercises 12 and 13, find the slope of the line passing through the two given points. State whether the line is increasing, decreasing, horizontal, or vertical.

12. $(-5, -2)$ and $(-1, -4)$

13. $(-4, -5)$ and $(-4, 3)$

14. Road A climbs steadily for 150 feet over a horizontal distance of 5000 feet. Road B climbs steadily for 95 feet over a horizontal distance of 3500 feet. Which road is steeper? Explain.

For Exercises 15–18, graph the equation by hand.

15. $y = -\dfrac{2}{3}x + 4$

16. $y = 2x - 3$

17. $x = -5$

18. $y = 3$

19. Find an equation of the line sketched in Fig. 42.

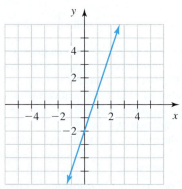

Figure 42 Exercise 19

20. A person's savings account balance decreased from $7500 to $2840 in 5 months. Find the approximate rate of change of the balance.

21. The number of deaths from patients' adverse reactions to drugs increased approximately steadily from 82.7 thousand deaths in 2010 to 123.9 thousand deaths in 2014 (Source: *FDA*). Find the approximate rate of change of the number of deaths.

22. The homeownership rate among people age 35 and under was 34.6% in 2015 and has decreased by about 0.8 percentage point per year since then (Source: *U.S. Census Bureau*). Let r be the homeownership rate at t years since 2015.
   a. Find the slope of the linear model that describes the situation. What does it mean in this situation?
   b. Find an equation of the model.
   c. Predict the homeownership rate among people age 35 and under in 2021.
   d. Predict when the homeownership rate among people age 35 and under will be 29%.

23. Some solutions of four equations are listed in Table 15. Which of the equations could be linear?

Table 15 Solutions of Four Equations

Equation 1		Equation 2		Equation 3		Equation 4	
x	y	x	y	x	y	x	y
0	11	0	56	3	35	1	1
1	13	1	53	4	44	2	1
2	16	2	50	5	53	3	1
3	20	3	47	6	62	4	1
4	25	4	44	7	71	5	1

24. The median sales prices of homes are shown in Table 16 for various years.

Table 16 Median Sales Prices of Homes

Year	Median Sales Price (thousands of dollars)
2010	221.8
2011	227.2
2012	245.2
2013	268.9
2014	282.8
2015	296.4

Source: *U.S. Census Bureau*

Let p be the median sales price (in thousands of dollars) of homes at t years since 2010. A model of the situation is $p = 16.1t + 216.8$.
   a. Use a graphing calculator to draw a scatterplot and the model in the same viewing window. Does the line come close to the data points?
   b. What is the slope? What does it mean in this situation?
   c. What is the p-intercept? What does it mean in this situation?
   d. When will the median sales price of homes be $400 thousand?
   e. Predict the median sales price of homes in 2020.
   f. The median household incomes in 2010 and 2015 were $54.4 thousand and $55.8 thousand, respectively. Compute the unit ratio of the median sales price of homes to the median household income for 2010. Also compute the unit ratio for 2015. Compare the two unit ratios. What does this comparison mean in this situation?

For Exercises 25–30, simplify the expression or solve the equation, as appropriate. Use a graphing calculator to verify your work.

25. $3r + 4 = 7r - 8$

26. $2(3x - 2) = 4(3x + 5) - 3x$

27. $4a - 5b + 6 - 2b - 7a$

28. $7 - 2(3p - 5) + 5(4p - 2)$

29. $\dfrac{2}{3}r - \dfrac{5}{6} = \dfrac{1}{2}$

30. $-(2a + 5) - (4a - 1)$

31. Solve $25.93 - 7.6(2.1x + 8.7) = 53.26$. Round the result to the second decimal place.

For Exercises 32 and 33, let x be a number. Translate the following an expression or an equation, as appropriate. Then simplify the expression or solve the equation.

32. The number, plus 9 times the quotient of the number and 3

33. 2 times the difference of 7 and twice the number is 87.

Solve the formula for the specified variable.

34. $A = 2\pi rh$, for h

35. $4x - 6y = 12$, for y

Linear Equations in Two Variables and Linear Inequalities in One Variable

5

Are many of the married couples you know happily married? The percentages of married persons who say they are "very happy" with their marriages are shown in Table 1 for various years. In Exercise 7 of Homework 5.4, you will predict when 59.5% of married persons will say they are very happy with their marriages.

In Chapter 4, we discussed how to simplify expressions and solve equations. In this chapter, we will use these skills to increase our effectiveness in graphing linear equations, finding equations of lines, and modeling authentic situations such as the Academy Award nominations and wins for best picture. Finally, we will discuss how to use a linear model to predict when a quantity will be more than (or less than) a certain amount.

Table 1 Percentages of Married Persons "Very Happy" with Their Marriages

Year	Percent
1970	67.0
1980	67.5
1990	64.6
2000	62.4
2010	63.0
2014	59.9

Source: *Institute for American Values*

5.1 Graphing Linear Equations in Two Variables

Objectives

» Graph a linear equation by solving for *y*.

» Graph a linear equation by finding the intercepts of its graph.

» Find coordinates of solutions of a linear equation.

» Describe the difference between equations in one variable and equations in two variables.

In this section, we will discuss how to graph linear equations by three methods: solving for *y*, finding intercepts, and finding three points.

Graphing Equations by Solving for *y*

In Section 3.4, we discussed how to use slope to graph an equation in the slope–intercept form $y = mx + b$. In Example 1, we will discuss how to graph equations that are not in this form, but that can be put into it.

Example 1   Graphing an Equation by Solving for *y*

Sketch the graph of $2x + 3y = 9$.

Solution

We will put the equation in $y = mx + b$ form so that we can use the *y*-intercept and the slope to help us graph the equation. To begin, we get *y* alone on one side of the equation:

$$2x + 3y = 9 \qquad \textit{Original equation}$$

$$2x + 3y - 2x = 9 - 2x \qquad \textit{Subtract 2x from both sides to get 3y alone on left-hand side.}$$

$$3y = -2x + 9 \qquad \textit{Combine like terms; rearrange right-hand side.}$$

$$\frac{3y}{3} = \frac{-2x}{3} + \frac{9}{3} \qquad \textit{Divide both sides by 3 to get y alone on left-hand side.}$$

$$y = -\frac{2}{3}x + 3 \qquad \textit{Simplify.}$$

Because $2x + 3y = 9$ can be put into the form $y = mx + b$ (as $y = -\frac{2}{3}x + 3$), we know

that the graph of $2x + 3y = 9$ is a line. The y-intercept is $(0, 3)$, and the slope is $-\frac{2}{3}$.

The graph is shown in Fig. 1.

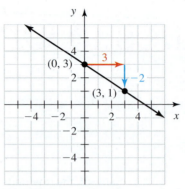

Figure 1   Graph of $2x + 3y = 9$

We can verify our result by checking that both $(0, 3)$ and $(3, 1)$ are solutions of $2x + 3y = 9$.

▶

Before we can use the y-intercept and the slope to graph a linear equation, we must solve for y to put the equation into the form $y = mx + b$.

WARNING   It is a common error to think the slope of an equation such as $2x + 3y = 9$ is 2

because 2 is the coefficient of x. However, we must solve for y (and get $y = -\frac{2}{3}x + 3$)

to determine that the slope is $-\frac{2}{3}$ (see Example 1).

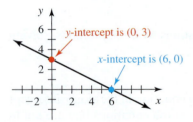

Figure 2   Intercepts of a line

Finding Intercepts

Another way to graph a linear equation is to use the intercepts of its graph. In Fig. 2, we show the x-intercept and y-intercept of a line.

Recall from Section 1.3 that for an x-intercept, the y-coordinate is 0 (see Fig. 2). For a y-intercept, the x-coordinate is 0 (see Fig. 2).

Intercepts of the Graph of an Equation

For an equation containing the variables x and y,

- To find the x-coordinate of each x-intercept, substitute 0 for y and solve for x.
- To find the y-coordinate of each y-intercept, substitute 0 for x and solve for y.

Example 2   Graphing an Equation by Using Intercepts

Consider the equation $5y - 2x = 10$.

1. Find the x-intercept.
2. Find the y-intercept.
3. Sketch the graph of the equation.

Solution

1. To find the x-intercept, we substitute 0 for y and solve for x:

$$5y - 2x = 10 \qquad \textit{Original equation}$$
$$5(0) - 2x = 10 \qquad \textit{Substitute 0 for y.}$$
$$-2x = 10 \qquad \textit{Simplify.}$$
$$\frac{-2x}{-2} = \frac{10}{-2} \qquad \textit{Divide both sides by } -2.$$
$$x = -5 \qquad \textit{Simplify.}$$

The x-intercept is $(-5, 0)$.

2. To find the y-intercept, we substitute 0 for x and solve for y:

$$5y - 2x = 10 \qquad \textit{Original equation}$$
$$5y - 2(0) = 10 \qquad \textit{Substitute 0 for x.}$$
$$5y = 10 \qquad \textit{Simplify.}$$
$$\frac{5y}{5} = \frac{10}{5} \qquad \textit{Divide both sides by 5.}$$
$$y = 2 \qquad \textit{Simplify.}$$

The y-intercept is $(0, 2)$.

3. The equation $5y - 2x = 10$ can be put into the form $y = mx + b$ (as $y = \frac{2}{5}x + 2$; try it). So, the graph of the equation $5y - 2x = 10$ is a line.

Before sketching the line, we find another solution to check our work. Here we substitute 3 for x and solve for y:

$$5y - 2x = 10 \qquad \textit{Original equation}$$
$$5y - 2(3) = 10 \qquad \textit{Substitute 3 for x.}$$
$$5y - 6 = 10 \qquad \textit{Multiply.}$$
$$5y = 16 \qquad \textit{Add 6 to both sides.}$$
$$y = \frac{16}{5} \qquad \textit{Divide both sides by 5.}$$

So, $\left(3, \dfrac{16}{5}\right)$ is a solution.

Finally, in Table 2, we list the three solutions of $5y - 2x = 10$ that we found, and in Fig. 3 we sketch the graph. We can verify our work by checking that the ordered pairs $(-5, 0)$ and $(0, 2)$ satisfy the equation $5y - 2x = 10$.

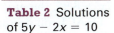

Table 2 Solutions of $5y - 2x = 10$

x	y
-5	0
0	2
3	$\dfrac{16}{5}$

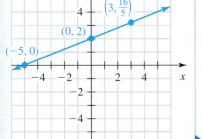

Figure 3 Graph of $5y - 2x = 10$

In Example 2, we used the intercepts of the line $5y - 2x = 10$ and another point on the line to graph the equation $5y - 2x = 10$.

Using Intercepts to Graph an Equation

To graph a linear equation whose graph has exactly two intercepts,

1. Find the intercepts.

2. Plot the intercepts and a third point on the line, and graph the line that contains the three points.

When working with models, we will sometimes want to find intercepts that have decimal coordinates.

Example 3   Finding Approximate Intercepts

Find the indicated intercept of the graph of $y = -2.67x + 8.95$. Round the coordinates to the second decimal place.

1. x-intercept   **2.** y-intercept

Solution

1. To find the x-intercept, we substitute 0 for y and solve for x:

$$
\begin{array}{ll}
y = -2.67x + 8.95 & \textit{Original equation} \\
0 = -2.67x + 8.95 & \textit{Substitute 0 for y.} \\
0 + 2.67x = -2.67x + 8.95 + 2.67x & \textit{Add 2.67x to both sides.} \\
2.67x = 8.95 & \textit{Combine like terms.} \\
\dfrac{2.67x}{2.67} = \dfrac{8.95}{2.67} & \textit{Divide both sides by 2.67.} \\
x \approx 3.35 & \textit{Round right-hand side to second decimal place.}
\end{array}
$$

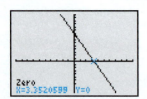

Figure 4 Verify the work

The approximate x-intercept is $(3.35, 0)$. We use "zero" on a graphing calculator to verify our work (see Fig. 4). See Appendix A.19 for graphing calculator instructions.

2. For an equation of the form $y = mx + b$, as is $y = -2.67x + 8.95$, the graph has y-intercept $(0, b)$. So, the y-intercept is $(0, 8.95)$. We use TRACE on a graphing calculator to verify our work (see Fig. 5). See Appendix A.5 for graphing calculator instructions.

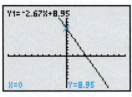

Figure 5 Verify the work

Finding Coordinates of Solutions

In Section 3.1, we graphed equations of the form $y = mx + b$ by plotting points. After learning how to solve equations in one variable in Chapter 4, we can now graph a linear equation in any form by plotting points.

Example 4   Graphing an Equation by Plotting Points

Complete the following steps to graph the equation $3x + 2y = 8$:

1. Find y when $x = 4$.
2. Find x when $y = 1$.
3. Sketch the graph of $3x + 2y = 8$.

Solution

1. We substitute 4 for x in the equation $3x + 2y = 8$ and solve for y:

$$
\begin{array}{ll}
3x + 2y = 8 & \textit{Original equation} \\
3(4) + 2y = 8 & \textit{Substitute 4 for x.} \\
12 + 2y = 8 & \textit{Multiply.} \\
12 + 2y - 12 = 8 - 12 & \textit{Subtract 12 from both sides.} \\
2y = -4 & \textit{Combine like terms.} \\
\dfrac{2y}{2} = \dfrac{-4}{2} & \textit{Divide both sides by 2.} \\
y = -2 & \textit{Simplify.}
\end{array}
$$

So, $y = -2$ when $x = 4$. The ordered pair $(4, -2)$ is a solution.

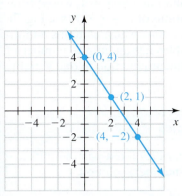

Figure 6 Graph of $3x + 2y = 8$

2. We substitute 1 for y in the equation $3x + 2y = 8$ and solve for x:

$$3x + 2y = 8 \qquad \textit{Original equation}$$
$$3x + 2(1) = 8 \qquad \textit{Substitute 1 for y.}$$
$$3x + 2 = 8 \qquad \textit{Multiply.}$$
$$3x + 2 - 2 = 8 - 2 \qquad \textit{Subtract 2 from both sides.}$$
$$3x = 6 \qquad \textit{Combine like terms.}$$
$$\frac{3x}{3} = \frac{6}{3} \qquad \textit{Divide both sides by 3.}$$
$$x = 2 \qquad \textit{Simplify.}$$

So, $x = 2$ when $y = 1$. The ordered pair $(2, 1)$ is a solution.

3. First, we find that the y-intercept is $(0, 4)$. (Try it.) Then we plot the points $(4, -2)$, $(2, 1)$, and $(0, 4)$ and sketch the line that contains them (see Fig. 6).

Using Three Solutions to Graph a Linear Equation

To graph any linear equation,

1. Find three solutions of the equation.
2. Plot the three solutions, and graph the line that contains them.

Examples 1, 2, and 4 show three ways to graph an equation in which y is not alone on one side of the equation. For an equation in which y is alone on one side, the y-intercept and the slope are usually best to use to graph the equation.

Comparing Equations in One and Two Variables

In Example 5, we will compare solutions of an equation in one variable with solutions of an equation in two variables.

Example 5   Comparing Equations in One and Two Variables

1. Describe the solution(s) of the equation $y = x + 1$, an equation in two variables.
2. Describe the solution(s) of the equation $3 = x + 1$, an equation in one variable.

Solution

1. The equation $y = x + 1$ is a linear equation in *two* variables, x and y. We can use the graph of the equation (a line) to describe the solutions (see Fig. 7). Because there is an infinite number of points on the line, there is an infinite number of solutions of the equation $y = x + 1$.

2. The equation $3 = x + 1$ is a linear equation in *one* variable, x. So, exactly one number is a solution:

$$3 = x + 1 \qquad \textit{Original equation}$$
$$3 - 1 = x + 1 - 1 \qquad \textit{Subtract 1 from both sides.}$$
$$2 = x \qquad \textit{Combine like terms.}$$

The solution is the number 2.

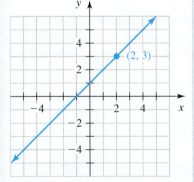

Figure 7 Graph of $y = x + 1$

There is a connection between the equations $y = x + 1$ and $3 = x + 1$. If we substitute 3 for y in the equation $y = x + 1$, the result is the equation $3 = x + 1$, whose solution is $x = 2$. So, $(2, 3)$ is one of an infinite number of solutions of $y = x + 1$ (see Fig. 7).

The graph of any linear equation in two variables is a line, which contains an infinite number of points. So, a linear equation in two variables has an infinite number of (ordered-pair) solutions. Recall from Section 4.4 that a linear equation in one variable has exactly one (real-number) solution.

Comparing Linear Equations in One and Two Variables

A linear equation in two variables has an infinite number of (ordered-pair) solutions. A linear equation in one variable has exactly one (real-number) solution.

When modeling, we use an equation in two variables to describe the relationship between two quantities. After substituting a value for one of the variables, we solve the equation in one variable to make an estimate or prediction.

Group Exploration

Section Opener: Graphing linear equations

1. a. Solve the equation $3x + 4y = 8$ for y.
   b. Use the result you found in part (a) to help you graph $3x + 4y = 8$ by hand.

2. a. Solve the equation $5x - 3y - 9 = 0$ for y.
   b. Use the result you found in part (a) to help you graph $5x - 3y - 9 = 0$ by hand.

Group Exploration

Comparing two graphing techniques

In this exploration, you will compare two methods of graphing an equation.

1. Consider the equation $4x + 5y = 20$.
   a. Graph the equation by finding the slope and the y-intercept.
   b. Graph the equation by finding both intercepts.
   c. Compare the graphs you sketched in parts (a) and (b). Decide which method you prefer for graphing the equation $4x + 5y = 20$.

2. Consider the equation $y = \frac{2}{5}x - 1$.

   a. Graph the equation by finding the slope and the y-intercept.
   b. Graph the equation by finding both intercepts.
   c. Compare the graphs you sketched in parts (a) and (b). Decide which method you prefer for graphing the equation $y = \frac{2}{5}x - 1$.

3. In general, which types of linear equations do you prefer to graph by using the slope and the y-intercept? Which types do you prefer to graph by finding the intercepts? Explain.

Tips for Success   Keep a Positive Attitude

Sometimes when students have difficulty in doing mathematics, they take this as evidence that they cannot learn it. However, getting stumped when trying to solve a math problem is something everyone has experienced—even your math instructor! Students who are successful at mathematics realize this is part of the process of learning the subject. When they have trouble learning something, they keep a positive attitude and continue working hard, knowing they will eventually learn the material.

Homework 5.1

For extra help ▶   **MyLab Math**   ⊞ Watch the videos in MyLab Math

Determine the slope and the y-intercept. Use the slope and the y-intercept to graph the equation by hand. Use a graphing calculator to verify your work.

1. $y = 2x - 3$ **2.** $y = 4x - 3$

3. $y = -3x + 5$ **4.** $y = -4x + 1$

5. $y = -\dfrac{3}{5}x - 2$ **6.** $y = -\dfrac{2}{3}x - 1$

7. $y = \dfrac{1}{2}x + 2$ **8.** $y = \dfrac{1}{3}x + 4$

9. $y = -3x$ **10.** $y = 2x$

11. $y = -4$ **12.** $y = 0$

13. $y + x = 3$ **14.** $y + x = 1$

15. $y + 2x = 4$ **16.** $y + 3x = -2$

17. $y - 2x = -1$ **18.** $y - 3x = 2$

19. $3y = 2x$ **20.** $2y = -5x$

21. $2y = -3x - 4$ **22.** $3y = -4x + 6$

23. $5y = 4x - 15$ **24.** $4y = 7x - 20$

25. $3x - 4y = 8$ **26.** $4x + 3y = 9$

27. $6x - 15y = 30$ **28.** $6x - 8y = 16$

29. $x + 4y = 4$ **30.** $-x + 5y = -20$

31. $-4 = x + 2y$ **32.** $9 = x + 3y$

33. $4x + y + 2 = 0$ **34.** $2x - y - 3 = 0$

35. $6x - 4y + 8 = 0$ **36.** $15x + 12y - 36 = 0$

37. $0 = 5x + 3y$ **38.** $0 = 4x - 6y$

39. $y - 3 = 0$ **40.** $y + 5 = 0$

Find the x-intercept and y-intercept. Then graph the equation by hand.

41. $x - 3y = 6$ **42.** $2x - y = 6$

43. $15 = 3x + 5y$ **44.** $20 = 4x - 5y$

45. $2x - 3y + 12 = 0$ **46.** $2x - 4y - 8 = 0$

47. $y = -3x + 6$ **48.** $y = x + 2$

49. $\dfrac{1}{2}x - \dfrac{1}{3}y = 2$ **50.** $\dfrac{1}{4}x - \dfrac{1}{2}y = -1$

51. $\dfrac{x}{3} + \dfrac{y}{5} = 1$ **52.** $\dfrac{x}{4} + \dfrac{y}{3} = 1$

Find the approximate x-intercept and the approximate y-intercept. Round the coordinates to the second decimal place.

53. $6.2x + 2.8y = 7.5$ **54.** $8.1x + 3.9y = 14.2$

55. $6.62x - 3.91y = -13.55$ **56.** $-7.29x + 4.72y = 26.36$

57. $y = -4.5x + 9.32$ **58.** $y = 3.5x - 4.8$

59. $y = -2.49x - 37.21$ **60.** $y = -8.79x - 92.58$

For Exercises 61–68, graph the equation by hand. Use any method.

61. $2x - y = 5$ **62.** $3x + y = 1$

63. $3y = 4x - 3$ **64.** $5y = 3x + 10$

65. $4y - 3x = 0$ **66.** $2y + 5x = 0$

67. $2x - 3y - 12 = 0$ **68.** $3x + 4y + 24 = 0$

69. Consider the equation $6x + 5y = -13$:
  a. Find y when $x = -3$.
  b. Find x when $y = -5$.
  c. Use the results you found in parts (a) and (b) to help you graph the equation $6x + 5y = -13$ by hand.

70. Consider the equation $y = -2x + 4$:
  a. Find y when $x = 3$.
  b. Find x when $y = 6$.
  c. Use the results you found in parts (a) and (b) to help you graph the equation $y = -2x + 4$ by hand.

Concepts

71. A student thinks the graph of $3y + 2x = 6$ has slope 2 because the coefficient of the $2x$ term is 2. Is the student correct? If yes, explain why. If no, find the slope.

72. A student says the y-intercept of the line $2y = 3x + 4$ is $(0, 4)$ because the constant term is 4. Is the student correct? Explain.

73. Recall that we can describe some or all the solutions of an equation in two variables with an equation, a table, a graph, or words.
  a. Describe the solutions of the equation $3x - 5y = 10$ by using a graph.
  b. Describe three solutions of the equation $3x - 5y = 10$ by using a table.
  c. Describe the solutions of the equation $3x - 5y = 10$ by using words.

74. Recall that we can describe some or all the solutions of an equation in two variables with an equation, a table, a graph, or words.
  a. Describe the solutions of the equation $3x + 4y = 8$ by using a graph.
  b. Describe three solutions of the equation $3x + 4y = 8$ by using a table.
  c. Describe the solutions of the equation $3x + 4y = 8$ by using words.

75. a. Find the intercepts of the line $\dfrac{x}{5} + \dfrac{y}{7} = 1$.

  b. Find the intercepts of the line $\dfrac{x}{4} + \dfrac{y}{6} = 1$.

  c. Find the intercepts of the line $\dfrac{x}{a} + \dfrac{y}{b} = 1$, where a and b are nonzero constants. [**Hint:** Do steps similar to the ones you did in parts (a) and (b).]

  d. Find an equation of a line whose x-intercept is $(2, 0)$ and whose y-intercept is $(0, 5)$. [**Hint:** See your work in part (c).]

76. a. Find the slope of the line $3x + 5y = 7$.
   b. Find the slope of the line $2x + 7y = 3$.
   c. Find the slope of the line $ax + by = c$, where a, b, and c are constants and b is nonzero. [**Hint:** Do steps similar to the ones you did in parts (a) and (b).]

77. For an equation containing the variables x and y, why do we substitute 0 for y to find the x-intercept?

78. a. Solve the equation $2y - 6 = 4x$ for y.
   b. Find two solutions of the equation $y = 2x + 3$.
   c. Show that the two ordered pairs you found in part (b) satisfy both of the equations $2y - 6 = 4x$ and $y = 2x + 3$.
   d. Explain why it makes sense that the graphs of $2y - 6 = 4x$ and $y = 2x + 3$ are the same.

79. If the graph of a linear equation has a defined slope, describe how to find the slope and the y-intercept, and use this information to graph the equation. (See page 9 for guidelines on writing a good response.)

80. If the graph of a linear equation has exactly two intercepts, describe how to find the intercepts, and use the intercepts to graph the equation. (See page 9 for guidelines on writing a good response.)

Related Review

For Exercises 81–88, refer to the graph sketched in Fig. 8.

81. Find y when $x = 6$.
82. Find y when $x = -2$.
83. Find x when $y = -1$.
84. Find x when $y = -5$.

85. Find the x-intercept.
86. Find the y-intercept.
87. Find the slope of the line.
88. Find the equation of the line.

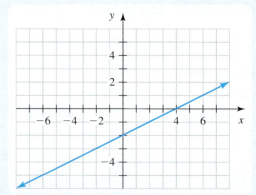

Figure 8 Exercises 81–88

Expressions, Equations, and Graphs

Perform the indicated instruction. Then use words such as linear, one variable, *and* two variables *to describe the expression or equation. For instance, to describe* $2x = 8$, *you could say "*$2x = 8$ *is a linear equation in one variable."*

89. Simplify $5 - 3(2x - 7)$.
90. Graph $y = -2x + 6$ by hand.
91. Solve $5 - 3(2x - 7) = 8$.
92. Solve $10 = -2x + 6$.

5.2 Finding Equations of Lines

Objectives

» Find an equation of a line by using the slope–intercept form of a linear equation.

» Find an equation of a line by using the *point–slope form* of a linear equation.

In this section, we will discuss two methods of finding an equation of a line. We will first use the slope–intercept form. Then we will use another form of an equation of a line called the *point–slope form*.

Method 1: Using the Slope–Intercept Form

In Example 1, we will use the concept that a point that lies on a line satisfies an equation of the line.

> **Example 1**   Using the Slope and a Point to Find a Linear Equation
>
> Find an equation of the line that has slope $m = 2$ and contains the point $(4, 3)$.

Solution

An equation of a nonvertical line can be put into the form $y = mx + b$. Because $m = 2$, we have

$$y = 2x + b$$

To find b, recall from Section 3.1 that any point on the graph of an equation represents a solution of that equation. In particular, the point $(4, 3)$ should satisfy the equation $y = 2x + b$:

$$y = 2x + b \qquad \text{Slope is } m = 2.$$
$$3 = 2(4) + b \qquad \text{Substitute 4 for } x \text{ and 3 for } y.$$
$$3 = 8 + b \qquad \text{Multiply.}$$
$$3 - 8 = 8 + b - 8 \qquad \text{Subtract 8 from both sides.}$$
$$-5 = b \qquad \text{Combine like terms.}$$

Now we substitute -5 for b in $y = 2x + b$:

$$y = 2x - 5$$

To verify our work, we check that the coefficient of the variable term $2x$ is 2 (the given slope) and we see whether $(4, 3)$ satisfies the equation $y = 2x - 5$:

$$y = 2x - 5 \quad \textit{The equation we found}$$
$$3 \overset{?}{=} 2(4) - 5 \quad \textit{Substitute 4 for x and 3 for y.}$$
$$3 \overset{?}{=} 8 - 5 \quad \textit{Multiply.}$$
$$3 \overset{?}{=} 3 \quad \textit{Subtract.}$$
$$\text{true}$$

▶

Finding an Equation of a Line by Using the Slope, a Point, and the Slope–Intercept Form

To find an equation of a line by using the slope and a point,

1. Substitute the given value of the slope m into the equation $y = mx + b$.
2. Substitute the coordinates of the given point into the equation you found in step 1 and solve for b.
3. Substitute the value of b you found in step 2 into the equation you found in step 1.
4. Check that the graph of your equation contains the given point.

Example 2   Using the Slope and a Point to Find a Linear Equation

Find an equation of the line that has slope $m = -\dfrac{2}{5}$ and contains the point $(-4, 1)$.

Solution

Because the slope is $-\dfrac{2}{5}$, the equation has the form $y = -\dfrac{2}{5}x + b$. To find b, we substitute the coordinates of the point $(-4, 1)$ into the equation $y = -\dfrac{2}{5}x + b$ and solve for b:

$$y = -\frac{2}{5}x + b \qquad \textit{Slope is } -\frac{2}{5}.$$

$$1 = -\frac{2}{5}(-4) + b \quad \textit{Substitute } -4 \textit{ for x and 1 for y.}$$

$$1 = \frac{8}{5} + b \qquad \textit{Simplify.}$$

$$5 = 8 + 5b \qquad \textit{Multiply both sides by LCD, 5, to clear equation of fractions.}$$

$$5 - 8 = 8 + 5b - 8 \qquad \textit{Subtract 8 from both sides.}$$

$$-3 = 5b \qquad \textit{Combine like terms.}$$

$$-\frac{3}{5} = b \qquad \textit{Divide both sides by 5.}$$

So, the equation is $y = -\dfrac{2}{5}x - \dfrac{3}{5}$. In Fig. 9, we use ZDecimal followed by TRACE on a graphing calculator to check that the line $y = -\dfrac{2}{5}x - \dfrac{3}{5}$ contains the point $(-4, 1)$.

▶

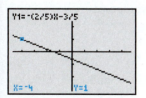

Figure 9 Check that the line contains $(-4, 1)$

In each of Examples 1 and 2, we found an equation of a line by using the slope of the line and a point. We can also find an equation of a line by using two points.

Example 3   Using Two Points to Find a Linear Equation

Find an equation of the line that passes through $(-1, 4)$ and $(2, -5)$.

Solution

First, we find the slope of the line:

$$m = \frac{y_2 - y_1}{x_2 - x_1} = \frac{-5 - 4}{2 - (-1)} = \frac{-5 - 4}{2 + 1} = \frac{-9}{3} = -3$$

So, we have $y = -3x + b$. Next, we will find the value of b. Because the line contains the point $(-1, 4)$, we substitute -1 for x and 4 for y:

$$\begin{array}{ll} y = -3x + b & \textit{Slope is m} = -3. \\ 4 = -3(-1) + b & \textit{Substitute } -1 \textit{ for x and 4 for y.} \\ 4 = 3 + b & \textit{Multiply.} \\ 4 - 3 = 3 + b - 3 & \textit{Subtract 3 from both sides.} \\ 1 = b & \textit{Combine like terms.} \end{array}$$

The equation is $y = -3x + 1$. To verify our equation, we check that both $(-1, 4)$ and $(2, -5)$ satisfy the equation $y = -3x + 1$:

Check that $(-1, 4)$ is a solution	**Check that $(2, -5)$ is a solution**
$y = -3x + 1$	$y = -3x + 1$
$4 \overset{?}{=} -3(-1) + 1$	$-5 \overset{?}{=} -3(2) + 1$
$4 \overset{?}{=} 3 + 1$	$-5 \overset{?}{=} -6 + 1$
$4 \overset{?}{=} 4$	$-5 \overset{?}{=} -5$
true	true

In Example 3, we substituted the coordinates of the given point $(-1, 4)$ into the equation $y = -3x + b$ to help us find the constant b. If we had used the other given point, $(2, -5)$, we would have found the same value of b. (Try it.)

Finding an Equation of a Line by Using Two Points and the Slope–Intercept Form

To find an equation of the line that passes through two given points whose x-coordinates are different,

1. Use the formula $m = \dfrac{y_2 - y_1}{x_2 - x_1}$ to find the slope of the line containing the two points.

2. Substitute the m value you found in step 1 into the equation $y = mx + b$.

3. Substitute the coordinates of one of the given points into the equation you found in step 2, and solve for b.

4. Substitute the b value you found in step 3 into the equation you found in step 2.

5. Check that the graph of your equation contains the two given points.

Example 4   Using Two Points to Find a Linear Equation

Find an equation of the line that passes through the points $(-9, -2)$ and $(-3, 7)$.

Solution

We begin by finding the slope of the line:

$$m = \frac{y_2 - y_1}{x_2 - x_1} = \frac{7 - (-2)}{-3 - (-9)} = \frac{9}{6} = \frac{3}{2}$$

So, we have $y = \frac{3}{2}x + b$. To find b, we substitute the coordinates of $(-3, 7)$ into the equation $y = \frac{3}{2}x + b$ and solve for b:

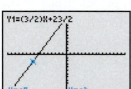

$$y = \frac{3}{2}x + b \qquad \textit{Slope is } \frac{3}{2}.$$

$$7 = \frac{3}{2}(-3) + b \qquad \textit{Substitute } -3 \textit{ for x and 7 for y.}$$

$$7 = -\frac{9}{2} + b \qquad \textit{Simplify.}$$

$$14 = -9 + 2b \qquad \textit{Multiply both sides by LCD, 2, to clear equation of fractions.}$$

$$14 + 9 = -9 + 2b + 9 \quad \textit{Add 9 to both sides.}$$

$$23 = 2b \qquad \textit{Combine like terms.}$$

$$\frac{23}{2} = b \qquad \textit{Divide both sides by 2.}$$

The equation is $y = \frac{3}{2}x + \frac{23}{2}$. We use ZStandard followed by ZSquare to check that the line $y = \frac{3}{2}x + \frac{23}{2}$ contains the points $(-9, -2)$ and $(-3, 7)$. See Fig. 10. For graphing calculator instructions, see Appendix A.3–A.6.

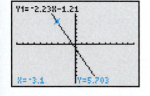

Figure 10 Check that the line contains both $(-9, -2)$ and $(-3, 7)$

In Example 5, we will work with decimal numbers to prepare us to find equations of models in Section 5.3.

Example 5 Finding an Approximate Equation of a Line

Find an approximate equation of the line that contains the points $(-3.1, 5.7)$ and $(1.6, -4.8)$.

Solution

First, we find the slope of the line:

$$m = \frac{y_2 - y_1}{x_2 - x_1} = \frac{-4.8 - 5.7}{1.6 - (-3.1)} = \frac{-10.5}{4.7} \approx -2.23$$

So, the equation has the form $y = -2.23x + b$. To find b, we substitute the coordinates of the point $(-3.1, 5.7)$ into the equation $y = -2.23x + b$ and solve for b:

$$y = -2.23x + b \qquad \textit{Slope is approximately } -2.23.$$

$$5.7 = -2.23(-3.1) + b \quad \textit{Substitute } -3.1 \textit{ for x and 5.7 for y.}$$

$$5.7 = 6.913 + b \qquad \textit{Multiply.}$$

$$5.7 - 6.913 = 6.913 + b - 6.913 \quad \textit{Subtract 6.913 from both sides.}$$

$$-1.21 \approx b \qquad \textit{Combine like terms.}$$

The approximate equation is $y = -2.23x - 1.21$.

We use a graphing calculator to check that the line $y = -2.23x - 1.21$ comes very close to the points $(-3.1, 5.7)$ and $(1.6, -4.8)$. See Fig. 11.

Figure 11 Check that the line comes very close to $(-3.1, 5.7)$ and $(1.6, -4.8)$

Example 6 Finding an Equation of a Vertical Line

Find an equation of the line that contains the points $(5, 1)$ and $(5, 3)$.

Solution

Because the x-coordinates of the given points are equal (both 5), the line that contains the points is vertical (see Fig. 12). An equation of the line is $x = 5$.

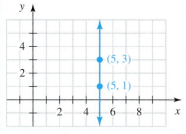

Figure 12 The line that contains $(5, 1)$ and $(5, 3)$

In Example 7, we will find an equation of a line that contains a given point and is perpendicular to a given line. Recall from Section 3.4 that if two nonvertical lines are perpendicular, then the slope of one line is the opposite reciprocal of the slope of the other line.

| Example 7 | Finding an Equation of a Line Perpendicular to a Given Line |

Find an equation of the line l that contains the point $(4, 2)$ and is perpendicular to the line $x + 3y = -6$.

Solution

First, we write $x + 3y = -6$ in slope–intercept form:

$$x + 3y = -6 \qquad \textit{Line perpendicular to line l}$$
$$x + 3y - x = -6 - x \qquad \textit{Subtract x from both sides.}$$
$$3y = -x - 6 \qquad \textit{Combine like terms.}$$
$$\frac{3y}{3} = \frac{-x}{3} - \frac{6}{3} \qquad \textit{Divide both sides by 3.}$$
$$y = -\frac{1}{3}x - 2 \qquad \textit{Simplify.}$$

For the line $y = -\frac{1}{3}x - 2$, the slope is $-\frac{1}{3}$. The slope of the line l is the opposite reciprocal of $-\frac{1}{3}$, or 3. An equation of line l has the form $y = 3x + b$. To find b, we substitute the coordinates of the given point $(4, 2)$ into $y = 3x + b$ and solve for b:

$$2 = 3(4) + b \qquad \textit{Substitute 4 for x and 2 for y.}$$
$$2 = 12 + b \qquad \textit{Multiply.}$$
$$2 - 12 = 12 + b - 12 \qquad \textit{Subtract 12 from both sides.}$$
$$-10 = b \qquad \textit{Combine like terms.}$$

An equation of line l is $y = 3x - 10$. We use ZStandard followed by ZSquare to verify our work (see Fig. 13).

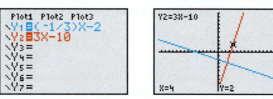

Figure 13 Check that the line contains $(4, 2)$ and is perpendicular to $x + 3y = -6$

Method 2: Using the Point–Slope Form

We can find an equation of a line by another method. Suppose a nonvertical line has slope m and contains the point (x_1, y_1). Then, if (x, y) represents a different point on the line, the slope of the line is

$$\frac{y - y_1}{x - x_1} = m$$

Multiplying both sides of the equation by $x - x_1$ gives

$$\frac{y - y_1}{x - x_1} \cdot (x - x_1) = m(x - x_1)$$
$$y - y_1 = m(x - x_1)$$

We say this linear equation is in **point–slope form**.*

---

*Although we assumed that (x, y) is different from (x_1, y_1), note that (x_1, y_1) is a solution of the equation $y - y_1 = m(x - x_1)$: $y_1 - y_1 = m(x_1 - x_1)$, or $0 = 0$, a true statement.

Point–Slope Form

If a nonvertical line has slope m and contains the point (x_1, y_1), then an equation of the line is

$$y - y_1 = m(x - x_1)$$

Example 8   Using the Point–Slope Form to Find an Equation of a Line

Use the point–slope form to find an equation of the line that has slope $m = -3$ and contains the point $(-4, 2)$. Then write the equation in slope–intercept form.

Solution

We begin by substituting $x_1 = -4$, $y_1 = 2$, and $m = -3$ into the point–slope form $y - y_1 = m(x - x_1)$:

$$\begin{aligned}
y - y_1 &= m(x - x_1) && \textit{Point–slope form} \\
y - 2 &= -3(x - (-4)) && \textit{Substitute } x_1 = -4, \ y_1 = 2, \text{ and } m = -3. \\
y - 2 &= -3(x + 4) && \textit{Simplify.} \\
y - 2 &= -3x - 12 && \textit{Distributive law} \\
y - 2 + 2 &= -3x - 12 + 2 && \textit{Add 2 to both sides.} \\
y &= -3x - 10 && \textit{Combine like terms.}
\end{aligned}$$

So, the equation is $y = -3x - 10$.

Example 9   Using the Point–Slope Form to Find an Equation of a Line

Use the point–slope form to find an equation of the line that contains the points $(2, -6)$ and $(5, -4)$. Then write the equation in slope–intercept form.

Solution

We begin by finding the slope of the line:

$$m = \frac{-4 - (-6)}{5 - 2} = \frac{2}{3}$$

Then we substitute $x_1 = 2$, $y_1 = -6$, and $m = \dfrac{2}{3}$ into the equation $y - y_1 = m(x - x_1)$:

$$\begin{aligned}
y - y_1 &= m(x - x_1) && \textit{Point–slope form} \\
y - (-6) &= \frac{2}{3}(x - 2) && \textit{Substitute } x_1 = 2, \ y_1 = -6, \text{ and } m = \frac{2}{3}. \\
y + 6 &= \frac{2}{3}x - \frac{4}{3} && \textit{Simplify; distributive law} \\
y + 6 - 6 &= \frac{2}{3}x - \frac{4}{3} - 6 && \textit{Subtract 6 from both sides.} \\
y &= \frac{2}{3}x - \frac{22}{3} && \textit{Combine like terms; } -\frac{4}{3} - 6 = -\frac{4}{3} - \frac{18}{3} = -\frac{22}{3}
\end{aligned}$$

So, the equation is $y = \dfrac{2}{3}x - \dfrac{22}{3}$.

Group Exploration

Finding equations of lines

The objective of this game is to earn 19 credits. You earn credits by finding equations of lines that pass through one or more of the following points:

$$(-3, 2), (-3, 0), (-2, -7), (-2, -1), (-1, 4), (0, 2),$$
$$(1, -1), (2, -2), (3, 1), (3, 3)$$

If a line passes through exactly one point, then you earn one credit. If a line passes through exactly two points,

then you earn three credits. If a line passes through exactly three points, then you earn five credits. You may use five equations. You may use points more than once. [**Hint:** First, construct a scatterplot. After finding your equations, use your graphing calculator to check that they are correct.]

Tips for Success   Show What You Know

Even if you don't know how to do one step of a problem, you can still show the instructor that you understand the other steps of the problem. Depending on how your instructor grades tests, you may earn partial credit even though you pick an incorrect number to be the result for a particular step, as long as you then show what you would do with that number in the remaining steps of the solution. Check with your instructor first.

For example, suppose you want to find an equation of the line that passes through the points $(1, 5)$ and $(2, 8)$, but you have forgotten how to find the slope. You could still write,

I've drawn a blank on finding slope. However, assuming that the slope is 2, then

$$y = 2x + b$$
$$5 = 2(1) + b$$
$$5 = 2 + b$$
$$5 - 2 = 2 + b - 2$$
$$3 = b$$

Therefore, $y = 2x + 3$.

You could point out that you know your result is incorrect because the graph of $y = 2x + 3$ does *not* pass through the point $(2, 8)$. Also, seeing your result (with the graph) may jog your memory about finding the slope and allow you to go back and do the problem correctly.

 # Homework 5.2

For extra help ▶ **MyLab Math**   Watch the videos in MyLab Math

Find an equation of the line that has the given slope and contains the given point. If possible, write your equation in slope–intercept form. Check that the ordered pair that represents the given point satisfies your equation.

1. $m = 2, (3, 5)$

2. $m = 3, (2, 4)$

3. $m = -3, (1, -2)$

4. $m = -5, (3, -8)$

5. $m = 2, (-4, -6)$

6. $m = 4, (-5, -1)$

7. $m = -6, (-2, -3)$

8. $m = -1, (-7, -4)$

9. $m = \dfrac{2}{5}, (3, 1)$

10. $m = \dfrac{1}{2}, (5, 3)$

11. $m = -\dfrac{3}{4}, (-2, -5)$

12. $m = -\dfrac{5}{3}, (-4, -2)$

13. $m = 0, (5, 3)$

14. $m = 0, (-1, -3)$

15. m is undefined, $(-2, 4)$

16. m is undefined, $(3, -2)$

Find an approximate equation of the line that has the given slope and contains the given point. Write your equation in slope–intercept form. Round the constant term to two decimal places. Use a graphing calculator to verify your equation.

17. $m = 2.1$, $(3.7, -5.9)$ **18.** $m = 5.8$, $(2.6, -4.3)$

19. $m = -5.6$, $(-4.5, 2.8)$ **20.** $m = -1.3$, $(-6.6, 3.8)$

21. $m = -6.59$, $(-2.48, -1.61)$

22. $m = -2.07$, $(-4.73, -9.60)$

Find an equation of the line that passes through the two given points. If possible, write your equation in slope–intercept form. Check that the graph of your equation contains the given points.

23. $(3, 2)$ and $(5, 6)$ **24.** $(1, 4)$ and $(2, 1)$

25. $(1, 6)$ and $(3, 2)$ **26.** $(4, 7)$ and $(6, 15)$

27. $(-1, -7)$ and $(2, 8)$ **28.** $(-2, -10)$ and $(3, 5)$

29. $(-5, -4)$ and $(-2, -7)$ **30.** $(-3, -2)$ and $(-1, -8)$

31. $(0, 0)$ and $(1, 1)$ **32.** $(0, 0)$ and $(1, -1)$

33. $(0, 9)$ and $(2, 1)$ **34.** $(-3, 1)$ and $(0, -5)$

35. $(3, 2)$ and $(5, 2)$ **36.** $(-5, -3)$ and $(-1, -3)$

37. $(-4, -1)$ and $(-4, 3)$ **38.** $(7, 1)$ and $(7, 6)$

39. $(4, 3)$ and $(8, 5)$ **40.** $(2, 3)$ and $(6, 1)$

41. $(-4, -3)$ and $(5, 3)$ **42.** $(-6, -2)$ and $(4, 6)$

43. $(-3, 2)$ and $(3, 1)$ **44.** $(2, -2)$ and $(6, -5)$

45. $(-2, 1)$ and $(5, -1)$ **46.** $(-6, 5)$ and $(-2, 2)$

47. $(-4, -2)$ and $(6, 4)$ **48.** $(-1, -2)$ and $(5, 6)$

49. $(-4, -8)$ and $(-2, -5)$ **50.** $(-6, -9)$ and $(-2, -4)$

Find an approximate equation of the line that passes through the two given points. Write your equation in slope–intercept form. Round the slope and the constant term to two decimal places. Use a graphing calculator to verify your equation.

51. $(2.3, 5.8)$ and $(4.5, 3.1)$

52. $(1.9, 5.6)$ and $(2.4, 8.3)$

53. $(-4.5, 2.2)$ and $(1.2, -7.5)$

54. $(-8.1, -5.3)$ and $(-3.3, -2.7)$

55. $(2.46, -1.84)$ and $(5.87, -5.29)$

56. $(3.57, -9.41)$ and $(5.98, -2.84)$

57. $(-4.57, -8.29)$ and $(7.17, -2.69)$

58. $(-8.99, 4.82)$ and $(-5.85, -3.92)$

Find an equation of the line that contains the given point and is parallel to the given line. Use a graphing calculator to verify your equation.

59. $(2, 3)$, $y = 2x - 7$ **60.** $(5, 1)$, $y = -3x + 4$

61. $(-3, 5)$, $4x + y = 1$ **62.** $(4, -2)$, $-2x + y = 3$

Find an equation of the line that contains the given point and is perpendicular to the given line. Use ZStandard followed by ZSquare with a graphing calculator to verify your equation.

63. $(3, 4)$, $y = \dfrac{1}{2}x + 5$ **64.** $(1, 2)$, $y = \dfrac{1}{4}x - 1$

65. $(-2, -5)$, $x + 4y = 12$ **66.** $(-3, -1)$, $x - 3y = 9$

67. Find an equation of the line sketched in Fig. 14. [**Hint:** Choose two points whose coordinates appear to be integers.]

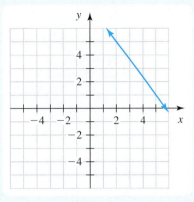

Figure 14 Exercise 67

68. Find an equation of the line sketched in Fig. 15. [**Hint:** Choose two points whose coordinates appear to be integers.]

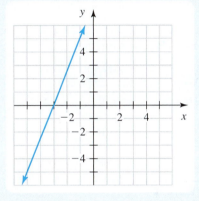

Figure 15 Exercise 68

Concepts

69. Consider the line that contains the points $(-2, -6)$ and $(4, 3)$.
 a. Find an equation of the line.
 b. Sketch the line on a coordinate system.
 c. Use a table to list five ordered pairs that correspond to points on the line.

70. Consider the line that contains the points $(-3, 4)$ and $(6, -2)$.
 a. Find an equation of the line.
 b. Sketch the line on a coordinate system.
 c. Use a table to list five ordered pairs that correspond to points on the line.

71. Decide whether it is possible for a line to have the indicated number of x-intercepts. If it is possible, find an equation of such a line. If it is not possible, explain why not.
 a. no x-intercepts
 b. exactly one x-intercept
 c. exactly two x-intercepts
 d. an infinite number of x-intercepts

72. Decide whether it is possible for a line to have the indicated number of y-intercepts. If it is possible, find an equation of such a line. If it is not possible, explain why not.
 a. no y-intercepts
 b. exactly one y-intercept
 c. exactly two y-intercepts
 d. an infinite number of y-intercepts

73. Let E be the enrollment (in thousands of students) at a college t years after the college opens. Some pairs of values of t and E are listed in Table 3.

Table 3 Enrollments

Age of College (years) t	Enrollment (thousands of students) E
2	9
4	13
7	19
9	23
12	29

Find an equation that describes the relationship between t and E.

74. Let s be a person's savings (in thousands of dollars) at t years since 2005. Some pairs of values of t and s are listed in Table 4.

Table 4 A Person's Savings

Years since 2005 t	Savings (thousands of dollars) s
1	5
4	14
5	17
7	23
10	32

Find an equation that describes the relationship between t and s.

75. a. Use each of the following forms to find an equation of the line that contains the points $(2, 1)$ and $(4, 7)$. Write each result in slope–intercept form.
   i. slope–intercept form
   ii. point–slope form
   b. Compare the results you found in parts (ai) and (aii).

76. a. Use each of the following forms to find an equation of the line that contains the points $(-4, 3)$ and $(2, -5)$. Write each result in slope–intercept form.
   i. slope–intercept form
   ii. point–slope form
   b. Compare the results you found in parts (ai) and (aii).

77. a. Find an equation of a line with slope -2. [**Hint:** There are *many* correct answers.]
   b. Find an equation of a line with y-intercept $(0, 4)$.
   c. Find an equation of a line that contains the point $(3, 5)$.
   d. Determine whether there is a line that has slope -2 and y-intercept $(0, 4)$ and contains the point $(3, 5)$. Explain.

78. Suppose you are trying to find an equation of a line that contains two given points, and you find the slope is undefined. What type of line is it? Explain.

79. Find an equation of the line that contains the points $(3, 7)$ and $(4, 9)$. After finding the slope, you used one of the points to complete the process. Now use the other point to complete the process. Did you get the same result? Why does this make sense?

80. Describe how to find an equation of a line that contains two given points. Also, explain how you can check that the graph of your equation contains the two points.

Related Review

81. Four sets of points are described in Table 5. For each set, find an equation of a line that contains the points.

Table 5 Four Sets of Points

Set 1		Set 2		Set 3		Set 4	
x	y	x	y	x	y	x	y
0	25	0	12	0	77	0	3
1	23	1	16	1	72	1	3
2	21	2	20	2	67	2	3
3	19	3	24	3	62	3	3
4	17	4	28	4	57	4	3

82. Sketch a vertical line on a coordinate system. Find an equation of the line. What is the slope of the line?

83. Sketch a decreasing line that is nearly horizontal on a coordinate system. Find an equation of the line.

84. Find an equation of a line that has no solutions in Quadrant I.

85. a. Graph $y = 2x - 3$ by hand.
   b. Choose two points that lie on the graph. Then use the two points to find an equation of the line that contains them. Compare your equation with the equation $y = 2x - 3$.

86. A line contains the points $(2, 4)$ and $(4, 1)$. Find three more points that lie on the line.

Expressions, Equations, and Graphs

Perform the indicated instruction. Then use words such as linear, one variable, *and* two variables *to describe the expression or equation. For instance, to describe* 2x + 7x, *you could say "2x + 7x is a linear expression in one variable."*

87. Graph $3x + 2y = 6$ by hand.

88. Simplify $\dfrac{5x}{8} + \dfrac{3}{4} - \dfrac{7x}{2}$.

89. Evaluate $3x + 2y$ for $x = 4$ and $y = -5$.

90. Solve $\dfrac{5x}{8} + \dfrac{3}{4} = \dfrac{7x}{2}$.

5.3 Finding Equations of Linear Models

Objectives

» Find an equation of a linear model by using data described in words.

» Find an equation of a linear model by using data displayed in a table.

In Section 5.2, we used two given points to find an equation of a line. In this section, we use this skill to find an equation of a linear model.

Finding a Model by Using Data Described in Words

In Example 1, we will use data described in words to find an equation of a linear model.

Example 1   Finding an Equation of a Linear Model

The percentage of American adults who want to lose weight has decreased approximately linearly from 59% in 2011 to 49% in 2015 (Source: *Gallup*). Let p be the percentage of American adults who want to lose weight at t years since 2000.

1. Find an equation of a linear model.
2. What is the slope? What does it mean in this situation?
3. What is the p-intercept? What does it mean in this situation?

Solution

1. Known values of t and p are shown in Table 6.

   A linear equation can be put into the form $y = mx + b$, where x affects (explains) y. Because the variables t and p are approximately linearly related and t affects (explains) p, we will find an equation of the form $p = mt + b$. We can use the data points $(11, 59)$ and $(15, 49)$ to find the values of m and b.

   First, we use $(11, 59)$ and $(15, 49)$ to find the slope:

$$m = \frac{49 - 59}{15 - 11} = -2.5$$

So, we can substitute -2.5 for m in the equation $p = mt + b$:

$$p = -2.5t + b$$

To find the constant b, we substitute the coordinates of the point $(11, 59)$ into the equation $p = -2.5t + b$ and then solve for b:

$59 = -2.5(11) + b$	*Substitute 11 for t and 59 for p.*
$59 = -27.5 + b$	*Multiply.*
$59 + 27.5 = -27.5 + b + 27.5$	*Add 27.5 to both sides.*
$86.5 = b$	*Combine like terms.*

Now we can substitute 86.5 for b in the equation $p = -2.5t + b$:

$$p = -2.5t + 86.5$$

We verify our equation by using a graphing calculator to check that our line comes very close to the points $(11, 59)$ and $(15, 49)$. See Fig. 16.

Table 6 Known Values of t and p

Years since 2000	Percent
t	p
11	59
15	49

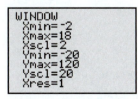 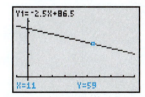 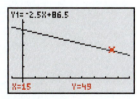

Figure 16 Checking that the model contains both $(11, 59)$ and $(15, 49)$

2. The slope is -2.5. According to the model, the percentage of American adults who want to lose weight decreases by 2.5 percentage points per year.
3. Because the model $p = -2.5t + 86.5$ is in slope–intercept form, the p-intercept is $(0, 86.5)$. So, the model estimates that 86.5% of American adults wanted to lose weight in 2000. A little research would show that 55% of American adults wanted to lose weight in 2000, so model breakdown has occurred.

Finding a Linear Model by Using Data Displayed in a Table

In Example 2, we will find an equation of a model by using data shown in a table.

Table 7 Total College Student Discretionary Spending

Year	Spending (billions of dollars)
2011	86
2012	120
2013	117
2014	163
2015	203

Source: *Refuel Agency*

Example 2   Finding an Equation of a Linear Model

College student *discretionary spending* consists of expenditures other than necessary ones such as tuition, boarding, and textbooks. Total college student discretionary spending is shown in Table 7 for various years. Let S be total college student discretionary spending (in billions of dollars) in the year that is t years since 2000. Find an equation of a model that comes close to the points in the scatterplot of the data.

Solution

We begin by viewing the positions of the data points in the scatterplot (see Fig. 17).

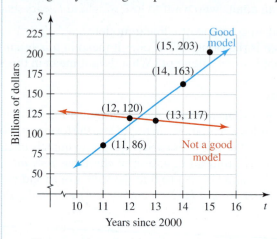

Figure 17 Discretionary spending scatterplot

To save time and improve the accuracy of plotting points, we can use a graphing calculator to view the scatterplot (see Fig. 18). For graphing calculator instructions on drawing scatterplots, see Appendix A.8, A.9, and A.12. For StatCrunch instructions, see Appendix B.1 and B.2.

It appears that t and S are approximately linearly related, so we will find an equation of a linear model. It is not necessary to use two *data* points to find an equation, although it is often convenient and satisfactory to do so.

The red line that contains the data points $(12, 120)$ and $(13, 117)$ does *not* come close to the other data points (see Fig. 17). However, the blue line that passes through the data points $(11, 86)$ and $(14, 163)$ appears to come close to the rest of the data points. We will find the equation of this line.

An equation of a line can be written in the form $y = mx + b$, where x is the explanatory variable. Because the variables t and S are approximately linearly related and t is the explanatory variable, we will find an equation of the form $S = mt + b$. To find the equation, we first use the points $(11, 86)$ and $(14, 163)$ to find the value of m (see Fig. 19):

$$m = \frac{163 - 86}{14 - 11} = \frac{77}{3} \approx 25.67$$

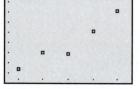

Figure 18 Graphing calculator scatterplot

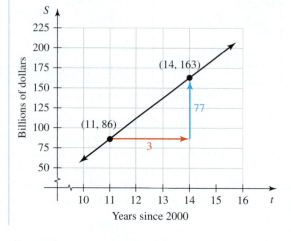

Figure 19 Points $(11, 86)$ and $(14, 163)$ of the discretionary spending scatterplot

Then we substitute 25.67 for m in the equation $S = mt + b$:

$$S = 25.67t + b$$

To find the constant b, we substitute the coordinates of the point $(14, 163)$ into the equation $S = 25.67t + b$ and then solve for b:

$$163 = 25.67(14) + b \qquad \text{\textit{Substitute 14 for t and 163 for S.}}$$
$$163 = 359.38 + b \qquad \text{\textit{Multiply.}}$$
$$163 - 359.38 = 359.38 + b - 359.38 \qquad \text{\textit{Subtract 359.38 from both sides.}}$$
$$-196.38 = b \qquad \text{\textit{Combine like terms.}}$$

Now we substitute -196.38 for b in the equation $S = 25.67t + b$:

$$S = 25.67t - 196.38$$

We check the correctness of our equation by using a graphing calculator to verify our line approximately contains the points $(11, 86)$ and $(14, 163)$ and comes close to the other data points (see Fig. 20). For graphing calculator instructions, see Appendix A.10 and A.11.

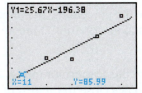 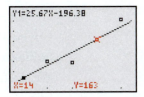

Figure 20 Checking that the model contains both $(11, 86)$ and $(14, 163)$

WARNING   It is a common error to skip constructing a scatterplot when we find an equation of a model. However, we benefit in many ways by viewing a scatterplot of data. First, we can determine whether the data are approximately linearly related. Second, if the data are approximately linearly related, viewing a scatterplot helps us choose two good points with which to find an equation of a linear model. Third, by graphing the model with the scatterplot, we can assess whether the model fits the data reasonably well.

Example 2 outlines four steps to find an equation of a linear model.

Finding an Equation of a Linear Model

To find an equation of a linear model, given some data,

1. Construct a scatterplot of the data.
2. Determine whether there is a line that comes close to the data points. If so, choose two points (not necessarily data points) that you can use to find an equation of a linear model.
3. Find an equation of the line.
4. Use a graphing calculator to verify that the graph of your equation contains the two chosen points and comes close to all the points of the scatterplot.

What should you do if you discover that a model does not fit a data set well? A good first step is to check for any graphing or calculation errors. If your work appears to be correct, then one option is to try using different points to find your equation. Another option is to increase or decrease the slope m and/or the constant term b until the fit is good. You can practice this "trial-and-error" process by completing the exploration in this section.

Example 3   Finding an Equation of a Linear Model

According to a poll performed by the National Consumers League, people are more concerned about privacy issues than about health care, education, crime, or taxes. The percentages of Americans of various age groups who consider their home address to be personal information are listed in Table 8.

Table 8 Americans Who Consider Their Home Address to Be Personal Information

Age Group	Age Used to Represent Age Group	Percent
18–24	21	84
25–34	29.5	80
35–44	39.5	74
45–54	49.5	74
55–64	59.5	67
over 64	75	60

Source: *American Demographics*

Let p be the percentage of Americans at age a years who consider their home address to be personal information. Find an equation that models the data well.

Solution

We see from the scatterplot in Fig. 21 that a line containing the data points $(21, 84)$ and $(59.5, 67)$ comes close to the rest of the data points.

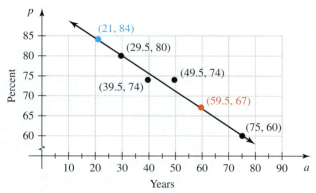

Figure 21 Personal information scatterplot and linear model

For an equation of the form $p = ma + b$, we first use the points $(21, 84)$ and $(59.5, 67)$ to find m:

$$m = \frac{67 - 84}{59.5 - 21} \approx -0.44$$

So, the equation has the form

$$p = -0.44a + b$$

To find b, we use the point $(21, 84)$ and substitute 21 for a and 84 for p:

$84 = -0.44(21) + b$	*Substitute* 21 *for a and* 84 *for p.*
$84 = -9.24 + b$	*Multiply.*
$84 + 9.24 = -9.24 + b + 9.24$	*Add* 9.24 *to both sides.*
$93.24 = b$	*Combine like terms.*

So, the equation is $p = -0.44a + 93.24$.

We use a graphing calculator to verify that the linear model approximately contains the points $(21, 84)$ and $(59.5, 67)$ and comes close to the other data points (see Fig. 22).

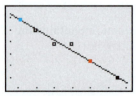

Figure 22 Checking that the model approximately contains both $(21, 84)$ and $(59.5, 67)$

Group Exploration

Adjusting the fit of a model

The winning times for the men's Olympic 100-meter freestyle swimming event are shown in Table 9 for various years.

Table 9 Winning Times for the Men's Olympic 100-Meter Freestyle

Year	Swimmer	Country	Winning Time (seconds)
1980	Jörg Woithe	E. Germany	50.40
1984	Rowdy Gaines	USA	49.80
1988	Matt Biondi	USA	48.63
1992	Aleksandr Popov	Unified Team	49.02
1996	Aleksandr Popov	Russia	48.74
2000	Pieter van den Hoogenband	Netherlands	48.30
2004	Pieter van den Hoogenband	Netherlands	48.17
2008	Alain Bernard	France	47.21
2012	Nathan Adrian	USA	47.52
2016	Kyle Chalmers	Australia	47.58

Source: *The New York Times Almanac*

Let w be the winning time (in seconds) at t years since 1980.

1. Use a graphing calculator to draw a scatterplot of the data. Do the variables t and w appear to be approximately linearly related?

2. The linear model $w = -0.0325t + 48.95$ can be found by using the data points $(20, 48.30)$ and $(24, 48.17)$. Draw the line and the scatterplot in the same viewing window. Check that the line contains these two points.

3. The model $w = -0.0325t + 48.95$ does not fit all the data points very well. Adjust the equation by increasing or decreasing the slope -0.0325 and/or the constant term 48.95 so that your new model will fit the data better. Keep adjusting the model until it fits the data points reasonably well.

4. Use your improved model to predict the winning time in the 2020 Olympics.

Tips for Success   Verify Your Work

Remember to use your graphing calculator to verify your work. In this section, for example, you can use your graphing calculator to check your equations. Checking your work increases your chances of catching errors and thus will likely improve your performance on homework assignments, quizzes, and tests.

 # Homework 5.3

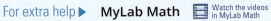

For extra help ▶ **MyLab Math**   Watch the videos in MyLab Math

1. The number of female members of Congress has increased approximately linearly from 85 members in 2005 to 108 members in 2015 (Source: *Congressional Research Service*). Let n be the number of female members of Congress at t years since 2000. Find an equation of a linear model to describe the data.

2. The average time an American adult spends daily with digital media has increased approximately linearly from 3.0 hours in 2009 to 5.6 hours in 2015 (Source: *eMarketer*). Let d be average time (in hours) an American adult spends daily with digital media at t years since 2000. Find an equation of a linear model to describe the data.

3. The number of inmates younger than 18 held in state prisons has decreased approximately linearly from 2717 inmates in 2008 to 1035 inmates in 2014 (Source: *Bureau of Justice Statistics*). Let n be the number of inmates younger than 18 held in state prisons at t years since 2000. Find an equation of a linear model to describe the data.

4. The percentage of children younger than 18 living with two married parents has decreased approximately linearly from 67.3% in 2005 to 64.7% in 2015 (Source: *U.S. Census Bureau*). Let p be the percentage of children younger than 18 living with two married parents at t years since 2000. Find an equation of a linear model to describe the data.

5. The average U.S. commute time to work (one way) has increased approximately linearly from 21.7 minutes in 1980 to 26.0 minutes in 2014 (Source: *Census Bureau*). Let c be the average U.S. commute time (in minutes) to work at t years since 1950.
   a. Find an equation of a linear model to describe the data.
   b. What is the slope? What does it mean in this situation?
   c. What is the c-intercept? What does it mean in this situation?

6. Light-vehicle sales in the United States have increased approximately linearly from 11.6 million vehicles in 2010 to 17.4 million vehicles in 2015 (Source: *WardsAuto*). Let s be the annual light-vehicle sales (in millions of vehicles) at t years since 2000.

a. Find an equation of a linear model to describe the data.
b. What is the slope? What does it mean in this situation?
c. What is the s-intercept? What does it mean in this situation?

7. The number of sexual-harassment charges filed has decreased approximately linearly from 7944 charges in 2010 to 6822 charges in 2015 (Source: *U.S. Equal Employment Opportunity Commission*). Let n be the number of sexual-harassment charges filed in the year that is t years since 2000.
a. Find an equation of a linear model to describe the data.
b. What is the slope? What does it mean in this situation?
c. What is the n-intercept? What does it mean in this situation?

8. The student-to-faculty ratio at the University of Tennessee at Martin has decreased approximately linearly from 18.1 in 2011 to 16.3 in 2015 (Source: *Common Data Set*). Let s be the student-to-faculty ratio at UT Martin at t years since 2000.
a. Find an equation of a linear model to describe the data.
b. What is the slope? What does it mean in this situation?
c. What is the s-intercept? What does it mean in this situation?

9. Find an equation of a line that comes close to the points listed in Table 10. Then use a graphing calculator to check that your line comes close to the points.

Table 10 Exercise 9

x	y
4	8
5	10
6	13
7	15
8	18

10. Find an equation of a line that comes close to the points listed in Table 11. Then use a graphing calculator to check that your line comes close to the points.

Table 11 Exercise 10

x	y
1	6
5	13
7	14
10	17
12	22

11. Find an equation of a line that comes close to the points listed in Table 12. Then use a graphing calculator to check that your line comes close to the points.

Table 12 Exercise 11

x	y
3	18
7	14
9	9
12	7
16	4

12. Find an equation of a line that comes close to the points listed in Table 13. Then use a graphing calculator to check that your line comes close to the points.

Table 13 Exercise 12

x	y
2	14
5	10
6	8
8	4
11	1

13. The percentages of all airline seats that are filled are shown in Table 14 for various years.

Table 14 Percentage of Airline Seats Filled

Year	Percent
2006	79.1
2008	80.0
2010	82.2
2012	83.4
2014	84.5
2016	84.7

Source: *Bureau of Transportation Statistics*

Let p be the percentage of airline seats that are filled in the year that is t years since 2000.
a. Use a graphing calculator to draw a scatterplot of the data.
b. Find an equation of a linear model to describe the data.
c. Draw your line and the scatterplot in the same viewing window. Verify that the line passes through the two points that you chose in finding the equation in part (b) and that it comes close to all the data points.

14. The percentages of Americans ages 25 or older who have earned a college degree are shown in Table 15 for various years.

Table 15 Percentages of Americans Ages 25 or Older Who Have Earned a College Degree

Year	Percent
1960	7.7
1970	10.7
1980	16.2
1990	21.3
2000	25.6
2010	29.9
2015	32.5

Source: *U.S. Census Bureau*

Let p be the percentage of Americans ages 25 or older who have earned a college degree at t years since 1960.

a. Use a graphing calculator to draw a scatterplot of the data.
b. Find an equation of a linear model to describe the data.
c. Draw your line and the scatterplot in the same viewing window. Verify that the line passes through the two points that you chose in finding the equation in part (b) and that it comes close to all the data points.

15. DATA. Table 16 shows how old a 20- to 50-pound dog is in relation to human years.

Table 16 Dog Years Compared with Human Years

Dog Years	Human Years
1	15
2	24
3	29
5	38
7	47
9	56
11	65
13	74
15	83

Source: *Fred Metzger, State College, PA*

Let H be the number of human years that is equivalent to d dog years.
a. Use a graphing calculator to draw a scatterplot of the data.
b. Find an equation of a linear model to describe the data.
c. Draw your line and the scatterplot in the same viewing window. Verify that the line passes through the two points that you chose in finding the equation in part (b) and that it comes close to all the data points.

16. The enrollments in an elementary algebra course at the College of San Mateo are shown in Table 17 for various Tuesdays leading up to the first day of class on Tuesday, January 17.

Table 17 Enrollments in an Elementary Algebra Course

Date	Number of Weeks since November 22	Enrollment
November 22	0	8
November 29	1	9
December 6	2	12
December 13	3	13
December 20	4	16
December 27	5	18
January 3	6	19

Source: *J. Lehmann*

Let E be the enrollment in the elementary algebra course at t weeks since November 22.
a. Use a graphing calculator to draw a scatterplot of the data.
b. Find an equation of a linear model to describe the data.
c. Draw your line and the scatterplot in the same viewing window. Verify that the line passes through the two points that you chose in finding the equation in part (b) and that it comes close to all the data points.

17. The numbers of Academy Award best-picture nominations and wins of movies up until 2016 are shown in Table 18 for various genres.

Table 18 Numbers of Academy Award Best-Picture Nominations and Wins of Movies of Various Genres

Genre	Nominations	Wins
Action	35	9
Comedy	108	16
Drama	462	80
Fantasy	31	2
Romance	215	36
Sci-Fi	11	0
Thriller	61	8
War	68	17

Source: *Collider, IMDb*

Let n be the number of nominations and w be the number of wins for movies of a certain genre.
a. Construct a scatterplot of the data.
b. Find an equation of a linear model to describe the data.
c. Draw your line and the scatterplot in the same viewing window. Verify that the line passes through the two points that you chose in finding the equation in part (b) and that it comes close to all the data points.
d. Find the total for the "Wins" column in Table 18. How is it possible for your result to be correct when there have been only 89 Academy Award ceremonies?

18. The Iraqi war death rates (number of deaths for every 100,000 people ages 18 to 54) are shown in Table 19 for various counties in the United States.

Table 19 Iraqi War Death Rates

County Population Group (thousands)	Population Used to Represent County Population Group (thousands)	Death Rate (number of deaths for every 100,000 people ages 18 to 54)
under 25	12.5	1.78
25–50	37.5	1.40
50–100	75.0	1.45
100–250	175.0	1.36
250–500	375.0	1.16
500–1000	750.0	0.85

Source: *The New York Times*

Let r be the death rate (number of deaths for every 100,000 people ages 18 to 54) of soldiers from counties of population p (in thousands).
a. Use a graphing calculator to draw a scatterplot of the data.
b. Find an equation of a linear model to describe the data.
c. Draw your line and the scatterplot in the same viewing window. Verify that the line passes through the two points you chose in finding the equation in part (b) and that it comes close to all the data points.

19. The average prices of battery packs for electric cars are shown in Table 20 for various years.

Table 20 Average Prices of Battery Packs for Electric Cars

Year	Average Price (dollars per kilowatt-hour)
2010	1000
2011	800
2012	642
2013	599
2014	540
2015	350
2016	273

Source: *Bloomberg New Energy Finance*

Let p be the average price (in dollars per kilowatt-hour) of battery packs for electric cars at t years since 2010.
a. Construct a scatterplot of the data.
b. Find an equation of a linear model to describe the data.
c. Draw your line and the scatterplot in the same viewing window. Verify that the line passes through the two points that you chose in finding the equation in part (b) and that it comes close to all the data points.
d. What is the slope? What does it mean in this situation?
e. What is the p-intercept? What does it mean in this situation?

20. The amounts of farmland in the United States are shown in Table 21 for various years.

Table 21 Farmland in the United States

Year	Amount of Farmland (millions of acres)
2008	919
2010	916
2012	915
2014	913
2016	911

Source: *Agricultural Statistics Board*

Let F be the amount of farmland (in millions of acres) in the United States at t years since 2000.
a. Use a graphing calculator to draw a scatterplot of the data.
b. Find an equation of a linear model to describe the data.
c. Draw your line and the scatterplot in the same viewing window. Verify that the line passes through the two points that you chose in finding the equation in part (b) and that it comes close to all the data points.
d. What is the slope? What does it mean in this situation?
e. What is the F-intercept? What does it mean in this situation?

21. ⬇DATA World land speed records are shown in Table 22 for various years.

Table 22 Land Speed Records

Year	Record (miles per hour)
1904	91
1914	124
1927	175
1935	301
1947	394
1960	407
1970	622
1983	633
1997	763

Source: *Fédération International de l'Automobile*

Let r be the land speed record (in miles per hour) at t years since 1900.
a. Use a graphing calculator to draw a scatterplot of the data.
b. Find an equation of a linear model to describe the data.
c. Draw your line and the scatterplot in the same viewing window. Verify that the line passes through the two points that you chose in finding the equation in part (b) and that it comes close to all the data points.
d. What is the slope? What does it mean in this situation?
e. What is the r-intercept? What does it mean in this situation?

22. PACE loans are designed to encourage homeowners to buy energy-efficient solar panels, window insulation, and air-conditioning units. The number of states with local governments that have set up PACE loans are shown in Table 23 for various years.

Table 23 Number of States with PACE Loans

Year	Number of States
2010	24
2011	26
2012	27
2013	30
2014	31
2015	34
2016	34

Source: *PACEnation*

Let n be the number of states with PACE loans at t years since 2010.
a. Construct a scatterplot of the data.
b. Find an equation of a linear model to describe the data.
c. Draw your line and the scatterplot in the same viewing window. Verify that the line passes through the two points that you chose in finding the equation in part (b) and that it comes close to all the data points.
d. What is the slope? What does it mean in this situation?
e. What is the n-intercept? What does it mean in this situation?

23. As a nurse's patient load increases, so does the chance that his or her patients will die. Table 24 lists the percent increase in patient mortality for various patient-to-nurse ratios compared with a patient-to-nurse ratio of 4 (that is, 4 to 1).

Table 24 Nurse Staffing and Patient Mortality Rates

Patient-to-Nurse Ratio	Percent Increase in Patient Mortality
4	0
5	7
6	14
7	23
8	31

Source: *University of Pennsylvania*

Let p be the percent increase in patient mortality if the patient-to-nurse ratio is r to 1 rather than 4 to 1.
  a. Use a graphing calculator to draw a scatterplot of the data.
  b. Find an equation of a linear model to describe the data.
  c. Draw your line and the scatterplot in the same viewing window. Verify that the line passes through the two points that you chose in finding the equation in part (b) and that it comes close to all the data points.

24. The numbers of firearm suicides are shown in Table 25 for various years.

Table 25 Numbers of Firearm Suicides

Year	Number of Firearm Suicides (thousands)
2008	18.2
2009	18.7
2010	19.4
2011	20.0
2012	20.7
2013	21.2
2014	21.3

Source: *Centers for Disease Control and Prevention*

Let n be the number (in thousands) of firearm suicides in the year that is t years since 2000.
  a. Construct a scatterplot of the data.
  b. Find an equation of a linear model to describe the data.
  c. Draw your line and the scatterplot in the same viewing window. Verify that the line passes through the two points that you chose in finding the equation in part (b) and that it comes close to all the data points.
  d. What is the slope? What does it mean in this situation?
  e. What is the n-intercept? What does it mean in this situation?

Concepts

25. DATA. Wind-energy capacities in the United States are shown in Table 26 for various years. Three students are to find a linear model to describe the situation. Student A uses the data for 2011 and 2012, Student B uses the data for 2010 and 2013, and Student C uses the data for 2012 and 2013. Which student made the best choice of points? Explain.

Table 26 U.S. Wind-Energy Capacities

Year	Capacity (gigawatts)
2008	25.4
2009	34.9
2010	40.2
2011	46.9
2012	60.0
2013	61.1
2014	65.9
2015	74.5

Source: *U.S. Department of Energy*

26. Levels of U.S. oat production are shown in Table 27 for various years. Three students are to find a linear model to describe the situation. Student A uses the data for 2003 and 2005, Student B uses the data for 2011 and 2013, and Student C uses the data for 2003 and 2014. Which student made the best choice of points? Explain.

Table 27 Levels of U.S. Oat Production

Year	Production (million metric tons)
2003	2.0
2005	2.0
2007	1.6
2009	1.5
2011	1.4
2013	1.1
2014	1.1

Source: *U.S. Agriculture Department*

For Exercises 27 and 28, consider the scatterplot of data and the graph of the model $y = mt + b$ in the indicated figure. Sketch the graph of a linear model that describes the data better, and then explain how you would adjust the values of m and b of the original model so that it would describe the data better.

27. See Fig. 23.

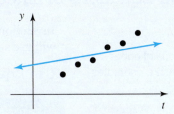

Figure 23 Exercise 27

28. See Fig. 24.

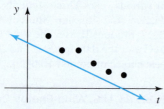

Figure 24 Exercise 28

29. Explain how to find an equation of a linear model for a given situation. Also, explain how you can verify that the line models the situation reasonably well. (See page 9 for guidelines on writing a good response.)

30. Describe the benefits of viewing a scatterplot of data. (See page 9 for guidelines on writing a good response.)

Related Review

31. In 2005, a stock is worth $10. Each year thereafter, the value of the stock increases by $2. Let V be the value (in dollars) of the stock at t years since 2005. Recall that we can describe a linear model using an equation, a table, a graph, or words.
   a. Find an equation of a linear model to describe the situation. Then perform a unit analysis of the equation.
   b. Use a graph to describe the linear model.
   c. Use a table of values of t and V to describe the linear model.

32. A person is on a car trip. At the start of the trip, there are 12 gallons of gasoline in the gasoline tank. The car consumes 3 gallons of gasoline per hour of driving. Let V be the volume of gasoline (in gallons) in the gasoline tank after t hours of driving. Recall that we can describe a linear model using an equation, a table, a graph, or words.
   a. Find an equation of a linear model to describe the situation. Then perform a unit analysis of the equation.
   b. Use a graph to describe the linear model.
   c. Use a table of values of t and V to describe the linear model.

33. The average attendance at college bowl games was 43.8 thousand in 2015 and has decreased by about 1.8 thousand spectators per year since then (Source: *NCAA*). Let n be the average attendance (in thousands) in the year that is t years since 2015.
   a. Find an equation of a linear model to describe the situation.
   b. What is the slope? What does it mean in this situation?
   c. Perform a unit analysis of the equation you found in part (a).

34. The number of U.S. employees who manufacture aerospace products and parts was 622 thousand in 2014 and has decreased by about 13 thousand employees per year since then (Source: *Deloitte*). Let n be the number (in thousands) of U.S. employees who manufacture aerospace products and parts at t years since 2014.
   a. Find an equation of a linear model to describe the data.
   b. What is the slope? What does it mean in this situation?
   c. Perform a unit analysis of the equation that you found in part (a).

35. To rent a 20-foot truck from Metro Truck Rental for one day, it costs $59.95 plus $0.69 per mile (Source: *Metro Truck Rental*). If the total cost to rent such a truck for one day was $119.29, how far was the truck driven?

36. To rent a 24-foot truck from Metro Truck Rental for one day, it costs $99.95 plus $0.89 per mile (Source: *Metro Truck Rental*). If the total cost to rent such a truck for one day was $166.70, how far was the truck driven?

Expressions, Equations, and Graphs

Perform the indicated instruction. Then use words such as linear, one variable, *and* two variables *to describe the expression or equation. For instance, to describe* 2x + 7 = 4, *you could say "*2x + 7 = 4 *is a linear equation in one variable."*

37. Solve $3 = -\frac{2}{5}x + 4$.

38. Simplify $4x - 3(2x - 5) + 1$.

39. Graph $y = -\frac{2}{5}x + 4$ by hand.

40. Solve $4x - 3(2x - 5) + 1 = 0$.

5.4 Using Equations of Linear Models to Make Estimates and Predictions

Objectives

» Use equations of linear models to make estimates and predictions.

» Determine a four-step modeling process.

» Use data described in words to make predictions.

In Section 5.3, we discussed how to find an equation of a linear model. In this section, we will use such an equation to make estimates and predictions.

Use Equations of Linear Models to Make Estimates and Predictions

Example 1   Making Estimates and Predictions

In Example 3 of Section 5.3, we found the equation $p = -0.44a + 93.24$, where p is the percentage of Americans at age a years who consider their home address to be personal information (see Table 28).

Table 28 Americans Who Consider Their Home Address to Be Personal Information

Age Group	Age Used to Represent Age Group	Percent
18–24	21	84
25–34	29.5	80
35–44	39.5	74
45–54	49.5	74
55–64	59.5	67
over 64	75	60

Source: *American Demographics*

1. Estimate the percentage of 19-year-old Americans who consider their home address to be personal information.
2. Estimate the age at which 70% of Americans consider their home address to be personal information.
3. Find the p-intercept. What does it mean in this situation?
4. Find the a-intercept. What does it mean in this situation?
5. What is the slope? What does it mean in this situation?

Solution

1. We substitute 19 for a in the equation $p = -0.44a + 93.24$:

$$p = -0.44(19) + 93.24 = 84.88$$

So, $p = 84.88$ when $a = 19$. According to the model, about 85% of 19-year-old Americans consider their home address to be personal information.

2. Here, we substitute 70 for p in the equation $p = -0.44a + 93.24$ and solve for a:

$70 = -0.44a + 93.24$	*Substitute 70 for p.*
$70 - 93.24 = -0.44a + 93.24 - 93.24$	*Subtract 93.24 from both sides.*
$-23.24 = -0.44a$	*Combine like terms.*
$\dfrac{-23.24}{-0.44} = \dfrac{-0.44a}{-0.44}$	*Divide both sides by -0.44.*
$52.82 \approx a$	*Divide; simplify.*

According to the model, 70% of 53-year-old Americans consider their home address to be personal information.

   We can verify our work in Problems 1 and 2 by using a graphing calculator graph (see Fig. 25). The values of a in these two problems are 19 and 52.82. To set up the window, we select a number (say, -10) that is less than 19 to be Xmin. We select a number (say, 80) that is more than 52.82 to be Xmax. Then we use ZoomFit to adjust Ymin and Ymax.

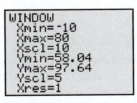
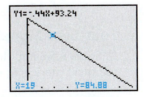
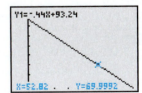

Figure 25 Verify the work

3. Because the model $p = -0.44a + 93.24$ is in slope–intercept form, we see that the p-intercept is $(0, 93.24)$. So, the model estimates that 93.24% of newborn Americans consider their home address to be personal information. Model breakdown has occurred.

Figure 26 Putting the table in "Ask" mode

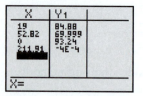

Figure 27 Verify the work

4. To find the *a*-intercept, we substitute 0 for *p* and solve for *a*:

$$0 = -0.44a + 93.24 \qquad \textit{Substitute 0 for p.}$$
$$0 + 0.44a = -0.44a + 93.24 + 0.44a \qquad \textit{Add 0.44a to both sides.}$$
$$0.44a = 93.24 \qquad \textit{Combine like terms.}$$
$$\frac{0.44a}{0.44} = \frac{93.24}{0.44} \qquad \textit{Divide both sides by 0.44.}$$
$$a \approx 211.91 \qquad \textit{Simplify; divide.}$$

The *a*-intercept is $(211.91, 0)$. So, the model estimates that no 212-year-old Americans consider their home address to be personal information. This estimate is correct, but only because no one lives to be 212 years old!

We can verify our work in Problems 1–4 by using a graphing calculator table (see Figs. 26 and 27). The expression "-4E-4" stands for the number -0.0004, which is close to 0. For graphing calculator instructions, see Appendix A.15.

5. The slope is -0.44. According to the model, the percentage of Americans who consider their home address to be personal information decreases by 0.44 percentage point per year of age.

Here we summarize how to use an equation of a model to make predictions (or estimates).

> **Using an Equation of a Linear Model to Make Predictions**
>
> - To make a prediction about the response variable of a linear model, substitute a chosen value for the explanatory variable in the model's equation. Then solve for the response variable.
> - To make a prediction about the explanatory variable of a linear model, substitute a chosen value for the response variable in the model's equation. Then solve for the explanatory variable.

We next describe in general how to use an equation of a model to find the intercepts of the model.

> **Using an Equation of a Linear Model to Find Intercepts**
>
> If an equation of the form $p = mt + b$, where $m \neq 0$, is used to model a situation, then
>
> - The *p*-intercept is $(0, b)$.
> - To find the *t*-coordinate of the *t*-intercept, substitute 0 for *p* in the model's equation and then solve for *t*.

Four-Step Modeling Process

In Section 5.3 and this section, we have discussed how to find linear models and how to use these models to make estimates and predictions. Here is a summary of this process.

> **Four-Step Modeling Process**
>
> To find a linear model and then make estimates and predictions,
>
> 1. Construct a scatterplot of the data to determine whether there is a nonvertical line that comes close to the points. If so, choose two points (not necessarily data points) that you can use to find an equation of a linear model.

2. Find an equation of your model.

3. Verify your equation by checking that the graph of your model contains the two chosen points and comes close to all the data points.

4. Use the equation of your model to make estimates, make predictions, and draw conclusions.

Using Data Described in Words to Make Predictions

In most application problems in this text so far, we have been provided variable names and their definitions. In Example 2, a key step will be to create variable names and define the variables.

Example 2   Making a Prediction

The percentage of Americans who favor legalizing marijuana has increased approximately linearly from 31% in 2002 to 60% in 2016 (Source: *Gallup*). Predict when 75% of Americans will favor legalizing marijuana.

Solution

Let p be the percentage of Americans who favor legalizing marijuana at t years since 2000. Known values of t and p are shown in Table 29. Because the variables t and p are approximately linearly related, we want an equation of the form $p = mt + b$. We can use the data points $(2, 31)$ and $(16, 60)$ to find the values of m and b.

First, we find the slope of the model:

$$m = \frac{60 - 31}{16 - 2} \approx 2.07$$

So, we can substitute 2.07 for m in the equation $p = mt + b$:

$$p = 2.07t + b$$

To find the constant b, we substitute the coordinates of the point $(2, 31)$ into the equation $p = 2.07t + b$ and solve for b:

$$31 = 2.07(2) + b \qquad \text{Substitute 2 for } t \text{ and 31 for } p.$$
$$31 = 4.14 + b \qquad \text{Multiply.}$$
$$31 - 4.14 = 4.14 + b - 4.14 \qquad \text{Subtract 4.14 from both sides.}$$
$$26.86 = b \qquad \text{Combine like terms.}$$

Now we substitute 26.86 for b in the equation $p = 2.07t + b$:

$$p = 2.07t + 26.86$$

Finally, we can predict when 75% of Americans will favor legalizing marijuana by substituting 75 for p in the equation $p = 2.07t + 26.86$ and solving for t:

$$75 = 2.07t + 26.86 \qquad \text{Substitute 75 for } p.$$
$$75 - 26.86 = 2.07t + 26.86 - 26.86 \qquad \text{Subtract 26.86 from both sides.}$$
$$48.14 = 2.07t \qquad \text{Combine like terms.}$$
$$\frac{48.14}{2.07} = \frac{2.07t}{2.07} \qquad \text{Divide both sides by 2.07.}$$
$$23.26 \approx t \qquad \text{Divide; simplify.}$$

The model predicts 75% of Americans will favor legalizing marijuana in $2000 + 23 = 2023$. We use a graphing calculator table to verify our work (see Fig. 28).

Table 29 Known Values of t and p

Years since 2000 t	Percent p
2	31
16	60

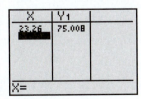

Figure 28 Verify the work

In Example 2, we defined t to be the number of years since *2000*. If we had defined t to be the number of years since *1900* (or any other year), we still would have obtained the same prediction of 2023, although the equation of our model would have been different. So, **if an exercise does not state a year from which to begin counting, choose any year.**

 Group Exploration

DATA Section Opener: Using a model's equation to make estimates and predictions

Consumer spendings for Halloween are shown in Table 30 for various years. Let s be the annual spending (in billions of dollars) for Halloween at t years since 2000.

1. Use a graphing calculator to draw a scatterplot of the data.
2. Find an equation of a model to describe the data.
3. Predict consumer spending for Halloween in 2021.
4. Predict in which year consumers will spend $10 billion for Halloween.
5. What is the slope? What does it mean in this situation?
6. What is the s-intercept? What does it mean in this situation?

Table 30 Halloween Spendings

Year	Spending (billions of dollars)
2007	5.1
2008	5.8
2009	4.7
2010	5.8
2011	6.9
2012	8.0
2013	7.0
2014	7.4
2015	6.9
2016	8.4

Source: *National Retail Federation*

Tips for Success   Stick with It

If you are having difficulty doing an exercise, don't panic! Reread the exercise and reflect on what you have already sorted out about the problem—what you know and where you need to go. Your solution to the problem may be just around the corner.

 # Homework 5.4

For extra help ▶   **MyLab Math**   ▦ Watch the videos in MyLab Math

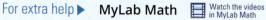

1. In Exercise 13 of Homework 5.3, you found an equation close to $p = 0.61t + 75.61$, where p is the percentage of all airline seats that are filled in the year that is t years since 2000 (see Table 31).

Table 31 Percentage of All Airline Seats Filled

Year	Percent
2006	79.1
2008	80.0
2010	82.2
2012	83.4
2014	84.5
2016	84.7

Source: *Bureau of Transportation Statistics*

a. Predict the percentage of seats that will be filled in 2022.
b. Estimate in which year all seats will be filled.
c. What is the p-intercept? What does it mean in this situation?
d. What is the slope? What does it mean in this situation?

2. In Exercise 14 of Homework 5.3, you found an equation close to $p = 0.46t + 7.06$, where p is the percentage of Americans ages 25 or older who have earned a college degree at t years since 1960 (see Table 32).

Table 32 Percentages of Americans Ages 25 or Older Who Have Earned a College Degree

Year	Percent
1960	7.7
1970	10.7
1980	16.2
1990	21.3
2000	25.6
2010	29.9
2015	32.5

Source: *U.S. Census Bureau*

a. Estimate the percentage of Americans ages 25 or older who will have earned a college degree in 2022.
b. Predict when 35% of Americans ages 25 or older will have earned a college degree.

c. What is the slope? What does it mean in this situation?

d. Find the *t*-intercept. What does it mean in this situation?

3. **DATA** In Exercise 17 of Homework 5.3, you found an equation close to $w = 0.17n - 0.50$, where w is the number of Academy Award best-picture wins for movies of a genre whose movies have had n nominations (see Table 33).

Table 33 Numbers of Academy Award Best-Picture Nominations and Wins

Genre	Nominations	Wins
Action	35	9
Comedy	108	16
Drama	462	80
Fantasy	31	2
Romance	215	36
Sci-Fi	11	0
Thriller	61	8
War	68	17

Source: *Collider, IMDb*

a. Crime movies have been nominated for best picture 57 times. Estimate the number of times crime movies have won best picture.

b. Adventure movies have won best picture 11 times. Estimate the number of times adventure movies have been nominated for best picture.

c. Find the slope. What does it mean in this situation?

d. Which data points are above the linear model? What does that mean in this situation?

4. In Exercise 18 of Homework 5.3, you found an equation close to $r = -0.001p + 1.57$, where r is the Iraqi war death rate (number of deaths for every 100,000 people ages 18 to 54) of soldiers from U.S. counties of population p (in thousands). See Table 34.

Table 34 Iraqi War Death Rates

County Population Group (thousands)	Population Used to Represent County Population Group (thousands)	Death Rate (number of deaths for every 100,000 people ages 18 to 54)
under 25	12.5	1.78
25–50	37.5	1.40
50–100	75.0	1.45
100–250	175.0	1.36
250–500	375.0	1.16
500–1000	750.0	0.85

Source: *The New York Times*

a. Estimate the Iraqi war death rate of St. Charles County, Missouri, which has a population of 361.6 thousand.

b. Estimate the population of a county with an Iraqi war death rate of 1.00 death for every 100,000 people ages 18 to 54.

c. What is the slope? What does it mean in this situation?

d. Assuming commanders are not putting soldiers from rural counties (smaller populations) at greater or less risk than soldiers from urban counties (larger populations), is a person from a rural county more or less likely to enlist in the armed forces, compared with a person from an urban county? [**Hint:** See Table 34.] Why might this be?

5. **DATA** In Exercise 15 of Homework 5.3, you found an equation close to $H = 4.5d + 15.3$, where H is the number of "human years" that is equivalent to d "dog years" for dogs that weigh between 20 pounds and 50 pounds, inclusive (see Table 35).

Table 35 Dog Years Compared with Human Years

Dog Years	Human Years
1	15
2	24
3	29
5	38
7	47
9	56
11	65
13	74
15	83

Source: *Fred Metzger, State College, PA*

a. The life expectancy of an American is about 79 years. Use the model to estimate the life expectancy of dogs.

b. The oldest dog ever was Bluey, an Australian cattle dog, who died at age 29 years. How long did Bluey live in human years?

c. The oldest American ever was Sarah Knauss, who died at age 119 years. How long did Knauss live in dog years?

d. What is the slope? What does it mean in this situation?

e. Although the information in Table 35 is much more accurate, it has long been believed that each dog year is equivalent to 7 human years. Use this old rule of thumb to find another model that describes the relationship between d and H.

f. What is the slope of the graph of the equation you found in part (e)? What does it mean in this situation? Compare this slope with the slope you found in part (d).

6. In Exercise 16 of Homework 5.3, you found an equation close to $E = 1.96t + 7.68$, where E is the enrollment in an elementary algebra course at the College of San Mateo at t weeks since November 22 (see Table 36).

Table 36 Enrollments in an Elementary Algebra Course

Date	Number of Weeks since November 22	Enrollment
November 22	0	8
November 29	1	9
December 6	2	12
December 13	3	13
December 20	4	16
December 27	5	18
January 3	6	19

Source: *J. Lehmann*

a. Find the *E*-intercept. What does it mean in this situation?

b. What is the slope? What does it mean in this situation?

c. Estimate when the enrollment reached 35 students, which is the maximum number of students allowed. The last day to add a class was Monday, January 31, which was 10 weeks after November 22. Has model breakdown occurred? Explain.

d. Estimate the enrollment on the first day of class, January 17, which was 8 weeks after November 22. The actual enrollment on the first day was 32 students. Explain why it is not surprising that your estimation is an underestimate.

7. The percentages of married persons who say they are "very happy" with their marriages are shown in Table 37 for various years.

Table 37 Percentages of Married Persons "Very Happy" with Their Marriages

Year	Percent
1970	67.0
1980	67.5
1990	64.6
2000	62.4
2010	63.0
2014	59.9

Source: *Institute for American Values*

Let p be the percentage of married persons who say they are "very happy" with their marriages at t years since 1970.
a. Use a graphing calculator to draw a scatterplot of the data.
b. Find an equation of a linear model to describe the data.
c. Use your model to predict when 59.5% of married persons will say they are "very happy" with their marriages.
d. Find the p-intercept. What does it mean in this situation?
e. Find the t-intercept. What does it mean in this situation?

8. The in-state tuitions at Austin Community College are shown in Table 38 for various years.

Table 38 In-State Tuitions at Austin Community College

Year	Tuition (dollars)
2007	4038
2009	4563
2011	6375
2013	7860
2015	9210

Source: *Graphiq Inc.*

Let T be the in-state tuition (in dollars) at t years since 2000.
a. Use a graphing calculator to draw a scatterplot of the data.
b. Find an equation of a linear model to describe the data.
c. Use the model to estimate the tuition in 2014.
d. Use the model to predict when the tuition will be $13,000.
e. What is the slope? What does it mean in this situation?

9. During the summer of 2003, a math professor was on a weight-loss program (see Table 39).

Table 39 Weights of the Math Professor

Number of Weeks on Weight-Loss Program	Weight (pounds)
0	159.0
1	157.0
2	157.0
3	155.0
4	154.5
5	152.0

Source: *J. Lehmann*

Let w be the weight (in pounds) of the math professor at t weeks since he started the weight-loss program.
a. Use a graphing calculator to draw a scatterplot of the data.
b. Find an equation of a linear model to describe the data.
c. What is the slope? What does it mean in this situation?
d. Estimate when the math professor reached his goal of 145 pounds.
e. Find the t-intercept. What does it mean in this situation?

10. The percentages of Americans who wear seat belts are shown in Table 40 for various years.

Table 40 Percentages of Americans Who Wear Seat Belts

Year	Percent
2003	79.2
2006	81.2
2009	84.1
2012	86.1
2015	88.5
2016	90.1

Source: *NOPUS*

Let p be the percentage of Americans who wear seat belts at t years since 2000.
a. Use a graphing calculator to draw a scatterplot of the data.
b. Find an equation of a linear model to describe the data.
c. What is the p-intercept? What does it mean in this situation?
d. Predict the percentage of Americans in 2022 who will wear seat belts.
e. Use your model to predict when every American will wear seat belts.

11. The percentages of high school students who dropped out of school are shown in Table 41 for various years.

Table 41 Percentages of High School Students Who Dropped Out of School

Year	Percent
2000	10.9
2003	9.9
2006	9.3
2009	8.1
2012	6.6
2014	6.5

Source: *U.S. Census Bureau*

Let p be the percentage of high school students who dropped out of school during the year that is t years since 2000.
a. Use a graphing calculator to draw a scatterplot of the data.
b. Find an equation of a linear model to describe the data.
c. Predict in which year 4% of high school students will drop out of school.
d. What is the p-intercept? What does it mean in this situation?
e. What is the t-intercept? What does it mean in this situation?

12. If you could stop time and live forever in good health at a particular age, what age would you choose? The average ideal ages chosen by various age groups are shown in Table 42.

Table 42 Ideal Ages

Age Group (years)	Age Used to Represent Age Group (years)	Average Ideal Age (years)
18–24	21	27
25–29	27	31
30–39	34.5	37
40–49	44.5	40
50–64	57	44
over 64	75	59

Source: *Harris Poll*

Let I be the average ideal age (in years) chosen by people whose actual age is a years.
a. Use a graphing calculator to draw a scatterplot of the data.
b. Find an equation of a linear model to describe the data.
c. Use your model to estimate the average ideal age chosen by 18-year-olds.
d. What is the slope? What does it mean in this situation?
e. What is the age of people whose ideal age is equal to their actual age?

13. In Exercise 23 of Homework 5.3, you found an equation close to $p = 7.8r - 31.8$, where p is the percent increase in patient mortality if the patient-to-nurse ratio were r to 1 rather than 4 to 1 (see Table 43).

Table 43 Nurse Staffing Patient Mortality Rates

Patient-to-Nurse Ratio	Percent Increase
4	0
5	7
6	14
7	23
8	31

Source: *University of Pennsylvania*

a. What is the slope? What does it mean in this situation?
b. In January 2004, California passed the first law in the United States that restricts the patient-to-nurse ratio. When the law was first being discussed, some people in the hospital industry felt that the ratio should be 10 to 1. Estimate the percent increase in patient mortality if the ratio were 10 to 1 rather than 4 to 1.
c. Beginning in 2005, the law requires a 5-to-1 ratio (the required ratio will be lower for intensive care and children's wards). Use the model to estimate the percent increase in patient mortality if the ratio is 5 to 1 rather than 4 to 1. Is your result an underestimate or an overestimate? Explain.
d. How does lowering the patient-to-nurse ratio affect the number of patients who can be admitted, the total labor costs for nurses, and the mortality rates of patients?

14. DATA. The prices of hot dogs and soft drinks at National League baseball stadiums are shown in Table 44. The two prices of the same color and row are from the same stadium.

Table 44 Prices of Hot Dogs and Soft Drinks at National League Baseball Stadiums

Price (dollars)			
Hot Dog	Soft Drink	Hot Dog	Soft Drink
5.50	4.00	4.25	5.25
3.50	2.50	3.25	3.25
1.00	1.00	3.75	4.00
6.25	5.00	5.00	5.00
6.00	4.50	4.75	4.75
5.25	4.50	5.50	5.75
4.75	3.25	4.00	4.00
2.75	1.50		

Source: *Team Marketing Report*

Let h be the price of hot dogs and s be the price of soft drinks, both in dollars, at a National League baseball stadium.
a. Use a graphing calculator to draw a scatterplot of the data. Treat the price of a hot dog as the explanatory variable and the price of a soft drink as the response variable.
b. Find an equation of a linear model to describe the data.
c. Estimate the price of a hot dog at Wrigley Field, home of the Chicago Cubs, where the price of a soft drink is $4.00.
d. What is the slope? What does it mean in this situation?
e. Which data point is the farthest above the linear model? What does this tell you about the situation?

15. The percentage of households with wireless telephones and no land-line telephones has increased approximately linearly from 22.7% in 2009 to 44.1% in 2014 (Source: *National Center for Health Statistics*). Predict when 78% of households will have wireless telephones and no land-line telephones.

16. The percentage of U.S. public libraries offering free access to e-books has increased approximately linearly from 38% in 2007 to 90% in 2014 (Source: *American Library Association*). Estimate when all U.S. public libraries offered free access to e-books.

17. The acreage in the Central Valley in California for growing grapes to produce raisins has decreased approximately linearly from 280 thousand acres in 2000 to 165 thousand acres in 2016 (Source: *California Agriculture Statistics Service*). Predict the acreage in 2021.

18. The percentage of Americans who say religion is very important in their lives has decreased approximately linearly from 61% in 2003 to 53% in 2016 (Source: *Gallup*). Predict when half of Americans will say religion is very important in their lives.

19. Revenue from home security systems has increased approximately linearly from $3.2 billion in 2007 to $4.4 billion in 2015 (Source: *Consumer Technology Association*). Predict the revenue in 2022.

20. The percentage of Americans who believe humans evolved, but God had no part in the process, has increased approximately linearly from 12% in 2001 to 19% in 2014 (Source: *Gallup*). Predict when the percentage will be 23%.

21. The percentage of mothers who smoke cigarettes during pregnancy has decreased approximately linearly from 13.9% in 1995 to 8.4% in 2014 (Source: *Centers for Disease Control and Prevention*).

a. Predict when 7% of mothers will smoke cigarettes during pregnancy.

b. The U.S. Department of Health and Human Services has set a goal that at most 1.4% of mothers smoke cigarettes during pregnancy in 2020. If the average rate of change of the percentage of mothers who smoke cigarettes during pregnancy continues to be the same as for the period 1995–2014, will this goal be met? Explain.

22. The population of East Coast's summer flounder are measured by the total weight of fish that are large enough to spawn. The total weight has decreased approximately linearly from 46 thousand metric tons in 2010 to 37 thousand metric tons in 2015 (Source: *Northeast Fisheries Science Center*).

a. Predict the total weight in 2021.

b. In 1989, the total weight was only 5 thousand metric tons due to overfishing. Predict when the total weight will return to 5 thousand metric tons.

Large Data Sets

23. DATA Access the data about the San Mateo Real Estate, which are available at MyLab Math and at the Pearson Downloadable Student Resources for Math & Stats website. Let p be the sales price (in dollars) of a home with x square feet.

a. Construct a scatterplot that describes the relationship between the square footages and the sales prices of the homes.

b. Find an equation of a linear model to describe the data.

c. Print the scatterplot and the graph of the model on the same coordinate system.

d. Estimate the sales price of a home with 2000 square feet.

e. Estimate the square footage of a home with sales price $2,900,000.

f. Find the slope of the graph of the model. What does it mean in this situation?

g. What is the square footage and the sales price of the home represented by the data point that is farthest from the model? How much is the sale price above what the model estimates?

24. DATA Access the data about wooden roller coasters, which are available at MyLab Math and at the Pearson Downloadable Student Resources for Math & Stats website. Let L be the length (in feet) of a wooden roller coaster with height H (in feet).

a. Construct a scatterplot that describes the relationship between the lengths and heights of the wooden roller coasters.

b. Find an equation of a linear model to describe the data.

c. Print the scatterplot and the graph of the model on the same coordinate system.

d. Estimate the length of a wooden roller coaster with height 30 feet.

e. Estimate the height of a wooden roller coaster with length 1600 feet.

f. Find the slope of the graph of the model. What does it mean in this situation?

g. Find the L-intercept. What does it mean in this situation?

Concepts

25. Describe in your own words the four-step modeling process.

26. For a linear model of the form $w = mt + b$, where m and b are constants, describe how to find the intercepts.

Related Review

27. The number of Americans who live alone was 35.0 million in 2015 and has increased by about 0.7 million each year since then (Source: *U.S. Census Bureau*). Let n be the number (in millions) of Americans who live alone at t years since 2015.

a. Find an equation of a linear model to describe the data.

b. What is the n-intercept? What does it mean in this situation?

c. Perform a unit analysis of the equation you found in part (a).

d. The population of California, the most populous state, is 38.8 million. Predict when 38.8 million people will live alone.

28. The number of immigrants in the United States was 42.4 million in 2014 and has increased by about 0.62 million per year since then (Source: *U.S. Census Bureau*). Let n be the number (in millions) of immigrants in the United States at t years since 2014.

a. Find an equation of a linear model to describe the data.

b. Perform a unit analysis of the equation you found in part (a).

c. The population of Spain is 48.1 million (Source: *CIA World Factbook*). Predict when there will be 48.1 million immigrants in the United States.

d. Predict the number of immigrants in 2021.

29. The percentage of Cuban-Americans in Miami-Dade who favored a U.S. trade embargo of Cuba was 87% in 1991 and has decreased by about 1.7 percentage points per year since then (Source: *Florida International University*). Let p be the percentage of Cuban-Americans in Miami-Dade who favored the embargo at t years since 1991.

a. Find an equation of a linear model to describe the data.

b. What is the p-intercept of the model? What does it mean in this situation?

c. Perform a unit analysis of the equation you found in part (a).

d. Estimate the percentage of Cuban-Americans in Miami-Dade who favored the embargo in 2016. Explain why it is not surprising that President Obama called for lifting the embargo in his State of the Union Address in 2016?

30. The unemployment rate in January was 5.7% in 2015 and has decreased by about 0.8 percentage point per year since then (Source: *Bureau of Labor Statistics*). Let r be the unemployment rate in January at t years since 2015.

a. Find an equation of a linear model to describe the data.

b. Perform a unit analysis of the equation you found in part (a).

c. Estimate the unemployment rate in 2017.

d. What is the t-intercept? What does it mean in this situation?

31. The company iRobot® Corporation builds products such as military and vacuum robots. The company's revenue was $560 million in 2015 and has increased by about $68.9 million per year since then (Source: *iRobot Corporation*). Predict when the revenue will be $900 million.

32. Revenue from herbal dietary supplements in the United States was $6.44 billion in 2014 and has increased by about $0.35 billion per year since then (Source: *Nutrition Business Journal*). Predict when the revenue will be $9 billion.

Expressions, Equations, and Graphs

Perform the indicated instruction. Then use words such as linear, one variable, *and* two variables *to describe the expression or equation.*

33. Solve $-4(3x - 5) = 3(2x + 1)$.

34. Graph $4x - 3y + 3 = 0$ by hand.

35. Simplify $-4(3x - 5) - 3(2x + 1)$.

36. Evaluate $4x - 3y + 3$ for $x = -2$ and $y = -3$.

5.5 Solving Linear Inequalities in One Variable

Objectives

» Describe the meaning of *inequality symbols* and *inequality*.

» Graph an inequality.

» Apply the properties of inequalities.

» Describe the meaning of *satisfy, solution,* and *solution set* for a *linear inequality in one variable*.

» Solve a linear inequality in one variable, and graph the solution set.

» Use linear inequalities to make predictions about authentic situations.

In Section 5.4, we predicted when a quantity would reach a certain amount. In this section, we will discuss how to use a model to predict when a quantity will be more than (or less than) a certain amount. For instance, in Example 11, we will predict the years when the annual U.S. revenue from music subscription and streaming services will be more than $5 billion.

Inequality Symbols and Inequalities

We use the **inequality symbols** $<, \leq, >,$ and $\geq$ to compare the sizes of two quantities. Here are the meanings of these symbols and some examples of *inequalities*:

Symbol	Meaning	Examples of Inequalities
$<$	is less than	$2 < 5, 0 < 5, -6 < -1$
$\leq$	is less than or equal to	$4 \leq 7, 2 \leq 2, -3 \leq 0$
$>$	is greater than	$9 > 2, -4 > -6, 2 > 0$
$\geq$	is greater than or equal to	$8 \geq 3, 5 \geq 5, -2 \geq -8$

An **inequality** contains one of the symbols $<, \leq, >,$ and $\geq$ with expressions on both sides. Here are some more examples of inequalities:

$$2x + 5 < 3x - 9 \qquad x \leq 5 \qquad 7 > 2 \qquad 4x - 1 \geq 6$$

Example 1   Inequalities

Decide whether the inequality statement is true or false.

1. $3 \leq 6$ **2.** $-5 > -2$ **3.** $8 \geq 8$ **4.** $9 < 9$

Solution

1. Because 3 is less than 6, the statement $3 \leq 6$ is true.
2. Because -5 lies to the left of -2 on the number line, -5 is less than -2. So, -5 is *not* greater than -2, and the statement $-5 > -2$ is false.
3. Because 8 is equal to itself, the statement $8 \geq 8$ is true.
4. Because 9 is not less than itself, the statement $9 < 9$ is false.

Graphing Inequalities

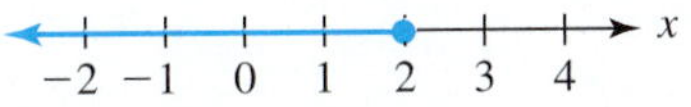

Figure 29 Graph of $x \leq 2$

Figure 30 Graph of $x < 2$

Consider the inequality $x \leq 2$. This inequality says the values of x are less than or equal to 2. We can represent these values graphically on a number line by shading the part of the number line that lies to the left of 2 (see Fig. 29). We draw a *filled-in* circle at 2 to indicate that 2 is a value of x, too.

To graph the inequality $x < 2$, we shade the part of the number line that lies to the left of 2, but draw an *open* circle at 2 to indicate that 2 is *not* a value of x (see Fig. 30).

We use **interval notation** to describe a set of numbers. For example, we describe the numbers greater than 3 by $(3, \infty)$. We describe the numbers greater than or equal to 3 by $[3, \infty)$. We describe the set of all real numbers by $(-\infty, \infty)$. More examples of inequalities and interval notation are shown in Fig. 31.

In Words	Inequality	Graph	Interval Notation
numbers less than 3	$x < 3$	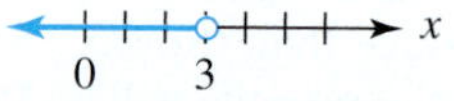	$(-\infty, 3)$
numbers less than or equal to 3	$x \leq 3$		$(-\infty, 3]$
numbers greater than 3	$x > 3$	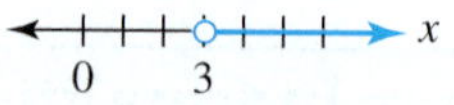	$(3, \infty)$
numbers greater than or equal to 3	$x \geq 3$	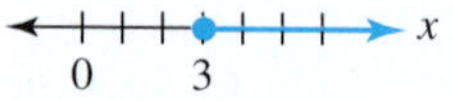	$[3, \infty)$

Figure 31 Words, inequalities, graphs, and interval notation

Example 2   Graphing an Inequality

Write the inequality $x > -2$ in interval notation, and graph the values of x.

Solution

Figure 32 Graph of $x > -2$

The inequality $x > -2$ means that the values of x are greater than -2. We describe these numbers in interval notation by $(-2, \infty)$. To graph the values of x, we shade the part of the number line that lies to the right of -2 and draw an open circle at -2 (see Fig. 32).

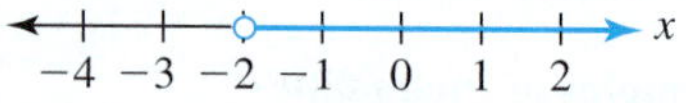

Addition Property of Inequalities

What happens if we add 3 to both sides of the inequality $4 < 6$?

$$4 < 6 \qquad \text{\textit{Original inequality}}$$
$$4 + 3 \overset{?}{<} 6 + 3 \qquad \text{\textit{Add 3 to both sides.}}$$
$$7 \overset{?}{<} 9 \qquad \text{\textit{Simplify.}}$$
$$\text{true}$$

What happens if we add -3 to both sides of the inequality $4 < 6$?

$$4 < 6 \qquad \text{\textit{Original inequality}}$$
$$4 + (-3) \overset{?}{<} 6 + (-3) \qquad \text{\textit{Add -3 to both sides.}}$$
$$1 \overset{?}{<} 3 \qquad \text{\textit{Simplify.}}$$
$$\text{true}$$

These examples suggest the following property:

Addition Property of Inequalities

$$\text{If } a < b, \text{ then } a + c < b + c.$$

Similar properties hold for $\leq$, $>$, and $\geq$.

Similar rules hold for subtraction because subtracting a number is the same as adding the opposite of the number.

We can use a number line to illustrate that if $a < b$, then $a + c < b + c$. From Figs. 33 and 34, we see that if a lies to the left of b ($a < b$), then $a + c$ lies to the left of $b + c$ ($a + c < b + c$). In Fig. 33, c is negative; in Fig. 34, c is positive.

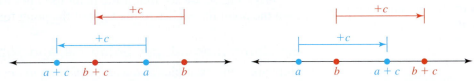

Figure 33 Adding c where c is negative   **Figure 34** Adding c where c is positive

Multiplication Property of Inequalities

What if we multiply both sides of the inequality $4 < 6$ by 3?

$$4 < 6 \qquad \textit{Original inequality}$$
$$4(3) \overset{?}{<} 6(3) \qquad \textit{Multiply both sides by 3.}$$
$$12 \overset{?}{<} 18 \qquad \textit{Simplify.}$$
$$\text{true}$$

Finally, what happens if we multiply both sides of $4 < 6$ by -3?

$$4 < 6 \qquad \textit{Original inequality}$$
$$4(-3) \overset{?}{<} 6(-3) \qquad \textit{Multiply both sides by } -3.$$
$$-12 \overset{?}{<} -18 \qquad \textit{Simplify.}$$
$$\text{false}$$

The result is the false statement $-12 < -18$. We can get a *true* statement if we *reverse the inequality symbol* when we multiply both sides of $4 < 6$ by -3:

$$4 < 6 \qquad \textit{Original inequality}$$
$$4(-3) \overset{?}{>} 6(-3) \qquad \textit{Reverse inequality symbol.}$$
$$-12 \overset{?}{>} -18 \qquad \textit{Simplify.}$$
$$\text{true}$$

So, when we multiply both sides of an inequality by a *negative* number, we *reverse* the inequality symbol.

Multiplication Property of Inequalities

- For a *positive* number c, if $a < b$, then $ac < bc$.
- For a *negative* number c, if $a < b$, then $ac > bc$.

Similar properties hold for $\leq$, $>$, and $\geq$. In words: When we multiply both sides of an inequality by a positive number, we keep the inequality symbol. When we multiply both sides by a negative number, we reverse the inequality symbol.

Similar rules apply for division because dividing by a nonzero number is the same as multiplying by its reciprocal. Therefore, **when we multiply or divide both sides of an inequality by a negative number, we reverse the inequality symbol.**

Consider multiplying both sides of $a < b$ by -1:

$$a < b \qquad \textit{Original inequality}$$
$$-1a > -1b \qquad \textit{Reverse inequality symbol.}$$
$$-a > -b \qquad -1a = -a$$

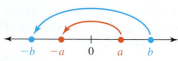

Figure 35 The points for *a*, *b*, −*a*, and −*b*

So, if $a < b$, then $-a > -b$. We can use a number line to illustrate this fact. To plot the point for $-a$, we move the point for a to the other side of the origin so that the points for $-a$ and a are the same distance from the origin (see Fig. 35).

From Fig. 35 we see that if the point for a lies to the *left* of the point for b ($a < b$), then the point for $-a$ lies to the *right* of the point for $-b$ ($-a > -b$).

Satisfying Inequalities, Solutions, and Solution Sets

Here are some examples of *linear inequalities in one variable*:

$$3x + 5 < 8, \qquad 2x \leq 5, \qquad x - 5 > 4 - 2x, \qquad 5(x - 3) \geq 1$$

Definition   Linear inequality in one variable

A **linear inequality in one variable** is an inequality that can be put into one of the forms

$$mx + b < 0, \qquad mx + b \leq 0, \qquad mx + b > 0, \qquad mx + b \geq 0$$

where m and b are constants and $m \neq 0$.

We say a number **satisfies** an inequality in one variable if the inequality becomes a true statement after we have substituted the number for the variable.

Example 3   Identifying Solutions of an Inequality

1. Does the number 4 satisfy the inequality $2x - 3 < 7$?
2. Does the number 6 satisfy the inequality $2x - 3 < 7$?

Solution

1. We substitute 4 for x in the inequality $2x - 3 < 7$:

$$2(4) - 3 \overset{?}{<} 7 \quad \textit{Substitute 4 for x.}$$
$$8 - 3 \overset{?}{<} 7 \quad \textit{Multiply.}$$
$$5 \overset{?}{<} 7 \quad \textit{Subtract.}$$
$$\text{true}$$

So, 4 satisfies the inequality $2x - 3 < 7$.

2. We substitute 6 for x in the inequality $2x - 3 < 7$:

$$2(6) - 3 \overset{?}{<} 7 \quad \textit{Substitute 6 for x.}$$
$$12 - 3 \overset{?}{<} 7 \quad \textit{Multiply.}$$
$$9 \overset{?}{<} 7 \quad \textit{Subtract.}$$
$$\text{false}$$

So, 6 does not satisfy the inequality $2x - 3 < 7$.

▶

Definition   *Solution*, *solution set*, and *solve* for an inequality in one variable

We say that a number is a **solution** of an inequality in one variable if it satisfies the inequality. The **solution set** of an inequality is the set of all solutions of the inequality. We **solve** an inequality by finding its solution set.

Solving a Linear Inequality

To solve a linear inequality in one variable, we apply properties of inequalities to get the variable alone on one side of the inequality. The steps are similar to solving a linear equation in one variable, but we must remember to reverse the inequality symbol when we multiply or divide both sides of an inequality by a negative number.

Example 4   Solving a Linear Inequality

Solve $2x - 3 < 7$. Describe the solution set as an inequality, in interval notation, and as a graph.

Solution

We get x alone on one side of the inequality:

$$2x - 3 < 7 \qquad \textit{Original inequality}$$
$$2x - 3 + 3 < 7 + 3 \qquad \textit{Add 3 to both sides.}$$
$$2x < 10 \qquad \textit{Combine like terms.}$$
$$\frac{2x}{2} < \frac{10}{2} \qquad \textit{Divide both sides by 2.}$$
$$x < 5 \qquad \textit{Simplify.}$$

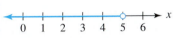

Figure 36 Graph of $x < 5$

The solution set is the set of all numbers less than 5, which we describe in interval notation as $(-\infty, 5)$. We graph the solution set on a number line in Fig. 36.

In Example 3, we found that 4 is a solution of the inequality $2x - 3 < 7$ but 6 is not. This checks with our work in Example 4 because 4 is on the graph in Fig. 36 and 6 is not.

Example 5   Solving a Linear Inequality

Solve the inequality $-3x \geq -12$. Describe the solution set as an inequality, in interval notation, and as a graph.

Solution

We divide both sides of the inequality by -3, a negative number:

$$-3x \geq -12 \qquad \textit{Original inequality}$$
$$\frac{-3x}{-3} \leq \frac{-12}{-3} \qquad \textit{Divide both sides by -3; reverse inequality symbol.}$$
$$x \leq 4 \qquad \textit{Simplify.}$$

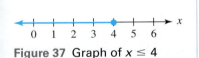

Figure 37 Graph of $x \leq 4$

Because we divided both sides of the inequality by a negative number, we reversed the inequality symbol. The solution set is $(-\infty, 4]$. We graph the solution set in Fig. 37.

WARNING   It is a common error to forget to reverse an inequality symbol when multiplying or dividing both sides of an inequality by a negative number. For instance, in Example 5, it is important that we reversed the inequality symbol $\geq$ when we divided both sides of the inequality $-3x \geq -12$ by -3.

Example 6   Solving a Linear Inequality

Solve the inequality $3x \geq -12$. Describe the solution set as an inequality, in interval notation, and as a graph.

Solution

We divide both sides of the inequality by 3, a positive number:

$$3x \geq -12 \qquad \textit{Original inequality}$$

$$\frac{3x}{3} \geq \frac{-12}{3} \qquad \textit{Divide both sides by 3.}$$

$$x \geq -4 \qquad \textit{Simplify.}$$

Because we divided both sides of the inequality by a positive number, we did *not* reverse the inequality symbol. We graph the solution set, $[-4, \infty)$, in Fig. 38.

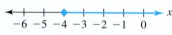

Figure 38 Graph of $x \geq -4$

Example 7   Solving a Linear Inequality

Solve $2x - 5 > 6x + 3$. Describe the solution set as an inequality, in interval notation, and as a graph.

Solution

$$2x - 5 > 6x + 3 \qquad \textit{Original inequality}$$

$$2x - 5 - 6x > 6x + 3 - 6x \qquad \textit{Subtract 6x from both sides.}$$

$$-4x - 5 > 3 \qquad \textit{Combine like terms.}$$

$$-4x - 5 + 5 > 3 + 5 \qquad \textit{Add 5 to both sides.}$$

$$-4x > 8 \qquad \textit{Combine like terms.}$$

$$\frac{-4x}{-4} < \frac{8}{-4} \qquad \textit{Divide both sides by }-4\textit{; reverse inequality symbol.}$$

$$x < -2 \qquad \textit{Simplify.}$$

Figure 39 Graph of $x < -2$

We graph the solution set, $(-\infty, -2)$, in Fig. 39.

To verify our result, we use a graphing calculator table to check that, for some values of x less than -2, the value of $2x - 5$ is greater than the value of $6x + 3$ (see Fig. 40). We do this by setting up the table so x begins at -2 and increases by 1. Then we scroll up 3 rows so we can view values of x that are less than -2 and values of x that are greater than -2.

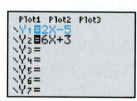

Figure 40 Verify the work

Example 8   Solving a Linear Inequality

Solve $3(x - 2) < 5x - 3$. Describe the solution set as an inequality, in interval notation, and as a graph.

Solution

$$3(x - 2) < 5x - 3 \qquad \textit{Original inequality}$$

$$3x - 6 < 5x - 3 \qquad \textit{Distributive law}$$

$$3x - 6 - 5x < 5x - 3 - 5x \qquad \textit{Subtract 5x from both sides.}$$

$$-2x - 6 < -3 \qquad \textit{Combine like terms.}$$

$$-2x - 6 + 6 < -3 + 6 \qquad \textit{Add 6 to both sides.}$$

$$-2x < 3 \qquad \textit{Combine like terms.}$$

$$\frac{-2x}{-2} > \frac{3}{-2} \qquad \textit{Divide both sides by }-2\textit{; reverse inequality symbol.}$$

$$x > -\frac{3}{2} \qquad \textit{Simplify.}$$

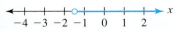

Figure 41 Graph of $x > -\dfrac{3}{2}$

We graph the solution set, $\left(-\dfrac{3}{2}, \infty\right)$, in Fig. 41.

To verify our result, we use a graphing calculator table to check that, for some values of x greater than $-\dfrac{3}{2}$, the value of $3(x - 2)$ is less than the value of $5x - 3$ (see Fig. 42). We do this by setting up the table so x begins at -1.5 and increases by 1. Then we scroll up 3 rows so we can view values of x that are less than -1.5 and values of x that are greater than -1.5.

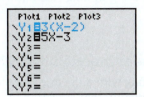
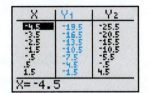

Figure 42 Verify the work

Example 9 Solving a Linear Inequality That Contains Decimals

Solve $-2.9x + 4.1 \le -6.05$. Describe the solution set as an inequality, in interval notation, and as a graph.

Solution

$$
\begin{aligned}
-2.9x + 4.1 &\le -6.05 && \textit{Original inequality} \\
-2.9x + 4.1 - 4.1 &\le -6.05 - 4.1 && \textit{Subtract 4.1 from both sides.} \\
-2.9x &\le -10.15 && \textit{Combine like terms.} \\
\frac{-2.9x}{-2.9} &\ge \frac{-10.15}{-2.9} && \textit{Divide both sides by } -2.9; \textit{ reverse inequality} \\
& && \textit{symbol.} \\
x &\ge 3.5 && \textit{Simplify.}
\end{aligned}
$$

Figure 43 Graph of $x \ge 3.5$

We graph the solution set, $[3.5, \infty)$, in Fig. 43.

We can use graphing calculator tables to check our work.

Example 10 Solving a Linear Inequality That Contains Fractions

Solve $\dfrac{2}{3}w + \dfrac{1}{2} \le \dfrac{5}{6}$. Describe the solution set as an inequality, in interval notation, and as a graph.

Solution

We begin by multiplying both sides by the LCD, 6, to clear the fractions:

$$
\begin{aligned}
\frac{2}{3}w + \frac{1}{2} &\le \frac{5}{6} && \textit{Original inequality} \\
6 \cdot \left(\frac{2}{3}w + \frac{1}{2}\right) &\le 6 \cdot \frac{5}{6} && \textit{Multiply both sides by 6.} \\
6 \cdot \frac{2}{3}w + 6 \cdot \frac{1}{2} &\le 6 \cdot \frac{5}{6} && \textit{Distributive law} \\
4w + 3 &\le 5 && \textit{Simplify.} \\
4w + 3 - 3 &\le 5 - 3 && \textit{Subtract 3 from both sides.} \\
4w &\le 2 && \textit{Simplify.} \\
\frac{4w}{4} &\le \frac{2}{4} && \textit{Divide both sides by 4.} \\
w &\le \frac{1}{2} && \textit{Simplify.}
\end{aligned}
$$

We graph the solution set, $\left(-\infty, \dfrac{1}{2}\right]$, in Fig. 44.

$$
\xleftarrow{\hspace{1cm}} \underset{\substack{-3\ \ -2\ \ -1\ \ \ 0\ \ \ 1\ \ \ 2\ \ \ 3}}{\rule{0pt}{0pt}} \xrightarrow{\hspace{1cm}} w
$$

Figure 44 Graph of $w \le \dfrac{1}{2}$

We can use graphing calculator tables to check our work.

Using Linear Inequalities to Make Predictions

When working with a linear model, we can make certain types of predictions by solving a linear inequality that is related to the linear model.

Example 11   Making a Prediction

Revenues from music subscription and streaming services in the United States are shown in Table 45 for various years.

Table 45 Annual U.S. Revenues from Music Subscription and Streaming Services

Year	Annual Revenue (billions of dollars)
2011	0.65
2012	1.03
2013	1.45
2014	1.87
2015	2.41

Source: *Recording Industry Association of America*

Let r be the annual revenue (in billions of dollars) at t years since 2010. A reasonable model is

$$r = 0.44t + 0.17$$

Predict the years when the annual revenue will be more than \$5 billion.

Solution

To predict when the annual revenue will be more than \$5 billion, we find the values of t such that the expression $0.44t + 0.17$ is more than 5:

$$0.44t + 0.17 > 5$$
$$0.44t + 0.17 - 0.17 > 5 - 0.17 \qquad \textit{Subtract 0.17 from both sides.}$$
$$0.44t > 4.83 \qquad \textit{Combine like terms.}$$
$$\frac{0.44t}{0.44} > \frac{4.83}{0.44} \qquad \textit{Divide both sides by 0.44.}$$
$$t > 10.98 \qquad \textit{Simplify; divide.}$$

So, the annual revenue will be more than \$5 billion after 2021 ($t > 11$).

Stage-fright control. It plays the entire solo of "Free Bird" in a pinch.

What's the button for?

Group Exploration

Section Opener: Properties of inequalities

The symbol "$<$" means *is less than*. For example, the statement $2 < 5$ means that 2 is less than 5, which is true. The statement $-5 < -3$ means that -5 is less than -3, which is also true.

The symbol "$>$" means *is greater than*. For example, the statement $6 > 1$ means that 6 is greater than 1, which is true.

Statements of the form $a < b$ or $a > b$ are examples of *inequalities*.

1. Decide whether each inequality is true.
   a. $4 < 9$ **b.** $3 > 8$ **c.** $-1 < -7$ **d.** $-2 > -6$

2. Decide whether performing the given operation on both sides of the true inequality $4 < 6$ will give a true statement.
   a. Add 2 to both sides.
   b. Add -2 to both sides.
   c. Multiply both sides by 2.
   d. Multiply both sides by -2.

3. For any true inequality (such as $3 < 7$), can you add the given type of number to both sides of the inequality and then still have a true inequality?
   a. positive number   **b.** negative number   **c.** zero

4. For any true inequality, can you multiply both sides of the inequality by the given type of number and then still have a true inequality?
   a. positive number   **b.** negative number   **c.** zero

5. Recall from Section 2.4 that subtracting a number is the same as adding the opposite of that number. What does this tell you about subtracting a number from both sides of an inequality?

6. Recall from Section 2.2 that dividing by a nonzero number is the same as multiplying by the reciprocal of that number. What does this tell you about dividing both sides of an inequality by a nonzero number?

Tips for Success   Scan Test Problems

When you take a test, scan the test problems quickly, pick the ones with which you feel most comfortable, and complete those problems first. By doing so, you will warm up and gain confidence, and you may do better on the rest of the test. Also, you will probably have a better idea of how to allot your time on the remaining problems.

Homework 5.5

For extra help ▶   **MyLab Math**   ▦ Watch the videos in MyLab Math

Decide whether the given inequality is true or false.

1. $-3 > -5$ **2.** $-6 \leq -2$

3. $4 \geq 4$ **4.** $-5 < -5$

For Exercises 5–12, sketch the graph of the given inequality.

5. $x < 4$ **6.** $x < -5$

7. $x \geq -1$ **8.** $x \geq 1$

9. $x \leq -2$ **10.** $x \leq 4$

11. $x > 6$ **12.** $x > -3$

13. Use words, inequalities, graphs, and interval notation to complete Fig. 45.

In Words	Inequality	Graph	Interval Notation
		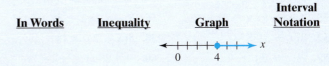	
numbers less than or equal to -2			
			$(-\infty, 1)$
	$x > -5$		

Figure 45 Exercise 13

14. Use words, inequalities, graphs, and interval notation to complete Fig. 46.

In Words	Inequality	Graph	Interval Notation
	$x \le -6$		
numbers greater than 1			
			$[-4, \infty)$
		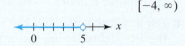	

Figure 46 Exercise 14

Which of the given numbers satisfy the given inequality?

15. $3x + 5 \ge 14$; $2, 3, 6$ **16.** $-4x - 7 > 1$; $-3, -2, 0$

17. $2x < x + 2$; $-4, 2, 3$ **18.** $x - 9 \le 4x$; $-3, 2, 5$

For Exercises 19–74, solve the inequality. Describe the solution set as an inequality, in interval notation, and as a graph. Then use a graphing calculator table to verify your result.

19. $x + 2 > 3$ **20.** $x + 5 \le 9$ **21.** $x - 1 < -4$

22. $x - 3 \ge -1$ **23.** $2x \le 6$ **24.** $3x > 9$

25. $4x \ge -8$ **26.** $2x < -10$ **27.** $-3t \ge 6$

28. $-2w \le 2$ **29.** $-5x < 20$ **30.** $-7x > 21$

31. $-2x > 1$ **32.** $-4x < -2$ **33.** $5x \le 0$

34. $-3x > 0$ **35.** $-x < 2$ **36.** $-x \le -1$

37. $\frac{1}{2}a < 3$ **38.** $\frac{1}{3}r > -2$ **39.** $-\frac{2}{3}x \ge 2$

40. $-\frac{5}{2}x \le 10$ **41.** $3x - 1 \ge 2$ **42.** $4x + 7 < 15$

43. $5 - 3x < -7$ **44.** $8 - 2x \le 6$

45. $-4x - 3 \ge 5$ **46.** $-9x + 7 > -11$

47. $3c - 6 \le 5c$ **48.** $7w + 4 > 3w$

49. $5x \ge x - 12$ **50.** $4x < 6 - 2x$

51. $-3.8x + 1.9 > -7.6$ **52.** $-2.4x + 5.8 \le 8.92$

53. $3b + 2 > 7b - 6$ **54.** $5k - 1 \le 2k + 8$

55. $4 - 3x < 9 - 2x$ **56.** $8 - x \ge 2 - 3x$

57. $2(x + 3) \le 8$ **58.** $5(x - 2) \ge 15$

59. $-(a - 3) > 4$ **60.** $-(t + 5) < -2$

61. $3(2x - 1) \le 2(2x + 1)$ **62.** $4(3x - 5) > 5(2x - 3)$

63. $4(2x - 3) + 1 \ge 3(4x - 5) - x$

64. $-2(5x + 3) - 2x < -3(2x - 4) + 2$

65. $4.3(1.5 - x) \ge 13.76$ **66.** $3.1(2.7 - x) > -1.55$

67. $\frac{1}{2}y + \frac{2}{3} \ge \frac{3}{2}$ **68.** $\frac{3}{4}t - \frac{1}{2} \le \frac{1}{4}$

69. $\frac{5}{3} - \frac{1}{6}x < \frac{1}{2}$ **70.** $\frac{1}{4} - \frac{2}{3}x > \frac{7}{12}$

71. $\frac{2(3 - x)}{3} > -4x$ **72.** $\frac{3(4 - x)}{2} \ge -6x$

73. $\frac{3r - 5}{4} \le \frac{2r - 3}{3}$ **74.** $\frac{4k + 6}{7} > \frac{3k + 4}{5}$

75. Total student loan amounts are shown in Table 46 for various years.

Table 46 Total Student Loan Amounts

Year	Total Student Loan Amount (trillions of dollars)
2011	0.96
2012	1.06
2013	1.15
2014	1.24
2015	1.32

Source: *Board of Governors of the Federal Reserve System*

Let L be the total student loan amount (in trillions of dollars) at t years since 2010. A reasonable model is $L = 0.09t + 0.88$.
a. What is the slope? What does it mean in this situation?
b. In which years will the total student loan amount be more than \$2 trillion?

76. The percentages of osteopathic doctors who are women are shown in Table 47 for various years.

Table 47 Percentages of Osteopathic Doctors Who Are Women

Year	Percent
1990	13.9
1995	17.9
2000	22.8
2005	26.7
2010	32.0
2015	40.0

Source: *American Medical Association*

Let P be the percentage of osteopathic doctors who are women at t years since 1990. A reasonable model is $P = 1.01t + 12.93$.
a. What is the slope? What does it mean in this situation?
b. In which years will more than 43% of osteopathic doctors be women?

77. The numbers of married-couple households (in millions) and the percentages of households that are married-couple households are shown in Table 48 for various years.

Table 48 Married-Couple Households

Year	Number of Married-Couple Households (millions)	Percentage of Households That Are Married-Couple Households
1975	47.6	66.9
1985	51.1	58.9
1995	54.9	55.5
2005	59.4	52.4
2015	60.0	48.2

Source: *U.S. Census Bureau*

Let p be the percentage of households that are married-couple households at t years since 1970.
a. Find an equation of a linear model to describe the data.
b. In which years were more than 45% of households married-couple households?
c. Although the number of married-couple households is *increasing*, the percentage of households that are married-couple households is *decreasing*. Explain how this is possible.

78. The percentages of American adults who have a great deal or a fair amount of trust in the mass media are shown in Table 49 for various years.

Table 49 Percentages of American Adults Who Have a Great Deal or a Fair Amount of Trust in the Mass Media

Year	Percent
2005	50
2007	47
2009	45
2011	44
2013	44
2015	40

Source: *Gallup*

Let p be the percentage of American adults who have a great deal or a fair amount of trust in the mass media at t years since 2000.
a. Find an equation of a linear model to describe the data.
b. Predict when one-third of American adults will have a great deal or a fair amount of trust in the mass media.
c. What is the p-intercept? What does it mean in this situation?
d. What is the t-intercept? What does it mean in this situation?
e. Predict when less than 36% of American adults will have a great deal or a fair amount of trust in the mass media.

79. The teenage birthrate in the United States has declined since 2010 (see Table 50).

Table 50 American Teenage Birthrate

Year	Birthrate (number of births per 1000 women ages 15–19)
2010	34.3
2011	31.3
2012	29.4
2013	26.5
2014	24.2
2015	22.3

Source: *U.S. National Center for Health Statistics*

Let B be the American teenage birthrate (number of births per 1000 women ages 15–19) at t years since 2010.
a. Find an equation of a linear model to describe the data.
b. Predict the teenage birthrate in 2020. Then predict the *number* of births to women ages 15–19 in 2020. Use the U.S. Census Bureau's prediction that there will be 11,007,000 women ages 15–19 in that year.

c. The American teenage birthrate is 2 to 7 times larger than that in most other Western countries. For example, the birthrate in Germany is 6.0 births per 1000 women ages 15–19. Predict in which years the American birthrate will be less than 6.0 births per 1000 women ages 15–19.

80. A dollar store is a no-frills retailer that sells many goods for one dollar. In 2002, there were 13,000 dollar stores in the United States—three times as many as in 1993. The percentages of households in various income groups that shop at dollar stores are shown in Table 51.

Table 51 Households That Shop at Dollar Stores

Household Income Group (thousands of dollars)	Income Used to Represent Income Group (thousands of dollars)	Percentage of Households That Shop at Dollar Stores
0–19.999	10	74
20–29.999	25	71
30–39.999	35	67
40–49.999	45	64
50–69.999	60	58
70+	100	45

Source: *ACNielsen Homescan Panel*

Let P be the percentage of households with an income of d thousand dollars that shop at dollar stores.
a. Find an equation of a linear model to describe the data.
b. What percentage of households with an income of $27 thousand shop at dollar stores?
c. At what incomes do more than half of households shop at dollar stores?

Concepts

81. A student tries to solve $-3x < 15$:

$$-3x < 15$$
$$\frac{-3x}{-3} < \frac{15}{-3}$$
$$x < -5$$

Describe any errors. Then solve the inequality correctly.

82. A student tries to solve $4x < -24$:

$$4x < -24$$
$$\frac{4x}{4} > \frac{-24}{4}$$
$$x > -6$$

Describe any errors. Then solve the inequality correctly.

83. a. List three numbers that satisfy $3x - 7 < 5$.
b. List three numbers that do not satisfy $3x - 7 < 5$.

84. a. List three numbers that satisfy $2(x + 3) > 17$.
b. List three numbers that do not satisfy $2(x + 3) > 17$.

85. Use the number line to show that if $a < b$, then $2a < 2b$.

86. Use the number line to show that if $a < b$, then $-2a > -2b$.

87. Give an example of a linear inequality of the form $mx + b \leq c$, where m, b, and c are constants. Then solve the inequality. Describe the solution set as an inequality, in interval notation, and as a graph.

88. Describe how to solve a linear inequality in one variable. Describe when you need to reverse an inequality symbol. Explain why it is necessary to reverse the symbol in this case. Finally, explain what you have accomplished by solving an inequality.

Related Review

Solve the equation or inequality. If the statement is an inequality, then describe the solution set as an inequality, in interval notation, and as a graph.

89. $-2x + 6 = 3x - 14$ **90.** $-3(2x - 5) = 21$

91. $-2x + 6 > 3x - 14$ **92.** $-3(2x - 5) \leq 21$

For Exercises 93–96, let x be a number. Translate the English sentence into an inequality. Then solve the inequality. Describe the solution set as an inequality, in interval notation, and as a graph.

93. The sum of the number and 5 is greater than 2.

94. The difference of 7 and the number is less than 3.

95. Twice the number is less than or equal to 5 times the number, minus 6.

96. 3 times the number is greater than or equal to the difference of the number and 8.

Expressions, Equations, and Graphs

Give an example of the following. Then solve, simplify, or graph, as appropriate.

97. Linear equation in one variable

98. Linear expression in one variable with four terms

99. Linear inequality in one variable

100. Linear equation in two variables

Taking It to the Lab

Climate Change Lab (continued from Chapter 3)

Recall from the Climate Change Lab in Chapter 1 that many scientists believe that an increase in average global temperature as small as 3.6°F could be a dangerous climate change (see Table 52). Recall from the Climate Change Lab in Chapter 2 that most scientists believe global warming is largely the result of carbon dioxide emissions from the burning of fossil fuels (see Table 53).

DATA. **Table 52** Average Surface Temperatures of Earth

Year	Average Temperature (degrees Fahrenheit)	Year	Average Temperature (degrees Fahrenheit)
1900	57.1	1960	57.2
1905	56.8	1965	57.0
1910	56.6	1970	57.3
1915	57.0	1975	57.1
1920	56.9	1980	57.6
1925	56.9	1985	57.3
1930	57.1	1990	57.8
1935	57.0	1995	57.9
1940	57.3	2000	57.8
1945	57.3	2005	58.3
1950	56.9	2010	58.3
1955	57.0	2015	59.0

Source: *NASA–GISS*

DATA. **Table 53** Carbon Dioxide Emissions from Burning of Fossil Fuels

Year	Carbon Dioxide Emissions (billions of metric tons) United States	World
1950	2.4	5.8
1955	2.7	7.2
1960	2.9	9.4
1965	3.5	11.2
1970	4.3	14.7
1975	4.4	16.5
1980	4.8	19.1
1985	4.6	19.4
1990	5.0	21.6
1995	5.3	22.2
2000	5.9	23.8
2005	6.0	28.4
2010	5.6	30.6
2014	5.4	35.7

Source: *U.S. Department of Energy*

Finally, recall from the Climate Change Lab in Chapter 2 that the goal of the Kyoto Protocol was to cut developed countries' carbon dioxide emissions to about 5% to 7% below 1990 levels by 2012. This was a crucial first goal, but the Intergovernmental Panel on Climate Change (IPCC) is calling for carbon dioxide emissions in 2050 to be 60% less than carbon dioxide emissions in 1990.

Many countries condemned the United States because the Bush and Obama administrations refused to ratify the Kyoto Protocol. Although the U.S. population in 2014 was only 4% of the world population, 15% of world carbon dioxide emissions that year were produced by the United States (see Tables 53 and 54).

DATA, Table 54 United States and World Populations

	Population (billions)	
Year	United States	World
1960	0.18	3.04
1970	0.21	3.71
1980	0.23	4.45
1990	0.25	5.29
2000	0.28	6.09
2010	0.31	6.85
2014	0.32	7.26

Source: *U.S. Census Bureau*

Critics of the Kyoto Protocol say a fairer pact would be for all countries to commit to the same level of carbon dioxide emissions *per person.* Using the IPCC's recommendation for 2050 carbon dioxide emissions, coupled with the United Nations' prediction of 9.7 billion people in 2050, carbon dioxide emissions should be about 0.9 metric ton per person that year.*

In 2014, annual carbon dioxide emissions were about 4.9 metric tons of carbon dioxide per person. Average annual carbon dioxide emissions for developing countries is 3.9 metric tons per person, which is more than four times IPCC's recommendation. Even worse, average annual carbon dioxide emissions for developed countries is 9.9 metric tons per person, more than ten times IPCC's recommendation (see Table 55).†

DATA, Table 55 GDP Ranks, per-Person GDPs, and per-Person Carbon Dioxide Emissions

Country	2014 GDP Rank	2014 per-Person GDP (thousands of international dollars)	2014 per-Person Carbon Emissions (metric tons)
Australia	12	61.2	16.0
Austria	28	51.3	7.4
Belgium	25	47.7	8.9
Denmark	34	60.6	6.8
France	6	44.5	5.0
Germany	4	47.6	9.8
Italy	8	35.8	5.5
Japan	3	36.3	9.7
Netherlands	17	51.4	9.1
Norway	27	97.0	8.7
Sweden	21	58.5	4.6
Switzerland	20	87.5	5.0
United Kingdom	5	45.7	6.7
United States	1	54.6	17.0

Sources: *Global Carbon Atlas, United Nations, International Monetary Fund*

* United Nations Population Division.
† Emissions Database for Global Atmospheric Research.

GDP, the gross domestic product, is a measure of a country's economic strength.

With annual carbon dioxide emissions of 17.0 metric tons per person, the United States would have to reduce emissions by 95% to meet the standard of 0.9 metric ton per person. This means Americans could emit only 5% of the carbon dioxide they currently emit. Imagine driving your car, heating and cooling your home, using your home appliances (including your refrigerator), using your computer, using your lights, and watching your television only 5% (one-twentieth) of the time you currently do.

Now that the Kyoto Protocol has expired, it is time for a new climate deal. Recall from the Climate Change Lab in Chapter 2 that all countries except the United States, Nicaragua, and Syria have either signed or ratified the Paris Climate Accord, which aims to limit the increase of global average temperature to well below 2°C above pre-industrial levels, to have global emissions peak as soon as possible, and to undertake rapid reductions thereafter.‡

Although President Obama ratified the accord in 2016, President Trump withdrew from the accord in 2017, claiming that it would do little to reduce global warming, is unfair to the United States, and would hurt the U.S. economy.§

Some experts, however, such as the engineer Alan Pears, codirector of the environmental consultancy Sustainable Solutions, believe that it is possible for emissions to be significantly reduced without harming a country's economy. Sweden, Switzerland, Denmark, and Norway, for instance, have a larger per-person GDP than the United States has, as well as significantly lower per-person carbon dioxide emissions. In fact, with the exception of Australia, all the countries listed in Table 55 have strong economies and significantly lower per-person carbon dioxide emissions than the United States has. A scatterplot of the data would show that countries with higher per-person GDP do not necessarily have higher per-person carbon dioxide emissions.

Many states have taken the matter into their own hands by adopting policies to reduce carbon dioxide emissions. And by using alternative sources of energy, many countries have slowed or reversed the growth of carbon dioxide emissions in recent years.¶

In addition to national, state, and even corporate actions, individuals can help lower carbon dioxide emissions by purchasing hybrid automobiles, major appliances with the Energy Star logo, solar thermal systems to help provide hot water, and compact fluorescent light bulbs. Individuals can also car pool or use public transportation.

‡ Center for Climate and Energy Solutions, "Outcomes of the U.N. Climate Change Conference in Paris," December 2015.
§ "Statement by President Trump on the Paris Climate Accord," www.whitehouse.gov, June 1, 2017.
¶ Pamela Person, "Reducing Greenhouse Gas Emissions," Maine Center for Economic Policy, *Choices*, VII(9), October. 11, 2001.

Analyzing the Situation

1. a. Let A be the average global temperature (in degrees Fahrenheit) at t years since 1900. Use a graphing calculator to draw a scatterplot of the data in Table 52, and then find an equation of a linear model to describe the data *for the years 1965 to 2010*. Does the model fit the data well for those years?

b. In Problem 2a of the Climate Change Lab in Chapter 3, you found the average global temperature from 1900 to 1965. If you didn't do this, do so now. [**Hint:** Divide the sum of the temperatures by the number of temperature readings.] Use your model to predict when the planet's average temperature will have increased by 3.6°F—a potentially dangerous climate change.

c. Use your model to estimate the average global temperature in 2015. Is your result an understimate or an overstimate? What *might* this mean about the prediction you made in part (b)?

2. Use Tables 53 and 54 to verify the claims that although the U.S. population in 2014 was only 4% of world population, 15% of annual world carbon dioxide emissions was produced by the United States in that year.

3. Use the United Nations' prediction that the world population will be 9.7 billion in 2050 to verify the claim that per-person carbon dioxide emissions that year should be about 0.9 metric ton per person for the IPCC recommendation of a 60% reduction by then.

4. Let P be the U.S. population (in billions) at t years since 1950. Use a graphing calculator to draw a scatterplot of the data and then find an equation of a linear model to describe the data. Finally, verify that your model fits the data well.

5. Let C be U.S. carbon dioxide emissions (in billions of metric tons) in the year that is t years since 1950. Use a graphing calculator to draw a scatterplot of the data and then find an equation of a linear model to describe the data. Finally, verify that your model fits the data well.

6. a. Use the U.S. population model you found in Problem 4 to estimate U.S. population in 2015.

b. Use the U.S. carbon dioxide emissions model you found in Problem 5 to estimate U.S. carbon dioxide emissions in 2015.

c. Use the results you found in parts (a) and (b) to estimate U.S. *per-person* carbon dioxide emissions in 2015.

d. The actual U.S. per-person carbon dioxide emissions in 2015 were 16.6 metric tons. Is the result you found in part (c) an underestimate or an overestimate? What does this suggest is starting to happen?

7. Let G be the per-person GDP (in international dollars) of a country with per-person carbon emissions c (in metric tons). Use a graphing calculator to draw a scatterplot of the data. Do countries with higher per-person GDPs always have higher per-person carbon emissions? Explain.

Rope Lab

In this lab, you will explore the relationship between the number of knots tied in a rope and the rope's length.

Check with your instructor whether you should collect your own data or use the data listed in Table 56.

Table 56 Lengths of a Rope with Diameter about 7 Millimeters

Number of Knots	Length of Rope (centimeters)
0	60.0
1	53.2
2	45.8
3	38.3
4	30.6

Source: *J. Lehmann*

Materials

1. A 60-centimeter-long piece of rope with diameter about 7 millimeters

2. A meterstick or other measuring device with units of millimeters

Recording of Data

Pull the rope taut and measure its length (in centimeters). Then tie a knot close to one end of the rope and measure the length of the rope again. Next, tie another knot next to the first one and measure the length of the rope. Continue tying knots, working your way along the rope and measuring the rope's length after you have tied each knot. Tie a total of four knots.

Analyzing the Situation

1. Display your data in a table or use the data in Table 56.

2. Let L be the length (in centimeters) of the rope with n knots. Use a graphing calculator to draw a scatterplot of the data. Copy the scatterplot by hand.

3. Find an equation of a linear model to describe the data.

4. Use a graphing calculator to draw a graph of your model and the scatterplot in the same viewing window. Also, graph the model and the scatterplot by hand. How well does the model fit the data points?

5. Is your line increasing or decreasing? What does that mean in this situation?

6. Find the L-intercept of your model. What does it mean in this situation?

7. Find the slope of your model. What does it mean in this situation?

8. Use your model to estimate the length of the rope with five knots.

9. Check whether the result you found in Problem 8 is an underestimate or an overestimate by tying a fifth knot in the rope and then measuring the rope's length.

10. Continue tying knots in the rope. When does model breakdown first occur? Explain.

11. Find the *n*-intercept of your model. What does it mean in this situation?

Shadow Lab

In this lab, you will explore the relationship between an object's height and the length of its shadow.

Check with your instructor whether you should collect your own data or use the data listed in Table 57.

Table 57 Heights of Objects and the Lengths of Their Shadows

Object	Height (inches)	Length of Shadow (inches)
Nothing	0	0
Wine bottle	15.5	10.3
Toy putter	20.8	13.0
Box	36.5	23.6
Mop	47.8	31.0
Person's shoulder	55.0	36.5
Person	63.0	41.5

Source: *J. Lehmann*

Materials

1. Six objects of various heights up to 7 feet
2. A building, pole, tree, or other tall object with height greater than 15 feet
3. A tape measure or other measuring device

Recording of Data

Run the experiment when the objects (including the tall object) have noticeable and measurable shadows. For each object, measure its height and the length of its shadow. Record the beginning and ending time of the experiment. Also, record the length of the shadow of the tall object. It is important that you record all the data quickly.

Analyzing the Situation

1. Display your data in a table similar to Table 57. Those data were collected from 2:10 P.M. to 2:20 P.M., and the tall object is a tree whose shadow has a length of 49.5 feet.

2. Let *h* be the height (in inches) of an object and *L* be the length (in inches) of the object's shadow. Use a graphing calculator to draw a scatterplot of the data. Copy the scatterplot by hand.

3. Find an equation of a linear model to describe the data.

4. Use a graphing calculator to draw the graph of your model and the scatterplot in the same viewing window. Also, graph the model and the scatterplot by hand. How well does the model fit the data points?

5. Is your line increasing or decreasing? What does that mean in this situation?

6. Find the *L*-intercept of your model. What does it mean in this situation?

7. Find the slope of your model. What does it mean in this situation?

8. Use the length of the shadow of the tall object to estimate the object's height.

9. Explain why it was important that you ran the experiment quickly.

10. Suppose you had run the experiment half an hour later. How would that have affected the slope of your model? Explain. (If the Sun would have set by then, describe the impact on the slope if the experiment had been performed half an hour earlier.) Would this change in time have resulted in a different estimate of the height of the tall object? Explain.

Linear Lab: Topic of Your Choice

Your objective in this lab is to use a linear model to describe some authentic situation. Find some data on two quantities that describe a situation that has not been discussed in this text. Almanacs, newspapers, magazines, scientific journals, and the Internet are good resources. Or you can conduct an experiment. Choose something that interests you!

Analyzing the Situation

1. What two quantities did you explore? Define variables for the quantities. Include units in your definitions.

2. Which variable is the explanatory variable? Which variable is the response variable? Explain.

3. Describe how you found your data. If you conducted an experiment, provide a careful description with specific details of how you ran your experiment. If you didn't conduct an experiment, state the source of your data.

4. Include a table of your data.

5. Use a graphing calculator to draw a scatterplot of your data. (If your data are not approximately linear, find some data that are.) Copy the scatterplot by hand.

6. Find an equation of a linear model to describe the data.

7. Use a graphing calculator to draw a graph of your model and the scatterplot in the same viewing window. Also, graph the model and the scatterplot by hand. How well does the model fit the data points?

8. What is the slope of your linear model? What does it mean in this situation?

9. Does it make sense that your variables are approximately linearly related in terms of the situation you chose to model? Explain.

10. Choose a value for your explanatory variable. On the basis of that chosen value, use your model to find a value for your response variable. Describe what your result means in the situation you are modeling.

11. Choose a value for your response variable. On the basis of that chosen value, use your model to find a value for your explanatory variable. Describe what your result means in the situation you are modeling.

12. Find the intercepts of your linear model. What do they mean in the situation you are modeling? Has model breakdown occurred at the intercepts?

13. Comment on your lab experience.
   a. For example, you might address whether the lab was enjoyable, insightful, and so on.
   b. Were you surprised by any of your findings? If so, which ones?
   c. How would you improve your process for this lab if you were to do it again?
   d. How would you improve your process if you had more time and money?

 # Chapter Summary

Key Points of Chapter 5

Section 5.1   Graphing Linear Equations in Two Variables

Solving for _y_ to graph	Before we can use the y-intercept and the slope to graph a linear equation, we must solve for y to put the equation into the form $y = mx + b$.
Intercepts of the graph of an equation	For an equation containing the variables x and y, • To find the x-coordinate of each x-intercept, substitute 0 for y and solve for x. • To find the y-coordinate of each y-intercept, substitute 0 for x and solve for y.
Using intercepts to graph an equation	To graph a linear equation whose graph has exactly two intercepts, **1.** Find the intercepts. **2.** Plot the intercepts and a third point on the line, and graph the line that contains the three points.
Using three solutions to graph a linear equation	To graph any linear equation, **1.** Find three solutions of the equation. **2.** Plot the three solutions, and graph the line that contains them.
Comparing linear equations in one and two variables	A linear equation in two variables has an infinite number of (ordered-pair) solutions. A linear equation in one variable has exactly one (real-number) solution.

Section 5.2   Finding Equations of Lines

Finding an equation of a line by using the slope, a point, and the slope-intercept form	To find an equation of a line by using the slope and a point, **1.** Substitute the given value of the slope m into the equation $y = mx + b$. **2.** Substitute the coordinates of the given point into the equation you found in step 1, and solve for b. **3.** Substitute the value of b you found in step 2 into the equation you found in step 1. **4.** Check that the graph of your equation contains the given point.
Finding an equation of a line by using two points and the slope-intercept form	To find an equation of the line that passes through two given points whose x-coordinates are different, **1.** Use the formula $m = \dfrac{y_2 - y_1}{x_2 - x_1}$, to find the slope of the line containing the two points. **2.** Substitute the m value you found in step 1 into the equation $y = mx + b$. **3.** Substitute the coordinates of one of the given points into the equation you found in step 2, and solve for b. **4.** Substitute the b value you found in step 3 into the equation you found in step 2. **5.** Check that the graph of your equation contains the two given points.
Point-slope form	If a nonvertical line has slope m and contains the point (x_1, y_1), then an equation of the line is $y - y_1 = m(x - x_1)$.

Section 5.3   Finding Equations of Linear Models

Finding an equation of a linear model	To find an equation of a linear model, given some data, 1. Construct a scatterplot of the data. 2. Determine whether there is a line that comes close to the data points. If so, choose two points (not necessarily data points) that you can use to find an equation of a linear model. 3. Find an equation of the line. 4. Use a graphing calculator to verify that the graph of your equation contains the two chosen points and comes close to all the points of the scatterplot.

Section 5.4   Using Equations of Linear Models to Make Estimates and Predictions

Using an equation of a linear model to make predictions	To make a prediction about the response variable of a linear model, substitute a chosen value for the explanatory variable in the model's equation. Then solve for the response variable. To make a prediction about the explanatory variable of a linear model, substitute a chosen value for the response variable in the model's equation. Then solve for the explanatory variable.
Using an equation of a linear model to find intercepts	If an equation of the form $p = mt + b$, where $m \neq 0$, is used to model a situation, then • The p-intercept is $(0, b)$. • To find the t-coordinate of the t-intercept, substitute 0 for p in the model's equation and solve for t.
Four-step modeling process	To find a linear model and then make estimates and predictions, 1. Construct a scatterplot of the data to determine whether there is a nonvertical line that comes close to the points. If so, choose two points (not necessarily data points) that you can use to find an equation of a linear model. 2. Find an equation of your model. 3. Verify your equation by checking that the graph of your model contains the two chosen points and comes close to all the data points. 4. Use the equation of your model to make estimates, make predictions, and draw conclusions.

Section 5.5   Solving Linear Inequalities in One Variable

Addition property of inequalities	If $a < b$, then $a + c < b + c$. Similar properties hold for $\leq$, $>$, and $\geq$.
Multiplication property of inequalities	• For a *positive* number c, if $a < b$, then $ac < bc$. • For a *negative* number c, if $a < b$, then $ac > bc$. Similar properties hold for $\leq$, $>$, and $\geq$.
Linear inequality in one variable	A **linear inequality in one variable** is an inequality that can be put into one of the forms $$mx + b < 0, \quad mx + b \leq 0, \quad mx + b > 0, \quad mx + b \geq 0$$ where m and b are constants and $m \neq 0$.
Satisfy for an inequality in one variable	We say a number **satisfies** an inequality in one variable if the inequality becomes a true statement after we have substituted the number for the variable.
Solution, solution set, and solve for an inequality in one variable	We say that a number is a **solution** of an inequality in one variable if it satisfies the inequality. The **solution set** of an inequality is the set of all solutions of the inequality. We **solve** an inequality by finding its solution set.

Chapter 5 Review Exercises

Determine the slope and the y-intercept. Use the slope and the y-intercept to graph the equation by hand. Use a graphing calculator to verify your work.

1. $y = 4x - 5$

2. $y = -3x + 4$

3. $y = \frac{1}{2}x + 1$

4. $y = -\frac{2}{3}x - 2$

5. $y = \frac{5}{3}x$

6. $y = -5$

7. $2x - y = 5$

8. $3x - 2y = -6$

9. $4x + 5y = 10$

10. $x + 3y = 6$

11. $2x + 5y - 20 = 0$

12. $y - 4 = 0$

Find the x-intercept and y-intercept. Then graph the equation by hand.

13. $4x - 5y = 20$ **14.** $3x + 2y = 6$

15. $3x + 4y + 12 = 0$ **16.** $y = 2x - 4$

17. $y = -x + 3$ **18.** $\frac{1}{3}x - \frac{1}{2}y = 1$

Find the approximate x-intercept and approximate y-intercept. Round the coordinates to the second decimal place.

19. $9.2x - 3.8y = 87.2$ **20.** $y = 2.56x + 97.25$

21. a. Find the intercepts of the graph of $y = 3x + 7$.
    b. Find the intercepts of the graph of $y = 2x + 9$.
    c. Find the intercepts of the graph of $y = mx + b$, where b is a constant and m is a nonzero constant.

22. Consider $2x - 4y = 8$.
    a. Graph the equation by finding the slope and the y-intercept.
    b. Graph the equation by finding the intercepts.
    c. Compare your graphs from parts (a) and (b). Decide which method you prefer for graphing the equation $2x - 4y = 8$.

Find an equation of the line that has the given slope and contains the given point. If possible, write your equation in slope-intercept form. Check that the graph of your equation contains the given point.

23. $m = -4, (2, -1)$ **24.** $m = 2, (-9, -3)$

25. $m = -3, (-4, 5)$ **26.** $m = -\frac{2}{5}, (-3, 6)$

27. $m = \frac{3}{7}, (2, 9)$ **28.** $m = -\frac{2}{3}, (-6, -4)$

29. m is undefined, $(2, 5)$ **30.** $m = 0, (-1, -4)$

Find an approximate equation of the line that has the given slope and contains the given point. Write your equation in slope-intercept form. Round the constant term to the second decimal place. Use a graphing calculator to verify your equation.

31. $m = -5.29, (-4.93, 8.82)$ **32.** $m = 1.45, (-2.79, -7.13)$

Find an equation of the line that passes through the two given points. If possible, write your equation in slope-intercept form. Check that the graph of your equation contains the given points.

33. $(2, 1)$ and $(5, 7)$ **34.** $(-2, 9)$ and $(1, 3)$

35. $(-2, -7)$ and $(1, 2)$ **36.** $(2, -5)$ and $(4, 5)$

37. $(-5, 8)$ and $(-1, 2)$ **38.** $(-8, -5)$ and $(4, -2)$

39. $(-3, 9)$ and $(6, -6)$ **40.** $(-4, -10)$ and $(-2, -7)$

41. $(5, -3)$ and $(5, 2)$ **42.** $(-4, -3)$ and $(-1, -3)$

For Exercises 43 and 44, find an approximate equation of the line that passes through the two given points. Write your equation in slope–intercept form. Round the slope and the constant term to two decimal places. Use a graphing calculator to verify your equation.

43. $(3.5, 9.2)$ and $(8.7, 4.8)$

44. $(-5.22, 2.49)$ and $(1.83, -3.99)$

45. Find an equation of the line sketched in Fig. 47.

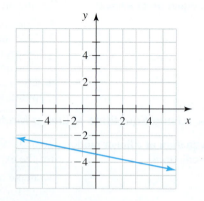

Figure 47 Exercise 45

46. Find an equation of a line that comes close to the points listed in Table 58. Then use a graphing calculator to check that your line comes close to those points.

Table 58 Exercise 46

x	y
1	28
4	23
6	16
9	13
10	8

47. The percentages of U.S. household income collected by the middle class are shown in Table 59 for various years.

Table 59 Percentages of U.S. Household Income Collected by the Middle Class

Year	Percent
1970	53
1980	52
1990	50
2000	47
2010	47
2014	46

Source: *Census Bureau*

Let p be the percentage of U.S. household income collected by the middle class at t years since 1970.
    a. Use a graphing calculator to draw a scatterplot of the data.
    b. Find an equation of a linear model to describe the data.
    c. Use your model to predict in which year 44% of U.S. household income will be collected by the middle class.
    d. Find the p-intercept. What does it mean in this situation?
    e. Find the t-intercept. What does it mean in this situation?

48. The percentages of Americans who support making smoking totally illegal are shown in Table 60 for various years.

Table 60 Percentages of Americans Who Support Making Smoking Totally illegal

Year	Percent
2005	16
2007	12
2009	17
2011	19
2013	22
2015	24

Source: *Gallup*

Let p be the percentage of Americans who support making smoking totally illegal at t years since 2000.
  a. Use a graphing calculator to draw a scatterplot of the data.
  b. Find an equation of a linear model to describe the data.
  c. What is the slope? What does it mean in this situation?
  d. Predict the percentage of Americans who will support making smoking totally illegal in 2020.
  e. Predict when 30% of Americans will support making smoking totally illegal.

49. The number of kidney transplants was 17.1 thousand transplants in 2014 and has increased by about 0.4 thousand transplants per year since then (Source: *U.S. Department of Health and Human Services*). Let n be the number (in thousands) of kidney transplants in the year that is t years since 2014.
  a. Find an equation of a linear model to describe the data.
  b. What is the slope? What does it mean in this situation?
  c. Predict in which year there will be 19.5 thousand kidney transplants.
  d. Predict the number of kidney transplants in 2022.

50. The average cost per 30-second ad slot during the Academy Awards has increased approximately linearly from $1.3 million in 2009 to $2.1 million in 2017 (Source: *Kantar Media*). Predict when the average cost will be $2.6 million.

Solve the inequality. Describe the solution set as an inequality, in interval notation, and as a graph. Then use a graphing calculator table to verify your result.

51. $x + 7 > 10$

52. $x - 3 \geq -4$

53. $3w \leq -15$

54. $-\frac{4}{3}p < -8$

55. $-4x < 8$

56. $5x - 3 > 3x - 9$

57. $-3(2a + 5) + 5a \geq 2(a - 3)$

58. $\frac{2b - 4}{3} \leq \frac{3b - 4}{4}$

59. Violent-crime rates in the United States are shown in Table 61 for various years.

Table 61 Violent-Crime Rates

Year	Number of Violent Crimes per 100,000 People
2000	507
2003	476
2006	479
2009	432
2012	388
2015	373

Source: *U.S. Department of Justice*

Let v be the violent-crime rate (number of violent crimes per 100,000 people) at t years since 2000.
  a. Find an equation of a linear model to describe the data.
  b. What is the slope? What does it mean in this situation?
  c. Predict the years when the violent-crime rate will be less than 320 violent crimes per 100,000 people.
  d. Find the t-intercept. What does it mean in this situation?

Chapter 5 Test

Determine the slope and the y-intercept. Use the slope and the y-intercept to graph the equation by hand.

1. $y = -3x - 1$ 2. $2x - 5y = 10$ 3. $y - 5 = 0$

4. Find the x-intercept and y-intercept of the graph of the equation $6x - 3y = 18$. Then graph the equation by hand.

5. Find the approximate x-intercept and approximate y-intercept of the graph of $5.93x - 4.81y = 43.79$. Round the coordinates to the second decimal place.

6. Find the x-intercept and the y-intercept of the graph of the equation $\frac{x}{2} + \frac{y}{7} = 1$.

7. A student thinks that the line $5x - 4y = 20$ has slope 5 because the coefficient of the $5x$ term is 5. Is the student correct? If yes, explain why. If no, then find the slope.

Find an equation of the line that has the given slope and contains the given point. If possible, write your equation in slope–intercept form.

8. $m = 7$, $(-2, -4)$

9. $m = -\frac{2}{3}$, $(6, -1)$

10. m is undefined, $(2, -4)$

11. Find an equation of the line that passes through the points $(-4, 6)$ and $(2, 3)$.

12. Find an approximate equation of the line that passes through the points $(-3.4, 2.9)$ and $(1.8, -7.1)$. Write your equation in slope-intercept form. Round the slope and the constant term to two decimal places.

13. Find an equation of the line sketched in Fig. 48.

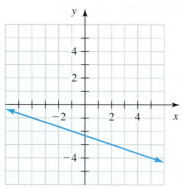

Figure 48 Exercise 13

14. Consider the scatterplot of data and the graph of the model $y = mt + b$ in Fig. 49. Sketch a linear model that describes the data better, and then explain how you would adjust the values of m and b of the original model to describe the data better.

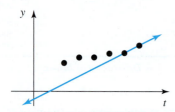

Figure 49 Exercise 14

15. Find an equation of a line that comes close to the points listed in Table 62. Then use a graphing calculator to check that your line comes close to those points.

Table 62 Exercise 15

x	y
1	7
3	11
7	16
12	20
14	23

16. The percentages of Nike's revenue that come from footwear are shown in Table 63 for various years.

Table 63 Percentages of Nike's Revenue That Come from Footwear

Year	Percent
2011	55
2012	56
2013	58
2014	58
2015	60
2016	60

Source: *Nike®*

Let p be the percentage of Nike's revenue that comes from footwear at t years since 2010.
a. Find an equation of a linear model to describe the data.
b. What is the slope? What does it mean in this situation?
c. What is the p-intercept? What does it mean in this situation?
d. When will two-thirds of Nike's revenue come from footwear?
e. Estimate in which years less than 65% of Nike's revenue came from footwear.

17. The percentage of U.S. consumers who own a single-cup coffee brewing system increased approximately linearly from 7% in 2011 to 29% in 2016 (Source: *National Coffee Association*). Predict when half of U.S. consumers will own a single-cup coffee brewing system.

Solve the inequality. Describe the solution set as an inequality, in interval notation, and as a graph.

18. $3(2x + 1) \leq 4(x + 2) - 1$

19. $\dfrac{5}{6} - \dfrac{2}{3}x > -\dfrac{1}{2}$

Systems of Linear Equations and Systems of Linear Inequalities

Did you know that although men's world record times are better than women's world record times for running and swimming events, women's record times are declining at a greater rate than men's record times? Table 1 shows the world record times for the 200-meter run. In Exercise 39 of Homework 6.1, you will predict when the women's record time for the 200-meter run will be equal to the men's record time.

Table 1 200-Meter Run World Record Times

Women		Men	
Year	Record Time (seconds)	Year	Record Time (seconds)
1973	22.38	1951	20.6
1974	22.21	1963	20.3
1978	22.06	1967	20.14
1984	21.71	1979	19.72
1988	21.34	1996	19.32
		2009	19.19

Source: *IAAF Statistics Handbook*

In Chapters 3–5, we took a close look at linear equations in two variables. In this chapter, we will work with *two* such equations together. This work will enable us to predict when two quantities will be equal, such as the number of women and the number of men who earned a bachelor's degree. We will also find when various quantities are greater (or less) than other quantities for a single authentic situation.

6.1 Using Graphs and Tables to Solve Systems

Objectives

» Describe a *system of linear equations in two variables*.

» Use a graphical approach to solve systems of linear equations.

» Use a graphing calculator to solve systems of linear equations.

» Use graphing to make predictions about situations that can be modeled by two linear models.

» Describe the three types of linear systems of two equations.

» Find the solution of a system of linear equations from tables of solutions of such equations.

In this section, we will work with graphs and tables that describe solutions of two linear equations in two variables. We will also work with situations that can be modeled by using two linear models.

System of Linear Equations in Two Variables

A **system of linear equations in two variables**, or a **linear system**, for short, consists of two or more linear equations in two variables. Here is an example of a system of two linear equations in two variables:

$$y = 3x - 1$$
$$y = -2x + 4$$

We will work with such systems throughout this chapter.

Recall from Section 3.1 that every point on the graph of an equation represents a solution of the equation and that every point *not* on the graph represents an ordered pair that is *not* a solution. Knowing the meaning of a graph will help us greatly in this section.

Example 1   Finding Ordered Pairs That Satisfy Both of Two Given Equations

Find all ordered pairs that satisfy both of the equations in the system

$$y = 3x - 1$$
$$y = -2x + 4$$

Solution

To begin, we graph each equation on the same coordinate system (see Fig. 1).

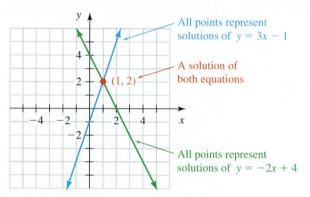

Figure 1 The intersection point is $(1, 2)$

For an ordered pair to be a solution of *both* equations, it must represent a point that lies on *both* lines. The intersection point $(1, 2)$ is the only point that lies on both lines. So, the ordered pair $(1, 2)$ is the only ordered pair that satisfies both equations.

We can verify that $(1, 2)$ satisfies both equations:

$$y = 3x - 1 \qquad\qquad y = -2x + 4$$
$$2 \overset{?}{=} 3(1) - 1 \qquad\quad 2 \overset{?}{=} -2(1) + 4$$
$$2 \overset{?}{=} 2 \qquad\qquad\quad 2 \overset{?}{=} 2$$
$$\text{true} \qquad\qquad\qquad \text{true}$$

Solving Systems by Graphing

In Example 1, we worked with the system

$$y = 3x - 1$$
$$y = -2x + 4$$

We found that the only point whose coordinates satisfy both equations is the intersection point $(1, 2)$. We call the set containing only $(1, 2)$ the *solution set of the system*.

Definition   *Solution, solution set,* and *solve* for a system

We say that an ordered pair (a, b) is a **solution** of a system of two equations in two variables if it satisfies both equations. The **solution set** of a system is the set of all solutions of the system. We **solve** a system by finding its solution set.

In general, **the solution set of a system of two linear equations can be found by locating any intersection point(s) of the graphs of the two equations.**

Example 2   Solving a System of Two Linear Equations by Graphing

Solve the system

$$y = 2x + 1$$
$$y = \frac{1}{2}x - 2$$

Solution

The graphs of the equations are sketched in Fig. 2.

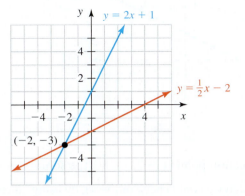

Figure 2 The intersection point is $(-2, -3)$

The intersection point is $(-2, -3)$. So, the solution is the ordered pair $(-2, -3)$.

We can check that $(-2, -3)$ satisfies both equations:

$$y = 2x + 1 \qquad\qquad y = \frac{1}{2}x - 2$$

$$-3 \overset{?}{=} 2(-2) + 1 \qquad -3 \overset{?}{=} \frac{1}{2}(-2) - 2$$

$$-3 \overset{?}{=} -3 \qquad\qquad -3 \overset{?}{=} -1 - 2$$

$$\text{true} \qquad\qquad\qquad -3 \overset{?}{=} -3$$

$$\text{true}$$

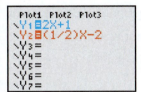

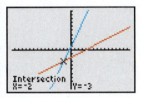

Figure 3 Verify that the intersection point is $(-2, -3)$

We can also check our work by using "intersect" on a graphing calculator (see Fig. 3). See Appendix A.17 for graphing calculator instructions.

WARNING
After solving a system of two linear equations, you commit a common error if you check that a result satisfies only one of the two equations. It is important to check that your result satisfies *both* equations.

In Example 2, we solved a system in which both equations are in $y = mx + b$ form. If any equations in a system are not in $y = mx + b$ form, we usually begin by writing them in $y = mx + b$ form.

Example 3   Solving a System of Two Linear Equations by Graphing

Solve the system

$$3x + 2y = -4 \qquad \textit{Equation (1)}$$
$$y = \frac{1}{2}x + 2 \quad \textit{Equation (2)}$$

Solution

First, we write equation (1) in slope–intercept form by solving the equation for y:

$$3x + 2y = -4 \qquad\qquad \textit{Equation (1)}$$
$$2y = -3x - 4 \qquad\qquad \textit{Subtract 3x from both sides.}$$
$$\frac{2y}{2} = \frac{-3x}{2} - \frac{4}{2} \qquad \textit{Divide both sides by 2.}$$
$$y = -\frac{3}{2}x - 2 \qquad\qquad \textit{Simplify.}$$

Next, we sketch a graph of the equations $y = -\frac{3}{2}x - 2$ and $y = \frac{1}{2}x + 2$ (see Fig. 4).

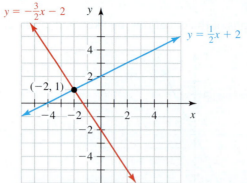

Figure 4 The solution is $(-2, 1)$

The intersection point is $(-2, 1)$. So, the solution is the ordered pair $(-2, 1)$.

Although we could check our work by using "intersect" on a graphing calculator with the equations $y = -\dfrac{3}{2}x - 2$ and $y = \dfrac{1}{2}x + 2$, this check would not verify that we solved $3x + 2y = -4$ for y correctly. To check all of our work, we check that $(-2, 1)$ satisfies both of the *original* equations:

$$3x + 2y = -4 \qquad\qquad y = \frac{1}{2}x + 2$$

$$3(-2) + 2(1) \overset{2}{=} -4 \qquad\qquad 1 \overset{2}{=} \frac{1}{2}(-2) + 2$$

$$-6 + 2 \overset{2}{=} -4 \qquad\qquad 1 \overset{2}{=} -1 + 2$$

$$-4 \overset{2}{=} -4 \qquad\qquad 1 \overset{2}{=} 1$$

$$\text{true} \qquad\qquad\qquad \text{true}$$

Using a Graphing Calculator to Solve a System

So far, we have solved systems with simple equations that can be accurately graphed by hand in a reasonable amount of time. In Example 4, we will use a graphing calculator to help us solve a system with equations that are more difficult to graph.

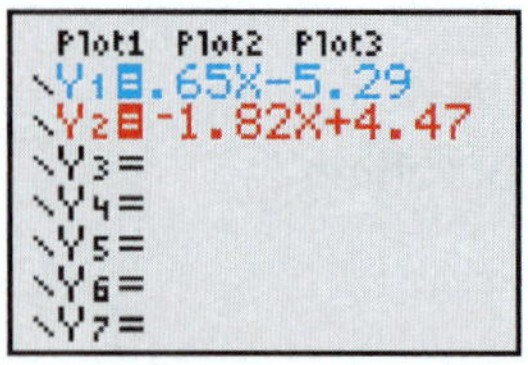

Plot1 Plot2 Plot3
\Y₁◼.65X-5.29
\Y₂◼-1.82X+4.47
\Y₃=
\Y₄=
\Y₅=
\Y₆=
\Y₇=

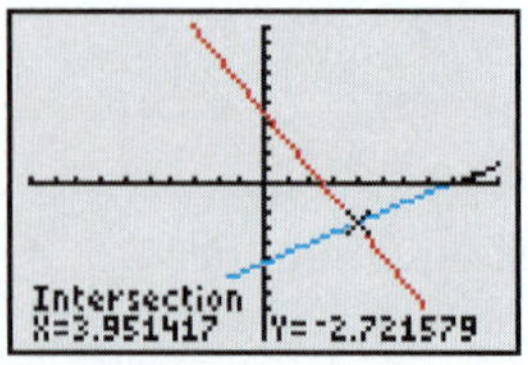

Intersection
X=3.951417   Y=-2.721579

Figure 5 The approximate solution is $(3.95, -2.72)$

> **Example 4**   Using a Graphing Calculator to Solve a System

Use a graphing calculator to find any approximate solutions of the system

$$y = 0.65x - 5.29$$
$$y = -1.82x + 4.47$$

Solution

We use "intersect" to find the approximate solution $(3.95, -2.72)$. See Fig. 5.
    We check that $(3.95, -2.72)$ approximately satisfies both of the original equations:

$$y = 0.65x - 5.29 \qquad\qquad y = -1.82x + 4.47$$

$$-2.72 = 0.65(3.95) - 5.29 \qquad -2.72 = -1.82(3.95) + 4.47$$

$$-2.72 \approx -2.7225 \qquad\qquad -2.72 \approx -2.719$$

Using Two Linear Models to Make a Prediction

We can use graphing to make estimates and predictions about some authentic situations that can be modeled by two linear equations.

| Example 5 | Making a Prediction |

To control costs during the past three decades, colleges have steadily increased the number of instructors that work part time (see Table 2).

Table 2 Numbers of Part-Time and Full-Time Instructors

| Year | Number of Instructors (thousands) | |
	Part Time	Full Time
1997	421	569
2001	495	618
2005	615	676
2009	710	729
2013	754	791

Source: *U.S. National Center for Education Statistics*

Let n be the number of college instructors (in thousands) at t years since 1990. Reasonable models for part-time instructors and full-time instructors are

$$n = 22.03t + 268.63 \quad \textit{Part-time instructors}$$
$$n = 13.88t + 468.48 \quad \textit{Full-time instructors}$$

Use graphing to estimate when the number of part-time instructors was equal to the number of full-time instructors. What was that number of instructors?

Solution

We begin by sketching graphs of the two models on the same coordinate system (see Fig. 6).

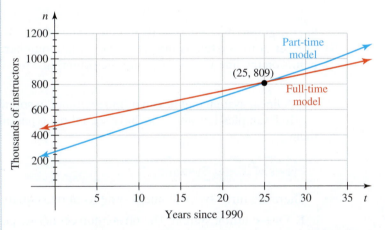

Figure 6 Part-time and full-time models

Note that the intersection point is about (25, 809). So, the models estimate there were the same number of part-time instructors as full-time instructors (809 thousand of each) in 1990 + 25 = 2015.

We verify our work by using "intersect" on a graphing calculator (see Fig. 7). See Appendix A.7 and A.17 for graphing calculator instructions.

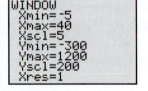

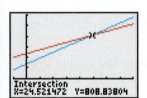

Figure 7 Verify the work

Intersection Point of the Graphs of Two Models

If the explanatory variable of two models represents time, then an intersection point of the graphs of the two models indicates when the quantities represented by the response variables were or will be equal.

Three Types of Linear Systems of Two Equations

Each of the systems in Examples 1–5 has one solution. Some systems do not have exactly one solution, however.

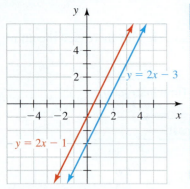

Figure 8 The solution set is the empty set

Example 6 | A System Whose Solution Is the Empty Set

Solve the system

$$y = 2x - 1$$
$$y = 2x - 3$$

Solution

Because the distinct lines have equal slopes, the lines are parallel (see Fig. 8). Parallel lines do not intersect, so there is no ordered pair that satisfies both equations. The solution set is the empty set.

A linear system whose solution is the empty set is called an **inconsistent system**.

Example 7 | A System That Has an Infinite Number of Solutions

Solve the system

$$y = 3x - 5 \quad \textit{Equation (1)}$$
$$6x - 2y = 10 \quad \textit{Equation (2)}$$

Solution

We write equation (2) in slope–intercept form:

$$6x - 2y = 10 \qquad \textit{Equation (2)}$$
$$-2y = -6x + 10 \qquad \textit{Subtract 6x from both sides.}$$
$$\frac{-2y}{-2} = \frac{-6x}{-2} + \frac{10}{-2} \qquad \textit{Divide both sides by −2.}$$
$$y = 3x - 5 \qquad \textit{Simplify.}$$

So, the graph of $6x - 2y = 10$ and the graph of $y = 3x - 5$ are the same line. The solution set of the system is the set of the infinite number of ordered pairs represented by points on the line $y = 3x - 5$ and the (same) line $6x - 2y = 10$.

A linear system that has an infinite number of solutions is called a **dependent system**.

In Examples 1, 6, and 7, we have seen three types of linear systems. We now describe them.

Types of Linear Systems

There are three types of linear systems of two equations:

1. **One-solution system:** The lines intersect in one point. The solution of the system is the ordered pair that corresponds to that point. See Fig. 9.

2. **Inconsistent system:** The lines are parallel. The solution set of the system is the empty set. See Fig. 10.

3. **Dependent system:** The lines are identical. The solution set of the system is the set of the infinite number of solutions represented by points on the same line. See Fig. 11.

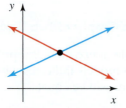

Figure 9 Graphs of the equations in a system with one solution

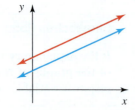

Figure 10 Graphs of the equations in an inconsistent system

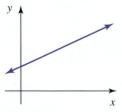

Figure 11 Graph of the equations in the dependent system

Finding the Solution of a System from Tables of Solutions

By the Rule of Four, we know we can use tables to describe some solutions of a linear equation in two variables. In Example 8, we will use such tables to help us find the solution of a system.

Example 8   Using Tables to Solve a System

Some solutions of two linear equations are shown in Tables 3 and 4. Find the solution of the system of the two equations.

Table 3 Solutions of One Equation

x	y
0	3
1	7
2	11
3	15
4	19

Table 4 Solutions of the Other Equation

x	y
0	21
1	19
2	17
3	15
4	13

Solution

Because the ordered pair $(3, 15)$ is listed in both tables, we know it is a solution of both equations. So, it is a solution of the system of equations.

The graphs of the two equations are two lines with slopes 4 and -2. Because the lines have different slopes, there is only one intersection point. So, the ordered pair $(3, 15)$ is the *only* solution of the system.

In Example 8, we used tables of solutions of two linear equations to help us find the solution of the linear system. **If an ordered pair is listed in both tables of solutions of two linear equations, then that ordered pair is a solution of the system.**

◈ Group Exploration

DATA Section Opener: Using a system of equations to model a situation

The average annual U.S. per-person consumptions of bottled water and soda are shown in Table 5 for various years. The annual per-person consumption (in gallons per person) W of bottled water and S of soda are modeled by the system

$$W = 1.5t + 27.8$$
$$S = -1.3t + 45.2$$

where t is the number of years since 2010.

Table 5 Average Annual U.S. Per-Person Consumptions of Bottled Water and Soda

Year	Average Annual Consumption (gallons per person)	
	Bottled Water	Soda
2010	28	45
2011	29	44
2012	31	43
2013	32	41
2014	34	40

Source: *Beverage Marketing Corporation*

1. Find the W-intercept of the bottled water model and the S-intercept of the soda model. What do these intercepts mean in this situation?

2. Find the slopes of both models. What do the slopes mean in this situation?

3. Your responses to Problems 1 and 2 should suggest an event that will happen in the future. Describe that event.

4. By using Zoom Out and "intersect" on a graphing calculator, estimate the coordinates of the point where the graphs of the models intersect. (See Appendix A.6 and A.17 for graphing calculator instructions.) In terms of consumption, what does it mean that the graphs intersect at this point? State the year and consumption for this event.

Group Exploration

Three types of systems

1. Solve the system

$$y = 2x + 1$$
$$y = -3x + 6$$

2. a. Graph by hand the equations in the system

$$y = 2x + 1$$
$$y = 2x + 3$$

   b. Are there any intersection points of the two lines? Explain.

   c. Are there any solutions of the system? If yes, find them. If no, explain why not.

3. a. Graph by hand the equations in the system

$$y = 2x + 1$$
$$y = 2x + 1$$

   b. Are there any intersection points of the two lines? If yes, describe them. If no, explain why not.

   c. Are there any solutions of the system? If yes, describe them. If no, explain why not.

4. Problems 1–3 suggest that there are three types of systems of two linear equations. Describe these three types.

5. Solve the system

$$y = -2x + 3$$
$$4x + 2y = 10$$

6. Solve the system

$$y = 3x - 1$$
$$9x - 3y = 3$$

Tips for Success   **Take a Break**

Have you ever had trouble solving a problem but returned to the problem hours later and found it easy to solve? By taking a break, you can return to the exercise with a different perspective and renewed energy. You've also given your unconscious mind a chance to reflect on the problem while you take your break. You can strategically take advantage of this phenomenon by allocating time to complete your homework assignment at two different points in your day.

Homework 6.1

For extra help ▶   **MyLab Math**   ▦ Watch the videos in MyLab Math

*Determine which of the given ordered pairs is a solution of the given system. [**Hint:** There is no need to graph.]*

1. $(2, 3), (1, -1), (-4, 6)$
$$y = 4x - 5$$
$$y = -2x + 1$$

2. $(-3, 5), (-7, 3), (3, -7)$
$$y = -x - 4$$
$$y = -3x + 2$$

3. $(-1, 8), (3, -2), (7, 1)$
$$5x + 2y = 11$$
$$3x - 4y = 17$$

4. $(3, -1), (-4, -2), (-2, -5)$
$$4x - 5y = 17$$
$$3x + 2y = -16$$

Find the solution of the system by graphing the equations by hand. Use "intersect" on a graphing calculator to verify your work.

5. $y = 3x - 5$
$$y = -2x + 5$$

6. $y = 4x - 2$
$$y = -2x + 4$$

7. $y = \dfrac{1}{2}x + 2$
$$y = -\dfrac{3}{2}x - 2$$

8. $y = -\dfrac{2}{3}x + 4$
$$y = \dfrac{5}{3}x - 3$$

Find the solution of the system by graphing the equations by hand. Verify any solution by checking that it satisfies both equations.

9. $y = -3x$
$$y = 4x$$

10. $y = x$
$$y = -x$$

11. $3x + y = 2$
$$2x - y = 8$$

12. $4x - y = -9$
$$2x + y = -3$$

13. $4y = -3x$
$$x - 2y = 10$$

14. $3y = 2x$
$$5x - 3y = -9$$

15. $4x + 3y = 6$
$$2x - 3y = 12$$

16. $x + 2y = -4$
$$-3x + 2y = 4$$

17. $6y + 15x = 12$
$$2y - x = -8$$

18. $6y + 10x = -18$
$$3y - x = 9$$

19. $x = 3$
$$y = -2$$

20. $x = 0$
$$y = 0$$

21. $x = \dfrac{1}{3}y$
$$y = -2x + 5$$

22. $y = x - 2$
$$x = \dfrac{1}{2}y$$

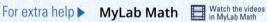

Use "intersect" on a graphing calculator to solve the system, with the coordinates of the solution rounded to the second decimal place. Check that your result approximately satisfies both equations.

23. $y = 2.18x - 5.34$
$y = -3.53x + 1.29$

24. $y = 4.95x + 7.51$
$y = -0.84x - 5.38$

25. $y = \frac{5}{4}x + 2$
$y = -\frac{1}{4}x - 5$

26. $y = -\frac{7}{3}x + 3$
$y = -\frac{2}{3}x - 2$

27. $y = \frac{3}{4}x - 8$
$5x + 3y = 6$

28. $y = -\frac{2}{3}x - 5$
$7y - 3x = 7$

29. $-2x + 5y = 15$
$6x + 14y = -14$

30. $12x + 9y = 18$
$-x - 4y = 20$

For Exercises 31–36, find the solution of the system by graphing the equations by hand. If the system is inconsistent or dependent, say so.

31. $y = -2x + 4$
$6x + 3y = 12$

32. $y = \frac{3}{5}x + 1$
$3x - 5y = -5$

33. $y = -2x - 1$
$3x - y = 6$

34. $y = 3x + 2$
$3x - 2y = 2$

35. $y = 3x - 4$
$6x - 2y = 6$

36. $y = \frac{2}{3}x - 2$
$2x - 3y = 9$

37. **DATA**. The numbers of women and men who earned a bachelor's degree are listed in Table 6 for various years.

Table 6 Women and Men Who Earned a Bachelor's Degree

| Year | Number of People Who Earned a Bachelor's Degree (thousands) | |
	Women	Men
1995	634	526
2000	712	532
2005	855	631
2010	982	734
2014	1082	813

Source: *U.S. National Center for Education Statistics*

Let n be the number of people (in thousands) who earned a bachelor's degree in the year that is t years since 1900. Reasonable models for women and men are

$$n = 24.27t - 1690.26 \quad \text{Women}$$
$$n = 16.09t - 1039.32 \quad \text{Men}$$

Use "intersect" on a graphing calculator with the window shown in Fig. 12 to estimate when the number of women who earned a bachelor's degree was equal to the number of men who earned a bachelor's degree. What is that number of people?

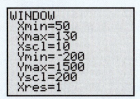

Figure 12 Exercise 37

38. **DATA**. The numbers of morning daily newspapers and evening daily newspapers are shown in Table 7 for various years.

Table 7 Numbers of Morning Dailies and Evening Dailies

| Year | Number of Daily Newspapers | |
	Morning	Evening
1980	387	1388
1985	482	1220
1990	559	1084
1995	656	891
2000	766	727
2005	817	645
2011	931	451
2015	972	389

Source: *Editor & Publisher Co.*

Let n be the number of daily newspapers at t years since 1980. Reasonable models for morning newspapers and evening newspapers are

$$n = 17t + 396.55 \quad \text{Morning newspapers}$$
$$n = -29t + 1360.43 \quad \text{Evening newspapers}$$

Use "intersect" on a graphing calculator with the window shown in Fig. 13 to estimate when the number of morning daily newspapers was equal to the number of evening daily newspapers. What is that number of newspapers?

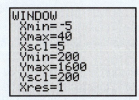

Figure 13 Exercise 38

39. **DATA**. World record times for the 200-meter run are listed in Table 8 for various years.

Table 8 200-Meter Run World Record Times

| Women | | Men | |
Year	Record Time (seconds)	Year	Record Time (seconds)
1973	22.38	1951	20.6
1974	22.21	1963	20.3
1978	22.06	1967	20.14
1984	21.71	1979	19.72
1988	21.34	1996	19.32
		2009	19.19

Source: *IAAF Statistics Handbook*

a. Let r be the women's record time (in seconds) at t years since 1900. Find an equation of a model to describe the data.

b. Let r be the men's record time (in seconds) at t years since 1900. Find an equation of a model to describe the data.

c. Use "intersect" on a graphing calculator with the window shown in Fig. 14 to predict when the women's record time will equal the men's record time. What is that record time?

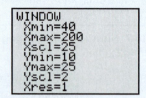

Figure 14 Exercise 39c

40. **DATA** World record times for the 500-meter race in speed skating are shown in Table 9 for various years.

Table 9 500-Meter Speed-Skating Record Times

| | Women | | Men |
Year	Record Time (seconds)	Year	Record Time (seconds)
1934	50.3	1900	45.2
1937	46.4	1914	43.4
1955	45.6	1938	41.8
1970	43.22	1953	40.9
1981	40.18	1970	39.09
1988	39.10	1987	36.55
2001	37.22	2000	34.63
2013	36.36	2015	33.98

Source: International Skating Union

a. Let r be the women's record time (in seconds) at t years since 1900. Find an equation of a model to describe the data.

b. Let r be the men's record time (in seconds) at t years since 1900. Find an equation of a model to describe the data.

c. Use "intersect" on a graphing calculator with the window shown in Fig. 15 to predict when the women's record time will be equal to the men's record time. What is that record time?

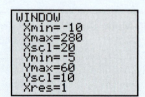

Figure 15 Exercise 40c

41. Figure 16 shows the graphs of two linear equations. Estimate, to the first decimal place, the coordinates of the solution of the system.

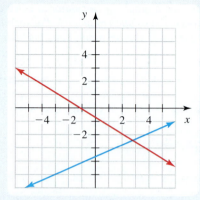

Figure 16 Exercise 41

42. Figure 17 shows the graphs of two linear equations. Estimate, to the first decimal place, the coordinates of the solution of the system.

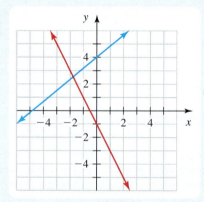

Figure 17 Exercise 42

43. Figure 18 shows the graphs of two linear equations. Estimate, to the nearest integer, the coordinates of the solution of the system. Explain. [**Hint:** Use the slope of each line.]

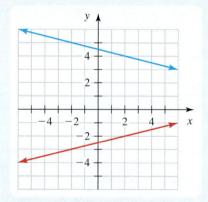

Figure 18 Exercise 43

44. Figure 19 shows the graphs of two linear equations. Estimate, to the nearest integer, the coordinates of the solution of the system. Explain. [**Hint:** Use the slope of each line.]

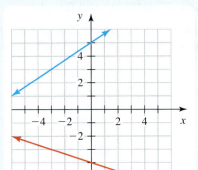

Figure 19 Exercise 44

45. Some solutions of two linear equations are shown in Tables 10 and 11. Find the solution of the system of the two equations.

Table 10 Solutions of One Equation

x	y
0	3
1	5
2	7
3	9
4	11

Table 11 Solutions of the Other Equation

x	y
0	21
1	17
2	13
3	9
4	5

46. Some solutions of two linear equations are shown in Tables 12 and 13. Find the solution of the system of the two equations.

Table 12 Solutions of One Equation

x	y
0	5
1	8
2	11
3	14
4	17

Table 13 Solutions of the Other Equation

x	y
0	21
1	16
2	11
3	6
4	1

Concepts

47. The graphs of $y = ax + b$ and $y = cx + d$ are sketched in Fig. 20.

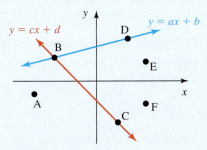

Figure 20 Exercise 47

Which of the points A, B, C, D, E, and F represent an ordered pair that
a. satisfies the equation $y = ax + b$?
b. satisfies the equation $y = cx + d$?
c. satisfies both equations?
d. does not satisfy either equation?

48. Consider the system

$$y = 4x - 5$$
$$y = -3x + 2$$

Find an ordered pair that
a. satisfies $y = 4x - 5$, but does not satisfy $y = -3x + 2$.
b. satisfies $y = -3x + 2$, but does not satisfy $y = 4x - 5$.
c. satisfies both equations.
d. does not satisfy either equation.

49. Create a system of two linear equations whose solution is $(1, 3)$. Verify your system graphically.

50. Create a system of two linear equations whose solution is $(-4, 2)$. Verify your system graphically.

51. Consider the system

$$y = 3x + 4$$
$$y = mx + b$$

For what values of m and b is the following true?
a. The system has exactly one solution.
b. The system has infinitely many solutions.
c. The solution set of the system is the empty set.

52. Is it possible for a system of linear equations to have exactly two solutions? Explain.

53. A student tries to solve the system

$$y = -x + 5$$
$$y = 2x - 4$$

After graphing the equations, the student believes that the solution is $(4, 1)$. The student then checks whether $(4, 1)$ satisfies $y = -x + 5$:

$$y = -x + 5$$
$$1 \stackrel{?}{=} -(4) + 5$$
$$1 \stackrel{?}{=} 1$$
$$\text{true}$$

The student concludes that $(4, 1)$ is the solution of the system. Is the student correct? Explain.

54. Explain why the solution(s) of a system of two linear equations can be found by locating intersection point(s) of the graphs of the two equations. (See page 9 for guidelines on writing a good response.)

Related Review

55. a. Determine whether the lines $y = 3x - 1$ and $y = 3x + 2$ are parallel.
b. Do parallel lines that are not the same line have an intersection point?
c. Solve the system

$$y = 3x - 1$$
$$y = 3x + 2$$

56. a. Determine whether the lines $3x - 7y = 14$ and $y = \dfrac{3}{7}x - 4$ are parallel.

b. Do parallel lines that are not the same line have an intersection point? Explain.

c. Solve the system

$$3x - 7y = 14$$
$$y = \frac{3}{7}x - 4$$

57. Create a system of two linear equations for which the slopes of both lines are negative and the solution of the system corresponds to a point in Quadrant IV.

58. Create a system of two linear equations for which the slopes of both lines are positive and the solution of the system corresponds to a point in Quadrant II.

Expressions, Equations, and Graphs

Perform the indicated instruction. Then use words such as linear, one variable, *and* two variables *to describe the expression, equation, or system.*

59. Solve the following:

$$y = 2x + 5$$
$$y = -3x - 5$$

60. Simplify $(2x + 5) - (-3x - 5)$.

61. Solve $2x + 5 = -3x - 5$.

62. Graph $y = -3x - 5$ by hand.

6.2 Using Substitution to Solve Systems

Objectives

» Use substitution to solve a system of two linear equations.

» Isolate a variable in an equation to help solve a system by substitution.

» Solve inconsistent systems and dependent systems by substitution.

In Section 6.1, we used graphs and tables to solve systems. In this section, we will use *equations* to solve systems.

The Substitution Method for Solving Systems

In Example 1, we will discuss how to solve a system by using a technique called *substitution*.

Example 1   Making a Substitution for *y*

Solve the system

$$y = x + 3 \quad \textit{Equation (1)}$$
$$2x + 3y = 14 \quad \textit{Equation (2)}$$

Solution

From equation (1), we know that for any solution of the system, the value of y is equal to the value of $x + 3$. So, we substitute $x + 3$ for y in equation (2):

$$2x + 3y = 14 \quad \textit{Equation (2)}$$
$$2x + 3(x + 3) = 14 \quad \textit{Substitute x + 3 for y.}$$

Note that, by making this substitution, we now have an equation in *one* variable. Next, we solve that equation for x:

$$\begin{aligned}
2x + 3(x + 3) &= 14 \\
2x + 3x + 9 &= 14 \quad &\textit{Distributive law} \\
5x + 9 &= 14 \quad &\textit{Combine like terms.} \\
5x &= 5 \quad &\textit{Subtract 9 from both sides.} \\
x &= 1 \quad &\textit{Divide both sides by 5.}
\end{aligned}$$

Thus, the x-coordinate of the solution is 1. To find the y-coordinate, we can substitute 1 for x in either of the original equations and solve for y:

$$\begin{aligned}
y &= x + 3 \quad &\textit{Equation (1)} \\
y &= 1 + 3 \quad &\textit{Substitute 1 for x.} \\
y &= 4 \quad &\textit{Add.}
\end{aligned}$$

So, the solution is $(1, 4)$. We check that $(1, 4)$ satisfies both of the system's equations:

$$y = x + 3 \qquad\qquad 2x + 3y = 14$$
$$4 \overset{?}{=} 1 + 3 \qquad\qquad 2(1) + 3(4) \overset{?}{=} 14$$
$$4 \overset{?}{=} 4 \qquad\qquad 2 + 12 \overset{?}{=} 14$$
$$\text{true} \qquad\qquad\qquad \text{true}$$

We can also verify that $(1, 4)$ is the solution by graphing equations (1) and (2) and checking that $(1, 4)$ is the intersection point of the two lines (see Fig. 21). To do so on a graphing calculator, we must first solve equation (2) for y:

$$2x + 3y = 14 \qquad \textit{Equation (2)}$$
$$3y = -2x + 14 \qquad \textit{Subtract 2x from both sides.}$$
$$\frac{3y}{3} = \frac{-2x}{3} + \frac{14}{3} \qquad \textit{Divide both sides by 3.}$$
$$y = -\frac{2}{3}x + \frac{14}{3} \qquad \textit{Simplify.}$$

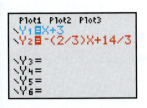

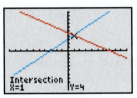

Figure 21 Verify that the solution is $(1, 4)$

Using Substitution to Solve a Linear System

To use **substitution** to solve a system of two linear equations,

1. Isolate a variable to one side of either equation.
2. Substitute the expression for the variable found in step 1 into the other equation.
3. Solve the equation in one variable found in step 2.
4. Substitute the solution found in step 3 into one of the original equations, and solve for the other variable.

   A system of equations can be solved by substitution as well as by graphing. The methods give the same result.
   In Example 1, we solved a system by making a substitution for y. In Example 2, we will solve a system by making a substitution for x.

Example 2   Making a Substitution for x

Solve the system

$$x = 2y - 8 \quad \textit{Equation (1)}$$
$$3x + 4y = 6 \qquad \textit{Equation (2)}$$

Solution

Because x is alone on one side of the equation $x = 2y - 8$, we begin by substituting $2y - 8$ for x in equation (2):

$$3x + 4y = 6 \qquad \textit{Equation (2)}$$
$$3(2y - 8) + 4y = 6 \qquad \textit{Substitute 2y − 8 for x.}$$
$$6y - 24 + 4y = 6 \qquad \textit{Distributive law}$$
$$10y - 24 = 6 \qquad \textit{Combine like terms.}$$
$$10y = 30 \quad \textit{Add 24 to both sides.}$$
$$y = 3 \quad \textit{Divide both sides by 10.}$$

The y-coordinate of the solution is 3. To find the x-coordinate, we substitute 3 for y in equation (1) and solve for x:

$$x = 2y - 8 \qquad \textit{Equation (1)}$$
$$x = 2(3) - 8 \qquad \textit{Substitute 3 for y.}$$
$$x = -2 \qquad \textit{Simplify.}$$

The solution is $(-2, 3)$. We could then check that $(-2, 3)$ satisfies *both* of the original equations.

▶

In Example 1, we made a substitution for y because y was alone on one side of one of the given equations. In Example 2, we made a substitution for x because x was alone on one side of one of the given equations.

In Example 3, we will solve a system of equations that have decimal coefficients. This will help prepare us for Section 6.4, where we will use systems to model situations.

Example 3   Solving a System of Equations

Solve the system

$$y = 1.72x - 4.38 \qquad \textit{Equation (1)}$$
$$y = -0.53x + 6.94 \qquad \textit{Equation (2)}$$

Solution

We start by substituting $1.72x - 4.38$ for y in equation (2):

$$y = -0.53x + 6.94 \qquad \textit{Equation (2)}$$
$$1.72x - 4.38 = -0.53x + 6.94 \qquad \textit{Substitute } 1.72x - 4.38 \textit{ for y.}$$
$$2.25x - 4.38 = 6.94 \qquad \textit{Add } 0.53x \textit{ to both sides.}$$
$$2.25x = 11.32 \qquad \textit{Add } 4.38 \textit{ to both sides.}$$
$$x \approx 5.03 \qquad \textit{Divide both sides by } 2.25.$$

Next, we substitute 5.03 for x in equation (1):

$$y = 1.72x - 4.38 \qquad \textit{Equation (1)}$$
$$y = 1.72(5.03) - 4.38 \qquad \textit{Substitute 5.03 for x.}$$
$$y \approx 4.27 \qquad \textit{Simplify.}$$

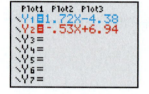

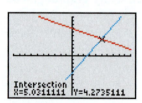

Figure 22 Verify that the approximate solution is (5.03, 4.27)

So, the approximate solution is (5.03, 4.27). We use "intersect" on a graphing calculator to check our work (see Fig. 22).

▶

Isolating a Variable

If no variable is alone on one side of either equation in a system of two linear equations, then we must first solve for a variable in one of the equations before we can make a substitution.

Example 4   Isolating a Variable and Then Using Substitution

Solve the system

$$3x + y = -7 \qquad \textit{Equation (1)}$$
$$2x - 5y = 1 \qquad \textit{Equation (2)}$$

Solution

We begin by solving for one of the variables in one of the equations. We can avoid fractions by choosing to solve equation (1) for y:

$$3x + y = -7 \qquad \textit{Equation (1)}$$
$$y = -3x - 7 \qquad \textit{Subtract 3x from both sides.}$$

Next, we substitute $-3x - 7$ for y in equation (2) and solve for x:

$$2x - 5y = 1 \qquad \textit{Equation (2)}$$
$$2x - 5(-3x - 7) = 1 \qquad \textit{Substitute } -3x - 7 \textit{ for y.}$$
$$2x + 15x + 35 = 1 \qquad \textit{Distributive law}$$
$$17x + 35 = 1 \qquad \textit{Combine like terms.}$$
$$17x = -34 \qquad \textit{Subtract 35 from both sides.}$$
$$x = -2 \qquad \textit{Divide both sides by 17.}$$

Finally, we substitute -2 for x in the equation $y = -3x - 7$ and solve for y:

$$y = -3(-2) - 7 \qquad \textit{Substitute } -2 \textit{ for x.}$$
$$y = 6 - 7 \qquad \textit{Multiply.}$$
$$y = -1 \qquad \textit{Subtract.}$$

The solution is $(-2, -1)$. We could verify our work by checking that $(-2, -1)$ satisfies both of the original equations.

Solving Inconsistent and Dependent Systems by Substitution

Each system in Examples 1–4 has one solution. What happens when we solve an inconsistent system (empty solution set) or a dependent system (infinitely many solutions) by substitution?

Example 5   Applying Substitution to an Inconsistent System

Consider the linear system

$$y = 3x + 2 \qquad \textit{Equation (1)}$$
$$y = 3x + 4 \qquad \textit{Equation (2)}$$

The graphs of the equations are distinct parallel lines (why?), so the system is inconsistent and the solution set is the empty set. What happens when we solve this system by substitution?

Solution

We substitute $3x + 2$ for y in equation (2) and solve for x:

$$y = 3x + 4 \qquad \textit{Equation (2)}$$
$$3x + 2 = 3x + 4 \qquad \textit{Substitute 3x + 2 for y.}$$
$$2 = 4 \qquad \textit{Subtract 3x from both sides.}$$
$$\text{false}$$

We get the *false* statement $2 = 4$.

Solving Inconsistent Systems by Substitution

If the result of applying substitution to a system of equations is a false statement, then the system is inconsistent; that is, the solution set is the empty set.

Example 6   Applying Substitution to a Dependent System

In Example 7 of Section 6.1, we found that the system

$$y = 3x - 5 \quad \text{Equation (1)}$$
$$6x - 2y = 10 \quad \text{Equation (2)}$$

is dependent and that the solution set is the infinite set of solutions of the equation $y = 3x - 5$. What happens when we solve this system by substitution?

Solution

We substitute $3x - 5$ for y in equation (2) and solve for x:

$$6x - 2y = 10 \quad \text{Equation (2)}$$
$$6x - 2(3x - 5) = 10 \quad \text{Substitute } 3x - 5 \text{ for } y.$$
$$6x - 6x + 10 = 10 \quad \text{Distributive law}$$
$$10 = 10 \quad \text{Combine like terms.}$$
$$\text{true}$$

We get the *true* statement $10 = 10$.

Solving Dependent Systems by Substitution

If the result of applying substitution to a linear system of two equations is a true statement that can be put into the form $a = a$, then the system is dependent; that is, the solution set is the set of ordered pairs represented by every point on the (same) line.

◆ Group Exploration

Comparing techniques for solving systems

1. Consider the system

$$y = -2x + 3$$
$$y = 4x + 3$$

   a. Solve the system by graphing the equations by hand.
   b. Now solve the system by substitution.
   c. Compare the results you found in parts (a) and (b).
   d. Decide which method you prefer for this system. Explain.

2. Consider the system

$$3x - 2y = -4$$
$$y = -2x + 9$$

   a. Solve the system by graphing the equations by hand.
   b. Now solve the system by substitution.
   c. Compare the results you found in parts (a) and (b).
   d. Decide which method you prefer for this system. Explain.

3. In general, are there any systems that you prefer to solve by graphing by hand? If yes, describe them and give an example. If no, explain why not.

4. In general, are there any systems that you prefer to solve by substitution? If yes, describe them and give an example. If no, explain why not.

Tips for Success   Write a Summary

Consider writing, after each class meeting, a summary of what you have learned. Your summaries will increase your understanding, as well as your memory, of concepts and procedures and will also serve as a good reference for quizzes and exams.

Homework 6.2

For extra help ▶   **MyLab Math**   Watch the videos in MyLab Math

Solve the system by substitution. Verify your solution by checking that it satisfies both equations of the system.

1. $y = 2x$
$3x + y = 10$

2. $y = 4x$
$2x + y = 18$

3. $x - 4y = -3$
$x = 2y - 1$

4. $x + 2y = 4$
$x = 3y - 1$

5. $2x + 3y = 5$
$y = x + 5$

6. $4x - 3y = 13$
$y = x - 4$

7. $-5x - 2y = 17$
$x = 4y + 1$

8. $3x + 2y = -9$
$x = 1 - 2y$

9. $2x - 5y - 3 = 0$
$y = 2x - 7$

10. $4x + 3y + 2 = 0$
$y = 1 - 3x$

11. $x = 2y + 6$
$-4x + 5y + 12 = 0$

12. $x = -2y + 5$
$7x + 2y + 13 = 0$

Solve the system by substitution. Verify your solution by using "intersect" on a graphing calculator.

13. $y = -2x - 1$
$y = 3x + 9$

14. $y = 4x + 2$
$y = -3x - 5$

15. $y = 2x$
$y = 3x$

16. $y = x$
$y = -x$

17. $y = 2(x - 4)$
$y = -3(x + 1)$

18. $y = 5(x - 1)$
$y = 2(x + 2)$

Use substitution to solve the system, with coordinates of solutions rounded to the second decimal place. Verify your work by using "intersect" on a graphing calculator.

19. $y = 2.57x + 7.09$
$y = -3.61x - 5.72$

20. $y = -1.45x - 6.18$
$y = 2.63x - 2.73$

21. $y = -3.17x + 8.92$
$y = 1.65x - 7.24$

22. $y = -0.51x - 2.64$
$y = -2.79x + 5.94$

23. $y = -1.82x + 3.95$
$y = 1.57x + 4.68$

24. $y = -2.49x - 6.17$
$y = 4.81x + 3.45$

Solve the system by substitution. Verify your solution by checking that it satisfies both equations of the system.

25. $2x + y = -9$
$5x - 3y = 5$

26. $-3x + y = -14$
$2x - 7y = 22$

27. $4x - 7y = 15$
$x - 3y = 5$

28. $2x + 3y = -1$
$x + 3y = -5$

29. $4x + 3y = 5$
$x - 2y = -7$

30. $6x - 5y = -8$
$x + 3y = 14$

31. $2x - y = 1$
$5x - 3y = 5$

32. $4x - 3y = -7$
$3x - y = -9$

33. $3x + 2y = -3$
$2x = y + 5$

34. $4x - 5y = -2$
$3x = y - 7$

Solve the system by substitution. If the system is inconsistent or dependent, say so.

35. $x = 4 - 3y$
$2x + 6y = 8$

36. $x = 2y - 1$
$4x - 8y = -4$

37. $5x - 2y = 18$
$y = -3x + 2$

38. $-3x - 2y = -10$
$y = 4 - 2x$

39. $y = -5x + 3$
$15x + 3y = 6$

40. $y = 2x - 5$
$-4x + 2y = -6$

41. $-4x + 12y = 4$
$x = 3y - 1$

42. $3x + 12y = 6$
$x = -4y + 2$

43. $y = 3x + 2$
$12x - 4y = 9$

44. $y = 4x - 1$
$-8x + 2y = -5$

Concepts

45. Some solutions of two linear equations are shown in Tables 14 and 15. Find the solution of the system of two equations. [**Hint:** You could begin by finding two equations.]

Table 14 Solutions of One Equation

x	y
0	5
1	8
2	11
3	14
4	17

Table 15 Solutions of the Other Equation

x	y
0	90
1	88
2	86
3	84
4	82

46. Some solutions of two linear equations are shown in Tables 16 and 17. Find the solution of the system of two equations. [**Hint:** You could begin by finding two equations.]

Table 16 Solutions of One Equation

x	y
-2	73
-1	67
0	61
1	55
2	49

Table 17 Solutions of the Other Equation

x	y
-2	-17
-1	-13
0	-9
1	-5
2	-1

47. Find the coordinates of the points A, B, C, and D as shown in Fig. 23. The equations of lines l_1 and l_2 are provided, but no attempt has been made to sketch the lines accurately except for showing the intersection points. Use TRACE and "intersect" on a graphing calculator to verify your work.

$l_1: y = -x + 8$
$l_2: y = 2x - 7$

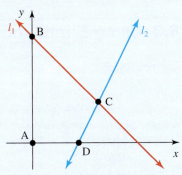

Figure 23 Exercise 47

48. Find the coordinates of the points A, B, C, and D as shown in Fig. 24. The equations of lines l_1 and l_2 are provided, but no attempt has been made to sketch the lines accurately except for showing the intersection points. Use TRACE and "intersect" on a graphing calculator to verify your work.

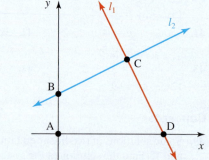

$l_1: y = -2x + 8$

$l_2: 2y - x = 6$

Figure 24 Exercise 48

49. Consider the system

$$x + y = 3$$
$$3x + 2y = 7$$

   a. Isolate x to one side of one of the equations, and then solve the system by substitution.

   b. Isolate y to one side of one of the equations, and then solve the system by substitution.

   c. Compare the results you found in parts (a) and (b).

50. Consider the system

$$2x + y = 8$$
$$x + 3y = 9$$

   a. Isolate x to one side of one of the equations, and then solve the system by substitution.

   b. Isolate y to one side of one of the equations, and then solve the system by substitution.

   c. Compare the results you found in parts (a) and (b).

51. Consider the system

$$y = 3x$$
$$2x + y = 5$$

A student tries to solve the system:

$$2x + 3x = 5$$
$$5x = 5$$
$$x = 1$$

Describe any errors. Then solve the system correctly.

52. **a.** When using substitution to solve a system, how can you tell if the solution set is the empty set?

   b. When using substitution to solve a system, how can you tell if there are an infinite number of solutions?

53. Describe how to solve a system by substitution.

54. Is it more reliable to use substitution or graphing by hand to find exact solutions of a system? Explain.

Related Review

55. **a.** Use "intersect" on a graphing calculator to solve the system

$$y = 2x - 3$$
$$y = -3x + 7$$

   b. Solve the same system by substitution.

   c. Solve the equation $2x - 3 = -3x + 7$.

   d. Compare the result you found in part (c) with the x-coordinate of the result you found in part (b).

   e. Use "intersect" to help you solve $x - 1 = -2x + 8$. Then solve this equation without using graphing. Compare your results.

   f. Summarize the main concepts addressed in this exercise. Explain how you can use "intersect" to verify your work.

56. **a.** Find an equation of the line that has slope 3 and contains the point $(3, 4)$.

   b. Find an equation of the line that has slope -2 and contains the point $(3, 4)$.

   c. Solve the system of the two linear equations you found in parts (a) and (b).

   d. Explain how your work in part (c) is related to your work in parts (a) and (b).

Expressions, Equations, and Graphs

Perform the indicated instruction. Then use words such as linear, one variable, *and* two variables *to describe the expression, equation, or system.*

57. Graph $y = -3x + 5$ by hand.

58. Solve $7 = -3x + 5$.

59. Solve the following:

$$9x + 2y = 1$$
$$y = -3x + 5$$

60. Simplify $9x + 2y - 3x + 5$.

6.3 Using Elimination to Solve Systems

Objectives

» Use elimination to solve a system of two linear equations.

» Solve inconsistent systems and dependent systems by elimination.

» Compare three ways to solve a system.

So far, we have solved systems by graphing and by substitution. In this section, we will use a third technique, called *elimination* or the *addition method*.

The Elimination Method for Solving Systems

To solve systems by elimination, we will need to use the following property:

Adding Left Sides and Right Sides of Two Equations

If $a = b$ and $c = d$, then

$$a + c = b + d$$

In words, the sum of the left sides of two equations is equal to the sum of the right sides.

For example, if we add the left sides and add the right sides of the equations $3 = 3$ and $5 = 5$, we obtain the true statement $3 + 5 = 3 + 5$.

Example 1   Solving a System by Elimination

Solve the system

$$3x + 2y = 9 \quad \text{\textit{Equation} (1)}$$
$$-3x + 5y = 12 \quad \text{\textit{Equation} (2)}$$

Solution

We begin by adding the left sides and adding the right sides of the two equations:

$$3x + 2y = 9 \quad \text{\textit{Equation} (1)}$$
$$\underline{-3x + 5y = 12} \quad \text{\textit{Equation} (2)}$$
$$0 + 7y = 21 \quad \text{\textit{Add left sides and add right sides; combine like terms.}}$$

By "eliminating" the variable x, we now have an equation in *one* variable. Next, we solve that equation for y:

$$0 + 7y = 21$$
$$7y = 21 \quad \text{\textit{0 + a = a}}$$
$$y = 3 \quad \text{\textit{Divide both sides by 7.}}$$

Then, we substitute 3 for y in either of the original equations and solve for x:

$$3x + 2y = 9 \quad \text{\textit{Equation} (1)}$$
$$3x + 2(3) = 9 \quad \text{\textit{Substitute 3 for y.}}$$
$$3x + 6 = 9 \quad \text{\textit{Multiply.}}$$
$$3x = 3 \quad \text{\textit{Subtract 6 from both sides.}}$$
$$x = 1 \quad \text{\textit{Divide both sides by 3.}}$$

The solution is $(1, 3)$. We check that $(1, 3)$ satisfies *both* equations (1) and (2):

$$3x + 2y = 9 \qquad\qquad -3x + 5y = 12$$
$$3(1) + 2(3) \stackrel{?}{=} 9 \qquad\qquad -3(1) + 5(3) \stackrel{?}{=} 12$$
$$3 + 6 \stackrel{?}{=} 9 \qquad\qquad -3 + 15 \stackrel{?}{=} 12$$
$$\text{true} \qquad\qquad\qquad \text{true}$$

Example 2   Solving a System by Elimination

Solve the system

$$-5x + 3y = -9 \quad \text{\textit{Equation} (1)}$$
$$3x + 6y = 21 \quad \text{\textit{Equation} (2)}$$

Solution

If we add the left sides and add the right sides of the equations as given to us, neither variable would be eliminated. Therefore, we first multiply both sides of equation $-5x + 3y = -9$ by -2, yielding the system

$$10x - 6y = 18 \quad \textit{Multiply both sides of equation (1) by } -2.$$
$$3x + 6y = 21 \quad \textit{Equation (2)}$$

Now that the coefficients of the y terms are equal in absolute value and opposite in sign, we add the left sides and add the right sides of the equations and solve for x:

$$
\begin{aligned}
10x - 6y &= 18 \\
\underline{3x + 6y} &= \underline{21} \\
13x + 0 &= 39 \quad \textit{Add left sides and add right sides; combine like terms.} \\
13x &= 39 \quad \textit{a + 0 = a} \\
x &= 3 \quad \textit{Divide both sides by 13.}
\end{aligned}
$$

We substitute 3 for x in equation (1) and solve for y:

$$
\begin{aligned}
-5x + 3y &= -9 \quad \textit{Equation (1)} \\
-5(3) + 3y &= -9 \quad \textit{Substitute 3 for x.} \\
-15 + 3y &= -9 \quad \textit{Simplify.} \\
3y &= 6 \quad \textit{Add 15 to both sides.} \\
y &= 2 \quad \textit{Divide both sides by 3.}
\end{aligned}
$$

The solution is $(3, 2)$. We could check that $(3, 2)$ satisfies both of the original equations.

Using Elimination to Solve a Linear System

To use **elimination** to solve a system of two linear equations,

1. Use the multiplication property of equality (Section 4.3) to get the coefficients of one variable to be equal in absolute value and opposite in sign.
2. Add the left sides and add the right sides of the equations to eliminate one of the variables.
3. Solve the equation in one variable found in step 2.
4. Substitute the solution found in step 3 into one of the original equations, and solve for the other variable.

Example 3 Solving a System by Elimination

Solve the system

$$5x - 4y = 13 \quad \textit{Equation (1)}$$
$$9x + 3y = 3 \quad \textit{Equation (2)}$$

Solution

To eliminate the y terms, we multiply both sides of equation (1) by 3 and multiply both sides of equation (2) by 4, yielding the system

$$15x - 12y = 39 \quad \textit{Multiply both sides of equation (1) by 3.}$$
$$36x + 12y = 12 \quad \textit{Multiply both sides of equation (2) by 4.}$$

The coefficients of the y terms are now equal in absolute value and opposite in sign. Next, we add the left sides and add the right sides of the equations and solve for x:

$$
\begin{aligned}
15x - 12y &= 39 \\
\underline{36x + 12y} &= \underline{12} \\
51x + 0 &= 51 \quad \textit{Add left sides and add right sides; combine like terms.}
\end{aligned}
$$

$$51x = 51 \quad \textit{a + 0 = a}$$
$$x = 1 \quad \textit{Divide both sides by 51.}$$

Substituting 1 for x in equation (1) gives

$$5x - 4y = 13 \quad \textit{Equation (1)}$$
$$5(1) - 4y = 13 \quad \textit{Substitute 1 for x.}$$
$$5 - 4y = 13 \quad \textit{Simplify.}$$
$$-4y = 8 \quad \textit{Subtract 5 from both sides.}$$
$$y = -2 \quad \textit{Divide both sides by} -4.$$

The solution is $(1, -2)$.

▶

In Example 3, we eliminated y by getting the coefficients of the y terms to be equal in absolute value and opposite in sign. Note that this process is similar to finding a least common multiple.

Solving Inconsistent Systems and Dependent Systems by Elimination

When we apply elimination to an inconsistent system or a dependent system, the results are similar to the results obtained from applying substitution to such systems.

Solving Inconsistent Systems and Dependent Systems by Elimination

If the result of applying elimination to a linear system of two equations is

- a false statement, then the system is inconsistent; that is, the solution set is the empty set.
- a true statement that can be put into the form $a = a$, then the system is dependent; that is, the solution set is the set of ordered pairs represented by every point on the (same) line.

Example 4   Using Elimination to Solve a Dependent System

Solve the system

$$2x - 7y = 5 \quad \textit{Equation (1)}$$
$$6x - 21y = 15 \quad \textit{Equation (2)}$$

Solution

To eliminate the x terms (and the y terms), we multiply both sides of equation (1) by -3, yielding the system

$$-6x + 21y = -15 \quad \textit{Multiply both sides of equation (1) by} -3.$$
$$6x - 21y = 15 \quad \textit{Equation (2)}$$

Now that the coefficients of the x terms (and those of the y terms) are equal in absolute value and opposite in sign, we add the left sides and add the right sides of the equations:

$$-6x + 21y = -15$$
$$\underline{6x - 21y = 15}$$
$$0 + 0 = 0 \quad \textit{Add left sides and add right sides; combine like terms.}$$
$$0 = 0 \quad \textit{Simplify.}$$

Because $0 = 0$ is a true statement of the form $a = a$, we conclude that the system is dependent and that the solution set of the system is the set of ordered pairs represented by the points on the line $2x - 7y = 5$ and the (same) line $6x - 21y = 15$.

▶

Comparing Three Ways to Solve a System

In Example 5, we will solve the same system that we solved by substitution in Example 3 of Section 6.2, but now we will solve it by elimination.

| Example 5 | Solving a System by Elimination |

Solve the system

$$y = 1.72x - 4.38 \quad \text{Equation (1)}$$
$$y = -0.53x + 6.94 \quad \text{Equation (2)}$$

Solution

First we multiply both sides of equation (1) by -1, yielding the system

$$-y = -1.72x + 4.38 \quad \text{Multiply equation (1) by } -1.$$
$$y = -0.53x + 6.94 \quad \text{Equation (2)}$$

Now that the coefficients of y are equal in absolute value and opposite in sign, we add the left sides and add the right sides of the equations and solve for x:

$$-y = -1.72x + 4.38$$
$$\underline{y = -0.53x + 6.94}$$
$$0 = -2.25x + 11.32 \quad \text{Add left sides and add right sides; combine like terms.}$$
$$2.25x = 11.32 \quad \text{Add 2.25x to both sides.}$$
$$x \approx 5.03 \quad \text{Divide both sides by 2.25.}$$

Then we substitute 5.03 for x in equation (1):

$$y = 1.72x - 4.38 \quad \text{Equation (1)}$$
$$y = 1.72(5.03) - 4.38 \quad \text{Substitute 5.03 for x.}$$
$$y \approx 4.27 \quad \text{Simplify.}$$

So, the approximate solution is $(5.03, 4.27)$, which is the same result that we obtained in Example 3 of Section 6.2.

▶

In both this section and Section 6.2, we solved the system

$$y = 1.72x - 4.38$$
$$y = -0.53x + 6.94$$

by graphing, substitution, and elimination. We obtained the same approximate solution by all three methods. In fact, **any linear system of two equations can be solved by graphing, substitution, or elimination. All three methods give the same result.**

How do we determine which method to use? When we solve a system of two equations, if a variable is alone or is easy to isolate to one side of either equation, substitution is often convenient. Otherwise, elimination is usually easiest. With most systems, we avoid solving by graphing or by using tables, except as a check, because these methods often lead only to approximate solutions.

 Group Exploration

Comparing techniques of solving systems

Consider the system

$$x + 2y = 4$$
$$y = 3x - 5$$

1. Use substitution to solve the system.

2. Use elimination to solve the system.

3. Solve the system by graphing the equations by hand.

4. Compare the results you found in Problems 1–3.

5. Give an example of a system that is easiest to solve by substitution. Also, give an example of a system that is easiest to solve by elimination. Finally, give an example of a system that is easiest to solve by graphing by hand. Explain. Solve your three systems.

Tips for Success   Cross-Checks

If you finish a quiz or an exam early, it pays to verify your answers with cross-checks. For example, suppose you determine by elimination that the solution of the system

$$5x + 2y = 9$$
$$3x + 4y = 11$$

is $(1, 2)$. There are several ways to verify your answer. You could check that $(1, 2)$ satisfies both equations. You could graph each equation and check that the intersection point is $(1, 2)$. Or you could solve the system by substitution.

Homework 6.3

For extra help ▶   **MyLab Math**   Watch the videos in MyLab Math

Solve the system by elimination. Verify your solution by checking that it satisfies both equations in the system.

1. $2x + 3y = 7$
$-2x + 5y = 1$

2. $-4x + 5y = 11$
$4x + 3y = 13$

3. $5x - 2y = 2$
$-3x + 2y = 2$

4. $3x + 5y = 19$
$2x - 5y = -4$

5. $x + 2y = -4$
$3x - 4y = 18$

6. $x - 4y = 5$
$5x - 2y = -11$

7. $2x - 3y = 8$
$5x + 6y = -7$

8. $3x + 4y = -5$
$-5x + 2y = 17$

9. $6x - 5y = 4$
$2x + 3y = -8$

10. $4x - 7y = 25$
$8x + 3y = -1$

11. $5x + 7y = -16$
$2x - 5y = 17$

12. $6x + 5y = -14$
$-4x - 7y = 2$

13. $-8x + 3y = 1$
$3x - 4y = 14$

14. $5x - 2y = 8$
$2x - 3y = 1$

15. $y = 3x - 6$
$y = -4x + 1$

16. $y = 2x + 7$
$y = -x - 8$

17. $3x + 6y - 18 = 0$
$17 = 7x + 9y$

18. $3 = -2x + 9y$
$5x - 5y - 10 = 0$

19. $\frac{2}{9}x + \frac{1}{3}y = 4$
$\frac{1}{2}x - \frac{2}{5}y = -\frac{5}{2}$

20. $\frac{1}{2}x - \frac{5}{4}y = \frac{9}{2}$
$\frac{3}{8}x + \frac{1}{2}y = \frac{1}{2}$

Solve the system by elimination. If the system is inconsistent or dependent, say so.

21. $4x - 7y = 3$
$8x - 14y = 6$

22. $4x + 6y = 10$
$-6x - 9y = -15$

23. $8x - 6y = 4$
$12x - 9y = 5$

24. $3x + 2y = 5$
$-12x - 8y = -17$

25. $3x - 2y = -14$
$6x + 5y = -19$

26. $8x - 3y = -2$
$5x + 12y = -29$

27. $6x - 15y = 7$
$-4x + 10y = -5$

28. $10x - 12y = 5$
$-15x + 18y = -8$

29. $3x - 9y = 12$
$-4x + 12y = -16$

30. $9x - 6y = 15$
$12x - 8y = 20$

Use elimination to solve the system, with coordinates of solutions rounded to the second decimal place. Verify your work by using "intersect" on a graphing calculator.

31. $y = 4.29x - 8.91$
$y = -1.26x + 9.75$

32. $y = 1.28x + 2.05$
$y = 3.94x - 8.83$

33. $y = -2.15x + 8.38$
$y = 1.67x + 2.57$

34. $y = 3.28x + 1.43$
$y = 0.56x + 6.72$

35. $y = -2.62x + 7.24$
$y = 1.89x - 6.44$

36. $y = 1.64x + 5.07$
$y = -2.57x - 3.39$

Solve the system by either elimination or substitution. Verify your work by using "intersect" on your graphing calculator or by checking that your result satisfies both equations of the system.

37. $4x - y = -12$
$3x + 5y = 14$

38. $6x + y = -13$
$3x - 2y = -19$

39. $-2x + 7y = -3$
$x = 3y + 2$

40. $4x - 5y = 23$
$y = 1 - 2x$

41. $2x + 7y = 13$
$3x - 4y = -24$

42. $5x - 2y = 7$
$4x - 3y = 7$

43. $2x - 7y = -1$
$-x - 3y = 7$

44. $3x - 8y = -4$
$x - 5y = -6$

45. $y = -2x - 3$
$y = 3x + 7$

46. $y = -4x - 9$
$y = 2x + 3$

47. $8x + 5y = 7$
$7y = -6x + 15$

48. $-3x + 2y = 19$
$5y = -4x - 10$

49. $3(2x - 5) + 4y = 11$
$5x - 2(3y + 1) = 1$

50. $4(2x - 7) + 3y = 7$
$7x - 5(2y + 3) = 3$

Concepts

For Exercises 51 and 52, solve the system of equations three times, once by each of the three methods: graphing by hand, substitution, and elimination. Decide which method you prefer for this system. Explain.

51. $3x + 2y = 8$
$2x - y = 3$

52. $y = -2x + 7$
$5x - 2y = 4$

53. Consider the system

$$5x + 3y = 11$$
$$2x - 4y = -6$$

a. Solve the system by eliminating the x terms.
b. Solve the system by eliminating the y terms.
c. Compare the results you found in parts (a) and (b).

54. Consider the system

$$4x - 7y = 15$$
$$5x + 3y = 7$$

a. Solve the system by eliminating the x terms.
b. Solve the system by eliminating the y terms.
c. Compare the results you found in parts (a) and (b).

55. Some solutions of two linear equations are shown in Tables 18 and 19. Find the solution of the system of the two equations. [**Hint:** You could begin by finding two equations.]

Table 18 Solutions of One Equation	
x	y
-2	99
-1	96
0	93
1	90
2	87

Table 19 Solutions of the Other Equation	
x	y
-2	-26
-1	-24
0	-22
1	-20
2	-18

56. Some solutions of two linear equations are shown in Tables 20 and 21. Find the solution of the system of the two equations. [**Hint:** You could begin by finding two equations.]

Table 20 Solutions of One Equation	
x	y
-2	-84
-1	-80
0	-76
1	-72
2	-68

Table 21 Solutions of the Other Equation	
x	y
-2	91
-1	88
0	85
1	82
2	79

57. Find the coordinates of the points A, B, C, and D as shown in Fig. 25. The equations of lines l_1 and l_2 are provided, but no attempt has been made to sketch the lines accurately except for showing the intersection points. Use TRACE and "intersect" on a graphing calculator to verify your work.

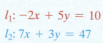

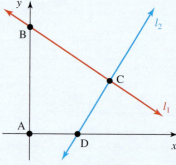

Figure 25 Exercise 57

58. Find the coordinates of the points A, B, C, and D as shown in Fig. 26. The equations of lines l_1 and l_2 are provided, but no attempt has been made to sketch the lines accurately except for showing the intersection points. Use TRACE and "intersect" on a graphing calculator to verify your work.

$l_1: -2x + 5y = 10$
$l_2: 7x + 3y = 47$

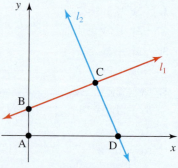

Figure 26 Exercise 58

59. When using elimination to solve a system, how can you tell if the solution set is the empty set?

60. When using elimination to solve a system, how can you tell if there are an infinite number of solutions?

61. Describe how to solve a system by elimination. (See page 9 for guidelines on writing a good response.)

62. Describe how to determine whether to solve a system by graphing, substitution, or elimination. (See page 9 for guidelines on writing a good response.)

Related Review

63. In this exercise, you will solve a system of equations to help find an equation of the line that contains the points $(2, 5)$ and $(4, 9)$.

a. Substitute 2 for x and 5 for y in the equation $y = mx + b$.
b. Substitute 4 for x and 9 for y in the equation $y = mx + b$.
c. Consider the system of equations formed by the results you found in parts (a) and (b). Solve this system to find values of m and b.
d. Substitute the values you found for m and b in the equation $y = mx + b$.
e. Verify your equation by using a graphing calculator.

64. In this exercise, you will solve a system of equations to help find an equation of the line that contains the points $(-1, 5)$ and $(2, -4)$.

   a. Substitute -1 for x and 5 for y in the equation $y = mx + b$.

   b. Substitute 2 for x and -4 for y in the equation $y = mx + b$.

   c. Consider the system of equations formed by the results you found in parts (a) and (b). Solve this system to find values of m and b.

   d. Substitute the values you found for m and b in the equation $y = mx + b$.

   e. Verify your equation by using a graphing calculator.

Expressions, Equations, and Graphs

Perform the indicated instruction. Then use words such as linear, one variable, *and* two variables *to describe the expression, equation, or system.*

65. Simplify $2(5x + 4) - 6(3x + 2)$.

66. Graph $5x + 4y = 8$ by hand.

67. Solve $2(5x + 4) - 6(3x + 2) = 0$.

68. Solve the following:

$$5x + 4y = 8$$
$$3x + 2y = 2$$

6.4 Using Systems to Model Data

Objectives

» Use substitution and elimination to make predictions about situations described by tables of data.

» Use substitution and elimination to make predictions about situations described by rates of change.

In Section 6.1, we used a graphical approach to find the intersection point of the graphs of two linear models. In this section, we will use substitution and elimination to find such an intersection point.

Using a Table of Data to Find a System for Modeling

In Example 1, we will work with a table of data and the two linear models we used in Section 6.1.

Example 1   Using a System to Make a Prediction

Let n be the number of college instructors (in thousands) at t years since 1990 (see Table 22). In Example 5 of Section 6.1, we used the following reasonable models for part-time instructors and full-time instructors:

$$n = 22.03t + 268.63 \quad \text{(1) } \textit{Part-time instructors}$$
$$n = 13.88t + 468.48 \quad \text{(2) } \textit{Full-time instructors}$$

Table 22 Numbers of Part-Time and Full-Time Instructors

Year	Number of Instructors (thousands)	
	Part Time	Full Time
1997	421	569
2001	495	618
2005	615	676
2009	710	729
2013	754	791

Source: *U.S. National Center for Education Statistics*

Use a nongraphical approach to estimate when the number of part-time instructors was equal to the number of full-time instructors. Also, find that number of instructors.

Solution

From Example 5 of Section 6.1, we saw that the intersection point of the graphs of the two models indicates the year the numbers of part-time instructors and full-time instructors were the same. Instead of graphing, we can find this intersection point by substitution (or elimination). To solve by substitution, we substitute $22.03t + 268.63$ for n in equation (2) and solve for t:

$$n = 13.88t + 468.48 \quad \textit{Equation (2)}$$
$$22.03t + 268.63 = 13.88t + 468.48 \quad \textit{Substitute } 22.03t + 268.63 \textit{ for n.}$$
$$8.15t + 268.63 = 468.48 \quad \textit{Subtract } 13.88t \textit{ from both sides.}$$
$$8.15t = 199.85 \quad \textit{Subtract } 268.63 \textit{ from both sides.}$$
$$t \approx 24.52 \quad \textit{Divide both sides by 8.15}$$

Next, we substitute 24.52 for t in equation (1):

$$n = 22.03(24.52) + 268.63 \approx 808.81$$

So, the approximate solution of the system is $(24.52, 808.81)$. According to the models, there were the same number of part-time instructors as full-time instructors (about 809 thousand of each) in 2015. This is the same conclusion we came to in Example 5 of Section 6.1.

▶

In Example 1, we used substitution to estimate when there were the same number of part-time instructors as full-time instructors. Graphically, this event is described by the intersection point of the graphs of the two instructor models. In general, **we can find an intersection point of the graphs of two linear models by graphing, substitution, or elimination.**

Example 2   Using a System to Make a Prediction

World record times for the 800-meter run are listed in Table 23.

Table 23 800-Meter Run Record Times

Women		Men	
Year	Record Time (seconds)	Year	Record Time (seconds)
1952	128.5	1938	108.4
1960	124.3	1939	106.6
1967	121.0	1955	105.7
1976	116.0	1968	104.3
1983	113.28	1976	103.5
		1997	101.11
		2012	100.91

Source: *Matti Koskimies/Peter Larrson*

Let r be the 800-meter record time (in seconds) at t years since 1900. Reasonable models for the women's record time and the men's record time are

$$r = -0.50t + 154.23 \quad \textit{Women}$$
$$r = -0.10t + 111.03 \quad \textit{Men}$$

1. Check that the models fit the data well.
2. Predict when the women's record time and the men's record time will be equal.

Solution

1. To keep track of the two types of data points, we use square "Marks" for the women's record scatterplot and use plus-sign "Marks" for the men's record scatterplot (see Fig. 27). See Appendix A.16 for graphing calculator instructions. The models appear to fit the data quite well.

2. We solve the system

$$r = -0.50t + 154.23 \quad \textit{Equation (1)}$$
$$r = -0.10t + 111.03 \quad \textit{Equation (2)}$$

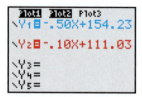

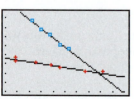

Figure 27 Check that the models fit the data well

by substitution. We do so by substituting $-0.50t + 154.23$ for r in equation (2):

$$r = -0.10t + 111.03 \quad \text{Equation (2)}$$
$$-0.50t + 154.23 = -0.10t + 111.03 \quad \text{Substitute } -0.50t + 154.23 \text{ for } r.$$
$$-0.40t + 154.23 = 111.03 \quad \text{Add } 0.10t \text{ to both sides.}$$
$$-0.40t = -43.20 \quad \text{Subtract } 154.23 \text{ from both sides.}$$
$$t = 108 \quad \text{Divide both sides by } -0.40.$$

Next, we substitute 108 for t in equation (1) and solve for r:

$$r = -0.50(108) + 154.23 = 100.23$$

So, according to the models, the record times for both women and men were 100.23 seconds in 2008. To verify our result, we use "intersect" on a graphing calculator (see Fig. 28).

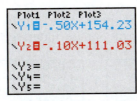 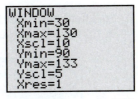 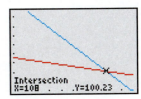

Figure 28 Verify that the intersection point is (108, 100.23)

Model breakdown has occurred. Currently, the men's record time is more than 12 seconds less than the women's record time (see Table 23).

Using Rate of Change to Find a System for Modeling

Recall from Section 3.5 that if the rate of change of the response variable with respect to the explanatory variable is constant, then there is a linear relationship between the variables. Also, for the linear model that describes that relationship, the slope is equal to the constant rate of change. We will use these ideas in Example 3.

Example 3   Using a System to Make a Prediction

The U.S. soda market share of Diet Pepsi® was 4.5% in 2014 and has decreased by about 0.26 percentage point per year since then. The U.S. soda market share of Sparkling Ice® was 1.4% in 2014 and has increased by about 0.25 percentage point per year since then (Source: *Euromonitor*).

1. Let s be Diet Pepsi's market share (in percent) at t years since 2014. Find an equation of a model to describe the data.
2. Let s be Sparkling Ice's market share (in percent) at t years since 2014. Find an equation of a model to describe the data.
3. When will Diet Pepsi's market share be equal to Sparkling Ice's market share? What is that market share?

Solution

1. Because Diet Pepsi's market share decreases by about 0.26 percentage point per year, we can model the situation by a linear model. The slope (rate of change) is -0.26 percentage point per year. Because the market share was 4.5% in 2014, the s-intercept is $(0, 4.5)$. So, a reasonable model is

$$s = -0.26t + 4.5$$

2. By similar reasoning, the model is $s = 0.25t + 1.4$.

3. To find when the two market shares will be equal, we solve the system

$$s = -0.26t + 4.5 \quad \text{(1) } \textit{Diet Pepsi}$$
$$s = 0.25t + 1.4 \quad \text{(2) } \textit{Sparkling Ice}$$

To solve the system by substitution, we substitute $-0.26t + 4.5$ for s in equation (2):

$$-0.26t + 4.5 = 0.25t + 1.4$$

Then we solve for t:

$$-0.51t + 4.5 = 1.4 \quad \textit{Subtract 0.25t from both sides.}$$
$$-0.51t = -3.1 \quad \textit{Subtract 4.5 from both sides.}$$
$$t \approx 6.08 \quad \textit{Divide both sides by } -0.51.$$

Next, we substitute 6.08 for t in equation (1):

$$s = -0.26(6.08) + 4.5 \approx 2.92$$

So, Diet Pepsi's market share and Sparkling Ice's market share will both be about 2.92% in 2020. To verify our result, we use "intersect" on a graphing calculator (see Fig. 29).

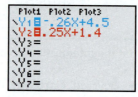 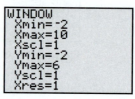 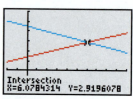

Figure 29 Verify the intersection point is about (6.08, 2.92)

◈ Group Exploration

Using a difference to make a prediction

1. Suppose that A and B are real numbers.
   a. If $A - B = 2$, which is larger, A or B? How much larger?
   b. If $A - B = -3$, which is larger, A or B? How much larger?
   c. If $A - B = 0$, what is true about A and B?

2. In Example 1, we worked with the models

$$n = 22.03t + 268.63 \quad \textit{Part-time instructors}$$
$$n = 13.88t + 468.48 \quad \textit{Full-time instructors}$$

Find the difference of the expressions $22.03t + 268.63$ and $13.88t + 468.48$. [**Hint:** Use parentheses and then simplify.]

3. Evaluate the result you found in Problem 2 for $t = 20$. What does your result mean in this situation?

4. Evaluate the result you found in Problem 2 for $t = 30$. What does your result mean in this situation?

5. For what values of t is the result you found in Problem 2 equal to 0? What does your result mean in this situation? [**Hint:** Solve an equation.]

6. Compare the result you found in Problem 5 with the result you found in Example 1.

Tips for Success   Reread a Problem

After you think you have solved a problem, reread it to make sure you have answered its question(s). Also, reread the problem with the solution in mind. If what you read makes sense, then you've provided another check of your result(s).

Homework 6.4

For extra help ▶ **MyLab Math** Watch the videos in MyLab Math

1. **DATA** Let n be the number of people (in thousands) who earned a bachelor's degree in the year that is t years since 1900 (see Table 24). In Exercise 37 of Homework 6.1, you worked with the following reasonable models for the numbers of women and men:

$$n = 24.27t - 1690.26 \quad \textit{Women}$$
$$n = 16.09t - 1039.32 \quad \textit{Men}$$

Table 24 Women and Men Who Earned a Bachelor's Degree

| Year | Number of People Who Earned a Bachelor's Degree (thousands) | |
	Women	Men
1995	634	526
2000	712	532
2005	855	631
2010	982	734
2014	1082	813

Source: *U.S. National Center for Education Statistics*

Use substitution or elimination to estimate when the number of women who earned a bachelor's degree was equal to the number of men who earned a bachelor's degree. What is that number of people?

2. **DATA** Let n be the number of daily newspapers at t years since 1980 (see Table 25). In Exercise 38 of Homework 6.1, you worked with the following reasonable models for the numbers of morning daily newspapers and evening daily newspapers:

$$n = 17t + 396.55 \quad \textit{Morning newspapers}$$
$$n = -29t + 1360.43 \quad \textit{Evening newspapers}$$

Table 25 Numbers of Morning Dailies and Evening Dailies

| Year | Number of Daily Newspapers | |
	Morning	Evening
1980	387	1388
1985	482	1220
1990	559	1084
1995	656	891
2000	766	727
2005	817	645
2011	931	451
2015	972	389

Source: *Editor & Publisher Co.*

Use substitution or elimination to estimate when the number of morning daily newspapers was equal to the number of evening daily newspapers. What is that number of newspapers?

3. **DATA** Let r be the record time (in seconds) for the 200-meter run at t years since 1900 (see Table 26). In Exercise 39 of Homework 6.1, you worked with the following reasonable models for the women's world record times and the men's world record times:

$$r = -0.064t + 27.00 \quad \textit{Women}$$
$$r = -0.026t + 21.86 \quad \textit{Men}$$

Table 26 200-Meter Run World Record Times

| Women | | Men | |
Year	Record Time (seconds)	Year	Record Time (seconds)
1973	22.38	1951	20.6
1974	22.21	1963	20.3
1978	22.06	1967	20.14
1984	21.71	1979	19.72
1988	21.34	1996	19.32
		2009	19.19

Source: *IAAF Statistics Handbook*

Use substitution or elimination to predict when the women's record time will equal the men's record time. What is that record time?

4. **DATA** Let r be the record time (in seconds) for the 500-meter race in speed skating at t years since 1900 (see Table 27). In Exercise 40 of Homework 6.1, you worked with the following reasonable models for the women's record times and the men's record times:

$$r = -0.17t + 54.32 \quad \textit{Women}$$
$$r = -0.10t + 45.38 \quad \textit{Men}$$

Table 27 500-Meter Speed-Skating Record Times

| Women | | Men | |
Year	Record Time (seconds)	Year	Record Time (seconds)
1934	50.3	1900	45.2
1937	46.4	1914	43.4
1955	45.6	1938	41.8
1970	43.22	1953	40.9
1981	40.18	1970	39.09
1988	39.10	1987	36.55
2001	37.22	2000	34.63
2013	36.36	2015	33.98

Source: *International Skating Union*

Use substitution or elimination to predict when the women's record time will be equal to the men's record time. What is that record time?

5. **DATA** Suppose you had to choose between never watching television again or never accessing the Internet again. Which would you choose to eliminate? The percentages for Americans giving each response are shown in Table 28 for various ages.

Table 28 Percentages of Americans Who Would Choose to Eliminate Watching Television and Percentages of Americans Who Would Choose to Eliminate Accessing the Internet

Age Group (years)	Age Used to Represent Age Group (years)	Percent	
		Eliminate Watching Television	Eliminate Accessing Internet
18–24	21.0	72	27
25–34	29.5	62	36
35–44	39.5	53	43
45–54	49.5	41	56
55–64	59.5	39	60
over 64	70.0	22	75

Sources: *Edison Research; Arbitron*

a. Let p be the percentage of Americans at age a years who would eliminate watching television. Find an equation of a model to describe the data.

b. Let p be the percentage of Americans at age a years who would eliminate accessing the Internet. Find an equation of a model to describe the data.

c. Use substitution or elimination to estimate at what age the percentage of Americans who would eliminate watching television equals the percentage of Americans who would eliminate accessing the Internet. What is that percentage of Americans?

6. **DATA** The numbers of people worldwide who access the Internet using desktop computers and mobile devices are shown in Table 29 for various years.

Table 29 Numbers of People Worldwide Who Access the Internet Using Desktop Computers and Mobile Devices

Year	Billions of People	
	Desktop	Mobile
2007	0.4	1.1
2009	0.7	1.3
2011	1.1	1.4
2013	1.5	1.6
2015	1.9	1.7

Source: *ComScore*

a. Let n be the number (in billions) of people who access the Internet using desktop computers at t years since 2000. Find an equation of a model to describe the data.

b. Let n be the number (in billions) of people who access the Internet using mobile devices at t years since 2000. Find an equation of a model to describe the data.

c. Use substitution or elimination to estimate when the number of people who accessed the Internet using desktop computers was equal to the number of people who accessed the Internet using mobile devices. What was that number of people?

d. On the basis of Table 29, a student concludes that a total of $1.9 + 1.7 = 3.6$ billion people accessed the Internet in 2015. What would you tell the student?

7. **DATA** The percentages of American adults who live in lower-income, middle-income, or upper-income households are shown in Table 30 for various years.

Table 30 Percentages of American Adults Who Live in Lower-Income, Middle-Income, or Upper-Income Households

Year	Percent		
	Lower Income	Middle Income	Upper Income
1971	25	61	14
1981	26	59	15
1991	27	56	17
2001	27	54	18
2011	29	51	20
2015	29	50	21

Source: *Pew Research Center*

a. Let p be the percentage of American adults who live in lower-income households at t years since 1900. Find an equation of a model to describe the data.

b. Let p be the percentage of American adults who live in middle-income households at t years since 1900. Find an equation of a model to describe the data.

c. Use substitution or elimination to predict when the percentage of American adults who live in lower-income households will equal the percentage of American adults who live in middle-income households. What is that percentage?

d. For the year you found in part (c), predict the percentage of American adults who will live in upper-income households.

8. **DATA** There are three age categories for U.S. marathon participants: juniors (under 20 years), open (20–39 years), and masters (at least 40 years). The percentages of U.S. marathon finishers in each age category are shown in Table 31 for various years.

Table 31 Percentage of U.S. Marathon Finishers in Age Categories

Year	Percent		
	Juniors	Open	Masters
1995	2	57	41
2000	2	54	44
2005	2	54	44
2010	2	52	46
2015	5	46	49

Source: *Running USA*

a. Let p be the percentage of marathon finishers who are in the open category at t years since 1990. Find an equation of a model to describe the data.

b. Let p be the percentage of marathon finishers who are in the masters category at t years since 1990. Find an equation of a model to describe the data.

c. Use substitution or elimination to predict when the percentage of marathon finishers who were in the open category was equal to the percentage of marathon finishers who were in the masters category. What is that percentage?

d. For the year you found in part (c), predict the percentage of marathon finishers who were in the junior category.

9. **DATA.** The numbers of women and men in the House of Representatives are shown in Table 32 for various years. Let n be the number of representatives at t years since 1980. Models for the numbers of female and male representatives are

$$n = 2.07t + 12.77 \quad \text{\textit{Women}}$$
$$n = -1.97t + 420.69 \quad \text{\textit{Men}}$$

Table 32 Numbers of Women and Men in the House of Representatives

Congress	Year	Number of Representatives	
		Women	Men
98th	1983	21	413
99th	1985	22	412
100th	1987	23	412
101st	1989	25	408
102nd	1991	28	407
103rd	1993	47	388
104th	1995	47	388
105th	1997	52	383
106th	1999	56	379
107th	2001	59	376
108th	2003	59	376
109th	2005	65	369
110th	2007	74	361
111th	2009	72	366
112th	2011	76	362
113th	2013	82	359
114th	2015	84	351
115th	2017	83	352

Source: *U.S. Census Bureau*

a. Use substitution or elimination to predict when the same number of women and men will be in the House of Representatives.

b. Find the greatest change in women from one Congress to the next. When did this change occur? Is the change about the same as the other changes in the number of women from one Congress to the next?

c. Here are models that fit the data well from the 103rd Congress to the 115th Congress:

$$n = 1.71t + 23.13 \quad \text{\textit{Women}}$$
$$n = -1.60t + 410.04 \quad \text{\textit{Men}}$$

Using these models, predict when the same number of women and men will be in the House of Representatives.

d. Explain why it makes sense that the year you predicted in part (c) is later than the year you predicted in part (a). Refer to the scatterplots of the data and/or the slopes of the models.

10. **DATA.** The numbers of women and men who live alone are shown in Table 33 for various years.

Table 33 Numbers of Women and Men Who Live Alone

Year	Number Living Alone (millions)	
	Women	Men
1980	11.3	7.0
1985	12.7	7.9
1990	14.0	9.0
1995	14.6	10.1
2000	15.6	11.2
2005	17.3	12.8
2010	17.4	14.0
2015	19.4	15.5

Source: *U.S. Census Bureau*

a. Let n be the number (in millions) of women who live alone at t years since 1980. Find an equation of a model to describe the data.

b. Let n be the number (in millions) of men who live alone at t years since 1980. Find an equation of a model to describe the data.

c. Compare the slopes of your two models. What do they mean in this situation?

d. Will the number of men who live alone ever equal the number of women who live alone, according to your models? If yes, explain why this event will not happen until far into the future. If no, explain why not.

11. The number of students who earned a bachelor's degree in biology was 104.6 thousand in 2014 and has since increased by about 4.7 thousand degrees each year. The number of students who earned a bachelor's degree in engineering was 92.2 thousand in 2014 and has since increased by about 5.2 thousand degrees each year (Source: *U.S. National Center for Education Statistics*).

a. Let d be the number (in thousands) of bachelor's degrees earned in biology in the year that is t years since 2014. Find an equation of a model to describe the data.

b. Let d be the number (in thousands) of bachelor's degrees earned in engineering in the year that is t years since 2014. Find an equation of a model to describe the data.

c. Use substitution or elimination to predict when the same number of bachelor's degrees will be earned in biology and engineering. What is that number of degrees?

12. The number of students who earned a bachelor's degree in parks, recreation, leisure, and fitness studies was 46.0 thousand in 2014 and has since increased by about 3.2 thousand degrees each year. The number of students who earned a bachelor's degree in multi/interdisciplinary studies was 48.3 thousand in 2014 and has since increased by about 2.6 thousand degrees each year (Source: *U.S. National Center for Education Statistics*).

a. Let d be the number (in thousands) of bachelor's degrees earned in parks, recreation, leisure, and fitness studies in the year that is t years since 2014. Find an equation of a model to describe the data.

b. Let d be the number (in thousands) of bachelor's degrees earned in multi/interdisciplinary studies in the year that is t years since 2014. Find an equation of a model to describe the data.

c. Use substitution or elimination to predict when the number of bachelor's degrees earned in parks, recreation, leisure, and fitness studies will equal the number of bachelor's degrees earned in multi/interdisciplinary studies. What is that number of degrees?

13. In 2014, the United States exported $16.5 billion worth of products to Chile, and the exports have since decreased by about $1.5 billion each year. In 2014, the United States imported $9.5 billion worth of products from Chile, and the imports have since decreased by about $0.6 billion each year (Source: *U.S. Census Bureau*).

a. Let v be the value (in billions of dollars) of U.S. exports to Chile in the year that is t years since 2014. Find an equation of a model to describe the data.

b. Let v be the value (in billions of dollars) of U.S. imports from Chile in the year that is t years since 2014. Find an equation of a model to describe the data.

c. Use substitution or elimination to predict when the value of U.S. exports to Chile will be equal to the value of U.S. imports from Chile. What is that value?

14. The Baltimore Orioles' payroll was $130 million in 2015 and has since increased by about $13.2 million each year. The Seattle Mariners' payroll was $137 million in 2015 and has since increased by about $16.4 million each year (Source: *Sportrac*).

a. Let p be the Orioles's payroll (in millions of dollars) in the year that is t years since 2015. Find an equation of a model to describe the data.

b. Let p be the Mariners' payroll (in millions of dollars) in the year that is t years since 2015. Find an equation of a model to describe the data.

c. Use substitution or elimination to estimate in which year the Orioles' and Mariners' payrolls were equal. What is that payroll?

15. In 2016, a 1-year-old Acura® RLX sedan was worth about $43,969, with a depreciation of about $4673 per year. A 1-year-old Honda® Accord sedan was worth about $19,985, with a depreciation of about $1848 per year (Source: *Edmunds.com*).

a. Let V be the value (in dollars) of the Acura RLX at t years since 2016. Find an equation of a model to describe the data.

b. Let V be the value (in dollars) of the Honda Accord at t years since 2016. Find an equation of a model to describe the data.

c. Use substitution or elimination to predict when the two cars will have the same value. What is that value?

16. In 2016, a 1-year-old Jeep® Patriot was worth about $16,644, with a depreciation of about $1145 per year. A 1-year-old Kia® Soul was worth about $13,920, with a depreciation of about $896 per year (Source: *Edmunds.com*).

a. Let V be the value (in dollars) of the Jeep Patriot at t years since 2016. Find an equation of a model to describe the data.

b. Let V be the value (in dollars) of the Kia Soul at t years since 2016. Find an equation of a model to describe the data.

c. Use substitution or elimination to predict when the two cars will have the same value. What is that value?

17. The average daily time people worldwide spend watching television was 175 minutes in 2015 and has since decreased by about 3.6 minutes per year. The average daily time people

worldwide are online was 128 minutes in 2015 and has since increased by about 9.7 minutes per year (Source: *Zenith via Recode*). Predict when the average daily time people worldwide spend watching television will be equal to the average daily time people worldwide are online. What is that amount of time?

18. The percentage of 30–34-year-old adults who are married was 53.4% in 2014 and has since decreased by about 1.8 percentage points each year. The percentage of 30–34-year-old adults who have never married was 36.1% in 2014 and has since increased by about 1.1 percentage points each year (Source: *U.S. Census Bureau*). Use substitution or elimination to predict when the percentage of 30–34-year-old adults who are married will equal the percentage of 30–34-year-old adults who have never married. What is that percentage?

Large Data Sets

19. DATA Access the data about head sizes and brain weights, which are available at MyLab Math and at the Pearson Downloadable Student Resources for Math & Stats website.

a. Let w be the weight (in grams) of a female brain for a head size of s cubic centimeters. Find an equation of a model to describe the data.

b. Let w be the weight (in grams) of a male brain for a head size of s cubic centimeters. Find an equation of a model to describe the data.

c. Estimate the female brain weight and the male brain weight for a head size of 3100 cubic centimeters.

d. Compare the slopes of the two models. What does your comparison tell you about this situation?

e. Explain why your work in parts (c) and (d) suggests that there may be a head size for which a female brain and a male brain have equal weight.

f. Estimate a head size for which a female brain and a male brain have equal weight. What is that brain weight?

g. The author plunged his head into a kitchen sink filled with water. The excess water drained into an adjacent sink. It took 18.4 cups of water to replace the displaced water. Estimate the weight of the author's brain in pounds. Round your result to the first decimal place. There are 236.6 cubic centimeters in 1 cup, and there is 0.002205 pound in 1 gram.

h. The data were collected for a study published in 1905. Do you have much faith in the result you found in part (g)? Explain.

20. DATA Access the data about automakers' market shares, which are available at MyLab Math and at the Pearson Downloadable Student Resources for Math & Stats website.

a. Let G be the market share (in percent) of General Motors in the year x. Find an equation of a model to describe the data.

b. Let H be the market share (in percent) of Honda in the year x. Find an equation of a model to describe the data.

c. Let T be the market share (in percent) of Toyota in the year x. Find an equation of a model to describe the data.

d. Predict when the market share of Toyota will be equal to the market share of General Motors. What is that market share?

e. Predict when the market share of Honda will be equal to the market share of General Motors. What is that market share?

f. Explain why it makes sense that the year you found in part (d) is before the year you found in part (e).

g. Compare the slopes of the Honda model and the Toyota model. What does your comparison tell you about this situation?

h. Find the sum of the slopes of all three models. What does it mean in this situation?

Concepts

21. Let v be the value (in dollars) of a car at age a years. Suppose a linear model describes the relationship between a and v for a certain car and another linear model describes the relationship between a and v for another car. If the slopes of the models are different, will their graphs definitely have an intersection point? If yes, explain why model breakdown might occur at the intersection point. If no, explain why the graphs might not have an intersection point.

22. Describe how you can find a system of linear equations to model a situation. Also, explain how you can use the system to make an estimate or prediction about the situation.

Related Review

23. The revenue from organic foods increased approximately linearly from $7.2 billion in 2012 to $10.8 billion in 2015. The revenue from non-genetically-modified foods increased approximately linearly from $2.9 billion in 2012 to $13.5 billion in 2015 (Source: *Nutrition Business Journal, non-GMO Project*). Estimate when the annual organic-food revenue was equal to the annual non-genetically-modified-food revenue. What was that revenue?

24. The percentage of high school students who smoke cigarettes decreased approximately linearly from 18.1% in 2011 to 10.8% in 2015. The percentage of high school students who use e-cigarettes increased approximately linearly from 1.5% in 2011 to 16.0% in 2015 (Source: *Centers for Disease Control and Prevention*). Estimate when the percentage of high school students who smoke cigarettes was equal to the percentage of high school students who use e-cigarettes. What was that percentage?

Expressions, Equations, and Graphs

Perform the indicated instruction. Then use words such as linear, one variable, *and* two variables *to describe the expression, equation, or system.*

25. Solve the following:

$$y = -2x + 6$$
$$y = 3x + 1$$

26. Evaluate $-2x + 6$ for $x = 1$.

27. Solve $-2x + 6 = 3x + 1$.

28. Simplify $(-2x + 6) + (3x + 1)$.

6.5 Perimeter, Value, Interest, and Mixture Problems

Objectives

» Describe a five-step problem-solving method.

» Solve perimeter, value, interest, and mixture problems.

In this section, we will use a five-step method to solve problems that involve the perimeter of a rectangle, the (dollar) value of an object, the interest from an investment, and the percentage of a substance in a mixture.

Five-Step Problem-Solving Method

To solve some problems in which we want to find two quantities, it is useful to perform the following five steps:

- **Step 1: Define each variable.** For each quantity we are trying to find, we usually define a variable to represent that unknown quantity.
- **Step 2: Write a system of two equations.** We find a system of two equations by using the variables from step 1. We can usually write each equation either by translating the information stated in the problem into mathematics or by making a substitution into a formula.
- **Step 3: Solve the system.** We solve the system of equations from step 2.
- **Step 4: Describe each result.** We use a complete sentence to describe the quantities we found.
- **Step 5: Check.** We reread the problem and check that the quantities we found agree with the given information.

Perimeter Problems

Recall from Section 4.6 that the formula of the perimeter P of a rectangle with length L and width W is $P = 2L + 2W$.

Figure 30 A
golden rectangle
drawn to scale

Example 1   Solving a Perimeter Problem

Throughout history, rectangles whose length is about 1.62 times their width, called *golden rectangles*, have been viewed as the most pleasing to the eye (see Fig. 30). Many famous structures, including the Parthenon in Athens and the Great Pyramid of Giza in Cairo, incorporate golden rectangles into their design. For Leonardo Da Vinci's *Mona Lisa*, the edges of the painting form a golden rectangle.

Suppose an artist wants a piece of canvas in the shape of a golden rectangle with a perimeter of 9 feet. Find the dimensions of the canvas.

Solution

Step 1: Define each variable.   Let W be the width (in feet) and L be the length (in feet). See Fig. 30.

Step 2: Write a system of two equations.   Because the length must be 1.62 times the width, our first equation is

$$L = 1.62W$$

Because the perimeter is 9 feet, we find our second equation by substituting 9 for P in the perimeter formula $P = 2L + 2W$:

$$9 = 2L + 2W$$

The system is

$$L = 1.62W \quad \textit{Equation (1)}$$
$$2L + 2W = 9 \quad \textit{Equation (2)}$$

Step 3: Solve the system.   We can solve the system by substitution. We substitute $1.62W$ for L in equation (2) and then solve for W:

$$2L + 2W = 9 \qquad \textit{Equation (2)}$$
$$2(1.62W) + 2W = 9 \qquad \textit{Substitute 1.62W for L.}$$
$$3.24W + 2W = 9 \qquad \textit{Simplify.}$$
$$5.24W = 9 \qquad \textit{Combine like terms.}$$
$$\frac{5.24W}{5.24} = \frac{9}{5.24} \qquad \textit{Divide both sides by 5.24.}$$
$$W \approx 1.72 \qquad \textit{Round } \frac{9}{5.24} \textit{ to the second decimal place.}$$

To find the approximate length, we substitute $\dfrac{9}{5.24}$ for W in equation (1):

$$L = 1.62\left(\frac{9}{5.24}\right) \approx 2.78$$

Step 4: Describe each result.   The approximate width is 1.72 feet, and the approximate length is 2.78 feet.

Step 5: Check.   We add the lengths of the four sides: $1.72 + 2.78 + 1.72 + 2.78 = 9$, which checks for a perimeter of 9 feet. We can also check that 2.78 is about 1.62 times 1.72 (try it).

Value Problems

Recall from Section 4.6 that if n objects each have value v, then their total value T is given by

$$T = vn$$

When some objects are sold, we refer to the total money collected as the **revenue** from selling the objects.

Example 2	Solving Value Problems

A vendor charges $5 per hamburger and $3 per hot dog.

1. What is the vendor's total revenue from selling 40 hamburgers and 85 hot dogs?
2. If the vendor sells a total of 135 hamburgers and hot dogs for a total revenue of $495, how many hamburgers and hot dogs did he sell?

Solution

1. The revenue from hamburgers is equal to the price per hamburger times the number of hamburgers sold: $5 \cdot 40 = 200$ dollars. The revenue from hot dogs is equal to the price per hot dog times the number of hot dogs sold: $3 \cdot 85 = 255$ dollars. We add the revenue of the hamburgers and that of the hot dogs to find the total revenue:

$$\underbrace{5}_{\substack{\text{dollars} \\ \text{hamburger}}} \cdot \underbrace{40}_{\text{hamburgers}} + \underbrace{3}_{\substack{\text{dollars} \\ \text{hot dog}}} \cdot \underbrace{85}_{\text{hot dogs}} = \underbrace{455}_{\substack{\text{total revenue} \\ \text{in dollars}}}$$

So, the total revenue from hamburgers and hot dogs is $455. The units of the expressions on both sides of the equation are dollars, which suggests that our work is correct.

2. ***Step 1: Define each variable.*** Let x be the number of hamburgers sold and y be the number of hot dogs sold.

   Step 2: Write a system of two equations. Our work in Problem 1 suggests that the formula of the total revenue T (in dollars) is

$$T = \underbrace{5}_{\substack{\text{dollars} \\ \text{hamburger}}} \cdot \underbrace{x}_{\text{hamburger}} + \underbrace{3}_{\substack{\text{dollars} \\ \text{hot dog}}} \cdot \underbrace{y}_{\text{hot dogs}}$$

   To obtain our first equation, we substitute 495 for T:

$$495 = 5x + 3y$$

   Because the vendor sells a total of 135 hamburgers and hot dogs, our second equation is

$$x + y = 135$$

   The system is

$$5x + 3y = 495 \quad \textit{Equation (1)}$$
$$x + y = 135 \quad \textit{Equation (2)}$$

   Step 3: Solve the system. We can use elimination to solve the system. To eliminate the y terms, we multiply both sides of equation (2) by -3, yielding the system

$$5x + 3y = 495 \quad \textit{Equation (1)}$$
$$-3x - 3y = -405 \quad \textit{Multiply both sides of equation (2) by } -3.$$

   Then we add the left sides and add the right sides of the equations and solve for x:

$$\begin{array}{ll} 5x + 3y = 495 & \\ \underline{-3x - 3y = -405} & \\ 2x \phantom{{}+ 0} + 0 = 90 & \textit{Add left sides and right sides; combine like terms.} \\ 2x = 90 & \textit{a + 0 = a} \\ x = 45 & \textit{Divide both sides by 2.} \end{array}$$

   Next, we substitute 45 for x in equation (2) and solve for y:

$$\begin{array}{ll} x + y = 135 & \textit{Equation (2)} \\ 45 + y = 135 & \textit{Substitute 45 for x.} \\ y = 90 & \textit{Subtract 45 from both sides.} \end{array}$$

Step 4: Describe each result. The vendor sold 45 hamburgers and 90 hot dogs.

Step 5: Check. First, we find the sum $45 + 90 = 135$, which is equal to the total number of hamburgers and hot dogs sold. Next, we find the total revenue from selling 45 hamburgers and 90 hot dogs: $5 \cdot 45 + 3 \cdot 90 = 495$, which checks.

▶

Notice that the arithmetic $5 \cdot 40 + 3 \cdot 85 = 455$ in Problem 1 of Example 2 suggests the equation $5x + 3y = 495$ that we formed in Problem 2. **If you have difficulty forming an equation, try making up numbers for unknown quantities and performing arithmetic to compute a related quantity. Your computation may suggest how to form the desired equation.**

Example 3   Solving a Value Problem

An auditorium has 500 balcony seats and 2100 main-level seats. If tickets for balcony seats cost \$18 less than tickets for main-level seats, what should the prices be for each type of ticket so that the total revenue from a sellout performance will be \$128,800?

Solution

Step 1: Define each variable. Let b be the price (in dollars) for balcony seats and m be the price (in dollars) for main-level seats.

Step 2: Write a system of two equations. Because tickets for balcony seats will cost \$18 less than tickets for main-level seats, our first equation is

$$\underset{b}{\underbrace{\text{balcony ticket price}}} \quad \underset{=}{\underbrace{\text{is}}} \quad \underset{m - 18}{\underbrace{\text{\$18 less than main-level ticket price}}}$$

Because the total revenue is \$128,800, our second equation is

$$\underset{b}{\underbrace{\frac{\text{dollars}}{\text{balcony ticket}}}} \cdot \underset{500}{\underbrace{\frac{\text{balcony}}{\text{tickets}}}} + \underset{m}{\underbrace{\frac{\text{dollars}}{\text{main-level ticket}}}} \cdot \underset{2100}{\underbrace{\frac{\text{main-level}}{\text{tickets}}}} = \underset{128{,}800}{\underbrace{\frac{\text{total revenue}}{\text{in dollars}}}}$$

The units of the expressions on both sides of the equation are dollars, which suggests that our work is correct. The system is

$$b = m - 18 \quad \textit{Equation (1)}$$
$$500b + 2100m = 128{,}800 \quad \textit{Equation (2)}$$

Step 3: Solve the system. We can use substitution to solve the system. We substitute $m - 18$ for b in equation (2) and solve for m:

$$
\begin{aligned}
500b + 2100m &= 128{,}800 && \textit{Equation (2)} \\
500(m - 18) + 2100m &= 128{,}800 && \textit{Substitute } m - 18 \textit{ for } b. \\
500m - 9000 + 2100m &= 128{,}800 && \textit{Distributive law} \\
2600m - 9000 &= 128{,}800 && \textit{Combine like terms.} \\
2600m &= 137{,}800 && \textit{Add } 9000 \textit{ to both sides.} \\
m &= 53 && \textit{Divide both sides by } 2600.
\end{aligned}
$$

Then we substitute 53 for m in equation (1) and solve for b:

$$b = 53 - 18 = 35$$

Step 4: Describe each result. Tickets for balcony seats should be priced at \$35 per ticket, and tickets for main-level seats should be priced at \$53 per ticket.

Step 5: Check. First, we find the difference in the ticket prices: $53 - 35 = 18$ dollars, which checks. Then we compute the total revenue from selling 500 of the \$35 tickets and 2100 of the \$53 tickets: $35 \cdot 500 + 53 \cdot 2100 = 128{,}800$ dollars, which checks.

▶

Interest Problems

Money deposited in an account, such as a savings account, certificate of deposit (CD), or mutual fund, is called **principal**. A person invests money in hopes of later getting back the principal plus additional money, called **interest**, which is a percentage of the principal (see Fig. 31). The **annual simple-interest rate** is the percentage of the principal that equals the interest earned per year. So, if we invest $100 and earn $5 per year, then the annual simple-interest rate is 5%.

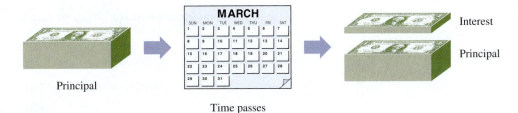

Time passes

Figure 31 Invest the principal; time passes; get back the principal plus interest

| **Example 4** | Finding the Interest from an Investment |

How much interest will a person earn by investing $2500 in an account at 6% annual interest for one year?

Solution

We find 6% of 2500:

$$0.06(2500) = 150$$

The person will earn $150 in interest.

Some people invest in a variety of mutual funds and stocks, including lower-risk investments and some higher-risk investments. These investors might earn large amounts of interest from the higher-risk investments, but will have a safety net of principal and interest from the lower-risk investments.

To have a specific interest rate to work with, we will assume the annual interest rate for a mutual fund in the next year will equal the previous five-year average. However, the future annual interest rate may turn out to be wildly different.

| **Example 5** | Solving Interest Problems |

1. How much interest will a person earn in one year from investing $6000 in a Mutual Financial Services fund that has a five-year average annual interest of 13.2% and $1000 in a John Hancock Preferred Income fund that has a five-year average annual interest of 9.7%?
2. A person plans to invest a total of $7000. She will invest in the two funds described in Problem 1. How much money should she invest in each fund to earn a total interest of $750 in one year?

Solution

1. We add the interest from both funds to compute the total interest:

interest from 13.2% fund		interest from 9.7% fund		total interest
0.132(6000)	+	0.097(1000)	=	889

2. **Step 1: Define each variable.**   Let x be the money (in dollars) invested at 13.2% annual interest, and let y be the money (in dollars) invested at 9.7% annual interest.

Step 2: Write a system of two equations.   Our work in Problem 1 suggests the first equation:

$$\underbrace{0.132x}_{\substack{\text{interest from} \\ \text{13.2\% fund}}} + \underbrace{0.097y}_{\substack{\text{interest from} \\ \text{9.7\% fund}}} = \underbrace{750}_{\text{total interest}}$$

Because the total investment is $7000, our second equation is

$$x + y = 7000$$

The system is

$$0.132x + 0.097y = 750 \qquad \textit{Equation (1)}$$
$$x + y = 7000 \qquad \textit{Equation (2)}$$

Step 3: Solve the system.   We can use elimination to solve the system. To eliminate the x terms, we multiply both sides of equation (2) by -0.132, yielding the system

$$0.132x + 0.097y = 750 \qquad \textit{Equation (1)}$$
$$-0.132x - 0.132y = -924 \qquad \textit{Multiply both sides of equation (2) by } -0.132.$$

Then we add the left sides and add the right sides of the equations and solve for y:

$$0.132x + 0.097y = 750$$
$$\underline{-0.132x - 0.132y = -924}$$
$$0 - 0.035y = -174 \qquad \textit{Add left sides and right sides; combine like terms.}$$
$$-0.035y = -174 \qquad \textit{0 + a = a}$$
$$y \approx 4971.43 \qquad \textit{Divide both sides by } -0.035.$$

Next, we substitute 4971.43 for y in equation (2) and solve for x:

$$x + y = 7000 \qquad \textit{Equation (2)}$$
$$x + 4971.43 = 7000 \qquad \textit{Substitute 4971.43 for y.}$$
$$x = 2028.57 \qquad \textit{Subtract 4971.43 from both sides.}$$

Step 4: Describe each result.   The person should invest $2028.57 at 13.2% annual interest and $4971.43 at 9.7% annual interest.

Step 5: Check.   First, we find the sum $2028.57 + 4971.43 = 7000$, which checks with the total money invested. Next, we calculate the total interest earned from investing $2028.57 at 13.2% and $4971.43 at 9.7%:

$$0.132(2028.57) + 0.097(4971.43) \approx 750$$

which checks.

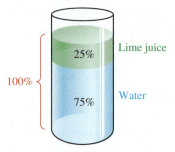

Figure 32 A 25% lime-juice solution

Mixture Problems

Many fields, including chemistry, cooking, and medicine, involve mixing different substances. Suppose that 2 ounces of lime juice is mixed with 6 ounces of water to make 8 ounces of unsweetened limeade. Note that $\frac{2}{8} = 0.25 = 25\%$ of the limeade is lime juice. We call the limeade a *25% lime-juice solution*.

Then, the remaining $\frac{6}{8} = 0.75 = 75\%$ of the limeade is water. The percentage of the solution that is lime juice plus the percentage of the solution that is water is equal to 100% (all) of the solution: $25\% + 75\% = 100\%$. See Fig. 32.

Therefore, in a 20% lime-juice solution, 20% of the solution is lime juice and $100\% - 20\% = 80\%$ of the solution is water. In general, **for an $x\%$ solution of two substances that are mixed, $x\%$ of the solution is one substance and $(100 - x)\%$ of the solution is the other substance.**

Example 6   Solving Mixture Problems

1. How much lime juice is in 6 ounces of a 15% lime-juice solution?
2. If 5 gallons of a 40% antifreeze solution is mixed with 3 gallons of a 75% antifreeze solution, how much pure antifreeze is in the mixture?

Solution

1. We find 15% of 6:

$$0.15(6) = 0.9$$

There is 0.9 ounce of lime juice in the 6 ounces of lime-juice solution.

2. The total amount of pure antifreeze doesn't change, regardless of how it is distributed in the two solutions. To find the amount of pure antifreeze in the mixture, we add the pure amounts of antifreeze in the 40% solution and the 75% solution:

number of gallons of pure antifreeze in 40% solution		number of gallons of pure antifreeze in 75% solution		number of gallons of pure antifreeze in mixture
$0.40(5)$	$+$	$0.75(3)$	$=$	4.25

There are 4.25 gallons of pure antifreeze in the mixture.

Our work in Problem 2 of Example 6 suggests how to find an equation for step 2 of our work in Example 7.

Example 7   Solving a Mixture Problem

A chemist needs 8 liters of a 20% alcohol solution, but she has only a 15% alcohol solution and a 35% alcohol solution. How many liters of each solution should she mix to make the desired 8 liters of 20% alcohol solution?

Solution

Step 1: Define each variable.   Let x be the number of liters of 15% alcohol solution, and let y be the number of liters of 35% alcohol solution.

Step 2: Write a system of two equations.   Because she wants 8 liters of the total mixture, our first equation is

$$x + y = 8$$

We find our second equation from the fact that the sum of the amounts of pure alcohol in the 15% alcohol solution and the 35% alcohol solution is equal to the amount of pure alcohol in the mixture:

pure alcohol in 15% solution		pure alcohol in 35% solution		pure alcohol in mixture
$0.15x$	$+$	$0.35y$	$=$	$0.20(8)$

The system is

$$x + y = 8 \qquad \textit{Equation (1)}$$
$$0.15x + 0.35y = 1.6 \qquad \textit{Equation (2)}$$

Step 3: Solve the system. We can solve the system by substitution. First, we solve equation (1) for y:

$$x + y = 8 \qquad \textit{Equation (1)}$$
$$y = 8 - x \quad \textit{Subtract x from both sides.}$$

Then, we substitute $8 - x$ for y in equation (2) and solve for x:

$$0.15x + 0.35y = 1.6 \qquad \textit{Equation (2)}$$
$$0.15x + 0.35(8 - x) = 1.6 \qquad \textit{Substitute } 8 - x \textit{ for y.}$$
$$0.15x + 2.8 - 0.35x = 1.6 \qquad \textit{Distributive law}$$
$$-0.20x + 2.8 = 1.6 \qquad \textit{Combine like terms.}$$
$$-0.20x = -1.2 \qquad \textit{Subtract 2.8 from both sides.}$$
$$x = 6 \qquad \textit{Divide both sides by } -0.20.$$

Next, we substitute 6 for x in the equation $y = 8 - x$ and solve for y:

$$y = 8 - 6 = 2$$

Step 4: Describe each result. Six liters of the 15% alcohol solution and 2 liters of the 35% alcohol solution are required.

Step 5: Check. First, we compute the total amount (in liters) of pure alcohol in 6 liters of the 15% alcohol solution and 2 liters of the 35% alcohol solution:

$$0.15(6) + 0.35(2) = 1.6$$

Next, we compute the amount (in liters) of pure alcohol in 8 liters of the 20% alcohol solution:

$$0.20(8) = 1.6$$

Because the two results are equal, they check. Also, $6 + 2 = 8$, which checks with the chemist wanting 8 liters of the 20% solution.

▶

Example 8   Solving a Mixture Problem

A chemist needs 6 ounces of a 5% acid solution, but has only a 40% acid solution. How much of the 40% solution and water should he mix to form the desired 6 ounces of the 5% solution?

Solution

Step 1: Define each variable. Let x be the number of ounces of the 40% solution, and let y be the number of ounces of water.

Step 2: Write a system of two equations. Because he wants 6 ounces of the total mixture, our first equation is

$$x + y = 6$$

There is no acid in pure water, so we find our second equation from the fact that the amount of pure acid in the 40% acid solution is equal to the amount of pure acid in the mixture:

$$\underbrace{0.40x}_{\substack{\text{amount of pure} \\ \text{acid in 40\% solution}}} = \underbrace{0.05(6)}_{\substack{\text{amount of pure} \\ \text{acid in mixture}}}$$

The system is

$$x + y = 6 \qquad \textit{Equation (1)}$$
$$0.40x = 0.3 \qquad \textit{Equation (2)}$$

Step 3: Solve the system. We begin by solving equation (2) for x:

$$0.40x = 0.3 \quad \textit{Equation (2)}$$
$$x = 0.75 \quad \textit{Divide both sides by 0.40.}$$

Next, we substitute 0.75 for x in equation (1):

$$x + y = 6 \quad \textit{Equation (1)}$$
$$0.75 + y = 6 \quad \textit{Substitute 0.75 for x.}$$
$$y = 5.25 \quad \textit{Subtract 0.75 from both sides.}$$

Step 4: Describe each result. The chemist needs to mix 0.75 ounce of the 40% acid solution with 5.25 ounces of water.

Step 5: Check. First, we find $0.75 + 5.25 = 6$, which checks with the chemist wanting 6 ounces of the 5% solution. Next, we compute the amount of pure acid in 0.75 ounce of the 40% solution: $0.40(0.75) = 0.3$ ounce. Finally, we compute the amount of pure acid in the 5% solution: $0.05(6) = 0.3$ ounce. The computed amounts of pure acid in the 40% solution and the 5% solution are equal, so they check.

▶

◈ Group Exploration

Describing an authentic situation for a given system

In many examples, we wrote a system of two equations that helped us find quantities for an authentic situation. For the following problems, you will work backward. For example, given the system

$$xy = 40$$
$$y = x + 3$$

We can relate the following to the system:

> A person wants to build a rectangular garden with an area of 40 square feet in which the length is 3 feet more than the width. Find the dimensions of the garden.

Describe an authentic situation for the given system. Find the unknown quantities for your situation.

1. $$y = 3x$$
   $$0.14x + 0.05y = 580$$

2. $$x + y = 12{,}000$$
   $$20x + 45y = 277{,}500$$

3. $$y = x - 5$$
   $$2x + 2y = 70$$

4. $$x + y = 12$$
   $$0.15x + 0.30y = 0.25(12)$$

Homework 6.5

For extra help ▶   **MyLab Math**   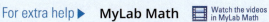 Watch the videos in MyLab Math

1. The length of a golden rectangle is equal to about 1.62 times the width. If an architect wants to design the base of a building to be a golden rectangle, what should be the dimensions of the base if the perimeter is to be 600 feet?

2. The length of a golden rectangle is equal to about 1.62 times the width. An architect wants to design a room so that the floor is a golden rectangle with a perimeter of 50 feet. What should be the dimensions of the floor?

3. A landscaper plans to dig a rectangular garden whose length is to be 5 feet more than the width. If the landscaper has 42 feet of fencing to enclose the garden, what should be the dimensions of the garden?

4. An official football field is a rectangle whose length (including end zones) is 2.25 times the width. If a little-league football field is to be constructed with a perimeter of 200 yards, what should be the width and length of the field?

5. An official tennis court is a rectangle whose length is 6 feet more than twice the width. The perimeter of the court is 228 feet. Find the dimensions of the court.

6. An official table tennis (ping pong) table is a rectangle whose length is 1 foot less than twice the width. The perimeter of the table is 28 feet. Find the dimensions of the table.

7. The length of a rectangle is 3 inches less than twice the width. If the perimeter is 108 inches, find the dimensions of the rectangle.

8. The length of a rectangle is 5 inches more than twice the width. If the perimeter is 112 inches, find the dimensions of the rectangle.

9. The length of a rectangle is 1 yard more than three times the width. If the perimeter is 146 yards, find the length and width of the rectangle.

10. The length of a rectangle is 2 yards less than three times the width. If the perimeter is 148 yards, find the length and width of the rectangle.

11. A 2000-seat theater has tickets for sale at $15 and $22. How many tickets should be sold at each price for a sellout performance to generate a total revenue of $33,500?

12. A 10,000-seat amphitheater has tickets for sale at $45 and $65. How many tickets should be sold at each price for a sellout performance to generate a total revenue of $560,000?

13. The album *A Moon Shaped Pool*, by Radiohead, has a list price of $14.99, and the album *The King of Limbs*, by Radiohead, has a list price of $9.99. If iTunes' total sales of both albums in one week is 253 albums and the total revenue from both albums in that week is $2722.47, how many of each album were sold?

14. The album *To Pimp a Butterfly*, by Kendrick Lamar, sells on iTunes for $13.99, and the album *good kid, m.A.A.d city*, by Kendrick Lamar, sells on iTunes for $15.99. If iTunes' total sales of both albums in one week is 373 recordings, and the total revenue from both albums in that week is $5402.27, how many of each album were sold?

15. *The Smaller Evil*, by Stephanie Kuehn, sells as a hardcover book with a list price of $17.99 and as an audio CD at the list price of $40.00. If a bookstore sold 1500 of these books and CDs at the list prices for a total revenue of $31,387, how many hardcover books and how many CDs were sold?

16. *You Killed Wesley Payne*, by Sean Beaudoin, sells as a hardcover book with a list price of $16.99 and as a paperback book with a list price of $8.99. If a bookstore sold 800 of these books at the list prices for a total revenue of $8792, how many of each type of book were sold?

17. An auditorium has 300 balcony seats and 1400 main-level seats. If tickets for balcony seats are priced at $12 less than the price of main-level seats, what should the prices be for each type of ticket so that the total revenue from a sellout performance will be $40,600?

18. An auditorium has 450 balcony seats and 1700 main-level seats. If tickets for balcony seats are priced at $15 less than the price of main-level seats, what should the prices be for each type of ticket so that the total revenue from a sellout performance will be $100,750?

19. An amphitheater has 8000 general seats and 4000 reserved seats. If tickets for general seats are priced at $25 less than the

price of reserved seats, what should the prices be for each type of ticket so that the total revenue from a sellout performance will be $544,000?

20. An amphitheater has 9500 general seats and 3200 reserved seats. If tickets for general seats are priced at $17 less than the price of reserved seats, what should the prices be for each type of ticket so that the total revenue from a sellout performance will be $435,400?

For Exercises 21–32, all interest rates are five-year averages.

21. A person invests the given amount of money in a BMO TCH Corporate Income fund at 5% annual interest. Find the interest in one year.
   a. 3500 dollars   **b.** 6500 dollars   **c.** *d* dollars

22. A person invests the given amount of money in a T. Rowe Price Blue Chip Growth fund at 13% annual interest. Find the interest in one year.
   a. 4000 dollars   **b.** 9000 dollars   **c.** *d* dollars

23. A person plans to invest in a Prudential QMA Strategic Value fund at 14% annual interest and a Summit Muni Intermed fund at 3% annual interest. Find the total interest earned in one year from the given investments.
   a. The person invests $5000 in the Prudential fund and $8000 in the Summit fund.
   b. The person invests $7000 in the Prudential fund and $12,000 in the Summit fund.
   c. The person invests *x* dollars in the Prudential fund and *y* dollars in the Summit fund.

24. A person plans to invest in a Nicholas fund at 13% annual interest and a Prudential Core Bond fund at 2% annual interest. Find the total interest earned in one year from the given investments.
   a. The person invests $3000 in the Nicholas fund and $6000 in the Prudential fund.
   b. The person invests $8000 in the Nicholas fund and $16,000 in the Prudential fund.
   c. The person invests *x* dollars in the Nicholas fund and 2*x* dollars in the Prudential fund.

25. A person plans to invest a total of $20,000 in a TFS Market Neutral fund at 2% annual interest and a ProFunds Biotechnology UltraSector fund at 38% annual interest. How much should she invest in each fund so that the total interest in one year will be $2560?

26. A person plans to invest a total of $11,000 in a Federated International Bond Strategy Pt fund at 4% annual interest and a Janus Venture fund at 17% annual interest. How much should she invest in each fund so that the total interest in one year will be $830?

27. A person plans to invest a total of $8500 in a Gibelli ABC AAA fund at 3.5% annual interest and a Becker Value Equity fund at 15% annual interest. How much should he invest in each fund so that the total interest in one year will be $600?

28. A person plans to invest a total of $9500 in a Calvert Income fund at 3.7% annual interest and a Turner Small Cap Growth fund at 15% annual interest. How much should he invest in each fund so that the total interest in one year will be $650?

29. A person plans to invest three times as much in a Vitrus Quality Small-Cap fund at 13% annual interest as in a Direxion Mthly S&P 500 Bull 2X fund at 30% annual interest. How much will the person have to invest in each fund to earn a total of $1725 in one year?

30. A person plans to invest three times as much in an Artisan International fund at 10% annual interest as in an SEI Large Cap Index fund at 16% annual interest. How much will the person have to invest in each fund to earn a total of $1840 in one year?

31. A person plans to invest twice as much in a Franklin Strategic Mortgage fund at 3.9% annual interest as in a TIAA-CREF Small-Cap Blend fund at 16% annual interest. How much will the person have to invest in each fund to earn a total of $1000 in one year?

32. A person plans to invest twice as much in a Putnam Global Equity fund at 11.9% annual interest as in a Bridgeway Small-Cap Growth fund at 20% annual interest. How much will the person have to invest in each fund to earn a total of $2800 in one year?

33. A salad dressing is a 65% oil solution. Determine the amount of oil in the given amount of oil solution.
 a. 2 ounces of oil solution
 b. 3 ounces of oil solution
 c. x ounces of oil solution

34. Some milk is a 2% butterfat solution. Determine the amount of butterfat in the given amount of 2% milk.
 a. 4 cups of milk
 b. 8 cups of milk
 c. M cups of milk

35. a. If 4 ounces of a 35% alcohol solution is mixed with 3 ounces of a 10% alcohol solution, how much pure alcohol is in the mixture?
 b. How many ounces of a 35% alcohol solution and a 10% alcohol solution need to be mixed to make 15 ounces of a 20% solution?

36. a. If 2 liters of a 24% alcohol solution is mixed with 5 liters of an 8% alcohol solution, how much pure alcohol is in the mixture?
 b. How many liters of a 24% alcohol solution and an 8% alcohol solution need to be mixed to make 8 liters of a 20% solution?

37. A chemist wants to mix a 20% acid solution and a 30% acid solution to make a 22% acid solution. How many quarts of each solution must be mixed to make 5 quarts of the 22% solution?

38. A chemist wants to mix a 30% acid solution and a 50% acid solution to make a 42% acid solution. How many quarts of each solution must be mixed to make 10 quarts of the 42% solution?

39. How many gallons of a 10% antifreeze solution and a 25% antifreeze solution need to be mixed to make 3 gallons of a 20% antifreeze solution?

40. How many liters of a 25% antifreeze solution and a 40% antifreeze solution need to be mixed to make 6 liters of a 30% antifreeze solution?

41. A chemist wants to mix a 15% acid solution and a 35% acid solution to make a 30% acid solution. How many quarts of each solution must be mixed to make 4 quarts of the 30% solution?

42. A chemist wants to mix a 5% acid solution and a 20% acid solution to make a 10% acid solution. How many quarts of each solution must be mixed to make 6 quarts of the 10% solution?

43. How many gallons of a 12% acid solution and a 24% acid solution need to be mixed to make 8 gallons of a 21% acid solution?

44. How many quarts of a 10% acid solution and a 45% acid solution need to be mixed to make 5 quarts of a 31% acid solution?

45. A chemist needs 5 ounces of a 12% alcohol solution, but has only a 20% alcohol solution. How much 20% solution and water should she mix to make the desired 5 ounces of 12% solution?

46. A chemist needs 6 liters of a 25% alcohol solution, but has only a 30% alcohol solution. How much 30% solution and water should he mix to make the desired 6 liters of 25% solution?

Concepts

47. For a sellout performance at a 300-seat theater,
 a. Find the total revenue if all tickets sell for $10.
 b. Find the total revenue if all tickets sell for $15.
 c. If some, none, or all tickets sell for $10 and the remaining tickets, if any, sell for $15, is it possible for the total revenue to be the amount that follows? If yes, how many of each ticket must be sold? If no, explain.
  i. $2500
  ii. $4875
  iii. $3250

48. A person plans to invest a total of $10,000 in an Appleseed Investor fund at 5% and a Schwab Health Care fund at 16% annual interest.
 a. Find the interest earned in one year from the following investments:
  i. The person invests all $10,000 in the Appleseed fund.
  ii. The person invests all $10,000 in the Schwab fund.
 b. Is it possible for her to earn the given amount of interest in one year? If yes, how much should she invest in each fund? If no, explain why not.
  i. $460
  ii. $840
  iii. $1650

49. If equal amounts of a 10% alcohol solution and a 20% alcohol solution are mixed, what percentage of the mixture is alcohol? Explain.

50. If a person invests equal amounts of money in an account at 5% annual interest and an account at 9% annual interest for one year, what percentage of the total money invested will be the total interest earned? Explain.

In many examples, we wrote a system of two equations that helped us find quantities for an authentic situation. For Exercises 51–54, you will work backward. For example, given the system

$$xy = 48$$
$$y = x - 2$$

we can relate the following to the system:

> A person wants to build a rectangular garden with an area of 48 square feet in which the width is 2 feet less than the length. Find the dimensions of the garden.

Describe an authentic situation for the given system. Find the unknown quantities for your situation.

51. $y = x + 3$
$2x + 2y = 50$

52. $x + y = 500$
$15x + 25y = 8300$

53. $y = 2x$
$0.03x + 0.07y = 340$

54. $x + y = 8$
$0.35x + 0.55y = 0.40(8)$

Related Review

Solve the equation for the specified variable.

55. $P = 2(L + W)$, for L

56. $P = 2(L + W)$, for W

57. $A = P(1 + rt)$, for r

58. $A = P(1 + rt)$, for P

Expressions, Equations, and Graphs

Perform the indicated instruction. Then use words such as linear, one variable, *and* two variables *to describe the expression, equation, or system.*

59. Simplify $-4(7x - 3)$.

60. Solve the following:

$$3x + 2y = 6$$
$$y = 2x - 11$$

61. Solve $-4(7x - 3) = 5x + 2$.

62. Graph $3x + 2y = 6$ by hand.

6.6 Linear Inequalities in Two Variables; Systems of Linear Inequalities in Two Variables

Objectives

» Graph a linear inequality in two variables.

» Solve a system of linear inequalities in two variables.

» Use a system of linear inequalities in two variables to make estimates.

Recall from Section 3.1 that a linear equation in two variables is an equation that can be put into the form $y = mx + b$ or $x = a$, where m, a, and b are constants. A **linear inequality in two variables** is an inequality that can be put into the form

$$y < mx + b \quad \text{or} \quad x < a$$

(or with $<$ replaced with $\leq$, $>$, or $\geq$), where m, a, and b are constants. Some examples are $y < 3x - 5$, $y \leq -2x + 4$, $5x + 2y > 10$, and $x \geq 2$. In the first part of this section, we will work with one linear inequality in two variables. In the second part, we will work with two or more such inequalities.

Graphing Linear Inequalities in Two Variables

We use the terms *satisfy*, *solution*, *solution set*, and *solve* for an inequality in two variables in much the same way that we have used them for equations in one or two variables and inequalities in one variable.

Definition   *Satisfy, solution, solution set,* and *solve* for an inequality in two variables

If an inequality in the two variables x and y becomes a true statement when a is substituted for x and b is substituted for y, we say the ordered pair (a, b) **satisfies** the inequality and call (a, b) a **solution** of the inequality. The **solution set** of an inequality is the set of all solutions of the inequality. We **solve** the inequality by finding its solution set.

We describe the solution set of an inequality in two variables by graphing all of the solutions.

Example 1   Graphing a Linear Inequality in Two Variables

Graph $y > x - 1$.

Solution

We begin by sketching the graph of $y = x - 1$ (see Fig. 33).

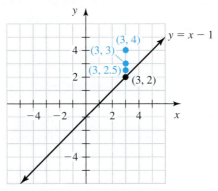

Figure 33 Some solutions of $y > x - 1$ (in blue)

To investigate how to solve $y > x - 1$, we choose a value of x (say, 3) and find several solutions with an x-coordinate of 3. For $y = x - 1$, if $x = 3$, then $y = (3) - 1 = 2$. So, the point $(3, 2)$ is on the line $y = x - 1$.

For $y > x - 1$, if $x = 3$, then

$$y > (3) - 1$$
$$y > 2$$

So, if $x = 3$, some possible values of y are $y = 2.5$, $y = 3$, and $y = 4$. Note that the points $(3, 2.5)$, $(3, 3)$, and $(3, 4)$ lie *above* the point $(3, 2)$, which is on the line $y = x - 1$ (see Fig. 33).

We could choose other values of x besides 3 and go through a similar argument. These investigations would suggest that the solutions of $y > x - 1$ lie *above* the graph of $y = x - 1$. This is, in fact, true. In Fig. 34, we shade the region that contains all the points that represent solutions of $y > x - 1$.

We make the line $y = x - 1$ dashed in Fig. 34 to indicate that its points are *not* solutions of $y > x - 1$. For example, the point $(3, 2)$ on the line $y = x - 1$ does *not* satisfy the inequality $y > x - 1$:

$$y > x - 1 \qquad \textit{Original inequality}$$
$$2 \overset{?}{>} (3) - 1 \qquad \textit{Substitute 3 for x and 2 for y.}$$
$$2 \overset{?}{>} 2 \qquad \textit{Subtract.}$$
$$\text{false}$$

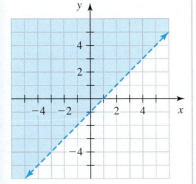

Figure 34 Graph of $y > x - 1$

We can draw a graph of $y > x - 1$ by using a graphing calculator (see Fig. 35), but we have to imagine that the border $y = x - 1$ is drawn with a dashed line. To shade above a line, press $\boxed{Y =}$ and then press $\boxed{\triangleleft}$ twice. Next press $\boxed{\text{ENTER}}$ as many times as necessary for the triangle shown in Fig. 35 to appear.

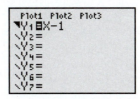 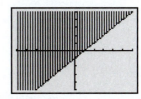

Figure 35 Graph of $y > x - 1$ (where we imagine that the border line is dashed)

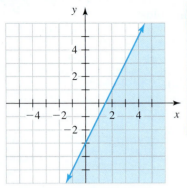

Figure 36 Graph of $y \le 2x - 3$

Example 2   Graphing a Linear Inequality in Two Variables

Sketch a graph of $y \le 2x - 3$.

Solution

The graph of $y \le 2x - 3$ is the line $y = 2x - 3$, as well as the region below that line (see Fig. 36). We use a *solid* line along the border $y = 2x - 3$ to indicate that the points on the line $y = 2x - 3$ are solutions of $y \le 2x - 3$.

Graph of a Linear Inequality in Two Variables

- The graph of a linear inequality of the form $y > mx + b$ is the region above the line $y = mx + b$. The graph of a linear inequality of the form $y < mx + b$ is the region below the line $y = mx + b$. For either inequality, we use a dashed line to show that $y = mx + b$ is not part of the graph.
- The graph of a linear inequality of the form $y \ge mx + b$ is the line $y = mx + b$, as well as the region above that line. The graph of a linear inequality of the form $y \le mx + b$ is the line $y = mx + b$, as well as the region below that line.

To sketch a graph of a linear inequality in two variables, if the variable y is not alone on one side of the inequality, we begin by isolating it. Recall from Section 5.5 that when we multiply or divide both sides of a linear inequality by a negative number, we must reverse the inequality symbol.

Example 3   Graphing a Linear Inequality in Two Variables

Sketch a graph of $-2x - 5y > 10$.

Solution

First, we get y alone on one side of the inequality:

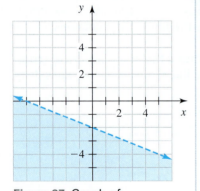

Figure 37 Graph of $-2x - 5y > 10$

$$-2x - 5y > 10 \qquad \textit{Original inequality}$$
$$-2x - 5y + 2x > 10 + 2x \qquad \textit{Add 2x to both sides.}$$
$$-5y > 2x + 10 \qquad \textit{Combine like terms; rearrange terms.}$$
$$\frac{-5y}{-5} < \frac{2x}{-5} + \frac{10}{-5} \qquad \textit{Divide both sides by −5; reverse inequality symbol.}$$
$$y < -\frac{2}{5}x - 2 \qquad \textit{Simplify.}$$

The graph of $y < -\frac{2}{5}x - 2$ is the region below the line $y = -\frac{2}{5}x - 2$ (see Fig. 37).

To verify our work, we choose a point on our graph—say, $(-4, -2)$—and check that it satisfies the original inequality:

$$-2x - 5y > 10 \quad \textit{Original inequality}$$
$$-2(-4) - 5(-2) \overset{?}{>} 10 \quad \textit{Substitute −4 for x and −2 for y.}$$
$$18 \overset{?}{>} 10 \quad \textit{Simplify.}$$
$$\text{true}$$

To check further, we could choose several other points on the graph and check that each point satisfies the inequality.

We could also choose several points that are *not* on the graph and check that each point does *not* satisfy the inequality. For example, note that $(0, 0)$ is not on the graph and that this point does not satisfy the original inequality $-2x - 5y > 10$:

$$-2x - 5y > 10 \quad \textit{Original inequality}$$
$$-2(0) - 5(0) \overset{?}{>} 10 \quad \textit{Substitute 0 for x and 0 for y.}$$
$$0 \overset{?}{>} 10 \quad \textit{Simplify.}$$
$$\text{false}$$

WARNING   It is a common error to think the graph of an inequality such as $-2x - 5y > 10$ is the region *above* the line $-2x - 5y = 10$ because the symbol ">" means "is greater than." However, we must first isolate y on the left side of a linear inequality before we can determine whether the graph includes the region that is above or below a line. In fact, in Example 3 we wrote the inequality $-2x - 5y > 10$ as $y < -\dfrac{2}{5}x - 2$ and concluded that the graph is the region *below* the line $y = -\dfrac{2}{5}x - 2$.

Example 4   Graphing Linear Inequalities in Two Variables.

Sketch the graph of the linear inequality in two variables.

1. $y \geq -2$ **2.** $x < 3$

Solution

1. The graph of $y \geq -2$ is the horizontal line $y = -2$, as well as the region above that line (see Fig. 38).
2. Ordered pairs with x-coordinates *less* than 3 are represented by points that lie to the *left* of the vertical line $x = 3$. So, the graph of $x < 3$ is the region to the left of $x = 3$ (see Fig. 39).

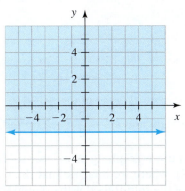

Figure 38 Graph of $y \geq -2$

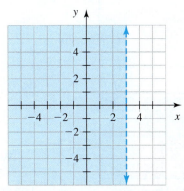

Figure 39 Graph of $x < 3$

Solving a System of Linear Inequalities in Two Variables

A **system of linear inequalities in two variables** consists of two or more linear inequalities in two variables. Here is an example:

$$y > x - 1$$
$$y \leq -\frac{2}{3}x - 2$$

Definition   *Solution, solution set,* and *solve* for a system of inequalities in two variables

We say an ordered pair (a, b) is a **solution** of a system of inequalities in two variables if it satisfies all the inequalities of the system. The **solution set** of a system is the set of all solutions of the system. We **solve** a system by finding its solution set.

Recall from Section 6.1 that we can find the solution set of a system of two linear *equations* in two variables by locating any intersection point(s) of the graphs of the two equations. Similarly, **the solution set of a system of inequalities in two variables can be found by locating the intersection of the graphs of all the inequalities.** This makes sense because a solution is an ordered pair that satisfies all the inequalities, which means the point that represents the ordered pair lies on the graphs of all the inequalities.

In this text, we use a graph to describe the solution set of a system of inequalities.

Example 5   Graphing the Solution Set of a System of Linear Inequalities

Graph the solution set of the system

$$y > x - 1$$

$$y \le -\frac{2}{3}x - 2$$

Solution

First, we sketch the graph of $y > x - 1$ (see Fig. 40, blue region) and the graph of $y \le -\frac{2}{3}x - 2$ (see Fig. 40, red region). The graph of the solution set of the system is the intersection of the graphs of the inequalities, which is shown in Fig. 41.

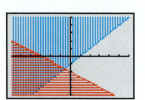

Figure 42 Verify our work (where we imagine that the border $y = x - 1$ is dashed)

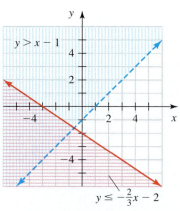

Figure 40 Graphs of $y > x - 1$ (in blue) and $y \le -\frac{2}{3}x - 2$ (in red)

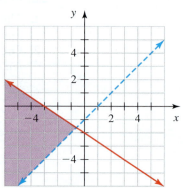

Figure 41 Graph of the solution set of the system

We can use a graphing calculator to draw a graph of the solution set of the system (see Fig. 42), where we imagine that the border $y = x - 1$ is drawn with a dashed line. The graph of the solution set is the region shaded by vertical lines (in blue) and horizontal lines (in red).

Example 6   Graphing the Solution Set of a System of Linear Inequalities

Graph the solution set of the system

$$y > -\frac{3}{2}x$$

$$-2x + 5y < 5$$

Solution

First, we sketch the graph of $y > -\frac{3}{2}x$ (see Fig. 43, blue region) and the graph of $-2x + 5y < 5$ (see Fig. 43, red region). The graph of the solution set of the system is the intersection of the graphs of the inequalities, which is shown in Fig. 44.

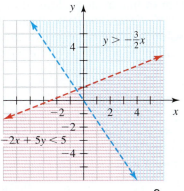

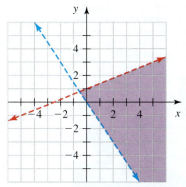

Figure 43 Graphs of $y > -\dfrac{3}{2}x$ (in blue) and $-2x + 5y < 5$ (in red)

Figure 44 Graph of the solution set of the system

Example 7   Graphing the Solution Set of a System of Linear Inequalities

Graph the solution set of the system of linear inequalities in two variables

$$-x + 3y \leq 9$$
$$-x + 3y \geq 3$$
$$x \geq 1$$
$$x \leq 4$$

Solution

In Fig. 45, we use arrows to indicate the graphs of each of the four inequalities. The solution set of the system is the intersection of the four graphs.

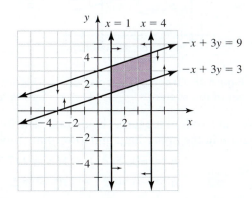

Figure 45 Graph of the solution set of the system

Using a Systems of Linear Inequalities to Make Estimates

We can model an authentic situation with a system of linear inequalities.

Example 8   Using a System of Linear Inequalities to Make Estimates

For you to get the most benefits, yet be safe, while exercising, your heart rate should be within a certain range, called the *target heart-rate zone*. Table 34 shows the lower and upper limits of the target heart-rate zones for various ages.

Let h be the heart rate (in beats per minute) for a person who is a years of age. Linear models for the lower and upper limits of the target heart-rate zones are

$$h = -0.75a + 165 \quad \textit{Upper limit}$$
$$h = -0.5a + 110 \quad \textit{Lower limit}$$

Table 34 Target Heart-Rate Zone

Age (years)	Target Heart-Rate Zone (beats per minute)	
	Lower Limit	Upper Limit
20	100	150.0
30	95	142.5
40	90	135.0
50	85	127.5
60	80	120.0
70	75	112.5
80	70	105.0
90	65	97.5

Source: *American Heart Association*

1. Find a system of linear inequalities that describes the target heart-rate zones for people between the ages of 10 years and 100 years.
2. Graph the solution set of the system of linear inequalities that you found in Problem 1.
3. What is the target heart-rate zone for a person who is 23 years old?

Solution

1. To be in the target heart-rate zone, a person's heart rate must be less than or equal to the upper limit ($h \le -0.75a + 165$) and greater than or equal to the lower limit ($h \ge -0.5a + 110$). We are seeking the zones for people between the ages of 10 years and 100 years, inclusive: $a \ge 10$ and $a \le 100$.

    So, the target heart-rate zones can be described by the system

    $$h \le -0.75a + 165$$
    $$h \ge -0.5a + 110$$
    $$a \ge 10$$
    $$a \le 100$$

2. In Fig. 46, we use arrows to indicate the graphs of each of the four inequalities. The solution set of the system is the intersection of the four graphs.
3. The target heart-rate zone for a 23-year-old person is represented by the blue vertical line segment above $a = 23$ on the a-axis in Fig. 47. From this line segment, we see that the target heart-rate zone for a 23-year-old person is between 98.5 beats per minute and 147.75 beats per minute, inclusive.

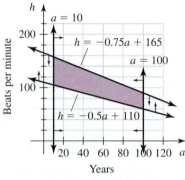

Figure 46 Graph of the solution set of the system

Figure 47 The target heart-rate zone for a 23-year-old person

We can also find the target heart-rate zone for a 23-year-old person by substituting 23 for a in the upper-limit and lower-limit models:

$$h = -0.75(23) + 165 = 147.75 \quad \textit{Upper limit}$$
$$h = -0.5(23) + 110 = 98.5 \quad \textit{Lower limit}$$

This means the target heart-rate zone is between 98.5 beats per minute and 147.75 beats per minute, inclusive, which checks.

◈ Group Exploration

Section Opener: Graphing a linear inequality in two variables

1. Graph $y = x + 1$ carefully by hand.
2. Find 4 points that lie above the line, 2 points that lie on the line, and 4 points that lie below the line. Choose your 10 points so that there are at least 2 points in each of the four quadrants. List the ordered pairs for these points so that it is clear which points are above, below, or on the line.
3. Consider the inequality $y > x + 1$. We say that the ordered pair $(2, 5)$ *satisfies* the inequality because the inequality becomes a true statement when we substitute 2 for x and 5 for y:

    $$y > x + 1$$
    $$5 \overset{?}{>} 2 + 1$$
    $$5 \overset{?}{>} 3$$
    $$\text{true}$$

Which of the ordered pairs you found in Problem 2 satisfy the given inequality or equation?

a. $y > x + 1$ **b.** $y < x + 1$ **c.** $y = x + 1$

4. Describe all the ordered pairs (not just those you found in Problem 2) that satisfy the given inequality or equation. Include in your description any of the three categories (above, below, or on the line) that are relevant.

a. $y > x + 1$ **b.** $y < x + 1$ **c.** $y = x + 1$

5. Draw a dashed line where the graph of $y = -\frac{1}{2}x - 1$ would be in a coordinate system. Then use shading to indicate points whose ordered pairs satisfy the inequality $y < -\frac{1}{2}x - 1$.

Tips for Success   The Value of Learning Mathematics

Imagine someone who is a math expert. Are you imagining a person wearing broken glasses that are taped together, a pocket protector, and wrinkled, unfashionable clothing—in other words, a math nerd? Of course, this stereotype does not accurately describe most mathematicians. But perhaps it is because of the stereotype of a math nerd that many students don't want to become "too good" at mathematics.

Learning mathematics will not transform you into a nerd. Rather, it will transform you into a more educated, well-rounded person. Learning mathematics will equip you to be more effective in many lines of work, as well as when you do math-intensive activities such as investing or filling out a tax return. Most people have a high respect for mathematicians due to their commitment to a useful and challenging subject.

Homework 6.6

For extra help ▶   **MyLab Math**   Watch the videos in MyLab Math

Which of the given ordered pairs satisfy the given inequality?

1. $y < 2x - 3$ $(4, 1), (-2, 3), (-3, -1)$
2. $y > -3x + 2$ $(4, -1), (-4, 3), (-2, -5)$
3. $2x - 5y \geq 10$ $(-3, -1), (-1, -4), (5, 0)$
4. $3y \leq -4x$ $(2, 1), (-3, 4), (-2, -1)$

Graph the linear inequality in two variables by hand.

5. $y > x - 3$
6. $y < 2x - 4$
7. $y \leq -2x + 3$
8. $y \geq -3x + 4$
9. $y < \frac{1}{3}x + 1$
10. $y > \frac{2}{3}x - 5$
11. $y \geq -\frac{3}{5}x - 1$
12. $y \leq -\frac{1}{2}x + 3$
13. $y > x$
14. $y < -x$
15. $y - 2x < 0$
16. $y + 3x \leq 0$
17. $3x + y \leq 2$
18. $2x + y \geq 3$
19. $4x - y > 1$
20. $x - y < 3$
21. $4x - 3y \geq 0$
22. $5x - 2y \leq 0$
23. $4x + 5y \geq 10$
24. $2x + 4y > 4$
25. $2x - 5y < 5$
26. $3x - 4y > 12$
27. $y > 3$
28. $y < -2$
29. $x \geq -3$
30. $x \leq 4$

31. $x < -2$
32. $y > 1$
33. $y \leq 0$
34. $x \geq 0$

For Exercises 35–58, graph the solution set of the system of linear inequalities in two variables by hand.

35. $y \geq \frac{1}{3}x - 3$
    $y \leq -\frac{3}{2}x + 2$

36. $y \leq \frac{2}{3}x + 1$
    $y \geq -\frac{4}{3}x - 2$

37. $y > 2x - 3$
    $y > -\frac{3}{4}x + 1$

38. $y < -3x + 5$
    $y < \frac{2}{5}x - 3$

39. $y \leq -\frac{2}{3}x - 3$
    $y > 2x + 1$

40. $y \leq \frac{4}{3}x - 2$
    $y > -x + 4$

41. $y \geq -2x - 1$
    $y > \frac{1}{3}x + 2$

42. $y \leq 3x - 3$
    $y < -\frac{1}{4}x + 1$

43. $y > -3x + 4$
    $y \leq 2x + 3$

44. $y > -4x + 5$
    $y \leq -x + 1$

45. $y \leq \frac{2}{3}x$
    $y < -\frac{2}{5}x$

46. $y \leq -\frac{1}{2}x$
    $y < \frac{4}{3}x$

47. $5x - 3y \leq 12$
    $-2y < x$

48. $x - 2y < 6$
    $3y \geq 6x$

49. $x - y \leq 2$
    $2x + y < 1$

50. $x + y < 1$
    $3x - y \geq 2$

51. $2x - 3y > 3$
    $3x + 5y \geq 10$

52. $-3x + 4y \leq 4$
    $5x - 2y < 2$

53. $y < 3$
    $x \geq -2$

54. $y \geq 2$
    $x < -1$

55. $y \leq 1$
    $y \geq -2$
    $x \geq -3$
    $x \leq 4$

56. $y \leq -1$
    $y \geq -4$
    $x \leq 2$
    $x \geq -5$

57. $2x - 5y \geq -5$
    $2x - 5y \leq 15$
    $x \geq -1$
    $x \leq 3$

58. $3x + 4y \leq 16$
    $3x + 4y \geq 8$
    $x \geq 1$
    $x \leq 4$

59. DATA. The lower and upper limits of ideal weights of men with a medium frame are listed in Table 35 for various heights. Assume that the men are wearing 5 pounds of clothing, including shoes with 1-inch heels. Let w be the ideal weight (in pounds, including clothes) of a man who has a medium frame and a height of h inches (including shoes). Linear models for the lower and upper limits of the ideal weights are

$$w = 3.50h - 80.97 \quad \textit{Upper limit}$$
$$w = 3.08h - 64.14 \quad \textit{Lower limit}$$

Table 35 Ideal Weights of Men with a Medium Frame

Height (inches)	Ideal Weight (pounds)	
	Lower Limit	Upper Limit
65	137	148
67	142	154
69	148	160
71	154	166
73	160	174
76	171	187

Source: *Metropolitan Life Insurance Company*

a. Find a system of linear inequalities that describes the ideal weights of men who have a medium frame and are between the heights of 63 inches and 78 inches, inclusive.

b. Graph by hand the solution set of the system of linear inequalities that you found in part (a).

c. What is the ideal weight range of men who have a medium frame and a height of 68 inches?

60. DATA. In Exercise 15 of Homework 5.3, you found an equation close to $H = 4.5d + 15.3$, where H is the number of "human years" that is equivalent to d "dog years" for dogs that weigh between 20 pounds and 50 pounds, inclusive (see Table 36). A reasonable model for 50–90-pound dogs is $H = 5.4d + 12.3$.

Assume that the model $H = 4.5d + 15.3$ is best suited for dogs that weigh 35 pounds (the mean of 20 pounds and 50 pounds) and that the model $H = 5.4d + 12.3$ is best suited for dogs that weigh 70 pounds (the mean of 50 pounds and 90 pounds).

Table 36 Dog Years Compared with Human Years

Dog Years	Human Years	
	20–50-Pound Dogs	50–90-Pound Dogs
1	15	14
2	24	22
3	29	29
5	38	40
7	47	50
9	56	61
11	65	72
13	74	82
15	83	93

Source: *Fred Metzger, State College, PA*

a. Find a system of linear inequalities that describes the numbers of human years that are equivalent to dog years between 10 dog years and 20 dog years, inclusive, for dog weights between 35 pounds and 70 pounds, inclusive.

b. Graph by hand the solution set of the system of linear inequalities that you found in part (a).

c. What are the numbers of human years that are equivalent to 15 dog years for dogs that weigh between 35 pounds and 70 pounds, inclusive?

Concepts

61. Find a linear inequality in two variables whose graph is shown in Fig. 48.

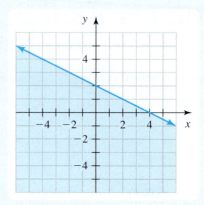

Figure 48 Exercise 61

62. Find a system of two linear inequalities whose solution set is shown in Fig. 49.

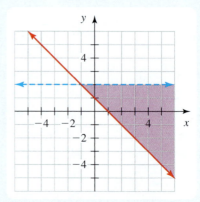

Figure 49 Exercise 62

63. A student believes that the graph of $5x - 2y < 6$ is the region below the line $5x - 2y = 6$. Is the student correct? Explain.

64. A student believes that the graph of $y \leq 3x - 1$ is the region below the line $y = 3x - 1$. Is the student correct? Explain.

65. The graphs of $y = ax + b$ and $y = cx + d$ are sketched in Fig. 50.

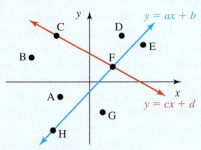

Figure 50 Exercise 65

Which of the points A, B, C, D, E, F, G, and H represent an ordered pair that

a. satisfies the inequality $y > ax + b$?

b. satisfies the inequality $y \leq cx + d$?

c. satisfies both $y > ax + b$ and $y \leq cx + d$?

d. satisfies neither $y > ax + b$ nor $y \leq cx + d$?

66. Consider the system of inequalities

$$y \geq 2x + 1$$
$$y < -x - 2$$

Find an ordered pair that

a. satisfies $y \geq 2x + 1$, but does not satisfy $y < -x - 2$.

b. satisfies $y < -x - 2$, but does not satisfy $y \geq 2x + 1$.

c. satisfies both inequalities.

d. does not satisfy either inequality.

67. Find three ordered pairs that are solutions of the inequality $y \geq 2x - 3$. Also, find three ordered pairs that are not solutions.

68. Find three ordered pairs that are solutions of the inequality $2x - y < 4$. Also, find three ordered pairs that are not solutions.

69. Why is the graph of a linear inequality of the form $y > mx + b$ the region above the line $y = mx + b$?

70. Why does the graph of a linear inequality of the form $y \geq mx + b$ include the line $y = mx + b$?

71. Explain how to graph a linear inequality in two variables by hand.

72. Explain how to solve a system of linear inequalities in two variables.

Related Review

73. Graph $y = 2x - 1$ by hand.

74. Graph $3x - 4y = 4$ by hand.

75. Graph the inequality $y \leq 2x - 1$ by hand.

76. Graph the inequality $3x - 4y > 4$ by hand.

77. Solve the inequality $3 \leq 2x - 1$, an inequality in *one* variable. Describe the solution set as an inequality, in interval notation, and as a graph.

78. Solve the inequality $3x - 4 > 4$, an inequality in *one* variable. Describe the solution set as an inequality, in interval notation, and as a graph.

Graph the solution set of the system of linear inequalities by hand.

79. $y \leq 2x - 1$
    $y \geq -x + 5$

80. $3x - 4y > 4$
    $x + 2y < 8$

Solve the system of linear equations.

81. $y = 2x - 1$
    $y = -x + 5$

82. $3x - 4y = 4$
    $x + 2y = 8$

Expressions, Equations, and Graphs

Give an example of the following, and then solve, simplify, or graph, as appropriate:

83. Linear equation in one variable

84. Linear equation in two variables

85. Linear inequality in one variable

86. Linear inequality in two variables

87. System of two linear equations in two variables

88. System of two linear inequalities in two variables

Taking It to the Lab

Climate Change Lab (continued from Chapter 5)

Recall from the Climate Change Lab in Chapter 5 that, to avoid a climate catastrophe, the IPCC has recommended that carbon dioxide emissions be lowered to 0.9 metric ton per person per year by 2050. Developed countries' carbon dioxide emissions are about 9.9 metric tons per person per year. So, these countries will have to reduce their per-person emissions significantly while trying to sustain their relatively strong economies. The United States, with carbon dioxide emissions of 17.0 metric tons per person per year, will have to reduce its emissions drastically.[*]

Developing countries' carbon dioxide emissions are about 3.9 metric tons per person per year. Due to their already low per-person emissions, it seems these countries will have an easier time meeting IPCC's goal than developed countries. However, by 2050, many such countries will have become *developed* countries, and their economies will have become much stronger. As a result of this growth, their carbon dioxide emissions will greatly increase without intervention.[†]

For example, China's carbon dioxide emissions were only 1.15 metric tons per person in 1990. Due to China's booming economy, however, the country's carbon dioxide emissions in 2050 may reach 41.5 metric tons per person per year — more than 36 times the 1990 per-person level. This large increase in per-person carbon dioxide emissions will be amplified by a population increase of 480 million people in those 60 years (from 0.98 billion to 1.46 billion).[‡]

An analysis of the data in Table 37 shows that developing countries' carbon dioxide emissions are generally increasing, but developed countries' carbon dioxide emissions are

Table 37 Carbon Dioxide Emissions

| Year | Carbon Dioxide Emissions (billion metric tons carbon) | | |
	Developed Countries	Developing Countries	International Transport
2007	14.4	16.0	1.1
2008	14.0	18.9	1.0
2009	13.1	18.3	1.0
2010	13.4	18.4	1.1
2011	13.4	19.6	1.1
2012	13.3	20.3	1.1
2013	13.3	20.9	1.1
2014	12.9	21.6	1.1

Source: *EDGAR*

generally decreasing. This is occurring not only because developing countries are becoming more industrialized, but also because their populations are increasing significantly. Developed countries' economies and populations are growing at a much slower rate; in fact, most developed countries' populations will begin to decline slowly after 2020.[§]

Many challenges lie ahead for developed and developing countries, both of which will need to develop efficient systems that rely on alternative energy sources whenever possible. Citizens of developed countries will have the extra challenges of foregoing certain conveniences that up until now have been taken for granted. Developing countries will have the extra challenge of harnessing large population growths.

Analyzing the Situation

1. Let c be carbon dioxide emissions (in billion metric tons) of developed countries in the year that is t years since 2000. Find an equation of a model to describe the data.

2. Let c be carbon dioxide emissions (in billion metric tons) of developing countries in the year that is t years since 2000. Find an equation of a model to describe the data.

3. Use a graphing calculator to draw the models and scatterplots of the data all in the same viewing window. Also, graph the models and the scatterplots in one coordinate system by hand. Make sure it is clear which data points are for the developing countries. How well do the models fit the data?

4. Compare the slopes of both models. What does the comparison tell you about the situation?

5. Explain why developing countries' carbon dioxide emissions are generally increasing, but developed countries' carbon dioxide emissions are generally decreasing.

6. Use substitution or elimination to estimate when developing countries' carbon dioxide emissions were equal to developed countries' carbon dioxide emissions. What is that carbon dioxide emission value?

7. Use the result you found in Problem 6 to estimate the per-person carbon dioxide emissions in the year when developing countries' carbon dioxide emissions were equal to developed countries' carbon dioxide emissions. Include carbon dioxide emissions from international transport in your estimate. Assume world population was 6.4 billion in that year.[¶] Given that the carbon dioxide emissions in 2000 were 4.2 metric tons per person, did per-person carbon dioxide emissions increase or decrease from 2000 to the year you found in Problem 6? Explain.

8. What are the challenges that lie ahead in trying to reduce carbon dioxide emissions? What challenges are unique to developed countries? To developing countries?

[*]International Energy Agency, "CO_2 Emissions from Fuel Combustion Highlights," 2011.

[†]Carbon Dioxide Information Analysis Center (CDIAC).

[‡]United Nations Population Division.

[§]U.S. Census Bureau.

[¶]U.S. Census Bureau.

Chapter Summary

Key Points of Chapter 6

Section 6.1   Using Graphs and Tables to Solve Systems

Solution, solution set, and solve for a system	We say that an ordered pair (a, b) is a **solution** of a system of two equations in two variables if it satisfies both equations. The **solution set** of a system is the set of all solutions of the system. We **solve** a system by finding its solution set.
Using graphing to solve a system	The solution set of a system of two linear equations can be found by locating any intersection point(s) of the graphs of the two equations.
Intersection point of the graphs of two models	If the explanatory variable of two models represents time, then an intersection point of the graphs of the two models indicates when the quantities represented by the response variables were or will be equal.
Types of linear systems	There are three types of linear systems of two equations: 1. **One-solution system:** The lines intersect in one point. The solution of the system is the ordered pair that corresponds to that point. 2. **Inconsistent system:** The lines are parallel. The solution set of the system is the empty set. 3. **Dependent system:** The lines are identical. The solution set of the system is the set of the infinite number of solutions represented by points on the same line.
Using tables to solve a linear system	If an ordered pair is listed in both tables of solutions of two linear equations, then that ordered pair is a solution of that system.

Section 6.2   Using Substitution to Solve Systems

Using substitution to solve a linear system	To use **substitution** to solve a system of two linear equations, 1. Isolate a variable to one side of either equation. 2. Substitute the expression for the variable found in step 1 into the other equation. 3. Solve the equation in one variable found in step 2. 4. Substitute the solution found in step 3 into one of the original equations, and solve for the other variable.
Solving inconsistent systems and dependent systems by substitution	If the result of applying substitution to a linear system of two equations is • a false statement, then the system is inconsistent; that is, the solution set is the empty set. • a true statement that can be put into the form $a = a$, then the system is dependent; that is, the solution set is the set of ordered pairs represented by every point on the (same) line.

Section 6.3   Using Elimination to Solve Systems

Using elimination to solve a linear system	To use **elimination** to solve a system of two linear equations, 1. Use the multiplication property of equality to get the coefficients of one variable to be equal in absolute value and opposite in sign. 2. Add the left sides and add the right sides of the equations to eliminate one of the variables. 3. Solve the equation in one variable found in step 2. 4. Substitute the solution found in step 3 into one of the original equations and solve for the other variable.
Solving inconsistent systems and dependent systems by elimination	If the result of applying elimination to a linear system of two equations is • a false statement, then the system is inconsistent; that is, the solution set is the empty set. • a true statement that can be put into the form $a = a$, then the system is dependent; that is, the solution set is the set of ordered pairs represented by every point on the (same) line.
Methods of solving a linear system	Any linear system of two equations can be solved by graphing, substitution, or elimination. All three methods give the same result.

Section 6.4 Using Systems to Model Data

Methods of finding an intersection point of two linear models	We can find an intersection point of the graphs of two linear models by graphing, substitution, or elimination.

Section 6.5 Perimeter, Value, Interest, and Mixture Problems

Five-step problem-solving method	To solve some problems in which we want to find two quantities, it is useful to perform the following five steps: • *Step 1: Define each variable.* • *Step 2: Write a system of two equations.* • *Step 3: Solve the system.* • *Step 4: Describe each result.* • *Step 5: Check.*
Using arithmetic to suggest how to form an equation	If you have difficulty forming an equation, try making up numbers for unknown quantities and performing arithmetic to compute a related quantity. Your computation may suggest how to form the desired equation.
Annual simple interest rate	The **annual simple interest rate** is the percentage of the **principal** that equals the **interest** earned per year.

Section 6.6 Linear Inequalities in Two Variables; Systems of Linear Inequalities in Two Variables

Linear inequality in two variables	A **linear inequality in two variables** is an inequality that can be put into the form $y < mx + b$ or $x < a$ (or with $<$ replaced with $\leq$, $>$, or $\geq$), where m, a, and b are constants.
Satisfy, solution, solution set, and solve for an inequality in two variables	If an inequality in the two variables x and y becomes a true statement when a is substituted for x and b is substituted for y, we say that the ordered pair (a, b) **satisfies** the inequality and call (a, b) a **solution** of the inequality. The **solution set** of an inequality is the set of all solutions of the inequality. We **solve** the inequality by finding its solution set.
Graph of a linear inequality in two variables	The graph of a linear inequality of the form $y > mx + b$ is the region above the line $y = mx + b$. The graph of a linear inequality of the form $y < mx + b$ is the region below the line $y = mx + b$. For either inequality, we use a dashed line to show that $y = mx + b$ is not part of the graph. The graph of a linear inequality of the form $y \geq mx + b$ is the line $y = mx + b$, as well as the region above that line. The graph of a linear inequality of the form $y \leq mx + b$ is the line $y = mx + b$, as well as the region below that line.
System of linear inequalities in two variables	A **system of linear inequalities in two variables** consists of two or more linear inequalities in two variables.
Solution, solution set, and solve for a system of inequalities in two variables	We say that an ordered pair (a, b) is a **solution** of a system of a linear inequalities in two variables if it satisfies all the inequalities of the system. The **solution set** of a system is the set of all solutions of the system. We **solve** a system by finding its solution set.
Finding the solution set of a system of inequalities	The solution set of a system of inequalities in two variables can be found by locating the intersection of the graphs of all the inequalities.

Chapter 6 Review Exercises

Find the solution of the system by graphing the equations by hand. Use "intersect" on a graphing calculator to verify your work.

1. $y = 2x - 3$
   $y = -3x + 7$

2. $y = \dfrac{3}{2}x + 4$
   $y = -\dfrac{1}{2}x - 4$

3. $y = \dfrac{2}{5}x$
   $y = -2x$

4. $4x + y = 3$
   $-3x + y = -4$

5. $-x + y = 4$
   $2x + y = -5$

6. $x - 3y = 3$
   $2x + 3y = -12$

Solve the system by substitution. Verify your solution by checking that it satisfies both equations of the system.

7. $3x - 2y = 11$
   $y = 5x - 16$

8. $4x - 3y - 5 = 0$
   $x = 4 - 2y$

9. $y = -5x$
   $y = 2x$

10. $y = -3(x + 2)$
   $y = 4(x - 5)$

11. $x + y = -1$
   $2x - y = 4$

12. $x + 2y = 5$
   $4x + 2y = -4$

Use substitution to solve the system, with coordinates of solutions rounded to the second decimal place. Verify your work by using "intersect" on a graphing calculator.

13. $y = -2.19x + 3.51$
$y = 1.54x - 6.22$

14. $y = -4.98x - 1.18$
$y = -0.57x + 4.08$

Solve the system by elimination. Verify your work by using "intersect" on a graphing calculator or by checking that your result satisfies both equations of the system.

15. $x - 2y = -1$
$3x + 5y = 19$

16. $2x - 5y = -3$
$4x + 3y = -19$

17. $3x + 8y = 2$
$5x - 2y = -12$

18. $4x - 3y = -6$
$-7x + 5y = 11$

19. $-2x - 5y = 2$
$3x + 6y = 0$

20. $y = 3x - 5$
$y = -2x + 5$

21. $2(x + 3) + y = 6$
$x - 3(y - 2) = -1$

22. $\frac{1}{2}x - \frac{2}{3}y = -\frac{5}{3}$
$\frac{1}{3}x - \frac{3}{2}y = -\frac{13}{6}$

Use elimination to solve the system, with coordinates of solutions rounded to the second decimal place. Verify your work by using "intersect" on a graphing calculator.

23. $y = 4.59x + 1.25$
$y = 0.52x + 4.39$

24. $y = 0.91x - 3.57$
$y = -3.58x + 6.05$

Solve the system by either substitution or elimination. Verify your work by using "intersect" on a graphing calculator or by checking that your result satisfies both equations of the system.

25. $2x - 7y = -13$
$5x + 3y = -12$

26. $4x + 7y = 8$
$x = 3 - 2y$

27. $-3x + 7y = 6$
$6x + 2y = -12$

28. $y = -x + 7$
$y = 2x - 5$

29. $4x + 5y = -6$
$2y = -3x - 8$

30. $y = x - 2$
$3x + 5y - 30 = 0$

31. $2x - 3y = 0$
$5x - 7y = -1$

32. $2(4x - 3) - 5y = 12$
$5(3x - 1) + 2y = 6$

For Exercises 33 and 34, solve the system by elimination. If the system is inconsistent or dependent, say so.

33. $2x - 6y = 4$
$-3x + 9y = -3$

34. $y = -4x + 3$
$8x + 2y = 6$

35. Consider the system

$$y = -2x + 7$$
$$y = 3x - 3$$

Find an ordered pair that

a. satisfies $y = -2x + 7$, but does not satisfy $y = 3x - 3$.
b. satisfies $y = 3x - 3$, but does not satisfy $y = -2x + 7$.
c. does not satisfy either equation.
d. satisfies both equations.

36. Figure 51 shows the graphs of two linear equations. To the first decimal place, estimate the coordinates of the solution of the system.

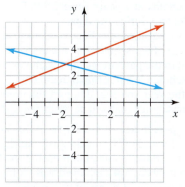

Figure 51 Exercise 36

37. Some solutions of two linear equations are shown in Tables 38 and 39. Find the solution of the system of the two equations.

Table 38 Solutions of One Equation		Table 39 Solutions of the Other Equation	
x	y	x	y
0	75	0	5
1	73	1	8
2	71	2	11
3	69	3	14
4	67	4	17

38. Solve the system of equations three times, once by each of the three methods: graphing by hand, substitution, and elimination. Decide which method you prefer for this system.

$$3x + 4y = 15$$
$$2y = -5x + 11$$

39. **DATA** In *non-family households*, the members are neither married nor related to each other. In *other-family households*, at least one member is related to another member, but none of the members are married to each other. The percentages of households that are married households, non-family households, or other-family households are shown in Table 40 for various years.

Table 40 Percentages of Households That Are Married Households, Non-Family Households, and Other-Family Households

Year	Percent		
	Married	Non-Family	Other-Family
2000	53	31	16
2004	51	32	17
2008	50	33	17
2012	49	34	17
2016	48	35	17

Source: *U.S. Census Bureau*

a. Let p be the percentage of households that are married households at t years since 2000. Find an equation of a model to describe the data.

b. Let p be the percentage of households that are non-family households at t years since 2000. Find an equation of a model to describe the data.

c. Use substitution or elimination to predict when the percentage of households that are married households will be equal to the percentage of households that are non-family households. What is that percentage?

d. For the year you found in part (c), predict the percentage of households that will be other-family households.

40. **DATA** The percentage of youths (ages 16–24 years) who were employed was 51.9% in July 2014 and has since increased by about 0.98 percentage point each year. The percentage of youths who were not employed was 48.1% in July 2014 and has since decreased by about 0.98 percentage point each year (Source: *U.S. Bureau of Labor Statistics*).

a. Let p be the percentage of youths who were employed in July t years since 2014. Find an equation of a model to describe the data.

b. Let p be the percentage of youths who were not employed in July t years since 2014. Find an equation of a model to describe the data.

c. Use substitution or elimination to estimate when the percentage of youths who were employed was equal to the percentage of youths who were not employed.

d. Estimate when the percentage of youths who were employed was equal to the percentage of youths who were not employed by substituting an appropriate value for p in one of your two models and then solving the resulting equation for t. Compare the result with the result you found in part (c).

e. Compare the slope of the employed model with the slope of the not-employed model. What do you notice about these slopes? Why does this make sense?

41. The length of a rectangle is 2 feet more than three times the width. If the perimeter is 44 feet, find the dimensions of the rectangle.

42. An 8000-seat amphitheater has tickets for sale at $22 and $39. How many tickets should be sold at each price for a sellout performance to generate a total revenue of $201,500?

Graph the linear inequality in two variables by hand.

43. $y \leq 3x - 5$

44. $y \geq -2x + 4$

45. $3x - 2y > 4$

46. $2y - 5x < 0$

47. $x \geq 3$

48. $y < -2$

For Exercises 49–52, graph the solution set of the system of linear inequalities in two variables by hand.

49. $y > x + 1$
$y \leq -2x + 5$

50. $y \geq \dfrac{3}{5}x + 1$
$x < -1$

51. $3x - 4y \geq 12$
$5y \leq -3x$

52. $x > 2$
$y \leq -1$

53. **DATA** The lower and upper limits of ideal weights of women with a medium frame are listed in Table 41 for various heights. Assume that the women are wearing 5 pounds of clothing, including shoes with 1-inch heels.

Table 41 Ideal Weights of Women with a Medium Frame

Height (inches)	Ideal Weight (pounds)	
	Lower Limit	Upper Limit
61	115	129
63	121	135
65	127	141
67	133	147
69	139	153
72	148	162

Source: *Metropolitan Life Insurance Company*

Let w be the ideal weight (in pounds, including clothes) of a woman who has a medium frame and a height of h inches (including shoes). Linear models for the lower and upper limits of the ideal weights are

$$w = 3h - 54 \quad \textit{Upper limit}$$
$$w = 3h - 68 \quad \textit{Lower limit}$$

a. Find a system of linear inequalities that describes the ideal weights of women who have a medium frame and are between the heights of 58 inches and 74 inches, inclusive.

b. Graph by hand the solution set of the system of linear inequalities that you found in part (a).

c. What is the ideal weight range of women who have a medium frame and a height of 70 inches?

54. Find three ordered pairs that are solutions of the inequality $4x - 3y > 9$. Also, find three ordered pairs that are not solutions.

Chapter 6 Test

1. Find the solution of this system by graphing the equations by hand:

$$y = -\frac{2}{5}x - 1$$
$$y = -2x + 7$$

2. Use "intersect" on a graphing calculator to solve this system, with the coordinates of the solution rounded to the second decimal place:

$$y = \frac{2}{3}x + 4$$
$$3x + 4y = -2$$

Solve the system by substitution.

3. $5x - 2y = 4$
$\qquad y = 3x - 1$

4. $3x + 4y = 9$
$\qquad x - 2y = -7$

Solve the system by elimination.

5. $-7x - 2y = -8$
$\qquad 5x + 4y = -2$

6. $2x - 5y = -18$
$\qquad 3x + 4y = -4$

For Exercises 7–10, solve the system by substitution or elimination. If the system is inconsistent or dependent, say so.

7. $3x - 5y = -21$
$\qquad x = 2(2 - y)$

8. $2x - 3y = 4$
$\qquad -4x + 6y = -8$

9. $\qquad x = 2y - 3$
$\qquad 3x - 6y = 12$

10. $4x - 7y = 6$
$\qquad -5x + 2y = -21$

11. Use substitution or elimination to solve this system, with coordinates of the solution rounded to the second decimal place:
$$y = -1.94x + 8.62$$
$$y = 1.25x - 2.38$$

12. The graphs of $y = ax + b$ and $y = cx + d$ are sketched in Fig. 52.

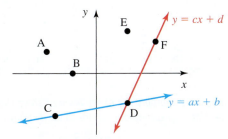

Figure 52 Exercise 12

Which of the points A, B, C, D, E, and F
a. satisfy the equation $y = ax + b$?
b. satisfy the equation $y = cx + d$?
c. satisfy both equations?
d. do not satisfy either equation?

13. Figure 53 shows the graphs of two linear equations. Find the solution of the system.

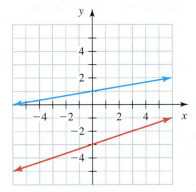

Figure 53 Exercise 13

14. Create a system of two linear equations whose solution is $(2, 4)$.

15. Find the coordinates of the points A, B, C, and D as shown in Fig. 54. The equations of the lines l_1 and l_2 are provided, but no attempt has been made to sketch the lines accurately except for showing the intersection points.

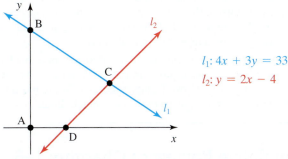

$l_1: 4x + 3y = 33$
$l_2: y = 2x - 4$

Figure 54 Exercise 15

16. <u>DATA</u> The number of U.S. House of Representatives in a region is based on the population of that region. The calculations are performed every 10 years. The percentages of House seats in the Northeast, Midwest, South, and West are shown in Table 42 for various years.

Table 42 Percentages of House Seats by Region

Year	Percent			
	Northeast	Midwest	South	West
1940	28	30	31	11
1950	26	30	31	14
1960	25	29	31	16
1970	24	28	31	17
1980	22	26	33	20
1990	20	24	34	21
2000	19	23	35	23
2010	18	22	37	23

Source: *U.S. Census Bureau*

a. Let p be the percentage of House seats in the Northeast at t years since 1900. Find an equation of a model to describe the data.
b. Let p be the percentage of House seats in the West at t years since 1900. Find an equation of a model to describe the data.
c. Use substitution or elimination to estimate when the percentage of House seats in the Northeast was equal to the percentage of House seats in the West. What is that percentage?
d. The percentages for the four regions in 1950 add to 101%, which is impossible. Explain how rounding can account for this. Give an example of what the actual percentages might be before rounding.
e. The percentages for the four regions in 1990 add to 99%, which is impossible. Explain how rounding can account for this. Give an example of what the actual percentages might be before rounding.

17. The number of guns in circulation in the United States was 369 million in 2014 and has since increased by about 12.7 million per year (Source: *Bureau of Alcohol, Tobacco, and Firearms*). The U.S. population was 319 million in 2014 and has since increased by about 2.4 million per year (Source: *U.S. Census Bureau*).

a. Let *n* be the number of guns in circulation (in millions) at *t* years since 2014. Find an equation of a model to describe the data.

b. Let *n* be the U.S. population (in millions) at *t* years since 2014. Find an equation of a model to describe the data.

c. Predict when there was, on average, one gun per person in the United States. Did everyone own a gun in that year? How many guns were in circulation?

18. A person plans to invest a total of $7000, part of it in a Homestead International Equity fund at 7.1% annual interest and the rest in an Evercore Equity fund at 14% annual interest. Both interest rates are averages. How much should the person

invest in each fund so that the total interest in one year will be $600?

19. A 20,000-seat theater has tickets for sale at $55 and $70. How many tickets should be sold at each price for a sellout performance to generate a total revenue of $1,197,500?

Graph the linear inequality in two variables by hand.

20. $5x - 2y \le 6$ **21.** $y < -3$

Graph the solution set of the system of linear inequalities in two variables by hand.

22.
$$y \le -3x + 4$$
$$x - 3y > 6$$

23. $y > 2$
$$x \ge -3$$

Cumulative Review of Chapters 1–6

1. The low temperatures in New York City for the first four days of March are 4°F, −2°F, −1°F, and 3°F. Let *F* be the temperature in Fahrenheit degrees. Use points on a number line to describe these low temperatures.

2. The temperature affects (explains) the rate at which a cricket chirps. Let *n* be the number of times a cricket chirps per minute when the temperature is *t* degrees Fahrenheit. What does the ordered pair (70, 129) represent?

For Exercises 3 and 4, perform the indicated operations. Use a calculator to verify your work.

3. $-\dfrac{26}{27} \cdot \dfrac{12}{13}$ **4.** $\dfrac{5}{7} - \left(-\dfrac{3}{5}\right)$

5. Evaluate $a^2 - bc + b^2$, where $a = -3$, $b = -2$, and $c = 4$.

6. Find the slope of the line that contains the points $(-3, -2)$ and $(5, -8)$. State whether the line is increasing, decreasing, horizontal, or vertical.

7. Graphs of four equations are shown in Fig. 55. State whether *m* and *b* are positive, negative, zero, or undefined for the $y = mx + b$ form of each equation.

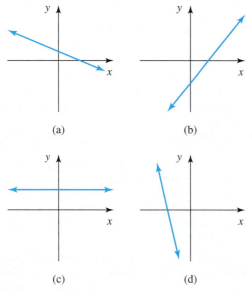

(a)   (b)

(c)   (d)

Figure 55 Exercise 7

8. Some solutions of four linear equations are provided in Table 43. Find an equation of each of the four lines.

Table 43 Solutions of Four Linear Equations

Equation 1		Equation 2		Equation 3		Equation 4	
x	*y*	*x*	*y*	*x*	*y*	*x*	*y*
0	49	0	11	3	39	2	14
1	41	1	15	4	37	4	20
2	33	2	19	5	35	5	23
3	25	3	23	6	33	8	32
4	17	4	27	7	31	9	35

For Exercises 9–12, simplify the expression or solve the equation, as appropriate. Use a graphing calculator to verify your work.

9. $2 - 5(4x + 8) = 3(2x - 7) + 3$

10. $-3(2p - w) - (7p + 2w) + 5$

11. $\dfrac{2}{3}(6w + 9y - 15)$

12. $\dfrac{3m}{5} - \dfrac{2}{3} = \dfrac{4}{5}$

13. Solve $ax + by = c$ for *y*.

Let x be a number. Translate the following into an expression or an equation, as appropriate, and then simplify the expression or solve the equation.

14. 6, plus 3 times the sum of 4 and the number

15. 4 minus the quotient of the number and 3 is 2.

Determine the slope and the y-intercept, and use them to graph the equation by hand. Use a graphing calculator to verify your work.

16. $y = 2x - 4$ **17.** $x - 2y = 6$

18. $5x + 2y - 12 = 0$ **19.** $y = -3$

For Exercises 20 and 21, find the x-intercept and y-intercept. Then graph the equation by hand.

20. $2x - 5y = 10$ **21.** $y = -2x + 4$

22. Graph the equation $x = 4$.

23. Determine whether the lines $2x + 5y = 7$ and $y = \frac{2}{5}x - 3$ are parallel.

24. Find an equation of the line that has slope $-\frac{2}{5}$ and contains the point $(3, -2)$. Write your equation in slope–intercept form. Check that the point $(3, -2)$ satisfies your equation.

Find an equation of the line that passes through the two given points. If possible, write your equation in slope–intercept form. If possible, verify your equation by using a graphing calculator.

25. $(-5, 1)$ and $(-2, -3)$ **26.** $(-2, 8)$ and $(-2, 1)$

For Exercises 27–30, refer to Fig. 56.

27. Find y when $x = -3$.   **28.** Find x when $y = -1$.
29. Find the x-intercept.   **30.** Find an equation of the line.

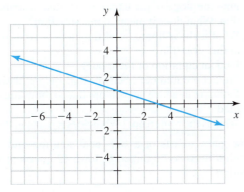

Figure 56 Exercise 27–30

31. Solve the inequality $-3(x - 5) > 18$, an inequality in one variable. Describe the solution set as an inequality, in interval notation, and as a graph.

32. Find the solution of this system by graphing the equations by hand:

$$y = 2x - 3$$
$$x + y = 3$$

33. Use "intersect" on a graphing calculator to solve this system, with the coordinates of the solution rounded to the second decimal place:

$$y = -2.9x + 7.8$$
$$y = 1.3x - 6.1$$

For Exercises 34 and 35, solve the system by substitution or elimination. Use "intersect" on a graphing calculator to verify your work.

34. $3x + 5y = -1$ **35.** $4x - y - 9 = 0$
$2x - 3y = 12$ $y = 5 - 3x$

36. Use elimination or substitution to solve the system, with coordinates of solutions rounded to the second decimal place. Verify your work by using "intersect" on a graphing calculator.

$$y = -2.9x + 97.8$$
$$y = 3.1x - 45.6$$

37. Solve the system three times, once by each of the three methods: graphing by hand, substitution, and elimination. Decide which method you prefer for this system.

$$2x - 3y = 19$$
$$y = 4x - 13$$

38. Some solutions of two linear equations are shown in Tables 44 and 45. Find the solution of the system of the two equations.

Table 44 Solutions of One Equation		Table 45 Solutions of the Other Equation	
x	y	x	y
0	97	0	7
1	93	1	9
2	89	2	11
3	85	3	13
4	81	4	15

39. Graph the inequality $y < -\frac{2}{3}x + 3$ by hand.

40. Graph the solution set of the following system by hand.

$$3x - 5y \leq 10$$
$$x > -4$$

41. DATA. The numbers of pedestrian fatalities and the percentages of traffic fatalities that are pedestrian fatalities are shown in Table 46 for various years.

Table 46 Pedestrian Fatalities

Year	Pedestrian Fatalities (thousands)	Percentage of Traffic Fatalities That Are Pedestrian Fatalities
2006	4.8	11
2007	4.7	11
2008	4.4	12
2009	4.1	12
2010	4.3	13
2011	4.5	14
2012	4.8	14
2013	4.8	15
2014	4.9	15
2015	5.4	15

Source: *Fatality Analysis Reporting System*

Let p be the percentage of traffic fatalities that are pedestrian fatalities at t years since 2000.
a. Did the value of p increase or decrease from 2007 to 2008? Explain how this is possible when the number of pedestrian fatalities decreased in that period.
b. Find an equation of a model to describe the values of t and p.
c. Find the p-intercept. What does it mean in this situation?
d. Predict when 18% of traffic fatalities will be pedestrian fatalities.
e. About 74% of pedestrian fatalities happen in the dark. Predict the *number* of pedestrian fatalities that will occur in the dark in 2022. Assume there will be 33.9 thousand traffic fatalities in that year.

42. <u>DATA</u> The percentages of children living with two, one, or no parents are shown in Table 47 for various years.

Table 47 Percentages of Children Living with Two, One, or No Parents

| Year | Percentage of Children Who Live with the Following Number of Parents | | |
	Two	One	None
1960	87	9	4
1970	81	14	4
1980	77	19	4
1990	76	20	4
2000	73	22	5
2010	70	25	5
2014	69	26	5

Source: *Pew Research Center*

a. Let p be the percentage of children who live with two parents at t years since 1960. Find an equation of a model to describe the data.

b. Let p be the percentage of children who live with one parent at t years since 1960. Find an equation of a model to describe the data.

c. Use substitution or elimination to predict when the percentage of children who live with two parents will equal the percentage of children who live with one parent. What is that percentage?

43. The revenue of department stores was $60 billion in 2014 and has decreased by about $1.7 billion per year since then. The revenue of off-price stores was $29 billion in 2014 and has increased by about $2.6 billion per year since then (Source: *Customer Growth Partners*).

a. Let r be the annual revenue (in billions of dollars) of department stores at t years since 2014. Find an equation of a model to describe the data.

b. Let r be the annual revenue (in billions of dollars) of off-price stores at t years since 2014. Find an equation of a model to describe the data.

c. Use substitution or elimination to predict when the annual revenues of department stores and off-price stores will be equal. What is that annual revenue?

44. The number of claims airline passengers filed against the Transportation Security Administration (TSA) was 8.9 thousand in 2014 and has since decreased by about 0.5 thousand claims per year (Source: *Transportation Security Administration*). Let n be the number (in thousands) of claims airline passengers filed against TSA in the year that is t years since 2014.

a. Find the slope of the linear model that describes the situation. What does it mean in this situation?

b. What is the n-intercept of the model? What does it mean in this situation?

c. Find an equation of the model.

d. Predict the number of claims that will be filed in 2021.

e. Estimate in which years more than 6 thousand claims were filed.

f. What is the t-intercept? What does it mean in this situation?

45. How many quarts of a 16% acid solution and a 28% acid solution need to be mixed to form 12 quarts of a 20% acid solution?

7 Polynomials and Properties of Exponents

How much time do you spend driving each day? Table 1 lists the average times Americans of various age groups spend driving each day. In Example 1 of Section 7.2, we will estimate the age of drivers who spend the most time driving.

In Chapters 1–6, we worked with linear expressions and equations. In Chapters 7–9, we will work with new expressions and equations called *power expressions*, *polynomial expressions*, and *polynomial equations*. This work will enable us to make estimates and predictions about a new type of authentic situation, such as the percentage of tenth graders who have used marijuana in the past 30 days.

Table 1 Average Times Spent Driving Each Day

Age Group	Age Used to Represent Age Group	Average Daily Driving Time (minutes)
15–19	17.0	25
20–24	22.0	52
25–54	39.5	64
55–64	59.5	58
over 64	70.0	39

Source: *U.S. Department of Transportation*

7.1 Graphing Quadratic Equations

Objectives

» Graph *quadratic equations in two variables*.

» Find any intercepts of a parabola.

In this section, we will graph *quadratic equations in two variables*.

Graphing Quadratic Equations in Two Variables

So far, we have graphed linear equations in two variables. These equations are one type of *polynomial equation in two variables*. Here are some other examples of polynomial equations in two variables:

$$y = 2x^2 - 18x + 36 \qquad y = x^3 - x^2 - 6x + 2 \qquad y = x^4 - 4x^2 - x$$

The graphs of these equations are shown in Fig. 1.

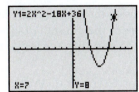
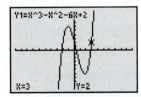
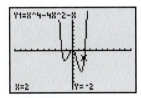

Figure 1 Graphs of three polynomial equations in two variables

Graphing most types of polynomial equations is beyond the scope of this course. After having graphed linear equations in two variables, we will now focus on graphing

365

polynomial equations such as $y = 2x^2 - 18x + 36$, which is called a *quadratic equation in two variables*. Here are some more equations of this type:

$$y = 3x^2 - 5x + 4 \qquad y = -2x^2 + 9 \qquad y = 4x^2 \qquad y = x^2 - 6x$$

Definition   Quadratic equation in two variables

A **quadratic equation in two variables** is an equation that can be put into the form

$$y = ax^2 + bx + c,$$

where $a, b,$ and c are constants and $a \neq 0$. This form is called **standard form**.

The equation $y = 3x^2 - 5x + 4$ is in the form $y = ax^2 + bx + c$, with $a = 3, b = -5,$ and $c = 4$. The equation $y = -2x^2 + 9$ is also in the form $y = ax^2 + bx + c$, with $a = -2, b = 0,$ and $c = 9$.

Table 2 Solutions of $y = x^2$

x	y
-3	$(-3)^2 = 9$
-2	$(-2)^2 = 4$
-1	$(-1)^2 = 1$
0	$0^2 = 0$
1	$1^2 = 1$
2	$2^2 = 4$
3	$3^2 = 9$

Example 1   Sketching a Graph

Sketch the graph of $y = x^2$.

Solution

First, we list some solutions of $y = x^2$ in Table 2. Then, in Fig. 2, we sketch a curve that contains the points corresponding to the solutions.

Recall from Section 3.1 that every point on the graph of an equation represents a solution of the equation. Every point *not* on the graph represents an ordered pair that is *not* a solution.

The graph of a quadratic equation in two variables is called a **parabola**. So, the curve sketched in Fig. 2 is a parabola. Two more examples of parabolas are sketched in Figs. 3 and 4.

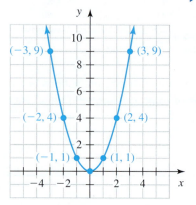

Figure 2 Graph of $y = x^2$

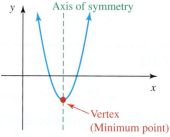

Figure 3 A parabola that opens upward

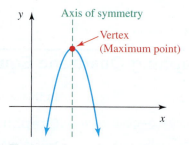

Figure 4 A parabola that opens downward

The lowest point of a parabola that *opens upward* (see Fig. 3) is called the **minimum point**. The highest point of a parabola that *opens downward* (see Fig. 4) is called the **maximum point**. The minimum point or the maximum point of a parabola is called the **vertex** of the parabola.

The vertical line that passes through a parabola's vertex is called the **axis of symmetry** (see Figs. 3 and 4). The part of the parabola that lies to the left of the axis of symmetry is the mirror reflection of the part that lies to the right.

Example 2   Sketching a Graph and Finding the Vertex

Sketch the graph of $y = 2x^2 - 8x + 6$, and give the coordinates of the vertex.

Solution

First, we list some solutions of $y = 2x^2 - 8x + 6$ in Table 3. Then, in Fig. 5, we sketch a curve that contains the points corresponding to the solutions.

Table 3 Solutions of
$y = 2x^2 - 8x + 6$

x	y
0	$2(0)^2 - 8(0) + 6 = 6$
1	$2(1)^2 - 8(1) + 6 = 0$
2	$2(2)^2 - 8(2) + 6 = -2$ ← Vertex
3	$2(3)^2 - 8(3) + 6 = 0$
4	$2(4)^2 - 8(4) + 6 = 6$

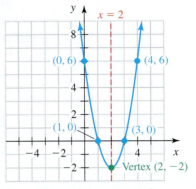

Figure 5 Graph of $y = 2x^2 - 8x + 6$

The vertex is $(2, -2)$, the lowest point of this parabola. The axis of symmetry is the line $x = 2$, the vertical line that contains the vertex.

To verify our work, we check that the part of the parabola that lies to the left of the axis of symmetry is the mirror reflection of the part that lies to the right. In particular, we check that the point $(3, 0)$ is the mirror reflection of the point $(1, 0)$ and that the point $(4, 6)$ is the mirror reflection of the point $(0, 6)$. We can also use a graphing calculator to verify our work (see Fig. 6).

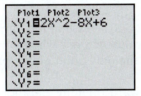 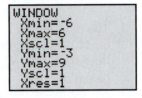 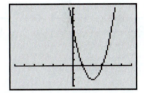

Figure 6 Verify the work

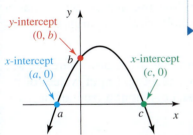

Figure 7 Intercepts of a curve

Intercepts of Curves

The definition of an intercept of a line is similar to the definition of an intercept of any type of curve. An **x-intercept of a curve** is a point where the curve and the x-axis intersect; a **y-intercept of a curve** is a point where the curve and the y-axis intersect (see Fig. 7).

Example 3   Sketching a Graph and Finding the Intercepts and the Vertex

Sketch the graph of $y = -x^2 - 2x + 3$, and give the coordinates of the intercepts and the vertex.

Solution

First, we list some solutions of $y = -x^2 - 2x + 3$ in Table 4. Then, in Fig. 8, we sketch a curve that contains the points corresponding to the solutions.

Table 4 Solutions of
$y = -x^2 - 2x + 3$

x	y
-4	$-(-4)^2 - 2(-4) + 3 = -5$
-3	$-(-3)^2 - 2(-3) + 3 = 0$
-2	$-(-2)^2 - 2(-2) + 3 = 3$
-1	$-(-1)^2 - 2(-1) + 3 = 4$ ← Vertex
0	$-(0)^2 - 2(0) + 3 = 3$
1	$-(1)^2 - 2(1) + 3 = 0$
2	$-(2)^2 - 2(2) + 3 = -5$

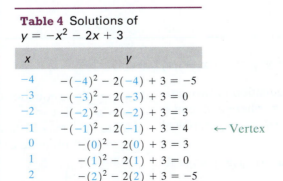

Figure 8 Graph of $y = -x^2 - 2x + 3$

The x-intercepts are $(-3, 0)$ and $(1, 0)$. The y-intercept is $(0, 3)$. The vertex is $(-1, 4)$, the highest point of the parabola. The axis of symmetry is the line $x = -1$, the vertical line that contains the vertex.

We use a graphing calculator to verify our work (see Fig. 9).

 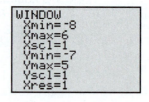 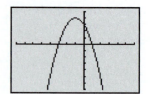

Figure 9 Verify the work

Example 4   Finding Coordinates of Points on a Parabola

A parabola is sketched in Fig. 10.

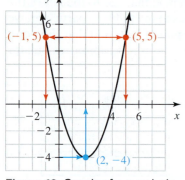

Figure 10 Graph of a parabola

1. Find x when $y = 5$.
3. Find x when $y = -5$.

2. Find x when $y = -4$.

Solution

1. The red arrows in Fig. 10 show that the output $y = 5$ originates from the two inputs $x = -1$ and $x = 5$. So, the values of x are -1 and 5 when $y = 5$.
2. The blue arrows in Fig. 10 show that the output $y = -4$ originates from the single input $x = 2$. So, the value of x is 2 when $y = -4$. [There is a single input because the vertex $(2, -4)$ is the only point on the parabola that has a y-coordinate equal to -4.]
3. No point of the upward-opening parabola is below the vertex, which has a y-coordinate of -4. So, there is no point on the parabola with a y-coordinate of -5.

By considering the shape of any parabola that opens upward or downward, we see that each (output) value of y originates from either two, one, or no (input) values of x.

Note that this situation is different from that for lines. For a nonhorizontal line, each value of y originates from *exactly* one value of x.

◈ Group Exploration ────────────────

Section Opener: Graphing quadratic equations in two variables

Graph the equation by hand. To begin, substitute the values $-3, -2, -1, 0, 1, 2,$ and 3 for x. Make other substitutions as necessary. Then, use a graphing calculator to verify your work.

1. $y = x^2$
3. $y = x^2 - 3$
5. $y = -x^2$

2. $y = x^2 + 1$
4. $y = 2x^2$

◈ Group Exploration ────────────────

Significance of a, h, and k for $y = a(x - h)^2 + k$

It is possible to write any quadratic equation in two variables in the form $y = a(x - h)^2 + k$, where $a, h,$ and k are constants and $a \neq 0$. You will explore the graphical significance of $a, h,$ and k.

1. Use ZStandard followed by ZSquare on a graphing calculator to draw a graph of $y = x^2$.

2. Graph these equations of the form $y = x^2 + k$ in order: $y = x^2 + 1, y = x^2 + 2, y = x^2 + 3,$ and $y = x^2 + 4$. State in terms of k how you could "move" the graph of $y = x^2$ to produce each graph. Do the same with the equations $y = x^2 - 1, y = x^2 - 2, y = x^2 - 3,$ and $y = x^2 - 4$.

3. Graph these equations of the form $y = (x - h)^2$ in order: $y = (x - 1)^2$, $y = (x - 2)^2$, $y = (x - 3)^2$ and $y = (x - 4)^2$. State in terms of h how you could "move" the graph of $y = x^2$ to produce each graph. Do the same with the equations $y = (x + 1)^2$, $y = (x + 2)^2$, $y = (x + 3)^2$, and $y = (x + 4)^2$.

4. Graph these equations of the form $y = ax^2$ in order: $y = 0.1x^2$, $y = 0.4x^2$, $y = x^2$, $y = 2x^2$, and $y = 5x^2$. State in terms of a how you could adjust the graph of $y = x^2$ to produce each graph. Do the same with the equations $y = -0.1x^2$, $y = -0.4x^2$, $y = -x^2$, $y = -2x^2$, and $y = -5x^2$.

5. a. Graph these equations in order: $y = x^2$, $y = 2x^2$, $y = -2x^2$, $y = -2(x + 1)^2$, and $y = -2(x + 1)^2 - 6$. Explain how these graphs relate to your observations in Problems 2–4.

b. Referring to your graph of $y = -2(x + 1)^2 - 6$ from part (a), find the coordinates of the vertex. Compare these coordinates with the equation $y = -2(x + 1)^2 - 6$. What do you notice?

6. Summarize your findings about a, h, and k in terms of how you could move or adjust the graph of $y = x^2$ to produce the graph of $y = a(x - h)^2 + k$. Also, discuss how the coordinates of the vertex are related to a, h, and k. If you are unsure, continue exploring.

Tips for Success   Examine Your Mistakes

When you make mistakes on a quiz or an exam, it is tempting to disregard these errors by calling them silly mistakes. Or perhaps you know you've made significant errors, but don't want to face them fully. This is understandable; no one enjoys examining his or her mistakes! However, your mistakes can be the gateway to your success. If you examine your errors closely, you can learn a lot about related concepts and perform better next time.

For each error that you made on a quiz or an exam, think carefully about what went wrong. It is a good idea to read about the related concepts and then do many related exercises to build your understanding and confidence in using those concepts. Discussing the concepts with other students or your instructor can help, too.

Ask your instructor or tutor to look for a pattern in your errors. It's possible that you made one kind of error five times rather than five different errors. Learning to correct that one type of error could save you many points on the next test.

Homework 7.1

For extra help ▶   **MyLab Math**   Watch the videos in MyLab Math

Graph the equation by hand. To begin, substitute the values $-3, -2, -1, 0, 1, 2,$ and 3 for x. Make other substitutions as necessary. Give the coordinate of the vertex. Then use a graphing calculator to verify your work.

1. $y = 2x^2$

2. $y = 3x^2$

3. $y = -x^2$

4. $y = -2x^2$

5. $y = x^2 + 1$

6. $y = x^2 - 4$

7. $y = 3x^2 - 5$

8. $y = -2x^2 + 4$

9. $y = x^2 + 4x$

10. $y = x^2 - 4x$

11. $y = -3x^2 - 6x$

12. $y = 2x^2 + 4x$

13. $y = x^2 - 4x + 3$

14. $y = x^2 - 2x - 1$

15. $y = -2x^2 + 4x - 2$

16. $y = -2x^2 + 8x - 3$

For Exercises 17–28, refer to the graph sketched in Fig. 11.

17. Find y when $x = -2$.

18. Find y when $x = -6$.

19. Find x when $y = 3$.

20. Find x when $y = -5$.

21. Find x when $y = 4$.

22. Find x when $y = 0$.

23. Find x when $y = 5$.

24. Find x when $y = 6$.

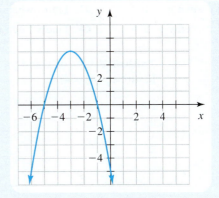

Figure 11 Exercises 17–28

25. Find the x-intercept(s).

26. Find the y-intercept(s).

27. Find the vertex.

28. Find the maximum point.

For Exercises 29–38, refer to the graph sketched in Fig. 12.

29. Find y when $x = 0$.

30. Find y when $x = 4$.

31. Find x when $y = 0$.

32. Find x when $y = 3$.

33. Find x when $y = -3$.   **34.** Find x when $y = -1$.

35. Find the y-intercept(s).   **36.** Find the x-intercept(s).

37. Find the minimum point.   **38.** Find the vertex.

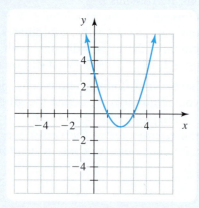

Figure 12 Exercises 29–38

39. Recall that we can describe some of or all the solutions of an equation in two variables with an equation, a table, a graph, or words.
   a. Describe the solutions of the equation $y = x^2 - 3$ by using a graph.
   b. Describe five solutions of the equation $y = x^2 - 3$ by using a table.
   c. Describe the solutions of the equation $y = x^2 - 3$ by using words.

40. Recall that we can describe some of or all the solutions of an equation in two variables with an equation, a table, a graph, or words.
   a. Describe the solutions of the equation $y = x^2 + 2x$ by using a graph.
   b. Describe five solutions of the equation $y = x^2 + 2x$ by using a table.
   c. Describe the solutions of the equation $y = x^2 + 2x$ by using words.

For Exercises 41–44, refer to Table 5.

41. Find y when $x = 1$.   **42.** Find y when $x = 2$.

43. Find x when $y = 3$.   **44.** Find x when $y = 4$.

Table 5 Values of x and y
(Exercises 41–44)

x	y
0	0
1	3
2	4
3	3
4	0

Concepts

45. a. What is the smallest possible value of x^2? Explain.
   b. What is the smallest possible value of $x^2 + 3$? Explain.
   c. What is the vertex of $y = x^2 + 3$? Explain how you can determine this without graphing the equation. Then graph the equation either by hand or by graphing calculator to verify your result.

46. Solve the system

$$y = x^2 + 2$$
$$y = -x^2 + 2$$

[**Hint:** Use graphing.]

47. a. For the equation $y = x^2 + 1$, a quadratic equation in two variables, find all outputs for the given input. State how many outputs there are for that single input.
      i. the input $x = 2$
      ii. the input $x = 4$
      iii. the input $x = -3$
   b. For $y = x^2 + 1$, how many outputs originate from any single input? Explain.
   c. Give an example of a quadratic equation in two variables. Using your equation, find all outputs for the given input. State how many outputs there are for that single input.
      i. the input $x = 1$
      ii. the input $x = 3$
      iii. the input $x = -2$
   d. For your equation, how many outputs originate from any single input? Explain.
   e. For *any* quadratic equation in two variables, how many outputs originate from a single input? Explain.

48. Use Fig. 13 to find x when
   a. $y = -1$.   **b.** $y = 2$.   **c.** $y = 3$.
   d. $y = 4$.   **e.** $y = 5$.
   f. For each of parts (a)–(e), you found two, one, or no values for x. Now describe all value(s) of y for which there are
      i. two values of x.
      ii. one value of x.
      iii. no values of x.

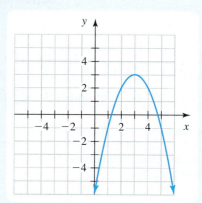

Figure 13 Exercise 48

49. Some students believe x^2 is greater than x for all real numbers x. Use ZDecimal on a graphing calculator to graph $y = x^2$ and $y = x$ on the same coordinate system. For which values of x is $x^2 < x$?

50. Find the y-intercept of the graph of $y = ax^2 + bx + c$.

Related Review

Graph the equation by hand. Then use a graphing calculator to verify your work.

51. $y = -2x - 1$ **52.** $y = 3x - 2$

53. $y = -2x^2 - 1$ **54.** $y = 3x^2 - 2$

55. **a.** Graph $y = 2x$.
    b. Graph $y = x^2$.
    c. Compare the y-intercepts of the graphs of $y = 2x$ and $y = x^2$.
    d. Which of the curves $y = 2x$ and $y = x^2$ appears to be "steeper" for large values of x? Explain.

56. **a.** Construct a table of values of x and y for the equation $y = 2x$. Use the values $1, 100, 200, 300, 400,$ and 500 for x.
    b. Construct a table of values of x and y for the equation $y = x^2$. Use the values $1, 100, 200, 300, 400,$ and 500 for x.
    c. As the values of x increase beyond 1, do the values of y increase at a greater rate for the equation $y = 2x$ or for the equation $y = x^2$? Explain.

Expressions, Equations, and Graphs

Perform the indicated instruction. Then use words such as linear, quadratic, one variable, *and* two variables *to describe the expression, equation, or system.*

57. Solve $8(3x - 2) = 4(x - 5)$.

58. Graph $y = 2x^2 - 4$ by hand.

59. Simplify $8(3x - 2) - 4(x - 5)$.

60. Graph $y = 2x - 4$ by hand.

7.2 Quadratic Models

Objectives

» Use a quadratic model to make estimates and predictions.

» Determine whether to model a situation by using a linear model, a quadratic model, or neither.

» Use a graphing calculator to make predictions with a quadratic model.

In Chapters 1–6, we used linear models to describe authentic situations. In Chapters 7–9, we will use *quadratic models* to describe other types of authentic situations.

Using a Quadratic Model to Make Estimates and Predictions

We begin by defining a quadratic model.

> **Definition   Quadratic model**
>
> A **quadratic model** is a quadratic equation in two variables that describes the relationship between two quantities for an authentic situation. We also refer to the quadratic equation's graph (a parabola) as a quadratic model.

> **Example 1**   Using a Parabola to Make Estimates

Table 6 lists the average times spent driving each day by various age groups.

Table 6 Average Times Spent Driving Each Day

Age Group	Age Used to Represent Age Group	Average Daily Driving Time (minutes)
15–19	17.0	25
20–24	22.0	52
25–54	39.5	64
55–64	59.5	58
over 64	70.0	39

Source: *U.S. Department of Transportation*

Let T be the average daily driving time (in minutes) for a person of age a years. A reasonable model is drawn from a scatterplot of the data in Fig. 14.

1. Estimate the average daily driving time of 22-year-old drivers.
2. What was the actual average daily driving time of 22-year-old drivers? Compute the error in the estimate from Problem 1.
3. Estimate the age(s) of drivers whose average daily driving time is 56 minutes.
4. Estimate the age of drivers who spend the most time driving.
5. Find the a-intercepts. What do they mean in this situation?

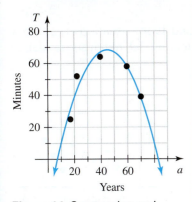

Figure 14 Scatterplot and model of daily driving times

Solution

1. The blue arrows in Fig. 15 show that the input $a = 22$ leads to the approximate output $T = 43$. So, according to the model, the average daily driving time for 22-year-old drivers is 43 minutes.

2. The estimated average daily driving time of 43 minutes for 22-year-old drivers is less than the actual time of 52 minutes (see Table 6 or Fig. 15). Because $52 - 43 = 9$ and we underestimated the time, the error is -9 minutes.

3. The red arrows in Fig. 16 show that the output $T = 56$ originates from the two approximate inputs $a = 29$ and $a = 61$. So, according to the model, both 29-year-old drivers and 61-year-old drivers have an average daily driving time of 56 minutes.

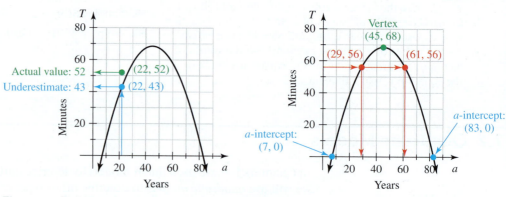

Figure 15 Estimated and actual daily driving times for 22-year-old drivers

Figure 16 Using the model to make more estimates

4. The maximum point on the parabola is the vertex, which is about $(45, 68)$. See Fig. 16. So, according to the model, 45-year-old drivers have an average daily driving time of 68 minutes, which is greater than the average for any other age.

5. From Fig. 16, it appears the a-intercepts are $(7, 0)$ and $(83, 0)$. According to the model, 7-year-old children do not drive, which is true. However, there is model breakdown for values of a near (but not equal to) 7; for example, the model suggests that 6-year-old children drive for negative amounts of time, which doesn't make sense, and that 8-year-old children drive briefly, which is not true.

The a-intercept $(83, 0)$ means 83-year-old adults do not drive. Model breakdown has occurred because some 83-year-old adults drive.

In Example 1, we found that $T = 56$ corresponds to *two* values of a. Recall from Section 7.1 that, for a parabola, each value of the response variable corresponds to two, one, or zero values of the explanatory variable.

Deciding Which Type of Model to Use

By constructing a scatterplot of some data, we can determine whether to use a linear model, a quadratic model, or neither.

Example 2   Selecting a Model

Consider the scatterplots of data shown in Figs. 17, 18, and 19 for situations 1, 2, and 3, respectively. For each situation, determine whether a linear model, quadratic model, or neither would describe the situation well.

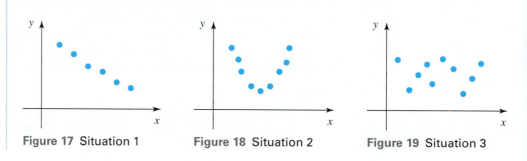

Figure 17 Situation 1

Figure 18 Situation 2

Figure 19 Situation 3

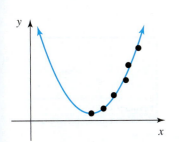

Figure 20 Scatterplot and a quadratic model

Solution

A linear model would describe the data for situation 1 well (see Fig. 17). It appears a quadratic model would fit the data for situation 2 well (see Fig. 18). Neither a linear model nor a quadratic model would fit the data well for situation 3 (see Fig. 19).

Throughout this course, we have seen that usually only a part of a model describes a situation well. For the scatterplot shown in Fig. 20, the sketched quadratic model fits the data well. At least part of the model that lies to the right of the vertex can be used to make reasonable estimates or predictions. The part of the model that lies to the left of the vertex may or may not describe the situation well. More data points could reveal whether this part of the parabola gives good estimates.

Using a Graphing Calculator to Make Predictions

When modeling a situation by using a quadratic equation, we can use TRACE on a graphing calculator to help us make estimates and predictions. We can also use "minimum" or "maximum" to locate the vertex of the model.

Table 7 Nike Clothing Revenues

Year	Clothing Revenue (billions of dollars)
2011	2.10
2012	2.99
2013	3.59
2014	3.94
2015	4.41
2016	4.75

Source: *Nike*

Example 3   Using a Graphing Calculator

The clothing revenues of Nike® are shown in Table 7 for various years. Let r be the annual clothing revenue (in billions of dollars) at t years since 2010. A linear model of the situation is

$$r = 0.51t + 1.84$$

A quadratic model of the situation is

$$r = -0.0584t^2 + 0.92t + 1.30$$

1. Which of the two models comes closer to the points in the scatterplot of the data?
2. Substitute a value for one of the variables in the quadratic model to predict the clothing revenue in 2021.
3. Use "maximum" on a graphing calculator to find the vertex of the quadratic model. What does it mean in this situation?
4. Which model would Nike hope gives better predictions?
5. Use "intersect" on a graphing calculator together with the quadratic model to estimate when the annual clothing revenue was $4 billion and when the annual clothing revenue will be $4 billion.

Solution

1. We use ZoomStat to get the window settings shown in Fig. 21. The linear model comes fairly close to the data points (see Fig. 22). The quadratic model comes quite close to the data points (see Fig. 23).

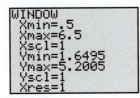

Figure 21 Window settings

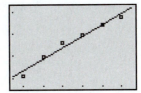

Figure 22 Linear model and scatterplot

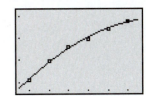

Figure 23 Quadratic model and scatterplot

2. Because 2021 is 11 years since 2010, we substitute the input 11 for t in the quadratic model and solve for r:

$$r = -0.0584(11)^2 + 0.92(11) + 1.30 \approx 4.35$$

The clothing revenue will be about $4.35 billion in 2021, according to the quadratic model.

3. Because the vertex is not displayed in Fig. 23, we Zoom Out on a graphing calculator to adjust the window settings so we can see the vertex (see Fig. 24). We also turn off the plotter so the scatterplot is no longer displayed. Then we use "maximum" to find the approximate vertex (see Fig. 24). See Appendix A.18 for graphing calculator instructions.

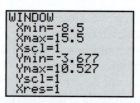 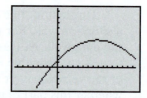 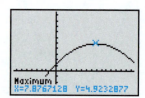

Figure 24 Use Zoom Out and "maximum" to find the vertex

The vertex is about $(7.88, 4.92)$. So, according to the quadratic model, the clothing revenue will be $4.92 billion in 2018, the largest annual clothing revenue for all time.

4. The part of the quadratic model to the right of its vertex predicts that the annual clothing revenue will decrease after 2018 (see Fig. 24). The linear model predicts that the annual clothing revenue will increase for all time (see Fig. 22). So, Nike would hope that the linear model gives better predictions.

5. We graph the quadratic model and the line $r = 4$ in the same coordinate system and use "intersect" to find the intersection points (see Fig. 25). We conclude that $r = 4$ when $t \approx 3.90$ and when $t \approx 11.85$. So, the clothing revenue was $4 billion in 2014 and will be $4 billion in 2022, according to the quadratic model.

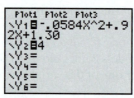 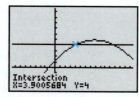 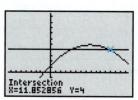

Figure 25 Find when the annual clothing revenue was or will be $4 billion

In Example 3, we used "maximum" on a graphing calculator to find the vertex of a quadratic model. In general, we can use "minimum" to find the vertex of a parabola that opens upward (see Fig. 26) and "maximum" to find the vertex of a parabola that opens downward (see Fig. 27). See Appendix A.18 for calculator instructions.

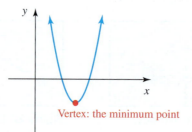

Figure 26 A parabola that opens upward

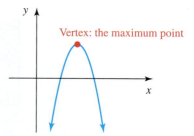

Figure 27 A parabola that opens downward

Group Exploration

Section Opener: Quadratic modeling

The percentages of car owners who owe more money on their vehicle than it is currently worth are shown in Table 8 for various years.

Table 8 Percentages of Car Owners Who Owe More Money on Their Vehicle Than It Is Currently Worth

Year	Percent
2004	37
2006	30
2008	28
2010	24
2012	24
2014	28
2016	31

Source: *J.D. Power*

Let p be the percentage of car owners who owe more money on their vehicle than it is currently worth at t years since 2000.

1. Use a graphing calculator to draw a scatterplot of the data.

2. Are the variables t and p approximately linearly related? Explain.

3. A *quadratic model* of the car-owner situation is $p = 0.262t^2 - 5.7t + 55.5$. Use a graphing calculator to draw the graph of the quadratic model and the scatterplot of the data in the same viewing window. Does the model fit the data well?

4. Use TRACE to estimate the percentage of car owners in 2013 who owed more money on their vehicle than it was worth.

5. Use TRACE to estimate in which years(s) 29% of car owners owed more money on their vehicle than it was worth.

6. Use TRACE to estimate in which year the smallest percentage of car owners owed more money on their vehicle than it was worth. What is that percentage?

Homework 7.2

For extra help ▶ MyLab Math   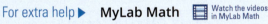 Watch the videos in MyLab Math

1. A scatterplot of a given situation is graphed in Fig. 28. Determine whether a linear model, a quadratic model, or neither type of model would be reasonable for modeling the data. Also, sketch a curve that you would use to model the data.

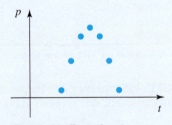

Figure 28 Exercise 1

2. A scatterplot of a given situation is graphed in Fig. 29. Determine whether a linear model, a quadratic model, or neither type of model would be reasonable for modeling the data. Also, sketch a curve that you would use to model the data.

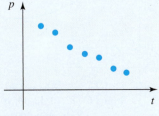

Figure 29 Exercise 2

3. A student believes the data listed in Table 9 suggest a quadratic relationship because the values of y increase and then decrease. Is the student correct? Explain. [**Hint:** Use a graphing calculator to draw a scatterplot of the data.]

Table 9 Is There a Quadratic Relationship?

x	y	x	y
0	1	6	18
1	2	7	7
2	3	8	3
3	7	9	2
4	18	10	1
5	30		

4. A student believes the data listed in Table 10 suggest a quadratic relationship because the values of y decrease and then increase. Is the student correct? Explain. [**Hint:** Use a graphing calculator to draw a scatterplot of the data.]

Table 10 Is There a Quadratic Relationship?

x	y	x	y
0	34	6	17
1	33	7	28
2	32	8	32
3	28	9	33
4	17	10	34
5	5		

5. DATA The numbers of traumatic brain injuries suffered by service members while serving in the U.S. military are shown in Table 11 for various years.

Table 11 Traumatic Brain Injuries

Year	Number of Traumatic Brain Injuries (thousands)
2006	17.0
2007	23.2
2008	28.5
2009	29.0
2010	29.4
2011	32.9
2012	30.8
2013	27.6
2014	25.0

Source: *Congressional Research Service*

Let n be the number (in thousands) of traumatic brain injuries in the year that is t years since 2000. A quadratic model is sketched in Fig. 30.

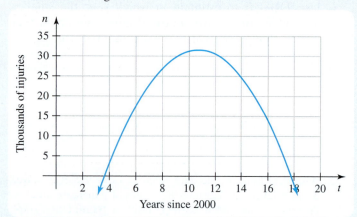

Figure 30 Exercise 5

a. Trace the graph in Fig. 30 carefully onto a piece of paper. Then, on the same graph, draw a scatterplot of the data. Does the model fit the data well?

b. Estimate the number of traumatic brain injuries in 2015.

c. Estimate in which year(s) there were 14 thousand traumatic brain injuries.

d. What is the vertex of the model? What does it mean in this situation?

e. What are the t-intercepts? What do they mean in this situation?

6. Table 12 lists the percentages of Americans of various age groups who listen to talk radio.

Table 12 Percentages of Americans Who Listen to Talk Radio

Age Group (years)	Age Used to Represent Age Group (years)	Percent
12–17	14.5	6
18–34	26.0	16
35–44	39.5	25
45–54	49.5	26
55–64	59.5	21
over 64	75.0	6

Sources: *Talkers Magazine; Mediamark*

Let p be the percentage of Americans of age a years who listen to talk radio. A quadratic model is graphed in Fig. 31.

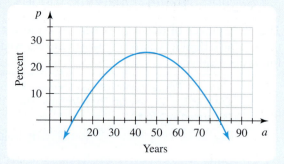

Figure 31 Exercise 6

a. Trace the graph in Fig. 31 carefully onto a piece of paper. Then, on the same graph, draw a scatterplot of the data. Does the model fit the data well?

b. Estimate the percentage of 20-year-old Americans who listen to talk radio.

c. Estimate the age(s) at which 20% of Americans listen to talk radio.

d. What is the vertex of the model? What does it mean in this situation?

e. What are the a-intercepts? What do they mean in this situation?

7. The U.S. market shares of General Motors (GM) are shown in Table 13 for various years.

Table 13 GM's U.S. Market Shares

Year	Market Share (percent)
2004	26.9
2006	23.9
2008	21.9
2010	18.8
2012	17.6
2015	17.7

Sources: *Bank of America; Merrill Lynch*

Let p be GM's U.S. market share (in percent) at t years since 2000. A quadratic model is graphed in Fig. 32.

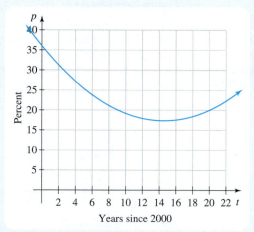

Figure 32 Exercise 7

a. Trace the graph in Fig. 32 carefully onto a piece of paper. Then, on the same graph, draw a scatterplot of the data. Does the model fit the data well?

b. Predict GM's U.S. market share in 2022.

c. Estimate when GM's U.S. market share was 20% or will be 20%.

d. What is the *p*-intercept? What does it mean in this situation?

e. Estimate when GM's U.S. market share was the least. What was that market share?

8. DATA. Percentages of 10th graders who have used marijuana in the past 30 days are shown in Table 14 for various years.

Table 14 Percentages of 10th Graders Who Have Used Marijuana in the Past 30 Days

Year	Percent
2008	13.8
2009	15.9
2010	16.7
2011	17.6
2012	17.0
2013	18.0
2014	16.6
2015	14.8

Source: *Monitoring the Future Study, University of Michigan*

Let *p* be the percentage of 10th graders who have used marijuana in the past 30 days at *t* years since 2000. A quadratic model is graphed in Fig. 33.

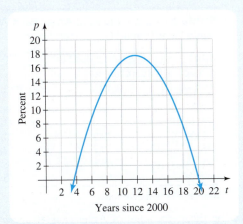

Figure 33 Exercise 8

a. Trace the graph in Fig. 33 carefully onto a piece of paper. Then, on the same graph, draw a scatterplot of the data. Does the model fit the data well?

b. Predict the percentage of 10th graders who will have used marijuana in the past 30 days in 2019.

c. Estimate in which year(s) 16% of 10th graders have used marijuana in the past 30 days.

d. What are the *t*-intercepts? What do they mean in this situation?

e. Estimate the year when the percentage was the greatest. What was that percentage?

9. The percentages of married people who reach major anniversaries are listed in Table 15.

Table 15 Percentages of Married People Who Reach Major Anniversaries

Anniversary	Percent
5th	82
10th	65
15th	52
25th	33
35th	20
50th	5

Source: *U.S. Census Bureau*

Let *p* be the percentage of married people who reach their *a*th anniversary. A quadratic model is graphed in Fig. 34.

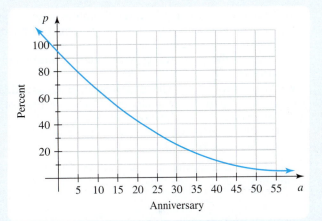

Figure 34 Exercise 9

a. Trace the graph in Fig. 34 carefully onto a piece of paper. Then, on the same graph, draw a scatterplot of the data. Does the model fit the data well?

b. What is the *p*-intercept? What does it mean in this situation?

c. Does the part of the parabola that lies to the right of the vertex describe the situation well? Explain.

d. What percentage of married people reach their 20th anniversary?

e. What anniversary do exactly half of marriages reach?

10. The average prices of smartphones worldwide are shown in Table 16 for various years.

Table 16 Average Prices of Smartphones

Year	Average Price (dollars)
2011	420
2012	381
2013	333
2014	310
2015	305

Source: *Consumer Technology*

Let p be the average price (in dollars) of smartphones at t years since 2010. A quadratic model is graphed in Fig. 35.

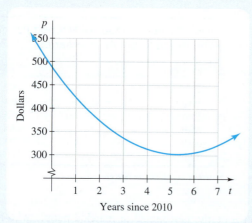

Figure 35 Exercise 10

a. Trace the graph in Fig. 35 carefully onto a piece of paper. Then, on the same graph, draw a scatterplot of the data. Does the model fit the data well?

b. What is the p-intercept? What does it mean in this situation?

c. What is the vertex? What does it mean in this situation?

d. Does the part of the model that lies to the right of the vertex describe the situation well? Explain.

e. Use the model to estimate when the average price of smartphones was $373.

11. DATA The number of wiretap authorizations has increased significantly since 9/11 (see Table 17).

Table 17 Number of Wiretap Authorizations

Year	Number of Wiretaps (thousands)
2004	1.7
2005	1.8
2006	1.8
2007	2.2
2008	1.9
2009	2.4
2010	3.2
2011	2.7
2012	3.4
2013	3.6
2014	3.6
2015	4.1

Source: *United States Courts*

Let n be the number (in thousands) of wiretap authorizations in the year that is t years since 2000. A scatterplot of the data and a quadratic model are graphed in Fig. 36.

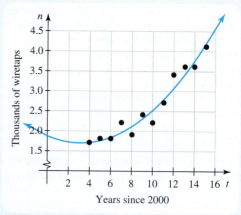

Figure 36 Exercise 11

a. Use the quadratic model to estimate the number of wiretaps in 2008.

b. What was the actual number of wiretaps in 2008?

c. Is the result you found in part (a) an underestimate or an overestimate? Explain how you can tell this from the graph of the scatterplot and the model. Compute the error in your estimate.

12. For this exercise, refer to Exercise 11, including Table 17 and Fig. 36.

a. Use the quadratic model to estimate the number of wiretaps in 2010.

b. What was the actual number of wiretaps in 2010?

c. Is the result you found in part (a) an underestimate or an overestimate? Explain how you can tell this from the graph of the scatterplot and the model. Compute the error in your estimate.

13. Table 18 shows the percentages of Americans of various age groups who have ever played organized soccer.

Table 18 Percentages of Americans Who Have Ever Played Organized Soccer

Age Group (years)	Age Group Used to Represent Age Group (years)	Percent
12–17	14.5	39.6
18–24	21	31.1
25–34	29.5	21.3
35–44	39.5	17.6
45–54	49.5	8.4
over 54	60	5.3

Source: *ESPN sports poll*

Let p be the percentage of Americans at age a years who have ever played organized soccer. A model of the situation is $p = 0.01a^2 - 1.5a + 58.7$.

a. Use a graphing calculator to draw the graph of the model and a scatterplot of the data in the same viewing window. Does the model fit the data well?

b. Substitute a value for one of the variables in the model's equation to estimate the percentage of 18-year-old Americans who have ever played organized soccer.

c. Use "minimum" on a graphing calculator, along with the quadratic model, to find the vertex of the model. What does it mean in this situation? Does the part of the parabola that lies to the right of the vertex describe the situation well? Explain.

d. Use TRACE on a graphing calculator to estimate at what age 12% of Americans have ever played organized soccer.

14. DATA The percentages of the U.S. population that are foreign born are shown in Table 19 for various years.

Table 19 Percentages of U.S. Population That Are Foreign Born

Year	Percent
1930	11.6
1940	8.8
1950	6.9
1960	5.4
1970	4.7
1980	6.2
1990	7.9
2000	11.1
2010	12.9
2015	13.7

Source: *U.S. Census Bureau*

Let p be the percentage of the U.S. population that is foreign born at t years since 1900. A model of the situation is $p = 0.00409t^2 - 0.554t + 24.34$.

a. Use a graphing calculator to draw the graph of the model and a scatterplot of the data in the same viewing window. Does the model fit the data well?

b. Substitute a value for one of the variables in the model's equation to predict the percentage of the U.S. population that will be foreign born in 2022.

c. Use "minimum" on a graphing calculator to estimate the percentage of the U.S. population that was foreign born when it was the lowest. Does that match what really happened?

d. Use TRACE on a graphing calculator to predict when 17% of the U.S. population will be foreign born.

15. DATA The numbers of international students in the United States are shown in Table 20 for various years.

Table 20 Numbers of International Students

Year	Number of Students (millions)
2009	0.67
2010	0.69
2011	0.72
2012	0.76
2013	0.82
2014	0.89
2015	0.97
2016	1.04

Source: *Institute of International Students*

Let n be the number (in millions) of international students at t years since 2000. A quadratic model of the situation is $n = 0.00536t^2 - 0.08t + 0.95$. A linear model of the situation is $n = 0.054t + 0.14$.

a. Use a graphing calculator to draw graphs of the quadratic and linear models and a scatterplot of the data in the same viewing window. Which model comes closer to the points in the scatterplot?

b. For each model, substitute a value for one of the variables in the model's equation to predict the number of international students in 2022. Refer to the models' graphs to explain why the quadratic model's prediction is so much larger than the linear model's prediction.

c. Which of the two models estimates that the number of international students was decreasing for years before 2007?

d. Which of the two models estimates that there was a negative number of international students in each year before 1997? Has model breakdown occurred for those years? Explain.

e. Use TRACE on a graphing calculator, along with the quadratic model, to predict when there will be 1.5 million international students.

16. DATA The numbers of pages in the IRS 1040 instruction booklet are shown in Table 21 for various years.

Table 21 Numbers of Pages in the IRS 1040 Instruction Booklet

Year	Number of Pages
1945	4
1955	16
1965	17
1975	39
1985	52
1995	84
2005	142
2016	241

Source: *National Taxpayers Union*

Let n be the number of pages in the IRS 1040 instruction booklet at t years since 1900. A quadratic model of the situation is $n = 0.064t^2 - 7.3t + 214.7$. A linear model of the situation is $n = 2.97t - 163.54$.

a. Use a graphing calculator to draw graphs of the quadratic and linear models and a scatterplot of the data in the same viewing window. Which model comes closer to the points in the scatterplot?

b. For each model, substitute a value for one of the variables in the model's equation to predict the number of pages in the booklet in 2021. Refer to the models' graphs to explain why the quadratic model's prediction is so much larger than the linear model's prediction.

c. Which of the two models estimates that the number of pages in the booklet was decreasing for years before 1957? Has model breakdown likely occurred for that model for those years? Explain.

d. Which of the two models estimates that there was a negative number of pages in the booklet in each year before 1955? Has model breakdown occurred for that model for those years? Explain.

e. Use TRACE on a graphing calculator, along with the quadratic model, to predict in which year there will be a total of 275 pages in the booklet.

Concepts

17. Suppose that a situation can be modeled well by a parabola, where t is the explanatory variable and p is the response variable. If a data point (c, d) is below the parabola, does the model underestimate or overestimate the value of p when $t = c$? Explain.

18. Suppose that a situation can be modeled well by a parabola, where t is the explanatory variable and p is the response variable. If a data point (c, d) is above the parabola, does the model underestimate or overestimate the value of p when $t = c$? Explain.

19. Which is more desirable, finding a quadratic model whose graph contains several (but not all) data points or finding a quadratic model whose graph does not contain any data points but comes close to all data points? Include in your discussion some sketches of scatterplots and quadratic models.

20. If a quantity increases from one year to the next, explain how in some cases it might be better to model the situation by a quadratic equation rather than a linear equation. Include an example of a scatterplot in your explanation.

21. Describe the meaning of *quadratic model* in your own words. (See page 9 for guidelines on writing a good response.)

22. Explain why the vertex and any intercepts of a quadratic model are often of special interest. (See page 9 for guidelines on writing a good response.)

Related Review

23. The revenues of Whole Foods are shown in Table 22 for various years.

Table 22 Whole Foods's Revenues

Year	Revenue (billions of dollars)
2006	5.6
2008	8.0
2010	9.0
2012	11.7
2014	14.2
2016	15.7

Source: *Whole Foods*

Let s be the annual revenue (in billions of dollars) of Whole Foods at t years since 2005.

a. Use a graphing calculator to draw a scatterplot of the data.

b. Find an equation of a model to describe the data.

c. What is the s-intercept? What does it mean in this situation?

d. Predict the revenue in 2020.

e. Predict when the annual revenue will be $22 billion.

24. The average personal incomes of Americans are shown in Table 23 for various years.

Table 23 Average Personal Incomes of Americans

Year	Average Personal Income (thousands of dollars)
1990	19.4
1995	23.3
2000	30.3
2005	35.4
2010	40.5
2015	47.7

Source: *U.S. Bureau of Economic Analysis*

Let I be the personal income (in thousands of dollars) of an American for the year that is t years since 1990.

a. Use a graphing calculator to draw a scatterplot of the data.

b. Find an equation of a model to describe the data.

c. What is the slope? What does it mean in this situation?

d. Predict the average personal income of Americans in 2020.

e. Predict the average personal income of Americans *per work hour* in 2022. Assume that a typical work schedule consists of 8 hours per day, 5 days per week, and 50 weeks per year.

25. Explain how constructing a scatterplot can help you determine whether to describe some data by a linear model, a quadratic model, or neither.

26. Suppose both a linear model and a quadratic model fit the points in a scatterplot of data well and the explanatory variable is time (in years). Will using the two models to make predictions for a certain year give similar results?

Expressions, Equations, and Graphs

Perform the indicated instruction. Then use words such as linear, quadratic, one variable, *and* two variables *to describe the expression, equation, or system.*

27. Solve the following:

$$3x - 2y = -4$$
$$4x + 5y = 33$$

28. Simplify $4x + 9(2x - 3)$.

29. Graph $3x - 2y = -4$ by hand.

30. Solve $4x + 9(2x - 3) = 0$.

7.3 Adding and Subtracting Polynomials

Objectives

» Describe the meaning of *term, monomial, polynomial, degree, coefficient,* and *like terms*.

» Combine like terms.

» Add polynomials.

» Subtract polynomials.

» Use a sum or a difference of two polynomials to model an authentic situation.

In Sections 7.1 and 7.2, we worked with quadratic equations in two variables, such as $y = 2x^2 - 8x + 6$, which is a special type of polynomial equation. We will now begin working with *polynomial expressions*, or *polynomials* for short. For example, $2x^2 - 8x + 6$ is a polynomial.

Polynomials

Recall from Section 4.1 that a **term** is a constant, a variable, or a product of a constant and one or more variables raised to powers. Here are some examples of terms:

$$5x^4 \qquad x \qquad 17 \qquad -6x^8y^2 \qquad -x^{-3}$$

A **monomial** is a constant, a variable, or a product of a constant and one or more variables raised to *counting-number* powers. Here are some examples of monomials:

$$-2x^7 \qquad y \qquad -3 \qquad \frac{2}{7}x^4y^9 \qquad x^3$$

A **polynomial**, or **polynomial expression**, is a monomial or a sum of monomials. Here are some examples of polynomials:

$$3x^2 + 5x + 2 \qquad 9x^5y^3 - 6x^3y^2 - xy + 5 \qquad -2x + 5 \qquad 7 \qquad 3x^4$$

The polynomial $7x^3 - 5x^2 + 8x - 1$ is a *polynomial in one variable*. It has four terms: $7x^3, -5x^2, 8x$, and -1. We usually write polynomials in one variable so that the exponents of the terms decrease from left to right, an arrangement called **decreasing order**. If a polynomial contains more than one variable, we usually write the polynomial so that the exponents of one of the variables decrease from left to right.

The **degree of a term in one variable** is the exponent on the variable. For example, the degree of the term $2x^7$ is 7. The **degree of a term in two or more variables** is the sum of the exponents on the variables. For example, the term $5x^4y^2$ has degree $4 + 2 = 6$. The **degree of a polynomial** is the largest degree of any nonzero term of the polynomial. For example, the polynomial $7x^4 - 2x^2 + 5$ has degree 4. A constant polynomial such as 5 has degree 0.

Polynomials with degrees 1, 2, or 3 have special names:

Degree	Name	Examples
1	linear (first-degree) polynomial	$-3x + 7, 2x$
2	quadratic (second-degree) polynomial	$8x^2 - 5x + 2, x^2 - 4$
3	cubic (third-degree) polynomial	$4x^3 + 2x^2 - 6x + 5, -6x^3 + x$

Example 1   Describing Polynomials

Use words such as *linear, quadratic, cubic, polynomial, degree, one variable,* and *two variables* to describe the expression.

1. $5x^2 - 3x + 7$ **2.** $-7x^3 + 8x^2 - 1$ **3.** $4a^3b^2 - 2a^2b^2 + 9ab^2$

Solution

1. The term $5x^2$ has degree 2, which is larger than the degrees of the other terms. So, $5x^2 - 3x + 7$ is a quadratic (or second-degree) polynomial in one variable.

2. The term $-7x^3$ has degree 3, which is larger than the degrees of the other terms. So, $-7x^3 + 8x^2 - 1$ is a cubic (or third-degree) polynomial in one variable.

3. The term $4a^3b^2$ has degree $3 + 2 = 5$, which is larger than the degrees of the other terms. So, $4a^3b^2 - 2a^2b^2 + 9ab^2$ is a fifth-degree polynomial in two variables.

Combining Like Terms

Recall from Section 4.2 that the **coefficient** of a term is the constant factor of the term. For the term $-3x^2$, the coefficient is -3. For the term x^3, the coefficient is 1 because $x^3 = 1x^3$. For the term 6, the coefficient is 6. The coefficients of a polynomial are the coefficients of the terms. For example, the coefficients of the polynomial $7x^3 - 5x^2 + 8x - 1$ are 7, -5, 8, and -1. The **leading coefficient** of a polynomial is the coefficient of the term with the largest degree. For $7x^3 - 5x^2 + 8x - 1$, the leading coefficient is 7.

Recall from Section 4.2 that **like terms** are either constant terms or variable terms that contain the same variable(s) raised to exactly the same power(s). For example, the terms $9x^2y^7$ and $4x^2y^7$ are like terms because both terms have an x with the exponent 2 and a y with the exponent 7. Recall from Section 4.2 that if terms are not like terms, we say they are **unlike terms**. For example, $3x^3$ and $5x^2$ are unlike terms because the exponents of x are different.

Recall also from Section 4.2 that we can combine like terms of linear polynomials by using the distributive law. For example,

$$3x + 5x = (3 + 5)x = 8x$$

Likewise, we can *combine like terms* of polynomials of higher degree, such as $4x^2 + 6x^2$, by using the distributive law:

$$4x^2 + 6x^2 = (4 + 6)x^2 = 10x^2$$

We can find $4x^2 + 6x^2$ in one step by adding the coefficients:

$$4x^2 + 6x^2 = 10x^2$$
$$4 + 6 = 10$$

We can also find $6x^3 - 2x^3$ by adding the coefficients:

$$6x^3 - 2x^3 = 4x^3$$
$$6 + (-2) = 4$$

However, we can't add the coefficients in $8x^4 + 3x^2$ because $8x^4$ and $3x^2$ are not like terms. There is no helpful way to use the distributive law for unlike terms.

Combining Like Terms

To combine like terms, add the coefficients of the terms.

Example 2   Combining Like Terms

Combine like terms when possible.

1. $7x^2 + 5x^2$
2. $-4x^3y^2 - 6x^3y^2$
3. $5a^2b^4 + 3a^4b^2$
4. $-6x^2 + x^2$

Solution

1. $7x^2 + 5x^2 = 12x^2$ *Add coefficients: $7 + 5 = 12$*
2. $-4x^3y^2 - 6x^3y^2 = -10x^3y^2$ *Add coefficients: $-4 + (-6) = -10$*
3. Although $5a^2b^4$ and $3a^4b^2$ have the same variables, the variables do not have the same exponents. They are not like terms, so we cannot combine them.
4. $-6x^2 + x^2 = -6x^2 + 1x^2 = -5x^2$ *Add coefficients: $-6 + 1 = -5$*

▶

Example 3   Combining Like Terms

Combine like terms: $8x - 7x^2 - 2x + 3x^2$.

Solution

$$8x - 7x^2 - 2x + 3x^2 = -7x^2 + 3x^2 + 8x - 2x \qquad \text{Rearrange terms.}$$
$$= -4x^2 + 6x \qquad \text{Combine like terms.}$$

▶

Addition of Polynomials

Now that we know how to combine like terms, we can add polynomials.

> **Adding Polynomials**
>
> To add polynomials, combine like terms.

Example 4   Adding Polynomials

Find the sum.

1. $\left(4x^2 + 3x\right) + \left(6x^2 - 2x\right)$
2. $\left(5x^3 - 3x^2 + 8x - 1\right) + \left(7x^3 - x^2 - 3x + 4\right)$
3. $\left(2x^3 - 5x\right) + \left(7x^2 - 4x\right)$
4. $\left(4x^2 + 3xy - 5y^2\right) + \left(3x^2 - 7xy + 9y^2\right)$

Solution

1. $\left(4x^2 + 3x\right) + \left(6x^2 - 2x\right) = 4x^2 + 6x^2 + 3x - 2x$ *Rearrange terms.*
$$= 10x^2 + x \qquad\qquad \textit{Combine like terms.}$$

2. $\left(5x^3 - 3x^2 + 8x - 1\right) + \left(7x^3 - x^2 - 3x + 4\right)$
$$= 5x^3 + 7x^3 - 3x^2 - x^2 + 8x - 3x - 1 + 4 \qquad \textit{Rearrange terms.}$$
$$= 12x^3 - 4x^2 + 5x + 3 \qquad\qquad\qquad \textit{Combine like terms.}$$

3. $\left(2x^3 - 5x\right) + \left(7x^2 - 4x\right) = 2x^3 + 7x^2 - 5x - 4x$ *Rearrange terms.*
$$= 2x^3 + 7x^2 - 9x \qquad \textit{Combine like terms.}$$

We use a graphing calculator table to verify the work (see Fig. 37).

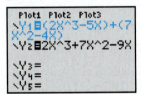 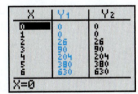

Figure 37 Verify the work

4. $\left(4x^2 + 3xy - 5y^2\right) + \left(3x^2 - 7xy + 9y^2\right)$
$$= 4x^2 + 3x^2 + 3xy - 7xy - 5y^2 + 9y^2 \qquad \textit{Rearrange terms.}$$
$$= 7x^2 - 4xy + 4y^2 \qquad\qquad\qquad \textit{Combine like terms.}$$

Subtraction of Polynomials

In Section 4.1, we discussed how to subtract expressions; those expressions were actually linear polynomials. In this section, we take similar steps to subtract polynomials of higher degree.

Example 5   Subtracting Polynomials

Find the difference.

1. $\left(8x^3 + 5x^2\right) - \left(6x^3 + 2x^2\right)$
2. $\left(2a^2 - 5ab + 7b^2\right) - \left(6a^2 - 4ab + 3b^2\right)$

Solution

1. To begin, we write $\left(8x^3 + 5x^2\right) - \left(6x^3 + 2x^2\right)$ as $\left(8x^3 + 5x^2\right) - 1\left(6x^3 + 2x^2\right)$ and then distribute the -1:

$$\left(8x^3 + 5x^2\right) - \left(6x^3 + 2x^2\right) = \left(8x^3 + 5x^2\right) - 1\left(6x^3 + 2x^2\right) \quad \textit{a − b = a − 1b}$$
$$= 8x^3 + 5x^2 - 6x^3 - 2x^2 \qquad \textit{Distributive law}$$
$$= 8x^3 - 6x^3 + 5x^2 - 2x^2 \qquad \textit{Rearrange terms.}$$
$$= 2x^3 + 3x^2 \qquad\qquad\qquad \textit{Combine like terms.}$$

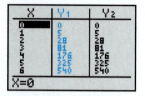

Figure 38 Verify the work

We use a graphing calculator table to verify the work (see Fig. 38).

2. $\left(2a^2 - 5ab + 7b^2\right) - \left(6a^2 - 4ab + 3b^2\right)$

$= \left(2a^2 - 5ab + 7b^2\right) - 1\left(6a^2 - 4ab + 3b^2\right)$ *a − b = a − 1b*

$= 2a^2 - 5ab + 7b^2 - 6a^2 + 4ab - 3b^2$ *Distributive law*

$= 2a^2 - 6a^2 - 5ab + 4ab + 7b^2 - 3b^2$ *Rearrange terms.*

$= -4a^2 - ab + 4b^2$ *Combine like terms.*

In Problem 1 of Example 5, a key step in finding $\left(8x^3 + 5x^2\right) - \left(6x^3 + 2x^2\right)$ is to write this difference as $\left(8x^3 + 5x^2\right) - 1\left(6x^3 + 2x^2\right)$, so that we can distribute the -1.

Subtracting Polynomials

To subtract polynomials, first distribute -1 and then combine like terms.

Using a Sum or Difference to Model an Authentic Situation

Suppose that A and B represent quantities. Then $A + B$ represents the sum of the two quantities.

The difference $A - B$ tells us how much more there is of one quantity than the other. For example, suppose $A = 7$ and $B = 2$. Then $A - B = 7 - 2 = 5$ tells us A is 5 more than B. Now suppose $A = 2$ and $B = 7$. Then $A - B = 2 - 7 = -5$ tells us A is 5 less than B.

The Meaning of the Sign of a Difference

If a difference $A - B$ is positive, then A is more than B. If a difference $A - B$ is negative, then A is less than B.

In Example 6, we will add and subtract polynomials that represent authentic quantities and will interpret the meanings of the sum and difference.

Example 6   Describing Quantities by Combining Polynomials

The numbers of commercial and educational FM radio stations are shown in Table 24 for various years.

Table 24 Numbers of Commercial and Educational FM Radio Stations

| Year | Number of FM Stations (thousands) | |
	Commercial	Educational
2005	6.2	2.6
2007	6.3	2.9
2009	6.5	3.1
2011	6.5	3.6
2013	6.6	4.0
2015	6.7	4.1

Source: *Federal Communications Commission*

Let n be the number of FM radio stations (in thousands) at t years since 2000. Reasonable models of the number of commercial and educational stations are

$$n = 0.049t + 5.98 \quad \textit{Commercial}$$
$$n = 0.16t + 1.77 \quad \textit{Educational}$$

1. Check that the models fit the data well.
2. Find the sum of the polynomials $0.049t + 5.98$ and $0.16t + 1.77$. What does your result represent?
3. Evaluate the result in Problem 2 for $t = 22$. What does your result mean in this situation?
4. Find the difference of the polynomials $0.049t + 5.98$ and $0.16t + 1.77$. What does your result mean in this situation?
5. Evaluate the result in Problem 4 for $t = 22$. What does your result mean in this situation?

Solution

1. To keep track of the two types of data points, we use square "Marks" for the commercial scatterplot and plus-sign "Marks" for the educational scatterplot (see Fig. 39). The models appear to fit the data quite well.

2. $$(0.049t + 5.98) + (0.16t + 1.77) = 0.209t + 7.75$$

   The polynomial $0.209t + 7.75$ represents the *total* number (in thousands) of commercial and educational FM stations at t years since 2000.

3. We substitute 22 for t in the polynomial $0.209t + 7.75$:

$$0.209(22) + 7.75 \approx 12.35$$

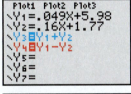

Figure 39 The models fit the data well

   The total number of commercial and educational FM radio stations in 2022 will be about 12.4 thousand stations.

4. $(0.049t + 5.98) - (0.16t + 1.77)$

$$
\begin{aligned}
&= (0.049t + 5.98) - 1(0.16t + 1.77) &&a - b = a - 1b\\
&= 0.049t + 5.98 - 0.16t - 1.77 &&\textit{Distributive law}\\
&= -0.111t + 4.21 &&\textit{Combine like terms.}
\end{aligned}
$$

   The polynomial $-0.111t + 4.21$ represents the *difference* (in thousands) of commercial and educational FM radio stations at t years since 2000.

5. We substitute 22 for t in the polynomial $-0.111t + 4.21$:

$$-0.111(22) + 4.21 \approx 1.77$$

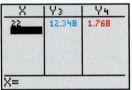

Figure 40 Verify the work

   In 2022, the difference in the number of commercial and educational FM radio stations will be about 1.8 thousand stations. In other words, there will be 1.8 thousand more commercial stations than educational stations.

   We can use Y_n references on a graphing calculator to verify our work in Problems 3 and 5 (see Fig. 40). See Appendix A.22 for graphing calculator instructions.

Group Exploration

DATA, Section Opener: Combining expressions that represent authentic quantities

In Example 1 of Section 6.4, we used the linear expression $22.03t + 268.63$ to model the number of part-time college instructors and the linear expression $13.88t + 468.48$ to model the number of full-time college instructors both, in thousands, at t years since 1990 (see Table 25).

1. Find the sum of the expressions $22.03t + 268.63$ and $13.88t + 468.48$. What does your result represent?

2. Evaluate the result you found in Problem 1 for $t = 31$. What does your result mean in this situation?

3. Find the difference of the expressions $22.03t + 268.63$ and $13.88t + 468.48$. What does your result represent?

4. Evaluate the result you found in Problem 3 for $t = 31$. What does your result mean in this situation? [**Hint:** The difference $7 - 2 = 5$ means that 7 is 5 more than 2.]

Table 25 Numbers of Part-Time and Full-Time Instructors

Year	Number of Instructors (thousands)	
	Part Time	Full Time
1997	421	569
2001	495	618
2005	615	676
2009	710	729
2013	754	791

Source: *U.S. National Center for Education Statistics*

Homework 7.3

For extra help ▶   MyLab Math   📷 Watch the videos in MyLab Math

Use words such as linear, quadratic, cubic, polynomial, degree, one variable, *and* two variables *to describe the expression.*

1. $3x^2 - 4x + 2$

2. $9x^2 + 5x^3 - 1$

3. $-7x^3 - 9x - 4$

4. $-4x^3 + 6x^2 - 9x + 8$

5. $3p^5q^2 - 5p^3q^3 + 7pq^4$

6. $8m^7n + 2m^4n^3 + 7m^2n^4$

Combine like terms when possible. Use a graphing calculator table to verify your work when possible.

7. $2x + 4x$

8. $3x + 7x$

9. $-4x - 9x$

10. $6x - 8x$

11. $3t^2 + 5t^2$

12. $-4a^2 + 9a^2$

13. $-8a^4b^3 - 3a^4b^3$

14. $9x^2y^5 - 2x^2y^5$

15. $4x^2 + x^2$

16. $-8x^2 - x^2$

17. $7x^2 - 3x$

18. $6x^2 + 2x$

19. $5b^3 - 8b^3$

20. $-4m^5 + 2m^5$

21. $-x^6 + 7x^6$

22. $-x^4 + 6x^4$

23. $2t^3w^5 + 4t^5w^3$

24. $4a^7b^2 - 7a^2b^7$

25. $-2.5p^4 + 9.9p^4$

26. $-3.05y^3 - 7.38y^3$

Perform the indicated operations. Use a graphing calculator table to verify your work.

27. $8x^2 + 2x - 3x^2 - 5x$

28. $-7x + 9x^2 + 4x - x^2$

29. $9x - 4x^2 + 5x - 2 + 3x^2 - 6$

30. $6x^2 - 5x - x^2 - 2x - 1$

31. $5x + 8x^3 - x^2 + 4x^3 + 2x - x^3$

32. $-5x^2 - 9x + 8x^3 - x + 1 - 7x^3$

33. $20.3t^2 - 5.4t - 45.1t^3 - 3.6t + 93.8t^2$

34. $-5.99k^2 + 2.35 - 6.911k^3 + 7.91k^2 - 1.99k^3$

Perform the addition. Use a graphing calculator table to verify your work when possible.

35. $\left(5x^2 - 4x - 2\right) + \left(-9x^2 - 3x + 8\right)$

36. $\left(-6x^2 + 2x - 5\right) + \left(4x^2 + 3x - 2\right)$

37. $\left(4x^2 - 3x + 2\right) + \left(-4x^2 + 3x - 2\right)$

38. $\left(-5x^2 + 2x - 7\right) + \left(5x^2 - 2x + 7\right)$

39. $\left(4x^3 - 7x^2 + 2x - 9\right) + \left(-7x^3 - 3x^2 + 5x - 2\right)$

40. $\left(-5x^3 + 4x^2 - 6x + 4\right) + \left(-2x^3 - 8x^2 + x - 5\right)$

41. $\left(5a^3 - a\right) + \left(8a^3 - 3a^2 - 1\right)$

42. $\left(2t^2 - 8t + 4\right) + \left(-7t^3 + 2t^2\right)$

43. $\left(5a^2 - 3ab + 7b^2\right) + \left(-4a^2 - 6ab + 2b^2\right)$

44. $\left(3m^2 + 6mn - 4n^2\right) + \left(6m^2 - 7mn - 2n^2\right)$

45. $\left(14.1x^3 - 7.9x^2 - 4.8x + 31.9\right) + \left(-8.2x^3 + 28.8x^2 - 9.5x + 32.2\right)$

46. $\left(-76.3x^3 + 8.2x^2 - 14.4x - 57.1\right) + \left(-19.3x^3 - 3.8x^2 - 3.2x + 80.9\right)$

For Exercises 47–58, find the difference. Use a graphing calculator table to verify your work when possible.

47. $(-3x + 8) - (4x - 3)$

48. $(2x - 4) - (-3x - 1)$

49. $\left(2x^2 + 5x - 1\right) - \left(4x^2 + 9x - 7\right)$

50. $\left(3x^2 - 7x + 2\right) - \left(2x^2 - x + 1\right)$

51. $\left(6y^3 - 2y^2 - 4y + 5\right) - \left(5y^3 - 8y^2 - 5y + 3\right)$

52. $\left(-3t^3 + 7t^2 - 2t + 4\right) - \left(4t^3 + 6t^2 - t - 4\right)$

53. $\left(5x^3 - 9x^2 + 2x - 4\right) - \left(5x^3 - 9x^2 + 2x - 4\right)$

54. $\left(-3x^3 + 2x^2 + 7x - 1\right) - \left(-3x^3 + 2x^2 + 7x - 1\right)$

55. $\left(3x^2 + 7xy - 2y^2\right) - \left(5x^2 - 4xy - 3y^2\right)$

56. $\left(-7a^2 - 4ab + 3b^2\right) - \left(-5a^2 + ab - 6b^2\right)$

57. $\left(2.54x^2 + 6.29x - 7.99\right) - \left(-4.21x^2 - 8.45x + 9.29\right)$

58. $\left(-7.72x^2 - 5.38x + 6.13\right) - \left(2.49x^2 - 3.36x - 8.83\right)$

59. Find the sum of the polynomials $2x^2 + 5x$ and $3x + 9$.

60. Find the sum of the polynomials $-4x^2 + x$ and $x^2 - 1$.

61. Find the difference of the polynomials $5x^3 - 5x + 8$ and $8x^2 - 3x - 3$.

62. Find the difference of the polynomials $4x^3 + 6x^2 - x$ and $7x^3 - 4x - 2$.

63. Subtract $3x^2 - 5x + 4$ from $2x^2 + 6x - 1$.

64. Subtract $4x^2 + 2x - 4$ from $7x^2 - 9x + 5$.

Perform the indicated operations. Use a graphing calculator to verify your work.

65. $\left(6x^3 - 2x\right) - \left(4x^2 - 7x + 2\right) + \left(3x^3 - x^2\right)$

66. $\left(4x^3 - 7x^2\right) + \left(5x^2 - 6x\right) - \left(7x^3 + 4x^2 - 5\right)$

67. 📊 In Exercise 1 of Homework 6.4, the number n (in thousands) of people who have earned a bachelor's degree is modeled by the system

$$n = 24.27t - 1690.26 \quad \textit{Women}$$
$$n = 16.09t - 1039.32 \quad \textit{Men}$$

where t is the number of years since 1900 (see Table 26).

Table 26 Women and Men Who Have Earned a Bachelor's Degree

Year	Number of People Who Have Earned a Bachelor's Degree (thousands)	
	Women	Men
1995	634	526
2000	712	532
2005	855	631
2010	982	734
2014	1082	813

Source: *U.S. National Center for Education Statistics*

a. Find the sum of the polynomials $24.27t - 1690.26$ and $16.09t - 1039.32$. What does your result represent?

b. Evaluate the result you found in part (a) for $t = 121$. What does your result mean in this situation?

c. Find the difference of the polynomials $24.27t - 1690.26$ and $16.09t - 1039.32$. What does your result represent?

d. Evaluate the result you found in part (c) for $t = 121$. What does your result mean in this situation?

68. DATA In Exercise 2 of Homework 6.4, the number n of daily newspapers is modeled by the system

$$n = 17t + 396.55 \quad \textit{Morning newspapers}$$
$$n = -29t + 1360.43 \quad \textit{Evening newspapers}$$

where t is the number of years since 1980 (see Table 27).

Table 27 Numbers of Morning Dailies and Evening Dailies

| | Number of Daily Newspapers | |
Year	Morning	Evening
1980	387	1388
1985	482	1220
1990	559	1084
1995	656	891
2000	766	727
2005	817	645
2011	931	451
2015	972	389

Source: *Editor & Publisher Co.*

a. Find the sum of the polynomials $17t + 396.55$ and $-29t + 1360.43$. What does your result represent?

b. Evaluate the result you found in part (a) for $t = 42$. What does your result mean in this situation?

c. Find the difference of the polynomials $17t + 396.55$ and $-29t + 1360.43$. What does your result represent?

d. Evaluate the result you found in part (c) for $t = 42$. What does your result mean in this situation?

69. DATA Sales of Hyundai Sonata and Lexus RX 400h are shown in Table 28 for various years.

Table 28 Sales of Sonata and RX 400h

| | Vehicle Sales (thousands) | |
Year	Sonata	RX 400h
2011	19.7	10.7
2012	20.8	12.2
2013	21.6	11.3
2014	21.1	9.4
2015	19.9	7.7

Source: *HybridCars.com*

Let S be vehicle annual sales (in thousands) at t years since 2010. Models of Sonata and RX 400h sales are

$$S = -0.421t^2 + 2.6t + 17.46 \quad \textit{Sonata}$$
$$S = -0.53t^2 + 2.29t + 9.2 \quad \textit{RX 400h}$$

a. Use a graphing calculator to check that the models fit the data well.

b. Find the sum of the polynomials $-0.421t^2 + 2.6t + 17.46$ and $-0.53t^2 + 2.29t + 9.2$. What does your result represent?

c. Evaluate the result you found in part (b) for $t = 6$. What does your result mean in this situation?

d. Find the difference of the polynomials $-0.421t^2 + 2.6t + 17.46$ and $-0.53t^2 + 2.29t + 9.2$. What does your result represent?

e. Evaluate the result you found in part (d) for $t = 6$. What does your result mean in this situation?

70. DATA The market shares of large-screen (40–49 inches) televisions and very-large-screen (greater than 49 inches) televisions are shown in Table 29 for various years.

Table 29 Market Shares of Large-Screen and Very-Large-Screen Televisions

| | Market Share (percent) | |
Year	Large	Very Large
2009	7	25
2010	7	32
2011	8	34
2012	10	41
2013	13	40
2014	18	45

Source: *IHS Technology*

Let p be the market share (in percent) of certain televisions at t years since 2000. Models of the market shares of large-screen televisions and very-large-screen televisions are

$$p = 0.59t^2 - 11.41t + 62.07 \quad \textit{Large-screen televisions}$$
$$p = -0.39t^2 + 12.78t - 57.69 \quad \textit{Very-large-screen televisions}$$

a. Use a graphing calculator to check that the models fit the data well.

b. Find the sum of the polynomials $0.59t^2 - 11.41t + 62.07$ and $-0.39t^2 + 12.78t - 57.69$. What does your result represent?

c. Evaluate the result you found in part (b) for $t = 16$. What does your result mean in this situation?

d. Find the difference of the polynomials $0.59t^2 - 11.41t + 62.07$ and $-0.39t^2 + 12.78t - 57.69$. What does your result represent?

e. Evaluate the result you found in part (d) for $t = 16$. What does your result mean in this situation?

Concepts

71. A student tries to find the difference $(5x^2 + 3x + 7) - (3x^2 + 2x + 1)$:

$$(5x^2 + 3x + 7) - (3x^2 + 2x + 1)$$
$$= 5x^2 + 3x + 7 - 3x^2 + 2x + 1$$
$$= 2x^2 + 5x + 8$$

Describe any errors. Then find the difference correctly.

72. A student tries to find the sum $3x^2 + 2x$:

$$3x^2 + 2x = 5x$$

Describe any errors. Then find the sum correctly.

73. Find two polynomials of degree 3 whose sum is the polynomial $6x^3 + 4x^2 + x - 2$.

74. Find two polynomials of degree 3 whose difference is the polynomial $6x^3 + 4x^2 + x - 2$.

75. Find two polynomials of degree 3 whose sum is $3x^2 - 5x + 4$.

76. Find two polynomials of degree 3 whose sum is 0.

77. Use the distributive law to explain why $2x^3 + 5x^3 = 7x^3$.

78. Use the distributive law to explain why $9x^4 - 3x^4 = 6x^4$.

79. If the difference $A - B$ is positive, then A is more than B. Explain why this is true.

80. Explain why you cannot have a simplified first-degree polynomial in one variable that has three terms.

Related Review

For Exercises 81–84, let x be a number. Translate the English phrase into a mathematical expression. Then simplify the expression.

81. The number squared, minus 3 times the number, plus 5 times the number squared

82. The number squared plus twice the number squared minus 3

83. 4 times the number squared, minus the sum of the number squared and the number

84. 6 times the number squared, plus the difference of the number squared and 5

85. DATA. The numbers of subscribers of pay-TV and Netflix are shown in Table 30 for various years.

Table 30 Numbers of Subscribers of Pay-TV and Netflix

Year	Number of Subscribers (millions)	
	Pay-TV	Netflix
2012	100.8	23.4
2013	99.3	29.2
2014	98.0	35.7
2015	97.2	41.4
2016	96.0	47.0
2017	94.6	50.9

Sources: *Netflix; TDG Research*

Let n be the number (in millions) of subscribers at t years since 2010. Models of the number of pay-TV and Netflix subscribers are

$$n = -1.19t + 103.01 \quad \textit{pay-TV}$$
$$n = 5.62t + 12.66 \quad \textit{Netflix}$$

a. Use a graphing calculator to check that the models fit the data well.

b. Find the difference of the polynomials $-1.19t + 103.01$ and $5.62t + 12.66$. What does your result represent?

c. Evaluate the result you found in part (b) for $t = 12$. What does your result mean in this situation?

d. Use elimination or substitution to predict when pay-TV and Netflix will have the same number of subscribers.

86. DATA. The percentages of women and men who use social networking sites are shown in Table 31 for various years.

Table 31 Percentages of Women and Men Who Use Social Networking Sites

	Percent	
Year	Women	Men
2010	50	42
2011	52	48
2012	59	51
2013	65	59
2014	63	60

Source: *Pew Research Center*

Let p be the percentage of adults who use social networking sites at t years since 2010. Models of the percentages of women and men who use social networking sites are

$$p = 3.9t + 50.0 \quad \textit{Women}$$
$$p = 4.7t + 42.6 \quad \textit{Men}$$

a. Use a graphing calculator to check that the models fit the data well.

b. Find the difference of the polynomials $3.9t + 50.0$ and $4.7t + 42.6$. What does your result represent?

c. Evaluate the result you found in part (b) for $t = 11$. What does your result mean in this situation?

d. Use elimination or substitution to estimate when the same percentage of women and men used social networking sites.

Expressions, Equations, and Graphs

Perform the indicated instruction. Then use words such as linear, quadratic, cubic, polynomial, degree, one variable, *and* two variables *to describe the expression, equation, or system.*

87. Simplify $3x^2 - 1 - 2x^2$.

88. Evaluate $3x^2 - 1 - 2x^2$ for $x = -2$.

89. Graph $y = 3x^2 - 1 - 2x^2$ by hand.

90. Solve $3x - 1 = -2x$.

7.4 Multiplying Polynomials

Objectives

» Describe the meaning of *binomial* and *trinomial*.

» Apply the product property for exponents.

» Multiply monomials.

» Multiply a monomial and a polynomial.

» Multiply two polynomials.

» Use a product of two polynomials to model an authentic situation.

In Section 7.3, we added and subtracted polynomials. How do we multiply polynomials?

Binomials and Trinomials

We refer to a polynomial as a monomial, a **binomial**, or a **trinomial**, depending on whether it has one, two, or three nonzero terms, respectively:

Name	Examples	Meaning
monomial	$8x^3$, x^2, $-4xy^2$, 9	one nonzero term
binomial	$7x^3 - 2x$, $-2x^2 + x$, $8y^2 + 5$	two nonzero terms
trinomial	$-x^3 + 8x - 7$, $4x^2 + 6x - 1$	three nonzero terms

Product Property for Exponents

Consider the product $x^2 \cdot x^3$. We can write this product as a single power:

$$
\begin{aligned}
x^2 \cdot x^3 &= (x \cdot x)(x \cdot x \cdot x) && \textit{Write factors without exponents.} \\
&= x \cdot x \cdot x \cdot x \cdot x && \textit{Remove parentheses.} \\
&= x^5 && \textit{Simplify.}
\end{aligned}
$$

We can find the same product by adding the exponents:

$$x^2 \cdot x^3 = x^{2+3} = x^5$$

Product Property for Exponents

If m and n are counting numbers, then

$$x^m x^n = x^{m+n}$$

In words: To multiply two powers of x, keep the base and add the exponents.

For example, $x^4 x^8 = x^{4+8} = x^{12}$. Also, $x^5 x = x^5 x^1 = x^{5+1} = x^6$.

Multiplication of Monomials

We can use the product property to help us multiply monomials.

Example 1   Finding Products of Monomials

Find the product.

1. $2x(7x^3)$

2. $3x^4(-5x^2)$

Solution

1. We rearrange factors so that the coefficients are adjacent and the powers of x are adjacent:

$$
\begin{aligned}
2x(7x^3) &= (2 \cdot 7)(x^1 x^3) && \textit{Rearrange factors; } x = x^1 \\
&= 14x^4 && \textit{Add exponents: } x^m x^n = x^{m+n}
\end{aligned}
$$

2.
$$
\begin{aligned}
3x^4(-5x^2) &= 3(-5)(x^4 x^2) && \textit{Rearrange factors.} \\
&= -15x^6 && \textit{Add exponents: } x^m x^n = x^{m+n}
\end{aligned}
$$

▶

WARNING   The polynomial $3x^4(-5x^2)$ is *not* the same as the polynomial $3x^4 - 5x^2$. The polynomial $3x^4(-5x^2)$ is a product, whereas $3x^4 - 5x^2$ is a difference. We can tell $3x^4(-5x^2)$ is a product because there is no operation symbol between the monomials $3x^4$ and $(-5x^2)$.

Example 2   Finding the Product of Two Monomials

Find the product $3a^2b\left(2a^3b^2\right)$.

Solution

We rearrange factors so the coefficients are adjacent, the powers of a are adjacent, and the powers of b are adjacent:

$$3a^2b\left(2a^3b^2\right) = (3 \cdot 2)\left(a^2a^3\right)\left(b^1b^2\right) \quad \text{Rearrange factors.}$$

$$= 6a^5b^3 \quad \text{Add exponents: } x^m x^n = x^{m+n}$$

Multiplication of a Monomial and a Polynomial

We can use the distributive law to help us multiply a monomial and a polynomial.

Example 3   Finding the Product of a Monomial and a Polynomial

Find the product.

1. $2x(3x + 5)$ **2.** $-4t\left(8t^2 + 3t\right)$ **3.** $3xy\left(2x^2 - 4xy + y^2\right)$

Solution

1. $2x(3x + 5) = 2x \cdot 3x + 2x \cdot 5 \quad \text{Distributive law}$

$$= 6x^2 + 10x \quad x \cdot x = x^2$$

2. $-4t\left(8t^2 + 3t\right) = -4t \cdot 8t^2 - 4t \cdot 3t \quad \text{Distributive law}$

$$= -32t^3 - 12t^2 \quad \text{Add exponents: } x^m x^n = x^{m+n}$$

3. $3xy\left(2x^2 - 4xy + y^2\right) = 3xy \cdot 2x^2 - 3xy \cdot 4xy + 3xy \cdot y^2 \quad \text{Distributive law}$

$$= 6x^3y - 12x^2y^2 + 3xy^3 \quad \text{Add exponents: } x^m x^n = x^{n}$$

Multiplication of Two Polynomials

We can also use the distributive law to help us find the product of two binomials. For instance, we can find the product $(a + b)(c + d)$ by using the distributive law three times:

$$(a + b)(c + d) = a(c + d) + b(c + d) \quad \text{Distribute } (c + d).$$

$$= ac + ad + bc + bd \quad \text{Distribute } a \text{ and distribute } b.$$

By examining the polynomial $ac + ad + bc + bd$, we can find the product $(a + b)(c + d)$ more directly by adding the four products formed by multiplying each term in the first sum by each term in the second sum:

$$(a + b)(c + d) = ac + ad + bc + bd$$

After using this technique, we combine like terms if possible.

Multiplying Two Polynomials

To multiply two polynomials, multiply each term in the first polynomial by each term in the second polynomial. Then combine like terms if possible.

Example 4   Finding Products of Binomials

Find the product.

1. $(x + 3)(x + 5)$
2. $(3x + 4y)(2x - 5y)$
3. $(-2.8x + 3.5)(4.1x + 9.7)$
4. $(x - 4)(x^2 - 3)$

Solution

1. $(x + 3)(x + 5) = x \cdot x + x \cdot 5 + 3 \cdot x + 3 \cdot 5$ *Multiply pairs of terms.*

   $= x^2 + 5x + 3x + 15$ *$x \cdot x = x^2$*

   $= x^2 + 8x + 15$ *Combine like terms.*

2. $(3x + 4y)(2x - 5y) = 3x \cdot 2x - 3x \cdot 5y + 4y \cdot 2x - 4y \cdot 5y$ *Multiply pairs of terms.*

   $= 6x^2 - 15xy + 8xy - 20y^2$ *$x \cdot x = x^2$*

   $= 6x^2 - 7xy - 20y^2$ *Combine like terms.*

3. $(-2.8x + 3.5)(4.1x + 9.7)$ *Multiply pairs of terms.*

   $= -2.8x(4.1x) - 2.8x(9.7) + 3.5(4.1x) + 3.5(9.7)$

   $= -11.48x^2 - 27.16x + 14.35x + 33.95$ *$x \cdot x = x^2$*

   $= -11.48x^2 - 12.81x + 33.95$ *Combine like terms.*

4. $(x - 4)(x^2 - 3) = x \cdot x^2 - x \cdot 3 - 4 \cdot x^2 + 4 \cdot 3$ *Multiply pairs of terms.*

   $= x^3 - 3x - 4x^2 + 12$ *Add exponents: $x^m x^n = x^{m+n}$*

   $= x^3 - 4x^2 - 3x + 12$ *Rearrange terms.*

We use a graphing calculator table to verify our work (see Fig. 41).

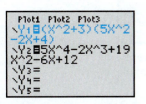

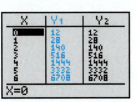

Figure 41 Verify the work

We can find the product of two polynomials of any degree in a similar fashion. The key idea is to multiply each term in the first polynomial by each term in the second polynomial. Then we combine like terms.

Example 5   Finding the Products of Polynomials

Find the product.

1. $(3a - 2)(a^2 + 4a + 5)$ 2. $(x^2 + 3)(5x^2 - 2x + 4)$

Solution

1. $(3a - 2)(a^2 + 4a + 5)$ *Multiply pairs of terms.*

   $= 3a \cdot a^2 + 3a \cdot 4a + 3a \cdot 5 - 2 \cdot a^2 - 2 \cdot 4a - 2 \cdot 5$

   $= 3a^3 + 12a^2 + 15a - 2a^2 - 8a - 10$ *Add exponents: $x^m x^n = x^{m+n}$*

   $= 3a^3 + 12a^2 - 2a^2 + 15a - 8a - 10$ *Rearrange terms.*

   $= 3a^3 + 10a^2 + 7a - 10$ *Combine like terms.*

2. $(x^2 + 3)(5x^2 - 2x + 4)$ *Multiply pairs of terms.*

   $= x^2 \cdot 5x^2 - x^2 \cdot 2x + x^2 \cdot 4 + 3 \cdot 5x^2 - 3 \cdot 2x + 3 \cdot 4$

   $= 5x^4 - 2x^3 + 4x^2 + 15x^2 - 6x + 12$ *Add exponents: $x^m x^n = x^{m+n}$*

   $= 5x^4 - 2x^3 + 19x^2 - 6x + 12$ *Combine like terms.*

Figure 42 Verify the work

We use a graphing calculator table to verify our work (see Fig. 42).

Using a Product of Two Polynomials to Model a Situation

In Section 7.3, we modeled authentic situations by using sums and differences of polynomials. Sometimes it is meaningful to find the product of two polynomials that represent authentic situations.

Example 6   Using a Product to Model a Situation

The average annual cost C of tuition (in 2014 dollars per student) at a public community college can be modeled by the equation

$$C = 70t + 1636$$

where t is the number of years since 1990 (see Table 32). The total enrollment E (in thousands of students) at all public community colleges can be modeled by the equation

$$E = 87t + 5194$$

where t is the number of years since 1990 (see Table 32).

Table 32 Average Tuition and Total Enrollment at All Public Community Colleges

Year	Average Annual Tuition (2014 dollars per student)	Total Enrollment (thousands of students)
1990	1611	5240
1995	2078	5492
2000	2264	5948
2005	2660	6488
2010	2997	7684
2014	3347	6714

Source: *U.S. National Center for Education Statistics*

1. Check that the models fit the data well.
2. Perform a unit analysis of the polynomial $(70t + 1636)(87t + 5194)$.
3. Find the product of the polynomials $70t + 1636$ and $87t + 5194$. What does your result represent?
4. Evaluate the result found in Problem 3 for $t = 31$. What does your result mean in this situation?

Solution

1. The tuition model fits the data quite well (see Fig. 43). The enrollment model fits the data fairly well (see Fig. 44).

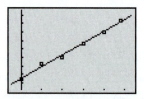

Figure 43 Tuition scatterplot and model

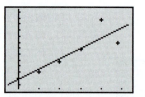

Figure 44 Enrollment scatterplot and model

2. Here is a unit analysis of the polynomial $(70t + 1636)(87t + 5194)$:

$$\underbrace{(70t + 1636)}_{\dfrac{\text{dollars}}{\text{student}}} \quad \underbrace{(87t + 5194)}_{\text{thousands of students}}$$

The units of the polynomial are thousands of dollars.

3. $(70t + 1636)(87t + 5194)$

$\quad = 70t(87t) + 70t(5194) + 1636(87t) + 1636(5194)$ *Multiply pairs of terms.*

$\quad = 6090t^2 + 363{,}580t + 142{,}332t + 8{,}497{,}384$ $x \cdot x = x^2$

$\quad = 6090t^2 + 505{,}912t + 8{,}497{,}384$ *Combine like terms.*

The polynomial represents the total money (in thousands of dollars) paid by all public community college students for annual tuition in the year that is t years since 1990.

4. We substitute 31 for t in the polynomial $6090t^2 + 505{,}912t + 8{,}497{,}384$:

$$6090(31)^2 + 505{,}912(31) + 8{,}497{,}384 = 30{,}033{,}146$$

In 2021, the total money paid by all public community college students for tuition will be $30,033,146 thousand, or about $30 billion.

Group Exploration

Section Opener: Product property for exponents

We can find the product x^3x^2 by the following work:

$$x^3x^2 = (x \cdot x \cdot x)(x \cdot x) = x \cdot x \cdot x \cdot x \cdot x = x^5$$

1. Find the product x^2x^4.

2. Find the product x^5x^3.

3. Find the product x^3x^4.

4. For each of Problems 1–3, compare the exponents of the original expression with the exponent of the result you found. What pattern do you notice? Refer to your work to explain why this pattern makes sense.

5. Use the pattern you described in Problem 4 to find the product x^mx^n, where m and n are counting numbers.

6. Find the product $x^{200}x^{400}$.

Tips for Success   "Work Out" by Solving Problems

Although there are many things you can do to enhance your learning, there is no substitute for solving problems. Your mathematical ability will respond to solving problems in much the same way that your muscles respond to lifting weights. Muscles increase greatly in strength when you work out intensely, frequently, and consistently.

     Just as with building muscles, to learn math, you must be an *active* participant. No amount of watching weight lifters lift, reading about weight-lifting techniques, or conditioning yourself psychologically can replace working out by lifting weights. The same is true of learning math: No amount of watching your instructor do problems, reading your text, or listening to a tutor can replace "working out" by solving problems.

Homework 7.4

For extra help ▶   **MyLab Math**   🎞 Watch the videos in MyLab Math

Find the product. Use a graphing calculator table to verify your work when possible.

1. x^4x^3 **2.** x^3x^5 **3.** w^8w **4.** pp^8

5. $9x^5(4x^7)$ **6.** $3x^8(5x^6)$

7. $-5x^4(-6x^3)$ **8.** $-6x^9(-2x^2)$

9. $x^6y^3(x^9y^4)$ **10.** $x^7y^5(x^6y^2)$

11. $3c^4d^6(8c^{10}d^2)$

12. $6a^8b^3(4ab^5)$

13. $4p^2t(-9p^3t^2)$

14. $2a^3b^2(-7ab^2)$

15. $\dfrac{4}{5}x^3\left(-\dfrac{7}{2}x^2\right)$

16. $-\dfrac{3}{8}x^5\left(-\dfrac{4}{5}x\right)$

17. $3w(w - 2)$

18. $6y(y + 1)$

19. $2.8x(9.4x - 7.3)$

20. $4.6x(5.7x - 1.4)$

21. $-4x(2x^2 + 3)$

22. $-5x(3x^2 - 7)$

23. $2mn^2(3m^2 + 5n)$

24. $4pq(3p + 2q^2)$

25. $2x(3x^2 - 2x + 7)$

26. $4x(x^2 + 3x - 1)$

27. $-3t^2(2t^2 + 4t - 2)$

28. $-2a^2(3a^2 - a - 4)$

29. $2xy^2(3x^2 - 4xy + 5y^2)$

30. $3p^2q(4p^2 - pq + 2q^2)$

Find the product. Use a graphing calculator table to verify your work when possible.

31. $(x + 2)(x + 4)$

32. $(x + 3)(x + 7)$

33. $(x - 2)(x + 5)$

34. $(x + 4)(x - 6)$

35. $(a - 3)(a - 2)$

36. $(p - 6)(p - 5)$

37. $(x + 6)(x - 6)$

38. $(x - 8)(x + 8)$

39. $(x - 5.3)(x - 9.2)$

40. $(x - 1.7)(x + 4.3)$

41. $(5y - 2)(3y + 4)$

42. $(6a + 3)(2a - 4)$

43. $(2x + 4)(2x + 4)$

44. $(5x + 2)(5x + 2)$

45. $(3x - 1)(3x - 1)$

46. $(4x - 3)(4x - 3)$

47. $(3x - 5y)(4x + y)$

48. $(2x + 7y)(3x - 2y)$

49. $(2a - 8b)(3a - 4b)$

50. $(3m - 7n)(2m - 4n)$

51. $(3x - 4)(3x + 4)$

52. $(2x + 7)(2x - 7)$

53. $(9x + 4y)(9x - 4y)$

54. $(6a - 5b)(6a + 5b)$

55. $(2.5x + 9.1)(4.6x - 7.7)$

56. $(4.8x - 2.5)(3.1x - 8.9)$

57. $(0.37x + 20.45)(-1.7x + 50.8)$

58. $(-8.2x + 143.7)(0.23x - 8.29)$

59. $(x + 6)(x^2 - 3)$

60. $(x - 4)(x^2 + 2)$

61. $(2t^2 - 5)(3t - 2)$

62. $(3a^2 - 3)(4a - 5)$

63. $(3a^2 + 5b^2)(2a^2 - 3b^2)$

64. $(4m^2 - 7n^2)(3m^2 - 2n^2)$

Find the product. Use a graphing calculator table to verify your work when possible.

65. $(x + 2)(x^2 + 3x + 5)$

66. $(x + 3)(x^2 + 2x + 4)$

67. $(x + 2)(x^2 - 2x + 4)$

68. $(x - 4)(x^2 + 4x + 16)$

69. $(2b^2 - 3b + 2)(b - 4)$

70. $(3y^2 - y + 5)(y - 2)$

71. $(2x^2 + 3)(3x^2 - x + 4)$

72. $(5x^2 + 2)(2x^2 - 2x - 3)$

73. $(4w^2 - 2w + 1)(3w^2 - 2)$

74. $(3t^2 + 3t - 4)(2t^2 - 1)$

75. $(2x^2 + 4x - 1)(3x^2 - x + 2)$

76. $(4x^2 - x + 3)(3x^2 + 2x + 1)$

77. $(a + b + c)(a + b - c)$

78. $(a - b - c)(a + b + c)$

79. **DATA** In Exercise 13 of Homework 1.4, you modeled average salaries for public school teachers. A reasonable model is $s = 1.1t + 31$, where s is the average salary (in thousands of dollars) at t years since 1990 (see Table 33). The number n (in millions) of teachers can be modeled by the equation $n = 0.034t + 2.9$, where t is the number of years since 1990.

Table 33 Average Salaries and Numbers of Public School Teachers

Year	Average Salary (thousands of dollars)	Number of Teachers (millions)
1990	31	2.8
1995	37	3.0
2000	42	3.4
2005	48	3.6
2010	55	3.5
2014	57	3.6

Sources: *National Center for Education Statistics; National Education Association*

a. Perform a unit analysis of the quadratic polynomial $(1.1t + 31)(0.034t + 2.9)$. [**Hint:** What is one thousand times one million?]

b. Find the product of the polynomials $1.1t + 31$ and $0.034t + 2.9$. Round the coefficients of your result to the second decimal place. What does your result represent?

c. Evaluate the result you found in part (b) for $t = 32$. What does your result mean in this situation?

80. **DATA** In Exercise 10 of Homework 1.4, you modeled the death rate from heart disease in the United States. A reasonable model is $r = -7.4t + 557$, where r is the death rate (number of deaths per 100,000 people) from heart disease for the year that is t years since 1960 (see Table 34). The U.S. population can be modeled by the equation $P = 26t + 1783$, where P is the population (in 100,000s of people) at t years since 1960.

Table 34 Death Rates from Heart Disease; U.S. Population

Year	Death Rate (number of deaths per 100,000 people)	Year	Population (in 100,000s)
1960	559	2000	2822
1970	492	2003	2901
1980	409	2006	2984
1990	317	2009	3068
2000	253	2012	3140
2010	177	2015	3209
2015	169		

Sources: *U.S. Center for Health Statistics; U.S. Census Bureau*

a. Perform a unit analysis of the quadratic polynomial $(-7.4t + 557)(26t + 1783)$.

b. Find the product of the polynomials $-7.4t + 557$ and $26t + 1783$. What does your result represent?

c. Evaluate the result you found in part (b) for $t = 61$. What does your result mean in this situation?

81. In Exercise 79 of Homework 5.5, you found an equation close to $r = -2.4t + 34$, where r is the American teenage birthrate (number of births per 1000 women ages 15–19) at t years since 2010 (see Table 35). The population (in thousands) of American teenage women ages 15–19 can be modeled by the equation $P = 21t^2 - 189t + 10{,}705$, where t is the number of years since 2010.

Table 35 American Teenage Birthrate; Numbers of Teenage Women

Year	Birthrate (number of births per 1000 women ages 15–19)	Year	Number of Women Ages 15–19 (thousands)
2010	34.3	2010	10,704
2011	31.3	2011	10,540
2012	29.4	2012	10,410
2013	26.5	2013	10,331
2014	24.2	2014	10,279
2015	22.3	2015	10,299
		2016	10,328

Sources: *U.S. National Center for Health Statistics; U.S. Census Bureau*

a. Perform a unit analysis of the quadratic polynomial $(-2.4t + 34)(21t^2 - 189t + 10{,}705)$.
b. Find the product of the polynomials $-2.4t + 34$ and $21t^2 - 189t + 10{,}705$. What does your result represent?
c. Evaluate the result you found in part (b) for $t = 11$. What does your result mean in this situation?

82. The numbers (in millions) of students who take AP exams can be modeled by the equation $n = 0.11t + 1.85$, where t is the number of years since 2010 (see Table 36). The average number of exams taken per student can be modeled by the equation $A = 0.0107t + 1.74$, where t is the number of years since 2010.

Table 36 Numbers of Students Who Take AP Exams; Average Numbers of AP Exams Taken per Student

Year	Number of Students (millions)	Average Number of Exams Taken per Student
2010	1.8	1.74
2011	2.0	1.75
2012	2.1	1.76
2013	2.2	1.77
2014	2.3	1.78
2015	2.4	1.80
2016	2.5	1.80

Source: *The College Board*

a. Perform a unit analysis of the quadratic polynomial $(0.11t + 1.85)(0.0107t + 1.74)$.
b. Find the product of the polynomials $0.11t + 1.85$ and $0.0107t + 1.74$. What does your result represent?
c. Evaluate the result you found in part (b) for $t = 12$. What does your result mean in this situation?

Concepts

83. A student tries to find the product $6x(-4x)$:

$$6x(-4x) = 6x - 4x = 2x$$

Describe any errors. Then find the product correctly.

84. A student tries to find the product $(x + 3)(x + 5)$:

$$(x + 3)(x + 5) = 2x + 5x + 3x + 8$$
$$= 10x + 8$$

Describe any errors. Then find the product correctly.

85. a. Find the given product. Then decide whether your result is a linear, quadratic, or cubic polynomial.
   i. $(2x + 3)(4x + 5)$
   ii. $(3x - 7)(5x + 2)$
   b. Create two linear polynomials and find their product. Is your result a linear, quadratic, or cubic polynomial?
   c. In general, is the product of two linear polynomials a linear, quadratic, or cubic polynomial? Explain.

86. a. Find the given product. Then decide whether your result is a linear, quadratic, or cubic polynomial.
   i. $(x + 2)(x^2 + 3x + 5)$
   ii. $(2x - 3)(3x^2 + 4x - 1)$
   b. Create a linear polynomial and a quadratic polynomial, and find their product. Is your result a linear, quadratic, or cubic polynomial?
   c. In general, is the product of a linear polynomial and a quadratic polynomial a linear, quadratic, or cubic polynomial? Explain.

87. a. Find the product $(x + 4)(x + 7)$.
   b. Find the product $(x + 7)(x + 4)$.
   c. Explain in terms of the laws of operations why it makes sense that $(x + 4)(x + 7)$ and $(x + 7)(x + 4)$ are equivalent polynomials.

88. Use the distributive law three times to help you show that $(x + 2)(x + 7)$ and $x^2 + 9x + 14$ are equivalent polynomials.

89. Which of the following polynomials are equivalent?

$$(x - 5)(x + 2) \quad x^2 - 3x + 10 \quad x(x - 3) - 10$$
$$(x - 2)(x + 5) \quad x^2 - 3x - 10 \quad (x + 2)(x - 5)$$

90. Which of the following polynomials are equivalent?

$$(x + 2)(x + 6) \quad x^2 + 8x + 8 \quad (x + 6)(x + 2)$$
$$x^2 + 8x + 12 \quad (x - 2)(x - 6) \quad x^2 + 4(2x + 3)$$

Related Review

Perform the indicated operation. Use a graphing calculator table to verify your work.

91. $(3x - 5)(2x^2 - 4x + 2)$

92. $(2x^2 - x - 8)(4x - 1)$

93. $(3x - 5) - (2x^2 - 4x + 2)$

94. $(2x^2 - x - 8) - (4x - 1)$

Simplify the right-hand side of the equation to help you decide whether the equation is linear or quadratic. State whether the graph of the equation is a line or a parabola. Use a graphing calculator graph to verify your decision.

95. $y = 3x(x - 2)$

96. $y = -4x(2x - 1)$

97. $y = (x - 3)(x - 4)$

98. $y = (x - 4)(x + 6)$

99. $y = (x + 2) - (3x + 5)$

100. $y = (4x - 3) + (-9x + 7)$

101. $y = (2x + 1)(5x - 2)$

102. $y = (3x - 5)(6x - 4)$

Expressions, Equations, and Graphs

Perform the indicated instruction. Then use words such as linear, quadratic, cubic, polynomial, degree, one variable, and two variables to describe the expression, equation, or system.

103. Solve $2x - 5 = 7x + 5$.

104. Graph $y = 2x - 5$ by hand.

105. Find the product $(2x - 5)(7x + 5)$.

106. Solve the following:

$$y = 2x - 5$$
$$y = 7x + 5$$

7.5 Powers of Polynomials; Product of Binomial Conjugates

Objectives

» Raise a product to a power.

» Find the power of a monomial.

» Simplify the square of a binomial.

» Put a quadratic equation into standard form, $y = ax^2 + bx + c$.

» Find the product of two *binomial conjugates*.

Recall from Section 2.6 that we refer to x^n as "the nth power of x." For most of this section, we will discuss how to find powers of polynomials.

Raising a Product to a Power

Consider the power $(xy)^3$. We can write this power as a product of monomials:

$$(xy)^3 = (xy)(xy)(xy) \qquad \textit{Write power without exponents.}$$
$$= (x \cdot x \cdot x)(y \cdot y \cdot y) \qquad \textit{Rearrange factors.}$$
$$= x^3 y^3 \qquad \textit{Simplify.}$$

We can get this same result by raising each factor of the base xy to the third power:

$$(xy)^3 = x^3 y^3$$

Raising a Product to a Power

If n is a counting number, then

$$(xy)^n = x^n y^n$$

In words: To raise a product to a power, raise each factor to the power.

For example, $(xy)^6 = x^6 y^6$.

Power of a Monomial

We can use the property of raising a product to a power to help us find the power of a monomial.

Example 1   Finding Powers of Monomials

Perform the indicated operation.

1. $(xy)^7$ **2.** $(7p)^2$ **3.** $(-2xy)^4$

Solution

1. $(xy)^7 = x^7 y^7$

2. $(7p)^2 = 7^2 p^2 = 49p^2$

3. $(-2xy)^4 = (-2)^4 x^4 y^4 = 16x^4 y^4$

▶

WARNING   The monomials $5x^2$ and $(5x)^2$ are *not* equivalent monomials:

$$5x^2 = 5x \cdot x$$
$$(5x)^2 = (5x)(5x) = 25x \cdot x$$

For $5x^2$, the base is the variable x. For $(5x)^2$, the base is the product $5x$.

Here we show a typical error and the correct work performed to find the power $(5x)^2$:

$$(5x)^2 = 5x^2 \quad \textit{Incorrect}$$
$$(5x)^2 = 5^2x^2 = 25x^2 \quad \textit{Correct}$$

Because the base $5x$ is a product, we need to raise *both* factors 5 and x to the second power.

In general, when finding a power of the form $(AB)^n$, remember to raise *both* factors A and B to the nth power.

Square of a Binomial

Next, we will discuss how to square a sum. The square of the binomial $x + 4$ is $(x+4)^2$:

$$(x+4)^2 = (x+4)(x+4) \quad y^2 = yy$$
$$= x^2 + 4x + 4x + 16 \quad \textit{Multiply pairs of terms.}$$
$$= x^2 + 8x + 16 \quad \textit{Combine like terms.}$$

We have simplified $(x+4)^2$ by writing it as $x^2 + 8x + 16$. We **simplify the square of a binomial** by writing it as a polynomial that does not have parentheses.

Now we generalize and simplify $(A+B)^2$:

$$(A+B)^2 = (A+B)(A+B) \quad y^2 = yy$$
$$= A^2 + AB + BA + B^2 \quad \textit{Multiply pairs of terms.}$$
$$= A^2 + 2AB + B^2 \quad \textit{Combine like terms.}$$

So, $(A+B)^2 = A^2 + 2AB + B^2$. We can use similar steps to find the property $(A-B)^2 = A^2 - 2AB + B^2$.

Squaring a Binomial

$$(A+B)^2 = A^2 + 2AB + B^2 \quad \textit{Square of a sum}$$
$$(A-B)^2 = A^2 - 2AB + B^2 \quad \textit{Square of a difference}$$

In words: The square of a binomial equals the first term squared, plus (or minus) twice the product of the two terms, plus the second term squared.

We can instead simplify $(x+4)^2$ by substituting x for A and 4 for B in the formula for the square of a sum:

$$(A+B)^2 = A^2 + 2 \cdot A \cdot B + B^2$$
$$(x+4)^2 = x^2 + 2 \cdot x \cdot 4 + 4^2 \quad \textit{Substitute.}$$
$$= x^2 + 8x + 16 \quad \textit{Simplify.}$$

The result is the same as our result from writing $(x+4)^2$ as $(x+4)(x+4)$ and then multiplying. So, there are two ways to simplify $(x+4)^2$. Similarly, there are two ways to simplify the square of any binomial. If you experiment with both methods on several exercises in the homework, you will be able to make an informed choice of methods for solving future problems.

Example 2   Simplifying Squares of Binomials

Simplify.

1. $(x + 3)^2$ **2.** $(x - 7)^2$

Solution

1. We substitute x for A and 3 for B:

$$(A + B)^2 = A^2 + 2 \cdot A \cdot B + B^2$$

$$(x + 3)^2 = x^2 + 2 \cdot x \cdot 3 + 3^2 \quad \textit{Substitute.}$$
$$= x^2 + 6x + 9 \quad \textit{Simplify.}$$

Another way to simplify $(x + 3)^2$ is to write it as $(x + 3)(x + 3)$ and then multiply each term in the first binomial by each term in the second binomial:

$$(x + 3)^2 = (x + 3)(x + 3) \qquad b^2 = bb$$
$$= x^2 + 3x + 3x + 9 \quad \textit{Multiply pairs of terms.}$$
$$= x^2 + 6x + 9 \quad \textit{Combine like terms.}$$

2. We substitute x for A and 7 for B:

$$(A - B)^2 = A^2 - 2 \cdot A \cdot B + B^2$$

$$(x - 7)^2 = x^2 - 2 \cdot x \cdot 7 + 7^2 \quad \textit{Substitute.}$$
$$= x^2 - 14x + 49 \quad \textit{Simplify.}$$

Another way to simplify $(x - 7)^2$ is to write it as $(x - 7)(x - 7)$ and then multiply each term in the first binomial by each term in the second binomial:

$$(x - 7)^2 = (x - 7)(x - 7) \qquad b^2 = bb$$
$$= x^2 - 7x - 7x + 49 \quad \textit{Multiply pairs of terms.}$$
$$= x^2 - 14x + 49 \quad \textit{Combine like terms.}$$

In Example 2, we found that $x^2 + 6x + 9$ and $x^2 - 14x + 49$ are *squares* of binomials. Both of these *trinomials* are called perfect-square trinomials. A **perfect-square trinomial** is a trinomial that is equivalent to the square of a binomial.

Example 3   Simplifying Squares of Binomials

Simplify.

1. $(5x - 4)^2$ **2.** $(3x + 5y)^2$

Solution

1. We substitute $5x$ for A and 4 for B:

$$(A - B)^2 = A^2 - 2 \cdot A \cdot B + B^2$$

$$(5x - 4)^2 = (5x)^2 - 2 \cdot 5x \cdot 4 + 4^2 \quad \textit{Substitute.}$$
$$= 25x^2 - 40x + 16 \quad \textit{Simplify.}$$

We use a graphing calculator table to verify our work (see Fig. 45).

2. We substitute $3x$ for A and $5y$ for B:

$$(A + B)^2 = A^2 + 2 \cdot A \cdot B + B^2$$

$$(3x + 5y)^2 = (3x)^2 + 2 \cdot 3x \cdot 5y + (5y)^2 \quad \textit{Substitute.}$$
$$= 9x^2 + 30xy + 25y^2 \quad \textit{Simplify.}$$

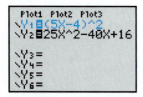

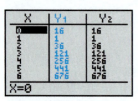

Figure 45 Verify the work

Here we compare squaring a product with squaring a sum:

$$(AB)^2 = A^2B^2 \qquad \text{\textit{Square of a product}}$$
$$(A + B)^2 = A^2 + 2AB + B^2 \qquad \text{\textit{Square of a sum}}$$

WARNING   It is a common error to omit the middle term in squaring a binomial. Here we show not only a typical error made in simplifying $(x + 6)^2$, but also the correct result:

$$(x + 6)^2 = x^2 + 36 \qquad \text{\textit{Incorrect}}$$
$$(x + 6)^2 = x^2 + 12x + 36 \quad \text{\textit{Correct}}$$

When simplifying $(A + B)^2$, don't omit the middle term, $2AB$, of $A^2 + 2AB + B^2$. Likewise, when simplifying $(A - B)^2$, don't omit the middle term, $-2AB$, of $A^2 - 2AB + B^2$.

Quadratic Equations Written in Standard Form

We can sometimes use our knowledge of squaring a binomial to help us write a quadratic equation in standard form, $y = ax^2 + bx + c$.

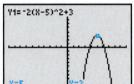

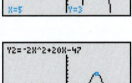

Figure 46 Verify the work

Example 4   Writing a Quadratic Equation in Standard Form

Write $y = -2(x - 5)^2 + 3$ in standard form.

Solution

We begin by simplifying $(x - 5)^2$ because we work with exponents before multiplying or adding:

$$
\begin{aligned}
y &= -2(x - 5)^2 + 3 && \text{\textit{Original equation}} \\
&= -2\left(x^2 - 10x + 25\right) + 3 && (A - B)^2 = A^2 - 2AB + B^2 \\
&= -2x^2 + 20x - 50 + 3 && \text{\textit{Distributive law}} \\
&= -2x^2 + 20x - 47 && \text{\textit{Combine like terms.}}
\end{aligned}
$$

We use graphing calculator graphs to verify our work (see Fig. 46).

Multiplication of Two Binomial Conjugates

We say that the sum of two terms and the difference of the same two terms are **binomial conjugates** of each other. For instance, $5x + 7$ and $5x - 7$ are binomial conjugates. To find the binomial conjugate of a binomial, we change the plus symbol to a minus symbol or vice versa.

How do we find the product of two binomial conjugates? To begin, we find the product of the binomial conjugates $x + 3$ and $x - 3$:

$$
\begin{aligned}
(x + 3)(x - 3) &= x^2 - 3x + 3x - 9 && \text{\textit{Multiply pairs of terms.}} \\
&= x^2 - 9 && \text{\textit{Combine like terms.}}
\end{aligned}
$$

Now we generalize and find the product of the binomial conjugates $A + B$ and $A - B$:

$$
\begin{aligned}
(A + B)(A - B) &= A^2 - AB + AB - B^2 && \text{\textit{Multiply pairs of terms.}} \\
&= A^2 - B^2 && \text{\textit{Combine like terms.}}
\end{aligned}
$$

We see that the product of $A + B$ and $A - B$ is the binomial $A^2 - B^2$, which is called a **difference of two squares**.

Product of Binomial Conjugates

$$(A + B)(A - B) = A^2 - B^2$$

In words: The product of two binomial conjugates is the difference of the square of the first term and the square of the second term.

By the commutative law of multiplication, we have

$$(A - B)(A + B) = (A + B)(A - B) = A^2 - B^2.$$

So, it is also true that

$$(A - B)(A + B) = A^2 - B^2$$

Example 5   Finding the Product of Binomial Conjugates

Find the product.

1. $(x + 5)(x - 5)$
2. $(3x - 8)(3x + 8)$
3. $(2a - 7b)(2a + 7b)$

Solution

1. We substitute x for A and 5 for B:

$$(A + B)(A - B) = A^2 - B^2$$
$$(x + 5)(x - 5) = x^2 - 5^2 \quad \textit{Substitute.}$$
$$= x^2 - 25 \quad \textit{Simplify.}$$

2. We substitute $3x$ for A and 8 for B:

$$(A - B)(A + B) = A^2 - B^2$$
$$(3x - 8)(3x + 8) = (3x)^2 - 8^2 \quad \textit{Substitute.}$$
$$= 9x^2 - 64 \quad \textit{Simplify.}$$

We use a graphing calculator table to verify our work (see Fig. 47).

3. We substitute $2a$ for A and $7b$ for B:

$$(A - B)(A + B) = A^2 - B^2$$
$$(2a - 7b)(2a + 7b) = (2a)^2 - (7b)^2 \quad \textit{Substitute.}$$
$$= 4a^2 - 49b^2 \quad \textit{Simplify.}$$

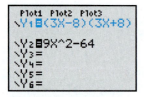

Figure 47 Verify the work

◈ Group Exploration

Section Opener: Product of binomial conjugates

1. Find the products.
   a. $(x + 2)(x - 2)$
   b. $(x + 3)(x - 3)$
   c. $(x + 4)(x - 4)$

2. What pattern do you notice in your work from Problem 1?

3. Use the pattern you described in Problem 2 to help you find each of the following products:
   a. $(x + 5)(x - 5)$
   b. $(x + 9)(x - 9)$
   c. $(x + k)(x - k)$

4. Compare the results of finding the product $(x + k)(x - k)$ with the results of finding the product $(x - k)(x + k)$. Explain in terms of one of the laws of operations why the results are the same.

Homework 7.5

For extra help ▶   **MyLab Math**   Watch the videos in MyLab Math

Perform the indicated operation. Use a graphing calculator table to verify your work when possible.

1. $(xy)^8$ **2.** $(xy)^5$ **3.** $(6x)^2$ **4.** $(9x)^2$

5. $(4p)^3$ **6.** $(2w)^2$ **7.** $(-8x)^2$ **8.** $(-5x)^2$

9. $(-3x)^3$ **10.** $(-2x)^5$ **11.** $(-a)^5$ **12.** $(-w)^4$

13. $(2mp)^4$ **14.** $(5bw)^2$ **15.** $(-3xy)^4$ **16.** $(-2xy)^3$

Simplify. Use a graphing calculator table to verify your work when possible.

17. $(x + 5)^2$ **18.** $(x + 2)^2$ **19.** $(x - 4)^2$

20. $(x - 3)^2$ **21.** $(2x + 3)^2$ **22.** $(3x + 7)^2$

23. $(5y - 2)^2$ **24.** $(4a - 9)^2$ **25.** $(3x + 6)^2$

26. $(4x + 3)^2$ **27.** $(9x - 2)^2$ **28.** $(6x - 3)^2$

29. $(2a + 5b)^2$ **30.** $(3t + 7w)^2$

31. $(8x - 3y)^2$ **32.** $(4m - 3n)^2$

Write the equation in standard form. Use graphing calculator graphs to verify your work.

33. $y = (x + 6)^2$ **34.** $y = (x - 8)^2$

35. $y = (x - 3)^2 + 1$ **36.** $y = (x + 4)^2 - 5$

37. $y = 2(x + 4)^2 - 3$ **38.** $y = 3(x - 1)^2 + 2$

39. $y = -3(x - 1)^2 - 2$ **40.** $y = -2(x - 2)^2 - 1$

Find the product. Use a graphing calculator table to verify your work when possible.

41. $(x + 4)(x - 4)$ **42.** $(x + 6)(x - 6)$

43. $(t - 7)(t + 7)$ **44.** $(a - 8)(a + 8)$

45. $(x + 1)(x - 1)$ **46.** $(x + 9)(x - 9)$

47. $(7a + 9)(7a - 9)$ **48.** $(5y + 4)(5y - 4)$

49. $(2x - 3)(2x + 3)$ **50.** $(4x - 7)(4x + 7)$

51. $(3x + 6y)(3x - 6y)$ **52.** $(5m + 8n)(5m - 8n)$

53. $(3t - 7w)(3t + 7w)$ **54.** $(2x - 5y)(2x + 5y)$

Perform the indicated operation. Use a graphing calculator table to verify your work when possible.

55. $(x + 11)(x - 11)$ **56.** $(x - 7)(x + 7)$

57. $(4x^2 - 5x) - (2x^3 - 8x)$ **58.** $(7x^3 - 2x) - (3x^2 - 6x)$

59. $(6m - 2n)^2$ **60.** $(2x - 7y)^2$

61. $5t(-2t^2)$ **62.** $-3w^2(-4w)$

63. $(3x + 4)(x^2 - x + 2)$ **64.** $(2x - 5)(x^2 + 4x - 1)$

65. $(2t - 3p)(2t + 3p)$ **66.** $(5a - 2b)(5a + 2b)$

67. $2xy^2(4x^2 - 8x - 5)$ **68.** $-3xy(8x^2 - 2xy + 9y^2)$

69. $(-6x^2 - 4x + 5) + (-2x^2 + 3x - 8)$

70. $(5x^3 - 2x + 1) + (-9x^3 - 5x^2 - 8x)$

71. $(x + 3)(x - 7)$ **72.** $(x - 2)(x - 4)$

73. $(4w - 8)^2$ **74.** $(6c - 4)^2$

75. $(x^2 + 2x - 5)(x - 3)$ **76.** $(x^2 - 2x + 3)(x + 2)$

77. $(3x - 7y)(2x + 3y)$ **78.** $(2a + 3b)(4a - 5b)$

79. $(6x - 7)(6x + 7)$ **80.** $(5x - 2)(5x + 2)$

81. $(2t + 7)^2$ **82.** $(5k + 4)^2$

Concepts

83. A student tries to simplify $(4x)^2$:
$$(4x)^2 = 4x^2$$
Describe any errors. Then simplify the power correctly.

84. A student tries to simplify $(-3x)^2$:
$$(-3x)^2 = -9x^2$$
Describe any errors. Then simplify the power correctly.

85. A student tries to simplify $(x + 7)^2$:
$$(x + 7)^2 = x^2 + 7^2$$
$$= x^2 + 49$$
Describe any errors. Then simplify the polynomial correctly.

86. A student tries to simplify $(x - 9)^2$:
$$(x - 9)^2 = x^2 - 9^2$$
$$= x^2 - 81$$
Describe any errors. Then simplify the polynomial correctly.

87. Use graphing calculator tables to show that $(x - 5)^2$ is not equivalent to $x^2 - 5^2$ and to show that $(x - 5)^2$ is not equivalent to $x^2 + 5^2$. Then simplify $(x - 5)^2$, and verify your work by using a graphing calculator table.

88. Use a graphing calculator table to show that $(x + 5)^2$ is not equivalent to $x^2 + 5^2$. Then simplify $(x + 5)^2$, and verify your work by using a graphing calculator table.

89. a. Use a graphing calculator table to show that $(x + 4)^2$ and $x^2 + 4^2$ are not equivalent.
  b. Simplify $(x + 4)^2$.
  c. Use a graphing calculator table to show that $(x + 4)^2$ and the result you found in part (b) are equivalent.

90. a. Use a graphing calculator table to show that $(x - 3)^2$ and $x^2 - 3^2$ are not equivalent.
  b. Simplify $(x - 3)^2$.
  c. Use a graphing calculator table to show that $(x - 3)^2$ and the result you found in part (b) are equivalent.

91. Show that the statement $(A + B)^2 = A^2 + B^2$ is false by substituting 2 for A and 3 for B. Explain why your work shows that the statement is false. Then write an important true statement that has the polynomial $(A + B)^2$ on the left side.

92. Show that the statement $(A - B)^2 = A^2 - B^2$ is false by substituting 5 for A and 3 for B. Explain why your work shows that the statement is false. Then write an important true statement that has the polynomial $(A - B)^2$ on the left side.

93. Which of the following polynomials are equivalent?

$(x - 2)^2$ $(x + 2)^2 - 4x$ $x(x - 4) + 4$ $(x - 2)(x + 2)$
$x^2 - 4x + 4$ $(x - 1)(x - 4)$

94. Which of the following polynomials are equivalent?

$(x + 3)^2$ $x^2 + 6x + 9$ $(x + 3)^2 + 6x$ $(x + 3)(x + 3)$
$(x - 3)^2 + 12x$ $(x + 3)(x - 3) + 6x$

95. Suppose that a student asks you, "Where did the term $14x$ come from?" in the equation $(x + 7)^2 = x^2 + 14x + 49$. How would you respond?

96. A student tries to simplify the polynomial $2(x + 4)^2$:

$$2(x + 4)^2 = (2x + 8)^2$$
$$= (2x)^2 + 2(2x)(8) + 8^2$$
$$= 4x^2 + 32x + 64$$

Describe any errors. Then simplify the polynomial correctly.

97. a. Simplify $(A + B + C)^2$. [**Hint:** $D^2 = DD$.]
b. Compare the property $(A + B)^2 = A^2 + 2AB + B^2$ with the result you found in part (a).

98. a. Is it possible for the product of two binomials to be equivalent to a trinomial? If yes, give an example. If no, why not?
b. Is it possible for the product of two binomials to be equivalent to a binomial? If yes, give an example. If no, why not?

99. In this exercise, you will explore a reasonable definition of b^0, where b is a nonzero real number.
a. **i.** Perform the exponentiation for 2^5, 2^4, 2^3, 2^2, and 2^1.
ii. What pattern do you notice in the results you found in part (i)?
iii. Based on the results you found in part (i), what would be a reasonable value for 2^0?
b. Perform the exponentiations for 3^4, 3^3, 3^2, and 3^1. Based on these values, what would be a reasonable value for 3^0?
c. If b is a nonzero real number, what would be a reasonable value for b^0?

100. a. Perform the exponentiations for $(-2)^2$, $(-2)^4$, $(-2)^6$, and $(-2)^8$. State whether each result is positive, negative, or 0.
b. Perform the exponentiations for $(-2)^3$, $(-2)^5$, $(-2)^7$, and $(-2)^9$. State whether each result is positive, negative, or 0.
c. For which counting numbers n is $(-2)^n$ positive? negative? Explain.
d. Let k be a negative number. For which counting numbers n is k^n positive? negative? Explain.

Related Review

Simplify the right-hand side of the equation to help you decide whether the equation is linear or quadratic. State whether the graph of the equation is a line or a parabola. Use a graphing calculator graph to verify your work.

101. $y = 2x(5x - 3)$ **102.** $y = -3(7x + 2)$

103. $y = (4x - 3)(6x - 5)$ **104.** $y = 2x(x + 5) - 2x^2$

105. $y = x^2 - (x - 3)^2$ **106.** $y = x - (x + 2)^2$

Expressions, Equations, and Graphs

Perform the indicated instruction. Then use words such as linear, quadratic, cubic, polynomial, degree, one variable, and two variables to describe the expression, equation, or system.

107. Solve the following:

$$2x - 5y = 15$$
$$y = 3x - 16$$

108. Simplify $(3a - 8)^2$.

109. Graph $2x - 5y = 15$ by hand.

110. Solve $2(3a - 8) = 2a + 7$.

7.6 Properties of Exponents

Objectives

» Apply properties of exponents.

» Determine whether a power expression is simplified.

» Describe the meaning of the exponent zero.

» Use combinations of properties of exponents to simplify power expressions.

In Sections 7.4 and 7.5, we discussed two properties of exponents. In this section, we will discuss several more properties of exponents.

Product Property and Raising a Product to a Power

We begin by reviewing the product property for exponents, from Section 7.4, and how to raise a product to a power, from Section 7.5.

If m and n are counting numbers, then

$$x^m x^n = x^{m+n} \quad \textit{Product property for exponents}$$
$$(xy)^n = x^n y^n \quad \textit{Raising a product to a power}$$

To multiply two powers of x, we keep the base and add the exponents. For example, $x^2 x^5 = x^{2+5} = x^7$. To raise a product to a power, we raise each factor to the power. For example, $(xy)^5 = x^5 y^5$.

Example 1   Performing Operations

Perform the indicated operation.

1. $\left(2x^3y^6\right)\left(3x^8y^7\right)$ 2. $(-2xy)^3$

Solution

1. We rearrange the factors so that the coefficients are adjacent, the powers of x are adjacent, and the powers of y are adjacent:

$$\left(2x^3y^6\right)\left(3x^8y^7\right) = (2\cdot3)\left(x^3x^8\right)\left(y^6y^7\right) \quad \textit{Rearrange factors.}$$
$$= 6x^{11}y^{13} \quad \textit{Add exponents: } x^m x^n = x^{m+n}$$

2. $(-2xy)^3 = (-2)^3 x^3 y^3$ *Raise each factor to third power:* $(xy)^n = x^n y^n$
$$= -8x^3y^3 \quad \textit{Simplify.}$$

▶

Determining whether a Power Expression Is Simplified

In Example 1, we worked with the power expressions $\left(2x^3y^6\right)\left(3x^8y^7\right)$ and $(-2xy)^3$. A **power expression** is an expression that contains one or more powers. Here are some more power expressions:

$$\left(w^2\right)^3 \qquad \left(9x^7\right)\left(5x^8\right) \qquad 3x^5(2xy)^3 \qquad \frac{8x^6y^9}{6x^2y^4} \qquad \left(\frac{7x^8}{x^3}\right)^2$$

We can use the two properties about exponents that we have discussed, as well as three more properties that we will discuss in this section, to *simplify a power expression*.

Simplifying a Power Expression

A power expression is simplified if

1. It includes no parentheses.

2. In any monomial, each variable or constant appears as a base at most once. For example, for nonzero x, we write x^3x^5 as x^8.

3. Each numerical expression (such as 5^2) has been calculated, and each numerical fraction has been simplified.

Quotient Property for Exponents

Consider the quotient $\dfrac{x^5}{x^2}$. We can simplify this quotient by first writing the expression without exponents:

$$\frac{x^5}{x^2} = \frac{x\cdot x\cdot x\cdot x\cdot x}{x\cdot x} \qquad \textit{Write quotient without exponents.}$$
$$= \frac{x\cdot x}{x\cdot x}\cdot\frac{x\cdot x\cdot x}{1} \qquad \frac{ac}{bd} = \frac{a}{b}\cdot\frac{c}{d}$$
$$= 1\cdot x\cdot x\cdot x \qquad \textit{Simplify: } \frac{x\cdot x}{x\cdot x} = 1$$
$$= x^3 \qquad \textit{Simplify.}$$

We can get the same result by subtracting the exponents 5 and 2:

$$\frac{x^5}{x^2} = x^{5-2} = x^3$$

> **Quotient Property for Exponents**
>
> If m and n are counting numbers and x is nonzero, then
>
> $$\frac{x^m}{x^n} = x^{m-n}$$
>
> In words: To divide two powers of x, keep the base and subtract the exponents.

For example, $\dfrac{x^9}{x^4} = x^{9-4} = x^5$.

Example 2   Quotient Property

Simplify.

1. $\dfrac{6x^9}{4x^5}$ **2.** $\dfrac{12x^6y^8}{4xy^7}$

Solution

1. $\dfrac{6x^9}{4x^5} = \dfrac{6}{4} \cdot \dfrac{x^9}{x^5}$ $\dfrac{ac}{bd} = \dfrac{a}{b} \cdot \dfrac{c}{d}$

$= \dfrac{3}{2} \cdot x^{9-5}$ *Simplify; subtract exponents:* $\dfrac{x^m}{x^n} = x^{m-n}$

$= \dfrac{3}{2} \cdot \dfrac{x^4}{1}$ *Simplify.*

$= \dfrac{3x^4}{2}$ *Multiply numerators; multiply denominators:* $\dfrac{A}{B} \cdot \dfrac{C}{D} = \dfrac{AC}{BD}$

2. $\dfrac{12x^6y^8}{4xy^7} = \dfrac{12}{4} \cdot \dfrac{x^6}{x^1} \cdot \dfrac{y^8}{y^7}$ $\dfrac{ac}{bd} = \dfrac{a}{b} \cdot \dfrac{c}{d}$

$= 3 \cdot x^{6-1} \cdot y^{8-7}$ *Simplify; subtract exponents:* $\dfrac{x^m}{x^n} = x^{m-n}$

$= 3x^5y^1$ *Simplify.*

$= 3x^5y$ $y^1 = y$

Zero as an Exponent

What is the meaning of 2^0? Computing powers of 2 can suggest the meaning:

$$2^4 = 16$$
$$2^3 = 8$$
The exponent decreases by 1. $\quad 2^2 = 4 \quad$ The value is divided by 2.
$$2^1 = 2$$
$$2^0 = 1$$

Each time we decrease the exponent by 1, the value is divided by 2. This pattern suggests that $2^0 = 1$.

Similar work with any other nonzero base x would suggest that $x^0 = 1$.

> **Definition**   Zero exponent
>
> For nonzero x,
>
> $$x^0 = 1$$

So, for nonzero x and y, $4^0 = 1$, $(xy)^0 = 1$, and $\left(\dfrac{5x^2}{y^4}\right)^0 = 1$.

Raising a Quotient to a Power

Consider the power expression $\left(\dfrac{x}{y}\right)^4$. We begin to write this expression in another form by writing it without exponents:

$$\left(\frac{x}{y}\right)^4 = \frac{x}{y}\cdot\frac{x}{y}\cdot\frac{x}{y}\cdot\frac{x}{y} \qquad \textit{Write power expression without exponents.}$$

$$= \frac{x\cdot x\cdot x\cdot x}{y\cdot y\cdot y\cdot y} \qquad \textit{Multiply numerators; multiply denominators: } \frac{A}{B}\cdot\frac{C}{D} = \frac{AC}{BD}$$

$$= \frac{x^4}{y^4} \qquad \textit{Simplify.}$$

We can get the same result by raising both the numerator and the denominator of the base $\dfrac{x}{y}$ to the fourth power:

$$\left(\frac{x}{y}\right)^4 = \frac{x^4}{y^4}$$

This method is called *raising a quotient to a power*.

Raising a Quotient to a Power

If n is a counting number and y is nonzero, then

$$\left(\frac{x}{y}\right)^n = \frac{x^n}{y^n}$$

In words: To raise a quotient to a power, raise both the numerator and the denominator to the power.

Example 3 — Raising a Quotient to a Power

Simplify.

1. $\left(\dfrac{x}{y}\right)^8$ **2.** $\left(\dfrac{x}{2}\right)^3$

Solution

1. $\left(\dfrac{x}{y}\right)^8 = \dfrac{x^8}{y^8}$ *Raise numerator and denominator to eighth power:* $\left(\dfrac{x}{y}\right)^n = \dfrac{x^n}{y^n}$

2. $\left(\dfrac{x}{2}\right)^3 = \dfrac{x^3}{2^3}$ *Raise numerator and denominator to third power:* $\left(\dfrac{x}{y}\right)^n = \dfrac{x^n}{y^n}$

$$= \frac{x^3}{8} \qquad \textit{Simplify.}$$

Raising a Power to a Power

Consider the power expression $\left(x^2\right)^3$. We can write this expression as the power x^6:

$$\left(x^2\right)^3 = x^2\cdot x^2\cdot x^2 \qquad \textit{Write expression without exponent 3.}$$
$$= x^{2+2+2} \qquad \textit{Add exponents: } x^m x^n = x^{m+n}$$
$$= x^6 \qquad \textit{Simplify.}$$

We can get the same result by multiplying the exponents 2 and 3:

$$\left(x^2\right)^3 = x^{2\cdot 3} = x^6$$

This method is called *raising a power to a power*.

Raising a Power to a Power

If m and n are counting numbers, then

$$\left(x^m\right)^n = x^{mn}$$

In words: To raise a power to a power, keep the base and multiply the exponents.

Example 4   Raising a Power to a Power

Simplify.

1. $\left(x^2\right)^7$

2. $\left(x^5\right)^8$

Solution

1. $\left(x^2\right)^7 = x^{2\cdot 7}$ *Multiply exponents:* $\left(x^m\right)^n = x^{mn}$
    $= x^{14}$ *Simplify.*

2. $\left(x^5\right)^8 = x^{5\cdot 8}$ *Multiply exponents:* $\left(x^m\right)^n = x^{mn}$
    $= x^{40}$ *Simplify.*

▶

Using Combinations of Properties of Exponents

Next, we use more than one property of exponents to simplify a power expression.

Example 5   Simplifying Power Expressions

Simplify.

1. $\left(2x^8y^5\right)^3$ **2.** $2x^7\left(4x^4\right)^2$ **3.** $\dfrac{8x^3x^4}{2x^7}$

Solution

1. $\left(2x^8y^5\right)^3 = 2^3\left(x^8\right)^3\left(y^5\right)^3$ *Raise each factor to third power:* $(xy)^n = x^ny^n$
    $= 8x^{24}y^{15}$ *Multiply exponents:* $\left(x^m\right)^n = x^{mn}$

2. $2x^7\left(4x^4\right)^2 = 2x^7[4^2\left(x^4\right)^2]$ *Raise each factor to second power:* $(xy)^n = x^ny^n$
    $= 2x^7[16x^8]$ *Multiply exponents:* $\left(x^m\right)^n = x^{mn}$
    $= (2\cdot 16)\left(x^7x^8\right)$ *Rearrange factors.*
    $= 32x^{15}$ *Add exponents:* $x^mx^n = x^{m+n}$

3. $\dfrac{8x^3x^4}{2x^7} = \dfrac{4x^7}{x^7}$ *Simplify; add exponents:* $x^mx^n = x^{m+n}$
    $= 4x^{7-7}$ *Subtract exponents:* $\dfrac{x^m}{x^n} = x^{m-n}$
    $= 4x^0$ *Simplify.*
    $= 4$ $x^0 = 1.$

▶

Example 6 Simplifying Power Expressions

Simplify.

1. $\left(\dfrac{2x^4}{3y^7}\right)^3$

2. $\dfrac{(2x^3y)^5}{x^8y}$

Solution

1. $\left(\dfrac{2x^4}{3y^7}\right)^3 = \dfrac{(2x^4)^3}{(3y^7)^3}$ *Raise numerator and denominator to third power:* $\left(\dfrac{x}{y}\right)^n = \dfrac{x^n}{y^n}$

$= \dfrac{2^3(x^4)^3}{3^3(y^7)^3}$ *Raise factors to third power:* $(xy)^n = x^ny^n$

$= \dfrac{8x^{12}}{27y^{21}}$ *Multiply exponents:* $(x^m)^n = x^{mn}$

2. $\dfrac{(2x^3y)^5}{x^8y} = \dfrac{2^5(x^3)^5y^5}{x^8y}$ *Raise factors to fifth power:* $(xy)^n = x^ny^n$

$= \dfrac{32x^{15}y^5}{x^8y^1}$ *Multiply exponents:* $(x^m)^n = x^{mn}$; $y = y^1$

$= 32x^{15-8}y^{5-1}$ *Subtract exponents:* $\dfrac{x^m}{x^n} = x^{m-n}$

$= 32x^7y^4$ *Simplify.*

Group Exploration

Section Opener: Properties of exponents

1. a. Write each expression as a product of fractions. Then simplify so that your result is a single fraction.

 i. $\left(\dfrac{x}{y}\right)^2$ **ii.** $\left(\dfrac{x}{y}\right)^3$ **iii.** $\left(\dfrac{x}{y}\right)^4$

 b. Write $\left(\dfrac{x}{y}\right)^n$ in another form. [**Hint:** Refer to the results you found in part (a).]

 c. Use the result you found in part (b) to write $\left(\dfrac{x}{y}\right)^{28}$ in another form in one step.

2. a. Write each expression as a product of powers. Then simplify so that your result is a single power.
 i. $(x^2)^3$ [**Hint:** Write as a product of three x^2 factors.]
 ii. $(x^5)^2$
 iii. $(x^3)^4$

 b. Write $(x^m)^n$ in another form. [**Hint:** Refer to part (a) and determine whether we add, subtract, multiply, or divide the exponents m and n.]

 c. Use the result you found in part (b) to write $(x^6)^9$ in another form in one step.

3. a. We can write the power x^3 as $x \cdot x \cdot x$ without using exponents. Write each expression without using exponents. Then simplify your result.

 i. $\dfrac{x^5}{x^3}$ **ii.** $\dfrac{x^6}{x^2}$ **iii.** $\dfrac{x^8}{x^5}$

 b. Write $\dfrac{x^m}{x^n}$ in another form. [**Hint:** Refer to your work in part (a) and determine whether we add, subtract, multiply, or divide the exponents m and n.]

 c. Use the result you found in part (b) to write $\dfrac{x^{25}}{x^{21}}$ in another form in one step.

Group Exploration

Properties of exponents

1. For the statement $x^2x^3 = x^5$, a student wants to know why there are two x's on the left-hand side of the equation and only one x on the right-hand side. What would you tell the student?

2. A student tries to simplify the expression $(3x^3)^2$:

$$(3x^3)^2 = (3 \cdot 2)x^{3 \cdot 2} = 6x^6$$

Describe any errors. Then simplify the expression correctly.

3. In simplifying power expressions, a student is confused about when to add exponents and when to multiply exponents. What would you tell the student?

4. A student tries to simplify $\dfrac{(2x^5)^4}{x^2}$:

$$\dfrac{(2x^5)^4}{x^2} = (2x^{5-2})^4 = (2x^3)^4 = 2^4(x^3)^4 = 16x^{12}$$

Describe any errors. Then simplify the expression correctly.

5. Simplify $x^3 + x^3$. Then simplify x^3x^3. Explain why your two results have different exponents.

Homework 7.6

For extra help ▶ **MyLab Math**   Watch the videos in MyLab Math

Simplify.

1. x^3x^5

2. x^2x^7

3. r^5r

4. yy^8

5. $(5x^4)(3x^5)$

6. $(-2x^7)(6x^8)$

7. $(-4b^3)(-8b^5)$

8. $(3t^5)(-7t^4)$

9. $(6a^2b^5)(9a^4b^3)$

10. $(10c^7d^2)(3c^5d)$

11. $(rt)^7$

12. $(ab)^4$

13. $(8x)^2$

14. $(2x)^4$

15. $(2xy)^5$

16. $(3xy)^3$

17. $(-2a)^4$

18. $(-3t)^3$

19. $(9xy)^0$

20. $(-3x)^0$

21. $\dfrac{a^5}{a^2}$

22. $\dfrac{w^8}{w^6}$

23. $\dfrac{6x^7}{3x^3}$

24. $\dfrac{8x^6}{4x^5}$

25. $\dfrac{15x^6y^8}{12x^3y}$

26. $\dfrac{14x^5y^9}{21x^3y^6}$

27. $\left(\dfrac{t}{w}\right)^7$

28. $\left(\dfrac{a}{b}\right)^4$

29. $\left(\dfrac{3}{t}\right)^3$

30. $\left(\dfrac{b}{7}\right)^2$

31. $\left(\dfrac{x}{3}\right)^0$

32. $\left(\dfrac{2}{x}\right)^0$

33. $(r^2)^4$

34. $(a^3)^8$

35. $(x^4)^9$

36. $(x^6)^9$

For Exercises 37–72, simplify.

37. $(6x^3)^2$

38. $(4x^5)^3$

39. $(-t^3)^4$

40. $(-a^2)^3$

41. $(2a^2a^7)^3$

42. $(2w^3w^4)^5$

43. $(x^2y^3)^4x^5y^8$

44. $(xy^6)^3x^2y^4$

45. $5x^4(3x^6)^2$

46. $4x^2(2x^5)^3$

47. $-3c^6(c^4)^5$

48. $-7w^4(w^3)^6$

49. $(xy^3)^5(xy)^4$

50. $(x^2y)^3(xy^3)^2$

51. $\dfrac{10t^5t^7}{8t^4}$

52. $\dfrac{15y^9y^2}{20y^8}$

53. $\dfrac{18x^{10}}{24x^4x^6}$

54. $\dfrac{12x^{15}}{21x^9x^6}$

55. $\left(\dfrac{y}{2x}\right)^3$

56. $\left(\dfrac{8x}{y}\right)^2$

57. $\left(\dfrac{x^2}{y^5}\right)^4$

58. $\left(\dfrac{x^8}{y^2}\right)^5$

59. $\left(\dfrac{r^6}{6}\right)^2$

60. $\left(\dfrac{2}{a^4}\right)^3$

61. $\left(\dfrac{2a^4}{3b^2}\right)^3$

62. $\left(\dfrac{4w^6}{7t^3}\right)^2$

63. $\left(\dfrac{3x^4}{5y^7}\right)^0$

64. $\left(\dfrac{-5y^2}{9x^4}\right)^0$

65. $\left(\dfrac{2a^6b}{3c^5}\right)^3$

66. $\left(\dfrac{5xy^4}{7w^3}\right)^2$

67. $\dfrac{(x^4y)^4}{x^5}$

68. $\dfrac{(x^3y^7)^5}{y^{10}}$

69. $\dfrac{(w^3)^4}{(2w)^5}$

70. $\dfrac{(3t)^3}{(t^2)^5}$

71. $\dfrac{(4x^5y^8)^2}{8x^8y^9}$

72. $\dfrac{(2x^4y^3)^3}{12x^5y^4}$

73. A person invests $5000 in an account at 3% interest compounded annually. The value V (in dollars) of the account after t years is described by $V = 5000(1.03)^t$. Find the value of the account after 10 years.

74. A person invests $8000 in an account at 7% interest compounded annually. The value V (in dollars) of the account after t years is described by $V = 8000(1.07)^t$. Find the value of the account after 6 years.

75. If an object falls from rest for t seconds, the distance d (in feet) the object will fall is described by the equation $d = 16t^2$, assuming no air resistance. Estimate how far a sky diver will have fallen 5 seconds after jumping out of an airplane.

76. The weight w (in ounces) of a thick-crust pizza with three toppings at Papa Del's in Champaign, Illinois, is described by the equation $w = 0.41d^2$, where d is the diameter (in inches) of the pizza. Estimate the weight of such a pizza with diameter 12 inches.

77. The power P (in watts) generated by a windmill is described by the equation $P = 0.8r^2v^3$, where r is the radius (in meters) of the windmill and v is the wind speed (in meters per second). Estimate the power a windmill with radius 0.57 meter can generate if the wind speed is 12.5 meters per second.

78. The intensity I (in watts per square meter) of a television signal at a distance of d kilometers is described by the equation $I = \dfrac{180}{d^2}$. Find the intensity of the signal at a distance of 5 kilometers.

79. The volume V (in cubic centimeters) of a cylinder is described by the equation $V = \pi r^2 h$, where h is the height (in centimeters) of the cylinder and r is the radius (in centimeters) of the bottom (see Fig. 48). Find the volume of a can of chili with height 11 centimeters and radius 3.5 centimeters.

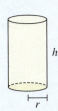

Figure 48 The height h and radius r of a cylinder— Exercise 79

80. The volume V (in cubic feet) of a rectangular box with a square base is described by the equation $V = L^2H$, where H is the height (in feet) of the box and L is the length (in feet) of the base (see Fig. 49). Find the volume of a 4-foot-tall cardboard box whose square base has length 2 feet.

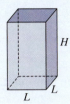

Figure 49 The height H and length L of a rectangular box with a square base— Exercise 80

Concepts

81. A student tries to simplify x^3x^5:

$$x^3x^5 = x^{3 \cdot 5} = x^{15}$$

Describe any errors. Then simplify the expression correctly.

82. A student tries to simplify $(x^4)^6$:

$$(x^4)^6 = x^{4+6} = x^{10}$$

Describe any errors. Then simplify the expression correctly.

83. A student tries to simplify $(5x^3)^2$:

$$(5x^3)^2 = 5(x^3)^2 = 5x^6$$

Describe any errors. Then simplify the expression correctly.

84. A student tries to simplify $2(3x)^2$:

$$2(3x)^2 = (6x)^2 = 36x^2$$

Describe any errors. Then simplify the expression correctly.

We can explain why $(xy)^3 = x^3y^3$ by writing

$$(xy)^3 = (xy)(xy)(xy) = (xxx)(yyy) = x^3y^3$$

For Exercises 85 and 86, use a similar approach to show that the given statement is true.

85. $\left(\dfrac{x}{y}\right)^3 = \dfrac{x^3}{y^3}$ **86.** $\dfrac{x^6}{x^4} = x^2$

87. A student tries to simplify $(2x^2)^4$:

$$(2x^2)^4 = (2 \cdot 4)x^{2 \cdot 4} = 8x^8$$

Describe any errors. Then simplify the expression correctly.

88. Describe what it means to use properties of exponents to simplify a power expression. Include several examples.

Related Review

Simplify.

89. x^3x^2 **90.** x^5x^3

91. $x^3 + x^2$ **92.** $x^5 + x^3$

93. $2x^4 + 3x^4$ **94.** $5x^3 - 2x^3$

95. $(2x^4)(3x^4)$ **96.** $(5x^3)(-2x^3)$

97. $(3x)^2$ **98.** $(6x)^2$

99. $(3 + x)^2$ **100.** $(6 + x)^2$

Expressions, Equations, and Graphs

Perform the indicated instruction. Then use words such as linear, quadratic, cubic, power, polynomial, degree, one variable, *and* two variables *to describe the expression, equation, or system.*

101. Solve $\dfrac{2}{3}x - \dfrac{5}{6} = \dfrac{1}{2}x$.

102. Find the product $(2p - 1)(3p^2 + p - 2)$.

103. Combine like terms: $\dfrac{2}{3}x - \dfrac{5}{6} - \dfrac{1}{2}x$.

104. Graph $y = x^2 - 4x + 3$ by hand.

7.7 Negative-Integer Exponents

Objectives

» Describe the meaning of a negative-integer exponent.

» Simplify power expressions containing negative-integer exponents.

» Work with models that have negative-integer exponents.

» Use scientific notation.

In Section 7.6, we worked with power expressions with nonnegative-integer exponents. In this section, we will extend our work with power expressions to include negative-integer exponents.

Definition of a Negative-Integer Exponent

We can begin to see the meaning of a negative-integer exponent by considering the expression $\dfrac{x^3}{x^5}$. We write this expression in two different forms:

$$\frac{x^3}{x^5} = x^{3-5} = x^{-2}$$

$$\frac{x^3}{x^5} = \frac{x \cdot x \cdot x}{x \cdot x \cdot x \cdot x \cdot x} = \frac{1}{x^2}$$

Equating the two results, we have

$$x^{-2} = \frac{1}{x^2}$$

Likewise, we could write $\frac{x^2}{x^5}$ in two different forms to show that

$$x^{-3} = \frac{1}{x^3}$$

Next, we describe this pattern in general.

Definition   Negative-integer exponent

If n is a counting number and x is nonzero, then

$$x^{-n} = \frac{1}{x^n}$$

In words: To find x^{-n}, take its reciprocal and change the sign of the exponent.

Simplifying Power Expressions Containing Negative-Integer Exponents

Simplifying a power expression includes writing it so that each exponent is positive.

Example 1   Simplifying Power Expressions

Simplify.

1. 5^{-2} **2.** x^{-6}

Solution

1. $5^{-2} = \dfrac{1}{5^2}$ *Write power so exponent is positive:* $x^{-n} = \dfrac{1}{x^n}$

$\phantom{5^{-2}} = \dfrac{1}{25}$ *Simplify.*

2. $x^{-6} = \dfrac{1}{x^6}$ *Write power so exponent is positive:* $x^{-n} = \dfrac{1}{x^n}$

▶

Next, we write $\dfrac{1}{x^{-n}}$ in another form, where x is nonzero and n is a counting number:

$$\frac{1}{x^{-n}} = 1 \div x^{-n} \quad \frac{a}{b} = a \div b$$

$$\phantom{\frac{1}{x^{-n}}} = 1 \div \frac{1}{x^n} \quad \textit{Write power so exponent is positive: } x^{-n} = \frac{1}{x^n}$$

$$\phantom{\frac{1}{x^{-n}}} = 1 \cdot \frac{x^n}{1} \quad \textit{Multiply by reciprocal of } \frac{1}{x^n}, \textit{ which is } \frac{x^n}{1}.$$

$$\phantom{\frac{1}{x^{-n}}} = x^n \quad \textit{Simplify.}$$

So, $\dfrac{1}{x^{-n}} = x^n$.

> **Negative-Integer Exponent in a Denominator**
>
> If x is nonzero and n is a counting number, then
>
> $$\frac{1}{x^{-n}} = x^n$$
>
> In words: To find $\frac{1}{x^{-n}}$, find its reciprocal and change the sign of the exponent.

Example 2   Simplifying Power Expressions

Simplify.

1. $\dfrac{1}{2^{-3}}$ 　　　　　　　　　　　　　　　**2.** $\dfrac{1}{x^{-7}}$

Solution

1. $\dfrac{1}{2^{-3}} = 2^3$ 　*Write power so exponent is positive:* $\dfrac{1}{x^{-n}} = x^n$

$\phantom{\dfrac{1}{2^{-3}}} = 8$ 　*Simplify.*

2. $\dfrac{1}{x^{-7}} = x^7$ 　*Write power so exponent is positive:* $\dfrac{1}{x^{-n}} = x^n$

In Example 3, we will simplify some quotients of two powers.

Example 3   Simplifying Power Expressions

Simplify.

1. $\dfrac{x^{-4}}{y^2}$ 　　　　　　　　　　　　　　　**2.** $\dfrac{x^3}{y^{-7}}$

Solution

1. $\dfrac{x^{-4}}{y^2} = x^{-4} \cdot \dfrac{1}{y^2}$ 　*Write quotient as a product:* $\dfrac{A}{B} = A \cdot \dfrac{1}{B}$

$\phantom{\dfrac{x^{-4}}{y^2}} = \dfrac{1}{x^4} \cdot \dfrac{1}{y^2}$ 　*Write powers so exponents are positive:* $x^{-n} = \dfrac{1}{x^n}$

$\phantom{\dfrac{x^{-4}}{y^2}} = \dfrac{1}{x^4 y^2}$ 　*Multiply numerators; multiply denominators:* $\dfrac{A}{B} \cdot \dfrac{C}{D} = \dfrac{AC}{BD}$

2. $\dfrac{x^3}{y^{-7}} = x^3 \cdot \dfrac{1}{y^{-7}}$ 　*Write quotient as a product:* $\dfrac{A}{B} = A \cdot \dfrac{1}{B}$

$\phantom{\dfrac{x^3}{y^{-7}}} = x^3 y^7$ 　*Write powers so exponents are positive:* $\dfrac{1}{x^{-n}} = x^n$

For Problem 1 of Example 3, we simplified $\dfrac{x^{-4}}{y^2}$. When simplifying such expressions, we usually skip the first two steps shown in the example and write

$$\frac{x^{-4}}{y^2} = \frac{1}{x^4 y^2}$$

Likewise, we can skip the first step of our work in Problem 2 of Example 3 by simplifying $\dfrac{x^3}{y^{-7}}$ as follows:

$$\frac{x^3}{y^{-7}} = x^3 y^7$$

Example 4   Simplifying Power Expressions

Simplify $\dfrac{4a^{-2}b^4}{7c^{-5}}$.

Solution

$$\frac{4a^{-2}b^4}{7c^{-5}} = \frac{4c^5 b^4}{7a^2}$$ *Write powers so exponents are positive: $x^{-n} = \dfrac{1}{x^n}$, $\dfrac{1}{x^{-n}} = x^n$*

In Section 7.6, we discussed various properties of counting-number exponents. It turns out that these properties are also true for all negative-integer exponents and the zero exponent.

Properties of Integer Exponents

If m and n are integers and x and y are nonzero, then

1. $x^m x^n = x^{m+n}$ Product property for exponents

2. $\dfrac{x^m}{x^n} = x^{m-n}$ Quotient property for exponents

3. $(xy)^n = x^n y^n$ Raising a product to a power

4. $\left(\dfrac{x}{y}\right)^n = \dfrac{x^n}{y^n}$ Raising a quotient to a power

5. $\left(x^m\right)^n = x^{mn}$ Raising a power to a power

We can use these properties to expand our rules for simplifying power expressions to include those that contain negative-integer exponents.

Simplifying Power Expressions

A power expression is simplified if

1. It includes no parentheses.

2. In any monomial, each variable or constant appears as a base at most once.

3. Each numerical expression (such as 7^2) has been calculated, and each numerical fraction has been simplified.

4. Each exponent is positive.

Example 5   Simplifying Power Expressions

Simplify.

1. $\left(x^{-4}\right)^3$ **2.** $\left(8t^6\right)\left(-3t^{-10}\right)$

Solution

1. $\left(x^{-4}\right)^3 = x^{-12}$ *Multiply exponents:* $\left(x^m\right)^n = x^{mn}$

   $= \dfrac{1}{x^{12}}$ *Write power so exponent is positive:* $x^{-n} = \dfrac{1}{x^n}$

2. $\left(8t^6\right)\left(-3t^{-10}\right) = 8(-3)t^6 t^{-10}$ *Rearrange factors.*

   $= -24t^{-4}$ *Add exponents:* $x^m x^n = x^{m+n}$

   $= -\dfrac{24}{t^4}$ *Write power so exponent is positive:* $x^{-n} = \dfrac{1}{x^n}$

In Example 6, we will discuss how to simplify the quotient of two powers in which the bases are the same.

| Example 6 | Simplifying Power Expressions |

Simplify.

1. $\dfrac{x^4}{x^9}$ 2. $\dfrac{4x^7}{3x^{-2}}$ 3. $\dfrac{6b^{-3}c^{-2}}{9b^4 c^{-5}}$

Solution

1. $\dfrac{x^4}{x^9} = x^{4-9}$ *Subtract exponents:* $\dfrac{x^m}{x^n} = x^{m-n}$

   $= x^{-5}$ *Subtract.*

   $= \dfrac{1}{x^5}$ *Write power so exponent is positive:* $x^{-n} = \dfrac{1}{x^n}$

2. $\dfrac{4x^7}{3x^{-2}} = \dfrac{4x^{7-(-2)}}{3}$ *Subtract exponents:* $\dfrac{x^m}{x^n} = x^{m-n}$

   $= \dfrac{4x^9}{3}$ *Simplify.*

3. $\dfrac{6b^{-3}c^{-2}}{9b^4 c^{-5}} = \dfrac{2b^{-3-4}c^{-2-(-5)}}{3}$ *Simplify; subtract exponents:* $\dfrac{x^m}{x^n} = x^{m-n}$

   $= \dfrac{2b^{-7}c^3}{3}$ *Simplify.*

   $= \dfrac{2c^3}{3b^7}$ *Write powers so exponents are positive:* $x^{-n} = \dfrac{1}{x^n}$

In the first step of Problem 2 of Example 6, we found that

$$\dfrac{4x^7}{3x^{-2}} = \dfrac{4x^{7-(-2)}}{3}$$

WARNING   Note that we need a subtraction symbol *and* a negative symbol in the expression on the right-hand side. It is a common error to omit writing one of these two symbols in problems of this type.

| Example 7 | Simplifying Power Expressions |

Simplify.

1. $\left(2x^{-5}\right)^{-3}$ 2. $\left(\dfrac{2x^{-4}}{y^6}\right)^{-5}$

Solution

1. $\left(2x^{-5}\right)^{-3} = 2^{-3}\left(x^{-5}\right)^{-3}$ *Raise factors to power* -3: $(xy)^n = x^n y^n$

$= 2^{-3}x^{15}$ *Multiply exponents:* $\left(x^m\right)^n = x^{mn}$

$= \dfrac{x^{15}}{2^3}$ *Write powers so exponents are positive:* $x^{-n} = \dfrac{1}{x^n}$

$= \dfrac{x^{15}}{8}$ *Simplify.*

2. $\left(\dfrac{2x^{-4}}{y^6}\right)^{-5} = \dfrac{\left(2x^{-4}\right)^{-5}}{\left(y^6\right)^{-5}}$ *Raise numerator and denominator to power* -5: $\left(\dfrac{x}{y}\right)^n = \dfrac{x^n}{y^n}$

$= \dfrac{2^{-5}\left(x^{-4}\right)^{-5}}{\left(y^6\right)^{-5}}$ *Raise each factor to power* -5: $(xy)^n = x^n y^n$

$= \dfrac{2^{-5}x^{20}}{y^{-30}}$ *Multiply exponents:* $\left(x^m\right)^n = x^{mn}$

$= \dfrac{x^{20}y^{30}}{2^5}$ *Write powers so exponents are positive:* $x^{-n} = \dfrac{1}{x^n}$; $\dfrac{1}{x^{-n}} = x^n$

$= \dfrac{x^{20}y^{30}}{32}$ *Simplify.*

Models That Have Negative-Integer Exponents

Many authentic situations can be modeled by equations that contain negative-integer exponents. Some examples of such situations are the sound level of a guitar, the intensity of illumination by a light bulb, and the gravitational force of the Sun acting on Earth.

| Example 8 | Using a Model That Has Negative-Integer Exponents |

The intensity of a television signal I (in watts per square meter) at a distance d kilometers from the transmitter is described by the equation

$$I = 250d^{-2}$$

1. Simplify the right-hand side of the equation.
2. Substitute 5 for d in the equation found in Problem 1, and solve for I. What does the result mean in this situation?

Solution

1. We use the definition of a negative-integer exponent to write the expression on the right-hand side without any negative exponents:

$I = 250d^{-2}$ *Original equation*

$I = \dfrac{250}{d^2}$ *Write power so exponent is positive:* $x^{-n} = \dfrac{1}{x^n}$

2. We substitute 5 for d in the equation $I = \dfrac{250}{d^2}$:

$$I = \dfrac{250}{5^2} = \dfrac{250}{25} = 10$$

So, the intensity is 10 watts per square meter at a distance of 5 kilometers from the transmitter.

Scientific Notation

Now that we know how to work with negative-integer exponents, we can use exponents to describe numbers in *scientific notation*. This will enable us to describe compactly a number whose absolute value is very large or very small. For example, the distance to Proxima Centauri, the nearest star other than the Sun, is 40,100,000,000,000 kilometers. Here we write 40,100,000,000,000 in scientific notation:

$$4.01 \times 10^{13}$$

The symbol "$\times$" stands for multiplication.

As another example, 1 square inch is approximately 0.00000016 acre. Here we write 0.00000016 in scientific notation:

$$1.6 \times 10^{-7}$$

Definition Scientific notation

A number is written in **scientific notation** if it has the form $N \times 10^k$, where k is an integer and the absolute value of N is between 1 and 10 or is equal to 1.

Here are more examples of numbers in scientific notation:

$$8.6 \times 10^{19} \qquad 2.159 \times 10^8 \qquad -4.23 \times 10^{-14} \qquad 7.94 \times 10^{-97}$$

The problems in Example 9 suggest how to convert a number from scientific notation $N \times 10^k$ to standard decimal notation.

Example 9 Writing Numbers in Standard Decimal Notation

Simplify.

1. 7×10^3 **2.** 7×10^{-3} **3.** 9.48×10^{-4}

Solution

1. We simplify $7 \times 10^3 = 7.0 \times 10^3$ by *multiplying* 7.0 by 10 three times and hence moving the decimal point three places to the *right*:

$$7.0 \times 10^3 = 7000.0 = 7000$$

three places to the right

2. Because

$$7 \times 10^{-3} = 7 \times \frac{1}{10^3} = \frac{7}{1} \times \frac{1}{10^3} = \frac{7}{10^3}$$

we see that we can simplify 7.0×10^{-3} by *dividing* 7.0 by 10 three times and hence moving the decimal point three places to the *left*:

$$7.0 \times 10^{-3} = 0.007$$

three places to the left

3. We *divide* 9.48 by 10 four times and hence move the decimal point of 9.48 four places to the *left*:

$$9.48 \times 10^{-4} = 0.000948$$

four places to the left

Converting from Scientific Notation to Standard Decimal Notation

To write the scientific notation $N \times 10^k$ in standard decimal notation, we move the decimal point of the number N as follows:

- If k is *positive*, we multiply N by 10 k times and hence move the decimal point k places to the *right*.
- If k is *negative*, we divide N by 10 k times and hence move the decimal point k places to the *left*.

The problems in Example 10 suggest how to convert a number from standard decimal notation to scientific notation.

Example 10   Writing Numbers in Scientific Notation

Write the number in scientific notation.

1. 845,000,000   **2.** 0.0000382

Solution

1. In scientific notation, we would have 8.45×10^k, but what is k? If we move the decimal point of 8.45 eight places to the right, the result is 845,000,000. So, $k = 8$ and the scientific notation is 8.45×10^8.
2. In scientific notation, we would have 3.82×10^k, but what is k? If we move the decimal point of 3.82 five places to the left, the result is 0.0000382. So, $k = -5$ and the scientific notation is 3.82×10^{-5}.

Converting from Standard Decimal Notation to Scientific Notation

To write a number in scientific notation, count the number k of places that the decimal point needs to be moved so that the absolute value of the new number N is between 1 and 10 or is equal to 1.

- If the decimal point is moved to the left, then the scientific notation is written as $N \times 10^k$.
- If the decimal point is moved to the right, then the scientific notation is written as $N \times 10^{-k}$.

Example 11   Writing Numbers in Scientific Notation

Write the number in scientific notation.

1. 778,000,000   (*Jupiter's average distance, in kilometers, from the Sun*)
2. 0.000012   (*the diameter, in meters, of a white blood cell*)

Solution

1. For 778,000,000, the decimal point needs to be moved eight places to the left so the new number N is between 1 and 10. Therefore, the scientific notation is 7.78×10^8.
2. For 0.000012, the decimal point needs to be moved five places to the right so the new number N is between 1 and 10. Therefore, the scientific notation is 1.2×10^{-5}.

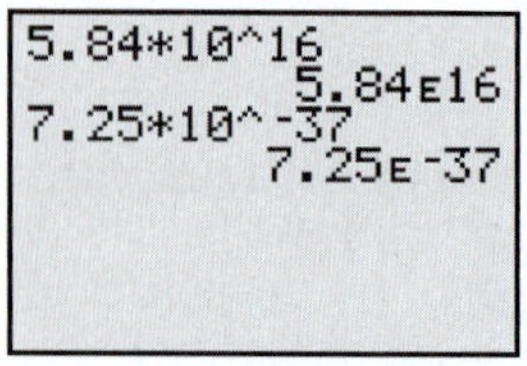

```
5.84*10^16
          5.84E16
7.25*10^-37
          7.25E-37
```

Figure 50 The numbers 5.84×10^{16} and 7.25×10^{-37}

Calculators express numbers in scientific notation so that the numbers "fit" on the screen. To represent 5.84×10^{16}, most calculators use the notation 5.84 E 16, where E stands for exponent (of 10). Calculators represent 7.25×10^{-37} as 7.25 E −37 (see Fig. 50).

Group Exploration

Properties of exponents

1. Because $x^{-5} = \dfrac{1}{x^5}$ for nonzero x, does it follow that $-5 = \dfrac{1}{5}$? Explain.

2. A student tries to simplify $7x^{-2}$:

$$7x^{-2} = \frac{1}{7x^2}$$

Describe any errors. Then simplify the expression correctly.

3. A student tries to simplify $\dfrac{x^8}{x^{-5}}$:

$$\frac{x^8}{x^{-5}} = x^{8-5} = x^3$$

Describe any errors. Then simplify the expression correctly.

4. A student tries to simplify $\left(5x^3\right)^{-2}$:

$$\left(5x^3\right)^{-2} = 5\left(x^3\right)^{-2} = 5x^{-6} = \frac{5}{x^6}$$

Describe any errors. Then simplify the expression correctly.

Tips for Success Complete the Rest of the Assignment

If you have spent a good amount of time trying to solve an exercise but can't, consider going on to the next exercise in the assignment. You may find that the next exercise involves a different concept or involves a more familiar situation. You may even find that, after completing the rest of the assignment, you are able to complete the exercise(s) you skipped. One explanation of this phenomenon is that you may have learned or remembered some concept in a later exercise that relates to the exercise with which you were struggling.

Homework 7.7

For extra help ▶ **MyLab Math** Watch the videos in MyLab Math

Simplify.

1. 6^{-2}

2. 9^{-2}

3. x^{-4}

4. x^{-2}

5. b^{-1}

6. t^{-8}

7. $\dfrac{1}{2^{-4}}$

8. $\dfrac{1}{9^{-2}}$

9. $\dfrac{1}{w^{-2}}$

10. $\dfrac{1}{a^{-5}}$

11. $\dfrac{x^{-3}}{y^5}$

12. $\dfrac{x^2}{y^{-7}}$

13. $\dfrac{a^{-4}}{b^{-2}}$

14. $\dfrac{r^{-4}}{t^{-8}}$

15. $\dfrac{2a^3b^{-5}}{5c^{-8}}$

16. $\dfrac{3x^{-1}y^6}{7w^{-2}}$

17. $\dfrac{4x^{-9}}{-6y^4w^{-1}}$

18. $\dfrac{-9x^{-1}}{6y^{-5}w^7}$

Simplify.

19. $\left(x^{-2}\right)^7$

20. $\left(x^5\right)^{-8}$

21. $\left(t^{-4}\right)^{-3}$

22. $\left(a^{-6}\right)^{-1}$

23. $\left(6t^{-4}\right)\left(5t^2\right)$

24. $\left(9w^4\right)\left(2w^{-6}\right)$

25. $\left(-4x^{-1}\right)\left(3x^{-8}\right)$

26. $\left(-5x^{-3}\right)\left(-8x^{-5}\right)$

27. $\left(-4x^3y^{-7}\right)\left(-x^{-5}y^4\right)$

28. $\left(-3b^{-2}c^{-6}\right)\left(-b^9c^{-1}\right)$

29. $\dfrac{x^2}{x^6}$

30. $\dfrac{x^5}{x^8}$

31. $\dfrac{a^{-3}}{a^5}$

32. $\dfrac{r^{-4}}{r^7}$

33. $\dfrac{a^3}{a^{-2}}$

34. $\dfrac{y^6}{y^{-1}}$

35. $\dfrac{7x^{-3}}{4x^{-9}}$

36. $\dfrac{4x^{-2}}{9x^{-5}}$

37. $\dfrac{2^{-1}}{2^4}$

38. $\dfrac{3^{-2}}{3^2}$

39. $\dfrac{5^{-6}}{5^{-4}}$

40. $\dfrac{7^{-5}}{7^{-6}}$

41. $\dfrac{3^4w^{-8}}{3^2w^{-3}}$

42. $\dfrac{2^5b^{-7}}{2^2b^{-2}}$

43. $\left(8c^{-3}\right)^2$

44. $\left(4t^{-7}\right)^3$

45. $\left(2x^{-1}\right)^{-5}$

46. $\left(2x^{-4}\right)^{-4}$

47. $\left(x^{-2}y^5\right)^{-6}$

48. $\left(x^4y^{-7}\right)^{-5}$

49. $\left(2a^{-6}b\right)^{-3}$

50. $\left(4r^3t^{-8}\right)^{-2}$

51. $\left(ab^2\right)^3\left(a^3\right)^{-2}$

52. $\left(t^3w\right)^5\left(w^2\right)^{-4}$

53. $\dfrac{1}{(xy)^{-3}}$

54. $\dfrac{1}{(xy)^{-5}}$

55. $\left(\dfrac{x^{-3}}{y^2}\right)^4$

56. $\left(\dfrac{x^5}{y^{-8}}\right)^3$

57. $\left(\dfrac{2}{c^{-7}}\right)^{-3}$

58. $\left(\dfrac{w^{-3}}{7}\right)^{-1}$

59. $\left(\dfrac{3r^{-5}}{t^9}\right)^3$

60. $\left(\dfrac{a^3}{5b^{-6}}\right)^2$

61. $\left(\dfrac{8a^{-3}b}{c^{-5}d^4}\right)^{-2}$

62. $\left(\dfrac{x^{-4}y^2}{tw^{-7}}\right)^{-3}$

63. $\dfrac{\left(2a^{-2}b\right)^{-3}}{\left(3cd^{-3}\right)^2}$

64. $\dfrac{\left(5xy^{-3}\right)^{-2}}{\left(2t^2w^{-1}\right)^{-3}}$

65. $\dfrac{6b^{-3}c^4}{8b^2c^{-3}}$

66. $\dfrac{9t^4w^{-6}}{3t^{-2}w^3}$

67. If an object is moving at a constant speed s (in miles per hour), then $s = dt^{-1}$, where d is the distance traveled (in miles) and t is the time (in hours).
   a. Simplify the right-hand side of the equation.
   b. Substitute 186 for d and 3 for t in the equation you found in part (a), and solve for s. What does your result mean in this situation?

68. The force F (in pounds) you must exert on a wrench handle of length L inches to loosen a bolt is described by the equation $F = 720L^{-1}$.
   a. Simplify the right-hand side of the equation.
   b. Substitute 12 for L in the equation you found in part (a), and solve for F. What does your result mean in this situation?

69. If a person plays an electric guitar outside, the sound level L (in decibels) at a distance d yards from the amplifier is described by the equation $L = 5760d^{-2}$.
   a. Simplify the right-hand side of the equation.
   b. Substitute 8 for d in the equation you found in part (a), and solve for L. What does your result mean in this situation?

70. The amount of light I (in milliwatts per square centimeter) of a 50-watt light bulb at a distance d centimeters is described by the equation $I = 8910d^{-2}$.
   a. Simplify the right-hand side of the equation.
   b. Substitute 80 for d in the equation you found in part (a), and solve for I. What does your result mean in this situation?

71. A person plans to invest in an account at 6% interest compounded annually. If the person wants the balance to be $10,000 in t years, the money P (in dollars) she must invest is described by $P = 10,000(1.06)^{-t}$.
   a. Simplify the right-hand side of the equation.
   b. Substitute 8 for t in the equation you found in part (a), and solve for P. What does your result mean in this situation?

72. A person plans to invest in an account at 4% interest compounded annually. If the person wants the balance to be $7000 in t years, the money P (in dollars) she must invest is described by $P = 7000(1.04)^{-t}$.
   a. Simplify the right-hand side of the equation.
   b. Substitute 6 for t in the equation you found in part (a), and solve for P. What does your result mean in this situation?

Write the number in standard decimal form.

73. 4.9×10^4

74. 8.31×10^6

75. 8.59×10^{-3}

76. 6.488×10^{-5}

77. 2.95×10^{-4}

78. 8.7×10^{-2}

79. -4.512×10^8

80. -9.46×10^{10}

Write the number in scientific notation.

81. 45,700,000

82. 280,000

83. 0.0000659

84. 0.000023

85. $-5,987,000,000,000$

86. $-308,000,000$

87. 0.000001

88. 0.0004

For each of Exercises 89 and 90, numbers are displayed in scientific notation in a graphing calculator table shown in the given figure. Write each of these numbers in the Y_1 column in standard decimal form.

89. See Fig. 51.

90. See Fig. 52.

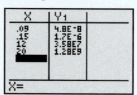

Figure 51 Exercise 89

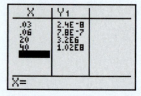

Figure 52 Exercise 90

For Exercises 91–96, the given sentence contains a number written in scientific notation. Write the number in standard decimal form.

91. The Milky Way galaxy contains about 4×10^{11} stars.

92. Light travels at 2.998×10^8 meters per second.

93. Bill Gates's net worth was 8.5×10^{10} dollars in September 2017 (Source: *Forbes*).

94. The intensity of the sound inside a running car is about 1×10^{-8} watt per square meter.

95. Orange juice has a hydrogen ion concentration of 6.3×10^{-4} mole per liter.

96. One second is 3.17×10^{-8} year.

For Exercises 97–102, the given sentence contains a number written in standard decimal form. Write the number in scientific notation.

97. The area of the United States is about 3,720,000 square miles.

98. The diameter of Earth is about 12,700 kilometers.

99. The temperature of the Sun's core is approximately 27,000,000°F.

100. The diameter of a human hair is about 0.00254 centimeter.

101. A human red blood cell has a diameter of about 0.0000075 meter.

102. One teaspoon is approximately 0.0013 gallon.

Concepts

103. A student tries to simplify $\dfrac{5x^{-3}y^2}{w}$:

$$\frac{5x^{-3}y^2}{w} = \frac{y^2}{5x^3w}$$

Describe any errors. Then simplify the expression correctly.

104. A student tries to simplify $\dfrac{-8x^5}{y^3}$:

$$\frac{-8x^5}{y^3} = \frac{x^5}{8y^3}$$

Describe any errors. Then simplify the expression correctly.

105. A student tries to simplify $\dfrac{x^6}{x^{-4}}$:

$$\frac{x^6}{x^{-4}} = x^{6-4} = x^2$$

Describe any errors. Then simplify the expression correctly.

106. A student thinks that 4^{-3} is a negative number. Is the student correct? Explain.

107. To write scientific notation $N \times 10^k$ in standard decimal notation, we move the decimal point of N to the right if k is positive. Why does this make sense?

108. What is an advantage of using scientific notation?

109. a. Find 1^5.
b. Find 1^0.
c. Find 1^{-3}.
d. Find 1^n, where n is an integer.

110. Explain how you can write a power expression with negative-integer exponents as an equivalent expression with only positive exponents. (See page 9 for guidelines on writing a good response.)

Related Review

Simplify.

111. $-3x^{-3} + \left(-3x^{-3}\right)$
112. $3x^3 + 3x^3$
113. $-3x^{-3}\left(-3x^{-3}\right)$
114. $3x^3\left(3x^3\right)$

115. $\left(-3x^{-3}\right)^3$
116. $\left(3x^3\right)^3$
117. $\left(-3x^3\right)^{-3}$
118. $\left(-3x^{-3}\right)^{-3}$

119. a. Simplify $\dfrac{x^4}{x^{-3}}$ by each of the following methods:
  i. Use the quotient property $\dfrac{x^m}{x^n} = x^{m-n}$.
  ii. Use the property $\dfrac{1}{x^{-n}} = x^n$.
b. Compare the results you found in part (a).

120. a. Simplify $\dfrac{x^6}{x^2}$ by each of the following methods:
  i. Use the quotient property $\dfrac{x^m}{x^n} = x^{m-n}$.
  ii. Write x^6 and x^2 without exponents.
b. Compare the results you found in part (a).

Expressions, Equations, and Graphs

Perform the indicated instruction. Then use words such as linear, quadratic, cubic, polynomial, degree, one variable, *and* two variables *to describe the expression, equation, or system.*

121. Simplify $3(2x - 5) - 4(3x + 2)$.
122. Evaluate $3(2x - 5) - 4(3x + 2)$ for $x = -2$.
123. Solve $3(2x - 5) - 4(3x + 2) = 0$.
124. Graph $7x - 4y = 12$.

7.8 Dividing Polynomials

Objectives

» Divide a polynomial by a monomial.

» Use long division to divide a polynomial by a binomial.

In this chapter, we have discussed how to add, subtract, and multiply polynomials. In this section, we will divide polynomials.

Dividing a Polynomial by a Monomial

First, we will discuss how to divide a polynomial by a monomial. Recall the following rule about adding fractions with a common denominator:

$$\frac{A}{B} + \frac{C}{B} = \frac{A + C}{B}, \quad \text{where } B \text{ is nonzero}$$

To divide by a monomial, we will go backward.

Dividing a Polynomial by a Monomial

If A, B, and C are monomials and B is nonzero, then

$$\frac{A + C}{B} = \frac{A}{B} + \frac{C}{B}$$

In words: To divide a polynomial by a monomial, divide each term of the polynomial by the monomial.

WARNING   It is a common error to divide only some of the terms of the polynomial by the monomial. Remember to divide *every* term of the polynomial by the monomial.

Example 1   Dividing by a Monomial

Find the quotient $\dfrac{15w^2 + 35w}{5w}$.

Solution

$$\dfrac{15w^2 + 35w}{5w} = \dfrac{15w^2}{5w^1} + \dfrac{35w^1}{5w^1} \qquad \textit{Divide each term by } 5w\text{: } \dfrac{A+C}{B} = \dfrac{A}{B} + \dfrac{C}{B}$$

$$= 3w^{2-1} + 7w^{1-1} \qquad \textit{Simplify; subtract exponents: } \dfrac{x^m}{x^n} = x^{m-n}$$

$$= 3w + 7 \qquad \textit{Simplify; } x^{1-1} = x^0 = 1$$

Because $\dfrac{6}{2} = 3$, it follows that $2 \cdot 3 = 6$. We can use this concept to verify our work in Example 1. To check that

$$\dfrac{15w^2 + 35w}{5w} = 3w + 7$$

we multiply $5w$ and $3w + 7$:

$$5w(3w + 7) = 15w^2 + 35w$$

which checks.

Example 2   Dividing by a Monomial

Find the quotient $\dfrac{6x^4 - 7x^3 + 8}{2x^3}$.

Solution

$$\dfrac{6x^4 - 7x^3 + 8}{2x^3} = \dfrac{6x^4}{2x^3} - \dfrac{7x^3}{2x^3} + \dfrac{8}{2x^3} \qquad \begin{array}{l}\textit{Divide each term by } 2x^3\text{:} \\ \dfrac{A - C + D}{B} = \dfrac{A}{B} - \dfrac{C}{B} + \dfrac{D}{B}\end{array}$$

$$= 3x^{4-3} - \dfrac{7x^{3-3}}{2} + \dfrac{4}{x^3} \qquad \textit{Simplify; subtract exponents: } \dfrac{x^m}{x^n} = x^{m-n}$$

$$= 3x - \dfrac{7}{2} + \dfrac{4}{x^3} \qquad \textit{Simplify.}$$

We use a graphing calculator table to verify our work (see Fig. 53).

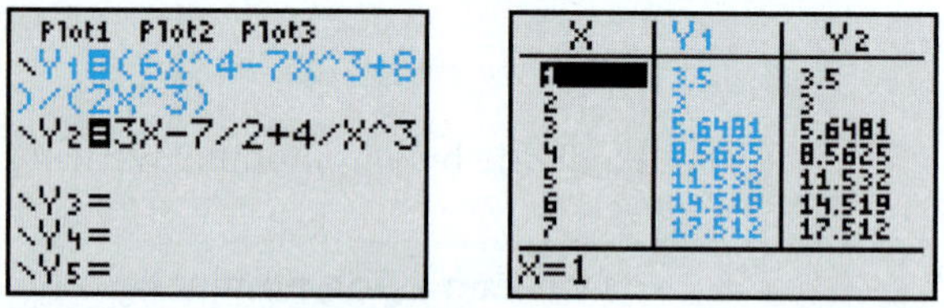

Figure 53 Verify the work

Example 3   Dividing by a Monomial

Find the quotient $\dfrac{9x^4y + 15x^3y^2 - 6x^2y^3}{-3x^2y}$.

Solution

$$\frac{9x^4y + 15x^3y^2 - 6x^2y^3}{-3x^2y}$$

$$= \frac{9x^4y}{-3x^2y} + \frac{15x^3y^2}{-3x^2y} - \frac{6x^2y^3}{-3x^2y}$$ *Divide each term by $-3x^2y$:* $\frac{A + C - D}{B} = \frac{A}{B} + \frac{C}{B} - \frac{D}{B}$

$$= -3x^{4-2}y^{1-1} - 5x^{3-2}y^{2-1} + 2x^{2-2}y^{3-1}$$ *Simplify; subtract exponents:* $\frac{x^m}{x^n} = x^{m-n}$

$$= -3x^2 - 5xy + 2y^2$$ *Simplify.*

We verify our work by finding the product $-3x^2y(-3x^2 - 5xy + 2y^2)$:

$$-3x^2y(-3x^2 - 5xy + 2y^2) = 9x^4y + 15x^3y^2 - 6x^2y^3$$

Using Long Division to Divide a Polynomial by a Binomial

We can use long division to divide a polynomial by a binomial. The steps are similar to performing long division with numbers. Here we review how to use long division to divide 3981 by 17:

$$
\begin{array}{r}
2\;3\;4 \quad \leftarrow \text{Quotient} \\
\text{Divisor} \to 17\overline{)3\;9\;8\;1} \quad \leftarrow \text{Dividend} \\
-3\;4\downarrow \quad\quad 2\cdot 17 = 34 \\
\overline{5\;8}\downarrow \quad \textit{Subtract and bring down next digit of dividend.} \\
-5\;1\downarrow \quad 3\cdot 17 = 51 \\
\overline{7\;1} \quad \textit{Subtract and bring down next digit of dividend.} \\
-6\;8 \quad 4\cdot 17 = 68 \\
\text{Remainder} \longrightarrow 3 \quad \textit{Subtract.}
\end{array}
$$

We conclude

$$\frac{3981}{17} = 234 + \frac{3}{17}$$

Recall that we can verify our work by checking that

Divisor	Quotient	Remainder	Dividend
↓	↓	↓	↓
17 ·	234 +	3 $\stackrel{?}{=}$	3981

which is true.

Example 4   Dividing by a Binomial

Divide: $\dfrac{3x^2 + 10x + 8}{x + 2}$.

Solution

The steps are similar to performing long division with numbers. To begin, we divide $3x^2$ (the first term of $3x^2 + 10x + 8$) by x (the first term of $x + 2$): $\dfrac{3x^2}{x} = 3x$.

$$
\begin{array}{r}
3x \quad\quad\quad\quad \dfrac{3x^2}{x} = 3x \\
x + 2\overline{)3x^2 + 10x + 8} \quad\quad \\
3x^2 + 6x \quad\quad 3x(x + 2) = 3x^2 + 6x
\end{array}
$$

To subtract $3x^2 + 6x$, we change the signs of $3x^2$ and $6x$ and add:

$$
\begin{array}{r}
3x \quad\quad\quad\quad \\
x + 2\overline{)3x^2 + 10x + 8} \\
\text{Change signs.} \quad {}^{-}3x^2 \mp 6x \\
\hline
4x \quad\quad\quad \textit{Add.}
\end{array}
$$

Next, we bring down the 8:

$$
\begin{array}{r}
3x \\
x + 2 \overline{)\, 3x^2 + 10x + 8} \\
\underline{-3x^2 - 6x} \qquad \downarrow \\
4x + 8
\end{array}
$$
Bring down the 8.

Then we repeat the process:

$$
\begin{array}{r}
3x + 4 \\
x + 2 \overline{)\, 3x^2 + 10x + 8} \\
\underline{-3x^2 - 6x} \\
4x + 8 \\
\underline{-4x \mp 8} \\
0
\end{array}
$$

Change signs. $\dfrac{4x}{x} = 4$ $4(x + 2) = 4x + 8$ *Add.*

We conclude

$$
\frac{3x^2 + 10x + 8}{x + 2} = 3x + 4
$$

We verify our work by checking that

$$
\underset{\substack{\downarrow \\ (x+2)}}{\text{Divisor}} \cdot \underset{\substack{\downarrow \\ (3x+4)}}{\text{Quotient}} + \underset{\substack{\downarrow \\ 0}}{\text{Remainder}} \overset{?}{=} \underset{\substack{\downarrow \\ 3x^2 + 10x + 8}}{\text{Dividend}}
$$

which is true.

▶

Example 5   Dividing by a Binomial

Divide: $\dfrac{6x^2 - 23x + 27}{3x - 4}$.

Solution

To begin, we divide $6x^2$ (the first term of $6x^2 - 23x + 27$) by $3x$ (the first term of $3x - 4$): $\dfrac{6x^2}{3x} = 2x$.

$$
\begin{array}{r}
2x \\
3x - 4 \overline{)\, 6x^2 - 23x + 27} \\
\underline{-6x^2 \mp 8x} \qquad \downarrow \\
-15x + 27
\end{array}
$$

Change signs. $\dfrac{6x^2}{3x} = 2x$ $2x(3x - 4) = 6x^2 - 8x$ *Add. Then bring down the 27.*

Then we repeat the process:

$$
\begin{array}{r}
2x - 5 \\
3x - 4 \overline{)\, 6x^2 - 23x + 27} \\
\underline{-6x^2 + 8x} \\
-15x + 27 \\
\underline{+15x \mp 20} \\
7
\end{array}
$$

Change signs. $\dfrac{-15x}{3x} = -5$ $-5(3x - 4) = -15x + 20$ *Add.*

We conclude

$$
\frac{6x^2 - 23x + 27}{3x - 4} = 2x - 5 + \frac{7}{3x - 4}
$$

To verify our work, we simplify $(3x - 4)(2x - 5) + 7$:

$$
\begin{array}{c}
\textcolor{blue}{\text{Divisor}} \qquad \textcolor{red}{\text{Quotient}} \qquad \textcolor{purple}{\text{Remainder}} \\
\downarrow \qquad\qquad \downarrow \qquad\qquad \downarrow
\end{array}
$$

$$
\begin{aligned}
(\textcolor{blue}{3x - 4}) \cdot (\textcolor{red}{2x - 5}) + \textcolor{purple}{7} &= 6x^2 - 15x - 8x + 20 + 7 \\
&= \textcolor{green}{6x^2 - 23x + 27} \leftarrow \text{Dividend}
\end{aligned}
$$

Because the result is the dividend, this checks.

▶

Example 6 Dividing a Polynomial with Missing Terms

Divide: $\dfrac{8x^3 - 27}{2x - 3}$.

Solution

Because the dividend, $8x^3 - 27$, doesn't have x^2 or x terms, we use $\textcolor{blue}{0x^2}$ and $\textcolor{blue}{0x}$ as placeholders:

$$8x^3 + \textcolor{blue}{0x^2} + \textcolor{blue}{0x} - 27$$

This way, like terms will line up when we perform long division:

$$
\begin{array}{l}
\,4x^2 + 6x + 9 \\
2x - 3 \overline{)8x^3 + 0x^2 + 0x - 27} \qquad \textit{Use } 0x^2 \textit{ and } 0x \textit{ as placeholders.}\\
\underline{8x^3 - 12x^2} \downarrow \qquad\quad 4x^2(2x - 3) = 8x^3 - 12x^2 \\
12x^2 + 0x \qquad\quad \textit{Add and bring down } 0x.\\
\underline{12x^2 - 18x} \downarrow \qquad\quad 6x(2x - 3) = 12x^2 - 18x \\
18x - 27 \qquad\quad \textit{Add and bring down } -27.\\
\underline{18x - 27} \qquad\quad 9(2x - 3) = 18x - 27 \\
0 \qquad\quad \textit{Add.}
\end{array}
$$

Change signs. (×3)

We conclude

$$\frac{8x^3 - 27}{2x - 3} = 4x^2 + 6x + 9$$

We use a graphing calculator table to verify our work (see Fig. 54).

 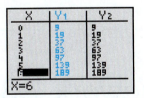

Figure 54 Verify the work

▶

Example 7 Dividing a Polynomial with Missing Terms

Divide: $\dfrac{4x^3 - 2x^2 + 7x + 8}{x^2 + 5}$.

Solution

Because the divisor, $x^2 + 5$, doesn't have an x term, we use $\textcolor{blue}{0x}$ as a placeholder:

$$x^2 + \textcolor{blue}{0x} + 5$$

Just like in Example 6, this will allow like terms to line up:

$$\begin{array}{r} 4x - 2 \\ x^2 + 0x + 5 \overline{)4x^3 - 2x^2 + 7x + 8} \end{array}$$

Change signs. $\underline{-4x^3 + 0x^2 + 20x}$ ↓ $4x(x^2 + 0x + 5) = 4x^3 + 0x^2 + 20x$

$-2x^2 - 13x + 8$ *Add and bring down the 8.*

Change signs. $\underline{+2x^2 + 0x + 10}$ $-2(x^2 + 0x + 5) = -2x^2 - 0x - 10$

$-13x + 18$ *Add.*

We conclude

$$\frac{4x^3 - 2x^2 + 7x + 8}{x^2 + 5} = 4x - 2 + \frac{-13x + 18}{x^2 + 5}$$

To verify our work, we simplify $(x^2 + 5)(4x - 2) + (-13x + 18)$:

Divisor	Quotient	Remainder
↓	↓	↓

$$(x^2 + 5) \cdot (4x - 2) + (-13x + 18) \quad = 4x^3 - 2x^2 + 20x - 10 - 13x + 1$$
$$= 4x^3 - 2x^2 + 7x + 8 \leftarrow \text{Dividend}$$

Because the result is the dividend, this checks.

▶

 Group Exploration

Section Opener: Dividing by a monomial

Recall the following rule about adding fractions with a common denominator:

$$\frac{A}{B} + \frac{C}{B} = \frac{A + C}{B}, \quad \text{where } B \text{ is nonzero}$$

To divide by a monomial, we will go backward:

$$\frac{A + C}{B} = \frac{A}{B} + \frac{C}{B}, \quad \text{where } B \text{ is nonzero}$$

Find the quotient. Then simplify your result.

1. $\dfrac{x^7 + x^4}{x^2}$

2. $\dfrac{4x^6 + 8x^3}{2x}$

3. $\dfrac{9x^5 - 12x^4 + 6x^3}{3x^3}$

4. $\dfrac{20x^7 + 8x^5 - 2x^4}{-4x^2}$

 # Homework 7.8

For extra help ▶   **MyLab Math**   📺 Watch the videos in MyLab Math

Divide. Use a graphing calculator table to verify your work.

1. $\dfrac{x^4 + x^2}{x}$

2. $\dfrac{x^5 + x^3}{x^2}$

3. $\dfrac{6x^3 + 8x^2}{2x}$

4. $\dfrac{10x^3 + 15x}{5x}$

5. $\dfrac{8m^4 - 4m^3 + 12m^2}{4m^2}$

6. $\dfrac{12b^5 + 9b^3 - 6b^2}{3b^2}$

7. $\dfrac{24x^5 - 5x^4 - 18x^3}{-6x^2}$

8. $\dfrac{21x^4 + 14x^3 - 2x^2}{-7x^2}$

Divide. Verify your work by multiplying.

9. $\dfrac{35x^4 - 7x^3 - 20x}{5x^3}$

10. $\dfrac{12x^5 - 13x^4 + 16x^2}{4x^4}$

11. $\dfrac{-6w^4 + 24w^3 - 16w}{-8w^2}$

12. $\dfrac{-10k^5 - 6k^4 + 12k}{-6k^3}$

13. $\dfrac{x^3y - x^2y^2 + xy^4}{xy}$

14. $\dfrac{x^4y^2 + x^2y^3 - xy^5}{xy}$

15. $\dfrac{6a^4b^2 + 3a^2b^3 + 9ab^4}{3ab^2}$

16. $\dfrac{2m^4p + 6m^3p^2 + 4m^2p^3}{2m^2p}$

17. $\dfrac{12x^5y^3 - 18x^4y^4 + 6x^2y^5}{-6x^2y^3}$ **18.** $\dfrac{15x^5y - 20x^4y^2 - 35x^3y^4}{-5x^3y}$

Divide. To verify your work, check that the product of the divisor and the quotient, plus the remainder is the dividend.

19. $\dfrac{2x^2 + 13x + 15}{x + 5}$ **20.** $\dfrac{3x^2 + 10x + 8}{x + 2}$

21. $\dfrac{4x^2 - 29x + 7}{x - 7}$ **22.** $\dfrac{5x^2 - 17x + 6}{x - 3}$

23. $\dfrac{4b^2 + 17b + 17}{b + 3}$ **24.** $\dfrac{6w^2 + 31w + 33}{w + 4}$

25. $\dfrac{15x^2 + x - 2}{3x + 2}$ **26.** $\dfrac{24x^2 - 2x - 21}{4x + 3}$

27. $\dfrac{6x^2 + x - 5}{2x - 1}$ **28.** $\dfrac{12x^2 + 11x - 7}{3x - 1}$

29. $\dfrac{8m^2 - 14m - 19}{2m - 5}$ **30.** $\dfrac{10k^2 - 11k - 4}{5k - 3}$

Divide. Use a graphing calculator table to verify your work.

31. $\dfrac{3x^3 + 7x^2 - 3x - 5}{x + 2}$ **32.** $\dfrac{2x^3 + 2x^2 - 11x + 7}{x + 3}$

33. $\dfrac{6x^3 - 5x^2 + 9x - 5}{2x - 1}$ **34.** $\dfrac{8x^3 - 22x^2 - 11x + 1}{4x - 1}$

35. $\dfrac{p^2 + 5}{p + 2}$ **36.** $\dfrac{m^2 + 4}{m + 3}$

37. $\dfrac{x^3 - 8}{x - 2}$ **38.** $\dfrac{x^3 - 27}{x - 3}$

39. $\dfrac{6x^3 - 11x^2 - 1}{2x - 1}$ **40.** $\dfrac{3x^3 - 10x^2 + 7}{3x - 1}$

41. $\dfrac{2w^3 - w^2 + 4w - 3}{w^2 - 1}$ **42.** $\dfrac{3y^3 + 2y^2 - y + 5}{y^2 + 2}$

43. $\dfrac{8x^3 - 2x^2 + 12x - 4}{2x^2 + 3}$ **44.** $\dfrac{6x^3 - 15x^2 + 4x - 15}{3x^2 + 2}$

Concepts

45. A student tries to find a quotient:

$$\frac{3x^2 + 5x}{3x^2} = \frac{3x^2}{3x^2} + 5x$$
$$= 1 + 5x$$

Describe any errors. Then find the quotient correctly.

46. A student tries to find a quotient:

$$\frac{2x^2 + 7x}{5x^2} = \frac{2x^2}{5x^2} + \frac{7x}{5x^2}$$
$$= \frac{2}{5} + \frac{7x}{5}$$

Describe any errors. Then find the quotient correctly.

47. A student tries to find the quotient $\dfrac{3x^2 + 11x + 10}{x + 2}$:

$$\begin{array}{r} -3x - 5 \\ x + 2 \overline{)\, 3x^2 + 11x + 10} \\ \underline{-3x^2 - 6x} \\ 5x + 10 \\ \underline{-5x - 10} \\ 0 \end{array}$$

$$\frac{3x^2 + 11x + 10}{x + 2} = -3x - 5$$

Describe any errors. Then find the quotient correctly.

48. a. Find the product $(x + 3)(x + 4)$.

b. Use long division to find the quotient $\dfrac{x^2 + 7x + 12}{x + 3}$.

c. Explain how your work in parts (a) and (b) is related.

49. Give an example of a polynomial and a monomial such that the quotient of the polynomial and the monomial is $3x^2 - 5x + 4$.

50. Give an example of a polynomial and a binomial such that the quotient of the polynomial and the binomial is $4x + 3$.

51. When a certain polynomial is divided by the binomial $x - 2$, the result is $3x - 4 + \dfrac{5}{x - 2}$. What is the polynomial?

52. If the division of two polynomials has remainder 0, what does the product of the divisor and quotient equal? Explain.

53. When we divide a polynomial by a monomial, we divide each term of the polynomial by the monomial. Explain why this makes sense.

54. Give several reasons why long division of polynomials is similar to long division of numbers.

55. Describe how to divide a polynomial by a monomial.

56. Describe how to divide a polynomial by a binomial.

57. After dividing a polynomial by a binomial, describe how to verify your work.

58. When dividing a polynomial with missing terms by a binomial, why do we write the missing terms with zero coefficients?

Related Review

Perform the indicated operation. Use a graphing calculator table to verify your work.

59. $(2x^2 + x - 4)(2x - 3)$

60. $(3x^2 - 5x - 7)(3x + 4)$

61. $(2x^2 + x - 4) - (2x - 3)$

62. $(3x^2 - 5x - 7) - (3x + 4)$

63. $\dfrac{2x^2 + x - 4}{2x - 3}$ **64.** $\dfrac{3x^2 - 5x - 7}{3x + 4}$

65. $(2x - 3)^2$ **66.** $(3x + 4)^2$

Expressions, Equations, and Graphs

Give an example of the following. Then solve, simplify, or graph, as appropriate.

67. A quadratic equation in two variables

68. A linear equation in two variables

69. A quadratic polynomial in one variable and with four terms

70. A linear equation in one variable

71. A cubic polynomial in one variable and with five terms

72. A system of two linear equations in two variables

Taking It to the Lab

Climate Change Lab (continued from Chapter 6)

Throughout our study of climate change, we have used the IPCC's yardstick that, by 2050, carbon dioxide emissions should be 0.9 metric ton per person per year. Yet it is difficult to imagine developed countries, with average annual carbon dioxide emissions of 9.9 metric tons per person today, reducing their emissions by 91% to meet the IPCC's recommendation. In particular, it is difficult to imagine that the United States will reduce its annual carbon dioxide emissions of 17.0 metric tons per person by 95% to reach that desired level.

Recall from the Climate Change Lab in Chapter 5 that the IPCC's yardstick is equivalent to recommending that 2050 carbon dioxide emissions be 60% less than the carbon dioxide emissions in 1990. We can reach this goal without reducing per-person carbon dioxide emissions at all—if we reduce world population.

Reducing world population seems impossible because it is currently growing by leaps and bounds. The United Nations (U.N.) predicts that, from 2000 to 2050, the world population will increase from 6.09 billion to 9.3 billion—a 53% increase (see Table 37).

Table 37 World Population

Year	Population (billions)	Year	Population (billions)
A.D. 1	0.17	1600	0.55
200	0.19	1800	0.81
400	0.19	2000	6.09
600	0.20	2025	7.9
800	0.22	2050	9.3
1000	0.25	2075	9.9
1200	0.36	2100	10.1
1400	0.35		

Sources: *The Cambridge Factfinder; U.N.*

Despite the large population growth in the first half of the century, the population will probably not grow nearly as quickly in the second half. The U.N. has predicted the population will increase from 9.3 billion in 2050 to 10.1 billion in 2100—a mere 9% increase. In fact, the U.N. has predicted that the rate of change in 2100 will be very close to zero and might even be negative.

The U.N. uses a complicated model to make these predictions. A simpler model that matches well with the U.N. model for the years 2000 to 2100 is the quadratic equation $p = -0.00046t^2 + 0.086t + 6.08$, where p is world population (in billions) at t years since 2000.

Using the yardstick of 2050 carbon dioxide emissions at 60% less than the 21.6 billion metric tons of carbon dioxide

emissions in 1990, we can calculate the largest carbon dioxide emissions that Earth can handle each year:

$$0.40(21.6) = 8.64 \text{ billion metric tons}$$

Next, we will consider two scenarios that stay within this limit.

For our first scenario, let's assume that in the future each person in the world emits 17.0 metric tons per year, the current per-person annual emissions rate for Americans (see Table 38). Then we could still be at the yearly limit of 8.64 billion metric tons of carbon dioxide if the world population were a mere 508 million. By modeling the data in Table 39, we could show that U.S. population alone will reach 508 million before 2090. Common sense dictates that countries would never voluntarily lower their populations enough to help reach this level.

Table 38 GDP Ranks, per-Person GDPs, and per-Person Carbon Dioxide Emissions

Country	2014 GDP Rank	2014 per-Person GDP (thousands of international dollars)	2014 per-Person Carbon Emissions (metric tons)
Sweden	21	58.5	4.6
Switzerland	20	87.5	5.0
France	6	44.5	5.0
Italy	8	35.8	5.5
United Kingdom	5	45.7	6.7
Denmark	34	60.6	6.8
Austria	28	51.3	7.4
Norway	27	97.0	8.7
Belgium	25	47.7	8.9
Japan	3	36.3	9.7
Germany	4	47.6	9.8
Australia	12	61.2	16.0
United States	1	54.6	17.0
Netherlands	17	51.4	9.1

Sources: *Global Carbon Atlas; United Nations; International Monetary Fund*

Table 39 U.S. Population

Year	U.S. Population (billions)
1960	0.18
1970	0.21
1980	0.23
1990	0.25
2000	0.28
2010	0.31
2014	0.32

Sources: *U.S. Census Bureau; The Cambridge Factfinder*

For our second scenario, let's make an assumption about how low carbon dioxide emissions could be. Although

0.9 metric ton per person per year seems unreachable, perhaps 5.0 metric tons per person per year is attainable. After all, Sweden, Switzerland, and France already have per-person annual carbon dioxide emissions at or below 5.0 metric tons (see Table 38). If all countries annually produced 5.0 metric tons of carbon dioxide per person, Earth could handle 1.7 billion people.

The first scenario reveals that we will not be able to get carbon dioxide emissions under control merely by reducing world population. Earlier, we observed that simply reducing per-person emissions will not be acceptable. So, as the second scenario suggests, it is only by reducing both world population and per-person emissions that we can reach the IPCC's goal.

Analyzing the Situation

1. Let w be world population (in billions of people) in the year t. For example, $t = 2005$ represents the year 2005.
   a. Use a graphing calculator to draw a scatterplot of world population data.
   b. Calculate the changes in world population for every 200 (or 199) years up until 2000.
   c. When did world population grow the most? Explain why it is surprising world population will increase by only about 0.2 billion from 2075 to 2100.
2. Now let w be world population (in billions of people) at t years *since 2000*.
   a. Use a graphing calculator to draw a scatterplot for the years 2000–2100 that are listed in Table 37. In the same viewing window, draw a graph of the scatterplot and the quadratic world population model provided earlier.
   b. Use the quadratic model to predict world population for the years 2000, 2025, 2050, 2075, and 2100.
   c. Does the quadratic model make predictions that are similar to the U.N. predictions listed in Table 37? Explain.
3. In Problem 4 of the Climate Change Lab in Chapter 5, you found a model of the U.S. population (in billions) at t years since 1950. If you didn't find such an equation, find it now. Then use it to predict when the U.S. population will reach 508 million. Finally, compare your result with the claim that this will happen before 2090.
4. Perform a unit analysis for the estimate of the amount of carbon dioxide emissions (in billions of tons of carbon dioxide) that Earth can withstand each year:

$$0.40(21.6) = 8.64 \text{ billion metric tons}$$

Explain why this computation gives the amount of carbon dioxide emissions that Earth can withstand each year.
5. Show the work for the claim that if everyone in the world emitted carbon dioxide as Americans do, then Earth could support only 508 million people.
6. Verify the claim that if worldwide annual carbon dioxide emissions were 5.0 metric tons of carbon dioxide per person, then Earth could support 1.7 billion people.

Projectile Lab

In this lab, you will estimate your vertical throwing speed.

Materials

You will need at least three people and the following items:
1. A baseball or other ball (a solid, heavy ball works best)
2. A digital stopwatch or some other timing device

Recording of Data

The first person should throw the ball straight up and say "Mark" at the moment the ball is released. The second person should hold his or her hand at the position of the ball's release and should say "Mark" when the ball returns to its initial height. The third person should begin timing when the first person says "Mark" and stop timing when the second person says "Mark."

Theory

Throughout this lab, we will use "height" to mean distance above the ground. Let h_0 be the ball's height (in feet) at the moment of its release. Let v_0 be the ball's speed (in feet per second) at that moment. We call h_0 the *initial height* and v_0 the *initial velocity*. The ball's height h (in feet) is given by

$$h = -16t^2 + v_0t + h_0$$

where t is the time (in seconds) since the ball was released. So, t and h are variables and v_0 and h_0 are constants.

Analyzing the Situation

1. Substitute 0 for t in the equation $h = -16t^2 + v_0t + h_0$ and solve for h. Explain why your work shows that h_0 represents the initial height.
2. What was the actual initial height of the ball in your experiment? Substitute this value for h_0 in the equation $h = -16t^2 + v_0t + h_0$.
3. How long did it take for the ball to return to its initial height? Substitute that time for t and the initial height for h in the equation you found in Problem 2. Then solve for v_0. What does your value for v_0 mean in this situation?
4. Substitute the value for v_0 that you found in Problem 3 into the equation you found in Problem 2. [**Hint:** Your equation should still have the variables t and h in it.]
5. Use a graphing calculator to draw a graph of your model. Use a pencil and paper to copy the graph.
6. What is the h-intercept of the model? What does it mean in this situation?
7. Use the model to estimate the height of the ball at 1 second.

Chapter Summary

Key Points of Chapter 7

Section 7.1   Graphing Quadratic Equations

Quadratic equation in two variables	A **quadratic equation in two variables** is an equation that can be put into the form $y = ax^2 + bx + c$, where a, b, and c are constants and $a \neq 0$. This form is called **standard form**.
Parabola, minimum point, maximum point, and vertex	A **parabola** is the graph of a quadratic equation in two variables. The lowest point of a parabola that *opens upward* is called the **minimum point**. The highest point of a parabola that *opens downward* is called the **maximum point**. The minimum point or the maximum point of a parabola is called the **vertex** of the parabola.
Intercepts of curves	An **x-intercept of a curve** is a point where the curve and the x-axis intersect; a **y-intercept of a curve** is a point where the curve and the y-axis intersect.

Section 7.2   Quadratic Models

Quadratic model	A **quadratic model** is a quadratic equation in two variables that describes the relationship between two quantities for an authentic situation. We also refer to the quadratic equation's graph (a parabola) as a quadratic model.
Deciding which type of model to use	By constructing a scatterplot of some data, we can determine whether to use a linear model, a quadratic model, or neither.

Section 7.3   Adding and Subtracting Polynomials

Term	A **term** is a constant, a variable, or a product of a constant and one or more variables raised to powers.
Monomial	A **monomial** is a constant, a variable, or a product of a constant and one or more variables raised to counting-number powers.
Polynomial or polynomial expression	A **polynomial**, or **polynomial expression**, is a monomial or a sum of monomials.
Degree of a term	The **degree of a term in one variable** is the exponent on the variable. The **degree of a term in two or more variables** is the sum of the exponents on the variables.
Degree of a polynomial	The **degree of a polynomial** is the largest degree of any nonzero term of the polynomial.
Like terms	**Like terms** are either constant terms or variable terms that contain the same variable(s) raised to exactly the same power(s).
Combining like terms	To combine like terms, add the coefficients of the terms.
Adding polynomials	To add polynomials, combine like terms.
Subtracting polynomials	To subtract polynomials, first distribute -1 and then combine like terms.
The meaning of the sign of a difference	If a difference $A - B$ is positive, then A is more than B. If a difference $A - B$ is negative, then A is less than B.

Section 7.4   Multiplying Polynomials

Binomial and trinomial	We refer to a polynomial as a **binomial** or a **trinomial**, depending on whether it has two or three nonzero terms, respectively.
Product property for exponents	If n and m are counting numbers, then $x^m x^n = x^{m+n}$.
Multiplying two polynomials	To multiply two polynomials, multiply each term in the first polynomial by each term in the second polynomial. Then combine like terms if possible.

Section 7.5   Powers of Polynomials; Product of Binomial Conjugates

Raising a product to a power	If n is a counting number, then $(xy)^n = x^n y^n$.
Square of a sum	$(A + B)^2 = A^2 + 2AB + B^2$
Square of a difference	$(A - B)^2 = A^2 - 2AB + B^2$
Product of binomial conjugates	$(A + B)(A - B) = A^2 - B^2$

Section 7.6 Properties of Exponents

Simplifying power expressions	A power expression is simplified if 1. It includes no parentheses. 2. In any monomial, each variable or constant appears as a base at most once. 3. Each numerical expression (such as 7^2) has been calculated, and each numerical fraction has been simplified.
Quotient property for exponents	If n and m are counting numbers and x is nonzero, then $\dfrac{x^m}{x^n} = x^{m-n}$.
Zero exponent	For nonzero x, $x^0 = 1$.
Raising a quotient to a power	If n is a counting number and y is nonzero, then $\left(\dfrac{x}{y}\right)^n = \dfrac{x^n}{y^n}$.
Raising a power to a power	If m and n are counting numbers, then $\left(x^m\right)^n = x^{mn}$.

Section 7.7 Negative-Integer Exponents

Negative-integer exponent	If n is a counting number and x is nonzero, then $x^{-n} = \dfrac{1}{x^n}$.
Negative-integer exponent in a denominator	If x is nonzero and n is a counting number, then $\dfrac{1}{x^{-n}} = x^n$.
Properties of integer exponents	If m and n are integers and x and y are nonzero, then • $x^m x^n = x^{m+n}$ **Product property for exponents** • $\dfrac{x^m}{x^n} = x^{m-n}$ **Quotient property for exponents** • $(xy)^n = x^n y^n$ **Raising a product to a power** • $\left(\dfrac{x}{y}\right)^n = \dfrac{x^n}{y^n}$ **Raising a quotient to a power** • $\left(x^m\right)^n = x^{mn}$ **Raising a power to a power**
Simplifying power expressions	A power expression is simplified if 1. It includes no parentheses. 2. In any monomial, each variable or constant appears as a base at most once. 3. Each numerical expression has been calculated, and each numerical fraction has been simplified. 4. Each exponent is positive.
Scientific notation	A number is written in **scientific notation** if it has the form $N \times 10^k$, where k is an integer and the absolute value of N is between 1 and 10 or is equal to 1.
Converting from scientific notation to standard decimal notation	To write the scientific notation $N \times 10^k$ in standard decimal notation, we move the decimal point of the number N as follows: • If k is *positive*, we multiply N by 10 k times and hence move the decimal point k places to the *right*. • If k is *negative*, we divide N by 10 k times and hence move the decimal point k places to the *left*.
Converting from standard decimal notation to scientific notation	To write a number in scientific notation, count the number k of places that the decimal point needs to be moved so that the absolute value of the new number N is either between 1 and 10 or equal to 1. • If the decimal point is moved to the left, then the scientific notation is written as $N \times 10^k$. • If the decimal point is moved to the right, then the scientific notation is written as $N \times 10^{-k}$.

Section 7.8 Dividing Polynomials

Dividing a polynomial by a monomial	To divide a polynomial by a monomial, use the property $\dfrac{A + C}{B} = \dfrac{A}{B} + \dfrac{C}{B}$, where B is nonzero, to divide each term of the polynomial by the monomial.
Dividing a polynomial by a binomial	To divide a polynomial by a binomial, use long division.

Chapter 7 Review Exercises

Graph the equation by hand. To begin, substitute the values $-3, -2, -1, 0, 1, 2,$ *and* 3 *for x. Make other substitutions as necessary. Then use a graphing calculator to verify your work.*

1. $y = -2x^2$ **2.** $y = 3x^2 - 4$ **3.** $y = 2x^2 - 8x + 4$

For Exercises 4–7, refer to Table 40.

4. Find y when $x = 3$.   **5.** Find y when $x = 2$.

6. Find x when $y = 2$.   **7.** Find x when $y = 1$.

Table 40 Values of x and y (Exercises 4–7)

x	y
0	5
1	2
2	1
3	2
4	5

8. Average annual expenditures by Americans are shown in Table 41 for various age groups.

Table 41 Average Annual Expenditures

Age Group (years)	Age Used to Represent Age Group (years)	Average Annual Expenditure (thousands of dollars)
18–24	21.0	24.3
25–34	29.5	40.3
35–44	39.5	48.3
45–54	49.5	48.7
55–64	59.5	44.3
65–74	69.5	32.2

Source: *Consumer Expenditure Survey*

Let E be the average annual expenditure (in thousands of dollars) by Americans at age a years. A quadratic model is graphed in Fig. 55.

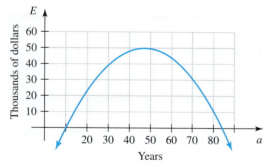

Figure 55 Exercise 8

a. Trace the graph in Fig. 55 carefully onto a piece of paper. Then, on the same graph, draw a scatterplot of the data. Does the model fit the data well?

b. Estimate the average annual expenditure for 19-year-old Americans.

c. Estimate the age(s) at which the average annual expenditure is $35 thousand.

d. What is the vertex of the model? What does it mean in this situation?

e. Find the a-intercepts. What do they mean in this situation?

9. DATA. The costs of a typical Thanksgiving dinner for ten people are shown in Table 42 for various years.

Table 42 Costs of a Typical Thanksgiving Dinner for Ten People

Year	Cost (dollars)
1986	28.74
1990	28.85
1995	29.64
2000	32.37
2005	36.78
2010	43.47
2015	50.11
2016	49.87

Source: *American Farm Bureau Federation*

Let c be the cost (in dollars) of a Thanksgiving dinner for ten people at t years since 1900. A quadratic model of the situation is $c = 0.0281t^2 - 4.93t + 244.8$. A linear model of the situation is $c = 0.78t - 41.85$.

a. Use a graphing calculator to draw graphs of the quadratic and linear models and, in the same window, the scatterplot of the data. Which model comes closer to the points in the scatterplot?

b. For each model, substitute a value for one of the variables in the model's equation to predict the cost of a typical Thanksgiving dinner for ten people in 2023. Refer to the models' graphs to explain why the quadratic model's prediction is larger than the linear model's prediction.

c. Use "minimum" on a graphing calculator to find the vertex of the quadratic model. What does the vertex represent in this situation?

d. Which of the two models estimates that the cost of a Thanksgiving dinner for ten people was negative for years before 1953?

e. Use TRACE on a graphing calculator, along with the quadratic model, to predict when the cost of a Thanksgiving dinner for ten people will be $60.

For Exercises 10 and 11, perform the indicated operation. Use a graphing calculator table to check your work.

10. $\left(-4x^3 + 7x^2 - x\right) + \left(-5x^3 - 9x^2 + 3\right)$

11. $\left(6x^3 - 2x^2 + 5\right) - \left(8x^3 - 4x^2 + 3x\right)$

12. Perform the indicated operations: $4x - 9x^2 - 6x + 2 - 7x^3 + x^2 - 8x$.

13. DATA. In Exercise 39 of the Review Exercises in Chapter 6, the percentages of households that are married households or non-family households is modeled by the system

$$p = -0.3t + 52.6 \quad \textit{Married households}$$
$$p = 0.25t + 31.0 \quad \textit{Non-family households}$$

where p is the percentage of households and t is the number of years since 2000 (see Table 43).

Table 43 Percentages of Households That Are Married Households, Non-Family Households, and Other-Family Households

Year	Percent Married	Non-Family	Other-Family
2000	53	31	16
2004	51	32	17
2008	50	33	17
2012	49	34	17
2016	48	35	17

Source: *U.S. Census Bureau*

a. Find the sum of the polynomials $-0.3t + 52.6$ and $0.25t + 31.0$. What does your result mean in this situation?
b. Evaluate the polynomial you found in part (a) for $t = 22$. What does your result mean in this situation?
c. Find the difference of the polynomials $-0.3t + 52.6$ and $0.25t + 31.0$. What does your result mean in this situation?
d. Evaluate the polynomial you found in part (c) for $t = 22$. What does your result mean in this situation?

Perform the indicated operation. Use a graphing calculator table to verify your work when possible.

14. $-3x^2(7x^5)$
15. $5x^3(2x^2 - 7x + 4)$
16. $(w - 3)(w - 9)$
17. $(2a + 5b)(3a - 8b)$
18. $(3x^2 - 4)(5x + 6)$
19. $(x + 4)(x^2 - 3x + 5)$
20. $(4b^2 - b + 3)(2b - 7)$
21. $(-2t)^3$
22. $(x + 7)^2$
23. $(x - 4)^2$
24. $(2p + 5)^2$
25. $-5(c + 2)^2$
26. $(x + 6)(x - 6)$
27. $(4m - 7n)(4m + 7n)$
28. $(3p - 2t)^2$
29. $(2a^2 - a + 3)(a^2 + 2a - 1)$
30. Is the product of a linear polynomial and a quadratic polynomial a linear, quadratic, or cubic polynomial? Give an example.

Write the equation in standard form. Use graphing calculator graphs to verify your work.

31. $y = (x - 5)^2 + 3$
32. $y = -2(x + 3)^2 - 6$

Simplify.

33. $(-x^5)^2$
34. $(2x^3)(6x^4)$
35. $(8a^2b^3)(-5a^4b^9)$
36. $\dfrac{8x^4y^8}{16x^3y^5}$

37. $\left(\dfrac{x}{2}\right)^3$
38. $(2x^9y^3)^5$
39. $3x^6(5x^4)^2$
40. $\dfrac{15c^2c^7}{10c^4}$
41. $\left(\dfrac{a^4}{9}\right)^2$
42. $\left(\dfrac{-9x^5}{5y^7}\right)^0$
43. $\dfrac{(3x^5y^4)^2}{6x^7y^3}$
44. $\left(\dfrac{3x^5}{4x^2}\right)^3$

Simplify.

45. 4^{-3}
46. $\dfrac{1}{x^{-5}}$
47. $\dfrac{r^{-7}}{r^3}$
48. $\dfrac{t^{-5}}{t^{-2}}$
49. $\dfrac{x^{-5}y^2}{w^{-7}}$
50. $(c^6)^{-4}$
51. $(-4x^3)(5x^{-9})$
52. $(2x^{-3})^{-5}$
53. $(3x^{-4}y^6)^{-3}$
54. $\dfrac{6a^{-2}b^3}{8a^{-5}b^6}$
55. $\dfrac{(3a^{-2}b)^2}{(2ab^{-3})^3}$
56. $\left(\dfrac{x^3y^{-2}}{w^{-4}}\right)^{-7}$

57. A student tries to simplify $\dfrac{x^9}{x^{-6}}$:

$$\dfrac{x^9}{x^{-6}} = x^{9-6}$$
$$= x^3$$

Describe any errors. Then simplify the expression correctly.

Write the number in standard decimal form.

58. 5.832×10^8
59. 3.17×10^{-4}

Write the number in scientific notation.

60. 74,200,000
61. 0.00008
62. Saturn is an average distance of about 1,426,000,000 kilometers from the Sun. Write 1,426,000,000 in scientific notation.

Divide. Verify your work by multiplying.

63. $\dfrac{12x^4 - 9x^3 + 15x^2}{3x^2}$
64. $\dfrac{-10b^4 + 7b^3 - 20b^2}{-5b^3}$
65. $\dfrac{8x^4y - 4x^3y^2 - 6x^2y^3}{2x^2y}$
66. $\dfrac{2x^2 - 13x + 15}{x - 5}$
67. $\dfrac{6w^2 - w - 11}{2w + 3}$
68. $\dfrac{4x^3 - 8x^2 + 5x - 3}{2x - 1}$
69. $\dfrac{27p^3 - 8}{3p - 2}$
70. $\dfrac{4x^3 - 3x^2 + x + 1}{x^2 + 2}$

Chapter 7 Test

1. Graph $y = -2x^2 + 4x + 1$ by hand. To begin, substitute the values $-3, -2, -1, 0, 1, 2,$ and 3 for x. Make other substitutions as necessary. Then use a graphing calculator to verify your work.

For Exercises 2–7, refer to the graph sketched in Fig. 56.

2. Find y when $x = -5$.   **3.** Find x when $y = -3$.

4. Find x when $y = -4$.   **5.** Find x when $y = -5$.

6. Find the x-intercepts.

7. Find the minimum point.

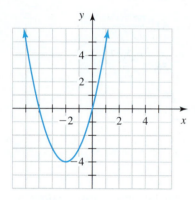

Figure 56 Exercises 2–7

8. DATA, The percentages of women and men who are obese are shown in Table 44 for various age groups.

Table 44 Percentages of Women and Men Who Are Obese

Age Group (years)	Age Used to Represent Age Group (years)	Percent	
		Women	Men
20–34	27.0	33.4	28.5
35–44	39.5	39.1	39.8
45–54	49.5	41.7	36.6
55–64	59.5	44.4	38.1
65–74	69.5	40.7	36.2
over 74	80.0	30.5	26.8

Source: *Centers for Disease Control and Prevention*

Let p be the percentage of women at age a years who are obese. A quadratic model is sketched in Fig. 57.

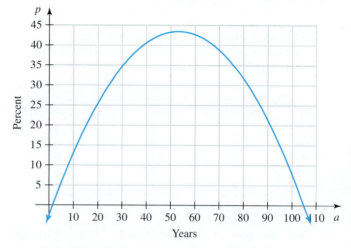

Figure 57 Exercise 8

a. Trace the graph in Fig. 57 carefully onto a piece of paper. Then, on the same graph, draw a scatterplot of the data. Does the model fit the data well?

b. Estimate the percentage of 30-year-old women who are obese.

c. Estimate the age(s) at which 37% of women are obese.

d. What is the vertex of the model? What does it mean in this situation?

e. What are the a-intercepts? What do they mean in this situation?

Perform the indicated operations.

9. $\left(2x^3 - 4x^2 + 7x\right) - \left(6x^3 - 3x^2 + 9x\right)$

10. $-5x^2\left(3x^2 - 8x + 2\right)$

11. $(3p - 2)(4p + 6)$

12. $(2k - 5)\left(3k^2 - 4k - 2\right)$

13. $(x - 6)^2$

14. $(4a + 7b)^2$

15. $\left(3w^2 - 2w + 3\right)\left(w^2 + w - 2\right)$

16. DATA, In Exercise 10 of Homework 6.4, the number of women and men who live alone is modeled by the system

$$n = 0.22t + 11.49 \quad \textit{Women}$$
$$n = 0.24t + 6.67 \quad \textit{Men}$$

where n is the number (in millions) of people and t is the number of years since 1980 (see Table 45).

Table 45 Numbers of Women and Men Who Live Alone

| Year | Number Living Alone (millions) | |
	Women	Men
1980	11.3	7.0
1985	12.7	7.9
1990	14.0	9.0
1995	14.6	10.1
2000	15.6	11.2
2005	17.3	12.8
2010	17.4	14.0
2015	19.4	15.5

Source: *U.S. Census Bureau*

a. Find the sum of the polynomials $0.22t + 11.49$ and $0.24t + 6.67$. What does your result mean in this situation?

b. Evaluate the polynomial you found in part (a) for $t = 41$. What does your result mean in this situation?

c. Find the difference of the polynomials $0.22t + 11.49$ and $0.24t + 6.67$. What does your result mean in this situation?

d. Evaluate the polynomial you found in part (c) for $t = 41$. What does your result mean in this situation?

17. Write $y = -3(x - 1)^2 + 5$ in standard form.

18. A student tries to simplify $(x + 4)^2$:

$$(x + 4)^2 = x^2 + 4^2$$
$$= x^2 + 16$$

Describe any errors. Then simplify the polynomial correctly.

Simplify.

19. $\dfrac{6x^7y^4}{8x^3y^9}$

20. $\left(4a^3b^5\right)^3 a^6 b$

21. $\left(\dfrac{x^3}{y^4}\right)^6$

22. $\left(7x^{-3}\right)^{-2}$

23. $\left(\dfrac{x^2y^{-6}}{w^{-3}}\right)^4$

24. $\dfrac{2p^{-5}t^2}{4p^{-2}t^{-3}}$

25. Write 0.000468 in scientific notation.

26. In 2002, college enrollment in the United States was about 1.65×10^7 students (Source: *U.S. Census Bureau*). Write 1.65×10^7 in standard decimal form.

Divide. Verify your work by multiplying.

27. $\dfrac{12x^3 - 6x^2 - 9x}{-3x^2}$

28. $\dfrac{4x^4y^2 + 6x^3y^3 - 8x^2y^4}{2xy^2}$

29. $\dfrac{6x^2 - 17x + 3}{3x - 1}$

30. $\dfrac{8w^2 - 5}{2w + 3}$

Factoring Polynomials and Solving Polynomial Equations

Do you or one of your friends own an iPhone? Worldwide sales of iPhones have increased greatly since 2007 (see Table 1). In Exercise 5 of Homework 8.6, you will estimate when worldwide iPhone sales were 46 million phones.

In Chapter 7, we discussed how to multiply polynomials. In this chapter, we will discuss how to do the reverse process, *factoring*, which will help us solve new types of equations. In turn, solving these equations will help us use quadratic models to make estimates about authentic situations, such as when the population of endangered manatees was as low as 1380 manatees.

Table 1 Worldwide iPhone Sales

Year	Sales (millions)
2007	1.4
2009	20.7
2011	72.3
2013	150.3
2015	231.2

Source: *Apple, Inc.*

8.1 Factoring Trinomials of the Form $x^2 + bx + c$ and Differences of Two Squares

Objectives

» Explain why multiplying and *factoring* are reverse processes.

» Factor a trinomial of the form $x^2 + bx + c$.

» Describe the meaning of a *prime* polynomial.

» Factor a difference of two squares.

We multiply 3 and 7 as follows: $3 \cdot 7 = 21$. We factor the number 21 as follows: $21 = 3 \cdot 7$. Thus, factoring 21 is the reverse process of multiplying 3 and 7. Next, we will make similar observations about polynomials.

Multiplying Polynomials versus Factoring Polynomials

In Chapter 7, we found products of polynomials—for example,

$$(x + 3)(x + 5) = x^2 + 5x + 3x + 15 \quad \textit{Multiply pairs of terms.}$$
$$= x^2 + 8x + 15 \quad \textit{Combine like terms.}$$

In this section, we will learn how to go backward. That is, we will learn how to write $x^2 + 8x + 15$ as a product. This process is called *factoring*.

We **factor** a polynomial by writing it as a product. We say that $(x + 3)(x + 5)$ is a **factored polynomial** and that both $(x + 3)$ and $(x + 5)$ are factors of the polynomial.

Multiplying versus Factoring

Multiplying and factoring are reverse processes. For example,

Multiplying

$$(x + 3)(x + 5) = x^2 + 8x + 15$$

Factoring

Factoring Trinomials of the Form $x^2 + bx + c$

To see how to factor $x^2 + 8x + 15$, let's take another look at how we find the product $(x + 3)(x + 5)$:

last terms

$$(x + 3)(x + 5) = x^2 + 5x + 3x + 3 \cdot 5$$
$$= x^2 + 8x + 15$$

sum of   product of
last terms   last terms
$3 + 5 = 8$ $3 \cdot 5 = 15$

For $x^2 + 8x + 15$, notice that the coefficient of x is 8, which is the sum of the *last terms* of $(x + 3)(x + 5)$, 3 and 5. Also, the constant term of $x^2 + 8x + 15$ is 15, which is the product of 3 and 5. Now we find the product $(x + p)(x + q)$:

$$(x + p)(x + q) = x^2 + qx + px + pq \qquad \text{Multiply pairs of terms.}$$
$$= x^2 + px + qx + pq \qquad \text{Rearrange terms.}$$
$$= x^2 + (p + q)x + pq \qquad \text{Distributive law}$$

In the result, we see that the coefficient of x is the sum of the last terms p and q and that the constant term is the product of the last terms p and q. This observation can help us factor some quadratic trinomials.

Factoring Trinomials with Positive Constant Terms

For the trinomial $x^2 + 8x + 15$, the constant term, 15, is positive. In Examples 1–3, we will factor three more trinomials whose constant term is positive.

Example 1   Factoring Trinomials of the Form $x^2 + bx + c$

Factor $x^2 + 6x + 8$.

Solution

To factor $x^2 + 6x + 8$, we need two integers whose product is 8 and whose sum is 6. We try only positive integers because both their product and their sum have to be positive. The only pairs of factors of 8 whose product is 8 are 1 and 8 or 2 and 4:

Product = 8	**Sum = 6?**
$1(8) = 8$	$1 + 8 = 9$
$2(4) = 8$	$2 + 4 = 6 \longleftarrow$ Success!

Because $2(4) = 8$ and $2 + 4 = 6$, we conclude that the last terms of the factors are 2 and 4:

$$x^2 + 6x + 8 = (x + 2)(x + 4)$$

We check the result by finding the product $(x + 2)(x + 4)$:

$$(x + 2)(x + 4) = x^2 + 4x + 2x + 8 = x^2 + 6x + 8$$

By the commutative law, $(x + 2)(x + 4) = (x + 4)(x + 2)$, so we can write the factors $x + 2$ and $x + 4$ in either order.

▶

We now summarize how to factor a trinomial of the form $x^2 + bx + c$.

Factoring $x^2 + bx + c$

To factor $x^2 + bx + c$, look for two integers p and q whose product is c and whose sum is b. That is, $pq = c$ and $p + q = b$. If such integers exist, the factored polynomial is

$$(x + p)(x + q)$$

Example 2   Factoring Trinomials of the Form $x^2 + bx + c$

Factor $x^2 + 8x + 12$.

Solution

To factor $x^2 + 8x + 12$, we need two integers whose product is 12 and whose sum is 8. We try only positive integers because both their product and their sum have to be positive. Here are the possibilities:

Product = 12	Sum = 8?
$1(12) = 12$	$1 + 12 = 13$
$2(6) = 12$	$2 + 6 = 8$ ⟵ Success!
$3(4) = 12$	$3 + 4 = 7$

Because $2(6) = 12$ and $2 + 6 = 8$, we conclude that the last terms of the factors are 2 and 6:

$$x^2 + 8x + 12 = (x + 2)(x + 6)$$

We check the result by finding the product $(x + 2)(x + 6)$:

$$(x + 2)(x + 6) = x^2 + 6x + 2x + 12 = x^2 + 8x + 12$$

▶

Example 3   Factoring a Trinomial of the Form $x^2 + bx + c$

Factor $x^2 - 12x + 36$.

Solution

To factor $x^2 - 12x + 36$, we need two integers whose product is 36 and whose sum is -12. Because the product 36 is positive, the two integers must have the same sign. Therefore, both integers must be negative because the sum -12 is negative. Here are the possibilities:

Product = 36	Sum = -12?
$-1(-36) = 36$	$-1 + (-36) = -37$
$-2(-18) = 36$	$-2 + (-18) = -20$
$-3(-12) = 36$	$-3 + (-12) = -15$
$-4(-9) = 36$	$-4 + (-9) = -13$
$-6(-6) = 36$	$-6 + (-6) = -12$ ⟵ Success!

Because $-6(-6) = 36$ and $-6 + (-6) = -12$, we conclude that the last terms of the factors are -6 and -6:

$$x^2 - 12x + 36 = (x - 6)(x - 6) = (x - 6)^2$$

Note that our result, $(x - 6)^2$, is the square of a binomial. So, the original polynomial $x^2 - 12x + 36$ is a perfect-square trinomial (Section 7.5).

We use a graphing calculator table to verify our work (see Fig. 1).

▶

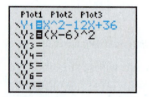

Figure 1 Verify the work

When the constant term of a trinomial is positive, we need to consider only certain possibilities for the factors of that constant term. In Examples 1 and 2, we worked with only positive factors of the positive constant term because the coefficient of the middle term was positive. In Example 3, we worked with only negative factors of the positive constant term because the coefficient of the middle term was negative.

Factoring $x^2 + bx + c$ with c Positive

To factor a trinomial of the form $x^2 + bx + c$ with a positive constant term c,

- If b is positive, look for two *positive* integers whose product is c and whose sum is b. For example,

$$x^2 + 10x + 21 = (x + 3)(x + 7)$$

Positive Positive   Both last
b c terms are
                            positive.

- If b is negative, look for two *negative* integers whose product is c and whose sum is b. For example,

$$x^2 - 11x + 28 = (x - 7)(x - 4)$$

Negative b Positive c Both last terms are negative.

Factoring Trinomials with Negative Constant Terms

How do we factor a quadratic trinomial whose constant term is negative?

Example 4   Factoring Trinomials of the Form $x^2 + bx + c$

Factor $w^2 - w - 20$.

Solution

To factor $w^2 - 1w - 20$, we need two integers whose product is -20 and whose sum is -1. Because the product -20 is negative, the two integers must have different signs. Here are the possibilities:

Product = -20	Sum = -1?
$1(-20) = -20$	$1 + (-20) = -19$
$2(-10) = -20$	$2 + (-10) = -8$
$4(-5) = -20$	$4 + (-5) = -1$ ⟵ Success!
$5(-4) = -20$	$5 + (-4) = 1$
$10(-2) = -20$	$10 + (-2) = 8$
$20(-1) = -20$	$20 + (-1) = 19$

Because $4(-5) = -20$ and $4 + (-5) = -1$, we conclude that the last terms of the factors are 4 and -5:

$$w^2 - w - 20 = (w + 4)(w - 5)$$

We use a graphing calculator table to verify our work (see Fig. 2).

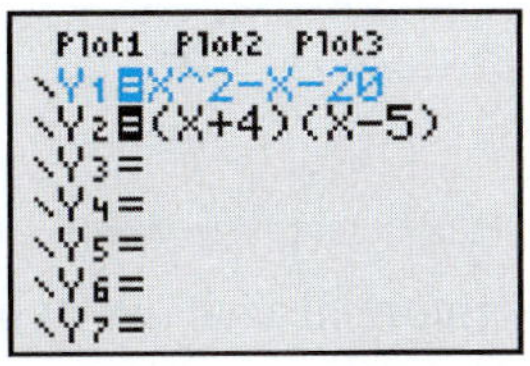

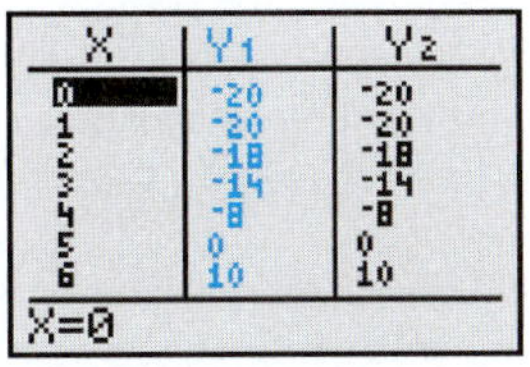

Figure 2 Verify the work

When the constant term of a trinomial is negative, any two integers whose product equals that negative constant term have different signs. For instance, in Example 4 we factored the trinomial $w^2 - w - 20$ by working with integers with different signs whose product is -20.

Factoring $x^2 + bx + c$ with c Negative

To factor a trinomial of the form $x^2 + bx + c$ with a negative constant term c, look for two integers with *different* signs whose product is c and whose sum is b. For example,

$$x^2 + 2x - 24 = (x - 4)(x + 6)$$

Negative c The last terms have different signs.

We can use a similar method to factor trinomials that have two variables.

Example 5   Factoring Trinomials with Two Variables

Factor $x^2 + 5xy + 6y^2$.

Solution

To help us find the last terms, we write the trinomial in the form $x^2 + (5y)x + 6y^2$. We need two monomials whose product is $6y^2$ and whose sum is $5y$. So, the last terms are $2y$ and $3y$:

$$x^2 + 5xy + 6y^2 = (x + 2y)(x + 3y)$$

We check by finding the product $(x + 2y)(x + 3y)$:

$$(x + 2y)(x + 3y) = x^2 + 3xy + 2xy + 6y^2 = x^2 + 5xy + 6y^2$$

Prime Polynomials

Just as a prime number has no positive factors other than itself and 1, a polynomial that cannot be factored is called **prime**.

For example, consider the polynomial $x^2 + 4x + 6$. To factor this polynomial, we need two integers whose product is 6 and whose sum is 4. We try only positive integers because both their product and sum have to be positive. Here are the possibilities:

Product = 6	Sum = 4?
$1(6) = 6$	$1 + 6 = 7$
$2(3) = 6$	$2 + 3 = 5$

None of the possible sums equal 4, so we conclude that the trinomial $x^2 + 4x + 6$ is prime.

Example 6   Identifying a Prime Polynomial

Factor $3x - 15 + x^2$.

Solution

First, we write $3x - 15 + x^2$ in descending order to avoid confusion about the coefficients:

$$x^2 + 3x - 15$$

To factor $x^2 + 3x - 15$, we need two integers whose product is -15 and whose sum is 3. Because the product -15 is negative, the two integers must have different signs. Here are the possibilities:

Product = -15	Sum = 3?
$1(-15) = -15$	$1 + (-15) = -14$
$3(-5) = -15$	$3 + (-5) = -2$
$5(-3) = -15$	$5 + (-3) = 2$
$15(-1) = -15$	$15 + (-1) = 14$

Because none of the sums equal 3, we conclude that the trinomial $x^2 + 3x - 15$ is prime. So, the original trinomial, $3x - 15 + x^2$, is prime.

Factoring the Difference of Two Squares

In Section 7.5, we found the product of two binomial conjugates by using the property $(A + B)(A - B) = A^2 - B^2$. The expression $A^2 - B^2$ is the difference of two squares. To factor a difference of two squares, we can use that property in reverse.

Difference of Two Squares

$$A^2 - B^2 = (A + B)(A - B)$$

In words: The difference of the squares of two terms is the product of the sum of the terms and the difference of the terms.

Example 7   Factoring Differences of Two Squares

Factor.

1. $x^2 - 16$ **2.** $9m^2 - 4$ **3.** $25p^2 - 49q^2$

Solution

1. Because $x^2 - 16 = (x)^2 - (4)^2$, we substitute x for A and 4 for B:

$$A^2 - B^2 = (A + B)(A - B)$$
$$x^2 - 16 = x^2 - 4^2 = (x + 4)(x - 4)$$

2. Because $9m^2 - 4 = (3m)^2 - (2)^2$, we substitute $3m$ for A and 2 for B:

$$A^2 - B^2 = (A + B)(A - B)$$
$$9m^2 - 4 = (3m)^2 - 2^2 = (3m + 2)(3m - 2)$$

3. Because $25p^2 - 49q^2 = (5p)^2 - (7q)^2$, we substitute $5p$ for A and $7q$ for B:

$$A^2 - B^2 = (A + B)(A - B)$$
$$25p^2 - 49q^2 = (5p)^2 - (7q)^2 = (5p + 7q)(5p - 7q)$$

WARNING   The binomial $x^2 + 16$ is prime. Some students think this polynomial can be factored as $(x + 4)^2$, but simplify $(x + 4)^2$; you'll see that the result is $x^2 + 8x + 16$, not $x^2 + 16$. In general, **a polynomial of the form $x^2 + k^2$, where $k \neq 0$, is prime.**

 ## Group Exploration

Section Opener: Factoring the difference of two squares

Note that $(x + 5)(x - 5) = x^2 - 25$. When we work backward and write $x^2 - 25 = (x + 5)(x - 5)$, we say we have *factored* $x^2 - 25$. So we factor a polynomial by writing it as a product.

1. Multiply: $(x + 4)(x - 4)$.   **2.** Factor $x^2 - 16$.

3. Factor $x^2 - 9$.   **4.** Factor $w^2 - 4$.

5. Use long division to find the quotient $\dfrac{w^2 - 4}{w - 2}$. To verify your work, check that the product of the divisor and

the quotient is the dividend. Explain how your work is related to your work in Problem 4.

6. Factor $25x^2 - 49$.

7. Use long division to find the quotient $\dfrac{25x^2 - 49}{5x + 7}$. To verify your work, check that the product of the divisor and the quotient is the dividend. Explain how your work is related to your work in Problem 6.

8. Factor $9w^2 - 4y^2$.

Homework 8.1

For extra help ▶   **MyLab Math**   Watch the videos in MyLab Math

Factor when possible. If a polynomial is prime, say so. Verify that you have factored correctly by finding the product of your factored polynomial.

1. $x^2 + 5x + 6$ **2.** $x^2 + 9x + 8$ **3.** $t^2 + 9t + 20$

4. $w^2 + 10w + 16$ **5.** $x^2 + 8x + 16$ **6.** $x^2 + 14x + 49$

7. $x^2 - 2x - 8$ **8.** $x^2 - 3x - 10$ **9.** $a^2 - 6a - 16$

10. $w^2 - 5w - 24$ **11.** $x^2 + 5x - 24$ **12.** $x^2 + 7x - 30$

13. $x^2 + 8x - 12$ **14.** $x^2 + 9x - 20$ **15.** $3t - 28 + t^2$

16. $y^2 - 14 + 5y$ **17.** $x^2 - 10x + 16$ **18.** $x^2 - 11x + 28$

19. $24 - 11x + x^2$ **20.** $36 + x^2 - 15x$ **21.** $x^2 - 3x + 10$

22. $x^2 - 2x + 8$ **23.** $r^2 - 10r + 25$ **24.** $a^2 - 6a + 9$

25. $x^2 + 36 - 12x$ **26.** $1 - 2x + x^2$

27. $x^2 + 10xy + 9y^2$ **28.** $a^2 + 8ab + 12b^2$

29. $m^2 - mn - 6n^2$ **30.** $r^2 + rt - 20t^2$

31. $a^2 - 7ab + 6b^2$ **32.** $m^2 - 7mn + 10n^2$

Factor when possible. Use a graphing calculator table to verify your work when possible.

33. $x^2 - 25$ **34.** $x^2 - 49$

35. $x^2 - 81$ **36.** $x^2 - 36$

37. $t^2 - 1$ **38.** $a^2 - 64$

39. $x^2 + 36$ **40.** $x^2 + 4$

41. $4x^2 - 25$ **42.** $9x^2 - 49$

43. $81r^2 - 1$ **44.** $25p^2 - 64$

45. $36x^2 + 49$ **46.** $25x^2 + 9$

47. $49p^2 - 100q^2$ **48.** $4a^2 - 9b^2$

49. $64m^2 - 9n^2$ **50.** $25b^2 - 9c^2$

Factor when possible. Use a graphing calculator table to verify your work when possible.

51. $x^2 - 3x - 18$ **52.** $x^2 - 8x - 20$

53. $x^2 + 14x + 49$ **54.** $x^2 + 10x + 25$

55. $a^2 - 4$ **56.** $m^2 - 100$

57. $x^2 + 4x + 12$ **58.** $x^2 + 12x + 24$

59. $x^2 - 8x + 12$ **60.** $x^2 - 9x + 18$

61. $-2w - 48 + w^2$ **62.** $-4t - 60 + t^2$

63. $x^2 - 8x + 16$ **64.** $x^2 - 2x + 1$

65. $w^2 + 49$ **66.** $b^2 + 25$

67. $m^2 - 6mn - 27n^2$ **68.** $a^2 - 19ab - 20b^2$

69. $32 - 18x + x^2$ **70.** $30 - 13x + x^2$

71. $100p^2 - 9t^2$ **72.** $64a^2 - 49b^2$

73. $p^2 + 12p + 36$ **74.** $m^2 + 18m + 81$

Concepts

75. A student tries to factor the polynomial $x^2 + 9$:

$$x^2 + 9 = (x + 3)(x + 3) = (x + 3)^2$$

Describe any errors. Then factor the polynomial correctly.

76. Two students try to factor the polynomial $x^2 + 14x + 48$:

Student A
$$x^2 + 14x + 48 = (x + 6)(x + 8)$$

Student B
$$x^2 + 14x + 48 = (x + 8)(x + 6)$$

Are both students, one student, or neither student correct? Explain.

77. Which of the following polynomials are equivalent?

$$(x - 3)(x + 7)   \quad x^2 - 21 \quad   x^2 - 4x - 21$$
$$x^2 + 4x - 21   \quad (x + 7)(x - 3)$$

78. Which of the following polynomials are equivalent?

$$x^2 - 2x - 24   \quad (x - 6)(x + 4) \quad   x^2 - 10x - 24$$
$$(x + 4)(x - 6)   \quad x^2 + 2x - 24$$

79. Factor the polynomial $x^2 - 5x - 24$. Then find the product of the result. What do you observe?

80. Find the product $(x - 2)(x + 8)$. Then factor the result. What do you observe?

81. Give three examples of quadratic polynomials in which $x + 5$ is a factor.

82. Give three examples of quadratic polynomials in which $x - 3$ is a factor.

83. Find all possible values of k so that $x^2 + kx + 12$ can be factored.

84. Find all possible values of k so that $x^2 + kx - 20$ can be factored.

85. Evaluate $x^2 - 6x + 8$ for $x = 2$ and for $x = 4$. Why does it make sense that both of your results are 0? [**Hint:** Factor $x^2 - 6x + 8$.]

86. Evaluate $x^2 + 5x + 6$ for $x = -3$ and for $x = -2$. Why does it make sense that both of your results are 0? [**Hint:** Factor $x^2 + 5x + 6$.]

87. Compare the process of factoring a polynomial with that of finding the product of some polynomials. (See page 9 for guidelines on writing a good response.)

88. Describe how to factor a difference of two squares. (See page 9 for guidelines on writing a good response.)

Related Review

If the polynomial is not factored, then factor it. If the polynomial is factored, then find the product.

89. $(x - 9)(x + 2)$ **90.** $(x - 1)(x - 8)$

91. $x^2 - 15x + 50$ **92.** $x^2 + 19x - 20$

93. $(3x - 7)(3x + 7)$

94. $(4x + 1)(4x - 1)$

95. $25x^2 - 36$

96. $49x^2 - 81$

99. Find the product $(5p + 7w)(2p - 4w)$.

100. Graph $y = 3(x - 5)$ by hand.

101. Factor $p^2 - 11pw + 18w^2$.

102. Solve the following:

$$y = 3(x - 5)$$
$$7x + 4y = -3$$

Expressions, Equations, and Graphs

Perform the indicated instruction. Then use words such as linear, *quadratic, cubic, polynomial, degree, one variable, and* two variables *to describe the expression, equation, or system.*

97. Find the difference $(5p + 7w) - (2p - 4w)$.

98. Solve $6 = 3(x - 5)$.

8.2 Factoring Out the GCF; Factoring by Grouping

Objectives

» Factor out the *greatest common factor (GCF)* of a polynomial.

» Factor polynomials completely.

» Factor out the opposite of the GCF of a polynomial.

» Factor a polynomial by grouping.

In Section 8.1, we discussed some ways to factor polynomials. In this section, we will discuss more factoring techniques.

Factoring Out the GCF

Consider the polynomial $2x + 10$. Note that 2 is a common factor of both $2x = 2 \cdot x$ and $10 = 2 \cdot 5$, so we have

$$2x + 10 = 2 \cdot x + 2 \cdot 5$$

We use the distributive law to factor out the common factor 2:

$$2x + 10 = 2 \cdot x + 2 \cdot 5 = 2(x + 5)$$

So, we have factored $2x + 10$ as $2(x + 5)$. We check the result by finding the product $2(x + 5)$:

$$2(x + 5) = 2x + 10$$

Example 1   Factoring Out a Common Factor

Factor.
1. $3x + 21$ **2.** $8x^2 - 6x$ **3.** $6x^3 + 12x^2$

Solution

1. The number 3 is a common factor of $3x = 3 \cdot x$ and $21 = 3 \cdot 7$. So, we use the distributive law to factor out 3:

$$3x + 21 = 3 \cdot x + 3 \cdot 7 \quad \textit{3 is a common factor.}$$
$$= 3(x + 7) \quad \textit{Factor out 3.}$$

We can verify our work by finding the product $3(x + 7)$:

$$3(x + 7) = 3x + 21$$

2. The monomial $2x$ is a common factor of $8x^2 = 2x \cdot 4x$ and $6x = 2x \cdot 3$. So, we use the distributive law to factor out $2x$:

$$8x^2 - 6x = 2x \cdot 4x - 2x \cdot 3 \quad \textit{2x is a common factor.}$$
$$= 2x(4x - 3) \quad \textit{Factor out 2x.}$$

3. The monomial $6x^2$ is a common factor of $6x^3 = 6x^2 \cdot x$ and $12x^2 = 6x^2 \cdot 2$. So, we use the distributive law to factor out $6x^2$:

$$6x^3 + 12x^2 = 6x^2 \cdot x + 6x^2 \cdot 2 \quad \textit{6x}^2 \textit{ is a common factor.}$$
$$= 6x^2(x + 2) \quad \textit{Factor out 6x}^2\textit{.}$$

In Problem 3 of Example 1, notice that $2x$ is also a common factor of $6x^3 = 2x \cdot 3x^2$ and $12x^2 = 2x \cdot 6x$. So, we could have factored $6x^3 + 12x^2$ by factoring out $2x$ rather than $6x^2$:

$$6x^3 + 12x^2 = 2x\left(3x^2 + 6x\right)$$

However, the resulting polynomial is not factored completely: We can still factor $3x^2 + 6x$ by factoring out $3x$:

$$6x^3 + 12x^2 = 2x\left(3x^2 + 6x\right) = 2x \cdot 3x(x + 2) = 6x^2(x + 2)$$

Although we have found the same final result, it was more efficient to factor out $6x^2$, which has a larger coefficient and a higher degree than $2x$. We call $6x^2$ the *greatest common factor* of $6x^3$ and $12x^2$.

Definition   Greatest common factor

The **greatest common factor (GCF)** of two or more terms is the monomial with the largest coefficient and the largest degree that is a factor of all the terms.

For each polynomial in Example 1, the common factor that we factored out of the polynomial was the GCF. In Example 2, we will factor some more polynomials by factoring out the GCF.

Example 2   Factoring Out the GCF

Factor.

1. $20x^2 + 35x$

2. $14p^3 - 21p^2$

Solution

1. We begin by factoring $20x^2$ and $35x$:

$$20x^2 = 2 \cdot 2 \cdot 5 \cdot x \cdot x$$
$$35x = 5 \cdot 7 \cdot x$$

Both 5 and x are common factors. So, the GCF is $5x$:

$$20x^2 + 35x = 5x \cdot 4x + 5x \cdot 7 \qquad \text{\color{blue}5x is the GCF.}$$
$$= 5x(4x + 7) \qquad \text{\color{blue}Factor out 5x.}$$

2. We begin by factoring $14p^3$ and $21p^2$:

$$14p^3 = 2 \cdot 7 \cdot p \cdot p \cdot p$$
$$21p^2 = 3 \cdot 7 \cdot p \cdot p$$

There are three common factors: $7, p$, and p. So, the GCF is $7p^2$:

$$14p^3 - 21p^2 = 7p^2 \cdot 2p - 7p^2 \cdot 3 \qquad \text{\color{blue}$7p^2$ is the GCF.}$$
$$= 7p^2(2p - 3) \qquad \text{\color{blue}Factor out $7p^2$.}$$

We use a graphing calculator table to verify our work (see Fig. 3).

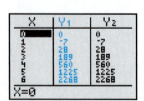

Figure 3 Verify the work

After you factor a polynomial, verify your work by finding the product of your result or by using a graphing calculator table.

So far, we have factored out the GCF for some binomials with one variable. We can also factor out the GCF for some polynomials with more than two terms and more than one variable.

Example 3   Factoring Out the GCF

Factor $12x^4y^2 - 6x^2y^3 + 15xy^2$.

Solution

We begin by factoring $12x^4y^2$, $6x^2y^3$, and $15xy^2$:

$$12x^4y^2 = 2 \cdot 2 \cdot 3 \cdot x \cdot x \cdot x \cdot x \cdot y \cdot y$$
$$6x^2y^3 = 2 \cdot 3 \cdot x \cdot x \cdot y \cdot y \cdot y$$
$$15xy^2 = 3 \cdot 5 \cdot x \cdot y \cdot y$$

There are four common factors: 3, x, y, and y. So, the GCF is $3xy^2$:

$$12x^4y^2 - 6x^2y^3 + 15xy^2 = 3xy^2 \cdot 4x^3 - 3xy^2 \cdot 2xy + 3xy^2 \cdot 5 \quad \textit{$3xy^2$ is the GCF.}$$
$$= 3xy^2(4x^3 - 2xy + 5) \quad \textit{Factor out $3xy^2$.}$$

Factoring Polynomials Completely

After we factor the GCF out of a polynomial, we must check whether the result can be further factored by using factoring techniques discussed in Section 8.1. If a result cannot be further factored, it is said to be **factored completely**.

Example 4   Factoring Polynomials Completely

Factor $4x^2 - 36$.

Solution

The GCF of $4x^2$ and 36 is 4, so:

$$4x^2 - 36 = 4(x^2 - 9) \quad \textit{Factor out GCF, 4.}$$
$$= 4(x + 3)(x - 3) \quad \textit{$A^2 - B^2 = (A + B)(A - B)$}$$

To factor $4x^2 - 36$ in Example 4, we first factored out the GCF, 4, and then factored the resulting difference of two squares, $x^2 - 9$. These steps require less work than first using the property for the difference of two squares:

$$4x^2 - 36 = (2x + 6)(2x - 6) \quad \textit{$A^2 - B^2 = (A + B)(A - B)$}$$
$$= 2(x + 3)(2)(x - 3) \quad \textit{Factor out GCF, 2.}$$
$$= 4(x + 3)(x - 3) \quad \textit{Simplify.}$$

Not only are there fewer steps in Example 4, but it is easier to use the property for the difference of squares to factor $x^2 - 9$ than $4x^2 - 36$. In general, **when the leading coefficient of a polynomial is positive and the GCF is not 1, we first factor out the GCF.** (We will soon discuss what to do when the leading coefficient of a polynomial is negative.)

Example 5   Factoring Polynomials Completely

Factor $2x^4y - 6x^3y - 20x^2y$.

Solution

The GCF of $2x^4y$, $6x^3y$, and $20x^2y$ is $2x^2y$:

$$2x^4y - 6x^3y - 20x^2y = 2x^2y(x^2 - 3x - 10) \quad \textit{Factor out GCF, $2x^2y$.}$$
$$= 2x^2y(x - 5)(x + 2) \quad \textit{Find two integers whose product is -10 and whose sum is -3.}$$

WARNING   It is a common error when factoring a polynomial to forget to factor it *completely*. In Example 5, we factored the GCF, $2x^2y$, out of $2x^4y - 6x^3y - 20x^2y$:

$$2x^4y - 6x^3y - 20x^2y = 2x^2y\left(x^2 - 3x - 10\right) \qquad \textit{Not factored completely}$$

However, we were not done factoring because we could still factor $x^2 - 3x - 10$:

$$2x^2y\left(x^2 - 3x - 10\right) = 2x^2y(x - 5)(x + 2) \qquad \textit{Factored completely}$$

When factoring a polynomial, always factor it *completely*.

Example 6   Factoring Polynomials Completely

Factor.

1. $12x^3 - 75x$ 2. $36x + 4x^3 - 24x^2$

Solution

1. The GCF of $12x^3$ and $75x$ is $3x$, so:

$$\begin{aligned} 12x^3 - 75x &= 3x\left(4x^2 - 25\right) && \textit{Factor out GCF, 3x.} \\ &= 3x(2x + 5)(2x - 5) && \textit{A}^2 - \textit{B}^2 = (\textit{A} + \textit{B})(\textit{A} - \textit{B}) \end{aligned}$$

2. We begin by writing $36x + 4x^3 - 24x^2$ in descending order:

$$\begin{aligned} 36x + 4x^3 - 24x^2 &= 4x^3 - 24x^2 + 36x && \textit{Write in descending order.} \\ &= 4x\left(x^2 - 6x + 9\right) && \textit{Factor out GCF, 4x.} \\ &= 4x(x - 3)(x - 3) && \textit{Find two integers whose product} \\ & && \textit{is 9 and whose sum is } -6. \\ &= 4x(x - 3)^2 && \textit{bb} = \textit{b}^2 \end{aligned}$$

Factoring Out the Opposite of the GCF of a Polynomial

How do we factor a polynomial in which the leading coefficient is negative? For example, consider the polynomial $-5x^3 + 30x^2 - 40x$, which has a negative leading coefficient, -5.

> **How to Factor when the Leading Coefficient Is Negative**
>
> When the leading coefficient of a polynomial is negative, we first factor out the opposite of the GCF.

Example 7   Factoring Out the Opposite of the GCF

Factor $-5x^3 + 30x^2 - 40x$.

Solution

For $-5x^3 + 30x^2 - 40x$, the GCF is $5x$. Because the leading coefficient, -5, is negative, we factor out the opposite of the GCF:

$$\begin{aligned} -5x^3 + 30x^2 - 40x &= -5x\left(x^2 - 6x + 8\right) && \textit{Factor out } -5x, \textit{ opposite of GCF.} \\ &= -5x(x - 2)(x - 4) && \textit{Find two integers whose product is} \\ & && \textit{8 and whose sum is } -6. \end{aligned}$$

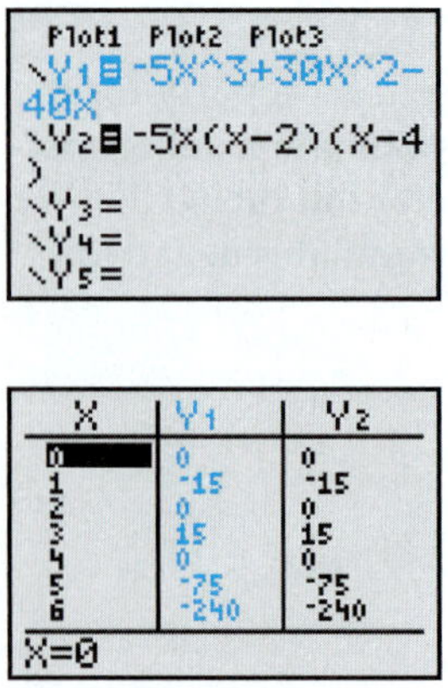

Figure 4 Verify the work

We use a graphing calculator table to verify our work (see Fig. 4).

| Example 8 | Factoring Out the Opposite of the GCF |

Factor $49 - w^2$.

Solution

First we write $49 - w^2$ in descending order: $-w^2 + 49$. The GCF of $-w^2 + 49$ is 1. Because the leading coefficient, -1, is negative, we factor out the opposite of the GCF:

$$-w^2 + 49 = -1(w^2 - 49) \qquad \textit{Factor out } -1, \textit{ opposite of GCF.}$$
$$= -1(w + 7)(w - 7) \qquad A^2 - B^2 = (A + B)(A - B)$$
$$= -(w + 7)(w - 7) \qquad -1a = -a$$

Factoring by Grouping

So far, we have factored out a GCF when the GCF is a monomial. For example, here we factor out the monomial p from the polynomial $x^2(p) + 2(p)$:

$$x^2(p) + 2(p) = (x^2 + 2)(p)$$

We can also factor out a GCF that is a binomial. For example, here we factor out the binomial $x + 3$ from the polynomial $x^2(x + 3) + 2(x + 3)$:

$$x^2(x + 3) + 2(x + 3) = (x^2 + 2)(x + 3)$$

We can factor some polynomials that contain four terms by first factoring the first two terms and the last two terms—for example,

$$\underbrace{x^3 + 3x^2}_{\text{factor}} + \underbrace{2x + 6}_{\text{factor}} = x^2(x + 3) + 2(x + 3) \qquad \textit{Factor both pairs of terms.}$$
$$= (x^2 + 2)(x + 3) \qquad \textit{Factor out GCF, x + 3.}$$

We call this method *factoring by grouping*.

| Example 9 | Factoring by Grouping |

Factor $2x^3 - 8x^2 - 3x + 12$.

Solution

$$2x^3 - 8x^2 - 3x + 12 = 2x^2(x - 4) - 3(x - 4) \qquad \textit{Factor both pairs of terms.}$$
$$= (2x^2 - 3)(x - 4) \qquad \textit{Factor out GCF, x - 4.}$$

WARNING   It is a common error to think a polynomial such as $2x^2(x - 4) - 3(x - 4)$ in Example 9 is factored. Even though both of the terms $2x^2(x - 4)$ and $3(x - 4)$ are factored, the entire expression $2x^2(x - 4) - 3(x - 4)$ is a difference, not a product. The polynomial $(2x^2 - 3)(x - 4)$ in Example 9 *is* factored because it is a product.

We now describe in general how to factor a polynomial by grouping.

Factoring by Grouping

For a polynomial with four terms, we **factor by grouping** (if it can be done) by

1. Factoring the first two terms and the last two terms.

2. Factoring out the binomial GCF.

When trying to factor a polynomial with four terms, consider trying to factor it by grouping.

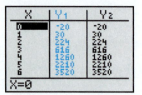

Figure 5 Verify the work

Example 10   Factoring by Grouping

Factor $9x^3 + 45x^2 - 4x - 20$.

Solution

$$\begin{aligned}
9x^3 + 45x^2 - 4x - 20 &= 9x^2(x + 5) - 4(x + 5) &&\textit{Factor both pairs of terms.}\\
&= (9x^2 - 4)(x + 5) &&\textit{Factor out GCF, } x + 5.\\
&= (3x + 2)(3x - 2)(x + 5) &&A^2 - B^2 = (A + B)(A - B)
\end{aligned}$$

We use a graphing calculator table to verify our work (see Fig. 5).

Example 11   Factoring by Grouping

Factor $2a - 2b + xa - xb$.

Solution

$$\begin{aligned}
2a - 2b + xa - xb &= 2(a - b) + x(a - b) &&\textit{Factor both pairs of terms.}\\
&= (2 + x)(a - b) &&\textit{Factor out GCF, } a - b.
\end{aligned}$$

In summary, there are two aspects to factoring that are important to remember:

1. If the leading coefficient of a polynomial is positive and the GCF is not 1, first factor out the GCF. If the leading coefficient is negative, first factor out the opposite of the GCF.

2. Always factor a polynomial *completely*.

 ## Group Exploration

Section Opener: Factoring out the greatest common factor

1. Simplify the following.
   a. $2(x + 3)$ b. $5(x - 4)$ c. $3(x^2 - 5x + 2)$

2. Factor the following. [**Hint:** Refer to your work in Problem 1.]
   a. $2x + 6$ b. $5x - 20$ c. $3x^2 - 15x + 6$

3. Factor the following.
   a. $3x + 12$ b. $2x - 10$ c. $5x^2 - 20x + 15$

4. Find the product of the following.
   a. $x(3x + 9)$ b. $2x^2(5x - 3)$ c. $7x(2x^2 + 5x - 4)$

5. Factor the following. [**Hint:** Refer to your work in Problem 4.]
   a. $3x^2 + 9x$ b. $10x^3 - 6x^2$ c. $14x^3 + 35x^2 - 28x$

6. Factor the following.
   a. $x^2 + 5x$ b. $6x^3 - 15x^2$ c. $2x^3 - 8x^2 - 10x$

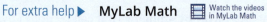

Homework 8.2

For extra help ▶   **MyLab Math**   🎞 Watch the videos in MyLab Math

Factor. Use a graphing calculator table to verify your work when possible.

1. $6x + 8$
2. $3x + 15$
3. $20w^2 + 35w$
4. $28p^2 + 21p$
5. $12x^3 - 30x^2$
6. $18x^4 - 12x^2$
7. $6a^2b - 9ab$
8. $8pq^3 - 6p^2q$
9. $8x^3y^2 + 12x^2y^3$
10. $27x^3y - 45x^2y^3$
11. $15x^3 - 10x - 30$
12. $28x^3 + 12x - 20$
13. $12t^4 + 8t^3 - 16t$
14. $6r^4 - 9r^2 - 12r$
15. $10a^4b - 15a^3b + 25ab$
16. $8p^4q + 6p^2q - 4pq$

Factor. Verify that you have factored correctly by finding the product of your factored polynomial.

17. $2x^2 - 18$ **18.** $3x^2 - 75$

19. $3m^2 + 21m + 30$ **20.** $5p^2 - 5p - 60$

21. $2x^2 - 18x + 36$ **22.** $4x^2 - 44x + 72$

23. $3x^3 - 27x$ **24.** $5x^3 - 20x$

25. $4r^3 - 16r^2 - 20r$ **26.** $2t^3 - 12t^2 - 32t$

27. $6x^4 - 24x^2$ **28.** $2x^4 - 50x^2$

29. $8m^4n - 18m^2n$ **30.** $75p^4y - 27p^2y$

31. $5x^4 + 10x^3 - 120x^2$ **32.** $8x^4 - 24x^3 + 16x^2$

33. $8x - 2x^3$ **34.** $24x - 6x^3$

35. $36t^2 + 32t + 4t^3$ **36.** $60y - 40y^2 + 5y^3$

37. $-12x^3 + 27x$ **38.** $-4x^4 + 4x^2$

39. $-3x^3 - 18x^2 + 48x$ **40.** $-4x^3 + 20x^2 - 24x$

41. $6a^4b + 36a^3b + 54a^2b$ **42.** $4m^4n - 40m^3n + 100m^2n$

Factor. Use a graphing calculator table to verify your work when possible.

43. $5x^2(x - 3) + 2(x - 3)$ **44.** $8x^2(x + 4) + 3(x + 4)$

45. $6x^2(2x + 5) - 7(2x + 5)$ **46.** $5x^2(4x - 1) - 3(4x - 1)$

47. $2p^3 + 6p^2 + 5p + 15$ **48.** $12r^3 + 3r^2 + 8r + 2$

49. $6x^3 - 2x^2 + 21x - 7$ **50.** $4x^3 - 16x^2 + 3x - 12$

51. $15w^3 + 5w^2 - 6w - 2$ **52.** $6b^3 - 10b^2 - 21b + 35$

53. $16x^3 - 12x^2 - 36x + 27$ **54.** $50x^3 + 125x^2 - 8x - 20$

55. $2b^3 - 5b^2 - 18b + 45$ **56.** $4t^3 - 7t^2 - 16t + 28$

57. $x^3 - x^2 - x + 1$ **58.** $x^3 + x^2 - x - 1$

59. $3x + 3y + ax + ay$ **60.** $5a + 5b + xa + xb$

61. $2xy - 8x + 3y - 12$ **62.** $5ab - 10a + 3b - 6$

Factor. Use a graphing calculator table to verify your work when possible.

63. $81x^2 - 25$ **64.** $9x^2 - 100$

65. $w^2 - 10w + 16$ **66.** $p^2 + 3p - 40$

67. $24 - 10x + x^2$ **68.** $-14 - 5x + x^2$

69. $20a^2b - 15ab^3$ **70.** $14xy^2 + 21x^2y$

71. $3r^2 + 30r + 75$ **72.** $2y^2 + 16y + 32$

73. $64x^3 - 49x$ **74.** $2x^3 - 162x$

75. $-m^2 + 6m - 9$ **76.** $-w^2 + 16w - 64$

77. $x^3 + 9x^2 - 4x - 36$ **78.** $28x^3 - 35x^2 + 8x - 10$

79. $2m^3n - 10m^2n^2 + 12mn^3$ **80.** $4p^3y + 8p^2y^2 - 12py^3$

Concepts

81. A student tries to factor $6x^3 + 8x^2 + 15x + 20$:

$$6x^3 + 8x^2 + 15x + 20 = 2x^2(3x + 4) + 5(3x + 4)$$

Describe any errors. Then factor the polynomial correctly.

82. Two students try to factor $15x^3 + 3x^2 - 35x - 7$:

Student A

$$15x^3 + 3x^2 - 35x - 7 = 3x^2(5x + 1) - 7(5x + 1)$$
$$= (3x^2 - 7)(5x + 1)$$

Student B

$$15x^3 + 3x^2 - 35x - 7 = 15x^3 - 35x + 3x^2 - 7$$
$$= 5x(3x^2 - 7) + 1(3x^2 - 7)$$
$$= (5x + 1)(3x^2 - 7)$$

Are both students, one student, or neither student correct?

83. A student tries to factor $4x^3 + 28x^2 + 40x$:

$$4x^3 + 28x^2 + 40x = 4x(x^2 + 7x + 10)$$

Describe any errors. Then factor the polynomial correctly.

84. A student tries to factor $5x^3 - 45x$:

$$5x^3 - 45x = 5x(x^2 - 9)$$

Describe any errors. Then factor the polynomial correctly.

85. A student tries to factor $4x^2 - 100$:

$$4x^2 - 100 = (2x + 10)(2x - 10)$$
$$= 2(x + 5)(2)(x - 5)$$
$$= 4(x + 5)(x - 5)$$

What would you tell the student?

86. A student tries to factor $64x^2 - 36$:

$$64x^2 - 36 = (8x + 6)(8x - 6)$$
$$= 2(4x + 3)(2)(4x - 3)$$
$$= 4(4x + 3)(4x - 3)$$

What would you tell the student?

87. Give three examples of cubic polynomials in which $2x$ is a factor.

88. Give three examples of quadratic polynomials in which $-5x$ is a factor.

89. A student tries to factor $2x^2 + 10x + 12$:

$$2x^2 + 10x + 12 = 2(x^2 + 5x + 6)$$

The student then checks that tables for $y = 2x^2 + 10x + 12$ and $y = 2(x^2 + 5x + 6)$ are the same. Explain why the student's work is incorrect even though the tables for the two equations are the same.

90. Explain why, when factoring a polynomial, it is a good idea to factor out the GCF first if it is not 1 and the leading coefficient of the polynomial is positive.

Related Review

If the polynomial is not factored, then factor it. If the polynomial is factored, then find the product.

91. $2x(x - 3)(x + 4)$ **92.** $-5x(x - 2)(x - 7)$

93. $5x^3 - 40x^2 + 80x$ **94.** $3x^3 + 12x^2 + 12x$

95. $6x^3 - 9x^2 - 4x + 6$ **96.** $8x^3 + 6x^2 - 20x - 15$

97. $(x - 3)(x^2 + 5)$ **98.** $(x^2 - 2)(x - 8)$

Expressions, Equations, and Graphs

Perform the indicated instruction. Then use words such as linear, *quadratic, cubic, polynomial, degree, one variable, and* two variables *to describe the expression, equation, or system.*

99. Graph $y = -4x + 1$ by hand.

100. Find the product $(3x - 2)(4x^2 - x + 5)$.

101. Solve $-4x + 1 = 2x - 5$.

102. Find the sum $(3x - 2) + (4x^2 - x + 5)$.

103. Solve the following:

$$y = -4x + 1$$
$$y = 2x - 5$$

104. Factor $36x^2 - 81$.

8.3 Factoring Trinomials of the Form $ax^2 + bx + c$

Objectives

» Factor a trinomial by trial and error.

» Factor out the GCF, and then factor by trial and error.

» Rule out possibilities when factoring by trial and error.

» Factor a trinomial by grouping.

In Section 8.1, we factored trinomials of the form $ax^2 + bx + c$, where $a = 1$. In this section, we will factor trinomials of the form $ax^2 + bx + c$, where $a \neq 1$. We will discuss two methods: factoring by trial and error and factoring by grouping. These two methods give equivalent results.

Method 1: Factoring Trinomials by Trial and Error

One way to factor trinomials of the form $ax^2 + bx + c$, where $a \neq 1$, is to make educated guesses at the factorization and then find the product of these guesses to see if any of them work. This method is called **factoring by trial and error**.

Example 1	Factoring by Trial and Error

Factor $3x^2 + 17x + 10$.

Solution

If we can factor $3x^2 + 17x + 10$, the result will be of the form

$$(3x + ?)(x + ?)$$

The product of the last terms has to be 10, so the last terms must be 1 and 10, or 2 and 5, where each pair of terms can be written in either order. We can rule out negative last terms in the factors because the middle term, $17x$, of $3x^2 + 17x + 10$ has the positive coefficient 17. We decide between the two pairs of possible last terms by multiplying:

$$(3x + 1)(x + 10) = 3x^2 + 30x + x + 10 = 3x^2 + 31x + 10$$
$$(3x + 10)(x + 1) = 3x^2 + 3x + 10x + 10 = 3x^2 + 13x + 10$$
$$(3x + 2)(x + 5) = 3x^2 + 15x + 2x + 10 = 3x^2 + 17x + 10 \longleftarrow \text{Success!}$$
$$(3x + 5)(x + 2) = 3x^2 + 6x + 5x + 10 = 3x^2 + 11x + 10$$

So, $3x^2 + 17x + 10 = (3x + 2)(x + 5)$. We use a graphing calculator table to verify our work (see Fig. 6).

Figure 6 Verify the work

In trying to factor a polynomial, once we find the factored polynomial, there is no need to multiply the other possibilities. In Example 1, we multiplied all possible factorizations of $3x^2 + 17x + 10$ only to show how to organize the work in case the last possibility is the correct one.

To use the method shown in Example 1, it is helpful to be able to multiply two binomials in one step. Consider the product of $2x + 3$ and $4x + 5$:

$$(2x + 3)(4x + 5) = 8x^2 + 10x + 12x + 15$$
$$= 8x^2 + 22x + 15$$

To find the product in one step, we must combine the like terms $10x$ and $12x$ mentally. Note that these like terms come from the product of the two *outer terms* and the product of the two *inner terms* of $(2x + 3)(4x + 5)$:

$$\overset{\text{outer terms}}{}\qquad\overset{\substack{\text{Add these}\\\text{terms mentally.}}}{}$$

$$(2x + 3)(4x + 5) = 8x^2 + 10x + 12x + 15$$
$$\underset{\text{inner terms}}{} = 8x^2 + 22x + 15$$

Example 2   Factoring by Trial and Error

Factor $2x^2 - 3x - 9$.

Solution

If we can factor $2x^2 - 3x - 9$, the result will be of the form
$$(2x + ?)(x + ?)$$

The product of the last terms is -9, so the last terms must be 1 and -9, 3 and -3, or -1 and 9, where each pair can be written in either order. We decide among the three pairs of possible last terms by multiplying:

$$(2x + 1)(x - 9) = 2x^2 - 17x - 9$$
$$(2x - 9)(x + 1) = 2x^2 - 7x - 9$$
$$(2x + 3)(x - 3) = 2x^2 - 3x - 9 \longleftarrow \text{Success!}$$
$$(2x - 3)(x + 3) = 2x^2 + 3x - 9$$
$$(2x - 1)(x + 9) = 2x^2 + 17x - 9$$
$$(2x + 9)(x - 1) = 2x^2 + 7x - 9$$

Therefore, $2x^2 - 3x - 9 = (2x + 3)(x - 3)$.

Factoring $ax^2 + bx + c$ by Trial and Error

To **factor a trinomial** of the form $ax^2 + bx + c$ **by trial and error**, if the trinomial can be factored as a product of two binomials, then the product of the coefficients of the first terms of the binomials is equal to a and the product of the last terms of the binomials is equal to c. For example,

$$\text{Coefficients of first terms:}$$
$$3 \cdot 2 = 6 = a$$
$$6x^2 + 23x + 20 = (3x + 4)(2x + 5)$$
$$a = 6 \quad b = 23 \quad c = 20 \qquad \text{Last terms:}$$
$$4 \cdot 5 = 20 = c$$

To find the correct factored polynomial, multiply the possible products and identify the one for which the coefficient of x is b.
For example,

$$\overset{\text{outer terms}}{}$$
$$(3x + 4)(2x + 5) = 6x^2 + 15x + 8x + 20$$
$$\underset{\text{inner terms}}{} = 6x^2 + 23x + 20$$
$$b = 23 \text{ is correct.}$$

Factoring Out the GCF and Then Factoring by Trial and Error

When factoring a polynomial, recall from Section 8.2 that if the GCF is not 1, then we first factor out the GCF (or its opposite) and continue factoring if possible. Always factor a polynomial completely.

Example 3   Factoring a Polynomial Completely

Factor $15x^4 - 39x^3 + 18x^2$.

Solution

To factor $15x^4 - 39x^3 + 18x^2$, we first factor out the GCF, $3x^2$:

$$3x^2\left(5x^2 - 13x + 6\right)$$

If we can factor further, the desired result is of the form

$$3x^2(5x + ?)(x + ?)$$

The product of the last terms has to be 6, so the last terms must be -1 and -6, or -2 and -3, where each pair can be written in either order. We can rule out positive last terms because the middle term, $-13x$, has a negative coefficient, -13. We decide by multiplying:

(We have temporarily put aside the GCF, $3x^2$.)

$$(5x - 1)(x - 6) = 5x^2 - 31x + 6$$
$$(5x - 6)(x - 1) = 5x^2 - 11x + 6$$
$$(5x - 2)(x - 3) = 5x^2 - 17x + 6$$
$$(5x - 3)(x - 2) = 5x^2 - 13x + 6 \longleftarrow \text{Success!}$$

So, $15x^4 - 39x^3 + 18x^2 = 3x^2\left(5x^2 - 13x + 6\right) = 3x^2(5x - 3)(x - 2)$.

WARNING   When we factor a polynomial by trial and error, we can easily forget about the GCF by the time we have found the other factors. **If there is more factoring to be done after factoring out the GCF, write a note several lines down that reminds you to include the GCF in your result.**

Ruling Out Possibilities When Factoring by Trial and Error

In Example 4, we will rule out possible factorizations to help speed up the process of factoring.

Example 4   Ruling Out Possibilities

Factor $6x^2 - 19x + 8$.

Solution

If we can factor $6x^2 - 19x + 8$, the result is of one of the following two forms:

$$(6x + ?)(x + ?) \qquad (3x + ?)(2x + ?)$$

The product of the last terms has to be 8, so the last terms must be -1 and -8, or -2 and -4, where each pair can be written in either order. We can rule out positive last terms because the middle term, $-19x$, has a negative coefficient, -19.

Because the terms of $6x^2 - 19x + 8$ do not have a common factor of 2, we can also rule out products that have a factor of 2. For example, we can rule out $(6x - 8)(x - 1)$ because it has a factor of 2:

$$(6x - 8)(x - 1) = 2(3x - 4)(x - 1)$$

We decide among the remaining possible last terms by multiplying:

$(6x - 1)(x - 8) = 6x^2 - 49x + 8$

Contains a factor of 2, rule out: $(6x - 2)(x - 4)$
Contains a factor of 2, rule out: $(6x - 4)(x - 2)$
Contains a factor of 2, rule out: $(3x - 1)(2x - 8)$
$(3x - 8)(2x - 1) = 6x^2 - 19x + 8 \longleftarrow \text{Success!}$
Contains a factor of 2, rule out: $(3x - 2)(2x - 4)$
Contains a factor of 2, rule out: $(3x - 4)(2x - 2)$

So, $6x^2 - 19x + 8 = (3x - 8)(2x - 1)$. We use a graphing calculator table to verify our work (see Fig. 7).

Figure 7 Verify the work

We can use a similar method to factor trinomials that have two variables.

Example 5   Factoring a Trinomial That Has Two Variables

Factor $10p^2 + 19pw + 6w^2$.

Solution

If we can factor $10p^2 + 19pw + 6w^2$, the result is of one of the following two forms:

$$(10p + ?)(p + ?)   (5p + ?)(2p + ?)$$

The product of the last terms has to be $6w^2$, so the last terms must be $6w$ and w, or $3w$ and $2w$, where each pair can be written in either order.

Because the terms of $10p^2 + 19pw + 6w^2$ do not have a common factor of 2, we can rule out products that have a factor of 2. We decide among the remaining possible last terms by multiplying:

Contains a factor of 2, rule out: $(10p + 6w)(p + w)$
$$(10p + w)(p + 6w) = 10p^2 + 61pw + 6w^2$$
$$(10p + 3w)(p + 2w) = 10p^2 + 23pw + 6w^2$$
Contains a factor of 2, rule out: $(10p + 2w)(p + 3w)$
$$(5p + 6w)(2p + w) = 10p^2 + 17pw + 6w^2$$
Contains a factor of 2, rule out: $(5p + w)(2p + 6w)$
Contains a factor of 2, rule out: $(5p + 3w)(2p + 2w)$
$$(5p + 2w)(2p + 3w) = 10p^2 + 19pw + 6w^2 \longleftarrow \text{Success!}$$

So, $10p^2 + 19pw + 6w^2 = (5p + 2w)(2p + 3w)$.

▶

Method 2: Factoring Trinomials by Grouping

Instead of using trial and error to factor a trinomial, we can factor by grouping.

To factor a trinomial of the form $x^2 + bx + c$, recall from Section 8.1 that we look for two integers whose product is c and whose sum is b. To factor a trinomial of the form $ax^2 + bx + c$, we must look for two integers whose product is ac and whose sum is b.

Factoring $ax^2 + bx + c$ by Grouping

To **factor a trinomial** of the form $ax^2 + bx + c$ **by grouping** (if it can be done),

1. Find pairs of numbers whose product is ac.

2. Determine which of the pairs of numbers from step 1 has the sum b. Call this pair of numbers m and n.

3. Write the bx term as $mx + nx$:

$$ax^2 + bx + c = ax^2 + mx + nx + c$$

4. Factor $ax^2 + mx + nx + c$ by grouping.

Another name for this technique is the **ac method**.

Example 6   Factoring a Trinomial by Grouping

Factor $3x^2 + 17x + 10$ by grouping.

Solution

Here, $a = 3$, $b = 17$, and $c = 10$.

Step 1:   Find the product ac: $ac = 3(10) = 30$.

Step 2: We want to find two numbers m and n that have the product $ac = 30$ and the sum $b = 17$:

Product = 30	Sum = 17?
$1(30) = 30$	$1 + 30 = 31$
$2(15) = 30$	$2 + 15 = 17$ ⟵ Success!
$3(10) = 30$	$3 + 10 = 13$
$5(6) = 30$	$5 + 6 = 11$

Because $2(15) = 30$ and $2 + 15 = 17$, we conclude that the two numbers m and n are 2 and 15.

Step 3: We write $3x^2 + 17x + 10 = 3x^2 + 2x + 15x + 10$.

Step 4: We factor $3x^2 + 2x + 15x + 10$ by grouping:

$$3x^2 + 2x + 15x + 10 = x(3x + 2) + 5(3x + 2) \quad \textit{Factor both pairs of terms.}$$
$$= (x + 5)(3x + 2) \quad \textit{Factor out GCF, } (3x + 2).$$

In Step 3 of Example 6, we could switch the mx and nx terms to get $3x^2 + 15x + 2x + 10$ and still be able to factor by grouping. (Try it.)

In Example 1, we used trial and error to factor $3x^2 + 17x + 10$ as $(3x + 2)(x + 5)$. In Example 6, we factored it as $(x + 5)(3x + 2)$ by using grouping. The two results are equivalent. In general, the results from factoring a trinomial by trial and error and factoring a trinomial by grouping are equivalent.

Example 7   Factoring Out the GCF and Then Factoring by Grouping

Factor $18x^3 + 70x^2 - 8x$ by grouping.

Solution

First, we factor out the GCF, $2x$:

$$18x^3 + 70x^2 - 8x = 2x(9x^2 + 35x - 4)$$

Next, we use grouping to try to factor $9x^2 + 35x - 4$. Here, $a = 9$, $b = 35$, and $c = -4$.

Step 1: Find the product ac: $ac = 9(-4) = -36$.

Step 2: We want to find two numbers m and n that have the product $ac = -36$ and the sum $b = 35$:

(We have temporarily put aside the GCF, $2x$.)

Product = -36	Sum = 35?
$1(-36) = -36$	$1 + (-36) = -35$
$2(-18) = -36$	$2 + (-18) = -16$
$3(-12) = -36$	$3 + (-12) = -9$
$4(-9) = -36$	$4 + (-9) = -5$
$6(-6) = -36$	$6 + (-6) = 0$
$9(-4) = -36$	$9 + (-4) = 5$
$12(-3) = -36$	$12 + (-3) = 9$
$18(-2) = -36$	$18 + (-2) = 16$
$36(-1) = -36$	$36 + (-1) = 35$ ⟵ Success!

Because $36(-1) = -36$ and $36 + (-1) = 35$, we conclude that the two numbers m and n are 36 and -1.

Step 3: We write $9x^2 + 35x - 4 = 9x^2 + 36x - 1x - 4$.

Step 4: We factor $9x^2 + 36x - 1x - 4$ by grouping:

$$9x^2 + 36x - 1x - 4 = 9x(x + 4) - 1(x + 4) \quad \textit{Factor both pairs of terms.}$$
$$= (9x - 1)(x + 4) \quad \textit{Factor out GCF, } (x + 4).$$

So, $18x^3 + 70x^2 - 8x = 2x(9x^2 + 35x - 4) = 2x(9x - 1)(x + 4)$. We use a graphing calculator table to verify our work (see Fig. 8).

Figure 8 Verify the work

Group Exploration

Factoring polynomials

1. A student tries to factor $2x^2 - 17x - 30$:

$$2x^2 - 17x - 30 = (2x - 5)(x - 6)$$

Multiply $(2x - 5)(x - 6)$ to show the work is incorrect. Then factor $2x^2 - 17x - 30$ correctly.

2. A student tries to factor $2x^2 + 10x + 12$:

$$2x^2 + 10x + 12 = (2x + 4)(x + 3)$$

Explain why the work is not correct. Then factor the polynomial correctly.

3. A student tries to factor $2x^2 - x - 6$. Because the product of -3 and 2 is -6 and the sum of -3 and 2 is -1, the student does the following work:

$$2x^2 - x - 6 = (2x - 3)(x + 2)$$

Find the product $(2x - 3)(x + 2)$ to show the work is incorrect. Explain what is wrong with the student's reasoning. Then factor the polynomial correctly.

Homework 8.3

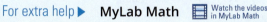

For extra help ▶ **MyLab Math** Watch the videos in MyLab Math

Factor if possible. Verify that you have factored correctly by finding the product of your factored polynomial.

1. $2x^2 + 7x + 3$

2. $3x^2 + 16x + 5$

3. $5x^2 + 11x + 2$

4. $4x^2 + 13x + 3$

5. $3x^2 + 8x + 4$

6. $5x^2 + 13x + 6$

7. $2t^2 + t - 6$

8. $3b^2 - 10b + 8$

9. $6x^2 - 13x + 6$

10. $8x^2 - 14x - 15$

11. $4x^2 + 20x + 25$

12. $9x^2 + 12x + 4$

13. $6x^2 - 37x + 6$

14. $4x^2 + 15x - 4$

15. $2r^2 + 5r + 4$

16. $3a^2 - 2a - 6$

17. $18x^2 + 21x - 4$

18. $12x^2 - 17x + 6$

19. $3m^2 - 22m + 24$

20. $10t^2 - 27t + 18$

21. $2x^2 - 21x + 40$

22. $4x^2 + 81x + 20$

23. $2a^2 + 5ab + 3b^2$

24. $3m^2 + 8mn + 4n^2$

25. $5x^2 + 18xy - 8y^2$

26. $2p^2 - 5pt - 12t^2$

27. $6b^2 - 15bc + 6c^2$

28. $8r^2 - 33rw + 4w^2$

Factor. Use a graphing calculator table to verify your work when possible.

29. $4x^2 + 26x + 30$

30. $6x^2 + 27x + 30$

31. $20a^2 - 40a + 15$

32. $12b^2 - 56b + 32$

33. $24x^2 + 15x - 9$

34. $12x^2 - 27x + 15$

35. $-20x^2 + 22x + 12$

36. $-24x^2 + 54x - 27$

37. $4w^4 - 6w^3 - 12w^2$

38. $9t^4 - 21t^3 - 12t^2$

39. $10x^4 - 5x^3 - 50x^2$

40. $36x^4 - 30x^3 + 6x^2$

41. $6a^2 - 34ab - 12b^2$

42. $6m^2 + 38mn + 12n^2$

43. $12r^3 + 40r^2w + 32rw^2$

44. $15x^3 + 36x^2y + 12xy^2$

Factor when possible. Use a graphing calculator table to verify your work when possible.

45. $x^2 - 6x - 27$

46. $x^2 - 4x - 45$

47. $-48x^2 + 40x$

48. $-35x^2 - 42x$

49. $5a^3 + 2a^2 - 15a - 6$

50. $10r^3 + 25r^2 - 6r - 15$

51. $x^2 + 9$

52. $4x^2 + 25$

53. $4x^2 - 12x + 9$

54. $25x^2 - 10x + 1$

55. $-17p^2 + 17$

56. $-5t^2 + 20$

57. $24 + 10x + x^2$

58. $-12 + x + x^2$

59. $b^2 - 3bc - 28c^2$

60. $p^2 + 4pt - 32t^2$

61. $8t^2 - 10t + 3$

62. $6a^2 - 19a + 15$

63. $7x^4 - 28x^2$

64. $2x^4 - 50x^2$

65. $12p^3 - 4p^2 - 27p + 9$

66. $50m^3 - 75m^2 - 2m + 3$

67. $x^2 - 6x + 12$

68. $x^2 - 3x - 8$

69. $3x^4 - 21x^3 + 30x^2$

70. $4x^4 - 12x^3 + 8x^2$

71. $20x^2 + 16x^4 - 42x^3$

72. $-14x^3 + 6x^4 - 40x^2$

73. $36a^2 - 49b^2$

74. $64m^2 - 25n^2$

75. $-2x^2y + 8xy + 24y$

76. $-6x^2y + 24xy - 24y$

77. $4y^3 - 9y^2 - 9y$

78. $12t^3 + 7t^2 - 5t$

Concepts

79. A student tries to factor the polynomial $2x^2 + 7x + 10$. Because the integers 2 and 5 have the product $2(5) = 10$ and the sum $2 + 5 = 7$, the student writes

$$2x^2 + 7x + 10 = (2x + 5)(x + 2)$$

Is the student correct? Explain.

80. A student tries to factor the polynomial $3x^2 + 12$:

$$3x^2 + 12 = 3(x^2 + 4) = 3(x + 2)(x + 2) = 3(x + 2)^2$$

Describe any errors. Then factor the polynomial correctly.

81. A student tries to factor the polynomial $8x^2 + 28x + 12$:

$$8x^2 + 28x + 12 = (4x + 2)(2x + 6)$$

The student then uses a graphing calculator table to verify that the polynomials $8x^2 + 28x + 12$ and $(4x + 2)(2x + 6)$ are equivalent. Explain why the student's work is incorrect even though the polynomials are equivalent.

82. A student tries to factor the polynomial $6x^2 + 28x - 10$. First, the student divides the polynomial by 2: $3x^2 + 14x - 5$. Then the student factors $3x^2 + 14x - 5$:

$$3x^2 + 14x - 5 = (3x - 1)(x + 5)$$

The student thinks that the factorization of $6x^2 + 28x - 10$ is $(3x - 1)(x + 5)$. What would you tell the student? Also, explain how using a graphing calculator table to check the work can help the student identify the error.

83. To factor $4x^2 + 28x + 48$, why is it a good idea to first factor out the GCF before using any other factoring technique?

84. Factor $x^2 + 9x + 20$ using grouping. Then factor $x^2 + 9x + 20$ by the method discussed in Section 8.1. Which method is easier?

85. Give three examples of quadratic polynomials in which $2x - 3$ is a factor.

86. Give three examples of cubic polynomials in which $3x + 1$ is a factor.

Related Review

If the polynomial is not factored, then factor it. If the polynomial is factored, then find the product.

87. $3x^2 + 16x - 12$

88. $5x^2 - 29x + 20$

89. $(4x - 7)(3x - 1)$

90. $(2x - 9)(5x + 3)$

91. $(x - 3)(2x^2 + 3x - 5)$

92. $(3x^2 - x - 2)(2x + 3)$

93. $6x^3 + 10x^2 - 4x$

94. $6x^4 - 33x^3 + 36x^2$

Expressions, Equations, and Graphs

Perform the indicated instruction. Then use words such as linear, quadratic, cubic, polynomial, degree, one variable, *and* two variables *to describe the expression, equation, or system.*

95. Graph $y = x^2 - 3$ by hand.

96. Solve the following:

$$7x + 2y = -6$$
$$3x - 4y = -22$$

97. Factor $x^2 - 2x - 3$.

98. Graph $7x + 2y = -6$ by hand.

99. Evaluate $x^2 - 2x - 3$ for $x = -5$.

100. Find the product $(7x + 2y)(3x - 4y)$.

8.4 Sums and Differences of Cubes; A Factoring Strategy

Objectives

» Factor a sum or difference of two cubes.

» Apply a factoring strategy.

In Section 8.1, we discussed how to factor a difference of two squares. Here, first we will discuss how to factor a sum or difference of two cubes. Then we will discuss how to sift through the many factoring techniques we have discussed in this chapter and select the best ones to factor a given polynomial completely.

Factoring a Sum or Difference of Two Cubes

To see how to factor the sum of two cubes, we begin by multiplying the polynomials $A + B$ and $A^2 - AB + B^2$:

$$(A + B)(A^2 - AB + B^2) = A \cdot A^2 - A \cdot AB + A \cdot B^2 + B \cdot A^2 - B \cdot AB + B \cdot B^2$$
$$= A^3 - A^2B + AB^2 + A^2B - AB^2 + B^3$$
$$= A^3 + B^3$$

So, $(A + B)(A^2 - AB + B^2) = A^3 + B^3$. Note that the right-hand side of the equation, $A^3 + B^3$, is a sum of two cubes. By similar work, we can also find a property for the difference of two cubes. You will do this in Exercise 85.

Sum or Difference of Two Cubes

$$A^3 + B^3 = (A + B)(A^2 - AB + B^2) \quad \textit{Sum of two cubes}$$
$$A^3 - B^3 = (A - B)(A^2 + AB + B^2) \quad \textit{Difference of two cubes}$$

We can use these two properties to factor any polynomial that is a sum or difference of two cubes. In order to use the properties, it will help to memorize the following cubes:

$$2^3 = 8 \qquad 3^3 = 27 \qquad 4^3 = 64 \qquad 5^3 = 125 \qquad 10^3 = 1000$$

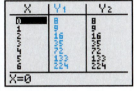

Figure 9 Verify the work

| **Example 1** | Factoring a Sum and a Difference of Two Cubes |

Factor.

1. $x^3 + 8$ **2.** $x^3 - 64$

Solution

1.
$$A^3 + B^3 = (A + B)(A^2 - A\,B + B^2)$$
$$x^3 + 8 = x^3 + 2^3 = (x + 2)(x^2 - x \cdot 2 + 2^2) \quad \textit{Factor.}$$
$$= (x + 2)(x^2 - 2x + 4) \qquad \textit{Simplify.}$$

The trinomial $x^2 - 2x + 4$ is prime, so we have factored $x^3 + 8$ completely. We use a graphing calculator table to verify our work (see Fig. 9).

2.
$$A^3 - B^3 = (A - B)(A^2 + A\,B + B^2)$$
$$x^3 - 64 = x^3 - 4^3 = (x - 4)(x^2 + x \cdot 4 + 4^2) \quad \textit{Factor.}$$
$$= (x - 4)(x^2 + 4x + 16) \qquad \textit{Simplify.}$$

The trinomial $x^2 + 4x + 16$ is prime, so we have factored $x^3 - 64$ completely.

| **Example 2** | Factoring a Sum and a Difference of Two Cubes |

Factor.

1. $27x^3 + 125$ **2.** $16p^3 - 2$

Solution

1. $27x^3 + 125 = (3x)^3 + 5^3$ *Write as a sum of cubes.*
$$= (3x + 5)((3x)^2 - 3x \cdot 5 + 5^2) \quad A^3 + B^3 = (A + B)(A^2 - AB + B^2)$$
$$= (3x + 5)(9x^2 - 15x + 25) \qquad \textit{Simplify.}$$

The trinomial $9x^2 - 15x + 25$ is prime, so we have factored $27x^3 + 125$ completely.

2. We begin by factoring out the GCF, 2:

$$16p^3 - 2 = 2(8p^3 - 1) \qquad\qquad\qquad\quad \textit{Factor out GCF, 2.}$$
$$= 2((2p)^3 - 1^3) \qquad\qquad\qquad \textit{Write as a difference of cubes.}$$
$$= 2(2p - 1)((2p)^2 + 2p \cdot 1 + 1^2) \quad A^3 - B^3 = (A - B)(A^2 + AB + B^2)$$
$$= 2(2p - 1)(4p^2 + 2p + 1) \qquad\qquad \textit{Simplify.}$$

The trinomial $4p^2 + 2p + 1$ is prime, so we have factored $16p^3 - 2$ completely.

Consider the properties for the sum of cubes and difference of cubes:

$$A^3 + B^3 = (A + B)(A^2 - AB + B^2) \qquad A^3 - B^3 = (A - B)(A^2 + AB + B^2)$$

Provided that we have first factored out the GCF (or its opposite), we can assume that the trinomials $A^2 - AB + B^2$ and $A^2 + AB + B^2$ are prime.

A Factoring Strategy

We will now discuss a five-step factoring strategy that will help us determine the best factoring techniques to use to factor a given polynomial completely.

> **Five-Step Factoring Strategy**
>
> The five steps that follow can be used to factor many polynomials. Steps 2–4 can be applied to the entire polynomial or to a factor of the polynomial.
>
> 1. If the leading coefficient is positive and the GCF is not 1, we factor out the GCF. If the leading coefficient is negative, we factor out the opposite of the GCF.
> 2. For a binomial, try using one of the properties for the difference of two squares, the sum of two cubes, or the difference of two cubes.
> 3. For a trinomial of the form $ax^2 + bx + c$,
>   a. If $a = 1$, try to find two integers whose product is c and whose sum is b.
>   b. If $a \neq 1$, try to factor by trial and error or by grouping.
> 4. For a polynomial with four terms, try factoring by grouping.
> 5. Continue applying steps 2–4 until the polynomial is factored completely.

Example 3   Factoring a Polynomial

Factor $x^3 - 7x^2y + 12xy^2$.

Solution

For $x^3 - 7x^2y + 12xy^2$, the GCF is x. First, we factor out x:

$$x^3 - 7x^2y + 12xy^2 = x\left(x^2 - 7xy + 12y^2\right)$$

Because the factor $x^2 - 7xy + 12y^2$ is a trinomial with a leading coefficient that is 1, we try to find two monomials whose product is $12y^2$ and whose sum is $-7y$. The monomials are $-3y$ and $-4y$, so we have

$$x^3 - 7x^2y + 12xy^2 = x\left(x^2 - 7xy + 12y^2\right) = x(x - 3y)(x - 4y)$$

Example 4   Factoring a Polynomial

Factor $32w^3 - 98w$.

Solution

For $32w^3 - 98w$, the GCF is $2w$. First, we factor out $2w$:

$$32w^3 - 98w = 2w\left(16w^2 - 49\right)$$

Because the factor $16w^2 - 49$ has two terms, we check to see whether it is the difference of two squares, which it is. So, we have

$$32w^3 - 98w = 2w\left(16w^2 - 49\right) = 2w(4w + 7)(4w - 7)$$

Example 5   Factoring a Polynomial

Factor $3x - 10 + 4x^2$.

Solution

First, we write $3x - 10 + 4x^2$ in descending order:

$$4x^2 + 3x - 10$$

Because $4x^2 + 3x - 10$ is a trinomial with a leading coefficient that is not 1, we try to factor it by grouping. Here, $a = 4$, $b = 3$, and $c = -10$.

Step 1: Find the product ac: $ac = 4(-10) = -40$.

Step 2: We want to find two numbers m and n that have the product $ac = -40$ and the sum $b = 3$. The integers are 8 and -5.

Step 3: We write $4x^2 + 3x - 10 = 4x^2 + 8x - 5x - 10$.

Step 4: We factor $4x^2 + 8x - 5x - 10$ by grouping:

$$4x^2 + 8x - 5x - 10 = 4x(x + 2) - 5(x + 2) \quad \textcolor{blue}{\textit{Factor both pairs of terms.}}$$
$$= (4x - 5)(x + 2) \quad \textcolor{blue}{\textit{Factor out GCF, } x + 2.}$$

So, $3x - 10 + 4x^2 = 4x^2 + 3x - 10 = (4x - 5)(x + 2)$.
Instead of using grouping, we could have used trial and error.

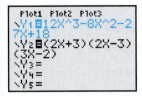

Example 6   Factoring a Polynomial

Factor $12x^3 - 8x^2 - 27x + 18$.

Solution

Because $12x^3 - 8x^2 - 27x + 18$ has four terms, we try to factor it by grouping:

$$12x^3 - 8x^2 - 27x + 18 = 4x^2(3x - 2) - 9(3x - 2) \quad \textcolor{blue}{\textit{Factor both pairs of terms.}}$$
$$= (4x^2 - 9)(3x - 2) \quad \textcolor{blue}{\textit{Factor out GCF, } 3x - 2.}$$
$$= (2x + 3)(2x - 3)(3x - 2) \quad \textcolor{blue}{A^2 - B^2 = (A + B)(A - B)}$$

We use a graphing calculator table to verify our work (see Fig. 10).

Figure 10 Verify the work

Example 7   Factoring a Polynomial

Factor $-5x^4 + 20x^3 - 30x^2$.

Solution

For $-5x^4 + 20x^3 - 30x^2$, the GCF is $5x^2$. Because the leading coefficient of the polynomial, -5, is negative, we factor out $-5x^2$:

$$-5x^4 + 20x^3 - 30x^2 = -5x^2(x^2 - 4x + 6)$$

Because the factor $x^2 - 4x + 6$ is a trinomial with a leading coefficient that is 1, we try to find two integers whose product is 6 and whose sum is -4. However, there is no such pair; the factor $x^2 - 4x + 6$ is prime. So, the completely factored result is $-5x^2(x^2 - 4x + 6)$.

◈ Group Exploration

Section Opener: Developing a factoring strategy

In this exploration, you will summarize what you have learned about factoring.

1. When factoring a polynomial, what should you try to do first? Give an example.

2. Describe various techniques that you can use to factor a polynomial with the given number of terms. For each technique, give an example of factoring a polynomial.

   a. two terms   **b.** three terms   **c.** four terms

3. Explain how you know when you are done factoring a polynomial.

Tips for Success   Solutions Manual

If you are having trouble solving a homework exercise, it can help to refer to the Solutions Manual, which provides step-by-step solutions to the exercises. It is a nice complement to other forms of support, such as your instructor, a tutor, or friends because you can refer to it at any time, day or night.

However, do not use the manual as a crutch. It is a common error to consult the manual before trying to complete an exercise without it. But if you have been trying to solve an exercise for a while and begin to feel frustrated, reach for the manual.

Once you have completed your homework assignment, assess how often you sought help from the manual. If you did so frequently, then continue to do more exercises until you can do any type of problem in the homework without referring to the manual. The ultimate goal is for you to understand the material, not just complete the assignment.

◆ Homework 8.4

For extra help ▶   **MyLab Math**   ▦ Watch the videos in MyLab Math

Factor. Use a graphing calculator table to verify your work when possible.

1. $x^3 + 27$ **2.** $x^3 + 64$ **3.** $x^3 + 125$

4. $x^3 + 1000$ **5.** $x^3 - 8$ **6.** $x^3 - 27$

7. $x^3 - 1$ **8.** $x^3 - 125$ **9.** $8t^3 + 27$

10. $27p^3 + 64$ **11.** $27x^3 - 8$ **12.** $8x^3 - 125$

13. $5x^3 + 40$ **14.** $2x^3 + 2$ **15.** $2x^3 - 54$

16. $3x^3 - 24$ **17.** $8x^3 + 27y^3$ **18.** $27m^3 + 64n^3$

19. $64a^3 - 27b^3$ **20.** $125p^3 - 8t^3$

Factor when possible. Use a graphing calculator table to verify your work when possible.

21. $x^2 - 64$ **22.** $x^2 - 1$ **23.** $m^2 + 11m + 28$

24. $a^2 + 5a - 14$ **25.** $2t^2 + 2t - 24$ **26.** $3a^2 - 21a + 30$

27. $x^2 + 49$ **28.** $x^2 + 4$ **29.** $-3x^2 + 24x - 45$

30. $-2x^2 + 22x - 60$ **31.** $1 + 15p^2 - 8p$

32. $-6t - 8 + 5t^2$ **33.** $x^2 - 2x + 1$

34. $x^2 - 8x + 16$ **35.** $24r^2 + 4r - 4$

36. $20y^3 - 12y^2 - 8y$ **37.** $-4ab^3 + 6a^2b^2$

38. $-6a^2b + 9ab^3$ **39.** $a^2 - ab - 20b^2$

40. $m^2 - 12mn + 32n^2$ **41.** $8x^3 - 20x^2 - 2x + 5$

42. $5x^3 - 2x^2 - 5x + 2$ **43.** $-12x^4 - 4x^3$

44. $-25x^4 + 35x^3$ **45.** $15a^4 + 25a^3 + 10a^2$

46. $6t^4 - 3t^3 - 30t^2$ **47.** $24 - 14x + x^2$

48. $9 - 6x + x^2$ **49.** $2w^4 + 4w^3 - 8w^2$

50. $3p^4 - 9p^3 + 24p^2$ **51.** $12x^4 - 27x^2$

52. $98x^4 - 18x^2$ **53.** $x^2 + 10x + 25$

54. $x^2 + 4x + 4$ **55.** $2x^2 + x - 21$

56. $4x^2 + 20x + 21$ **57.** $m^3 - 13m^2n + 36mn^2$

58. $p^2q + 5pq^2 - 12q^3$ **59.** $4x - 5 + 2x^2$

60. $7x - 4 + 3x^2$ **61.** $100x^2 - 9y^2$

62. $49b^2 - 36c^2$ **63.** $4x^2 + 12x + 9$

64. $16x^2 - 8x + 1$ **65.** $18x^3 + 27x^2 - 8x - 12$

66. $12x^3 + 9x^2 + 8x + 6$ **67.** $3a^3 - 10a^2b + 8ab^2$

68. $5p^2t + 18pt^2 - 8t^3$ **69.** $x^2 - 9x - 20$ **70.** $x^2 - 13x - 30$

71. $x^3 - 1000$ **72.** $x^3 + 1$ **73.** $64a^3 + 27$

74. $27r^3 + 8$ **75.** $3x^3 + 24$ **76.** $7x^3 - 7$

Concepts

For Exercises 77–80, discuss which technique(s) you should consider using when factoring the given type of polynomial. Also, refer to an exercise in Homework 8.4 that illustrates this technique.

77. binomial

78. trinomial of the form $x^2 + bx + c$

79. trinomial of the form $ax^2 + bx + c$, where $a \neq 1$

80. polynomial with four terms

81. A student tries to factor the polynomial $x^3 - 27$:
$$x^3 - 27 = (x - 3)\left(x^2 + 6x + 9\right)$$
Describe any errors. Then factor the polynomial correctly.

82. A student tries to factor the polynomial $x^3 + 8$:
$$x^3 + 8 = (x + 2)\left(x^2 + 2x + 4\right)$$
Describe any errors. Then factor the polynomial correctly.

83. Is the following statement true? Explain.
$$A^3 + B^3 = (A + B)\left(A^2 - 2AB + B^2\right)$$

84. When factoring a polynomial with a negative leading coefficient, what should you do? Give an example.

85. Show that $A^3 - B^3 = (A - B)\left(A^2 + AB + B^2\right)$ by finding the product $(A - B)\left(A^2 + AB + B^2\right)$.

86. Show that $A^2 + AB + B^2$ is prime when $A = x$ and $B = y$.

87. When factoring a polynomial, does it matter whether you factor out the GCF in the first step or in the last step? Explain.

88. Describe a factoring strategy in your own words.

Related Review

If the polynomial is not factored, then factor it. If the polynomial is factored, then find the product.

89. $(5x - 7)(5x + 7)$ **90.** $(2x + 5)\left(x^2 - x + 4\right)$

91. $81x^2 - 16$ **92.** $x^2 - 5x - 36$

93. $3x^3 + 9x^2 - 12x$ **94.** $30x^2 - 24$

95. $-2x\left(7x^2 - 5x + 1\right)$ **96.** $(4x - 3)(2x + 9)$

Expressions, Equations, and Graphs

Perform the indicated instruction. Then use words such as linear, quadratic, cubic, polynomial, degree, one variable, *and* two variables *to describe the expression, equation, or system.*

97. Graph $y = 4x - 5$ by hand.

98. Factor $x^3 - 3x^2 - 4x + 12$.

99. Solve $4x - 5 = 3x + 2$.

100. Simplify $(2x - 7)^2$.

101. Find the product $(4x - 5)(3x + 2)$.

102. Graph $y = 3x^2$ by hand.

8.5 Using Factoring to Solve Polynomial Equations

Objectives

» Apply the *zero factor property*.

» Use factoring to solve *quadratic equations in one variable*.

» Use graphing to solve a polynomial equation in one variable.

» Combine like terms to help solve quadratic equations in one variable.

» Solve quadratic equations in one variable that contain fractions.

» Find any *x*-intercept(s) of a parabola.

» Use factoring to solve *cubic equations in one variable*.

» Compare solving polynomial equations with factoring polynomials.

In this section, we will discuss how to use factoring to help solve equations. In Section 8.6, we will use this skill to make efficient predictions with some quadratic models.

Zero Factor Property

In Section 7.1, we graphed quadratic equations in *two* variables. In this section, we will solve quadratic equations in *one* variable, such as

$$x^2 - 4x - 21 = 0 \qquad 8x^2 - 50 = 0 \qquad 4x^2 = 20x - 25$$

A **quadratic equation in one variable** is an equation that can be put into the form

$$ax^2 + bx + c = 0$$

where a, b, and c are constants and $a \neq 0$. The connection between solving a quadratic equation and factoring a quadratic polynomial lies in the *zero factor property*.

Zero Factor Property

Let A and B be real numbers:

$$\text{If } AB = 0, \text{ then } A = 0 \text{ or } B = 0.$$

In words: If the product of two numbers is zero, then at least one of the numbers must be zero.

Solving Quadratic Equations by Factoring

In the next several examples, we will use the zero factor property to solve some quadratic equations in one variable.

Example 1   Solving a Quadratic Equation

Solve $(x - 3)(x - 8) = 0$.

Solution

$$
\begin{array}{ll}
(x - 3)(x - 8) = 0 & \textit{Original equation} \\
x - 3 = 0 \quad \text{or} \quad x - 8 = 0 & \textit{Zero factor property} \\
x = 3 \quad \text{or} \qquad x = 8 & \textit{Add 3 to both sides./Add 8 to both sides.}
\end{array}
$$

So, the solutions are 3 and 8. We check that both 3 and 8 satisfy the original equation:

$$
\begin{array}{cc}
\textbf{Check } x = \textbf{3} & \textbf{Check } x = \textbf{8} \\
(x - 3)(x - 8) = 0 & (x - 3)(x - 8) = 0 \\
(3 - 3)(3 - 8) \stackrel{?}{=} 0 & (8 - 3)(8 - 8) \stackrel{?}{=} 0 \\
0(-5) \stackrel{?}{=} 0 & 5(0) \stackrel{?}{=} 0 \\
0 \stackrel{?}{=} 0 & 0 \stackrel{?}{=} 0 \\
\text{true} & \text{true}
\end{array}
$$

Example 2   Solving a Quadratic Equation by Factoring

Solve $x^2 - 4x - 21 = 0$.

Solution

The zero factor property tells us about the *product* of two numbers that equals zero. So, to solve $x^2 - 4x - 21 = 0$ we begin by factoring the left side:

$$
\begin{array}{ll}
x^2 - 4x - 21 = 0 & \textit{Original equation} \\
(x + 3)(x - 7) = 0 & \textit{Factor left side.} \\
x + 3 = 0 \quad \text{or} \quad x - 7 = 0 & \textit{Zero factor property} \\
x = -3 \quad \text{or} \qquad x = 7 & \textit{Subtract 3 from both sides./Add 7 to both sides.}
\end{array}
$$

So, the solutions are -3 and 7. We check that both -3 and 7 satisfy the original equation:

Check $x = -3$	Check $x = 7$
$x^2 - 4x - 21 = 0$	$x^2 - 4x - 21 = 0$
$(-3)^2 - 4(-3) - 21 \overset{?}{=} 0$	$7^2 - 4(7) - 21 \overset{?}{=} 0$
$9 + 12 - 21 \overset{?}{=} 0$	$49 - 28 - 21 \overset{?}{=} 0$
$0 \overset{?}{=} 0$	$0 \overset{?}{=} 0$
true	true

In Examples 1 and 2, each original equation has one side that is 0. If neither side of a quadratic equation in one variable is 0, we must first use properties of equality to get one side to be 0 and then apply the zero factor property.

Example 3   Solving a Quadratic Equation by Factoring

Solve $4x^2 = 20x - 25$.

Solution

$$
\begin{array}{ll}
4x^2 = 20x - 25 & \textit{Original equation} \\
4x^2 - 20x + 25 = 0 & \textit{Write in } ax^2 + bx + c = 0 \textit{ form.} \\
(2x - 5)(2x - 5) = 0 & \textit{Factor left side.} \\
2x - 5 = 0 & \textit{Zero factor property} \\
2x = 5 & \textit{Add 5 to both sides.} \\
x = \dfrac{5}{2} & \textit{Divide both sides by 2.}
\end{array}
$$

So, the solution is $\dfrac{5}{2}$. We use a graphing calculator table to check that if $\dfrac{5}{2}$ is substituted for x in the polynomials $4x^2$ and $20x - 25$, the two results are equal (see Fig. 11).

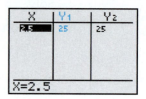

Figure 11 Verify the work

Example 4   Solving a Quadratic Equation by Factoring

Solve $12w^2 - 75 = 0$.

Solution

$$
\begin{array}{ll}
12w^2 - 75 = 0 & \textit{Original equation} \\
3\left(4w^2 - 25\right) = 0 & \textit{Factor out GCF, 3.} \\
3(2w + 5)(2w - 5) = 0 & \textit{Factor left side completely.}
\end{array}
$$

Now we can apply a variation on the zero factor property: If $3AB = 0$, then $A = 0$ or $B = 0$. Here we take $(2w + 5)$ to be A and $(2w - 5)$ to be B:

$$
\begin{array}{lll}
2w + 5 = 0 & \text{or} \quad 2w - 5 = 0 & \textit{Zero factor property} \\
2w = -5 & \text{or} \quad\quad 2w = 5 & \\
w = -\dfrac{5}{2} & \text{or} \quad\quad\; w = \dfrac{5}{2} &
\end{array}
$$

So, the solutions are $-\dfrac{5}{2}$ and $\dfrac{5}{2}$. We can check that the solutions satisfy the original equation.

Example 5   Solving a Quadratic Equation by Factoring

Solve $6x^2 = 18x$.

Solution

$$6x^2 = 18x \qquad \textit{Original equation}$$
$$6x^2 - 18x = 0 \qquad \textit{Write in ax}^2 + \textit{bx} = 0 \textit{ form.}$$
$$6x(x - 3) = 0 \qquad \textit{Factor left side.}$$
$$6x = 0 \quad \text{or} \quad x - 3 = 0 \quad \textit{Zero factor property}$$
$$x = 0 \quad \text{or} \quad x = 3$$

So, the solutions are 0 and 3. We can check that the solutions satisfy the original equation. (Try it.)

WARNING   In Example 5, we solved the equation $6x^2 = 18x$ by first writing it in the form $ax^2 + bx = 0$ (there is no c term) and then factoring the left side of the equation. It is a common error to try to solve the equation $6x^2 = 18x$ by dividing both sides by $6x$ to get the result $x = 3$. But note that this incorrect method gives only one of the two solutions 0 and 3.

In general, when solving quadratic equations, we can't divide by x because we don't know if the value of x is zero. Recall from Section 2.2 that division by zero is undefined.

Solving Quadratic Equations by Graphing

We can solve quadratic equations by graphing in much the same way we solved linear equations by graphing in Sections 4.3 and 4.4.

Example 6   Solving a Quadratic Equation in One Variable by Graphing

The graphs of $y = x^2 - 10x + 28$, $y = 7$, $y = 3$, and $y = 1$ are shown in Fig. 12. Use these graphs to solve the given equation.

1. $x^2 - 10x + 28 = 7$ **2.** $x^2 - 10x + 28 = 3$ **3.** $x^2 - 10x + 28 = 1$

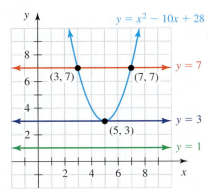

Figure 12 Solving three quadratic equations in one variable

Solution

1. The graphs of $y = x^2 - 10x + 28$ and $y = 7$ intersect only at the points $(3, 7)$ and $(7, 7)$, which have x-coordinates 3 and 7. So, 3 and 7 are the solutions of $x^2 - 10x + 28 = 7$.
2. The graphs of $y = x^2 - 10x + 28$ and $y = 3$ intersect only at the point $(5, 3)$, which has x-coordinate 5. So, 5 is the solution of $x^2 - 10x + 28 = 3$.
3. The graphs of $y = x^2 - 10x + 28$ and $y = 1$ do not intersect. So, no real number is a solution of $x^2 - 10x + 28 = 1$.

Our results in Example 6 suggest how many real-number solutions a quadratic equation in one variable can have.

Number of Solutions of a Quadratic Equation in One Variable

A quadratic equation in one variable has either two real-number solutions, one real-number solution, or no real-number solutions.

We will solve quadratic equations that have no real-number solutions in Chapter 9.

Combining Like Terms to Help Solve Quadratic Equations

After getting one side of an equation equal to zero, it is sometimes necessary to combine like terms to write the equation in $ax^2 + bx + c = 0$ form.

Example 7   Solving a Quadratic Equation by Factoring

Solve $12x^2 - 31x = 3x - 24$.

Solution

$$12x^2 - 31x = 3x - 24 \qquad \text{\textit{Original equation}}$$
$$12x^2 - 31x - 3x + 24 = 3x - 24 - 3x + 24 \qquad \text{\textit{Subtract 3x from both sides; add}}$$
$$\text{\textit{24 to both sides.}}$$
$$12x^2 - 34x + 24 = 0 \qquad \text{\textit{Combine like terms.}}$$
$$6x^2 - 17x + 12 = 0 \qquad \text{\textit{Divide both sides by 2.}}$$
$$(2x - 3)(3x - 4) = 0 \qquad \text{\textit{Factor left side.}}$$
$$2x - 3 = 0 \quad \text{or} \quad 3x - 4 = 0 \qquad \text{\textit{Zero factor property}}$$
$$2x = 3 \quad \text{or} \quad 3x = 4$$
$$x = \frac{3}{2} \quad \text{or} \quad x = \frac{4}{3}$$

Figure 13 Verify the work

So, the solutions are $\frac{3}{2}$ and $\frac{4}{3}$. For each solution, we check that if it is substituted for x in the polynomials $12x^2 - 31x$ and $3x - 24$, the two results are equal (see Fig. 13).

Example 8   Solving a Quadratic Equation by Factoring

Solve $(x - 2)(x + 5) = 8$.

Solution

Although the left-hand side of $(x - 2)(x + 5) = 8$ is factored, the right-hand side is not zero. First, we multiply the left-hand side and then write the equation in the form $ax^2 + bx + c = 0$:

$$(x - 2)(x + 5) = 8 \qquad \text{\textit{Original equation}}$$
$$x^2 + 3x - 10 = 8 \qquad \text{\textit{Multiply left-hand side.}}$$
$$x^2 + 3x - 18 = 0 \qquad \text{\textit{Write in } ax^2 + bx + c = 0 \text{ form.}}$$
$$(x + 6)(x - 3) = 0 \qquad \text{\textit{Factor left-hand side.}}$$
$$x + 6 = 0 \quad \text{or} \quad x - 3 = 0 \qquad \text{\textit{Zero factor property}}$$
$$x = -6 \quad \text{or} \quad x = 3$$

The solutions are -6 and 3. Recall from Section 4.3 that we can use "intersect" on a graphing calculator to solve an equation in one variable. In Fig. 14, we graph the equations $y = (x - 2)(x + 5)$ and $y = 8$ and find the intersection points $(-6, 8)$ and $(3, 8)$, which have x-coordinates -6 and 3, respectively. So, the solutions of the original equation are -6 and 3, which checks.

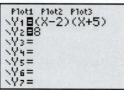

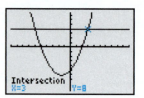

Figure 14 Verify the work

Solving Quadratic Equations That Contain Fractions

Recall from Section 4.4 that if an equation contains fractions, we multiply both sides of the equation by the least common denominator (LCD) of all the fractions.

| Example 9 | Solving a Quadratic Equation That Contains Fractions |

Solve $\frac{1}{2}x^2 + 4 = \frac{11}{3}x$.

Solution

To find the LCD of the two fractions in the equation, we list the multiples of 2 and the multiples of 3:

$$\text{Multiples of 2:}\quad 2, 4, 6, 8, 10, \ldots$$
$$\text{Multiples of 3:}\quad 3, 6, 9, 12, 15, \ldots$$

The LCD is 6. To clear the equation of fractions, we multiply both sides of the equation by the LCD, 6:

$$\frac{1}{2}x^2 + 4 = \frac{11}{3}x \qquad \text{\textit{Original equation}}$$
$$6 \cdot \frac{1}{2}x^2 + 6 \cdot 4 = 6 \cdot \frac{11}{3}x \qquad \text{\textit{Multiply both sides by LCD, 6.}}$$
$$3x^2 + 24 = 22x \qquad \text{\textit{Simplify.}}$$
$$3x^2 - 22x + 24 = 0 \qquad \text{\textit{Write in } } ax^2 + bx + c = 0 \text{ \textit{form.}}$$
$$(3x - 4)(x - 6) = 0 \qquad \text{\textit{Factor left side.}}$$
$$3x - 4 = 0 \quad \text{or} \quad x - 6 = 0 \qquad \text{\textit{Zero factor property}}$$
$$3x = 4 \quad \text{or} \qquad x = 6$$
$$x = \frac{4}{3} \quad \text{or} \qquad x = 6$$

So, the solutions are $\frac{4}{3}$ and 6. To verify our work, we use ZStandard followed by ZoomFit to graph the equations $y = \frac{1}{2}x^2 + 4$ and $y = \frac{11}{3}x$ and then use "intersect" to find the approximate intersection points (1.33, 4.89) and (6, 22), which have x-coordinates 1.33 and 6 (see Fig. 15). So, the approximate solutions of the original equation are 1.33 and 6, which checks.

  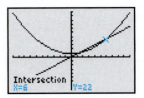

Figure 15 Verify the work

Finding x-intercept(s) of a Parabola

Recall from Section 5.1 that for an equation containing the variables x and y, we can find the x-intercept(s) of the graph by substituting 0 for y and then solving for x. In Example 10, we use this strategy to find the x-intercepts of a parabola.

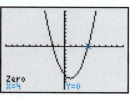

Figure 16 Verify the work

Example 10   Finding *x*-Intercepts

Find the *x*-intercepts of the parabola $y = x^2 - 2x - 8$.

Solution

We substitute 0 for y and solve for x:

$$0 = x^2 - 2x - 8 \qquad \text{\textit{Substitute 0 for y.}}$$
$$0 = (x + 2)(x - 4) \qquad \text{\textit{Factor right-hand side.}}$$
$$x + 2 = 0 \quad \text{or} \quad x - 4 = 0 \qquad \text{\textit{Zero factor property}}$$
$$x = -2 \quad \text{or} \quad x = 4$$

So, the *x*-intercepts are $(-2, 0)$ and $(4, 0)$. We use "zero" on a graphing calculator to verify our work (see Fig. 16). See Appendix A.19 for graphing calculator instructions.

▶

Solving Cubic Equations by Factoring

So far in this section, we have solved quadratic equations. We will now solve some cubic equations, such as

$$3x^3 - 21x^2 + 30x = 0 \qquad x^3 - 9x = 4x^2 - 36$$

A **cubic equation in one variable** is an equation that can be put into the form

$$ax^3 + bx^2 + cx + d = 0$$

where $a, b, c,$ and d are constants and $a \neq 0$. We can solve some cubic equations by using the zero factor property for three factors:

If $ABC = 0$, then $A = 0, B = 0,$ or $C = 0$.

Example 11   Solving a Cubic Equation by Factoring

Solve $3x^3 - 21x^2 + 30x = 0$.

Solution

$$3x^3 - 21x^2 + 30x = 0 \qquad \text{\textit{Original equation}}$$
$$3x\left(x^2 - 7x + 10\right) = 0 \qquad \text{\textit{Factor out GCF, 3x.}}$$
$$3x(x - 2)(x - 5) = 0 \qquad \text{\textit{Factor } x^2 - 7x + 10.}$$
$$3x = 0 \quad \text{or} \quad x - 2 = 0 \quad \text{or} \quad x - 5 = 0 \quad \text{\textit{Zero factor property}}$$
$$x = 0 \quad \text{or} \quad x = 2 \quad \text{or} \quad x = 5$$

So, the solutions are $0, 2,$ and 5. We can check that the solutions satisfy the original equation. (Try it.)

▶

Note that solving a cubic equation in one variable is similar to solving a quadratic equation in one variable. We now summarize the steps used to solve either type of equation.

Solving Quadratic and Cubic Equations by Factoring

If an equation can be solved by factoring, we solve it by the following steps:

1. Write the equation so that one side of it is equal to zero.

2. Factor the nonzero side of the equation.

3. Apply the zero factor property.

4. Solve each equation that results from using the zero factor property.

Example 12   Solving a Cubic Equation by Factoring

Solve $x^3 - 9x = 4x^2 - 36$

Solution

$x^3 - 9x = 4x^2 - 36$	*Original equation*
$x^3 - 4x^2 - 9x + 36 = 0$	*Write in $ax^3 + bx^2 + cx + d = 0$ form.*
$x^2(x - 4) - 9(x - 4) = 0$	*Factor both pairs of terms.*
$(x^2 - 9)(x - 4) = 0$	*Factor out GCF, $(x - 4)$.*
$(x + 3)(x - 3)(x - 4) = 0$	$A^2 - B^2 = (A + B)(A - B)$
$x + 3 = 0$ or $x - 3 = 0$ or $x - 4 = 0$	*Zero factor property*
$x = -3$ or $x = 3$ or $x = 4$	

So, the solutions are -3, 3, and 4. We can check that the solutions satisfy the original equation. (Try it.)

▶

WARNING   It is a common error to try to apply the zero factor property to an equation such as $x^2(x - 4) - 9(x - 4) = 0$ and incorrectly conclude that the solutions are 0 and 4. However, the polynomial $x^2(x - 4) - 9(x - 4)$ is *not* factored because it is a difference, not a product. Only after we factor the left side of the equation $x^2(x - 4) - 9(x - 4) = 0$ can we apply the zero factor property.

Solving Polynomial Equations versus Factoring Polynomials

In Section 4.5, we compared solving a linear equation with simplifying a linear expression. **When we solve any equation, our objective is to find all *numbers* that satisfy the equation. When we factor a polynomial, our objective is to write the polynomial as a product of polynomials, which is an *expression*.**

Here we compare solving a polynomial equation with factoring a polynomial:

Solving the Equation $x^2 - 5x + 6 = 0$

$x^2 - 5x + 6 = 0$	*Original equation*
$(x - 2)(x - 3) = 0$	*Factor left side.*
$x - 2 = 0$ or $x - 3 = 0$	*Zero factor property*
$x = 2$ or $x = 3$	

The solutions, 2 and 3, are numbers.

Factoring the Polynomial $x^2 - 5x + 6$

$$x^2 - 5x + 6 = (x - 2)(x - 3)$$

The result, $(x - 2)(x - 3)$, is an expression.

◈ Group Exploration

Section Opener: Zero factor property

1. What can you say about A or B if $AB = 0$?

2. What can you say about A or B if $A(B - 3) = 0$?

3. What can you say about x if $x(x - 3) = 0$?

4. Solve the equation $x^2 - 3x = 0$. [**Hint:** Does this have something to do with Problem 3?]

5. Solve $3x^2 - 6x = 0$.

6. Solve $x^2 - 7x + 10 = 0$.

Group Exploration

Solving systems of quadratic equations

1. Use "intersect" on a graphing calculator to solve the system

$$y = x^2 - 6x$$
$$y = x - 10$$

2. Now solve the system by substitution.

3. Solve the equation $x^2 - 6x = x - 10$.

4. Compare the results you found in Problem 3 with the x-coordinates of the results you found in Problem 2.

5. Use "intersect" to help you solve $x^2 - 2x = x + 4$. Then solve this equation without using graphing. Compare your results.

6. Explain why you cannot solve the equation $x^2 - 3x = x + 2$ by using factoring. Then use "intersect" to find approximate solutions of the equation.

7. Summarize what you have learned in this exploration. Explain how you can use "intersect" to verify work, as well as to solve equations that you cannot solve by factoring.

 # Homework 8.5

For extra help ▶ **MyLab Math**   Watch the videos in MyLab Math

Solve. Verify any results by checking that they satisfy the equation.

1. $(x - 2)(x - 7) = 0$
2. $(x + 3)(x - 1) = 0$
3. $x^2 + 7x + 10 = 0$
4. $x^2 + 5x + 6 = 0$
5. $w^2 + 3w - 28 = 0$
6. $t^2 - 4t - 45 = 0$
7. $x^2 - 14x + 49 = 0$
8. $x^2 - 2x + 1 = 0$
9. $r^2 + 3r = 4$
10. $p^2 + 2p = 24$
11. $5x = x^2 + 6$
12. $9x = x^2 + 14$
13. $k^2 - 64 = 0$
14. $y^2 - 9 = 0$
15. $36x^2 - 49 = 0$
16. $4x^2 - 25 = 0$
17. $x^2 = 49$
18. $x^2 = 4$
19. $w^2 = w$
20. $r^2 = -r$

Solve. Use a graphing calculator to verify your work.

21. $(3x - 7)(2x + 5) = 0$
22. $(4x + 3)(9x - 1) = 0$
23. $4x^2 - 8x + 3 = 0$
24. $3x^2 - 11x + 10 = 0$
25. $10k^2 + 3k - 4 = 0$
26. $6p^2 - 5p - 6 = 0$
27. $25x^2 + 20x + 4 = 0$
28. $9p^2 - 6p + 1 = 0$
29. $x^2 - 2x = 12 - 6x$
30. $x^2 - 10x = 6x - 15$
31. $3t^2 - 4t = 6t - 8$
32. $2r^2 + 5r = -15 - 6r$
33. $(x - 2)(x - 5) = 4$
34. $(x - 3)(x - 1) = 8$
35. $4x(x + 1) = 15$
36. $-9x(x - 1) = 2$
37. $(2w - 3)(w - 2) = 3$
38. $(3y + 4)(y - 1) = -2$
39. $(x - 2)^2 - 3x = -6$
40. $(x + 3)^2 + 2x = -3$

Solve. Use a graphing calculator to verify your work.

41. $\frac{1}{4}x^2 + 2x + 3 = 0$
42. $\frac{1}{5}x^2 + \frac{6}{5}x + 1 = 0$
43. $\frac{3}{8}m^2 + m - 2 = 0$
44. $\frac{5}{2}t^2 + 9t - 4 = 0$
45. $-\frac{1}{3}x^2 + \frac{1}{3}x + 10 = 6$
46. $-\frac{1}{2}x^2 - \frac{5}{2}x + 12 = 5$

For Exercises 47–50, use the graph of $y = x^2 - 6x + 5$ shown in Fig. 17 to solve the given equation.

47. $x^2 - 6x + 5 = -3$
48. $x^2 - 6x + 5 = 0$
49. $x^2 - 6x + 5 = -4$
50. $x^2 - 6x + 5 = -5$

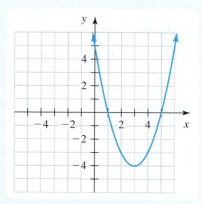

Figure 17 Exercises 47–50

For Exercises 51–54, use the graph of $y = -x^2 - 4x - 1$ shown in Fig. 18 to solve the given equation.

51. $-x^2 - 4x - 1 = 2$ **52.** $-x^2 - 4x - 1 = -1$

53. $-x^2 - 4x - 1 = 4$ **54.** $-x^2 - 4x - 1 = 3$

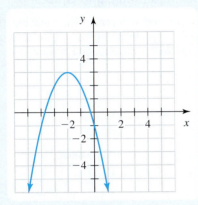

Figure 18 Exercises 51–54

Use "intersect" on a graphing calculator to solve the equation. Round any solutions to the second decimal place.

55. $x^2 - 5x = -2x + 3$ **56.** $x^2 + 3x = x + 5$

57. $-x^2 - 2x + 4 = -x - 5$ **58.** $-2x^2 + 4x + 8 = 2x - 1$

For Exercises 59–62, solve the given equation by referring to the solutions of $y = 2x^2 - 8x + 10$ shown in Table 2.

59. $2x^2 - 8x + 10 = 4$ **60.** $2x^2 - 8x + 10 = 10$

61. $2x^2 - 8x + 10 = 2$ **62.** $2x^2 - 8x + 10 = -3$

Table 2 Exercises 59–62

x	y
−1	20
0	10
1	4
2	2
3	4
4	10
5	20

Find all x-intercepts. Verify your work by using a graphing calculator.

63. $y = x^2 - 3x - 28$ **64.** $y = x^2 - 9x + 18$

65. $y = x^2 - 8x + 16$ **66.** $y = x^2 + 6x + 9$

67. $y = 2x^2 - x - 3$ **68.** $y = 3x^2 + x - 10$

Solve. Use a graphing calculator to verify your work.

69. $2x^3 - 5x^2 - 18x + 45 = 0$

70. $x^3 - x^2 - x + 1 = 0$

71. $3x^3 - 4x^2 - 12x + 16 = 0$

72. $8x^3 + 12x^2 - 18x - 27 = 0$

73. $18y^3 - 27y^2 = 8y - 12$

74. $12w^3 - 75w = 50 - 8w^2$

Solve. Use a graphing calculator to verify your work.

75. $5x^2 + 10x - 40 = 0$ **76.** $4x^2 + 16x - 20 = 0$

77. $2x^3 - 50x = 0$ **78.** $3x^3 - 12x = 0$

79. $3p^3 + 15p^2 = -18p$ **80.** $3r^3 + 27r^2 = -60r$

81. $2x^3 = 4x^2 + 16x$ **82.** $2x^3 = 4x - 2x^2$

Concepts

83. A student tries to solve the equation $4x^2 = 8x$:

$$4x^2 = 8x$$
$$\frac{4x^2}{4x} = \frac{8x}{4x}$$
$$x = 2$$

Describe any errors. Then solve the equation correctly.

84. A student tries to solve the equation $x^2 = 25$:

$$x^2 = 25$$
$$x = 5$$

Describe any errors. Then solve the equation correctly.

85. A student tries to solve the equation $2x^2 - 26x + 80 = 0$:

$$2x^2 - 26x + 80 = 0$$
$$2(x^2 - 13x + 40) = 0$$
$$2(x - 5)(x - 8) = 0$$
$$x = 2, x = 5, \text{ or } x = 8$$

Describe any errors. Then solve the equation correctly.

86. A student tries to solve the equation $x^3 + 3x^2 - 4x - 12 = 0$:

$$x^3 + 3x^2 - 4x - 12 = 0$$
$$x^2(x + 3) - 4(x + 3) = 0$$
$$x + 3 = 0$$
$$x = -3$$

Describe any errors. Then solve the equation correctly.

87. Give an example of a quadratic equation in one variable whose solutions are 2 and 5.

88. Give an example of a cubic equation in one variable whose solutions are −4, 1, and 3.

89. Consider the equation $y = x^2 - 6x + 8$.
  a. Find x when $y = 3$.   **b.** Find x when $y = -1$.
  c. Use the results you found in parts (a) and (b) to help you graph the equation $y = x^2 - 6x + 8$ by hand. [**Hint:** Your work in part (b) should help you determine the vertex of the graph. (Why?)]

90. Here is the work done in solving the equation $x^2 = 5x - 6$:

$$x^2 = 5x - 6$$
$$x^2 - 5x + 6 = 0$$
$$(x - 2)(x - 3) = 0$$
$$x - 2 = 0 \quad \text{or} \quad x - 3 = 0$$
$$x = 2 \quad \text{or} \quad x = 3$$

Show that *each line* of the work becomes a true statement when 2 or 3 is substituted for x.

91. If $(x + 5)(x + 6) = 0$, then $x + 5 = 0$ or $x + 6 = 0$. Explain why.

92. If an equation in one variable has exactly one solution, must the equation be linear? If yes, explain why. If no, give an example to show why not.

93. Compare factoring a quadratic polynomial with solving a quadratic equation. (See page 9 for guidelines on writing a good response.)

94. Explain in your own words how to solve a quadratic equation in one variable. (See page 9 for guidelines on writing a good response.)

Related Review

Solve.

95. $x^2 + 3x = 7x + 12$

96. $3x(x - 2) = 24$

97. $x + 3 = 7x + 12$

98. $3(x - 2) = 24$

99. Solve the formula $A = P + PRT$ for P. [**Hint:** Begin by factoring the right-hand side of the formula.]

100. Solve the formula $S = 2WL + 2WH + 2LH$ for L. [**Hint:** You will need to use factoring eventually.]

Expressions, Equations, and Graphs

Perform the indicated instruction. Then use words such as linear, quadratic, cubic, polynomial, degree, one variable, *and* two variables *to describe the expression, equation, or system.*

101. Factor $x^2 - 4x - 5$.

102. Factor $x^2 - 4$.

103. Solve $x^2 - 4x - 5 = 0$.

104. Solve $x^2 - 4 = 0$.

105. Graph $y = x^2 - 4x - 5$ by hand.

106. Graph $y = x^2 - 4$ by hand.

8.6 Using Factoring to Make Predictions with Quadratic Models

Objectives

» Use an equation of a quadratic model to make estimates and predictions.

» Model projectile motion.

» Model the area of rectangular objects.

In Section 7.2, we used the *graph* of a quadratic model to make predictions. In this section, we will use an *equation* of such a model to make similar predictions.

Using a Quadratic Model's Equation to Make Predictions

In Example 1, we will check that a specific quadratic equation models a situation well and then use the model to make estimates.

| Example 1 | Making Predictions |

The enrollments in an intermediate algebra course at the College of San Mateo are shown in Table 3 for various Mondays leading up to Monday, August 6, which was one week before the first day of class.

Table 3 Enrollments in an Intermediate Algebra Course

Date	Number of Weeks since July 2	Enrollment
July 2	0	11
July 9	1	12
July 16	2	13
July 30	4	25
August 6	5	35

Source: *J. Lehmann*

Let E be the enrollment in the intermediate algebra course at t weeks since July 2. A model of the situation is

$$E = t^2 - \frac{1}{3}t + 11$$

1. Use a graphing calculator to draw the graph and, in the same viewing window, the scatterplot of the data. Does the model fit the data well?

2. Estimate when the enrollment was 19 students.

3. Use a graphing calculator to find the vertex of the model. What does the vertex mean in this situation?

4. The maximum enrollment for the class is 35 students. Estimate the number of students who were wait-listed as of the first day of class (August 13), which was 6 weeks after July 2.

Solution

1. The graph of the model and the scatterplot of the data are shown in Fig. 19. The model appears to fit the data well.

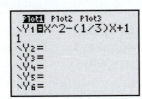 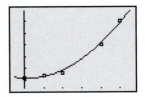

Figure 19 Check the fit

2. To estimate when the enrollment was 19 students, we substitute the output 19 for E in the equation $E = t^2 - \frac{1}{3}t + 11$ and solve for t:

$$19 = t^2 - \frac{1}{3}t + 11 \qquad \textit{Substitute 19 for E.}$$

$$0 = t^2 - \frac{1}{3}t - 8 \qquad \textit{Write in } ax^2 + bx + c = 0 \textit{ form.}$$

$$0 = 3t^2 - t - 24 \qquad \textit{Multiply both sides by LCD, 3.}$$

$$0 = (3t + 8)(t - 3) \qquad \textit{Factor right-hand side.}$$

$$3t + 8 = 0 \quad \text{or} \quad t - 3 = 0 \qquad \textit{Zero factor property}$$

$$3t = -8 \quad \text{or} \qquad t = 3$$

$$t = -\frac{8}{3} \quad \text{or} \qquad t = 3$$

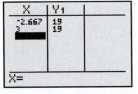

Figure 20 Verify the work

The negative value, $t = -\frac{8}{3}$, does not make sense, but the positive value, $t = 3$, is reasonable. According to the model, there were 19 students 3 weeks after July 2, which was July 23. We can use a graphing calculator to verify our work (see Fig. 20).

3. To get a better view of the vertex, we Zoom Out (see Fig. 21). Because the parabola opens upward, we use "minimum" to find the approximate vertex (see Fig. 21). See Appendix A.18 for graphing calculator instructions.

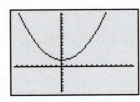 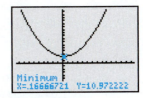

Figure 21 Finding the approximate vertex

The vertex is about $(0.17, 10.97)$. This means enrollment was a minimum on July 2 (11 students), according to the model. Although the enrollment really was 11 students on July 2, the *minimum* enrollment was 0 students whenever the course first became available, so the part of the model that lies to the left of the vertex is not accurate (model breakdown).

4. We substitute the input 6 for t in $E = t^2 - \frac{1}{3}t + 11$:

$$E = (6)^2 - \frac{1}{3}(6) + 11 = 45$$

So, 45 students signed up for the course, according to the model. Because the maximum enrollment was 35 students, this would mean $45 - 35 = 10$ students were wait-listed.

Making Predictions about the Response and Explanatory Variables

- To make a prediction about the response variable of a quadratic model, substitute a chosen value for the explanatory variable in the model's equation. Then solve for the response variable.
- To make a prediction about the explanatory variable of a quadratic model, substitute a chosen value for the response variable in the model's equation. Then solve for the explanatory variable, which involves writing the equation so one side of it is zero and then factoring the nonzero side.

Projectile Motion

Any thrown object is called a *projectile*. If you launch a projectile such as a baseball or stone into the air, the relationship between time and the height of the projectile can be described well by a quadratic model. Such a model will not work well for an object that encounters a lot of air resistance—a feather, for example.

On the basis of my assumption of the initial speed of the ball, I've got six seconds to get into position.

Example 2   Making Estimates

A batter hits a baseball into the air. The height h (in feet) of the baseball at t seconds after the ball is hit is given by

$$h = -16t^2 + 96t + 4$$

1. When is the baseball at a height of 84 feet? Explain why there are two such times.
2. When is the baseball at a height of 148 feet? Explain why there is one such time.
3. When is the baseball at a height of 150 feet?

Solution

1. We substitute the output 84 for h in the equation $h = -16t^2 + 96t + 4$ and solve for t:

$$
\begin{aligned}
84 &= -16t^2 + 96t + 4 &&\textit{Substitute 84 for h.}\\
0 &= -16t^2 + 96t - 80 &&\textit{Write in } 0 = ax^2 + bx + c \textit{ form.}\\
0 &= -16(t^2 - 6t + 5) &&\textit{Factor out } -16 \textit{ (on right-hand side).}\\
0 &= -16(t - 1)(t - 5) &&\textit{Factor right-hand side completely.}
\end{aligned}
$$

$$
\begin{aligned}
t - 1 = 0 \quad &\text{or} \quad t - 5 = 0 &&\textit{Zero factor property}\\
t = 1 \quad &\text{or} \qquad\; t = 5
\end{aligned}
$$

So, 1 second after the baseball is hit, and again 5 seconds after it is hit, the baseball is at a height of 84 feet. It makes sense that there are two times because the baseball reaches 84 feet both on its way up and on its way down (see Fig. 22).

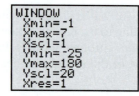
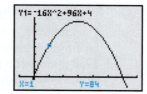
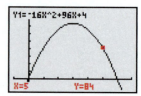

Figure 22 Finding when the baseball is at a height of 84 feet

2. We substitute the output 148 for h in the equation $h = -16t^2 + 96t + 4$ and solve for t:

$$148 = -16t^2 + 96t + 4 \qquad \textit{Substitute 148 for h.}$$
$$0 = -16t^2 + 96t - 144 \qquad \textit{Write in } 0 = ax^2 + bx + c \textit{ form.}$$
$$0 = -16\left(t^2 - 6t + 9\right) \qquad \textit{Factor out } -16 \textit{ (on right-hand side).}$$
$$0 = -16(t - 3)(t - 3) \qquad \textit{Factor right-hand side completely.}$$
$$t - 3 = 0 \qquad\qquad\qquad \textit{Zero factor property}$$
$$t = 3$$

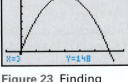

Figure 23 Finding when the baseball is at a height of 148 feet

So, 3 seconds after the baseball is hit, it is at a height of 148 feet. It makes sense that there is only one time the baseball reaches a height of 148 feet if when it reaches this height it is at the top of its climb—in other words, if the vertex of the parabola is (3, 148). Figure 23 confirms the situation.

3. Because the maximum height of the baseball is 148 feet, the baseball never reaches a height of 150 feet (see Fig. 23).

Consider the following observations about a quadratic model, a parabola, and a quadratic equation in one variable:

- Our work in Example 2 suggests that the baseball reaches each height either two times, one time, or never.

- In Sections 7.1 and 7.2, we found that, for a parabola opening upward or downward, each value of y corresponds to two, one, or no values of x.

- In Section 8.5, we learned that a quadratic equation in one variable has either two, one, or no real-number solutions.

These observations suggest an important common thread between quadratic models, parabolas, and quadratic equations in one variable. For the purposes of this section, that common thread means that when we use a quadratic model to predict the value of an explanatory variable for a specific value of the response variable, there will be either two, one, or no real-number values.

Area of Rectangular Objects

Recall from Section 4.6 that the area A of a rectangle is given by the formula $A = LW$, where L is the length and W is the width.

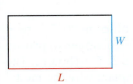

Figure 24 The length L and width W of a rectangular garden

Example 3   Solving an Area Problem

A rectangular garden has an area of 40 square feet. If the length is 3 feet more than the width, find the dimensions of the garden.

Solution

We use the five-step problem-solving method of Section 6.5.

Step 1: Define each variable. Let L be the length (in feet) and W be the width (in feet). See Fig. 24.

Step 2: Write a system of two equations. Because the length is 3 feet more than the width, our first equation is $L = W + 3$. Because the area is 40 square feet, we find our second equation by substituting 40 for A in the area formula $A = LW$: $40 = LW$. The system is

$$L = W + 3 \qquad \textit{Equation (1)}$$
$$LW = 40 \qquad \textit{Equation (2)}$$

Step 3: *Solve the system.* We can solve the system by substitution. We substitute $W + 3$ for L in equation (2) and then solve for W:

$$LW = 40 \qquad \textit{Equation (2)}$$
$$(W + 3)W = 40 \qquad \textit{Substitute } W + 3 \textit{ for L.}$$
$$W^2 + 3W = 40 \qquad \textit{Distributive law}$$
$$W^2 + 3W - 40 = 0 \qquad \textit{Subtract 40 from both sides.}$$
$$(W + 8)(W - 5) = 0 \qquad \textit{Factor left side.}$$
$$W + 8 = 0 \quad \text{or} \quad W - 5 = 0 \qquad \textit{Zero factor property}$$
$$W = -8 \quad \text{or} \qquad W = 5$$

The result $W = -8$ means that the width is negative; model breakdown has occurred because a width must be positive. So, the width is 5 feet (our other result). To find the length, we substitute 5 for W in equation (1):

$$L = W + 3 = 5 + 3 = 8$$

Step 4: *Describe each result.* The width is 5 feet and the length is 8 feet.

Step 5: *Check.* The product of 5 and 8 is 40, which checks for an area of 40 square feet. Also, 8 is indeed 3 more than 5.

▶

 Group Exploration

Section Opener: Using a quadratic model to make a prediction

The U.S. government's shares of outstanding student debt are shown in Table 4 for various years.

Table 4 U.S. Government's Shares of Outstanding Student Debt

Year	Percent
2010	88.7
2011	90.1
2012	91.0
2013	91.6
2014	92.1
2015	92.4
2016	92.6

Sources: *MeasureOne; U.S. Department of Education*

Let p be the U.S. government's share (in percent) of outstanding student debt at t years since 2010.

1. Use a graphing calculator to draw a scatterplot of the data. Can the data be better described by a linear model or a quadratic model? Explain.

2. Use a graphing calculator to draw the graph of the model $p = -\frac{1}{10}t^2 + \frac{6}{5}t + 89$ and, in the same viewing window, the scatterplot of the data. Does the model fit the data well?

3. Use "maximum" on a graphing calculator to estimate when the government's share of outstanding student debt was the highest. What is that share?

4. Substitute a value for one of the variables in the model's equation to estimate the government's share of outstanding student debt in 2017.

5. The total outstanding student debt in 2017 was $1.4 trillion. How much of this money is owed to the government?

6. Predict when the government's share of outstanding student debt will be 91%. [**Hint:** At some point after finding an equation in one variable, multiply both sides by the LCD.]

7. President Trump, who came into office in 2017, is seen as making the student-loan sector friendlier to private lenders such as J.P. Morgan Chase & Co. (Source: The Wall Street Journal, *12/27/16*). Explain why the model suggests the government's share of outstanding student debt would have decreased no matter who came into office.

Tips for Success   Create an Example

When learning a definition or property, try to create an example. For example, while studying the material in Sections 8.1–8.4, you could create polynomials that might be factored by the following methods:

- Factor out the GCF.
- Factor a cubic polynomial of the form $ax^3 + bx^2 + cx + d$ by grouping.
- Factor a difference of two squares.
- Factor a sum of two cubes.
- Factor a difference of two cubes.
- Factor a quadratic trinomial of the form $x^2 + bx + c$.
- Factor a quadratic trinomial of the form $ax^2 + bx + c$, where $a \neq 1$, by using trial and error or by grouping.
- Factor a polynomial completely by using a combination of methods.

You could create these examples by finding products of polynomials. For instance, create a trinomial of the form $ax^2 + bx + c$ by finding the product $(2x + 5)(3x - 4)$. After creating many polynomials, let time pass so that you won't remember their factored forms. Then practice factoring them. While factoring each polynomial, reflect on the polynomial's type and on why you chose to use a particular method to factor that type.

Creating examples will shed light on many details of a concept and will personalize the information. It will also help you understand the similarities and differences between related concepts.

Homework 8.6

For extra help ▶ **MyLab Math**   Watch the videos in MyLab Math

1. Worldwide sales of "super-premium" vehicles with prices starting at about $100,000 are shown in Table 5 for various years.

Table 5 Worldwide Super-Premium Vehicle Sales

Years	Sales (thousands)
2010	40
2011	41
2012	43
2013	47
2014	51
2015	60
2016	73

Source: *LMC Automotive*

Let s be the worldwide annual sales (in thousands) of super-premium vehicles at t years since 2010. A quadratic model of the situation is $s = t^2 - t + 41$.
a. Use a graphing calculator to draw the graph of the model and, in the same viewing window, the scatterplot of the data. Does the model fit the data well?
b. Predict the worldwide sales of super-premium vehicles in 2022.
c. Predict when the worldwide annual sales of super-premium vehicles will be 113 thousand vehicles.

2. The percentages of mothers who use Twitter are shown in Table 6 for various years.

Table 6 Percentages of Mothers Who Use Twitter

Year	Percent
2012	10
2013	13
2014	18
2015	17
2016	15

Source: *Edison Research*

Let p be the percentage of mothers who use Twitter at t years since 2010.
a. Use a graphing calculator to draw a scatterplot of the data. Can the data be described better by a linear model or a quadratic model? Explain.
b. Use a graphing calculator to draw the graph of the model $p = -t^2 + 9t - 3$ and, in the same viewing window, the scatterplot of the data. Does the model fit the data well?
c. Estimate the percentage of mothers who used Twitter in 2017.
d. Predict when 5% of mothers will use Twitter.

3. Florida's manatee populations are shown in Table 7 for various years.

Table 7 Florida's Manatee Populations

Year	Population
1991	1267
1995	1823
2000	1646
2005	3143
2010	5077
2015	6063
2017	6620

Source: *Florida Fish and Wildlife Conservation Commission*

Let p be Florida's manatee population at t years since 1990. A quadratic model of the situation is $p = 6t^2 + 42t + 1200$.

a. Use a graphing calculator to draw the graph of the model and, in the same viewing window, the scatterplot of the data. Does the model fit the data well?

b. Predict Florida's manatee population in 2021.

c. Estimate when there were 1380 manatees in Florida.

d. The manatee is classified as endangered. The group Save Crystal River is petitioning the federal wildlife agency to reclassify the manatee to the less dire status of threatened, so the Crystal River economy does not suffer. Find the unit ratio of Florida's manatee population in 2017 to Florida's manatee population in 1991. Explain why Save Crystal River might refer to the unit ratio.

4. The revenues of LinkedIn are shown in Table 8 for various years.

Table 8 LinkedIn Revenues

Year	Revenue (millions of dollars)
2009	120
2010	243
2011	522
2012	972
2013	1529
2014	2219
2015	2991

Source: *LinkedIn*

Let r be the annual revenue (in millions of dollars) at t years since 2009. A quadratic model of the situation is $r = 60t^2 + 120t + 120$.

a. Use a graphing calculator to draw the graph of the model and, in the same viewing window, the scatterplot of the data. Does the model fit the data well?

b. Predict the revenue in 2022.

c. Estimate when the annual revenue was $3900 million ($3.9 billion).

5. Table 9 shows worldwide iPhone sales for various years.

Table 9 Worldwide iPhone Sales

Year	Sales (millions)
2007	1.4
2009	20.7
2011	72.3
2013	150.3
2015	231.2

Source: *Apple, Inc.*

Let S be annual worldwide sales (in millions) of iPhones at t years since 2005. A linear model of the situation is $S = 29.46t - 81.58$. A quadratic model of the situation is $S = 3t^2 - 6t + 1$.

a. Use a graphing calculator to draw a scatterplot of the data and both models in the same viewing window. Which model comes closer to the points in the scatterplot?

b. Use first the linear model and then the quadratic model to estimate iPhone sales in 2011. Which is the better estimate? Explain.

c. Use the quadratic model to predict iPhone sales in 2021.

d. Use the quadratic model to estimate when iPhone annual sales were 46 million phones.

6. The enrollments at Thomas A. Edison High School in Elmira Heights, New York, are shown in Table 10 for various years.

Table 10 Enrollments at Thomas A. Edison High School

Year	Enrollment (number of students)
2008	337
2010	346
2012	340
2014	320
2015	302

Source: *New York State Education Department*

Let E be the enrollment at t years since 2005. A linear model of the situation is $E = -5t + 363$. A quadratic model of the situation is $E = -2t^2 + 21t + 292$.

a. Use a graphing calculator to draw a scatterplot of the data and both models in the same viewing window. Which model comes closer to the points in the scatterplot?

b. Use first the linear model and then the quadratic model to estimate the enrollment in 2012. Which is the better estimate? Explain.

c. Use the quadratic model to predict the enrollment in 2020.

d. Use the quadratic model to estimate when the enrollment was 281 students.

e. Use "maximum" on a graphing calculator to find the vertex of the model. What does it mean in this situation?

7. The numbers of U.S. overdose deaths are shown in Table 11 for various years.

Table 11 Numbers of U.S. Overdose Deaths

Year	Number of Deaths (thousands)
1980	5
1990	9
2000	17
2010	42
2015	52

Source: *Centers for Disease Control and Prevention*

Let n be the number (in thousands) of U.S. overdose deaths in the year that is t years since 1980. A quadratic model of the situation is $n = \frac{1}{20}t^2 - \frac{2}{5}t + 6$.

a. Use a graphing calculator to draw the graph of the model and, in the same viewing window, the scatterplot of the data. Does the model fit the data well?

b. Estimate in which year there were 15 thousand overdose deaths.

c. Estimate the number of overdose deaths in 2016.

d. All the data for 2016 have not been collected yet, but *The New York Times* estimates that there were between 59 thousand and 65 thousand overdose deaths in that year. Does the result you found in part (c) fall in this range? Does this suggest that overdose deaths are climbing faster than ever or slower than ever? Explain.

8. The U.S. costs of the "War on Terror" are shown in Table 12 for various years.

Table 12 U.S. Annual Costs of the War on Terror

Year	Cost (billions of dollars)
2003	75
2005	109
2007	170
2009	156
2011	158
2013	99
2015	74

Source: *Mercatus Center*

Let c be the annual U.S. cost (in billions of dollars) of the War on Terror at t years since 2000.

a. Use a graphing calculator to draw a scatterplot of the data. Can the data be described better by a linear model or a quadratic model? Explain.

b. Use a graphing calculator to draw the graph of the model $c = -\frac{5}{2}t^2 + 45t - 41$ and, in the same viewing window, the scatterplot of the data. Does the model fit the data well?

c. Estimate the U.S. cost of the War on Terror in 2012.

d. Estimate when the annual U.S. cost of the War on Terror was $39 billion.

e. Determine which is costlier, the War on Terror or the Vietnam war, which cost about $686 billion in current dollars. Explain.

9. A company's annual profit can be modeled by $p = t^2 - 3t + 5$, where p is the annual profit (in thousands of dollars) at t years since 2015. Estimate when the annual profit was or will be $23 thousand.

10. A company's annual profit can be modeled by $p = t^2 - 6t + 17$, where p is the annual profit (in thousands of dollars) at t years since 2015. Estimate when the annual profit was or will be $24 thousand.

11. A company's annual revenue can be modeled by $r = 2t^2 - 13t + 25$, where r is the annual revenue (in millions of dollars) at t years since 2010. Estimate when the annual revenue was or will be $10 million.

12. A company's annual revenue can be modeled by $r = 3t^2 - 19t + 35$, where r is the annual revenue (in millions of dollars) at t years since 2010. Estimate when the annual revenue was or will be $29 million.

13. A batter hits a baseball into the air. The height h (in feet) of the baseball after t seconds is given by the quadratic equation $h = -16t^2 + 64t + 3$.

a. When is the baseball at a height of 3 feet? Explain why there are two such times. [**Hint:** Use a graphing calculator to graph the model.]

b. When is the baseball at a height of 51 feet? Explain why there are two such times.

c. When is the baseball at a height of 67 feet? Explain why there is just one such time.

14. A person throws a baseball into the air. The height h (in feet) of the baseball after t seconds is given by $h = -16t^2 + 80t + 4$.

a. When is the baseball at a height of 68 feet? Explain why there are two such times. [**Hint:** Use a graphing calculator to graph the model.]

b. When is the baseball at a height of 100 feet? Explain why there are two such times.

c. When is the baseball at a height of 104 feet? Explain why there is just one such time.

15. A rectangular boardroom table has an area of 60 square feet. If the length is 7 feet more than the width, find the dimensions of the table.

16. A rectangular rug has an area of 80 square feet. If the length is 2 feet more than the width, find the dimensions of the rug.

17. A rectangular office floor has an area of 98 square feet. If the length is twice the width, find the dimensions of the floor.

18. A rectangular garden has an area of 162 square feet. If the length is twice the width, find the dimensions of the garden.

Concepts

19. For a given set of data, describe how you can determine whether to model the situation by a linear model, a quadratic model, or neither.

20. Explain how to make a prediction for the response variable of a quadratic model. Explain how to make a prediction for the explanatory variable.

Related Review

21. The percentages of tax returns that were filed online are shown in Table 13 for various years.

Table 13 Percentages of Tax Returns That Were Filed Online

Year	Percent
2009	67.2
2010	69.8
2011	77.2
2012	82.7
2013	84.7
2014	86.0
2015	91.0

Source: *efile.com*

Let p be the percentage of tax returns that were filed online at t years since 2000.

a. Use a graphing calculator to draw a scatterplot of the data. Can the data be described better by a linear model or a quadratic model? Explain.

b. Find an equation of a model to describe the data.

c. Estimate in which year all tax returns were filed online.

d. Estimate the *number* of tax returns that were filed online in 2008. The total number of tax returns in 2008 was 153.7 million.

22. The number of newborn boys for every 100 newborn girls is called the *sex ratio at birth*. The *birth order* of a first child is 1, the birth order of a second child is 2, and so on. The U.S. sex ratios at birth are shown in Table 14 for various birth orders.

Table 14 Birth Orders versus U.S. Sex Ratios at Birth

Birth Order	Sex Ratio at Birth
1	105.8
2	105.1
3	104.8
4	104.4
5	104.0
6	103.4

Source: *U.S. Centers for Disease Control and Prevention*

Let r be the U.S. sex ratio at birth for newborns whose birth order is n.

a. Use a graphing calculator to draw a scatterplot of the data. Can the data be described better by a linear model or a quadratic model? Explain.

b. Find an equation of a model to describe the data.

c. Estimate the U.S. sex ratio for children with birth order 7.

d. China limits the number of babies a couple can have to one or two babies. Chinese parents tend to want to have boys rather than girls (Source: *NBC News*). The sex ratio at birth in China in 2014 was 116 (Source: *Xinhua News Agency*). Give a possible reason why the sex ratio at birth might have been so high.

23. The number of law school applicants was 54.5 thousand in 2015 and decreased approximately linearly by about 2.4 thousand students per year since then (Source: *Law School Admission Council*). Let n be the number (in thousands) of law school applicants at t years since 2015.

a. Find an equation of a model to describe the data.

b. Predict when there will be 40 thousand applicants.

24. The number of new scripted TV shows aired online in the United States was 57 shows in 2016 and has increased approximately linearly by about 14 shows per year since then (Source: *IHS Markit*). Let n be the number of new scripted TV shows aired online at t years since 2016.

a. Find an equation of a model to describe the data.

b. Predict when 110 new scripted TV shows will be aired online in the United States.

25. The average cost of room and board at public universities has increased approximately linearly from \$4900 for the school year ending in 2001 to \$9455 for the school year ending in 2016 (Source: *The College Board*). Predict the average cost of room and board for the school year ending in 2021.

26. About 91% of 21-year-old Americans attend movies, and about 43% of 65-year-old Americans attend movies (Source: *Motion Picture Association of America*). Let p be the percentage of Americans who attend movies at age a years. There is an approximate linear relationship between a and p for values of a between 21 and 65. Estimate the percentage of 30-year-old Americans who attend movies.

Expressions, Equations, and Graphs

Give an example of each. Then solve, simplify, or graph, as appropriate.

27. cubic polynomial in one variable with five terms

28. quadratic polynomial in one variable with four terms

29. linear polynomial in one variable with four terms

30. quadratic equation in one variable

31. system of two linear equations in two variables

32. linear equation in one variable

33. quadratic equation in two variables

34. linear equation in two variables

Chapter Summary

Key Points of Chapter 8

Section 8.1 Factoring Trinomials of the Form $x^2 + bx + c$ and Differences of Two Squares

Factor	We **factor** a polynomial by writing it as a product.
Multiplying versus factoring	Multiplying and factoring are reverse processes.
Factoring $x^2 + bx + c$	To factor $x^2 + bx + c$, look for two integers p and q whose product is c and whose sum is b. That is, $pq = c$ and $p + q = b$. If such integers exist, the factored polynomial is $(x + p)(x + q)$.
Factoring $x^2 + bx + c$ with c positive	To factor a trinomial of the form $x^2 + bx + c$ with a positive constant term c, • If b is positive, look for two *positive* integers whose product is c and whose sum is b. • If b is negative, look for two *negative* integers whose product is c and whose sum is b.
Factoring $x^2 + bx + c$ with c negative	To factor a trinomial of the form $x^2 + bx + c$ with a negative constant term c, look for two integers with *different* signs whose product is c and whose sum is b.
Prime polynomial	A polynomial that cannot be factored is called **prime**.
Difference of two squares	$A^2 - B^2 = (A + B)(A - B)$
Polynomial $x^2 + k^2$, where $k \neq 0$	A polynomial of the form $x^2 + k^2$, where $k \neq 0$, is prime.

Section 8.2 Factoring Out the GCF; Factoring by Grouping

GCF	The **greatest common factor (GCF)** of two or more terms is the monomial with the largest coefficient and the largest degree that is a factor of all the terms.
Factor out GCF	When the leading coefficient of a polynomial is positive and the GCF is not 1, we first factor out the GCF.
Factor completely	When factoring a polynomial, always factor it *completely*.
Factoring when the leading coefficient is negative	When the leading coefficient of a polynomial is negative, we first factor out the opposite of the GCF.
Factor by grouping	For a polynomial with four terms, **factor by grouping** (if it can be done) by 1. Factoring the first two terms and the last two terms. 2. Factoring out the binomial GCF.

Section 8.3 Factoring Trinomials of the Form $ax^2 + bx + c$

Factor $ax^2 + bx + c$ by trial and error	To **factor a trinomial** of the form $ax^2 + bx + c$ **by trial and error**, if the trinomial can be factored as a product of two binomials, then the product of the coefficients of the first terms of the binomials is equal to a and the product of the last terms of the binomials is equal to c. To find the correct factored polynomial, multiply the possible products and identify the one for which the coefficient of x is b.
Factoring $ax^2 + bx + c$ by grouping (*ac* method)	To **factor a trinomial** of the form $ax^2 + bx + c$ **by grouping** (if it can be done), 1. Find pairs of numbers whose product is ac. 2. Determine which of the pairs of numbers from step 1 has the sum b. Call this pair of numbers m and n. 3. Write the bx term as $mx + nx$: $$ax^2 + bx + c = ax^2 + mx + nx + c$$ 4. Factor $ax^2 + mx + nx + c$ by grouping. Another name for this technique is the ***ac* method**.

Section 8.4   Sums and Differences of Cubes; A Factoring Strategy

Sum of two cubes	$A^3 + B^3 = (A + B)(A^2 - AB + B^2)$
Difference of two cubes	$A^3 - B^3 = (A - B)(A^2 + AB + B^2)$
Factoring strategy	The five steps that follow can be used to factor many polynomials. Steps 2–4 can be applied to the entire polynomial or to a factor of the polynomial.

  1. If the leading coefficient is positive and the GCF is not 1, factor out the GCF. If the leading coefficient is negative, factor out the opposite of the GCF.
  2. For a binomial, try using one of the properties for the difference of two squares, the sum of two cubes, or the difference of two cubes.
  3. For a trinomial of the form $ax^2 + bx + c$,
     a. If $a = 1$, try to find two integers whose product is c and whose sum is b.
     b. If $a \neq 1$, try to factor by trial and error or by grouping.
  4. For a polynomial with four terms, try factoring by grouping.
  5. Continue applying steps 2–4 until the polynomial is factored completely.

Section 8.5   Using Factoring to Solve Polynomial Equations

Quadratic equation in one variable	A **quadratic equation in one variable** is an equation that can be put into the form $ax^2 + bx + c = 0$, where a, b, and c are constants and $a \neq 0$.
Zero factor property	Let A and B be real numbers: If $AB = 0$, then $A = 0$ or $B = 0$.
Number of solutions: quadratic equation	A quadratic equation in one variable has two real-number solutions, one real-number solution, or no real-number solutions.
Cubic equation in one variable	A **cubic equation in one variable** is an equation that can be put into the form $$ax^3 + bx^2 + cx + d = 0$$ where a, b, c, and d are constants and $a \neq 0$.
Solving quadratic or cubic equations by factoring	If an equation can be solved by factoring, we solve it by the following steps:

  1. Write the equation so one side of the equation is equal to zero.
  2. Factor the nonzero side of the equation.
  3. Apply the zero factor property.
  4. Solve each equation that results from applying the zero factor property.

Section 8.6   Using Factoring to Make Predictions with Quadratic Models

Making a prediction about the response variable	To make a prediction about the response variable of a quadratic model, substitute a chosen value for the explanatory variable in the model's equation. Then solve for the response variable.
Making a prediction about the explanatory variable	To make a prediction about the explanatory variable of a quadratic model, substitute a chosen value for the response variable in the model's equation. Then solve for the explanatory variable, which involves writing the equation so that one side of it is zero and then factoring the nonzero side.

Chapter 8 Review Exercises

For Exercises 1–32, factor when possible. Use a graphing calculator table to verify your work when possible.

1. $x^2 + 9x + 20$
2. $6x^2 - 2x - 8$
3. $81x^2 - 49$
4. $x^2 + 14x + 49$
5. $-18t^4 - 33t^3 + 30t^2$
6. $p^2 - 3pq - 54q^2$
7. $32 - 12x + x^2$
8. $-9x^2 + 4$
9. $4w^2 + 25$
10. $20m^2n - 45mn^3$
11. $16x^2 + 14x + 2x^3$
12. $16x^3 - 32x^2 + 16x$
13. $24x^3 - 32x^2$
14. $5x^4y - 35x^3y + 60x^2y$
15. $-m^2 - 2m + 35$
16. $4r^3 - 10r^2 + 6r - 15$
17. $2p^2 + 7p + 3$
18. $x^2 - 9x + 20$
19. $6t^2 + 11ty - 10y^2$
20. $2x^3 - 50x$
21. $x^2 - 10x + 25$
22. $p^2 - 81$
23. $8w^2 - 12w + 3$
24. $x^2 + 4$
25. $4x^3 - 4x^2 - 9x + 9$
26. $4x^2 + 20x + 25$
27. $12w^3 - 50w^2 + 8w$
28. $49a^2 - 9b^2$
29. $x^2 - 7x - 12$
30. $x^3 + 3x^2 - 4x - 12$
31. $r^3 + 8$
32. $8t^3 - 27$

33. A student tries to factor $x^2 + 25$:

$$x^2 + 25 = (x + 5)(x + 5) = (x + 5)^2$$

Describe any errors. Then factor the polynomial correctly.

34. A student tries to factor $5x^3 + 35x^2 + 60x$:

$$5x^3 + 35x^2 + 60x = 5x(x^2 + 7x + 12)$$

Describe any errors. Then factor the polynomial correctly.

35. Find all possible values of k so that $x^2 + kx + 24$ can be factored.

For Exercises 36–51, solve the given equation. Verify your results by checking that they satisfy the equation.

36. $x^2 + 9x + 14 = 0$

37. $2x^3 + 16x^2 = -24x$

38. $(m - 3)(m + 2) = -4$

39. $t^2 - 6t + 9 = 0$

40. $x^2 - 3x = 5x - 15$

41. $25x^2 - 81 = 0$

42. $2x^3 - 7x^2 - 2x + 7 = 0$

43. $6x^2 + x - 2 = 0$

44. $3x^2 = 15x$

45. $8r^2 - 18r + 9 = 0$

46. $a^2 = 2a + 35$

47. $x(x - 4) = 12$

48. $\frac{1}{3}x^2 - \frac{1}{3}x - 4 = 6$

49. $3x^3 - 2x^2 = 27x - 18$

50. $a^2 = 4$

51. $5p^2 + 20p - 60 = 0$

52. Use "intersect" on a graphing calculator to solve the equation $2x^2 + 4x - 5 = 2x + 3$. Round the solution(s) to the second decimal place.

For Exercises 53–56, solve the given equation by referring to the solutions of $y = -3x^2 + 6x + 20$ shown in Table 15.

53. $-3x^2 + 6x + 20 = 23$

54. $-3x^2 + 6x + 20 = -4$

55. $-3x^2 + 6x + 20 = 27$

56. $-3x^2 + 6x + 20 = 20$

Table 15 Exercises 53–56

x	y
−2	−4
−1	11
0	20
1	23
2	20
3	11
4	−4

For Exercises 57 and 58, find all x-intercepts. Verify your work by using your graphing calculator.

57. $y = x^2 - 49$

58. $y = 8x^2 - 14x - 15$

59. Give an example of a quadratic equation in one variable whose solutions are 3 and −6.

60. Give an example of a cubic equation in one variable whose solutions are −2, 0, and 1.

61. The numbers of worldwide malaria deaths are shown in Table 16 for various years.

Table 16 Numbers of Worldwide Malaria Deaths

Year	Number of Deaths (thousands)
2000	882
2005	830
2010	639
2013	584
2015	438

Source: *World Health Organization (WHO)*

Let n be the number (in thousands) of worldwide malaria deaths in the year that is t years since 2000.

a. Use a graphing calculator to draw the graph of the model $n = -\frac{3}{2}t^2 - 6t + 885$ and, in the same viewing window, the scatterplot of the data. Does the model fit the data well?

b. Predict the number of malaria deaths in 2021.

c. Estimate in which year(s) the number of malaria deaths was 867 thousand deaths.

d. An article in the British health journal *The Lancet* reports 1.24 million (1240 thousand) people died of malaria in 2010. Is this estimate close to WHO's? (See Table 16.) *The Lancet* collected its data by interviewing people about the symptoms of relatives and friends who had died possibly from malaria. WHO collected its data by inspecting medical records. Explain why the two estimates might be so different.

62. A rectangular banner has area 30 square feet. If the length is 4 feet more than twice the width, find the dimensions of the banner.

Chapter 8 Test

For Exercises 1–9, factor.

1. $x^2 - 3x - 40$

2. $24 + x^2 - 10x$

3. $8m^2n^3 - 10m^3n$

4. $p^2 - 14pq + 40q^2$

5. $25p^2 - 36y^2$

6. $3x^4y - 21x^3y + 36x^2y$

7. $8x^3 + 20x^2 - 18x - 45$

8. $8x^2 - 26x + 15$

9. $64x^3 - 1$

10. Which of the following polynomials are equivalent?

$$x^2 - 7x - 10 \quad (x - 5)(x + 2) \quad x^2 - 3x - 10$$
$$x^2 + 3x - 10 \quad (x + 2)(x - 5)$$

11. A student tries to factor $5x^3 + 3x^2 - 20x - 12$:

$$5x^3 + 3x^2 - 20x - 12 = x^2(5x + 3) - 4(5x + 3)$$

Describe any errors. Then factor the polynomial correctly.

Solve.

12. $x^2 - 13x + 36 = 0$

13. $49x^2 - 9 = 0$

14. $t(t + 14) = 2(t - 18)$

15. $\frac{1}{4}p^2 - \frac{1}{2}p - 6 = 0$

16. $3x^3 - 12x = 8 - 2x^2$

17. $2x^3 = 8x^2 + 10x$

18. $(x - 4)(x - 2) = -1$

19. $6x^2 - 19x = -15$

For Exercises 20–23, use the graph of $y = x^2 + 6x + 7$ *shown in Fig. 25 to solve the given equation.*

20. $x^2 + 6x + 7 = 2$ **21.** $x^2 + 6x + 7 = -1$

22. $x^2 + 6x + 7 = -2$ **23.** $x^2 + 6x + 7 = -4$

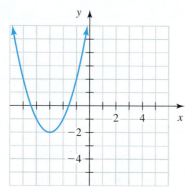

Figure 25 Exercises 20–23

24. Find the x-intercepts of the graph of $y = 10x^2 - 11x - 6$.

25. Give an example of a quadratic equation in one variable whose solutions are -3 and 8.

26. ᴰᴬᵀᴬ The percentages of workers who have saved money for retirement are shown in Table 17.

Table 17 Percentages of Workers Who Have Saved Money for Retirement

Year	Percent
2009	75
2010	69
2011	68
2012	66
2013	66
2014	64
2015	67
2016	69

Source: *IC3*

Let p be the percentage of workers who have saved money for retirement at t years since 2005.

a. Use a graphing calculator to draw a scatterplot of the data. Can the data be described better by a linear model or a quadratic model? Explain.

b. Use a graphing calculator to draw the graph of the model $p = \frac{1}{2}t^2 - 8t + 97$ and, in the same viewing window, the scatterplot of the data. Does the model fit the data well?

c. Predict the percentage of workers who will have saved money for retirement in 2020.

d. Estimate in which year(s) 73% of workers had saved money for retirement.

27. A company's revenue can be modeled by $r = t^2 - 3t + 28$, where r is the revenue (in millions of dollars) for the year that is t years since 2015. Estimate when the annual revenue was or will be $68 million.

Cumulative Review of Chapters 1–8

1. Let p be the percentage of Americans of age a years who work. What is the explanatory variable? What is the response variable?

2. Suppose that an explanatory variable t and a response variable p are approximately linearly related. If a data point (c, d) lies above a linear model, does the model underestimate or overestimate the value of p when $t = c$? Explain.

3. An hour ago, the temperature was 4°F. The temperature is now -2°F. What is the change in temperature?

4. Find the slope of the line that passes through the points $(-3, -1)$ and $(5, 3)$. State whether the line is increasing, decreasing, horizontal, or vertical.

5. For the equation $y = -4x + 9$, describe the change in the value of y as the value of x is increased by 1.

6. Find an equation of the line that has slope $\frac{2}{3}$ and contains the point $(-2, -6)$. Write your equation in slope–intercept form.

7. Find an equation of the line that contains the points $(-4, 7)$ and $(2, -3)$. Write your equation in slope–intercept form.

8. Find an equation of the line sketched in Fig. 26.

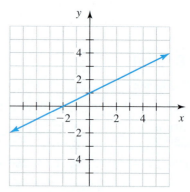

Figure 26 Exercise 8

Let x be a number. Translate each statement into an expression or an equation. Then simplify the expression or solve the equation, as appropriate.

9. 3 times the sum of the number and 5 is 9.

10. 7, minus 4 times the quotient of the number and 2

For Exercises 11 and 12, solve the system by either substitution or elimination. Verify your work by using "intersect" on a graphing calculator or by checking that your result satisfies both equations in the system.

11. $2x - 3y = 7$
$y = 4x - 9$

12. $3x + 4y = 4$
$7x - 5y = 38$

13. Solve the system of equations three times, once by each of the three methods: elimination, substitution, and graphing by hand. Decide which method you prefer for this system.

$$4x - 5y = -22$$
$$x + 2y = 1$$

14. Figure 27 shows the graphs of two linear equations. Find the solution of the system.

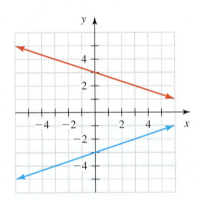

Figure 27 Exercise 14

Perform the indicated operations.

15. $5(-2) - (-6)^2 + 4$

16. $9 - (6 - 8)^3 + 4 \div (-2)$

Evaluate the given expression for $a = -4$, $b = -2$, and $c = 5$.

17. $b^2 - 4ac$

18. $\dfrac{c + a^2}{a - b^2}$

For Exercises 19–24, refer to the graph sketched in Fig. 28.

19. Find y when $x = 4$.
20. Find x when $y = 4$.
21. Find x when $y = 3$.
22. Find x when $y = 5$.
23. Find the x-intercepts.
24. Find the maximum point.

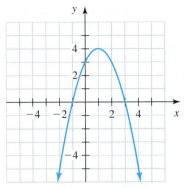

Figure 28 Exercises 19–24

25. Graph $x = 2$ by hand.

For Exercises 26–43, perform the indicated instruction. Then use words such as linear, quadratic, cubic, polynomial, degree, one variable, *and* two variables *to describe the expression, equation, or system.*

26. Graph $y = 2x^2$ by hand.

27. Factor $4m^2 - 49n^2$.

28. Simplify $(3p - 7q)^2$.

29. Factor $w^2 + 5w - 14$.

30. Solve $5x^2 + 18x - 8 = 0$.

31. Solve $x^2 = 2x + 35$.

32. Factor $6x^3y + x^2y - 15xy$.

33. Find the product $(8p - 3)(2p + 3)$.

34. Find the product $-4xy(5x^2 + 2xy - 3y^2)$.

35. Factor $a^2 - 3ab - 40b^2$.

36. Solve $5(t - 2) = 4 - 7t$.

37. Find the sum $(3x^2 - 5x) + (-7x^2 - 2x + 9)$.

38. Factor $3x^2 - 33x + 54$.

39. Graph $2x - 4y = 8$ by hand.

40. Find the product $(2m - 5)(3m^2 - 2m + 4)$.

41. Find the difference $(x^2 - x) - (2x^3 - 4x^2 + 5x)$.

42. Factor $4r^3 + 8r^2 - 9r - 18$.

43. Factor $8x^3 - 27$.

44. Solve the formula $S = 2\pi r^2 + rh$ for h.

45. Solve the inequality $2(x - 1) \le 5(x + 2)$. Describe the solution set as an inequality, in interval notation, and as a graph.

46. Graph the inequality $4x - 5y \ge 20$ by hand.

47. Graph the solution set of the system.

$$y > -\frac{1}{2}x + 2$$
$$y < 3$$

48. Which of the following polynomials are equivalent?

$$(x - 2)(x + 4) \qquad 8 - 2x + x^2 \qquad x^2 + 2x - 8$$
$$(x + 2)(x - 4) \qquad (x + 4)(x - 2) \qquad x(x + 2) - 8$$

49. Write the equation $y = -3(x - 2)^2$ in standard form.

Find all x-intercepts. Verify your work by using a graphing calculator.

50. $y = x^2 + 4x - 21$

51. $2x - 5y = 20$

For Exercises 52–55, simplify.

52. $\left(-2ab^2\right)^3$

53. $\dfrac{2^2 r^{-2}}{2^5 r^{-7}}$

54. $\left(4x^{-8}y^5\right)\left(2x^3y^{-2}\right)$

55. $\left(\dfrac{x^2 y^{-6}}{w^{-4}}\right)^3$

56. A 6000-seat theater has tickets for sale at $20 and $35. How many tickets should be sold at each price for a sellout performance to generate a total revenue of $147,000?

57. Worldwide sales of Nokia/Microsoft cell phones are shown in Table 18 for various years.

Table 18 Worldwide Sales of Nokia/Microsoft Cell Phones

Year	Sales (millions)
2010	110
2011	108
2012	83
2013	63
2014	50
2015	33

Sources: *Gartner, Microsoft*

Let s be the worldwide annuals sales (in millions) of Nokia/Microsoft cell phones at t years since 2010.

a. Use a graphing calculator to draw a scatterplot of the data.

b. Find an equation of a model to describe the data.

c. What is the slope? What does it mean in this situation?

d. Use the model to estimate the sales in 2011. Is your result an underestimate or an overestimate? Explain.

e. Estimate when the sales were 17 million cell phones.

f. Find the s-intercept. What does it mean in this situation?

g. Find the t-intercept. What does it mean in this situation?

58. The median household income in North Dakota was $59,099 in 2014 and has since increased by about $1530 per year. The median household income in Minnesota was $61,554 in 2014 and has since increased by about $770 per year (Source: *U.S. Census Bureau*).

a. Let I be the median household income (in dollars) in North Dakota at t years since 2014. Find an equation of a model to describe the situation.

b. Let I be the median household income (in dollars) in Minnesota at t years since 2014. Find an equation of a model to describe the situation.

c. Use substitution or elimination to estimate in which year the median household income in the two states was equal. What is that median income?

59. Table 19 lists the percentages of Americans of various age groups who visit online trading sites.

Table 19 Americans Who Visit Online Trading Sites

Age Group (years)	Age Used to Represent Age Group (years)	Percent
2–17	9.5	4.7
18–24	21.0	9.7
25–34	29.5	22.9
35–44	39.5	22.2
45–54	49.5	20.6
55–64	59.5	14.6
65+	70.0	5.3

Source: *comScore Media Metrix*

Let p be the percentage of Americans of age t years who visit online trading sites. A quadratic model is sketched in Fig. 29.

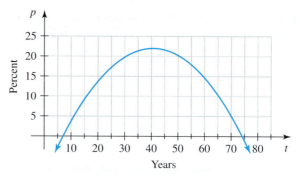

Figure 29 Exercise 59

a. Carefully trace the graph in Fig. 29 onto a piece of paper. Then, on the same graph, draw a scatterplot of the data. Does the model fit the data well?

b. What are the t-intercepts? What do they mean in this situation?

c. What is the vertex of the model? What does it mean in this situation?

d. Estimate the percentage of 19-year-old Americans who visit online trading sites.

e. Estimate the age(s) at which 18% of Americans visit online trading sites.

60. A company's profit can be modeled by $p = t^2 - 2t + 8$, where p is the profit (in millions of dollars) for the year that is t years since 2015. Estimate when the annual profit was or will be $32 million.

61. A rectangular rug has an area of 84 square feet. If the length is 8 feet more than the width, find the dimensions of the rug.

Solving Quadratic Equations

9

Have you ever run a red light? Younger people are more likely to admit to running red lights than older people (see Table 1). In Exercise 5 of Homework 9.6, you will estimate at what age half of all drivers admit to running red lights.

In Chapter 8, we used factoring to solve quadratic equations in one variable. However, most quadratic equations cannot be solved by factoring. In this chapter, we will discuss other ways to solve quadratic equations; some of those ways can be used to solve *any* quadratic equation. This work will enable us to make estimates and predictions about any situation that can be described well by a quadratic model, including the red-light data just presented.

Table 1 Percentages of Drivers Who Admit to Running Red Lights

Age Group (years)	Age Used to Represent Age Group (years)	Percent
18–25	21.5	75
26–35	30.5	73
36–45	40.5	63
46–55	50.5	56
over 55	65.0	35

Source: *The Social Science Research Center*

9.1 Simplifying Radical Expressions

Objectives

» Describe the meaning of *square root* and *principal square root.*

» Approximate a principal square root.

» Apply the *product property for square roots.*

» Use the product property for square roots to simplify radicals.

» Simplify a square root whose radicand contains x^2.

So far, we have worked with polynomial expressions. In this section and Section 9.2, we will work with a new type of expression called a *radical expression*. This work will prepare us to solve more quadratic equations in Sections 9.3–9.6.

Principal Square Roots

What number squared is equal to 9? The numbers that "work" are -3 and 3:

$$(-3)^2 = 9 \qquad 3^2 = 9$$

We say that -3 and 3 are *square roots* of 9. A **square root** of a number a is the number we square to get a.

Of the square roots of 9, only the nonnegative square root, 3, is the *principal square root* of 9. Using symbols, we write $\sqrt{9} = 3$.

Definition   Principal square root

If a is a nonnegative number, then $\sqrt{a}$ is the nonnegative number we square to get a. We call $\sqrt{a}$ the **principal square root** of a.

For example, $\sqrt{25} = 5$ because $5^2 = 25$. Also, $\sqrt{64} = 8$ because $8^2 = 64$.

The symbol "$\sqrt{}$" is called a **radical sign**. An expression under a radical sign is called a **radicand**. For $\sqrt{4x - 7}$, the radicand is $4x - 7$. A radical sign together with a radicand is called a **radical**. Here we label the radical sign and radicand of the radical $\sqrt{2x + 5}$:

$$\underbrace{\sqrt{2x + 5}}_{\text{Radical}} \xleftarrow{} \text{Radical sign} \\ \xleftarrow{} \text{Radicand}$$

Here are some more radicals: $\sqrt{5}$, $\sqrt{x}$, $\sqrt{9x + 4}$.

An expression that contains a radical is called a **radical expression**. Here are some radical expressions:

$$\sqrt{5}, \quad \sqrt{x}, \quad \sqrt{2x - 5}, \quad 4\sqrt{x + 8} - 8x, \quad \left(3\sqrt{x} - 2\right)\left(\sqrt{x} + 7\right)$$

WARNING   Is a square root of a negative number a real number? Consider $\sqrt{-9}$. Note that $\sqrt{-9} \neq -3$ because $(-3)^2$ is equal to 9, not -9. Because any number squared is non-negative, we see that $\sqrt{-9}$ is not a real number. In general, a square root of a negative number is not a real number.

Example 1   Finding Square Roots

Find the square root.

1. $\sqrt{49}$ 2. $\sqrt{-49}$ 3. $-\sqrt{49}$ 4. $-\sqrt{-49}$

Solution

1. $\sqrt{49} = 7$ because $7^2 = 49$.
2. $\sqrt{-49}$ is not a real number because the radicand -49 is negative.
3. $-\sqrt{49} = -7$.
4. $-\sqrt{-49}$ is not a real number because the radicand -49 is negative.

Rational and Irrational Square Roots

Recall from Section 1.1 that a rational number is a number that can be written in the form $\dfrac{n}{d}$, where n and d are integers and d is nonzero. A **perfect square** is a number whose principal square root is rational. For example, 25 is a perfect square because $\sqrt{25} = 5 = \dfrac{5}{1}$ is rational. By squaring the integers from 0 to 15, we can find the integer perfect squares between 0 and 225, inclusive (see Table 2). You should memorize the perfect squares shown in this table because you will work with them again and again.

For a number that is not a perfect square, any square root of the number is not rational. Recall from Section 1.1 that we call such a number *irrational*. For example, $\sqrt{7}$ is irrational. We know that $\sqrt{7}$ is a number between 2 and 3 because $2^2 = 4$ and $3^2 = 9$. To use a calculator to get the estimate $\sqrt{7} \approx 2.645751311$, press $\boxed{\text{2nd}}$ $[\sqrt{}]$ $\boxed{7}$ $\boxed{)}$ $\boxed{\text{ENTER}}$ (see Fig. 1).

Table 2 Perfect Squares

x	Perfect Square x^2
0	0
1	1
2	4
3	9
4	16
5	25
6	36
7	49
8	64
9	81
10	100
11	121
12	144
13	169
14	196
15	225

```
√(7)
        2.645751311
```

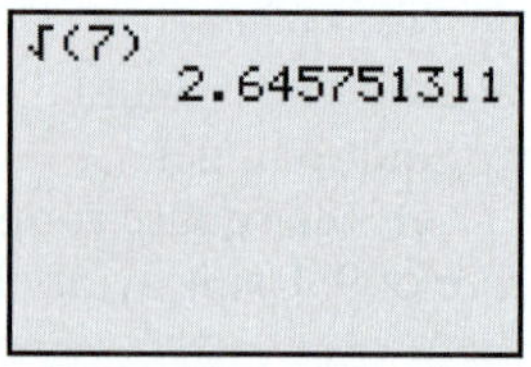

Figure 1 Estimating $\sqrt{7}$

Example 2   Approximating Square Roots

State whether the square root is rational or irrational. If the square root is rational, find the (exact) value. If the square root is irrational, estimate its value by rounding to the second decimal place.

1. $\sqrt{19}$ 2. $\sqrt{169}$

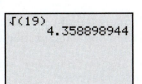

Figure 2 Estimating $\sqrt{19}$

Solution

1. The number 19 is not a perfect square, so $\sqrt{19}$ is irrational. We use a calculator to compute $\sqrt{19} \approx 4.36$ (see Fig. 2).
2. The number 169 is a perfect square, so $\sqrt{169}$ is rational. In fact, $\sqrt{169} = 13$ because $13^2 = 169$.

▶

Product Property for Square Roots

Consider the following computations with radicals:

$$\sqrt{4 \cdot 9} = \sqrt{36} = 6$$
$$\sqrt{4}\sqrt{9} = 2 \cdot 3 = 6$$

Because both results are equal, we can write

$$\sqrt{4 \cdot 9} = \sqrt{4}\sqrt{9}$$

This equation suggests the following **product property for square roots**, which can help us work with the principal square root of a product.

Product Property for Square Roots

For nonnegative numbers a and b,

$$\sqrt{ab} = \sqrt{a}\sqrt{b}$$

In words: The square root of a product is the product of the square roots.

For example, $\sqrt{16 \cdot 49} = \sqrt{16}\sqrt{49} = 4 \cdot 7 = 28$.

Using the Product Property for Square Roots to Simplify Radicals

We can sometimes write a radical as an expression with a smaller radicand. For example, consider $\sqrt{24}$. Note that the perfect square 4 is a factor of 24: $4 \cdot 6 = 24$. We can use the product property for square roots to write

$$\sqrt{24} = \sqrt{4 \cdot 6} = \sqrt{4}\sqrt{6} = 2\sqrt{6}$$

When writing an expression such as $2\sqrt{6}$, we write the "2" to the left of "$\sqrt{6}$." If we were to write the expression as $\sqrt{6}2$, the radicand might appear to be 62, not 6.

A square root is **simplified** when the radicand does not have any perfect-square factors other than 1.

Example 3   Simplifying Radicals

Simplify.

1. $\sqrt{12}$ 2. $\sqrt{32x}$ 3. $7\sqrt{45x}$

Solution

1. Note that 4 is the largest perfect-square factor of 12. We write 12 as a product of 4 and 3 and then apply the product property for square roots:

$$\sqrt{12} = \sqrt{4 \cdot 3} \qquad \text{4 is the largest perfect-square factor.}$$
$$= \sqrt{4}\sqrt{3} \qquad \text{Product property: } \sqrt{ab} = \sqrt{a}\sqrt{b}$$
$$= 2\sqrt{3} \qquad \sqrt{4} = 2$$

We verify our work by using a calculator to compare approximations for $\sqrt{12}$ and $2\sqrt{3}$ (see Fig. 3).

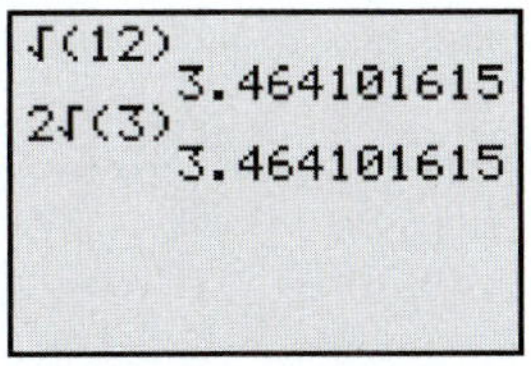

Figure 3 Verify the work

2. Both 4 and 16 are perfect-square factors of $32x$. Because 16 is the largest perfect-square factor, we write $32x$ as the product of 16 and $2x$ and then apply the product property for square roots:

$$\sqrt{32x} = \sqrt{16 \cdot 2x} \quad \text{16 is the largest perfect-square factor.}$$
$$= \sqrt{16}\sqrt{2x} \quad \text{Product property: } \sqrt{ab} = \sqrt{a}\sqrt{b}$$
$$= 4\sqrt{2x} \quad \sqrt{16} = 4$$

We use a graphing calculator table with nonnegative values of x to verify our work (see Fig. 4).

 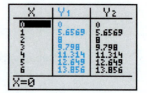

Figure 4 Verify the work

3. $7\sqrt{45x} = 7\sqrt{9 \cdot 5x} \quad$ 9 *is the largest perfect-square factor.*
$\qquad\quad = 7\sqrt{9}\sqrt{5x} \quad$ *Product property:* $\sqrt{ab} = \sqrt{a}\sqrt{b}$
$\qquad\quad = 7 \cdot 3 \cdot \sqrt{5x} \quad \sqrt{9} = 3$
$\qquad\quad = 21\sqrt{5x} \quad$ *Multiply.*

In Problem 2 of Example 3, we used the largest perfect-square factor, 16, of the radicand $32x$ to simplify $\sqrt{32x}$. What would happen if we used the smaller perfect-square factor, 4?

$$\sqrt{32x} = \sqrt{4 \cdot 8x} = \sqrt{4}\sqrt{8x} = 2\sqrt{8x}$$

The radicand, $8x$, has a perfect-square factor of 4, so we continue to simplify:

$$\sqrt{32x} = 2\sqrt{8x} = 2\sqrt{4 \cdot 2x} = 2\sqrt{4}\sqrt{2x} = 2 \cdot 2\sqrt{2x} = 4\sqrt{2x}$$

The result is the same as our result in Example 3. This exploration suggests two things:

- The most efficient way to simplify a square root is to use the *largest* perfect-square factor of the radicand.
- If we don't use the largest perfect-square factor of the radicand, we can simplify the square root by continuing to use perfect-square factors until the radicand has no perfect-square factors other than 1.

Simplifying a Square Root with a Small Radicand

To simplify a square root in which it is easy to determine the largest perfect-square factor of the radicand,

1. Write the radicand as the product of the *largest* perfect-square factor and another number.

2. Apply the product property for square roots.

In Example 4, we will discuss how to simplify a radical that has a large radicand.

Example 4   Simplifying a Radical with a Large Radicand

Simplify $\sqrt{72}$.

Solution

We begin by finding the prime factorization of the radicand, 72:

$$\begin{aligned}
\sqrt{72} &= \sqrt{2 \cdot 2 \cdot 2 \cdot 3 \cdot 3} && \textit{Find prime factorization of 72.} \\
&= \sqrt{2 \cdot 2 \cdot 3 \cdot 3 \cdot 2} && \textit{Rearrange factors to highlight pairs of identical factors.} \\
&= \sqrt{2 \cdot 2} \cdot \sqrt{3 \cdot 3} \cdot \sqrt{2} && \textit{Product property: } \sqrt{ab} = \sqrt{a}\sqrt{b} \\
&= 2 \cdot 3 \cdot \sqrt{2} && \sqrt{2 \cdot 2} = \sqrt{4} = 2; \sqrt{3 \cdot 3} = \sqrt{9} = 3 \\
&= 6\sqrt{2} && \textit{Multiply.}
\end{aligned}$$

Simplifying a Square Root with a Large Radicand

To simplify a square root in which it is difficult to determine the largest perfect-square factor of the radicand,

1. Find the prime factorization of the radicand.
2. Look for pairs of identical factors of the radicand, and rearrange the factors to highlight them.
3. Use the product property for square roots.

Simplifying a Square Root Whose Radicand Contains x^2

If x is nonnegative, then $\sqrt{x^2} = x$ because squaring x gives x^2. In this course, we will not discuss the case in which x is negative.

Simplifying $\sqrt{x^2}$

If x is nonnegative, then

$$\sqrt{x^2} = x$$

Example 5   Simplifying a Radical

Simplify. Assume that any variables are nonnegative.

1. $\sqrt{21t^2w^2}$

2. $3\sqrt{20x^2y}$

Solution

1.
$$\begin{aligned}
\sqrt{21t^2w^2} &= \sqrt{21}\sqrt{t^2}\sqrt{w^2} && \textit{Product property: } \sqrt{ab} = \sqrt{a}\sqrt{b} \\
&= \left(\sqrt{21}\right)tw && \sqrt{x^2} = x \textit{ for nonnegative } x \\
&= tw\sqrt{21} && \textit{Rearrange factors.}
\end{aligned}$$

2.
$$\begin{aligned}
3\sqrt{20x^2y} &= 3\sqrt{4 \cdot 5 \cdot x^2 \cdot y} && \textit{4 and } x^2 \textit{ are perfect squares.} \\
&= 3\sqrt{4 \cdot x^2 \cdot 5y} && \textit{Rearrange factors.} \\
&= 3\sqrt{4}\sqrt{x^2}\sqrt{5y} && \textit{Product property: } \sqrt{ab} = \sqrt{a}\sqrt{b} \\
&= 3 \cdot 2 \cdot x \cdot \sqrt{5y} && \sqrt{4} = 2, \sqrt{x^2} = x \textit{ for nonnegative } x \\
&= 6x\sqrt{5y} && \textit{Multiply.}
\end{aligned}$$

Group Exploration

Section Opener: Product property for square roots

1. If a is a nonnegative number, then $\sqrt{a}$ is the nonnegative number we square to get a. We call $\sqrt{a}$ the *principal square root* of a. So, $\sqrt{9} = 3$ because $3^2 = 9$. Find the principal square root or the product of principal square roots.
   a. $\sqrt{25}$ b. $\sqrt{2 \cdot 8}$
   c. $\sqrt{25} \cdot \sqrt{49}$ d. $\sqrt{13}$ [**Hint:** Use a calculator.]

2. Find the principal square root $\sqrt{4 \cdot 9}$ and the product $\sqrt{4} \cdot \sqrt{9}$. Then compare your results.

3. Find the principal square root $\sqrt{4 \cdot 25}$ and the product $\sqrt{4} \cdot \sqrt{25}$. Then compare your results.

4. Write $\sqrt{ab}$ as a product of two principal square roots. [**Hint:** See Problems 2 and 3.] This is called the *product property for square roots*.

5. Use a calculator to estimate $\sqrt{4 + 9}$. Then find the sum $\sqrt{4} + \sqrt{9}$. Finally, compare your results.

6. Use a calculator to estimate $\sqrt{16 + 25}$. Then find the sum $\sqrt{16} + \sqrt{25}$. Finally, compare your results.

7. Is the statement $\sqrt{a + b} = \sqrt{a} + \sqrt{b}$ true for all real numbers a and b? Explain.

Tips for Success   Use 3-by-5 Cards

Do you have trouble memorizing definitions and properties? If so, try writing a word or phrase on one side of a 3-by-5 card, and, on the other side, put its definition or state its property and how it can be applied. For example, you could write "product property for square roots" on one side of a card and "For nonnegative numbers a and b, $\sqrt{ab} = \sqrt{a}\sqrt{b}$" on the other side. You could also describe, in your own words, the meaning of the property and how you can apply it.

Once you have completed a card for each definition and property, shuffle the cards and quiz yourself until you are confident you know the definitions and properties and how to apply them. Quiz yourself again later to make sure you have retained the information.

In addition to memorizing definitions and properties, it is important you strive to understand their meanings and how to apply them.

 # Homework 9.1

For extra help ▶   **MyLab Math**   Watch the videos in MyLab Math

Find the square root.

1. $\sqrt{4}$ 2. $\sqrt{16}$ 3. $\sqrt{81}$ 4. $\sqrt{36}$
5. $\sqrt{121}$ 6. $\sqrt{100}$ 7. $\sqrt{144}$ 8. $\sqrt{225}$
9. $-\sqrt{16}$ 10. $-\sqrt{9}$ 11. $-\sqrt{81}$ 12. $-\sqrt{169}$
13. $\sqrt{-9}$ 14. $\sqrt{-4}$ 15. $-\sqrt{-25}$ 16. $-\sqrt{-16}$

State whether the square root is rational or irrational. If the square root is rational, find the (exact) value. If the square root is irrational, estimate its value by rounding to the second decimal place.

17. $\sqrt{30}$ 18. $\sqrt{62}$ 19. $\sqrt{78}$
20. $\sqrt{256}$ 21. $\sqrt{196}$ 22. $\sqrt{95}$

Simplify. Use a calculator to verify your work.

23. $\sqrt{20}$ 24. $\sqrt{28}$ 25. $\sqrt{45}$ 26. $\sqrt{18}$
27. $\sqrt{27}$ 28. $\sqrt{8}$ 29. $\sqrt{50}$ 30. $\sqrt{32}$
31. $\sqrt{300}$ 32. $\sqrt{75}$ 33. $-\sqrt{98}$ 34. $-\sqrt{80}$
35. $4\sqrt{72}$ 36. $5\sqrt{52}$ 37. $3\sqrt{120}$ 38. $2\sqrt{135}$

Simplify. Assume that any variables are nonnegative.

39. $\sqrt{9x}$ 40. $\sqrt{25x}$ 41. $\sqrt{64t}$
42. $\sqrt{49a}$ 43. $\sqrt{81x^2}$ 44. $\sqrt{36x^2}$
45. $\sqrt{225t^2w^2}$ 46. $\sqrt{169a^2b^2}$ 47. $\sqrt{5x^2}$
48. $\sqrt{17x^2}$ 49. $7\sqrt{39x^2y}$ 50. $4\sqrt{23x^2y}$
51. $\sqrt{12p}$ 52. $\sqrt{24t}$ 53. $2\sqrt{63x}$
54. $5\sqrt{40x}$ 55. $\sqrt{60a^2b^2}$ 56. $\sqrt{72t^2w^2}$
57. $3\sqrt{125xy^2}$ 58. $5\sqrt{48xy^2}$

59. Suppose that, while traveling on a dry road, a driver slams on the brakes. Let D be the distance (in feet) that the car will skid and S be the speed (in miles per hour) before braking. The relationship between D and S is described by the model $S = \sqrt{30FD}$, where F is a drag factor, which is a measure of the roughness of the surface of the road. The drag factor on new concrete is 0.95, and the drag factor on polished concrete or asphalt is 0.75.

   A motorist who was involved in an accident claims he was driving at the posted speed limit of 60 miles per hour. A police officer measures the car's skid marks on the asphalt road to be 210 feet long. Assuming the motorist applied the brakes suddenly, estimate the speed at which the motorist was traveling before braking.

60. The size of a rectangular television screen is usually described by the length of its diagonal, given (in inches) by $d = \sqrt{w^2 + h^2}$, where w is the screen's width and h is the screen's height, both in inches. Find the length of the diagonal of a television screen whose width is 17 inches and height is 11 inches.

Concepts

61. A student thinks that $\sqrt{-25} = -5$. Is the student correct? Explain.

62. A student thinks that $-\sqrt{-9} = 3$. Is the student correct? Explain.

Without using a calculator, find the two consecutive integers nearest to the given radical. Verify your result by using a calculator.

63. $\sqrt{22}$ 64. $\sqrt{15}$ 65. $\sqrt{71}$ 66. $\sqrt{43}$

67. **a.** Which of the two numbers is larger?
   i. 2, $\sqrt{2}$ **ii.** 5, $\sqrt{5}$ **iii.** 8, $\sqrt{8}$
   b. Describe any patterns in the results you found in part (a).
   c. Which of the two numbers is larger? [**Hint:** Use a calculator to be sure.]
   i. 0.2, $\sqrt{0.2}$ **ii.** 0.5, $\sqrt{0.5}$ **iii.** 0.8, $\sqrt{0.8}$
   d. Describe any patterns in the results you found in part (c).
   e. For which type of number does finding the principal square root give a smaller result? A larger result?

68. **a.** Square the given number. Then find the principal square root of the result.
   i. 3   **ii.** 5   **iii.** 6
   b. Find the square root.
   i. $\sqrt{3^2}$ **ii.** $\sqrt{5^2}$ **iii.** $\sqrt{6^2}$
   c. Describe any patterns from your work in parts (a) and (b).
   d. Explain why $\sqrt{x^2} = x$, where x is nonnegative.

69. Is $\sqrt{10}$ twice as big as $\sqrt{5}$? Explain. Include the product rule for square roots in your discussion.

70. Is $2\sqrt{3}$ equal to $\sqrt{6}$? Explain. Include the product rule for square roots in your discussion.

Sketch the graph of the equation. Verify your result by using a graphing calculator.

71. $y = \sqrt{x}$ 72. $y = -\sqrt{x}$

73. Explain why the square root of a negative number is not a real number.

74. Choose nonnegative numbers for a and b, and show $\sqrt{ab} = \sqrt{a}\sqrt{b}$.

Related Review

Simplify. Assume that x is nonnegative.

75. $(7x)^2$ 76. $(5x)^2$ 77. $\sqrt{49x^2}$ 78. $\sqrt{25x^2}$

Expressions, Equations, and Graphs

Perform the indicated instruction. Then use words such as linear, quadratic, cubic, power, radical, polynomial, degree, one variable, and two variables to describe the expression, equation, or system.

79. Graph $y = -2x^2$ by hand.

80. Factor $4r^3 - 12r^2 - r + 3$.

81. Solve $x^2 = 6x - 8$.

82. Simplify $(m + 2n)^2$.

83. Simplify $\sqrt{68x}$.

84. Solve the following:
$$6x + 5y = 8$$
$$2x - 3y = -16$$

9.2 Simplifying More Radical Expressions

Objectives

» Apply the *quotient property for square roots.*

» Use the quotient property for square roots to simplify a radical.

» *Rationalize the denominator* of a radical expression.

» Simplify a radical quotient.

» Summarize how to simplify a radical expression.

In Section 9.1, we discussed how to use the product property for square roots to help us simplify the principal square root of a product. In this section, we will discuss a property that will help us simplify the principal square root of a quotient.

Quotient Property for Square Roots

Here we compute the square root of a quotient and a quotient of square roots:

Square root of a quotient: $\sqrt{\dfrac{16}{49}} = \dfrac{4}{7}$ because $\left(\dfrac{4}{7}\right)^2 = \dfrac{16}{49}$

Quotient of square roots: $\dfrac{\sqrt{16}}{\sqrt{49}} = \dfrac{4}{7}$

Because the two results are equal, we can write

$$\sqrt{\dfrac{16}{49}} = \dfrac{\sqrt{16}}{\sqrt{49}}$$

This equation suggests the **quotient property for square roots**.

> **Quotient Property for Square Roots**
>
> For a nonnegative number a and a positive number b,
>
> $$\sqrt{\frac{a}{b}} = \frac{\sqrt{a}}{\sqrt{b}}$$
>
> In words: The square root of a quotient is the quotient of the square roots.

Using the Quotient Property for Square Roots to Simplify Radicals

If a radical has a fractional radicand, we **simplify the radical** by writing it as an expression whose radicand is not a fraction.

Example 1   Simplifying Radicals

Simplify. Assume that x and y are positive.

1. $\sqrt{\dfrac{5}{x^2}}$ **2.** $\sqrt{\dfrac{50}{81}}$ **3.** $\sqrt{\dfrac{3x^2y}{49}}$

Solution

1. $\sqrt{\dfrac{5}{x^2}} = \dfrac{\sqrt{5}}{\sqrt{x^2}}$ *Quotient property:* $\sqrt{\dfrac{a}{b}} = \dfrac{\sqrt{a}}{\sqrt{b}}$

      $= \dfrac{\sqrt{5}}{x}$ $\sqrt{x^2} = x$ *for nonnegative* x

We use a graphing calculator table with positive values of x to verify our work (see Fig. 5).

 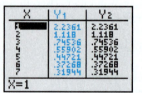

Figure 5 Verify the work

2. $\sqrt{\dfrac{50}{81}} = \dfrac{\sqrt{50}}{\sqrt{81}}$ *Quotient property:* $\sqrt{\dfrac{a}{b}} = \dfrac{\sqrt{a}}{\sqrt{b}}$

      $= \dfrac{\sqrt{25 \cdot 2}}{9}$ *25 is a perfect square;* $\sqrt{81} = 9$

      $= \dfrac{\sqrt{25}\sqrt{2}}{9}$ *Product property:* $\sqrt{ab} = \sqrt{a}\sqrt{b}$

      $= \dfrac{5\sqrt{2}}{9}$ $\sqrt{25} = 5$

3. $\sqrt{\dfrac{3x^2y}{49}} = \dfrac{\sqrt{3x^2y}}{\sqrt{49}}$ *Quotient property:* $\sqrt{\dfrac{a}{b}} = \dfrac{\sqrt{a}}{\sqrt{b}}$

      $= \dfrac{\sqrt{x^2}\sqrt{3y}}{7}$ *Product property:* $\sqrt{ab} = \sqrt{a}\sqrt{b}$; $\sqrt{49} = 7$

      $= \dfrac{x\sqrt{3y}}{7}$ $\sqrt{x^2} = x$ *for nonnegative* x

Rationalizing the Denominator of a Radical Expression

We simplify an expression of the form $\dfrac{p}{\sqrt{q}}$ by leaving no denominator as a radical expression. We call this process **rationalizing the denominator**.

Example 2   Rationalizing the Denominator

Simplify.

1. $\dfrac{5}{\sqrt{7}}$ **2.** $\dfrac{2}{\sqrt{x}}$ **3.** $\dfrac{14}{\sqrt{24}}$

Solution

1. $\dfrac{5}{\sqrt{7}} = \dfrac{5}{\sqrt{7}} \cdot 1$ *$a = a \cdot 1$*

$= \dfrac{5}{\sqrt{7}} \cdot \dfrac{\sqrt{7}}{\sqrt{7}}$ *Rationalize denominator.*

$= \dfrac{5\sqrt{7}}{\sqrt{49}}$ *Product property: $\sqrt{a}\sqrt{b} = \sqrt{ab}$*

$= \dfrac{5\sqrt{7}}{7}$ *$\sqrt{49} = 7$*

2. $\dfrac{2}{\sqrt{x}} = \dfrac{2}{\sqrt{x}} \cdot \dfrac{\sqrt{x}}{\sqrt{x}}$ *Rationalize denominator.*

$= \dfrac{2\sqrt{x}}{\sqrt{x^2}}$ *Product property: $\sqrt{a}\sqrt{b} = \sqrt{ab}$*

$= \dfrac{2\sqrt{x}}{x}$ *$\sqrt{x^2} = x$ for nonnegative x*

3. $\dfrac{14}{\sqrt{24}} = \dfrac{14}{2\sqrt{6}}$ *$\sqrt{24} = \sqrt{4 \cdot 6} = \sqrt{4}\sqrt{6} = 2\sqrt{6}$*

$= \dfrac{7}{\sqrt{6}}$ *Simplify.*

$= \dfrac{7}{\sqrt{6}} \cdot \dfrac{\sqrt{6}}{\sqrt{6}}$ *Rationalize denominator.*

$= \dfrac{7\sqrt{6}}{\sqrt{36}}$ *Product property: $\sqrt{a}\sqrt{b} = \sqrt{ab}$*

$= \dfrac{7\sqrt{6}}{6}$ *$\sqrt{36} = 6$*

▶

As shown in Example 2, **to rationalize the denominator of a fraction of the form** $\dfrac{p}{\sqrt{q}}$, **where q is positive, we multiply the fraction by 1 in the form** $\dfrac{\sqrt{q}}{\sqrt{q}}$.

Example 3   Rationalizing the Denominator

Simplify.

1. $\sqrt{\dfrac{7}{3}}$

2. $\sqrt{\dfrac{3t}{20}}$

Solution

1. We first use the quotient property for square roots and then rationalize the denominator:

$$\sqrt{\dfrac{7}{3}} = \dfrac{\sqrt{7}}{\sqrt{3}} \qquad \textit{Quotient property: } \sqrt{\dfrac{a}{b}} = \dfrac{\sqrt{a}}{\sqrt{b}}$$

$$= \dfrac{\sqrt{7}}{\sqrt{3}} \cdot \dfrac{\sqrt{3}}{\sqrt{3}} \qquad \textit{Rationalize denominator.}$$

$$= \dfrac{\sqrt{21}}{\sqrt{9}} \qquad \textit{Product property: } \sqrt{a}\sqrt{b} = \sqrt{ab}$$

$$= \dfrac{\sqrt{21}}{3} \qquad \sqrt{9} = 3$$

2. $$\sqrt{\dfrac{3t}{20}} = \dfrac{\sqrt{3t}}{\sqrt{20}} \qquad \textit{Quotient property: } \sqrt{\dfrac{a}{b}} = \dfrac{\sqrt{a}}{\sqrt{b}}$$

$$= \dfrac{\sqrt{3t}}{2\sqrt{5}} \qquad \sqrt{20} = \sqrt{4 \cdot 5} = \sqrt{4}\sqrt{5} = 2\sqrt{5}$$

$$= \dfrac{\sqrt{3t}}{2\sqrt{5}} \cdot \dfrac{\sqrt{5}}{\sqrt{5}} \qquad \textit{Rationalize denominator.}$$

$$= \dfrac{\sqrt{15t}}{2\sqrt{25}} \qquad \textit{Product property: } \sqrt{a}\sqrt{b} = \sqrt{ab}$$

$$= \dfrac{\sqrt{15t}}{10} \qquad 2\sqrt{25} = 2 \cdot 5 = 10$$

Simplifying a Radical Quotient

Next we will **simplify a radical quotient**. We will use this skill when solving quadratic equations in Section 9.5.

Example 4   Simplifying a Radical Quotient

Simplify $\dfrac{6 + 3\sqrt{2}}{12}$.

Solution

We first factor the numerator and the denominator, and then simplify the expression:

$$\dfrac{6 + 3\sqrt{2}}{12} = \dfrac{3\left(2 + \sqrt{2}\right)}{3(4)} \qquad \textit{Factor out 3.}$$

$$= \dfrac{3}{3} \cdot \dfrac{2 + \sqrt{2}}{4} \qquad \dfrac{ac}{bd} = \dfrac{a}{b} \cdot \dfrac{c}{d}$$

$$= \dfrac{2 + \sqrt{2}}{4} \qquad \dfrac{3}{3} = 1; \textit{ simplify.}$$

Example 5   Simplifying a Radical Quotient

Simplify $\dfrac{8 - \sqrt{28}}{10}$.

Solution

We begin by simplifying the radical $\sqrt{28}$:

$$\dfrac{8 - \sqrt{28}}{10} = \dfrac{8 - 2\sqrt{7}}{10} \qquad \sqrt{28} = \sqrt{4 \cdot 7} = \sqrt{4}\sqrt{7} = 2\sqrt{7}$$

$$= \dfrac{2\left(4 - \sqrt{7}\right)}{2(5)} \qquad \textit{Factor out 2.}$$

$$= \dfrac{2}{2} \cdot \dfrac{4 - \sqrt{7}}{5} \qquad \dfrac{ac}{bd} = \dfrac{a}{b} \cdot \dfrac{c}{d}$$

$$= \dfrac{4 - \sqrt{7}}{5} \qquad \dfrac{2}{2} = 1; \textit{ simplify.}$$

We verify our work by using a calculator to compare approximations for $\dfrac{8 - \sqrt{28}}{10}$ and $\dfrac{4 - \sqrt{7}}{5}$ (see Fig. 6). When we perform such a check, it is important that all parentheses be entered as shown.

Figure 6 Verify the work

```
(8-√(28))/10
        .2708497378
(4-√(7))/5
        .2708497378
```

In Example 5, we obtained the result $\dfrac{4 - \sqrt{7}}{5}$. Keep in mind this radical expression is a number. From our check in Fig. 6, we see that $\dfrac{4 - \sqrt{7}}{5} \approx 0.27$.

Summary of Simplifying a Radical Expression

We have discussed various ways to simplify a radical expression. What follows is a summary of these methods.

Simplifying a Radical Expression

To simplify a radical expression,

1. Use the quotient property for square roots so that no radicand is a fraction.
2. Use the product property for square roots so that no radicands have perfect-square factors other than 1.
3. Rationalize denominators so that no denominator is a radical expression.
4. Use the property $\dfrac{a}{a} = 1$, where a is nonzero, to simplify.
5. Continue applying steps 1–4 until the radical expression is simplified completely.

◆ Group Exploration

Section Opener: Quotient property for square roots

1. Find $\sqrt{\dfrac{4}{9}}$ and $\dfrac{\sqrt{4}}{\sqrt{9}}$. Then compare your results.

2. Find $\sqrt{\dfrac{16}{25}}$ and $\dfrac{\sqrt{16}}{\sqrt{25}}$. Then compare your results.

3. Find $\sqrt{\dfrac{49}{81}}$ and $\dfrac{\sqrt{49}}{\sqrt{81}}$. Then compare your results.

4. Write $\sqrt{\dfrac{a}{b}}$ as a quotient of two square roots. [**Hint:** See Problems 1, 2, and 3.]

5. Write each expression as a radical expression in which no radicand is a fraction. Assume that x is positive.

   a. $\sqrt{\dfrac{5}{16}}$ **b.** $\sqrt{\dfrac{20}{x^2}}$

Homework 9.2

Simplify. Assume that any variables are positive. Use a graphing calculator table to verify your work when possible.

1. $\sqrt{\dfrac{25}{36}}$ **2.** $\sqrt{\dfrac{9}{49}}$ **3.** $\sqrt{\dfrac{121}{x^2}}$ **4.** $\sqrt{\dfrac{196}{x^2}}$

5. $\sqrt{\dfrac{7x}{25}}$ **6.** $\sqrt{\dfrac{3x}{16}}$ **7.** $\sqrt{\dfrac{19}{a^2}}$ **8.** $\sqrt{\dfrac{23}{t^2}}$

9. $\sqrt{\dfrac{5}{x^2 y^2}}$ **10.** $\sqrt{\dfrac{3}{a^2 b^2}}$ **11.** $-\sqrt{\dfrac{8}{49}}$ **12.** $-\sqrt{\dfrac{27}{16}}$

13. $\sqrt{\dfrac{20}{81}}$ **14.** $\sqrt{\dfrac{45}{4}}$ **15.** $\sqrt{\dfrac{75w}{36}}$ **16.** $\sqrt{\dfrac{32y}{9}}$

17. $\sqrt{\dfrac{4a}{b^2}}$ **18.** $\sqrt{\dfrac{25x}{y^2}}$ **19.** $\sqrt{\dfrac{80}{x^2}}$ **20.** $\sqrt{\dfrac{200}{x^2}}$

21. $\sqrt{\dfrac{7r^2 t}{81}}$ **22.** $\sqrt{\dfrac{5ab^2}{49}}$

Simplify. Assume that any variables are positive. Use a graphing calculator table to verify your work when possible.

23. $\dfrac{2}{\sqrt{3}}$ **24.** $\dfrac{5}{\sqrt{2}}$ **25.** $\dfrac{a}{\sqrt{13}}$ **26.** $\dfrac{t}{\sqrt{17}}$

27. $\dfrac{7}{\sqrt{x}}$ **28.** $\dfrac{9}{\sqrt{w}}$ **29.** $\dfrac{15}{\sqrt{27}}$ **30.** $\dfrac{14}{\sqrt{40}}$

31. $\sqrt{\dfrac{2}{7}}$ **32.** $\sqrt{\dfrac{3}{5}}$ **33.** $\sqrt{\dfrac{11}{2}}$ **34.** $\sqrt{\dfrac{5}{6}}$

35. $\sqrt{\dfrac{7}{p}}$ **36.** $\sqrt{\dfrac{2}{b}}$ **37.** $\sqrt{\dfrac{3}{8}}$ **38.** $\sqrt{\dfrac{5}{18}}$

39. $\sqrt{\dfrac{3x}{50}}$ **40.** $\sqrt{\dfrac{7x}{32}}$ **41.** $\sqrt{\dfrac{5w^2}{12}}$ **42.** $\sqrt{\dfrac{3c^2}{28}}$

43. $\sqrt{\dfrac{7x^2 y}{5}}$ **44.** $\sqrt{\dfrac{3xy^2}{7}}$

Simplify. Use a calculator to verify your work.

45. $\dfrac{9 + 3\sqrt{2}}{6}$ **46.** $\dfrac{10 + 6\sqrt{5}}{8}$ **47.** $\dfrac{8 - 4\sqrt{7}}{4}$

48. $\dfrac{15 - 5\sqrt{2}}{5}$ **49.** $\dfrac{8 + 12\sqrt{13}}{6}$ **50.** $\dfrac{21 + 6\sqrt{17}}{9}$

51. $\dfrac{4 + \sqrt{12}}{8}$ **52.** $\dfrac{6 + \sqrt{20}}{4}$ **53.** $\dfrac{10 - \sqrt{50}}{20}$

54. $\dfrac{6 - \sqrt{27}}{12}$ **55.** $\dfrac{9 - \sqrt{45}}{6}$ **56.** $\dfrac{25 - \sqrt{75}}{10}$

57. Let T be the amount of time (in seconds) it takes a baseball to fall to the ground when the ball is dropped from h feet above the ground. A good model is $T = \sqrt{\dfrac{h}{16}}$.

    a. Simplify the right-hand side of the equation.
    b. How long would it take for a baseball to reach the ground if it were dropped from the top of Chicago's Willis Tower, which is 1450 feet tall?

58. The distance d (in miles) to the horizon at an altitude of h feet above sea level is given by the equation $d = \sqrt{\dfrac{3h}{2}}$.

    a. Simplify the right-hand side of the equation.
    b. New York City's Empire State Building is 1250 feet tall. What would be the distance from the top of the building to the horizon? Assume that the base of the building is at sea level.
    c. If an airplane flies at an altitude of 34,000 feet over the skyscraper, what is the distance from the airplane to the horizon?

Concepts

59. A student tries to simplify $\dfrac{5}{\sqrt{3}}$:

$$\frac{5}{\sqrt{3}} = \left(\frac{5}{\sqrt{3}}\right)^2$$
$$= \frac{5^2}{(\sqrt{3})^2}$$
$$= \frac{25}{3}$$

Describe any errors. Then simplify the expression correctly.

60. A student thinks that $\dfrac{\sqrt{14}}{7} = 2$. Is the student correct? Explain.

61. A student tries to simplify $\dfrac{3}{\sqrt{20}}$:

$$\frac{3}{\sqrt{20}} = \frac{3}{\sqrt{20}} \cdot \frac{\sqrt{20}}{\sqrt{20}}$$
$$= \frac{3\sqrt{20}}{20}$$
$$= \frac{3\sqrt{4 \cdot 5}}{20}$$
$$= \frac{3(2)\sqrt{5}}{2(10)}$$
$$= \frac{3\sqrt{5}}{10}$$

Is the work correct? Is there an easier approach? Explain.

62. Choose a nonnegative number for a and a positive number for b, and show $\sqrt{\dfrac{a}{b}} = \dfrac{\sqrt{a}}{\sqrt{b}}$.

63. To simplify $\dfrac{1}{\sqrt{7}}$, why don't we square the fraction?

64. Describe how to simplify a radical expression.

Related Review

Factor.

65. $12 + 8x$ **66.** $10 - 35x$

67. $12 + 8\sqrt{3}$ **68.** $10 - 35\sqrt{2}$

69. $14 - 21x$ **70.** $18 + 27x$

71. $14 - 21\sqrt{7}$ **72.** $18 + 27\sqrt{5}$

Expressions, Equations, and Graphs

Perform the indicated instruction. Then use words such as linear, quadratic, cubic, power, radical, polynomial, degree, one variable, *and* two variables *to describe the expression, equation, or system.*

73. Solve $x^2 - 4x = 0$.

74. Find the difference $(2x + 4) - (5x - 3)$.

75. Factor $x^2 - 4x$.

76. Find the product $(2x + 4)(5x - 3)$.

77. Graph $y = x^2 - 4x$ by hand.

78. Solve $(2x + 4)(5x - 3) = 0$.

9.3 Solving Quadratic Equations by the Square Root Property; The Pythagorean Theorem

Objectives

» Use the *square root property* to solve a quadratic equation of the form $x^2 = k$.

» Use the square root property to solve a quadratic equation of the form $(x + p)^2 = k$.

» Describe the *Pythagorean theorem*.

» Use the Pythagorean theorem to find the length of a side of a right triangle.

» Use the Pythagorean theorem to model a situation.

In Section 8.5, we discussed how to use factoring to help solve some quadratic equations in one variable. In this section, we will discuss another way to solve some quadratic equations.

Solving Quadratic Equations of the Form $x^2 = k$

Consider the equation $x^2 = 9$. We first use factoring to solve this equation:

$$x^2 = 9 \qquad \textit{Original equation}$$
$$x^2 - 9 = 0 \qquad \textit{Subtract 9 from both sides.}$$
$$(x + 3)(x - 3) = 0 \qquad \textit{Factor left side.}$$
$$x + 3 = 0 \quad \text{or} \quad x - 3 = 0 \qquad \textit{Zero factor property}$$
$$x = -3 \quad \text{or} \quad x = 3$$

So, the solutions are -3 and 3. We can use the notation ± 3 to stand for the numbers -3 and 3.

It is not necessary to use factoring to solve the equation $x^2 = 9$, however. Notice that a solution of this equation is any number that, when squared, is equal to 9. So, the solutions are the square roots of 9; in symbols, we write $x = \pm \sqrt{9} = \pm 3$.

Here we solve $x^2 = 16$:

$$x^2 = 16$$
$$x = \pm \sqrt{16}$$
$$x = \pm 4$$

This work suggests the *square root property*.

Square Root Property

Let k be a nonnegative constant. Then $x^2 = k$ is equivalent to

$$x = \pm \sqrt{k}$$

Example 1   Using the Square Root Property to Solve Equations

Solve.

1. $x^2 = 25$ **2.** $x^2 = 3$ **3.** $x^2 = -9$

Solution

1. $x^2 = 25$ *Original equation*
   $ x = \pm \sqrt{25}$ *Square root property*
   $ x = \pm 5$ *Simplify.*

2. $x^2 = 3$ *Original equation*
   $ x = \pm \sqrt{3}$ *Square root property*

3. Because the square of a number is nonnegative, we conclude that $x^2 = -9$ has no real-number solutions.

WARNING   It is a common error to confuse solving an equation such as $x^2 = 25$ with computing a principal square root such as $\sqrt{25}$. In Problem 1 of Example 1, we found that the solutions of $x^2 = 25$ are the *two* numbers -5 and 5. Recall from Section 9.1 that $\sqrt{25}$ is the *one* number 5.

Example 2   Using the Square Root Property to Solve Equations

Solve.

1. $x^2 - 24 = 0$ **2.** $3x^2 - 4 = 3$

Solution

1. First, we isolate x^2 to (get x^2 alone on) one side of the equation:

$$x^2 - 24 = 0 \qquad \textit{Original equation}$$
$$x^2 = 24 \qquad \textit{Add 24 to both sides.}$$
$$x = \pm\sqrt{24} \qquad \textit{Square root property}$$
$$x = \pm 2\sqrt{6} \qquad \textit{Simplify.}$$

2. First, we isolate x^2 to one side of the equation:

$$3x^2 - 4 = 3 \qquad \textit{Original equation}$$
$$3x^2 = 7 \qquad \textit{Add 4 to both sides.}$$
$$x^2 = \frac{7}{3} \qquad \textit{Divide both sides by 3.}$$
$$x = \pm\sqrt{\frac{7}{3}} \qquad \textit{Square root property}$$
$$x = \pm\frac{\sqrt{7}}{3} \qquad \textit{Quotient property: } \sqrt{\frac{a}{b}} = \frac{\sqrt{a}}{\sqrt{b}}$$
$$x = \pm\frac{\sqrt{7}}{\sqrt{3}} \cdot \frac{\sqrt{3}}{\sqrt{3}} \qquad \textit{Rationalize denominator.}$$
$$x = \pm\frac{\sqrt{21}}{3} \qquad \textit{Simplify.}$$

To verify our work, we use a graphing calculator to graph $y = 3x^2 - 4$ and $y = 3$ in the same coordinate system and then use "intersect" to find the approximate intersection points $(-1.53, 3)$ and $(1.53, 3)$, which have respective x-coordinates -1.53 and 1.53 (see Fig. 7). So, the approximate solutions of the original equation are -1.53 and 1.53, which checks.

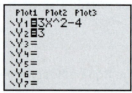 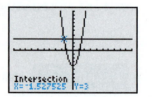 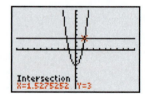

Figure 7 Verify the work

From Example 2, we see that **to solve a quadratic equation of the form $ax^2 + c = k$, we isolate x^2 to one side of the equation before using the square root property.**

Solving Quadratic Equations of the Form $(x + p)^2 = k$

We can also use the square root property to solve equations of the form $(x + p)^2 = k$.

Example 3   Using the Square Root Property to Solve Equations

Solve.

1. $(x + 4)^2 = 36$ **2.** $(x - 7)^2 = 50$

Solution

1. Note that the base of $(x + 4)^2$ is $x + 4$. We can still use the square root property to solve the equation $(x + 4)^2 = 36$.

$$
\begin{aligned}
(x + 4)^2 &= 36 && \textit{Original equation} \\
x + 4 &= \pm\sqrt{36} && \textit{Square root property} \\
x + 4 &= \pm 6 && \textit{Simplify.} \\
x + 4 = -6 \quad &\text{or} \quad x + 4 = 6 && \textit{Write as two equations.} \\
x = -10 \quad &\text{or} \qquad\quad x = 2
\end{aligned}
$$

2.
$$
\begin{aligned}
(x - 7)^2 &= 50 && \textit{Original equation} \\
x - 7 &= \pm\sqrt{50} && \textit{Square root property} \\
x - 7 &= \pm 5\sqrt{2} && \textit{Simplify.} \\
x &= 7 \pm 5\sqrt{2} && \textit{Add 7 to both sides.}
\end{aligned}
$$

So, the solutions are $7 - 5\sqrt{2}$ and $7 + 5\sqrt{2}$. We use graphing calculator tables to check that if $7 - 5\sqrt{2}$ or $7 + 5\sqrt{2}$ is substituted for x in the polynomial $(x - 7)^2$, the result is 50 (see Fig. 8).

 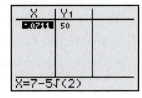 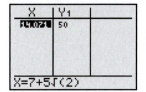

Figure 8 Verify the work

The Pythagorean Theorem

Now that we know the square root property, we can use it to help us work with a special type of triangle called a *right triangle*. An angle of 90° is called a *right angle*. If one angle of a triangle measures 90°, the triangle is a **right triangle** (see Fig. 9).

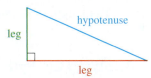

Figure 9 A right triangle

The side opposite the right angle is the longest side. We call that side the **hypotenuse**, and we call the two shorter sides the **legs**. The **Pythagorean theorem** describes the relationship between the lengths of the legs and the hypotenuse of a right triangle.

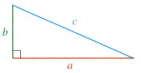

Figure 10 The Pythagorean theorem: $a^2 + b^2 = c^2$

> **Pythagorean Theorem**
>
> If a and b are the lengths of the legs of a right triangle and c is the length of the hypotenuse, then
>
> $$a^2 + b^2 = c^2$$
>
> In words: The sum of the squares of the lengths of the legs is equal to the square of the length of the hypotenuse (see Fig. 10).

Finding the Length of a Side of a Right Triangle

If we know the lengths of two of the three sides of a right triangle, how can we use the Pythagorean theorem to find the length of the third side?

Example 4   Finding the Length of a Side of a Right Triangle

The lengths of two sides of a right triangle are given. Find the length of the third side.

1.

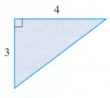

2.

Solution

1. Because the lengths of the legs are given, we are to find the length of the hypotenuse. We substitute $a = 3$ and $b = 4$ into the equation $c^2 = a^2 + b^2$ and solve for c:

$$c^2 = a^2 + b^2 \quad \textit{Pythagorean theorem}$$
$$c^2 = 3^2 + 4^2 \quad \textit{Substitute 3 for a and 4 for b.}$$
$$c^2 = 9 + 16 \quad \textit{Find the squares.}$$
$$c^2 = 25 \quad \textit{Add.}$$
$$c = \sqrt{25} \quad \textit{c is nonnegative, so disregard } -\sqrt{25}.$$
$$c = 5 \quad \textit{Simplify.}$$

The length of the hypotenuse is 5 units.

2. The length of the hypotenuse is 8, and the length of one of the legs is 4. We substitute $a = 4$ and $c = 8$ into the equation $a^2 + b^2 = c^2$ and solve for b:

$$a^2 + b^2 = c^2 \quad \textit{Pythagorean theorem}$$
$$4^2 + b^2 = 8^2 \quad \textit{Substitute 4 for a and 8 for c.}$$
$$16 + b^2 = 64 \quad \textit{Find the squares.}$$
$$b^2 = 48 \quad \textit{Isolate } b^2\text{: Subtract 16 from both sides.}$$
$$b = \sqrt{48} \quad \textit{b is nonnegative, so disregard } -\sqrt{48}.$$
$$b = 4\sqrt{3} \quad \textit{Simplify.}$$

The length of the second leg is $4\sqrt{3}$ units (about 6.93 units).

Applying the Pythagorean Theorem to Model a Situation

The Pythagorean theorem can be used to model many situations. It is used by surveyors, architects, pilots, navigators, engineers, and many other professionals. Several applications are included in the Homework exercises and in Example 5.

Example 5   Modeling a Situation

A person is flying a kite. If the person has let out 102.9 feet of string and the horizontal distance from the person to the kite is 61.5 feet, find the vertical distance between the bottom of the kite and the ground.

Solution

First, we define h to be the vertical distance (in feet) between the bottom of the kite and the ground. Then, we draw a sketch that describes the situation (see Fig. 11).

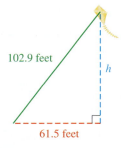

Figure 11 Flying a kite

The triangle in Fig. 11 is a right triangle in which the hypotenuse has length 102.9 feet and one of its legs has length 61.5 feet. We use the Pythagorean theorem to find h.

$$61.5^2 + h^2 = 102.9^2 \qquad \textit{Pythagorean theorem}$$
$$3782.25 + h^2 = 10{,}588.41 \qquad \textit{Find the squares.}$$
$$h^2 = 6806.16 \qquad \textit{Isolate } h^2\text{: Subtract } 3782.25 \textit{ from both sides.}$$
$$h = \sqrt{6806.16} \qquad \textit{h is nonnegative, so disregard } -\sqrt{6806.16}.$$
$$h \approx 82.50 \qquad \textit{Find approximate square root.}$$

So, the bottom of the kite is approximately 82.5 feet above the ground.

 Group Exploration ────────────────────────────────────

Pythagorean theorem and its converse

For this exploration, you will need a ruler, scissors, and paper to help you work with right triangles and other triangles. It will be helpful to have a tool for drawing right angles, such as a protractor or graph paper, although a corner of a piece of paper will suffice. For each triangle, assume c is the length of the *longest* side (in some triangles, more than one side can be equally the longest) and a and b are the lengths of the other sides.

1. Sketch three right triangles of different sizes. Measure the sides. Show that, for each right triangle, $a^2 + b^2 = c^2$.

2. Sketch three triangles of different sizes that are *not* right triangles. For these triangles, check whether $a^2 + b^2 = c^2$.

3. Sketch a triangle that has an angle close, but not equal, to 90°. Check whether $a^2 + b^2 \approx c^2$. If you cannot

show this for your triangle, repeat the problem with a triangle that has an angle even closer to 90°.

4. Note that if $a = 3$, $b = 5$, and $c = \sqrt{34}$, then $a^2 + b^2 = c^2$. Cut three thin strips of paper that are about $3, 5,$ and $\sqrt{34}$ inches in length. Form a triangle with the three strips of paper. Is the triangle a right triangle?

5. Find values of a, b, and c, other than the ones in Problem 4, such that $a^2 + b^2 = c^2$. Then repeat Problem 4, using your values.

6. Find three more values of a, b, and c, other than those you used in Problems 4 and 5, such that $a^2 + b^2 = c^2$. Then repeat Problem 4, using your values.

7. Summarize at least three concepts addressed in this exploration.

Tips for Success   Plan for Final Exams

Don't wait until the last minute to begin studying for your final exams. Look at your finals schedule, and decide how you will allocate your time to prepare for each final.

It is important to be well rested during your finals so you can concentrate fully during your exams. Plan to do some fun activities that involve exercise, as that is a great way to neutralize stress.

Homework 9.3

For extra help ▶   **MyLab Math**   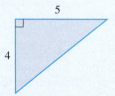 Watch the videos in MyLab Math

Solve. Use a graphing calculator to verify your work.

1. $x^2 = 4$ **2.** $x^2 = 16$ **3.** $x^2 = 196$

4. $x^2 = 169$ **5.** $x^2 = 0$ **6.** $x^2 = 1$

7. $w^2 = 15$ **8.** $r^2 = 17$ **9.** $x^2 = 20$

10. $x^2 = 18$ **11.** $x^2 = 27$ **12.** $x^2 = 50$

13. $x^2 = -49$ **14.** $x^2 = -25$ **15.** $b^2 - 28 = 0$

16. $c^2 - 40 = 0$ **17.** $x^2 + 17 = 0$ **18.** $x^2 + 8 = 0$

19. $4x^2 = 5$ **20.** $9x^2 = 11$ **21.** $5x^2 = 7$

22. $7x^2 = 3$ **23.** $8m^2 = 5$ **24.** $12y^2 = 11$

25. $2x^2 + 4 = 7$ **26.** $3x^2 + 1 = 8$ **27.** $5x^2 - 3 = 11$

28. $7x^2 - 8 = -3$

Solve. Use a graphing calculator to verify your work.

29. $(x + 2)^2 = 16$ **30.** $(x + 4)^2 = 25$ **31.** $(p - 3)^2 = 36$

32. $(t - 6)^2 = 49$ **33.** $(x - 7)^2 = 13$ **34.** $(x - 4)^2 = 35$

35. $(x + 3)^2 = -16$ **36.** $(x - 1)^2 = -9$ **37.** $(x + 2)^2 = 18$

38. $(x + 7)^2 = 45$ **39.** $(r - 6)^2 = 24$ **40.** $(p - 1)^2 = 75$

41. $(x - 5)^2 = 0$ **42.** $(x + 9)^2 = 0$

Use factoring or the square root property to solve the equation. Use a graphing calculator to verify your work.

43. $x^2 = 4x + 12$ **44.** $x^2 = -12x - 20$

45. $y^2 - 81 = 0$ **46.** $m^2 - 49 = 0$

47. $(x - 2)^2 = 24$ **48.** $(x + 8)^2 = 32$

49. $3x^2 + 4 = 15$ **50.** $25x^2 - 1 = 16$

51. $2x^2 - 15 = -7x$ **52.** $3x^2 + 2 = 5x$

53. $(t - 1)^2 = 1$ **54.** $(y - 7)^2 = 36$

The lengths of two sides of a right triangle are given. Find the length of the third side. (The triangle is not drawn to scale.)

55. See Fig. 12.   **56.** See Fig. 13.

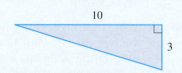

Figure 12 Exercise 55   **Figure 13 Exercise 56**

57. See Fig. 14.   **58.** See Fig. 15.

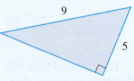

Figure 14 Exercise 57   **Figure 15 Exercise 58**

59. See Fig. 16.   **60.** See Fig. 17.

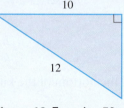

Figure 16 Exercise 59   **Figure 17 Exercise 60**

Let a and b be the lengths of the legs of a right triangle, and let c be the length of the hypotenuse. Values for two of the three lengths are given. Find the third length.

61. $a = 6$ and $b = 8$ **62.** $a = 5$ and $b = 12$

63. $a = 2$ and $b = 3$ **64.** $a = 2$ and $b = 7$

65. $a = 4$ and $b = 8$ **66.** $a = 3$ and $b = 6$

67. $a = 6$ and $c = 10$ **68.** $a = 12$ and $c = 13$

69. $a = 2$ and $c = 5$ **70.** $a = 8$ and $c = 11$

71. $b = 2$ and $c = 7$ **72.** $b = 6$ and $c = 8$

For Exercises 73–86, round approximate results to the first decimal place.

73. A student drives 8 miles west from home and then 17 miles north to get to school. What would be the length of the trip if it were possible to drive along a straight line from home to school?

74. A commuter drives 5 miles south from home and then 9 miles east to get to work. What would be the length of the trip if it were possible to drive along a straight line from home to work?

75. A 12-foot-long ladder is leaning against a building. The bottom of the ladder is 4 feet from the base of the building. Will the ladder reach the bottom of a window that is 11 feet above the ground?

76. A 20-foot-long ladder is leaning against a building. The bottom of the ladder is 5 feet from the base of the building. Will the ladder reach the bottom of a window that is 19 feet above the ground?

77. The size of a rectangular television is usually described in terms of the length of its diagonal. If a 62-inch television screen has a height of 30 inches, what is the width of the screen?

78. The size of a rectangular television is usually described in terms of the length of its diagonal. If a 26-inch television has a height of 15 inches, what is the width of the screen?

79. A rectangular painting has a width of 24 inches. The length of the diagonal is 37 inches. Find the length of the painting.

80. A rectangular painting has a length of 42 inches. The length of the diagonal is 53 inches. Find the width of the painting.

81. A surveyor wants to estimate the distance (in miles) across a lake from point A to point B, as shown in Fig. 18. Find that distance.

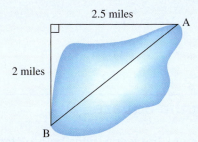

Figure 18 Exercise 81

82. A surveyor wants to estimate the distance (in miles) across a lake from point A to point B, as shown in Fig. 19. Find that distance.

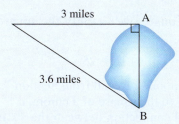

Figure 19 Exercise 82

83. Not including the end zones, an official football field is rectangular with a length of 100 yards and a width of 53 yards, 1 foot. If a player runs diagonally across this rectangle, how far will he run? [**Hint:** Use units of feet to perform your calculations.]

84. A soccer field is rectangular. For international matches, the length must be between 110 and 120 yards, inclusive, and the width must be between 70 and 80 yards, inclusive.
   a. Find the length of the diagonal for the largest permitted field for international matches.
   b. Find the length of the diagonal for the smallest permitted field for international matches.

85. Sioux Falls, South Dakota, is about 266 miles almost directly south of Fargo, North Dakota, and is almost directly west of Madison, Wisconsin. The distance between Fargo and Madison is about 514 miles. What is the approximate distance between Sioux Falls and Madison?

86. Jackson, Mississippi, is about 406 miles almost directly east of Dallas, Texas, and is almost directly south of Memphis, Tennessee. The distance between Dallas and Memphis is about 460 miles. What is the approximate distance between Jackson and Memphis?

Concepts

87. **a.** Use factoring to solve the equation $x^2 - 36 = 0$.
   b. Use the square root property to solve the equation $x^2 - 36 = 0$.
   c. Compare the results you found in parts (a) and (b).
   d. In your opinion, is it easier to solve $x^2 - 36 = 0$ by factoring or by using the square root property? Explain.

88. **a.** Use factoring to solve the equation $(x - 3)^2 = 25$.
   b. Use the square root property to solve the equation $(x - 3)^2 = 25$.
   c. Compare the results you found in parts (a) and (b).
   d. In your opinion, is it easier to solve $(x - 3)^2 = 25$ by factoring or by using the square root property? Explain.

89. **a.** Can the equation $(x - 2)^2 = 7$ be solved by means of the square root property? If so, solve the equation by this method.
   b. Can the equation $(x - 2)^2 = 7$ be solved by factoring? If so, solve the equation by this method.
   c. Can all equations that can be solved by means of the square root property be solved by factoring? Explain.

90. A student tries to solve the equation $x^2 - 2x - 24 = 0$:

$$x^2 - 2x - 24 = 0$$
$$x^2 = 2x + 24$$
$$x = \pm\sqrt{2x + 24}$$

Describe any errors. Then solve the equation correctly.

91. Why does $x^2 = 9$ have two solutions, even though $\sqrt{9}$ is equal to only one of those solutions?

92. For the Pythagorean theorem $a^2 + b^2 = c^2$, does it matter which leg of the triangle is a? Explain.

93. Explain how to solve a quadratic equation by using the square root property.

94. Describe the Pythagorean theorem in your own words. Give an example of how to use it.

Related Review

95. **a.** Solve $x^2 = 49$.
   b. Find $\sqrt{49}$.
   c. Explain why the results you found in parts (a) and (b) are different.

96. A student says 3 is a solution of the equation $-x^2 = 9$ because $-3^2 = 9$. Is the student correct? Explain.

97. A rectangle has an area of 16 square inches. Let W be the width, L be the length, D be the length of the diagonal (all in inches), and A be the area (in inches squared).
   a. Sketch three possible rectangles, including one that is a square.
   b. For each of your three rectangles, find the approximate length D of the diagonal. Round each result to the second decimal place.
   c. Which of your three rectangles has the smallest diagonal length D?
   d. Which of the symbols W, L, D, and A are variables? Explain.
   e. Which of the symbols W, L, D, and A are constants? Explain.

98. A rectangle has a perimeter of 20 inches. Let W be the width, L be the length, D be the length of the diagonal, and P be the perimeter, all with units in feet.

   a. Sketch three possible rectangles, including one that is a square.

   b. For each of your three rectangles, find the approximate length D of the diagonal. Round each result to the second decimal place.

   c. Which of your three rectangles has the smallest diagonal length D?

   d. Which of the symbols W, L, D, and P are variables? Explain.

   e. Which of the symbols W, L, D, and P are constants? Explain.

Expressions, Equations, and Graphs

Perform the indicated instruction. Then use words such as linear, quadratic, cubic, power, radical, polynomial, degree, one variable, *and* two variables *to describe the expression, equation, or system.*

  99. Find the product $(3x^2 - 2)(4x^2 + x - 5)$.

100. Solve $5 - 2(3w - 7) = 4 + 2w$.

101. Factor $3x^4 - 12x^3 - 63x^2$.

102. Solve $4x^2 - 13 = 0$.

103. Graph $3x + 5y = 15$ by hand.

104. Simplify $\sqrt{\dfrac{5}{x}}$.

9.4 Solving Quadratic Equations by Completing the Square

Objectives

» Describe the relationship between b and c for a perfect-square trinomial of the form $x^2 + bx + c$.

» Solve quadratic equations of the form $x^2 + bx + c = 0$ by completing the square.

» Solve quadratic equations of the form $ax^2 + bx + c = 0$ by completing the square.

» Explain why any quadratic equation can be solved by completing the square.

So far, we have discussed how to solve quadratic equations both by factoring and by using the square root property. In this section, we will discuss yet another way, called *completing the square*.

Perfect-Square Trinomials

To begin our study of completing the square, we simplify the square of a sum. For example,

$$(x + 5)^2 = x^2 + 10x + 25$$

So, $x^2 + 10x + 25$ is a perfect-square trinomial. Recall from Section 7.5 that a *perfect-square trinomial* is a trinomial equivalent to the square of a binomial.

For $x^2 + 10x + 25$, there is a special connection between the 10 and the 25. If we divide the 10 by 2 and then square the result, we get 25:

$$x^2 + 10x + 25$$

$$\left(\frac{10}{2}\right)^2 = 5^2 = 25$$

This is no coincidence. Consider simplifying $(x - 3)^2$: $(x - 3)^2 = x^2 - 6x + 9$. If we divide -6 by 2 and then square the result, we get 9:

$$x^2 - 6x + 9$$

$$\left(\frac{-6}{2}\right)^2 = (-3)^2 = 9$$

For the general case, we simplify $(x + k)^2$, where k is a constant:

$$(x + k)^2 = x^2 + (2k)x + k^2$$

$$\left(\frac{2k}{2}\right)^2 = k^2$$

For $x^2 + (2k)x + k^2$, we see that if we divide the coefficient of x by 2 and then square the result, we get the constant term k^2.

Perfect-Square Trinomial Property

For a perfect-square trinomial of the form $x^2 + bx + c$, dividing b by 2 and then squaring the result gives c:

$$x^2 + bx + c$$

$$\left(\frac{b}{2}\right)^2 = c$$

Note that this property is for a perfect-square trinomial $ax^2 + bx + c$, where $a = 1$.

Example 1   Factoring Perfect-Square Trinomials

Find the value of c such that the polynomial is a perfect-square trinomial. Then factor the perfect-square trinomial.

1. $x^2 + 12x + c$ **2.** $x^2 - 18x + c$

Solution

1. We divide $b = 12$ by 2 and then square the result:

$$\left(\frac{12}{2}\right)^2 = 6^2 = 36 = c$$

The polynomial is $x^2 + 12x + 36$, with factored form $(x + k)^2$, for some positive integer k. So, we have

$$x^2 + 12x + 36 = (x + k)^2 = x^2 + 2kx + k^2$$

The constant terms
36 and k^2 are equal.

Here $k^2 = 36$, or $k = 6$ (k is positive). So, the factored form of $x^2 + 12x + 36$ is $(x + 6)^2$.

2. We divide $b = -18$ by 2 and then square the result:

$$\left(\frac{-18}{2}\right)^2 = (-9)^2 = 81 = c$$

The polynomial is $x^2 - 18x + 81$, with factored form $(x - k)^2$, for some positive integer k. So, we have

$$x^2 - 18x + 81 = (x - k)^2 = x^2 - 2kx + k^2$$

The constant terms
81 and k^2 are equal.

Here $k^2 = 81$, or $k = 9$ (k is positive). So, the factored form of $x^2 - 18x + 81$ is $(x - 9)^2$.

▶

Solving $x^2 + bx + c = 0$ by Completing the Square

In Example 2, we will begin to solve a quadratic equation by forming a perfect-square trinomial on one side of the equation.

Example 2   Solving by Completing the Square

Solve $x^2 + 8x = 3$.

Solution

Because $\left(\dfrac{8}{2}\right)^2 = 4^2 = 16$, we add 16 to both sides of $x^2 + 8x = 3$ so the left side will be a perfect-square trinomial:

$$
\begin{aligned}
x^2 + 8x &= 3 && \text{\textit{Original equation}} \\
x^2 + 8x + 16 &= 3 + 16 && \text{\textit{Add 16 to both sides.}} \\
(x + 4)^2 &= 19 && \text{\textit{Factor the left side.}} \\
x + 4 &= \pm\sqrt{19} && \text{\textit{Square root property}} \\
x &= -4 \pm \sqrt{19} && \text{\textit{Subtract 4 from both sides.}}
\end{aligned}
$$

We use graphing calculator tables to check that if $-4 - \sqrt{19}$ or $-4 + \sqrt{19}$ is substituted for x in the polynomial $x^2 + 8x$, the result is 3 (see Fig. 20).

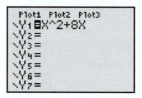 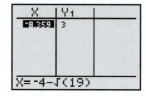 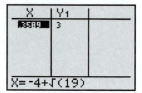

Figure 20 Verify the work

In Example 2, we added 16 to both sides of $x^2 + 8x = 3$ and then factored the left side to get $(x + 4)^2$. By adding 16 to both sides of the equation, we say we *completed the square* for $x^2 + 8x$.

WARNING   In solving a quadratic equation such as $x^2 + 8x = 3$ by completing the square, it is a common error to add a number (in this case 16) to only the left side of the equation. Remember to add the number to *both* sides of the equation.

Example 3   Solving by Completing the Square

Solve $p^2 - 14p + 4 = 0$.

Solution

We're trying to get a perfect-square trinomial on the left side of the equation $p^2 - 14p + 4 = 0$, so we subtract 4 from both sides:

$$p^2 - 14p = -4$$

Because $\left(\dfrac{-14}{2}\right)^2 = (-7)^2 = 49$, we add 49 to both sides of $p^2 - 14p = -4$ to complete the square for $p^2 - 14p$:

$$
\begin{aligned}
p^2 - 14p + 49 &= -4 + 49 && \text{\textit{Add 49 to both sides.}} \\
(p - 7)^2 &= 45 && \text{\textit{Factor left side.}} \\
p - 7 &= \pm\sqrt{45} && \text{\textit{Square root property}} \\
p - 7 &= \pm 3\sqrt{5} && \text{\textit{Simplify.}} \\
p &= 7 \pm 3\sqrt{5} && \text{\textit{Add 7 to both sides.}}
\end{aligned}
$$

Recall from Section 8.5 that some quadratic equations do not have any real-number solutions. We will see such an equation in Example 4.

Example 4   A Quadratic Equation with No Real-Number Solutions

Solve $x^2 + 2x = -7$.

Solution

Because $\left(\dfrac{2}{2}\right)^2 = 1^2 = 1$, we add 1 to both sides of $x^2 + 2x = -7$ to complete the square for $x^2 + 2x$:

$$
\begin{aligned}
x^2 + 2x &= -7 && \text{\textit{Original equation}}\\
x^2 + 2x + 1 &= -7 + 1 && \text{\textit{Add 1 to both sides.}}\\
(x + 1)^2 &= -6 && \text{\textit{Factor left side.}}
\end{aligned}
$$

Because the square of a number is nonnegative, we conclude that $x^2 + 2x = -7$ has no real-number solutions.

Solving $ax^2 + bx + c = 0$ by Completing the Square

So far, we have worked with trinomials only of the form $ax^2 + bx + c$, where $a = 1$. Consider simplifying $(2x + 3)^2$:

$$(2x + 3)^2 = 4x^2 + 12x + 9$$

Dividing $b = 12$ by 2 and then squaring the result does *not* give 9:

$$\left(\frac{12}{2}\right)^2 = 6^2 = 36 \neq 9$$

So, the perfect-square trinomial property given for $ax^2 + bx + c$, where $a = 1$, does not extend to trinomials with $a \neq 1$. However, **when solving a quadratic equation of the form $ax^2 + bx + c = 0$ with $a \neq 1$, we can first divide both sides by a to obtain an equation involving $1x^2$—one to which we *can* apply the property.**

Example 5   Solving by Completing the Square

Solve $2x^2 + 12x - 22 = 0$.

Solution

$$
\begin{aligned}
2x^2 + 12x - 22 &= 0 && \text{\textit{Original equation}}\\
2x^2 + 12x &= 22 && \text{\textit{Add 22 to both sides.}}\\
x^2 + 6x &= 11 && \text{\textit{Divide both sides by 2 so that coefficient of x^2 will be 1.}}\\
x^2 + 6x + 9 &= 11 + 9 && \left(\tfrac{6}{2}\right)^2 = 3^2 = 9;\ \textit{add 9 to both sides.}\\
(x + 3)^2 &= 20 && \text{\textit{Factor left side.}}\\
x + 3 &= \pm\sqrt{20} && \text{\textit{Square root property}}\\
x + 3 &= \pm 2\sqrt{5} && \text{\textit{Simplify.}}\\
x &= -3 \pm 2\sqrt{5} && \text{\textit{Subtract 3 from both sides.}}
\end{aligned}
$$

Which Equations Can Be Solved by Completing the Square?

Any quadratic equation can be solved by completing the square. Here is a summary of this method:

> **Solving a Quadratic Equation by Completing the Square**
>
> To solve a quadratic equation by **completing the square,**
>
> 1. Write the equation in the form $ax^2 + bx = k$, where $a, b,$ and k are constants.
> 2. If $a \neq 1$, divide both sides of the equation by a.
> 3. Complete the square for the polynomial on the left side of the equation.
> 4. Solve the equation by using the square root property.

WARNING   To solve an equation of the form $ax^2 + bx = k$ with $a \neq 1$, we must divide both sides of the equation by a before completing the square for the left side of the equation.

Group Exploration

Identifying errors in solving by completing the square

1. A student tries to solve the equation $x^2 + 6x - 5 = 0$:

$$x^2 + 6x - 5 = 0$$
$$x^2 + 6x = 5$$
$$x^2 + 6x + 9 = 5$$
$$(x + 3)^2 = 5$$
$$x + 3 = \pm\sqrt{5}$$
$$x = -3 \pm \sqrt{5}$$

Describe any errors. Then solve the equation correctly.

2. A student tries to solve the equation $4x^2 - 8x = 12$:

$$4x^2 - 8x = 12$$
$$4x^2 - 8x + 16 = 12 + 16$$
$$(2x - 4)^2 = 28$$
$$2x - 4 = \pm\sqrt{28}$$
$$2x - 4 = \pm 2\sqrt{7}$$
$$2x = 4 \pm 2\sqrt{7}$$
$$x = 2 \pm \sqrt{7}$$

Describe any errors. Then solve the equation correctly.

Homework 9.4

For extra help ▶   MyLab Math   ▦ Watch the videos in MyLab Math

Find the value of c for which the polynomial is a perfect-square trinomial. Then factor the perfect-square trinomial.

1. $x^2 + 6x + c$

2. $x^2 + 16x + c$

3. $x^2 + 14x + c$

4. $x^2 + 4x + c$

5. $x^2 + 2x + c$

6. $x^2 + 20x + c$

7. $x^2 - 8x + c$

8. $x^2 - 12x + c$

9. $x^2 - 10x + c$

10. $x^2 - 2x + c$

11. $x^2 - 20x + c$

12. $x^2 - 16x + c$

Solve by completing the square. Use a graphing calculator to verify your work.

13. $x^2 + 6x = 5$

14. $x^2 + 8x = 1$

15. $x^2 + 14x = -20$

16. $x^2 + 4x = 13$

17. $x^2 - 10x = -4$

18. $x^2 - 16x = 3$

19. $t^2 - 2t = 5$

20. $p^2 - 12p = -3$

21. $x^2 + 12x = -4$

22. $x^2 + 10x = 3$

23. $x^2 - 4x = 14$

24. $x^2 - 14x = -9$

25. $x^2 - 8x = 8$

26. $x^2 - 2x = 19$

27. $x^2 - 16x = -70$

28. $x^2 - 6x = -10$

29. $y^2 + 20y = -40$

30. $r^2 + 18r = -6$

31. $x^2 + 6x + 1 = 0$

32. $x^2 + 12x - 3 = 0$

33. $x^2 - 4x + 1 = 0$

34. $x^2 - 16x + 9 = 0$

35. $x^2 - 10x - 7 = 0$

36. $x^2 - 14x - 11 = 0$

37. $y^2 + 20y + 120 = 0$

38. $m^2 + 18m + 100 = 0$

39. $x^2 + 8x + 4 = 0$

40. $x^2 + 2x - 17 = 0$

Solve by completing the square. Use a graphing calculator to verify your work.

41. $2x^2 - 16x = 6$

42. $3x^2 - 18x = 6$

43. $5x^2 + 10x = 35$

44. $2x^2 + 8x = 16$

45. $3w^2 - 30w = -21$

46. $5t^2 - 60t = -20$

47. $6x^2 + 12x - 6 = 0$

48. $2x^2 + 20x + 6 = 0$

49. $4x^2 - 24x + 4 = 0$

50. $3x^2 - 6x - 33 = 0$

51. $5x^2 + 20x - 20 = 0$

52. $2x^2 + 16x - 8 = 0$

Solve by the method of your choice. Use a graphing calculator to verify your work.

53. $x^2 - 9 = 0$

54. $x^2 - 16 = 0$

55. $r^2 = 11r - 30$

56. $p^2 = 6p + 27$

57. $(x - 5)^2 = 32$

58. $(x + 3)^2 = 45$

59. $3x^2 + 5x = 12$

60. $2x^2 - 3x = 5$

61. $x^2 = 13$

62. $x^2 = 21$

63. $t^2 - 6t - 3 = 0$

64. $m^2 + 8m - 4 = 0$

Concepts

65. a. Solve the equation $x^2 - 6x + 8 = 0$ by factoring.
 b. Solve the equation $x^2 - 6x + 8 = 0$ by completing the square.
 c. In your opinion, is it easier to solve $x^2 - 6x + 8 = 0$ by factoring or by completing the square? Explain.

66. a. Solve the equation $x^2 + 4x = 12$ by factoring.
 b. Solve the equation $x^2 + 4x = 12$ by completing the square.
 c. In your opinion, is it easier to solve $x^2 + 4x = 12$ by factoring or by completing the square? Explain.

67. a. Can the equation $x^2 + 4x = 7$ be solved by completing the square? If so, solve the equation by using this method.
 b. Can the equation $x^2 + 4x = 7$ be solved by factoring? If so, solve the equation by using this method.
 c. Explain how to decide whether to solve a quadratic equation by completing the square or by factoring.

68. a. Simplify $(x + k)^2$.
 b. Consider the polynomial $x^2 + 2kx + k^2$, where k is a constant. Show that, if you divide the coefficient of x by 2 and square the result, you get the constant term.

69. Compare the methods of solving a quadratic equation by factoring, by using the square root property, and by completing the square. Describe the methods, as well as their advantages and disadvantages. (See page 9 for guidelines on writing a good response.)

70. Describe how to solve a quadratic equation by completing the square. (See page 9 for guidelines on writing a good response.)

71. For an equation of the form $ax^2 + bx = c$, where $a \neq 1$, explain why it is necessary to divide both sides of the equation by a when we solve by completing the square.

72. To solve $x^2 + 6x = 5$ by completing the square, we begin by computing $\left(\dfrac{6}{2}\right)^2 = 3^2 = 9$. Why must we add 9 to *both* sides of the equation $x^2 + 6x = 5$?

Related Review

73. Simplify $(x + 7)^2$.

74. Simplify $(x - 5)^2$.

75. Factor $x^2 - 16x + 64$.

76. Factor $x^2 + 12x + 36$.

Expressions, Equations, and Graphs

Perform the indicated instruction. Then use words such as linear, quadratic, cubic, power, radical, polynomial, degree, one variable, *and* two variables *to describe the expression, equation, or system.*

77. Simplify $(x - 4)^2$.

78. Solve the following:
$$y = -3x + 1$$
$$5x + 2y = 1$$

79. Solve $(x - 4)^2 = 3$.

80. Solve $4(2x - 5) = -3x + 1$.

81. Factor $p^2 - 8pq + 16q^2$.

82. Graph $y = -3x + 1$ by hand.

9.5 Solving Quadratic Equations by the Quadratic Formula

Objectives

» Solve quadratic equations by using the *quadratic formula.*

» Find approximate solutions of a quadratic equation.

» Find approximate x-intercepts of a parabola.

» Determine whether to solve a quadratic equation by factoring, by using the square root property, by completing the square, or by using the quadratic formula.

Any quadratic equation can be solved by completing the square. However, it is difficult to use this method with most quadratic equations. An easier option is to use an important equation called the *quadratic formula*, which can also be used to solve *any* quadratic equation.

The Quadratic Formula

We will find the quadratic formula by solving the general quadratic equation
$$ax^2 + bx + c = 0$$
by completing the square. For now, we assume that a is positive:

$$ax^2 + bx + c = 0 \qquad \textit{General quadratic equation}$$
$$ax^2 + bx = -c \qquad \textit{Subtract c from both sides.}$$
$$x^2 + \frac{b}{a}x = -\frac{c}{a} \qquad \textit{Divide both sides by (nonzero) a so that the coefficient of } x^2 \textit{ will be 1.}$$

To begin completing the square, we divide $\dfrac{b}{a}$ by 2:

$$\frac{b}{a} \div 2 = \frac{b}{a} \cdot \frac{1}{2} = \frac{b}{2a}$$

Then we square the result:

$$\left(\frac{b}{2a}\right)^2 = \frac{b^2}{4a^2}$$

Next, we add $\dfrac{b^2}{4a^2}$ to both sides of the equation $x^2 + \dfrac{b}{a}x = -\dfrac{c}{a}$:

$$x^2 + \frac{b}{a}x + \frac{b^2}{4a^2} = -\frac{c}{a} + \frac{b^2}{4a^2} \qquad \text{Add } \frac{b^2}{4a^2} \text{ to both sides.}$$

$$\left(x + \frac{b}{2a}\right)^2 = -\frac{c}{a} \cdot \frac{4a}{4a} + \frac{b^2}{4a^2} \qquad \text{Factor left-hand side; find a common denominator, } 4a^2, \text{ for right-hand side.}$$

$$\left(x + \frac{b}{2a}\right)^2 = \frac{b^2 - 4ac}{4a^2} \qquad \text{Add fractions.}$$

$$x + \frac{b}{2a} = \pm\sqrt{\frac{b^2 - 4ac}{4a^2}} \qquad \text{Square root property}$$

$$x + \frac{b}{2a} = \pm\frac{\sqrt{b^2 - 4ac}}{\sqrt{4a^2}} \qquad \sqrt{\frac{A}{B}} = \frac{\sqrt{A}}{\sqrt{B}}$$

$$x = -\frac{b}{2a} \pm \frac{\sqrt{b^2 - 4ac}}{2a} \qquad \text{Subtract } \frac{b}{2a} \text{ from both sides;}$$
$$\sqrt{4a^2} = \sqrt{4}\sqrt{a^2} = 2a \text{ for a nonnegativ}$$

$$x = \frac{-b \pm \sqrt{b^2 - 4ac}}{2a} \qquad \text{Add and subtract the fractions.}$$

So, we have found a formula (the last line) for the solutions of a quadratic equation $ax^2 + bx + c = 0$, where a is positive. In a similar way, we could derive the same formula for a quadratic equation in which a is negative.

Quadratic Formula

The solutions of a quadratic equation $ax^2 + bx + c = 0$ are given by the **quadratic formula**:

$$x = \frac{-b \pm \sqrt{b^2 - 4ac}}{2a}$$

WARNING   For the fraction in the quadratic formula, notice that the term $-b$ is part of the numerator:

$$\frac{-b \pm \sqrt{b^2 - 4ac}}{2a} \qquad \text{Correct}$$

$$-b \pm \frac{\sqrt{b^2 - 4ac}}{2a} \qquad \text{Incorrect}$$

Example 1   Solving by Using the Quadratic Formula

Solve $x^2 - 7x + 10 = 0$ by using the quadratic formula.

Solution

Comparing $x^2 - 7x + 10$ with $ax^2 + bx + c$, we see that $a = 1$, $b = -7$, and $c = 10$. We substitute these values for a, b, and c into the quadratic formula:

$$x = \frac{-b \pm \sqrt{b^2 - 4ac}}{2a} \qquad \textit{Quadratic formula}$$

$$x = \frac{-(-7) \pm \sqrt{(-7)^2 - 4(1)(10)}}{2(1)} \qquad \textit{Substitute into quadratic formula.}$$

$$x = \frac{7 \pm \sqrt{49 - 40}}{2} \qquad \textit{Simplify.}$$

$$x = \frac{7 \pm \sqrt{9}}{2} \qquad \textit{Subtract.}$$

$$x = \frac{7 \pm 3}{2} \qquad \sqrt{9} = 3$$

$$x = \frac{7 - 3}{2} \quad \text{or} \quad x = \frac{7 + 3}{2} \qquad \textit{Write as two equations.}$$

$$x = \frac{4}{2} \quad \text{or} \quad x = \frac{10}{2} \qquad \textit{Subtract./Add.}$$

$$x = 2 \quad \text{or} \quad x = 5$$

The solutions are 2 and 5. We can check that the solutions satisfy the original equation. (Try it.)

▶

Note that instead of using the quadratic formula, we can solve $x^2 - 7x + 10 = 0$ by factoring:

$$x^2 - 7x + 10 = 0 \qquad \textit{Original equation}$$
$$(x - 2)(x - 5) = 0 \qquad \textit{Factor left side.}$$
$$x - 2 = 0 \quad \text{or} \quad x - 5 = 0 \qquad \textit{Zero factor property}$$
$$x = 2 \quad \text{or} \qquad x = 5$$

A benefit of the quadratic formula is that we can use it to solve equations that are difficult or impossible to solve by factoring. In Example 2, we will solve an equation that would be impossible to solve by factoring.

Example 2   Solving by Using the Quadratic Formula

Solve $5x^2 - x = 2$.

Solution

First, we write the equation in the form $ax^2 + bx + c = 0$:

$$5x^2 - x = 2 \qquad \textit{Original equation}$$
$$5x^2 - x - 2 = 0 \qquad \textit{Subtract 2 from both sides.}$$

Then we substitute $a = 5$, $b = -1$, and $c = -2$ into the quadratic formula:

$$x = \frac{-b \pm \sqrt{b^2 - 4ac}}{2a} \qquad \textit{Quadratic formula}$$

$$x = \frac{-(-1) \pm \sqrt{(-1)^2 - 4(5)(-2)}}{2(5)} \qquad \textit{Substitute into quadratic formula.}$$

$$x = \frac{1 \pm \sqrt{1 + 40}}{10} \qquad \textit{Simplify.}$$

$$x = \frac{1 \pm \sqrt{41}}{10} \qquad \textit{Add.}$$

We use graphing calculator tables to check that if $\dfrac{1 - \sqrt{41}}{10}$ or $\dfrac{1 + \sqrt{41}}{10}$ is substituted for x in the polynomial $5x^2 - x$, the result is 2 (see Fig. 21).

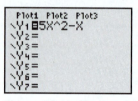 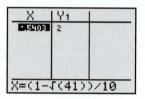 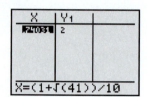

Figure 21 Verify the work

By the check in Fig. 21, we see that $\dfrac{1 - \sqrt{41}}{10}$ is about -0.54. So, $\dfrac{1 - \sqrt{41}}{10}$ is simply a number. Likewise, $\dfrac{1 + \sqrt{41}}{10}$ is a number (about 0.74).

Example 3   Solving by Using the Quadratic Formula

Solve $-2x^2 - 6x + 3 = 0$.

Solution

First, we multiply both sides of the equation by -1 so that we can avoid having a negative denominator after we use the quadratic formula:

$$-2x^2 - 6x + 3 = 0 \qquad \textit{Original equation}$$
$$2x^2 + 6x - 3 = 0 \qquad \textit{Multiply both sides by } -1.$$

(Another benefit is that we will have fewer negative numbers to substitute into the quadratic formula.)

Then we substitute $a = 2$, $b = 6$, and $c = -3$ into the quadratic formula:

$$x = \frac{-b \pm \sqrt{b^2 - 4ac}}{2a} \qquad \textit{Quadratic formula}$$

$$x = \frac{-(6) \pm \sqrt{(6)^2 - 4(2)(-3)}}{2(2)} \qquad \textit{Substitute into quadratic formula.}$$

$$x = \frac{-6 \pm \sqrt{36 + 24}}{4} \qquad \textit{Simplify.}$$

$$x = \frac{-6 \pm \sqrt{60}}{4} \qquad \textit{Add.}$$

$$x = \frac{-6 \pm 2\sqrt{15}}{4} \qquad \textit{$\sqrt{60} = \sqrt{4 \cdot 15} = \sqrt{4}\sqrt{15} = 2\sqrt{15}$}$$

$$x = \frac{2\left(-3 \pm \sqrt{15}\right)}{2(2)} \qquad \textcolor{blue}{\textit{Factor out 2.}}$$

$$x = \frac{2}{2} \cdot \frac{-3 \pm \sqrt{15}}{2} \qquad \textcolor{blue}{\frac{ab}{cd} = \frac{a}{c} \cdot \frac{b}{d}}$$

$$x = \frac{-3 \pm \sqrt{15}}{2} \qquad \textcolor{blue}{\frac{2}{2} = 1; \textit{ simplify.}}$$

Recall from Section 9.1 that the square root of a negative number is not a real number. For instance, $\sqrt{-23}$ is not a real number. We will deal with such a situation in Example 4.

Example 4   Solving by Using the Quadratic Formula

Solve $\frac{3}{2}x^2 = x - \frac{5}{2}$.

Solution

We clear the equation of fractions by multiplying by the LCD, 2, and then write the equation in the form $ax^2 + bx + c = 0$:

$$\frac{3}{2}x^2 = x - \frac{5}{2} \qquad \textcolor{blue}{\textit{Original equation}}$$

$$2 \cdot \frac{3}{2}x^2 = 2 \cdot x - 2 \cdot \frac{5}{2} \qquad \textcolor{blue}{\textit{Multiply both sides by the LCD, 2.}}$$

$$3x^2 = 2x - 5 \qquad \textcolor{blue}{\textit{Simplify.}}$$

$$3x^2 - 2x + 5 = 0 \qquad \textcolor{blue}{\textit{Write in } ax^2 + bx + c = 0 \textit{ form.}}$$

Next, we substitute $a = 3, b = -2$, and $c = 5$ into the quadratic formula:

$$x = \frac{-(-2) \pm \sqrt{(-2)^2 - 4(3)(5)}}{2(3)} \qquad \textcolor{blue}{\textit{Substitute into quadratic formula.}}$$

$$x = \frac{2 \pm \sqrt{4 - 60}}{6} \qquad \textcolor{blue}{\textit{Simplify.}}$$

$$x = \frac{2 \pm \sqrt{-56}}{6} \qquad \textcolor{blue}{\textit{Subtract.}}$$

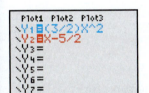

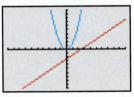

Figure 22 Verify the work

Because the square root of a negative number is not a real number, we conclude that there are no real-number solutions. To verify our work, we use a graphing calculator to graph $y = \frac{3}{2}x^2$ and $y = x - \frac{5}{2}$ (see Fig. 22). Because the curves do not intersect, the original equation has no real-number solutions.

If the result of solving a quadratic equation is a radical expression in which a radicand is negative, such as $\frac{2 \pm \sqrt{-56}}{6}$, then there are no real-number solutions.

Finding Approximate Solutions of a Quadratic Equation

When we use a model to make predictions, it is not necessary to find exact solutions of an equation. To prepare us for working with quadratic models in Section 9.6, we will use a calculator to help us find approximate solutions in Example 5.

> **Example 5**   Finding Approximate Solutions

Find approximate solutions of $7.2w^2 + 4.8w + 2.3 = 3.7$. Round the results to the second decimal place.

Solution

$7.2w^2 + 4.8w + 2.3 = 3.7$ *Original equation*

$7.2w^2 + 4.8w - 1.4 = 0$ *Subtract 3.7 from both sides.*

$$w = \frac{-4.8 \pm \sqrt{4.8^2 - 4(7.2)(-1.4)}}{2(7.2)}$$

Substitute $a = 7.2$, $b = 4.8$, and $c = -1.4$ into quadratic formula.

$$w = \frac{-4.8 \pm \sqrt{23.04 + 40.32}}{14.4}$$

Simplify.

$$w = \frac{-4.8 \pm \sqrt{63.36}}{14.4}$$

Add.

$$w \approx \frac{-4.8 \pm 7.9599}{14.4}$$

Approximate square root.

$$w \approx \frac{-4.8 - 7.9599}{14.4} \quad \text{or} \quad w \approx \frac{-4.8 + 7.9599}{14.4}$$

Write as two approximations.

$$w \approx \frac{-12.7599}{14.4} \quad \text{or} \quad w \approx \frac{3.1599}{14.4}$$

Subtract./Add.

$$w \approx -0.89 \quad \text{or} \quad w \approx 0.22$$

Finding Approximate x-Intercepts of a Parabola

Recall from Section 8.5 that we can find the x-intercepts of the graph of a quadratic equation $y = ax^2 + bx + c$ by substituting 0 for y and then solving for x.

> **Example 6**   Finding Approximate x-Intercepts

Find approximate x-intercepts of the parabola $y = -1.9x^2 + 25.7x - 68.2$. Round the coordinates of the results to the second decimal place.

Solution

First, we substitute 0 for y. Then, we use the quadratic formula to solve for x:

$0 = -1.9x^2 + 25.7x - 68.2$ *Substitute 0 for y.*

$$x = \frac{-25.7 \pm \sqrt{25.7^2 - 4(-1.9)(-68.2)}}{2(-1.9)}$$

Substitute $a = -1.9$, $b = 25.7$, and $c = -68.2$ into quadratic formula.

$$x = \frac{-25.7 \pm \sqrt{660.49 - 518.32}}{-3.8}$$

Simplify.

$$x = \frac{-25.7 \pm \sqrt{142.17}}{-3.8}$$

Subtract.

$$x \approx \frac{-25.7 \pm 11.9235}{-3.8}$$

Approximate square root.

$$x \approx 9.90 \quad \text{or} \quad x \approx 3.63$$

Calculate.

So, the x-intercepts are about $(3.63, 0)$ and $(9.90, 0)$. We use "zero" on a graphing calculator to verify our work (see Fig. 23).

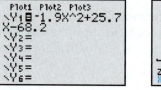 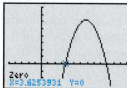 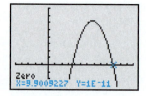

Figure 23 Verify the work

Deciding Which Method to Use to Solve a Quadratic Equation

In Chapter 8 and in this chapter, we have discussed four ways to solve quadratic equations: factoring, the square root property, completing the square, and the quadratic formula. How do we know which method to use?

Remember that **any quadratic equation can be solved by the quadratic formula**. Although any equation can also be solved by completing the square, that method is much more difficult to use than the quadratic formula.

Not many quadratic equations can be solved by factoring because almost all polynomials are prime. However, if an equation can be solved by simple factoring techniques, then it is easier to solve the equation by factoring than by any of the other three methods.

Here are some guidelines on deciding which method to use to solve a quadratic equation:

Method:	When to Use:
Factoring	For equations that can easily be put into the form $ax^2 + bx + c = 0$ and where $ax^2 + bx + c$ can easily be factored
Square root property	For equations that can easily be put into the form $x^2 = k$ or $(x + p)^2 = k$
Completing the square	When the directions require it
Quadratic formula	For all equations except those that can easily be solved by factoring or by the square root property

Example 7   Deciding Which Method to Use

Solve.

1. $(x - 5)^2 = 7$ **2.** $x^2 + 5x + 3 = 0$ **3.** $x^2 + 4x - 21 = 0$

Solution

1. The equation is of the form $(x + p)^2 = k$, so we solve it by using the square root property:

$$(x - 5)^2 = 7 \qquad \textit{Original equation}$$
$$x - 5 = \pm\sqrt{7} \qquad \textit{Square root property}$$
$$x = 5 \pm \sqrt{7} \qquad \textit{Add 5 to both sides.}$$

2. The trinomial $x^2 + 5x + 3$ is prime, so we can't solve the equation $x^2 + 5x + 3 = 0$ by factoring. The equation $x^2 + 5x + 3 = 0$ can't be put into the form $x^2 = k$, and it can't easily be put into the form $(x + p)^2 = k$, so we don't try to solve it by using the square root property. Instead, we substitute $a = 1, b = 5$, and $c = 3$ into the quadratic formula:

$$x = \frac{-5 \pm \sqrt{5^2 - 4(1)(3)}}{2(1)} \qquad \textit{Quadratic formula}$$
$$x = \frac{-5 \pm \sqrt{25 - 12}}{2} \qquad \textit{Simplify.}$$
$$x = \frac{-5 \pm \sqrt{13}}{2} \qquad \textit{Subtract.}$$

3. The trinomial $x^2 + 4x - 21$ is factorable, so we solve the equation $x^2 + 4x - 21 = 0$ by factoring:

$$x^2 + 4x - 21 = 0 \qquad \textit{Original equation}$$
$$(x + 7)(x - 3) = 0 \qquad \textit{Factor left side.}$$
$$x + 7 = 0 \quad \text{or} \quad x - 3 = 0 \quad \textit{Zero factor property}$$
$$x = -7 \quad \text{or} \qquad x = 3$$

Group Exploration

Comparing methods of solving quadratic equations

1. Solve the equation $x^2 + 6x + 8 = 0$ by factoring.

2. Solve the equation $x^2 + 6x + 8 = 0$ by completing the square.

3. Solve the equation $x^2 + 6x + 8 = 0$ by using the quadratic formula.

4. Compare the results you found in Problems 1, 2, and 3. In your opinion, which method was easiest to use?

5. Try solving the equation $x^2 + 2x - 5 = 0$ by each of the three methods. What did you find?

6. Compare the methods of solving quadratic equations by factoring, by completing the square, and by using the quadratic formula. What are the advantages and disadvantages of each method?

 Homework 9.5

For extra help ▶   MyLab Math   Watch the videos in MyLab Math

Solve by using the quadratic formula.

1. $2x^2 + 5x + 3 = 0$
2. $2x^2 + 3x + 1 = 0$
3. $4x^2 + 7x + 2 = 0$
4. $3x^2 + 5x + 1 = 0$
5. $x^2 + 3x - 5 = 0$
6. $x^2 + 7x - 2 = 0$
7. $3w^2 - 5w - 3 = 0$
8. $2p^2 - 7p - 2 = 0$
9. $5x^2 + 2x - 1 = 0$
10. $2x^2 + 4x - 3 = 0$
11. $-3m^2 + 6m - 2 = 0$
12. $-5t^2 + 8t - 2 = 0$
13. $4x^2 + 2x + 3 = 0$
14. $5x^2 + 3x + 2 = 0$
15. $2x^2 + 5x = 4$
16. $3x^2 - 2x = 7$
17. $-r^2 = r - 1$
18. $-2y^2 = 3y - 4$
19. $5x^2 + 3 = 2x$
20. $7x^2 + 4 = x$
21. $x^2 = \frac{7}{3}x - 1$
22. $2x^2 = \frac{5}{2}x + 1$
23. $3x(3x - 1) = 1$
24. $5x(5x - 1) = 1$
25. $(t + 4)(t - 2) = 3$
26. $(p - 3)(p - 5) = 2$

Find approximate solutions of the equation. Round your results to the second decimal place. Use a graphing calculator table to verify your work.

27. $4x^2 - 3x - 9 = 0$
28. $8x^2 + 2x - 5 = 0$
29. $2.1r^2 + 6.8r - 17.1 = 0$
30. $0.5m^2 - 3.5m - 14.5 = 0$
31. $-1.2x^2 = 2.8x - 12.9$
32. $-1.8x^2 = 3.8x - 52.9$
33. $0.4x^2 - 3.4x + 17.4 = 54.9$
34. $2.7x^2 + 3.8x - 39.7 = 5.6$

Find approximate x-intercept(s). Round the coordinates of your results to the second decimal place. Use a graphing calculator to verify your work.

35. $y = 2x^2 - 5x + 1$
36. $y = 7x^2 + 4x - 3$
37. $y = -5x^2 + 3x + 4$
38. $y = -4x^2 - 2x + 5$

39. $y = 3.7x^2 + 5.2x - 7.5$
40. $y = 1.2x^2 - 5.4x - 3.9$
41. $y = -2.9x^2 - 1.9x + 8.4$
42. $y = -4.1x^2 + 9.8x + 6.6$

Solve by the method of your choice.

43. $x^2 + 11x = -18$
44. $x^2 - 3x = 28$
45. $(x + 4)^2 = 13$
46. $(x - 7)^2 = 5$
47. $3r^2 + 5r = 1$
48. $5t^2 - 6t = 7$
49. $36x^2 - 49 = 0$
50. $25x^2 - 9 = 0$
51. $14x^2 = 21x$
52. $4x^2 = 6x$
53. $7x^2 - 3 = 2$
54. $4x^2 + 5 = 8$
55. $6x^2 = 7x - 2$
56. $4x^2 = -4x - 3$
57. $6m^2 - 2m + 5 = 0$
58. $3p^2 + 4p + 6 = 0$
59. $\frac{7}{2}x^2 = x + \frac{3}{2}$
60. $\frac{2}{7}x^2 = x + \frac{3}{7}$
61. $(w + 3)^2 - 2w = 7$
62. $(r - 2)^2 + 3r = 7$

For Exercises 63–66, use the graph of $y = -x^2 + 2x + 3$ shown in Fig. 24 to solve the given equation.

63. $-x^2 + 2x + 3 = 0$
64. $-x^2 + 2x + 3 = 3$
65. $-x^2 + 2x + 3 = 4$
66. $-x^2 + 2x + 3 = 5$

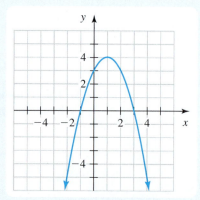

Figure 24 Exercises 63–66

For Exercises 67–70, use the graph of $y = x^2 + 6x + 7$ shown in Fig. 25 to solve the given equation.

67. $x^2 + 6x + 7 = 2$

68. $x^2 + 6x + 7 = -1$

69. $x^2 + 6x + 7 = -4$

70. $x^2 + 6x + 7 = -2$

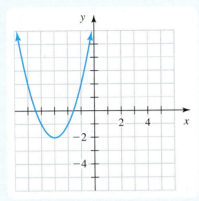

Figure 25 Exercises 67–70

Use "intersect" on a graphing calculator to solve the equation. Round the solution(s) to the second decimal place.

71. $x^2 - 5 = 2x - 1$

72. $7x - 2x^2 = -2x + 5$

73. $2x^2 - 5x + 1 = -x^2 + 3x + 3$

74. $-2x^2 + 6x - 4 = x^2 - 5x + 1$

For Exercises 75–78, solve the given equation by referring to the solutions of $y = -2x^2 + 4x + 4$ shown in Table 3.

75. $-2x^2 + 4x + 4 = -12$

76. $-2x^2 + 4x + 4 = -2$

77. $-2x^2 + 4x + 4 = 6$

78. $-2x^2 + 4x + 4 = 7$

Table 3 Exercises 75–78

x	y
-2	-12
-1	-2
0	4
1	6
2	4
3	-2
4	-12

Concepts

79. A student tries to solve the equation $x^2 + 5x = 3$:

$$x^2 + 5x = 3$$

$$x = \frac{-5 \pm \sqrt{5^2 - 4(1)(3)}}{2(1)}$$

$$x = \frac{-5 \pm \sqrt{25 - 12}}{2}$$

$$x = \frac{-5 \pm \sqrt{13}}{2}$$

Describe any errors. Then solve the equation correctly.

80. A student tries to solve the equation $3x^2 + 2x - 7 = 0$:

$$3x^2 + 2x - 7 = 0$$

$$x = \frac{-2 \pm \sqrt{2^2 - 4(3)(-7)}}{2(3)}$$

$$x = \frac{-2 \pm \sqrt{4 + 84}}{6}$$

$$x = \frac{-2 \pm \sqrt{88}}{6}$$

$$x = \frac{-2 \pm 2\sqrt{22}}{6}$$

$$x = \frac{-1 \pm 2\sqrt{22}}{3}$$

Describe any errors. Then solve the equation correctly.

81. a. Solve $x^2 + 2x - 3 = 0$ by factoring.

b. Solve $x^2 + 2x - 3 = 0$ by using the quadratic formula.

c. Solve $x^2 + 2x - 3 = 0$ by completing the square.

d. Compare the results you found in parts (a), (b), and (c).

e. In your opinion, which method was easiest to use? Explain.

82. A student tries to solve the equation $7x^2 + 3x - 2 = 0$:

$$7x^2 + 3x - 2 = 0$$

$$x = -3 \pm \frac{\sqrt{3^2 - 4(7)(-2)}}{2(7)}$$

$$x = -3 \pm \frac{\sqrt{9 + 56}}{14}$$

$$x = -3 \pm \frac{\sqrt{65}}{14}$$

Describe any errors. Then solve the equation correctly.

83. Find nonzero values of m and b so the equation $x^2 = mx + b$ has no real-number solutions. [**Hint:** Think about the graphs of $y = x^2$ and $y = mx + b$.]

84. Refer to the quadratic formula to explain why the graph of an equation of the form $y = ax^2 + bx + c$ can have at most two x-intercepts.

85. Describe how to solve a quadratic equation by using the quadratic formula.

86. Compare the methods of solving a quadratic equation by factoring, by the square root property, by completing the square, and by using the quadratic formula. Describe the methods, as well as their advantages and disadvantages.

Related Review

Solve.

87. $2x - 6 = 1$

88. $3x - 8 = -2$

89. $2x^2 - 6 = 1$

90. $3x^2 - 8 = -2$

91. $2x^2 - 6x = 1$

92. $3x^2 - 8x = -2$

Find approximate x-intercept(s). Round the coordinates of your result(s) to the second decimal place. Use a graphing calculator to verify your work.

93. $y = 3x^2 - 2x - 4$

94. $y = -2x^2 + 4x + 7$

95. $y = 3x - 2$

96. $y = -2x + 4$

Expressions, Equations, and Graphs

Perform the indicated instruction. Then use words such as linear, quadratic, cubic, power, radical, polynomial, degree, one variable, *and* two variables *to describe the expression, equation, or system.*

97. Solve $x^2 = 5 - 3x$.

98. Graph $y = x^2 - 3$ by hand.

99. Factor $6x^2 - x - 1$.

100. Solve $x^2 - 3 = 0$.

101. Simplify $3x - 4(7x - 5) + 3$.

102. Find the product $(x^2 - 3)(x^3 + 2)$.

9.6 More Quadratic Models

Objectives

» Use quadratic models to make estimates and predictions.

In Section 8.6, we used quadratic models to describe authentic situations. The situations and models were carefully chosen so we would be able to use factoring to help make estimates and predictions about the explanatory variable. Recall from Section 9.5 that factoring is rarely helpful because almost all polynomials are prime. *Any* quadratic equation can be solved by means of the quadratic formula. So, now that we know it, **we can use the quadratic formula to make estimates and predictions about the explanatory variable for any situation that is described well by a quadratic model.**

| Example 1 | Using a Quadratic Model to Make Estimates |

Table 4 lists the percentages of Americans of various age groups who watched prime-time television during the prior week.

Table 4 Percentages of Americans Who Watched Prime-Time Television during the Prior Week

Age Group (years)	Age Used to Represent Age Group (years)	Percent
18–24	21	71.9
25–34	29.5	77.5
35–44	39.5	83.5
45–54	49.5	85.4
55–64	59.5	87.4
over 64	70	90.3

Source: *GfK Mediamark Research & Intelligence, LLC*

Let p be the percentage of Americans at age t years who watched prime-time television during the prior week. A model of the situation is

$$p = -0.0061t^2 + 0.91t + 55.95$$

1. Use a graphing calculator to draw the graph of the model and, in the same viewing window, the scatterplot of the data. Does the model fit the data well?
2. Estimate the percentage of 19-year-old Americans who watched prime-time television during the prior week.
3. Use "maximum" on a graphing calculator to estimate the age of Americans who are most likely to have watched prime-time television during the prior week. According to the model, what percentage of Americans at this age watched prime-time television during the prior week?
4. Estimate the age at which 86% of Americans watched prime-time television during the prior week.
5. Find the p-intercept. What does it mean in this situation?
6. Find the t-intercepts. What do they mean in this situation?

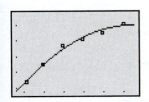

Figure 26 Check how well the model fits the data

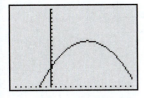

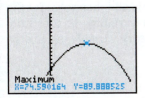

Figure 27 Finding the approximate vertex

Solution

1. The graph of the model and the scatterplot of the data are shown in Fig. 26. The model appears to fit the data quite well.
2. We substitute the input 19 for t in the equation $p = -0.0061t^2 + 0.91t + 55.95$ and solve for p:

$$p = -0.0061(19)^2 + 0.91(19) + 55.95 \approx 71.04$$

So, about 71.0% of 19-year-old Americans watched prime-time television during the prior week.

3. To get a better view of the vertex, we Zoom Out (see Fig. 27). Because the parabola opens downward, we use "maximum" to find the approximate vertex (see Fig. 27). The vertex is about $(74.59, 89.89)$. So, according to the model, about 89.9% of 75-year-old Americans watched prime-time TV during the prior week, the highest percentage for any age group. However, the actual maximum percentage is larger than 89.9%. From Table 4, we see 90.3% of Americans over 64 watched prime-time TV during the prior week.

   Note that the part of the model that lies to the right of the vertex is decreasing. This means Americans older than 75 years are less likely to watch prime-time TV than 75-year-olds, but this doesn't make sense because as people become more sedentary, they are more likely to watch TV. So, model breakdown occurs for the part of the model that lies to the right of the vertex.

4. To find the age, we substitute the output 86 for p in the equation $p = -0.0061t^2 + 0.91t + 55.95$ and solve for t:

$$86 = -0.0061t^2 + 0.91t + 55.95 \qquad \text{Substitute 86 for } p.$$
$$0 = -0.0061t^2 + 0.91t - 30.05 \qquad \text{Subtract 86 from both sides.}$$
$$t = \frac{-0.91 \pm \sqrt{0.91^2 - 4(-0.0061)(-30.05)}}{2(-0.0061)} \qquad \text{Substitute into the quadratic formula.}$$
$$t = \frac{-0.91 \pm \sqrt{0.09488}}{-0.0122} \qquad \text{Simplify.}$$
$$t \approx \frac{-0.91 \pm 0.3080}{-0.0122} \qquad \text{Approximate square root.}$$
$$t \approx 99.84 \quad \text{or} \quad t \approx 49.34 \qquad \text{Calculate.}$$

The values of t we found represent the approximate ages 49 years and 100 years. The point $(99.84, 86)$ lies approximately on the part of the parabola to the right of the vertex, where model breakdown occurs. So, model breakdown occurs for the estimate 100 years. The model does make a reasonable estimate that 86% of 49-year-old Americans watched prime-time television during the prior week.

5. To find the p-intercept, we substitute 0 for t in the equation $p = -0.0061t^2 + 0.91t + 55.95$ and solve for p:

$$p = -0.0061(0)^2 + 0.91(0) + 55.95 = 55.95$$

The p-intercept is $(0, 55.95)$. So, the model estimates about 56.0% of newborns watched prime-time TV during the week prior; model breakdown has occurred.

6. To find the t-intercepts, we substitute 0 for p in the equation $p = -0.0061t^2 + 0.91t + 55.95$ and solve for t:

$$0 = -0.0061t^2 + 0.91t + 55.95 \qquad \text{Substitute 0 for } p.$$
$$t = \frac{-0.91 \pm \sqrt{0.91^2 - 4(-0.0061)(55.95)}}{2(-0.0061)} \qquad \text{Substitute into the quadratic formula.}$$
$$t = \frac{-0.91 \pm \sqrt{2.19328}}{-0.0122} \qquad \text{Simplify.}$$
$$t \approx \frac{-0.91 \pm 1.4810}{-0.0122} \qquad \text{Approximate square root.}$$
$$t \approx 195.98 \quad \text{or} \quad t \approx -46.80 \qquad \text{Calculate.}$$

The *t*-intercepts are approximately $(-46.80, 0)$ and $(195.98, 0)$. The ages -47 years and 196 years don't make sense. So, model breakdown has occurred at both *t*-intercepts.

We use a graphing calculator to verify our work in Problems 2, 4, 5, and 6 (see Fig. 28).

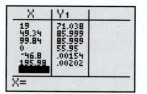

Figure 28 Verify the work

Recall from Section 8.6 that, to make a prediction about the response variable of a quadratic model, we substitute a value for the explanatory variable and then solve for the response variable. To make a prediction about the explanatory variable, we substitute a value for the response variable and then solve for the explanatory variable, usually by using the quadratic formula.

Example 2   Comparing a Linear Model and a Quadratic Model

Worldwide sales of electric cars are shown in Table 5 for various years.

Table 5 Worldwide Sales of Electric Cars

Year	Sales (thousands of cars)
2010	6.8
2011	47.6
2012	118.1
2013	203.7
2014	323.4
2015	547.1
2016	753.0

Source: *International Energy Agency*

Let *s* be annual electric-car sales (in thousands of cars) at *t* years since 2010. A linear model of the situation is

$$s = 122.96t - 83.21$$

A quadratic model of the situation is

$$s = 19.76t^2 + 4.41t + 15.58$$

1. Which model describes the situation better?
2. Use first the linear model and then the quadratic model to estimate electric-car sales in 2013. Which is the better estimate? Explain.
3. Use first the linear model and then the quadratic model to predict the worldwide electric-car sales in 2020. Refer to the graphs of the models to explain why it makes sense that the quadratic model's prediction is so much larger than the linear model's prediction.
4. Predict when worldwide electric-car sales will reach 3 million cars.

Solution

1. We see how well each model fits the data (see Figs. 29 and 30). It appears that the quadratic model fits the data points better than the linear model does.

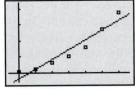
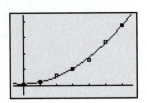

Figure 29 Linear model   **Figure 30** Quadratic model

2. First we substitute the input 3 for t in the linear model:

$$s = 122.96(3) - 83.21 = 285.67$$

The linear model estimates that about 285.7 thousand electric cars were sold in 2013. Now we substitute the input 3 in the quadratic model to make another estimate of electric-car sales in 2013:

$$s = 19.76(3)^2 + 4.41(3) + 15.58 = 206.65$$

The quadratic model estimates that about 206.7 thousand electric cars were sold in 2013. The actual sales in 2013 were 203.7 thousand electric cars (see Table 5). So, the quadratic model's estimate is much better than the linear model's estimate.

3. First we substitute the input 10 for t in the linear model:

$$s = 122.96(10) - 83.21 = 1146.39$$

The linear model predicts that about 1146 thousand (1.146 million) electric cars will be sold in 2020. Now we substitute the input 10 in the quadratic model:

$$s = 19.76(10)^2 + 4.41(10) + 15.58 = 2035.68$$

The quadratic model predicts that about 2036 thousand (2.036 million) electric cars will be sold in 2020. It makes sense that the quadratic model's prediction is so much greater than the linear model's prediction because the parabola is so much higher than the line at $t = 10$ (see Fig. 31).

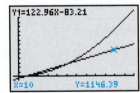
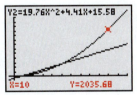

Figure 31 The parabola is well above the line at $x = 10$

4. Because the quadratic model fits the data points better than the linear model, we will use the quadratic model to predict when the worldwide electric-car sales will reach 3 million (3000 thousand) cars. So, we substitute the output 3000 for s in the equation $s = 19.76t^2 + 4.41t + 15.58$ and solve for t:

$3000 = 19.76t^2 + 4.41t + 15.58$	*Substitute* 3000 *for s.*
$0 = 19.76t^2 + 4.41t - 2984.42$	*Subtract* 3000 *from both sides.*
$t = \dfrac{-4.41 \pm \sqrt{(4.41)^2 - 4(19.76)(-2984.42)}}{2(19.76)}$	*Substitute into the quadratic formula.*
$t = \dfrac{-4.41 \pm \sqrt{235,908.0049}}{39.52}$	*Simplify.*
$t \approx \dfrac{-4.41 \pm 485.70}{39.52}$	*Approximate square root.*
$t \approx -12.40 \quad \text{or} \quad t \approx 12.18$	*Calculate.*

The solutions represent the years 1998 and 2022. A little research would show that worldwide electric-car sales in 1998 were less than 1 thousand cars, so model breakdown has occurred for that year. Furthermore, we set out to make a prediction, so we disregard the faulty estimate of 1998 and predict that worldwide electric-car sales will reach 3 million cars in 2022. We verify our work in Fig. 32.

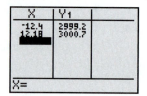

Figure 32 Verify the work

Tips for Success   Create a Mind Map for the Final Exam

In preparing for a final exam, consider how all the concepts you have learned are interconnected. One way to help yourself do this is to make a *mind map*. Put the main topic in the middle. Around it, attach concepts that relate to it, then concepts that relate to those concepts, and so on.

A portion of a mind map that describes this course is illustrated in Fig. 33. Many more "concept rectangles" could be added to it. You could make one mind map showing an overview of the course and several mind maps for the components of the overview map.

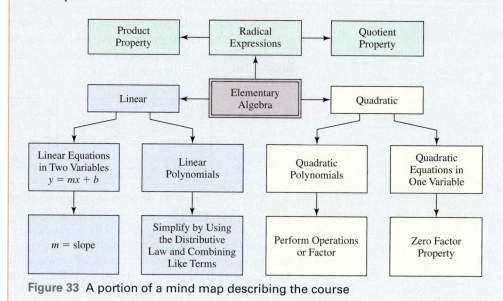

Figure 33 A portion of a mind map describing the course

◈ Group Exploration

Section Opener: Using a quadratic model to make estimates

A total of 1011 Australians participated in a study about sleep. The percentages of the subjects who used prescription sleep medication are shown in Table 6 for various age groups.

Let p be the percentage of subjects at age t years who used prescription sleep medication. A model of the situation is $p = -0.0176t^2 + 1.45t - 17.76$.

Table 6 Percentages of Subjects Who Used Prescription Sleep Medication

Age Group (years)	Age Used to Represent Age Group (years)	Percent
18–24	21.0	5.0
25–34	29.5	9.8
35–44	39.5	12.3
45–54	49.5	11.0
55–64	59.5	6.5

Sources: *The University of Adelaide; The Adelaide Institute for Sleep Health*

1. Use a graphing calculator to draw the graph of the model and, in the same viewing window, the scatterplot of the data. Does the model fit the data well?

2. Use "maximum" on a graphing calculator to find the vertex of the quadratic model. What does it mean in this situation?

3. Estimate the percentage of 19-year-old subjects who used prescription sleep medication.

4. Find the t-intercepts. What do they mean in this situation?

5. Use the model to estimate the percentage of 70-year-old subjects who used prescription sleep medication. Explain how your result relates to the results you found in Problem 4.

6. In reality, 9.1% of the subjects older than 64 years used prescription sleep medication. On the basis of the data shown in Table 6, explain why this is surprising.

Homework 9.6

For extra help ▶ **MyLab Math**   Watch the videos in MyLab Math

1. **DATA** The cocaine cultivations and production potentials in Columbia are shown in Table 7 for various years.

Table 7 Columbia's Cocaine Cultivations and Production Potentials

Year	Cultivation (thousand of hectares)	Production Potential (metric tons)
2009	116	315
2010	100	280
2011	83	220
2012	78	210
2013	81	235
2014	112	320
2015	159	520
2016	188	710

Source: *U.S. Office of National Drug Control Policy*

Let p be the annual cocaine production potential (in metric tons) in Columbia at t years since 2000. A model of the situation is $p = 24.6t^2 - 564t + 3425$.

   a. Use a graphing calculator to draw the graph of the model and, in the same viewing window, the scatterplot of the data. Does the model fit the data well?

   b. Predict Columbia's cocaine production potential in 2020.

   c. Predict when Columbia's annual cocaine production potential will be 3000 metric tons.

   d. Find the unit ratio of production potential to cultivation for cocaine in Columbia for each of the years 2014, 2015, and 2016. What do your results suggest is happening?

2. **DATA** Tiffany's revenues are shown in Table 8 for various years. Let r be the annual revenue (in billions of dollars) at t years since 2010. A model of the situation is the quadratic equation $r = -0.054t^2 + 0.585t + 2.56$.

Table 8 Tiffany's Revenues

Year	Revenue (billions of dollars)
2010	2.6
2011	3.0
2012	3.6
2013	3.8
2014	4.0
2015	4.2
2016	4.1
2017	4.0

Source: *Tiffany*

   a. Use a graphing calculator to draw the graph of the model and, in the same viewing window, the scatterplot of the data. Does the model fit the data well?

   b. Predict the revenue in 2022.

   c. Predict when the annual revenue will be $3 billion.

3. The percentages of lawyers who are women are shown in Table 9 for various years.

Table 9 Percentages of Lawyers Who Are Women

Year	Percent
1970	5
1980	14
1990	24
2000	30
2010	33
2017	35

Sources: *U.S. Census Bureau; American Bar Association*

Let p be the percentage of lawyers who are women at t years since 1970. A model of the situation is the quadratic equation $p = -0.011t^2 + 1.16t + 4.6$.

a. Use a graphing calculator to draw the graph of the model and, in the same viewing window, the scatterplot of the data. Does the model fit the data well?

b. Predict the percentage of lawyers who will be women in 2020.

c. Estimate when 31% of lawyers were women.

d. Use "maximum" on a graphing calculator to find the vertex. What does it mean in this situation?

e. If the percentage of lawyers who are women continues to increase in the future, for what years in the future is there model breakdown for certain?

4. DATA The percentages of people working at age 65 or older are shown in Table 10 for various years.

Table 10 Percentages of People Working at Age 65 or Older

Year	Percent
1970	17
1975	14
1980	13
1985	11
1990	12
1995	12
2000	13
2005	15.1
2010	17.4
2016	18.8

Source: *Bureau of Labor Statistics*

Let p be the percentage of people working at age 65 or older at t years since 1970. A model of the situation is $p = 0.0117t^2 - 0.46t + 16.4$.

a. Use a graphing calculator to draw the graph of the model and, in the same viewing window, the scatterplot of the data. Does the model fit the data well?

b. Predict the percentage of people working at age 65 or older in 2020.

c. Estimate when the percentage of people working at age 65 or older was or will be 24%.

d. Use "minimum" on a graphing calculator to estimate when the lowest percentage of people worked at age 65 or older. According to the model, what was that percentage?

5. The percentages of drivers who admit to running red lights are shown in Table 11 for various age groups.

Table 11 Percentages of Drivers Who Admit to Running Red Lights

Age Group (years)	Age Used to Represent Age Group (years)	Percent
18–25	21.5	75
26–35	30.5	73
36–45	40.5	63
46–55	50.5	56
over 55	65.0	35

Source: *The Social Science Research Center*

Let p be the percentage of drivers at age t years who admit to running red lights. A model of the situation is the equation $p = -0.015t^2 + 0.38t + 74$.

a. Use a graphing calculator to draw the graph of the model and, in the same viewing window, the scatterplot of the data. Does the model fit the data well?

b. Estimate the percentage of 37-year-old drivers who admit to running red lights.

c. At what age do half of all drivers admit to running red lights?

d. Find the t-intercepts. What do they mean in this situation?

e. Use "maximum" on a graphing calculator to estimate the age at which drivers are most likely to admit to running red lights. According to the model, what was that percentage?

6. Table 12 lists the percentages of U.S. households, by income groups, that have Internet access.

Table 12 Percentages of U.S. Households That Have Internet Access

Household Income Group (in thousands of dollars)	Income Used to Represent Household Income Group (in thousands of dollars)	Percent
0–15	7.5	39.6
15–25	20	52.6
25–35	30	63.3
35–50	42.5	77.9
50–75	62.5	87.1
over 75	100	96.4

Source: *U.S. Census Bureau*

For U.S. households with income d thousand dollars, let p be the percentage that have Internet access. A model of the situation is $p = -0.0073d^2 + 1.41d + 28.7$.

a. Use a graphing calculator to draw the graph of the model and, in the same viewing window, the scatterplot of the data. Does the model fit the data well?

b. For households with income $23 thousand, estimate the percentage that have Internet access.

c. Estimate the income(s) at which 70% of households have Internet access.

7. DATA The average sizes of U.S. households are shown in Table 13 for various years.

Table 13 Average Sizes of U.S. Households

Year	Average Size (number of people)
1930	4.01
1940	3.68
1950	3.38
1960	3.29
1970	3.11
1980	2.80
1990	2.63
2000	2.62
2010	2.59
2016	2.57

Source: *U.S. Census Bureau*

Let s be the average size (in number of people) of a U.S. household at t years since 1900. A model of the situation is $s = 0.000172t^2 - 0.042t + 5.11$.

  a. Use a graphing calculator to draw the graph of the model and, in the same viewing window, the scatterplot of the data. Does the model fit the data well?

  b. Predict the average size of a U.S. household in 2020.

  c. Estimate when the average U.S. household size was 3 people.

  d. Find the s-intercept. What does it mean in this situation?

8. The percentages of Americans who have high cholesterol are shown in Table 14 for various years.

Table 14 Percentages of Americans Who Have High Cholesterol

Age (years)	Percent
50	29
60	43
70	50
80	47
90	39

Source: *Gallup Organization*

Let p be the percentage of Americans at age t years who have high cholesterol. A model of the situation is $p = -0.0386t^2 + 5.64t - 156$.

  a. Use a graphing calculator to draw the graph of the model and, in the same viewing window, the scatterplot of the data. Does the model fit the data well?

  b. Estimate the percentage of 55-year-old Americans who have high cholesterol.

  c. At what ages do 40% of Americans have high cholesterol?

  d. Use "maximum" on a graphing calculator to estimate at what age Americans are most likely to have high cholesterol. What is that percentage?

  e. Find the t-intercepts. What do they mean in this situation?

9. The percentages of Americans who are concerned about joblessness are shown in Table 15 for various years.

Table 15 Percentages of Americans Who Are Concerned about Joblessness

Age Group (years)	Age Used to Represent Age Group (years)	Percent
18–34	26	43
35–44	39.5	41
45–54	49.5	39
55–64	59.5	38
over 64	70	32

Source: *Edward Jones survey*

Let p be the percentage of Americans at age t years who are concerned about joblessness. A model of the situation is $p = -0.0047t^2 + 0.22t + 40.2$.

  a. Use a graphing calculator to draw the graph of the model and, in the same viewing window, the scatterplot of the data. Does the model fit the data well?

  b. Estimate the percentage of 23-year-old Americans who are concerned about joblessness.

  c. At what age are 36% of Americans concerned about joblessness?

  d. Find the t-intercepts. What do they mean in this situation?

10. The percentages of Americans who have ever played organized basketball are shown in Table 16 for various age groups.

Table 16 Percentages of Americans Who Have Ever Played Organized Basketball

Age Group (years)	Age Used to Represent Age Group (years)	Percent
12–17	14.5	48.1
18–24	21	34.6
25–34	29.5	36.9
35–44	39.5	30.8
45–54	49.5	27.4
over 55	60	25.0

Source: *ESPN sports poll*

Let p be the percentage of Americans at age t years who have ever played organized basketball. A model of the situation is $p = 0.0082t^2 - 1.04t + 58.44$.

  a. Use a graphing calculator to draw the graph of the model and, in the same viewing window, the scatterplot of the data. Does the model fit the data well?

  b. What percentage of 17-year-old Americans have ever played organized basketball?

  c. Estimate at what age 28% of Americans have ever played organized basketball.

  d. Find the p-intercept. What does it mean in this situation?

11. The numbers of women who have run in the New York City (NYC) Marathon are shown in Table 17 for various years.

Table 17 Numbers of Women Who Have Run in the NYC Marathon

Year	Number of Women
1970	1
1980	1962
1990	5204
2000	8641
2010	16,253
2016	21,457

Source: *New York Road Runners Club*

Let n be the number of women who have run in the NYC Marathon at t years since 1970. A linear model of the situation is $n = 459t - 2253$. A quadratic model of the situation is $n = 9t^2 + 33.5t + 316$.

  a. Use a graphing calculator to draw the graphs of both models and, in the same viewing window, the scatterplot of the data. Which model describes the situation better?

  b. Use first the linear model and then the quadratic model to estimate the number of women who ran in the marathon in 2000. Which is the better estimate? Explain.

  c. Use first the linear model and then the quadratic model to predict in which year 25,700 women will run in the marathon—when about half of the runners will be women. Use the graphs of the models to explain why the quadratic model's result is earlier than the linear model's result.

12. The numbers of assaults in New York City are shown in Table 18 for various years.

Table 18 Assaults in New York City

Year	Number of Assaults (in thousands)
2010	27.3
2011	29.8
2012	31.2
2013	31.8
2014	31.5
2015	30.5

Source: *City-data.com*

Let n be the number (in thousands) of assaults in New York City at t years since 2010. A linear model of the situation is $n = 0.62t + 28.8$. A quadratic model of the situation is $n = -0.43t^2 + 2.79t + 27.35$.

a. Use a graphing calculator to draw the graphs of both models and, in the same viewing window, the scatterplot of the data. Which model describes the situation better?

b. Use first the linear model and then the quadratic model to estimate the number of assaults in 2012. Which is the better estimate? Explain.

c. Find the t-intercepts of the quadratic model. What do they mean in this situation?

d. Use "maximum" on a graphing calculator to find the vertex of the quadratic model. What does it mean in this situation?

e. For the years after 2014, which model would the mayor of New York City hope would describe the situation better? Explain.

13. Table 19 shows that people with more education tend to earn more money.

Table 19 Education versus Median Annual Income

Grade-Level Completion or Degree	Full-Time Equivalent Years in School	Median Annual Income (thousands of dollars)
9th to 12th Grade, no diploma	10	25.6
High school	12	35.3
Some college	13	38.4
Associate's	14	41.5
Bachelor's	16	59.1
Master's	18	69.7
Doctoral	20	84.4

Source: *U.S. Bureau of Labor Statistics*

Let I be the median annual income (in thousands of dollars) for people with a full-time equivalent of t years of education. A model for the situation is $I = 0.199t^2 - 0.02t + 5.7$.

a. Use a graphing calculator to draw the graph of the model and, in the same viewing window, the scatterplot of the data. Does the model fit the data well?

b. Estimate the full-time equivalent number of years of education for people whose median annual income is $50 thousand.

c. In August 2016, a high school graduate is trying to decide whether to get a job or to go to a public, in-state, 4-year college to get a bachelor's degree. For the 2016–2017 academic year, average tuition and fees for a public, in-state, 4-year college are $9650. In parts (i)–(iv), assume tuition and fees, as well as incomes, are constant.

  i. If the student doesn't go to college, estimate the student's total earnings in 4 years.

  ii. If the student goes to college, estimate the total cost of 4 years of college.

  iii. If the student goes to college, estimate after how many years of work will the student be in the same financial position (total earnings minus total costs) as if the student hadn't gone to college and had gotten a job instead.

  iv. Assume the student earns a bachelor's degree in 4 years and then works for 33 years until retirement. How much more will the student's lifetime earnings minus total college tuition and fees be than if the person had not gone to college (and had worked for 37 years)?

  v. During the past 20 years, those with larger incomes have had larger growth in incomes. Is the estimate you made in part (iv) an underestimate or an overestimate? Explain.

14. Sales of downloaded albums are shown in Table 20 for various years.

Table 20 Downloaded Album Sales

Year	Sales (millions of downloaded albums)
2010	85.8
2011	103.9
2012	116.7
2013	118.0
2014	117.6
2015	109.4

Source: *Recording Industry Association of America*

Let s be the annual sales (in millions) of downloaded albums at t years since 2010. A model of the situation is $s = -3.3t^2 + 21t + 86.1$.

a. Use a graphing calculator to draw the graph of the model and, in the same viewing window, the scatterplot of the data. Does the model fit the data well?

b. Estimate the sales in 2016.

c. Predict when the annual sales will be 40 million downloaded albums.

d. Find the t-intercepts.

e. In which years is there model breakdown for certain?

Large Data Sets

15. ^{DATA} Access the data about power cleans and shot puts, which are available at MyLab Math and at the Pearson Downloadable Student Resources for Math & Stats website. The data describe the maximum power cleans (weight lifting) and best shot puts by 28 female college shot-putters. Let s be the best shot put (in meters) by a shot-putter whose maximum power clean is p kilograms. A model of the situation is $s = -0.00104p^2 + 0.283p - 1.53$.

a. Use a graphing calculator to draw the graph of the model and, in the same viewing window, the scatterplot of the data. Does the model fit the data well?

b. Estimate the best shot put by a shot-putter whose maximum power clean is 100 kilograms.

c. Find the p-intercepts. What do they mean in this situation?

d. Find the vertex. What does it mean in this situation?

e. For which values of p is there probably model breakdown?

f. For shot-putters with a certain maximum power clean, which have the better technique, the ones represented by data points above the model or the ones represented by data points below the model?

g. A student says the shot-putter with maximum power clean 102 kilograms and best shot put 14.8 meters has better technique than the shot-putter with maximum power clean 63.6 kilograms and best shot put 13.2 meters because 14.8 meters is greater than 13.2 meters. What would you tell the student?

16. DATA Access the data about the clarity of Lake Tahoe, which are available at MyLab Math and at the Pearson Downloadable Student Resources for Math & Stats website. The clarity of Lake Tahoe in California is measured by the depth at which a white disk with diameter 10 inches remains visible when lowered beneath the water's surface. Let c be the clarity (in meters) of Lake Tahoe at t years since 1960.

a. Use a graphing calculator to draw the graph of the model $c = 0.00907t^2 - 0.733t + 38.14$ and, in the same viewing window, the scatterplot of the winter data. Does the model fit the data well?

b. Find an equation of a linear model that describes the relationship between t and c for the summer data.

c. Use a graphing calculator to draw the graph of the model $c = 0.00617t^2 - 0.559t + 34.12$ and, in the same viewing window, the scatterplot of the annual data. Does the model fit the data well.

d. In a given year, does there tend to be more clarity in the winter or in the summer? Explain. Is this true in every year?

e. The graph of which of the three models is between the graphs of the other two? Why does this make sense?

f. Which of the three models is most suggestive that the clarity of Lake Tahoe is improving? Which of the three models is most suggestive that the clarity is getting worse?

g. Federal and state regulators have set a clarity restoration target of 29.7 meters. Use the annual-data model to predict when the target will be reached.

Concepts

17. Discuss how to make predictions about the explanatory or response variable of a quadratic model.

18. Discuss how to find intercepts of a quadratic model.

Related Review

19. The percentages of consumers who will try a product if it is sponsoring an event for an artist they like are shown in Table 21 for various age groups.

Table 21 Percentages of Consumers Who Will Try a Product if It Is Sponsoring an Event for an Artist They Like

Age Group (years)	Age Used to Represent Age Group (years)	Percent
13–17	15.0	31
18–24	21.0	24
25–34	29.5	26
35–44	39.5	18
45–54	49.5	13
over 54	60.0	10

Source: *Nielsen*

Let p be the percentage of consumers at age a years who will try a product if it is sponsoring an event for an artist they like.

a. Find an equation of a model to describe the data.

b. Estimate the percentage of 19-year-old consumers who will try a product if it is sponsoring an event for an artist they like.

c. Estimate at what age 20% of consumers will try a product if it is sponsoring an event for an artist they like.

20. DATA The rate of illnesses and injuries suffered in the workplace has declined since 2008 (see Table 22).

Table 22 Illness and Injury Rate in the Workplace

Year	Illness and Injury Rate (cases per 100 full-time workers per year)
2008	3.9
2009	3.6
2010	3.5
2011	3.4
2012	3.4
2013	3.3
2014	3.2
2015	3.0

Source: *Bureau of Labor Statistics surveys*

Let r be the rate of illnesses and injuries (cases per 100 full-time workers per year) suffered in the workplace at t years since 2000.

a. Find an equation of a model to describe the data.

b. Predict when the illness and injury rate will be 2 cases per 100 full-time workers.

c. Predict the illness and injury rate in 2021.

d. In 2015, Wal-Mart® was the largest U.S. company, with 2.3 million employees. Estimate the number of illness and injury cases at Wal-Mart in 2015.

21. In 2015, 71% of farms owned or leased computers. The percentage has increased about 2 percentage points per year since then (Source: *National Agricultural Statistics Service*). Let p be the percentage of farms that own or lease computers at t years since 2015.

a. Find an equation of a model to describe the data.

b. Predict the percentage of farms that will own or lease computers in 2021.

c. Predict when 87% of farms will own or lease computers.

22. In 2015, 13.0% of Americans ages 18–24 smoked cigarettes. The percentage has decreased about 1.5 percentage points per year since then (Source: *Centers for Disease Control and Prevention*). Let p be the percentage of Americans ages 18–24 who smoke at t years since 2015.
    a. Find an equation of a model to describe the data.
    b. Predict what percentage of Americans ages 18–24 will smoke cigarettes in 2020.
    c. Predict when 3% of Americans ages 18–24 will smoke cigarettes.

23. The percentage of Apple's revenue that is from the iPad has decreased approximately linearly from 19.8% in 2012 to 9.6% in 2016 (Source: *Apple*). Predict when 2% of Apple's revenue will be from the iPad.

24. The number of malware specimens such as computer viruses, worms, and ransomware increased approximately linearly from 2.09 million specimens in 2010 to 6.83 million specimens in 2016 (Source: *G DATA, AV-TEST*). Predict in which year there will be 12 million malware specimens.

Expressions, Equations, and Graphs

Give an example of each. Then solve, simplify, or graph, as appropriate.

25. Quadratic equation in one variable that can be solved by factoring

26. Linear equation in one variable

27. Radical expression in which a radicand is a fraction

28. Quadratic polynomial with four terms

29. Linear equation in two variables

30. System of two linear equations in two variables

31. Quadratic equation in one variable that can't be solved by factoring

32. Cubic polynomial with five terms

33. Linear polynomial with four terms

34. Quadratic equation in two variables

 # Taking It to the Lab

Climate Change Lab (continued from Chapter 7)

Some developing countries' carbon dioxide emissions are increasing greatly as their economies continue to grow. In particular, China's future carbon dioxide emissions will affect the planet considerably because of that country's large population, large population growth, and increasing per-person carbon dioxide emissions (see Table 23). Unless China takes measures to slow the growth of carbon dioxide emissions, its per-person emissions in 2050 could reach 37.2 metric tons—about 19 times the 1990 per-person emissions.

Although the United States has the largest gross domestic product (GDP) in the world, China's GDP is likely to surpass U.S. GDP in 2030.[*] So, it is not surprising China has large carbon dioxide emissions. In fact, China has already overtaken the United States as the top emitter of carbon dioxide in the world; in 2014, China's emissions were 10.4 billion metric tons and U.S. emissions were 5.4 billion metric tons.[†]

Even though China has larger carbon dioxide emissions than the United States, its 2014 per-person emissions are only about 45% of the United States' (see Tables 23 and 24). If U.S. per-person emissions were to hold steady at 17.0 metric tons, China's per-person emissions could reach that level as early as 2029.

DATA. Table 24 U.S. Population and Per-Person Carbon Dioxide Emissions

Year	Population (billions)	Per-Person Carbon Emissions (metric tons per person)
1950	0.15	16.0
1960	0.18	16.1
1970	0.21	20.5
1980	0.23	20.9
1990	0.25	20.0
2000	0.28	21.1
2010	0.31	18.1
2014	0.32	17.0

Sources: *U.S. Census Bureau; The Cambridge Factfinder*

Recall from the Climate Change Lab in Chapter 7 that if world population were to decline to 1.7 billion, then the IPCC's 2050 carbon dioxide emissions goal could be met if

DATA. Table 23 China's Population and Per-Person Carbon Dioxide Emissions

Year	Population (billions)	Per-Person Carbon Emissions (metric tons per person)
1990	1.15	1.98
1995	1.21	2.38
2000	1.27	2.86
2005	1.31	4.21
2010	1.34	6.15
2014	1.37	7.60

Sources: *U.S. Census Bureau; U.S. Energy Information Administration*

---

[*]*The New York Times*

[†]*Carbon Dioxide Information Analysis Center*

carbon dioxide emissions were limited to 5.0 metric tons per person each year. This per-person limit would create room for most developing countries to expand their economies.

Such a population–emissions strategy would require some developing countries to learn to expand their economies in new ways that would not increase per-person carbon dioxide emissions so much. Other developing countries such as China and oil-rich countries such as Qatar would actually have to reduce their per-person emissions. It would also require developed countries to learn to increase their economies while reducing their per-person carbon dioxide emissions. Finally, it would require that world population be reduced by 77% from its 2016 level of 7.35 billion. All requirements would be great challenges.

Analyzing the Situation

1. Let p be China's population (in billions) at t years since 1950. Construct a scatterplot for China's population data. Then find a linear equation of a model that describes the situation well.

2. Let y be China's *per-person* annual carbon dioxide emissions (in metric tons per person) at t years since 1950. A reasonable model is $y = 0.00973t^2 - 0.775t + 17.4$. Verify the claim that China's per-person emissions in 2050 could be as high as 37.2 metric tons of carbon dioxide per person.

3. Let C be China's annual carbon dioxide emissions (in billions of metric tons) at t years since 1950. Find an equation that describes the relationship between t and C. [**Hint:** Add, subtract, multiply, or divide the right-hand sides of the results you found in Problems 1 and 2.]

4. Perform a unit analysis of the model you found in Problem 3.

5. In Problem 5 of the Climate Change Lab in Chapter 5, you found a model of U.S. annual carbon dioxide emissions C (in billions of metric tons) at t years since 1950. If, however, you didn't find such an equation, find it now. Then use "intersect" on a graphing calculator to estimate when China and the United States had the same annual amount of carbon dioxide emissions. What is that annual amount of carbon dioxide emissions?

6. Verify the claim that China's *per-person* annual carbon dioxide emissions could reach the United States' 2014 level of 17.0 metric tons of carbon dioxide per person as early as 2029.

7. Explain why the year you found in Problem 5 is so much earlier than the year you found in Problem 6.

Projectile Lab (continued from Chapter 7)

In the Projectile Lab in Chapter 7, you found an equation of a quadratic model that describes the times and heights of a ball that you threw straight up. In this lab, you will analyze that situation further.

Analyzing the Situation

1. What model did you work with in the Projectile Lab in Chapter 7? Provide the equation and define the variables.

2. Use the model to estimate both times when the ball was at a height of 10 feet.

3. Find the average of the two times you found in Problem 2.

4. Use the model to estimate the height of the ball at the time you found in Problem 3.

5. Use a graphing calculator to draw a graph of the model. Copy the graph and indicate on the graph the two points that represent when the ball was at a height of 10 feet and the point that represents when the ball was at the height you found in Problem 4.

6. By referring to your graph in Problem 5, explain why the result you found in Problem 4 is the maximum height of the ball.

7. Use "maximum" on a graphing calculator to estimate the maximum height of the ball. Compare this result with the result you found in Problem 6.

8. Find the average speed of the ball on its way up. That is, find the ball's average speed from the moment it was released to the moment it reached its maximum height.

9. Explain why it makes sense that the result you found in Problem 8 is less than the initial speed you found in the Projectile Lab in Chapter 7.

Chapter Summary

Key Points of Chapter 9

Section 9.1 Simplifying Radical Expressions

Principal square root	If a is a nonnegative number, then $\sqrt{a}$ is the nonnegative number we square to get a. We call $\sqrt{a}$ the **principal square root** of a.
Perfect square	A **perfect square** is a number whose principal square root is rational.

Section 9.1   Simplifying Radical Expressions (*Continued*)

Product property for square roots	For nonnegative numbers a and b, $$\sqrt{ab} = \sqrt{a}\sqrt{b}$$
Simplified square root	A square root is **simplified** when the radicand does not have any perfect-square factors other than 1.
Simplifying a square root with a small radicand	To simplify a square root in which it is easy to determine the largest perfect-square factor of the radicand, 1. Write the radicand as the product of the *largest* perfect-square factor and another number. 2. Apply the product property for square roots.
Simplifying a square root with a large radicand	To simplify a square root in which it is difficult to determine the largest perfect-square factor of the radicand, 1. Find the prime factorization of the radicand. 2. Look for pairs of identical factors of the radicand, and rearrange the factors to highlight them. 3. Use the product property for square roots.
Simplifying $\sqrt{x^2}$	If x is nonnegative, then $\sqrt{x^2} = x$.

Section 9.2 Simplifying More Radical Expressions

Quotient property for square roots	For a nonnegative number a and a positive number b, $$\sqrt{\frac{a}{b}} = \frac{\sqrt{a}}{\sqrt{b}}$$
Rationalize the denominator	To **rationalize the denominator** of a fraction of the form $\dfrac{p}{\sqrt{q}}$, where q is positive, we multiply the fraction by 1 in the form $\dfrac{\sqrt{q}}{\sqrt{q}}$.
Simplifying a radical expression	To simplify a radical expression, 1. Use the quotient property for square roots so that no radicand is a fraction. 2. Use the product property for square roots so that no radicands have perfect-square factors other than 1. 3. Rationalize denominators so that no denominator is a radical expression. 4. Use the property $\dfrac{a}{a} = 1$, where a is nonzero, to simplify. 5. Continue applying steps 1–4 until the radical expression is simplified completely.

Section 9.3 Solving Quadratic Equations by the Square Root Property; The Pythagorean Theorem

Square root property	Let k be a nonnegative constant. Then $x^2 = k$ is equivalent to $x = \pm\sqrt{k}$.
Pythagorean theorem	If a and b are the lengths of the legs of a right triangle and c is the length of the hypotenuse, then $a^2 + b^2 = c^2$.

Section 9.4 Solving Quadratic Equations by Completing the Square

Perfect-square trinomial property	For a perfect-square trinomial of the form $x^2 + bx + c$, dividing b by 2 and then squaring the result gives c: $$x^2 + bx + c$$ $$\left(\frac{b}{2}\right)^2 = c$$

Section 9.4   Solving Quadratic Equations by Completing the Square (*Continued*)

When completing the square can be used	Any quadratic equation can be solved by completing the square.
Solving a quadratic equation by completing the square	To solve a quadratic equation by **completing the square,** 1. Write the equation in the form $ax^2 + bx = k$, where $a, b,$ and k are constants. 2. If $a \neq 1$, divide both sides of the equation by a. 3. Complete the square for the polynomial on the left side of the equation. 4. Solve the equation by using the square root property.

Section 9.5 Solving Quadratic Equations by the Quadratic Formula

Quadratic formula	The solutions of a quadratic equation $ax^2 + bx + c = 0$ are given by the **quadratic formula**: $$x = \frac{-b \pm \sqrt{b^2 - 4ac}}{2a}$$
When the quadratic formula can be used	Any quadratic equation can be solved by the quadratic formula.
Guidelines on solving quadratic equations	Here are some guidelines on deciding which method to use to solve a quadratic equation:

Method:	When to Use:
Factoring	For equations that can easily be put into the form $ax^2 + bx + c = 0$ and where $ax^2 + bx + c$ can easily be factored
Square root property	For equations that can easily be put into the form $x^2 = k$ or $(x + p)^2 = k$
Completing the square	When the directions require it
Quadratic formula	For all equations except those that can easily be solved by factoring or by the square root property

Section 9.6 More Quadratic Models

Using the quadratic formula to make estimates and predictions	We can use the quadratic formula to make estimates and predictions about the explanatory variable for any situation that is described well by a quadratic model.

Chapter 9 Review Exercises

Simplify.

1. $\sqrt{196}$ **2.** $-\sqrt{64}$

3. $\sqrt{-25}$ **4.** $-\sqrt{-81}$

Find an approximate value of the radical. Round your result to the second decimal place.

5. $\sqrt{95}$ **6.** $-7.29\sqrt{38.36}$

Simplify. Assume that any variables are nonnegative.

7. $\sqrt{84}$ **8.** $\sqrt{36x}$

9. $\sqrt{18ab^2}$ **10.** $\sqrt{98m^2n^2}$

11. Without using a calculator, find the two nearest consecutive integers to $\sqrt{39}$.

Simplify. Assume that any variables are positive.

12. $\sqrt{\dfrac{5}{9}}$ **13.** $\sqrt{\dfrac{50y}{x^2}}$ **14.** $\dfrac{4}{\sqrt{x}}$ **15.** $\sqrt{\dfrac{5b^2}{32}}$

16. A student tries to simplify $\dfrac{3}{\sqrt{7}}$:

$$\frac{3}{\sqrt{7}} = \left(\frac{3}{\sqrt{7}}\right)^2$$
$$= \frac{3^2}{(\sqrt{7})^2}$$
$$= \frac{9}{7}$$

Describe any errors. Then simplify the expression correctly.

Solve. Use a graphing calculator to verify your work.

17. $p^2 = 45$ **18.** $5t^2 = 7$

19. $5x^2 + 4 = 12$ **20.** $(x + 4)^2 = 27$

21. $(w - 6)^2 = -15$ **22.** $(m - 8)^2 = 0$

Let a and b be the lengths of the legs of a right triangle, and let c be the length of the hypotenuse. Values for two of the three lengths are given. Find the third length.

23. $a = 4$ and $b = 8$ **24.** $b = 3$ and $c = 6$

For Exercises 25 and 26, the lengths of two sides of a right triangle are given. Find the length of the third side. (The triangle is not drawn to scale.)

25. See Fig. 34.

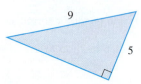

Figure 34 Exercise 25

26. See Fig. 35.

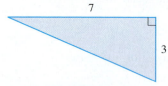

Figure 35 Exercise 26

27. A 12-foot-long ladder is leaning against a building. The bottom of the ladder is 3.5 feet from the base of the building. Will the ladder reach the bottom of a window that is 11 feet above the ground?

Solve by completing the square.

28. $x^2 + 6x = 2$ **29.** $x^2 + 2x = 17$

30. $w^2 - 8w - 4 = 0$ **31.** $t^2 - 12t + 40 = 0$

32. $2x^2 + 8x = 12$ **33.** $3x^2 - 18x - 27 = 0$

Solve by using the quadratic formula.

34. $3x^2 + 7x + 1 = 0$ **35.** $2x^2 - 5x - 4 = 0$

36. $-5y^2 = 3 - 2y$ **37.** $-p^2 + 3p = -5$

38. $2x^2 - x = \dfrac{3}{2}$ **39.** $2x(x - 1) = 5$

Find approximate solutions of the equation. Round your results to the second decimal place. Use a graphing calculator table to verify your work.

40. $6w^2 = 8w + 5$ **41.** $-1.9t^2 - 5.4t + 27.9 = 14.1$

42. A student tries to solve the equation $2x^2 + 3x - 7 = 0$:

$$2x^2 + 3x - 7 = 0$$

$$x = -3 \pm \frac{\sqrt{3^2 - 4(2)(-7)}}{2(2)}$$

$$= -3 \pm \frac{\sqrt{65}}{4}$$

Describe any errors. Then solve the equation correctly.

Solve by the method of your choice.

43. $x^2 = 5x + 14$ **44.** $3x^2 - 4x = 1$

45. $r^2 + 13 = 0$ **46.** $(t - 4)^2 = 24$

47. $x^2 = 2x + 35$ **48.** $5x^2 + 8 = 20$

49. $(x - 2)^2 - 3x = 1$ **50.** $3x^2 + 2x - 8 = 0$

51. $(w + 7)^2 = 27$ **52.** $-2p^2 = 7p - 3$

53. $x^2 - 2x = \dfrac{5}{2}$ **54.** $4x(x - 2) = -3$

For Exercises 55–58, use the graph of $y = \dfrac{1}{2}x^2 - 2x - 1$ shown in Fig. 36 to solve the given equation.

55. $\dfrac{1}{2}x^2 - 2x - 1 = -3$ **56.** $\dfrac{1}{2}x^2 - 2x - 1 = -4$

57. $\dfrac{1}{2}x^2 - 2x - 1 = -1$ **58.** $\dfrac{1}{2}x^2 - 2x - 1 = 5$

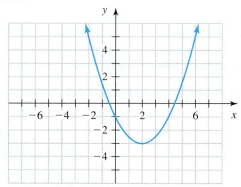

Figure 36 Exercises 55–58

59. Use "intersect" on a graphing calculator to solve the equation $7 - 2x^2 = x^2 - x - 8$. Round the solution(s) to the second decimal place.

Find approximate x-intercepts. Round the coordinates of your results to the second decimal place. Use a graphing calculator to verify your results.

60. $y = 3x^2 - 8x + 2$ **61.** $y = -3.9x^2 + 7.1x + 54.9$

62. a. Solve $x^2 - 2x - 15 = 0$ by factoring.
    b. Solve $x^2 - 2x - 15 = 0$ by using the quadratic formula.
    c. Solve $x^2 - 2x - 15 = 0$ by completing the square.
    d. Compare the results you found in parts (a), (b), and (c).

63. A person is said to have moderate or severe memory impairment if the person can recall 4 or fewer words out of 20 on combined immediate and delayed recall tests. The percentages of Americans with moderate or severe memory impairment are shown in Table 25 for various age groups.

Table 25 Percentages of Americans with Moderate or Severe Memory Impairment

Age Group (years)	Age Used to Represent Age Group (years)	Percent
65–69	67	1.1
70–74	72	2.5
75–79	77	4.5
80–84	82	6.4
over 84	90	12.9

Source: *Health and Retirement Study*

Let p be the percentage of Americans with moderate or severe memory impairment at age t years. A model of the situation is $p = 0.016t^2 - 2t + 63$.

  a. Use a graphing calculator to draw the graph of the model and, in the same viewing window, the scatterplot of the data. Does the model fit the data well?

  b. Estimate the percentage of Americans at age 70 that have moderate or severe memory impairment.

  c. Estimate at what age 10% of Americans have moderate or severe memory impairment.

64. The percentages of medical degrees earned by women has increased greatly since 1970 (see Table 26). Let p be the percentage of medical degrees earned by women at t years since 1970. A model of the situation is the equation $p = -0.019t^2 + 1.72t + 9.3$.

  a. Use a graphing calculator to draw the graph of the model and, in the same viewing window, the scatterplot of the data. Does the model fit the data well?

Table 26 Medical Degrees Earned by Women

Year	Percent
1970	9.2
1980	24.9
1990	36.0
2000	43.2
2010	48.4
2015	47.6

Source: *Association of American Medical Colleges*

  b. Predict the percentage of medical degrees that will be earned by women in 2022.

  c. Predict when women will earn 47% of medical degrees.

  d. Find the t-intercepts. What do they mean in this situation?

Chapter 9 Test

Simplify.

1. $-\sqrt{121}$ **2.** $-\sqrt{-9}$

Simplify. Assume that any variables are positive.

3. $\sqrt{48}$ **4.** $\sqrt{64x}$ **5.** $\sqrt{45a^2b}$

6. $\dfrac{3}{\sqrt{7}}$ **7.** $\sqrt{\dfrac{2m^2}{n}}$ **8.** $\sqrt{\dfrac{3}{20}}$

9. The lengths of two sides of the triangle shown in Fig. 37 are given. Find the length of the third side. (The triangle is not drawn to scale.)

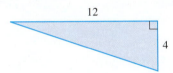

12

4

Figure 37 Exercise 9

10. The width of a picture frame is 11 inches, and the diagonal has a length of 17 inches. Estimate the length of the picture frame to the second decimal place.

Solve by the method of your choice.

11. $3x^2 = 5$ **12.** $-5x^2 + 4 = -28$

13. $(t - 5)^2 = 18$ **14.** $2w^2 + 3w - 6 = 0$

15. $x(x - 3) = 40$ **16.** $\dfrac{3}{2}x^2 = x + 2$

17. Solve $5p^2 - 20p - 35 = 0$ by completing the square.

18. Find approximate solutions of the quadratic equation $1.4x^2 - 2.3x - 38.5 = 7.4$. Round your results to the second decimal place.

19. Find approximate x-intercepts of the graph of $y = -1.2x^2 + 37.9x - 50.4$. Round the coordinates of your results to the second decimal place.

For Exercises 20–23, solve the given equation by referring to the solutions of $y = -x^2 + 6x + 5$ shown in Table 27.

20. $-x^2 + 6x + 5 = 13$ **21.** $-x^2 + 6x + 5 = 10$

22. $-x^2 + 6x + 5 = 14$ **23.** $-x^2 + 6x + 5 = 15$

Table 27 Exercises 20–23

x	y
0	5
1	10
2	13
3	14
4	13
5	10
6	5

24. A student tries to solve the equation $2x^2 + 4x - 3 = 0$:

$$2x^2 + 4x - 3 = 0$$

$$x = \frac{-4 \pm \sqrt{4^2 - 4(2)(-3)}}{2(2)}$$

$$x = \frac{-4 \pm \sqrt{16 + 24}}{4}$$

$$x = \frac{-4 \pm \sqrt{40}}{4}$$

$$x = \frac{-4 \pm 2\sqrt{10}}{4}$$

$$x = -1 \pm 2\sqrt{10}$$

Describe any errors. Then solve the equation correctly.

25. The U.S. median asking rents (in 2015 dollars) are shown in Table 28 for various years.

Table 28 Median Asking Rents

Year	Median Asking Rent (2015 dollars)
2009	708
2010	698
2011	694
2012	717
2013	734
2014	762
2015	813

Source: *U.S. Census Bureau*

Let r be the median asking rent (in 2015 dollars) at t years since 2000. A model of the situation is $r = 5.39t^2 - 112t + 1280$.

a. Use a graphing calculator to draw the graph of the model and, in the same viewing window, the scatterplot of the data. Does the model fit the data well?

b. Use "minimum" on a graphing calculator to estimate when the median asking rent was the least. According to the model, what was the median asking rent in that year? Compare your results with the actual values.

c. Predict the median asking rent in 2022.

d. In 2015, the median asking rent in the West was $1081 (in 2015 dollars). Predict when the U.S. median asking rent will be $1081 (in 2015 dollars).

Rational Expressions and Equations

10

Do you use online banking? The numbers of households in the United States that use online banking are shown in Table 1 for various years. In Example 6 of Section 10.5, we will predict when 90% of households will use online banking.

In Chapters 1–6, we worked with linear expressions and equations. In Chapters 7–9, we worked with polynomial expressions and equations. In this chapter, we will work with yet another type of expressions and equations called *rational* expressions and equations. This work will enable us to make estimates and predictions about authentic situations, such as the percentage of recent high school graduates who will enroll in college in 2021.

Table 1 Households That Use Online Banking

Year	Households That Use Online Banking (millions)	Total Number of Households (millions)
2008	56.3	116.8
2010	65.0	117.5
2012	76.2	121.1
2014	81.9	123.2
2016	90.9	125.8

Sources: *Nielsen Scarborough; U.S. Census Bureau*

10.1 Simplifying Rational Expressions

Objectives

» Describe the meaning of *rational expression.*

» Find all *excluded values* of a rational expression.

» Simplify a rational expression.

» Describe a rational equation in two variables.

» Make estimates and predictions with a *rational model.*

In this section, we will discuss the meaning of a *rational expression* and how to simplify it. We will also discuss how to use a *rational model* to make estimates and predictions about an authentic situation.

Rational Expressions

Throughout this course, we have worked with polynomials. If P and Q are polynomials with Q nonzero, we call the ratio $\dfrac{P}{Q}$ a **rational expression**. The name "rational" refers to a ratio. Here are some examples:

$$\frac{x^2 + 5x - 3}{x - 7} \qquad \frac{4x^2 - 7}{x^3} \qquad \frac{5x^4}{x^2 - 9}$$

Excluded Values

Consider the rational expression $\dfrac{3}{x}$. Note that substituting 0 for x in $\dfrac{3}{x}$ leads to $\dfrac{3}{0}$, which is undefined because we cannot divide by 0 (Section 2.2). We say that 0 is an excluded value of $\dfrac{3}{x}$.

> **Definition**   **Excluded value**
>
> A number is an **excluded value** of a rational expression if substituting the number into the expression leads to a division by 0.

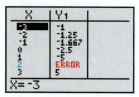

Figure 1 Verify the work

Example 1   Finding Excluded Values

Find any excluded values of the given expression.

1. $\dfrac{5}{x-2}$ **2.** $\dfrac{x}{3}$

Solution

1. The number 2 is an excluded value because $\dfrac{5}{2-2}$ has a division by 0. No other value of x leads to a division by 0, so 2 is the only excluded value. To verify our work, we construct a graphing calculator table for $y = \dfrac{5}{x-2}$. See Fig. 1. The "ERROR" message across from $x = 2$ in the table supports the idea that the value 2 for x leads to a division by 0. (The TI-83 and TI-84 display either "ERROR" or "ERR:".)

2. The denominator is the constant 3, so no substitution for x will lead to a division by 0. Thus, there are no excluded values.

WARNING   For the expression $\dfrac{5}{x-2}$, 0 itself is *not* an excluded value because $\dfrac{5}{0-2} = \dfrac{5}{-2}$ does not contain a division by 0.

For the expression $\dfrac{x}{3}$, again, 0 is not an excluded value because $\dfrac{0}{3}$ does not have a division by 0. In fact, $\dfrac{0}{3}$ is defined and $\dfrac{0}{3} = 0$.

Example 2   Finding Excluded Values

Find any excluded values of the given expression.

1. $\dfrac{x+1}{4x-7}$ **2.** $\dfrac{x-4}{2x^2-5x-3}$

Solution

1. To find which values of x lead to a division by 0, we set the denominator of $\dfrac{x+1}{4x-7}$ equal to 0 and solve for x:

$$4x - 7 = 0 \quad \textit{Set denominator of } \dfrac{x+1}{4x-7} \textit{ equal to 0.}$$
$$4x = 7 \quad \textit{Add 7 to both sides.}$$
$$x = \dfrac{7}{4} \quad \textit{Divide both sides by 4.}$$

So, the only excluded value is $\dfrac{7}{4}$.

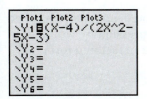

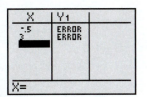

Figure 2 Verify the work

2. We set the denominator of $\dfrac{x-4}{2x^2-5x-3}$ equal to 0 and solve for x:

$$2x^2 - 5x - 3 = 0 \qquad \textcolor{blue}{\text{Set denominator of } \dfrac{x-4}{2x^2-5x-3} \text{ equal to } 0.}$$

$$(2x+1)(x-3) = 0 \qquad \textcolor{blue}{\text{Factor left side.}}$$

$$2x+1 = 0 \quad \text{or} \quad x - 3 = 0 \qquad \textcolor{blue}{\text{Zero factor property}}$$

$$2x = -1 \quad \text{or} \qquad x = 3$$

$$x = -\frac{1}{2} \quad \text{or} \qquad x = 3$$

So, the excluded values are $-\dfrac{1}{2}$ and 3. We use a graphing calculator table to verify our work (see Fig. 2).

Finding Excluded Values

To find any excluded values of a rational expression $\dfrac{P}{Q}$,

1. Set the denominator Q equal to 0.
2. Solve the equation $Q = 0$.

Simplifying a Rational Expression

In Sections 9.2 and 9.5, we simplified radical quotients by factoring the numerator and the denominator and then using the property $\dfrac{a}{a} = 1$, where $a \neq 0$. Here we use similar steps to simplify $\dfrac{15}{35}$:

$$\frac{15}{35} = \frac{5 \cdot 3}{5 \cdot 7} = \frac{5}{5} \cdot \frac{3}{7} = 1 \cdot \frac{3}{7} = \frac{3}{7}$$

We simplify the rational expression $\dfrac{3x+6}{x^2-4}$ in a similar manner:

$$\frac{3x+6}{x^2-4} = \frac{3(x+2)}{(x-2)(x+2)} \qquad \textcolor{blue}{\text{Factor numerator and denominator.}}$$

$$= \frac{3}{x-2} \cdot \frac{x+2}{x+2} \qquad \textcolor{blue}{\dfrac{AC}{BD} = \dfrac{A}{B} \cdot \dfrac{C}{D}}$$

$$= \frac{3}{x-2} \cdot 1 \qquad \textcolor{blue}{\text{Simplify: } \dfrac{x+2}{x+2} = 1}$$

$$= \frac{3}{x-2} \qquad \textcolor{blue}{a \cdot 1 = a}$$

Figure 3 Verify the work

The number 2 is the only excluded value of our result, $\dfrac{3}{x-2}$. The numbers -2 and 2 are the only excluded values of the original expression, $\dfrac{3x+6}{x^2-4}$. (Try it.) When we write

$$\frac{3x+6}{x^2-4} = \frac{3}{x-2}$$

we mean that the two expressions yield the same number for each real number substituted for x, except for any excluded values of either expression (-2 and 2). In Fig. 3, we use a graphing calculator table to verify our work.

A rational expression is in **lowest terms** if the numerator and denominator have no common factors other than 1 or −1. We **simplify a rational expression** by writing it in lowest terms. We also write the numerator and the denominator in factored form.

Simplifying a Rational Expression

To simplify a rational expression,

1. Factor the numerator and the denominator.

2. Use the property

$$\frac{AB}{AC} = \frac{A}{A} \cdot \frac{B}{C} = 1 \cdot \frac{B}{C} = \frac{B}{C}$$

where A and C are nonzero, so that the expression is in lowest terms.

Throughout the rest of this chapter, you may assume that the form $\frac{A}{B}$ represents a rational expression.

For the expression $\frac{AB}{AC}$, note that the polynomial A is a factor of both the numerator and the denominator. The expression $\frac{2x}{5x}$ can be simplified to $\frac{2}{5}$ when $x \neq 0$ because x is a factor of both the numerator and the denominator.

WARNING   The expression

$$\frac{2x + 7}{5x}$$

is in lowest terms already. Although x is a factor of the term $2x$, it is not a factor of $2x + 7$, the (entire) *numerator*.

Likewise, the expression

$$\frac{(x + 3)(x + 5) + 4}{7(x + 3)}$$

is in lowest terms already. The binomial $x + 3$ is not a factor of $(x + 3)(x + 5) + 4$, the (entire) numerator. To see that the rational expression is in lowest terms, we simplify the numerator to get $\frac{x^2 + 8x + 19}{7(x + 3)}$. (Try it.)

Example 3   Simplifying Rational Expressions

Simplify.

1. $\dfrac{20x^2}{12x^4}$

2. $\dfrac{5t + 15}{t^2 + 5t + 6}$

Solution

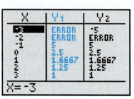

Figure 4 Verify the work

1.
$$\frac{20x^2}{12x^4} = \frac{2 \cdot 2 \cdot 5 \cdot x \cdot x}{2 \cdot 2 \cdot 3 \cdot x \cdot x \cdot x \cdot x} \qquad \textit{Factor numerator and denominator.}$$

$$= \frac{5}{3x^2} \qquad \textit{Simplify: } \frac{2 \cdot 2 \cdot x \cdot x}{2 \cdot 2 \cdot x \cdot x} = 1$$

2.
$$\frac{5t + 15}{t^2 + 5t + 6} = \frac{5(t + 3)}{(t + 2)(t + 3)} \qquad \textit{Factor numerator and denominator.}$$

$$= \frac{5}{t + 2} \qquad \textit{Simplify: } \frac{t + 3}{t + 3} = 1$$

We use a graphing calculator table to verify our work (see Fig. 4). There are "ERROR" messages across from $x = -3$ and $x = -2$ in the table because both −3 and −2 are excluded values of the original expression and −2 is an excluded value of our result.

Example 4   Simplifying Rational Expressions

Simplify.

1. $\dfrac{2a^2 - 50}{2a^2 - 6a - 20}$

2. $\dfrac{x^2 - 5x + 6}{x^3 + 3x^2 - 4x - 12}$

Solution

1.

$$\dfrac{2a^2 - 50}{2a^2 - 6a - 20} = \dfrac{2(a^2 - 25)}{2(a^2 - 3a - 10)}$$

$$= \dfrac{2(a - 5)(a + 5)}{2(a - 5)(a + 2)}$$

Factor numerator and denominator.

$$= \dfrac{a + 5}{a + 2}$$

Simplify: $\dfrac{2(a - 5)}{2(a - 5)} = 1$

2.

$$\dfrac{x^2 - 5x + 6}{x^3 + 3x^2 - 4x - 12} = \dfrac{(x - 2)(x - 3)}{x^2(x + 3) - 4(x + 3)}$$

$$= \dfrac{(x - 2)(x - 3)}{(x^2 - 4)(x + 3)}$$

Factor numerator and denominator.

$$= \dfrac{(x - 2)(x - 3)}{(x - 2)(x + 2)(x + 3)}$$

$$= \dfrac{x - 3}{(x + 2)(x + 3)}$$

Simplify: $\dfrac{x - 2}{x - 2} = 1$

Rational Equations in Two Variables

A **rational equation in two variables** is an equation in two variables in which each side can be written as a rational expression. Here are some examples of such equations:

$$y = \dfrac{1}{x} \qquad y = \dfrac{x + 2}{x - 4} \qquad y = \dfrac{3x}{x^2 - x - 6}$$

We show graphing calculator graphs of these equations in Fig. 5.

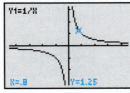 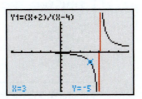 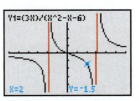

Figure 5 Graphs of three rational equations in two variables; the vertical lines (in red) displayed in the middle and right-hand screens are not part of the graph

Graphing rational equations in two variables by hand is beyond the scope of this course.

Rational Models

A **rational model** is a rational equation in two variables that describes the relationship between two quantities for an authentic situation. Rational models can help us describe equal shares of a quantity. For example, a rational model can describe the relationship between the number of people who attend a party and the equal per-person cost of the party.

Table 2 Number of Students and Per-Person Bus Cost

Number of Students n	Per-Person Bus Cost (dollars per person) p
1	$\frac{350}{1}$
2	$\frac{350}{2}$
3	$\frac{350}{3}$
4	$\frac{350}{4}$
n	$\frac{350}{n}$

Example 5   Finding a Rational Model

For a ski trip, the ski club at a college plans to charter a bus for a total cost of $350. Let p be the per-person bus cost (in dollars per person) if n students go on the trip.

1. Use a table to help find an equation for n and p. Show the arithmetic to help you see a pattern.
2. Perform a unit analysis of the equation you found in Problem 1.
3. Substitute 16 for n in the equation. What does the result mean in this situation?
4. Use a graphing calculator graph to verify the result found in Problem 3.
5. Is the graph of the model in Quadrant I increasing, decreasing, or neither? What does that mean in this situation?

Solution

1. We create Table 2. From the last row of the table, we see that the per-person cost (in dollars per person) can be represented by $\frac{350}{n}$. So, $p = \frac{350}{n}$.

2. Here is a unit analysis of the equation $p = \frac{350}{n}$:

$$\underbrace{p}_{\text{dollars per person}} = \underbrace{\frac{350}{n}}_{\text{dollars per person}} \begin{matrix} \leftarrow \text{dollars} \\ \leftarrow \text{number of students} \end{matrix}$$

The units of the expressions on both sides of the equation are dollars per person, which suggests that our equation is correct.

3. We substitute 16 for n in the equation $p = \frac{350}{n}$:

$$p = \frac{350}{16} = 21.875$$

So, the per-person cost is about $21.88 if 16 students go on the trip. For the ski club to pay exactly $350, eight members could pay $21.87 each and the other eight members could pay $21.88 each.

4. In Fig. 6, we verify our work in Problem 3.

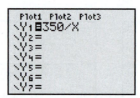
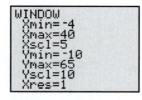
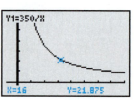

Figure 6 Verify the work

5. The graph appears to be decreasing in Quadrant I. That means the more people who go on the trip, the smaller the per-person cost will be, which makes sense.

In Example 5, we used a rational equation in two variables to model equal shares of a quantity. Rational equations in two variables can also be helpful in modeling percentages of a population or the average value of some quantity. For instance, a rational model can describe the relationship between the year and the percentage of households that are burglarized. A rational model can also describe the relationship between the year and the average household debt.

| Example 6 | Using a Rational Model to Make Estimates |

The number of people between the ages of 16 and 24 years who graduated from high school and the number of those students who then enrolled in college within 12 months are shown in Table 3 for various years.

Table 3 College Enrollment of Recent High School Graduates

Year	Number of High School Graduates (millions)	Number of High School Graduates Who Enrolled in College within 12 Months (millions)
1990	2.36	1.42
1995	2.60	1.61
2000	2.76	1.75
2005	2.68	1.84
2010	3.16	2.15
2015	2.97	2.06

Source: *U.S. National Center for Education Statistics*

Let p be the percentage of high school graduates who enrolled in college at t years since 1990 and within 12 months of receiving their high school diploma. A reasonable rational model is

$$p = \frac{2.8t + 145}{0.027t + 2.42}$$

1. Use the model to estimate the percentage of recent high school graduates who enrolled in college in 2015.
2. Find the *actual* percentage of recent high school graduates who enrolled in college in 2015.
3. Is the result you found in Problem 1 an underestimate or an overestimate? Explain.
4. Use the model to predict the percentage of recent high school graduates who will enroll in college in 2021.

Solution

1. Because $2015 - 1990 = 25$, we substitute 25 for t in the equation $p = \dfrac{2.8t + 145}{0.027t + 2.42}$ and solve for p:

$$p = \frac{2.8(25) + 145}{0.027(25) + 2.42} = \frac{215}{3.095} \approx 69.47$$

According to the model, about 69.5% of recent high school graduates enrolled in college in 2015.

2. From Table 3, we see that, of the 2.97 million recent high school graduates, 2.06 million enrolled in college. We begin to find the percentage by dividing the number enrolled in college by the number of recent high school graduates:

$$\frac{2.06}{2.97} \approx 0.694$$

By moving the decimal point of 0.694 two places to the right, we see that about 69.4% of recent high school graduates enrolled in college in 2015.

3. The result we found in Problem 1 is a slight overestimate: 69.5% > 69.4%.

4. Because $2021 - 1990 = 31$, we substitute 31 for t in the equation $p = \dfrac{2.8t + 145}{0.027t + 2.42}$ and solve for p:

$$p = \frac{2.8(31) + 145}{0.027(31) + 2.42} = \frac{231.8}{3.257} \approx 71.17$$

According to the model, about 71.2% of recent high school graduates will enroll in college in 2021.

Group Exploration

Simplifying rational expressions

We can write $\dfrac{2x}{5x} = \dfrac{2}{5}$ because x is a common factor of the numerator and the denominator. For each problem, use this concept to help you determine whether the work is correct. If you are unsure, it may help to decide whether you can show work similar to the following:

$$\frac{2x}{5x} = \frac{2}{5} \cdot \frac{x}{x} = \frac{2}{5} \cdot 1 = \frac{2}{5}$$

1. a. $\dfrac{4x}{9x} = \dfrac{4}{9}$ **b.** $\dfrac{4+x}{9x} = \dfrac{4+1}{9} = \dfrac{5}{9}$

2. a. $\dfrac{3x \cdot 5}{7x} = \dfrac{3 \cdot 5}{7} = \dfrac{15}{7}$ **b.** $\dfrac{3x+5}{7x} = \dfrac{3+5}{7} = \dfrac{8}{7}$

3. a. $\dfrac{8+x^5}{6+x^2} = \dfrac{4+x^3}{3}$ **b.** $\dfrac{8x^5}{6x^2} = \dfrac{4x^3}{3}$

4. a. $\dfrac{(x+2)(x-4)}{5(x-4)} = \dfrac{x+2}{5}$

b. $\dfrac{(x+2)(x-4)+1}{5(x-4)} = \dfrac{(x+2)+1}{5} = \dfrac{x+3}{5}$

Homework 10.1

For extra help ▶   **MyLab Math**   ▤ Watch the videos in MyLab Math

Find all excluded values.

1. $\dfrac{7}{x}$ **2.** $\dfrac{3}{x}$ **3.** $\dfrac{x}{4}$ **4.** $\dfrac{x}{6}$

5. $\dfrac{3}{x-4}$ **6.** $\dfrac{8}{x-1}$ **7.** $-\dfrac{x}{x+9}$ **8.** $-\dfrac{x}{x+2}$

9. $\dfrac{2p}{3p-12}$ **10.** $\dfrac{8t}{5t-20}$

11. $\dfrac{x-4}{6x+8}$ **12.** $\dfrac{x+1}{12x+9}$

13. $\dfrac{2}{x^2+5x+6}$ **14.** $\dfrac{5}{x^2+8x+15}$

15. $\dfrac{r}{r^2-2r-35}$ **16.** $\dfrac{2y}{y^2-10y+24}$

17. $\dfrac{x-9}{x^2-10x+25}$ **18.** $\dfrac{x+7}{x^2+6x+9}$

19. $\dfrac{3x-1}{x^2-16}$ **20.** $\dfrac{x+8}{x^2-25}$

21. $\dfrac{c+4}{25c^2-49}$ **22.** $\dfrac{k-3}{36k^2-1}$

23. $\dfrac{2x-5}{2x^2+13x+15}$ **24.** $\dfrac{7x-4}{3x^2-7x+2}$

25. $\dfrac{3x}{6x^2-13x+6}$ **26.** $\dfrac{x+1}{4x^2+4x-3}$

27. $\dfrac{w-5}{w^3+2w^2-4w-8}$ **28.** $\dfrac{t+2}{t^3+3t^2-9t-27}$

Simplify.

29. $\dfrac{4x}{6}$ **30.** $\dfrac{12}{15x}$ **31.** $\dfrac{12t^3}{15t}$ **32.** $\dfrac{21y^4}{14y^2}$

33. $\dfrac{18x^3y}{27x^2y^4}$ **34.** $\dfrac{35x^2y^3}{25x^5y}$ **35.** $\dfrac{3x-6}{5x-10}$

36. $\dfrac{7x+21}{4x+12}$ **37.** $\dfrac{2x+12}{3x+18}$ **38.** $\dfrac{5x-25}{2x-10}$

39. $\dfrac{a^3+4a^2}{7a^2+28a}$ **40.** $\dfrac{4r^2-24r}{r^3-6r^2}$ **41.** $\dfrac{x^2-y^2}{3x+3y}$

42. $\dfrac{x^2-xy}{2x-2y}$ **43.** $\dfrac{4x+8}{x^2+7x+10}$ **44.** $\dfrac{3x+12}{x^2+7x+12}$

45. $\dfrac{5x-35}{x^2-9x+14}$ **46.** $\dfrac{4x-4}{x^2-9x+8}$ **47.** $\dfrac{t^2+5t+4}{t^2+9t+20}$

48. $\dfrac{b^2+3b+2}{b^2+8b+12}$ **49.** $\dfrac{x^2-9x+14}{x^2-8x+7}$ **50.** $\dfrac{x^2-x-30}{x^2-4x-12}$

51. $\dfrac{x^2-4}{x^2-3x-10}$ **52.** $\dfrac{x^2-9}{x^2+5x-24}$

53. $\dfrac{x^3+8x^2+16x}{x^3-16x}$ **54.** $\dfrac{x^3-14x^2+49x}{x^3-49x}$

55. $\dfrac{6x-16}{9x^2-64}$ **56.** $\dfrac{27x+21}{81x^2-49}$

57. $\dfrac{-4w+8}{w^2+2w-8}$ **58.** $\dfrac{-6a+30}{a^2-6a+5}$

59. $\dfrac{4x^2-25}{2x^2+x-15}$ **60.** $\dfrac{9x^2-4}{3x^2-8x+4}$

61. $\dfrac{3x^2+9x+6}{6x^2+5x-1}$ **62.** $\dfrac{2x^2+6x-20}{4x^2-5x-6}$

63. $\dfrac{a^2+2ab+b^2}{a^2-b^2}$ **64.** $\dfrac{a^2+3ab+2b^2}{a^2+4ab+3b^2}$

65. $\dfrac{5x+10}{x^3+2x^2-3x-6}$ **66.** $\dfrac{5x-15}{x^3-3x^2-4x+12}$

67. $\dfrac{t^2 + 2t + 1}{t^3 + t^2 - t - 1}$

68. $\dfrac{a^2 + 4a + 4}{a^3 + 2a^2 - 4a - 8}$

69. $\dfrac{x^3 + 8}{x^2 + 7x + 10}$

70. $\dfrac{x^3 + 27}{x^2 - 2x - 15}$

71. $\dfrac{x^3 - 64}{x^2 - 16}$

72. $\dfrac{x^3 - 8}{x^2 - 4}$

73. Some students have a party and share equally in the expense, which is \$60 for both soft drinks and snacks. Let p be the per-person cost (in dollars per person) if n students go to the party.
   a. Complete Table 4 to help find an equation for n and p. Show the arithmetic to help you see a pattern.

Table 4 Per-Person Cost for the Party

Number of Students n	Cost per Student (dollars per student) p
10	
20	
30	
40	
n	

   b. Perform a unit analysis of the equation you found in part (a).
   c. Substitute 25 for n in the equation, and solve for p. What does your result mean in this situation?
   d. For positive values of n, what happens to the value of p as the value of n is increased? What does that trend mean in this situation?

74. A student has taken a total of 70 units (hours or credits) of courses at a college. Let A be the average number of units the student has taken per semester and n be the number of semesters the student has been enrolled at the college.
   a. Complete Table 5 to help find an equation for n and A. Show the arithmetic to help you see a pattern.

Table 5 Average Number of Units per Semester

Number of Semesters n	Average Number of Units per Semester A
4	
5	
6	
7	
n	

   b. Perform a unit analysis of the equation you found in part (a).
   c. Substitute 8 for n in the equation, and solve for A. What does your result mean in this situation?
   d. For positive values of n, what happens to the value of A as the value of n is increased? What does that trend mean in this situation?

75. DATA The numbers of U.S. households that have been burglarized are shown in Table 6 for various years.

Table 6 Numbers of Households Burglarized

Year	Number of Households Burglarized (millions)	Total Number of Households (millions)
2011	1.63	118.7
2012	1.56	121.1
2013	1.43	122.5
2014	1.27	123.2
2015	1.13	124.6

Sources: *FBI; U.S. Census Bureau*

Let p be the percentage of households that were burglarized in the year that is t years since 2010. A reasonable model is

$$p = \frac{-13t + 179}{1.4t + 117.9}$$

   a. Use the model to estimate the percentage of households that were burglarized in 2012.
   b. Now find the *actual* percentage of households that were burglarized in 2012.
   c. Is the result you found in part (a) an underestimate or an overestimate?
   d. Predict the percentage of households that will be burglarized in 2021.

76. DATA The numbers of households that stream Netflix are shown in Table 7 for various years.

Table 7 Numbers of Households That Stream Netflix

Year	Households That Stream Netflix (millions)	Total Number of Households (millions)
2012	23.4	121.1
2013	29.2	122.5
2014	35.7	123.2
2015	41.4	124.6
2016	47.0	125.8

Sources: *Netflix; U.S. Census Bureau*

Let p be the percentage of households that stream Netflix at t years since 2010. A reasonable model is

$$p = \frac{594t + 1158}{1.15t + 118.84}$$

   a. Use the model to estimate the percentage of households that streamed Netflix in 2016.
   b. Now find the *actual* percentage of households that streamed Netflix in 2016.
   c. Is the result you found in part (a) an underestimate or an overestimate? Explain.
   d. Predict the percentage of households that will stream Netflix in 2022.

Concepts

77. A student thinks that 0 is an excluded value of the expression $\dfrac{x}{2}$. Is the student correct? Explain.

78. A student thinks that the numbers 3 and 5 are excluded values of the expression $\dfrac{x - 3}{x - 5}$. Is the student correct? Explain.

79. It is a common error to think that 0 is an excluded value of the expression $\frac{5}{x+2}$. Substitute 0 for x in the expression $\frac{5}{x+2}$ to show that 0 is not an excluded value.

80. It is a common error to think that 4 is an excluded value of the expression $\frac{x-4}{x+6}$. Substitute 4 for x in the expression $\frac{x-4}{x+6}$ to show that 4 is not an excluded value.

81. A student tries to simplify the expression $\frac{4x+3}{2x}$:

$$\frac{4x+3}{2x} = \frac{4+3}{2} = \frac{7}{2}$$

Describe any errors. Then simplify the expression correctly.

82. A student tries to simplify the expression

$$\frac{(x-2)(x+4)}{x^2(x-2)+5(x-2)}:$$

$$\frac{(x-2)(x+4)}{x^2(x-2)+5(x-2)} = \frac{x+4}{x^2+5(x-2)}$$

$$= \frac{x+4}{x^2+5x-10}$$

Describe any errors. Then simplify the expression correctly.

83. Give three examples of rational expressions, each of whose only excluded value is 7.

84. Give three examples of rational expressions, each of whose only excluded values are -3 and 5.

85. Give three examples of rational expressions, each equivalent to the expression $\frac{x}{x-4}$.

86. Give three examples of rational expressions, each equivalent to the expression $\frac{5}{x+2}$.

87. Explain why we can simplify the expression $\frac{3x}{7x}$ but the expression $\frac{3+x}{7x}$ is in lowest terms already.

88. Does $\frac{(x+2)(x+3)}{x+2}$ have the same value as $x+3$ for all real numbers? Explain.

89. Describe how to find all excluded values of a rational expression. (See page 9 for guidelines on writing a good response.)

90. Describe how to simplify a rational expression. (See page 9 for guidelines on writing a good response.)

Related Review

Simplify.

91. $\frac{x^{-4}y^9}{5w^{-2}}$ **92.** $\frac{2w^{-3}}{x^{-5}y^{-1}}$ **93.** $\frac{4x-12}{x^2-7x+12}$

94. $\frac{x^2-4}{x^2+5x-14}$ **95.** $\frac{6-\sqrt{32}}{8}$ **96.** $\frac{10+\sqrt{50}}{35}$

Expressions, Equations, and Graphs

Perform the indicated instruction. Then use words such as linear, quadratic, cubic, power, radical, rational, polynomial, degree, one variable, *and* two variables *to describe the expression, equation, or system.*

97. Factor $2p^3 - 3p^2 - 18p + 27$.

98. Simplify $3(2x-5) - (5x+2) + 2(x-3)$.

99. Graph $y = x^2 - 2x$ by hand.

100. Simplify $-2(4x-6)^2$.

101. Solve $x(3x-2) = 4$.

102. Graph $y = 3x^2$ by hand.

10.2 Multiplying and Dividing Rational Expressions; Converting Units

Objectives

» Multiply rational expressions.

» Divide rational expressions.

» Convert units of quantities.

In Section 10.1, we simplified rational expressions. In this section, we will multiply and divide rational expressions.

Multiplication of Rational Expressions

Recall from Section 2.2 that we multiply fractions by using the property $\frac{a}{b} \cdot \frac{c}{d} = \frac{ac}{bd}$, where b and d are nonzero. For example,

$$\frac{2}{5} \cdot \frac{3}{7} = \frac{2 \cdot 3}{5 \cdot 7} = \frac{6}{35}$$

We multiply more complicated rational expressions in a similar way.

Multiplying Rational Expressions

If $\frac{A}{B}$ and $\frac{C}{D}$ are rational expressions and B and D are nonzero, then

$$\frac{A}{B} \cdot \frac{C}{D} = \frac{AC}{BD}$$

In words: To multiply two rational expressions, write the numerators as a product and write the denominators as a product.

Example 1   Finding the Product of Two Rational Expressions

Find the product $\dfrac{4x}{5} \cdot \dfrac{7}{6x^4}$. Simplify the result.

Solution

$$\frac{4x}{5} \cdot \frac{7}{6x^4} = \frac{4x \cdot 7}{5 \cdot 6x^4}$$

Write numerators as a product and denominators as a product: $\dfrac{A}{B} \cdot \dfrac{C}{D} = \dfrac{AC}{BD}$

$$= \frac{2 \cdot 2 \cdot x \cdot 7}{5 \cdot 2 \cdot 3 \cdot x \cdot x \cdot x \cdot x}$$

Factor numerator and denominator.

$$= \frac{14}{15x^3}$$

Simplify: $\dfrac{2 \cdot x}{2 \cdot x} = 1$

To find the product of rational expressions, we usually begin by factoring the numerators and the denominators if possible. That will put us in a good position to simplify after we have found the product.

Example 2   Finding the Product of Two Rational Expressions

Find the product $\dfrac{2x + 6}{3} \cdot \dfrac{6x}{5x + 15}$. Simplify the result.

Solution

$$\frac{2x + 6}{3} \cdot \frac{6x}{5x + 15} = \frac{2(x + 3)}{3} \cdot \frac{2 \cdot 3 \cdot x}{5(x + 3)}$$

Factor numerators and denominators.

$$= \frac{2(x + 3) \cdot 2 \cdot 3 \cdot x}{3 \cdot 5(x + 3)}$$

Write both numerators and denominators as a product: $\dfrac{A}{B} \cdot \dfrac{C}{D} = \dfrac{AC}{BD}$

$$= \frac{4x}{5}$$

Simplify: $\dfrac{(x + 3) \cdot 3}{(x + 3) \cdot 3} = 1$

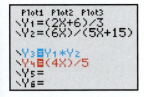

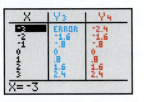

Figure 7 Verify the work

We verify our work by constructing a graphing calculator table for $y = \dfrac{2x + 6}{3} \cdot \dfrac{6x}{5x + 15}$ and $y = \dfrac{4x}{5}$. See Fig. 7. To enter an equation by using Y_n references, see Appendix A.22. There is an "ERROR" message across from $x = -3$ in the table because -3 is an excluded value of the original expression.

Example 3   Finding the Product of Two Rational Expressions

Find the product $\dfrac{x^2 - 49}{5x - 10} \cdot \dfrac{x^2 - 2x}{x^2 - 9x + 14}$. Simplify the result.

Solution

$$\frac{x^2 - 49}{5x - 10} \cdot \frac{x^2 - 2x}{x^2 - 9x + 14} = \frac{(x - 7)(x + 7)}{5(x - 2)} \cdot \frac{x(x - 2)}{(x - 7)(x - 2)}$$

Factor numerators and denominators.

$$= \frac{(x - 7)(x + 7) \cdot x(x - 2)}{5(x - 2)(x - 7)(x - 2)}$$

Multiply numerators; multiply denominators.

$$= \frac{x(x + 7)}{5(x - 2)}$$

Simplify: $\dfrac{(x - 7)(x - 2)}{(x - 7)(x - 2)} = 1$

> **How to Multiply Rational Expressions**
>
> To multiply two rational expressions,
>
> **1.** Factor the numerators and the denominators.
>
> **2.** Multiply by using the property $\dfrac{A}{B} \cdot \dfrac{C}{D} = \dfrac{AC}{BD}$, where B and D are nonzero.
>
> **3.** Simplify the result.

Example 4   Finding the Product of Two Rational Expressions

Find the product $\dfrac{4x^2 - 9}{2x - 10} \cdot \dfrac{x^2 - 10x + 25}{2x^2 - 7x + 6}$. Simplify the result.

Solution

$$
\begin{aligned}
\frac{4x^2 - 9}{2x - 10} \cdot \frac{x^2 - 10x + 25}{2x^2 - 7x + 6} &= \frac{(2x - 3)(2x + 3)}{2(x - 5)} \cdot \frac{(x - 5)(x - 5)}{(2x - 3)(x - 2)} \qquad \textit{Factor numerators and denominators.} \\[2mm]
&= \frac{(2x - 3)(2x + 3)(x - 5)(x - 5)}{2(x - 5)(2x - 3)(x - 2)} \qquad \textit{Multiply numerators; multiply denominators.} \\[2mm]
&= \frac{(x - 5)(2x + 3)}{2(x - 2)} \qquad \textit{Simplify: } \frac{(2x - 3)(x - 5)}{(2x - 3)(x - 5)} = 1
\end{aligned}
$$

Division of Rational Expressions

Recall from Section 2.2 that, to divide two fractions, we use the property $\dfrac{a}{b} \div \dfrac{c}{d} = \dfrac{a}{b} \cdot \dfrac{d}{c}$, where b, c, and d are nonzero. For example,

$$\frac{2}{5} \div \frac{3}{7} = \frac{2}{5} \cdot \frac{7}{3} = \frac{14}{15}$$

Note that dividing by $\dfrac{3}{7}$ is the same as multiplying by the reciprocal of $\dfrac{3}{7}$, which is $\dfrac{7}{3}$. We divide more complicated rational expressions in a similar way.

> **Dividing Rational Expressions**
>
> If $\dfrac{A}{B}$ and $\dfrac{C}{D}$ are rational expressions and B, C, and D are nonzero, then
>
> $$\frac{A}{B} \div \frac{C}{D} = \frac{A}{B} \cdot \frac{D}{C}$$
>
> In words: To divide by a rational expression, multiply by its reciprocal.

Example 5   Finding the Quotient of Two Rational Expressions

Find the quotient $\dfrac{15x^3}{2} \div \dfrac{35x^7}{4}$.

Solution

$$
\begin{aligned}
\frac{15x^3}{2} \div \frac{35x^7}{4} &= \frac{15x^3}{2} \cdot \frac{4}{35x^7} \qquad \textit{Multiply by reciprocal of } \frac{35x^7}{4}, \textit{ which is } \frac{4}{35x^7}. \\[2mm]
&= \frac{3 \cdot 5x^3}{2} \cdot \frac{2 \cdot 2}{5 \cdot 7x^7} \qquad \textit{Factor numerators and denominators.} \\[2mm]
&= \frac{3 \cdot 5 \cdot x^3 \cdot 2 \cdot 2}{2 \cdot 5 \cdot 7 \cdot x^7} \qquad \textit{Multiply numerators; multiply denominators.} \\[2mm]
&= \frac{6}{7x^4} \qquad \textit{Simplify: } \frac{2 \cdot 5 \cdot x^3}{2 \cdot 5 \cdot x^3} = 1
\end{aligned}
$$

How to Divide Rational Expressions

To divide two rational expressions,

1. Write the quotient as a product by using the property $\dfrac{A}{B} \div \dfrac{C}{D} = \dfrac{A}{B} \cdot \dfrac{D}{C}$, where B, C, and D are nonzero.
2. Find the product.
3. Simplify.

Example 6   Finding the Quotient of Two Rational Expressions

Find the quotient.

1. $\dfrac{x+1}{x^2 - x - 6} \div \dfrac{3x+3}{x^2-9}$

2. $\dfrac{3x^2 + 11x + 10}{x^2 + 2x - 24} \div \dfrac{x^2 - 4}{3x^2 + 18x}$

Solution

1. $\dfrac{x+1}{x^2 - x - 6} \div \dfrac{3x+3}{x^2-9}$

$= \dfrac{x+1}{x^2 - x - 6} \cdot \dfrac{x^2 - 9}{3x + 3}$ *Multiply by reciprocal of $\dfrac{3x+3}{x^2-9}$.*

$= \dfrac{x+1}{(x-3)(x+2)} \cdot \dfrac{(x-3)(x+3)}{3(x+1)}$ *Factor numerators and denominators.*

$= \dfrac{(x+1)(x-3)(x+3)}{(x-3)(x+2) \cdot 3(x+1)}$ *Multiply numerators; multiply denominators.*

$= \dfrac{x+3}{3(x+2)}$ *Simplify: $\dfrac{(x+1)(x-3)}{(x+1)(x-3)} = 1$*

We verify our work by constructing a graphing calculator table for the equations $y = \dfrac{x+1}{x^2 - x - 6} \div \dfrac{3x+3}{x^2-9}$ and $y = \dfrac{x+3}{3(x+2)}$. See Fig. 8. There are "ERROR" messages across from the values $x = -3$, $x = -2$, $x = -1$, and $x = 3$ because each of these values is an excluded value of either the original expression or our result.

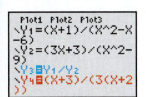

Figure 8 Verify the work

2. $\dfrac{3x^2 + 11x + 10}{x^2 + 2x - 24} \div \dfrac{x^2 - 4}{3x^2 + 18x}$

$= \dfrac{3x^2 + 11x + 10}{x^2 + 2x - 24} \cdot \dfrac{3x^2 + 18x}{x^2 - 4}$ *Multiply by reciprocal of $\dfrac{x^2 - 4}{3x^2 + 18x}$.*

$= \dfrac{(3x+5)(x+2)}{(x-4)(x+6)} \cdot \dfrac{3x(x+6)}{(x-2)(x+2)}$ *Factor numerators and denominators.*

$= \dfrac{(3x+5)(x+2) \cdot 3x(x+6)}{(x-4)(x+6)(x-2)(x+2)}$ *Multiply numerators; multiply denominators.*

$= \dfrac{3x(3x+5)}{(x-4)(x-2)}$ *Simplify: $\dfrac{(x+2)(x+6)}{(x+2)(x+6)} = 1$*

Example 7   Finding the Quotient of Two Rational Expressions

Find the quotient $\dfrac{x^2 + 6x + 9}{x - 4} \div \left(x^2 + x - 6 \right)$.

Solution

$$\dfrac{x^2 + 6x + 9}{x - 4} \div \left(x^2 + x - 6 \right) = \dfrac{x^2 + 6x + 9}{x - 4} \cdot \dfrac{1}{x^2 + x - 6} \qquad \textit{Multiply by reciprocal of } \dfrac{x^2 + x -}{1}$$

$$= \dfrac{(x + 3)(x + 3)}{x - 4} \cdot \dfrac{1}{(x + 3)(x - 2)} \qquad \textit{Factor numerators and denominators.}$$

$$= \dfrac{(x + 3)(x + 3)}{(x - 4)(x + 3)(x - 2)} \qquad \textit{Multiply numerators; multiply denominators.}$$

$$= \dfrac{x + 3}{(x - 4)(x - 2)} \qquad \textit{Simplify: } \dfrac{x + 3}{x + 3} = 1$$

Converting Units of Quantities

Suppose we're ordering a rug measured in feet, but we measured the width and the length of the floor in inches. There is no need to remeasure: We can convert the units of a quantity to equivalent units by multiplying by ratios of units, where each ratio is equal to 1. For example, because there are 12 inches in 1 foot, the following ratio is equal to 1:

$$\dfrac{12 \text{ inches}}{1 \text{ foot}} = 1$$

The reciprocal is also equal to 1:

$$\dfrac{1 \text{ foot}}{12 \text{ inches}} = 1$$

Suppose the width of a floor is 110 inches. Here we convert from units of inches to feet:

$$\dfrac{110 \text{ inches}}{1} \cdot \dfrac{1 \text{ foot}}{12 \text{ inches}} \approx 9.2 \text{ feet}$$

When converting, we can eliminate "inches" because $\dfrac{\text{inches}}{\text{inches}} = 1$.

We can eliminate a pair of the same units even if one is in singular form and the other is in plural form. For example, $\dfrac{\text{feet}}{\text{foot}} = 1$. However, $\dfrac{\text{inch}^2}{\text{inch}} = \text{inch}$, not 1.

Some equivalent units are shown in the margin.

Example 8   Converting Units

Make the indicated unit conversions. Round the results to two decimal places for Problems 2–4.

1. The official height of a basketball hoop is 10 feet. What is its height in yards?
2. An electric 1974 Fender® Jazz Bass® is 46.25 inches long. What is the length in centimeters?
3. In 2015, Americans consumed an average of 214 pounds of meat (Source: *U.S. Department of Agriculture*). What is this average in ounces per day?
4. A person walks at a speed of 4 miles per hour. What is the person's speed in feet per second?

Equivalent Units

Length

1 inch = 2.54 centimeters
1 foot = 12 inches
1 yard = 3 feet
1 mile = 5280 feet
1 mile ≈ 1.61 kilometers

Volume

1 cup = 8 ounces
1 quart = 4 cups
1 quart ≈ 0.946 liter
1 gallon = 4 quarts

Weight

1 gram = 1000 milligrams
1 pound = 16 ounces

Time

1 year = 365 days

Solution

1. Because there are 3 feet in 1 yard, we can multiply 10 feet by $\dfrac{1 \text{ yard}}{3 \text{ feet}} = 1$. By doing so, we can eliminate "feet" because $\dfrac{\text{feet}}{\text{feet}} = 1$:

$$\frac{10 \text{ feet}}{1} \cdot \frac{1 \text{ yard}}{3 \text{ feet}} = \frac{10}{3} \text{ yards}$$

The official height of the hoop is $3\dfrac{1}{3}$ yards.

2. Because there are 2.54 centimeters in 1 inch, we multiply 46.25 inches by $\dfrac{2.54 \text{ centimeters}}{1 \text{ inch}} = 1$ so that the inches are eliminated:

$$\frac{46.25 \text{ inches}}{1} \cdot \frac{2.54 \text{ centimeters}}{1 \text{ inch}} \approx 117.48 \text{ centimeters}$$

The bass is approximately 117.48 centimeters long.

3. There are 16 ounces in 1 pound and approximately 365 days in 1 year. To convert, we multiply by ratios equal to 1 or approximately equal to 1. We arrange the ratios so that the units we want to eliminate appear in one numerator and one denominator:

$$\frac{214 \text{ pounds}}{1 \text{ year}} \cdot \frac{16 \text{ ounces}}{1 \text{ pound}} \cdot \frac{1 \text{ year}}{365 \text{ days}} \approx 9.38 \frac{\text{ounces}}{\text{day}}$$

On average, Americans consume about 9.38 ounces of meat per day.

4. There are 5280 feet in 1 mile, 60 minutes in 1 hour, and 60 seconds in 1 minute. To convert, we multiply by ratios equal to 1. Again, we arrange the ratios so that the units we want to eliminate appear in one numerator and one denominator:

$$\frac{4 \text{ miles}}{1 \text{ hour}} \cdot \frac{5280 \text{ feet}}{1 \text{ mile}} \cdot \frac{1 \text{ hour}}{60 \text{ minutes}} \cdot \frac{1 \text{ minute}}{60 \text{ seconds}} \approx 5.87 \frac{\text{feet}}{\text{second}}$$

So, the person is walking at a speed of about 5.87 feet per second.

Converting Units

To convert the units of a quantity,

1. Write the quantity in the original units.
2. Multiply by fractions equal to 1 so that the units you want to eliminate appear in one numerator and one denominator.

◈ Group Exploration

Section Opener: Multiplying and dividing rational expressions

Find the indicated product.

1. $\dfrac{3}{x} \cdot \dfrac{2}{x}$

2. $\dfrac{x}{4} \cdot \dfrac{5}{x+3}$

3. $\dfrac{x-3}{5} \cdot \dfrac{x+2}{x}$

4. $\dfrac{x^2+5x+6}{x-5} \cdot \dfrac{x+1}{x^2+7x+10}$

5. $\dfrac{x^2-25}{x^2-x-12} \cdot \dfrac{(x-4)^2}{4x+20}$

6. $\dfrac{x-2}{3} \div \dfrac{x+3}{x}$

7. $\dfrac{x^2+7x+12}{x-2} \div \dfrac{x+3}{x^2-7x+10}$

8. $\dfrac{x^2-9}{x^7} \div \dfrac{3x+9}{4x^3}$

Homework 10.2

Perform the indicated operation. Simplify the result.

1. $\dfrac{3}{x} \cdot \dfrac{5}{x}$

2. $\dfrac{x}{2} \cdot \dfrac{x}{7}$

3. $\dfrac{x}{6} \div \dfrac{3}{2x}$

4. $\dfrac{8}{x} \div \dfrac{10x}{7}$

5. $\dfrac{6a^2}{7} \cdot \dfrac{21}{5a^8}$

6. $\dfrac{15}{3t^4} \cdot \dfrac{2t^9}{40}$

7. $\dfrac{2}{x-3} \cdot \dfrac{x-4}{x+5}$

8. $\dfrac{x-1}{x+7} \cdot \dfrac{x-6}{5}$

9. $\dfrac{k-2}{k-6} \div \dfrac{k+6}{k+4}$

10. $\dfrac{y-4}{y+8} \div \dfrac{y+3}{y+4}$

11. $\dfrac{6}{7x-14} \cdot \dfrac{5x-10}{9}$

12. $\dfrac{3x+15}{8} \cdot \dfrac{3}{4x+20}$

13. $\dfrac{3x+18}{x-6} \div \dfrac{x+6}{2x-12}$

14. $\dfrac{x+4}{7x-7} \div \dfrac{3x+12}{x-1}$

15. $\dfrac{4w^6}{w+3} \cdot \dfrac{w+5}{2w^2}$

16. $\dfrac{b-1}{6b^3} \cdot \dfrac{9b^7}{b+4}$

17. $\dfrac{(x-4)(x+1)}{(x-7)(x+2)} \cdot \dfrac{5(x+2)}{3(x-4)}$

18. $\dfrac{(x+3)(x+9)}{(x-6)(x-1)} \cdot \dfrac{x(x-6)}{6(x+3)}$

19. $\dfrac{4(x-4)^2}{(x+5)^2} \div \dfrac{14(x-4)}{15(x+5)}$

20. $\dfrac{9(x+2)^2}{8(x-1)^2} \div \dfrac{3(x+2)}{10(x-1)}$

21. $\dfrac{4t^7}{3t-9} \cdot \dfrac{5t-15}{8t^3}$

22. $\dfrac{10w+10}{18w^5} \cdot \dfrac{27w^3}{5w+5}$

23. $\dfrac{8x^2}{x^2-49} \div \dfrac{4x^5}{3x+21}$

24. $\dfrac{33x^6}{5x-30} \div \dfrac{11x^{10}}{x^2-36}$

25. $\dfrac{15a^4b}{8ab^5} \div \dfrac{25ab^3}{4a}$

26. $\dfrac{4xy^3}{9x^3y} \div \dfrac{10y}{3xy^5}$

27. $\dfrac{x^2-3x-28}{x^2+4x-45} \cdot \dfrac{x+9}{x-7}$

28. $\dfrac{x^2+x-42}{x^2-x-20} \cdot \dfrac{x-5}{x-6}$

29. $\dfrac{t^2-36}{t^2-81} \div \dfrac{t+6}{t-9}$

30. $\dfrac{r^2-1}{r^2-49} \div \dfrac{r-1}{r+7}$

31. $\dfrac{x^2+6x+8}{x^2-5x} \cdot \dfrac{4x-20}{3x+6}$

32. $\dfrac{6x-12}{5x+35} \cdot \dfrac{x^2+10x+21}{2x^2-4x}$

33. $\dfrac{x^2-4}{x^2-7x+12} \div \dfrac{x^2-8x+12}{x^2-2x-3}$

34. $\dfrac{x^2-11x+28}{x^2-x} \div \dfrac{x^2-5x-14}{x^2-1}$

35. $\dfrac{a^2+6a+9}{a^2+6a} \cdot \dfrac{a^2+11a+30}{4a^2-36}$

36. $\dfrac{c^2-10c+25}{3c^2-12c} \cdot \dfrac{c^2-6c+8}{2c^2-50}$

37. $\dfrac{x^2+x-6}{x^2+9x+8} \cdot \dfrac{x^2+5x-24}{x^2-7x+10}$

38. $\dfrac{x^2-x-30}{x^2-x-12} \cdot \dfrac{x^2-6x+8}{x^2+3x-10}$

39. $\dfrac{x^2+16}{x^2+8x+16} \div \dfrac{4x}{x^2-16}$

40. $\dfrac{3x}{x^2-6x+9} \div \dfrac{x^2+9}{x^2-9}$

41. $\dfrac{9x^2-25}{x^2-8x+16} \cdot \dfrac{x^2-4x}{3x^2-2x-5}$

42. $\dfrac{x^2+5x}{2x^2-x-21} \cdot \dfrac{4x^2-49}{x^2+10x+25}$

43. $\dfrac{3p^2+15p+12}{4p^2} \div \dfrac{9p+36}{6p^4}$

44. $\dfrac{15w^6}{2w^2+20w+48} \div \dfrac{5w^2}{8w+48}$

45. $\dfrac{x^2-25}{x^3-3x^2-4x} \div \dfrac{x^2-10x+25}{x^2-4x}$

46. $\dfrac{3x^3+18x^2+24x}{x^2-36} \div \dfrac{6x^2+24x}{5x-30}$

47. $\dfrac{-6x+12}{x^2+10x+21} \cdot \dfrac{x^2-9}{-3x+6}$

48. $\dfrac{x^2-2x+1}{-10x+40} \cdot \dfrac{-6x+24}{x^2-3x+2}$

49. $(b^2-25) \cdot \dfrac{2b}{b^2-6b+5}$

50. $(t^2-11t+24) \cdot \dfrac{t+1}{4t-32}$

51. $\dfrac{x^2+8x}{x-4} \div (x^2+16x+64)$

52. $\dfrac{4x+28}{x+1} \div (x^2-49)$

53. $\dfrac{a^2-b^2}{4a^2-9b^2} \cdot \dfrac{2a-3b}{a+b}$

54. $\dfrac{a^2b-ab^2}{a+3b} \cdot \dfrac{3a^2+9ab}{a^2-b^2}$

55. $\dfrac{x^2-y^2}{3x+6y} \div \dfrac{x^2+2xy+y^2}{x^2+2xy}$

56. $\dfrac{2x^2+3xy+y^2}{5x-10y} \div \dfrac{4x^2-y^2}{xy-2y^2}$

57. $\dfrac{x^3-2x^2+3x-6}{x^2-3x} \cdot \dfrac{3x-9}{x^2-4}$

58. $\dfrac{x^3+3x^2-2x-6}{x^2+2x} \cdot \dfrac{5x+10}{x^2-9}$

59. $\dfrac{x^3+27}{6x-3} \div \dfrac{2x^2-6x+18}{2x^2-x}$

60. $\dfrac{x^3-8}{x^2-9} \cdot \dfrac{5x+15}{3x-6}$

61. Perform the indicated operation. Simplify the result.

a. $\dfrac{x^2 - 1}{x^2 - 4} \cdot \dfrac{x^2 + 3x + 2}{x^2 - 3x + 2}$

b. $\dfrac{x^2 - 1}{x^2 - 4} \div \dfrac{x^2 + 3x + 2}{x^2 - 3x + 2}$

62. Perform the indicated operation. Simplify the result.

a. $\dfrac{x^2 + x - 6}{x^2 - 4x - 5} \cdot \dfrac{x^2 - x - 2}{x^2 - 2x - 15}$

b. $\dfrac{x^2 + x - 6}{x^2 - 4x - 5} \div \dfrac{x^2 - x - 2}{x^2 - 2x - 15}$

For Exercises 63–72, round approximate results to two decimal places. Refer to the list of equivalent units in the margin of page 546 as needed.

63. On an official tennis court, the height of the net is 3 feet at the center. What is that height in inches?

64. An NBA basketball court is 94 feet long. What is its length in yards?

65. A person races in a 10-kilometer run. How long is the race in miles?

66. A person buys 13.9 gallons of gasoline. How many liters of gasoline is that?

67. In 2015, Americans consumed an average of 15.5 pounds of fish and shellfish per year (Source: *National Oceanic and Atmospheric Administration*). What is this average in ounces per day?

68. A person drives at a speed of 68 miles per hour. What is the driving speed in feet per second?

69. A Porsche® 911 Turbo® Coupe has a gas mileage equal to 6.29 kilometers per liter. What is the car's gas mileage in miles per gallon?

70. There are 2250 milligrams of salt in a 10-ounce bag of Snyder's® pretzels. How many grams of salt are there in a pound of Snyder's pretzels?

71. A 12-ounce can of Jolt® has 71.2 milligrams of caffeine (Source: *National Soft Drink Association*). How many grams of caffeine are in 1 gallon of Jolt?

72. A 1-cup serving of Coca-Cola® has 26 grams of sugar. How many milligrams of sugar are in 1 ounce of Coca-Cola?

73. Find the product of $\dfrac{x^3}{12}$ and $\dfrac{3}{x}$.

74. Find the product of $\dfrac{15}{x^2}$ and $\dfrac{x^6}{20}$.

75. Find the quotient of $\dfrac{x - 2}{x^2 - 3x - 18}$ and $\dfrac{x + 4}{x^2 + 4x + 3}$.

76. Find the quotient of $\dfrac{x^2 - 4}{6x - 42}$ and $\dfrac{5x - 10}{3x - 21}$.

Concepts

77. A student tries to find the product $\dfrac{x + 2}{x + 4} \cdot \dfrac{x + 4}{x + 6}$:

$$\dfrac{x + 2}{x + 4} \cdot \dfrac{x + 4}{x + 6} = \dfrac{(x + 2)(x + 4)}{(x + 4)(x + 6)} = \dfrac{x^2 + 6x + 8}{x^2 + 10x + 24}$$

Describe any errors. Then find the product correctly.

78. A student tries to find the quotient $\dfrac{x + 4}{x - 5} \div \dfrac{x - 5}{x + 8}$:

$$\dfrac{x + 4}{x - 5} \div \dfrac{x - 5}{x + 8} = \dfrac{x + 4}{x + 8}$$

Describe any errors. Then find the quotient correctly.

79. A student tries to find the product $\dfrac{x}{3} \cdot \dfrac{x + 4}{x - 7}$:

$$\dfrac{x}{3} \cdot \dfrac{x + 4}{x - 7} = \dfrac{x^2 + 4}{3x - 7}$$

Find a value to substitute for x to show that the student's work is incorrect. Then perform the multiplication correctly. Use a graphing calculator table to verify your result.

80. A student tries to find the product $\dfrac{x + 3}{x + 7} \cdot \dfrac{x + 3}{x - 7}$:

$$\dfrac{x + 3}{x + 7} \cdot \dfrac{x + 3}{x - 7} = \dfrac{x^2 + 9}{x^2 - 49}$$

Find a value to substitute for x to show that the student's work is incorrect. Then perform the multiplication correctly. Use a graphing calculator table to verify your result.

81. Give an example of two rational expressions whose product is 1.

82. Give an example of two rational expressions whose quotient is 1.

83. Explain how dividing rational expressions is similar to dividing rational numbers.

84. When converting units, why can we only multiply by ratios equal to 1?

85. Describe how to multiply two rational expressions. Include an example.

86. Describe how to divide two rational expressions. Include an example.

Related Review

Find the quotient twice, first by using factoring and then by using long division. Compare your results.

87. $\dfrac{x^2 - 7x + 12}{x - 3}$

88. $\dfrac{x^2 - 49}{x + 7}$

Expressions, Equations, and Graphs

Perform the indicated instruction. Then use words such as linear, quadratic, cubic, power, radical, rational, polynomial, degree, one variable, *and* two variables *to describe the expression, equation, or system.*

89. Simplify $-3(x - 5)^2$.

90. Graph $y = 3x^2$ by hand.

91. Solve $-3(x - 5)^2 = -24$.

92. Find the difference $\left(3x^2 - 2x\right) - \left(7x^2 + 4x\right)$.

93. Evaluate $-3(x - 5)^2$ for $x = 2$.

94. Solve $3x^2 - 2x = 4$.

10.3 Adding Rational Expressions

Objectives

» Add rational expressions that have a common denominator.

» Add rational expressions that have different denominators.

In Section 10.2, we multiplied and divided rational expressions. How do we add rational expressions?

Addition of Rational Expressions That Have a Common Denominator

Recall from Section 2.2 that we add fractions that have a common denominator by using the property $\frac{a}{b} + \frac{c}{b} = \frac{a + c}{b}$, where b is nonzero. For example,

$$\frac{2}{7} + \frac{3}{7} = \frac{2 + 3}{7} = \frac{5}{7}$$

We add more complicated rational expressions that have a common denominator in a similar way.

Adding Rational Expressions That Have a Common Denominator

If $\frac{A}{B}$ and $\frac{C}{B}$ are rational expressions, where B is nonzero, then

$$\frac{A}{B} + \frac{C}{B} = \frac{A + C}{B}$$

In words: To add two rational expressions that have a common denominator, add the numerators and keep the common denominator.

Example 1   Adding Rational Expressions That Have a Common Denominator

Perform the addition $\frac{4x + 3}{x} + \frac{2}{x}$.

Solution

$$\frac{4x + 3}{x} + \frac{2}{x} = \frac{4x + 3 + 2}{x} \qquad \textit{Add numerators and keep common denominator:} \frac{A}{B} + \frac{C}{B} = \frac{A + C}{B}$$

$$= \frac{4x + 5}{x} \qquad \textit{Combine like terms.}$$

After we add rational expressions, we may be able to simplify the result.

Example 2   Adding Rational Expressions That Have a Common Denominator

Perform the addition $\frac{x^2 - 10x}{x - 5} + \frac{2x + 15}{x - 5}$.

Solution

$$\frac{x^2 - 10x}{x - 5} + \frac{2x + 15}{x - 5} = \frac{x^2 - 10x + 2x + 15}{x - 5} \qquad \textit{Add numerators and keep common denominator:} \frac{A}{B} + \frac{C}{B} = \frac{A + C}{B}$$

$$= \frac{x^2 - 8x + 15}{x - 5} \qquad \textit{Combine like terms.}$$

$$= \frac{(x - 5)(x - 3)}{x - 5} \qquad \textit{Factor numerator.}$$

$$= x - 3 \qquad \textit{Simplify:} \frac{x - 5}{x - 5} = 1$$

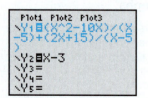

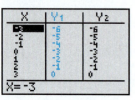

Figure 9 Verify the work

We use a graphing calculator table to verify our work (see Fig. 9).

Addition of Rational Expressions That Have Different Denominators

When adding two fractions that have different denominators in Section 2.2, we listed multiples of the denominators to help us find the least common denominator (LCD) of the fractions. In this section, we discuss another way to find the LCD of two fractions.

Suppose that two brothers, John and Paul, own the following numbers and types of musical instruments:

John	Paul
3 guitars	1 guitar
1 bass guitar	2 bass guitars
1 sitar	1 sitar

The brothers will not give their instruments to each other, but each brother wants to own the same numbers and types of musical instruments that the other brother has.

Because Paul has one more bass than John, John wants one more bass guitar. Because John has two more guitars than Paul, Paul wants two more guitars. John and Paul have the same number of sitars, so neither brother wants another sitar.

Let's compare this situation with finding the LCD of the sum $\frac{7}{24} + \frac{5}{18}$. First, we find the prime factorization of each denominator:

$$24 = 2 \cdot 2 \cdot 2 \cdot 3$$
$$18 = 2 \cdot 3 \cdot 3$$

What factors does each of the denominators need so that the denominators can become the same? Because 18 has one more 3 factor than 24, the denominator 24 needs one 3 factor. Because 24 has two more 2 factors than 18, the denominator 18 needs two 2 factors.

We use the fact that $\frac{A}{A} = 1$ when $A \neq 0$ to introduce the one 3 factor for 24 and the two 2 factors for 18:

$$\frac{7}{24} + \frac{5}{18} = \frac{7}{2 \cdot 2 \cdot 2 \cdot 3} + \frac{5}{2 \cdot 3 \cdot 3}$$ *Find prime factorization of each denominator.*

$$= \frac{7}{2 \cdot 2 \cdot 2 \cdot 3} \cdot \frac{3}{3} + \frac{5}{2 \cdot 3 \cdot 3} \cdot \frac{2 \cdot 2}{2 \cdot 2}$$ *Introduce missing factors.*

$$= \frac{7 \cdot 3}{2 \cdot 2 \cdot 2 \cdot 3 \cdot 3} + \frac{5 \cdot 2 \cdot 2}{2 \cdot 3 \cdot 3 \cdot 2 \cdot 2}$$ *Find products.*

$$= \frac{21}{72} + \frac{20}{72}$$ *Simplify.*

$$= \frac{41}{72}$$ *Add numerators and keep common denominator: $\frac{A}{B} + \frac{C}{B} = \frac{A + C}{B}$*

Note that this method not only helps us find the LCD, but also suggests what forms of $\frac{A}{A}$ to use to introduce missing factors.

Example 3 Adding Rational Expressions That Have Different Denominators

Find the sum $\frac{1}{6} + \frac{9}{2w}$.

Solution

We begin by factoring the denominators:

$$6 = 2 \cdot 3$$
$$2w = 2 \cdot w$$

Because $2w$ has a w factor and 6 has no w factors, the denominator 6 needs a w factor. Because 6 has a 3 factor and $2w$ has no 3 factors, the denominator $2w$ needs a 3 factor.

We use the fact that $\dfrac{A}{A} = 1$, where A is nonzero, to introduce the missing factors:

$$\frac{1}{6} + \frac{9}{2w} = \frac{1}{2 \cdot 3} + \frac{9}{2 \cdot w} \qquad \textcolor{blue}{\textit{Factor denominators.}}$$

$$= \frac{1}{2 \cdot 3} \cdot \frac{w}{w} + \frac{9}{2 \cdot w} \cdot \frac{3}{3} \qquad \textcolor{blue}{\textit{Introduce missing factors.}}$$

$$= \frac{w}{6w} + \frac{27}{6w} \qquad \textcolor{blue}{\textit{Find products.}}$$

$$= \frac{w + 27}{6w} \qquad \textcolor{blue}{\textit{Add numerators and keep common denominator: } \frac{A}{B} + \frac{C}{B} = \frac{A + C}{B}}$$

The result is in lowest terms.

▶

Example 4   Adding Rational Expressions That Have Different Denominators

Find the sum $\dfrac{7}{12x} + \dfrac{5}{8x^3}$.

Solution

We begin by factoring the denominators:

$$12x = 2 \cdot 2 \cdot 3 \cdot x$$
$$8x^3 = 2 \cdot 2 \cdot 2 \cdot x \cdot x \cdot x$$

Because $8x^3$ has one more 2 factor and two more x factors than $12x$, the denominator $12x$ needs one more 2 factor and two more x factors. Because $12x$ has a 3 factor and $8x^3$ does not have any 3 factors, the denominator $8x^3$ needs a 3 factor.

We use the fact that $\dfrac{A}{A} = 1$, where A is nonzero, to introduce the missing factors:

$$\frac{7}{12x} + \frac{5}{8x^3} = \frac{7}{2 \cdot 2 \cdot 3 \cdot x} + \frac{5}{2 \cdot 2 \cdot 2 \cdot x \cdot x \cdot x} \qquad \textcolor{blue}{\textit{Factor denominators.}}$$

$$= \frac{7}{2 \cdot 2 \cdot 3 \cdot x} \cdot \frac{2 \cdot x \cdot x}{2 \cdot x \cdot x} + \frac{5}{2 \cdot 2 \cdot 2 \cdot x \cdot x \cdot x} \cdot \frac{3}{3} \qquad \textcolor{blue}{\textit{Introduce missing factors.}}$$

$$= \frac{14x^2}{24x^3} + \frac{15}{24x^3} \qquad \textcolor{blue}{\textit{Find products.}}$$

$$= \frac{14x^2 + 15}{24x^3} \qquad \textcolor{blue}{\textit{Add numerators and keep common denominator.}}$$

The result is in lowest terms.

▶

Example 5   Adding Rational Expressions That Have Different Denominators

Find the sum $\dfrac{2}{x - 4} + \dfrac{6}{x + 7}$.

Solution

The denominator $x - 4$ needs an $x + 7$ factor, and the denominator $x + 7$ needs an $x - 4$ factor. We use the fact that $\dfrac{A}{A} = 1$, where A is nonzero, to introduce the missing factors:

$$\frac{2}{x - 4} + \frac{6}{x + 7} = \frac{2}{x - 4} \cdot \frac{x + 7}{x + 7} + \frac{6}{x + 7} \cdot \frac{x - 4}{x - 4} \qquad \text{Introduce missing factors.}$$

$$= \frac{2(x + 7)}{(x - 4)(x + 7)} + \frac{6(x - 4)}{(x + 7)(x - 4)} \qquad \text{Find products.}$$

$$= \frac{2(x + 7) + 6(x - 4)}{(x - 4)(x + 7)} \qquad \text{Add numerators and keep common denominator:} \; \frac{A}{B} + \frac{C}{B} = \frac{A + C}{B}$$

$$= \frac{2x + 14 + 6x - 24}{(x - 4)(x + 7)} \qquad \text{Distributive law}$$

$$= \frac{8x - 10}{(x - 4)(x + 7)} \qquad \text{Combine like terms.}$$

$$= \frac{2(4x - 5)}{(x - 4)(x + 7)} \qquad \text{Factor numerator.}$$

The result is in lowest terms.

▶

For the sum of two expressions that have different denominators, we first factor the denominators, if possible, to help us find the LCD.

Example 6   Adding Rational Expressions That Have Different Denominators

Find the sum $\dfrac{5}{6} + \dfrac{7}{2x - 8}$.

Solution

First, we factor the denominators:

$$6 = 2 \cdot 3$$
$$2x - 8 = 2(x - 4)$$

Because $2x - 8$ has an $x - 4$ factor and 6 does not, the denominator 6 needs an $x - 4$ factor. Because 6 has a 3 factor and $2x - 8$ does not, the denominator $2x - 8$ needs a 3 factor. Thus,

$$\frac{5}{6} + \frac{7}{2x - 8} = \frac{5}{2 \cdot 3} + \frac{7}{2(x - 4)} \qquad \text{Factor denominators.}$$

$$= \frac{5}{2 \cdot 3} \cdot \frac{x - 4}{x - 4} + \frac{7}{2(x - 4)} \cdot \frac{3}{3} \qquad \text{Introduce missing factors.}$$

$$= \frac{5(x - 4)}{6(x - 4)} + \frac{21}{6(x - 4)} \qquad \text{Find products.}$$

$$= \frac{5(x - 4) + 21}{6(x - 4)} \qquad \text{Add numerators and keep common denominator:} \; \frac{A}{B} + \frac{C}{B} = \frac{A + C}{B}$$

$$= \frac{5x - 20 + 21}{6(x - 4)} \qquad \text{Distributive law}$$

$$= \frac{5x + 1}{6(x - 4)} \qquad \text{Combine like terms.}$$

▶

Example 7   Adding Rational Expressions That Have Different Denominators

Find the sum $\dfrac{x}{2x-4} + \dfrac{-4}{x^2-4}$.

Solution

First, we factor the denominators:

$$2x - 4 = 2(x - 2)$$
$$x^2 - 4 = (x - 2)(x + 2)$$

Because $x^2 - 4$ has an $x + 2$ factor and $2x - 4$ does not, the denominator $2x - 4$ needs an $x + 2$ factor. Because $2x - 4$ has a 2 factor and $x^2 - 4$ does not, the denominator $x^2 - 4$ needs a 2 factor. Thus,

$$\frac{x}{2x-4} + \frac{-4}{x^2-4} = \frac{x}{2(x-2)} + \frac{-4}{(x-2)(x+2)} \qquad \text{\textit{Factor denominators.}}$$

$$= \frac{x}{2(x-2)} \cdot \frac{x+2}{x+2} + \frac{-4}{(x-2)(x+2)} \cdot \frac{2}{2} \qquad \text{\textit{Introduce missing facto}}$$

$$= \frac{x(x+2)}{2(x-2)(x+2)} + \frac{-8}{2(x-2)(x+2)} \qquad \text{\textit{Find products.}}$$

$$= \frac{x(x+2) - 8}{2(x-2)(x+2)} \qquad \text{\textit{Add numerators and ke}}$$
$$\qquad\qquad \text{\textit{common denominator.}}$$

$$= \frac{x^2 + 2x - 8}{2(x-2)(x+2)} \qquad \text{\textit{Distributive law}}$$

$$= \frac{(x-2)(x+4)}{2(x-2)(x+2)} \qquad \text{\textit{Factor numerator.}}$$

$$= \frac{x+4}{2(x+2)} \qquad \text{\textit{Simplify: } \dfrac{x-2}{x-2} = 1}$$

Figure 10 Verify the work ▶

We use a graphing calculator table to verify our work (see Fig. 10). There are "ERROR" messages across from $x = -2$ and $x = 2$ in the table because -2 and 2 are excluded values of either the original expression or our result.

How to Add Rational Expressions That Have Different Denominators

To add rational expressions that have different denominators,

1. Factor the denominators of the expressions if possible. Determine which factors are missing.
2. Use the property $\dfrac{A}{A} = 1$, where A is nonzero, to introduce missing factors.
3. Add the expressions by using the property $\dfrac{A}{B} + \dfrac{C}{B} = \dfrac{A+C}{B}$, where B is nonzero.
4. Simplify.

| Example 8 | Adding Rational Expressions That Have Different Denominators |

Find the sum $\dfrac{2x}{x^2 - x - 2} + \dfrac{x + 4}{x + 1}$.

Solution

First, we factor the denominators:

$$x^2 - x - 2 = (x - 2)(x + 1)$$
$$x + 1 = x + 1$$

The denominator $x^2 - x - 2$ does not need any factors. Because $x^2 - x - 2$ has an $x - 2$ factor and $x + 1$ does not, the denominator $x + 1$ needs an $x - 2$ factor. Thus,

$$\dfrac{2x}{x^2 - x - 2} + \dfrac{x + 4}{x + 1} = \dfrac{2x}{(x - 2)(x + 1)} + \dfrac{x + 4}{x + 1} \qquad \textit{Factor denominator.}$$

$$= \dfrac{2x}{(x - 2)(x + 1)} + \dfrac{x + 4}{x + 1} \cdot \dfrac{x - 2}{x - 2} \qquad \textit{Introduce missing factor.}$$

$$= \dfrac{2x}{(x - 2)(x + 1)} + \dfrac{(x + 4)(x - 2)}{(x + 1)(x - 2)} \qquad \textit{Find product.}$$

$$= \dfrac{2x + (x + 4)(x - 2)}{(x - 2)(x + 1)} \qquad \textit{Add numerators and keep common denominator.}$$

$$= \dfrac{2x + x^2 + 2x - 8}{(x - 2)(x + 1)} \qquad \textit{Find product.}$$

$$= \dfrac{x^2 + 4x - 8}{(x - 2)(x + 1)} \qquad \textit{Combine like terms.}$$

The result is in lowest terms.

▶

 Group Exploration

Performing operations with rational expressions

For Problems 1–3, a student tries to perform an operation and then simplify the result. If the result is correct, decide whether there is a more efficient way to do the problem. If the result is incorrect, describe any errors and do the problem correctly.

1. $\dfrac{x + 2}{x + 5} + \dfrac{6}{(x + 2)(x + 5)} = \dfrac{1}{x + 5} + \dfrac{6}{x + 5}$

$$= \dfrac{7}{x + 5}$$

2. $\dfrac{1}{x^2 + 6x + 8} + \dfrac{1}{x^2 + 5x + 6}$

$$= \dfrac{1}{x^2 + 6x + 8} \cdot \dfrac{x^2 + 5x + 6}{x^2 + 5x + 6} + \dfrac{1}{x^2 + 5x + 6} \cdot \dfrac{x^2 + 6x + 8}{x^2 + 6x + 8}$$

$$= \dfrac{\left(x^2 + 5x + 6\right) + \left(x^2 + 6x + 8\right)}{\left(x^2 + 5x + 6\right)\left(x^2 + 6x + 8\right)}$$

$$= \dfrac{2x^2 + 11x + 14}{\left(x^2 + 5x + 6\right)\left(x^2 + 6x + 8\right)}$$

3. $\dfrac{4}{x + 2} \cdot \dfrac{3}{x + 1} = \left(\dfrac{4}{x + 2} \cdot \dfrac{x + 1}{x + 1}\right) \cdot \left(\dfrac{3}{x + 1} \cdot \dfrac{x + 2}{x + 2}\right)$

$$= \dfrac{4(x + 1)}{(x + 2)(x + 1)} \cdot \dfrac{3(x + 2)}{(x + 1)(x + 2)}$$

$$= \dfrac{12(x + 1)(x + 2)}{(x + 1)^2(x + 2)^2}$$

$$= \dfrac{12}{(x + 1)(x + 2)}$$

Homework 10.3

For extra help ▶   **MyLab Math**   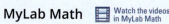 Watch the videos in MyLab Math

Find the sum. Simplify the result.

1. $\dfrac{7}{x} + \dfrac{2}{x}$

2. $\dfrac{3}{x^2} + \dfrac{5}{x^2}$

3. $\dfrac{2x}{x-1} + \dfrac{6x}{x-1}$

4. $\dfrac{-4x}{x+5} + \dfrac{x}{x+5}$

5. $\dfrac{3x-2}{x+3} + \dfrac{5x+4}{x+3}$

6. $\dfrac{6x+3}{x-4} + \dfrac{9x-8}{x-4}$

7. $\dfrac{t^2}{t+5} + \dfrac{7t+10}{t+5}$

8. $\dfrac{a^2+9a}{a+3} + \dfrac{18}{a+3}$

9. $\dfrac{x}{x^2-4} + \dfrac{2}{x^2-4}$

10. $\dfrac{x}{x^2-16} + \dfrac{-4}{x^2-16}$

11. $\dfrac{x^2-5x}{x^2+5x+6} + \dfrac{4x-12}{x^2+5x+6}$

12. $\dfrac{x^2-4x}{x^2+3x-4} + \dfrac{6x-8}{x^2+3x-4}$

13. $\dfrac{3r^2-5r}{r^2+6r+9} + \dfrac{-2r^2+r-21}{r^2+6r+9}$

14. $\dfrac{5w^2-2w+12}{w^2-12w+36} + \dfrac{-4w^2-6w}{w^2-12w+36}$

Find the sum. Simplify the result.

15. $\dfrac{3}{x} + \dfrac{5}{2x}$

16. $\dfrac{8}{3x} + \dfrac{2}{x}$

17. $\dfrac{3}{2w} + \dfrac{5}{6}$

18. $\dfrac{3}{8} + \dfrac{7}{4a}$

19. $\dfrac{5x}{6} + \dfrac{3}{4x}$

20. $\dfrac{4}{9x} + \dfrac{7x}{6}$

21. $\dfrac{5}{8x^3} + \dfrac{3}{10x}$

22. $\dfrac{7}{6x} + \dfrac{1}{3x^3}$

23. $\dfrac{a}{b} + \dfrac{b}{a}$

24. $\dfrac{a}{3b} + \dfrac{b}{2a}$

25. $\dfrac{5}{4x} + \dfrac{2}{x+3}$

26. $\dfrac{7}{x-5} + \dfrac{4}{3x}$

27. $\dfrac{3}{r+2} + \dfrac{4}{r-5}$

28. $\dfrac{2}{b-1} + \dfrac{6}{b+7}$

29. $\dfrac{6}{x-1} + \dfrac{3}{5x}$

30. $\dfrac{7}{2x} + \dfrac{4}{x+6}$

31. $\dfrac{5x}{x-4} + \dfrac{2}{x+2}$

32. $\dfrac{3}{x+5} + \dfrac{4x}{x-3}$

33. $\dfrac{1}{a+b} + \dfrac{1}{a-b}$

34. $\dfrac{1}{x+y} + \dfrac{1}{x+2y}$

35. $\dfrac{x}{x+4} + \dfrac{2}{5x+20}$

36. $\dfrac{3}{x-6} + \dfrac{7x}{4x-24}$

37. $\dfrac{p}{3p-9} + \dfrac{-1}{p-3}$

38. $\dfrac{c}{2c-4} + \dfrac{-1}{c-2}$

39. $\dfrac{2x}{5x-25} + \dfrac{4}{3x-15}$

40. $\dfrac{3x}{4x+12} + \dfrac{2}{6x+18}$

41. $\dfrac{6}{x^2-1} + \dfrac{3}{x+1}$

42. $\dfrac{1}{x+2} + \dfrac{4}{x^2-4}$

43. $\dfrac{t^2+2t}{t^2+11t+18} + \dfrac{4}{t+9}$

44. $\dfrac{r^2+3r}{r^2+7r+12} + \dfrac{2}{r+4}$

45. $\dfrac{x}{2x-6} + \dfrac{3}{x^2-9}$

46. $\dfrac{2x}{x^2-4} + \dfrac{5}{3x+6}$

47. $\dfrac{x}{x^2-4} + \dfrac{1}{x^2+2x}$

48. $\dfrac{1}{x^2+x} + \dfrac{2x}{x^2-1}$

49. $\dfrac{4}{(a-3)(a+1)} + \dfrac{2}{(a+1)(a+4)}$

50. $\dfrac{7}{(w-5)(w-2)} + \dfrac{4}{(w-3)(w-2)}$

51. $\dfrac{3}{x^2-16} + \dfrac{5}{x^2+5x+4}$

52. $\dfrac{2}{x^2+5x-14} + \dfrac{6}{x^2-49}$

53. $\dfrac{4x}{x^2+3x-18} + \dfrac{2}{x^2+10x+24}$

54. $\dfrac{5}{x^2+3x+2} + \dfrac{7x}{x^2-x-2}$

55. $\dfrac{x}{x^2+9x+20} + \dfrac{-4}{x^2+8x+15}$

56. $\dfrac{6}{x^2-9} + \dfrac{-5}{x^2-x-6}$

57. $3 + \dfrac{w-2}{w+5}$

58. $2 + \dfrac{a+3}{a-4}$

Find the sum. Simplify the result.

59. $\dfrac{x-3}{x+5} + \dfrac{x+2}{x-4}$

60. $\dfrac{x+7}{x-2} + \dfrac{x-5}{x+1}$

61. $\dfrac{y-2}{y-3} + \dfrac{y+3}{y+2}$

62. $\dfrac{a+5}{a+1} + \dfrac{a-1}{a-5}$

63. $\dfrac{x-6}{x-5} + \dfrac{2x}{x^2-2x-15}$

64. $\dfrac{x+4}{x+3} + \dfrac{7}{x^2+5x+6}$

65. $\dfrac{5}{b^2-7b+12} + \dfrac{b+2}{b-4}$

66. $\dfrac{3t}{t^2-4t-5} + \dfrac{t-3}{t+1}$

67. $\dfrac{x-2}{4x+12} + \dfrac{5}{x^2-9}$

68. $\dfrac{3}{x^2-4} + \dfrac{x-3}{5x-10}$

69. $\dfrac{2x}{x-y} + \dfrac{2xy}{x^2-2xy+y^2}$

70. $\dfrac{3y}{x+y} + \dfrac{-9xy}{3x^2+4xy+y^2}$

71. $\dfrac{2x}{x^3-4x^2+2x-8} + \dfrac{5}{3x^2+6}$

72. $\dfrac{2}{x^3+2x^2-5x-10} + \dfrac{3x}{2x^2-10}$

73. $\dfrac{5x}{x^3 - 8} + \dfrac{4}{x^2 - 4}$

74. $\dfrac{3x}{x^3 + 27} + \dfrac{6}{x^2 - 9}$

75. There are two 100-watt lights in a room. A person is twice as far from one light as from the other. The illumination (the brightness of the light) can be measured in watts per square meter $\left(\text{W/m}^2 \right)$. The illumination (in W/m^2) from the closer light source is

$$\frac{18}{d^2}$$

where d is the distance (in meters) to that light. The illumination (in W/m^2) from the other light source is

$$\frac{18}{(2d)^2}$$

where $2d$ is the distance (in meters) to that light.

a. Find an expression for the total illumination (in W/m^2) from the two light sources.

b. Write the result you found in part (a) as a single fraction.

c. Evaluate the expression you found in part (b) for $d = 1.2$. What does your result mean in this situation?

76. A student plans to drive from El Camino College in Torrance, California, to Chandler-Gilbert Community College in Sun Lakes, Arizona. The trip involves 231 miles in California, followed by 174 miles in Arizona. The speed limit is 70 mph in California and 75 mph in Arizona. The student plans to drive at a mph above the speed limits. The driving time (in hours) in California is

$$\frac{231}{a + 70}$$

The driving time (in hours) in Arizona is

$$\frac{174}{a + 75}$$

a. Find an expression for the total driving time (in hours).

b. Write the result you found in part (a) as a single fraction.

c. Evaluate the expression you found in part (b) for $a = 5$. What does your result mean in this situation?

Concepts

77. A student tries to find the sum $\dfrac{3}{x + 2} + \dfrac{5}{x + 3}$:

$$\frac{3}{x + 2} + \frac{5}{x + 3} = \frac{3}{x + 2} \cdot \frac{1}{x + 3} + \frac{5}{x + 3} \cdot \frac{1}{x + 2}$$

$$= \frac{3}{(x + 2)(x + 3)} + \frac{5}{(x + 3)(x + 2)}$$

$$= \frac{8}{(x + 2)(x + 3)}$$

Describe any errors. Then find the sum correctly.

78. A student tries to find the sum $\dfrac{x}{3} + \dfrac{2}{x}$:

$$\frac{x}{3} + \frac{2}{x} = \frac{x + 2}{3 + x} = \frac{x + 2}{x + 3}$$

Describe any errors. Then find the sum correctly.

79. A student tries to find the sum $\dfrac{2}{x^2 + 2x} + \dfrac{3}{x + 2}$:

$$\frac{2}{x^2 + 2x} + \frac{3}{x + 2} = \frac{2}{x^2 + 2x} \cdot \frac{x + 2}{x + 2} + \frac{3}{x + 2} \cdot \frac{x^2 + 2x}{x^2 + 2x}$$

$$= \frac{2x + 4}{(x^2 + 2x)(x + 2)} + \frac{3x^2 + 6x}{(x + 2)(x^2 + 2x)}$$

$$= \frac{3x^2 + 8x + 4}{(x^2 + 2x)(x + 2)}$$

$$= \frac{(3x + 2)(x + 2)}{x(x + 2)(x + 2)}$$

$$= \frac{3x + 2}{x(x + 2)}$$

Is the work correct? Is there a better way to find the sum? Explain.

80. a. Add $\dfrac{2}{5} + \dfrac{4}{7}$. **b.** Add $\dfrac{7}{2} + \dfrac{5}{3}$.

c. Add $\dfrac{a}{b} + \dfrac{c}{d}$, where b and d are nonzero.

d. Use the result you found in part (c) to help you find the sum $\dfrac{2}{5} + \dfrac{4}{7}$. Compare your result with the result you found in part (a).

81. Explain how adding rational expressions is similar to adding rational numbers.

82. Why do we factor the denominators of rational expressions before finding the LCD?

83. Describe how to add two rational expressions that have a common denominator. Include an example.

84. Describe how to add two rational expressions that have different denominators. Include an example.

Related Review

Find the sum. Simplify the result.

85. $\left(4x^2 - 7x + 2 \right) + \left(-3x^2 + 2x + 4 \right)$

86. $\left(-5x^2 + 8x - 17 \right) + \left(6x^2 - 5x - 23 \right)$

87. $\dfrac{4x^2 - 7x + 2}{x - 3} + \dfrac{-3x^2 + 2x + 4}{x - 3}$

88. $\dfrac{-5x^2 + 8x - 17}{x - 5} + \dfrac{6x^2 - 5x - 23}{x - 5}$

Expressions, Equations, and Graphs

Perform the indicated instruction. Then use words such as linear, quadratic, cubic, power, radical, rational, polynomial, degree, one variable, *and* two variables *to describe the expression, equation, or system.*

89. Solve the following:

$$y = 3x - 2$$
$$y = 5x + 4$$

90. Factor $t^2 - t - 20$.

91. Graph $y = 3x - 2$ by hand.

92. Solve $t^2 - t - 20 = 0$.

93. Solve $3x - 2 = 5x + 4$.

94. Find the sum $\dfrac{t}{t^2 - t - 20} + \dfrac{3}{t^2 + 7t + 12}$.

10.4 Subtracting Rational Expressions

Objectives

» Subtract rational expressions that have a common denominator.

» Subtract rational expressions that have different denominators.

In Sections 10.2 and 10.3, we multiplied, divided, and added rational expressions. In this section, we subtract rational expressions.

Subtraction of Rational Expressions That Have a Common Denominator

Recall from Section 2.2 that we subtract fractions that have a common denominator by using the property $\dfrac{a}{b} - \dfrac{c}{b} = \dfrac{a-c}{b}$, where b is nonzero. For example,

$$\frac{5}{7} - \frac{2}{7} = \frac{5-2}{7} = \frac{3}{7}$$

We subtract more complicated rational expressions that have a common denominator in a similar way.

> **Subtracting Rational Expressions That Have a Common Denominator**
>
> If $\dfrac{A}{B}$ and $\dfrac{C}{B}$ are rational expressions, where B is nonzero, then
>
> $$\frac{A}{B} - \frac{C}{B} = \frac{A-C}{B}$$
>
> In words: To subtract two rational expressions that have a common denominator, subtract the numerators and keep the common denominator.

After subtracting two rational expressions, it may be possible to simplify the result.

Example 1   Subtracting Two Rational Expressions That Have a Common Denominator

Find the difference $\dfrac{x}{x^2 - 4} - \dfrac{2}{x^2 - 4}$.

Solution

$$\frac{x}{x^2 - 4} - \frac{2}{x^2 - 4} = \frac{x-2}{x^2-4} \qquad \text{Subtract numerators and keep common denominator: } \frac{A}{B} - \frac{C}{B} = \frac{A-C}{B}$$

$$= \frac{x-2}{(x-2)(x+2)} \qquad \text{Factor denominator.}$$

$$= \frac{1}{x+2} \qquad \text{Simplify: } \frac{x-2}{x-2} = 1$$

Example 2   Subtracting Two Rational Expressions That Have a Common Denominator

Find the difference $\dfrac{x^2}{x+4} - \dfrac{3x+28}{x+4}$.

Solution

$$\frac{x^2}{x+4} - \frac{3x+28}{x+4} = \frac{x^2 - (3x+28)}{x+4} \qquad \text{Subtract numerators and keep common denominator: } \frac{A}{B} - \frac{C}{B} = \frac{A-C}{B}$$

$$= \frac{x^2 - 3x - 28}{x+4} \qquad \text{Simplify.}$$

$$= \frac{(x-7)(x+4)}{x+4} \qquad \text{Factor numerator.}$$

$$= x - 7 \qquad \text{Simplify: } \frac{x+4}{x+4} = 1$$

WARNING   It is a common error to write

$$\frac{x^2}{x+4} - \frac{3x+28}{x+4} = \frac{x^2 - 3x + 28}{x+4} \qquad \textit{Incorrect}$$

This work is incorrect. **When subtracting rational expressions, be sure to subtract the** *entire* **numerator:**

$$\frac{x^2}{x+4} - \frac{3x+28}{x+4} = \frac{x^2 - (3x+28)}{x+4} = \frac{x^2 - 3x - 28}{x+4}$$

See Example 2 for the rest of the work.

Subtraction of Rational Expressions That Have Different Denominators

When subtracting rational expressions that have different denominators, we use the method discussed in Section 10.3 to find the LCD.

Example 3   Subtracting Two Rational Expressions That Have Different Denominators

Find the difference $\dfrac{2}{9a} - \dfrac{5}{6a^2}$.

Solution

$$\frac{2}{9a} - \frac{5}{6a^2} = \frac{2}{3 \cdot 3 \cdot a} - \frac{5}{2 \cdot 3 \cdot a \cdot a} \qquad \textit{Factor denominators.}$$

$$= \frac{2}{3 \cdot 3 \cdot a} \cdot \frac{2 \cdot a}{2 \cdot a} - \frac{5}{2 \cdot 3 \cdot a \cdot a} \cdot \frac{3}{3} \qquad \textit{Introduce missing factors.}$$

$$= \frac{4a}{18a^2} - \frac{15}{18a^2} \qquad \textit{Find products.}$$

$$= \frac{4a - 15}{18a^2} \qquad \textit{Subtract numerators and keep common denominator: } \frac{A}{B} - \frac{C}{B} = \frac{A-C}{B}$$

Example 4   Subtracting Two Rational Expressions That Have Different Denominators

Find the difference $\dfrac{x}{x^2-9} - \dfrac{2}{5x+15}$.

Solution

$$\frac{x}{x^2-9} - \frac{2}{5x+15} = \frac{x}{(x-3)(x+3)} - \frac{2}{5(x+3)} \qquad \textit{Factor denominators.}$$

$$= \frac{x}{(x-3)(x+3)} \cdot \frac{5}{5} - \frac{2}{5(x+3)} \cdot \frac{x-3}{x-3} \qquad \textit{Introduce missing factors.}$$

$$= \frac{5x}{5(x-3)(x+3)} - \frac{2(x-3)}{5(x+3)(x-3)} \qquad \textit{Find products.}$$

$$= \frac{5x - 2(x-3)}{5(x-3)(x+3)} \qquad \frac{A}{B} - \frac{C}{B} = \frac{A-C}{B}$$

$$= \frac{5x - 2x + 6}{5(x-3)(x+3)} \qquad \textit{Distributive law}$$

$$= \frac{3x + 6}{5(x-3)(x+3)} \qquad \textit{Combine like terms.}$$

$$= \frac{3(x+2)}{5(x-3)(x+3)} \qquad \textit{Factor numerator.}$$

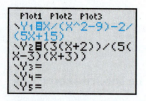

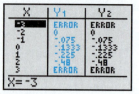

Figure 11 Verify the work

Note that $\dfrac{3(x + 2)}{5(x - 3)(x + 3)}$ is in lowest terms. We use a graphing calculator table to verify our work (see Fig. 11). There are "ERROR" messages across from $x = -3$ and $x = 3$ in the table because -3 and 3 are excluded values of either the original expression or our result.

How to Subtract Two Rational Expressions That Have Different Denominators

To subtract two rational expressions that have different denominators,

1. Factor the denominators of the expressions if possible. Determine which factors are missing.

2. Use the property $\dfrac{A}{A} = 1$, where A is nonzero, to introduce missing factors.

3. Subtract the expressions by using the property $\dfrac{A}{B} - \dfrac{C}{B} = \dfrac{A - C}{B}$, where B is nonzero.

4. Simplify.

Example 5   Subtracting Two Rational Expressions That Have Different Denominators

Find the difference $\dfrac{x + 2}{x - 3} - \dfrac{x - 4}{x + 1}$.

Solution

$$\frac{x + 2}{x - 3} - \frac{x - 4}{x + 1} = \frac{x + 2}{x - 3} \cdot \frac{x + 1}{x + 1} - \frac{x - 4}{x + 1} \cdot \frac{x - 3}{x - 3}$$ *Introduce missing factors.*

$$= \frac{(x + 2)(x + 1)}{(x - 3)(x + 1)} - \frac{(x - 4)(x - 3)}{(x + 1)(x - 3)}$$ *Find products.*

$$= \frac{(x + 2)(x + 1) - (x - 4)(x - 3)}{(x - 3)(x + 1)}$$ *Subtract numerators and keep com denominator:* $\dfrac{A}{B} - \dfrac{C}{B} = \dfrac{A - C}{B}$

$$= \frac{x^2 + 3x + 2 - (x^2 - 7x + 12)}{(x - 3)(x + 1)}$$ *Find products.*

$$= \frac{x^2 + 3x + 2 - x^2 + 7x - 12}{(x - 3)(x + 1)}$$ *Subtract.*

$$= \frac{10x - 10}{(x - 3)(x + 1)}$$ *Combine like terms.*

$$= \frac{10(x - 1)}{(x - 3)(x + 1)}$$ *Factor numerator.*

◆ Group Exploration

Performing operations with rational expressions

For Problems 1–3, a student tries to perform an operation and then simplify the result. If the result is correct, decide whether there is a more efficient way to do the problem. If the result is incorrect, describe any errors and do the problem correctly.

1. $\dfrac{7x}{x - 3} - \dfrac{5x + 9}{x - 3} = \dfrac{7x - 5x + 9}{x - 3}$

$= \dfrac{2x + 9}{x - 3}$

2. $\dfrac{5}{4x} - \dfrac{3}{2x^2} = \dfrac{5}{4x} \cdot \dfrac{2x^2}{2x^2} - \dfrac{3}{2x^2} \cdot \dfrac{4x}{4x}$

$\qquad = \dfrac{10x^2 - 12x}{8x^3}$

$\qquad = \dfrac{2x(5x - 6)}{8x^3}$

$\qquad = \dfrac{5x - 6}{4x^2}$

3. $\dfrac{2}{x+3} \cdot \dfrac{5}{x} = \left(\dfrac{2}{x+3} \cdot \dfrac{x}{x} \right) \cdot \left(\dfrac{5}{x} \cdot \dfrac{x+3}{x+3} \right)$

$\qquad = \dfrac{2x}{x(x+3)} \cdot \dfrac{5(x+3)}{x(x+3)}$

$\qquad = \dfrac{10x(x+3)}{x^2(x+3)^2}$

$\qquad = \dfrac{10}{x(x+3)}$

Tips for Success   Consider Various Uses of a Technique

At several points during the term, think about how a technique can be used in various contexts. For example, consider how we have used factoring in this course. In Sections 8.1–8.4, we learned how to factor polynomials. In Section 8.5, we used factoring to help us solve polynomial equations. In Section 8.6, we used factoring to help us solve quadratic equations that lead to estimates and predictions about authentic situations. In Section 9.2, we used factoring to help us simplify radical expressions, which also helped us simplify solutions of quadratic equations in Section 9.5. In Section 9.4, we used factoring to help us complete the square. In Sections 10.1–10.4, we used factoring to simplify rational expressions and to perform operations with rational expressions.

Considering the various ways we have used a technique will help you see the big picture. If you do that for all the techniques that are used frequently in this course, you will be more likely to recognize what technique to use to solve a particular exercise and you will increase your understanding of how the topics in this course are connected. This will improve your performance on your upcoming final exam, as well as in the next mathematics course you may take.

Homework 10.4

For extra help ▶ **MyLab Math**   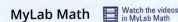 Watch the videos in MyLab Math

Find the difference. Simplify the result.

1. $\dfrac{6}{x} - \dfrac{4}{x}$

2. $\dfrac{8}{x^3} - \dfrac{3}{x^3}$

3. $\dfrac{9x}{x-2} - \dfrac{2x}{x-2}$

4. $\dfrac{4x}{x+7} - \dfrac{6x}{x+7}$

5. $\dfrac{x}{x^2-9} - \dfrac{3}{x^2-9}$

6. $\dfrac{x}{x^2-25} - \dfrac{5}{x^2-25}$

7. $\dfrac{3r}{r+6} - \dfrac{7r-4}{r+6}$

8. $\dfrac{2t}{t-4} - \dfrac{3t-5}{t-4}$

9. $\dfrac{x^2}{x+1} - \dfrac{2x+3}{x+1}$

10. $\dfrac{x^2}{x-7} - \dfrac{3x+28}{x-7}$

11. $\dfrac{x^2+7x}{x^2-2x-8} - \dfrac{3x+32}{x^2-2x-8}$

12. $\dfrac{x^2-2x}{x^2+4x-5} - \dfrac{5x-6}{x^2+4x-5}$

13. $\dfrac{4a^2-5a-12}{a^2+8a+16} - \dfrac{3a^2-6a}{a^2+8a+16}$

14. $\dfrac{2k^2-4k}{k^2-10k+25} - \dfrac{k^2-2k+15}{k^2-10k+25}$

Find the difference. Simplify the result.

15. $\dfrac{3}{4x} - \dfrac{2}{x}$

16. $\dfrac{7}{x} - \dfrac{3}{5x}$

17. $\dfrac{5}{2b} - \dfrac{3}{8}$

18. $\dfrac{2}{3t} - \dfrac{5}{6}$

19. $\dfrac{5x}{8} - \dfrac{1}{6x}$

20. $\dfrac{5x}{9} - \dfrac{7}{12x}$

21. $\dfrac{5}{10x^4} - \dfrac{3}{15x^2}$

22. $\dfrac{1}{8x} - \dfrac{5}{4x^3}$

23. $\dfrac{a}{2b} - \dfrac{b}{3a}$

24. $\dfrac{a}{b} - \dfrac{b}{a}$

25. $\dfrac{3}{p-2} - \dfrac{5}{4p}$

26. $\dfrac{4}{w+6} - \dfrac{2}{3w}$

27. $\dfrac{7}{x-1} - \dfrac{3}{x+4}$

28. $\dfrac{3}{x+6} - \dfrac{4}{x-5}$

29. $\dfrac{3x}{x-2} - \dfrac{5}{x+3}$

30. $\dfrac{4}{x-4} - \dfrac{3x}{x+2}$

31. $\dfrac{1}{a+b} - \dfrac{1}{a-b}$

32. $\dfrac{2}{t+2r} - \dfrac{3}{t+3r}$

33. $\dfrac{c}{2c-8} - \dfrac{3}{c-4}$

34. $\dfrac{y}{4y-20} - \dfrac{6}{y-5}$

35. $\dfrac{4x}{6x-24} - \dfrac{7}{4x-16}$

36. $\dfrac{5x}{6x+42} - \dfrac{3}{8x+56}$

37. $\dfrac{x}{x-1} - \dfrac{2}{x^2-1}$

38. $\dfrac{3}{x-5} - \dfrac{5x-16}{x^2-25}$

39. $\dfrac{3x-1}{x^2+2x-15} - \dfrac{2}{x+5}$

40. $\dfrac{6x}{x^2-8x+15} - \dfrac{9}{x-3}$

41. $\dfrac{t}{3t-21} - \dfrac{4}{t^2-49}$

42. $\dfrac{3}{4a+4} - \dfrac{a}{a^2-1}$

43. $\dfrac{4x}{x^2-25} - \dfrac{2}{x^2+5x}$

44. $\dfrac{2x}{x^2-16} - \dfrac{3}{x^2-4x}$

45. $\dfrac{3}{(x-5)(x+2)} - \dfrac{4}{(x+2)(x+4)}$

46. $\dfrac{5}{(x-2)(x+3)} - \dfrac{2}{(x-6)(x-2)}$

47. $\dfrac{7}{x^2-5x+6} - \dfrac{2}{x^2-3x}$

48. $\dfrac{2}{x^2-4x-32} - \dfrac{6}{x^2+4x}$

49. $\dfrac{5b}{b^2+3b-10} - \dfrac{3}{b^2+4b-12}$

50. $\dfrac{4}{t^2-5t-24} - \dfrac{3t}{t^2+2t-3}$

51. $\dfrac{2x}{x^2+11x+18} - \dfrac{5}{x^2-5x-14}$

52. $\dfrac{3}{x^2-12x+35} - \dfrac{8}{x^2-3x-10}$

53. $\dfrac{x+3}{x-6} - 4$

54. $\dfrac{x-1}{x+5} - 2$

Find the difference. Simplify the result.

55. $\dfrac{x+2}{x-4} - \dfrac{x-3}{x+1}$

56. $\dfrac{x-6}{x-3} - \dfrac{x-4}{x+5}$

57. $\dfrac{x+2}{x-4} - \dfrac{4}{x^2-9x+20}$

58. $\dfrac{x-7}{x-2} - \dfrac{3}{x^2+3x-10}$

59. $\dfrac{5t}{t^2-10t+21} - \dfrac{t+4}{t-7}$

60. $\dfrac{4r}{r^2-9r+8} - \dfrac{r-3}{r-8}$

61. $\dfrac{x-4}{3x+3} - \dfrac{6}{x^2-1}$

62. $\dfrac{x-2}{3x-12} - \dfrac{5}{x^2-16}$

63. $\dfrac{3x}{x+y} - \dfrac{3xy}{2x^2+3xy+y^2}$

64. $\dfrac{2y}{x-y} - \dfrac{6xy}{3x^2-2xy-y^2}$

65. $\dfrac{2}{x^3-6x^2-3x+18} - \dfrac{3x}{5x^2-15}$

66. $\dfrac{4x}{x^3-2x^2+2x-4} - \dfrac{5}{4x^2+8}$

67. $\dfrac{3x}{x^3+1} - \dfrac{2}{x^2+2x+1}$

68. $\dfrac{2x}{x^3-64} - \dfrac{4}{x^2-8x+16}$

Concepts

69. A student tries to find the difference $\dfrac{3x}{x+2} - \dfrac{5x+7}{x+2}$:

$$\dfrac{3x}{x+2} - \dfrac{5x+7}{x+2} = \dfrac{3x-5x+7}{x+2}$$
$$= \dfrac{-2x+7}{x+2}$$

Describe any errors. Then find the difference correctly.

70. A student tries to find the difference $\dfrac{5x}{x-7} - \dfrac{2x+4}{x-7}$:

$$\dfrac{5x}{x-7} - \dfrac{2x+4}{x-7} = \dfrac{5x-(2x+4)}{x-7}$$
$$= \dfrac{5x-2x+4}{x-7}$$
$$= \dfrac{3x+4}{x-7}$$

Describe any errors. Then find the difference correctly.

71. When we subtract two rational expressions, we subtract the *entire* numerator of the second rational expression. Give an example that illustrates this, where the numerator of the second rational expression is a binomial.

72. Explain how subtracting rational expressions is similar to subtracting rational numbers.

73. Describe how to subtract two rational expressions that have a common denominator.

74. Describe how to subtract two rational expressions that have different denominators.

Related Review

75. Find the quotient of $\dfrac{x+4}{x+3}$ and $\dfrac{x-7}{x+3}$.

76. Find the product of $\dfrac{x+4}{x+3}$ and $\dfrac{x-7}{x+3}$.

77. Find the difference of $\dfrac{x+4}{x+3}$ and $\dfrac{x-7}{x+3}$.

78. Find the sum of $\dfrac{x+4}{x+3}$ and $\dfrac{x-7}{x+3}$.

Perform the indicated operation. Simplify the result.

79. $\dfrac{x^2 - 9}{x^2 + 10x + 25} \cdot \dfrac{x^2 + 5x}{x^2 - 4x - 21}$

80. $\dfrac{x^2 - 2x - 8}{x^2 + 10x + 9} \div \dfrac{x^2 - 16}{x^2 + 18x + 81}$

81. $\dfrac{2x}{x^2 - 4x + 4} + \dfrac{4}{x^2 - 9x + 14}$

82. $\dfrac{7}{x^2 - 11x + 30} - \dfrac{3x}{x^2 - 7x + 10}$

83. $\dfrac{x^2 - 10x + 16}{5x^2 - 3x} \div \dfrac{x^2 - 3x - 40}{25x^2 - 9}$

84. $\dfrac{4x^2 - 49}{4x - 12} \cdot \dfrac{x^2 - 9}{2x^2 + 5x - 7}$

85. $\dfrac{3}{x^2 + 4x - 21} - \dfrac{5x}{2x - 6}$

86. $\dfrac{2x}{5x - 20} + \dfrac{4}{x^2 - 11x + 28}$

87. Perform the indicated operation. Simplify the result.

  a. $\dfrac{x}{2} + \dfrac{4}{x}$ **b.** $\dfrac{x}{2} - \dfrac{4}{x}$ **c.** $\dfrac{x}{2} \cdot \dfrac{4}{x}$ **d.** $\dfrac{x}{2} \div \dfrac{4}{x}$

88. Perform the indicated operation. Simplify the result.

  a. $\dfrac{x + 2}{x + 3} + \dfrac{x + 4}{x + 5}$ **b.** $\dfrac{x + 2}{x + 3} - \dfrac{x + 4}{x + 5}$

  c. $\dfrac{x + 2}{x + 3} \cdot \dfrac{x + 4}{x + 5}$ **d.** $\dfrac{x + 2}{x + 3} \div \dfrac{x + 4}{x + 5}$

Expressions, Equations, and Graphs

Perform the indicated instruction. Then use words such as linear, quadratic, cubic, power, radical, rational, polynomial, degree, one variable, and two variables to describe the expression, equation, or system.

89. Solve $3x^2 + 2x = 8$.

90. Solve $5x^2 = 8$.

91. Factor $3x^2 + 2x - 8$.

92. Find the product $(4w - 1)(3w^2 + 2w - 5)$.

93. Find the difference $\dfrac{5}{3x^2 + 2x - 8} - \dfrac{3x}{x^2 - 4}$.

94. Simplify $\sqrt{40ab^2}$. Assume that a and b are nonnegative.

10.5 Solving Rational Equations

Objectives

» Solve *rational equations in one variable.*

» Compare solving rational equations with simplifying rational expressions.

» Use a rational model to make estimates and predictions.

In Sections 10.1–10.4, we worked mostly with rational expressions. In this section, we will solve rational *equations* in one variable.

Solving Rational Equations in One Variable

Here are some examples of rational equations in one variable:

$$\frac{5}{x - 2} = 7 \qquad \frac{2}{3x} + \frac{x}{4} = \frac{5}{6x} \qquad \frac{2x}{x - 3} - \frac{4}{x + 5} = \frac{1}{x^2 + 2x - 10}$$

A **rational equation in one variable** is an equation in one variable in which each side can be written as a rational expression. Recall from Section 4.4 that we can clear an equation of fractions by multiplying both sides of the equation by the LCD.

With rational equations, it is possible to take the usual steps for solving equations, yet arrive at x values that are excluded values for one or more of the fractions in the equation. These values of x are *not* solutions. We call these values **extraneous solutions**.

Example 1   Solving a Rational Equation

Solve $7 - \dfrac{4}{x} = 2 + \dfrac{6}{x}$.

Solution

We note that 0 is an excluded value. We clear the equation of fractions by multiplying both sides of the equation by x, which is the LCD of both of the fractions $\dfrac{4}{x}$ and $\dfrac{6}{x}$:

$$7 - \frac{4}{x} = 2 + \frac{6}{x} \qquad \textit{Original equation}$$

$$x \cdot \left(7 - \frac{4}{x}\right) = x \cdot \left(2 + \frac{6}{x}\right) \qquad \textit{Multiply both sides by LCD, x.}$$

$$x \cdot 7 - x \cdot \frac{4}{x} = x \cdot 2 + x \cdot \frac{6}{x} \qquad \textit{Distributive law}$$

$$7x - 4 = 2x + 6 \qquad \textit{Simplify.}$$

$$5x = 10 \qquad \textit{Subtract 2x on both sides; add 4 to both sides.}$$

$$x = 2 \qquad \textit{Divide both sides by 5.}$$

Because 2 is not an excluded value, we conclude that 2 is the solution of the equation. We check that 2 satisfies the original equation:

$$7 - \frac{4}{x} = 2 + \frac{6}{x}$$

$$7 - \frac{4}{2} \overset{?}{=} 2 + \frac{6}{2}$$

$$7 - 2 \overset{?}{=} 2 + 3$$

$$5 \overset{?}{=} 5$$

$$\text{true}$$

Example 2   Solving a Rational Equation

Solve $\dfrac{x}{4} - \dfrac{3}{2x} = \dfrac{5}{4}$.

Solution

We note that 0 is an excluded value. We clear the equation of fractions by multiplying both sides of the equation by $4x$, which is the LCD of all of the fractions:

$$\frac{x}{4} - \frac{3}{2x} = \frac{5}{4} \qquad \textit{Original equation}$$

$$4x \cdot \left(\frac{x}{4} - \frac{3}{2x} \right) = 4x \cdot \frac{5}{4} \qquad \textit{Multiply both sides by the LCD, 4x.}$$

$$4x \cdot \frac{x}{4} - 4x \cdot \frac{3}{2x} = 4x \cdot \frac{5}{4} \qquad \textit{Distributive law}$$

$$x^2 - 6 = 5x \qquad \textit{Simplify.}$$

$$x^2 - 5x - 6 = 0 \qquad \textit{Write in } ax^2 + bx + c = 0 \textit{ form.}$$

$$(x + 1)(x - 6) = 0 \qquad \textit{Factor left side.}$$

$$x + 1 = 0 \quad \text{or} \quad x - 6 = 0 \qquad \textit{Zero factor property}$$

$$x = -1 \quad \text{or} \qquad x = 6$$

Because neither of our results, -1 and 6, is an excluded value, we conclude that the solutions are -1 and 6. We use "intersect" on a graphing calculator to check our work (see Fig. 12).

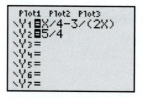 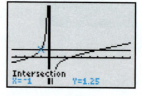 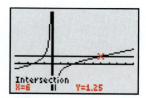

Figure 12 Verify the work

In Example 3, we will see the importance of keeping track of excluded values.

Example 3   Solving a Rational Equation

Solve $\dfrac{4}{w - 3} + 2 = \dfrac{w + 1}{w - 3}$.

Solution

We note that 3 is an excluded value. We clear the equation of fractions by multiplying both sides of the equation by the LCD, $w - 3$:

$$\frac{4}{w-3} + 2 = \frac{w+1}{w-3} \qquad \text{Original equation}$$

$$(w-3) \cdot \left(\frac{4}{w-3} + 2\right) = (w-3) \cdot \left(\frac{w+1}{w-3}\right) \qquad \text{Multiply both sides by LCD, } w-3.$$

$$(w-3) \cdot \frac{4}{w-3} + (w-3) \cdot 2 = (w-3) \cdot \frac{w+1}{w-3} \qquad \text{Distributive law}$$

$$4 + (w-3) \cdot 2 = w + 1 \qquad \text{Simplify.}$$

$$4 + 2w - 6 = w + 1 \qquad \text{Distributive law}$$

$$2w - 2 = w + 1 \qquad \text{Combine like terms.}$$

$$w = 3 \qquad \text{Subtract } w \text{ from both sides; add 2 to both sides.}$$

Our result, 3, is *not* a solution because 3 is an excluded value. Because the only possibility is not a solution of the original equation, we conclude that no number is a solution. We say that the solution set is the *empty set*.

In Example 3, we multiplied both sides of the rational equation $\frac{4}{w-3} + 2 = \frac{w+1}{w-3}$ by the LCD and simplified both sides. We got the equation $4 + (w-3) \cdot 2 = w + 1$, which has the solution 3. However, 3 is *not* a solution of the original rational equation.

To see where we introduced the extraneous solution 3, notice that 3 does not satisfy the equation

$$(w-3) \cdot \frac{4}{w-3} + (w-3) \cdot 2 = (w-3) \cdot \frac{w+1}{w-3}$$

because 3 is an excluded value of the expression $\frac{4}{w-3}$ and $\frac{w+1}{w-3}$. However, 3 does satisfy the next equation, $4 + (w-3) \cdot 2 = w + 1$:

$$4 + (w-3) \cdot 2 = w + 1$$
$$4 + (3-3) \cdot 2 \stackrel{?}{=} 3 + 1$$
$$4 \stackrel{?}{=} 4$$
$$\text{true}$$

Because multiplying both sides of a rational equation by the LCD and then simplifying both sides may introduce extraneous solutions, we must always check that any proposed solution is not an excluded value.

Example 4 Solving a Rational Equation

Solve $\frac{x+4}{x-3} = \frac{x-6}{x+1}$.

Solution

We note that 3 and -1 are excluded values. We clear the equation of fractions by multiplying both sides of the equation by the LCD, $(x-3)(x+1)$:

$$\frac{x+4}{x-3} = \frac{x-6}{x+1} \qquad \text{Original equation}$$

$$(x-3)(x+1) \cdot \left(\frac{x+4}{x-3}\right) = (x-3)(x+1) \cdot \left(\frac{x-6}{x+1}\right) \qquad \text{Multiply both sides by LCD, } (x-3)(x+1).$$

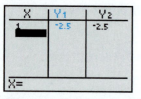

Figure 13 Verify the work

$$(x + 1)(x + 4) = (x - 3)(x - 6) \qquad \textit{Simplify.}$$
$$x^2 + 5x + 4 = x^2 - 9x + 18 \qquad \textit{Find the products.}$$
$$5x + 4 = -9x + 18 \qquad \textit{Subtract } x^2 \textit{ from both si}$$
$$\qquad\qquad\qquad \textit{Add } 9x \textit{ to both sides; su}$$
$$14x = 14 \qquad\qquad \textit{4 from both sides.}$$
$$x = 1 \qquad\qquad \textit{Divide both sides by } 14.$$

Because 1 is not an excluded value, the solution is 1. We use a graphing calculator table to check our work (see Fig. 13).

To solve a rational equation, we factor the denominators of fractions to help us determine any excluded values, to find the LCD, and, later, to help us simplify rational expressions.

Example 5   Solving a Rational Equation

Solve $\dfrac{x}{x - 5} + \dfrac{2}{x - 6} = \dfrac{2}{x^2 - 11x + 30}$.

Solution

We begin by factoring the denominator of the expression on the right-hand side of the equation:

$$\frac{x}{x - 5} + \frac{2}{x - 6} = \frac{2}{x^2 - 11x + 30} \qquad \textit{Original equation}$$

$$\frac{x}{x - 5} + \frac{2}{x - 6} = \frac{2}{(x - 6)(x - 5)} \qquad \textit{Factor denominator.}$$

The excluded values are 5 and 6. Next, we clear the equation of fractions by multiplying both sides by the LCD, $(x - 6)(x - 5)$:

$$(x - 6)(x - 5) \cdot \left(\frac{x}{x - 5} + \frac{2}{x - 6} \right) = (x - 6)(x - 5) \cdot \frac{2}{(x - 6)(x - 5)}$$

On the left-hand side, we use the distributive law. On the right-hand side, we simplify:

$$(x - 6)(x - 5) \cdot \frac{x}{x - 5} + (x - 6)(x - 5) \cdot \frac{2}{x - 6} = 2$$

$$(x - 6) \cdot x + (x - 5) \cdot 2 = 2 \qquad \textit{Simplify.}$$
$$x^2 - 6x + 2x - 10 = 2 \qquad \textit{Distributive law}$$
$$x^2 - 4x - 10 = 2 \qquad \textit{Combine like terms.}$$
$$x^2 - 4x - 12 = 0 \qquad \textit{Write in } ax^2 + bx + c = 0 \textit{ for}$$
$$(x - 6)(x + 2) = 0 \qquad \textit{Factor left-hand side.}$$
$$x - 6 = 0 \quad \text{or} \quad x + 2 = 0 \qquad \textit{Zero factor property}$$
$$x = 6 \quad \text{or} \qquad x = -2$$

Because 6 is an excluded value, it is *not* a solution. The only solution is -2. We use a graphing calculator table to check our work (see Fig. 14).

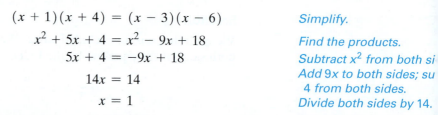

Figure 14 Verify the work

Solving a Rational Equation in One Variable

To solve a rational equation in one variable,

1. Factor the denominator(s) if possible.
2. Identify any excluded values.
3. Find the LCD of all of the fractions.

4. Multiply both sides of the equation by the LCD, which gives a simpler equation to solve.

5. Solve the simpler equation.

6. Discard any proposed solutions that are excluded values.

Solving Equations versus Simplifying Expressions

Throughout this course, we have solved equations and simplified expressions. In solving an equation, our objective is to find any *numbers* that satisfy the equation. In simplifying an expression, our objective is to find a simpler, yet equivalent, *expression*.

Solving a Rational Equation versus Simplifying a Rational Expression

To solve a rational equation, clear the fractions in it by multiplying both sides of the equation by the LCD. To simplify a rational expression, do *not* multiply it by the LCD. The only multiplication permissible is multiplication by 1, usually in the form $\dfrac{A}{A}$, where A is a nonzero polynomial.

Here we compare solving a rational equation with simplifying a rational expression:

Solving the Equation $\dfrac{3}{2} = \dfrac{6}{x}$

The number 0 is an excluded value.

$$\dfrac{3}{2} = \dfrac{6}{x} \qquad \textit{Original equation}$$

$$2x \cdot \dfrac{3}{2} = 2x \cdot \dfrac{6}{x} \qquad \begin{array}{l}\textit{Multiply both}\\ \textit{sides by LCD, 2x.}\end{array}$$

$$3x = 12 \qquad \textit{Simplify.}$$

$$x = 4 \qquad \textit{The result is a number.}$$

Simplifying the Expression $\dfrac{3}{2} + \dfrac{6}{x}$

$$\dfrac{3}{2} + \dfrac{6}{x} = \dfrac{3}{2} \cdot \dfrac{x}{x} + \dfrac{6}{x} \cdot \dfrac{2}{2} \qquad \begin{array}{l}\textit{Introduce missing}\\ \textit{factors.}\end{array}$$

$$= \dfrac{3x}{2x} + \dfrac{12}{2x} \qquad \textit{Find products.}$$

$$= \dfrac{3x + 12}{2x} \qquad \begin{array}{l}\textit{The result is an}\\ \textit{expression.}\end{array}$$

For the equation $\dfrac{3}{2} = \dfrac{6}{x}$, the solution is the *number* 4. By contrast, we simplify the expression $\dfrac{3}{2} + \dfrac{6}{x}$ by writing it as the *expression* $\dfrac{3x + 12}{2x}$. In general, **the result of solving a rational equation is the empty set or a set of one or more numbers. The result of simplifying a rational expression is an expression.**

Rational Models

When using a rational model to describe a situation, we can make predictions about the explanatory variable by substituting a value for the response variable and then solving the equation by using techniques discussed in this section.

Example 6   Using a Rational Model to Make a Prediction

The numbers of households in the United States that use online banking are shown in Table 8 for various years.

Honey, you do understand that those checks you've been feeding your computer are going nowhere, right?

Table 8 Households That Use Online Banking

Year	Households That Use Online Banking (millions)	Total Number of Households (millions)
2008	56.3	116.8
2010	65.0	117.5
2012	76.2	121.1
2014	81.9	123.2
2016	90.9	125.8

Sources: *Nielsen Scarborough; U.S. Census Bureau*

Let p be the percentage of households that use online banking at t years since 2000. A reasonable model is

$$p = \frac{431t + 2240}{1.19t + 106.67}$$

1. Predict the percentage of households that will use online banking in 2022.
2. Predict when 90% of households will use online banking.

Solution

1. We substitute 22 for t in the equation $p = \dfrac{431t + 2240}{1.19t + 106.67}$ and solve for p:

$$p = \frac{431(22) + 2240}{1.19(22) + 106.67} \approx 88.23$$

So, about 88.2% of households will use online banking in 2022, according to the model.

2. We substitute the output 90 for p in the equation $p = \dfrac{431t + 2240}{1.19t + 106.67}$ and solve for t:

$$90 = \frac{431t + 2240}{1.19t + 106.67} \qquad \text{Substitute 90 for } p.$$

$$(1.19t + 106.67) \cdot 90 = (1.19t + 106.67) \cdot \frac{431t + 2240}{1.19t + 106.67} \qquad \begin{array}{l}\textit{Multiply both sides by LC}\\ \textit{1.19t + 106.67.}\end{array}$$

$$107.1t + 9600.3 = 431t + 2240 \qquad \textit{Distributive law; simplify.}$$

$$107.1t - 431t = 2240 - 9600.3 \qquad \begin{array}{l}\textit{Subtract 431t from both si}\\ \textit{subtract 9600.3 from both}\end{array}$$

$$-323.9t = -7360.3 \qquad \textit{Combine like terms.}$$

$$t = 22.72 \qquad \textit{Divide both sides by } -32\text{.}$$

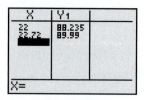

Figure 15 Verify the work

The value of t found represents the year $2000 + 23 = 2023$. So, 90% of households will use online banking in 2023, according to the model. We use a graphing calculator table to verify our work in Problems 1 and 2 (see Fig. 15).

◈ Group Exploration

Section Opener: Solving rational equations

1. Solve $\dfrac{1}{2} + \dfrac{x}{3} = \dfrac{5}{6}$.

2. Solve $\dfrac{1}{x - 2} + \dfrac{2}{3} = \dfrac{5}{x - 2}$. [**Hint:** Multiply both sides by the LCD of all three fractions.]

3. Solve $\dfrac{x}{x - 2} - 5 = \dfrac{2}{x - 2}$. Check whether your result satisfies the equation. What is the solution set? Explain.

4. The equations in Problems 1–3 are called *rational equations in one variable*. What does your work in Problem 3 suggest you should always do when solving a rational equation in one variable?

Homework 10.5

For extra help ▶   **MyLab Math**   📹 Watch the videos in MyLab Math

Solve.

1. $\dfrac{3}{x} - 2 = \dfrac{7}{x}$

2. $\dfrac{9}{x} - 3 = \dfrac{3}{x}$

3. $5 - \dfrac{4}{x} = 3 + \dfrac{2}{x}$

4. $6 - \dfrac{8}{x} = 3 + \dfrac{4}{x}$

5. $\dfrac{5}{p - 1} = \dfrac{2p + 1}{p - 1}$

6. $\dfrac{7}{y + 3} = \dfrac{3y + 4}{y + 3}$

7. $\dfrac{8x + 4}{x + 2} = \dfrac{5x - 2}{x + 2}$

8. $\dfrac{3x - 2}{x - 5} = \dfrac{x + 8}{x - 5}$

9. $\dfrac{w + 2}{w - 4} + 3 = \dfrac{2}{w - 4}$

10. $\dfrac{t - 5}{t + 1} + 2 = \dfrac{3}{t + 1}$

11. $\dfrac{2}{x} + \dfrac{5}{4} = \dfrac{3}{x}$

12. $\dfrac{3}{x} - \dfrac{7}{2} = \dfrac{6}{x}$

13. $\dfrac{5}{6x} - \dfrac{1}{2} = \dfrac{3}{4x}$

14. $\dfrac{3}{8x} - \dfrac{3}{4} = \dfrac{1}{6}$

15. $\dfrac{4}{x - 2} = \dfrac{2}{x + 3}$

16. $\dfrac{3}{x + 4} = \dfrac{4}{x - 1}$

17. $\dfrac{2r + 7}{4r} = \dfrac{5}{3}$

18. $\dfrac{5p - 2}{3p} = \dfrac{6}{7}$

19. $\dfrac{5}{x + 3} + \dfrac{3}{4} = 2$

20. $\dfrac{4}{x - 3} + \dfrac{5}{2} = 3$

21. $\dfrac{2}{x - 3} + \dfrac{1}{x + 3} = \dfrac{5}{x^2 - 9}$

22. $\dfrac{3}{x + 5} + \dfrac{2}{x - 5} = \dfrac{1}{x^2 - 25}$

23. $\dfrac{4}{x + 2} + \dfrac{3}{x + 1} = \dfrac{3}{x^2 + 3x + 2}$

24. $\dfrac{2}{x - 3} + \dfrac{4}{x + 5} = \dfrac{16}{x^2 + 2x - 15}$

25. $\dfrac{5}{x^2 - 4} + \dfrac{2}{x + 2} = \dfrac{4}{x - 2}$

26. $\dfrac{3}{x^2 - 1} + \dfrac{7}{x - 1} = \dfrac{4}{x + 1}$

27. $\dfrac{3}{y - 4} - \dfrac{4}{y - 3} = \dfrac{3}{y^2 - 7y + 12}$

28. $\dfrac{2}{r + 2} - \dfrac{3}{r - 6} = \dfrac{-16}{r^2 - 4r - 12}$

29. $\dfrac{4}{x^2 - 3x} - \dfrac{5}{x} = \dfrac{7}{x - 3}$

30. $\dfrac{2}{x^2 + 6x} - \dfrac{4}{x} = \dfrac{5}{x + 6}$

31. $\dfrac{3}{2x - 6} + \dfrac{5x}{6x - 18} = \dfrac{2}{4x - 12}$

32. $\dfrac{5}{4x + 8} - \dfrac{3x}{2x + 4} = \dfrac{2}{5x + 10}$

33. $\dfrac{2}{x^2 - x - 6} - \dfrac{4}{x + 2} = \dfrac{3}{2x - 6}$

34. $\dfrac{2}{x^2 + 4x + 3} - \dfrac{3}{x + 1} = \dfrac{5}{3x + 9}$

35. $\dfrac{3}{x} = \dfrac{4}{x^2} - 1$

36. $\dfrac{2}{x} = \dfrac{24}{x^2} - 2$

37. $1 = \dfrac{15}{t^2} - \dfrac{2}{t}$

38. $1 = \dfrac{-14}{w^2} + \dfrac{9}{w}$

39. $\dfrac{2}{x^2} + \dfrac{3}{x} = 5$

40. $\dfrac{4}{x^2} - \dfrac{1}{x} = 3$

41. $\dfrac{x - 3}{x + 2} = \dfrac{x + 5}{x - 1}$

42. $\dfrac{x + 6}{x + 2} = \dfrac{x - 7}{x - 3}$

43. $\dfrac{x}{x + 1} = \dfrac{3}{x - 1} + \dfrac{2}{x^2 - 1}$

44. $\dfrac{x}{x + 3} = \dfrac{2}{x - 3} - \dfrac{12}{x^2 - 9}$

45. $\dfrac{r}{r - 3} - \dfrac{2}{r + 5} = \dfrac{10}{r^2 + 2r - 15}$

46. $\dfrac{4}{p + 2} - \dfrac{p}{p + 1} = \dfrac{5}{p^2 + 3p + 2}$

47. $\dfrac{1}{x - 2} = \dfrac{2x}{x + 1} - \dfrac{6}{x^2 - x - 2}$

48. $\dfrac{x}{x + 5} = \dfrac{3}{x + 4} + \dfrac{5}{x^2 + 9x + 20}$

49. $\dfrac{2}{x^2 - 9} = \dfrac{x}{x - 3} - \dfrac{x - 5}{x + 3}$

50. $\dfrac{-1}{x^2 - 4} = \dfrac{x}{x - 2} - \dfrac{x - 4}{x + 2}$

51. $\dfrac{3}{x - 4} + \dfrac{7}{x - 3} = \dfrac{x + 4}{x - 3}$

52. $\dfrac{5}{x + 5} + \dfrac{4}{x - 2} = \dfrac{x - 5}{x + 5}$

53. $\dfrac{2}{m - 3} + \dfrac{5m}{m^2 - 9} = \dfrac{4}{m + 3} - \dfrac{3}{m^2 - 9}$

54. $\dfrac{6w}{w^2 - 16} - \dfrac{3}{w + 4} = \dfrac{5}{w^2 - 16} + \dfrac{2}{w - 4}$

55. **DATA** In Exercise 75 of Homework 10.1, you worked with the model

$$p = \dfrac{-13t + 179}{1.4t + 117.9}$$

where p is the percentage of households that were burglarized in the year that is t years since 2010 (see Table 9).

Table 9 Numbers of Households Burglarized

Year	Number of Households Burglarized (millions)	Total Number of Households (millions)
2011	1.63	118.7
2012	1.56	121.1
2013	1.43	122.5
2014	1.27	123.2
2015	1.13	124.6

Sources: *FBI; U.S. Census Bureau*

Predict in which year 0.4% of households will be burglarized.

56. <u>DATA</u> In Exercise 76 of Homework 10.1, you worked with the model

$$p = \frac{594t + 1158}{1.15t + 118.84}$$

where p is the percentage of households that stream Netflix at t years since 2010 (see Table 10).

Table 10 Numbers of Households That Stream Netflix

Year	Households That Stream Netflix (millions)	Total Number of Households (millions)
2012	23.4	121.1
2013	29.2	122.5
2014	35.7	123.2
2015	41.4	124.6
2016	47.0	125.8

Sources: *Netflix; U.S. Census Bureau*

Predict when 62% of households will stream Netflix.

57. <u>DATA</u> The numbers of households with broadband access to the Internet are shown in Table 11 for various years.

Table 11 Numbers of Households with Broadband Access to the Internet

Year	Number of Households with Broadband Access to the Internet (millions)	Total Number of Households (millions)
2008	77.1	116.8
2010	84.5	117.5
2012	92.5	121.1
2014	97.8	123.2
2016	106.1	125.8

Sources: *ITU; Federal Communications Commission; U.S. Census Bureau*

Let p be the percentage of households with broadband access at t years since 2000. A reasonable model is

$$p = \frac{357t + 4882}{1.19t + 106.7}$$

a. Use the model to estimate the percentage of households that had broadband access in 2014. Then compute the actual percentage. Is your result from the model an underestimate or an overestimate?

b. Predict the percentage of households that will have broadband access in 2021.

c. Estimate when 96% of households had broadband access.

58. <u>DATA</u> Total household debt in the United States is shown in Table 12 for various years. Totals include debt from mortgages, auto loans, credit cards, student loans, and other types of loans.

Table 12 Household Debt

Year	Total Household Debt (trillions of dollars)	Number of Households (millions)
2008	12.6	116.8
2010	11.9	117.5
2012	11.4	121.1
2014	11.6	123.2
2016	12.3	125.8

Sources: *New York Fed Consumer Credit Panel; Equifax; U.S. Census Bureau*

Let D be the average household debt (in *thousands* of dollars) at t years since 2000. A reasonable model of the situation is

$$D = \frac{62.5t^2 - 1545t + 21{,}000}{1.19t + 106.7}$$

a. Estimate the average household debt in 2016. Then compute the actual average household debt. [**Hint:** Pay careful attention to the units.] Is your result from the model an underestimate or an overestimate?

b. Predict the average household debt in 2022.

c. Predict in what year the average household debt will be $120 thousand.

Concepts

59. A student tries to simplify $\dfrac{5}{x + 2} + \dfrac{3}{x}$:

$$\frac{5}{x + 2} + \frac{3}{x} = x(x + 2) \cdot \left(\frac{5}{x + 2} + \frac{3}{x} \right)$$

$$= x(x + 2) \cdot \frac{5}{x + 2} + x(x + 2) \cdot \frac{3}{x}$$

$$= 5x + 3(x + 2)$$

$$= 5x + 3x + 6$$

$$= 8x + 6$$

Describe any errors. Then simplify the expression correctly.

60. A student tries to solve the equation $\dfrac{x}{x - 5} + \dfrac{1}{2} = \dfrac{2}{x - 5}$:

$$\frac{x}{x - 5} + \frac{1}{2} = \frac{2}{x - 5}$$

$$\frac{2}{2} \cdot \frac{x}{x - 5} + \frac{x - 5}{x - 5} \cdot \frac{1}{2} = \frac{2}{2} \cdot \frac{2}{x - 5}$$

$$\frac{2x + x - 5}{2(x - 5)} = \frac{4}{2(x - 5)}$$

$$\frac{3x - 5}{2(x - 5)} = \frac{4}{2(x - 5)}$$

$$3x - 5 = 4$$

$$3x = 9$$

$$x = 3$$

Is the work correct? Is there a better way to solve the equation? Explain.

61. A student tries to solve $\dfrac{7}{x^2 + x - 20} = \dfrac{4}{x + 5} - \dfrac{2}{x - 4}$:

$$\dfrac{7}{x^2 + x - 20} = \dfrac{4}{x + 5} - \dfrac{2}{x - 4}$$

$$= \dfrac{4}{x + 5} \cdot \dfrac{x - 4}{x - 4} - \dfrac{2}{x - 4} \cdot \dfrac{x + 5}{x + 5}$$

$$= \dfrac{4(x - 4) - 2(x + 5)}{(x - 4)(x + 5)}$$

$$= \dfrac{4x - 16 - 2x - 10}{(x - 4)(x + 5)}$$

$$= \dfrac{2x - 26}{(x - 4)(x + 5)}$$

$$= \dfrac{2(x - 13)}{(x - 4)(x + 5)}$$

Describe any errors. Then solve the equation correctly.

62. A student tries to solve a rational equation. The student's result is $\dfrac{x + 1}{x - 4}$. What would you tell the student?

63. Why must we check whether any proposed solution of a rational equation is an excluded value?

64. Why do we multiply both sides of a rational equation by the LCD of all the fractions?

65. When simplifying a rational expression, can we multiply it by the LCD? Explain. When solving a rational equation, can we multiply both sides of the equation by the LCD? Explain. (See page 9 for guidelines on writing a good response.)

66. Describe how to solve a rational equation. (See page 9 for guidelines on writing a good response.)

Related Review

Solve or simplify, whichever is appropriate.

67. $\dfrac{7}{x} + \dfrac{3}{x}$

68. $\dfrac{4}{x + 5} - \dfrac{x + 1}{x + 5} = 2$

69. $\dfrac{7}{x} + \dfrac{3}{x} = 1$

70. $\dfrac{4}{x + 5} - \dfrac{x + 1}{x + 5}$

71. $\dfrac{2}{x - 3} + \dfrac{3x}{x + 2} = \dfrac{2}{x^2 - x - 6}$

72. $\dfrac{5}{x - 4} - \dfrac{2}{x^2 - 16}$

73. $\dfrac{2}{x - 3} + \dfrac{3x}{x + 2}$

74. $\dfrac{5}{x - 4} - \dfrac{2}{x^2 - 16} = \dfrac{3}{x + 4}$

Solve.

75. $3x^2 - 2x = 4$

76. $5(2x - 3) = 4x - 7$

77. $3x^3 + 8 = 2x^2 + 12x$

78. $x^2 = 4x + 12$

79. $2x^2 + 1 = 8$

80. $\dfrac{2x}{x - 1} + \dfrac{5}{x + 1} = \dfrac{4x - 3}{x^2 - 1}$

Expressions, Equations, and Graphs

Perform the indicated instruction. Then use words such as linear, quadratic, cubic, power, radical, rational, polynomial, degree, one variable, *and* two variables *to describe the expression, equation, or system.*

81. Simplify $3(5m - 2n) - 4(m + 3n) + n$.

82. Find the quotient $\dfrac{x^2 - 9}{x^3 + 3x^2 - 28x} \div \dfrac{5x + 15}{x^3 - 16x}$.

83. Factor $2x^3 - 18x^2 + 28x$.

84. Graph $2x - 5y = 10$ by hand.

85. Solve $\dfrac{2}{x - 4} - \dfrac{3}{x + 4} = \dfrac{5}{x^2 - 16}$.

86. Solve $x(x - 2) = 48$.

10.6 Proportions; Similar Triangles

Objectives

» Use *proportions* to make estimates.

» Find the length of a side of a *similar triangle*.

In Section 2.5, we worked with ratios. In this section, we will discuss an equation called a *proportion*, which contains ratios. We will use proportions to make estimates of quantities and to find the lengths of the sides of special pairs of triangles called *similar triangles*.

Throughout this chapter, we have been working with rational expressions and rational equations. Recall from Section 10.1 that the word "rational" refers to a ratio. Recall from Section 2.5 that the ratio of a to b can be written as the fraction $\dfrac{a}{b}$ or as $a : b$. For example, if there are 9 women and 5 men on a coed softball team, then the ratio of the number of women to the number of men is

$$\dfrac{9 \text{ women}}{5 \text{ men}}$$

Proportions

A **proportion** is a statement of the equality of two ratios. For example,

$$\frac{4}{6} = \frac{2}{3}$$

is a proportion that says the ratios $\frac{4}{6}$ and $\frac{2}{3}$ are equal.

There are many situations in which proportions can describe equal ratios. For example, suppose that a certain type of pen costs \$2. Then the ratio of the total cost of some pens to the number of pens is constant for any group of pens. Here is the ratio for a group of 3 pens:

$$\begin{array}{l} \text{Total cost} \longrightarrow \\ \text{Number of pens} \longrightarrow \end{array} \frac{\$6}{3 \text{ pens}} = \$2 \text{ per pen}$$

Here is the ratio for a group of 5 pens:

$$\begin{array}{l} \text{Total cost} \longrightarrow \\ \text{Number of pens} \longrightarrow \end{array} \frac{\$10}{5 \text{ pens}} = \$2 \text{ per pen}$$

The proportion

$$\frac{\$6}{3 \text{ pens}} = \frac{\$10}{5 \text{ pens}}$$

correctly states that the ratio of the total cost to the number of pens is the same for a group of 3 pens and a group of 5 pens. We say the total cost and the number of pens are *proportional*.

In general, if the ratio of two related quantities is constant, we say the two quantities are **proportional**.

When we work with a proportion of the form

$$\frac{a}{b} = \frac{c}{d}$$

if we know the values of three of the four variables, we can solve for the fourth variable.

Example 1   Using a Proportion to Make an Estimate

While commuting to work and running errands, a person travels 576 miles in a 3-week period.

1. Estimate over what period the person will travel 1000 miles while commuting to work and running errands.
2. Discuss any assumptions you made in Problem 1. Describe a scenario in which the result you found in Problem 1 is an overestimate.

Solution

1. We let t be the period (in weeks) required for the person to travel 1000 miles while commuting to work and running errands. Assuming that the ratio of the total miles traveled to the number of weeks is constant, we set up a proportion:

$$\text{Total miles traveled} \longrightarrow \frac{576}{3} = \frac{1000}{t} \longleftarrow \text{Total miles traveled} \\ \text{Number of weeks} \longrightarrow \qquad\qquad \longleftarrow \text{Number of weeks}$$

Next, we multiply both sides of the equation by the LCD, $3t$:

$$\frac{576}{3} \cdot \frac{3t}{1} = \frac{1000}{t} \cdot \frac{3t}{1} \qquad \textit{Multiply both sides by 3t.}$$

$$576t = 3000 \qquad \textit{Simplify.}$$

$$t \approx 5.21 \qquad \textit{Divide both sides by 576.}$$

So, the person will drive 1000 miles over about a 5.2-week period.

2. We assumed that the ratio of the total miles traveled to the number of weeks is constant. This will not be the case if the person's driving patterns over the course of the 1000 miles are significantly different from those for the 576 miles. For example, if the person is transferred to a work location that is farther from home and the person's driving patterns while running errands remained the same, our estimate in Problem 1 would be an overestimate.

In Example 1, the person traveled 576 miles in a 3-week period. For this period, the unit ratio of total miles traveled to number of weeks is

$$\frac{576 \text{ miles}}{3 \text{ weeks}} = \frac{192 \text{ miles}}{1 \text{ week}}$$

To estimate over what period the person would travel 1000 miles, we assumed the ratio of total miles to number of weeks would remain $\frac{192 \text{ miles}}{1 \text{ week}}$. If that ratio were not close to $\frac{192 \text{ miles}}{1 \text{ week}}$, then our result would not be accurate.

If we use a proportion to estimate the value of one of two quantities, the accuracy of our result depends on how much the ratio of the two quantities varies. If the ratio is constant, the estimate will be accurate. If the ratio is approximately constant, the estimate will likely be reasonable. If the ratio varies a great deal, the estimate will likely be inaccurate.

Example 2   Using a Proportion to Make an Estimate

In a poll of 2209 adults, 1679 adults said professional baseball players who use steroids should be banned from the National Baseball Hall of Fame (Source: *Harris Interactive, Inc.*).

1. Estimate how many of the 20,000 students at Cincinnati State Technical and Community College would say that professional baseball players who use steroids should be banned from the National Baseball Hall of Fame.
2. Discuss any assumptions you made in Problem 1. Describe a scenario in which the result you found in Problem 1 is an underestimate.

Solution

1. We let n be the number of students at Cincinnati State Technical and Community College who would say professional baseball players who use steroids should be banned from the National Baseball Hall of Fame. Assuming the ratio of people who support the ban to the total number of people is the same at the university as in the poll, we set up a proportion:

Number of students in favor of ban $\longrightarrow$ $\dfrac{n}{20,000} = \dfrac{1679}{2209}$ $\longleftarrow$ Number of polled adults in favor of ban

Total number of students $\longrightarrow$ $\longleftarrow$ Total number of polled adults

Next, we multiply both sides of the equation by 20,000:

$$\frac{n}{20,000} \cdot 20,000 = \frac{1679}{2209} \cdot 20,000 \qquad \textit{Multiply both sides by 20,000.}$$

$$n \approx 15,201 \qquad \textit{Simplify.}$$

So, about 15,201 of the 20,000 students would say that professional baseball players who use steroids should be banned from the National Baseball Hall of Fame.

2. We assumed the ratio of people who support the ban to the total number of people is the same at the college as in the poll. This may not be the case. For example, if students at the college are more disapproving of steroid use than the adults in the survey, our estimate would be an underestimate.

Similar Triangles

Two triangles are called **similar triangles** if their corresponding angles are equal in measure. For example, the two triangles in Fig. 16 are similar triangles because the measures of angles A and X are equal, the measures of angles B and Y are equal, and the measures of angles C and Z are equal.

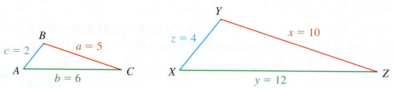

Figure 16 Two similar triangles

Here we find the ratio of the lengths of corresponding sides of the two triangles in Fig. 16:

$$\frac{c}{z} = \frac{2}{4} = \frac{1}{2}, \qquad \frac{a}{x} = \frac{5}{10} = \frac{1}{2}, \qquad \frac{b}{y} = \frac{6}{12} = \frac{1}{2}$$

Notice that all three ratios are equal.

Side Lengths of Similar Triangles

The lengths of the corresponding sides of two similar triangles are proportional.

Similar triangles have the same shape, but not necessarily the same size.

Example 3   Finding the Length of the Side of a Triangle

The two triangles shown in Fig. 17 are similar. Find the length of the side labeled "x feet" on the larger triangle.

Figure 17 Two similar triangles

Solution

Because the triangles are similar, the lengths of corresponding sides are proportional. Thus,

$$\frac{2}{x} = \frac{7}{9} \qquad \text{\textit{Use a proportion.}}$$

$$\frac{2}{x} \cdot 9x = \frac{7}{9} \cdot 9x \qquad \text{\textit{Multiply both sides by LCD, 9x.}}$$

$$18 = 7x \qquad \text{\textit{Simplify.}}$$

$$\frac{18}{7} = x \qquad \text{\textit{Divide both sides by 7.}}$$

$$2.57 \approx x \qquad \text{\textit{Round to second decimal place.}}$$

So, the length of the side labeled "x feet" is approximately 2.57 feet.

Group Exploration

Section Opener: Proportions

Recall from Section 2.5 that the ratio of a to b is the quotient $\frac{a}{b}$.

1. a. The price of a notebook is \$3. For the given number of notebooks, find the ratio of the total cost to the number of notebooks. *Do not simplify the ratio.*
  i. 2 notebooks [**Hint:** First find the total cost of 2 notebooks. Then divide your result by 2. Include units in your result.]
  ii. 5 notebooks
 b. Compare the results you found in parts (i) and (ii). In particular, describe how the following equation is related to your results:

$$\frac{6}{2} = \frac{15}{5}$$

When an equation says that two ratios are equal, we call the equation a *proportion*.

 c. For any number of notebooks, what is the ratio of the total cost to the number of notebooks? Explain.

2. a. A student earns \$8 per hour. For the given number of hours worked, find the ratio of the total income to the number of hours worked. *Do not simplify the ratio.*
  i. 3 hours
  ii. 5 hours
 b. Write a proportion that is related to the results you found in parts (i) and (ii).
 c. For any number of hours worked, what is the ratio of the total income to the number of hours worked? Explain how this ratio is related to the rate of change of income.

 # Homework 10.6

For extra help ▶ **MyLab Math**   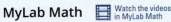 Watch the videos in MyLab Math

1. A band pays \$720 for 6 hours of recording time in a studio. How much will the band pay for 8 hours of recording time?

2. A household pays \$146.25 for 3 months' use of cable TV. How much will the household pay for 5 months' use of cable TV?

3. If 0.75 cup of Post Grape-Nuts Flakes® cereal contains 4 grams of sugar, what amount, in cups, of that cereal contains 7 grams of sugar?

4. If 0.5 cup of Chef Boyardee Mini Ravioli® contains 4.5 grams of fat, what amount, in cups, of that ravioli would contain 11.25 grams of fat?

5. A 10-ounce solution contains 4 ounces of acid. How many ounces of acid are in 6 ounces of that solution?

6. A 7-ounce solution contains 3 ounces of lemon juice. How many ounces of lemon juice are in 5 ounces of the solution?

7. For the fall semester 2016, 2 out of 5 undergraduate students at Boston University were male (Source: *Boston University*). If there were 7173 male undergraduate students, how many undergraduate students were there?

8. For the fall semester 2016, 3 out of 8 new freshmen at Adelphi University lived in the residence halls (Source: *Adelphi University*). If 3144 new freshmen lived in the residence halls, how many new freshmen were there?

9. If 1.25 inches on a map represent 50 miles, how many inches on the map represent 270 miles?

10. If 0.75 inch on a map represents 3 miles, how many inches on the map represent 10 miles?

11. If 5 U.S. dollars can be exchanged for 4.78 European euros, how many U.S. dollars can be exchanged for 260 euros?

12. If 15 U.S. dollars can be exchanged for 309 Mexican pesos, how many U.S. dollars can be exchanged for 4000 pesos?

13. The weight of an object on Earth and the weight of the same object on the Moon are proportional. An astronaut who weighs 180 pounds on Earth weighs 29.8 pounds on the Moon. What is the weight of a person on Earth if she weighs 28.5 pounds on the Moon?

14. The weight of an object on Earth and the weight of the same object on Jupiter are proportional. An astronaut who weighs 150.0 pounds on Earth would weigh 379.9 pounds on Jupiter. What is the weight of a person on Earth if he would weigh 450 pounds on Jupiter?

15. An American pays \$175.50 for 3 months of phone service.
 a. Estimate how much the person will pay for 7 months of phone service.
 b. Discuss any assumptions you made in part (a). If the person begins to make a lot of international calls during the 7-month period and the person's U.S.-call patterns remain the same, is the result you found in part (a) an underestimate or an overestimate? Explain.

16. A family spends \$475 on groceries in a 5-month period.
 a. Estimate the family's total money for groceries during a 3-month period.
 b. Discuss any assumptions you made in part (a). If hungry children were away during part of the 5-month period but were home during the 3-month period, is the result you found in part (a) an underestimate or an overestimate? Explain.

17. In a poll of 1017 adults in the United States, 478 adults believed the U.S. government should redistribute wealth by heavy taxes on the rich (Source: *Gallup*).
   a. Estimate how many of the 12,758 adults in Deerfield, Illinois, believe the U.S. government should redistribute wealth by heavy taxes on the rich.
   b. Discuss any assumptions you made in part (a). The 2015 median household income was $144,621 in Deerfield and $56,516 in the United States overall (Source: *U.S. Census Bureau*). Do you think the result you found in part (a) is an underestimate or an overestimate? Explain.

18. In a poll of 2242 adults in the United States, 1143 of the adults believed the Olympics should be restricted to amateur athletes (Source: *Harris Interactive*).
   a. Estimate how many of the 14,953 students at the Rockville Campus of Montgomery College believe the Olympics should be restricted to amateur athletes.
   b. A much higher percentage of adults over age 40 believe in the restriction than do adults under age 40. Do you think that the result you found in part (a) is an underestimate or an overestimate? Explain.

19. When a 4000-pound car travels at 40 miles per hour and then hits a stationary object, such as a concrete wall, the force of impact is 26 tons. When the speed is 60 miles per hour instead, the force of impact is 55 tons. Are the speed and the force of impact proportional for a 4000-pound car? Explain.

20. If you park your car with the windows closed in a sunny location, the temperature inside the car will increase by about 29°F in 20 minutes and by about 34°F in 30 minutes (Source: *Jan Null*). Are the time elapsed and the increase in temperature proportional? Explain.

21. In fall 2003, the full-time equivalent (FTE) enrollment at Louisiana State University (LSU) was 27,764 students and the number of FTE faculty was 1294. The ratio of the FTE enrollment to the number of FTE faculty at 23 peer institutions is 17:1 (Source: *LSU; U.S. News & World Report*).
   a. Find the ratio of the FTE enrollment to the number of FTE faculty at LSU in fall 2003. Is the ratio greater than, equal to, or less than 17:1?
   b. The LSU deans and department chairs met to determine the cost of reducing their ratio of FTE enrollment to the number of FTE faculty to 17:1. How many FTE faculty would need to be hired to do that? The average cost (salary plus benefits) of adding an FTE faculty member at LSU in 2003 was $73,125. What would be the total cost of hiring enough faculty to reduce the ratio to 17:1?
   c. How much would the FTE enrollment at LSU have to be reduced to reduce the ratio of the FTE enrollment to the number of FTE faculty to 17:1? The revenue (tuition plus an academic excellence fee) from an FTE student was $3345 at LSU in 2003. What would be the total cost of reducing the FTE enrollment to reduce the ratio to 17:1?
   d. Which is the cheaper way to reduce the ratio, hiring FTE faculty or reducing the FTE enrollment?

22. The ratio of 2 and a number is equal to the ratio of the number and 8. Find the number. (The result is called the *geometric mean* of 2 and 8.)

For Exercises 23–28, the given figure shows two similar triangles. Find the length of the side labeled "x inches," "x feet," or "x meters." Round your result to the second decimal place.

23. See Fig. 18.

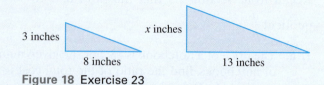

Figure 18 Exercise 23

24. See Fig. 19.

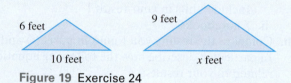

Figure 19 Exercise 24

25. See Fig. 20.

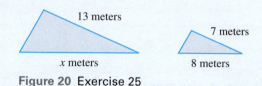

Figure 20 Exercise 25

26. See Fig. 21.

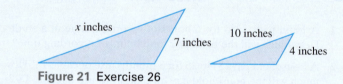

Figure 21 Exercise 26

27. See Fig. 22.

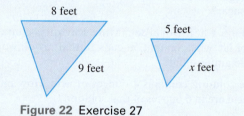

Figure 22 Exercise 27

28. See Fig. 23.

Figure 23 Exercise 28

Concepts

29. Compare the meaning of *ratio* with the meaning of *proportion*.

30. Give an example of an authentic situation in which two quantities are proportional. Why are the quantities proportional?

Related Review

31. If x and y are proportional, then the ratio of x and y is constant. That is,

$$\frac{y}{x} = k$$

where k is a constant.

a. For the equation $\frac{y}{x} = k$, get y alone on one side of the equation. What is the y-intercept of the graph of your result?

b. Bicycle shop A charges \$15 per hour to rent a bicycle. Let C be the total cost (in dollars) of renting a bicycle for t hours. Find an equation to describe this situation. Refer to the result you found in part (a) to help you decide whether t and C are proportional. Explain.

c. Bicycle shop B charges a flat fee of \$25, plus \$10 per hour, to rent a bicycle. Let C be the total cost (in dollars) of renting a bicycle for t hours. Find an equation to describe this situation. Refer to the result you found in part (a) to help you decide whether t and C are proportional. Explain.

d. Find the total cost of renting a bicycle for 2 hours at bicycle shop B. Next, find the total cost of renting a bicycle for 3 hours. Finally, use your work to find two unit ratios to help you determine whether t and C are proportional. Is your conclusion the same as your conclusion in part (c)?

32. A student earns \$72 for working 8 hours at a clothing store.

a. Use a proportion to find the student's total earnings for working 40 hours.

b. What is the rate of change of the student's total earnings?

c. Let E be the student's total earnings (in dollars) for working t hours. Find an equation to describe the situation.

d. Use your linear model to find the student's total earnings for working 40 hours. Compare your result with the result you found in part (a).

33. Consider the line shown in Fig. 24.

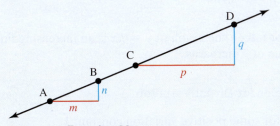

Figure 24 Exercise 33

a. Use the points A and B to find the slope of the line in terms of m and n.

b. Use the points C and D to find the slope of the line in terms of p and q.

c. Use the concept of similar triangles to help you explain why the results you found in parts (a) and (b) are equal.

d. Explain why your work in parts (a), (b), and (c) shows that no matter which two distinct points on a line are used to find the slope of the line, the result will be the same.

34. Two similar right triangles are shown in Fig. 25. Find the length of the side labeled "x feet." [**Hint:** First use the Pythagorean theorem, and then set up a proportion.]

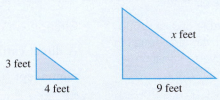

Figure 25 Exercise 34

Expressions, Equations, and Graphs

Perform the indicated instruction. Then use words such as linear, quadratic, cubic, power, radical, rational, polynomial, degree, one variable, *and* two variables *to describe the expression, equation, or system.*

35. Solve $2x^2 + 7x - 15 = 0$.

36. Find the sum $\dfrac{2x}{x-4} + \dfrac{5}{x+3}$.

37. Factor $2x^2 + 7x - 15$.

38. Find the difference $\dfrac{2x}{x-4} - \dfrac{5}{x+3}$.

39. Evaluate $2x^2 + 7x - 15$ for $x = -2$.

40. Find the quotient $\dfrac{2x}{x-4} \div \dfrac{5}{x+3}$.

10.7 Variation

Objectives

» Describe the meaning of *direct variation* and *inverse variation*.

» In direct variation and inverse variation, describe how a

(continued)

In this section, we will discuss two simple, yet important, types of equations in two variables: *direct variation equations* and *inverse variation equations*.

Direct Variation

Recall from Section 10.6 that two related quantities are proportional if the ratio of the two quantities is constant. For example, x and y are proportional if

$$\frac{y}{x} = k$$

where k is a constant. If we multiply both sides of this equation by x, we have

$$y = kx$$

Definition   Direct variation

If $y = kx$ for some constant k, we say that **y varies directly as x** or that **y is proportional to x.** We call k the **variation constant** or the **constant of proportionality.** The equation $y = kx$ is called a **direct variation equation.**

For $y = 3x$, we say y varies directly as x with variation constant 3. For $w = 8t$, we say w varies directly as t with variation constant 8.

Changes in Values for Direct Variation

A direct variation equation $y = kx$ is a linear equation whose graph has slope k and y-intercept $(0, 0)$. In Fig. 26, we sketch three such equations, with $k = \dfrac{1}{2}, k = 1$, and $k = 2$.

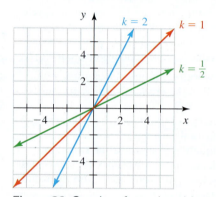

Figure 26 Graphs of $y = kx$ with $k = \dfrac{1}{2}, k = 1$, and $k = 2$

If k is positive (positive slope), then the graph of $y = kx$ is an increasing line. So, if the value of x increases, the value of y increases.

Changes in Values of Variables for Direct Variation

Assume y varies directly as x for some positive variation constant k.

• If the value of x increases, then the value of y increases.
• If the value of x decreases, then the value of y decreases.

Using One Point to Find a Direct Variation Equation

If one variable varies directly as another variable and we know one point that lies on the graph, we can find the variation constant k as well as the direct variation equation.

Example 1   Finding a Direct Variation Equation

The variable y varies directly as x with positive variation constant k.

1. What happens to the value of y as the value of x increases?
2. If $y = 7$ when $x = 2$, find an equation for x and y.

Solution

1. Because y varies directly as x with the positive variation constant k, the value of y increases as the value of x increases.
2. The equation is of the form $y = kx$. We find the constant k by substituting 2 for x and 7 for y:

$$y = kx \qquad \textit{y varies directly as x.}$$
$$7 = k(2) \quad \textit{Substitute 2 for x and 7 for y.}$$
$$\frac{7}{2} = k \qquad \textit{Divide both sides by 2.}$$

The equation is $y = \dfrac{7}{2}x$. We use a graphing calculator graph to check that the curve $y = \dfrac{7}{2}x$ contains the point $(2, 7)$. See Fig. 27.

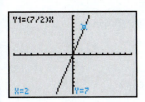

Figure 27 Verify the work

Example 2   Using a Direct Variation Equation

The variable p varies directly as t. If $p = 6$ when $t = 2$, find the value of t when $p = 10$.

Solution

Because p varies directly as t, we can describe the relationship between t and p by the equation $p = kt$. We find the constant k by substituting 2 for t and 6 for p:

$$p = kt \qquad \textit{p varies directly as t.}$$
$$6 = k(2) \quad \textit{Substitute 2 for t and 6 for p.}$$
$$3 = k \qquad \textit{Divide both sides by 2.}$$

The equation is $p = 3t$. We find the value of t when $p = 10$ by substituting 10 for p in the equation $p = 3t$ and then solving for t:

$$10 = 3t \quad \textit{Substitute 10 for p.}$$
$$\frac{10}{3} = t \quad \textit{Divide both sides by 3.}$$

So, $t = \dfrac{10}{3}$ when $p = 10$.

Direct Variation Models

Many authentic situations can be modeled well by direct variation equations. For example, when a person drives at a constant speed, the driving time and the distance traveled can be modeled by such an equation.

Example 3   Using a Direct Variation Model

For California residents, the cost of tuition at Solano Community College varies directly as the number of units (hours or credits) a student takes. For fall 2017, the cost of 6 units of classes is \$276. Let C be the total cost (in dollars) of u units.

1. As the value of u increases, what happens to the value of C? What does that pattern mean in this situation?
2. Find an equation for u and C.
3. Find the total cost of 12 units of classes.
4. A student can afford to pay at most \$700 for classes. How many units of classes can the student enroll in?

Solution

1. Because C varies directly as u, as u increases, the value of C increases. This means the greater the number of units taken, the greater the total cost will be.

2. Because C varies directly as u, we can model the situation with the equation $C = ku$. We substitute 6 for u and 276 for C to find the constant k:

$$C = ku \qquad \textit{C varies directly as u.}$$
$$276 = k(6) \qquad \textit{Substitute 6 for u and 276 for C.}$$
$$46 = k \qquad \textit{Divide both sides by 6.}$$

The equation is $C = 46u$.

3. We substitute 12 for u into the equation $C = 46u$ and solve for C:

$$C = 46(12) = 552$$

The cost of 12 units is $552.

4. We substitute 700 for C in the equation $C = 46u$ and solve for u:

$$700 = 46u \qquad \textit{Substitute 700 for C.}$$
$$\frac{700}{46} = u \qquad \textit{Divide both sides by 46.}$$
$$15.22 \approx u \qquad \textit{Compute.}$$

The student can afford 15 units of classes. We use a graphing calculator table to verify our work (see Fig. 28).

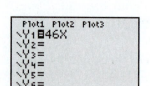

Figure 28 Verify the work

Finding and Using a Direct Variation Model

Assume a quantity p varies directly as a quantity t. To make estimates about an authentic situation,

1. Substitute the values of a data point into the equation $p = kt$ and then solve for k.
2. Substitute the value of k into the equation $p = kt$.
3. Use the equation found in step 2 to make estimates of quantity t or quantity p.

A Variable Varying Directly as an Expression

So far, we have described equations in which the response variable varies directly as the explanatory *variable*. Here, for constant k, we list some examples of equations in which the response variable varies directly as an *expression*:

$$y = kx^2 \qquad \textit{y varies directly as } x^2.$$
$$w = kt^3 \qquad \textit{w varies directly as } t^3.$$
$$H = k(p + 5) \qquad \textit{H varies directly as } p + 5.$$

So, we use "varies directly" to mean that the response variable is equal to a constant times an expression in terms of the explanatory variable.

Inverse Variation

In Sections 10.1 and 10.5, we worked with rational equations in two variables. Now we will focus on a simple type of rational equation in two variables: an inverse variation equation.

Definition   Inverse variation

If $y = \dfrac{k}{x}$ for some constant k, we say that **y varies inversely as x** or that **y is inversely proportional to x**. We call k the **variation constant** or the **constant of proportionality**. The equation $y = \dfrac{k}{x}$ is called an **inverse variation equation**.

For $y = \dfrac{7}{x}$, we say y varies inversely as x with variation constant 7. For $r = \dfrac{4}{t}$, we say r varies inversely as t with variation constant 4.

Changes in Values for Inverse Variation

In Fig. 29, we graph three equations of the form $y = \dfrac{k}{x}$, with $k = 1$, $k = 2$, and $k = 4$.

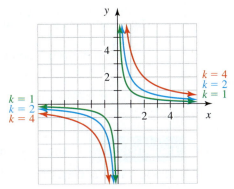

Figure 29 Graphs of $y = \dfrac{k}{x}$ with $k = 1$, $k = 2$, and $k = 4$

These graphs suggest that, if k is positive, then the graph of $y = \dfrac{k}{x}$ is a decreasing curve for positive values of x, which is true. So, if the value of x increases, the value of y decreases.

Changes in Values of Variables for Inverse Variation

Assume y varies inversely as x for some positive variation constant k. For positive values of x,

• If the value of x increases, then the value of y decreases.
• If the value of x decreases, then the value of y increases.

Use One Point to Find an Inverse Variation Equation

If one variable varies inversely as another variable and we know one point that lies on the graph, we can find the variation constant k, as well as the inverse variation equation.

Example 4   Finding an Inverse Variation Equation

The variable y varies inversely as x with the positive variation constant k.

1. For positive values of x, what happens to the value of y as the value of x increases?
2. If $y = 3$ when $x = 5$, find an equation for x and y.

Solution

1. Because y varies inversely as x with positive variation constant k and the values of x are positive, the value of y decreases as the value of x increases.

2. The equation is of the form $y = \dfrac{k}{x}$. We find the constant k by substituting 5 for x and 3 for y in the equation $y = \dfrac{k}{x}$:

$$y = \frac{k}{x} \qquad \text{\textit{y varies inversely as x.}}$$

$$3 = \frac{k}{5} \qquad \text{\textit{Substitute 5 for x and 3 for y.}}$$

$$15 = k \qquad \text{\textit{Multiply both sides by the LCD, 5.}}$$

The equation is $y = \dfrac{15}{x}$. We use a graphing calculator graph to check that the curve $y = \dfrac{15}{x}$ contains the point $(5, 3)$. See Fig. 30.

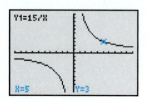

Figure 30 Verify the work

Example 5   Using an Inverse Variation Equation

The variable B varies inversely as P. If $B = 7$ when $P = 4$, find the value of P when $B = 5$.

Solution

Because B varies inversely as P, we can describe the relationship between P and B by the equation $B = \dfrac{k}{P}$. We find the constant k by substituting 4 for P and 7 for B:

$$B = \frac{k}{P} \qquad \text{\textit{B varies inversely as P.}}$$

$$7 = \frac{k}{4} \qquad \text{\textit{Substitute 4 for P and 7 for B.}}$$

$$28 = k \qquad \text{\textit{Multiply both sides by LCD, 4.}}$$

The equation is $B = \dfrac{28}{P}$. We find the value of P when $B = 5$ by substituting 5 for B in the equation $B = \dfrac{28}{P}$ and then solving for P:

$$5 = \frac{28}{P} \qquad \text{\textit{Substitute 5 for B.}}$$

$$5P = 28 \qquad \text{\textit{Multiply both sides by LCD, P.}}$$

$$P = \frac{28}{5} \qquad \text{\textit{Divide both sides by 5.}}$$

So, $P = \dfrac{28}{5}$ when $B = 5$.

Inverse Variation Models

Many authentic situations can be modeled well by inverse variation equations. For example, the length of a wrench handle and the force you must exert on the handle to loosen a bolt can be modeled by such an equation.

Example 6   Using an Inverse Variation Model

When a stone is tied to a string and whirled in a circle at constant speed, the tension T (in newtons) of the string varies inversely as the radius r (in centimeters) of the circle. If the radius is 50 centimeters, the tension is 96 newtons.

1. Find an equation for r and T.
2. For positive values of r, as the value of r increases, what happens to the value of T? What does that pattern mean in this situation?
3. Find the tension if the radius is 60 centimeters.

Solution

1. Because T varies inversely as r, we can model the situation with the equation $T = \dfrac{k}{r}$. We substitute 50 for r and 96 for T to find the constant k:

$$T = \frac{k}{r} \qquad \textit{T varies inversely as r.}$$

$$96 = \frac{k}{50} \qquad \textit{Substitute 50 for r and 96 for T.}$$

$$4800 = k \qquad \textit{Multiply both sides by LCD, 50.}$$

The equation is $T = \dfrac{4800}{r}$.

2. Because T varies inversely as r and k is positive, as the value of r increases, the value of T decreases. This means the larger the radius, the smaller the tension will be.

3. We substitute 60 for r in the equation $T = \dfrac{4800}{r}$ and solve for T:

$$T = \frac{4800}{60} = 80$$

So, the tension is 80 newtons when the radius is 60 centimeters. We use a graphing calculator table to verify our work (see Fig. 31).

Figure 31 Verify the work

Finding and Using an Inverse Variation Model

Assume a quantity p varies inversely as a quantity t. To make estimates about an authentic situation,

1. Substitute the values of a data point into the equation $p = \dfrac{k}{t}$ and then solve for k.

2. Substitute the value of k into the equation $p = \dfrac{k}{t}$.

3. Use the equation found in step 2 to make estimates of quantity t or quantity p.

A Variable Varying Inversely as an Expression

Here, for constant k, we list some examples of equations in which the response variable varies inversely as an *expression*:

$$y = \frac{k}{x^2} \qquad \textit{y varies inversely as } x^2.$$

$$w = \frac{k}{r^3} \qquad \textit{w varies inversely as } r^3.$$

$$p = \frac{k}{t - 2} \qquad \textit{p varies inversely as } t - 2.$$

So, we use "varies inversely" to mean the response variable is equal to a constant divided by an expression in terms of the explanatory variable.

Example 7   Making an Estimate by Using an Inverse Variation Model

Radiation can be used to treat a tumor. The intensity of radiation varies inversely as the square of the distance from the machine that produces the radiation. If the intensity is 60 milliroentgens per hour (mr/h) at a distance of 3 meters, at what distance is the intensity 40 mr/h?

Solution

We let I be the intensity (in mr/h) at distance d meters from the machine. Our desired model has the form

$$I = \frac{k}{d^2}$$

The denominator of the fraction on the right-hand side is d^2 because the intensity varies inversely as the *square* of the distance d.

Next, we substitute 3 for d and 60 for I and solve for k:

$$60 = \frac{k}{3^2} \qquad \textit{Substitute 3 for d and 60 for I.}$$

$$60 = \frac{k}{9} \qquad 3^2 = 9$$

$$540 = k \qquad \textit{Multiply both sides by LCD, 9.}$$

The model is $I = \dfrac{540}{d^2}$. Next, we substitute 40 for I and solve for d:

$$40 = \frac{540}{d^2} \qquad \textit{Substitute 40 for I.}$$

$$40d^2 = 540 \qquad \textit{Multiply both sides by } d^2.$$
$$d^2 = 13.5 \qquad \textit{Divide both sides by 40.}$$
$$d = \sqrt{13.5} \qquad \textit{d is nonnegative, so disregard } -\sqrt{13.5}.$$
$$d \approx 3.67 \qquad \textit{Approximate the square root.}$$

So, the intensity is 40 mr/h at a distance of about 3.67 meters.

▶

Use a Table of Data to Find an Equation of a Model

If we assume that one quantity varies directly as another quantity, then we need only one point to find the direct variation equation. A similar idea applies to finding an inverse variation equation. However, if we do not make any assumptions about the relationship between two quantities, then we need to have *several* data points so we can construct a scatterplot and determine which type of equation will best model the situation.

Table 13 The Lengths of the Telescopes and the Diameters of Vision

Length of Telescope (centimeters) L	Diameter of Vision (centimeters) D
10	70
15	46
20	36
25	29
30	24
35	21
40	19
45	16

Source: *J. Lehmann*

Example 8   Using a Table of Data to Find an Equation of a Model

A person constructs "telescopes" by rolling and taping pieces of card stock. He makes eight telescopes that vary in length but have the same radius (4.8 centimeters). He then attaches a tape measure to a wall and extends the tape along the wall. While standing 1.5 meters away from the wall, he records how much of the tape measure he can see (the diameter of vision) through each telescope (see Table 13). Let D be the diameter of vision (in centimeters) at a distance of 1.5 meters for a telescope that is L centimeters long.

1. Construct a scatterplot of the data. Are the data modeled better by a direct variation equation or an inverse variation equation?
2. Find an equation of a model to describe the data.
3. What is the length of a telescope that has a diameter of vision equal to 26 centimeters (at a distance of 1.5 meters)?

Solution

1. A scatterplot is shown in Fig. 32. It appears that the data are better modeled by an inverse variation model.

2. An inverse variation model is of the form $D = \dfrac{k}{L}$. We need only one point to find k. It seems all the points in the scatterplot fit about the same pattern, so any of the points will likely generate a reasonable model. We choose the point $(30, 24)$ arbitrarily:

$$D = \frac{k}{L} \qquad \textit{Inverse variation equation}$$

$$24 = \frac{k}{30} \qquad \textit{Substitute 30 for L and 24 for D.}$$

$$720 = k \qquad \textit{Multiply both sides by 30.}$$

The equation is $D = \dfrac{720}{L}$. We see in Fig. 33 that the model fits the data quite well.

3. To estimate the length of a telescope with a diameter of vision equal to 26 centimeters (at a distance of 1.5 meters), we substitute 26 for D in the equation $D = \dfrac{720}{L}$ and solve for L:

$$26 = \frac{720}{L} \qquad \textit{Substitute 26 for D.}$$

$$26L = 720 \qquad \textit{Multiply both sides by LCD, L.}$$

$$L = \frac{720}{26} \qquad \textit{Divide both sides by 26.}$$

$$L \approx 27.69 \qquad \textit{Compute.}$$

So, the length of the telescope is about 27.7 centimeters.

Figure 32 The telescope scatterplot

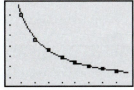

Figure 33 Check the fit

◈ Group Exploration

Section Opener: Inverse variation

1. Use ZDecimal on a graphing calculator to graph
$$y = \frac{1}{x}, \quad y = \frac{2}{x}, \quad y = \frac{3}{x}$$
in that order, and describe any patterns you observe.

2. By viewing any of the graphs on the calculator screen, determine what happens to the value of y as the value of x increases, for positive values of x.

3. Explain why the pattern you described in Problem 2 makes sense by referring to any of the three equations.

   [**Hint:** Are the values $\dfrac{1}{2}, \dfrac{1}{3}, \dfrac{1}{4}, \ldots$ getting larger or smaller?]

4. Let P be the pressure (in atmospheres) of air in a sealed syringe and V be the volume of air (in cubic centimeters). Here is a model that describes the relationship between P and V for a specific syringe:

$$P = \frac{7.4}{V}$$

Describe what happens to the pressure as the volume is increased. Explain in terms of the graph of the model and the model's equation.

5. For the situation described in Problem 4, describe what happens to the pressure as the volume is decreased. Explain in terms of the graph of the model and the model's equation.

6. In your opinion, is it possible to squeeze the syringe so hard (without breaking the seal) that the volume of air in the syringe becomes 0 cubic centimeters? Does the model support your theory? Explain. [**Hint:** In addition to substituting 0 for V in the model's equation, substitute a small value, such as 0.000001, for V.]

Group Exploration

Comparing methods of finding a direct variation equation

In this exploration, you will use three different methods to find an equation for a direct variation model.

1. Suppose that an employee's total earnings E (in dollars) varies directly as the number t of hours worked. The employee earns $72 for an 8-hour workday.

    a. Use the method discussed in this section to find an equation of a direct variation model to describe the data.

    b. What does the ordered pair $(0, 0)$ mean in this situation? Does that make sense? Use the slope formula $m = \dfrac{E_2 - E_1}{t_2 - t_1}$ with the ordered pairs $(0, 0)$

and $(8, 72)$ to help you find an equation of a model to describe the data.

    c. Use the information that the employee earns $72 for an 8-hour workday to compute the rate of change of earnings per hour. Then use the fact that the slope is a rate of change to help you find an equation of a model to describe the data.

2. Compare the three equations you found in Problem 1. Which method was easiest for you to use?

3. If a variable y varies directly as a variable x, does it follow that the rate of change of y with respect to x is constant? Explain.

 # Homework 10.7

For extra help ▶ **MyLab Math**   Watch the videos in MyLab Math

Translate the sentence into an equation.

1. w varies directly as t.

2. p varies directly as r.

3. F varies inversely as r.

4. H varies inversely as x.

5. c varies directly as s squared.

6. y varies directly as the square root of w.

7. B varies inversely as t cubed.

8. z varies inversely as $d + 3$.

Translate the equation into a sentence by using the phrase "varies directly" or "varies inversely."

9. $H = ku$

10. $d = kt$

11. $P = \dfrac{k}{V}$

12. $w = \dfrac{k}{t}$

13. $E = kc^2$

14. $p = k\sqrt{x}$

15. $F = \dfrac{k}{r^2}$

16. $T = \dfrac{k}{d^3}$

Find an equation that meets the given conditions.

17. y varies directly as x, and $y = 8$ when $x = 4$.

18. y varies directly as x, and $y = 15$ when $x = 3$.

19. F varies inversely as r, and $F = 2$ when $r = 6$.

20. T varies inversely as d, and $T = 7$ when $d = 3$.

21. H varies directly as $\sqrt{u}$, and $H = 6$ when $u = 4$.

22. p varies directly as t^3, and $p = 40$ when $t = 2$.

23. C varies inversely as x^2, and $C = 2$ when $x = 3$.

24. M varies inversely as $\sqrt{r}$, and $M = 7$ when $r = 25$.

For Exercises 25–32, find the requested value of the variable.

25. If y varies directly as x, and $y = 20$ when $x = 5$, find y when $x = 3$.

26. If T varies directly as W, and $T = 28$ when $W = 4$, find T when $W = 6$.

27. If P varies inversely as V, and $P = 6$ when $V = 3$, find P when $V = 2$.

28. If H varies inversely as t, and $H = 8$ when $t = 4$, find H when $t = 16$.

29. If d varies directly as t^2, and $d = 12$ when $t = 2$, find d when $t = 4$.

30. If y varies directly as w^2, and $y = 32$ when $w = 4$, find w when $y = 32$.

31. If F varies inversely as r^2, and $F = 1$ when $r = 4$, find r when $F = 4$.

32. If I varies inversely as d^2, and $I = 9$ when $d = 2$, find I when $d = 3$.

33. The variable w varies directly as t with positive variation constant k. Describe what happens to the value of w as the value of t increases.

34. The variable p varies directly as r with positive variation constant k. Describe what happens to the value of p as the value of r decreases.

35. The variable z varies inversely as u with positive variation constant k. For positive values of u, what happens to the value of z as the value of u increases?

36. The variable B varies inversely as x with positive variation constant k. For positive values of x, what happens to the value of B as the value of x decreases?

37. The conduction of electricity in a strand of DNA varies directly with humidity for the positive variation constant k. If the humidity is increased, what happens to the conduction of electricity in a strand of DNA?

38. The intensity of the emission of longwave radiation from a cloud varies directly as the temperature for the positive variation constant k. If the temperature increases, what happens to the intensity of the emission of longwave radiation from a cloud?

39. The demand for refinancing home loans varies inversely with the rate of interest for the positive variation constant k. If the interest rate increases, what happens to the demand for refinancing home loans?

40. Relative humidity varies inversely with air temperature for the positive variation constant k. As the air warms, what happens to the relative humidity?

41. One version of "Murphy's law" is that if anything can go wrong, it will. A spinoff is "The number of doors left open varies inversely with the outside temperature." What does this spinoff mean? Why is it a version of Murphy's law?

42. The famous mathematician Bertrand Russell once said, "The degree of one's emotions varies inversely with one's knowledge of the facts." What does this quotation mean?

43. For a Massachusetts resident, the total cost of tuition and fees at MassBay Community College varies directly as the number of credits hours a student takes. For fall 2017, the total cost of 12 credit hours is $2544. What is the total cost of 15 credit hours?

44. For a resident of Jackson County, Michigan, the total cost of tuition at Jackson Community College varies directly as the number of billing contact hours (units or credits) a student takes. For the academic year 2016–2017, the total cost of 9 billing contact hours is $1215. What is the total cost of 12 billing contact hours?

45. The time it takes to fill up a car's gasoline tank varies directly as the total volume of gasoline being pumped. A Chevron® pump takes 81 seconds to pump 10 gallons of gasoline. How much gasoline can be pumped in 104 seconds?

46. A car is traveling at speed s (in mph) on a dry asphalt road, and the brakes are applied forcefully. The braking distance d (in feet) varies directly as the square of the speed s. If a car traveling at 50 mph has a braking distance of 157.5 feet, what is the braking distance of a car traveling at 65 mph?

47. The weight of a pepperoni, mushroom, and garlic pizza varies directly as the square of the radius of the pizza. At Amici's East Coast Pizzeria in San Mateo, California, a pepperoni, mushroom, and garlic pizza with a diameter of 15 inches weighs 32 ounces. What is the weight of a 13-inch-diameter pizza with the same toppings?

48. The distance d that an object falls varies directly as the square of the time the object is in free fall. If an object falls freely for 2 seconds, it will fall 64.4 feet. To estimate the height of a cliff, a person drops a stone at the edge of the cliff and times how long it takes the stone to reach the base of the cliff. If it takes the stone 2.9 seconds to reach the base, what is the height of the cliff?

49. The force you must exert on a wrench handle to loosen a bolt on a bike varies inversely as the length of the handle.
  a. Must a greater force be exerted by a wrench with a short handle or a wrench with a long handle? Explain in terms of inverse variation.

  b. If a force of 40 pounds is needed when the handle is 6 inches long, what force is needed when the handle is 8 inches long?

50. A person plays an electric guitar outside. The sound level (in decibels) varies inversely as the square of the distance from the amplifier. If the sound level is 114 decibels 30 feet from the amplifier, what is the sound level at 40 feet?

51. The brightness of light, in milliwatts per square centimeter $\left(\text{mW/cm}^2\right)$, of a 25-watt light bulb varies inversely as the square of the distance, in centimeters, from the light bulb. If the brightness of light is 0.55 mW/cm^2 at a distance of 90 centimeters, at what distance is the brightness of light equal to 0.26 mW/cm^2?

52. The intensity of a television signal varies inversely as the square of the distance from the transmitter. If the intensity of a television signal is 30 watts per square meter $\left(\text{W/m}^2\right)$ at a distance of 2.6 km, at what distance is the intensity 20 W/m^2?

53. **DATA** The more you stretch a spring with your hands, the more *force* the spring exerts on your hands. A person ran an experiment comparing the amount of stretch with the forces exerted by the spring (see Table 14).

Table 14 Stretches and Forces of a Spring

Stretch (meters)	Force (newtons)	Stretch (meters)	Force (newtons)
0.000	0.0	0.104	3.0
0.018	0.5	0.121	3.5
0.035	1.0	0.139	4.0
0.052	1.5	0.156	4.5
0.069	2.0	0.173	5.0
0.087	2.5		

Source: *Richard Taylor, The Hockaday School*

Let F be the force (in newtons) of the spring when the spring is stretched x meters.
  a. Use a graphing calculator to draw a scatterplot of the data.
  b. Find an equation of a model to describe the data.
  c. In a sentence that uses the phrase "varies directly" or "varies inversely," describe how the stretch and force of the spring are related. (This law for springs is known as *Hooke's law*.)
  d. Use the model you found in part (b) to estimate at what stretch the force will be 4.2 newtons.

54. A person finds the total weights of various numbers of identical textbooks (see Table 15).

Table 15 Total Weights of Identical Textbooks

Number of Textbooks	Total Weight (pounds)
0	0
1	3.5
2	7.1
3	10.7
4	14.1
5	17.6

Source: *J. Lehmann*

Let w be the total weight (in pounds) of n textbooks.
a. Use a graphing calculator to draw a scatterplot of the data.
b. Find an equation of a model to describe the data.
c. In a sentence that uses the phrase "varies directly" or "varies inversely," describe how the number of textbooks and the total weight are related.
d. Use the model you found in part (b) to estimate the total weight of six textbooks.
e. Explain how you can tell that not all the points of the scatterplot lie on the graph of your model. What does this tell you about the books and/or the scale the person used to weigh the books?

55. **DATA** When a sealed syringe is filled with air, it gets harder and harder to squeeze it further the more you squeeze. Some volumes in cubic centimeters $\left(\text{cm}^3\right)$ and corresponding pressures in atmospheres (atm) for a sealed syringe are given in Table 16.

Table 16 Volumes and Pressures in a Syringe

Volume $\left(\text{cm}^3\right)$	Pressure (atm)	Volume $\left(\text{cm}^3\right)$	Pressure (atm)
3	2.23	13	0.56
5	1.46	15	0.48
7	1.05	17	0.42
9	0.83	19	0.37
11	0.67	20	0.35

Source: *J. Lehmann*

Let P be the pressure (in atm) in the syringe at volume V (in cm^3).
a. Use a graphing calculator to draw a scatterplot of the data.
b. Find an equation of a model to describe the data.
c. In a sentence that uses the phrase "varies directly" or "varies inversely," describe how the volume and pressure are related.
d. Use the model you found in part (b) to estimate at what volume the pressure will be 0.9 atm.

56. **DATA** When a guitarist picks the thickest string on a six-string guitar (the "open E" string), the string vibrates at a frequency of 82.4 hertz, which means that it vibrates an average of 82.4 times per second. By using a finger to press the E string firmly against the fret board, the guitarist shortens the effective length of the string. As a result, the frequency increases (and so does the pitch of the note).

We use some of the letters of the alphabet to refer to these notes. The effective lengths for the open E string and seven other notes for a Fender Squire® guitar and the corresponding frequencies are listed in Table 17.

Table 17 Effective Lengths and Frequencies of Eight Notes on the E String

Note	Effective Length of the E String (inches)	Frequency (in hertz)
E	25.50	82.4
F	24.07	87.3
G	21.44	98.0
A	19.10	110.0
B	17.02	123.5
C	16.06	130.8
D	14.31	146.8
E	12.75	164.8

Source: *J. Lehmann*

Let F be the frequency (in hertz) of the E string when the effective length is L inches.
a. Use a graphing calculator to draw a scatterplot of the data.
b. Find an equation of a model to describe the data.
c. In a sentence that uses the phrase "varies directly" or "varies inversely," describe how the effective length and the frequency are related.
d. When the E string is vibrating, what is its frequency if its effective length is 7.58 inches?

Concepts

57. a. Give an example of an equation in which y varies inversely as x. Using your equation, find any outputs for the given input. State how many outputs there are for that single input.
   i. the input $x = 3$
   ii. the input $x = 5$
   iii. the input $x = -2$
b. For your equation, how many outputs originate from a single input? Explain.
c. For *any* equation in which y varies inversely as x, how many outputs originate from a single input? Explain.

58. Assume y varies directly as x for some *positive* variation constant k. Recall that if the value of x increases, then the value of y increases. What can you say if k is negative? Explain. [**Hint:** Sketch a graph.]

59. Six inches is the same distance as 15.24 centimeters. One student finds the model $y = 2.54x$, and another student finds the model $y = 0.39x$. Even though their results are different, both students are correct. How is this possible? What does 2.54 of the model $y = 2.54x$ represent? What does 0.39 of the model $y = 0.39x$ represent?

60. Describe the meanings of *direct variation equation* and *inverse variation equation*. Compare these two types of equations. Compare some properties of these types of equations.

Related Review

61. a. If y varies directly as x, are x and y linearly related? Explain.
b. If w and t are linearly related, does w vary directly as t? Explain.

62. a. If a quantity y varies directly as a quantity x, how many data points do you need to find an equation of a model to describe the situation?

  b. If quantities w and t are linearly related, how many data points do you need to find an equation of a model to describe the situation?

63. The total cost of 4 pillows is $50.

  a. Use a proportion to find the total cost of 6 pillows.

  b. Let c be the total cost (in dollars) of n pillows. Find a variation equation to describe this situation.

  c. Use the variation equation you found in part (b) to determine the total cost of 6 pillows. Is your result equal to the result you found in part (a)?

64. A person's earnings E (in dollars) vary directly as the amount of time t (in hours) that he has worked. The person earns $540 for working 30 hours. What is the slope of the model that describes the situation? What does the slope mean in this situation?

Expressions, Equations, and Graphs

Perform the indicated instruction. Then use words such as linear, quadratic, cubic, power, radical, rational, polynomial, degree, one variable, *and* two variables *to describe the expression, equation, or system.*

65. Solve $2w^2 - 3w - 5 = 0$.

66. Solve $\dfrac{3}{x-1} + \dfrac{1}{x+2} = \dfrac{5}{x-1}$.

67. Evaluate $2w^2 - 3w - 5$ for $w = -1$.

68. Find the sum $\dfrac{3}{x-1} + \dfrac{1}{x+2}$.

69. Factor $2w^2 - 3w - 5$.

70. Find the difference $\dfrac{3}{x-1} - \dfrac{1}{x+2}$.

10.8 Simplifying Complex Rational Expressions

Objective

» Simplify *complex rational expressions.*

In this section, we will work with complex rational expressions. Here are some examples of such expressions:

$$\dfrac{\dfrac{8}{x}}{\dfrac{3}{x^2}} \qquad \dfrac{\dfrac{5x}{x+4}}{\dfrac{x^2}{x-1}} \qquad \dfrac{\dfrac{x}{7} - \dfrac{6}{x}}{\dfrac{x}{5} + \dfrac{3}{x}}$$

A **complex rational expression** is a rational expression whose numerator or denominator (or both) is a rational expression.

Here we find the values of two numerical complex rational expressions:

$$\dfrac{\dfrac{3}{3}}{3} = \dfrac{3}{1} = 3 \qquad \dfrac{3}{\dfrac{3}{3}} = \dfrac{1}{3}$$

From these two examples, we see that it is important to keep track of the main fraction bar (the longest one, in bold) of the complex fraction.

We will discuss two methods for simplifying complex rational expressions. Ask your instructor whether you are required to know method 1, method 2, or both methods. If the choice is yours, compare the use of method 1 in Examples 1 and 2 with the use of method 2 in Examples 3 and 4. The complex rational expressions in these examples are simplified by both methods, so you can get a sense of the advantages and disadvantages of each one.

Method 1: Writing a Complex Rational Expression as a Quotient of Two Rational Expressions

Recall from Section 2.2 that an expression in the form $\dfrac{R}{S}$, where R and S are expressions, can be written in the form $R \div S$. We use this idea to help simplify a complex rational expression:

$$\dfrac{\dfrac{2}{5}}{\dfrac{3}{7}} = \dfrac{2}{5} \div \dfrac{3}{7} \qquad \color{blue}{\dfrac{R}{S} = R \div S}$$

$$= \dfrac{2}{5} \cdot \dfrac{7}{3} \qquad \color{blue}{\text{Multiply by the reciprocal of } \dfrac{3}{7}, \text{ which is } \dfrac{7}{3}.}$$

$$= \dfrac{14}{15} \qquad \color{blue}{\text{Multiply numerators; multiply denominators.}}$$

We **simplify a complex rational expression** by writing it as a rational expression $\dfrac{P}{Q}$, with $\dfrac{P}{Q}$ in lowest terms.

Example 1 Simplifying Complex Rational Expressions by Method 1

Simplify by method 1.

1. $\dfrac{\dfrac{6}{x^3}}{\dfrac{10}{x}}$

2. $\dfrac{\dfrac{x}{x+5}}{\dfrac{3}{x^2-25}}$

Solution

1. $\dfrac{\dfrac{6}{x^3}}{\dfrac{10}{x}} = \dfrac{6}{x^3} \div \dfrac{10}{x}$ $\quad \dfrac{R}{S} = R \div S$

$\qquad = \dfrac{6}{x^3} \cdot \dfrac{x}{10}$ *Multiply by reciprocal of $\dfrac{10}{x}$, which is $\dfrac{x}{10}$.*

$\qquad = \dfrac{2 \cdot 3}{x^3} \cdot \dfrac{x}{2 \cdot 5}$ *Factor numerator and denominator.*

$\qquad = \dfrac{2 \cdot 3x}{2 \cdot 5x^3}$ *Multiply numerators; multiply denominators.*

$\qquad = \dfrac{3}{5x^2}$ *Simplify: $\dfrac{2x}{2x} = 1$*

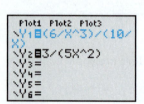

Figure 34 Verify the work

We use a graphing calculator table to verify our work (see Fig. 34). There are "ERROR" messages across from the value $x = 0$ because 0 is an excluded value of the numerator (and the denominator) of the original expression and 0 is an excluded value of the result.

2. $\dfrac{\dfrac{x}{x+5}}{\dfrac{3}{x^2-25}} = \dfrac{x}{x+5} \div \dfrac{3}{x^2-25}$ $\quad \dfrac{R}{S} = R \div S$

$\qquad = \dfrac{x}{x+5} \cdot \dfrac{x^2-25}{3}$ *Multiply by reciprocal of $\dfrac{3}{x^2-25}$.*

$\qquad = \dfrac{x}{x+5} \cdot \dfrac{(x-5)(x+5)}{3}$ *Factor numerator.*

$\qquad = \dfrac{x(x-5)(x+5)}{3(x+5)}$ *Multiply numerators; multiply denominators.*

$\qquad = \dfrac{x(x-5)}{3}$ *Simplify: $\dfrac{x+5}{x+5} = 1$*

Next, we use method 1 to simplify a complex rational expression that has two rational expressions in the numerator and two rational expressions in the denominator.

Example 2 Simplifying a Complex Rational Expression by Method 1

Simplify $\dfrac{\dfrac{1}{a} + \dfrac{2}{a^2}}{\dfrac{3}{4} - \dfrac{1}{a}}$ by method 1.

Solution

We write both the numerator and the denominator as a fraction and then simplify as before:

$$\frac{\dfrac{1}{a}+\dfrac{2}{a^2}}{\dfrac{3}{4}-\dfrac{1}{a}}=\frac{\left.\dfrac{1}{a}\cdot\dfrac{a}{a}+\dfrac{2}{a^2}\right\}}{\left.\dfrac{3}{4}\cdot\dfrac{a}{a}-\dfrac{1}{a}\cdot\dfrac{4}{4}\right\}}$$

Introduce missing factors to get a common denominator, a^2.
Introduce missing factors to get a common denominator, $4a$.

$$=\frac{\dfrac{a}{a^2}+\dfrac{2}{a^2}}{\dfrac{3a}{4a}-\dfrac{4}{4a}}$$

Find products.

$$=\frac{\dfrac{a+2}{a^2}}{\dfrac{3a-4}{4a}}$$

$\dfrac{A}{B}+\dfrac{C}{B}=\dfrac{A+C}{B}$

$$=\frac{a+2}{a^2}\div\frac{3a-4}{4a}$$

$\dfrac{R}{S}=R\div S$

$$=\frac{a+2}{a^2}\cdot\frac{4a}{3a-4}$$

Multiply by reciprocal of $\dfrac{3a-4}{4a}$.

$$=\frac{4a(a+2)}{a^2(3a-4)}$$

Multiply numerators; multiply denominators.

$$=\frac{4(a+2)}{a(3a-4)}$$

Simplify: $\dfrac{a}{a}=1$

Because our result is in lowest terms, we are done.

▶

Using Method 1 to Simplify a Complex Rational Expression

To simplify a complex rational expression by method 1,

1. Write both the numerator and the denominator as fractions.

2. To write the complex rational expression as the quotient of two rational expressions, use the property

$$\frac{\dfrac{A}{B}}{\dfrac{C}{D}}=\frac{A}{B}\div\frac{C}{D}, \text{ where } B, C, \text{ and } D \text{ are nonzero}$$

3. Divide the rational expressions.

Method 2: Multiplying by $\dfrac{\text{LCD}}{\text{LCD}}$

Instead of using method 1, we can simplify a complex rational expression by first finding the LCD of all the fractions appearing in the numerator and denominator. Then we multiply by 1 in the form $\dfrac{\text{LCD}}{\text{LCD}}$.

In Example 3, we will simplify the same complex rational expressions that we simplified in Example 1, but now we use method 2.

| Example 3 | Simplifying Complex Rational Expressions by Method 2 |

Simplify by method 2.

1. $\dfrac{\dfrac{6}{x^3}}{\dfrac{10}{x}}$ **2.** $\dfrac{\dfrac{x}{x+5}}{\dfrac{3}{x^2-25}}$

Solution

1. The LCD of $\dfrac{6}{x^3}$ and $\dfrac{10}{x}$ is x^3. So, we multiply the complex rational expression by 1 in the form $\dfrac{x^3}{x^3}$:

$$\dfrac{\dfrac{6}{x^3}}{\dfrac{10}{x}} = \dfrac{\dfrac{6}{x^3}}{\dfrac{10}{x}} \cdot \dfrac{\dfrac{x^3}{1}}{\dfrac{x^3}{1}} \qquad \textit{Multiply by } \dfrac{LCD}{LCD}, \dfrac{x^3}{x^3} = 1.$$

$$= \dfrac{\dfrac{6x^3}{x^3}}{\dfrac{10x^3}{x}} \qquad \textit{Simplify.}$$

$$= \dfrac{6}{10x^2} \qquad \textit{Simplify fractions in numerator and in denominator.}$$

$$= \dfrac{2 \cdot 3}{2 \cdot 5x^2} \qquad \textit{Factor numerator and denominator.}$$

$$= \dfrac{3}{5x^2} \qquad \textit{Simplify: } \dfrac{2}{2} = 1$$

The result is the same as the result we found in Problem 1 of Example 1.

2. $\dfrac{\dfrac{x}{x+5}}{\dfrac{3}{x^2-25}} = \dfrac{\dfrac{x}{x+5}}{\dfrac{3}{(x-5)(x+5)}}$ *Factor numerators and denominators of fractions.*

$$= \dfrac{\dfrac{x}{x+5}}{\dfrac{3}{(x-5)(x+5)}} \cdot \dfrac{\dfrac{(x-5)(x+5)}{1}}{\dfrac{(x-5)(x+5)}{1}} \qquad \textit{Multiply by } \dfrac{LCD}{LCD}, \dfrac{(x-5)(x+5)}{(x-5)(x+5)}.$$

$$= \dfrac{\dfrac{x(x-5)(x+5)}{x+5}}{\dfrac{3(x-5)(x+5)}{(x-5)(x+5)}} \qquad \dfrac{A}{B} \cdot \dfrac{C}{D} = \dfrac{AC}{BD}$$

$$= \dfrac{x(x-5)}{3} \qquad \textit{Simplify: } \dfrac{x+5}{x+5} = 1,$$
$$\dfrac{(x-5)(x+5)}{(x-5)(x+5)} = 1$$

The result is the same as the result we found in Problem 2 of Example 1.

▶

In comparing our work in Examples 1 and 3, we see that methods 1 and 2 required about the same number of steps. One advantage that method 1 has over method 2 for *these* complex rational expressions is that method 1 does not require us to find the LCD.

In Example 4, we will simplify the same expression as in Example 2, but now we use method 2.

> **Example 4**   Simplifying a Complex Rational Expression by Method 2
>
> Simplify $\dfrac{\dfrac{1}{a} + \dfrac{2}{a^2}}{\dfrac{3}{4} - \dfrac{1}{a}}$ by method 2.

Solution

The LCD of the rational expressions in the numerator and the denominator is $4a^2$. To simplify, we multiply by $\dfrac{4a^2}{4a^2}$:

$$\frac{\dfrac{1}{a} + \dfrac{2}{a^2}}{\dfrac{3}{4} - \dfrac{1}{a}} = \frac{\dfrac{1}{a} + \dfrac{2}{a^2}}{\dfrac{3}{4} - \dfrac{1}{a}} \cdot \frac{4a^2}{4a^2} \qquad \textit{Multiply by } \frac{LCD}{LCD}, \frac{4a^2}{4a^2}.$$

$$= \frac{\dfrac{1}{a} \cdot \dfrac{4a^2}{1} + \dfrac{2}{a^2} \cdot \dfrac{4a^2}{1}}{\dfrac{3}{4} \cdot \dfrac{4a^2}{1} - \dfrac{1}{a} \cdot \dfrac{4a^2}{1}} \qquad \textit{Distributive law}$$

$$= \frac{4a + 8}{3a^2 - 4a} \qquad \textit{Simplify.}$$

$$= \frac{4(a + 2)}{a(3a - 4)} \qquad \textit{Factor numerator and denominator.}$$

Note that the result is the same as the result we found in Example 2.

WARNING   For the first step in Example 4, it would be incorrect to multiply by the fraction

$$\frac{\text{LCD of the numerator}}{\text{LCD of the denominator}} = \frac{a^2}{4a}:$$

$$\frac{\dfrac{1}{a} + \dfrac{2}{a^2}}{\dfrac{3}{4} - \dfrac{1}{a}} = \frac{\dfrac{1}{a} + \dfrac{2}{a^2}}{\dfrac{3}{4} - \dfrac{1}{a}} \cdot \frac{a^2}{4a} \qquad \textit{Incorrect}$$

This is incorrect because the expression $\dfrac{a^2}{4a}$ is not equivalent to 1. It *is* correct to multiply by $\dfrac{4a^2}{4a^2} = 1$.

In comparing our work in Examples 2 and 4, we see that method 2 required fewer steps than method 1. In general, when the numerator, denominator, or both contain two or more rational expressions, method 2 is more efficient.

Using Method 2 to Simplify a Complex Rational Expression

To simplify a complex rational expression by method 2,

1. Find the LCD of all the fractions appearing in the numerator and denominator.

2. Multiply by 1 in the form $\dfrac{LCD}{LCD}$.

3. Simplify the numerator and the denominator to polynomials.

4. Simplify the rational expression.

| Example 5 | Simplifying a Complex Rational Expression by Method 2 |

Simplify $\dfrac{3 - \dfrac{12}{x^2}}{2 - \dfrac{4}{x}}$ by method 2.

Solution

The LCD of the rational expressions in the numerator and the denominator is x^2. To simplify, we multiply by $\dfrac{x^2}{x^2}$:

$$\frac{3 - \dfrac{12}{x^2}}{2 - \dfrac{4}{x}} = \frac{3 - \dfrac{12}{x^2}}{2 - \dfrac{4}{x}} \cdot \frac{x^2}{x^2} \qquad \textit{Multiply by } \frac{LCD}{LCD}, \frac{x^2}{x^2}.$$

$$= \frac{3x^2 - \dfrac{12}{x^2} \cdot \dfrac{x^2}{1}}{2x^2 - \dfrac{4}{x} \cdot \dfrac{x^2}{1}} \qquad \textit{Distributive law}$$

$$= \frac{3x^2 - 12}{2x^2 - 4x} \qquad \textit{Simplify.}$$

$$= \frac{3\left(x^2 - 4\right)}{2x(x - 2)} \qquad \textit{Factor numerator and denominator.}$$

$$= \frac{3(x - 2)(x + 2)}{2x(x - 2)} \qquad \textit{Factor numerator.}$$

$$= \frac{3(x + 2)}{2x} \qquad \textit{Simplify: } \frac{x - 2}{x - 2} = 1$$

◀

Group Exploration

Section Opener: Simplifying complex rational expressions

A **complex rational expression** is a rational expression whose numerator or denominator (or both) is a rational expression. Here we simplify a complex rational expression:

$$\frac{\dfrac{a}{b}}{\dfrac{c}{d}} = \frac{a}{b} \div \frac{c}{d}$$

$$= \frac{a}{b} \cdot \frac{d}{c}$$

$$= \frac{ad}{bc}$$

Simplify the complex rational expressions.

1. $\dfrac{\dfrac{2}{x}}{\dfrac{3}{y}}$

2. $\dfrac{\dfrac{8}{x^2}}{\dfrac{6}{x^5}}$

3. $\dfrac{\dfrac{3x - 6}{6x^6}}{\dfrac{5x - 10}{4x^4}}$

4. $\dfrac{\dfrac{x^2 - 25}{x^2 - 2x - 3}}{\dfrac{2x + 10}{x - 3}}$

Homework 10.8

For extra help ▶ **MyLab Math**   Watch the videos in MyLab Math

Simplify.

1. $\dfrac{\dfrac{4}{5}}{\dfrac{8}{3}}$

2. $\dfrac{\dfrac{3}{4}}{\dfrac{5}{6}}$

3. $\dfrac{\dfrac{x}{4}}{\dfrac{x}{7}}$

4. $\dfrac{\dfrac{3}{x}}{\dfrac{5}{x}}$

5. $\dfrac{\dfrac{6}{5x}}{\dfrac{3}{7x}}$

6. $\dfrac{\dfrac{4}{3x}}{\dfrac{8}{5x}}$

7. $\dfrac{\dfrac{w}{6}}{\dfrac{w^2}{9}}$

8. $\dfrac{\dfrac{10}{y}}{\dfrac{12}{y^2}}$

9. $\dfrac{\dfrac{15x^3}{8}}{\dfrac{25x^5}{12}}$

10. $\dfrac{\dfrac{7}{6x^8}}{\dfrac{21}{4x^7}}$

11. $\dfrac{\dfrac{14}{x^2-9}}{\dfrac{21}{x-3}}$

12. $\dfrac{\dfrac{8}{x^2-16}}{\dfrac{20}{x+4}}$

13. $\dfrac{\dfrac{5a-10}{4}}{\dfrac{3a-6}{2}}$

14. $\dfrac{\dfrac{3t-12}{14}}{\dfrac{4t-16}{21}}$

15. $\dfrac{\dfrac{x^2-2x-15}{6x}}{\dfrac{x-5}{10}}$

16. $\dfrac{\dfrac{x^2+3x-28}{16x}}{\dfrac{x+7}{12}}$

17. $\dfrac{\dfrac{4}{x}+\dfrac{2}{x}}{\dfrac{9}{x}-\dfrac{7}{x}}$

18. $\dfrac{\dfrac{5}{x}-\dfrac{2}{x}}{\dfrac{3}{x}+\dfrac{4}{x}}$

19. $\dfrac{\dfrac{3}{8}+\dfrac{1}{4}}{\dfrac{1}{2}+\dfrac{5}{8}}$

20. $\dfrac{\dfrac{2}{3}+\dfrac{5}{6}}{\dfrac{7}{12}-\dfrac{1}{6}}$

21. $\dfrac{\dfrac{7}{4x}+\dfrac{1}{x}}{\dfrac{3}{2x}+\dfrac{5}{x}}$

22. $\dfrac{\dfrac{2}{x}+\dfrac{5}{3x}}{\dfrac{8}{5x}-\dfrac{1}{x}}$

23. $\dfrac{\dfrac{5}{3r}-\dfrac{3}{2r}}{\dfrac{1}{2r}-\dfrac{4}{3r}}$

24. $\dfrac{\dfrac{7}{2t}+\dfrac{3}{5t}}{\dfrac{3}{5t}+\dfrac{5}{2t}}$

25. $\dfrac{\dfrac{2}{3}-\dfrac{4}{3x}}{\dfrac{5}{6x}}$

26. $\dfrac{\dfrac{5}{4}-\dfrac{7}{4x}}{\dfrac{3}{8x}}$

27. $\dfrac{2+\dfrac{5}{x}}{4-\dfrac{1}{x}}$

28. $\dfrac{3-\dfrac{7}{x}}{5-\dfrac{6}{x}}$

29. $\dfrac{\dfrac{2}{b}-\dfrac{3}{b^2}}{\dfrac{4}{b}+\dfrac{5}{b^2}}$

30. $\dfrac{\dfrac{8}{w}+\dfrac{5}{w^2}}{\dfrac{6}{w}-\dfrac{3}{w^2}}$

31. $\dfrac{\dfrac{1}{x}-\dfrac{4}{x^3}}{\dfrac{1}{x}-\dfrac{2}{x^2}}$

32. $\dfrac{\dfrac{1}{x}-\dfrac{3}{x^2}}{\dfrac{1}{x}-\dfrac{9}{x^3}}$

33. $\dfrac{\dfrac{3}{4x}+\dfrac{1}{x^2}}{\dfrac{5}{2x}+\dfrac{1}{x^2}}$

34. $\dfrac{\dfrac{4}{3x}-\dfrac{1}{x^3}}{\dfrac{3}{2x}+\dfrac{2}{x^2}}$

35. $\dfrac{\dfrac{r}{3}-\dfrac{3}{r}}{\dfrac{r}{2}-\dfrac{2}{r}}$

36. $\dfrac{\dfrac{a}{2}-\dfrac{2}{a}}{\dfrac{a}{5}-\dfrac{5}{a}}$

37. $\dfrac{2-\dfrac{8}{x^2}}{1-\dfrac{2}{x}}$

38. $\dfrac{1-\dfrac{9}{x^2}}{2-\dfrac{6}{x}}$

Concepts

39. A student tries to simplify a complex rational expression:

$$\frac{x}{\dfrac{1}{x}+\dfrac{1}{3}}=x\div\left(\frac{1}{x}+\frac{1}{3}\right)$$

$$=x\cdot\left(\frac{x}{1}+\frac{3}{1}\right)$$

$$=x\cdot(x+3)$$

$$=x^2+3x$$

Describe any errors. Then simplify the expression correctly.

40. A student tries to simplify a complex rational expression:

$$\frac{\dfrac{2}{x}-\dfrac{7}{x^2}}{\dfrac{4}{3}+\dfrac{5}{x}}=\frac{\dfrac{2}{x}-\dfrac{7}{x^2}}{\dfrac{4}{3}+\dfrac{5}{x}}\cdot\frac{x^2}{3x}$$

$$=\frac{\dfrac{2}{x}\cdot\dfrac{x^2}{1}-\dfrac{7}{x^2}\cdot\dfrac{x^2}{1}}{\dfrac{4}{3}\cdot\dfrac{3x}{1}+\dfrac{5}{x}\cdot\dfrac{3x}{1}}$$

$$=\frac{2x-7}{4x+15}$$

Describe any errors. Then simplify the expression correctly.

41. Simplify $\dfrac{\dfrac{2}{x}+\dfrac{1}{x^2}}{\dfrac{3}{x}+\dfrac{4}{x^3}}$ first using method 1 and then using method 2. Which method was easier?

42. The equation $\dfrac{6}{2}=3$ implies that $3\cdot2=6$, which is true. What does the equation $\dfrac{4}{\frac{1}{2}}=8$ imply? Is your result true?

43. Describe a complex rational expression. Give an example. Then simplify your complex rational expression.

44. Describe how to simplify a complex rational expression.

Related Review

Simplify.

45. $\dfrac{x^2-9}{3x+15}\div\dfrac{x-3}{2x+10}$

46. $\dfrac{3x-6}{x^2+x-12}\div\dfrac{4x-8}{x+4}$

47. $\dfrac{\dfrac{x^2-9}{3x+15}}{\dfrac{x-3}{2x+10}}$

48. $\dfrac{\dfrac{3x-6}{x^2+x-12}}{\dfrac{4x-8}{x+4}}$

49. $\dfrac{2x^{-1}}{5\cdot3^{-1}}$

50. $\dfrac{4\cdot2^{-1}}{3x^{-1}}$

51. $\dfrac{2+x^{-1}}{5+3^{-1}}$ [**Hint:** First write as a complex rational expression.]

52. $\dfrac{4+2^{-1}}{3+x^{-1}}$ [**Hint:** First write as a complex rational expression.]

53. a. Solve the equation $\dfrac{3}{5}x=\dfrac{7}{2}$ by multiplying both sides by $\dfrac{5}{3}$.

    b. Solve the equation $\dfrac{3}{5}x=\dfrac{7}{2}$ by dividing both sides by $\dfrac{3}{5}$.

    c. Explain why it makes sense that the methods in parts (a) and (b) give the same result.

54. Find the slope of the line that contains the points $\left(\dfrac{2}{3},\dfrac{5}{6}\right)$ and $\left(\dfrac{7}{2},\dfrac{1}{4}\right)$.

Expressions, Equations, and Graphs

Give an example of the following. Then solve, simplify, or graph, as appropriate.

55. Rational equation in one variable

56. Linear equation in one variable

57. System of two linear equations in two variables

58. Quadratic equation in two variables

59. Quadratic equation in one variable that can be solved by factoring

60. Linear equation in two variables

61. Quadratic equation in one variable that can't be solved by factoring

62. Rational expression in one variable that is not in lowest terms

63. Linear polynomial with four terms

64. Quadratic polynomial with four terms

◆◇ Taking It to the Lab

Climate Change Lab (continued from Chapter 9)

In previous Climate Change Labs, we found various models related to carbon dioxide emissions and human populations. We can use these models to find meaningful ratios and percentages of quantities.

As a warm-up activity, suppose that a student takes a quiz, correctly answering 15 out of 20 questions. To compute the percentage of questions answered correctly, we divide the number of questions answered correctly by the total number of questions and multiply the result by 100:

$$\text{Percentage of correct answers}=\frac{15}{20}\cdot100=75\%$$

We will now take similar steps to find a model that describes the percentage of the world population that lives in the United States. Let u be the U.S. population and w be the world population (both in billions), both at t years since 1950. Here are some reasonable models:

$$u = 0.0026t + 0.154 \qquad \textit{United States}$$
$$w = -0.00046t^2 + 0.132t + 0.63 \quad \textit{World}$$

The world population model describes United Nations predictions well only for the years 2000–2100.

To find a model of the percentage p of the world population that is in the United States at t years since 1950, we divide the right-hand side of the U.S. population model by the right-hand side of the world population model and multiply the result by 100:

$$p = \frac{0.0026t + 0.154}{-0.00046t^2 + 0.132t + 0.63} \cdot 100$$
$$= \frac{0.26t + 15.4}{-0.00046t^2 + 0.132t + 0.63}$$

We can use this model to help us understand the situation better, which you will do in Problem 1 of this lab.

Analyzing the Situation

1. Use a graphing calculator to construct a table of values of t and p for the percentage model

$$p = \frac{0.26t + 15.4}{-0.00046t^2 + 0.132t + 0.63}$$

Your table should contain the values 50, 60, 70, 80, 90, and 100 for t. What do your results tell you about this situation?

2. A model of U.S. carbon dioxide emissions C (in billions of metric tons) in the year that is t years since 1950 is $c = 0.054t + 2.73$. A model of the U.S. population u (in billions) at t years since 1950 is $u = 0.0026t + 0.154$. Find a model of the U.S. per-person carbon dioxide emissions y (in metric tons per person) in the year that is t years since 1950. [**Hint:** Find a quotient. You do *not* have to multiply the fraction by 100 because you are not trying to describe a percentage.] Perform a unit analysis of that model.

3. Use a graphing calculator to construct a table of values of t and y for the model you found in Problem 2. Your table should contain the values 50, 60, 70, 80, 90, and 100 for t. What do your results tell you about this situation?

4. A student takes a quiz, answering 24 out of 30 questions correctly. What percentage of the questions did the student get right?

5. Let u be U.S. carbon dioxide emissions and w be world carbon dioxide emissions (both in billions of metric tons), both in the year that is t years since 1950. Here are some reasonable models:

$$u = 0.054t + 2.73 \quad \textit{United States}$$
$$w = 0.42t + 5.25 \quad \textit{World}$$

Find a model of the percentage p of worldwide carbon dioxide emissions emitted in the United States in the year that is t years since 1950.

6. Use a graphing calculator to construct a table of values for t and p for the percentage model you found in Problem 5. Your table should contain the values 50, 60, 70, 80, 90, and 100 for t. What do your results tell you about this situation?

7. Explain how it is possible for the U.S. share of annual worldwide carbon dioxide emissions to decline even if U.S. per-person annual carbon dioxide emissions increase. [**Hint:** There may be two reasons. Consider the result you found in Problem 1. Consider also what is happening in developing countries.]

Estimating π Lab

One of the great moments in early mathematics was the discovery of π and its usefulness. Long before there were calculators, many great mathematicians tried to estimate π. In this lab, you will build on your work from the Volume Lab in Chapter 1 to find an estimate of π. If you haven't done that lab yet, do it now. Then respond to the questions that follow.

Analyzing the Situation

1. Recall from the Volume Lab in Chapter 1 that V is the volume of water (in ounces) in a cylinder when the height of the water in the cylinder is h centimeters (cm). In the Volume Lab, you likely sketched a linear model that contains the origin $(0, 0)$. If not, sketch a new linear model that does.

2. Find an equation of a model to describe the data.

3. What does the slope mean in this situation? Make sure that you include units in your description.

4. There are 29.57353 cm^3 in one ounce. Use this fact to help you find the slope of the model, in units of cm^2.

5. The formula for the volume of a cylinder is $V = \pi r^2 h$, where r is the cylinder's radius and h is the cylinder's height. In Problem 2, you found an equation of the form $V = mh$, where m is the slope. Explain why we can conclude that $m = \pi r^2$.

6. Substitute the slope of your model for m and the radius of the cylinder for r in the equation $m = \pi r^2$. Then treat π as if it is an unknown constant and solve for it. If you are using the data provided in the Volume Lab, the base radius of the cylinder is 4.45 cm. What is your estimate of π?

7. Use a calculator to help you find the error in your estimate of π.

Chapter Summary

Key Points of Chapter 10

Section 10.1 Simplifying Rational Expressions

	Throughout these key points, assume $A, B, C,$ and D are polynomials.
Rational expression	If P and Q are polynomials with Q nonzero, we call the ratio $\dfrac{P}{Q}$ a **rational expression**.
Excluded value	A number is an **excluded value** of a rational expression if substituting the number into the expression leads to a division by 0.
Finding excluded values	To find any excluded values of a rational expression $\dfrac{P}{Q}$, 1. Set the denominator Q equal to 0. 2. Solve the equation $Q = 0$.
Lowest terms and simplify a rational expression	A rational expression is in **lowest terms** if the numerator and denominator have no common factors other than 1 or -1. We **simplify a rational expression** by writing it in lowest terms.
Simplifying a rational expression	To simplify a rational expression, 1. Factor the numerator and the denominator. 2. Use the property $\dfrac{AB}{AC} = \dfrac{A}{A} \cdot \dfrac{B}{C} = 1 \cdot \dfrac{B}{C} = \dfrac{B}{C}$, where A and C are nonzero, so the expression is in lowest terms.
Rational equation in two variables	A **rational equation in two variables** is an equation in two variables in which each side can be written as a rational expression.
Rational model	A **rational model** is a rational equation in two variables that describes the relationship between two quantities for an authentic situation.

Section 10.2 Multiplying and Dividing Rational Expressions; Converting Units

Multiplying rational expressions	If $\dfrac{A}{B}$ and $\dfrac{C}{D}$ are rational expressions and B and D are nonzero, then $\dfrac{A}{B} \cdot \dfrac{C}{D} = \dfrac{AC}{BD}$.
How to multiply rational expressions	To multiply two rational expressions, 1. Factor the numerators and the denominators. 2. Multiply by using the property $\dfrac{A}{B} \cdot \dfrac{C}{D} = \dfrac{AC}{BD}$, where B and D are nonzero. 3. Simplify the result.
Dividing rational expressions	If $\dfrac{A}{B}$ and $\dfrac{C}{D}$ are rational expressions and $B, C,$ and D are nonzero, then $\dfrac{A}{B} \div \dfrac{C}{D} = \dfrac{A}{B} \cdot \dfrac{D}{C}$.
How to divide rational expressions	To divide two rational expressions, 1. Write the quotient as a product by using the property $\dfrac{A}{B} \div \dfrac{C}{D} = \dfrac{A}{B} \cdot \dfrac{D}{C}$, where $B, C,$ and D are nonzero. 2. Find the product. 3. Simplify.
Converting units	To convert the units of a quantity, 1. Write the quantity in the original units. 2. Multiply by fractions equal to 1 so that the units you want to eliminate appear in one numerator and one denominator.

Section 10.3 Adding Rational Expressions

Adding rational expressions that have a common denominator	If $\dfrac{A}{B}$ and $\dfrac{C}{B}$ are rational expressions, where B is nonzero, then $\dfrac{A}{B} + \dfrac{C}{B} = \dfrac{A + C}{B}$.
How to add rational expressions that have different denominators	To add rational expressions that have different denominators, 1. Factor the denominators of the expressions if possible. Determine which factors are missing. 2. Use the property $\dfrac{A}{A} = 1$, where A is nonzero, to introduce missing factors. 3. Add the expressions by using the property $\dfrac{A}{B} + \dfrac{C}{B} = \dfrac{A + C}{B}$, where B is nonzero. 4. Simplify.

Section 10.4 Subtracting Rational Expressions

Subtracting rational expressions that have a common denominator	If $\dfrac{A}{B}$ and $\dfrac{C}{B}$ are rational expressions, where B is nonzero, then $\dfrac{A}{B} - \dfrac{C}{B} = \dfrac{A - C}{B}$.
Subtract entire numerator	When subtracting rational expressions, be sure to subtract the *entire* numerator.
How to subtract two rational expressions that have different denominators	To subtract two rational expressions that have different denominators, 1. Factor the denominators of the expressions if possible. Determine which factors are missing. 2. Use the property $\dfrac{A}{A} = 1$, where A is nonzero, to introduce missing factors. 3. Subtract the expressions by using the property $\dfrac{A}{B} - \dfrac{C}{B} = \dfrac{A - C}{B}$, where B is nonzero. 4. Simplify.

Section 10.5 Solving Rational Equations

Rational equation in one variable	A **rational equation in one variable** is an equation in one variable in which each side can be written as a rational expression.
Solving a rational equation in one variable	To solve a rational equation in one variable, 1. Factor the denominator(s) if possible. 2. Identify any excluded values. 3. Find the LCD of all the fractions. 4. Multiply both sides of the equation by the LCD, which gives a simpler equation to solve. 5. Solve the simpler equation. 6. Discard any proposed solutions that are excluded values.
Solving a rational equation versus simplifying a rational expression	To solve a rational equation, clear the fractions in it by multiplying both sides of the equation by the LCD. To simplify a rational expression, do *not* multiply it by the LCD. The only multiplication permissible is multiplication by 1, usually in the form $\dfrac{A}{A}$, where A is a nonzero polynomial.
Results of solving rational equations and simplifying rational expressions	The result of solving a rational equation is the empty set or a set of one or more numbers. The result of simplifying a rational expression is an expression.

Section 10.6 Proportions; Similar Triangles

Proportion	A **proportion** is a statement of the equality of two ratios.
Proportional	If the ratio of two related quantities is constant, we say the two quantities are **proportional**.
Similar triangles	Two triangles are called **similar triangles** if their corresponding angles are equal in measure.
Side lengths of similar triangles	The lengths of the corresponding sides of two similar triangles are proportional.

Section 10.7 Variation

Direct variation	If $y = kx$ for some constant k, we say **y varies directly as x** or **y is proportional to x**. We call k the **variation constant** or the **constant of proportionality**. The equation $y = kx$ is called a **direct variation equation**.
Changes in values of variables for direct variation	Assume y varies directly as x for some positive variation constant k. • If the value of x increases, then the value of y increases. • If the value of x decreases, then the value of y decreases.
Finding and using a direct variation model	Assume a quantity p varies directly as a quantity t. To make estimates about an authentic situation, 1. Substitute the values of a data point into the equation $p = kt$, and then solve for k. 2. Substitute the value of k into the equation $p = kt$. 3. Use the equation found in step 2 to make estimates of quantity t or quantity p.
Inverse variation	If $y = \dfrac{k}{x}$ for some constant k, we say **y varies inversely as x** or **y is inversely proportional to x**. We call k the **variation constant** or the **constant of proportionality**. The equation $y = \dfrac{k}{x}$ is called an **inverse variation equation**.

Section 10.7   Variation (*Continued*)

Changes in values of variables for inverse variation	Assume y varies inversely as x for some positive variation constant k. For positive values of x, • If the value of x increases, then the value of y decreases. • If the value of x decreases, then the value of y increases.
Finding and using an inverse variation model	Assume a quantity p varies inversely as a quantity t. To make estimates about an authentic situation, **1.** Substitute the values of a data point into the equation $p = \dfrac{k}{t}$, and then solve for k. **2.** Substitute the value of k into the equation $p = \dfrac{k}{t}$. **3.** Use the equation found in step 2 to make estimates of quantity t or quantity p.

Section 10.8   Simplifying Complex Rational Expressions

Complex rational expression	A **complex rational expression** is a rational expression whose numerator or denominator (or both) is a rational expression.
Using method 1 to simplify a complex rational expression	To simplify a complex rational expression by method 1, **1.** Write both the numerator and the denominator as fractions. **2.** To write the complex rational expression as the quotient of two rational expressions, use the property $$\frac{\dfrac{A}{B}}{\dfrac{C}{D}} = \frac{A}{B} \div \frac{C}{D}, \text{ where } B, C, \text{ and } D \text{ are nonzero}$$ **3.** Divide the rational expressions.
Using method 2 to simplify a complex rational expression	To simplify a complex rational expression by method 2, **1.** Find the LCD of all the fractions appearing in the numerator and denominator. **2.** Multiply by 1 in the form $\dfrac{\text{LCD}}{\text{LCD}}$. **3.** Simplify the numerator and the denominator to polynomials. **4.** Simplify the rational expression.

Chapter 10 Review Exercises

Find any excluded values.

1. $\dfrac{4}{x}$

2. $\dfrac{x}{7}$

3. $\dfrac{9}{3x - 5}$

4. $\dfrac{x - 5}{x^2 - 6x + 8}$

5. $\dfrac{5t - 6}{t^2 + 6t + 9}$

6. $\dfrac{7w}{4w^2 - 9}$

7. $\dfrac{2x + 3}{5x^2 + 18x - 8}$

8. $\dfrac{x + 9}{2x^3 + 5x^2 - 18x - 45}$

Simplify.

9. $\dfrac{28x^3y^5}{35x^7y^2}$

10. $\dfrac{7x + 21}{x^2 + x - 6}$

11. $\dfrac{w^2 + 7w + 12}{w^2 - 16}$

12. $\dfrac{2a^2 - 2ab}{a^2 - b^2}$

13. $\dfrac{x^2 - 8x + 12}{3x^2 - 16x - 12}$

14. $\dfrac{x^2 + 10x + 25}{2x^3 - 3x^2 - 50x + 75}$

Perform the indicated operation. Simplify the result.

15. $\dfrac{m - 4}{m + 3} \cdot \dfrac{m + 2}{m - 3}$

16. $\dfrac{25b^3}{b^2 - b} \cdot \dfrac{b^2 - 1}{35b}$

17. $\dfrac{x^3 + 8x^2 + 16x}{x^2 - 10x + 21} \cdot \dfrac{4x - 12}{3x + 12}$

18. $\dfrac{-3x + 12}{x^2 - 1} \cdot \dfrac{-2x - 2}{x^2 - 16}$

19. $\dfrac{5x - 15}{6x^5} \div \dfrac{2x - 6}{4x^2}$

20. $\dfrac{9x^2 - 36}{3x + 3} \div \dfrac{2x^2 - 6x - 20}{2x^2 + 7x + 5}$

21. $\dfrac{7t + 14}{t - 7} \div \left(3t^2 + 2t - 8\right)$

22. $\dfrac{a^2 - 9b^2}{4a^2 - 8ab} \div \dfrac{a^2 - 4ab - 21b^2}{a^2 - 4ab + 4b^2}$

23. $\dfrac{x^2}{x + 7} + \dfrac{5x - 14}{x + 7}$

24. $\dfrac{3}{4x} + \dfrac{5}{6x^3}$

25. $\dfrac{5}{x^2 + 7x + 6} + \dfrac{2x}{x^2 - 3x - 4}$

26. $\dfrac{x + 4}{x - 2} + \dfrac{2}{x^2 + 4x - 12}$

27. $\dfrac{p^2}{p^2 - 4} - \dfrac{4p + 12}{p^2 - 4}$

28. $\dfrac{x - 4}{x^2 + 2x - 3} - \dfrac{x + 2}{x^2 - 6x + 5}$

29. $\dfrac{2xy}{x^2 - 10xy + 25y^2} - \dfrac{2y}{x - 5y}$

30. $\dfrac{y}{y^3 + 64} - \dfrac{3}{y^2 + 4y}$

31. Find the product of $\dfrac{8x^2 + 4x}{9x^4}$ and $\dfrac{12x^9}{10x + 5}$.

32. Find the quotient of $\dfrac{x^2 - 36}{x^2 + 13x + 36}$ and $\dfrac{x^2 + 12x + 36}{9x + 36}$.

33. Find the sum of $\dfrac{5m}{m^2 + 4m - 45}$ and $\dfrac{3}{m^2 + 6m - 27}$.

34. Find the difference of $\dfrac{c - 5}{c + 2}$ and $\dfrac{c + 3}{c - 1}$.

35. A student tries to find the difference $\dfrac{2x}{x + 4} - \dfrac{7x - 3}{x + 4}$:

$$\frac{2x}{x + 4} - \frac{7x - 3}{x + 4} = \frac{2x - 7x - 3}{x + 4}$$
$$= \frac{-5x - 3}{x + 4}$$

Describe any errors. Then find the difference correctly.

For Exercises 36 and 37, round approximate results to two decimal places. Refer to the list of equivalent units in the margin of page 546 as needed.

36. A TI-84 graphing calculator weighs 0.62 pound. What is the calculator's weight in ounces?

37. Americans consume an average of 121 gallons of water per year (Source: *Wirthlin Worldwide*). What is this average in cups per day?

Solve.

38. $\dfrac{3x - 2}{x + 4} = \dfrac{9}{x + 4}$

39. $\dfrac{5}{2x} - \dfrac{4}{x} = \dfrac{3}{2}$

40. $\dfrac{3}{r - 1} + \dfrac{2}{4r - 4} = \dfrac{7}{4}$

41. $\dfrac{3}{t - 2} + \dfrac{2}{t + 2} = \dfrac{4}{t^2 - 4}$

42. $\dfrac{5}{x - 4} - \dfrac{2}{x - 7} = \dfrac{4}{x^2 - 11x + 28}$

43. $\dfrac{3}{x + 3} - \dfrac{2}{x - 3} = \dfrac{-12}{x^2 - 9}$

44. $\dfrac{5}{x} = \dfrac{3}{x^2} - 2$

45. $\dfrac{4}{x + 6} + \dfrac{2}{x + 1} = \dfrac{x - 2}{x + 6}$

46. DATA The numbers and revenues of print newspapers (excluding online and other media newspapers) are shown in Table 18 for various years.

Table 18 Numbers and Revenues of Print Newspapers

Year	Number of Print Newspapers	Revenue (millions of dollars)
2002	1457	11,030
2005	1452	10,740
2008	1408	10,090
2011	1382	9990
2014	1331	10,740

Sources: *Editor and Publisher Co.; Pew Research Center*

Let r be the average annual revenue (in millions of dollars) per print newspaper at t years since 2000. A reasonable model is

$$r = \frac{21t^2 - 378t + 11{,}833}{-10.7t + 1492}$$

a. Use the model to estimate the average revenue per print newspaper in 2014. Then find the actual average revenue. Is your result from using the model an underestimate or an overestimate?

b. Predict the average revenue per print newspaper in 2022.

c. Predict when the average annual revenue per print newspaper will be $10 million.

47. If a recipe for 4 servings of chicken cacciatore calls for 14 ounces of diced tomatoes, how many ounces of diced tomatoes should be used to make 7 servings?

48. Two similar triangles are shown in Fig. 35. Find the length of the side labeled "x yards." Round your result to the second decimal place.

Figure 35 Exercise 48

Translate the sentence into an equation.

49. A varies directly as w.

50. F varies inversely as r squared.

Translate the equation into a sentence by using the phrase "varies directly" or "varies inversely."

51. $V = kr^3$

52. $I = \dfrac{k}{d^2}$

For Exercises 53 and 54, find the requested value of the variable.

53. If w varies directly as t^2, and $w = 18$ when $t = 3$, find t when $w = 50$.

54. If H varies inversely as d, and $H = 4$ when $d = 3$, find H when $d = 6$.

55. The variable w varies directly as p with positive variation constant k. What happens to the value of w as the value of p decreases?

56. The variable C varies inversely as t with positive variation constant k. For positive values of t, what happens to the value of C as the value of t increases?

57. An author's total royalties vary directly as the number of books sold. If the total royalties are $47,500 from selling 5000 books, what would be the total royalties from selling 7500 books?

58. The time it takes a commuter to drive to work varies inversely as the average driving speed. If it takes the commuter 40 minutes when traveling at an average speed of 45 mph, how long will the commute take if the average speed is 50 mph?

59. DATA As you move away from an object, it appears to decrease in size. To describe this relationship, a person stood 4 feet away from a painting, held a yardstick at arm's length (2 feet), and measured the image of the painting. The image had a height of 14 inches. For this exercise, we call the height of such an image the *apparent height* of the painting. Apparent heights of the painting were collected at various distances from it (see Table 19).

Table 19 Apparent Heights of a Painting

Distance away from Painting (feet)	Apparent Height (inches)
4	14.0
5	10.3
6	8.8
7	7.8
8	7.0
9	6.0
10	5.5
11	5.0

Source: *J. Lehmann*

Let A be the apparent height (in inches) of the painting when the person is D feet away.

a. Use a graphing calculator to draw a scatterplot of the data.

b. Find an equation of a model to describe the data.

c. In a sentence that uses the phrase "varies directly" or "varies inversely," describe how the distance from the painting and the apparent height are related.

d. Find the apparent height of the painting when the person is standing 12 feet away from it.

e. Estimate the actual height of the painting. [**Hint:** Take into account the fact that the person held the yardstick at arm's length (2 feet).]

Simplify.

60. $\dfrac{\dfrac{12}{x^2}}{\dfrac{9}{x^3}}$

61. $\dfrac{\dfrac{x^2 - 6x - 16}{25x}}{\dfrac{x - 8}{35}}$

62. $\dfrac{5 - \dfrac{2}{w}}{1 - \dfrac{3}{w}}$

63. $\dfrac{\dfrac{3}{2b} + \dfrac{1}{b^2}}{\dfrac{1}{3b} - \dfrac{2}{b^2}}$

Chapter 10 Test

For Exercises 1–3, find any excluded values.

1. $\dfrac{x}{2}$

2. $\dfrac{w - 2}{w + 9}$

3. $\dfrac{x - 5}{x^2 + 3x - 54}$

4. Give three examples of rational expressions, each of whose only excluded value is -4.

Simplify.

5. $\dfrac{p^2 - 16}{p^2 - 9p + 20}$

6. $\dfrac{5m^2 + 10m}{3m^3 - 2m^2 - 12m + 8}$

Perform the indicated operation. Simplify the result.

7. $\dfrac{x^2 + 5x + 6}{x^2 - 10x + 25} \cdot \dfrac{x^2 - 25}{x^2 - 2x - 8}$

8. $\dfrac{x^2 - 2xy}{20x^6 y} \div \dfrac{xy - 2y^2}{45x^4 y^2}$

9. $\dfrac{5a}{a^2 - a - 20} + \dfrac{2}{a^2 - 3a - 10}$

10. $\dfrac{t}{t^2 - 49} - \dfrac{5}{2t - 14}$

11. Find the quotient of $\dfrac{9x^2 - 4}{-2x - 8}$ and $\dfrac{6x^2 - x - 2}{-3x - 12}$.

12. Find the difference of $\dfrac{x - 1}{x + 4}$ and $\dfrac{x + 2}{x - 7}$.

For Exercises 13 and 14, solve.

13. $\dfrac{4}{p - 2} + \dfrac{3}{5} = 1$

14. $\dfrac{3}{w^2 - 4w - 21} - \dfrac{5}{w - 7} = \dfrac{2}{3w + 9}$

15. The speed limit is 130 kilometers per hour on motorways in France. Describe this speed limit in miles per hour. Round your result to two decimal places. Refer to the list of equivalent units in the margin of page 546 as needed.

16. DATA The numbers of prisoners and the numbers of releases from prisons are shown in Table 20 for various years.

Table 20 Prisoners and Releases from Prisons

Year	Number of Prisoners Released from Prison (thousands)	Total Number of Prisoners (thousands)
2002	634	1440
2005	702	1526
2008	734	1608
2011	691	1599
2014	636	1562

Source: *Bureau of Justice Statistics*

Let p be the percentage of prisoners who are released in the year that is t years since 2000. A reasonable model of the situation is

$$p = \dfrac{-255t^2 + 4053t + 56{,}408}{-2.67t^2 + 53.4t + 1339}$$

a. Use the model to estimate the percentage of prisoners who were released in 2014. Then find the actual percentage of prisoners who were released. Is your result from using the model an underestimate or an overestimate?

b. Predict the percentage of prisoners who will be released in 2022.

c. Predict in which year 20% of prisoners will be released.

17. A 2017 Toyota Prius Two Eco® uses 3 gallons of gasoline to travel 159 miles on highways. Estimate how many gallons of gasoline are required to travel 400 miles on highways.

18. Two similar triangles are shown in Fig. 36. Find the length of the side labeled "x meters." Round your result to the second decimal place.

14 meters

8 meters

x meters

9 meters

Figure 36 Exercise 18

19. If H varies directly as t, and $H = 12$ when $t = 3$, find H when $t = 6$.

20. If p varies inversely as w^2, and $p = 9$ when $w = 2$, find w when $p = 4$.

21. The variable c varies inversely as u for the positive variation constant k. For positive values of u, describe what happens to the value of c as the value of u increases.

22. The total Michigan out-of-district tuition at Kalamazoo Valley Community College varies directly as the number of contact hours (units or credits). For the summer term 2017, the total tuition is $2064 for 12 contact hours. If a student can afford $2600 for tuition, in how many contact hours can she enroll?

23. Simplify $\dfrac{\dfrac{4}{3x} - \dfrac{1}{x^2}}{\dfrac{1}{2x} - \dfrac{5}{x^2}}$.

11 More Radical Expressions and Equations

Has a McDonald's® restaurant opened in your neighborhood in the last year? From 2012 to 2016, the number of McDonald's restaurants in the world increased by about 2.4 thousand restaurants (see Table 1). In Exercise 53 of Homework 11.3, you will predict when there will be 38 thousand McDonald's restaurants in the world.

Table 1 Numbers of McDonald's Restaurants

Year	Number of Restaurants (thousands)
2012	34.5
2013	35.4
2014	36.3
2015	36.5
2016	36.9

Source: *McDonald's*

In Sections 9.1 and 9.2, we simplified radical expressions. In this chapter, we will discuss how to add, subtract, and multiply radical expressions. We will also discuss how to solve equations that have radicals, and we will use this skill to make predictions about authentic situations such as when the number of 3D movie screens in the United States will reach 56 thousand screens.

11.1 Adding and Subtracting Radical Expressions

Objectives

» Add and subtract radical expressions.

Recall from Section 4.2 that we can use the distributive law to add like terms, such as $3x$ and $6x$:

$$3x + 6x = (3 + 6)x = 9x$$

How do we add (or subtract) radical expressions? We can again use the distributive law if the radicals are like radicals. We say $3\sqrt{x}$ and $6\sqrt{x}$ are like radicals because they have the same radicand. In general, square root radicals that have the same radicand are called **like radicals**.

We add the like radicals $3\sqrt{x}$ and $6\sqrt{x}$ as follows:

$$3\sqrt{x} + 6\sqrt{x} = (3 + 6)\sqrt{x} = 9\sqrt{x}$$

To add or subtract like radicals, we use the distributive law. When we add or subtract like radicals, we say we *combine like radicals*.

Example 1   Adding and Subtracting Radical Expressions

Perform the indicated operation.

1. $3\sqrt{2} + 5\sqrt{2}$
2. $7\sqrt{x} + \sqrt{x}$
3. $9\sqrt{5x} - 2\sqrt{5x}$
4. $2\sqrt{3} + 6\sqrt{7}$
5. $5t\sqrt{2} + 3t\sqrt{2}$

Solution

1. $3\sqrt{2} + 5\sqrt{2} = (3 + 5)\sqrt{2} = 8\sqrt{2}$
2. $7\sqrt{x} + \sqrt{x} = 7\sqrt{x} + 1\sqrt{x} = (7 + 1)\sqrt{x} = 8\sqrt{x}$
3. $9\sqrt{5x} - 2\sqrt{5x} = (9 - 2)\sqrt{5x} = 7\sqrt{5x}$
4. Because the radicals $2\sqrt{3}$ and $6\sqrt{7}$ have different radicands, they are not like radicals and we cannot use the distributive law. The expression $2\sqrt{3} + 6\sqrt{7}$ is already in simplified form.
5. $5t\sqrt{2} + 3t\sqrt{2} = (5t + 3t)\sqrt{2} = 8t\sqrt{2}$

▶

Example 2   Adding and Subtracting Radical Expressions

Perform the indicated operations: $5\sqrt{x} + 3\sqrt{2} - 7\sqrt{x}$.

Solution

$$
\begin{aligned}
5\sqrt{x} + 3\sqrt{2} - 7\sqrt{x} &= 5\sqrt{x} - 7\sqrt{x} + 3\sqrt{2} & \textit{Rearrange terms.} \\
&= (5 - 7)\sqrt{x} + 3\sqrt{2} & \textit{Distributive law} \\
&= -2\sqrt{x} + 3\sqrt{2} & \textit{Subtract.}
\end{aligned}
$$

▶

Sometimes, simplifying radicals will allow us to combine like radicals.

Example 3   Adding and Subtracting Radical Expressions

Perform the indicated operations.

1. $\sqrt{8} + 5\sqrt{2}$ 2. $5\sqrt{27} - 2\sqrt{75}$ 3. $6\sqrt{4w} - 7\sqrt{9w} + 5\sqrt{w}$

Solution

1. $$
\begin{aligned}
\sqrt{8} + 5\sqrt{2} &= \sqrt{4 \cdot 2} + 5\sqrt{2} & \textit{4 is a perfect square.} \\
&= \sqrt{4}\sqrt{2} + 5\sqrt{2} & \sqrt{ab} = \sqrt{a}\sqrt{b} \\
&= 2\sqrt{2} + 5\sqrt{2} & \sqrt{4} = 2 \\
&= 7\sqrt{2} & \textit{Combine like radicals.}
\end{aligned}
$$

2. $$
\begin{aligned}
5\sqrt{27} - 2\sqrt{75} &= 5\sqrt{9 \cdot 3} - 2\sqrt{25 \cdot 3} & \textit{9 and 25 are perfect squares.} \\
&= 5\sqrt{9}\sqrt{3} - 2\sqrt{25}\sqrt{3} & \sqrt{ab} = \sqrt{a}\sqrt{b} \\
&= 5 \cdot 3 \cdot \sqrt{3} - 2 \cdot 5 \cdot \sqrt{3} & \sqrt{9} = 3; \sqrt{25} = 5 \\
&= 15\sqrt{3} - 10\sqrt{3} & \textit{Multiply.} \\
&= 5\sqrt{3} & \textit{Combine like radicals.}
\end{aligned}
$$

We use a graphing calculator to verify our work (see Fig. 1).

3. $$
\begin{aligned}
6\sqrt{4w} - 7\sqrt{9w} + 5\sqrt{w} &= 6\sqrt{4}\sqrt{w} - 7\sqrt{9}\sqrt{w} + 5\sqrt{w} & \sqrt{ab} = \sqrt{a}\sqrt{b} \\
&= 6 \cdot 2 \cdot \sqrt{w} - 7 \cdot 3 \cdot \sqrt{w} + 5\sqrt{w} & \sqrt{4} = 2; \sqrt{9} = 3 \\
&= 12\sqrt{w} - 21\sqrt{w} + 5\sqrt{w} & \textit{Multiply.} \\
&= -4\sqrt{w} & \textit{Combine like radicals.}
\end{aligned}
$$

We use a graphing calculator table to verify our work (see Fig. 2).

▶

Recall from Section 9.1 that if x is nonnegative, then $\sqrt{x^2} = x$.

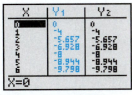

Figure 1 Verify the work

Figure 2 Verify the work

Example 4   Adding and Subtracting Radical Expressions

Perform the indicated operations. Assume that x is nonnegative.

1. $5\sqrt{3x^2} - x\sqrt{12}$ 2. $x\sqrt{20} - 6\sqrt{2x} + 4\sqrt{45x^2}$

Solution

1.
$$5\sqrt{3x^2} - x\sqrt{12} = 5\sqrt{x^2 \cdot 3} - x\sqrt{4 \cdot 3}$$ *Rearrange factors; x^2 and 4 are perfect squ*
$$= 5\sqrt{x^2}\sqrt{3} - x\sqrt{4}\sqrt{3}$$ $\sqrt{ab} = \sqrt{a}\sqrt{b}$
$$= 5x\sqrt{3} - 2x\sqrt{3}$$ $\sqrt{x^2} = x$ *for nonnegative x;* $\sqrt{4} = 2$
$$= 3x\sqrt{3}$$ *Combine like radicals.*

2.
$$x\sqrt{20} - 6\sqrt{2x} + 4\sqrt{45x^2}$$
$$= x\sqrt{4 \cdot 5} - 6\sqrt{2x} + 4\sqrt{9x^2 \cdot 5}$$ *4, 9, and x^2 are perfect squares.*
$$= x\sqrt{4}\sqrt{5} - 6\sqrt{2x} + 4\sqrt{9}\sqrt{x^2}\sqrt{5}$$ $\sqrt{ab} = \sqrt{a}\sqrt{b}$
$$= x \cdot 2 \cdot \sqrt{5} - 6\sqrt{2x} + 4 \cdot 3 \cdot x \cdot \sqrt{5}$$ $\sqrt{4} = 2; \sqrt{9} = 3; \sqrt{x^2} = x$ *for nonnegative x*
$$= 2x\sqrt{5} - 6\sqrt{2x} + 12x\sqrt{5}$$ *Multiply.*
$$= 2x\sqrt{5} + 12x\sqrt{5} - 6\sqrt{2x}$$ *Rearrange terms.*
$$= 14x\sqrt{5} - 6\sqrt{2x}$$ *Combine like radicals.*

Group Exploration

Section Opener: Combining like radicals

Recall from Section 4.2 that we can use the disributive law to combine like terms. For example, here we add the like terms $5x$ and $3x$: $5x + 3x = (5 + 3)x = 8x$. We can also use the distributive law to "combine" some radicals.

1. Use the distributive law to perform the operations.
   a. $5\sqrt{x} + 3\sqrt{x}$ b. $9\sqrt{2} - 6\sqrt{2}$
   c. $4\sqrt{7} + \sqrt{7} - 2\sqrt{7}$

2. Can you use the distributive law to "combine" $4\sqrt{3}$ and $6\sqrt{7}$ in the sum $4\sqrt{3} + 6\sqrt{7}$? If yes, show how. If no, explain why not.

3. Can you use the distributive law to "combine" $8\sqrt{2}$ and $3\sqrt{5}$ in the difference $8\sqrt{2} - 3\sqrt{5}$? If yes, show how. If no, explain why not.

4. Describe in general when you can use the distributive law to "combine" radicals in a sum or difference of two radicals.

5. Use the distributive law to perform the operations. [**Hint:** To begin, simplify the radicals.]
   a. $\sqrt{8} + \sqrt{18}$
   b. $\sqrt{27} - \sqrt{12}$

Homework 11.1

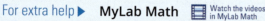
For extra help ▶   **MyLab Math**   ▤ Watch the videos in MyLab Math

Perform the indicated operations. Use a graphing calculator table to verify your work.

1. $4\sqrt{3} + 5\sqrt{3}$
2. $2\sqrt{7} + 6\sqrt{7}$
3. $-3\sqrt{5} + 9\sqrt{5}$
4. $-5\sqrt{6} + 3\sqrt{6}$
5. $4\sqrt{x} + \sqrt{x}$
6. $\sqrt{x} + 7\sqrt{x}$
7. $\sqrt{7} + \sqrt{7}$
8. $\sqrt{3} + \sqrt{3}$
9. $2\sqrt{6} - 9\sqrt{6}$
10. $5\sqrt{10} - 8\sqrt{10}$
11. $-4\sqrt{5t} - 7\sqrt{5t}$
12. $-8\sqrt{3a} - 5\sqrt{3a}$
13. $7\sqrt{5} - 3\sqrt{7}$
14. $8\sqrt{2} - 5\sqrt{6}$
15. $5\sqrt{2} - \sqrt{2} + 2\sqrt{2}$
16. $8\sqrt{3} - \sqrt{3} + 5\sqrt{3}$
17. $-4\sqrt{7} + 8\sqrt{7} - 3\sqrt{5}$
18. $-3\sqrt{2} + 5\sqrt{3} - 7\sqrt{2}$
19. $7\sqrt{p} - 4\sqrt{p} - 3\sqrt{p}$
20. $9\sqrt{b} - 5\sqrt{b} - 4\sqrt{b}$

21. $\sqrt{18} + 4\sqrt{2}$

22. $\sqrt{12} + 6\sqrt{3}$

23. $5\sqrt{6} - \sqrt{24}$

24. $7\sqrt{5} - \sqrt{45}$

25. $-4\sqrt{5} - \sqrt{20}$

26. $8\sqrt{3} - \sqrt{27}$

27. $\sqrt{28} + \sqrt{63}$

28. $\sqrt{32} + \sqrt{8}$

29. $2\sqrt{27} + 5\sqrt{12}$

30. $3\sqrt{20} + 4\sqrt{45}$

31. $5\sqrt{8} - 5\sqrt{18}$

32. $2\sqrt{32} - 3\sqrt{50}$

33. $\sqrt{20} - 4\sqrt{5} + \sqrt{45}$

34. $\sqrt{32} - 3\sqrt{2} + \sqrt{18}$

35. $5\sqrt{12} + 4\sqrt{75} - 2\sqrt{3}$

36. $3\sqrt{8} + 2\sqrt{18} - 8\sqrt{2}$

37. $\sqrt{4x} + \sqrt{81x}$

38. $\sqrt{64x} + \sqrt{16x}$

39. $2\sqrt{9x} - 3\sqrt{4x}$

40. $4\sqrt{25x} - 5\sqrt{9x}$

41. $\sqrt{50x} - \sqrt{20} - \sqrt{32x}$

42. $\sqrt{98x} - \sqrt{28} - \sqrt{72x}$

43. $3\sqrt{12t} - 2\sqrt{44} + 5\sqrt{27t}$

44. $4\sqrt{5a} - 5\sqrt{63} + 3\sqrt{45a}$

Perform the indicated operations. Assume that any variables are nonnegative. Use a graphing calculator table to verify your work.

45. $x\sqrt{20} + x\sqrt{45}$

46. $x\sqrt{18} + x\sqrt{50}$

47. $x\sqrt{40} - 5\sqrt{10x^2}$

48. $x\sqrt{32} - 3\sqrt{2x^2}$

49. $\sqrt{5w^2} - \sqrt{45w^2}$

50. $\sqrt{7b^2} - \sqrt{28b^2}$

51. $2\sqrt{75x^2} + 3\sqrt{12x^2}$

52. $4\sqrt{27x^2} + 8\sqrt{300x^2}$

53. $\sqrt{3x^2} - 2x\sqrt{3} + 5\sqrt{4x}$

54. $\sqrt{8x^2} - 5x\sqrt{2} + 6\sqrt{3x}$

55. $5\sqrt{48y^2} - 3\sqrt{20y} - 2y\sqrt{12}$

56. $\sqrt{54t^2} - 9\sqrt{24t} - t\sqrt{600}$

57. The time it takes a planet to make one revolution around the Sun is called the planet's *period*. The period T (in years) of a planet whose average distance from the Sun is d million kilometers is modeled by the equation $T = 0.0005443d\sqrt{d}$.

   a. What is the period of Saturn, whose average distance from the Sun is 1427.0 million kilometers?

   b. If a person is 59 years old in "Earth years," approximately how old is the person in "Saturn years"? [**Hint:** 1 "Earth year" is the amount of time it takes Earth to revolve around the Sun once.]

58. A person builds a *pendulum* by tying one end of some thread to a washer (a metal ring) and attaching the other end to the bottom of a table so that the washer is suspended and can swing freely. The *period* of the pendulum is the amount of time it takes the washer to swing forward and backward once. Let T be the period (in seconds) of the pendulum, where L is the length (in feet) of the thread. A model of the situation is $T = \dfrac{\pi}{4}\sqrt{2L}$. Estimate the period of the pendulum if its length is 2 feet.

59. **DATA.** The percentages of incomes unreported to the IRS are shown in Table 2 for various incomes.

Table 2 Percentages of Incomes Unreported to the IRS

Income (thousands of dollars)	Income Used to Represent Income Group (thousands of dollars)	Percent
20–25	22.5	5
25–30	27.5	6
30–40	35	7
40–50	45	7
50–75	62.5	8
75–100	87.5	8
100–200	150	13
200–500	350	20

Source: The Distribution of Income Tax Noncompliance *by Johns and Slemrod*

Let p be the percentage of income unreported to the IRS by taxpayers with income d (in thousands of dollars). A model of the situation is $p = 1.04\sqrt{d}$.

   a. Use a graphing calculator to draw the graph of the model and, in the same viewing window, the scatterplot of the data. Does the model fit the data well?

   b. Substitute a value for one of the variables in the model's equation to estimate the percentage of income unreported by taxpayers whose income is $250 thousand.

   c. Use TRACE on a graphing calculator to estimate the income of taxpayers who do not report 11% of their income.

60. The percentages of mothers who use social networking websites several times per day are shown in Table 3 for various years.

Table 3 Percentages of Mothers Who Use Social Networking Websites Several Times per Day

Year	Percent
2008	11
2010	32
2012	46
2014	49
2016	56
2017	62

Source: *Edison Research*

Let p be the percentage of mothers who use social networking websites several times per day at t years since 2008. A model of the situation is $p = 16.3\sqrt{t} + 11$.

   a. Use a graphing calculator to draw the graph of the model and, in the same viewing window, the scatterplot of the data. Does the model fit the data well?

   b. Substitute a value for one of the variables in the model's equation to predict the percentage of mothers who will use social networking websites several times per day in 2021.

   c. Use TRACE to predict when 72% of mothers will use social networking websites several times per day.

Concepts

61. A student tries to simplify $2\sqrt{5} + 4\sqrt{5}$:

$$2\sqrt{5} + 4\sqrt{5} = 6\sqrt{10}$$

Describe any errors. Then simplify the expression correctly.

62. A student tries to simplify $6\sqrt{2} + 7\sqrt{3}$:

$$6\sqrt{2} + 7\sqrt{3} = 13\sqrt{5}$$

Describe any errors. Then simplify the expression correctly.

63. Explain how the distributive law can be used to show that the statement $4\sqrt{7} + 5\sqrt{7} = 9\sqrt{7}$ is true.

64. Explain how the distributive law can be used to show that the statement $9\sqrt{2} - 6\sqrt{2} = 3\sqrt{2}$ is true.

Related Review

Simplify. Assume x is nonnegative.

65. $5x - 7x$ **66.** $2x + 4x$ **67.** $5\sqrt{x} - 7\sqrt{x}$

68. $2\sqrt{x} + 4\sqrt{x}$ **69.** $\sqrt{12x^2} + \sqrt{3x^2}$ **70.** $\sqrt{2x^2} - \sqrt{18x^2}$

71. $12x^2 + 3x^2$ **72.** $2x^2 - 18x^2$

Expressions, Equations, and Graphs

Perform the indicated instruction. Then use words such as linear, quadratic, cubic, power, radical, rational, polynomial, degree, one variable, *and* two variables *to describe the expression, equation, or system.*

73. Solve $t(t - 2) = 7$.

74. Factor $10p^2 - 13pq - 3q^2$.

75. Find the difference $\dfrac{3w}{w^2 - 9} - \dfrac{1}{4w + 12}$.

76. Solve $t^2 - 3t - 18 = 0$.

77. Factor $x^3 - 10x^2 + 25x$.

78. Graph $y = x^2 - 1$ by hand.

11.2 Multiplying Radical Expressions

Objectives

» Multiply two radical expressions.

» Find the product of two *radical conjugates*.

» Simplify the square of a radical expression that has two terms.

In Section 11.1, we added and subtracted radical expressions. How do we multiply two radical expressions?

Multiplication of Radical Expressions

In Section 9.1, we discussed the product property for square roots, namely, $\sqrt{ab} = \sqrt{a}\sqrt{b}$, where a and b are nonnegative. We can use this property to help us multiply radical expressions. Here we multiply $\sqrt{2}$ and $\sqrt{3}$:

$$\sqrt{2} \cdot \sqrt{3} = \sqrt{2 \cdot 3} = \sqrt{6}$$

Next, we find the power $\left(\sqrt{x}\right)^2$, where x is nonnegative:

$$\left(\sqrt{x}\right)^2 = \sqrt{x}\sqrt{x} = \sqrt{x^2} = x$$

Power Property for Square Roots

If x is nonnegative, then

$$\left(\sqrt{x}\right)^2 = x$$

It is good practice to check whether the product of radical expressions can be simplified.

Example 1   Finding the Product of Radical Expressions

Find the product.
1. $\sqrt{5} \cdot \sqrt{7}$ 2. $5\sqrt{3} \cdot \sqrt{6}$
3. $2\sqrt{x} \cdot 3\sqrt{x}$ 4. $\sqrt{3t} \cdot \sqrt{5t}$

Solution

1. $\sqrt{5} \cdot \sqrt{7} = \sqrt{5 \cdot 7}$ $\sqrt{a}\sqrt{b} = \sqrt{ab}$

$\qquad\qquad = \sqrt{35}$ *Multiply.*

2. $5\sqrt{3} \cdot \sqrt{6} = 5\sqrt{18}$ *$\sqrt{a}\sqrt{b} = \sqrt{ab}$*

$= 5\sqrt{9 \cdot 2}$ *9 is a perfect square.*

$= 5\sqrt{9}\sqrt{2}$ *$\sqrt{ab} = \sqrt{a}\sqrt{b}$*

$= 5 \cdot 3\sqrt{2}$ *$\sqrt{9} = 3$*

$= 15\sqrt{2}$ *Multiply.*

3. $2\sqrt{x} \cdot 3\sqrt{x} = 2 \cdot 3 \cdot \sqrt{x} \cdot \sqrt{x}$ *Rearrange factors.*

$= 6x$ *Multiply; $\sqrt{x}\sqrt{x} = x$ for $x \geq 0$*

4. $\sqrt{3t} \cdot \sqrt{5t} = \sqrt{15t^2}$ *$\sqrt{a}\sqrt{b} = \sqrt{ab}$*

$= \sqrt{15}\sqrt{t^2}$ *$\sqrt{ab} = \sqrt{a}\sqrt{b}$*

$= t\sqrt{15}$ *$\sqrt{x^2} = x$ for $x \geq 0$*

We can use the distributive law to help us find products of radical expressions.

Example 2   Finding the Product of Radical Expressions

Find the product.

1. $\sqrt{5}\left(3 - \sqrt{2}\right)$ **2.** $\sqrt{x}\left(\sqrt{x} - \sqrt{5}\right)$

3. $2\sqrt{3}\left(4\sqrt{3} + 6\sqrt{7}\right)$

Solution

1. $\sqrt{5}\left(3 - \sqrt{2}\right) = \sqrt{5} \cdot 3 - \sqrt{5} \cdot \sqrt{2}$ *Distributive law*

$= 3\sqrt{5} - \sqrt{10}$ *Rearrange factors; $\sqrt{a}\sqrt{b} = \sqrt{ab}$*

2. $\sqrt{x}\left(\sqrt{x} - \sqrt{5}\right) = \sqrt{x} \cdot \sqrt{x} - \sqrt{x} \cdot \sqrt{5}$ *Distributive law*

$= x - \sqrt{5x}$ *$\sqrt{x}\sqrt{x} = x$ for $x \geq 0$; $\sqrt{a}\sqrt{b} = \sqrt{ab}$*

3. $2\sqrt{3}\left(4\sqrt{3} + 6\sqrt{7}\right) = 2\sqrt{3} \cdot \left(4\sqrt{3}\right) + 2\sqrt{3} \cdot \left(6\sqrt{7}\right)$ *Distributive law*

$= 2 \cdot 4\sqrt{3}\sqrt{3} + 2 \cdot 6\sqrt{3}\sqrt{7}$ *Rearrange factors.*

$= 8 \cdot 3 + 12\sqrt{21}$ *$\sqrt{x}\sqrt{x} = x$ for $x \geq 0$; $\sqrt{a}\sqrt{b} = \sqrt{ab}$*

$= 24 + 12\sqrt{21}$ *Multiply.*

Next, we find the product of two radical expressions for which each factor has two terms.

Example 3   Finding the Product of Radical Expressions

Find the product.

1. $\left(5 - \sqrt{3}\right)\left(4 + \sqrt{3}\right)$ **2.** $\left(\sqrt{x} + \sqrt{3}\right)\left(\sqrt{x} + \sqrt{5}\right)$

Solution

1. We begin by multiplying each term in the first radical expression by each term in the second radical expression:

$\left(5 - \sqrt{3}\right)\left(4 + \sqrt{3}\right) = 5 \cdot 4 + 5 \cdot \sqrt{3} - \sqrt{3} \cdot 4 - \sqrt{3} \cdot \sqrt{3}$ *Multiply pairs of terms.*

$= 20 + 5\sqrt{3} - 4\sqrt{3} - \sqrt{3}\sqrt{3}$ *Simplify.*

$= 20 + \sqrt{3} - 3$ *Combine like radicals; $\sqrt{x}\sqrt{x} = x$ for $x \geq 0$*

$= 17 + \sqrt{3}$ *Simplify.*

Figure 3 Verify the work

2. Again, we begin by multiplying each term in the first radical expression by each term in the second radical expression:

$$\left(\sqrt{x} + \sqrt{3}\right)\left(\sqrt{x} + \sqrt{5}\right) = \sqrt{x}\sqrt{x} + \sqrt{x}\sqrt{5} + \sqrt{3}\sqrt{x} + \sqrt{3}\sqrt{5} \quad \textit{Multiply pairs of term}$$

$$= x + \sqrt{5x} + \sqrt{3x} + \sqrt{15} \qquad \begin{array}{l}\sqrt{x}\sqrt{x} = x \text{ for } x \geq \\ \sqrt{a}\sqrt{b} = \sqrt{ab}\end{array}$$

We cannot combine the terms $\sqrt{5x}$ and $\sqrt{3x}$ because they are not like radicals. So, our result is simplified. We use a graphing calculator table to verify our work (see Fig. 3).

> ### Product of Two Radical Expressions, Each with Two Terms
>
> To find the product of two radical expressions in which each factor has two terms,
>
> **1.** Multiply each term in the first radical expression by each term in the second radical expression.
>
> **2.** Combine like radicals.

Example 4 Finding the Product of Radical Expressions

Find the product $\left(5 - 2\sqrt{7}\right)\left(3 - 4\sqrt{7}\right)$.

Solution

We begin by multiplying each term in the first radical expression by each term in the second radical expression:

$$\left(5 - 2\sqrt{7}\right)\left(3 - 4\sqrt{7}\right) = 5 \cdot 3 - 5\left(4\sqrt{7}\right) - \left(2\sqrt{7}\right)3 + \left(2\sqrt{7}\right)\left(4\sqrt{7}\right) \quad \textit{Multiply pairs of terms.}$$

$$= 15 - 5 \cdot 4 \cdot \sqrt{7} - 2 \cdot 3 \cdot \sqrt{7} + 2 \cdot 4 \cdot \sqrt{7}\sqrt{7} \quad \textit{Rearrange factors.}$$

$$= 15 - 20\sqrt{7} - 6\sqrt{7} + 8 \cdot 7 \qquad \sqrt{x}\sqrt{x} = x \text{ for } x \geq 0$$

$$= 15 - 26\sqrt{7} + 56 \qquad \begin{array}{l}\textit{Combine like radicals;}\\ \textit{multiply.}\end{array}$$

$$= 71 - 26\sqrt{7} \qquad \textit{Simplify.}$$

Multiplication of Two Radical Conjugates

Recall from Section 7.5 that we call binomials such as $3x + 5$ and $3x - 5$ conjugates of each other. Similarly, we call the radical expressions $3 + 5\sqrt{x}$ and $3 - 5\sqrt{x}$ radical conjugates of each other. The sum of two radical terms and the difference of the same two radical terms are **radical conjugates** of each other.

> ### Product of Two Radical Conjugates
>
> To find the product of two radical conjugates, use the property
>
> $$(A + B)(A - B) = A^2 - B^2$$

In Example 5, we will find products of two radical conjugates.

| Example 5 | Finding the Product of Two Radical Conjugates |

Find the product.

1. $\left(\sqrt{7} + \sqrt{2} \right)\left(\sqrt{7} - \sqrt{2} \right)$ **2.** $\left(5\sqrt{a} - 4\sqrt{b} \right)\left(5\sqrt{a} + 4\sqrt{b} \right)$

Solution

1. We substitute $\sqrt{7}$ for A and $\sqrt{2}$ for B in the property for the product of two radical conjugates:

$$(A + B)\ (A - B) = A^2 - B^2$$
$$\left(\sqrt{7} + \sqrt{2} \right)\left(\sqrt{7} - \sqrt{2} \right) = \left(\sqrt{7} \right)^2 - \left(\sqrt{2} \right)^2 \quad \textit{Substitute.}$$
$$= 7 - 2 \quad \left(\sqrt{x} \right)^2 = x \textit{ for } x \geq 0$$
$$= 5 \quad \textit{Subtract.}$$

2. We substitute $5\sqrt{a}$ for A and $4\sqrt{b}$ for B in the property for the product of two radical conjugates:

$$(A - B)\ (A + B) = A^2 - B^2$$
$$\left(5\sqrt{a} - 4\sqrt{b} \right)\left(5\sqrt{a} + 4\sqrt{b} \right) = \left(5\sqrt{a} \right)^2 - \left(4\sqrt{b} \right)^2 \quad \textit{Substitute.}$$
$$= 5^2\left(\sqrt{a} \right)^2 - 4^2\left(\sqrt{b} \right)^2 \quad \begin{array}{l}\textit{Raise each factor to second}\\ \textit{power: } (xy)^n = x^n y^n\end{array}$$
$$= 25a - 16b \quad \left(\sqrt{x} \right)^2 = x \textit{ for } x \geq 0$$

Simplifying the Square of a Radical Expression That Has Two Terms

Recall from Section 7.5 the properties for the square of a binomial:

$$(A + B)^2 = A^2 + 2AB + B^2 \quad \textit{Square of a sum}$$
$$(A - B)^2 = A^2 - 2AB + B^2 \quad \textit{Square of a difference}$$

For each problem in Example 6, we will first simplify a radical expression without using these properties. We will then simplify the expression again, using one of the properties.

| Example 6 | Simplifying the Square of a Radical Expression That Has Two Terms |

Simplify.

1. $\left(3 + \sqrt{5} \right)^2$ **2.** $\left(\sqrt{x} - \sqrt{3} \right)^2$

Solution

1. We begin by using the fact that $C^2 = CC$ and then multiply pairs of terms:

$$\left(3 + \sqrt{5} \right)^2 = \left(3 + \sqrt{5} \right)\left(3 + \sqrt{5} \right) \quad C^2 = CC$$
$$= 3 \cdot 3 + 3 \cdot \sqrt{5} + \sqrt{5} \cdot 3 + \sqrt{5} \cdot \sqrt{5} \quad \textit{Multiply pairs of terms.}$$
$$= 9 + 3\sqrt{5} + 3\sqrt{5} + \sqrt{5}\sqrt{5} \quad \textit{Simplify.}$$
$$= 9 + 6\sqrt{5} + 5 \quad \begin{array}{l}\textit{Combine like radicals;}\\ \sqrt{x}\sqrt{x} = x \textit{ for } x \geq 0\end{array}$$
$$= 14 + 6\sqrt{5} \quad \textit{Simplify.}$$

Another way to simplify $(3 + \sqrt{5})^2$ is to substitute 3 for A and $\sqrt{5}$ for B in the property for the square of a sum:

$$(A \; + \; B)^2 = A^2 \; + \; 2\,A \;\; B \; + \;\; B^2$$

$$(3 + \sqrt{5})^2 = 3^2 + 2(3)\sqrt{5} + (\sqrt{5})^2 \qquad \textit{Substitute.}$$
$$= 9 + 6\sqrt{5} + 5 \qquad (\sqrt{x})^2 = x \textit{ for } x \geq 0$$
$$= 14 + 6\sqrt{5} \qquad \textit{Simplify.}$$

2. We begin by using the fact that $C^2 = CC$ and then multiply pairs of terms:

$$(\sqrt{x} - \sqrt{3})^2 = (\sqrt{x} - \sqrt{3})(\sqrt{x} - \sqrt{3}) \qquad C^2 = CC$$
$$= \sqrt{x}\sqrt{x} - \sqrt{x}\sqrt{3} - \sqrt{3}\sqrt{x} + \sqrt{3}\sqrt{3} \qquad \textit{Multiply pairs of terms.}$$
$$= x - \sqrt{3x} - \sqrt{3x} + 3 \qquad \begin{array}{l}\sqrt{x}\sqrt{x} = x \textit{ for } x \geq 0; \\ \sqrt{a}\sqrt{b} = \sqrt{ab};\ ab = ba\end{array}$$
$$= x - 2\sqrt{3x} + 3 \qquad \textit{Combine like radicals.}$$

Another way to simplify $(\sqrt{x} - \sqrt{3})^2$ is to substitute $\sqrt{x}$ for A and $\sqrt{3}$ for B in the property for the square of a difference:

$$(A \; - \; B)^2 = \;\; A^2 \; - \; 2\,A \;\; B \; + \;\; B^2$$

$$(\sqrt{x} - \sqrt{3})^2 = (\sqrt{x})^2 - 2\sqrt{x}\sqrt{3} + (\sqrt{3})^2 \quad \textit{Substitute.}$$
$$= x - 2\sqrt{3x} + 3 \qquad (\sqrt{x})^2 = x \textit{ for } x \geq 0;\ \sqrt{a}\sqrt{b} = \sqrt{ab}$$

In Example 6, we showed two ways to simplify the square of a radical expression that has two terms: (1) using the definition of "square" and (2) using the property for the square of a sum or difference. When simplifying similar expressions in the Homework, experiment with both methods to help you decide which one works better for you. Many students are more successful with writing the square as a product of two identical expressions and then multiplying pairs of terms.

Simplifying the Square of a Radical Expression

To simplify the square of a radical expression that has two terms,

- Use $C^2 = CC$ to write the square as a product of two identical expressions, and then multiply each term in the first radical expression by each term in the second radical expression,

or

- Use the square-of-a-sum property $(A + B)^2 = A^2 + 2AB + B^2$ or the square-of-a-difference property $(A - B)^2 = A^2 - 2AB + B^2$.

WARNING   In simplifying $(x + k)^2$, it is important to remember the middle term of $x^2 + 2kx + k^2$. Likewise, in simplifying $(x - k)^2$, it is important to remember the middle term of $x^2 - 2kx + k^2$. Do not make the following typical error in simplifying $(\sqrt{x} + \sqrt{7})^2$:

$$(\sqrt{x} + \sqrt{7})^2 = (\sqrt{x})^2 + (\sqrt{7})^2 = x + 7 \qquad \textit{Incorrect}$$
$$(\sqrt{x} + \sqrt{7})^2 = (\sqrt{x})^2 + 2\sqrt{7}\sqrt{x} + (\sqrt{7})^2 = x + 2\sqrt{7x} + 7 \quad \textit{Correct}$$

Group Exploration

Section Opener: Multiplying two radical expressions

In Section 9.1, we discussed the product property for square roots:

$$\sqrt{ab} = \sqrt{a}\sqrt{b} \quad \text{where } a \text{ and } b \text{ are nonnegative numbers}$$

We can use this property to multiply two radical expressions:

$$\sqrt{2}\sqrt{7} = \sqrt{2\cdot 7} = \sqrt{14}$$

Perform the indicated operation.

1. $\sqrt{3}\cdot\sqrt{5}$
2. $6\sqrt{5}\cdot\sqrt{2}$
3. $\sqrt{3}\left(\sqrt{2} - \sqrt{5}\right)$
4. $\left(x + \sqrt{5}\right)\left(x + \sqrt{7}\right)$
5. $\left(x - \sqrt{2}\right)\left(x + \sqrt{3}\right)$
6. $\left(x + \sqrt{3}\right)\left(x - \sqrt{3}\right)$
7. $\left(x - \sqrt{5}\right)^2$

 Group Exploration

Rationalizing the denominator

In Section 9.2, we "rationalized the denominator" of fractions of the form $\dfrac{1}{\sqrt{a}}$ by finding an equivalent expression that does not have a radical in any denominator. Here you will explore how to rationalize the denominator of a fraction when that denominator is a sum or a difference involving radicals.

1. Find the product.
   a. $\left(x + \sqrt{3}\right)\left(x - \sqrt{3}\right)$ b. $\left(x - \sqrt{7}\right)\left(x + \sqrt{7}\right)$
   c. $\left(2 + \sqrt{x}\right)\left(\left(2 - \sqrt{x}\right)\right)$ d. $\left(6 - \sqrt{x}\right)\left(6 + \sqrt{x}\right)$

2. What patterns do you notice from your work in Problem 1?

3. Rationalize the denominator of the expression $\dfrac{1}{5 - \sqrt{x}}$ by performing the multiplication

$$\frac{1}{5 - \sqrt{x}}\cdot\frac{5 + \sqrt{x}}{5 + \sqrt{x}}$$

Use a graphing calculator table to verify your work.

4. Rationalize the denominator of the expression $\dfrac{1}{8 + \sqrt{x}}$.

5. Describe how to rationalize the denominator of a radical expression.

 # Homework 11.2

For extra help ▶ **MyLab Math**   📺 Watch the videos in MyLab Math

Perform the indicated operation. Assume that any variables are nonnegative. Use a graphing calculator to verify your work when possible.

1. $\sqrt{2}\cdot\sqrt{5}$
2. $\sqrt{7}\cdot\sqrt{3}$
3. $-\sqrt{6}\cdot\sqrt{2}$
4. $-\sqrt{2}\cdot\sqrt{10}$
5. $\sqrt{17}\sqrt{17}$
6. $\sqrt{19}\sqrt{19}$
7. $2\sqrt{5}\cdot\sqrt{10}$
8. $2\sqrt{15}\cdot\sqrt{3}$
9. $7\sqrt{2}\left(-3\sqrt{14}\right)$
10. $4\sqrt{10}\left(-2\sqrt{5}\right)$
11. $5\sqrt{t}\cdot 7\sqrt{t}$
12. $2\sqrt{a}\cdot 8\sqrt{a}$
13. $\sqrt{7t}\sqrt{3t}$
14. $\sqrt{5b}\sqrt{2b}$
15. $\left(8\sqrt{x}\right)^2$
16. $\left(6\sqrt{x}\right)^2$
17. $\sqrt{ab}\sqrt{bc}$
18. $\sqrt{rw}\sqrt{rtw}$
19. $\sqrt{5}\left(1 + \sqrt{7}\right)$
20. $\sqrt{2}\left(4 + \sqrt{3}\right)$

21. $-\sqrt{7}\left(\sqrt{2} + \sqrt{5}\right)$
22. $-\sqrt{3}\left(\sqrt{5} + \sqrt{7}\right)$
23. $5\sqrt{c}\left(9 - \sqrt{c}\right)$
24. $3\sqrt{w}\left(8 - \sqrt{w}\right)$
25. $-4\sqrt{5}\left(3\sqrt{2} - \sqrt{5}\right)$
26. $-3\sqrt{2}\left(6\sqrt{7} - 5\right)$

Perform the indicated operation. Use a graphing calculator table to verify your work when possible.

27. $\left(4 + \sqrt{5}\right)\left(2 - \sqrt{5}\right)$
28. $\left(6 + \sqrt{2}\right)\left(3 - \sqrt{2}\right)$
29. $\left(\sqrt{3} - \sqrt{5}\right)\left(\sqrt{3} + \sqrt{2}\right)$
30. $\left(\sqrt{5} - \sqrt{2}\right)\left(\sqrt{5} + \sqrt{3}\right)$
31. $\left(\sqrt{x} - \sqrt{7}\right)\left(\sqrt{x} - \sqrt{2}\right)$
32. $\left(\sqrt{x} - \sqrt{5}\right)\left(\sqrt{x} - \sqrt{3}\right)$
33. $\left(y + \sqrt{2}\right)\left(y - \sqrt{11}\right)$
34. $\left(a + \sqrt{5}\right)\left(a - \sqrt{2}\right)$
35. $\left(4 - 2\sqrt{5}\right)\left(2 - 3\sqrt{5}\right)$
36. $\left(1 - 3\sqrt{2}\right)\left(3 - 4\sqrt{2}\right)$

37. $\left(\sqrt{3} + \sqrt{7}\right)\left(\sqrt{3} - \sqrt{7}\right)$

38. $\left(\sqrt{2} + \sqrt{5}\right)\left(\sqrt{2} - \sqrt{5}\right)$

39. $\left(r + \sqrt{3}\right)\left(r - \sqrt{3}\right)$

40. $\left(t + \sqrt{5}\right)\left(t - \sqrt{5}\right)$

41. $\left(5\sqrt{2} - 2\sqrt{3}\right)\left(5\sqrt{2} + 2\sqrt{3}\right)$

42. $\left(4\sqrt{3} - 3\sqrt{5}\right)\left(4\sqrt{3} + 3\sqrt{5}\right)$

43. $\left(3\sqrt{a} - 5\sqrt{b}\right)\left(3\sqrt{a} + 5\sqrt{b}\right)$

44. $\left(4\sqrt{m} - 7\sqrt{n}\right)\left(4\sqrt{m} + 7\sqrt{n}\right)$

45. $\left(4 + \sqrt{7}\right)^2$ **46.** $\left(6 + \sqrt{2}\right)^2$

47. $\left(\sqrt{3} - \sqrt{5}\right)^2$ **48.** $\left(\sqrt{2} - \sqrt{7}\right)^2$

49. $\left(b + \sqrt{2}\right)^2$ **50.** $\left(w + \sqrt{6}\right)^2$

51. $\left(\sqrt{x} - \sqrt{6}\right)^2$ **52.** $\left(\sqrt{x} - \sqrt{5}\right)^2$

53. $\left(2\sqrt{5} - 3\sqrt{2}\right)^2$ **54.** $\left(3\sqrt{3} - 4\sqrt{5}\right)^2$

Concepts

55. A student tries to simplify $\left(x + \sqrt{3}\right)^2$:

$$\left(x + \sqrt{3}\right)^2 = x^2 + \left(\sqrt{3}\right)^2 = x^2 + 3$$

Describe any errors. Then simplify the expression correctly.

56. A student tries to simplify $\left(x - \sqrt{5}\right)^2$:

$$\left(x - \sqrt{5}\right)^2 = x^2 - \left(\sqrt{5}\right)^2 = x^2 - 5$$

Describe any errors. Then simplify the expression correctly.

57. A student tries to find the product $3\left(2\sqrt{5}\right)$:

$$3\left(2\sqrt{5}\right) = 6\sqrt{15}$$

Describe any errors. Then find the product correctly.

58. A student tries to find the product $\left(4\sqrt{7}\right)\left(5\sqrt{7}\right)$:

$$\left(4\sqrt{7}\right)\left(5\sqrt{7}\right) = 20\sqrt{7}$$

Describe any errors. Then find the product correctly.

59. Why does the product of two radical conjugates contain no radicals?

60. Why can we write the product of $\sqrt{2}$ and $\sqrt{3}$ as one radical but we cannot write the sum of $\sqrt{2}$ and $\sqrt{3}$ as one radical?

61. a. Decide whether each of the following is true or false:

   i. $\sqrt{ab} = \sqrt{a}\sqrt{b}$ **ii.** $\sqrt{a + b} = \sqrt{a} + \sqrt{b}$

   iii. $(ab)^2 = a^2 b^2$ **iv.** $(a + b)^2 = a^2 + b^2$

   v. $\dfrac{1}{ab} = a^{-1} b^{-1}$ **vi.** $\dfrac{1}{a + b} = a^{-1} + b^{-1}$

b. Compare the types of equations that are true and the types of equations that are false in part (a). What patterns do you notice?

62. Describe how to simplify a radical expression. (See page 9 for guidelines on writing a good response.)

Related Review

Perform the indicated operation.

63. $3\sqrt{x} + 2\sqrt{x}$ **64.** $5\sqrt{x} - 7\sqrt{x}$

65. $\left(3\sqrt{x}\right)\left(2\sqrt{x}\right)$ **66.** $\left(5\sqrt{x}\right)\left(-7\sqrt{x}\right)$

67. $\left(3 + \sqrt{x}\right)\left(2 + \sqrt{x}\right)$ **68.** $\left(5 + \sqrt{x}\right)\left(7 - \sqrt{x}\right)$

69. $(3 + x)(2 + x)$ **70.** $(5 + x)(7 - x)$

71. $\left(3\sqrt{x}\right)^2$ **72.** $\left(-5\sqrt{x}\right)^2$

73. $\left(3 + \sqrt{x}\right)^2$ **74.** $\left(5 - \sqrt{x}\right)^2$

75. $(3 + x)^2$ **76.** $(5 - x)^2$

Expressions, Equations, and Graphs

Perform the indicated instruction. Then use words such as linear, quadratic, cubic, power, radical, rational, polynomial, degree, one variable, *and* two variables *to describe the expression, equation, or system.*

77. Graph $y = -\dfrac{2}{5}x - 1$ by hand.

78. Solve $-3x = 7x - 4(2x - 1)$.

79. Solve $\dfrac{p}{p + 4} - \dfrac{4}{p - 4} = \dfrac{p^2 + 16}{p^2 - 16}$.

80. Simplify $\left(\sqrt{a} + 2\right)^2$.

81. Find the quotient $\dfrac{x^2 - 1}{x^2 + 5x + 6} \div \dfrac{x^2 - 3x - 4}{x^2 + 3x}$.

82. Simplify $3\sqrt{75} - 5\sqrt{12}$.

11.3 Solving Square Root Equations

Objectives

» Solve a *square root equation* in one variable.

» Graph a square root equation in two variables.

» Use a *square root model* to make predictions.

In Sections 9.1, 9.2, 11.1, and 11.2, we worked with radical *expressions*—specifically, square root expressions. In this section, we will solve square root *equations*.

Solving Square Root Equations in One Variable

A square root equation is an equation that contains at least one square root expression. Here are some examples of a square root equation in one variable:

$$\sqrt{x} = 7 \qquad 2\sqrt{4x - 7} = 12 \qquad \sqrt{x - 5} = x - 8$$

Consider the square root equation

$$\sqrt{x} = 5$$

If x is negative, then $\sqrt{x}$ cannot be 5 because $\sqrt{x}$ is not a real number. If x is nonnegative, then $\left(\sqrt{x}\right)^2 = x$. To get the left side of the equation $\sqrt{x} = 5$ to be x, we square both sides:

$$\left(\sqrt{x}\right)^2 = 5^2$$
$$x = 25$$

So, the solution is 25. This checks out because $\sqrt{25} = 5$.

This work suggests using the **squaring property of equality** to help us solve square root equations.

Squaring Property of Equality

If A and B are expressions, then all solutions of the equation $A = B$ are *among* the solutions of the equation $A^2 = B^2$. That is, the solutions of an equation are among the solutions of the equation obtained by squaring both sides.

Recall from Section 10.5 that if we clear a rational equation of fractions and arrive at a value of x that is an excluded value, then we call that result an extraneous solution. In general, if a proposed solution of any type of equation is *not* a solution, we call it an **extraneous solution**.

Squaring both sides of an equation can introduce extraneous solutions. Consider the simple equation $x = 3$, whose only solution is 3. Here we square both sides of $x = 3$ and then solve by using the square root property (Section 9.3):

$$\begin{aligned} x &= 3 && \text{\textit{The only solution is 3.}} \\ x^2 &= 3^2 && \text{\textit{Square both sides.}} \\ x^2 &= 9 && 3^2 = 9 \\ x &= \pm\sqrt{9} && x^2 = k \text{ \textit{is equivalent to} } x = \pm\sqrt{k} \\ &&& \quad \text{\textit{for nonnegative constant} } k. \\ x &= \pm 3 && \sqrt{9} = 3 \end{aligned}$$

Squaring both sides of the equation $x = 3$ introduced the extraneous solution -3 (which is *not* a solution of the original equation).

Checking Proposed Solutions

Because squaring both sides of a square root equation may introduce extraneous solutions, it is essential to check that each proposed solution satisfies the original equation.

Example 1   Solving a Square Root Equation

Solve $\sqrt{6x - 2} = 4$.

Solution

First, we square both sides of the equation:

$$\begin{aligned} \sqrt{6x - 2} &= 4 && \text{\textit{Original equation}} \\ \left(\sqrt{6x - 2}\right)^2 &= 4^2 && \text{\textit{Square both sides.}} \\ 6x - 2 &= 16 && \left(\sqrt{x}\right)^2 = x \text{ \textit{for} } x \geq 0 \\ 6x &= 18 && \text{\textit{Add 2 to both sides.}} \\ x &= 3 && \text{\textit{Divide both sides by 6.}} \end{aligned}$$

We check that 3 satisfies the original equation:

$$\sqrt{6x - 2} = 4 \qquad \textit{Original equation}$$
$$\sqrt{6(3) - 2} \stackrel{?}{=} 4 \qquad \textit{Substitute 3 for x.}$$
$$\sqrt{16} \stackrel{?}{=} 4 \qquad \textit{Simplify.}$$
$$4 \stackrel{?}{=} 4 \qquad \sqrt{16} = 4$$
$$\text{true}$$

So, the solution is 3.

▶

Example 2   Solving a Square Root Equation

Solve $\sqrt{x + 5} = -3$.

Solution

First, we square both sides of the equation:

$$\sqrt{x + 5} = -3 \qquad \textit{Original equation}$$
$$\left(\sqrt{x + 5}\right)^2 = (-3)^2 \qquad \textit{Square both sides.}$$
$$x + 5 = 9 \qquad \textit{$(\sqrt{x})^2 = x$ for $x \geq 0$}$$
$$x = 4 \qquad \textit{Subtract 5 from both sides.}$$

We check that 4 satisfies the original equation:

$$\sqrt{x + 5} = -3 \qquad \textit{Original equation}$$
$$\sqrt{4 + 5} \stackrel{?}{=} -3 \qquad \textit{Substitute 4 for x.}$$
$$\sqrt{9} \stackrel{?}{=} -3 \qquad \textit{Add.}$$
$$3 \stackrel{?}{=} -3 \qquad \sqrt{9} = 3$$
$$\text{false}$$

The result, 4, is an extraneous solution—it is *not* a solution. The solution set is the empty set.

▶

In Example 2, we introduced an extraneous solution by squaring both sides of an equation. Remember that if you square both sides of an equation, you must check that each proposed solution satisfies the original equation.

Example 3   Solving a Square Root Equation

Solve $\sqrt{6x - 5} = \sqrt{4x + 2}$.

Solution

First, we square both sides of the equation:

$$\sqrt{6x - 5} = \sqrt{4x + 2} \qquad \textit{Original equation}$$
$$\left(\sqrt{6x - 5}\right)^2 = \left(\sqrt{4x + 2}\right)^2 \qquad \textit{Square both sides.}$$
$$6x - 5 = 4x + 2 \qquad \textit{$(\sqrt{x})^2 = x$ for $x \geq 0$}$$
$$2x = 7 \qquad \textit{Subtract 4x from both sides; add 5 to both sides.}$$
$$x = \frac{7}{2} \qquad \textit{Divide both sides by 2.}$$

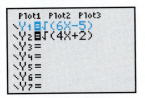

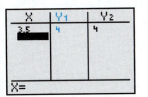

Figure 4 Verify the work

We check that $\dfrac{7}{2}$ satisfies the original equation:

$$\sqrt{6x - 5} = \sqrt{4x + 2} \qquad \textit{Original equation}$$

$$\sqrt{6\left(\dfrac{7}{2}\right) - 5} \stackrel{?}{=} \sqrt{4\left(\dfrac{7}{2}\right) + 2} \qquad \textit{Substitute } \dfrac{7}{2} \textit{ for x.}$$

$$\sqrt{21 - 5} \stackrel{?}{=} \sqrt{14 + 2} \qquad \textit{Simplify.}$$

$$\sqrt{16} \stackrel{?}{=} \sqrt{16} \qquad \textit{Simplify.}$$

$$\text{true}$$

The solution is $\dfrac{7}{2}$. We also use a graphing calculator table to verify our work (see Fig. 4).

Example 4   Isolating a Square Root to One Side of the Equation

Solve $2 + 3\sqrt{y} = 23$.

Solution

We isolate $\sqrt{y}$ to one side of the equation and then square both sides:

$$2 + 3\sqrt{y} = 23 \qquad \textit{Original equation}$$

$$3\sqrt{y} = 21 \qquad \textit{Subtract 2 from both sides.}$$

$$\sqrt{y} = 7 \qquad \textit{Divide both sides by 3.}$$

$$\left(\sqrt{y}\right)^2 = 7^2 \qquad \textit{Square both sides.}$$

$$y = 49 \qquad \left(\sqrt{x}\right)^2 = x \textit{ for } x \geq 0$$

We check that 49 satisfies the original equation:

$$2 + 3\sqrt{y} = 23 \qquad \textit{Original equation}$$

$$2 + 3\sqrt{49} \stackrel{?}{=} 23 \qquad \textit{Substitute 49 for y.}$$

$$2 + 3(7) \stackrel{?}{=} 23 \qquad \sqrt{49} = 7$$

$$23 \stackrel{?}{=} 23 \qquad \textit{Simplify.}$$

$$\text{true}$$

The solution is 49.

We see from Example 4 that to solve a square root equation, we isolate a square root term to one side of the equation before squaring both sides.

Solving a Square Root Equation in One Variable

We solve a square root equation in one variable by following these steps:

1. Isolate a square root term to one side of the equation.
2. Square both sides.
3. Solve the new equation.
4. Check that each proposed solution satisfies the original equation.

Example 5   Isolating a Square Root to One Side of the Equation

Solve $\sqrt{3x - 5} + 4 = x + 1$.

Solution

We isolate $\sqrt{3x - 5}$ to one side of the equation and then square both sides:

$$\sqrt{3x - 5} + 4 = x + 1 \qquad \textit{Original equation}$$
$$\sqrt{3x - 5} = x - 3 \qquad \textit{Subtract 4 from both sides.}$$
$$\left(\sqrt{3x - 5}\right)^2 = (x - 3)^2 \qquad \textit{Square both sides.}$$
$$3x - 5 = x^2 - 6x + 9 \qquad \left(\sqrt{x}\right)^2 = x \text{ for } x \geq 0; (A - B)^2 = A^2 - 2AB + B^2$$
$$0 = x^2 - 9x + 14 \qquad \textit{Write in } 0 = ax^2 + bx + c \textit{ form.}$$
$$0 = (x - 2)(x - 7) \qquad \textit{Factor right-hand side.}$$
$$x - 2 = 0 \quad \text{or} \quad x - 7 = 0 \qquad \textit{Zero factor property}$$
$$x = 2 \quad \text{or} \quad x = 7$$

We check that both 2 and 7 satisfy the original equation:

Check $x = 2$	Check $x = 7$
$\sqrt{3x - 5} + 4 = x + 1$	$\sqrt{3x - 5} + 4 = x + 1$
$\sqrt{3(2) - 5} + 4 \overset{?}{=} 2 + 1$	$\sqrt{3(7) - 5} + 4 \overset{?}{=} 7 + 1$
$\sqrt{1} + 4 \overset{?}{=} 3$	$\sqrt{16} + 4 \overset{?}{=} 8$
$5 \overset{?}{=} 3$	$8 \overset{?}{=} 8$
false	true

So, the only solution is 7.

In Example 5, we simplified the right-hand side of the equation

$$\left(\sqrt{3x - 5}\right)^2 = (x - 3)^2$$

by using the property for the square of a difference, namely, $(A - B)^2 = A^2 - 2AB + B^2$:

$$(x - 3)^2 = x^2 - 6x + 9 \quad \textit{Correct}$$

WARNING   Remember that, in general, $(A - B)^2$ is *not* equal to $A^2 - B^2$, so it is incorrect to say

$$(x - 3)^2 = x^2 - 9 \quad \textit{Incorrect}$$

If an equation contains two square root terms, we may need to use the squaring property of equality twice.

Example 6   Using the Squaring Property of Equality Twice

Solve $\sqrt{x + 21} = \sqrt{x} + 3$.

Solution

$$\sqrt{x + 21} = \sqrt{x} + 3 \qquad \textit{Original equation}$$
$$\left(\sqrt{x + 21}\right)^2 = \left(\sqrt{x} + 3\right)^2 \qquad \textit{Square both sides.}$$
$$x + 21 = \left(\sqrt{x}\right)^2 + 2\left(\sqrt{x}\right)3 + 3^2 \qquad \left(\sqrt{x}\right)^2 = x \text{ for } x \geq 0; (A + B)^2 = A^2 + 2AB + B^2$$
$$x + 21 = x + 6\sqrt{x} + 9 \qquad \left(\sqrt{x}\right)^2 = x \text{ for } x \geq 0; \textit{simplify.}$$
$$12 = 6\sqrt{x} \qquad \textit{Subtract } x \textit{ from both sides; subtract 9 from both sides.}$$
$$2 = \sqrt{x} \qquad \textit{Divide both sides by 6.}$$
$$2^2 = \left(\sqrt{x}\right)^2 \qquad \textit{Square both sides.}$$
$$4 = x \qquad \textit{Simplify.}$$

We check that 4 satisfies the original equation:

$$\sqrt{x + 21} = \sqrt{x} + 3 \quad \textit{Original equation}$$
$$\sqrt{4 + 21} \overset{?}{=} \sqrt{4} + 3 \quad \textit{Substitute 4 for } x.$$
$$5 \overset{?}{=} 5 \quad \textit{Simplify.}$$
$$\text{true}$$

The solution is 4.

Graphs of Square Root Equations in Two Variables

So far in this section, we have solved square root equations in *one* variable. We can describe the solutions of square root equations in *two* variables with a graph. Here are some examples of square root equations in two variables:

$$y = \sqrt{x} \qquad y = \sqrt{x + 5} \qquad y = 2\sqrt{x} + 3$$

Example 7   Graphing a Square Root Equation

Sketch the graph of $y = \sqrt{x}$.

Solution

We list some solutions in Table 4. We choose perfect squares as values of x because we can find their principal square roots mentally. Because the radicand of $\sqrt{x}$ must be nonnegative, we cannot choose any negative numbers as values of x. Then we plot the points that correspond to the solutions we found and sketch a curve that contains these points (see Fig. 5).

Table 4 Solutions of $y = \sqrt{x}$

x	y
0	$\sqrt{0} = 0$
1	$\sqrt{1} = 1$
4	$\sqrt{4} = 2$
9	$\sqrt{9} = 3$
16	$\sqrt{16} = 4$

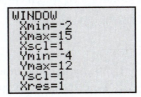

Figure 5 Graph of $y = \sqrt{x}$

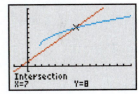

Figure 6 Graphs of two square root equations in two variables

In Fig. 6, we show the graphs of $y = \sqrt{x + 5}$ and $y = 2\sqrt{x} + 3$.

In Example 5, we found that 7 is the solution of the equation $\sqrt{3x - 5} + 4 = x + 1$. We can verify our work by using "intersect" on a graphing calculator (see Fig. 7).

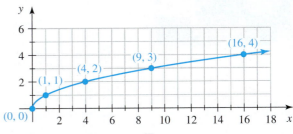

Figure 7 Verify the work

Example 8   Solving a Radical Equation in One Variable by Graphing

The graphs of $y = \sqrt{x + 2} + 1$ and $y = 3$ are shown in Fig. 8. Use these graphs to solve the equation $\sqrt{x + 2} + 1 = 3$.

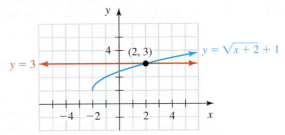

Figure 8 Solving a radical equation in one variable

Solution

The graphs of $y = \sqrt{x + 2} + 1$ and $y = 3$ intersect only at the point $(2, 3)$, which has x-coordinate 2. So, 2 is the solution of $\sqrt{x + 2} + 1 = 3$.

Square Root Models

Just as we have used linear equations, quadratic equations, and rational equations to model authentic situations, we can use square root equations in two variables to model such situations. A **square root model** is a square root equation in two variables that is used to describe an authentic situation.

| **Example 9** | Using a Square Root Model |

The percentages of American adults who watch cable television are shown in Table 5 for various income groups.

Table 5 Percentages of American Adults Who Watch Cable Television

| | Thousands of Dollars | |
Annual Income Group	Income Used to Represent Income Group	Percent
0–9.999	5.0	55.9
10–19.999	15.0	62.3
20–29.999	25.0	67.8
30–34.999	32.5	72.2
35–39.999	37.5	74.2
40–49.999	45.0	77.2
50+	70.0	84.8

Source: *Mediamark Research, Inc.*

Let p be the percentage of American adults with an annual income of I thousand dollars who watch cable television. A square root model of the situation is

$$p = 4.7\sqrt{I} + 45$$

1. Verify that the model fits the data well.
2. Find the p-intercept. What does it mean in this situation?
3. Estimate the percentage of adults with an income of $18 thousand who watch cable television.
4. Estimate the income at which 73% of adults watch cable television.

Solution

1. We draw the graph of the model and, in the same viewing window, the scatterplot of the data (see Fig. 9). It appears that the model fits the data well.
2. To find the p-intercept, we substitute 0 for I in the equation $p = 4.7\sqrt{I} + 45$ and solve for p:

$$p = 4.7\sqrt{0} + 45 = 45$$

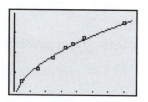

Figure 9 Check how well the model fits the data

The p-intercept is $(0, 45)$. According to the model, 45% of adults who have no income watch cable television. Model breakdown has likely occurred because it is hard to imagine that as many as 45% of American adults with no income could afford to subscribe to cable television.

3. We substitute 18 for I in the equation $p = 4.7\sqrt{I} + 45$ and solve for p:

$$p = 4.7\sqrt{18} + 45 \approx 64.94$$

So, about 64.9% of adults with an income of $18 thousand watch cable television.

4. We substitute 73 for p in the equation $p = 4.7\sqrt{I} + 45$ and solve for I:

$73 = 4.7\sqrt{I} + 45$	*Substitute 73 for p.*
$28 = 4.7\sqrt{I}$	*Subtract 45 from both sides.*
$\dfrac{28}{4.7} = \sqrt{I}$	*Divide both sides by 4.7.*
$\left(\dfrac{28}{4.7}\right)^2 = \left(\sqrt{I}\right)^2$	*Square both sides.*
$35.49 \approx I$	*Calculate; $\left(\sqrt{x}\right)^2 = x$ for $x \geq 0$*

Our result 35.49 does approximately satisfy $73 = 4.7\sqrt{I} + 45$. (Try it.) So, 73% of Americans whose income is about $35.5 thousand watch cable television, according to the model.

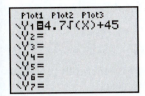

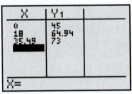

Figure 10 Verify the work

We use a graphing calculator table to verify our work in Problems 2–4 (see Fig. 10).

▶

◈ Group Exploration

Solving square root equations

1. A student tries to solve $\sqrt{x} + 3 = 5$:

$$\sqrt{x} + 3 = 5$$
$$\left(\sqrt{x}\right)^2 + 3^2 = 5^2$$
$$x + 9 = 25$$
$$x = 16$$

Describe any errors. Then solve the equation correctly.

2. A student tries to solve $\sqrt{2x - 5} = -3$:

$$\sqrt{2x - 5} = -3$$
$$\left(\sqrt{2x - 5}\right)^2 = (-3)^2$$
$$2x - 5 = 9$$
$$2x = 14$$
$$x = 7$$

Describe any errors. Then solve the equation correctly.

3. A student tries to solve $3\sqrt{x} - 2 = 5$:

$$3\sqrt{x} - 2 = 5$$
$$3\sqrt{x} = 7$$
$$\left(3\sqrt{x}\right)^2 = 7^2$$
$$3x = 49$$
$$x = \frac{49}{3}$$

Describe any errors. Then solve the equation correctly.

> **Tips for Success** **Retake Quizzes and Exams to Prepare for the Final Exam**
>
> To study for your final exam, consider retaking your quizzes and other exams. These quizzes and exams can reveal your weak areas. If you have difficulty with a certain concept, you can refer to Homework exercises that address that concept. Reflect on *why* you are having such difficulty, rather than just doing more Homework exercises that address the concept.

Homework 11.3

For extra help ▶   **MyLab Math**   📹 Watch the videos in MyLab Math

Solve.

1. $\sqrt{x} = 6$

2. $\sqrt{x} = 2$

3. $\sqrt{5x + 1} = 4$

4. $\sqrt{2x - 6} = 2$

5. $\sqrt{x} = -7$

6. $\sqrt{x} = -3$

7. $\sqrt{3t - 2} = 2$

8. $\sqrt{4w - 3} = 5$

9. $\sqrt{x + 4} = -2$

10. $\sqrt{x - 8} = -5$

11. $\sqrt{5x - 2} = \sqrt{3x + 8}$

12. $\sqrt{7x - 1} = \sqrt{4x + 17}$

13. $\sqrt{25p^2 + 4p - 8} = 5p$

14. $\sqrt{9r^2 - 7r + 21} = 3r$

15. $\sqrt{x^2 - 6x} = 4$

16. $\sqrt{x^2 + 5x} = 6$

17. $\sqrt{x} + 3 = 7$

18. $\sqrt{x} - 5 = 1$

19. $3\sqrt{x} = 6$

20. $4\sqrt{x} = 20$

21. $5\sqrt{w} + 3 = 23$

22. $4\sqrt{m} - 1 = 11$

23. $-3\sqrt{x} + 7 = 2$

24. $-2\sqrt{x} + 10 = 3$

25. $4\sqrt{x} - 3 = -5$

26. $5\sqrt{x} - 2 = -12$

27. $5 + \sqrt{a - 3} = 8$

28. $7 + \sqrt{b + 1} = 10$

29. $\sqrt{x - 2} = x - 2$

30. $\sqrt{x + 4} = x - 2$

31. $\sqrt{x + 3} = x + 1$

32. $\sqrt{x - 5} = x - 7$

33. $\sqrt{4r - 3} - r = -2$

34. $\sqrt{5t - 1} - t = 1$

35. $\sqrt{(x + 5)(x + 2)} = x + 3$

36. $\sqrt{(x - 1)(x + 4)} = x + 1$

37. $\sqrt{x + 5} = \sqrt{x} + 1$

38. $\sqrt{x - 4} = \sqrt{x} - 2$

39. $\sqrt{m - 5} + 2 = \sqrt{m + 3}$

40. $\sqrt{p + 2} - 3 = \sqrt{p - 7}$

Graph the equation by hand. Then use a graphing calculator to verify your work.

41. $y = 1 + \sqrt{x}$

42. $y = 2\sqrt{x}$

For Exercises 43–46, use the graph of $y = \sqrt{x + 4} - 1$ shown in Fig. 11 to solve the given equation.

43. $\sqrt{x + 4} - 1 = -1$

44. $\sqrt{x + 4} - 1 = 0$

45. $\sqrt{x + 4} - 1 = -2$

46. $\sqrt{x + 4} - 1 = 1$

Use "intersect" on a graphing calculator to solve the equation. Round the solution(s) to the second decimal place.

47. $\sqrt{x} - 4 = 3 - x$

48. $\sqrt{x} + 2 = 2x - 5$

49. $\sqrt{x} + 3 = x^2 - 5$

50. $\sqrt{x} - 4 = 3 - x^2$

51. The amounts of land worldwide used for genetically modified crops are shown in Table 6 for various years.

Table 6 Genetically Modified Crops

Year	Genetically Modified Crops (millions of acres)
2010	365.0
2011	395.0
2012	420.8
2013	433.2
2014	448.0
2015	444.0
2016	457.4

Sources: *Clive James; Department of Agriculture*

Let c be the amount (in millions of acres) of land used for genetically modified crops at t years since 2010. A model of the situation is $c = 37.9\sqrt{t} + 365$.

a. Use a graphing calculator to draw the graph of the model and, in the same viewing window, the scatterplot of the data. Does the model fit the data well?

b. What is the c-intercept? What does it mean in this situation?

c. Predict the number of acres of genetically modified crops in 2022.

d. Predict when there will be 500 million acres of genetically modified crops.

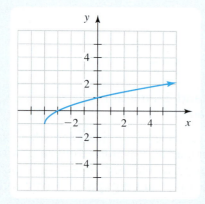

Figure 11 Exercises 43–46

52. The numbers of 3D movie screens in the United States are shown in Table 7 for various years.

Table 7 Numbers of 3D Movie Screens

Year	Number of Screens (thousands)
2010	14.7
2011	25.6
2012	33.1
2013	36.8
2014	38.4

Source: *IHS*

Let n be the number (in thousands) of 3D movie screens at t years since 2010. A square root model of the situation is $n = 12.3\sqrt{t} + 14.7$.
a. Use a graphing calculator to draw the graph of the model and, in the same viewing window, the scatterplot of the data. Does the model fit the data well?
b. What is the n-intercept? What does it mean in this situation?
c. Predict how many 3D movie screens there will be in 2023.
d. Predict when there will be 56 thousand 3D movie screens.

53. The numbers of McDonald's restaurants around the world are shown in Table 8 for various years.

Table 8 Numbers of McDonald's Restaurants

Year	Number of Restaurants (thousands)
2012	34.5
2013	35.4
2014	36.3
2015	36.5
2016	36.9

Source: *McDonald's*

Let n be the number (in thousands) of McDonald's restaurants at t years since 2012. A model of the situation is $n = 1.2\sqrt{t} + 34.5$.
a. Use a graphing calculator to draw the graph of the model and, in the same viewing window, the scatterplot of the data. Does the model fit the data well?
b. Predict the number of McDonald's restaurants in 2022.
c. There are 192 countries in the world. What will be the average number of McDonald's restaurants per country in 2023?
d. Predict when there will be 38 thousand McDonald's restaurants.

54. The percentages of convicts released from a state prison who have subsequently been arrested for a new crime are shown in Table 9 for various numbers of years since their release.

Table 9 Released Convicts Who Have Been Arrested for a New Crime

Number of Years since Release	Percent
0	0
0.5	30
1	44
2	59
3	68

Source: *U.S. Department of Justice*

Let p be the percentage of convicts released from a state prison who have been arrested for a new crime after being out of prison for t years. A model of the situation is $p = 41\sqrt{t}$.
a. Use a graphing calculator to draw the graph of the model and, in the same viewing window, the scatterplot of the data. Does the model fit the data well?
b. Estimate the percentage of released convicts who have been arrested for a new crime after being out of prison for 5 years.
c. Estimate the number of years since their release when all convicts will have been arrested for a new crime.

55. Tire dumps are unsightly and sometimes catch fire, releasing hazardous chemicals. In 2003 alone, Americans threw out 300 million tires, about one per person in the United States. Table 10 shows the percentages of tires that were reused or recycled for various years.

Table 10 Reused or Recycled Tires

Year	Percent
1990	11
1994	55
1998	67
2001	78
2003	83
2007	89
2008	90

Source: *Rubber Manufacturers Association*

Let p be the percentage of tires that have been reused or recycled at t years since 1990. A model of the situation is $p = 19.5\sqrt{t} + 11$.
a. Use a graphing calculator to draw the graph of the model and, in the same viewing window, the scatterplot of the data. Does the model fit the data well?
b. Estimate when all tires were first reused or recycled.
c. In 2006, there were two U.S. plants that burned tires to create energy. A third plant had been proposed that would have burned 10 million tires per year and produced enough energy for up to 20 thousand homes. However, there were concerns about the release of harmful chemicals from the

plant. Should the plant have been built? To decide, make the following assumptions:

- Construction would have begun in 2006, and it would have taken two years for the plant to be built.
- It is better to reuse or to recycle a tire than to burn it for energy.
- If a tire can't be reused or recycled, it is better to burn it for energy than to dump it in a landfill or tire pile.

[**Hint:** Refer to the result you found in part (b).]

56. The numbers of shopping malls are shown in Table 11 for various years.

Table 11 Numbers of Shopping Malls

Year	Number of Malls
1977	576
1987	874
1997	1043
2007	1165
2017	1211

Source: *CoStar Group*

Let n be the number of malls at t years since 1977. A model of the situation is $n = 103\sqrt{t} + 576$.

a. Use a graphing calculator to draw the graph of the model and, in the same viewing window, the scatterplot of the data. Does the model fit the data well?

b. Predict the number of malls in 2020.

c. Predict the average number of malls per state in 2022.

d. Predict when there will be 1275 malls.

Large Data Sets

57. **DATA.** Access the data about GDPs and happiness ratings, which are available at MyLab Math and at the Pearson Downloadable Student Resources for Math & Stats website. For the World Happiness Report 2016, people from 162 countries rated their happiness on a scale from 0 to 10. Let H be the average happiness rating for residents of a country with per-person GDP G international dollars. A model of the situation is $H = 0.024\sqrt{G} + 2.26$.

a. Use a graphing calculator to draw the graph of the model and, in the same viewing window, the scatterplot of the data. Describe how well the model fits the data.

b. Residents of Barbados were not surveyed. Estimate what the average happiness rating would be for Barbados, which has a per-person GDP of 4663 international dollars.

c. Estimate the average happiness rating for the United States, which has a per-person GDP of 52,118 international dollars. Is your result an underestimate or an overestimate?

d. As the inputs of the model increase, do the outputs increase or decrease? What does that mean in this situation?

e. Experts divide happiness into two components, *evaluative* and *affective*. Evaluative happiness is a sense that your life is good. Affective happiness means that a person experiences positive emotions more often than negative emotions. Research has shown that affective happiness increases as income increases but only up to a certain income. Affective happiness is unaffected by income increases above that amount (Source: "*Can Money Buy*

Happiness?" Wall Street Journal, *11/10/14*). On the basis of this research and the result you found in part (d), which component of happiness do you think $f(G)$ measures?

58. **DATA.** Access the data about health spending and life expectancy, which are available at MyLab Math and at the Pearson Downloadable Student Resources for Math & Stats website. Let L be the life expectancy at birth (in years) for residents of a country in which the 2014 average per-person spending on health care is S international dollars. A model of the situation is $L = 0.4\sqrt{S} + 60$.

a. Use a graphing calculator to draw the graph of the model and, in the same viewing window, the scatterplot of the data. Does the model fit the data well?

b. As the inputs of the model increase, do the outputs increase or decrease? What does that mean in this situation?

c. Estimate the life expectancy for residents of a country in which the average per-person spending on health care is $4000.

d. Estimate the average per-person spending on health care of a country in which the life expectancy at birth is 65 years.

e. The U.S. average per-person spending on health care is $9403. Use the model to estimate the life expectancy of Americans. Is your result an underestimate or an overestimate? List some possible reasons why the data point for the U.S. data does not fit the overall pattern of the other data points in the scatterplot.

Concepts

59. A student tries to solve $\sqrt{x^2 + x + 6} = x + 4$:

$$\sqrt{x^2 + x + 6} = x + 4$$
$$\left(\sqrt{x^2 + x + 6}\right)^2 = (x + 4)^2$$
$$x^2 + x + 6 = x^2 + 16$$
$$x = 10$$

Describe any errors. Then solve the equation correctly.

60. A student tries to solve $\sqrt{x - 1} = x - 3$:

$$\sqrt{x - 1} = x - 3$$
$$\left(\sqrt{x - 1}\right)^2 = (x - 3)^2$$
$$x - 1 = x^2 - 6x + 9$$
$$0 = x^2 - 7x + 10$$
$$0 = (x - 2)(x - 5)$$
$$x - 2 = 0 \quad \text{or} \quad x - 5 = 0$$
$$x = 2 \quad \text{or} \quad x = 5$$

The student says that the solutions are 2 and 5. Is the student correct? Explain.

61. Solve the system; then use "intersect" on a graphing calculator to verify your work:

$$y = 3\sqrt{x} - 4$$
$$y = -2\sqrt{x} + 6$$

62. For the equation $\sqrt{x} + 3 = 7$, why do we first subtract 3 from both sides rather than first square both sides?

63. Find nonzero values of m and b so the equation $\sqrt{x} = mx + b$ has no real-number solutions. [**Hint:** Think about the graphs of $y = \sqrt{x}$ and $y = mx + b$.]

64. Describe how to solve a square root equation. Is checking your answers for extraneous solutions optional?

Related Review

Solve.

65. $2x^2 + 4x = 7$

66. $2(3x - 5) - 4(2x + 1) = 8$

67. $\sqrt{x + 2} = x - 4$

68. $x^2 - 5x = 24$

69. $\dfrac{3}{x - 2} - \dfrac{5x}{x^2 - 4} = \dfrac{7}{x + 2}$ **70.** $(x + 3)^2 = 40$

Expressions, Equations, and Graphs

Give an example of each. Then solve, simplify, or graph, as appropriate.

71. Sum of two rational expressions

72. Radical equation in one variable

73. Linear equation in two variables

74. System of two linear equations in two variables

75. Quadratic equation in two variables

76. Linear equation in one variable

77. Rational equation in one variable

78. Quadratic polynomial with four terms

79. Radical expression with four terms

80. Quadratic equation in one variable

Chapter Summary

Key Points of Chapter 11

Section 11.1   Adding and Subtracting Radical Expressions

Like radicals	Square root radicals that have the same radicand are called **like radicals**.
Adding or subtracting like radicals	To add or subtract like radicals, we use the distributive law.

Section 11.2   Multiplying Radical Expressions

Power property for square roots	If x is nonnegative, then $(\sqrt{x})^2 = x$.
Product of two radical expressions, each with two terms	To find the product of two radical expressions in which each factor has two terms, **1.** Multiply each term in the first radical expression by each term in the second radical expression. **2.** Combine like radicals.
Radical conjugates	The sum of two radical terms and the difference of the same two radical terms are **radical conjugates** of each other.
Product of two radical conjugates	To find the product of two radical conjugates, use the property $(A + B)(A - B) = A^2 - B^2$.
Simplifying the square of a radical expression	To simplify the square of a radical expression that has two terms, • Use $C^2 = CC$ to write the square as a product of two identical expressions, and then multiply each term in the first radical expression by each term in the second radical expression, or • Use the square-of-a-sum property $(A + B)^2 = A^2 + 2AB + B^2$ or the square-of-a-difference property $(A - B)^2 = A^2 - 2AB + B^2$.

Section 11.3   Solving Square Root Equations

Square root equation	A **square root equation** is an equation that contains at least one square root expression.
Squaring property of equality	If A and B are expressions, then all solutions of the equation $A = B$ are *among* the solutions of the equation $A^2 = B^2$. That is, the solutions of an equation are among the solutions of the equation obtained by squaring both sides.
Checking proposed solutions	Because squaring both sides of a square root equation may introduce extraneous solutions, it is essential to check that each proposed solution satisfies the original equation.
Solving a square root equation in one variable	We solve a square root equation in one variable by following these steps: **1.** Isolate a square root term to one side of the equation. **2.** Square both sides. **3.** Solve the new equation. **4.** Check that each proposed solution satisfies the original equation.
Square root model	A **square root model** is a square root equation in two variables that is used to describe an authentic situation.

Chapter 11 Review Exercises

Perform the indicated operation(s). Assume that any variables are nonnegative. Use a graphing calculator to verify your work when possible.

1. $6\sqrt{3} + 9\sqrt{3}$

2. $-4\sqrt{2x} + 3\sqrt{2x}$

3. $7\sqrt{5x} - 9\sqrt{5x}$

4. $8\sqrt{3} + 4\sqrt{6}$

5. $2\sqrt{5} - 3\sqrt{7} - 8\sqrt{5}$

6. $\sqrt{27} + 8\sqrt{3}$

7. $-3\sqrt{6} - \sqrt{24}$

8. $\sqrt{32} + \sqrt{18}$

9. $\sqrt{45} - \sqrt{20}$

10. $5\sqrt{8} + 3\sqrt{50}$

11. $2\sqrt{40} - 5\sqrt{90}$

12. $\sqrt{32} - 5\sqrt{3} - \sqrt{72}$

13. $7\sqrt{5} + 4\sqrt{20} - 8\sqrt{24}$

14. $\sqrt{64x} + \sqrt{81x}$

15. $2\sqrt{25x} - 3\sqrt{49x}$

16. $\sqrt{12x} + \sqrt{8} - \sqrt{75x}$

17. $w\sqrt{28} - w\sqrt{63}$

18. $\sqrt{7a^2} + \sqrt{28a^2}$

19. $5\sqrt{3x^2} - 3x\sqrt{48}$

20. $2\sqrt{8x} - 4\sqrt{45x^2} + x\sqrt{20}$

21. $\sqrt{7a}\sqrt{5b}$

22. $4\sqrt{3}\left(-2\sqrt{6}\right)$

23. $\left(5\sqrt{x}\right)^2$

24. $\sqrt{2}\left(\sqrt{3} + \sqrt{7}\right)$

25. $-\sqrt{3}\left(\sqrt{7} - \sqrt{5}\right)$

26. $\sqrt{x}\left(\sqrt{x} - \sqrt{2}\right)$

27. $2\sqrt{7}\left(5\sqrt{3} + \sqrt{7}\right)$

28. $\left(\sqrt{5} - 3\right)\left(\sqrt{5} + 6\right)$

29. $\left(\sqrt{2} - 9\right)\left(\sqrt{2} - 8\right)$

30. $\left(\sqrt{2} - \sqrt{7}\right)\left(\sqrt{3} - \sqrt{5}\right)$

31. $\left(\sqrt{b} + 8\right)\left(\sqrt{b} - 1\right)$

32. $\left(t + \sqrt{3}\right)\left(t + \sqrt{5}\right)$

33. $\left(3\sqrt{7} - 1\right)\left(2\sqrt{7} - 2\right)$

34. $\left(\sqrt{5} - \sqrt{7}\right)\left(\sqrt{5} + \sqrt{7}\right)$

35. $\left(4\sqrt{a} + 3\sqrt{b}\right)\left(4\sqrt{a} - 3\sqrt{b}\right)$

36. $\left(b - \sqrt{3}\right)\left(b + \sqrt{3}\right)$

37. $\left(4 - \sqrt{5}\right)^2$

38. $\left(\sqrt{x} + 6\right)^2$

39. $\left(x + \sqrt{3}\right)^2$

40. $\left(3\sqrt{5} - 4\sqrt{2}\right)^2$

For Exercises 41–54, solve.

41. $\sqrt{9x^2 - 5x + 20} = 3x$

42. $\sqrt{2x - 8} = 6$

43. $\sqrt{3r + 8} = -7$

44. $\sqrt{4p + 3} = \sqrt{6p - 2}$

45. $\sqrt{x^2 - 3x} = 2$

46. $\sqrt{x} + 2 = 9$

47. $2\sqrt{x} - 5 = 11$

48. $-3\sqrt{x} + 4 = 16$

49. $3 + \sqrt{y - 2} = 5$

50. $\sqrt{p + 5} = p + 3$

51. $\sqrt{(x + 3)(x - 4)} = x - 2$

52. $1 + \sqrt{2x - 3} = x$

53. $\sqrt{5w + 1} - w = -1$

54. $\sqrt{t + 5} = \sqrt{t} + 1$

55. Use "intersect" on a graphing calculator to solve the equation $-2\sqrt{x} + 3 = \dfrac{1}{2}x^2 - 6$. Round the solution(s) to the second decimal place.

56. The percentages of adults who got a flu shot are shown in Table 12 for various years.

Table 12 Percentages of Adults Who Got a Flu Shot

Year	Percent
2005	27
2007	35
2009	40
2011	44
2013	44
2015	47

Source: *Harris Interactive*

Let p be the percentage of adults who got a flu shot in the year that is t years since 2005. A model of the situation is $p = 6.3\sqrt{t} + 27$.

a. Use a graphing calculator to draw the graph of the model and, in the same viewing window, the scatterplot of the data. Does the model fit the data well?

b. Data are not available for 2006. Estimate the percentage of adults who got flu shots in 2006.

c. Predict when 53% of adults will get a flu shot.

Chapter 11 Test

Perform the indicated operation(s). Assume that any variables are nonnegative.

1. $5\sqrt{3} - 3\sqrt{2} - 7\sqrt{3}$

2. $-3\sqrt{20x} - 2\sqrt{45x}$

3. $5\sqrt{45} - 5\sqrt{18x} - 2\sqrt{32x}$

4. $3\sqrt{24b^2} - 7\sqrt{6b^2}$

5. $-8\sqrt{14} \cdot 5\sqrt{2}$

6. $\sqrt{x}\left(\sqrt{x} - 3\right)$

7. $\left(\sqrt{5} - 2\right)\left(\sqrt{5} + 4\right)$

8. $\left(\sqrt{2} - \sqrt{5}\right)\left(\sqrt{2} + \sqrt{7}\right)$

9. $\left(5\sqrt{a} - 2\sqrt{b}\right)\left(5\sqrt{a} + 2\sqrt{b}\right)$

10. $\left(4 + \sqrt{3}\right)^2$

11. $\left(\sqrt{x} - \sqrt{5}\right)^2$

12. $\left(3\sqrt{2} + 2\sqrt{3}\right)^2$

Solve.

13. $\sqrt{3x - 5} = 4$

14. $\sqrt{2x + 7} = \sqrt{5x - 8}$

15. $4\sqrt{t} - 3 = 5$

16. $3 + \sqrt{w + 5} = 9$

17. $\sqrt{x - 3} = x - 5$

18. $\sqrt{x + 8} = \sqrt{x} + 2$

For Exercises 19–22, use the graph of $y = \sqrt{x + 5} - 3$ shown in Fig. 12 to solve the given equation.

19. $\sqrt{x + 5} - 3 = 0$

20. $\sqrt{x + 5} - 3 = -1$

21. $\sqrt{x + 5} - 3 = -3$

22. $\sqrt{x + 5} - 3 = -4$

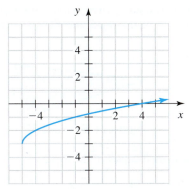

Figure 12 Exercises 19–22

Let p be the percentage of viewers who watch their programs within d days. A model of the situation is $p = 21\sqrt{d} + 46$.

a. Use a graphing calculator to draw the graph of the model and, in the same viewing window, the scatterplot of the data. Does the model fit the data well?

b. Find the p-intercept. What does it mean in this situation?

c. Estimate the percentage of viewers who watch their programs within 6 days.

d. After how many days have 88% of viewers watched their programs?

23. Because of DVRs and other technologies, many people are postponing viewing their television programs (see Table 13).

Table 13 Percentages of Viewers Who Watch Their Programs within the Specified Numbers of Days

Number of Days	Percent
0	46
1	62
2	76
3	84
5	95
7	100

Source: *Nielsen Media Research*

Cumulative Review of Chapters 1–11

For Exercises 1–16, perform the indicated operation(s) and simplify the result. State whether the expression is a linear polynomial, a quadratic polynomial, a cubic polynomial, a fourth-degree polynomial, a rational expression, or a radical expression.

1. $-3ab^2(5a - 2b)$

2. $7 - 2(4p + 3) - 5p$

3. $(3x^2 - 5x + 1) + (6x^2 - 2x - 4)$

4. $(4x^3 - x^2 + x) - (5x^3 - 2x + 1)$

5. $(2m - 5n)(6m + 7n)$

6. $(3x + 2)(x^2 - 4x + 3)$

7. $(2p + 3q)^2$

8. $\dfrac{5m - 10}{m^3 - 2m^2 - 15m} \cdot \dfrac{3m^2 + 9m}{m^2 - 4}$

9. $\dfrac{2w^2 + 5w - 3}{w^2 - 10w + 25} \div \dfrac{4w^2 - 1}{2w - 10}$

10. $\dfrac{2x}{x^2 - 36} + \dfrac{4}{5x - 30}$

11. $\dfrac{3}{x^2 - x - 6} - \dfrac{5x}{x^2 + 6x + 8}$

12. $\sqrt{\dfrac{3}{7}}$

13. $2\sqrt{27} - 5\sqrt{12}$

14. $\left(\sqrt{x} - \sqrt{2}\right)\left(\sqrt{x} + \sqrt{5}\right)$

15. $\left(2\sqrt{b} - 5\sqrt{c}\right)\left(2\sqrt{b} + 5\sqrt{c}\right)$

16. $\left(\sqrt{3} - \sqrt{7}\right)^2$

17. Write $y = -3(x + 2)^2 + 1$ in standard form.

18. Simplify $\left(3x^4y^8\right)^3\left(2xy^4\right)$.

Simplify. Assume that any variables are nonnegative.

19. $\sqrt{50x^2y}$

20. $\sqrt{\dfrac{t}{7}}$

Evaluate the expression for $a = -2$, $b = 8$, and $c = -4$.

21. $ac^2 - \dfrac{b}{c}$

22. $\dfrac{a + bc}{ab - c}$

For Exercises 23–30, factor the expression.

23. $16p^2 - 81q^2$

24. $a^2 - 6a - 27$

25. $w^2 + 14w + 49$

26. $6m^2n + 13mn^2 + 6n^3$

27. $2x^2 - 16x + 32$

28. $x^3 + 5x^2 - 9x - 45$

29. $12x^4 + 4x^3 - 40x^2$

30. $x^3 + 27$

31. Solve the inequality $-4(x - 2) > 20$. Describe the solution set as an inequality, in interval notation, and in a graph.

32. Graph the inequality $3x - 2y \geq 2$.

33. Graph the solution set of the system

$$y > \frac{1}{4}x - 3$$
$$y \geq -2x$$

For Exercises 34–45, solve. State whether the equation is linear, quadratic, rational, or radical.

34. $2x + 5 = 7x - 3$

35. $6(2x - 3) = 4(3x + 1) - 3(2x - 5)$

36. $w^2 = 5w + 24$

37. $(2r - 1)(3r - 2) = 1$

38. $3x^2 - 7 = 13$

39. $(x + 4)^2 = 60$

40. $3x^2 - 5x - 4 = 0$

41. $2x(x + 3) = -3$

42. $\dfrac{6}{p - 2} = \dfrac{5}{p - 3}$

43. $\dfrac{5}{m - 3} = \dfrac{m}{m - 2} + \dfrac{m}{m^2 - 5m + 6}$

44. $5\sqrt{x} - 2 = 1$

45. $\sqrt{x} = x - 2$

46. Solve the formula $a = \dfrac{v - v_0}{t}$ for t.

47. Solve by completing the square: $3x^2 - 12x = 42$.

48. Give examples of three equations that are equivalent to $2(x - 3) = 10$.

Solve the system.

49. $2x - 7y = 3$
$-5x + 3y = 7$

50. $3x - 4y = 35$
$y = 2x - 5$

Graph the equation.

51. $4x - 3y = 6$

52. $y = 2x^2 - 5$

For Exercises 53 and 54, find all x-intercepts.

53. $2x - 5y = 20$

54. $y = x^2 + 2x - 48$

55. Ski run A declines steadily for 130 yards over a horizontal distance of 610 yards. Ski run B declines steadily for 165 yards over a horizontal distance of 700 yards. Which run is steeper? Explain.

56. Find the slope of the line that contains the points $(-5, -4)$ and $(-1, -2)$. State whether the line is increasing, decreasing, horizontal, or vertical.

57. Find an equation of the line that has slope $m = -\dfrac{2}{5}$ and contains the point $(-3, 4)$.

58. Find an equation of the line that passes through the points $(-2, 4)$ and $(6, -3)$.

59. Find an equation of the line sketched in Fig. 13.

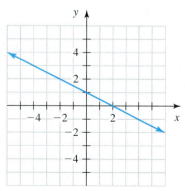

Figure 13 Exercise 59

60. The lengths of two sides of a right triangle are given (see Fig. 14). Find the length of the third side. (The triangle is not drawn to scale.)

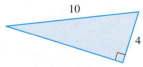

Figure 14 Exercise 60

61. Find any excluded values of the expression $\dfrac{5}{2x^2 - 7x + 5}$.

62. Simplify $\dfrac{-7x + 14}{x^2 - 7x + 10}$.

63. Two hours ago, the temperature was 6°F. Now the temperature is −3°F. What is the change in temperature over the past two hours?

64. A recipe for 8 servings of bread pudding calls for 3 tablespoons of sugar. How many tablespoons of sugar are required for 6 servings?

65. If p varies directly as w^2, and $p = 12$ when $w = 2$, find p when $w = 3$.

66. If c varies inversely as r, and $c = 10$ when $r = 2$, find r when $c = 4$.

67. Simplify $\dfrac{\dfrac{1}{4x} + \dfrac{3}{x^2}}{\dfrac{1}{2x} - \dfrac{5}{x^2}}$.

68. A person plans to invest a total of $12,000. She will invest in both an Amana Growth Fund at 12% annual interest and a Vanguard Primecap Inv Fund at 18% annual interest. How much should she invest in each account so that the total interest in one year will be $1710?

69. Just before beginning its descent, an airplane is at an altitude of 27,500 feet. The airplane then descends steadily at a rate of 1350 feet per minute. Let A be the airplane's altitude (in feet) at t minutes after the plane has begun its descent.
   a. Find an equation of a model to describe the data.
   b. Estimate the airplane's altitude 7 minutes after the plane has begun its descent.
   c. Estimate when the airplane reached an altitude of 1200 feet.
   d. What is the slope? What does it mean in this situation?
   e. What is the t-intercept? What does it mean in this situation?

70. Table 14 shows the percentages of American adults of various age groups who say that they are interested in soccer.

Age (in years)	Age Used to Represent Age Group (in years)	Percent
18–24	21.0	32.8
25–34	29.5	26.9
35–44	39.5	25.1
45–54	49.5	23.1
55–64	59.5	19.0
over 64	70	16.2

Table 14 American Adults Who Say That They Are Interested in Soccer

Source: *ESPN Sports Poll*

Let p be the percentage of American adults at age a years who say that they are interested in soccer.

a. Use a graphing calculator to draw a scatterplot of the data. Can the data be modeled better by a linear equation or a quadratic equation? Explain.

b. Find an equation of a model to describe the data.

c. What is the slope? What does it mean in this situation?

d. Estimate what percentage of 25-year-old Americans say that they are interested in soccer.

e. Estimate at what age 21% of American adults say that they are interested in soccer.

f. Find the a-intercept. What does it mean in this situation?

71. Average daily oil production, both onshore and offshore, in the United States is shown in Table 15 for various years.

Table 15 Onshore and Offshore Oil Production

Year	Average Daily Oil Production (millions of barrels a day)	
	Onshore	Offshore
2012	5.17	1.31
2013	6.16	1.31
2014	7.32	1.45
2015	7.87	1.55
2016	7.25	1.63

Source: *Energy Information Administration*

Let p be the average daily oil production (in millions of barrels a day) in the year that is t years since 2010. Reasonable models for onshore and offshore production are

$$p = 0.587t + 4.41 \quad \textit{onshore}$$
$$p = 0.088t + 1.10 \quad \textit{offshore}$$

a. Find the sum of the expressions $0.587t + 4.41$ and $0.088t + 1.10$. What does your result represent?

b. Evaluate the result you found in part (a) for $t = 11$. What does your result mean in this situation?

c. Estimate the rate of change of total average daily oil production (both onshore and offshore).

d. Use substitution or elimination to estimate when onshore and offshore average daily oil production were equal. Find that average daily oil production.

72. A student drives 12 miles south from home and then 3 miles east to go to school. How many miles would the trip take if she could drive along a straight line from home to school?

73. The revenue from selling Dr. Grip® pens varies directly as the number of pens sold. The revenue from selling 1275 pens is $10,378.50. If the revenue is $12,405.36, how many pens were sold?

74. The percentages of male drivers who were speeding when they became involved in fatal crashes are shown in Table 16 for various age groups.

Table 16 Speeding Male Drivers in Fatal Crashes

Age Group (years)	Age Used to Represent Age Group (years)	Percent
15–24	19.5	38
25–34	29.5	28
35–44	39.5	20
45–54	49.5	15
55–64	59.5	13
65–74	69.5	10
75+	80.0	7

Source: *National Center for Statistics & Analysis*

Let p be the percentage of male drivers at age a years who were speeding when they became involved in a fatal crash.

a. Use a graphing calculator to draw a scatterplot of the data. Can the data be modeled better by a linear equation or a quadratic equation? Explain.

b. A model of the situation is $p = 0.0078t^2 - 1.26t + 58.80$. Use a graphing calculator to draw the graph of the model and, in the same viewing window, the scatterplot of the data. Does the model fit the data well?

c. Estimate the percentage of 23-year-old male drivers who were speeding when they became involved in a fatal crash.

d. Estimate at what age 25% of male drivers were speeding when they became involved in a fatal crash.

75. The average percentages of Americans who regularly attend church in a typical week are shown in Table 17 for various age groups.

Table 17 Percentages of Americans Who Regularly Attend Church in a Typical Week

Age Group (years)	Age Used to Represent Age Group (years)	Percent
20s	25	31
30s	35	42
40s	45	47
50s	55	48
60 or more	65	53

Source: *Barna Research Group*

Let p be the average percentage of Americans at the age that is a years more than 25 years who regularly attend church in a typical week. (So, $a = 5$ represents the age $25 + 5 = 30$ years.) A model of the situation is $p = 3.4\sqrt{a} + 31$.

a. Use a graphing calculator to draw the graph of the model and, in the same viewing window, the scatterplot of the data. Does the model fit the data well?

b. Estimate what percentage of 30-year-old Americans go to church in a typical week. [**Hint:** Recall that a is the age that is a years more than 25 years.]

c. Estimate at what age half of all Americans go to church in a typical week.

Using a TI-83 or TI-84 Graphing Calculator

The more you experiment with a graphing calculator, the more comfortable and efficient you will become with it.

A TI graphing calculator can detect several types of errors and display an error message. When this occurs, refer to Appendix A.23 for explanations of some common error messages and how to fix these types of mistakes. Errors do not hurt the calculator. In fact, you can't hurt the calculator regardless of the order in which you press its keys. So, the more you experiment with the calculator, the better off you will be.

To access a TI-83 command written in yellow above a key, first press 2nd, and then the key. Whenever a key must follow the 2nd key, this appendix will use brackets for the key. For example, "Press 2nd [OFF]" means to press 2nd and then press ON (because "OFF" is written in yellow above the ON key). The same applies for TI-84 commands written in blue above a key.

Aside from having different-colored keys, the TI-83 and TI-84 are similar calculators: Virtually all of the key combinations for a TI-83 and a TI-84 are the same.

Instructions for using a TI-85 and TI-86 (as well as a TI-82 and a TI-83) graphing calculator are available at the website www.prenhall.com/divisions/esm/app/calc_v2/. This site also can serve as a cross-reference for TI-83 graphing calculator instructions. Because the TI-83 and TI-84 are so similar, TI-84 users will find the site helpful even though the TI-84 is not mentioned.

A.1 Turning a Graphing Calculator On or Off

To turn a graphing calculator on, press ON. To turn it off, first press 2nd. Then press [OFF].

A.2 Making the Screen Lighter or Darker

To make the screen darker, first press 2nd (then release it); then hold the △ key down for a while. To make the screen lighter, first press 2nd (then release it); then hold the ▽ key down for a while.

A.3 Entering an Equation

When we show two or more buttons in a row, press them one at a time and in order.

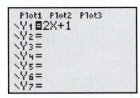

Figure 1 Entering an equation

To enter the equation $y = 2x + 1$,

1. Press $\boxed{Y=}$.
2. If necessary, press $\boxed{CLEAR}$ to erase a previously entered equation.
3. Press 2 $\boxed{X, T, \Theta, n}$ $\boxed{+}$ $\mathbf{1}$. The screen will look like the one displayed in Fig. 1.
4. If you want to enter another equation, press $\boxed{ENTER}$. Then type in the next equation.
5. Use the $\boxed{\triangle}$ or $\boxed{\triangledown}$ key to get from one equation to another.

A.4 Graphing an Equation

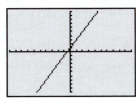

Figure 2 Graphing an Equation

To graph the equation $y = 2x + 1$,

1. Enter the equation $y = 2x + 1$; see Appendix A.3.
2. Press $\boxed{ZOOM}$ $\mathbf{6}$ to draw a graph of your equation between the values of -10 and 10 for both x and y. The screen will look like the one displayed in Fig. 2.
3. See Appendix A.6 if you want to zoom in or zoom out to get another part of the graph to appear on the calculator screen. Or see Appendix A.7 to change the window format manually; then press $\boxed{GRAPH}$.

A.5 Tracing a Curve without a Scatterplot

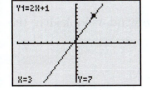

Figure 3 Tracing a curve

To *trace a curve*, we find coordinates of points on the curve. To trace the line $y = 2x + 1$,

1. Graph $y = 2x + 1$ (see Appendix A.4).
2. Press $\boxed{TRACE}$.
3. If you see a flashing "$\times$" on the curve, the coordinates of that point will be listed at the bottom of the screen. If you don't see the flashing "$\times$", press $\boxed{ENTER}$, and your calculator will adjust the viewing window so you can see it.
4. To find coordinates of points on your curve that are off to the right, press $\boxed{\triangleright}$.
5. To find coordinates of points on your curve that are off to the left, press $\boxed{\triangleleft}$.
6. Find the y-coordinate of a point by entering the x-coordinate. For example, to find the y-coordinate of the point that has x-coordinate 3, press 3 $\boxed{ENTER}$. The screen will look like the one displayed in Fig. 3. This feature works for values of x between Xmin and Xmax, inclusive (see Appendix A.7).
7. If more than one equation has been graphed, press $\boxed{\triangledown}$ to trace the second equation. Continue pressing $\boxed{\triangledown}$ to trace the third equation, and so on. Press $\boxed{\triangle}$ to return to the previous equation. Notice that the equation of the curve being traced is listed in the upper left corner of the screen.

A.6 Zooming

The $\boxed{ZOOM}$ menu has several features that allow you to adjust the viewing window. Some of the features adjust the values of x that are used in tracing.

- **Zoom In** magnifies the graph around the cursor location. The following instructions are for zooming in on the graph of $y = 2x + 1$:
  1. Graph $y = 2x + 1$ (see Appendix A.4).
  2. Press $\boxed{ZOOM}$ $\mathbf{2}$.

3. Use ◁, ▷, △, and ▽ to position the cursor on the portion of the line that you want to zoom in on.

If you lose sight of the line, you can always press TRACE ENTER.

4. To zoom in, press ENTER.
5. To zoom in on the graph again, you have two options:
   a. To zoom in at the same point, press ENTER.
   b. To zoom in at a new point, move the cursor to the new point; then press ENTER.

When zooming out, you will return to the original graph only if you did not move the cursor while zooming in.

6. To return to your original graph, press ZOOM 6. Or zoom out (see the next instruction) the same number of times you zoomed in.

- **Zoom Out** does the reverse of Zoom In: It allows you to see *more* of a graph. To zoom out, follow the preceding instructions, but press ZOOM 3 instead of ZOOM 2 in step 2.
- **ZStandard** will change your viewing screen so both x and y will go from -10 to 10. To use ZStandard, press ZOOM 6.
- **ZDecimal** lets you trace a curve by using the numbers 0, ± 0.1, ± 0.2, ± 0.3, ... for x. ZDecimal will change your viewing screen so x will go from -4.7 to 4.7 and y will go from -3.1 to 3.1. To use ZDecimal, press ZOOM 4.
- **ZInteger** allows you to trace a curve by using the numbers 0, ± 1, ± 2, ± 3, ... for x. ZInteger can be used for any viewing window, although it will change the view slightly. To use ZInteger, press ZOOM 8 ENTER.
- **ZSquare** will change your viewing window so the spacing of the tick marks on the x-axis is the same as that on the y-axis. To use ZSquare, press ZOOM 5.
- **ZoomStat** will change your viewing window so you can see a scatterplot of points that you have entered in the statistics editor. To use ZoomStat, press ZOOM 9.
- **ZoomFit** will adjust the dimensions of the y-axis to display as much of a curve as possible. The dimensions of the x-axis will remain unchanged. To use ZoomFit, press ZOOM 0.

A.7 Setting the Window Format

To graph the equation $y = 2x + 1$ between the values of -2 and 3 for x and between the values of -5 and 7 for y,

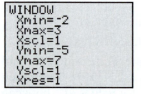

Figure 4 Window settings

1. Enter the equation $y = 2x + 1$ (see Appendix A.3).
2. Press WINDOW. Then change the window settings so the window looks like the one displayed in Fig. 4 after you have used steps 3–8.
3. Press (−) 2 ENTER to set the smallest value of x to -2.
4. Press 3 ENTER to set the largest value of x to 3.
5. Press 1 ENTER to set the scaling for the x-axis to increments of 1.
6. Press (−) 5 ENTER to set the smallest value of y to -5.
7. Press 7 ENTER to set the largest value of y to 7.
8. Press 1 ENTER to set the scaling for the y-axis to increments of 1.

If you press ZOOM 6 or ZOOM 9 or zoom in or zoom out, your window settings will change accordingly.

9. Press GRAPH to view the graph of $y = 2x + 1$. The screen will look like the graph drawn in Fig. 5.

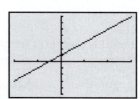

Figure 5 Graph of $y = 2x + 1$

A.8 Drawing a Scatterplot

Table 1 Drawing a Scatterplot

x	y
2	4
3	7
4	10
5	11

Make sure you press CLEAR rather than DEL. If you press DEL, the column will vanish. If you ever do this by mistake, press STAT 5 ENTER to get back the missing column.

If Plot 1 is off, your points will be saved in columns L_1 and L_2, but they will not be plotted.

To draw a scatterplot of the data displayed in Table 1,

1. To enter the data, press STAT 1.
2. If there are numbers listed in the first column (list L_1), clear the column by pressing ◁ as many times as necessary to get to column L_1. Next, press △ once to get to the top of column L_1. Then press CLEAR ENTER.
3. If there are numbers listed in the second column (list L_2), clear the column by pressing ▷ to move the cursor to column L_2. Then press △ CLEAR ENTER.
4. To return to the first entry position of list L_1, press ◁.
5. Press 2 ENTER 3 ENTER 4 ENTER 5 ENTER to enter the data in column L_1. (If you make a mistake, you can delete an entry by pressing DEL; then insert an entry by pressing 2nd [DEL].)
6. Press ▷ to move to the first entry position of list L_2.
7. Press 4 ENTER 7 ENTER 10 ENTER 11 ENTER to enter the elements of L_2.
8. Press 2nd [STAT PLOT].
9. Press 1 to select Plot 1.
10. Press ENTER to turn Plot 1 on.
11. Press ▽ ENTER to choose the scatterplot mode.
12. Press ▽ so the cursor is at "Xlist." Then press 2nd [L_1].
13. Press ▽ so the cursor is at "Ylist." Then press 2nd [L_2].
14. Use squares, plus signs, or dots to represent the points plotted on the scatterplot. These three symbols are called "Marks." Press ▽ once so the cursor is on one of the three Mark symbols. Next, press ▷ and/or ◁ to select a symbol. Then press ENTER. If you selected the square symbol, the screen will look like the one displayed in Fig. 6.
15. Press ZOOM 9. If you selected the square symbol, the screen will look like the one displayed in Fig. 7.

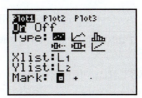

Figure 6 Setting up Plot 1

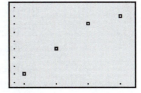

Figure 7 Drawing a scatterplot

A.9 Tracing a Scatterplot

To see the coordinates of a point in a scatterplot,

1. Draw a scatterplot (see Appendix A.8).
2. Press TRACE.
3. Notice the flashing "×" on one of the points of the scatterplot. The coordinates of this point are listed at the bottom of the screen.
4. To find the coordinates of the next point to the right, press ▷.
5. To find the coordinates of the next point to the left, press ◁.

A.10 Graphing Equations with a Scatterplot

Table 2 Drawing a Scatterplot

x	y
2	4
3	7
4	10
5	11

To graph the equation $y = 2x + 1$ with a scatterplot of the data displayed in Table 2,

1. Enter the equation $y = 2x + 1$ (see Appendix A.3).
2. Follow the instructions in Appendix A.8 to draw the scatterplot. (The graph of the equation will also be drawn because you turned the equation on.) The screen will look like the one displayed in Fig. 8.

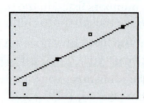

Figure 8 Graphing an equation and a scatterplot

A.11 Tracing a Curve with a Scatterplot

When tracing a curve, if you do not see the flashing "×," press ENTER, and the calculator will adjust the window settings so you can see the curve. Press TRACE and then keep pressing ▽ until you see the flashing "×."

To trace a curve with a scatterplot,

1. Graph an equation with a scatterplot (see Appendix A.10).
2. Press TRACE to trace points that make up the scatterplot. Then press ▽ to trace points that lie on the curve. If other equations are graphed, continue pressing ▽ to trace the second equation, and so on. Press △ to begin to return to the scatterplot. Notice that the label "P1: L_1, L_2" is in the upper left corner of the screen when Plot 1's points are being traced and that the equation entered in the Y= mode is listed in the upper left corner of the screen when the curve is being traced.

A.12 Turning a Plotter On or Off

To change the on/off status of the plotter,

1. Press Y=.
2. Press △. A flashing rectangle will be on "Plot 1."
3. Press ▷ if necessary to move the flashing rectangle to the plotter you wish to turn on or off.
4. Press ENTER to turn your plotter on or off. The plotter is on if the plotter icon is highlighted.

A.13 Constructing a Table

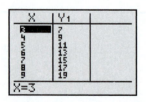

Figure 9 Table of ordered pairs for $y = 2x + 1$

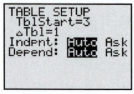

Figure 10 Table setup

To construct a table of ordered pairs for the equation $y = 2x + 1$, where the values of x are 3, 4, 5, . . . (see Fig. 9),

1. Enter the equation $y = 2x + 1$ for Y_1 (see Appendix A.3).
2. Press 2nd [TBLSET].
3. Press 3 ENTER to tell the calculator that the first x value in your table is 3.
4. Press 1 ENTER to tell the calculator that the x values in your table increase by 1.
5. Press ENTER ▽ ENTER to highlight "Auto" for both "Indpnt" and "Depend." The screen will now look like the one displayed in Fig. 10.
6. Press 2nd [TABLE] to construct the table shown in Fig. 9.
7. Hold the ▽ key down for a while to scroll down.
8. Hold the △ key down for a while to scroll up.

A.14 Constructing a Table for Two Equations

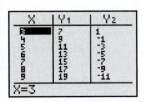

Figure 11 Table for two equations

To construct a table of ordered pairs for the equations $y = 2x + 1$ and $y = -2x + 7$, where the values of x are 3, 4, 5, . . . (see Fig. 11),

1. Enter the equation $y = 2x + 1$ for Y_1, and enter the equation $y = -2x + 7$ for Y_2 (see Appendix A.3).
2. Follow steps 2–8 of Appendix A.13.

A.15 Using "Ask" in a Table

Table 3 Using "Ask" in a Table with $y = 2x + 1$

x	y
2	
2.9	
5.354	
7	
100	

To use the "Ask" option in the Table Setup mode to complete Table 3 for $y = 2x + 1$,

1. Enter the equation $y = 2x + 1$ for Y_1 (see Appendix A.3).
2. Press 2nd [TBLSET].
3. Press ENTER twice. Next, press ▷. Then press ENTER. The "Ask" option for "Indpnt" will now be highlighted. Make sure the Auto option for "Depend" is highlighted.
4. Press 2nd [TABLE].
5. Press 2 ENTER 2.9 ENTER 5.354 ENTER 7 ENTER 100 ENTER. The screen will now look like the one displayed in Fig. 12.

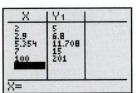

Figure 12 Using "Ask" for a table with $y = 2x + 1$

A.16 Drawing Two Scatterplots

Table 4 Drawing the First of Two Scatterplots

x	y
2	4
3	7
4	10
5	11

Make sure you press CLEAR rather than DEL. If you press DEL, the column will vanish. If you ever do this by mistake, press STAT 5 ENTER to get back the missing column.

Table 5 Drawing the Second of Two Scatterplots

x	y
2	11
2	9
3	6
5	4

It is possible to draw two scatterplots on the same calculator screen and use different markings for the two sets of points. To begin, follow the instructions in Appendix A.8 to draw a scatterplot of the data values in Table 4.

These data are stored in columns L_1 and L_2. The points are plotted by the plotter called "Plot 1."

You will now draw a scatterplot of the data values in Table 5.

These data will be stored in columns L_3 and L_4. The points will be plotted by the plotter called "Plot 2." To do this,

1. To enter the data, press STAT 1.
2. To clear list L_3, press ▷ and/or ◁ to move the cursor to column L_3. Then press △ CLEAR ENTER.
3. To clear list L_4, press ▷ to move the cursor to column L_4. Then press △ CLEAR ENTER.
4. To return to the first entry position of list L_3, press ◁.
5. Press 2 ENTER 2 ENTER 3 ENTER 5 ENTER to enter the elements of L_3.
6. Press ▷ to move to the first entry position of list L_4.
7. Press 11 ENTER 9 ENTER 6 ENTER 4 ENTER to enter the elements of L_4.
8. Press 2nd [STAT PLOT].
9. Press 2 to select "Plot 2."
10. Press ENTER to turn Plot 2 on.

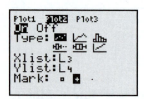

Figure 13 Setting up Plot 2

11. Press $\boxed{\bigtriangledown}$ twice so the cursor is at "Xlist." Then press $\boxed{2nd}$ $[L_3]$.

12. Press $\boxed{ENTER}$ so the cursor is at "Ylist." Then press $\boxed{2nd}$ $[L_4]$.

13. Press $\boxed{\bigtriangledown}$ once so the cursor is on one of the three choices for "Mark." Next, press $\boxed{\triangleright}$ and/or $\boxed{\triangleleft}$ to select a symbol different from the one you used for the first scatterplot. Then press $\boxed{ENTER}$. If you selected the plus symbol, the screen will look like the one in Fig. 13.

14. Press $\boxed{ZOOM}$ **9** to obtain the two scatterplots with different symbols. If you selected the square symbol for Plot 1 and the plus symbol for Plot 2, the screen will look like the one displayed in Fig. 14.

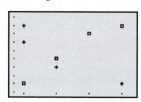

Figure 14 Drawing two scatterplots

A.17 Finding the Intersection Point(s) of Two Curves

To find the intersection point of the lines $y = 2x + 1$ and $y = -2x + 7$,

1. Enter the equation $y = 2x + 1$ for Y_1, and enter the equation $y = -2x + 7$ for Y_2 (see Appendix A.3).

2. By zooming in or out or by changing the window settings, draw a graph of both curves so you can see an intersection point. For our example, press $\boxed{ZOOM}$ **6**.

3. Press $\boxed{2nd}$ [CALC]. The screen will look like the one displayed in Fig. 15.

4. Press **5** to select "intersect."

5. You will now see a flashing cursor on your first curve. If there is more than one intersection point on your display screen, move the cursor by pressing $\boxed{\triangleright}$ or $\boxed{\triangleleft}$ so it is much closer to the intersection point you want to find. The screen will look something like the one displayed in Fig. 16.

6. Press $\boxed{ENTER}$ to put the cursor on the second curve. Press $\boxed{ENTER}$ again to display "Guess?" Press $\boxed{ENTER}$ once more. The screen will look like the one displayed in Fig. 17. The intersection point is (1.5, 4).

Figure 15 Menu of choices

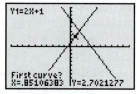

Figure 16 Put cursor near intersection point

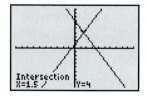

Figure 17 Location of intersection point

A.18 Finding the Minimum Point(s) or Maximum Point(s) of a Curve

To find the minimum point of the curve $y = x^2 - 3x + 1$,

1. Enter the equation $y = x^2 - 3x + 1$ (see Appendix A.3).

2. Use ZDecimal to draw a graph of the equation (see Appendix A.6).

3. Press $\boxed{2nd}$ [CALC].

4. Press **3** to select "minimum."

5. Move the flashing cursor by pressing $\boxed{\triangleleft}$ or $\boxed{\triangleright}$ so it is to the left of the minimum point, and press $\boxed{ENTER}$. Or type a number between Xmin and the x-coordinate of the minimum point, and press $\boxed{ENTER}$.

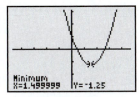

Figure 18 Finding the minimum point of $y = x^2 - 3x + 1$

6. Move the flashing cursor by pressing ▷ so it is to the right of the minimum point, and press ENTER. Or type a number between the x-coordinate of the minimum point and Xmax, and press ENTER.

7. Press ENTER. The calculator will display the coordinates of the minimum point — about $(1.50, -1.25)$. See Fig. 18.

You can find the maximum point of a curve in a similar fashion, but press **4** to select the "maximum" option, rather than the "minimum" option, in step 4.

A.19 Finding Any *x*-Intercepts of a Curve

To find the x-intercept of the line $y = x - 2$,

1. Enter the equation $y = x - 2$ (see Appendix A.3).
2. Use ZDecimal to draw a graph of the equation (see Appendix A.6).
3. Press 2nd [CALC].
4. Press **2** to choose the "zero" option.
5. Move the flashing cursor by pressing ◁ or ▷ so it is to the left of the x-intercept, and press ENTER. Or type a number between Xmin and the x-coordinate of the x-intercept, and press ENTER.

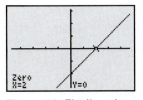

Figure 19 Finding the x-intercept of $y = x - 2$

6. Move the flashing cursor by pressing ▷ so it is to the right of the x-intercept, and press ENTER. Or type a number between the x-coordinate of the x-intercept and Xmax, and press ENTER.

7. Press ENTER. The screen will look like the one displayed in Fig. 19. The x-intercept is $(2, 0)$.

A.20 Turning an Equation On or Off

You can graph an equation only if its equals sign is highlighted. (The equation is then "on"). Up to 10 equations can be graphed at one time. To change the on–off status of an equation,

1. Press Y=.
2. Move the cursor to the equation whose status you want to change.
3. Use ◁ to place the cursor over the "=" sign of the equation.
4. Press ENTER to change the status.

A.21 Finding Coordinates of Points

To find the coordinates of particular points,

1. Press GRAPH to get into graphing mode.
2. Press ▷ to get a cursor to appear on the screen. (If you cannot see the cursor, it is probably on one or both of the axes. If it is on an axis, you should still be able to see a small flashing dot.) Notice that the coordinates of the point where the cursor is currently positioned are at the bottom of the screen.
3. Use ◁, ▷, △, or ▽ to move the cursor left, right, up, or down, respectively.

A.22 Entering an Equation by Using Y_n References

To enter the complicated equation $y = \dfrac{x+1}{x-3} \div \dfrac{x-2}{x+5}$ by using Y_n references,

1. Enter $Y_1 = \dfrac{x+1}{x-3}$ and $Y_2 = \dfrac{x-2}{x+5}$ (see Appendix A.3).

2. Turn both equations off (see Appendix A.20).

3. Move the flashing cursor to the right of "$Y_3 = .$"

4. Press $\boxed{\text{VARS}}$ $\boxed{\triangleright}$ $\boxed{\text{ENTER}}$ $\boxed{\text{ENTER}}$. "Y_1" will now appear to the right of "$Y_3 = $" in the $\boxed{\text{Y=}}$ window.

5. Press $\boxed{\div}$.

6. Press $\boxed{\text{VARS}}$ $\boxed{\triangleright}$ $\boxed{\text{ENTER}}$.

7. Move the cursor to "$2:Y_2$" and press $\boxed{\text{ENTER}}$. "Y_1/Y_2" will now appear to the right of "$Y_3 = $" in the $\boxed{\text{Y=}}$ window. The screen will look like the one displayed in Fig. 20.

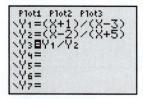

Figure 20 Using Y_n references

A.23 Responding to Error Messages

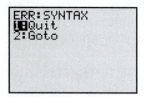

Figure 21 "Syntax" error message

Here are several common error messages and how to respond to them:

- The **Syntax** error (see Fig. 21) means you have misplaced one or more parentheses, operations, or commas. The calculator will find this type of error if you choose "Goto" by pressing $\boxed{\nabla}$ $\boxed{\text{ENTER}}$. Your error will be highlighted by a flashing black rectangle.

   The most common "Syntax" error is pressing $\boxed{(-)}$ when you should have pressed $\boxed{-}$, or vice versa:

   1. Press the $\boxed{(-)}$ key when you want to take the opposite of a number or are working with negative numbers. To compute $-5(-2)$, press $\boxed{(-)}$ **5** $\boxed{(}$ $\boxed{(-)}$ **2** $\boxed{)}$ $\boxed{\text{ENTER}}$.

   2. Press the $\boxed{-}$ key when you want to subtract two numbers. To compute $5 - 2$, press **5** $\boxed{-}$ **2** $\boxed{\text{ENTER}}$.

- The **Invalid** error (see Fig. 22) means you have tried to enter an inappropriate number, expression, or command. The most common "Invalid" error is to try to enter a number that is not between Xmin and Xmax, inclusive, when you use a command such as $\boxed{\text{TRACE}}$, "minimum," or "maximum."

- The **Invalid dimension** error (see Fig. 23) means you have the plotter turned on (see Fig. 24) but have not entered any data points in the STAT list editor (see Fig. 25). In this case, first press $\boxed{\text{ENTER}}$ to exit the error message display; then either turn the plotter off or enter data in the STAT list editor.

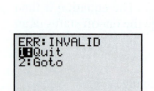

Figure 22 "Invalid" error message

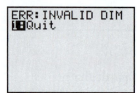

Figure 23 "Invalid dimension" error message

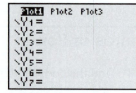

Figure 24 Plotter is on

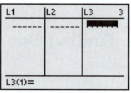

Figure 25 STAT list editor's columns are empty

Figure 26 "Dimension mismatch" error message

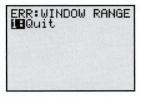

Figure 28 "Window range" error message

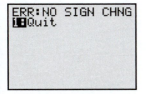

Figure 29 "No sign change" error message

- The **Dimension mismatch** error (see Fig. 26) is fixed in two ways:
  1. In the STAT list editor, one column you are using to plot has more numbers than the other column has (see Fig. 27). In this case, first press ENTER to exit the error message display; then add or delete numbers so the two columns have the same length.

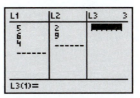

Figure 27 Columns of unequal length in STAT list editor

  2. In the STAT list editor, one column you are using to plot has more numbers than the other column has, but you didn't notice the difference in length because you deleted one or both of the columns by mistake. You can find the missing column(s) by pressing STAT 5 ENTER.

- The **Window range** error (see Fig. 28) means one of two things:
  1. You made an error in setting up your window. This usually means you entered a larger number for *Xmin* than for *Xmax* or you entered a larger number for *Ymin* than for *Ymax*. In this case, first press ENTER to exit the error message display; then change your window settings accordingly (see Appendix A.7).
  2. You pressed ZOOM 9 when only one data-point pair was entered in the STAT list editor. (On some TI graphing calculators, the command ZoomStat works only when you have two or more pairs of data points in the STAT list editor.) In this case, first press ENTER to exit the error message display; then either add more points to the STAT list editor or avoid pressing ZOOM 9 and set up your window settings manually (see Appendix A.7).

- The **No sign change** error (see Fig. 29) means one of two things:
  1. You are trying to locate a point that does not appear on the screen. For example, you may be trying to find an intersection point of two curves that intersect off-screen. Or you may be trying to find a zero of an equation that does not appear on the screen. In this case, press ENTER and change your window settings so the point you are trying to locate is on the screen.
  2. You are trying to locate a point that does not exist. For example, you may be trying to find an intersection point of two parallel lines. Or you may be trying to find a zero of an equation that does not have one. In this case, press ENTER and stop looking for the point that doesn't exist!

- The **Nonreal answer** error (see Fig. 30) means your computation did not yield a real number. For example, $\sqrt{-4}$ is not a real number. The calculator will locate this computation if you choose "Goto" by pressing ▽, then ENTER.

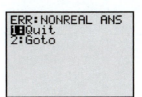

Figure 30 "Nonreal answer" error message

- The **Divide by 0** error (see Fig. 31) means you asked the calculator to perform a calculation that involves a division by zero. For example, $3 \div (5 - 5)$ will yield the error message shown in Fig. 31.

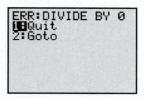

Figure 31 "Divide by zero" error message

The calculator will locate the division by zero if you choose "Goto" by pressing ▽, then ENTER.

Using StatCrunch

StatCrunch produces excellent statistical diagrams and does not require much time to learn how to use. The more you experiment with StatCrunch, the more comfortable and efficient you will become with it.

B.1 Entering Data

To manually enter data, click on the cell at the top of the first column, delete "var1," and enter the name of the variable. Next, enter the values of the variable below the name of the variable in the first column. If you have data for other variables, enter those observations in other columns similarly.

To perform a cut and paste of data from other spreadsheet software such as Excel, copy the data, click on the top of the first column of StatCrunch's spreadsheet, and paste the data. In some cases, you can add data by simply dragging and dropping a file onto StatCrunch's spreadsheet.

There are many other ways to import data, including a feature called "StatCrunch-This," which imports data tables from the Internet. See the instructions in StatCrunch or ask your professor for details.

B.2 Constructing Scatterplots

To construct a scatterplot,

1. Click on the button Graph. Then click on "Scatter Plot" in the drop-down menu.
2. In the drop-down menu under the heading "**X variable**," select a variable to be the explanatory variable.
3. In the drop-down menu under the heading "**Y variable**," select a variable to be the response variable.
4. Under the heading "**Display**," select "Points."
5. To construct a scatterplot, click on the button Compute! at the bottom of the window. To then make adjustments, click on the button Options at the top of the window, and in the drop-down menu click on "Edit." To take advantage of more features, do some or all of the following steps.
6. Suppose that you chose the variables Height and Weight of students for Steps 2 and 3, respectively, and that a column of your data set consists of ethnicities with the word "Ethnicity" in the cell at the top of the column. To construct a scatterplot comparing heights and weights for each ethnicity, you would select "Ethnicity" in the drop-down menu under the heading "**Group by**."

7. To adjust the size of the dots, select a size in the drop-down menu under the heading "**Point size**."

8. StatCrunch will automatically display the variable names on the axes, but you can enter something more complete such as the variable names and the units by making the appropriate entries in the boxes labeled "X-axis label" and "Y-axis label," which are under the heading "**Graph properties**."

9. To title the scatterplot, enter the title in the box labeled "Title," which is under the heading "**Graph properties**."

10. To show light horizontal and/or vertical lines, click on one or both of the boxes labeled "Horizontal lines" and "Vertical lines," which are under the heading "**Graph properties**."

11. To display the scatterplot, click on the button Compute! at the bottom of the window.

12. You can adjust the labels on the axes, the scaling on the axes, and the title by clicking on the small icon consisting of three horizontal lines near the bottom left corner of the window.

13. Click on the button Options at the top of the window and select from the dropdown menu to edit, save, copy, print, or download the scatterplot.

Answers to Odd-Numbered Exercises

Answers to most discussion exercises and to exercises in which answers may vary have been omitted.

Chapter 1

Homework 1.1   **1.** 25 thousand fans attended the concert.   **3.** In 2016, 199 million Americans had smartphones.   **5.** In 2015, 54.9 million iPads were sold.   **7.** In 2015, BP lost $6.5 billion.   **9.** The statement $t = 9$ represents the year 2019.   **11.** The statement $t = -3$ represents the year 2002.   **13.** h; 67, 72; -5, 0; answers may vary.   **15.** p; 50, 60; -2, -8; answers may vary.   **17.** T; 15, 40; 240, -10; answers may vary.   **19.** s; 25, 32; -15, -9; answers may vary.

21. a. The rectangles are not drawn to scale. Answers may vary.   4 inches ☐ 6 inches   3 inches ☐ 8 inches   1 inch ▭ 24 inches   **b.** W, L **c.** A

23. a. The rectangles are not drawn to scale. Answers may vary.   5 feet ☐ 5 feet   3 feet ☐ 7 feet   1 foot ▭ 9 feet   **b.** W, L **c.** P

25. a. The rectangles are not drawn to scale. Answers may vary.   1 inch ▭ 4 inches   2 inches ☐ 5 inches   3 inches ☐ 6 inches   **b.** W, L, A **c.** None

27. a. The rectangles are not drawn to scale. Answers may vary.   2 yards ☐ 2 yards   2 yards ☐ 3 yards   2 yards ☐ 4 yards   **b.** L, P **c.** W

29. 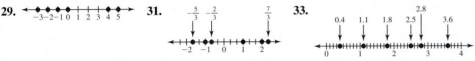   **31.** [number line with $-\frac{5}{3}, -\frac{2}{3}, \frac{7}{3}$]   **33.** [number line with 0.4, 1.1, 1.8, 2.5, 2.8, 3.6]

35. [number line with -1.8, 0.5, 1.2, 3.1]   **37.** 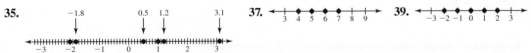   **39.** [number line -3 to 3]

41. [number line -3 to 5]   **43.** [number line -4 to 0]   **45.** 3, 356   **47.** -4 **49.** $\sqrt{7}, \pi$ **51.** $-2, -5, -7$; answers may vary.

53. $-8, -9, -27$; answers may vary.   **55.** $-2, -5, -40$; answers may vary.   **57.** $\frac{5}{4}, \frac{3}{2}, \frac{7}{4}$; answers may vary.   **59.** $-2.1, -2.3, -2.8$; answers may vary.   **61.** [number line, Average: 11 units; 9, 15; 4 6 8 10 12 14 16; u; Number of units]   **63.** [number line, Average: 12.8%; 10, 14, 16; 9 11 13 15 17; p; Percent]   **65.** 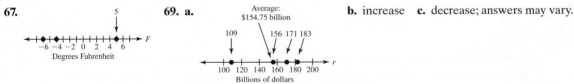 [22 24, 30 32; 21 23 25 27 29 31 33; L; Hours]

67. [number line, 5; -6 -4 -2 0 2 4 6; F; Degrees Fahrenheit]   **69. a.** [number line, Average: $154.75 billion; 109, 156 171 183; 100 120 140 160 180 200; r; Billions of dollars]   **b.** increase   **c.** decrease; answers may vary.

71. a. 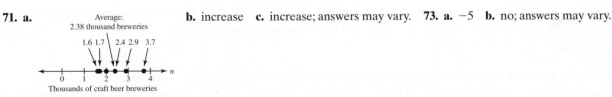 [number line, Average: 2.38 thousand breweries; 1.6 1.7, 2.4 2.9 3.7; 0 1 2 3 4; n; Thousands of craft beer breweries]   **b.** increase   **c.** increase; answers may vary.   **73. a.** -5 **b.** no; answers may vary.

75. a. i. 8 [number line 6–10]   **ii.** 3 [number line 0–6]   **iii.** 5 [number line 1–9]   **b.** Answers may vary.   **c.** infinitely many; answers may vary.   **77.** Answers may vary.   **79.** Answers may vary.

Homework 1.2   **1–15 odd.** 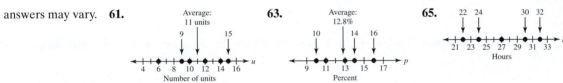 [graph with points $(-5, 4)$, $(-3, 0)$, $(0, 2)$, $(5, 1)$, $(4, -2)$, $(-1.3, -3.9)$, $(2.5, -4.5)$, $(-3, -6)$]   **17.** 2   **19.** explanatory: n; response: s **21.** explanatory: a; response: h

23. explanatory: *c*; response: *T*   **25.** explanatory: *A*; response: *n*   **27.** explanatory: *t*; response: *h*   **29.** A telemarketer who works 32 hours per week will sell an average of 43 magazine subscriptions per week.   **31.** In 2016, 70% of U.S. consumers binge watched an average of 5 episodes of television at a time.   **33.** $598.5 billion was spent on defense in 2015.   **35.** In 2013, 70% of Americans felt that a college education was very important.

37.
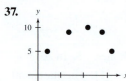
39. A$(-4, -3)$, B$(-5, 0)$, C$(-2, 4)$, D$(1, 3)$, E$(0, -2)$, F$(5, -4)$

41. a.
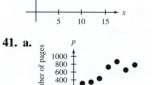
b. The fifth book   **c.** The third book to the fourth book; answers may vary.

43. a.
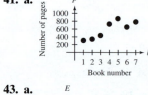
b. 2014; $0.1 million   **c.** 2007; $10.9 million   **d.** no; answers may vary.

45. a.
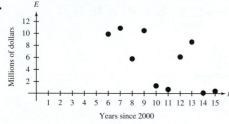
b. increase   **c.** increase; answers may vary.   **d.** Answers may vary.

47. a.
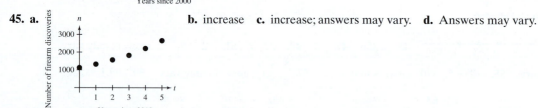
b. drivers older than 74 years   **c.** 16–20-year-old drivers   **d.** decrease   **e.** Answers may vary.

49. a.
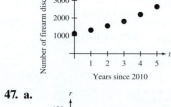
b. It has taken less time for recent inventions to reach mass use; answers may vary.   **c.** No; it took longer for the microwave to reach mass use than it did for several other earlier inventions.
d. Answers may vary.   **e.** It took longer for the automobile to reach mass use than it did for the earlier inventions of electricity and the telephone; answers may vary.

51. a. engineering; $65 thousand   **b.** education; $35 thousand   **c.** $55 thousand   **53.** Answers may vary; answers may vary; the points lie on the same vertical line; answers may vary.   **55.** Three possibilities; one possibility: $(6, 1)$ and $(6, 5)$; another possibility: $(-2, 1)$ and $(-2, 5)$ **57. a.** *x*-coordinate: positive; *y*-coordinate: positive   **b.** *x*-coordinate: negative; *y*-coordinate: positive
c. *x*-coordinate: negative; *y*-coordinate: negative   **d.** *x*-coordinate: positive; *y*-coordinate: negative   **59.** The points on the *x*-axis
61. Answers may vary.

Homework 1.3   **1.** 2   **3.** 6   **5.** $(2, 0)$ **7.** -2 **9.** -6 **11.** $(0, -1)$ **13. a. and b.**
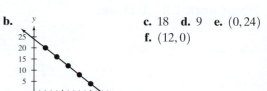
c. 18   **d.** 9   **e.** $(0, 24)$
f. $(12, 0)$

15. a. 18 thousand gallons   **b.** 4.2 hours   **c.** 30 thousand gallons   **d.** 5 hours   **17.** No; answers may vary.

19. a. 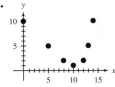   **b.** No; answers may vary.   **21. a.** 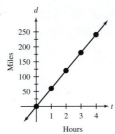   **b.** 150 miles   **c.** 3.5 hours

23. a. 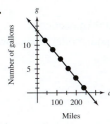   **b.** 17 thousand students   **c.** 7 years   **25. a.** 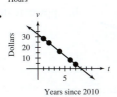   **b.** 2015   **c.** $(8, 0)$; the stock will have no value in 2018.   **d.** $(0, 32)$; the value of the stock was \$32 in 2010.

27. a. 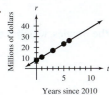   **b.** 6 gallons   **c.** 220 miles   **d.** $(260, 0)$; the gasoline tank will be empty after 260 miles of driving (if no refueling takes place).   **e.** $(0, 13)$; the car has a 13-gallon gasoline tank.

29. a. 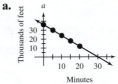   **b.** \$38 million   **c.** 2012   **d.** $(0, 8)$; the revenue was \$8 million in 2010.   **31. a.** **b.** 22 thousand feet   **c.** 30 minutes   **d.** underestimate; answers may vary.

33. a. 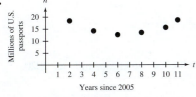   **b.** No; answers may vary.   **35.** No; the y-coordinate (not the y-intercept) of $(2, 5)$ is 5.

37. No; the y-coordinate of an x-intercept must be 0.   **39.** No; an x-intercept is a point that corresponds to an ordered pair with two coordinates (not a single number).   **41.** Answers may vary.   **43. a. i.** Answers may vary; one.   **ii.** Answers may vary; one.
iii. Answers may vary; one.   **b.** One; answers may vary.   **c.** One; answers may vary.   **45.** Answers may vary.   **47.** Answers may vary.

Homework 1.4

Throughout this section, answers may vary.

1. a. and c. 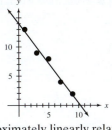   **b.** approximately linearly related   **d.** $(8, 3.2)$ **e.** $(5.9, 6)$ **f.** $(0, 14)$ **g.** $(10.3, 0)$

3. a. approximately linearly related   **b.**    **c.** \$3.74   **d.** \$5.53   **e.** up; answers may vary.

5. a. and b.
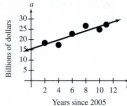
c. $25.3 billion   **d.** 2010

7. a. and c.
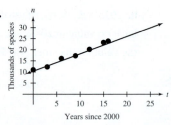
b. approximately linearly related   **d.** 2014   **e.** 28 thousand species

9. a. and b.
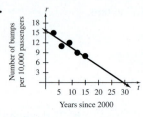
c. $(0, 16)$; in 2000, the voluntary bumping rate was 16 bumps per 10,000 passengers.
d. 2022   **e.** $(30, 0)$; in 2030, no passengers will volunteer to be bumped; model breakdown has likely occurred.

11. a. and b.
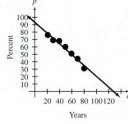
c. 80%   **d.** 59 years   **e.** $(128, 0)$; no 128-year-old Americans go to the movies, which is true because no American has lived that long.

13. a. and b.
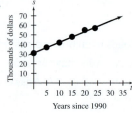
c. $(0, 31)$; the average salary was $31 thousand in 1990.   **d.** $67 thousand   **e.** 2020.

15. a. and b.
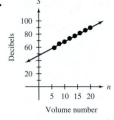
c. 88 decibels   **d.** volume number 11   **e.** $(0, 48)$; the sound level is 48 decibels when the volume number is 0; model breakdown has likely occurred.

17. a. 1.8 thousand injuries   **b.** 1.6 thousand injuries   **c.** overestimate; the line is above the data point; 0.2 thousand injuries

19. a. and b.

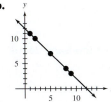

c. (3.7, 0); for a 3.7-hundred-foot-long cruise ship, there are no crew members; model breakdown has occurred.   **d.** 7.1 hundred feet   **e.** 11.1 hundred (1.11 thousand) crew members   **f.** 4.6 hundred crew members; less   **g.** Answers may vary.

21. overestimate; answers may vary.   **23.** Answers may vary.   **25.** Answers may vary.   **27.** Not necessarily; model breakdown may occur for some points on the line.

Chapter 1 Review Exercises

1. The total box office gross was $11.13 billion in 2015.   **2.** 2021   **3.** p; 60, 70; -12, 107; answers may vary.

4. a. The rectangles are not drawn to scale. Answers may vary.

b. W, L **c.** P

5.

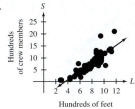

6. ◄─┼─●─┼─●─┼─●─┼─●─┼─►
        -6 -5 -4 -3 -2 -1 0

7. ◄─┼─●─┼─●─┼─┼─●─┼─●─┼─►p
        -5 -4 -3 -2 -1 0 1 2 3 4 5
        Millions of dollars

8.

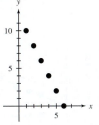

9. -6 **10.** -4

11. explanatory: a; response: p **12.** explanatory: t; response: a **13.** There were 540 U.S. billionaires in 2016.

14. In 2014, the revenue from ADHD drugs was $11 billion.   **15.**

16. a.

b. 2010   **c.** 1980

17. a. France; 76%   **b.** Belgium, Slovenia; 38%   **c.** 57%   **18.** -1 **19.** -5 **20.** 4   **21.** -6 **22.** $(0, -2)$ **23.** $(-4, 0)$

24. a. and b.

c. 1   **d.** 7   **e.** $(12, 0)$ **f.** $(0, 12)$ **25. a.**

b. linearly related

26. a.

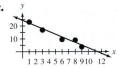

b. $2 million   **c.** 2012   **d.** $(0, 22)$; in 2010, the profit was $22 million.   **e.** $(11, 0)$; in 2021, the profit will be 0 dollars.

27. 0   **28. a. and c.**

b. approximately linearly related   **d.** $(5, 13.5)$ **e.** $(2, 20)$ **f.** $(0, 24.3)$ **g.** $(11.2, 0)$

29. a. and b. 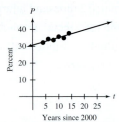 **c.** 40%   **d.** 2023   **e.** Answers may vary.   **30. a. and b.**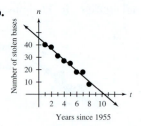

c. $(0, 45.2)$; Mays stole 45 bases in 1955, according to the model.   **d.** $(10.4, 0)$; Mays did not steal any bases in 1965, according to the model.   **e.** overestimate; yes; answers may vary.   **f.** underestimate; yes; answers may vary.

Chapter 1 Test

1. a. The rectangles are not drawn to scale. Answers may vary.   6 feet ☐ 4 feet ☐ 2 feet ▭   **b.** W, L **c.** A
6 feet   9 feet   18 feet

2. 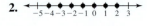   **3.** 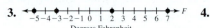   **4.**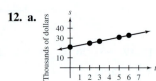

5. c; answers may vary.   **6.** In 2016, Joe Mauer's salary was $23 million.

7. a. 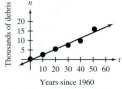   **b.** $(30, 18.5)$; Americans in the age group 26–34 years are the most likely to be without health insurance.   **c.** $(70, 1.6)$; Americans who are older than 64 years are the least likely to be without health insurance.

8. -3 **9.** 4   **10.** $(0, -1)$ **11.** $(2, 0)$ **12. a.** 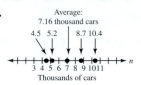   **b.** $29 thousand   **c.** 7 years   **d.** $(0, 21)$; when the person was hired, her salary was $21 thousand.

13. Answers may vary.   **14. a. and b.**   **c.** 1992   **d.** 14.2 thousand debris   **e.** 1.8 thousand debris

15. Answers may vary; answers may vary; no; answers may vary.

Chapter 2

Homework 2.1   **1.** 8   **3.** 3   **5.** 42   **7.** 2   **9.** 12   **11.** 36   **13.** 43.96; the total cost of 4 albums is $43.96.   **15.** $102.69

17. a.

Tuition (dollars)	Total Cost (dollars)
400	400 + 20
401	401 + 20
402	402 + 20
403	403 + 20
t	$t + 20$

$t + 20$ **b.** 437; if the tuition is $417, then the total cost is $437.

19. a.

Number of Shares	Total Value (dollars)
1	16.30 · 1
2	16.30 · 2
3	16.30 · 3
4	16.30 · 4
n	$16.30n$

$16.30n$ **b.** 114.1; the total value of 7 shares is $114.10.

21. a.

Number of Credit Hours	Total Cost (dollars)
1	$153 \cdot 1$
2	$153 \cdot 2$
3	$153 \cdot 3$
4	$153 \cdot 4$
n	$153n$

$153n$ **b.** 2295; the total cost for 15 credit hours of classes is $2295.   **23.** $x + 4$; 12   **25.** $x \div 2$; 4   **27.** $x - 5$; 3   **29.** $7x$; 56   **31.** $16 \div x$; 2   **33.** The number divided by 2   **35.** 7 minus the number   **37.** The number plus 5   **39.** The product of 9 and the number   **41.** The difference of the number and 7   **43.** The number times 2   **45.** 9   **47.** 3   **49.** 18   **51.** xy; 27   **53.** $x - y$; 6   **55.** 186; the car traveled 186 miles when driven for 3 hours at 62 mph.   **57.** 20; if a car can travel 240 miles on 12 gallons of gasoline, the car's gas mileage is 20 miles per gallon.

59. $170,000   **61. a.** 5, 10, 15, 20; the person earns $5, $10, $15, and $20 for working 1, 2, 3, and 4 hours, respectively.
b. $5 per hour   **c.** Answers may vary.   **63. a.** 50, 100, 150, 200; the person drives 50, 100, 150, and 200 miles in 1, 2, 3, and 4 hours, respectively.   **b.** 50 mph   **c.** Answers may vary.   **65.** Answers may vary.   **67.** Answers may vary.   **69.** Answers may vary.

Homework 2.2

1. 7   **3.** $2 \cdot 2 \cdot 5$ **5.** $2 \cdot 2 \cdot 3 \cdot 3$ **7.** $3 \cdot 3 \cdot 5$ **9.** $2 \cdot 3 \cdot 13$ **11.** $\frac{3}{4}$ **13.** $\frac{1}{4}$ **15.** $\frac{3}{5}$ **17.** $\frac{2}{5}$ **19.** $\frac{1}{5}$ **21.** $\frac{5}{6}$
23. $\frac{2}{15}$ **25.** $\frac{3}{10}$ **27.** $\frac{5}{3}$ **29.** $\frac{5}{6}$ **31.** $\frac{2}{3}$ **33.** $\frac{2}{15}$ **35.** $\frac{5}{7}$ **37.** $\frac{3}{4}$ **39.** $\frac{1}{5}$ **41.** $\frac{1}{3}$ **43.** $\frac{3}{4}$ **45.** $\frac{19}{12}$ **47.** $\frac{14}{3}$ **49.** $\frac{1}{9}$ **51.** $\frac{17}{63}$
53. $\frac{11}{5}$ **55.** 1   **57.** 599   **59.** undefined   **61.** 0   **63.** 1   **65.** 0   **67.** $\frac{1}{3}$ **69.** $\frac{9}{5}$ **71.** $\frac{1}{3}$ **73.** 0.17   **75.** 1.33   **77.** 0.43
79. Answers may vary.   **81.** $\frac{1}{10}$ square mile   **83.** $\frac{1}{4}$ of the course points   **85.** the quotient of the number and 3

87.

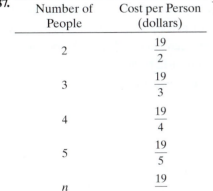

Number of People	Cost per Person (dollars)
2	$\frac{19}{2}$
3	$\frac{19}{3}$
4	$\frac{19}{4}$
5	$\frac{19}{5}$
n	$\frac{19}{n}$

$\frac{19}{n}$ **89. a. i.** $\frac{5}{9}$ **ii.** $\frac{5}{4}$ **iii.** $\frac{3}{2}$ **iv.** $\frac{1}{6}$ **b.** Answers may vary.   **91.** Answers may vary.
93. Answers may vary.   **95. a. and b.**   **c.** 0.04 inch

d. 2016   **e.** $(87, 0)$; there will be no grass on putting surfaces in 2037; this prediction is highly unlikely.   **97. a. i.** -6 **ii.** -8 **iii.** -7 **b.** The results are all negative.
c. -9 **d.** Answers may vary.   **99.** Answers may vary.

Homework 2.3

1. 4   **3.** -7 **5.** 3   **7.** 8   **9.** -4 **11.** -7 **13.** -5 **15.** -5 **17.** 2   **19.** -3 **21.** -10 **23.** -3 **25.** 0   **27.** 0
29. -13 **31.** -22 **33.** -1145 **35.** 0   **37.** -6.7 **39.** -4.8 **41.** -97.3 **43.** $\frac{2}{7}$ **45.** $-\frac{1}{4}$ **47.** $-\frac{3}{4}$ **49.** $\frac{7}{12}$ **51.** 6221.4
53. $-97,571.14$ **55.** -0.11 **57.** -1 **59.** -6 **61.** $x + 2$; -4 **63.** $-5 + x$; -11 **65.** $175
67.

Check No.	Date	Description of Transaction	Payment	Deposit	Balance
					-89.00
	7/18	Transfer		300.00	211.00
3021	7/22	State Farm	91.22		119.78
3022	7/22	MCI	44.26		75.52
	7/31	Paycheck		870.00	945.52

69. -2871 dollars   **71.** -1633 dollars
73. 4°F

75. a.

Weight before Diet (pounds)	Weight after Diet (pounds)
160	160 + (−20)
165	165 + (−20)
170	170 + (−20)
175	175 + (−20)
B	B + (−20)

$B + (-20)$ **b.** 149; the person's current weight is 149 pounds.

77. a.

Deposit (dollars)	New Balance (dollars)
50	−80 + 50
100	−80 + 100
150	−80 + 150
200	−80 + 200
d	−80 + d

$-80 + d$ **b.** 45; the new balance is \$45.   **79.** negative   **81.** The numbers are equal in absolute value and opposite in sign, or the numbers are both 0.

83. a. 3   **b.** 4   **c.** 6   **d.** no; answers may vary.

Homework 2.4   **1.** −2   **3.** −6   **5.** 9   **7.** −1   **9.** −3   **11.** 11   **13.** −6   **15.** −79   **17.** 420   **19.** −5.4   **21.** −11.3   **23.** 5.7   **25.** 15.98   **27.** −1   **29.** $\frac{1}{2}$ **31.** $\frac{3}{4}$ **33.** $-\frac{13}{24}$ **35.** 2   **37.** −2   **39.** $-\frac{1}{4}$ **41.** −2.7   **43.** −7   **45.** −2   **47.** −3128.17   **49.** 112,927.91   **51.** −0.95   **53.** −12°F   **55.** 11°F   **57. a.** −12°F   **b.** −6°F   **c.** Answers may vary.   **59.** 20,602 feet   **61. a.** 0.7 percentage point, −2.5 percentage points, 4.5 percentage points, −7.7 percentage points, 0.5 percentage point, 9.1 percentage points, −0.2 percentage point, −6.1 percentage points, 2.5 percentage points   **b.** 9.1 percentage points   **c.** no; answers may vary.   **63. a.** 180 thousand cars   **b.** From 2009 to 2010, from 2011 to 2012   **c.** From 2008 to 2009, from 2010 to 2011, from 2012 to 2015

65. a.

Score on the Second Exam (points)	Change in Score (points)
80	80 − 87
85	85 − 87
90	90 − 87
95	95 − 87
p	p − 87

$p - 87$ **b.** −6; the score decreased by 6 points.

67. a.

Change in Enrollment	Current Enrollment
100	100 + 24,500
200	200 + 24,500
300	300 + 24,500
400	400 + 24,500
c	c + 24,500

$c + 24,500$ **b.** 23,800; the current enrollment is 23,800 students due to a decrease in enrollment of 700 students in the past year.

69. −3   **71.** −7   **73.** 9   **75.** −3 − x; 2   **77.** x − 8; −13   **79.** x − (−2); −3   **81.** Answers may vary.   **83. a. i.** 2   **ii.** 8   **iii.** 5   **b.** Answers may vary.   **85. a. i.** −10   **ii.** −24   **iii.** −63   **b.** Answers may vary.   **c.** −21   **d.** Answers may vary.

87. a. 3   **b.** −3   **c.** They are equal in absolute value and opposite in sign.   **d.** −6, 6; they are equal in absolute value and opposite in sign.   **e.** Answers may vary.   **f.** They are equal in absolute value and opposite in sign.   **89.** Positive

Homework 2.5   **1.** 0.63   **3.** 0.09   **5.** 8%   **7.** 0.073   **9.** 5.2%   **11.** 2450 cars   **13.** \$13.80   **15.** 11,523 female undergraduates   **17.** −12   **19.** 18   **21.** −1   **23.** −8   **25.** −5   **27.** 8   **29.** 555   **31.** −39   **33.** 0.08   **35.** −0.975   **37.** −0.3   **39.** −9   **41.** 4   **43.** $-\frac{1}{10}$ **45.** $\frac{1}{15}$ **47.** $-\frac{9}{14}$ **49.** $\frac{15}{14}$ **51.** −3   **53.** 13   **55.** 6   **57.** −100   **59.** $-\frac{1}{4}$ **61.** $\frac{4}{3}$ **63.** $-\frac{11}{12}$ **65.** $-\frac{9}{20}$ **67.** $-\frac{4}{5}$ **69.** $\frac{3}{4}$ **71.** $-\frac{1}{2}$ **73.** 1   **75.** $-\frac{1}{24}$ **77.** 10,252.84   **79.** −6.78   **81.** 0.48   **83.** −8.07   **85.** −24   **87.** $-\frac{3}{2}$ **89.** −48   **91.** $\frac{1}{2}$ **93.** $\frac{w}{2}$; −4   **95.** w(−5); 40

97. $\frac{3}{4}$ **99.** $\frac{2.25}{1}$; One World Trade Center is 2.25 times taller than the John Hancock Tower.   **101.** $\frac{1.31}{1}$; the number of U.S.

billionaires in 2016 is 1.31 times the number of U.S. billionaires in 2011.   **103. a.** $\frac{0.8 \text{ red bell pepper}}{1 \text{ black olive}}$; for each black olive used,

0.8 red bell pepper is needed.   **b.** $\frac{1.25 \text{ black olives}}{1 \text{ red bell pepper}}$; for each red bell pepper used, 1.25 black olives are required.   **105. a.** $\frac{16.10}{1}$; the

FTE enrollment at Texas A&M University is 16.10 times larger than that at St. Olaf College.   **b.** $\frac{3.24}{1}$; the number of FTE faculty at

University of Massachusetts Amherst is 3.24 times greater than that at Butler University.   **c.** Butler University: $\frac{11.37}{1}$; St. Olaf College:

$\frac{11.93}{1}$; Stonehill College: $\frac{12.14}{1}$; University of Massachusetts Amherst: $\frac{18.08}{1}$; Texas A&M University: $\frac{23.00}{1}$ **d.** Texas A&M University;

Butler University   **e.** No; answers may vary.   **107. a.** $\frac{2.39}{1}$ **b.** For each \$1 the person pays to her MasterCard account, she should pay

about \$2.39 to her Discover account.   **109.** -3162 dollars   **111.** -29.52 dollars   **113. a.** -6 **b.** 8   **c.** A negative number times a

negative number is equal to a positive number.   **d.** Answers may vary.   **115.** $\frac{a}{b} = \frac{-a}{-b}, \frac{-a}{b} = \frac{a}{-b} = -\frac{a}{b} = -\frac{-a}{-b}$ **117.** Answers may

vary.   **119.** One number is positive and one number is negative.   **121.** a or b is zero.   **123. a.** 16   **b.** 1   **c.** yes; answers may vary.

125. a. Answers may vary.   **b.** Answers may vary.   **c.** 0

Homework 2.6   **1.** 64   **3.** 32   **5.** -64 **7.** 64   **9.** $\frac{36}{49}$ **11.** 12   **13.** -18 **15.** 10   **17.** -3 **19.** $-\frac{5}{2}$ **21.** $-\frac{2}{3}$ **23.** 10   **25.** -17

27. -50 **29.** -10 **31.** -14 **33.** 20   **35.** 15   **37.** -27 **39.** -57 **41.** $\frac{1}{2}$ **43.** 27   **45.** -48 **47.** 1   **49.** -17 **51.** 5   **53.** 2

55. -41 **57.** 3   **59.** $-\frac{9}{7}$ **61.** -9 **63.** 48   **65.** -613.37 **67.** -1.54 **69.** -1.33 **71.** -14 **73.** -8 **75.** -5 **77.** 40   **79.** $\frac{5}{4}$

81. $-\frac{13}{7}$ **83.** $-\frac{5}{2}$ **85.** $\frac{3}{5}$ **87.** -27 **89.** -12 **91.** 32   **93.** $5 + (-6)x; 29$ **95.** $\frac{x}{-2} - 3; -1$

97. a.

Years since 2000	Congressional Pay (thousands of dollars)
0	$3.6 \cdot 0 + 141.3$
1	$3.6 \cdot 1 + 141.3$
2	$3.6 \cdot 2 + 141.3$
3	$3.6 \cdot 3 + 141.3$
4	$3.6 \cdot 4 + 141.3$
t	$3.6t + 141.3$

$3.6t + 141.3$ **b.** 166.5; congressional pay was \$166.5 thousand in 2007.
c. \$175.4 thousand

99. a.

Years since 1980	Population (thousands)
0	$-2.1 \cdot 0 + 152$
1	$-2.1 \cdot 1 + 152$
2	$-2.1 \cdot 2 + 152$
3	$-2.1 \cdot 3 + 152$
4	$-2.1 \cdot 4 + 152$
t	$-2.1 \cdot t + 152$

$-2.1t + 152$ **b.** 63.8; Gary's population will be 63.8 thousand in 2022.   **101.** 19 attacks
103. 49.0 million people   **105.** 2.6 million subscribers
107. 35.6 million people   **109.** 8 cubic feet   **111.** Answers may vary; 25.
113. no; answers may vary.   **115. a.** -2 **b.** 3   **c.** Answers may vary; -2.
117. a. 24   **b.** 24   **c.** The results are equal.   **d.** $-40; -40$; the results
are equal.   **e.** Answers may vary.   **f.** yes; answers may vary.
g. Answers may vary.
119. a. $4; 1; 0; 1; 4$ **b.** nonnegative   **c.** nonnegative
121. a. Answers may vary.   **b.** Answers may vary.   **c.** $0, 1$

Chapter 2 Review Exercises

1. 6   **2.** -12 **3.** -3 **4.** 10   **5.** -16 **6.** -4 **7.** -3 **8.** 12   **9.** -1 **10.** $-\frac{3}{2}$ **11.** -11 **12.** -13 **13.** -16 **14.** -4 **15.** -20

16. -26 **17.** -12 **18.** 4   **19.** 14   **20.** 4   **21.** 10.9   **22.** $-\frac{2}{15}$ **23.** $\frac{5}{6}$ **24.** $\frac{7}{9}$ **25.** $\frac{1}{24}$ **26.** $-\frac{1}{6}$ **27.** 81   **28.** -81 **29.** 16   **30.** $\frac{27}{64}$

31. -54 **32.** 3   **33.** 0   **34.** $\frac{2}{3}$ **35.** $-\frac{8}{11}$ **36.** -19 **37.** -3 **38.** 58   **39.** $\frac{3}{4}$ **40.** $-\frac{4}{5}$ **41.** -8.68 **42.** 13.79   **43.** $2\frac{1}{6}$ yards

44. −4095.49 dollars   **45.** −4700 feet   **46. a.** −12°F   **b.** −4°F   **c.** Answers may vary.   **47. a.** \$193 million   **b.** −61 million dollars

c. from 2000 to 2004; \$193 million   **d.** from 2008 to 2012; \$102 million   **48.** $\dfrac{1.41}{1}$; the average number of texts sent or received per

day by female teen cell-phone users is 1.41 times larger than the average number of texts sent or received per day by male teen

cell-phone users.   **49.** 0.75   **50.** 0.029   **51.** 0.18 million (180 thousand) Americans   **52.** 4986 people   **53.** −4394.4 dollars   **54.** −10

55. 57   **56.** −2   **57.** $-\dfrac{11}{4}$ **58.** 55   **59.** $-\dfrac{1}{2}$ **60.** $x+5; 2$ **61.** $-7-x; -4$ **62.** $2-x(4); 14$ **63.** $1+\dfrac{-24}{x}; 9$ **64.** 50; if the total

cost is \$650 and there are 13 players on the team, the cost is \$50 per player.

65. a.

Time (hours)	Volume of Water (cubic feet)
0	−50·0 + 400
1	−50·1 + 400
2	−50·2 + 400
3	−50·3 + 400
4	−50·4 + 400
t	−50t + 400

−50t + 400   **b.** 50; there will be 50 cubic feet of water in the basement after 7 hours of pumping.   **66.** 11.9 million Americans

Chapter 2 Test

1. −13   **2.** 63   **3.** −6   **4.** −8   **5.** $\dfrac{1}{2}$ **6.** 3   **7.** −25   **8.** −0.08   **9.** $-\dfrac{45}{4}$ **10.** $\dfrac{13}{40}$ **11.** 81   **12.** −16   **13.** 6   **14.** −17   **15.** $-\dfrac{21}{4}$

16. −4°F **17. a.** 3.2 audits per 1000 tax returns   **b.** −1.2 audits per 1000 tax returns   **c.** from 2003 to 2005; 3.2 audits per 1000 tax

returns   **18.** $\dfrac{2.74}{1}$; the number of background checks conducted for gun purchases in 2015 was 2.74 times larger than the number of

background checks conducted for gun purchases in 2000.   **19.** −33   **20.** $-\dfrac{2}{3}$ **21.** 11   **22.** 124   **23.** $2x-3x; 5$ **24.** $\dfrac{-10}{x}-6; -4$

25. a.

Years since 2014	First-Class Mail Volume (billions of pieces)
0	−2.1(0) + 63.8
1	−2.1(1) + 63.8
2	−2.1(2) + 63.8
3	−2.1(3) + 63.8
4	−2.1(4) + 63.8
t	−2.1t + 63.8

−2.1t + 63.8   **b.** 49.1; the first-class mail volume will be 49.1 billion pieces in 2021.

26. 888 bats

Cumulative Review of Chapters 1 and 2

1. a. The rectangles are not drawn to scale. Answers may vary.   **b.** W, L **c.** P

2.   **3.**   **4.** −5   **5.** explanatory variable: t; response variable: V

6. a.

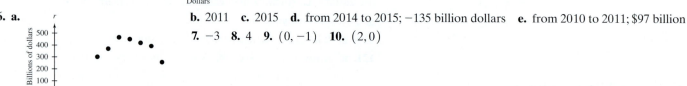

b. 2011   **c.** 2015   **d.** from 2014 to 2015; −135 billion dollars   **e.** from 2010 to 2011; \$97 billion

7. −3   **8.** 4   **9.** $(0, -1)$ **10.** $(2, 0)$

11. a.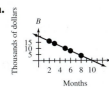
b. $12 thousand   **c.** 7 months after the person was laid off   **d.** $(0, 20)$; when the person was laid off, the balance was $20 thousand.   **e.** $(10, 0)$; there was no money in the account 10 months after the person was laid off.

12. a.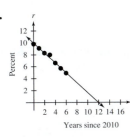
b. 7.5%; −0.5 percentage point   **c.** $(0, 10.0)$; in January 2010, the unemployment rate was 10.0%.
d. $(12, 0)$; in January 2022, the unemployment rate will be 0%; model breakdown has occurred.

13. $\frac{3}{4}$ **14.** -4 **15.** $\frac{18}{25}$ **16.** $-\frac{11}{24}$ **17.** -17 **18.** $-\frac{2}{9}$ **19.** $-25°F$ **20.** -1865 dollars   **21.** $-\frac{1}{2}$

22. 49   **23.** $x - \frac{-12}{x}$; -7 **24.** $7 + (-2)x$; 15   **25.** 7.14; the percent growth of the investment was 7.14%.

26. a.

Years since 2013	Enrollment (thousands of students)
0	$-11.3 \cdot 0 + 67$
1	$-11.3 \cdot 1 + 67$
2	$-11.3 \cdot 2 + 67$
3	$-11.3 \cdot 3 + 67$
4	$-11.3 \cdot 4 + 67$
t	$-11.3t + 67$

$-11.3t + 67$ **b.** 10.5; in 2018, the enrollment will be 10.5 thousand students.

Chapter 3

Homework 3.1   **1.** $(-3, -10), (2, 0)$ **3.** $(0, 7), (4, -5)$

5. $(0, 2)$ **7.** $(0, -4)$ **9.** $(0, 0)$ **11.** $(0, 0)$ **13.** $(0, 0)$ **15.** $(0, 0)$ 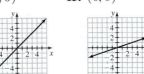   **17.** $(0, 0)$

19. $(0, 1)$ **21.** $(0, -3)$ **23.** $(0, 5)$ **25.** $(0, -3)$ **27.** $(0, -3)$ **29.** $(0, 1)$

31.   **33.**   **35.**   **37.**   **39.**
$x = 0$

41. a.

x	y
0	-3
1	-1
2	1

Answers may vary.   **b.**   **c.** For each solution, the y-coordinate is 3 less than twice the x-coordinate.

43. a. i. 7; one   **ii.** 13; one   **iii.** -5; one   **b.** one; answers may vary.   **c. i.** Answers may vary; one   **ii.** Answers may vary; one
iii. Answers may vary; one   **d.** one; answers may vary.   **e.** one; answers may vary.

45. a. i.   **ii.**   **iii.**   **b.** *x*-intercept: $(0,0)$; *y*-intercept: $(0,0)$

x-intercept: $(0,0)$;   *x*-intercept: $(0,0)$;   *x*-intercept: $(0,0)$;
y-intercept: $(0,0)$ *y*-intercept: $(0,0)$ *y*-intercept: $(0,0)$

47. Answers may vary. **49.** 3 **51.** 0 **53.** 4 **55.** −2 **57.** C, D, E **59.** Answers may vary; infinitely many **61.** $y = x + 3$ **63.** $y = x$

65. $x = -3$ **67. a.** Answers may vary. **b.** $y = 3x$ **69.**   **71.** Answers may vary.

73. vertical line; answers may vary. **75.** no; answers may vary. **77.** Answers may vary. **79.** Answers may vary.

Homework 3.2

1. a.

Drink Cost (dollars)	Total Cost (dollars)
d	*T*
2	2 + 3
3	3 + 3
4	4 + 3
5	5 + 3
d	*d* + 3

$T = d + 3$ **b.** The units for both of the expressions T and $d + 3$ are dollars.

c.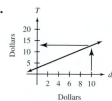

d. $(0, 3)$; if the person does not buy any drinks, then the total cost is $3. **e.** $13

3. a.

Number of Credits	Total Cost (dollars)
c	*T*
3	94 · 3
6	94 · 6
9	94 · 9
12	94 · 12
c	94 · *c*

$T = 94c$ **b.** The units for both of the expressions T and $94c$ are dollars.

c.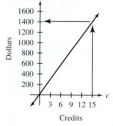

d. $1410

5. a.

Time at Company (years)	Salary (thousands of dollars)
t	*s*
0	3 · 0 + 24
1	3 · 1 + 24
2	3 · 2 + 24
3	3 · 3 + 24
4	3 · 4 + 24
t	3 · *t* + 24

$s = 3t + 24$ **b.** The units for both of the expressions s and $3t + 24$ are thousands of dollars. **c.**

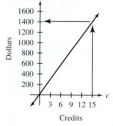

d. $(0, 24)$; the starting salary is $24 thousand. **e.** 6 years

7. a.

Years Since 2014	Sales (millions of games)
t	s
0	$80 - 12 \cdot 0$
1	$80 - 12 \cdot 1$
2	$80 - 12 \cdot 2$
3	$80 - 12 \cdot 3$
4	$80 - 12 \cdot 4$
t	$80 - 12t$

$s = 80 - 12t$ **b.** The units for both of the expressions s and $80 - 12t$ are millions of games.

c.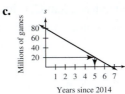

d. $(0, 80)$, in 2014, the sales were 80 million games.   **e.** 2019

9. a. $a = r - 5$ **b.** The units for both of the expressions a and $r - 5$ are minutes.   **c.** 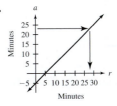   **d.** 28 minutes

11. a. $d = 60t$ **b.** The units for both of the expressions d and $60t$ are miles.

c. 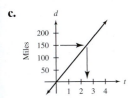   **d.** $(0, 0)$; the person will not travel any distance in 0 hours of driving.   **e.** 2.5 hours

13. a. $T = 46u + 36$ **b.** The units for both of the expressions T and $46u + 36$ are dollars.   **c.** \$726

15. a. $p = 3n + 62$, where p is the pressure (in psi) after n pumps; answers may vary.   **b.** The units for both of the expressions p and $3n + 62$ are psi; answers may vary.   **c.** 113 psi   **17. a.** $g = 11 - 2t$ **b.** The units for both of the expressions g and $11 - 2t$ are gallons.

c. 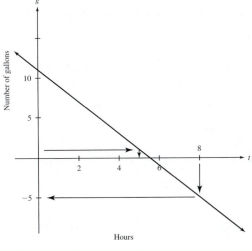   **d.** 5 hours   **e.** -5 gallons; model breakdown has occurred.

19. a.

Years Since 2014	Sales (millions of pounds)
t	s
0	167.1
1	180.7
2	194.3
3	207.9
4	221.5

Answers may vary.   **b.** $s = 13.6t + 167.1$ **c.**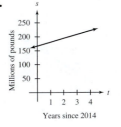

21. $n = 45$;

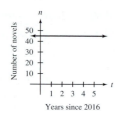

23. a. 2014   **b.** 0.46 thousand independent CD and record stores   **c.** $(0, 1.98)$; in 2010, there were 1.98 thousand independent CD and record stores.   **d.** $(15.6, 0)$; there will be neither independent CD stores nor independent record stores in 2026; model breakdown has likely occurred.

25. a. i. 4   **ii.** 0   **b. i.** -4 **ii.** 0   **27. a. i.** 5   **ii.** 3   **b. i.** -5 **ii.** -3

29. a.

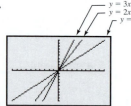

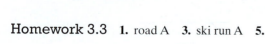

b. $y = x, y = 2x, y = 3x$ **c.** Answers may vary.   **d.** Answers may vary.   **31.** Answers may vary.

Homework 3.3   **1.** road A   **3.** ski run A   **5.**

run: 2; rise: 3; slope: $\frac{3}{2}$ **7.**

run: 2; rise: -4; slope: -2

9.

run: 6; rise: 4; slope: $\frac{2}{3}$ **11.**

run: 2; rise: -4; slope: -2 **13.** 2; increasing   **15.** -4; decreasing

17. $\frac{1}{2}$; increasing   **19.** $-\frac{1}{3}$; decreasing   **21.** -1; decreasing   **23.** $-\frac{1}{2}$; decreasing   **25.** 2; increasing   **27.** $\frac{3}{2}$; increasing

29. $-\frac{4}{5}$; decreasing   **31.** $\frac{2}{3}$; increasing   **33.** $-\frac{1}{2}$; decreasing   **35.** 0; horizontal   **37.** undefined slope; vertical   **39.** -9.25; decreasing

41. 1.14; increasing   **43.** -0.21; decreasing   **45.** 0.71; increasing   **47.** $\frac{2}{3}$ **49.** -3 **51. a.** negative   **b.** positive   **c.** undefined

d. zero   **53.** Answers may vary.   **55.** Answers may vary.   **57.** Answers may vary.   **59.** Answers may vary.   **61.** Answers may vary; $\frac{4}{3}$

63. Answers may vary; $\frac{13}{4}$ **65.** Answers may vary; yes   **67. a.** Answers may vary.   **b.** no   **c.** $|-2| = 2; |-3| = 3$; yes

d. Answers may vary.   **69.** Answers may vary.   **71.** Answers may vary.   **73.** Answers may vary.   **75.** not necessarily; answers may vary.

Homework 3.4

1. 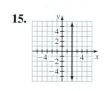   **3.**   **5.** 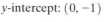   **7.**   **9.**   **11.**   **13.**

15.   **17.** slope: $\frac{2}{3}$;   **19.** slope: $-\frac{1}{3}$;   **21.** slope: $\frac{4}{3}$;   **23.** slope: $-\frac{4}{5}$;

y-intercept: $(0, -1)$ y-intercept: $(0, 4)$ y-intercept: $(0, 2)$ y-intercept: $(0, -1)$

25. slope: $\frac{1}{2}$;
y-intercept: $(0, 0)$

27. slope: $-\frac{5}{3}$;
y-intercept: $(0, 0)$

29. slope: 4;
y-intercept: $(0, -2)$

31. slope: -2;
y-intercept: $(0, 4)$

33. slope: -4;
y-intercept: $(0, -1)$

35. slope: 1;
y-intercept: $(0, 1)$

37. slope: -1;
y-intercept: $(0, 3)$

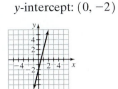

39. slope: -3;
y-intercept: $(0, 0)$

41. slope: 1;
y-intercept: $(0, 0)$

43. slope: 0;
y-intercept: $(0, -3)$

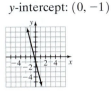

45. slope: 0;
y-intercept: $(0, 0)$

$y = 0$

47. a.

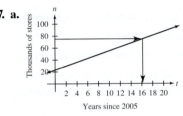

Years since 2005

b. 2021

49. a.

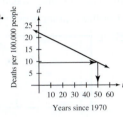

Years since 1970

b. 2020

51. a. m is positive; b is negative **b.** m is zero; b is negative **c.** m is negative; b is positive **d.** m is positive; b is positive

53. Answers may vary. **55.** Answers may vary. **57.** Answers may vary. **59.** $y = 3x - 4$ **61.** $y = -\frac{6}{5}x + 3$ **63.** $y = -\frac{2}{7}x$

65. $y = 2x + 3$ **67.** perpendicular **69.** neither **71.** parallel **73.** neither **75.** parallel **77.** perpendicular **79.** no; answers may vary.

81. -2 **83.** 3 **85.** $\frac{1}{3}$ **87.** no; answers may vary. **89.**   Answers may vary. **91. a.**   **b.**

x	y
-2	1
0	2
2	3

Answers may vary. **c.** For each solution, the y-coordinate is two more than half the x-coordinate. **93.** Answers may vary;

95. a.   **b.** $y = 2x + 3$ **97. a.** The slope of each line is zero. **b.** 0 **99. a.** k; answers may vary.

b. b; answers may vary. **101.** Answers may vary.

Homework 3.5

1. 300 gallons per hour **3.** -1650 feet per minute **5.** 1.30 shark attacks per year **7.** -1.3 million visits per year **9.** -2.67 percentage points per year **11.** \$113 per credit hour **13.** \$6630 per person **15. a.** yes; 4; the revenue increases by \$4 million per year.

b. i. $r = 4t + 3$ **ii.**

Years since 2010 t	Revenue (millions of dollars) r
0	3
1	7
2	11
3	15
4	19

Answers may vary.   **iii.**

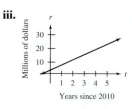

17. a. yes; -12; the number of sites is decreasing by about 12 sites per year.   **b.** $(0, 348)$; in 2014, there were 348 sites.
c. $n = -12t + 348$ **d.** The units for both of the expressions n and $-12t + 348$ are sites.   **e.** 264 sites   **19. a.** -650; the balance
declines by \$650 per month.   **b.** $(0, 4700)$; the balance is \$4700 on September 1.   **c.** $B = -650t + 4700$ **d.** The units for both of
the expressions B and $-650t + 4700$ are dollars.   **e.** \$800   **21. a.** $T = 592c + 16$, where T is the total one-semester cost (in dollars)
of c credits of classes plus the part-time student fee; answers may vary.   **b.** The units for both of the expressions T and $592c + 16$ are
dollars; answers may vary.   **c.** 592; the tuition increases by \$592 per credit.   **d.** \$5344   **23. a.** -0.02; the car uses 0.02 gallon of gasoline per mile.   **b.** $(0, 11.9)$; there were 11.9 gallons of gasoline in the tank at the start of the trip.   **c.** $G = -0.02d + 11.9$ **d.** The
units for both of the expressions G and $-0.02d + 11.9$ are gallons.   **e.** 10.5 gallons   **25. a.** -21.1; worldwide digital camera sales are
decreasing by about 21.1 million cameras per year.   **b.** $(0, 130.6)$; in 2010, worldwide digital camera sales were 130.6 million cameras.
c. 4 million cameras   **27. a.**

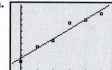

yes   **b.** 3.23; the percentage of adult Internet users who use social networking sites
is increasing by 3.23 percentage points per year.   **c.** $3.5, 3.5, 2$ (all in percentage points per year); answers may vary.   **d.** $(0, 61.10)$; in 2010, 61.1% of
adult Internet users used social networking sites.   **e.** 93.4%.

29. a.

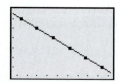

yes   **b.** 3.55; 0   **c.** -0.0025; the qualifying core GPA decreases by 0.0025 for an increase of 1 point
on the SAT.   **d.** $(0, 4.55)$; the qualifying core GPA is 4.55 for an SAT score of 0; model breakdown has occurred because the highest possible GPA is 4.0 and the lowest possible SAT score is
400 points.   **e.** no

31. a. yes; answers may vary.   **b.** 60 miles per hour   **33. a.** yes; answers may vary.   **b.** -1.5 gallons per hour   **35.** 1.75; worldwide
revenue increases by \$1.75 million as U.S. revenue increases by \$1 million.   **37.** -0.03; a dog breed that has a maximum weight
1 pound greater than another dog breed tends to have an average life expectancy 0.03 year less than the other dog breed.   **39.** yes; 30;
the total charge increases by \$30 per hour.   **41.** equations 1 and 3   **43.**

Equation 1		Equation 2		Equation 3		Equation 4	
x	y	x	y	x	y	x	y
0	3	0	99	21	16	43	17
1	8	1	92	22	14	44	20
2	13	2	85	23	12	45	23
3	18	3	78	24	10	46	26
4	23	4	71	25	8	47	29

45. Set 1: $y = 2x + 5$; Set 2: $y = -3x + 20$; Set 3: $y = 8x + 21$;
Set 4: $y = -5x + 9$

47.   a. and d.

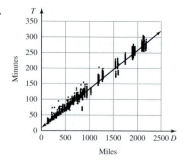

b. Each vertically aligned clump of data points might represent flights that travel the
same route (in either direction).   **c.** Delta probably offers more routes less than
1000 miles; some vertically aligned clumps of data points might represent several
routes.   **e.** 0.12; a route that is 1 mile longer than another route tends to have an
airborne time that is 0.12 minute greater.   **f.** 500 miles per hour   **g.** 784 miles
per hour; it is unlikely an airplane world fly so much faster than the typical speed of
500 miles per hour because much more fuel would be consumed.

49. The value of y increases by 7.   **51.** Answers may vary.   **53. a.** The slope 3 is the number multiplied by x.   **b.** If the run is 1, the
rise is 3.   **c.** As the value of x increases by 1, the value of y increases by 3.   **d.** Answers may vary.   **55.** $2m$; answers may vary.
57. Answers may vary.

Chapter 3 Review Exercises

1. $(-3, 9), (4, -5)$ **2.** -1 **3.** -3 **4.** -2 **5.** -2 **6.** -4 **7.** 4   **8.** airplane A   **9.** 2; increasing   **10.** -1; decreasing   **11.** $\frac{1}{2}$; increasing   **12.** $-\frac{1}{2}$; decreasing   **13.** -1; decreasing   **14.** $-\frac{1}{3}$; decreasing   **15.** $-\frac{9}{8}$; decreasing   **16.** $-\frac{2}{3}$; decreasing   **17.** undefined slope; vertical   **18.** 0; horizontal   **19.** -3.62; decreasing   **20.** 0.94; increasing   **21.** Answers may vary.

22.

23.

24.

25. slope: $\frac{3}{4}$; y-intercept: $(0, -1)$

26. slope: $-\frac{1}{2}$; y-intercept: $(0, 3)$

27. slope: $-\frac{2}{5}$; y-intercept: $(0, -1)$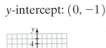

28. slope: $\frac{2}{3}$; y-intercept: $(0, 0)$

29. slope: -4; y-intercept: $(0, 0)$

30. slope: 2; y-intercept: $(0, -4)$

31. slope: -3; y-intercept: $(0, 1)$

32. slope: 1; y-intercept: $(0, 2)$

33. slope: 0; y-intercept: $(0, -5)$

34.

35.

36. a.

x	y
-1	3
0	1
1	-1

Answers may vary.   **b.**   **c.** For each solution, the y-coordinate is 1 more than -2 times the x-coordinate.

37. a. $B = -3t + 19$ **b.** 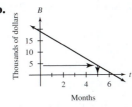   **c.** $(0, 19)$; when the person first lost his job, the balance was \$19 thousand.

d. 5 months   **38.** $x = 5$ **39.** $y = -\frac{2}{3}x + 4$ **40.** neither   **41.** perpendicular   **42.** perpendicular   **43.** parallel   **44.** -1.5°F per hour   **45.** -1.99 thousand nuns per year   **46. a.** $c = 2d + 3$ **b.** The units for both of the expressions c and $2d + 3$ are dollars.   **c.** \$37   **47. a.** -0.3; the budget decreases by \$0.3 billion per year.   **b.** $A = -0.3t + 11.3$ **c.** \$8.9 billion   **48. a.** 0.5; the total amount Americans spend on Father's Day increases by \$0.5 billion per year.   **b.** $(0, 12.5)$; Americans spent a total of \$12.5 billion on Father's Day in 2014.   **c.** $A = 0.5t + 12.5$ **d.** \$16 billion   **49.** yes; 119.99; the cost is \$119.99 per calculator.   **50. a.** 22.2%   **b.** \$109 thousand   **c.** -0.13; the percentage of American adults who are confident they will retire ahead of their schedule is 0.13 percentage point less for American adults who earn \$1 thousand more than other American adults.   **51.** equations 1, 3, and 4

52.

Equation 1		Equation 2		Equation 3		Equation 4	
x	y	x	y	x	y	x	y
0	50	0	12	61	25	26	−4
1	41	1	16	62	23	27	−1
2	32	2	20	63	21	28	2
3	23	3	24	64	19	29	5
4	14	4	28	65	17	30	8

53. The value of y decreases by 6.

Chapter 3 Test

1. 3   **2.** 3   **3.** $(0, 1)$ **4.** $(1.5, 0)$ **5.** ski run B   **6.** 3; increasing   **7.** $-\dfrac{1}{2}$; decreasing   **8.** 0; horizontal   **9.** undefined slope; vertical

10. 1.29; increasing   **11.**   **12.** slope: $-\dfrac{3}{2}$,   **13.** slope: $\dfrac{5}{6}$,   **14.** slope: 3;

y-intercept: $(0, 2)$ y-intercept: $(0, 0)$ y-intercept: $(0, -4)$

15. slope: 0;   **16.** slope: −2;   **17.** $y = \dfrac{1}{2}x + 1$ **18. a.** $v = -2t + 17$ **b.**

y-intercept: $(0, 2)$ y-intercept: $(0, 3)$

c. $(0, 17)$; the used car is currently worth $17 thousand.   **d.** 6 years from now   **19. a.** m is positive; b is positive   **b.** m is zero; b is negative   **c.** m is negative; b is positive   **d.** m is negative; b is negative   **20.** neither   **21.** parallel   **22.** −2.4°F per hour   **23.** −0.2 billion dollars per year   **24. a.** 30; the number of television shows has increased by 30 shows per year.   **b.** $n = 30t + 409$ **c.** 589 shows   **25. a.**    yes

b. 0.24; the cooking time increases by 0.24 hour per pound of turkey.   **c.** $(0, 1.64)$; the cooking time of a 0-pound turkey is 1.64 hours; model breakdown has occurred.   **d.** 6.2 hours   **26. a.** yes   **b.** 8 miles per hour   **27.** Set 1: $y = -3x + 25$; Set 2: $y = 4x + 2$; Set 3: $y = -5x + 12$; Set 4: $y = 6x + 47$ **28.** The value of y increases by 3.

Chapter 4

Homework 4.1   **1.** $x + 5$ **3.** $7 + 2p$ **5.** yx **7.** $x(-2)$ **9.** $15 + m \cdot 4$ **11.** $x + (4 + y)$ **13.** $4(bc)$ **15.** $(x + y) + 3$
17. $(ab)c$ **19.** $10x$ **21.** $p + 7$ **23.** $36x$ **25.** $3b + 11$ **27.** $-4x$ **29.** $\dfrac{7x}{4}$ **31.** $-3x + 2$ **33.** $2k + 8p - 7$ **35.** $x + y - 2$
37. $3x + 27$ **39.** $2x - 10$ **41.** $-2t - 10$ **43.** $10x - 30$ **45.** $-24x - 42$ **47.** $6x - 10y$ **49.** $-25x - 15y + 40$
51. $-0.3x - 0.06$ **53.** $2x + 4$ **55.** $21x - 27$ **57.** $2x + 5$ **59.** $4a + 19$ **61.** $-12x + 9$ **63.** $-9a + 19$ **65.** $10.5x - 38.56$
67. $-t - 2$ **69.** $-8x + 9y$ **71.** $-5x - 8y + 1$ **73.** $-3x + 3$ **75.** $-x + 5$ **77.** $-2t + 3$ **79.** $-8x + 14y - 3$ **81.** $8x; 16$;
answers may vary.   **83.** $5x - 35; -25$; answers may vary.   **85.** $-3x - 7; -13$; answers may vary.   **87.** $3(x + 2); 3x + 6$
89. $-4(2x - 5); -8x + 20$ **91.** $5 - 2(x - 4); -2x + 13$ **93.** $-2 + 7(2x + 1); 14x + 5$ **95. a.** $2x - 2$ **b.** 8   **c.** 8

d. The results are equal; the result in part (a) might be correct.   **97.** 18; 10; the work is incorrect.   **99. a.** 24; 48   **b.** Answers may vary.
c. $(ab)c$ **101.** $y = 2x - 6$ **103.** $y = -3x - 3$ **105.** $y = -6x$ **107.** $(7 + 3 + 6)10 = 160$ dollars; $7(10) + 3(10) + 6(10) = 160$ dollars; answers may vary.   **109.** Answers may vary.   **111.** $3 = 1$; answers may vary.   **113.** commutative law for addition; associative law for addition; commutative law for

addition; associative law for addition   **115.** commutative law for multiplication; distributive law; commutative law for multiplication   **117.** $-a = -1 \cdot a$; associative law for multiplication; the product of two real numbers with different signs is negative; $-1 \cdot a = -a$; $-(-a) = a$ **119.** Answers may vary.

Homework 4.2   **1.** $7x$ **3.** $5x$ **5.** $-13w$ **7.** $4t$ **9.** $-0.5x$ **11.** $\frac{7}{3}x$ **13.** $-3x - 3$ **15.** $-2p - 7$ **17.** $3x + y + 1$
19. $-9.9x + 1.1y + 2.1$ **21.** $-a + 15$ **23.** $43.16x + 23.08$ **25.** $7a - 8b$ **27.** $-x + 2$ **29.** $2x - 2y$ **31.** $-13t - 5$
33. $-2x - 4y + 6$ **35.** $-8x - 14$ **37.** $-24x - 38y$ **39.** 0   **41.** $-4x + 7y - 21$ **43.** $7x - 13y - 2$ **45.** $-\frac{2}{7}a + \frac{6}{7}$ **47.** $3x - 3$
49. $x + 5x$; $6x$ **51.** $4(x - 2)$; $4x - 8$ **53.** $x + 3(x - 7)$; $4x - 21$ **55.** $2x - 4(x + 6)$; $-2x - 24$ **57.** Twice the number plus 6 times the number (answers may vary); $8x$ **59.** 7 times the difference of the number and 5 (answers may vary); $7x - 35$
61. The number, plus 5 times the sum of the number and 1 (answers may vary); $6x + 5$ **63.** Twice the number, minus 3 times the difference of the number and 9 (answers may vary); $-x + 27$ **65.** $8x - 5$ **67.** $-3x + 9$ **69.** $5x + 2$; 22; answers may vary.
71. $3x + 11$; 23; answers may vary.   **73. a.** $8x + 12$ **b.** 28   **c.** 28   **d.** The results are equal; the result in part (a) might be correct.   **75.** $-2(x - 3), 2(3 - x), -3(x - 2) + x, -2x + 6$ **77.** 5, 14, 14; 14; $9x - 13$ **79.** Answers may vary.
81. $y = -2x$ **83.** $y = 2x - 4$ **85.** $y = 3x - 4$ **87. a.** Answers may vary.   **b.** Answers may vary.   **c.** $2(x - 3) - x + 6$; x **89.** Answers may vary.

Homework 4.3   **1.** yes   **3.** no   **5.** yes   **7.** 5   **9.** -14 **11.** 24   **13.** -28 **15.** -3 **17.** 3   **19.** 3   **21.** -4 **23.** 5   **25.** $\frac{4}{3}$ **27.** $\frac{6}{5}$
29. 0   **31.** 15   **33.** $\frac{21}{2}$ **35.** -12 **37.** $-\frac{10}{3}$ **39.** 6   **41.** -3 **43.** $\frac{1}{2}$ **45.** -11.1 **47.** 41.76   **49.** 2.3   **51.** \$820   **53.** 40 points
55. 1865 thousand (1.865 million) students   **57.** \$2631 billion (\$2.631 trillion)   **59.** -4 **61.** 4   **63.** 5   **65.** 4   **67.** 3   **69.** -3 **71.** 3
73. -2 **75.** 2   **77.** 2   **79.** Answers may vary; 5   **81.** $4(3)$ is equal to 12; $\frac{4(3)}{4} = 3$ is equal to $\frac{12}{4} = 3$; 3 is equal to 3.   **83.** yes; answers may vary.   **85.** Answers may vary.   **87.** Answers may vary.   **89.** no   **91.** Answers may vary.   **93.** Answers may vary.   **95.** yes   **97. a.** 5   **b.** 4   **c.** $k - b$ **99.** Answers may vary.   **101.** Answers may vary.

Homework 4.4   **1.** 5   **3.** -5 **5.** $-\frac{4}{3}$ **7.** 12   **9.** 2   **11.** 6   **13.** -5 **15.** 2   **17.** 3   **19.** $\frac{3}{4}$ **21.** -1 **23.** $\frac{5}{4}$ **25.** $\frac{5}{2}$ **27.** $\frac{8}{5}$ **29.** 6
31. $-\frac{41}{3}$ **33.** $-\frac{27}{10}$ **35.** $\frac{12}{13}$ **37.** -7 **39.** $\frac{11}{3}$ **41.** 1.67   **43.** 6.34   **45.** 21.85   **47. a.** $w = 0.4t + 30.4$ **b.** 33.2 pounds   **c.** 2020
49. a. $v = 3.36t + 8.16$ **b.** \$18.24   **c.** 6 months after Stewart was sentenced.
51. a. 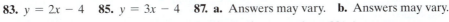 yes   **b.** 1.59; the percentage of freshmen at Boise State University whose GPA in high school was at least 3.5 is increasing by 1.59 percentage points per year.   **c.** 55.1%   **d.** 2017   **e.** Answers may vary.
53. a. $F = 2.25d + 3.25$, where F is the fare (in dollars) for d miles; answers may vary.   **b.** 13 miles
55. a. $V = -3281t + 31,681$, where V is the value (in dollars) of the car at t years since 2016; answers may vary.   **b.** 2022   **57.** 12 weeks   **59.** 2021   **61.** \$1.27 trillion   **63.** 33.0 pounds   **65.** \$15.0 billion   **67.** 225 square feet   **69.** 3
71. -4 **73.** -4 **75.** -7 **77.** Twice the number, minus 3, is 7 (answers may vary); 5   **79.** 6 times the number, minus 3, is equal to

8 times the number, minus 4 (answers may vary); $\frac{1}{2}$ **81.** Twice the difference of the number and 4 is 10 (answers may vary); 9

83. 4, minus 7 times the sum of the number and 1 is 2 (answers may vary); $-\frac{5}{7}$ **85.** -1 **87.** -2 **89.** 2.83   **91.** 3.45   **93.** -5.43

95. 2   **97.** -2 **99.** 4   **101.** -1 **103.** 2   **105.** -3 **107.** set of all real numbers; identity   **109.** empty set; inconsistent equation

111. 2; conditional equation   **113.** set of all real numbers; identity   **115.** empty set; inconsistent equation

117. a. and b.

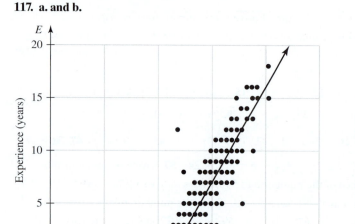

Age (years)

yes   **c.** 7 years   **d.** 34 years   **e.** 0.90; a player has 0.90 year more experience than a player who is 1 year younger.   **f.** The ordered pair $(24, 12)$ has at least one incorrect coordinate. If a 24-year-old player had 12 years of uninterrupted experience, he would have started with the NFL at age 12 years, which would mean the oldest age he could have graduated from high school was age 9 years. The ordered pair $(25, 8)$ probably has at least one incorrect coordinate. If a 25-year-old player had 8 years of uninterrupted experience, he would have started with the NF at age 17, which would mean the oldest age he could have graduated from high school is 14 years.   **119.** Answers may vary; answers may vary; 7   **121.** Answers may vary; 7   **123.** Answers may vary.   **125.** 3; answers may vary.   **127.** 3; 3; answers may vary.   **129. a.** m can be any real number except 7; b can be any real number.   **b.** $m = 7$; $b = 4$ **c.** $m = 7$; b can be any real number except 4.

Homework 4.5   **1.** linear equation   **3.** linear expression   **5.** linear expression   **7.** linear equation   **9.** 2   **11.** $7x$ **13.** $-4b + 5$

15. $\frac{5}{4}$ **17.** $\frac{7}{19}$ **19.** $19x - 7$ **21.** $-\frac{3}{4}$ **23.** $-16x - 12$ **25.** $5.9p - 4.5$ **27.** 2   **29.** 3.5   **31.** $-6.1x + 35.28$ **33.** $-\frac{3w}{4}$ **35.** -2

37. $\frac{x}{12} + \frac{1}{2}$ **39.** -6 **41.** $\frac{14}{33}$ **43.** $\frac{17}{4}x - \frac{7}{6}$ **45.** $3 + 2x = -10$; $-\frac{13}{2}$ **47.** $4 - 6(x - 2)$; $-6x + 16$ **49.** $-9x = x - 5$; $\frac{1}{2}$

51. $\frac{x}{2} = 3(x - 5)$; 6   **53.** $x + x(6)$; $7x$ **55.** $x + \frac{x}{2}$; $\frac{3x}{2}$ **57.** Answers may vary.   **59.** Answers may vary.   **61.** no; answers may vary.

63. Answers may vary; $\frac{7}{12}x$ **65.** Answers may vary.   **67.** Answers may vary.   **69.** Answers may vary.   **71.** 1; answers may vary.

73. Answers may vary.

Homework 4.6   **1.** $P = 4S$ **3.** $P = 2H + 2S + B$ **5.** $P = 2A + 2B + 2C + 2D$ **7.** $V = 5$ **9.** $B = 4$ **11.** $t \approx 1.86$

13. $L = 3$ **15.** $b = 7$ **17.** 14.5 feet   **19.** 16 inches   **21.** 59 feet   **23. a.** $A = 3x$ **b.** $P = 2x + 6$ **c.** The units for both of the expressions P and $2x + 6$ are inches.   **25. a.** Answers may vary.   **b.** Answers may vary.   **c.** Answers may vary.   **27. a.** 30 cents **b.** 40 cents   **c.** $T = 10d$ **d.** The units for both of the expressions T and $10d$ are cents.   **29.** $T = 725n$ **31. a.** $T = 15x$ **b.** 2000; if 2000 people buy tickets in advance, the total sales will be \$30,000.   **33. a.** $C = 77k$ **b.** $E = 250n$ **c.** $T = 77k + 250n$ **d.** 90; for 65 \$77 tickets and 90 \$250 tickets, the total cost is \$27,505.   **35. a.** $V = LWH$ **b.** 8 feet   **37. a.** $I = Prt$ **b.** \$600   **c.** $B = P + Prt$

d. \$2400   **39.** 82 points   **41.** $W = \frac{A}{L}$ **43.** $T = \frac{PV}{nR}$ **45.** $M = -\frac{Ur}{Gm}$ **47.** $B = \frac{2A}{H}$ **49.** $t = \frac{v - v_0}{g}$ **51.** $r = \frac{A - P}{Pt}$

53. $b = 3A - a - c$ **55.** $x = \frac{y - k + mh}{m}$ **57.** $y = a - x$ **59.** $y = -\frac{3}{4}x + 4$ **61.** $y = -\frac{1}{2}x + 2$ **63.** $y = \frac{5}{2}x - 3$

65. $y = -\frac{3}{7}x - \frac{5}{7}$ **67. a.**

yes   **b.** $t = \frac{n - 0.34}{0.45}$ **c.** 2016   **d.** 2017, 2018, 2019, 2020, and 2021

69. a. $s = 44.6t + 626$ **b.** $t = \frac{s - 626}{44.6}$ **c.** 2017   **d.** 2018, 2019, 2020, 2021, and 2022   **71. a.** $p = -1.4t + 39.1$ **b.** 2021   **c.** $t = \frac{p - 39.1}{-1.4}$ **d.** 2021

e. The results are the same; $t = \dfrac{p - 39.1}{-1.4}$; answers may vary.   **f.** $p = -1.4t + 39.1$; answers may vary; 29.3% **73. a.** The entries in the third column are $50 \cdot 4, 70 \cdot 3, 65 \cdot 2, 55 \cdot 5, st$, all in miles; $d = st$ **b.** The units for both of the expressions d and st are miles.   **c.** $t = \dfrac{d}{s}$ **d.** 4.5; it takes 4.5 hours to travel 315 miles at 70 miles per hour.   **e.** 5.56 hours   **75.** The units of s are miles per hour and the units of dt are miles-hours. Because the units are different, the formula is incorrect.   **77. a.** It doubles the perimeter.   **b.** The area is multiplied by 4.   **79.** Answers may vary.   **81.** Answers may vary.

Chapter 4 Review Exercises

1. $9 + 5w$ **2.** $8 + pw$ **3.** $(2 + k) + y$ **4.** $b(xw)$ **5.** $-20x$ **6.** $-24x - 12$ **7.** $12y - 28$ **8.** $-3x + 6y + 8$ **9.** $\dfrac{7}{9}x$
10. $4a - 9b + 2$ **11.** $-18x - 8y$ **12.** $-8.06x - 20.2$ **13.** $-5m - 4$ **14.** $-3a - 40b$ **15.** $-4(x - 7); -4x + 28$
16. $-7 + 2(x + 8); 2x + 9$ **17.** Answers may vary; $a(b + c) = ab + ac$ **18.** Answers may vary.   **19.** $-5(x - 4), 5(4 - x)$,
$-2(x - 10) - 3x, -5x + 20$ **20.**   **21.**   **22.** no   **23.** 7   **24.** -5 **25.** 3   **26.** -6 **27.** 2.7   **28.** $-\dfrac{7}{2}$
29. $\dfrac{5}{11}$ **30.** 1   **31.** $\dfrac{15}{8}$ **32.** $\dfrac{38}{3}$ **33.** $\dfrac{44}{3}$ **34.** Answers may vary; 7   **35.** Answers may vary.   **36.** -8.31
37. a. $n = -74t + 1231$ **b.** 2022   **c.** 861 gun seizures   **38.** $\dfrac{39}{4}$ **39.** -39 **40.** -1.64 **41.** 3   **42.** 1   **43.** 2   **44.** 0   **45.** the set of all real numbers; identity   **46.** $\dfrac{22}{15}$; conditional equation   **47.** empty set; inconsistent equation   **48.** Answers may vary; 12
49. linear expression   **50.** linear equation   **51.** $-2t$ **52.** 3   **53.** -5 **54.** $0.2a + 0.1$ **55.** $-3p - 54$ **56.** $-\dfrac{11}{32}$ **57.** -18
58. $\dfrac{13}{3}r - \dfrac{11}{12}$ **59.** no; answers may vary.   **60.** Answers may vary; $\dfrac{2}{3}x + \dfrac{7}{5}$ **61.** $4(6 - x) = 17; \dfrac{7}{4}$ **62.** $x - \dfrac{x}{2}; \dfrac{x}{2}$
63. $P = 2A + 2B + 2C + 2D + 2E$ **64. a.** $T = 15n + 25w$ **b.** 220; if the total cost is \$11,050 and 370 \$15 tickets were sold, 220 \$25 tickets were sold.   **65.** $r = \dfrac{C}{2\pi}$ **66.** $c = P - a - b$ **67.** $y = \dfrac{1}{2}x - 3$ **68.** $T = \dfrac{2A - HB}{H}$ **69. a.** $n = 1.5t + 23.0$
b. $t = \dfrac{n - 23.0}{1.5}$ **c.** 2021   **d.** 2021   **e.** The results are the same; $t = \dfrac{n - 23.0}{1.5}$; answers may vary.   **f.** $n = 1.5t + 23.0$; answers may vary; 33.5 thousand stores   **70.** 2022   **71.** 5.09 million people

Chapter 4 Test

1. $3p + 4$ **2.** $(3x)y$ **3.** $-4x + 6$ **4.** $-4.08x + 0.96$ **5.** $-22w + 53$ **6.** $-11a - 3b - 2$ **7.** Answers may vary; $-2x - 10$
8.   **9.** $\dfrac{11}{3}$ **10.** 10   **11.** 7   **12.** $\dfrac{10}{13}$ **13.** $-\dfrac{41}{4}$ **14.** $-\dfrac{32}{25}$ **15.** 1.18   **16.** -32 **17.** $23x + 24$ **18.** $-\dfrac{12}{11}$ **19.** no; answers may vary.   **20.** Answers may vary.   **21.** $5(x - 2) = 29; \dfrac{39}{5}$ **22.** $2 + 4(3 + x); 4x + 14$ **23.** 2   **24.** 4   **25.** -2
26. 4   **27. a.** $n = 22.8t + 630$ **b.** 766.8 thousand patent applications   **c.** 2020   **28.** 11 months   **29.** 269 thousand complaints   **30.** 18 feet   **31.** $a = 2A - b$

Cumulative Review of Chapters 1–4

1. $n; 275, 300; 0, -150$; answers may vary.   **2.**   **3.** In 2015, the average number of unique monthly visitors was 170.7 million visitors.   **4.** -2 **5.** 4   **6.** $-\dfrac{3}{10}$ **7.** $-\dfrac{47}{40}$ **8.** $-\dfrac{3}{5}$
9. $\dfrac{19}{6}$ feet   **10.** -7 points   **11.** -27 **12.** $-\dfrac{1}{2}$; decreasing   **13.** undefined slope; vertical   **14.** road A

15.   **16.**   **17.**   **18.**   **19.** $y = 3x - 2$ **20.** -932 dollars per month

21. 10.3 thousand deaths per year   **22. a.** -0.8; the home ownership rate among people age 35 and under decreases by 0.8 percentage point per year.   **b.** $r = -0.8t + 34.6$

c. 29.8%   **d.** 2022

23. equations 2, 3, and 4   **24. a.** 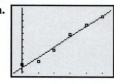 yes   **b.** 16.1; the median sales price of homes is increasing by $16.1 thousand per year.   **c.** $(0, 216.8)$; in 2010, the median sales price of homes was $216.8 thousand.   **d.** 2021   **e.** $377.8 thousand   **f.** $\frac{4.1}{1}; \frac{5.3}{1}$; the unit ratio for 2015 is larger than the unit ratio for 2010; answers may vary.   **25.** 3   **26.** -8 **27.** $-3a - 7b + 6$ **28.** $14p + 7$ **29.** 2

30. $-6a - 4$ **31.** -5.86 **32.** $x + 9\left(\frac{x}{3}\right); 4x$ **33.** $2(7 - 2x) = 87; -\frac{73}{4}$ **34.** $h = \frac{A}{2\pi r}$ **35.** $y = \frac{2}{3}x - 2$

Chapter 5

Homework 5.1

1. slope: 2;   **3.** slope: -3;   **5.** slope: $-\frac{3}{5}$;   **7.** slope: $\frac{1}{2}$;

y-intercept: $(0, -3)$ y-intercept: $(0, 5)$ y-intercept: $(0, -2)$ y-intercept: $(0, 2)$

9. slope: -3;   **11.** slope: 0;   **13.** slope: -1;   **15.** slope: -2;   **17.** slope: 2;

y-intercept: $(0, 0)$ y-intercept: $(0, -4)$ y-intercept: $(0, 3)$ y-intercept: $(0, 4)$ y-intercept: $(0, -1)$

19. slope: $\frac{2}{3}$;   **21.** slope: $-\frac{3}{2}$;   **23.** slope: $\frac{4}{5}$;   **25.** slope: $\frac{3}{4}$;   **27.** slope: $\frac{2}{5}$;

y-intercept: $(0, 0)$ y-intercept: $(0, -2)$ y-intercept: $(0, -3)$ y-intercept: $(0, -2)$ y-intercept: $(0, -2)$

  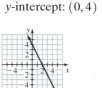   

29. slope: $-\frac{1}{4}$;   **31.** slope: $-\frac{1}{2}$;   **33.** slope: -4;   **35.** slope: $\frac{3}{2}$;   **37.** slope: $-\frac{5}{3}$;

y-intercept: $(0, 1)$ y-intercept: $(0, -2)$ y-intercept: $(0, -2)$ y-intercept: $(0, 2)$ y-intercept: $(0, 0)$

  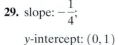 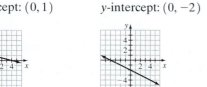 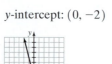

39. slope: 0;   **41.** *x*-intercept: $(6,0)$;   **43.** *x*-intercept: $(5,0)$;   **45.** *x*-intercept: $(-6,0)$;   **47.** *x*-intercept: $(2,0)$;
  y-intercept: $(0,3)$ *y*-intercept: $(0,-2)$ *y*-intercept: $(0,3)$ *y*-intercept: $(0,4)$ *y*-intercept: $(0,6)$

49. *x*-intercept: $(4,0)$;   **51.** *x*-intercept: $(3,0)$;   **53.** *x*-intercept: $(1.21,0)$;   **55.** *x*-intercept: $(-2.05,0)$;
  y-intercept: $(0,-6)$ *y*-intercept: $(0,5)$ *y*-intercept: $(0,2.68)$ *y*-intercept: $(0,3.47)$

57. *x*-intercept: $(2.07,0)$; *y*-intercept: $(0,9.32)$ **59.** *x*-intercept: $(-14.94,0)$; *y*-intercept: $(0,-37.21)$ **61.**

63.   **65.**   **67.**   **69. a.** 1 **b.** 2 **c.**   **71.** no; $-\dfrac{2}{3}$

73. a.   **b.** Answers may vary.   **c.** For each solution, the difference of three times the *x*-coordinate and five times the
   y-coordinate is equal to 10.

75. a. *x*-intercept: $(5,0)$; *y*-intercept: $(0,7)$ **b.** *x*-intercept: $(4,0)$; *y*-intercept: $(0,6)$ **c.** *x*-intercept: $(a,0)$; *y*-intercept: $(0,b)$

d. $\dfrac{x}{2} + \dfrac{y}{5} = 1$ **77.** Answers may vary.   **79.** Answers may vary.   **81.** 1   **83.** 2   **85.** $(4,0)$ **87.** $\dfrac{1}{2}$ **89.** $-6x + 26$; linear expression

in one variable   **91.** 3; linear equation in one variable

Homework 5.2   **1.** $y = 2x - 1$ **3.** $y = -3x + 1$ **5.** $y = 2x + 2$ **7.** $y = -6x - 15$ **9.** $y = \dfrac{2}{5}x - \dfrac{1}{5}$ **11.** $y = -\dfrac{3}{4}x - \dfrac{13}{2}$

13. $y = 3$ **15.** $x = -2$ **17.** $y = 2.1x - 13.67$ **19.** $y = -5.6x - 22.4$ **21.** $y = -6.59x - 17.95$ **23.** $y = 2x - 4$

25. $y = -2x + 8$ **27.** $y = 5x - 2$ **29.** $y = -x - 9$ **31.** $y = x$ **33.** $y = -4x + 9$ **35.** $y = 2$ **37.** $x = -4$ **39.** $y = \dfrac{1}{2}x + 1$

41. $y = \dfrac{2}{3}x - \dfrac{1}{3}$ **43.** $y = -\dfrac{1}{6}x + \dfrac{3}{2}$ **45.** $y = -\dfrac{2}{7}x + \dfrac{3}{7}$ **47.** $y = \dfrac{3}{5}x + \dfrac{2}{5}$ **49.** $y = \dfrac{3}{2}x - 2$ **51.** $y = -1.23x + 8.62$

53. $y = -1.70x - 5.46$ **55.** $y = -1.01x + 0.65$ **57.** $y = 0.48x - 6.11$ **59.** $y = 2x - 1$ **61.** $y = -4x - 7$ **63.** $y = -2x + 10$

65. $y = 4x + 3$ **67.** $y = -\dfrac{4}{3}x + \dfrac{23}{3}$ **69. a.** $y = \dfrac{3}{2}x - 3$ **b.**   **c.** Answers may vary.

71. a. possible; answers may vary.   **b.** possible; answers may vary.   **c.** not possible; answers may vary.   **d.** possible; $y = 0$

73. $E = 2t + 5$ **75. a. i.** $y = 3x - 5$ **ii.** $y = 3x - 5$ **b.** The results are the same.   **77. a.** Answers may vary.   **b.** Answers may

vary.   **c.** Answers may vary.   **d.** no such line; answers may vary.   **79.** $y = 2x + 1$; $y = 2x + 1$; yes; answers may vary.

81. Set 1: $y = -2x + 25$; Set 2: $y = 4x + 12$; Set 3: $y = -5x + 77$; Set 4: $y = 3$ **83.** Answers may vary.

85. a.   **b.** $y = 2x - 3$; it is the same equation.   **87.**   linear equation in two variables

89. 2; linear expression in two variables

Homework 5.3   **1.** $n = 2.3t + 73.5$ **3.** $n = -280.33t + 4959.67$ **5. a.** $c = 0.13t + 17.91$ **b.** 0.13; the average U.S. commute time to work is increasing by 0.13 minute per year.   **c.** $(0, 17.91)$; in 1950, the average U.S. commute time to work was 17.91 minutes.

7. a. $n = -224.4t + 10,188$ **b.** -224.4; the number of sexual-harassment charges filed is decreasing by 224.4 charges per year.

c. $(0, 10,188)$; in 2000, 10,188 sexual-harassment charges were filed.   **9.** $y = 2.5x - 2.2$; answers may vary.

11. $y = -1.11x + 20.83$; answers may vary.   **13. a.** 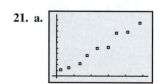   **b.** $p = 0.61t + 75.61$; answers may vary.

c. 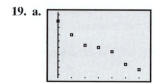   **15. a.** 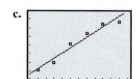   **b.** $H = 4.5d + 15.3$; answers may vary.   **c.**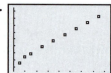

17. a. 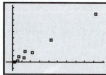   **b.** $w = 0.17n - 0.50$; answers may vary.   **c.** 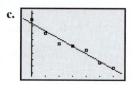   **d.** 168; answers may vary.

19. a. 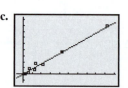   **b.** $p = -113.68t + 941.61$; answers may vary.   **c.** 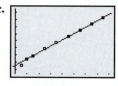   **d.** -113.68; the average price is decreasing by \$113.68 per kilowatt-hour per year.
e. $(0, 941.61)$; in 2010, the average price was about \$942 per kilowatt-hour.

21. a. 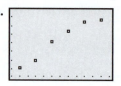   **b.** $r = 7.54t + 23.99$; answers may vary.   **c.** 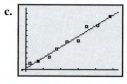   **d.** 7.54; the land speed record is increasing by 7.54 mph each year.
e. $(0, 23.99)$; in 1900, the land speed record was about 24 mph.

23. a. 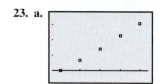   **b.** $p = 7.8r - 31.8$; answers may vary.   **c.**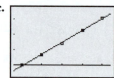

25. Student B   **27.**    increase m and decrease b

29. Answers may vary.   **31. a.** $V = 2t + 10$; the units for both of the expressions V and $2t + 10$ are dollars.

b.

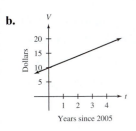

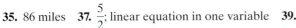

c. Answers may vary.   **33. a.** $n = -1.8t + 43.8$ **b.** -1.8; the average attendance is decreasing by 1.8 thousand spectators per year.   **c.** The units for both of the expressions n and $-1.8t + 43.8$ are thousands of spectators.

35. 86 miles   **37.** $\dfrac{5}{2}$; linear equation in one variable   **39.**

linear equation in two variables

Homework 5.4   1. a. 89.0%   **b.** 2040   **c.** $(0, 75.61)$; in 2000, about 75.6% of seats were filled.   **d.** 0.61; the percentage of seats that are filled increases by 0.61 percentage point per year.   **3. a.** 9 times   **b.** 68 times   **c.** 0.17; movies of a certain genre would have 0.17 win more than movies of a genre that have had 1 less nomination.   **d.** the data points for the genres action, drama, and war; for nominated movies, action, drama, and war movies are more likely to win than movies of a typical genre.   **5. a.** 14 dog years
b. 146 human years   **c.** 23 dog years   **d.** 4.5; a dog aging 1 year is equivalent to a human aging 4.5 years.   **e.** $H = 7d$ **f.** 7; a dog aging 1 year is equivalent to a human aging 7 years; the slope of the graph of the equation found in part (e) is greater than the slope found in part (d).

7. a.

b. $p = -0.16t + 67.81$; answers may vary.   **c.** 2022   **d.** $(0, 67.81)$; in 1970, about 67.8% of married persons said they were "very happy" with their marriages.   **e.** $(423.81, 0)$; in 2394, no married people will say they are "very happy" with their marriages; model breakdown has occurred.

9. a.

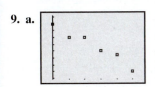

b. $w = -1.27t + 158.93$; answers may vary.   **c.** -1.27; the math professor lost 1.27 pounds per week.
d. 11 weeks after the math professor started the weight-loss program   **e.** $(125.14, 0)$; the math professor was weightless about 125 weeks after he started the weight-loss program; model breakdown has occurred.

11. a.

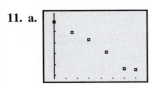

b. $p = -0.33t + 10.99$; answers may vary.   **c.** 2021   **d.** $(0, 10.99)$; in 2000, about 11.0% of high school students dropped out of high school.   **e.** $(33.30, 0)$; in 2033, no high school students will drop out of high school; model breakdown has occurred.

13. a. 7.8; the percent increase in patients who would die increases by 7.8 percentage points if the patient-to-nurse ratio is increased by 1.
b. 46.2%   **c.** 7.2%; overestimate; answers may vary.   **d.** Answers may vary.   **15.** 2022   **17.** 129 thousand acres   **19.** \$5.5 billion
21. a. 2019   **b.** no; answers may vary.

23. a. and c.

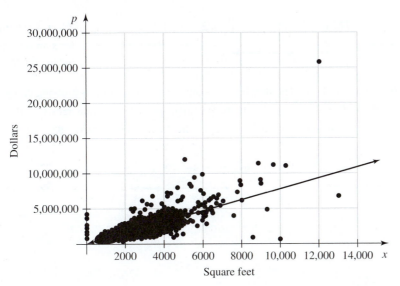

b. $p = 800.16x - 288,387$; answers may vary.   **d.** \$1,311,933   **e.** 3985 square feet
f. 800.16; the sales price of a home is \$800.16 more than the sales price for a home with 1 less square foot.   **g.** 12,180 square feet; \$26,000,000; \$16,542,438

25. Answers may vary.   **27. a.** $n = 0.7t + 35.0$ **b.** $(0, 35.0)$; in 2015, 35.0 million Americans lived alone.   **c.** The units for both of the expressions n and $0.7t + 35.0$ are millions of Americans.   **d.** 2020   **29. a.** $p = -1.7t + 87$ **b.** $(0, 87)$; in 1991, 87% of Cuban-Americans in Miami-Dade favored a U.S. trade embargo of Cuba.   **c.** The units for both of the expressions p and $-1.7t + 87$ are percentage points.   **d.** 44.5%; answers may vary.   **31.** 2020   **33.** $\frac{17}{18}$; linear equation in one variable   **35.** $-18x + 17$; linear expression in one variable

Homework 5.5

1. true   **3.** true   **5.** 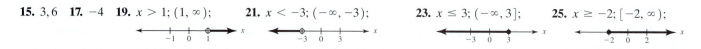   **7.**   **9.**   **11.**

13.

In Words	Inequality	Graph	Interval Notation
numbers greater than or equal to 4	$x \geq 4$		$[4, \infty)$
numbers less than or equal to -2	$x \leq -2$		$(-\infty, -2]$
numbers less than 1	$x < 1$		$(-\infty, 1)$
numbers greater than -5	$x > -5$		$(-5, \infty)$

15. 3, 6   **17.** -4 **19.** $x > 1$; $(1, \infty)$;   **21.** $x < -3$; $(-\infty, -3)$;   **23.** $x \leq 3$; $(-\infty, 3]$;   **25.** $x \geq -2$; $[-2, \infty)$;

27. $t \leq -2$; $(-\infty, -2]$;   **29.** $x > -4$; $(-4, \infty)$;   **31.** $x < -\frac{1}{2}$; $\left(-\infty, -\frac{1}{2}\right)$;   **33.** $x \leq 0$; $(-\infty, 0]$;   **35.** $x > -2$; $(-2, \infty)$;

37. $a < 6$; $(-\infty, 6)$;   **39.** $x \leq -3$; $(-\infty, -3]$;   **41.** $x \geq 1$; $[1, \infty)$;   **43.** $x > 4$; $(4, \infty)$;   **45.** $x \leq -2$; $(-\infty, -2]$;

47. $c \geq -3$; $[-3, \infty)$;   **49.** $x \geq -3$; $[-3, \infty)$;   **51.** $x < 2.5$; $(-\infty, 2.5)$;   **53.** $b < 2$; $(-\infty, 2)$;   **55.** $x > -5$; $(-5, \infty)$;

57. $x \leq 1$; $(-\infty, 1]$;   **59.** $a < -1$; $(-\infty, -1)$;   **61.** $x \leq \frac{5}{2}$; $\left(-\infty, \frac{5}{2}\right]$;   **63.** $x \leq \frac{4}{3}$; $\left(-\infty, \frac{4}{3}\right]$;   **65.** $x \leq -1.7$; $(-\infty, -1.7]$;

67. $y \geq \frac{5}{3}$; $\left[\frac{5}{3}, \infty\right)$;   **69.** $x > 7$; $(7, \infty)$;   **71.** $x > -\frac{3}{5}$; $\left(-\frac{3}{5}, \infty\right)$;   **73.** $r \leq 3$; $(-\infty, 3]$;   **75. a.** 0.09; total student loan amount is increasing by \$0.09 trillion per year.

b. after 2022   **77. a.** $p = -0.44t + 67.36$ **b.** before 2021   **c.** The number of nonmarried households is growing at a greater rate than the number of married households.   **79. a.** $B = -2.41t + 34.01$ **b.** 9.9 births per 1000 women ages $15-19$; 108,969 births   **c.** after 2022   **81.** Answers may vary; $x > -5$ **83. a.** Answers may vary.   **b.** Answers may vary.   **85.** Answers may vary.   **87.** Answers may vary.

89. 4   **91.** $x < 4$; $(-\infty, 4)$;   **93.** $x + 5 > 2$; $x > -3$; $(-3, \infty)$;   **95.** $2x \leq 5x - 6$; $x \geq 2$; $[2, \infty)$;   **97.** Answers may vary.   **99.** Answers may vary.

Chapter 5 Review Exercises

1. slope: 4;
  y-intercept: $(0, -5)$

2. slope: -3;
  y-intercept: $(0, 4)$

3. slope: $\frac{1}{2}$;
  y-intercept: $(0, 1)$

4. slope: $-\frac{2}{3}$;
  y-intercept: $(0, -2)$

5. slope: $\frac{5}{3}$;
  y-intercept: $(0, 0)$

6. slope: 0;
  y-intercept: $(0, -5)$

7. slope: 2;
  y-intercept: $(0, -5)$

8. slope: $\frac{3}{2}$;
  y-intercept: $(0, 3)$

9. slope: $-\frac{4}{5}$;
  y-intercept: $(0, 2)$

10. slope: $-\frac{1}{3}$;
  y-intercept: $(0, 2)$

11. slope: $-\frac{2}{5}$;
  y-intercept: $(0, 4)$

12. slope: 0;
  y-intercept: $(0, 4)$

13. x-intercept: $(5, 0)$;
  y-intercept: $(0, -4)$

14. x-intercept: $(2, 0)$;
  y-intercept: $(0, 3)$
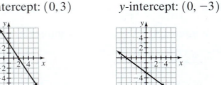

15. x-intercept: $(-4, 0)$;
  y-intercept: $(0, -3)$

16. x-intercept: $(2, 0)$;
  y-intercept: $(0, -4)$

17. x-intercept: $(3, 0)$;
  y-intercept: $(0, 3)$

18. x-intercept: $(3, 0)$;
  y-intercept: $(0, -2)$

19. x-intercept: $(9.48, 0)$;
  y-intercept: $(0, -22.95)$

20. x-intercept: $(-37.99, 0)$; y-intercept: $(0, 97.25)$

21. a. x-intercept: $\left(-\frac{7}{3}, 0\right)$; y-intercept: $(0, 7)$

b. x-intercept: $\left(-\frac{9}{2}, 0\right)$; y-intercept: $(0, 9)$

c. x-intercept: $\left(-\frac{b}{m}, 0\right)$; y-intercept: $(0, b)$ **22. a.** slope: $\frac{1}{2}$;
  y-intercept: $(0, -2)$ **b.** x-intercept: $(4, 0)$;
  y-intercept: $(0, -2)$ **c.** The graphs are the same; answers may vary.

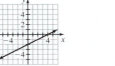

23. $y = -4x + 7$ **24.** $y = 2x + 15$

25. $y = -3x - 7$ **26.** $y = -\frac{2}{5}x + \frac{24}{5}$

27. $y = \frac{3}{7}x + \frac{57}{7}$ **28.** $y = -\frac{2}{3}x - 8$ **29.** $x = 2$ **30.** $y = -4$ **31.** $y = -5.29x - 17.26$ **32.** $y = 1.45x - 3.08$ **33.** $y = 2x - 3$

34. $y = -2x + 5$ **35.** $y = 3x - 1$ **36.** $y = 5x - 15$ **37.** $y = -\frac{3}{2}x + \frac{1}{2}$ **38.** $y = \frac{1}{4}x - 3$ **39.** $y = -\frac{5}{3}x + 4$ **40.** $y = \frac{3}{2}x - 4$

41. $x = 5$ **42.** $y = -3$ **43.** $y = -0.85x + 12.16$ **44.** $y = -0.92x - 2.31$ **45.** $y = -\frac{1}{5}x - \frac{17}{5}$ **46.** $y = -2.13x + 30.38$; answers

may vary.   **47. a.**
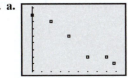
b. $p = -0.17t + 53.16$; answers may vary.   **c.** 2024   **d.** $(0, 53.16)$; in 1970, the middle class collected about 53% of U.S. household income.   **e.** $(312.71, 0)$; in 2283, the middle class will not have any income; model breakdown has occurred.

48. a. 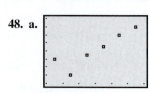   **b.** $p = 1.03t + 8.05$; answers may vary.   **c.** 1.03; the percentage of Americans who support making smoking totally illegal is increasing by 1.03 percentage points per year.   **d.** 29%   **e.** 2021

49. a. $n = 0.4t + 17.1$ **b.** 0.4; the number of kidney transplants is increasing by 0.4 thousand (400) transplants per year.   **c.** 2020   **d.** 20.3 thousand transplants   **50.** 2022

51. $x > 3$; $(3, \infty)$;

52. $x \geq -1$; $[-1, \infty)$;

53. $w \leq -5$; $(-\infty, -5]$;

54. $p > 6$; $(6, \infty)$;

55. $x > -2$; $(-2, \infty)$;

56. $x > -3$; $(-3, \infty)$;

57. $a \leq -3$; $(-\infty, -3]$;

58. $b \geq -4$; $[-4, \infty)$

59. a. $v = -9.34t + 512.57$; answers may vary.   **b.** -9.34; the violent-crime rate is decreasing by about 9 violent crimes per 100,000 people per year.   **c.** after 2021   **d.** $(54.88, 0)$; the violent-crime rate will be 0 violent crimes per 100,000 people in 2055; model breakdown has likely occurred.

Chapter 5 Test

1. slope: -3;   y-intercept: $(0, -1)$

2. slope: $\frac{2}{5}$;   y-intercept: $(0, -2)$

3. slope: 0;   y-intercept: $(0, 5)$

4. x-intercept: $(3, 0)$;   y-intercept: $(0, -6)$

5. x-intercept: $(7.38, 0)$;   y-intercept: $(0, -9.10)$

6. x-intercept: $(2, 0)$;   y-intercept: $(0, 7)$

7. no; $\frac{5}{4}$ **8.** $y = 7x + 10$

9. $y = -\frac{2}{3}x + 3$ **10.** $x = 2$ **11.** $y = -\frac{1}{2}x + 4$ **12.** $y = -1.92x - 3.64$ **13.** $y = -\frac{1}{3}x - \frac{7}{3}$ **14.**

decrease m and increase b **15.** $y = 1.15x + 6.88$; answers may vary.   **16. a.** $p = 1.06t + 54.13$; answers may vary.   **b.** 1.06; the percentage of Nike's revenue that comes from footwear is increasing by 1.06 percentage points per year.   **c.** $(0, 54.13)$; in 2010, about 54% of Nike's revenue came from footwear.   **d.** 2022   **e.** before 2020   **17.** 2021   **18.** $x \leq 2$; $(-\infty, 2]$;   **19.** $x < 2$; $(-\infty, 2)$;

Chapter 6

Homework 6.1   **1.** $(1, -1)$ **3.** $(3, -2)$ **5.** $(2, 1)$ **7.** $(-2, 1)$ **9.** $(0, 0)$ **11.** $(2, -4)$ **13.** $(4, -3)$ **15.** $(3, -2)$ **17.** $(2, -3)$ **19.** $(3, -2)$ **21.** $(1, 3)$ **23.** $(1.16, -2.81)$ **25.** $(-4.67, -3.83)$ **27.** $(4.14, -4.90)$ **29.** $(-4.83, 1.07)$ **31.** all solutions of the line $y = -2x + 4$; dependent system   **33.** $(1, -3)$ **35.** empty set; inconsistent system   **37.** 1980; 241 thousand graduates of each gender   **39. a.** $r = -0.064t + 27.00$ **b.** $r = -0.026t + 21.86$ **c.** 2035; 18.34 seconds   **41.** $(2.8, -2.4)$ **43.** $(14, 1)$ **45.** $(3, 9)$ **47. a.** B, D   **b.** B, C   **c.** B   **d.** A, E, F   **49.** Answers may vary.   **51. a.** m can be any real number except 3; b can be any real number.   **b.** $m = 3$; $b = 4$ **c.** $m = 3$; b can be any real number except 4.   **53.** no; answers may vary.   **55. a.** The lines are parallel.   **b.** no   **c.** empty set   **57.** Answers may vary.   **59.** $(-2, 1)$; system of two linear equations in two variables   **61.** -2; linear equation in one variable

Homework 6.2   **1.** $(2, 4)$ **3.** $(1, 1)$ **5.** $(-2, 3)$ **7.** $(-3, -1)$ **9.** $(4, 1)$ **11.** $(-2, -4)$ **13.** $(-2, 3)$ **15.** $(0, 0)$
17. $(1, -6)$ **19.** $(-2.07, 1.76)$ **21.** $(3.35, -1.71)$ **23.** $(-0.22, 4.34)$ **25.** $(-2, -5)$ **27.** $(2, -1)$ **29.** $(-1, 3)$ **31.** $(-2, -5)$
33. $(1, -3)$ **35.** infinite number of solutions of the equation $x = 4 - 3y$; dependent system   **37.** $(2, -4)$ **39.** empty set;
inconsistent system   **41.** infinite number of solutions of the equation $x = 3y - 1$; dependent system   **43.** empty set; inconsistent
system   **45.** $(17, 56)$ **47.** A: $(0, 0)$; B: $(0, 8)$; C: $(5, 3)$; D: $\left(\dfrac{7}{2}, 0\right)$ **49. a.** $(1, 2)$ **b.** $(1, 2)$ **c.** The results are the same.
51. Answers may vary; $(1, 3)$ **53.** Answers may vary.   **55.**   **a.** $(2, 1)$ **b.** $(2, 1)$ **c.** 2   **d.** They are the same.   **e.** 3; 3; they are
the same.   **f.** Answers may vary.   **57.**

linear equation in two variables   **59.** $(-3, 14)$; system of two linear equations
in two variables

Homework 6.3   **1.** $(2, 1)$ **3.** $(2, 4)$ **5.** $(2, -3)$ **7.** $(1, -2)$ **9.** $(-1, -2)$ **11.** $(1, -3)$ **13.** $(-2, -5)$ **15.** $(1, -3)$
17. $(-4, 5)$ **19.** $(3, 10)$ **21.** the infinite number of solutions of the equation $4x - 7y = 3$; dependent system   **23.** empty
set; inconsistent system   **25.** $(-4, 1)$ **27.** empty set; inconsistent system   **29.** the infinite number of solutions of the equation
$3x - 9y = 12$; dependent system   **31.** $(3.36, 5.51)$ **33.** $(1.52, 5.11)$ **35.** $(3.03, -0.71)$ **37.** $(-2, 4)$ **39.** $(5, 1)$ **41.** $(-4, 3)$
43. $(-4, -1)$ **45.** $(-2, 1)$ **47.** $(-1, 3)$ **49.** $(3, 2)$ **51.** $(2, 1)$; answers may vary.   **53. a.** $(1, 2)$ **b.** $(1, 2)$ **c.** The results
are the same.   **55.** $(23, 24)$ **57.** A: $(0, 0)$; B: $(0, 4)$; C: $(3, 2)$; D: $\left(\dfrac{9}{5}, 0\right)$ **59.** Answers may vary.   **61.** Answers may vary.
63. a. $5 = 2m + b$ **b.** $9 = 4m + b$ **c.** $m = 2; b = 1$ **d.** $y = 2x + 1$ **e.** Answers may vary.   **65.** $-8x - 4$; linear expression in
one variable   **67.** $-\dfrac{1}{2}$; linear equation in one variable

Homework 6.4   **1.** 1980; 241 thousand graduates of each gender   **3.** 2035; 18.34 seconds   **5. a.** $p = -0.96a + 91.17$
b. $p = 0.94a + 7.24$ **c.** 44 years; 49%   **7. a.** $p = 0.091t + 18.57$ **b.** $p = -0.25t + 79.17$ **c.** 2078; 35%   **d.** 31%
9. a. 2081   **b.** 19 women; from the 102nd Congress to the 103rd Congress; no   **c.** 2097   **d.** Answers may vary.   **11. a.** $d = 4.7t + 104.6$
b. $d = 5.2t + 92.2$ **c.** 2039; 221.2 thousand degrees   **13. a.** $v = -1.5t + 16.5$ **b.** $v = -0.6t + 9.5$ **c.** 2022; $4.8 billion
15. a. $V = -4673t + 43{,}969$ **b.** $V = -1848t + 19{,}985$ **c.** 2024; $4296   **17.** 2019; 162 minutes   **19. a.** $w = 0.2666s + 307.94$
b. $w = 0.2374s + 430.30$ **c.** 1134 grams; 1166 grams   **d.** The slope of the female model (0.2666) is greater than the slope of the
male model (0.2374). When comparing a head size to a larger one, the female brain weight will increase more than the male brain
weight.   **e.** For a head size of 3100 cubic centimeters, a typical female brain is lighter than a typical male brain, but when comparing
a head size to a larger one, the female brain weight will increase more than the male brain weight. So, for a certain head size (greater
than 3100 cubic centimeters), the female brain weight is equal to the male brain weight.   **f.** 4190 cubic centimeters; 1425 grams
g. 3.2 pounds   **h.** no; answers may vary.   **21.** Answers may vary.   **23.** 2014; $9.4 billion   **25.** $(1, 4)$; system of two linear equations
in two variables   **27.** 1; linear equation in one variable

Homework 6.5   **1.** width: 114.50 feet, length: 185.50 feet   **3.** width: 8 feet, length: 13 feet   **5.** width: 36 feet, length: 78 feet
7. width: 19 inches, length: 35 inches   **9.** width: 18 yards, length: 55 yards   **11.** 1500 $15 tickets, 500 $22 tickets   **13.** 39 *A Moon Shaped
Pool* albums, 214 *The King of Limbs* albums   **15.** 1300 hardcover books, 200 audio CDs   **17.** main level: $26, balcony: $14   **19.** general:
$37, reserved: $62   **21. a.** $175   **b.** $325   **c.** $0.05d$ **23. a.** $940   **b.** $1340   **c.** $0.14x + 0.03y$ **25.** TFS fund: $14,000, ProFunds fund;
$6000   **27.** Gibelli fund: $5869.57, Becker fund: $2630.43   **29.** Vitrus fund: $7500, Direxion fund: $2500   **31.** Franklin fund: $8403.36,
TIAA-CREF fund: $4201.68   **33. a.** 1.3 ounces   **b.** 1.95 ounces   **c.** $0.65x$ ounces   **35. a.** 1.7 ounces   **b.** 35% solution: 6 ounces,
10% solution: 9 ounces   **37.** 20% solution: 4 quarts, 30% solution: 1 quart   **39.** 10% solution: 1 gallon, 25% solution: 2 gallons
41. 15% solution: 1 quart, 35% solution: 3 quarts   **43.** 12% solution: 2 gallons, 24% solution: 6 gallons   **45.** 20% solution:
3 ounces, water: 2 ounces   **47. a.** $3000   **b.** $4500   **c. i.** no; answers may vary.   **ii.** no; answers may vary.   **iii.** yes; 250 $10 tickets,
50 $15 tickets   **49.** 15%; answers may vary.   **51.** Answers may vary.   **53.** Answers may vary.   **55.** $L = \dfrac{P - 2W}{2}$ **57.** $r = \dfrac{A - P}{Pt}$
59. $-28x + 12$; linear expression in one variable   **61.** $\dfrac{10}{33}$; linear equation in one variable

Homework 6.6   **1.** $(4,1)$ **3.** $(-1,-4),(5,0)$ **5.**   **7.**   **9.**   **11.**

13.   **15.**   **17.**   **19.**   **21.**   **23.**

25.   **27.**   **29.**   **31.**   **33.**   **35.**

37.   **39.**   **41.**   **43.**   **45.**   **47.**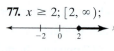

49.   **51.**   **53.**   **55.**   **57.**

59. a. $w \le 3.50h - 80.97, w \ge 3.08h - 64.14, h \ge 63, h \le 78$ **b.**   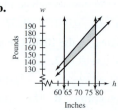   **c.** The ideal weights are between 145 pounds and 157 pounds, inclusive.

61. $y \le -\dfrac{1}{2}x + 2$; answers may vary.   **63.** no; answers may vary.   **65. a.** A, B, C, D   **b.** A, B, C, F, G, H   **c.** A, B, C   **d.** E
67. Answers may vary.   **69.** Answers may vary.   **71.** Answers may vary.   **73.**   **75.**   **77.** $x \ge 2; [2, \infty);$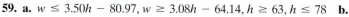

79.   **81.** $(2,3)$ **83.** Answers may vary.   **85.** Answers may vary.   **87.** Answers may vary.

Chapter 6 Review Exercises

1. $(2,1)$ **2.** $(-4,-2)$ **3.** $(0,0)$ **4.** $(1,-1)$ **5.** $(-3,1)$ **6.** $(-3,-2)$ **7.** $(3,-1)$ **8.** $(2,1)$ **9.** $(0,0)$ **10.** $(2,-12)$
11. $(1,-2)$ **12.** $(-3,4)$ **13.** $(2.61,-2.20)$ **14.** $(-1.19,4.76)$ **15.** $(3,2)$ **16.** $(-4,-1)$ **17.** $(-2,1)$ **18.** $(-3,-2)$ **19.** $(4,-2)$
20. $(2,1)$ **21.** $(-1,2)$ **22.** $(-2,1)$ **23.** $(0.77,4.79)$ **24.** $(2.14,-1.62)$ **25.** $(-3,1)$ **26.** $(-5,4)$ **27.** $(-2,0)$ **28.** $(4,3)$
29. $(-4,2)$ **30.** $(5,3)$ **31.** $(-3,-2)$ **32.** $(1,-2)$ **33.** empty set; inconsistent system   **34.** all solutions of the equation
$y = -4x + 3$; dependent system   **35. a.** Answers may vary.   **b.** Answers may vary.   **c.** Answers may vary.   **d.** $(2,3)$
36. $(-1.4,2.8)$ **37.** $(14,47)$ **38.** $(1,3)$; answers may vary.   **39. a.** $p = -0.3t + 52.6$ **b.** $p = 0.25t + 31.0$ **c.** 2039; 41%
d. 18%   **40. a.** $p = 0.98t + 51.9$ **b.** $p = -0.98t + 48.1$ **c.** 2012   **d.** 2012   **e.** The slope of the employed model is 0.98, and the
slope of the not-employed model is -0.98; the slopes are equal in absolute value and opposite in sign; answers may vary.

41. width: 5 feet, length: 17 feet   **42.** 6500 $22 tickets, 1500 $39 tickets   **43.**   **44.**   **45.**

46.   **47.**   **48.**   **49.**   **50.**   **51.**

52.   **53. a.** $w \leq 3h - 54, w \geq 3h - 68, h \geq 58, h \leq 74$ **b.** **c.** The ideal weight range is between 142 pounds and 156 pounds, inclusive.

54. Answers may vary.

Chapter 6 Test

1. $(5, -3)$ **2.** $(-3.18, 1.88)$ **3.** $(-2, -7)$ **4.** $(-1, 3)$ **5.** $(2, -3)$ **6.** $(-4, 2)$ **7.** $(-2, 3)$ **8.** all solutions of the equation $2x - 3y = 4$; dependent system **9.** empty set; inconsistent system **10.** $(5, 2)$ **11.** $(3.45, 1.93)$ **12. a.** C, D **b.** D, F **c.** D

d. A, B, E **13.** $(24, 5)$ **14.** Answers may vary. **15.** A: $(0, 0)$; B: $(0, 11)$; C: $\left(\frac{9}{2}, 5\right)$; D: $(2, 0)$ **16. a.** $p = -0.15t + 33.64$

b. $p = 0.18t + 5.0$ **c.** $1987; 21\%$ **d.** Answers may vary. **e.** Answers may vary. **17. a.** $n = 12.7t + 369$ **b.** $n = 2.4t + 319$

c. 2009; no; 307 million guns **18.** Homestead fund: $5507.25, Evercore fund: $1492.75 **19.** 13,500 tickets at $55, 6500 tickets at $70

20.   **21.**   **22.**   **23.**

Cumulative Review of Chapters 1–6

1.   Fahrenheit degrees   **2.** A cricket chirps 129 times per minute when the temperature is $70°$F. **3.** $-\frac{8}{9}$ **4.** $\frac{46}{35}$ **5.** 21

6. $-\frac{3}{4}$; decreasing **7. a.** m is negative; b is positive. **b.** m is positive; b is negative. **c.** m is zero; b is positive. **d.** m is negative; b is neg-

ative. **8.** equation (1): $y = -8x + 49$; equation (2): $y = 4x + 11$; equation (3): $y = -2x + 45$; equation (4): $y = 3x + 8$ **9.** $-\frac{10}{13}$

10. $-13p + w + 5$ **11.** $4w + 6y - 10$ **12.** $\frac{22}{9}$ **13.** $y = \frac{c - ax}{b}$ **14.** $6 + 3(4 + x); 3x + 18$ **15.** $4 - \frac{x}{3} = 2; 6$

16. slope: 2;   **17.** slope: $\frac{1}{2}$;   **18.** slope: $-\frac{5}{2}$;   **19.** slope: 0;   **20.** x-intercept: $(5, 0)$;

y-intercept: $(0, -4)$ y-intercept: $(0, -3)$ y-intercept: $(0, 6)$ y-intercept: $(0, -3)$ y-intercept: $(0, -2)$

21. x-intercept: $(2, 0)$;   **22.**   **23.** not parallel   **24.** $y = -\frac{2}{5}x - \frac{4}{5}$ **25.** $y = -\frac{4}{3}x - \frac{17}{3}$ **26.** $x = -2$ **27.** 2 **28.** 6

y-intercept: $(0, 4)$

29. $(3, 0)$ **30.** $y = -\dfrac{1}{3}x + 1$ **31.** $x < -1$; $(-\infty, -1)$ **32.** $(2, 1)$ **33.** $(3.31, -1.80)$ **34.** $(3, -2)$ **35.** $(2, -1)$

36. $(23.90, 28.49)$ **37.** $(2, -5)$ **38.** $(15, 37)$ **39.**   **40.**   **41. a.** increase; answers may vary.

b. $p = 0.52t + 7.73$ **c.** $(0, 7.73)$; in 2000, about 8% of traffic fatalities were pedestrian fatalities.   **d.** 2020   **e.** 4.8 thousand pedestrian fatalities   **42.**   **a.** $p = -0.31t + 85.07$ **b.** $p = 0.29t + 10.80$ **c.** 2084; 47%   **43. a.** $r = -1.7t + 60$
b. $r = 2.6t + 29$ **c.** 2021; \$47.7 billion   **44. a.** -0.5; the number of claims filed annually is decreasing by 0.5 thousand claims per year.   **b.** $(0, 8.9)$; in 2014, 8.9 thousand claims were field.   **c.** $n = -0.5t + 8.9$ **d.** 5.4 thousand claims   **e.** before 2020
f. $(17.8, 0)$; in 2032, no one will file a claim; model breakdown has occurred.   **45.** 16% solution: 8 quarts, 28% solution: 4 quarts

Chapter 7

Homework 7.1

1. $(0, 0)$ **3.** $(0, 0)$ **5.** $(0, 1)$ **7.** $(0, -5)$ **9.** $(-2, -4)$

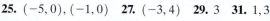

11. $(-1, 3)$ **13.** $(2, -1)$ **15.** $(1, 0)$ **17.** 3   **19.** $-4, -2$ **21.** -3 **23.** no such value
25. $(-5, 0), (-1, 0)$ **27.** $(-3, 4)$ **29.** 3   **31.** 1, 3
33. no such value   **35.** $(0, 3)$ **37.** $(2, -1)$

39. a. 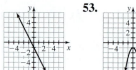   **b.** Answers may vary.   **c.** For each solution, the y-coordinate is 3 less than the square of the x-coordinate.
41. 3   **43.** 1, 3   **45. a.** 0   **b.** 3   **c.** $(0, 3)$; answers may vary.   **47. a. i.** 5; one   **ii.** 17; one   **iii.** 10; one
b. one; answers may vary.   **c. i.** Answers may vary; one   **ii.** Answers may vary; one   **iii.** Answers may vary; one   **d.** one; answers may vary.   **e.** one; answers may vary.   **49.** the real numbers between 0 and 1

51.   **53.**   **55. a.**   **b.**   **c.** The y-intercept is $(0, 0)$ for both equations.
d. $y = x^2$; answers may vary.   **57.** $-\dfrac{1}{5}$; linear equation in one variable   **59.** $20x + 4$; linear expression in one variable

Homework 7.2   **1.** quadratic model   **3.** no; answers may vary.

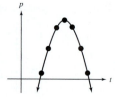

5. a.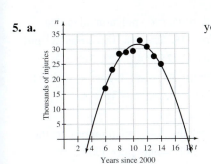

yes   **b.** 20.0 thousand traumatic brain injuries   **c.** 2005, 2016
d. $(10.7, 31.6)$; in 2011, there were 31.6 thousand traumatic brain injuries, the most in any year.   **e.** $(3.6, 0)$, $(17.8, 0)$; there were no traumatic brain injuries in 2004 and 2018; model breakdown has occurred.

7. a.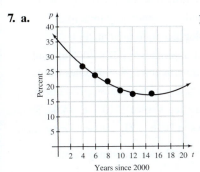

yes   **b.** 22.4%   **c.** 2009, 2020   **d.** $(0, 35.9)$; in 2000, GM's market share was 35.9%.
e. 2015; 17.4%

9. a.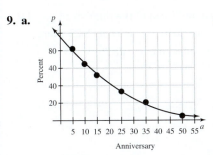

yes   **b.** $(0, 95)$; 95% of married people are still married when they first get married; model breakdown has occurred.   **c.** no; answers may vary.   **d.** 43%   **e.** 17th anniversary.
11. a. 2.3 thousand wiretaps   **b.** 1.9 thousand wiretaps   **c.** overestimate; the data point for 2008 is below the model's graph; 0.4 thousand wiretaps

13. a.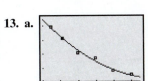

yes   **b.** 34.9%   **c.** $(75.00, 2.45)$; the percentage of 75-year-old Americans who have ever played organized soccer is 2.45%—the smallest percentage for any age, according to the model; no; answers may vary.   **d.** 44 years

15. a.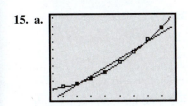

quadratic model   **b.** linear: 1.33 million students; quadratic: 1.78 million students; answers may vary.
c. quadratic model   **d.** linear model; yes; answers may vary.   **e.** 2020

17. overestimate; answers may vary.   **19.** It is more desirable to find a quadratic model whose graph does not contain any data points but comes close to all data points; answers may vary.   **21.** Answers may vary.

23. a.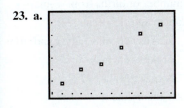

b. $s = 1.03t + 4.55$ **c.** $(0, 4.55)$; in 2005, Whole Foods's revenue was $4.55 billion.   **d.** $20 billion
e. 2022   **25.** Answers may vary.   **27.** $(2, 5)$; system of two linear equations in two variables

29. 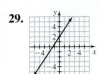   linear equation in two variables

Homework 7.3   **1.** quadratic (or second-degree) polynomial in one variable   **3.** cubic (or third-degree) polynomial in one variable   **5.** seventh-degree polynomial in two variables   **7.** $6x$ **9.** $-13x$ **11.** $8t^2$ **13.** $-11a^4b^3$ **15.** $5x^2$ **17.** The terms $7x^2$ and $-3x$ cannot be combined.   **19.** $-3b^3$ **21.** $6x^6$ **23.** The terms $2t^3w^5$ and $4t^5w^3$ cannot be combined.   **25.** $7.4p^4$ **27.** $5x^2 - 3x$ **29.** $-x^2 + 14x - 8$ **31.** $11x^3 - x^2 + 7x$ **33.** $-45.1t^3 + 114.1t^2 - 9t$ **35.** $-4x^2 - 7x + 6$ **37.** 0 **39.** $-3x^3 - 10x^2 + 7x - 11$ **41.** $13a^3 - 3a^2 - a - 1$ **43.** $a^2 - 9ab + 9b^2$ **45.** $5.9x^3 + 20.9x^2 - 14.3x + 64.1$ **47.** $-7x + 11$ **49.** $-2x^2 - 4x + 6$ **51.** $y^3 + 6y^2 + y + 2$ **53.** 0 **55.** $-2x^2 + 11xy + y^2$ **57.** $6.75x^2 + 14.74x - 17.28$ **59.** $2x^2 + 8x + 9$ **61.** $5x^3 - 8x^2 - 2x + 11$ **63.** $-x^2 + 11x - 5$ **65.** $9x^3 - 5x^2 + 5x - 2$ **67. a.** $40.36t - 2729.58$; the polynomial represents the total number (in thousands) of women and men who have earned bachelor's degrees in the year that is t years since 1900.   **b.** 2153.98; there will be a total of about 2.15 million women and men who will earn a bachelor's degree in 2021.   **c.** $8.18t - 650.94$; the polynomial represents the difference (in thousands) in the number of bachelor's degrees earned by women and men in the year that is t years since 1900.   **d.** 338.84; women will earn about 339 thousand more bachelor's degrees than men in 2021.

69. a. 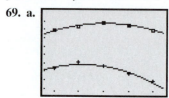   **b.** $-0.951t^2 + 4.89t + 26.66$; the polynomial represents the total of Sonata and RX 400h annual sales (in thousands) at t years since 2010.   **c.** 21.76; the total of Sonata and RX 400h sales were about 21.76 thousand vehicles in 2016.   **d.** $0.109t^2 + 0.31t + 8.26$; the polynomial represents the difference of Sonata and RX 400h annual sales (in thousands) at t years since 2010.   **e.** 14.04; Sonata sales were about 14.0 thousand vehicles more than RX 400h sales in 2016.   **71.** Answers may vary; $2x^2 + x + 6$ **73.** Answers may vary.   **75.** Answers may vary.   **77.** $2x^3 + 5x^3 = (2 + 5)x^3 = 7x^3$ **79.** Answers may vary.   **81.** $x^2 - 3x + 5x^2$; $6x^2 - 3x$ **83.** $4x^2 - (x^2 + x)$; $3x^2 - x$

85. a. 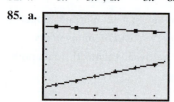   **b.** $-6.81t + 90.35$; the polynomial represents the difference (in millions) of the number of pay-TV subscribers and the number of Netflix subscribers at t years since 2010.   **c.** 8.63; in 2022, there will be about 8.6 million more pay-TV subscribers than Netflix subscribers.   **d.** 2023   **87.** $x^2 - 1$; quadratic (or second-degree) polynomial in one variable

89.

quadratic equation in two variables

Homework 7.4   **1.** x^7 **3.** w^9 **5.** $36x^{12}$ **7.** $30x^7$ **9.** $x^{15}y^7$ **11.** $24c^{14}d^8$ **13.** $-36p^5t^3$ **15.** $-\dfrac{14}{5}x^5$ **17.** $3w^2 - 6w$ **19.** $26.32x^2 - 20.44x$ **21.** $-8x^3 - 12x$ **23.** $6m^3n^2 + 10mn^3$ **25.** $6x^3 - 4x^2 + 14x$ **27.** $-6t^4 - 12t^3 + 6t^2$ **29.** $6x^3y^2 - 8x^2y^3 + 10xy^4$ **31.** $x^2 + 6x + 8$ **33.** $x^2 + 3x - 10$ **35.** $a^2 - 5a + 6$ **37.** $x^2 - 36$ **39.** $x^2 - 14.5x + 48.76$ **41.** $15y^2 + 14y - 8$ **43.** $4x^2 + 16x + 16$ **45.** $9x^2 - 6x + 1$ **47.** $12x^2 - 17xy - 5y^2$ **49.** $6a^2 - 32ab + 32b^2$ **51.** $9x^2 - 16$ **53.** $81x^2 - 16y^2$ **55.** $11.5x^2 + 22.61x - 70.07$ **57.** $-0.629x^2 - 15.969x + 1038.86$ **59.** $x^3 + 6x^2 - 3x - 18$ **61.** $6t^3 - 4t^2 - 15t + 10$ **63.** $6a^4 + a^2b^2 - 15b^4$ **65.** $x^3 + 5x^2 + 11x + 10$ **67.** $x^3 + 8$ **69.** $2b^3 - 11b^2 + 14b - 8$ **71.** $6x^4 - 2x^3 + 17x^2 - 3x + 12$ **73.** $12w^4 - 6w^3 - 5w^2 + 4w - 2$ **75.** $6x^4 + 10x^3 - 3x^2 + 9x - 2$ **77.** $a^2 + b^2 - c^2 + 2ab$ **79. a.** billions of dollars   **b.** $0.04t^2 + 4.24t + 89.9$; the polynomial represents the total money (in billions of dollars) paid for teacher salaries in the year that is t years since 1990.   **c.** 266.54; the total money paid for teacher salaries will be about \$267 billion in 2022.   **81. a.** number of births   **b.** $-50.4t^3 + 1167.6t^2 - 32{,}118t + 363{,}970$; the polynomial represents the number of births from women ages 15–19 in the year that is t years since 2010.   **c.** $84{,}869.2$; there will be about 84,869 births from women ages 15–19 in 2021.   **83.** Answers may vary; $-24x^2$ **85. a. i.** $8x^2 + 22x + 15$; quadratic   **ii.** $15x^2 - 29x - 14$; quadratic   **b.** Answers may vary; quadratic polynomial   **c.** quadratic polynomial; answers may vary.   **87. a.** $x^2 + 11x + 28$ **b.** $x^2 + 11x + 28$ **c.** Answers may vary.   **89.** $(x - 5)(x + 2)$, $x(x - 3) - 10$, $x^2 - 3x - 10$, $(x + 2)(x - 5)$ **91.** $6x^3 - 22x^2 + 26x - 10$ **93.** $-2x^2 + 7x - 7$ **95.** $y = 3x^2 - 6x$; quadratic; parabola   **97.** $y = x^2 - 7x + 12$; quadratic; parabola   **99.** $y = -2x - 3$; linear; line   **101.** $y = 10x^2 + x - 2$; quadratic; parabola   **103.** -2; linear equation in one variable   **105.** $14x^2 - 25x - 25$; quadratic (or second-degree) polynomial in one variable

Homework 7.5   **1.** x^8y^8 **3.** $36x^2$ **5.** $64p^3$ **7.** $64x^2$ **9.** $-27x^3$ **11.** $-a^5$ **13.** $16m^4p^4$ **15.** $81x^4y^4$ **17.** $x^2 + 10x + 25$
19. $x^2 - 8x + 16$ **21.** $4x^2 + 12x + 9$ **23.** $25y^2 - 20y + 4$ **25.** $9x^2 + 36x + 36$ **27.** $81x^2 - 36x + 4$ **29.** $4a^2 + 20ab + 25b^2$
31. $64x^2 - 48xy + 9y^2$ **33.** $y = x^2 + 12x + 36$ **35.** $y = x^2 - 6x + 10$ **37.** $y = 2x^2 + 16x + 29$ **39.** $y = -3x^2 + 6x - 5$
41. $x^2 - 16$ **43.** $t^2 - 49$ **45.** $x^2 - 1$ **47.** $49a^2 - 81$ **49.** $4x^2 - 9$ **51.** $9x^2 - 36y^2$ **53.** $9t^2 - 49w^2$ **55.** $x^2 - 121$
57. $-2x^3 + 4x^2 + 3x$ **59.** $36m^2 - 24mn + 4n^2$ **61.** $-10t^3$ **63.** $3x^3 + x^2 + 2x + 8$ **65.** $4t^2 - 9p^2$ **67.** $8x^3y^2 - 16x^2y^2 - 10xy^2$
69. $-8x^2 - x - 3$ **71.** $x^2 - 4x - 21$ **73.** $16w^2 - 64w + 64$ **75.** $x^3 - x^2 - 11x + 15$ **77.** $6x^2 - 5xy - 21y^2$ **79.** $36x^2 - 49$
81. $4t^2 + 28t + 49$ **83.** Answers may vary; $16x^2$ **85.** Answers may vary; $x^2 + 14x + 49$ **87.** Answers may vary;
$x^2 - 10x + 25$; answers may vary.   **89. a.** Answers may vary.   **b.** $x^2 + 8x + 16$ **c.** Answers may vary.   **91.** $25 = 13$;
answers may vary; $(A + B)^2 = A^2 + 2AB + B^2$ **93.** $(x - 2)^2, x(x - 4) + 4, x^2 - 4x + 4$ **95.** Answers may vary.
97. a. $A^2 + B^2 + C^2 + 2AB + 2AC + 2BC$ **b.** Answers may vary.   **99. a. i.** $32, 16, 8, 4, 2$ **ii.** Answers may vary.
iii. 1   **b.** $81, 27, 9, 3; 1$ **c.** 1   **101.** $y = 10x^2 - 6x$; quadratic; parabola   **103.** $y = 24x^2 - 38x + 15$; quadratic; parabola
105. $y = 6x - 9$; linear; line   **107.** $(5, -1)$; system of two linear equations in two variables
109.

linear equation in two variables

Homework 7.6   **1.** x^8 **3.** r^6 **5.** $15x^9$ **7.** $32b^8$ **9.** $54a^6b^8$ **11.** r^7t^7 **13.** $64x^2$ **15.** $32x^5y^5$ **17.** $16a^4$ **19.** 1   **21.** a^3 **23.** $2x^4$
25. $\dfrac{5x^3y^7}{4}$ **27.** $\dfrac{t^7}{w^7}$ **29.** $\dfrac{27}{t^3}$ **31.** 1   **33.** r^8 **35.** x^{36} **37.** $36x^6$ **39.** t^{12} **41.** $8a^{27}$ **43.** $x^{13}y^{20}$ **45.** $45x^{16}$ **47.** $-3c^{26}$ **49.** x^9y^{19}
51. $\dfrac{5t^8}{4}$ **53.** $\dfrac{3}{4}$ **55.** $\dfrac{y^3}{8x^3}$ **57.** $\dfrac{x^8}{y^{20}}$ **59.** $\dfrac{r^{12}}{36}$ **61.** $\dfrac{8a^{12}}{27b^6}$ **63.** 1   **65.** $\dfrac{8a^{18}b^3}{27c^{15}}$ **67.** $x^{11}y^4$ **69.** $\dfrac{w^7}{32}$ **71.** $2x^2y^7$ **73.** \$6719.58
75. 400 feet   **77.** 507.66 watts   **79.** 423.33 cubic centimeters   **81.** Answers may vary; x^8 **83.** Answers may vary; $25x^6$
85. $\left(\dfrac{x}{y}\right)^3 = \dfrac{x}{y} \cdot \dfrac{x}{y} \cdot \dfrac{x}{y} = \dfrac{x \cdot x \cdot x}{y \cdot y \cdot y} = \dfrac{x^3}{y^3}$ **87.** Answers may vary; $16x^8$ **89.** x^5 **91.** $x^3 + x^2$ **93.** $5x^4$ **95.** $6x^8$ **97.** $9x^2$
99. $x^2 + 6x + 9$ **101.** 5; linear equation in one variable   **103.** $\dfrac{1}{6}x - \dfrac{5}{6}$; linear (or first-degree) polynomial in one variable

Homework 7.7   **1.** $\dfrac{1}{36}$ **3.** $\dfrac{1}{x^4}$ **5.** $\dfrac{1}{b}$ **7.** 16   **9.** w^2 **11.** $\dfrac{1}{x^3y^5}$ **13.** $\dfrac{b^2}{a^4}$ **15.** $\dfrac{2a^3c^8}{5b^5}$ **17.** $-\dfrac{2w}{3x^9y^4}$ **19.** $\dfrac{1}{x^{14}}$ **21.** t^{12} **23.** $\dfrac{30}{t^2}$
25. $-\dfrac{12}{x^9}$ **27.** $\dfrac{4}{x^2y^3}$ **29.** $\dfrac{1}{x^4}$ **31.** $\dfrac{1}{a^8}$ **33.** a^5 **35.** $\dfrac{7x^6}{4}$ **37.** $\dfrac{1}{32}$ **39.** $\dfrac{1}{25}$ **41.** $\dfrac{9}{w^5}$ **43.** $\dfrac{64}{c^6}$ **45.** $\dfrac{x^5}{32}$ **47.** $\dfrac{x^{12}}{y^{30}}$ **49.** $\dfrac{a^{18}}{8b^3}$ **51.** $\dfrac{b^6}{a^3}$
53. x^3y^3 **55.** $\dfrac{1}{x^{12}y^8}$ **57.** $\dfrac{1}{8c^{21}}$ **59.** $\dfrac{27}{r^{15}t^{27}}$ **61.** $\dfrac{a^6d^8}{64b^2c^{10}}$ **63.** $\dfrac{a^6d^6}{72b^3c^2}$ **65.** $\dfrac{3c^7}{4b^5}$ **67. a.** $s = \dfrac{d}{t}$ **b.** 62; an object that travels 186 miles
in 3 hours at a constant speed is traveling at a speed of 62 miles per hour.   **69. a.** $L = \dfrac{5760}{d^2}$ **b.** 90; the sound level is 90 decibels at
a distance of 8 yards from the amplifier.   **71. a.** $P = \dfrac{10{,}000}{1.06^t}$ **b.** 6274.12; the person needs to invest \$6274.12 for the balance to be
\$10,000 in 8 years.   **73.** 49,000   **75.** 0.00859   **77.** 0.000295   **79.** $-451{,}200{,}000$ **81.** 4.57×10^7 **83.** 6.59×10^{-5}
85. -5.987×10^{12} **87.** 1×10^{-6} **89.** from top to bottom of the second column: 0.000000048; 0.0000017; 35,800,000; 1,280,000,000
91. 400,000,000,000   **93.** 85,000,000,000   **95.** 0.00063   **97.** 3.72×10^6 **99.** 2.7×10^7 **101.** 7.5×10^{-6} **103.** Answers may vary;
$\dfrac{5y^2}{x^3w}$ **105.** Answers may vary; x^{10} **107.** Answers may vary.   **109. a.** 1   **b.** 1   **c.** 1   **d.** 1   **111.** $-\dfrac{6}{x^3}$ **113.** $\dfrac{9}{x^6}$ **115.** $-\dfrac{27}{x^9}$
117. $-\dfrac{1}{27x^9}$ **119. a. i.** x^7 **ii.** x^7 **b.** They are the same.   **121.** $-6x - 23$; linear (or first-degree) polynomial in one variable
123. $-\dfrac{23}{6}$; linear equation in one variable

Homework 7.8   **1.** $x^3 + x$ **3.** $3x^2 + 4x$ **5.** $2m^2 - m + 3$ **7.** $-4x^3 + \dfrac{5}{6}x^2 + 3x$ **9.** $7x - \dfrac{7}{5} - \dfrac{4}{x^2}$ **11.** $\dfrac{3}{4}w^2 - 3w + \dfrac{2}{w}$

13. $x^2 - xy + y^3$ **15.** $2a^3 + ab + 3b^2$ **17.** $-2x^3 + 3x^2y - y^2$ **19.** $2x + 3$ **21.** $4x - 1$ **23.** $4b + 5 + \dfrac{2}{b + 3}$

25. $5x - 3 + \dfrac{4}{3x + 2}$ **27.** $3x + 2 - \dfrac{3}{2x - 1}$ **29.** $4m + 3 - \dfrac{4}{2m - 5}$ **31.** $3x^2 + x - 5 + \dfrac{5}{x + 2}$ **33.** $3x^2 - x + 4 - \dfrac{1}{2x - 1}$

35. $p - 2 + \dfrac{9}{p + 2}$ **37.** $x^2 + 2x + 4$ **39.** $3x^2 - 4x - 2 - \dfrac{3}{2x - 1}$ **41.** $2w - 1 + \dfrac{6w - 4}{w^2 - 1}$ **43.** $4x - 1 - \dfrac{1}{2x^2 + 3}$

45. Answers may vary; $1 + \dfrac{5}{3x}$ **47.** Answers may vary; $3x + 5$ **49.** Answers may vary.   **51.** $3x^2 - 10x + 13$ **53.** Answers may

vary.   **55.** Answers may vary.   **57.** Answers may vary.   **59.** $4x^3 - 4x^2 - 11x + 12$ **61.** $2x^2 - x - 1$ **63.** $x + 2 + \dfrac{2}{2x - 3}$

65. $4x^2 - 12x + 9$ **67.** Answers may vary.   **69.** Answers may vary.   **71.** Answers may vary.

Chapter 7 Review Exercises

1.   **2.**   **3.**   **4.** 2   **5.** 1   **6.** 1, 3   **7.** 2   **8. a.** 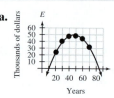   yes   **b.** $21 thousand

c. 27 years, 67 years

d. $(47, 50)$; the

average annual

expenditure for

47-year-old Americans is $50 thousand, the largest average annual expenditure for any age.   **e.** $(10, 0)$, $(84, 0)$; both 10-year-old and

84-year-old Americans do not spend any money; model breakdown has occurred.

9. a. 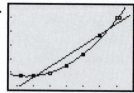   quadratic model   **b.** linear model; $54.09, quadratic model; $63.53; answers may vary.

c. $(87.72, 28.56)$; in 1988, the cost of a Thanksgiving dinner for ten people was $28.56, the least in any

year; model breakdown has occurred.   **d.** linear model   **e.** 2021   **10.** $-9x^3 - 2x^2 - x + 3$

11. $-2x^3 + 2x^2 - 3x + 5$ **12.** $-7x^3 - 8x^2 - 10x + 2$ **13. a.** $-0.05t + 83.6$; the polynomial

represents the percentage of households that are married households or non-family households at

t years since 2000.   **b.** 82.5; in 2022, 82.5% of households will be married households or non-family households.   **c.** $-0.55t + 21.6$;

the polynomial represents the difference of the percentage of households that are married households and the percentage of

households that are non-family households at t years since 2000.   **d.** 9.5; in 2022, the percentage of households that are married

households will be 9.5 percentage points greater than the percentage of households that are non-family households.   **14.** $-21x^7$

15. $10x^5 - 35x^4 + 20x^3$ **16.** $w^2 - 12w + 27$ **17.** $6a^2 - ab - 40b^2$ **18.** $15x^3 + 18x^2 - 20x - 24$ **19.** $x^3 + x^2 - 7x + 20$

20. $8b^3 - 30b^2 + 13b - 21$ **21.** $-8t^3$ **22.** $x^2 + 14x + 49$ **23.** $x^2 - 8x + 16$ **24.** $4p^2 + 20p + 25$ **25.** $-5c^2 - 20c - 20$

26. $x^2 - 36$ **27.** $16m^2 - 49n^2$ **28.** $9p^2 - 12pt + 4t^2$ **29.** $2a^4 + 3a^3 - a^2 + 7a - 3$ **30.** cubic polynomial; answers may vary.

31. $y = x^2 - 10x + 28$ **32.** $y = -2x^2 - 12x - 24$ **33.** x^{10} **34.** $12x^7$ **35.** $-40a^6b^{12}$ **36.** $\dfrac{xy^3}{2}$ **37.** $\dfrac{x^3}{8}$ **38.** $32x^{45}y^{15}$

39. $75x^{14}$ **40.** $\dfrac{3c^5}{2}$ **41.** $\dfrac{a^8}{81}$ **42.** 1   **43.** $\dfrac{3x^3y^5}{2}$ **44.** $\dfrac{27x^9}{64}$ **45.** $\dfrac{1}{64}$ **46.** x^5 **47.** $\dfrac{1}{r^{10}}$ **48.** $\dfrac{1}{t^3}$ **49.** $\dfrac{w^7y^2}{x^5}$ **50.** $\dfrac{1}{c^{24}}$ **51.** $-\dfrac{20}{x^6}$

52. $\dfrac{x^{15}}{32}$ **53.** $\dfrac{x^{12}}{27y^{18}}$ **54.** $\dfrac{3a^3}{4b^3}$ **55.** $\dfrac{9b^{11}}{8a^7}$ **56.** $\dfrac{y^{14}}{x^{21}w^{28}}$ **57.** Answers may vary; x^{15} **58.** 583,200,000   **59.** 0.000317

60. 7.42×10^7 **61.** 8×10^{-5} **62.** 1.426×10^9 **63.** $4x^2 - 3x + 5$ **64.** $2b - \dfrac{7}{5} + \dfrac{4}{b}$ **65.** $4x^2 - 2xy - 3y^2$ **66.** $2x - 3$

67. $3w - 5 + \dfrac{4}{2w + 3}$ **68.** $2x^2 - 3x + 1 - \dfrac{2}{2x - 1}$ **69.** $9p^2 + 6p + 4$ **70.** $4x - 3 + \dfrac{-7x + 7}{x^2 + 2}$

Chapter 7 Test

1. 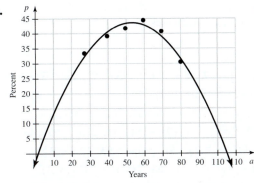   **2.** 5   **3.** $-3, -1$ **4.** -2 **5.** no such value   **6.** $(-4, 0), (0, 0)$ **7.** $(-2, -4)$

8. a.   yes   **b.** 34.8%   **c.** 33 years, 73 years   **d.** $(53.1, 43.5)$; the percentage of 53-year-old women who are obese is 43.5%, which is the largest percentage for any age.   **e.** $(1.4, 0), (104.9, 0)$; no 1-year-old babies are obese and no 105-year-old adults are obese.   **9.** $-4x^3 - x^2 - 2x$ **10.** $-15x^4 + 40x^3 - 10x^2$ **11.** $12p^2 + 10p - 12$ **12.** $6k^3 - 23k^2 + 16k + 10$ **13.** $x^2 - 12x + 36$ **14.** $16a^2 + 56ab + 49b^2$ **15.** $3w^4 + w^3 - 5w^2 + 7w - 6$ **16. a.** $0.46t + 18.16$; the polynomial represents the total number of people who live alone at t years since 1980.   **b.** 37.02; in 2021, about 37.0 million people will live alone.   **c.** $-0.02t + 4.82$; the polynomial represents the difference in the number (in millions) of women and men who live alone at t years since 1980.   **d.** 4; in 2021, there will be 4 million more women than men living alone.   **17.** $y = -3x^2 + 6x + 2$

18. Answers may vary; $x^2 + 8x + 16$ **19.** $\dfrac{3x^4}{4y^5}$ **20.** $64a^{15}b^{16}$ **21.** $\dfrac{x^{18}}{y^{24}}$ **22.** $\dfrac{x^6}{49}$ **23.** $\dfrac{x^8 w^{12}}{y^{24}}$ **24.** $\dfrac{t^5}{2p^3}$ **25.** 4.68×10^{-4}

26. 16,500,000   **27.** $-4x + 2 + \dfrac{3}{x}$ **28.** $2x^3 + 3x^2 y - 4xy^2$ **29.** $2x - 5 - \dfrac{2}{3x - 1}$ **30.** $4w - 6 + \dfrac{13}{2w + 3}$

Chapter 8

Homework 8.1   **1.** $(x + 2)(x + 3)$ **3.** $(t + 4)(t + 5)$ **5.** $(x + 4)^2$ **7.** $(x - 4)(x + 2)$ **9.** $(a - 8)(a + 2)$ **11.** $(x + 8)(x - 3)$ **13.** prime   **15.** $(t - 4)(t + 7)$ **17.** $(x - 8)(x - 2)$ **19.** $(x - 8)(x - 3)$ **21.** prime   **23.** $(r - 5)^2$ **25.** $(x - 6)^2$ **27.** $(x + y)(x + 9y)$ **29.** $(m - 3n)(m + 2n)$ **31.** $(a - 6b)(a - b)$ **33.** $(x - 5)(x + 5)$ **35.** $(x - 9)(x + 9)$ **37.** $(t - 1)(t + 1)$ **39.** prime   **41.** $(2x - 5)(2x + 5)$ **43.** $(9r - 1)(9r + 1)$ **45.** prime   **47.** $(7p - 10q)(7p + 10q)$ **49.** $(8m - 3n)(8m + 3n)$ **51.** $(x - 6)(x + 3)$ **53.** $(x + 7)^2$ **55.** $(a - 2)(a + 2)$ **57.** prime   **59.** $(x - 6)(x - 2)$ **61.** $(w - 8)(w + 6)$ **63.** $(x - 4)^2$ **65.** prime   **67.** $(m - 9n)(m + 3n)$ **69.** $(x - 16)(x - 2)$ **71.** $(10p - 3t)(10p + 3t)$ **73.** $(p + 6)^2$ **75.** Answers may vary; the polynomial is prime.   **77.** $(x - 3)(x + 7), x^2 + 4x - 21, (x + 7)(x - 3)$ **79.** $(x - 8)(x + 3); x^2 - 5x - 24$; answers may vary.   **81.** Answers may vary.   **83.** $-13, -8, -7, 7, 8, 13$ **85.** $0; 0$; answers may vary.   **87.** Answers may vary.   **89.** $x^2 - 7x - 18$ **91.** $(x - 5)(x - 10)$ **93.** $9x^2 - 49$ **95.** $(5x - 6)(5x + 6)$ **97.** $3p + 11w$; linear (or first-degree) polynomial in two variables   **99.** $10p^2 - 6pw - 28w^2$; quadratic (or second-degree) polynomial in two variables   **101.** $(p - 9w)(p - 2w)$; quadratic (or second-degree) polynomial in two variables

Homework 8.2   **1.** $2(3x + 4)$ **3.** $5w(4w + 7)$ **5.** $6x^2(2x - 5)$ **7.** $3ab(2a - 3)$ **9.** $4x^2 y^2(2x + 3y)$ **11.** $5(3x^3 - 2x - 6)$ **13.** $4t(3t^3 + 2t^2 - 4)$ **15.** $5ab(2a^3 - 3a^2 + 5)$ **17.** $2(x - 3)(x + 3)$ **19.** $3(m + 2)(m + 5)$ **21.** $2(x - 6)(x - 3)$ **23.** $3x(x - 3)(x + 3)$ **25.** $4r(r - 5)(r + 1)$ **27.** $6x^2(x - 2)(x + 2)$ **29.** $2m^2 n(2m - 3)(2m + 3)$ **31.** $5x^2(x - 4)(x + 6)$ **33.** $-2x(x - 2)(x + 2)$ **35.** $4t(t + 1)(t + 8)$ **37.** $-3x(2x - 3)(2x + 3)$ **39.** $-3x(x - 2)(x + 8)$ **41.** $6a^2 b(a + 3)^2$ **43.** $(x - 3)(5x^2 + 2)$ **45.** $(2x + 5)(6x^2 - 7)$ **47.** $(p + 3)(2p^2 + 5)$ **49.** $(3x - 1)(2x^2 + 7)$ **51.** $(3w + 1)(5w^2 - 2)$ **53.** $(2x - 3)(2x + 3)(4x - 3)$ **55.** $(b - 3)(b + 3)(2b - 5)$ **57.** $(x - 1)^2(x + 1)$ **59.** $(3 + a)(x + y)$ **61.** $(2x + 3)(y - 4)$

63. $(9x - 5)(9x + 5)$ **65.** $(w - 8)(w - 2)$ **67.** $(x - 6)(x - 4)$ **69.** $5ab(4a - 3b^2)$ **71.** $3(r + 5)^2$ **73.** $x(8x - 7)(8x + 7)$
75. $-(m - 3)^2$ **77.** $(x - 2)(x + 2)(x + 9)$ **79.** $2mn(m - 2n)(m - 3n)$ **81.** Answers may vary; $(2x^2 + 5)(3x + 4)$
83. Answers may vary; $4x(x + 2)(x + 5)$ **85.** Answers may vary.   **87.** Answers may vary.   **89.** Answers may vary.
91. $2x^3 + 2x^2 - 24x$ **93.** $5x(x - 4)^2$ **95.** $(2x - 3)(3x^2 - 2)$ **97.** $x^3 - 3x^2 + 5x - 15$ **99.**   linear equation in two
variables

101. 1; linear equation in one variable   **103.** $(1, -3)$; system of two linear equations in two variables

Homework 8.3   **1.** $(x + 3)(2x + 1)$ **3.** $(x + 2)(5x + 1)$ **5.** $(x + 2)(3x + 2)$ **7.** $(t + 2)(2t - 3)$ **9.** $(2x - 3)(3x - 2)$
11. $(2x + 5)^2$ **13.** $(x - 6)(6x - 1)$ **15.** prime   **17.** $(3x + 4)(6x - 1)$ **19.** $(m - 6)(3m - 4)$ **21.** $(x - 8)(2x - 5)$
23. $(2a + 3b)(a + b)$ **25.** $(5x - 2y)(x + 4y)$ **27.** $3(2b - c)(b - 2c)$ **29.** $2(x + 5)(2x + 3)$ **31.** $5(2a - 3)(2a - 1)$
33. $3(x + 1)(8x - 3)$ **35.** $-2(2x - 3)(5x + 2)$ **37.** $2w^2(2w^2 - 3w - 6)$ **39.** $5x^2(x + 2)(2x - 5)$ **41.** $2(a - 6b)(3a + b)$
43. $4r(r + 2w)(3r + 4w)$ **45.** $(x - 9)(x + 3)$ **47.** $-8x(6x - 5)$ **49.** $(a^2 - 3)(5a + 2)$ **51.** prime   **53.** $(2x - 3)^2$
55. $-17(p + 1)(p - 1)$ **57.** $(x + 4)(x + 6)$ **59.** $(b - 7c)(b + 4c)$ **61.** $(4t - 3)(2t - 1)$ **63.** $7x^2(x - 2)(x + 2)$
65. $(2p + 3)(2p - 3)(3p - 1)$ **67.** prime   **69.** $3x^2(x - 5)(x - 2)$ **71.** $2x^2(8x - 5)(x - 2)$ **73.** $(6a + 7b)(6a - 7b)$
75. $-2y(x - 6)(x + 2)$ **77.** $y(4y + 3)(y - 3)$ **79.** no; answers may vary.   **81.** Answers may vary.   **83.** Answers may vary.
85. Answers may vary.   **87.** $(x + 6)(3x - 2)$ **89.** $12x^2 - 25x + 7$ **91.** $2x^3 - 3x^2 - 14x + 15$ **93.** $2x(3x - 1)(x + 2)$
95. quadratic equation in two variables

97. $(x - 3)(x + 1)$; quadratic (or second-degree) polynomial in one variable   **99.** 32; quadratic (or second-degree) polynomial in
one variable

Homework 8.4   **1.** $(x + 3)(x^2 - 3x + 9)$ **3.** $(x + 5)(x^2 - 5x + 25)$ **5.** $(x - 2)(x^2 + 2x + 4)$ **7.** $(x - 1)(x^2 + x + 1)$
9. $(2t + 3)(4t^2 - 6t + 9)$ **11.** $(3x - 2)(9x^2 + 6x + 4)$ **13.** $5(x + 2)(x^2 - 2x + 4)$ **15.** $2(x - 3)(x^2 + 3x + 9)$
17. $(2x + 3y)(4x^2 - 6xy + 9y^2)$ **19.** $(4a - 3b)(16a^2 + 12ab + 9b^2)$ **21.** $(x - 8)(x + 8)$ **23.** $(m + 4)(m + 7)$
25. $2(t - 3)(t + 4)$ **27.** prime   **29.** $-3(x - 5)(x - 3)$ **31.** $(3p - 1)(5p - 1)$ **33.** $(x - 1)^2$ **35.** $4(2r + 1)(3r - 1)$
37. $-2ab^2(2b - 3a)$ **39.** $(a - 5b)(a + 4b)$ **41.** $(2x - 5)(2x - 1)(2x + 1)$ **43.** $-4x^3(3x + 1)$ **45.** $5a^2(a + 1)(3a + 2)$
47. $(x - 2)(x - 12)$ **49.** $2w^2(w^2 + 2w - 4)$ **51.** $3x^2(2x - 3)(2x + 3)$ **53.** $(x + 5)^2$ **55.** $(x - 3)(2x + 7)$
57. $m(m - 9n)(m - 4n)$ **59.** prime   **61.** $(10x - 3y)(10x + 3y)$ **63.** $(2x + 3)^2$ **65.** $(2x + 3)(3x - 2)(3x + 2)$
67. $a(a - 2b)(3a - 4b)$ **69.** prime   **71.** $(x - 10)(x^2 + 10x + 100)$ **73.** $(4a + 3)(16a^2 - 12a + 9)$ **75.** $3(x + 2)(x^2 - 2x + 4)$
77. Answers may vary.   **79.** Answers may vary.   **81.** Answers may vary; $x^3 - 27 = (x - 3)(x^2 + 3x + 9)$ **83.** no; answers may vary.
85. $(A - B)(A^2 + AB + B^2) = A^3 + A^2B + AB^2 - A^2B - AB^2 - B^3 = A^3 - B^3$.   **87.** Answers may vary.   **89.** $25x^2 - 49$
91. $(9x - 4)(9x + 4)$ **93.** $3x(x - 1)(x + 4)$ **95.** $-14x^3 + 10x^2 - 2x$ **97.** linear equation in two variables

99. 7; linear equation in one variable   **101.** $12x^2 - 7x - 10$; quadratic (or second-degree) polynomial in one variable

Homework 8.5   **1.** $2, 7$ **3.** $-5, -2$ **5.** $-7, 4$ **7.** 7 **9.** $-4, 1$ **11.** $2, 3$ **13.** $-8, 8$ **15.** $-\dfrac{7}{6}, \dfrac{7}{6}$ **17.** $-7, 7$ **19.** $0, 1$ **21.** $-\dfrac{5}{2}, \dfrac{7}{3}$
23. $\dfrac{3}{2}, \dfrac{1}{2}$ **25.** $-\dfrac{4}{5}, \dfrac{1}{2}$ **27.** $-\dfrac{2}{5}$ **29.** $-6, 2$ **31.** $2, \dfrac{4}{3}$ **33.** $1, 6$ **35.** $-\dfrac{5}{2}, \dfrac{3}{2}$ **37.** $\dfrac{1}{2}, 3$ **39.** $2, 5$ **41.** $-6, -2$ **43.** $-4, \dfrac{4}{3}$ **45.** $-3, 4$

47. $2, 4$ **49.** 3 **51.** $-3, -1$ **53.** no real-number solutions   **55.** $-0.79, 3.79$ **57.** $-3.54, 2.54$ **59.** $1, 3$ **61.** 2 **63.** $(-4, 0), (7, 0)$

65. $(4, 0)$ **67.** $(-1, 0), \left(\dfrac{3}{2}, 0\right)$ **69.** $-3, 3, \dfrac{5}{2}$ **71.** $-2, 2, \dfrac{4}{3}$ **73.** $-\dfrac{2}{3}, \dfrac{2}{3}, \dfrac{3}{2}$ **75.** $-4, 2$ **77.** $-5, 0, 5$ **79.** $-3, -2, 0$ **81.** $-2, 0, 4$

83. Answers may vary; $0, 2$ **85.** Answers may vary; $5, 8$ **87.** Answers may vary.   **89. a.** $1, 5$ **b.** 3 **c.**

91. Answers may vary.   **93.** Answers may vary.   **95.** $-2, 6$ **97.** $-\dfrac{3}{2}$ **99.** $P = \dfrac{A}{1 + RT}$ **101.** $(x - 5)(x + 1)$; quadratic (or second-degree)

polynomial in one variable   **103.** $-1, 5$; quadratic equation in one variable   **105.** 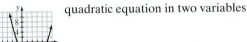   quadratic equation in two variables

Homework 8.6

1. a. 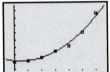   yes   **b.** 173 thousand vehicles   **c.** 2019

3. a.   yes   **b.** 8268 manatees   **c.** $1980, 1993$ **d.** $\dfrac{5.22}{1}$; answers may vary.

5. a.   quadratic model   **b.** 95.2 million iPhones; 73.0 million iPhones; 73.0 million iPhones is the better estimate; answers may vary.   **c.** 673 million iPhones   **d.** 2010

7. a.   **b.** 1998   **c.** 56 thousand overdose deaths   **d.** no; climbing faster than ever; answers may vary.

9. $2012, 2021$ **11.** $2012, 2015$ **13. a.** 0 seconds, 4 seconds; answers may vary.   **b.** 1 second, 3 seconds; answers may vary.
c. 2 seconds; answers may vary.   **15.** width: 5 feet, length: 12 feet   **17.** width: 7 feet, length: 14 feet   **19.** Answers may vary.
21. a. 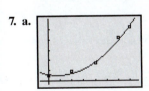   linear model   **b.** $p = 3.98t + 32.1$ **c.** 2017   **d.** 98.3 million tax returns   **23. a.** $n = -2.4t + 54.5$
b. 2021   **25.** $10,973   **27.** Answers may vary.   **29.** Answers may vary.   **31.** Answers may vary.
33. Answers may vary.

Chapter 8 Review Exercises

1. $(x + 4)(x + 5)$ **2.** $2(3x - 4)(x + 1)$ **3.** $(9x - 7)(9x + 7)$ **4.** $(x + 7)^2$ **5.** $-3t^2(2t + 5)(3t - 2)$ **6.** $(p - 9q)(p + 6q)$
7. $(x - 8)(x - 4)$ **8.** $-(3x - 2)(3x + 2)$ **9.** prime   **10.** $5mn(4m - 9n^2)$ **11.** $2x(x + 7)(x + 1)$ **12.** $16x(x - 1)^2$
13. $8x^2(3x - 4)$ **14.** $5x^2y(x - 3)(x - 4)$ **15.** $-(m - 5)(m + 7)$ **16.** $(2r^2 + 3)(2r - 5)$ **17.** $(2p + 1)(p + 3)$

18. $(x - 5)(x - 4)$ **19.** $(3t - 2y)(2t + 5y)$ **20.** $2x(x - 5)(x + 5)$ **21.** $(x - 5)^2$ **22.** $(p - 9)(p + 9)$ **23.** prime   **24.** prime
25. $(x - 1)(2x - 3)(2x + 3)$ **26.** $(2x + 5)^2$ **27.** $2w(6w - 1)(w - 4)$ **28.** $(7a - 3b)(7a + 3b)$ **29.** prime
30. $(x - 2)(x + 2)(x + 3)$ **31.** $(r + 2)(r^2 - 2r + 4)$ **32.** $(2t - 3)(4t^2 + 6t + 9)$ **33.** Answers may vary; the polynomial is
prime.   **34.** Answers may vary; $5x(x + 3)(x + 4)$ **35.** $-25, -14, -11, -10, 10, 11, 14, 25$ **36.** $-7, -2$ **37.** $-6, -2, 0$ **38.** $-1, 2$
39. 3   **40.** 3, 5   **41.** $-\dfrac{9}{5}, \dfrac{9}{5}$ **42.** $-1, 1, \dfrac{7}{2}$ **43.** $-\dfrac{2}{3}, \dfrac{1}{2}$ **44.** 0, 5   **45.** $\dfrac{3}{2}, \dfrac{3}{4}$ **46.** $-5, 7$ **47.** $-2, 6$ **48.** $-5, 6$ **49.** $-3, 3, \dfrac{2}{3}$ **50.** $-2, 2$
51. $-6, 2$ **52.** $-2.56, 1.56$ **53.** 1   **54.** $-2, 4$ **55.** no real-number solutions   **56.** 0, 2   **57.** $(-7, 0), (7, 0)$ **58.** $\left(-\dfrac{3}{4}, 0\right), \left(\dfrac{5}{2}, 0\right)$
59. Answers may vary.   **60.** Answers may vary.   **61. a.** 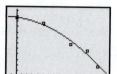   yes   **b.** 98 thousand deaths   **c.** 1994, 2002   **d.** no;
answers may vary.   **62.** width: 3 feet, length: 10 feet.

Chapter 8 Test

1. $(x - 8)(x + 5)$ **2.** $(x - 6)(x - 4)$ **3.** $2m^2n(4n^2 - 5m)$ **4.** $(p - 10q)(p - 4q)$ **5.** $(5p - 6y)(5p + 6y)$
6. $3x^2y(x - 4)(x - 3)$ **7.** $(2x - 3)(2x + 3)(2x + 5)$ **8.** $(4x - 3)(2x - 5)$ **9.** $(4x - 1)(16x^2 + 4x + 1)$
10. $(x - 5)(x + 2), x^2 - 3x - 10, (x + 2)(x - 5)$ **11.** Answers may vary; $(x - 2)(x + 2)(5x + 3)$ **12.** 4, 9   **13.** $-\dfrac{3}{7}, \dfrac{3}{7}$
14. -6 **15.** $-4, 6$ **16.** $-2, -\dfrac{2}{3}, 2$ **17.** $-1, 0, 5$ **18.** 3   **19.** $\dfrac{3}{2}, \dfrac{5}{3}$ **20.** $-5, -1$ **21.** $-4, -2$ **22.** -3 **23.** no real-number solution
24. $\left(-\dfrac{2}{5}, 0\right), \left(\dfrac{3}{2}, 0\right)$ **25.** Answers may vary.   **26. a.**    quadratic model; answers may vary.

b. 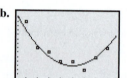   yes **c.** 90%   **d.** 2009, 2017   **27.** 2010, 2023

Cumulative Review of Chapters 1–8

1. explanatory: a, response: p **2.** underestimate   **3.** $-6°F$ **4.** $\dfrac{1}{2}$; increasing   **5.** The value of y decreases by 4 as the value of x is
increased by 1.   **6.** $y = \dfrac{2}{3}x - \dfrac{14}{3}$ **7.** $y = -\dfrac{5}{3}x + \dfrac{1}{3}$ **8.** $y = \dfrac{1}{2}x + 1$ **9.** $3(x + 5) = 9; -2$ **10.** $7 - 4\left(\dfrac{x}{2}\right); -2x + 7$
11. $(2, -1)$ **12.** $(4, -2)$ **13.** $(-3, 2)$ **14.** $(9, 0)$ **15.** -42 **16.** 15   **17.** 84   **18.** $-\dfrac{21}{8}$ **19.** -5 **20.** 1   **21.** 0, 2   **22.** no such
value   **23.** $(-1, 0), (3, 0)$ **24.** $(1, 4)$

25.   **26.**   quadratic equation in two variables

27. $(2m - 7n)(2m + 7n)$; quadratic (or second-degree) polynomial in two variables   **28.** $9p^2 - 42pq + 49q^2$; quadratic (or second-
degree) polynomial in two variables   **29.** $(w + 7)(w - 2)$; quadratic (or second-degree) polynomial in one variable   **30.** $-4, \dfrac{2}{5}$;
quadratic equation in one variable   **31.** $-5, 7$; quadratic equation in one variable   **32.** $xy(2x - 3)(3x + 5)$; fourth-degree polynomial in two variables   **33.** $16p^2 + 18p - 9$; quadratic (or second-degree) polynomial in one variable   **34.** $-20x^3y - 8x^2y^2 + 12xy^3$;
fourth-degree polynomial in two variables   **35.** $(a - 8b)(a + 5b)$; quadratic (or second-degree) polynomial in two variables
36. $\dfrac{7}{6}$; linear equation in one variable   **37.** $-4x^2 - 7x + 9$; quadratic (or second-degree) polynomial in one variable

38. $3(x - 9)(x - 2)$; quadratic (or second-degree) polynomial in one variable   **39.**   linear equation in two variables

40. $6m^3 - 19m^2 + 18m - 20$; cubic (or third-degree) polynomial in one variable   **41.** $-2x^3 + 5x^2 - 6x$; cubic (or third-degree) polynomial in one variable   **42.** $(r + 2)(2r - 3)(2r + 3)$; cubic (or third-degree) polynomial in one variable

43. $(2x - 3)(4x^2 + 6x + 9)$; cubic (or third-degree) polynomial in one variable   **44.** $h = \dfrac{S - 2\pi r^2}{r}$ **45.** $x \geq -4$; $[-4, \infty)$

46. 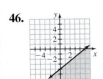   **47.**   **48.** $(x - 2)(x + 4), x^2 + 2x - 8, (x + 4)(x - 2), x(x + 2) - 8$ **49.** $y = -3x^2 + 12x - 12$

50. $(-7, 0), (3, 0)$ **51.** $(10, 0)$ **52.** $-8a^3b^6$ **53.** $\dfrac{r^5}{8}$ **54.** $\dfrac{8y^3}{x^5}$ **55.** $\dfrac{x^6 w^{12}}{y^{18}}$ **56.** 4200 \$20 tickets, 1800 \$35 tickets

57. a. 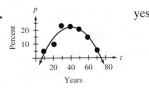   **b.** $s = -16.54t + 115.86$ **c.** -16.54; worldwide sales of Nokia/Microsoft cell phones are decreasing by 16.54 million cell phones per year.   **d.** 99 million cell phones; underestimate; answers may vary.
e. 2016   **f.** $(0, 115.86)$; in 2010, worldwide sales of Nokia/Microsoft cell phones were 116 million cell phones.   **g.** $(7.00, 0)$; in 2017, no Nokia/Microsoft cell phones were sold.

58. a. $I = 1530t + 59,099$ **b.** $I = 770t + 61,554$ **c.** 2017; 64,041   **59. a.**   yes

b. $(7.0, 0), (74.3, 0)$; no 7-year-old children visit online trading sites. Also, no 74-year-old adults visit online trading sites; model breakdown has occurred.   **c.** $(40.7, 22.0)$; this means that 22% of 41-year-old adults visit online trading sites, the highest percentage for any age group, according to the model. In reality, 22.9% of Americans between the ages of 25 and 34 years, inclusive, visit online trading sites.   **d.** 12.9%   **e.** 26-year-old and 55-year-old Americans   **60.** 2011, 2021   **61.** width: 6 feet, length: 14 feet

Chapter 9

Homework 9.1   **1.** 2   **3.** 9   **5.** 11   **7.** 12   **9.** -4 **11.** -9 **13.** not a real number   **15.** not a real number   **17.** irrational; 5.48
19. irrational; 8.83   **21.** rational; 14   **23.** $2\sqrt{5}$ **25.** $3\sqrt{5}$ **27.** $3\sqrt{3}$ **29.** $5\sqrt{2}$ **31.** $10\sqrt{3}$ **33.** $-7\sqrt{2}$ **35.** $24\sqrt{2}$ **37.** $6\sqrt{30}$
39. $3\sqrt{x}$ **41.** $8\sqrt{t}$ **43.** $9x$ **45.** $15tw$ **47.** $x\sqrt{5}$ **49.** $7x\sqrt{39y}$ **51.** $2\sqrt{3p}$ **53.** $6\sqrt{7x}$ **55.** $2ab\sqrt{15}$ **57.** $15y\sqrt{5x}$
59. 68.7 miles per hour   **61.** no; answers may vary.   **63.** 4 and 5   **65.** 8 and 9   **67. a. i.** 2   **ii.** 5   **iii.** 8   **b.** Each of the numbers 2, 5, and 8 is larger than its principal square root.   **c. i.** $\sqrt{0.2}$ **ii.** $\sqrt{0.5}$ **iii.** $\sqrt{0.8}$ **d.** Each of the numbers 0.2, 0.5, and 0.8 is smaller than its principal square root.   **e.** a number greater than 1; a positive number less than 1   **69.** no; answers may vary.
71.   **73.** Answers may vary.   **75.** $49x^2$ **77.** $7x$ **79.**   quadratic equation in two variables

81. 2, 4; quadratic equation in one variable   **83.** $2\sqrt{17x}$; radical expression in one variable

Homework 9.2   **1.** $\dfrac{5}{6}$ **3.** $\dfrac{11}{x}$ **5.** $\dfrac{\sqrt{7x}}{5}$ **7.** $\dfrac{\sqrt{19}}{a}$ **9.** $\dfrac{\sqrt{5}}{xy}$ **11.** $-\dfrac{2\sqrt{2}}{7}$ **13.** $\dfrac{2\sqrt{5}}{9}$ **15.** $\dfrac{5\sqrt{3w}}{6}$ **17.** $\dfrac{2\sqrt{a}}{b}$ **19.** $\dfrac{4\sqrt{5}}{x}$ **21.** $\dfrac{r\sqrt{7t}}{9}$

23. $\dfrac{2\sqrt{3}}{3}$ **25.** $\dfrac{a\sqrt{13}}{13}$ **27.** $\dfrac{7\sqrt{x}}{x}$ **29.** $\dfrac{5\sqrt{3}}{3}$ **31.** $\dfrac{\sqrt{14}}{7}$ **33.** $\dfrac{\sqrt{22}}{2}$ **35.** $\dfrac{\sqrt{7p}}{p}$ **37.** $\dfrac{\sqrt{6}}{4}$ **39.** $\dfrac{\sqrt{6x}}{10}$ **41.** $\dfrac{w\sqrt{15}}{6}$ **43.** $\dfrac{x\sqrt{35y}}{5}$

45. $\dfrac{3+\sqrt{2}}{2}$ **47.** $2-\sqrt{7}$ **49.** $\dfrac{4+6\sqrt{13}}{3}$ **51.** $\dfrac{2+\sqrt{3}}{4}$ **53.** $\dfrac{2-\sqrt{2}}{4}$ **55.** $\dfrac{3-\sqrt{5}}{2}$ **57. a.** $T=\dfrac{\sqrt{h}}{4}$ **b.** 9.5 seconds

59. Answers may vary; $\dfrac{5\sqrt{3}}{3}$ **61.** yes; answers may vary.   **63.** Answers may vary.   **65.** $4(2x+3)$ **67.** $4(3+2\sqrt{3})$

69. $-7(3x-2)$ **71.** $7(2-3\sqrt{7})$ **73.** $0,4$; quadratic equation in one variable   **75.** $x(x-4)$; quadratic (or second-degree)

polynomial in one variable   **77.**   quadratic equation in two variables

Homework 9.3   **1.** ±2 **3.** ±14 **5.** 0 **7.** $\pm\sqrt{15}$ **9.** $\pm2\sqrt{5}$ **11.** $\pm3\sqrt{3}$ **13.** no real-number solutions   **15.** $\pm2\sqrt{7}$

17. no real-number solutions   **19.** $\pm\dfrac{\sqrt{5}}{2}$ **21.** $\pm\dfrac{\sqrt{35}}{5}$ **23.** $\pm\dfrac{\sqrt{10}}{4}$ **25.** $\pm\dfrac{\sqrt{6}}{2}$ **27.** $\pm\dfrac{\sqrt{70}}{5}$ **29.** $-6,2$ **31.** $-3,9$ **33.** $7\pm\sqrt{13}$

35. no real-number solutions   **37.** $-2\pm3\sqrt{2}$ **39.** $6\pm2\sqrt{6}$ **41.** 5 **43.** $-2,6$ **45.** ±9 **47.** $2\pm2\sqrt{6}$ **49.** $\pm\dfrac{\sqrt{33}}{3}$ **51.** $-5,\dfrac{3}{2}$

53. $0,2$ **55.** $\sqrt{41}$ **57.** $2\sqrt{14}$ **59.** $2\sqrt{11}$ **61.** 10 **63.** $\sqrt{13}$ **65.** $4\sqrt{5}$ **67.** 8 **69.** $\sqrt{21}$ **71.** $3\sqrt{5}$ **73.** 18.8 miles   **75.** yes
77. 54.3 inches   **79.** 28.2 inches   **81.** 3.2 miles   **83.** 340 feet (113 yards and 1 foot)   **85.** 439.8 miles   **87. a.** ±6 **b.** ±6 **c.** They
are the same results.   **d.** Answers may vary.   **89. a.** yes; $2\pm\sqrt{7}$ **b.** no   **c.** no; answers may vary.   **91.** Answers may vary.
93. Answers may vary.   **95. a.** ±7 **b.** 7 **c.** Answers may vary.   **97. a.** Answers may vary.   **b.** Answers may vary.   **c.** the square
d. W, L, D **e.** A **99.** $12x^4+3x^3-23x^2-2x+10$; fourth-degree polynomial in one variable   **101.** $3x^2(x-7)(x+3)$;
fourth-degree polynomial in one variable   **103.**   linear equation in two variables

Homework 9.4   **1.** $9; (x+3)^2$ **3.** $49; (x+7)^2$ **5.** $1; (x+1)^2$ **7.** $16; (x-4)^2$ **9.** $25; (x-5)^2$ **11.** $100; (x-10)^2$
13. $-3\pm\sqrt{14}$ **15.** $-7\pm\sqrt{29}$ **17.** $5\pm\sqrt{21}$ **19.** $1\pm\sqrt{6}$ **21.** $-6\pm4\sqrt{2}$ **23.** $2\pm3\sqrt{2}$ **25.** $4\pm2\sqrt{6}$ **27.** no real-number
solutions   **29.** $-10\pm2\sqrt{15}$ **31.** $-3\pm2\sqrt{2}$ **33.** $2\pm\sqrt{3}$ **35.** $5\pm4\sqrt{2}$ **37.** no real-number solutions   **39.** $-4\pm2\sqrt{3}$
41. $4\pm\sqrt{19}$ **43.** $-1\pm2\sqrt{2}$ **45.** $5\pm3\sqrt{2}$ **47.** $-1\pm\sqrt{2}$ **49.** $3\pm2\sqrt{2}$ **51.** $-2\pm2\sqrt{2}$ **53.** ±3 **55.** $5,6$ **57.** $5\pm4\sqrt{2}$

59. $-3,\dfrac{4}{3}$ **61.** $\pm\sqrt{13}$ **63.** $3\pm2\sqrt{3}$ **65. a.** $2,4$ **b.** $2,4$ **c.** Answers may vary.   **67. a.** yes; $-2\pm\sqrt{11}$ **b.** no

c. Answers may vary.   **69.** Answers may vary.   **71.** Answers may vary.   **73.** $x^2+14x+49$ **75.** $(x-8)^2$ **77.** $x^2-8x+16$;
quadratic (or second-degree) polynomial in one variable   **79.** $4\pm\sqrt{3}$; quadratic equation in one variable   **81.** $(p-4q)^2$;
quadratic (or second-degree) polynomial in two variables

Homework 9.5   **1.** $-1,-\dfrac{3}{2}$ **3.** $\dfrac{-7\pm\sqrt{17}}{8}$ **5.** $\dfrac{-3\pm\sqrt{29}}{2}$ **7.** $\dfrac{5\pm\sqrt{61}}{6}$ **9.** $\dfrac{-1\pm\sqrt{6}}{5}$ **11.** $\dfrac{3\pm\sqrt{3}}{3}$ **13.** no real-number

solutions   **15.** $\dfrac{-5\pm\sqrt{57}}{4}$ **17.** $\dfrac{-1\pm\sqrt{5}}{2}$ **19.** no real-number solutions   **21.** $\dfrac{7\pm\sqrt{13}}{6}$ **23.** $\dfrac{1\pm\sqrt{5}}{6}$ **25.** $-1\pm2\sqrt{3}$

27. $-1.17, 1.92$ **29.** $-4.90, 1.66$ **31.** $-4.65, 2.31$ **33.** $-6.32, 14.82$ **35.** $(0.22, 0), (2.28, 0)$ **37.** $(-0.64, 0), (1.24, 0)$

39. $(-2.29, 0), (0.89, 0)$ **41.** $(-2.06, 0), (1.41, 0)$ **43.** $-9, -2$ **45.** $-4\pm\sqrt{13}$ **47.** $\dfrac{-5\pm\sqrt{37}}{6}$ **49.** $\pm\dfrac{7}{6}$ **51.** $0,\dfrac{3}{2}$ **53.** $\pm\dfrac{\sqrt{35}}{7}$

55. $\dfrac{1}{2},\dfrac{2}{3}$ **57.** no real-number solutions   **59.** $\dfrac{1\pm\sqrt{22}}{7}$ **61.** $-2\pm\sqrt{2}$ **63.** $-1,3$ **65.** 1 **67.** $-5,-1$ **69.** no real-number solutions

71. $-1.24, 3.24$ **73.** $-0.23, 2.90$ **75.** $-2, 4$ **77.** 1 **79.** Answers may vary; $\dfrac{-5 \pm \sqrt{37}}{2}$ **81. a.** $-3, 1$ **b.** $-3, 1$ **c.** $-3, 1$

d. The results are the same. **e.** Answers may vary. **83.** Answers may vary. **85.** Answers may vary. **87.** $\dfrac{7}{2}$ **89.** $\pm\dfrac{\sqrt{14}}{2}$

91. $\dfrac{3 \pm \sqrt{11}}{2}$ **93.** $(1.54, 0), (-0.87, 0)$ **95.** $(0.67, 0)$ **97.** $\dfrac{-3 \pm \sqrt{29}}{2}$; quadratic equation in one variable **99.** $(3x + 1)(2x - 1)$;

quadratic (or second-degree) polynomial in one variable **101.** $-25x + 23$; linear (or first-degree) polynomial in one variable

Homework 9.6 **1. a.**  yes **b.** 1985 metric tons **c.** 2022 **d.** $\dfrac{2.86}{1}; \dfrac{3.27}{1}; \dfrac{3.78}{1}$; answers may vary.

3. a. 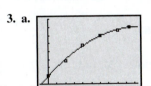 yes **b.** 35.1% **c.** 2003 **d.** $(52.73, 35.18)$; in 2023, the percentage of lawyers who are women will be 35% — the largest percentage for the past and future, according to the model. **e.** after 2023

5. a. 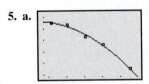 yes **b.** 68% **c.** 55 years **d.** $(-58.70, 0), (84.04, 0)$; no -59-year-old drivers and no 84-year-old drivers admit to running red lights; model breakdown has occurred. **e.** 13-year-old drivers; 76%; model breakdown has occurred.

7. a. 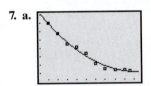 yes **b.** 2.55 people **c.** 1971 **d.** $(0, 5.11)$; in 1900, the average household size was 5.11 people.

9. a. 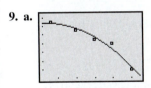 yes **b.** 43% **c.** 61 years **d.** $(-71.99, 0), (118.80, 0)$; no -72-year-old Americans and no 119-year-old Americans are worried about joblessness; model breakdown has occurred.

11. a. 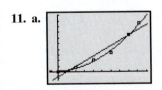 quadratic model **b.** 11,517 women; 9421 women; 9421 women; answers may vary. **c.** 2031; 2021; answers may vary.

13. a. 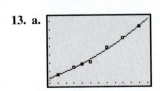 yes **b.** 15 years **c. i.** $141,200 **ii.** $38,600 **iii.** 7.55 years **iv.** $605,600 **v.** underestimate; answers may vary.

15. a. 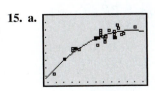 yes **b.** 16.4 meters **c.** $(5.52, 0), (266.60, 0)$; a shot-putter with maximum power clean of about 5.5 kilograms or 266.6 kilograms has a best shot put of 0 meters; model breakdown has occurred. **d.** $(136.06, 17.72)$; a shot-putter with maximum power clean of about 136.1 kilograms has a best shot put of about 17.7 meters, which is the best shot put of all the shot-putters.

e. $p \le 5.52$ or $p > 136.06$ **f.** above the model   **g.** Answers may vary.   **17.** Answers may vary.   **19. a.** $p = -0.45a + 36.56$
b. 28%   **c.** 37 years   **21. a.** $p = 2t + 71$ **b.** 83%   **c.** 2023   **23.** 2019   **25.** Answers may vary.   **27.** Answers may vary.
29. Answers may vary.   **31.** Answers may vary.   **33.** Answers may vary.

Chapter 9 Review Exercises

1. 14   **2.** -8 **3.** not a real number   **4.** not a real number   **5.** 9.75   **6.** -45.15 **7.** $2\sqrt{21}$ **8.** $6\sqrt{x}$ **9.** $3b\sqrt{2a}$ **10.** $7mn\sqrt{2}$
11. 6 and 7   **12.** $\dfrac{\sqrt{5}}{3}$ **13.** $\dfrac{5\sqrt{2y}}{x}$ **14.** $\dfrac{4\sqrt{x}}{x}$ **15.** $\dfrac{b\sqrt{10}}{8}$ **16.** Answers may vary; $\dfrac{3\sqrt{7}}{7}$ **17.** $\pm 3\sqrt{5}$ **18.** $\pm\dfrac{\sqrt{35}}{5}$ **19.** $\pm\dfrac{2\sqrt{10}}{5}$
20. $-4 \pm 3\sqrt{3}$ **21.** no real-number solutions   **22.** 8   **23.** $4\sqrt{5}$ **24.** $3\sqrt{3}$ **25.** $\sqrt{58}$ **26.** $2\sqrt{14}$ **27.** yes   **28.** $-3 \pm \sqrt{11}$
29. $-1 \pm 3\sqrt{2}$ **30.** $4 \pm 2\sqrt{5}$ **31.** no real-number solutions   **32.** $-2 \pm \sqrt{10}$ **33.** $3 \pm 3\sqrt{2}$ **34.** $\dfrac{-7 \pm \sqrt{37}}{6}$ **35.** $\dfrac{5 \pm \sqrt{57}}{4}$
36. no real-number solutions   **37.** $\dfrac{3 \pm \sqrt{29}}{2}$ **38.** $\dfrac{1 \pm \sqrt{13}}{4}$ **39.** $\dfrac{1 \pm \sqrt{11}}{2}$ **40.** $-0.46, 1.80$ **41.** $-4.47, 1.63$ **42.** Answers may
vary; $\dfrac{-3 \pm \sqrt{65}}{4}$ **43.** $-2, 7$ **44.** $\dfrac{2 \pm \sqrt{7}}{3}$ **45.** no real-number solutions   **46.** $4 \pm 2\sqrt{6}$ **47.** $-5, 7$ **48.** $\pm\dfrac{2\sqrt{15}}{5}$ **49.** $\dfrac{7 \pm \sqrt{37}}{2}$
50. $-2, \dfrac{4}{3}$ **51.** $-7 \pm 3\sqrt{3}$ **52.** $\dfrac{-7 \pm \sqrt{73}}{4}$ **53.** $\dfrac{2 \pm \sqrt{14}}{2}$ **54.** $\dfrac{1}{2}, \dfrac{3}{2}$ **55.** 2   **56.** no real-number solutions   **57.** $0, 4$ **58.** $-2, 6$
59. $-2.08, 2.41$ **60.** $(0.28, 0), (2.39, 0)$ **61.** $(-2.95, 0), (4.77, 0)$ **62. a.** $-3, 5$ **b.** $-3, 5$ **c.** $-3, 5$ **d.** The results are the same.
63. a.    yes   **b.** 1.4%   **c.** 87 years   **64. a.** 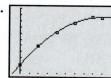   yes   **b.** 47.4%   **c.** 2023

d. $(-5.12, 0), (95.64, 0)$; no women earned medical degrees in 1965 and no women will earn medical degrees in 2066; model
breakdown has occurred.

Chapter 9 Test

1. -11 **2.** not a real number   **3.** $4\sqrt{3}$ **4.** $8\sqrt{x}$ **5.** $3a\sqrt{5b}$ **6.** $\dfrac{3\sqrt{7}}{7}$ **7.** $\dfrac{m\sqrt{2n}}{n}$ **8.** $\dfrac{\sqrt{15}}{10}$ **9.** $4\sqrt{10}$ **10.** 12.96 inches
11. $\pm\dfrac{\sqrt{15}}{3}$ **12.** $\pm\dfrac{4\sqrt{10}}{5}$ **13.** $5 \pm 3\sqrt{2}$ **14.** $\dfrac{-3 \pm \sqrt{57}}{4}$ **15.** $-5, 8$ **16.** $\dfrac{1 \pm \sqrt{13}}{3}$ **17.** $2 \pm \sqrt{11}$ **18.** $-4.96, 6.61$
19. $(1.39, 0), (30.19, 0)$ **20.** $2, 4$ **21.** $1, 5$ **22.** 3   **23.** no real-number solutions   **24.** Answers may vary; $\dfrac{-2 \pm \sqrt{10}}{2}$

25. a.    yes   **b.** 2010; $698 (in 2015 dollars); actually, the minimum median asking rent was $694 (in 2015
dollars), which occurred in 2011.   **c.** $1425   **d.** 2019

Chapter 10

Homework 10.1   **1.** 0   **3.** no excluded values   **5.** 4   **7.** -9 **9.** 4   **11.** $-\dfrac{4}{3}$ **13.** $-3, -2$ **15.** $-5, 7$ **17.** 5   **19.** ± 4
21. $\pm\dfrac{7}{5}$ **23.** $-5, -\dfrac{3}{2}$ **25.** $\dfrac{2}{3}, \dfrac{3}{2}$ **27.** ± 2 **29.** $\dfrac{2x}{3}$ **31.** $\dfrac{4t^2}{5}$ **33.** $\dfrac{2x}{3y^3}$ **35.** $\dfrac{3}{5}$ **37.** $\dfrac{2}{3}$ **39.** $\dfrac{a}{7}$ **41.** $\dfrac{x-y}{3}$ **43.** $\dfrac{4}{x+5}$ **45.** $\dfrac{5}{x-2}$
47. $\dfrac{t+1}{t+5}$ **49.** $\dfrac{x-2}{x-1}$ **51.** $\dfrac{x-2}{x-5}$ **53.** $\dfrac{x+4}{x-4}$ **55.** $\dfrac{2}{3x+8}$ **57.** $-\dfrac{4}{\omega+4}$ **59.** $\dfrac{2x+5}{x+3}$ **61.** $\dfrac{3(x+2)}{6x-1}$ **63.** $\dfrac{a+b}{a-b}$

65. $\dfrac{5}{x^2 - 3}$ **67.** $\dfrac{1}{t - 1}$ **69.** $\dfrac{x^2 - 2x + 4}{x + 5}$ **71.** $\dfrac{x^2 + 4x + 16}{x + 4}$ **73. a.**

$p = \dfrac{60}{n}$

Number of Students n	Cost per Student (dollars per student) p
10	$\dfrac{60}{10}$
20	$\dfrac{60}{20}$
30	$\dfrac{60}{30}$
40	$\dfrac{60}{40}$
n	$\dfrac{60}{n}$

b. The units on both sides of the equation are dollars per student.   **c.** 2.4; the per-person cost is \$2.40 if 25 students go to the party.
d. The value of p decreases; the greater the number of students at the party, the lower is the per-person cost.   **75. a.** 1.27% **b.** 1.29%

c. underestimate **d.** 0.27%   **77.** no; answers may vary.   **79.** $\dfrac{5}{2}$ **81.** Answers may vary; the expression is already simplified.

83. Answers may vary.   **85.** Answers may vary.   **87.** Answers may vary.   **89.** Answers may vary.   **91.** $\dfrac{y^9 w^2}{5x^4}$ **93.** $\dfrac{4}{x - 4}$

95. $\dfrac{3 - 2\sqrt{2}}{4}$ **97.** $(p - 3)(p + 3)(2p - 3)$; cubic (or third-degree) polynomial in one variable   **99.** quadratic

equation in two variables   **101.** $\dfrac{1 \pm \sqrt{13}}{3}$; quadratic equation in one variable

Homework 10.2   **1.** $\dfrac{15}{x^2}$ **3.** $\dfrac{x^2}{9}$ **5.** $\dfrac{18}{5a^6}$ **7.** $\dfrac{2(x - 4)}{(x - 3)(x + 5)}$ **9.** $\dfrac{(k - 2)(k + 4)}{(k - 6)(k + 6)}$ **11.** $\dfrac{10}{21}$ **13.** 6 **15.** $\dfrac{2w^4(w + 5)}{w + 3}$

17. $\dfrac{5(x + 1)}{3(x - 7)}$ **19.** $\dfrac{30(x - 4)}{7(x + 5)}$ **21.** $\dfrac{5t^4}{6}$ **23.** $\dfrac{6}{x^3(x - 7)}$ **25.** $\dfrac{3a^3}{10b^7}$ **27.** $\dfrac{x + 4}{x - 5}$ **29.** $\dfrac{t - 6}{t + 9}$ **31.** $\dfrac{4(x + 4)}{3x}$ **33.** $\dfrac{(x + 1)(x + 2)}{(x - 6)(x - 4)}$

35. $\dfrac{(a + 3)(a + 5)}{4a(a - 3)}$ **37.** $\dfrac{(x - 3)(x + 3)}{(x - 5)(x + 1)}$ **39.** $\dfrac{(x^2 + 16)(x - 4)}{4x(x + 4)}$ **41.** $\dfrac{x(3x + 5)}{(x - 4)(x + 1)}$ **43.** $\dfrac{p^2(p + 1)}{2}$ **45.** $\dfrac{x + 5}{(x - 5)(x + 1)}$

47. $\dfrac{2(x - 3)}{x + 7}$ **49.** $\dfrac{2b(b + 5)}{b - 1}$ **51.** $\dfrac{x}{(x - 4)(x + 8)}$ **53.** $\dfrac{a - b}{2a + 3b}$ **55.** $\dfrac{x(x - y)}{3(x + y)}$ **57.** $\dfrac{3(x^2 + 3)}{x(x + 2)}$ **59.** $\dfrac{x(x + 3)}{6}$

61. a. $\dfrac{(x + 1)^2}{(x - 2)^2}$ **b.** $\dfrac{(x - 1)^2}{(x + 2)^2}$ **63.** 36 inches **65.** 6.21 miles **67.** 0.68 ounce per day **69.** 14.78 miles per gallon **71.** 0.76 gram

73. $\dfrac{x^2}{4}$ **75.** $\dfrac{(x - 2)(x + 1)}{(x - 6)(x + 4)}$ **77.** Answers may vary; $\dfrac{x + 2}{x + 6}$ **79.** Answers may vary; $\dfrac{x(x + 4)}{3(x - 7)}$ **81.** Answers may vary.

83. Answers may vary.   **85.** Answers may vary.   **87.** $x - 4$; $x - 4$; the results are the same.   **89.** $-3x^2 + 30x - 75$; quadratic (or
second-degree) polynomial in one variable   **91.** $5 \pm 2\sqrt{2}$; quadratic equation in one variable   **93.** -27; quadratic (or second-degree) polynomial in one variable

Homework 10.3   **1.** $\dfrac{9}{x}$ **3.** $\dfrac{8x}{x - 1}$ **5.** $\dfrac{8x + 2}{x + 3}$ **7.** $t + 2$ **9.** $\dfrac{1}{x - 2}$ **11.** $\dfrac{x - 4}{x + 2}$ **13.** $\dfrac{r - 7}{r + 3}$ **15.** $\dfrac{11}{2x}$ **17.** $\dfrac{5w + 9}{6w}$

19. $\dfrac{10x^2 + 9}{12x}$ **21.** $\dfrac{12x^2 + 25}{40x^3}$ **23.** $\dfrac{a^2 + b^2}{ab}$ **25.** $\dfrac{13x + 15}{4x(x + 3)}$ **27.** $\dfrac{7(r - 1)}{(r - 5)(r + 2)}$ **29.** $\dfrac{3(11x - 1)}{5x(x - 1)}$ **31.** $\dfrac{5x^2 + 12x - 8}{(x - 4)(x + 2)}$

33. $\dfrac{2a}{(a + b)(a - b)}$ **35.** $\dfrac{5x + 2}{5(x + 4)}$ **37.** $\dfrac{1}{3}$ **39.** $\dfrac{2(3x + 10)}{15(x - 5)}$ **41.** $\dfrac{3}{x - 1}$ **43.** $\dfrac{t + 4}{t + 9}$ **45.** $\dfrac{x^2 + 3x + 6}{2(x - 3)(x + 3)}$ **47.** $\dfrac{x - 1}{x(x - 2)}$

49. $\dfrac{2(3a+5)}{(a-3)(a+1)(a+4)}$ **51.** $\dfrac{8x-17}{(x-4)(x+1)(x+4)}$ **53.** $\dfrac{2(2x^2+9x-3)}{(x-3)(x+4)(x+6)}$ **55.** $\dfrac{x^2-x-16}{(x+3)(x+4)(x+5)}$

57. $\dfrac{4w+13}{w+5}$ **59.** $\dfrac{2(x^2+11)}{(x-4)(x+5)}$ **61.** $\dfrac{2y^2-13}{(y-3)(y+2)}$ **63.** $\dfrac{x^2-x-18}{(x-5)(x+3)}$ **65.** $\dfrac{b^2-b-1}{(b-4)(b-3)}$ **67.** $\dfrac{x^2-5x+26}{4(x-3)(x+3)}$

69. $\dfrac{2x^2}{(x-y)^2}$ **71.** $\dfrac{11x-20}{3(x-4)(x^2+2)}$ **73.** $\dfrac{9x^2+18x+16}{(x-2)(x+2)(x^2+2x+4)}$ **75. a.** $\dfrac{18}{d^2}+\dfrac{18}{(2d)^2}$ **b.** $\dfrac{45}{2d^2}$ **c.** 15.63; the total illumi-

nation is 15.63 W/m² when the person is 1.2 meters away from the closer light and 2.4 meters away from the other light. **77.** Answers

may vary; $\dfrac{8x+19}{(x+2)(x+3)}$ **79.** yes; yes **81.** Answers may vary. **83.** Answers may vary. **85.** x^2-5x+6 **87.** $x-2$

89. $(-3,-11)$; system of two linear equations in two variables **91.**   linear equation in two variables **93.** -3; linear

equation in one variable

Homework 10.4 **1.** $\dfrac{2}{x}$ **3.** $\dfrac{7x}{x-2}$ **5.** $\dfrac{1}{x+3}$ **7.** $-\dfrac{4(r-1)}{r+6}$ **9.** $x-3$ **11.** $\dfrac{x+8}{x+2}$ **13.** $\dfrac{a-3}{a+4}$ **15.** $-\dfrac{5}{4x}$ **17.** $\dfrac{-3b+20}{8b}$

19. $\dfrac{15x^2-4}{24x}$ **21.** $\dfrac{-2x^2+5}{10x^4}$ **23.** $\dfrac{3a^2-2b^2}{6ab}$ **25.** $\dfrac{7p+10}{4p(p-2)}$ **27.** $\dfrac{4x+31}{(x-1)(x+4)}$ **29.** $\dfrac{3x^2+4x+10}{(x-2)(x+3)}$ **31.** $\dfrac{-2b}{(a+b)(a-b)}$

33. $\dfrac{c-6}{2(c-4)}$ **35.** $\dfrac{8x-21}{12(x-4)}$ **37.** $\dfrac{x+2}{x+1}$ **39.** $\dfrac{1}{x-3}$ **41.** $\dfrac{t^2+7t-12}{3(t-7)(t+7)}$ **43.** $\dfrac{2(2x^2-x+5)}{x(x-5)(x+5)}$ **45.** $\dfrac{-x+32}{(x-5)(x+2)(x+4)}$

47. $\dfrac{5x+4}{x(x-3)(x-2)}$ **49.** $\dfrac{5b^2+27b-15}{(b-2)(b+5)(b+6)}$ **51.** $\dfrac{2x^2-19x-45}{(x-7)(x+2)(x+9)}$ **53.** $-\dfrac{3(x-9)}{x-6}$ **55.** $\dfrac{10(x-1)}{(x-4)(x+1)}$

57. $\dfrac{x^2-3x-14}{(x-5)(x-4)}$ **59.** $-\dfrac{(t-6)(t+2)}{(t-7)(t-3)}$ **61.** $\dfrac{(x-7)(x+2)}{3(x-1)(x+1)}$ **63.** $\dfrac{6x^2}{(2x+y)(x+y)}$ **65.** $\dfrac{-3x^2+18x+10}{5(x-6)(x^2-3)}$

67. $\dfrac{x^2+5x-2}{(x+1)^2(x^2-x+1)}$ **69.** Answers may vary; $\dfrac{-2x-7}{x+2}$ **71.** Answers may vary. **73.** Answers may vary. **75.** $\dfrac{x+4}{x-7}$

77. $\dfrac{11}{x+3}$ **79.** $\dfrac{x(x-3)}{(x-7)(x+5)}$ **81.** $\dfrac{2(x^2-5x-4)}{(x-7)(x-2)^2}$ **83.** $\dfrac{(x-2)(5x+3)}{x(x+5)}$ **85.** $\dfrac{-5x^2-35x+6}{2(x-3)(x+7)}$ **87. a.** $\dfrac{x^2+8}{2x}$

b. $\dfrac{x^2-8}{2x}$ **c.** 2 **d.** $\dfrac{x^2}{8}$ **89.** $-2,\dfrac{4}{3}$; quadratic equation in one variable **91.** $(3x-4)(x+2)$; quadratic (or second-degree)

polynomial in one variable **93.** $\dfrac{-9x^2+17x-10}{(x-2)(x+2)(3x-4)}$; rational expression in one variable

Homework 10.5 **1.** -2 **3.** 3 **5.** 2 **7.** empty set **9.** 3 **11.** $\dfrac{4}{5}$ **13.** $\dfrac{1}{6}$ **15.** -8 **17.** $\dfrac{3}{2}$ **19.** 1 **21.** $\dfrac{2}{3}$ **23.** empty set

25. $-\dfrac{7}{2}$ **27.** empty set **29.** $\dfrac{19}{12}$ **31.** $-\dfrac{6}{5}$ **33.** 2 **35.** $-4,1$ **37.** $-5,3$ **39.** $-\dfrac{2}{5},1$ **41.** $-\dfrac{7}{11}$ **43.** 5 **45.** $-4,1$ **47.** $\dfrac{7}{2}$

49. $\dfrac{17}{11}$ **51.** 7 **53.** -7 **55.** 2020 **57. a.** 80.1%; 79.4%; overestimate **b.** 94.0% **c.** 2022 **59.** Answers may vary; $\dfrac{2(4x+3)}{x(x+2)}$

61. Answers may vary; $\dfrac{33}{2}$ **63.** Answers may vary. **65.** Answers may vary. **67.** $\dfrac{10}{x}$ **69.** 10 **71.** $2,\dfrac{1}{3}$ **73.** $\dfrac{(x-1)(3x-4)}{(x-3)(x+2)}$

75. $\dfrac{1\pm\sqrt{13}}{3}$ **77.** $\pm2,\dfrac{2}{3}$ **79.** $\pm\dfrac{\sqrt{14}}{2}$ **81.** $11m-17n$; linear (or first-degree) polynomial in two variables **83.** $2x(x-7)(x-2)$;

cubic (or third-degree) polynomial in one variable **85.** 15; rational equation in one variable

Homework 10.6   **1.** \$960   **3.** 1.31 cups   **5.** 2.4 ounces   **7.** 17,933 students   **9.** 6.75 inches   **11.** 271.97 U.S. dollars   **13.** 172.1 pounds
15. a. \$409.50   **b.** Answers may vary; underestimate   **17. a.** 5996 adults   **b.** Answers may vary; overestimate   **19.** no; answers may vary.
21. a. 21.5:1; greater   **b.** 339 FTE faculty; \$24.8 million   **c.** 5766 students; \$19.3 million   **d.** reduce FTE enrollment   **23.** 4.88 inches
25. 14.86 meters   **27.** 5.63 feet   **29.** Answers may vary.   **31. a.** $y = kx$; $(0, 0)$ **b.** $C = 15t$; the variables t and C are proportional;
answers may vary.   **c.** $C = 10t + 25$; the variables t and C are not proportional; answers may vary.   **d.** \$45; \$55; the variables t and
C are not proportional; yes   **33. a.** $\dfrac{n}{m}$ **b.** $\dfrac{q}{p}$ **c.** Answers may vary.   **d.** Answers may vary.   **35.** $-5, \dfrac{3}{2}$; quadratic equation in one
variable   **37.** $(2x - 3)(x + 5)$; quadratic (or second-degree) polynomial in one variable   **39.** -21; quadratic (or second-degree)
polynomial in one variable

Homework 10.7   **1.** $w = kt$ **3.** $F = \dfrac{k}{r}$ **5.** $c = ks^2$ **7.** $B = \dfrac{k}{t^3}$ **9.** H varies directly as u.   **11.** P varies inversely as V.

13. E varies directly as c^2.   **15.** F varies inversely as r^2.   **17.** $y = 2x$ **19.** $F = \dfrac{12}{r}$ **21.** $H = 3\sqrt{u}$ **23.** $C = \dfrac{18}{x^2}$ **25.** 12   **27.** 9

29. 48   **31.** ± 2 **33.** increases   **35.** decreases   **37.** increases   **39.** decreases   **41.** Answers may vary.   **43.** \$3180   **45.** 12.8 gallons
47. 24 ounces   **49. a.** short handle   **b.** 30 pounds   **51.** 130.9 centimeters   **53. a.**
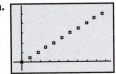
b. $F = 28.92x$

c. The force varies directly
as the stretch.

d. 0.145 meter   **55. a.**
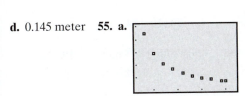
b. $P = \dfrac{7.18}{V}$ **c.** The pressure varies inversely as the volume.   **d.** 8.0 cm³

57. a. i. Answers may vary; one   **ii.** Answers may vary; one   **iii.** Answers may vary; one   **b.** one; answers may vary.   **c.** one; answers
may vary.   **59.** One student let x be the length in inches and y be the length in centimeters, which gives the model $y = 2.54x$; the other
student let x be the length in centimeters and y be the length in inches, which gives the model $y = 0.39x$; 2.54 centimeters is equal to
1 inch; 0.39 inch is (approximately) equal to 1 centimeter.   **61. a.** yes; answers may vary.   **b.** not necessarily; answers may vary.

63. a. \$75   **b.** $c = 12.5n$ **c.** \$75; yes   **65.** $\dfrac{5}{2}, -1$; quadratic equation in one variable   **67.** 0; quadratic (or second-degree)
polynomial in one variable   **69.** $(2w - 5)(w + 1)$; quadratic (or second-degree) polynomial in one variable

Homework 10.8   **1.** $\dfrac{3}{10}$ **3.** $\dfrac{7}{4}$ **5.** $\dfrac{14}{5}$ **7.** $\dfrac{3}{2w}$ **9.** $\dfrac{9}{10x^2}$ **11.** $\dfrac{2}{3(x + 3)}$ **13.** $\dfrac{5}{6}$ **15.** $\dfrac{5(x + 3)}{3x}$ **17.** 3   **19.** $\dfrac{5}{9}$ **21.** $\dfrac{11}{26}$

23. $-\dfrac{1}{5}$ **25.** $\dfrac{4(x - 2)}{5}$ **27.** $\dfrac{2x + 5}{4x - 1}$ **29.** $\dfrac{2b - 3}{4b + 5}$ **31.** $\dfrac{x + 2}{x}$ **33.** $\dfrac{3x + 4}{2(5x + 2)}$ **35.** $\dfrac{2(r - 3)(r + 3)}{3(r - 2)(r + 2)}$ **37.** $\dfrac{2(x + 2)}{x}$ **39.** Answers

may vary; $\dfrac{3x^2}{x + 3}$ **41.** $\dfrac{x(2x + 1)}{3x^2 + 4}$ **43.** Answers may vary.   **45.** $\dfrac{2(x + 3)}{3}$ **47.** $\dfrac{2(x + 3)}{3}$ **49.** $\dfrac{6}{5x}$ **51.** $\dfrac{3(2x + 1)}{16x}$ **53. a.** $\dfrac{35}{6}$

b. $\dfrac{35}{6}$ **c.** Answers may vary.   **55.** Answers may vary.   **57.** Answers may vary.   **59.** Answers may vary.   **61.** Answers may vary.

63. Answers may vary.

Chapter 10 Review Exercises

1. 0   **2.** no excluded values   **3.** $\dfrac{5}{3}$ **4.** 2, 4   **5.** -3 **6.** $\pm\dfrac{3}{2}$ **7.** $-4, \dfrac{2}{5}$ **8.** $-\dfrac{5}{2}, \pm 3$ **9.** $\dfrac{4y^3}{5x^4}$ **10.** $\dfrac{7}{x - 2}$ **11.** $\dfrac{w + 3}{w - 4}$ **12.** $\dfrac{2a}{a + b}$

13. $\dfrac{x - 2}{3x + 2}$ **14.** $\dfrac{x + 5}{(x - 5)(2x - 3)}$ **15.** $\dfrac{(m - 4)(m + 2)}{(m - 3)(m + 3)}$ **16.** $\dfrac{5b(b + 1)}{7}$ **17.** $\dfrac{4x(x + 4)}{3(x - 7)}$ **18.** $\dfrac{6}{(x - 1)(x + 4)}$ **19.** $\dfrac{5}{3x^3}$

20. $\dfrac{3(x - 2)(2x + 5)}{2(x - 5)}$ **21.** $\dfrac{7}{(t - 7)(3t - 4)}$ **22.** $\dfrac{(a - 3b)(a - 2b)}{4a(a - 7b)}$ **23.** $x - 2$ **24.** $\dfrac{9x^2 + 10}{12x^3}$ **25.** $\dfrac{2x^2 + 17x - 20}{(x - 4)(x + 1)(x + 6)}$

26. $\dfrac{x^2 + 10x + 26}{(x - 2)(x + 6)}$ **27.** $\dfrac{p - 6}{p - 2}$ **28.** $-\dfrac{14}{(x - 5)(x + 3)}$ **29.** $\dfrac{10y^2}{(x - 5y)^2}$ **30.** $-\dfrac{2(y^2 - 6y + 24)}{y(y + 4)(y^2 - 4y + 16)}$ **31.** $\dfrac{16x^6}{15}$

32. $\dfrac{9(x - 6)}{(x + 6)(x + 9)}$ **33.** $\dfrac{5m^2 - 12m - 15}{(m - 5)(m - 3)(m + 9)}$ **34.** $\dfrac{-11c - 1}{(c - 1)(c + 2)}$ **35.** Answers may vary; $\dfrac{-5x + 3}{x + 4}$ **36.** 9.92 ounces

37. 5.30 cups per day **38.** $\dfrac{11}{3}$ **39.** -1 **40.** 3 **41.** $\dfrac{2}{5}$ **42.** $\dfrac{31}{3}$ **43.** empty set **44.** $-3, \dfrac{1}{2}$ **45.** $-2, 9$ **46. a.** \$7.9 million;

\$8.1 million; underestimate **b.** \$10.9 million **c.** 2020 **47.** 24.5 ounces **48.** 27.5 yards **49.** $A = kw$ **50.** $F = \dfrac{k}{r^2}$ **51.** V varies

directly as r^3. **52.** I varies inversely as d^2. **53.** ± 5 **54.** 2 **55.** decreases **56.** decreases **57.** \$71,250 **58.** 36 minutes

59. a.

b. $A = \dfrac{54.36}{D}$ **c.** The apparent height varies inversely as the distance. **d.** 4.5 inches **e.** 27.2 inches

60. $\dfrac{4x}{3}$ **61.** $\dfrac{7(x + 2)}{5x}$ **62.** $\dfrac{5w - 2}{w - 3}$ **63.** $\dfrac{3(3b + 2)}{2(b - 6)}$

Chapter 10 Test

1. no excluded values **2.** -9 **3.** $-9, 6$ **4.** Answers may vary. **5.** $\dfrac{p + 4}{p - 5}$ **6.** $\dfrac{5m}{(m - 2)(3m - 2)}$ **7.** $\dfrac{(x + 3)(x + 5)}{(x - 5)(x - 4)}$

8. $\dfrac{9}{4x}$ **9.** $\dfrac{5a^2 + 12a + 8}{(a - 5)(a + 2)(a + 4)}$ **10.** $\dfrac{-3t - 35}{2(t - 7)(t + 7)}$ **11.** $\dfrac{3(3x + 2)}{2(2x + 1)}$ **12.** $\dfrac{-14x - 1}{(x - 7)(x + 4)}$ **13.** 12 **14.** $-\dfrac{22}{17}$

15. 80.75 miles per hour **16. a.** 40.4%; 40.7%; underestimate **b.** 18.1% **c.** 2022 **17.** 7.55 gallons **18.** 15.75 meters **19.** 24

20. ± 3 **21.** decreases **22.** 15 contact hours **23.** $\dfrac{2(4x - 3)}{3(x - 10)}$

Chapter 11

Homework 11.1   **1.** $9\sqrt{3}$ **3.** $6\sqrt{5}$ **5.** $5\sqrt{x}$ **7.** $2\sqrt{7}$ **9.** $-7\sqrt{6}$ **11.** $-11\sqrt{5t}$ **13.** $7\sqrt{5} - 3\sqrt{7}$ **15.** $6\sqrt{2}$ **17.** $4\sqrt{7} - 3\sqrt{5}$
19. 0 **21.** $7\sqrt{2}$ **23.** $3\sqrt{6}$ **25.** $-6\sqrt{5}$ **27.** $5\sqrt{7}$ **29.** $16\sqrt{3}$ **31.** $-5\sqrt{2}$ **33.** $\sqrt{5}$ **35.** $28\sqrt{3}$ **37.** $11\sqrt{x}$ **39.** 0 **41.** $\sqrt{2x} - 2\sqrt{5}$
43. $21\sqrt{3t} - 4\sqrt{11}$ **45.** $5x\sqrt{5}$ **47.** $-3x\sqrt{10}$ **49.** $-2w\sqrt{5}$ **51.** $16x\sqrt{3}$ **53.** $-x\sqrt{3} + 10\sqrt{x}$ **55.** $16y\sqrt{3} - 6\sqrt{5}y$
57. a. 29.3 years **b.** 2 "Saturn years" **59. a.**

yes **b.** 16% **c.** \$112 thousand **61.** Answers may vary; $6\sqrt{5}$

63. $4\sqrt{7} + 5\sqrt{7} = (4 + 5)\sqrt{7} = 9\sqrt{7}$ **65.** $-2x$ **67.** $-2\sqrt{x}$ **69.** $3x\sqrt{3}$ **71.** $15x^2$ **73.** $1 \pm 2\sqrt{2}$; quadratic equation in one

variable **75.** $\dfrac{11w + 3}{4(w - 3)(w + 3)}$; rational expression in one variable **77.** $x(x - 5)^2$; cubic (or third-degree) polynomial in

one variable

Homework 11.2   **1.** $\sqrt{10}$ **3.** $-2\sqrt{3}$ **5.** 17 **7.** $10\sqrt{2}$ **9.** $-42\sqrt{7}$ **11.** $35t$ **13.** $t\sqrt{21}$ **15.** $64x$ **17.** $b\sqrt{ac}$ **19.** $\sqrt{5} + \sqrt{35}$
21. $-\sqrt{14} - \sqrt{35}$ **23.** $45\sqrt{c} - 5c$ **25.** $-12\sqrt{10} + 20$ **27.** $3 - 2\sqrt{5}$ **29.** $3 + \sqrt{6} - \sqrt{15} - \sqrt{10}$ **31.** $x - \sqrt{2x} - \sqrt{7x} + \sqrt{14}$
33. $y^2 - y\sqrt{11} + y\sqrt{2} - \sqrt{22}$ **35.** $38 - 16\sqrt{5}$ **37.** -4 **39.** $r^2 - 3$ **41.** 38 **43.** $9a - 25b$ **45.** $23 + 8\sqrt{7}$ **47.** $8 - 2\sqrt{15}$
49. $b^2 + 2b\sqrt{2} + 2$ **51.** $x - 2\sqrt{6x} + 6$ **53.** $-12\sqrt{10} + 38$ **55.** Answers may vary; $x^2 + 2x\sqrt{3} + 3$ **57.** Answers may vary; $6\sqrt{5}$
59. Answers may vary. **61. a. i.** true **ii.** false **iii.** true **iv.** false **v.** true **vi.** false **b.** Answers may vary. **63.** $5\sqrt{x}$
65. $6x$ **67.** $x + 5\sqrt{x} + 6$ **69.** $x^2 + 5x + 6$ **71.** $9x$ **73.** $x + 6\sqrt{x} + 9$ **75.** $x^2 + 6x + 9$ **77.**

linear equation in

two variables **79.** empty set; rational equation in one variable

81. $\dfrac{x(x - 1)}{(x - 4)(x + 2)}$; rational expression in one variable

Homework 11.3   **1.** 36   **3.** 3   **5.** empty set   **7.** 2   **9.** empty set   **11.** 5   **13.** 2   **15.** $-2, 8$ **17.** 16   **19.** 4   **21.** 16

23. $\dfrac{25}{9}$ **25.** empty set   **27.** 12   **29.** 2, 3   **31.** 1   **33.** 7   **35.** -1 **37.** 4   **39.** 6   **41.**   **43.** -4 **45.** empty set

47. 4.81   **49.** 3.13   **51. a.** 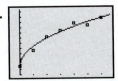 yes   **b.** $(0, 365)$; in 2010, there were 365 million acres of genetically modified crops.

c. 496.3 million acres   **d.** 2023   **53. a.** 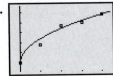 yes   **b.** 38.3 thousand restaurants   **c.** 200 restaurants   **d.** 2021

55. a. yes   **b.** 2011   **c.** Since all tires could have been reused or recycled by 2011, the plant would have been needed to be in operation for only 3 years. This would not have been enough time to warrant building such a plant.

57. a. fairly well   **b.** 3.90   **c.** 7.74; overestimate   **d.** increase; the higher the GDP of a country, the higher is the average happiness rating for the residents of the country.   **e.** evaluative   **59.** Answers may vary; $-\dfrac{10}{7}$

61. $(4, 2)$ **63.** Answers may vary.   **65.** $\dfrac{-2 \pm 3\sqrt{2}}{2}$ **67.** 7   **69.** $\dfrac{20}{9}$ **71.** Answers may vary.   **73.** Answers may vary.

75. Answers may vary.   **77.** Answers may vary.   **79.** Answers may vary.

Chapter 11 Review Exercises

1. $15\sqrt{3}$ **2.** $-\sqrt{2x}$ **3.** $-2\sqrt{5x}$ **4.** $8\sqrt{3} + 4\sqrt{6}$ **5.** $-6\sqrt{5} - 3\sqrt{7}$ **6.** $11\sqrt{3}$ **7.** $-5\sqrt{6}$ **8.** $7\sqrt{2}$ **9.** $\sqrt{5}$ **10.** $25\sqrt{2}$
11. $-11\sqrt{10}$ **12.** $-2\sqrt{2} - 5\sqrt{3}$ **13.** $15\sqrt{5} - 16\sqrt{6}$ **14.** $17\sqrt{x}$ **15.** $-11\sqrt{x}$ **16.** $-3\sqrt{3x} + 2\sqrt{2}$ **17.** $-w\sqrt{7}$
18. $3a\sqrt{7}$ **19.** $-7x\sqrt{3}$ **20.** $4\sqrt{2x} - 10x\sqrt{5}$ **21.** $\sqrt{35ab}$ **22.** $-24\sqrt{2}$ **23.** $25x$ **24.** $\sqrt{6} + \sqrt{14}$ **25.** $-\sqrt{21} + \sqrt{15}$
26. $x - \sqrt{2x}$ **27.** $10\sqrt{21} + 14$ **28.** $3\sqrt{5} - 13$ **29.** $-17\sqrt{2} + 74$ **30.** $\sqrt{6} - \sqrt{10} - \sqrt{21} + \sqrt{35}$ **31.** $b + 7\sqrt{b} - 8$
32. $t^2 + t\sqrt{5} + t\sqrt{3} + \sqrt{15}$ **33.** $-8\sqrt{7} + 44$ **34.** -2 **35.** $16a - 9b$ **36.** $b^2 - 3$ **37.** $-8\sqrt{5} + 21$ **38.** $x + 12\sqrt{x} + 36$
39. $x^2 + 2x\sqrt{3} + 3$ **40.** $-24\sqrt{10} + 77$ **41.** 4   **42.** 22   **43.** empty set   **44.** $\dfrac{5}{2}$ **45.** $-1, 4$ **46.** 49   **47.** 64
48. empty set   **49.** 6   **50.** -1 **51.** $\dfrac{16}{3}$ **52.** 2   **53.** 7   **54.** 4   **55.** 3.28   **56. a.** yes   **b.** 33%   **c.** 2022

Chapter 11 Test

1. $-2\sqrt{3} - 3\sqrt{2}$ **2.** $-12\sqrt{5x}$ **3.** $-23\sqrt{2x} + 15\sqrt{5}$ **4.** $-b\sqrt{6}$ **5.** $-80\sqrt{7}$ **6.** $x - 3\sqrt{x}$ **7.** $2\sqrt{5} - 3$
8. $2 + \sqrt{14} - \sqrt{10} - \sqrt{35}$ **9.** $25a - 4b$ **10.** $8\sqrt{3} + 19$ **11.** $x - 2\sqrt{5x} + 5$ **12.** $12\sqrt{6} + 30$ **13.** 7
14. 5   **15.** 4   **16.** 31   **17.** 7   **18.** 1   **19.** 4   **20.** -1 **21.** -5 **22.** empty set   **23. a.** 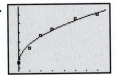 yes

b. $(0, 46)$; 46% of viewers watch their programs the day the program airs.   **c.** 97%   **d.** 4 days

Cumulative Review of Chapters 1–11

1. $-15a^2b^2 + 6ab^3$; fourth-degree polynomial   **2.** $-13p + 1$; linear polynomial   **3.** $9x^2 - 7x - 3$; quadratic polynomial

4. $-x^3 - x^2 + 3x - 1$; cubic polynomial   **5.** $12m^2 - 16mn - 35n^2$; quadratic polynomial   **6.** $3x^3 - 10x^2 + x + 6$; cubic polynomial

7. $4p^2 + 12pq + 9q^2$; quadratic polynomial   **8.** $\dfrac{15}{(m - 5)(m + 2)}$; rational expression   **9.** $\dfrac{2(w + 3)}{(w - 5)(2w + 1)}$; rational expression

10. $\dfrac{2(7x + 12)}{5(x - 6)(x + 6)}$; rational expression   **11.** $\dfrac{-5x^2 + 18x + 12}{(x - 3)(x + 2)(x + 4)}$; rational expression   **12.** $\dfrac{\sqrt{21}}{7}$; radical expression

13. $-4\sqrt{3}$; radical expression   **14.** $x - \sqrt{2x} + \sqrt{5x} - \sqrt{10}$; radical expression   **15.** $4b - 25c$; radical expression   **16.** $10 - 2\sqrt{21}$;

radical expression   **17.** $y = -3x^2 - 12x - 11$ **18.** $54x^{13}y^{28}$ **19.** $5x\sqrt{2y}$ **20.** $\dfrac{\sqrt{7t}}{7}$ **21.** -30 **22.** $\dfrac{17}{6}$ **23.** $(4p - 9q)(4p + 9q)$

24. $(a - 9)(a + 3)$ **25.** $(w + 7)^2$ **26.** $n(2m + 3n)(3m + 2n)$ **27.** $2(x - 4)^2$ **28.** $(x - 3)(x + 3)(x + 5)$

29. $4x^2(x + 2)(3x - 5)$ **30.** $(x + 3)(x^2 - 3x + 9)$ **31.** $x < -3$; $(-\infty, -3)$ 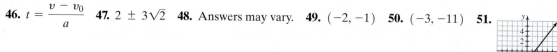   **32.**

33. 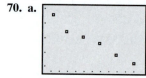   **34.** $\dfrac{8}{5}$; linear equation   **35.** $\dfrac{37}{6}$; linear equation   **36.** $-3, 8$; quadratic equation   **37.** $\dfrac{1}{6}, 1$; quadratic equation

38. $\pm\dfrac{2\sqrt{15}}{3}$; quadratic equation   **39.** $-4 \pm 2\sqrt{15}$; quadratic equation   **40.** $\dfrac{5 \pm \sqrt{73}}{6}$; quadratic equation   **41.** $\dfrac{-3 \pm \sqrt{3}}{2}$; qua-

dratic equation   **42.** 8; rational equation   **43.** 5; rational equation   **44.** $\dfrac{9}{25}$; radical equation   **45.** 4; radical equation

46. $t = \dfrac{v - v_0}{a}$ **47.** $2 \pm 3\sqrt{2}$ **48.** Answers may vary.   **49.** $(-2, -1)$ **50.** $(-3, -11)$ **51.** 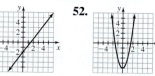   **52.**

53. $(10, 0)$ **54.** $(-8, 0), (6, 0)$ **55.** ski run B   **56.** $\dfrac{1}{2}$; increasing   **57.** $y = -\dfrac{2}{5}x + \dfrac{14}{5}$ **58.** $y = -\dfrac{7}{8}x + \dfrac{9}{4}$ **59.** $y = -\dfrac{1}{2}x + 1$

60. $2\sqrt{21}$ **61.** $1, \dfrac{5}{2}$ **62.** $-\dfrac{7}{x - 5}$ **63.** $-9°F$ **64.** 2.25 tablespoons   **65.** 27   **66.** 5   **67.** $\dfrac{x + 12}{2(x - 10)}$ **68.** Amana account: $7500,

Vanguard account: $4500   **69. a.** $A = -1350t + 27{,}500$ **b.** 18,050 feet   **c.** 19.48 minutes   **d.** -1350; the airplane

descends by 1350 feet per minute.   **e.** $(20.37, 0)$; the airplane will land about 20.4 minutes after beginning its descent.

70. a. 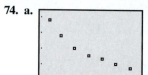   linear equation   **b.** $p = -0.31a + 37.92$ **c.** -0.31; the percentage of American adults who say that

they are interested in soccer decreases by 0.31 percentage point per year of age.   **d.** 30.2%

e. 55 years   **f.** $(122.32, 0)$; no 122-year-old Americans are interested in soccer. A little research would

show that the oldest American to date was 119 years old.   **71. a.** $0.675t + 5.51$; the expression

represents the total average daily oil production (in millions of barrels a day) in the year that is t years since 2010.   **b.** 12.94;

in 2021, total average daily oil production will be 12.94 million barrels per day.   **c.** Each year, total average daily oil

production increases by about 0.68 million barrels.   **d.** 2003; 0.52 million barrels per day   **72.** 12.4 miles   **73.** 1524 pens

74. a. 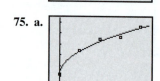   quadratic equation; answers may vary.   **b.** 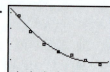   yes   **c.** 34%   **d.** 34 years

75. a. 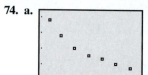   yes   **b.** 39%   **c.** 56 years

Index